Inderbir Singh's

TEXTBOOK OF
ANATOMY

Sixth Edition

Volume III

Inderbir Singh's

TEXTBOOK OF
ANATOMY

Sixth Edition

Revised and Edited by

Sudha Seshayyan MS PDHM

Director and Professor
Upgraded Institute of Anatomy
Vice Principal, Madras Medical College
Chennai

JAYPEE *The Health Sciences Publisher*

New Delhi | London | Philadelphia | Panama

 **Jaypee Brothers Medical Publishers (P) Ltd**

Headquarters

Jaypee Brothers Medical Publishers (P) Ltd
4838/24, Ansari Road, Daryaganj
New Delhi 110 002, India
Phone: +91-11-43574357
Fax: +91-11-43574314
Email: jaypee@jaypeebrothers.com

Overseas Offices

J.P. Medical Ltd
83 Victoria Street, London
SW1H 0HW (UK)
Phone: +44 20 3170 8910
Fax: +44 (0)20 3008 6180
Email: info@jpmedpub.com

Jaypee Medical Inc
The Bourse
111 South Independence Mall East
Suite 835, Philadelphia, PA 19106, USA
Phone: +1 267-519-9789
Email: jpmed.us@gmail.com

Jaypee Brothers Medical Publishers (P) Ltd
Bhotahity, Kathmandu, Nepal
Phone: +977-9741283608
Email: kathmandu@jaypeebrothers.com

Jaypee-Highlights Medical Publishers Inc
City of Knowledge, Bld. 237, Clayton
Panama City, Panama
Phone: +1 507-301-0496
Fax: +1 507-301-0499
Email: cservice@jphmedical.com

Jaypee Brothers Medical Publishers (P) Ltd
17/1-B Babar Road, Block-B, Shaymali
Mohammadpur, Dhaka-1207
Bangladesh
Mobile: +08801912003485
Email: jaypeedhaka@gmail.com

Website: www.jaypeebrothers.com
Website: www.jaypeedigital.com

Textbook of Anatomy

First Edition   : 1996
Second Edition : 1999
Third Edition   : 2003
Fourth Edition : 2007
Fifth Edition   : 2011
Sixth Edition   : **2016**

ISBN 978-93-5152-986-6

Printed at Replika Press Pvt. Ltd.

Late Professor Inderbir Singh

(1930–2014)

Tribute to a Legend

Professor Inderbir Singh, a legendary anatomist, is renowned for being a pillar in the education of generations of medical graduates across the globe. He was one of the greatest teachers of his times. He was a passionate writer who poured his soul into his work. His eagle's eye for details and meticulous way of writing made his books immensely popular amongst students. He managed to become enmeshed in millions of hearts in his lifetime. He was conferred the title of Professor Emeritus by Maharishi Dayanand University, Rohtak.

On 12th May, 2014, he was awarded posthumously with* Emeritus Teacher Award *by* National Board of Examination *for making invaluable contribution in teaching of Anatomy. This award is given to honour legends who have made tremendous contribution in the field of medical graduate. He was a visionary for his times, and the legacies he left behind are his various textbooks on *Gross Anatomy*, *Histology*, *Neuroanatomy* and *Embryology*. Although his mortal frame is not present amongst us, his genius will live on forever.

Preface

Castles of all medical wisdom are anchored to the knowledge of anatomy. Both the learning and the teaching of anatomy have undergone masterly changes. Though the limits of human anatomy appear to be confined to the boundaries of the human body, newer frontiers have constantly appeared due to two primary factors—one, expanding basic medical and clinical research and two, larger understanding of hitherto unexplained areas.

The preparation of a textbook on Anatomy should have the scope to adequately accommodate the growing changes. At the same time, it also cannot become disproportionately large, considering the time span within which an average undergraduate medical student would have to acquire this knowledge.

This edition of Inderbir Singh's *Textbook of Anatomy* has been prepared keeping the twin factors of the restructuring of medical curriculum and the knowledge expansion in mind. Many of the chapters have been completely revised and rewritten. Clinical Correlation has been clearly laid out. Embryological and Histological details have been added so as to give the reader a comprehensive picture. Newer features like Multiple Choice Questions and Clinical Problem-solving have been appended to each chapter in order to provide the reader with the opportunity of self-assessment.

A student entering the medical curriculum is faced with a completely new atmosphere. In an attempt to familiarize the student not only with Anatomy but also with the nuances of the medical world, new sections on *General Anatomy* and *Genetics* have been added. Professor Inderbir Singh's eye for details and meticulous writing style have always been popular amongst generations of medical students. Though many areas of the book have been revisited, the basic spirit and nature of the book have been retained. Additional features like *Added Information* and *Clinical Correlation* in any chapter will be of much help not only to the undergraduate students but also to the postgraduates.

At this juncture, I would like to place on record my appreciation and gratitude to Dr Hannah Sugirthabai Rajila Rajendran, Professor, Department of Anatomy, Chettinad Hospital and Research Institute, Kanchipuram District, Tamil Nadu, India; Dr M Nirmaladevi, Associate Professor, PSGIMS & R, Coimbatore, Tamil Nadu, India and Dr J Sreevidya, Assistant Professor-cum-Civil Surgeon, Madras Medical College, Chennai, Tamil Nadu, India for their painstaking editorial assistance. I would like to thank Dr Indumathi. S, Professor and HOD, Department of Anatomy, Chettinad Hospital and Research Institute, Dr T Anitha, Dr Elamathi Bose and Dr Bhuvaneswari, Assistant Professors of Anatomy, Madras Medical College, Chennai for their help during the preparation and review of the manuscripts and formulation of chapters.

I would be failing in my duty if I do not acknowledge the contributions of Dr Lakshmi, Dr Kanagavalli, Dr Arrchana, Assistant Professors, Department of Anatomy, Madras Medical College, Chennai and Dr Dharani, Assistant Professor, Villupuram Government Medical College, Villupuram, Tamil Nadu, India towards the completion of this edition. Shri RAC Mathews, Shri Ranganathan and Shri Sashikumar were instrumental in providing the necessary assistance, and Shri E Senthilkumar provided some of the illustrations for the book and I would like to extend my thanks to each of them.

Special thanks to Shri Jitendar P Vij (Group Chairman) and Mr Ankit Vij (Group President), Jaypee Brothers Medical Publishers (P) Ltd., without whom this edition would not have seen the light of the day. I am extremely thankful to them for reposing their confidence in me and providing the opportunity to revise Inderbir Singh's *Textbook of Anatomy*. Dr Sakshi Arora (Director, Content and Strategy) has been the driving force behind all efforts and deserves a very special thanks. She has provided insights and innovative ideas which have gone a long way in consolidating this book to best meet the needs of the taught and the teacher alike. We are thankful to her entire Development and Content Strategy team consisting of Ms Nitasha Arora (Project Manager), Mr Bunty Kashyap, Mr Phool Kumar, Mr Puneet Kumar Das, Mr Vikas Kumar, Mr Prabhat Ranjan, Mr Neeraj Choudhary, Mr Sanjeev Kumar and Mr Sandeep Kumar (Designers and Operators), and Ms Ankita Singh, Ms Sonal Jain, Ms Neelam Kakariya, Mr Prashant Soni (Editorial) for their constant technical support throughout the project.

This book is the combined effort of a number of people who have contributed in myriad ways and it may not be humanly possible to list down the many; however, I take this opportunity to extend my thanks to all of them.

Sudha Seshayyan

Contents

Section 6

Head and Neck

Chapter 1

Overview of Head and Neck

Head and neck are both important regions of the body. It is hard to separate the two of them because structures of the head are in anatomical and functional continuity with structures of the neck and vice versa.

The human head can be described as the superior part of the body that houses the brain and gives accommodation for organs of special senses. The term *cephalon* is used to denote the head region in several lower organisms. In humans, the term is not directly used; however, *cephalic*, a derivative of cephalon (Old French.cephalique; Latin. cephalicus; Greek.kephele=head, related to head), is regularly used. Brain and parts of brain are indicated by the term *encephalon* (in the head).

The neck is that portion which attaches the head to the rest of the body. It actually serves as a conduit transmitting various structures from and to the head.

Presence of head and neck can be attributed to an evolutionary and developmental process called *cephalisation*. Several lower organisms do not have a distinct head; even those which seem to have a separate and identifiable head do not possess a distinct neck. However, organisms and animals belonging to more and more of higher levels of evolution do have distinct and specialised heads and necks. Development of a distinct head followed the development of bilateral symmetry in animals. As bilateral symmetry made the organism more and more compact, the nervous elements of the organism's body got concentrated in the anterior region. Structures related to

impulse reception also developed in the same region so that coordination between senses and the neural elements would be maximal. As selection of food is related to the senses, organs of ingestion also developed in the anterior part of the body. Compact spatial arrangement of all the structures resulted in cephalisation (concentration of important structures in the head region).

A distinct neck was not required as long as the organisms lived in water or on the surface of land. When they had to move their heads up to some levels above the surface (as in birds and mammals unlike the reptiles), 'neck' became a necessity. The head had to move/turn/rotate separate from the movements of the rest of the body; neck provided 'freedom' to the head, all the time keeping the latter in contact with the rest of the body.

Head and neck, which are anterior in the quadrupeds, ascended up to become superior in humans.

REGIONS OF HEAD AND NECK

Though the head appears a compact 'whole' and the neck looks like a stalk holding the head up, several regions are described for purposes of study. The regions are:

- ❏ *Scalp (regio epicranius):* It is the skin that covers the 'top' of the head. Extending from the forehead to the back of head, this region forms an important landmark in the external morphology of the individual.
- ❏ *Face (regio facies):* It is the anterior portion of the head that serves as a communication tool. The main receptor organs of the special senses (eyes, ears, mouth and nose) are located on the face making the latter both a region of reception and a region of expression.
- ❏ *Parotid region (regio parotideus):* It is the region on the sides of the face, near and a little lower to the ear (hence the name; parotid=beside ear). It consists of a major salivary gland called the *parotid gland*.

Important vessels and nerves of head and neck traverse through this gland making it a zone of neurovascular significance.

❑ *Temporal region (regio temporale):* It is the region on the sides of the face, behind the eye and above the ear. It is also the lateral region of the scalp.

❑ *Infratemporal region (regio infratemporale):* It is the area that is deep to the parotid region and the mandible. It is not seen from the exterior and houses important neurovascular structures.

❑ *Submandibular region (regio submandibulare):* It is the area beneath the lower jaw. Since the area is below the maxillary bone, in the olden times, it was referred to as the submaxillary region. The submandibular triangle and the submental triangle form part of this region.

❑ *Suboccipital region (regio suboccipitale):* It is the area at the back of the head below the prominence of the bone. It extends inferiorly into the back of neck.

❑ *Anterior cervical region (regio antecollum* or *regio colli anterior):* It is the anterior portion of the neck and consists of the structures contained within the neck anterior to the larynx and trachea. The muscular triangle and the carotid triangle form part of this region.

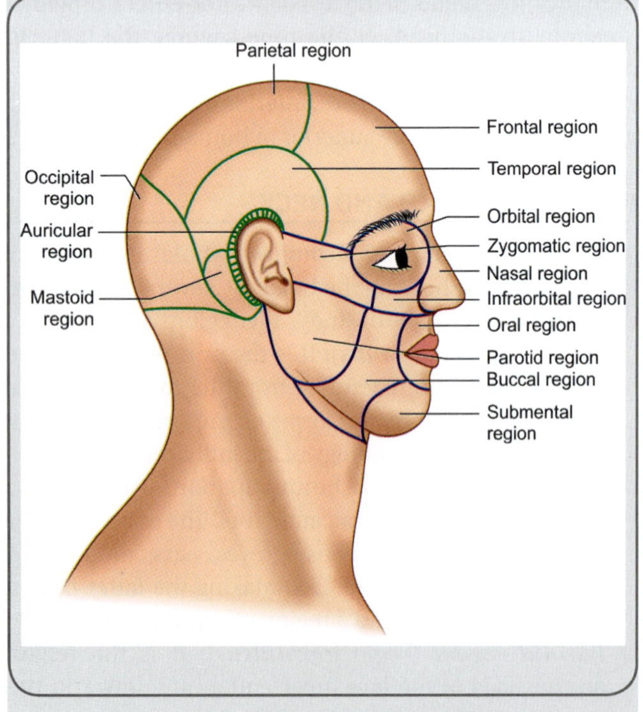

Fig. 1.1: Regions of head and neck

contd...

❑ *Posterior cervical region (regio postcollum* or *regio colli posterior):* It is the posterior portion of the neck consisting mainly of the fasciae, ligaments and muscles on the posterior aspect of the cervical vertebral column.

❑ *Lateral cervical region (regio sternomastoidale* or *regio colli lateralis):* It is the lateral portion of the neck consisting mainly of the posterior triangle of the neck.

❑ *Prevertebral region (regio prevertebralis):* It is the area consisting of structures which lie on the anterior aspect of the cervical vertebral column. The prevertebral muscles form the major bulk of this region.

❑ *Tracheo-oesophageal region (regio tracheo-oesophagialis):* It is the area related to the trachea and the oesophagus and the space between them. Important neurovascular structures are related to this area.

Several other regions like the sternocleidomastoid region, temporozygomatic region, parotideomasseteric region, temporoparietal region, frontonasal region, oromaxillary region and orbital region have been described. These are the areas which form part of the major regions mentioned above.

SURFACE LANDMARKS AND FEATURES

It is essential to know some of the important surface landmarks of head and neck before a detailed study of the internal structures is commenced.

Surface Features of Cranium

- *Frontal bossing:* These are the right and left prominences seen at the upper part of the forehead, a little below where the forehead curves to join the hairy part of the scalp. These are also called the *frontal tubera*.
- *Pinnae and ear apertures:* These are the external ear auricles and the openings of the external acoustic meati.
- *Mastoid processes:* These are the prominences seen and well felt behind the auricles. They are well formed in adults but are not present in infants.
- *External occipital protuberance:* This is the prominent point that can be felt in the midline on the occipital region at the point where the neck commences. It may sometimes not be felt clearly, especially in females.

Surface Features of Face

- *Supercilia:* These are the eyebrows, one on either side.
- *Superciliary arches:* These are the arched bony prominences seen above the eyebrows.
- *Glabella:* This is the hairless midline area between the eyebrows.
- *Palpebrae:* These are the eyelids which are fascio-fibro-muscular folds overlying the eyeball and protecting the latter.
- *Palpebral fissures:* These are the openings of the eyes on either side.
- *Superior and inferior palpebral sulci:* These are the shallow depressions seen above and below the eyes on both sides.
- *Nose or external nose:* This is the pyramidal structure that projects anteriorly and downward in the median part of the face.
- *Apex of nose:* This is the pointed anterior angular tip of the nose.
- *Bridge of nose:* This is the slightly depressed area that overlies the junction of the nose and the forehead. It is also called the *root of nose*.
- *Dorsum of nose:* This is the anterior border of nose that runs from the root to the apex.
- *Alae of nose:* These are the rounded wing-like projections, one on either side, each around the anterior nasal opening of its corresponding side.
- *Philtrum:* This is the vertical shallow groove seen in the median region of the upper lip.
- *Oral fissure:* This is the opening of the mouth.
- *Angles of the mouth:* These are the right and left edges of the oral fissure.

- *Mental protuberance:* This is the prominence of the chin. It is separated from the lower lip by a shallow transverse groove called the *mentolabial sulcus*.
- *Nasolabial sulci:* These are the right and left grooves which run from the alae of the nose to the angles of the mouth and become prominent while smiling or laughing.

Surface Features of Neck

- *Sternocleidomastoid:* This muscle can be made prominent in a living individual by asking the individual to rotate the face towards the contralateral side and elevate the chin. The muscle stands out as a broad band and its anterior and posterior borders can be well made out.
- *Jugular notch:* This depression can be palpated in the median plane of the root of neck between the sternal heads of the two sternocleidomastoid muscles.
- *Spine of the vertebra prominens:* This is the spine of the seventh cervical vertebra which can be palpated at the junction of the nape of the neck with the back.
- *Greater and lesser supraclavicular fossae:* These are two depressions found above the clavicle. The lesser fossa is between the clavicular and sternal heads of sternocleidomastoid and the greater fossa is between the clavicular head and the trapezius.

FASCIAE

The head and the neck accommodate important structures. In order to give protection to these structures, the fasciae of both head and neck have undergone specialisation.

In the scalp, the subcutaneous connective tissue forms a richly vascularised and a richly innervated layer beneath the skin. Another layer of loose connective tissue lies beneath the aponeurotic layer of the occipitofrontalis muscle; this layer has potential spaces which may distend with fluid.

Several fascial layers and planes exist in the face; these layers merge and separate in varied combinations to create spaces of different dimensions. Details of these fasciae are dealt with in chapters on face, parotid and temporal regions. The superficial fascia of face gains importance by the presence of the muscles of facial expression within and attached to it.

The superficial fascia of neck has no special features. However, the deep fascia of the neck forms clear sheets. One of these sheets forms a complete sleeve around the deeper structures of the neck and is called the *investing layer of deep cervical fascia*. Another sheet forms the pretracheal fascia while the third forms the prevertebral fascia.

CUTANEOUS INNERVATION

The cutaneous innervation of the head can be conveniently divided into that of the face and that of the scalp. The sensory supply of the skin of both these regions is predominantly subserved by the trigeminal nerve and its branches. The cutaneous innervations of the neck can be studied as a whole.

Cutaneous Innervation of Face

The trigeminal nerve has three branches—***ophthalmic***, ***maxillary*** and ***mandibular divisions***. These three divisions take care of the cutaneous supply of the face in three broad areas with posterosuperior wings, one below the other.

Branches of Ophthalmic Division

The cutaneous branches of the ophthalmic division of the trigeminal nerve take care of the supply of the upper eyelid, forehead and the anterior part of scalp. The external nasal nerve supplies the strip along the dorsum of the nose. The infratrochlear nerve supplies the glabella and the medial part of the upper eyelid. The supratrochlear nerve supplies the region above the glabella. The supraorbital nerve supplies a wide area on the forehead and scalp extending almost till the level of lambdoid suture. The lacrimal nerve supplies a small area in the lateral part of the upper eyelid.

Branches of Maxillary Division

The cutaneous branches of the maxillary division take care of the lower eyelid, upper lip, cheeks and some parts of the nose. The infraorbital nerve supplies the skin from near the lateral part of the root of nose, lower eyelid, cheek, upper lip and ala of nose (including the skin of the vestibule). The zygomaticofacial nerve supplies the prominence of the cheek. The zygomaticotemporal nerve supplies the hairless skin of the anterior aspect of temporal region.

Branches of Mandibular Division

These branches take care of the lower lip, lower jaw and a band of skin that passes up from the lower jaw through the auricular area to the posterior aspect of the temporal region. The mental branch supplies the skin of chin, mentolabial sulcus and the overhanging lower lip. The buccal nerve (buccal branch of mandibular division) supplies the inferolateral aspect of the cheek. The auriculotemporal nerve supplies the skin of preauricular area, auricle, external acoustic meatus and posterior part of temporal region.

Cutaneous Innervation of Scalp

The skin of scalp is supplied by the branches of trigeminal nerve and branches of cervical spinal nerves.

The skin of forehead and that of the anterior portion of scalp till the level of the lambdoid suture are supplied by the supraorbital and supratrochlear branches of the ophthalmic division of trigeminal nerve. The skin of the temporal region is supplied by the zygomaticotemporal branch of the maxillary division and the auriculotemporal branch of the mandibular division.

The posterior part of scalp is taken care of by the cervical spinal nerves.

The great auricular nerve, a branch of the cervical plexus and deriving fibres from the ventral rami of C2 and C3 spinal nerves, supplies the skin over the angle of mandible, inferior lobe of auricle and the parotid sheath. The lesser occipital nerve, again a branch of the cervical plexus and deriving fibres from the ventral rami of C2 and C3 spinal nerves, supplies the skin of the scalp posterior to the auricle.

The median portion of the skin of scalp is supplied by branches from the dorsal rami of C2 and C3 spinal nerves. The greater occipital nerve supplies the skin of the occipital region. The third occipital nerve supplies the skin of the lower occipital and suboccipital regions.

Cutaneous Innervation of Neck

The skin of neck derives its supply from the branches of the cervical plexus and from the cutaneous branches of the posterior rami of cervical spinal nerves.

Three cutaneous branches of the cervical plexus supply the anterior and lateral portions of the neck. The lesser occipital nerve supplies the small area of skin of neck behind the auricle. The great auricular nerve supplies the skin over the mastoid, skin in the space between the mastoid and the angle of mandible and a small area below it. The transverse cervical nerve supplies the skin covering the anterior cervical region.

The supraclavicular nerves, which are also branches of the cervical plexus, supply the skin over the lower part of the neck and above the clavicle.

The skin of the posterior aspect of the neck is supplied by the cutaneous branches of the posterior rami of C4 to C8 spinal nerves.

DERMATOMAL MAP OF HEAD AND NECK

The ***dermatomal map of the face and scalp*** reveals the interesting coordination that can be witnessed in the cutaneous supply of these areas. The cutaneous distribution of the face follows the developmental distribution of the latter. Viewed from front, the distribution can be likened to three cups fitted in the sequence of the upper cup pushed into the next lower cup. The uppermost cup is actually full to the brim and also has a linear central stalk that runs along the dorsum of the nose.

The cutaneous distribution of the scalp follows a clear division between the branches of the ventral and dorsal rami of spinal nerves. The posterior part of the scalp, as should be anticipated, is supplied by branches from the dorsal rami (greater occipital from C2 and third occipital from C3). The lateral part of the scalp is supplied by branches from the ventral rami (lesser occipital and great auricular nerves from C2,C3).

The ***dermatomal map of the neck*** recapitulates the segmental pattern of nervous distribution. The upper part of neck derives supply from C2,C3 segments while the lower part from C3,C4 segments.

LYMPH NODES

Lymph nodes which drain the structures of head and neck are distributed in ***terminal*** and ***outlying*** groups.

- ❑ The ***terminal group*** is related to the carotid sheath and lies along the internal jugular vein; it is called the ***deep cervical group of lymph nodes***. All the lymph vessels of the head and neck drain into these nodes either directly or indirectly through the outlying group.
- ❑ The ***outlying group*** is, in turn, arranged into an outer circle and an inner circle. The outer circle of lymph nodes extends from the chin to the occiput as a cervical collar at the junction of head and neck and is superficially placed. The inner circle surrounds the upper part of the respiratory and alimentary passages. Deep to the inner circle, a submucosal ring of lymphoid tissue, the ***Waldeyer's ring***, is present.

Lymphatics from the lower part of deep cervical lymph nodes are collected together to form the ***descending jugular lymph trunks***. The left jugular lymph trunk terminates at the junction of the left subclavian and internal jugular veins after joining with the thoracic duct. The right jugular trunk opens at the junction of the right subclavian and right internal jugular veins either directly or after joining with the right lymphatic duct.

Outer Circle of the Outlying Group

The outer circle of the outlying group is arranged around the junction of head and neck from the occiput posteriorly to the chin anteriorly. The constituent lymph nodal groups are as follows:

- ❑ The nodes of the ***occipital group*** lie along the attachment of trapezius to the occipital bone.
- ❑ The nodes of the ***retroauricular group*** (also called the ***mastoid group***) lie superficial to the upper attachment of sternocleidomastoid muscle.
- ❑ The parotid lymph nodes are in two groups, ***superficial*** and ***deep***. The nodes of the ***superficial parotid group*** (also called the ***preauricular group***) lie over the parotid gland and those of the ***deep parotid group*** are embedded within the gland.
- ❑ The nodes of the ***submandibular group*** lie over the submandibular gland. Some of them are embedded within the gland.
- ❑ The ***submental nodes*** lie below the chin overlying the mylohyoid muscle, between the anterior bellies of the right and left digastric muscles.

Apart from these lymph nodes, the following nodes are also part of the outer circle:

- ❑ The ***buccal nodes*** which lie along the facial vein,
- ❑ The ***superficial cervical nodes*** which lie along the external jugular vein, superficial to the sternocleidomastoid muscle, and
- ❑ The ***anterior cervical nodes*** which lie along the anterior jugular vein.

The lymphatics from these lymph nodes drain into the deep cervical lymph nodes either directly or through the inner circle of outlying group of lymph nodes (Fig. 1.2).

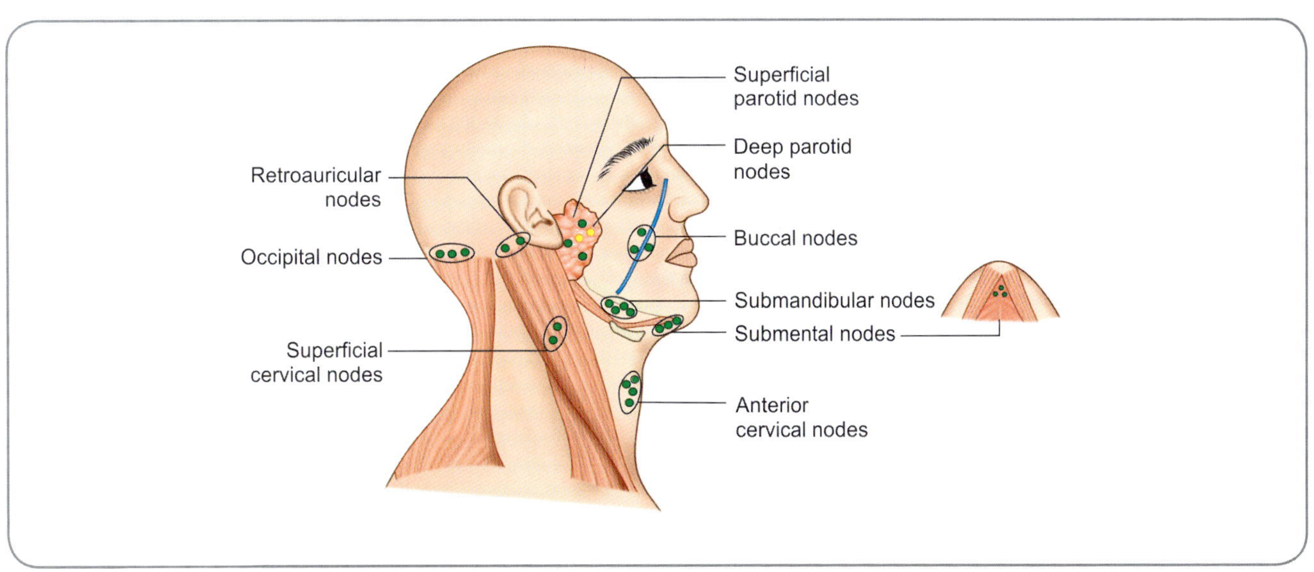

Fig. 1.2: Lymph nodes draining superficial tissues of the head and neck

Inner Circle of the Outlying Group

The lymph nodes of the inner circle consist of the prelaryngeal, pretracheal, paratracheal, retropharyngeal, lingual and infrahyoid nodes.

- The **prelaryngeal** nodes are located in front of the cricovocal membrane.
- The **pretracheal** nodes are in front of the trachea, above the isthmus of thyroid gland.
- The **paratracheal** nodes are present on either side of the trachea and oesophagus.

All these nodes drain lymph from the larynx below the vocal folds, the trachea, the oesophagus and the thyroid gland.

- The **retropharyngeal** nodes occupy the retropharyngeal space and receive afferents from the pharynx, palate, palatine tonsils, posterior part of nasal cavity, auditory tube, tympanic cavity, sphenoidal and posterior ethmoidal air sinuses.
- The **infrahyoid** nodes are located over the thyrohyoid membrane.
- The lingual nodes are situated over the hyoglossus.

Deep Cervical Nodes

The deep cervical lymph nodes are located along the internal jugular vein and lie mostly under cover of sterno-cleidomastoid. The deeper tissues of the head and neck drain into these nodes. They also receive afferents from the outlying group of lymph nodes—both inner and outer circles. They are further subdivided into the **superior** and **inferior groups (Fig. 1.3)**.

The **superior group of deep cervical lymph nodes** are located near the upper part of internal jugular vein. Though most of the nodes lie deep to the sternocleidomastoid, one subgroup of nodes lie in a triangle bounded behind by the internal jugular vein, above and in front by the posterior belly of the digastric muscle, and below and in front by

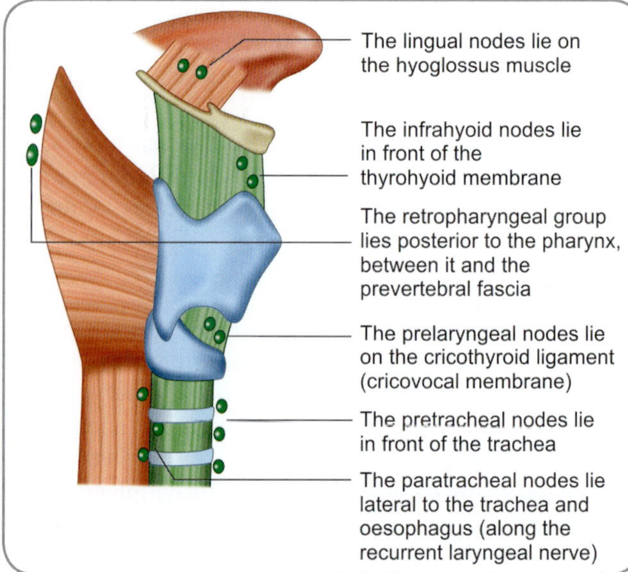

Fig. 1.4: Outlying members of the deep cervical group of lymph nodes

the facial vein. These nodes are called the **jugulodigastric nodes** and are mainly associated with the drainage of the tongue and also the palatine tonsils. Efferents from the upper deep cervical group either drain into the lower deep cervical or directly into the jugular lymph trunk.

The **inferior deep cervical lymph nodes** lie along the lower part of internal jugular vein and are also located deep to the sternocleidomastoid. One node of the inferior group lies just above the intermediate tendon of the omohyoid muscle and is called the **jugulo-omohyoid node**. It is also associated with the drainage of the tongue. A few nodes of the deep cervical group also extend in front of scalenus anterior muscle. Enlargement of the left scalene node is a common finding in carcinoma of stomach **(Virchow's node)**. Efferents from the lower deep cervical group drain into the jugular lymph trunks (Fig. 1.4).

Clinical Correlation

- Lymph nodes in the head and neck may be enlarged in various diseases. The most important of these are tuberculosis and malignancy. Physical examination of the neck involves systematic palpation of the nodes. The surgeon stands behind the patient whose neck is slightly flexed to relax the muscles to palpate the lymph nodes.
- The lymph nodes in the neck may be involved in carcinoma at various sites. The primary growth may lie in the thyroid, larynx, base of the tongue, laryngopharynx or paranasal sinuses. Rarely secondaries from carcinoma of the breast, the bronchi, the stomach or testis can also reach these nodes.
- Block dissection of the neck is sometimes done for removal of cervical lymph nodes along with some adjacent structures in malignancies to arrest the metastasis. Usually, the submandibular gland and a segment of the internal jugular

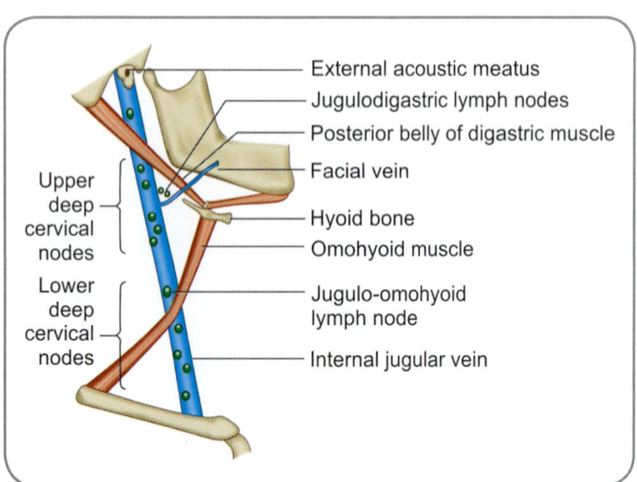

Fig. 1.3: Scheme to show the deep cervical lymph nodes

contd...

✎ Clinical Correlation *contd...*

vein are also removed in block dissection. Removal of the vein on one side is compensated by drainage through the vein of the other side. However, if bilateral removal is required, an interval of a few weeks is given between operations on the two sides to allow collateral venous channels to open up.

❑ In block dissection, special care is taken not to injure the carotid arteries, the vagus nerve, spinal accessory nerve, the mandibular branch of the facial nerve, phrenic nerve,hypoglosssal nerve and the cervical part of sympathetic trunk. However, sometimes the vagus, spinal accessory and hypoglossal nerves may have to be removed depending on the extent of the malignancy. Nerves lying deep to the prevertebral fascia namely the cervical plexus, brachial plexus and their branches usually remain intact. If necessary, the sternocleidomastoid muscle is divided for better access.

❑ Various other tumours may also be seen in the neck.
  ○ In infancy, a swelling of the sternocleidomastoid may be seen which later leads to torticollis.
  ○ Tumours may form in remnants of the thyroglossal duct or in the carotid body.

❑ Enlarged lymph nodes (submental or suprasternal nodes) may cause midline swellings which forms a differential diagnosis for other midline swellings like thyroglossal cysts, enlargements of thyroid gland and carcinoma of the larynx.

Multiple Choice Questions

1. The lingual nodes of the inner outlying group are related to:
   a. Mylohyoid
   b. Hyoglossus
   c. Genioglossus
   d. Styloglossus
2. The skin over the angle of mandible is supplied by the:
   a. Great auricular nerve
   b. Lesser occipital nerve
   c. Greater occipital nerve
   d. Third occipital nerve
3. The bony prominence not found in infants is:
   a. External occipital protuberance
   b. Parietal tuberosity
   c. Frontal tuber
   d. Mastoid process
4. The medial boundary of the lesser supraclavicular fossa is:
   a. Clavicular head of sternocleidomastoid
   b. Sternal head of sternocleidomastoid
   c. Anterior margin of trapezius
   d. Anterior midline of neck
5. The muscular triangle is a component of:
   a. Anterior cervical region
   b. Posterior cervical region
   c. Lateral cervical region
   d. Suboccipital region

ANSWERS

1. b   2. a   3. d   4. b   5. a

Clinical Problem-solving

Case Study 1: A 67-year-old woman complained of altered sensations over the dorsum of her nose.
❑ Which nerve is affected?
❑ What is the disposition of the dermatomal innervation of the face?
❑ Can you relate this disposition to developmental factors?

Case Study 2: The jugulodigastric lymph nodes are enlarged in an 11-year-old boy.
❑ Which organ/structure is likely to be involved/affected?
❑ Where do the efferents of this group of nodes go to?

(For solutions see Appendix).

Chapter 2

Bones and Joints of Head and Neck

Frequently Asked Questions

❏ Write notes on: (a) Norma basalis, (b) Mandible, (c) Occiput bone, (d) Atypical cervical vertebrae.
❏ Write in detail about the mandible. Give a neat diagram.
❏ Explain very briefly the following terms: (a) Jugum sphenoidale, (b) Sella turcica, (c) Basiocciput, (d) Mastoid.
❏ Write notes on: (a) Hyoid bone, (b) Hypophyseal fossa, (c) Foramen ovale, (d) Foramen lacerum.
❏ Write notes on: (a) Jugular foramen, (b) Foramen magnum, (c) Superior orbital fissure, (d) Cribriform plate.
❏ Write notes on: (a) Crista galli, (b) Pterygoid processes, (c) Vidian canal, (d) Pterion, (e) Fontanelles.

BONES OF HEAD AND NECK

The bones of the head and neck comprise the **skull**, **mandible**, **hyoid** and the **cervical vertebrae**.

SKULL

Whole Skull

Skull forms the skeleton of the head. It forms a protective frame for the brain, organs of special senses and the upper parts of the aerodigestive system. It also gives attachments to several muscles. The skull consists of 28 individual bones, many of which are paired. In general, bones in the median plane are single. The individual skull bones are of varying shapes and sizes. Bones which are covered by muscles tend to be thinner. Flat skull bones have two plates of compact bone—**inner** and **outer plates**. Between the two plates is a narrow layer of cancellous bone that houses bone marrow. The layer of cancellous bone is called **diploe** (Greek.diplous=double).

The bone forming the lower jaw is called the **mandible**. Other bones of the skull are firmly united to each other at joints called **sutures**. These bones collectively form the **cranium** (Cranium=skull minus mandible; Greek. kranion=head).

The parts of the cranium are:

❏ **Cranial cavity:** The large space that contains the brain;
❏ **Cranial vault:** The upper dome like part (also called the **skull cap** or **calvaria**);
❏ **Cranial base:** The portion that forms the floor of the cranial cavity;
❏ **Facial skeleton:** The portion that forms the frame for the face.

The usual way of classifying the parts of skull is to group them into two—the **neurocranium** and the **viscerocranium**. Neurocranium (otherwise called **cranium cerebrale**) is that part of the skull that houses the brain and organs of special senses. Viscerocranium (otherwise called **cranium viscerale**) is that part of the skull that is associated with face and organs of aerodigestive tract.

Added Information

The correct way of examining the skull would be to do so with the skull in the **Frankfort plane** (or **Frankfort horizontal plane**). The decision was taken at a convention of anthropologists at Frankfort in 1882 and so the name. In this plane, the lower margins of the orbital apertures and the upper margins of the external acoustic meati lie on the same horizontal plane (also called **orbitomeatal plane**). This is almost the same position in which the skull would be when an individual stands in the normal anatomical position. To achieve this in an isolated skull, place a two-inch block under the foramen magnum. The chin should rest on the table. The elevation of the posterior part of the skull takes it to the required plane.

Superior View (Skull seen from Above) (Fig. 2.1)

Other names: Norma verticalis, norma superior.

❏ The contour is modified ovoid and the greatest width is near the occipital end. Four bones are seen separated by three prominent sutures.
❏ The bone forming the anterior part of the vault is the **frontal bone**. The greater part of the roof and side walls

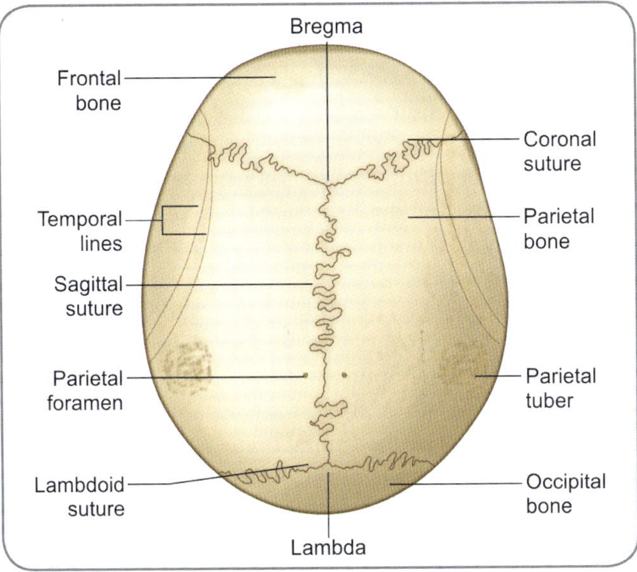

Fig. 2.1: Skull viewed from above

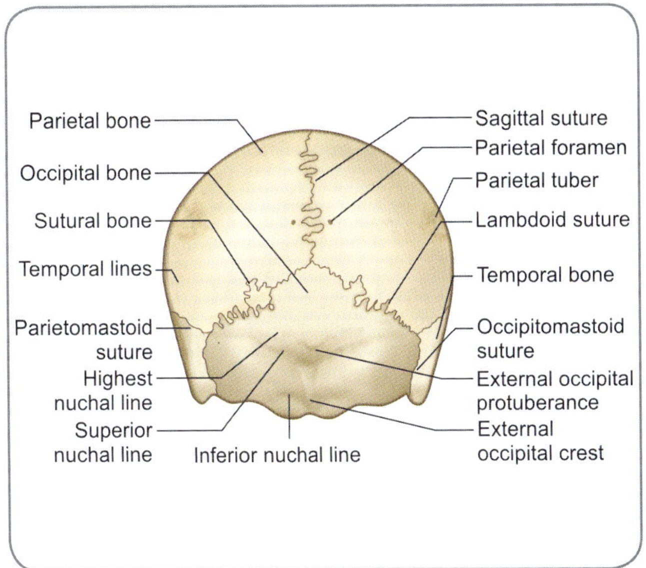

Fig. 2.2: Features seen on the skull when viewed from behind

of the cranial cavity are formed by the right and left *parietal bones*.

- ○ The two parietal bones meet in the midline at the *sagittal suture*.
- ○ Their anterior margins join the frontal bone at the *coronal suture* that runs transversely across the vault.
- ❑ The posterior part of the vault is formed by the *occipital bone*. The occipital bone is joined to the parietal bone at the *lambdoid suture*. It is so called because it is shaped like the Greek letter 'lambda' (that is like an inverted 'Y').
- ❑ The point where the coronal and sagittal sutures meet is called the *bregma* (Greek.bregma=forepart), while the point where the sagittal suture meets the lambdoid suture is called the *lambda*. In the foetal skull (and for a few months after birth), there are gaps in the bones of the skull at bregma and lambda. Membranes fill the gaps. These gaps are called the *anterior* and *posterior fontanelles* (Latin.fontaine=fountain) respectively.
- ❑ One area of the parietal bone is more convex than the rest of the bone. This area is called the *parietal tuber* (or *parietal eminence*). The convexities of both sides appear prominent.

Added Information

- ❑ When the skull is viewed from above, the face (or some part of it except the nose) is usually not seen. If any part of the face, especially the lower jaw, is seen in this view, such a skull is termed the *prognathic* (Greek.gnathion=jaw) skull.
- ❑ The zygomatic arches are not seen in many races. In some, they may be seen. Skulls where the zygomatic arches are hidden (from the superior view) are called *cryptozygous skulls*.

contd...

Added Information *contd...*

- ❑ Cephalic index is calculated by the measurements of the superior view. The proportion of the maximum width to maximum length of the norma verticalis gives this index. When the width is less than 75% of the length, it is 'long head'—dolicocephaly. When the width is more than 80% of the length, it is 'broad head'—brachycephaly. When the width is between 75 to 80% of the length, it is 'mid head'—mesaticephaly.

Posterior View (Skull seen from behind) (Fig. 2.2)

Other names: Norma occipitalis, norma posterior.

- ❑ The occipital bone (Latin.occipitis=back of head) is the most prominently seen bone in this view. Small parts of the parietal bones are seen superolaterally and those of the temporal bones inferolaterally. Near the middle of the occipital bone, a median projection called the *external occipital protuberance* is present. Extending laterally from the protuberance, on either side, is a curved ridge called the *superior nuchal line* (French. nuque=nape of neck). This line runs to a point above the mastoid process. Extending downwards (and forwards) from the protuberance is a median ridge called the *external occipital crest* (also called the *median nuchal line*). Extending laterally from the crest is the *inferior nuchal line*, which runs parallel to but below the superior nuchal line. A little above the superior nuchal lines (and parallel to them), the *highest nuchal lines* (or supreme nuchal lines) may sometimes be noted.
- ❑ The parietal bones articulate with the occipital bone at the lambdoid suture. The lambdoid suture meets the parietomastoid and the occipitomastoid sutures on the lateral aspect.

Frontal/Anterior View (Skull seen from front) (Fig. 2.3)

Other names: Norma frontalis, norma anterior, norma facialis.

> **Foramina seen in norma frontalis and structures passing through:**
> - Supraorbital notch (and foramen)—supraorbital vessels and nerve
> - Orbital opening
> - Anterior nasal aperture
> - Infraorbital foramen—infraorbital vessels and nerve
> - Zygomaticofacial foramen—zygomaticofacial vessels and nerve

- The region of the forehead is formed by the anterior part of the frontal bone. A small part of the parietal bone can be seen lateral to the frontal bone. Just below the frontal bone (on either side of the midline), the large openings of the orbit (***aditus orbitae***) are seen. Lateral to the orbit a part of the temporal bone and the zygomatic bone are seen. Inferior to the orbit is the maxilla (upper jaw) bearing the upper teeth. At the midline is the nasal aperture (***piriform aperture***) that leads into the nasal cavity. Just above the nasal aperture are the right and left nasal bones.

- On either side a little above the orbit, the frontal bone is more convex than elsewhere. This area is the ***frontal eminence*** (also called the ***frontal tuber*** or ***frontal tuberosity***). Superomedial to the orbit, a rounded prominence is seen; it is well marked in males; it is the ***superciliary arch***. There is a small smooth elevation in the midline (also the point where the two superciliary arches meet); it is called the ***glabella*** (Latin.glaber=smooth). Sometimes, the remnants of a midline suture may be found at the glabella. If present, the midline suture is called the ***interfrontal*** or ***metopic suture*** (Greek.metopon=forehead).

- The upper margin of the orbit (supraorbital margin) is formed entirely by the frontal bone. Near its medial end the margin shows the ***supraorbital notch***. Medial to it, there is a smaller ***frontal notch*** (or foramen). The lateral margin of the orbit is formed by the ***zygomatic process of the frontal bone*** above; and by the ***frontal process of the zygomatic bone*** below. The medial margin of the orbit is formed by the ***nasal process of the frontal bone*** above; and by the ***frontal process of the maxilla*** inferiorly. The inferior margin of the orbit (infraorbital margin) is formed by the zygomatic bone that is joined by the ***zygomatic process of the maxilla***.

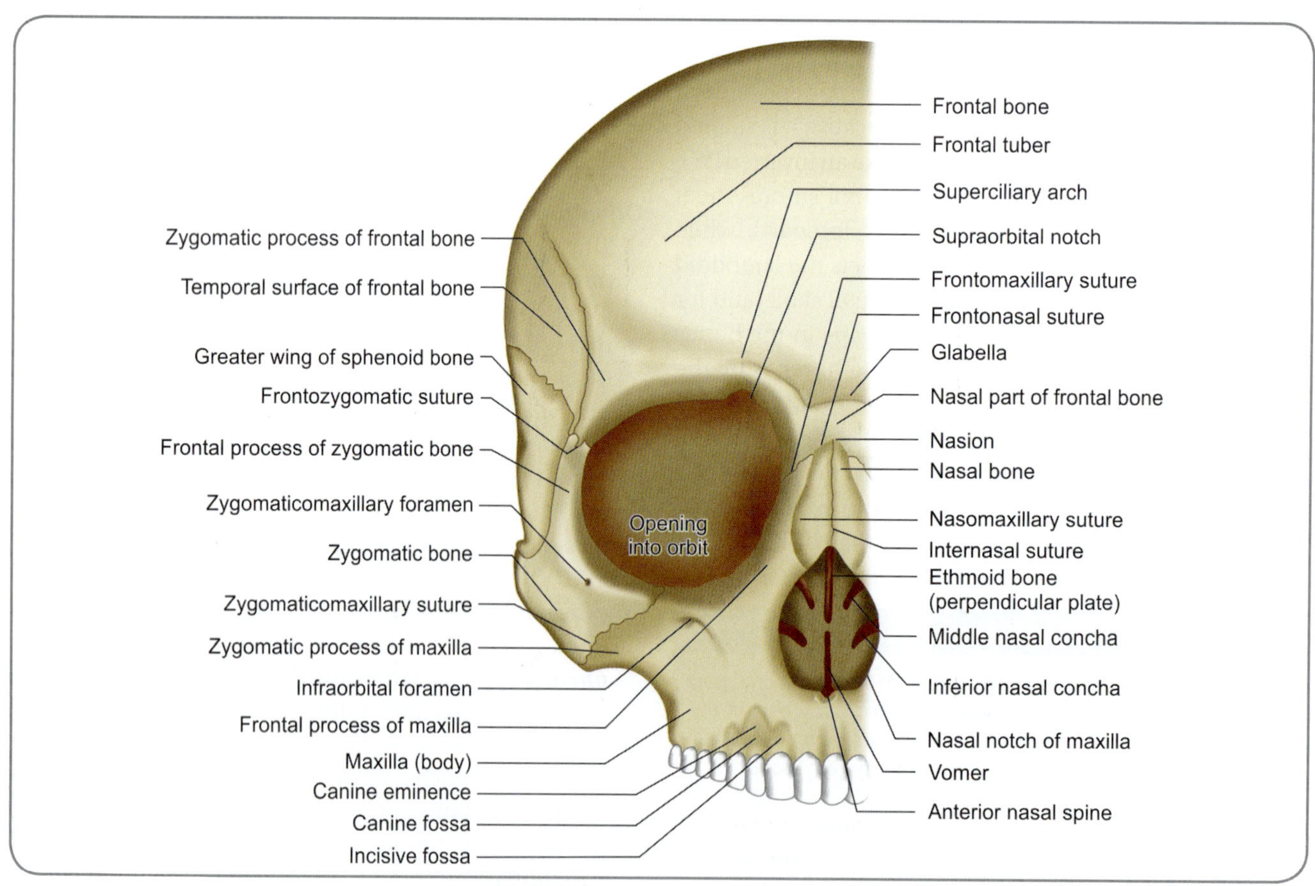

Fig. 2.3: Skull as seen from the front

Development

❑ Developmentally, the norma frontalis is bisected into the right and the left halves. The bisection is seen at birth. The parietal, nasal and maxillary bones remain separate and paired. The two halves of the mandible fuse by the second year of life (symphysis menti). The two halves of the frontal bone fuse by the second year (interfrontal or metopic suture). The interparietal (sagittal) suture usually tends to fuse by the fourth decade of life.

❑ The bones of the skull ossify in membrane. Ossification first commences in the frontal and parietal bone areas at points of their greatest fullness during the second intrauterine month. These points develop into the frontal and parietal tuberosities.

❑ A little below the orbital margin, the anterior surface of the maxilla shows the *infraorbital foramen*. On the lateral surface of the zygomatic bone, a *zygomaticofacial foramen* is found.

❑ Below the glabella, the frontal bone and the two nasal bones articulate at the frontonasal suture. The (midline) suture between the two nasal bones is the internasal suture. The junction of the frontonasal and the internasal sutures is the *nasion*.

❑ The anterior nasal aperture is the prominent feature of the midface. The lower border and the lower part of the lateral border of the anterior nasal aperture is formed by the nasal notch of the maxilla. The intermaxillary suture is marked by the anterior nasal spine at the lower margin of the anterior nasal aperture. The anterior nasal aperture is piriform in shape and its boundaries are superiorly, the nasal bones, laterally and inferiorly, the maxillae. Through the nasal aperture, some bones which lie within the nasal cavity are noted. In the midline are the *ethmoid bone* and the *vomer*. These form part of the *nasal septum*. Laterally two curved plates, the *middle and inferior nasal conchae* projecting into the nasal cavity are seen.

❑ Each maxilla bears eight *teeth*. Beginning from the midline these are the two incisors, one canine, two premolars and three molars. The part of the maxilla that bears teeth is called the *alveolar process*.

Added Information

The three foramina—*supraorbital*, *infraorbital* and *mental* foramina are in almost the same vertical line that passes between the two premolar teeth.

Lateral View (Skull seen from Lateral Side) (Fig. 2.4)

Other names: Norma lateralis, norma temporalis.

❑ The frontal bone forms the region of the forehead. The parietal bone is seen forming the vault behind the frontal bone. The occipital bone is at the posterior end. The maxilla forms the upper jaw and bears the upper teeth. A fan like portion of the temporal bone is seen below the parietal bone. In front of the temporal bone, the greater wing of sphenoid can be made out. Below the orbital opening and above the upper jaw, the quadrangular zygomatic bone can be seen. This prominent bone makes up most of the cheek's elevation.

❑ The temporal and infratemporal fossae are the most marked features of the lateral view. Temporal fossa is related to the temple of the head (Latin.tempus=time; named after the fact that greying of hair first occurs at this place and indicates passage of time). It is defined by the temporal lines.

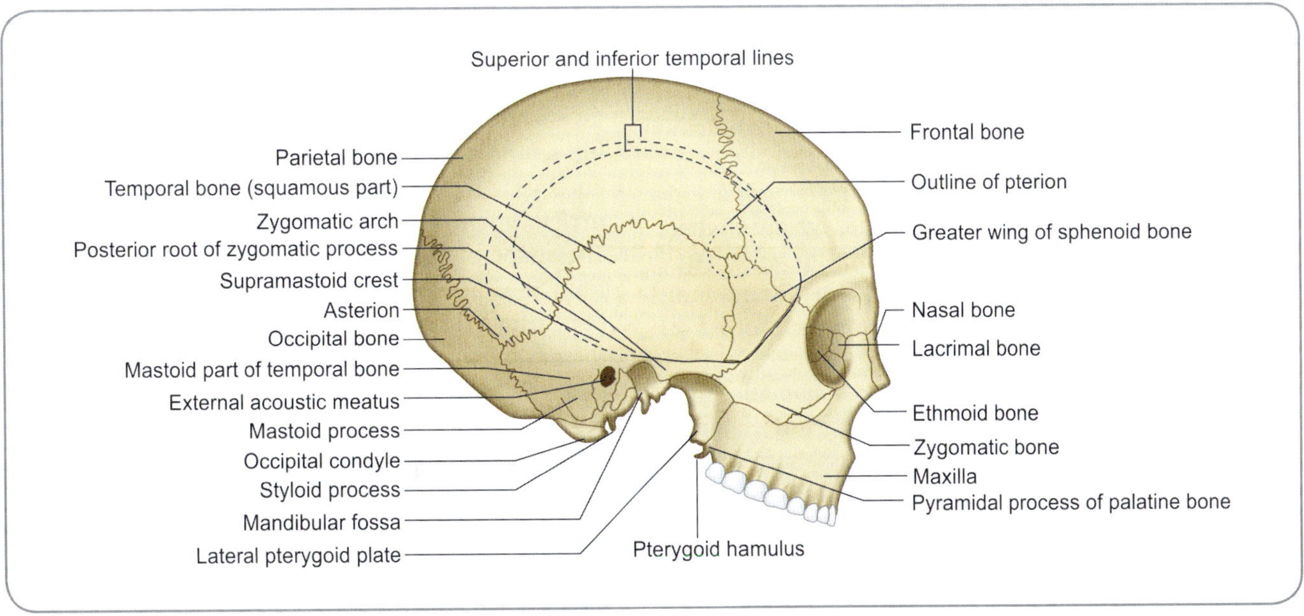

Fig. 2.4: Skull as seen from lateral side

- Running across the frontal, parietal and temporal bones are the two C-shaped *temporal lines*. Anteriorly, there is only one line, but over the parietal bone, two lines, namely the superior and inferior lines can be distinguished. At the anterior end, the unified (single) temporal line becomes continuous with the sharp lateral edge of the zygomatic process of the frontal bone. As they arch across, the two lines are prominently seen. The *superior temporal line*, however, fades away over the posterior part of the parietal bone. The *inferior temporal line* becomes more distinct as it curves down the squamous part of temporal bone, forms the supramastoid crest at the base of the mastoid process and runs forward to join the zygomatic arch.
- The *temporal fossa* is actually the area enclosed by the temporal line(s). Its floor is formed by:
  - Squamous part of temporal bone
  - Parietal bone
  - Frontal bone and
  - Greater wing of sphenoid.
- These four bones meet in the shape of 'H' within a small area. This junctional area is called the *pterion* and forms an important landmark on the side of skull. It lies over two important intracranial structures, the anterior branch of the middle meningeal artery and the lateral cerebral fissure. The lateral cerebral fissure is otherwise called the *Sylvian fissure*, for which reason the pterion is also called the *Sylvian point*. Pterion corresponds to the anteroinferior or sphenoidal fontanelle of the foetal and neonatal skull; this fontanelle closes about three months after birth.
- The parietal and the temporal bones articulate at the curved suture called the *squamosal suture*. The anteroinferior extension of the squamosal suture is the sphenosquamosal suture where the greater wing of sphenoid and squamous part of temporal bone articulate with each other.
- The *zygomatic arch* is a bar of bone lying horizontally over the lateral aspect of the skull. There is a gap between it and the floor of the temporal fossa. The temporal and the infratemporal fossae communicate with each other through this gap. The posterior part of the zygomatic arch is formed by the zygomatic process of the temporal bone and the anterior part by the temporal process of the zygomatic bone.
- Just below the posterior end of the zygomatic arch is a large oval aperture. This is the *external acoustic meatus*. It leads into the ear. The meatus is surrounded by a plate of bone with an irregular surface. This plate belongs to the tympanic part of the temporal bone.
- Just behind the external acoustic meatus is a thick downward projection called the *mastoid process* (Greek.masto=nipple, eidos=resembling; named after the resemblance of the process to a nipple). It forms the mastoid part of the temporal bone and is also called the *temporal apophysis* or the *mastoid bone*. A little below the external acoustic meatus, but a little internal is a pin-like process directed downwards and forwards. This is the *styloid process* (Latin.stylus=pen; indicating a pointed structure), which is also a part of the temporal bone. The base of the styloid process is sheathed by the tympanic plate of the temporal bone but the length of the process is very variable. Running medially into the base of the skull (seen from below), yet another part of the temporal bone is seen. This is called the *petrous part*, as it is stone like (Latin.petrosus=rock). The greater part of the ear lies within the petrous part of the temporal bone (it can be noted that the temporal bone has a squamous part, a petromastoid part, a tympanic part and the styloid process).

Foramina seen in the norma lateralis and structures passing through:
- External acoustic meatus
- Zygomaticotemporal foramen—zygomaticotemporal vessels and nerve
- Mastoid foramen—emissary vein from sigmoid sinus
- Parietal foramen—emissary vein from superior sagittal sinus
- Pterygomaxillary fissure—pterygopalatine fossa and infratemporal fossa communicate through this.

- The mastoid process articulates superiorly with the posteroinferior angle of the parietal bone at the parietomastoid suture and posteriorly with the occipital bone at the occipitomastoid suture. The lateral end of the lambdoid suture meets these two sutures. The junctional point of the three sutures is called the *asterion* (Greek. asterios=starry). Asterion is the place of the posterolateral or the mastoidal fontanelle of the neonatal skull; this fontanelle closes by the second year of life.
- A little in front of the external acoustic meatus is a depression called the *mandibular fossa* (or the *glenoid fossa*), into which the head of the mandible fits, to form the *temporomandibular joint*. The fossa is bounded anteriorly by a small inferior projection called the *articular eminence*.
- The shape of the *zygomatic bone* is best appreciated from the lateral side. It articulates with the frontal bone, the temporal bone and the maxilla. In addition to the lateral surface the bone also has a temporal surface directed towards the temporal fossa. The term *zygomatic arch* is usually restricted to the bony rod formed by the temporal process of the zygomatic bone and the zygomatic process of the temporal bone.
- Some parts of the *sphenoid bone* are also visible in this view. The greater part of this bone lies in the base of skull. However, the *greater wing* forms part of the floor of the temporal fossa. Another part of the sphenoid bone that is seen from the lateral side is the *pterygoid*

process, which is made up of the medial and lateral ***pterygoid plates*** (Greek.pteryx=wing). The pterygoid process comes into contact with the posterior aspect of the maxilla (because of the general resemblance of the sphenoid bone to a bird with wings, several of its parts have been given names which mean wings and wing-like; it is the same reason that the pterion has also been so named).

❑ The ***infratemporal fossa,*** another marked area seen in the lateral view, is a space that lies lateral to the pterygoid process. Its roof is formed by the infratemporal surface of the greater wing of sphenoid and its anterior wall by the posterior surface of the maxilla. The fossa communicates with the temporal fossa through the gap between the zygomatic arch and the side of the skull. A gap can be seen between the lateral pterygoid plate and the posterior aspect of maxilla. This is the ***pterygomaxillary fissure***.

Added Information

❑ On a cursory glance, three contour lines can be defined on the norma lateralis. The three lines are concentric—outermost line is the outline of the skull in this view, middle is the temporal line and the innermost is the line of squamosal suture.
❑ The cerebrum lies above the level of the inion (which can be seen in this view as the most prominent point of the external occipital protuberance). The cerebellum lies below the level of the inion.

Inferior View (Skull seen from below) (Fig. 2.5)

Other names: Norma basalis, norma basilaris, norma ventralis, norma inferior, base of skull, basis crania externa, norma basalis externa.

❑ Bones seen in an inferior view of the skull are the maxilla (pink), sphenoid (purple), temporal (green) and occipital (blue), zygomatic bone, vomer and palatine bones.

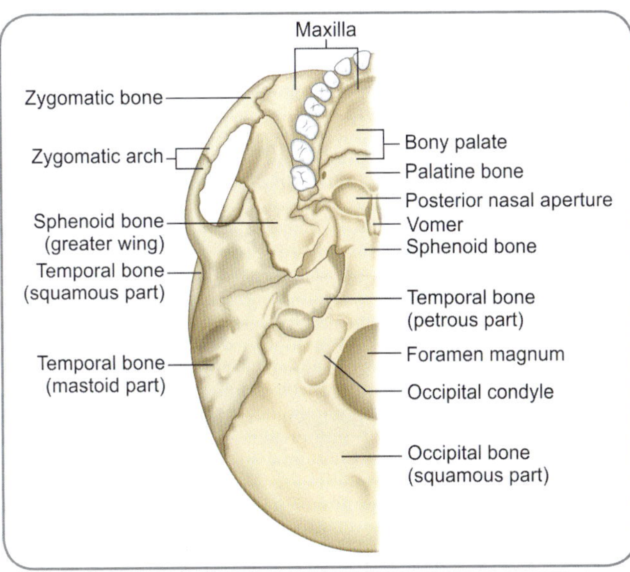

Fig. 2.5: Skull as seen from below

❑ The inferior surface of the skull can be subdivided into the anterior, middle and posterior parts. The posterior limit of the anterior part is the posterior border of the hard palate. The posterior limit of the middle part is the anterior margin of the foramen magnum.

Anterior Part of Cranial Base

The anterior part consists of the hard palate and the upper teeth. It lies at a lower level than the rest of the norma basalis. The alveolar processes of the right and the left maxillae bear the upper teeth and form the ***superior alveolar arch***.

❑ Within the superior alveolar arch is the hard palate (or the bony palate). Anterior part of the bony palate is formed by the palatine processes of maxillae and the posterior part by the horizontal plates of the right and left ***palatine bones***. The articulation between the two palatine processes is the intermaxillary suture and that between the two horizontal plates is the interpalatine suture. The articulation between the palatine process of one side and the horizontal plate of the palatine bone of the same side is the palatomaxillary suture.

❑ The alveolar process of the maxilla projects downwards and provides attachment to the upper teeth. The posterior end of each alveolar process forms a backward projection called the ***maxillary tuberosity***. The part of the alveolar process bearing the incisor teeth and including the adjoining part of the palate is called the ***pre-maxilla***. The incisive fossa lies behind the central incisor teeth; lateral incisive foramina open into the lateral walls of the incisive fossa. The lateral incisive foramina lead to the incisive canals which pass to the nasal cavity. If the median incisive foramina are present, they open into the anterior and posterior walls of the incisive fossa. Lateral to the alveolar arch, the inferior aspect of the ***zygomatic process*** of the maxilla is seen as it passes laterally to meet the zygomatic bone.

❑ The posterior borders of the horizontal plates of the palatine bones are free and form the posterior margin of the hard palate. A little in front of the posterior border is seen a curved ridge called the ***palatine crest***. The part of the palate formed by the palatine bone shows the ***greater*** and ***lesser palatine foramina***. The greater palatine foramen lies on the most lateral part of the horizontal plate, just medial to the last molar tooth. It is the lower opening of the canal of the same name. The lesser palatine foramina, usually two, are present just behind the greater palatine foramen.

❑ Just above the posterior margin of the hard palate are the two ***posterior nasal apertures***. Each aperture (also called ***choana*** or ***postnaris*** or ***posterior naris***; Greek. choane=funnel) is bounded, below, by the posterior edge of the horizontal plate of the palatine bone;

laterally, by the perpendicular plate of the palatine bone; and medially, by the vomer.

Foramina in anterior norma basalis and structures passing through:
- Incisive fossa—nasopalatine nerve and the termination of the greater palatine vessels
- Median incisive foramina (if present)—left nasopalatine nerve through the anterior and right nasopalatine nerve through the posterior
- Greater palatine foramen—greater palatine nerve and vessels
- Lesser palatine foramina—lesser palatine nerves and vessels
- Posterior nasal apertures
- Palatovaginal canal—pharyngeal branch of pterygopalatine ganglion and pharyngeal branch of maxillary artery
- Vomerovaginal canal (if present)—pharyngeal branch of the maxillary artery to the anterior part of the palatovaginal canal

---

- The ***vomer*** is a flat plate of bone that forms part of the nasal septum. The superior border of vomer is applied to the inferior aspect of the sphenoid. At this point, the vomer gives out an extension on either side that is also closely applied to the inferior aspect of sphenoid. This wing-like expansion of the vomer is called the **ala of vomer**. A similar expansion projects medially from the medial pterygoid plate towards the ala of vomer. This expansion is called the **vaginal process**. The vomerine ala and the vaginal process may touch each other or overlap one another.
- Looking through the posterior nasal aperture, certain additional intricate details may be noticed. Overlapping the vaginal process anteriorly is an expansion that projects medially from the perpendicular plate of the palatine bone. This expansion is called the **sphenoidal process**. A slender anteroposterior groove runs on the inferior aspect of the vaginal process; a reciprocal groove runs on the superior aspect of the sphenoidal process. The two grooves together form a canal called the ***palatovaginal canal***. If the vaginal process overlaps the vomerine ala, a canal may be formed between the two of them and is called the ***vomerovaginal canal***.

Middle Part of Cranial Base

- The middle part of the cranial base consists of the sphenoid, temporal and occipital bones. Immediately behind the vomer, the sphenoid which is an unpaired bone is seen. It has a median part, the body. On either side of the body, is the greater wing (that is seen partly on the base of skull and partly on the lateral wall).
- Posteriorly, the body of sphenoid is continuous with the basilar part of the occipital bone. The part of the body of sphenoid seen in the norma basalis is sometimes referred to as the basisphenoid. The basilar part of the occipital bone is the basiocciput. The basisphenoid and the basiocciput meet at the spheno-occipital synchondrosis. This is a primary cartilaginous joint in the young and growing skull. It contributes largely to the anteroposterior growth of the skull and ossifies by 25 to 30 years of age.

- The ***pterygoid process*** projects down from the junction of the body and the greater wing of sphenoid and consists of the medial and lateral pterygoid plates. The two plates are fused anteriorly except inferiorly where the pyramidal process of the palatine bone intervenes between the two. The two plates are separated posteriorly by the ***pterygoid fossa***. On its lateral aspect, the lateral pterygoid plate is separated from the maxilla by the pterygomaxillary fissure (which was already seen in the norma lateralis). The posterior border of the medial pterygoid plate is sharp. Superiorly, the plate expands to enclose the ***scaphoid fossa***. Inferiorly, the plate has a rounded projection called the ***pterygoid hamulus***.
- On the lateral aspect, the posterior surface of the maxilla is separated from the greater wing of the sphenoid by the ***inferior orbital fissure***. The greater wing of sphenoid, squamous and tympanic parts of temporal bones constitute the lateral aspect of the midcranial base. The petrous part of the temporal bone wedges itself between the sphenoid and the occipital bones. The mandibular fossa and the articular eminence (already noted in the norma lateralis) are clearly seen in this view. The zygomatic arch is also seen from below.
- The petrous part of the temporal bone is shaped like a pyramid and therefore, is called the ***petrous process***. Each petrous process meets the basiocciput at the petro-occipital suture. This suture is incomplete posteriorly and the resultant gap forms the ***jugular foramen***. The petrosphenoidal suture is formed between the anterior margin of the petrous process and the greater wing of sphenoid. The ***groove for the pharyngotympanic tube*** is also seen at this junction. The apex of the petrous process does not touch the spheno-occipital junction and the gap so produced forms the foramen lacerum. Though called a ***foramen***, the opening is actually a short vertical canal of about 1 cm length.
- The ***foramen lacerum*** is bounded:
  - Anteriorly by the body of sphenoid, root of pterygoid process and greater wing of sphenoid;
  - Medially by the basilar part of occipital bone and
  - Posterolaterally by the apex of petrous process.

Foramina and fissures in the middle norma basalis and structures passing through:
- Inferior orbital fissure—infraorbital and zygomatic branches of maxillary nerve and accompanying vessels.
- Pterygomaxillary fissure—leads into the pterygopalatine fossa.
- Foramen lacerum—no major structure through and through; internal carotid artery in the upper part; meningeal branches of the ascending pharyngeal artery and emissary veins from the cavernous sinus.

---

contd...

contd...

> ❑ Foramen ovale—mandibular nerve, lesser petrosal nerve, accessory meningeal branch of maxillary artery, emissary vein from cavernous sinus
> ❑ Foramen spinosum—middle meningeal artery and meningeal branch of mandibular nerve
> ❑ Sphenoidal emissary foramen of Vesalius—emissary vein
> ❑ Pterygoid canal—nerve of pterygoid canal and accompanying vessels
> ❑ Petrosphenoidal groove—pharyngotympanic tube
> ❑ Petrotympanic fissure—chorda tympani branch of facial nerve.

❑ Lying lateral to the foramen lacerum, but in the greater wing of sphenoid are the foramen ovale and the foramen spinosum. The foramen ovale is near the root of the lateral pterygoid plate. Posterolateral to the foramen ovale is the smaller and rounded foramen spinosum and posterolateral to the latter is the spine of sphenoid. If the groove for the pharyngotympanic tube already noted in the petrosphenoidal junction is traced posterolaterally, it will be seen that the spine of sphenoid also forms part of its anterolateral wall. The groove, when traced further posterolaterally, leads into the bony part of the pharyngotympanic tube that is located within the petrous temporal bone.

❑ Yet another smaller foramen called the ***sphenoidal emissary foramen of Vesalius*** may occasionally be seen between the foramen ovale and the scaphoid fossa.

❑ A large rounded foramen is found posterolateral to the foramen lacerum on the inferior surface of the petrous temporal bone. This is the inferior opening of the carotid canal. If a malleable probe is inserted into the inferior opening of the carotid canal, it can be pushed through the carotid canal that ascends up initially, then turns horizontal and runs anteromedially to reach the posterior wall of the foramen lacerum.

❑ Running forward from the anterior margin of the foramen lacerum above and between the two pterygoid plates is the pterygoid canal (or Vidian canal). The pterygoid canal leads to the pterygopalatine fossa.

❑ A midline tubercle is seen on the inferior aspect of the basiocciput; it is called the ***pharyngeal tubercle***.

Posterior Part of Cranial Base

❑ The posterior part of the cranial base consists of the occipital and temporal bones.

❑ Immediately behind the basilar part of the occipital bone is a large midline foramen called the ***foramen magnum***. The cranial cavity communicates with the vertebral canal through this foramen. Posterior to the foramen magnum, the occipital bone forms a large part of the base of skull.

> ***Foramina in the posterior norma basalis and structures passing through:***
> ❑ Foramen magnum—medulla vertebral arteries, spinal accessory nerve
> ❑ Hypoglossal canal—hypoglossal nerve, meningeal branch of ascending pharyngeal artery, emissary vein from basilar plexus
> ❑ Posterior condylar canal—emissary vein from sigmoid sinus
> ❑ Jugular foramen—inferior petrosal sinus, Glossopharyngeal, vagus and accessory nerves, Internal jugular vein
> ❑ Mastoid canaliculus—auricular branch of vagus nerve.
> ❑ Canaliculus for tympanic nerve—tympanic branch of the glossopharyngeal nerve to the middle ear
> ❑ Stylomastoid foramen—facial nerve and stylomastoid artery
> ❑ Mastoid foramen—emissary vein from sigmoid sinus.

❑ A large bony mass overlaps the anterolateral aspect of the foramen magnum on either side of the midline. This is the ***occipital condyle***. The condyle projects down to articulate with the superior articular facet of the lateral mass of atlas. Traversing the condyle and running laterally and forwards is the ***hypoglossal canal*** (also called the ***anterior condylar canal***). Behind the condyle is a depression called the ***condylar fossa***. An opening may sometimes be present in the condylar fossa. If present it is called the (***posterior***) ***condylar canal***.

❑ The posteromedial border of the petrous process articulates with the occipital bone at the petro-occipital suture and with the mastoid process at the petromastoid suture. At the posterior end of the petro-occipital suture is the ***jugular foramen***. The jugular foramen is bounded anteriorly by the ***jugular fossa*** of the petrous temporal bone and posteriorly by the ***jugular process*** of the occipital bone. A small opening called the ***mastoid canaliculus*** is seen on the lateral wall of the jugular fossa. Yet another tiny opening is seen on the ridge between the carotid canal and the jugular fossa. This opening is the ***canaliculus for the tympanic nerve***.

❑ The ***stylomastoid foramen*** is seen between the styloid and mastoid processes of the temporal bone. A groove called the ***mastoid notch*** is seen on the medial aspect of the mastoid process. The mastoid and the occipital bone articulate at the occipitomastoid suture. A small ***mastoid foramen*** may be present near the occipitomastoid suture. The squamous part of the occipital bone, the external occipital protuberance and the nuchal lines are also seen in this view.

Cranial Fossae

When the top of the skull (skull cap) is removed by a transverse cut, the floor of the cranial cavity can be seen. It is divided into three depressions called the ***cranial fossae—anterior***, ***middle*** and ***posterior***.

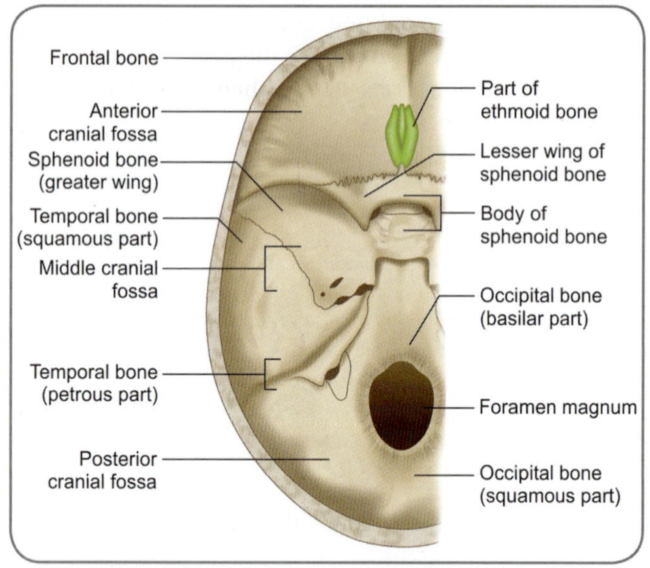

Fig. 2.6: Cranial fossae as seen from above

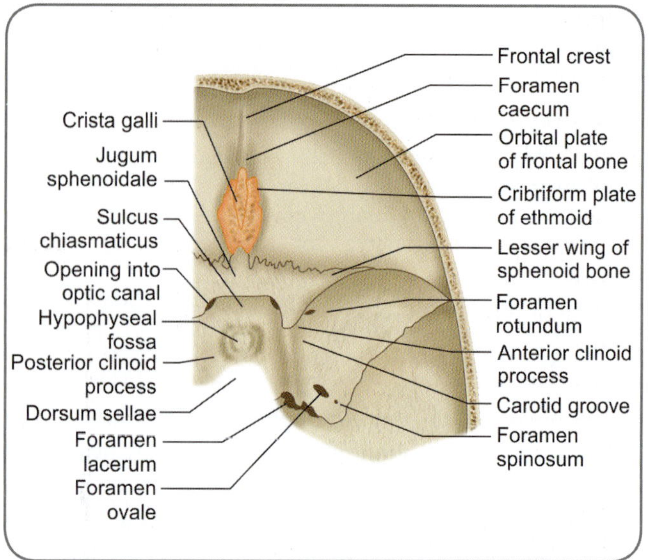

Fig. 2.7: Parts of the anterior and middle cranial fossae

Anterior Cranial Fossa

❏ The floor of the anterior cranial fossa (Fig. 2.6) is formed mainly by the orbital plates (right and left) of the frontal bone. Anteriorly, the right and left halves of the frontal bone are separated by a median projection called the *frontal crest*. Just behind the crest, there is a depression called the *foramen caecum*. Between the right and left orbital plates of the frontal bone, there is a notch occupied by the *cribriform plate of the ethmoid bone*. This plate has numerous foramina and bears a median vertical projection called the *crista galli* that lies immediately behind the foramen caecum.

❏ The posterior part of the floor of the anterior cranial fossa is formed by the sphenoid bone. This part which is smooth is called the *jugum sphenoidale*. The superior surface of the body of sphenoid and the superior surfaces of the lesser wings form the jugum sphenoidale. The lesser wing also forms the sharp posterior edge of the floor of the anterior cranial fossa. The medial edge of each lesser wing projects backwards as the *anterior clinoid process*.

Middle Cranial Fossa

❏ The middle cranial fossa (Figs 2.7 and 2.8) has a raised median part formed by the body of the sphenoid bone and two large deep hollow areas on either side.

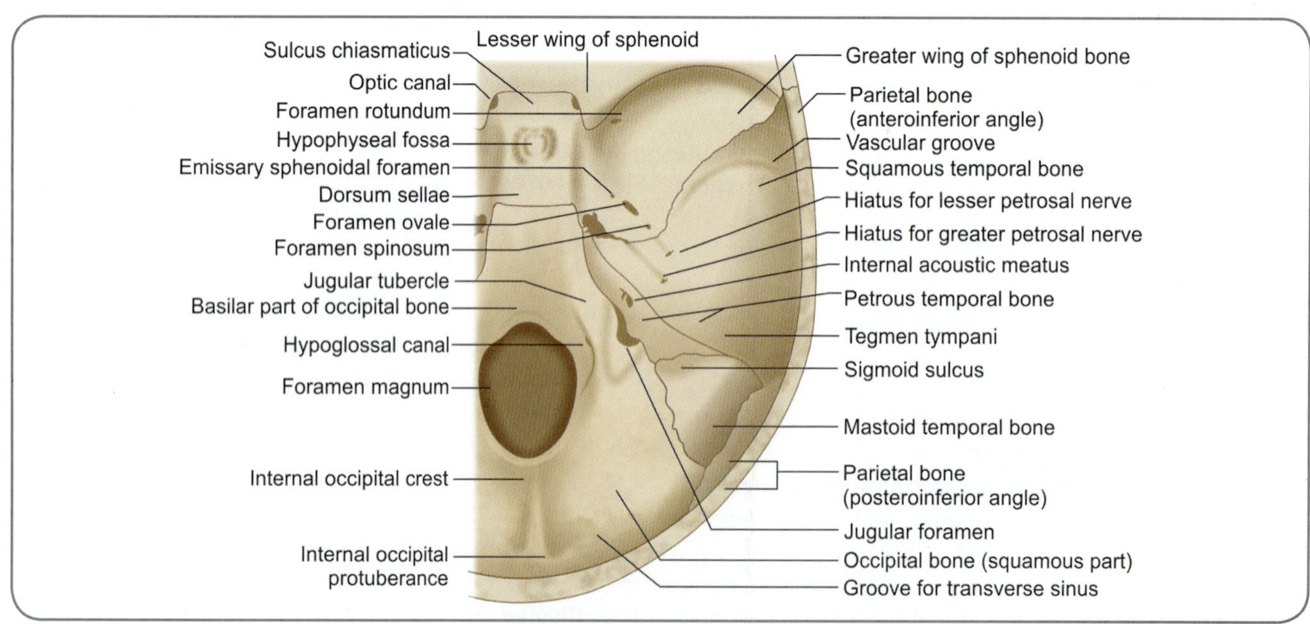

Fig. 2.8: Floor of middle and posterior cranial fossae

Immediately behind the jugum sphenoidale, the body of sphenoid is crossed by a transverse shallow groove that connects the two optic canals. It is called the ***sulcus chiasmaticus*** (even though the optic chiasma does not lie exactly over the sulcus). Behind the sulcus the superior surface of the body of sphenoid shows a median elevation, the ***tuberculum sellae***; and behind the tuberculum, there is a depression called the ***hypophyseal fossa***. Posterior to the hypophyseal fossa is a vertical plate of bone called the ***dorsum sellae***. The deep hollow bounded anteriorly by the tuberculum sellae and posteriorly by the dorsum sellae is the ***sella turcica***.

❑ The superolateral angles of the dorsum sellae project as the ***posterior clinoid processes***. On each side, the body of the sphenoid slopes downwards into the floor of the deep lateral part of the middle cranial fossa. At this junction of the body and the greater wing, is a shallow groove for the internal carotid artery, called the ***carotid groove***. The posterior extent of the carotid groove is at the foramen lacerum. Anteriorly, the groove turns upwards, medial to the anterior clinoid process.

❑ On either side, the anterior wall of the middle cranial fossa is formed by the greater and lesser wings of the sphenoid. The lesser wings are attached to the sides of the body of the sphenoid by two roots; anterior (or upper), and posterior (or lower). The optic canal passes forwards and laterally between the body of the sphenoid and the two roots of the lesser wing. The greater and lesser wings are separated by the ***superior orbital fissure*** (also called the ***foramen lacerum anterius***) that leads into the orbit.

❑ Just below the medial end of the superior orbital fissure and just lateral to the carotid groove, is the ***foramen rotundum*** (round foramen or upper sphenotic foramen). This foramen opens anteriorly into the pterygopalatine fossa.

❑ The posterior wall of the middle cranial fossa is formed, on either side, by the anterior sloping surface of the petrous temporal bone. The apex of the bone is separated from the body of the sphenoid by the ***foramen lacerum*** (or lacerated foramen or foramen lacerum medium or lower sphenotic foramen). A little above and lateral to the foramen, the surface of the petrous temporal bone shows a shallow depression called the ***trigeminal impression***. Lateral to this impression, an elevation called the ***arcuate eminence*** is seen. It is produced by the underlying anterior semicircular canal.

❑ The anterior surface of the petrous temporal bone is formed by a thin plate of bone that separates the middle cranial fossa from the cavities of the middle ear, the auditory tube and the mastoid antrum. This plate is called the ***tegmen tympani***. The floor of the deep lateral part of the middle cranial fossa is formed by the greater wing of the sphenoid, medially, and by the squamous part of the temporal bone, laterally. Near the posterior margin of the greater wing are the ***foramen ovale*** (oval foramen or the intermediate sphenotic foramen) and the ***foramen spinosum***.

❑ The lateral wall of the middle cranial fossa is formed, anteriorly, by the greater wing of sphenoid, and posteriorly, by the squamous temporal bone.

Posterior Cranial Fossa

❑ The most prominent landmark in the posterior cranial fossa (Fig. 2.8) is the ***foramen magnum*** (great foramen). Anterior to the foramen magnum the wall of the fossa is formed by the basilar part of the occipital bone that is continuous above with the posterior surface of the body of the sphenoid. The lateral margin of the basilar part is separated from the petrous temporal bone by a fissure that ends below in the ***jugular foramen*** (also called the ***foramen lacerum posterius***). Between the jugular foramen, laterally, and the anterior part of the foramen magnum, medially, there is a rounded elevation called the ***jugular tubercle***. In the interval between the jugular tubercle and the foramen magnum, there is a fossa. The ***hypoglossal canal*** opens into this fossa.

❑ When present, the ***posterior condylar canal*** opens just lateral to the jugular tubercle immediately behind the jugular foramen. The lateral part of the anterior wall of the posterior cranial fossa is formed by the posterior surface of the petrous temporal bone. A little above the jugular foramen this surface presents the opening of the ***internal acoustic meatus***.

❑ The floor and lateral walls of the posterior cranial fossa are formed, posteriorly, by the squamous part of the occipital bone; and in the anterolateral part by the mastoid part of the temporal bone.

❑ Behind the foramen magnum is a midline ridge called the ***internal occipital crest***. Posteriorly, the crest ends in an elevation called the ***internal occipital protuberance***. Running laterally from the protuberance, in the transverse plane, is a prominent wide groove (***transverse sulcus***) in which the transverse venous sinus is lodged. The groove first lies on the occipital bone, and near its lateral (or anterior) end it crosses the posteroinferior angle of the parietal bone. It then runs downwards and medially with an S-shaped curve to reach the jugular foramen. This S-shaped part of the groove is called the ***sigmoid sulcus***. The terminal part of the groove lies on the occipital bone just behind the jugular foramen.

Added Information

❑ The middle cranial fossa is likened to a bed mounted on four bed posts. The anterior and posterior clinoid processes are the bed posts (Greek.kline-bed). The tentorium is considered the guy tent for the bed posts, for to each of them the tent is attached.

Added Information contd...

❑ A crescent of foramina is noted on either side of the median part of the middle cranial fossa. The crescent faces laterally. From before backwards, the foramina which make up the crescent are the lacrimal foramen (inconstant foramen present near the apex of the superior orbital fissure, transmitting a communicating branch to the lacrimal artery), superior orbital fissure, foramen rotundum, sphenoidal emissary foramen (called the **foramen of Vesalius**, situated anteromedial to foramen ovale and transmitting an emissary vein), foramen ovale, canaliculus innominatus (inconstant foramen transmitting the lesser petrosal nerve if present) and foramen spinosum. The foramen lacerum is not part of the crescent.

❑ Foramen lacerum is actually an artifact of the dried skull. It is short canal closed by a cartilaginous plate in life. No major structure passes through and through the canal except for an occasional emissary vein.

Age Changes in the Skull and Face

❑ In an infant and small child, the viscerocranium is much smaller when compared to the cerebrocranium. Expansion of the cerebrocranium occurs during infancy due to rapid growth of brain. Bones of the calvaria grow and expand resulting in an overall increase in the volume of the cranial box. Alongside these changes, the mandible and maxillae also grow in tune with eruption of the teeth. This causes a slight increase in the length of the viscerocranium.

❑ In the first and second decades of life, marked changes are seen. Eruption of permanent teeth and the increase in the number of teeth cause posterior growth of the maxillae and mandible. Resorption of bone in the anterior ramus-body angle and deposition in the posterior ramus-body angle result in change of the lower contour of the face. The jaw 'gets set'. Deposition of dento alveolar bone leads to increase in the height of the lower face. Therefore, the lower face undergoes dimensional changes in the anteroposterior and vertical aspects.

❑ In the second decade of life, the upper and midface also undergo changes. The paranasal sinuses, which are either small or rudimentary at birth, grow and expand rapidly at and around puberty. This causes the orbital-oral interspace to increase (due to growth of maxillary air sinuses) and the supraorbital area to expand (due to growth of frontal air sinuses). Anteroposterior increase of the cranial base also occurs. Eruption of the posterior molars (wisdom teeth) produces changes in the jaw angle. The basisphenoid-basiocciput junction remains cartilaginous and growing until mid to late 20s allowing for these changes.

❑ Cranial sutures start fusing by 30 to 40 years of age. Fusion starts in the internal aspect and proceeds outwards. Fusion can be seen on the external aspect at around mid 50s. Obliteration of sutures starts at bregma and proceeds towards the coronal suture through the sagittal suture. Lambdoid suture obliterates late.

❑ In old age, cranial bones thin out; diploe gets filled with grey gelatinous material. The maxillae and mandible become edentulous (lacking in teeth) and the alveolar borders get resorbed.

Individual Skull Bones

The individual skull bones can be classified into two groups—the cranial bones (ossa cranii) and the facial bones (ossa faciei).

Cranial bones

These are the bones of the cerebral cranium. They surround the brain and afford protection. These bones are:

Unpaired—Frontal	**Paired**—Parietal
Occipital	Temporal
Sphenoid	
Ethmoid	

$$4 + 4 = 8 \text{ bones}$$

Facial bones

These are the bones of the visceral cranium. They form the orbits and surround the mouth and nose. These bones are:

Unpaired—Vomer	**Paired**—Nasal
(Mandible)	Lacrimal
(Hyoid)	Palatine
	Inferior nasal concha
	Zygomatic bone
	maxilla

$$3 + 12 = 15 \text{ bones}$$

The term **skull** is not well defined. It may include the mandible or not. Very often, it does not include the mandible. However, mandible forms part of the facial skeleton. Hyoid is often considered a bone of the neck and not included in the list of skull bones. Again, hyoid forms an integral part of the facial skeleton. Therefore, the number of individual skull bones may vary.

It is customary to describe that the skull has 28 individual bones. The list is as follows:

Bones of the cerebrocranium	– 08
Bones of the viscerocranium	– 14 (excluding the hyoid)
Ear ossicles	3 + 3 = 06
Totalling to	28

Frontal Bone

Other names: Frontalis, frontis, metopon, os frontale.

The frontal bone (Latin.frontis=forehead) forms the forehead portion of the skull. It is unpaired and has three parts, namely the squamous part, nasal part and the orbital part.

Squamous Part

❑ The plate like squamous part forms the major portion of the bone. The important features of the squamous portion are:

○ Two frontal tubera (singular, tuber), one on either side, a little above the supraorbital margins; the bone takes a backward curve at the tubera;

○ Two superciliary arches, one on either side, just above the supraorbital margins;

- ○ The smooth glabella in the midline, where the two superciliary arches meet;
- ○ The supraorbital margin (one on either side) that forms the upper margin of the orbital opening;
- ○ The supraorbital notch (or foramen) at the junction of the medial third with the lateral two thirds of the supraorbital margin;
- ○ The large zygomatic process at the lateral end of the supraorbital margin that articulates with the zygomatic bone.
- ❑ The squamous part has an internal surface which is concave and shows a median groove for the superior sagittal sinus. The edges of the groove unite below to form the frontal crest which gives attachment to the falx cerebri. The crest ends in a small notch. This notch is completed posteriorly by the ethmoid bone to form a small opening called the *foramen caecum*. Impressions of cerebral sulci and gyri can be seen on the internal surface. Along the median groove, impressions of the arachnoid granulations can also be made out.

Nasal Part

The nasal part of the bone lies between the supraorbital margins on the inferior aspect. This part forms the bridge of nose and has a small nasal spine. It articulates with the nasal bones, maxillae and the lacrimal bones.

Orbital Part

The orbital part is in the form of two (right and left)plates. The two plates are very thin and translucent, project backward and are separated by the ethmoidal notch. The orbital surface of the orbital plate is concave and forms the roof of the orbital cavity. On its anterolateral aspect is a shallow depression called the *lacrimal fossa* (for the lacrimal gland) and on the anteromedial aspect is the trochlear fovea that gives attachment to the fibrous trochlea for the superior oblique tendon. The convex surface of the orbital plate has impressions for sulci and gyri.

Posterior Border of the Frontal Bone

This border is also called the *parietal border* because it articulates with the parietal bones forming the coronal suture. It is serrated and bevelled. Inferiorly it expands into a rough triangular surface for articulation with the greater wing of sphenoid.

Ethmoidal Notch

- ❑ This is the rectangular gap seen between the orbital plates and is occupied by the cribriform plate of ethmoid bone in the articulated skull.
- ❑ The frontal air sinuses are located within the frontal bone, one on either side of the midline, above the medial ends of the supraorbital margins. They ascend between the two laminae of the bone and are separated by a midline septum which usually is deflected to one side.

Parietal Bone

Other name: Os parietale.

- ❑ The two parietal bones (Latin.parie=wall) form the cranial roof and the sides of the skull. Each parietal bone is four sided, has four borders, four angles and two surfaces.
- ❑ The external surface is smooth and convex. It has a central prominence called the *parietal tuber*. The temporal lines can be seen running on this surface. Close to the sagittal border, a parietal foramen that transmits an emissary vein from the superior sagittal sinus and a branch of the occipital artery is sometimes seen.
- ❑ The internal surface is marked by impressions of the cerebral sulci and gyri and groove for the middle meningeal artery. The artery ascends posterosuperiorly. A groove for the superior sagittal sinus is seen at the superior (sagittal) border; this groove is complemented by a similar groove on the opposite bone.
- ❑ The superior border is the sagittal border which is dentate. It articulates with the fellow of the opposite side at the sagittal suture. The inferior border is called the *squamosal border*. It articulates with (from before backwards) the greater wing of sphenoid, squamous part of temporal and mastoid part of temporal. Most of it is bevelled externally. The anterior border is the frontal border. It articulates with the frontal bone to form the coronal suture. The *posterior border* is called the *occipital border*, *dentate* and articulates with the occipital bone to form the lambdoid suture.
- ❑ All the four angles of the parietal bone have a fontanelle each in the foetal or neonatal skull. The anterosuperior angle is the frontal angle; it is at the bregma and is the site of the anterior fontanelle. The anteroinferior angle is the sphenoidal angle; it is at the pterion and is the site of the sphenoidal fontanelle. The posterosuperior angle is the occipital angle; it is at the lambda and is the site of the posterior fontanelle. The posteroinferior angle is the mastoid angle; it is at the asterion and is the site of the mastoidal fontanelle. The internal surface of the anteroinferior angle shows the groove for the middle meningeal artery and that of the posteroinferior angle has a wide groove for the transverse sinus.

Occipital Bone

Other name: Os occipitale.

- ❑ The unpaired occipital bone (Greek.occipitus= back of head) forms the back and base of the skull. It has four parts—a squamous part, two lateral parts and a basilar part.

 **Development**

Development of Skull

The development of skull is quite complicated. However, it can be seen that some of the individual bones develop from membranous ossification and some from cartilaginous ossification. Thus there are bones formed in four categories—cartilaginous neurocranium, membranous neurocranium, cartilaginous viscerocranium and membranous viscerocranium.

1. Cartilaginous neurocranium—part of occipital, sphenoid, part of ethmoid, otic capsule and part of temporal.
2. Membranous neurocranium—frontal, parietals, part of occipital, part of temporal and part of ethmoid.
3. Cartilaginous viscerocranium:
   1st arch cartilage—mandible, malleus and incus, part of maxillae.
   2nd arch cartilage—styloid process, part of hyoid, stapes
   3rd arch cartilage—part of hyoid.
4. Membranous viscerocranium—premaxilla, part of maxillae, palatine bone, zygomatic bone, nasal, lacrimal, vomer.

Squamous Part

❑ The squamous part is the large plate like portion that forms the posterior aspect of the bone. It is convex externally and concave internally. The external surface has an external occipital protuberance. Two curved lines extend laterally from the protuberance on either side. The upper line is faint and is the supreme nuchal line. The medial part of the supreme nuchal line gives attachment to the epicranial aponeurosis and the lateral part to the occipital belly of the occipitofrontalis. The lower is the superior nuchal line. A midline crest called the ***median external occipital crest*** descends from the external occipital protuberance to the posterior margin of the foramen magnum. From the centre of the crest, the inferior nuchal line extends laterally on either side.

❑ On the internal surface, the internal occipital protuberance is seen. Sagittal and horizontal extensions from the protuberance form a cruciform impression which divides the entire surface into two superior and two inferior fossae. The superior fossae are triangular in shape and lodge the occipital poles of the cerebral hemispheres. The inferior fossae are quadrangular in shape and accommodate the cerebellar hemispheres. A wide groove runs up from the internal protuberance to the superior angle of the bone. It lodges the posterior part of the superior sagittal sinus and its lips give attachment to the falx cerebri. The internal occipital crest runs down from the protuberance to the foramen magnum. The crest gives attachment to the falx cerebelli and bifurcates at the foramen magnum. The triangular area enclosed by the bifurcation is called the ***vermian fossa*** and lodges the inferior cerebellar vermis. A wide groove extends laterally from the protuberance on each side, lodges the transverse sinus and provides attachment to the tentorium cerebelli. The internal occipital protuberance is slightly depressed on one side and lodges the confluence of sinuses.

❑ The posterior border of the squamous part is shaped like an inverted 'V' and has a median superior angle and two lateral angles. It is called the ***lambdoid border*** because it forms the lambdoid suture by articulating with the posterior borders of the parietal bones. The superior angle is at the lambda and is the site of the posterior fontanelle and the lateral angles are at the asteria and are sites of the mastoidal fontanelles.

Lateral Part

This is the part of the occipital bone which is on either side of the foramen magnum. It is also called the ***condylar*** or the ***jugular*** or the ***exoccipital part***. The inferior surface of this part has the occipital condyle that articulates with the superior articular facet of the atlas vertebra. The hypoglossal canal starts a little above the anterolateral part of the foramen magnum and runs anterolaterally. It transmits the hypoglossal nerve and a meningeal branch of the ascending pharyngeal artery. A condylar fossa is seen on the inferior surface behind the occipital condyle. It lodges the superior facet of the atlas in full extension of the skull. It may have a (posterior) condylar canal that transmits an emissary vein from the sigmoid sinus. A jugular process projects laterally from the condyle and forms the posterior boundary of the jugular foramen. The jugular notch indents the anterior aspect of the jugular process. The jugular process articulates with the temporal bone laterally. A jugular tubercle is present on the superior surface of the occipital condyle just above the hypoglossal canal.

Basilar Part

❑ The basilar part of the occipital bone is that part which lies in front of the foramen magnum. It articulates with the basisphenoid through the spheno-occipital synchondrosis. This plate of cartilage closes by around 25-30 years of age. The inferior surface of the basilar part has the pharyngeal tubercle which gives attachment to the fibrous pharyngeal raphe. The prevertebral muscles are attached to this surface. The anterior margin of the foramen magnum gives attachment to the anterior atlanto-occipital membrane.

❑ The superior surface of the basilar part is slightly concave and forms part of the clivus. Clivus (or Blumenbach's clivus; Latin.klivon=slope) is the slope that ascends up from the anterior margin of the foramen magnum and lodges the pons and the medulla. Sulci for the inferior petrosal sinuses are present on the lateral margins of this surface.

Sphenoid Bone

Other names: Os sphenoidale, bird bone, wedge bone of the skull.

The unpaired sphenoid bone (Greek.sphene=wedge) is large, extending across the entire width of the base of the skull and extending also onto the lateral wall of the vault. It is made up of several parts which are:

- Body (that is median in position);
- Right and left greater and lesser wings; and
- Right and left pterygoid processes.

Body

- The body of sphenoid is cuboidal. The superior surface is the cerebral surface and articulates in front with the ethmoid bone. From before backwards, this surface shows
  - a smooth portion called the ***jugum sphenoidale*** (Latin.juga=yoke),
  - a transverse groove called the ***sulcus chiasmaticus***,
  - a small prominence called the ***tuberculum sellae*** and
  - a deep depression called the ***sella turcica*** (or ***pars sellaris***; Latin.sella=saddle, turcica=Turkish).
- The sella turcica contains the hypophyseal fossa and the hypophysis cerebri in it. The anterior edge of sella turcica has a pair of tubercles called the ***middle clinoid processes*** at its angles. The posterior boundary of the sella turcica is made up by a square plate of bone called the ***dorsum sellae***. The lateral angles of the dorsum sellae have the posterior clinoid processes. Diaphragma sella and tentorium cerebelli are attached to the clinoid processes. The petrosal part of the bone is below the dosum sellae on each side and articulates with the petrous temporal bone. Posterior to the dorsum sellae, the body of sphenoid slopes down to join the basiocciput.
- The lateral aspects of the body have the greater wings attached to themselves. A broad groove is found on the superior aspect of this region and is the groove for the internal carotid artery. The lateral margin of the groove is sharp and forms a projection called the ***lingula***. The lingula extends posteriorly to overhang the pterygoid canal.

> Posteromedial to the foramina ovale and spinosum, and to the spine of the sphenoid, the posterior margin of the greater wing forms the anterior wall of a prominent groove. The posterior wall of this groove is formed by the petrous temporal bone. The two bones meet in the floor of the groove that is meant for the cartilaginous part of the ***auditory tube***. Traced laterally, the groove ends in relation to the opening of the bony part of the auditory tube.

- A broad crest is found on the anterior surface of the body; it forms part of the nasal septum. The openings of the sphenoidal sinuses are seen on either side of the crest. The anterior surface closes the posterior ethmoidal sinus and forms part of the posterior nasal roof.

- The inferior surface has a midline projection called the ***sphenoidal rostrum***. The rostrum fits into the alae of the vomer. On either side of the rostrum, the vaginal processes projecting from the medial pterygoid plates can be seen.

Posteriorly, the body of sphenoid is directly continuous with the basilar part (or body) of the occipital bone.

Dissection

Take specimens of a full skull (unopened), skull cap, base of skull (skull cap cut off) and large individual bones like frontal, occipital, parietal and sphenoid. Make a complete study of the various parts and sutures. Locate places of the fontanelles. Try to mark the most important foramina. Send in thin probes through the foramina to see their lines of passage.

Try to superimpose the individual bones over the skull regions.

Greater Wings

Each greater wing is a wide superolateral extension. Most important of its features are the three surfaces—***cerebral***, ***lateral*** and ***orbital surfaces***. The ***cerebral surface*** is concave and forms part of the middle cranial fossa. It shows undulations which fit the sulci and gyri of the temporal lobe of the cerebrum. Close to where the greater wing spans out of the body and lateral to the carotid groove, is the ***foramen rotundum***. It transmits the maxillary nerve. Posterolateral to the foramen rotundum is the ***foramen ovale*** that transmits the mandibular nerve and accessory meningeal artery. The lesser petrosal nerve may sometimes travel through the foramen ovale. Very often, a small opening is found medial to foramen spinosum. It is called the ***canaliculus innominatus*** and transmits the lesser petrosal nerve. Posterolateral to the foramen ovale is the ***foramen spinosum*** which transmits the middle meningeal artery and the meningeal branch of mandibular nerve. A small ***sphenoidal foramen*** (foramen of Vesalius) may be found anteromedial to the foramen ovale and if present, will transmit an emissary vein from the cavernous sinus. The ***lateral surface*** of the greater wing is vertical and forms part of the temporal and the infratemporal fossae. The temporal part has the attachment of temporalis and the infratemporal part the attachment of lateral pterygoid. The foramina ovale and spinosum open on this surface. The ***spine of sphenoid*** can be seen as a small projection posterior to the foramen of the same name. The relationship of the spine to the pharyngotympanic tube (the spine appears on the tube's lateral wall) and the chorda tympani nerve (the nerve grooves the medial aspect of the spine) should be remembered. The ***orbital surface*** is four-sided and forms part of the lateral wall of the orbit. Its medial margin forms the inferior boundary of the superior orbital fissure and the inferior margin forms the superior boundary of the inferior orbital fissure. The anteromedial aspect of this surface forms the posterior wall of the pterygopalatine fossa and has the anterior opening of the foramen rotundum. The posterior opening of the pterygoid canal can be seen on the posterior margin of the greater wing.

Posteriorly, the greater wing meets the anterior margin of the petrous temporal bone.

Lesser Wings

Each lesser wing is a triangular plate that projects laterally from the upper part of the body of the bone. Its posterior border forms the posterior boundary of the anterior cranial fossa and in life, fits into the lateral sulcus of the cerebrum. The superior surface of the lesser wing is smooth, related to the frontal lobe of the cerebrum and is called the ***jugum sphenoidale*** (Latin.jugum=yoke; connecting two parts). The lesser wing is connected to the body by two roots and between the roots is the **optic canal**. The medial end of the lesser wing has a projection called the **anterior clinoid process**. The internal carotid artery ascends up medial to the anterior clinoid process. The anterior and the middle clinoid processes may be connected by a bony spicule resulting in a ***caroticoclinoid foramen***.

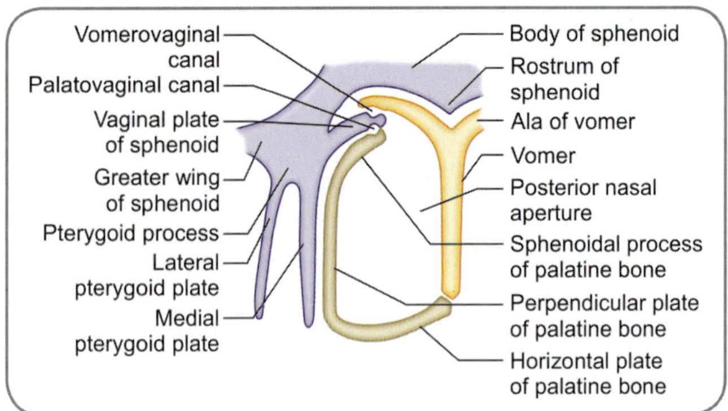

Fig. 2.9: Schematic coronal section to show relationship of pterygoid process to the rest of the sphenoid bone

Pterygoid Processes

Each pterygoid process projects downwards from the junction of the body of sphenoid with the greater wing. It consists of **medial** and **lateral pterygoid plates**. These plates meet anteriorly, but posteriorly they are free. The space between them is called the **pterygoid fossa**. Anteriorly, the pterygoid process is fused to the posterior aspect of the maxilla in its middle part. Higher up, it is separated from the maxilla by the **pterygomaxillary fissure**.

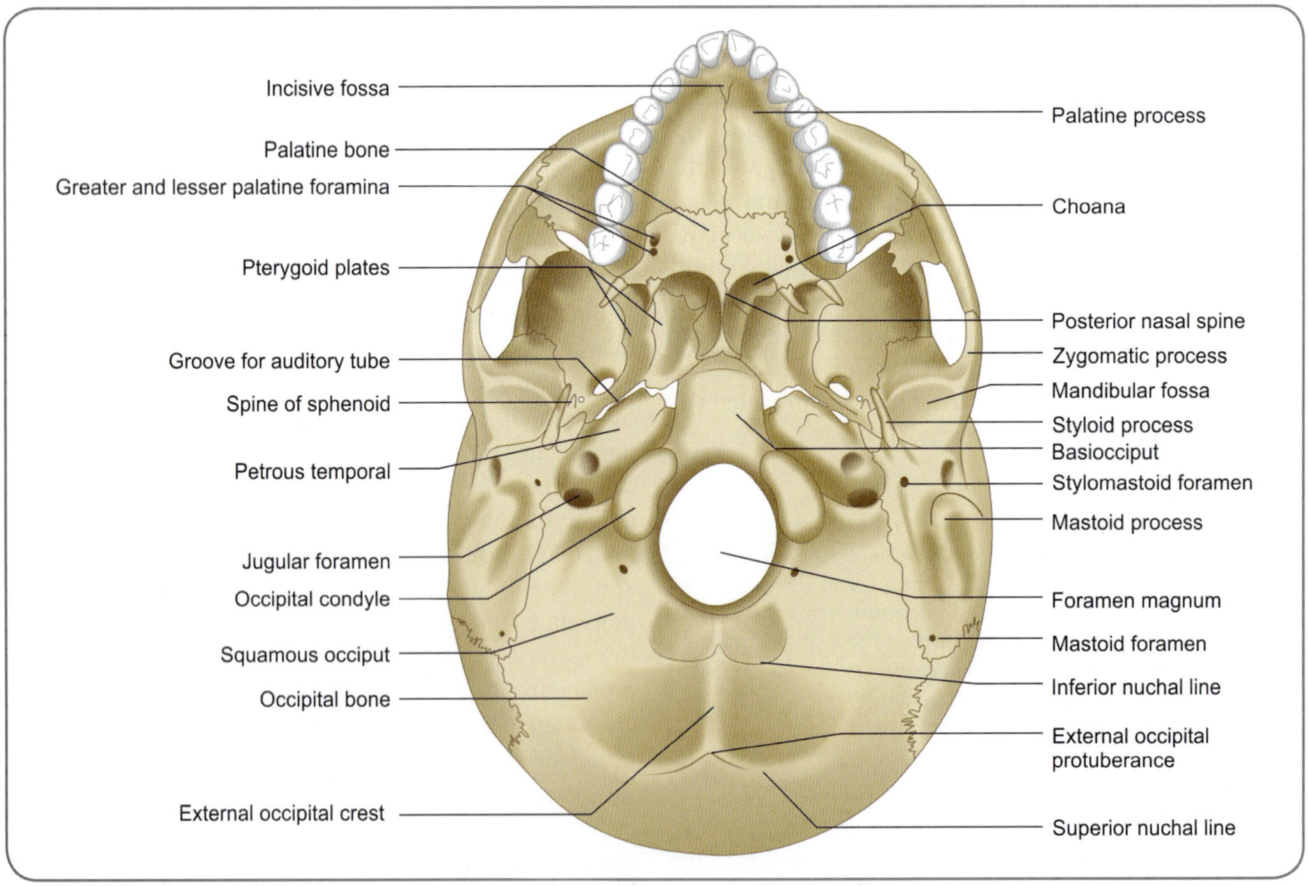

Fig. 2.10: Primary features of base of skull

The medial pterygoid plate is directed backwards so that it has medial and lateral surfaces, and a free posterior border. The lower end of the posterior border is prolonged downwards and laterally to form the ***pterygoid hamulus***. The lateral pterygoid plate projects backwards and laterally. It has medial and lateral surfaces. At its upper end, its lateral surface becomes continuous with the infratemporal surface of the greater wing (Fig. 2.9).

Temporal Bone (Figs 2.10 and 2.11)

Other name: Os temporale.

The paired temporal bone is an irregular bone occupying the sides of the skull (region of the temple) and comprising four parts—***squamous part***, ***petromastoid part***, ***tympanic part*** and ***styloid process***.

Squamous part

The squamous part of the temporal bone has a ***temporal surface*** that forms part of the temporal fossa and an infratemporal surface that roofs the infratemporal fossa

(along with the infratemporal surface of the greater wing of the sphenoid).

Behind its infratemporal surface, the squamous part bears the ***mandibular fossa***. This fossa is bounded anteriorly by a rounded eminence called the ***articular tubercle***. The ***articular area for the mandible*** extends onto the tubercle. Behind the articular fossa is the ***external acoustic meatus*** which is completed anteroinferiorly by the tympanic part. The ***supramastoid crest*** is a prominent ridge on the posterior part of the squamous part. The ridge is a posterior continuation of the zygomatic process. It is an important landmark for the mastoid antrum and forms a boundary of the suprameatal triangle. The ***zygomatic process*** articulates with the temporal process of the zygomatic bone to form the zygomatic arch. A small ***squamosal foramen*** may be present above the posterior end of the zygomatic process and when present, will transmit the petrosquamous sinus.

Between the mandibular fossa and the tympanic part is the ***squamotympanic fissure*** (which is seen clearly on

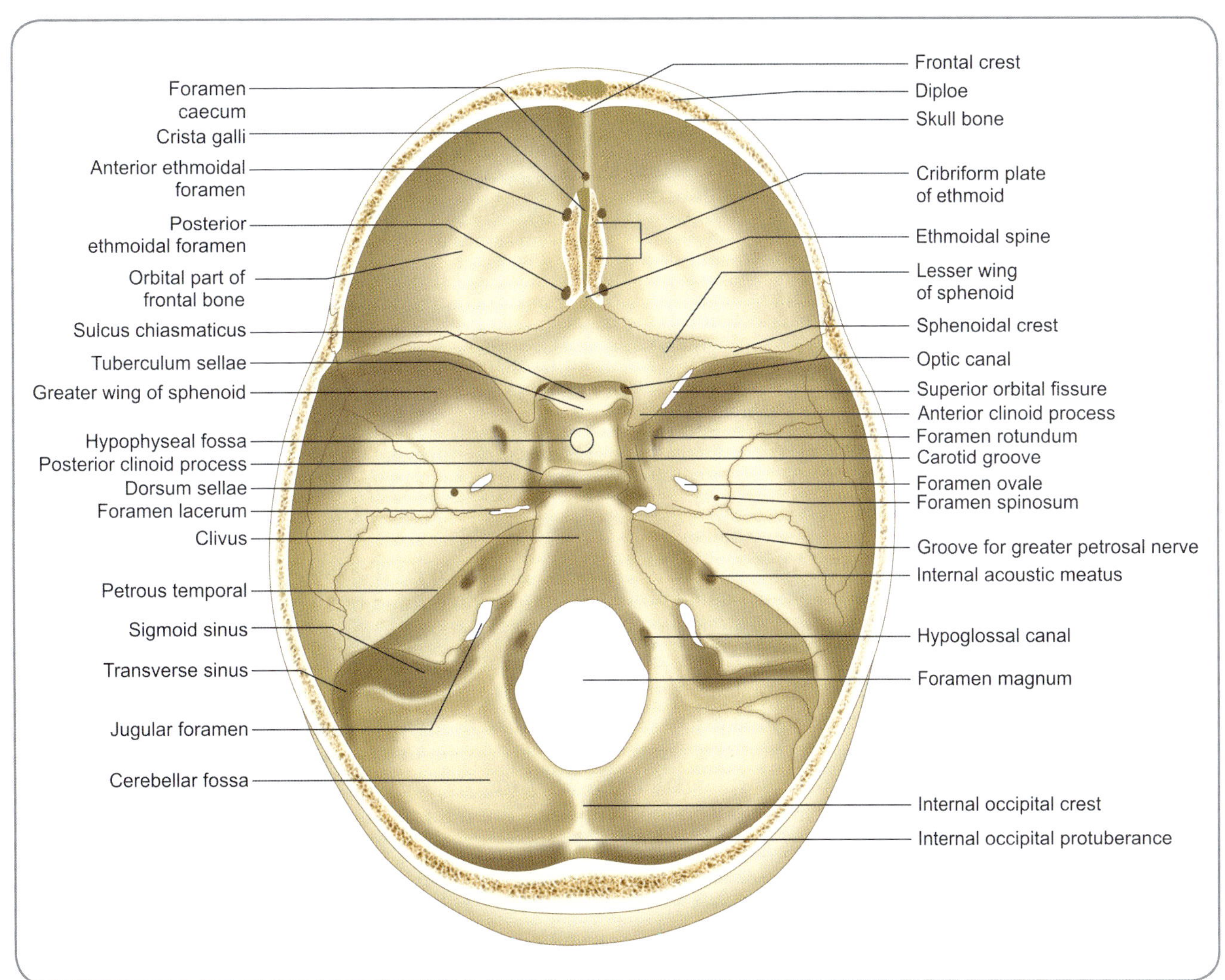

Fig. 2.11: Primary features of cranial fossae

the inferior aspect). A thin projection of a bone from above lies in this fissure; this bony projection is the downturned edge of the tegmen tympani (which is a portion of the petromastoid part). The squamotympanic fissure is therefore subdivided into an anterior ***petrosquamous fissure*** and a posterior ***petrotympanic fissure***. A probe passed into the petrotympanic fissure will lead into the tympanic cavity. The fissure lodges the anterior malleolar ligament and the anterior tympanic branch of the maxillary artery. The anterior canaliculus for the chorda tympani nerve opens in the fissure.

Petromastoid part

The petromastoid part is usually described in two parts—the ***mastoid*** and the ***petrous parts***.

The ***mastoid part*** has an outer surface roughened by the attachments of occipitofrontalis muscle. The ***mastoid process*** is a large conical projection. Sternocleidomastoid, splenius capitis and longissimus capitis muscles are attached to its external surface. On its medial aspect is a ***mastoid notch*** to which the posterior belly of digastric is attached. A shallow ***occipital groove*** is also present on the medial aspect and lodges the occipital artery. The internal surface of the mastoid process (the internal surface and medial aspect are not the same; the former is the inner surface of the cranial box and the latter, the outer surface of the cranial box) has a curved ***sigmoid sulcus*** for the venous sinus of the same name. ***Mastoid air cells*** lie within the process.

The ***petrous part*** is wedge shaped and lies between the sphenoid (anterolaterally) and the occipital (posteromedially) bones in the skull base. It contains the ***internal ear***. It has a base, an apex, three surfaces (anterior, posterior and inferior) and three borders (superior, posterior and anterior). The ***apex*** is well identified; it is blunt and irregular; it contains the ***anterior opening of the carotid canal*** and forms the posterior boundary of the foramen lacerum. The ***anterior surface*** is part of the middle cranial fossa and is wavy because of the sulci and gyri of the temporal lobe. This surface has the ***trigeminal impression*** (for the trigeminal ganglion) and roofs the carotid canal, the internal acoustic meatus, cochlea, vestibule and the facial canal. A thin strip of this surface is called the ***tegmen tympani*** (Latin.tego=roof; tegmen tympani = roof of tympanic cavity). This strip roofs the mastoid antrum, the tympanic cavity and the canal for tensor tympani. It is this strip that folds down anterior to the tensor canal and reaches the squamotympanic fissure to divide the fissure into the petrosquamous and the petrotympanic parts. A small groove can be seen on the tegmen; this groove starts in an opening and runs anteromedially to reach the foramen lacerum. It is the ***groove for the greater petrosal nerve***. A similar but smaller opening and groove are found lateral to this and are related to the lesser petrosal nerve. The ***groove for***

the lesser petrosal nerve leads to the foramen ovale. The ***posterior surface*** forms part of the posterior cranial fossa. It has the ***internal acoustic meatus*** near its centre. The ***inferior surface*** is part of the norma basalis and is irregular. A large circular opening is found on this surface; it is the ***inferior opening of the carotid canal***. Posterolateral to this opening is the ***jugular fossa*** that lodges the superior jugular bulb. This fossa leads posteriorly into the ***jugular foramen*** which is bounded posteriorly and below by the occipital bone, and opens into the posterior cranial fossa. Anteromedial to the fossa is the depression for the inferior ganglion of the glossopharyngeal nerve.

The ***superior border*** is grooved by the ***superior petrosal sinus*** and the attached margin of tentorium cerebelli is attached to it. The ***posterior border*** forms (along with the occipital bone) a groove for the ***inferior petrosal sinus.*** The ***anterior border*** articulates with the sphenoid.

Tympanic part

This is a curved plate of bone that forms the walls of the external acoustic meatus. It also forms a sheath for the styloid process (called the ***vaginal process of the styloid***).

Styloid process

The styloid process is a pointed projection from the inferior aspect of the temporal bone. It has two parts—the ***tympanohyal***—the portion sheathed by the tympanic plate and the ***stylohyal***—the portion that is pointed and gives attachments to ligaments and muscles. The styloid process has important relations. It is covered laterally by the parotid gland; crossed by the facial nerve at the base; crossed by the external carotid artery (embedded in the parotid gland) at its tip; separated from the internal jugular vein by the stylopharyngeus muscle.

Between the mastoid and the styloid processes is the ***stylomastoid foramen*** which transmits the facial nerve and the stylomastoid artery.

Ethmoid Bone

Other names: Os ethmoidale, sieve bone.

The unpaired ethmoid bone (Greek.ethmos=sieve) forms part of the anterior cranial fossa. It can be likened to a modified cross. It has a median perpendicular plate, a horizontal plate and two labyrinths. The horizontal plate cuts across the perpendicular plate. The labyrinths are suspended at the two lateral ends of the horizontal plate. The horizontal plate is the cribriform plate (sieve like) that fits into the ethmoidal notch of the frontal bone. From the superior surface of the cribriform plate, a triangular process projects in the median plane. This is the crista galli (Latin.crista=ridge, gallus=cock; cock's comb; named after its shape) and gives attachment to the falx cerebri. The anterior part of the crista galli spreads into two alae which, along with the frontal bone, form the foramen caecum. The perforations of the cribriform plate transmit the olfactory nerves. On either side of the crista, is a foramen that transmits the anterior ethmoidal nerve and vessels to the nasal cavity. The perpendicular plate descends in the median plane and forms part of the nasal septum. The ethmoidal labyrinth is a cuboidal structure that has two vertical plates. The lateral vertical plate is the orbital plate and forms part of the medial orbital wall. The medial vertical plate is not plainly vertical but has several ups and downs. It is a thin plate that forms the lateral nasal wall. Its medial surface has two convoluted projections—the *superior* and the *middle nasal conchae*. The ethmoidal air cells lie between the two vertical plates. Sphenoid, palatine, frontal, lacrimal and maxillary bones close the air cells at different aspects. The medial surface of the medial vertical plate shows various projections and depressions that form the bulla ethmoidalis and the superior and middle meati of the lateral nasal wall.

Zygomatic Bone

Other names: Yoke bone, jugal bone, malar bone, os malare, cheek bone, zygoma.

The paired zygomatic bones occupy the prominences of the cheek on either side. It forms part of the orbital walls, temporal and infratemporal fossae and completes the zygomatic arch. It is a quadrilateral bone. A *frontal process* runs up to articulate with the zygomatic process of the frontal bone. A tubercle called the **Whitnall's tubercle** is present on the orbital aspect of the frontal process. This tubercle gives attachment to the lateral palpebral ligament and the aponeurosis of levator palpebrae superioris. The *temporal process* runs backwards to articulate with the zygomatic process of the temporal bone to form the *zygomatic arch*.

Maxilla

Other names: Os maxillare.

The paired maxillary bones are the largest of the facial bones except the mandible. Together, they form the upper jaw. They also contribute to the walls of the nasal cavity and the floor of the orbit. The maxillae also are responsible for the formation of two important fossae of the cranial region—the *infratemporal* and the *pterygopalatine fossae*.

Each maxilla has a body and four processes—*zygomatic*, *frontal*, *alveolar* and *palatine processes*. The *body* is roughly pyramidal and houses the *maxillary air sinus*. Its external aspect gives attachments to some of the facial muscles. On its anterior aspect is a sharp margin that forms part of the *infraorbital margin*; below this margin is the *infraorbital foramen* that transmits the infraorbital nerve and vessels. The inferior aspect shows the dental sockets. Posterior to the third molar socket is a rounded prominence of the bone called the *maxillary tuberosity*. The bone above the tuberosity forms the smooth anterior boundary of the pterygopalatine fossa. The superior aspect of the

body has a smooth surface that forms the floor of the orbit. This surface has an *infraorbital groove* that runs into the *infraorbital canal*. As can be known from the names, the groove and the canal transmit the nerve and vessels of the same name. On the medial aspect of the body is the nasal surface that forms part of the lateral nasal wall. Air sinuses which are completed by ethmoid and lacrimal bones and features which contribute to the lateral nasal wall are seen. With keen observation, the greater palatine groove and the nasolacrimal groove can be made out.

The *zygomatic process* is a triangular pyramid that projects laterally and articulates with the zygomatic bone. The *frontal process* projects superiorly to articulate with the frontal bone. The *alveolar process* projects inferiorly and has the sockets for the upper teeth. The *palatine process* projects medially to complete the hard palate.

Foramina of the Skull

There are numerous foramina in the roof, floor and walls of the skull box. Most of them transmit nerves and vessels to and from the skull box. A list of important foramina and the contents they transmit is as follows:

- ❑ ***Foramen magnum:*** The lower end of the medulla oblongata passes to become continuous with the spinal cord. Other important structures passing through this foramen are the vertebral arteries and the spinal part of the accessory nerve.
- ❑ Inferior opening of the carotid canal—internal carotid artery enters the skull.
- ❑ ***Jugular foramen:*** The termination of sigmoid sinus and the commencement of the internal jugular vein lie in this foramen.
- ❑ ***Foramina in the cribriform plate of ethmoid:*** Bundles of nerve fibres that constitute the olfactory nerve pass through these minute apertures.
- ❑ ***Optic canal:*** Optic nerve passes from the middle cranial fossa into the orbit.
- ❑ ***Superior orbital fissure:*** Oculomotor, trochlear and abducent nerves enter the orbit.
- ❑ ***Foramina for divisions of trigeminal nerve:*** The trigeminal nerve has three divisions, each of which leaves the middle cranial fossa through a different foramen.
  - ○ ***Superior orbital fissure:*** Ophthalmic division enters the orbit.
  - ○ ***Foramen rotundum:*** Maxillary division passes into the pterygopalatine fossa.
  - ○ ***Foramen ovale:*** Mandibular division passes into the infratemporal region.
- ❑ ***Internal acoustic meatus:*** Facial and the vestibulocochlear nerves, nervus intermedius and labyrinthine vessels leave the posterior cranial fossa by passing through this opening.
- ❑ ***Stylomastoid foramen:*** Facial nerve emerges out of the cranial cavity; the stylomastoid branch of the posterior auricular artery enters (from outside the cranium) into this foramen to supply the facial nerve, tympanic cavity and mastoid air cells.
- ❑ ***Jugular foramen:*** Glossopharyngeal, vagus and accessory nerves leave the posterior cranial fossa to enter the neck.
- ❑ ***Hypoglossal canal:*** Hypoglossal nerve leaves the posterior cranial fossa.

Nasal Bone

The paired nasal bones are small triangular pieces which form the upper part of the bridge of the nose.

Lacrimal Bone

The paired lacrimal bones (or Ossa unguis) are two thin plates of bone. Each of these forms part of the medial wall of the ipsilateral orbit.

Palatine Bone

The paired palatine bones are L-shaped. Each bone has a vertical plate, a horizontal plate and three processes—*pyramidal*, *orbital* and *sphenoidal processes*. Between the orbital and the sphenoidal processes is a notch called the *sphenopalatine notch*. The notch is converted into a foramen by the sphenoid. The sphenopalatine foramen is the gateway to the nasal cavity.

Inferior Nasal Concha

The inferior nasal concha or turbinate is a scroll like bone that forms one of the processes of the lateral nasal wall.

Vomer

The vomer or the plowshare bone is unpaired and forms the posteroinferior part of the nasal septum.

Mandible

Other names: Mandibulum, mandibula, jaw bone.
The unpaired mandible (Latin.mansus=to chew) is the bone of the lower jaw and bears the lower teeth (Figs 2.12 to 2.15). It consists of an anterior U-shaped ***body***. Two ***rami*** (right and left) project upwards from the posterior parts of the body.

The body of the bone (mandibular corpus) has the external and internal surfaces and the upper and lower borders. The ***external surface*** shows, on the anterior midline, a faint ridge that indicates the position of the foetal symphysis menti. Near the inferior border, this ridge encloses a triangular prominence called the ***mental protuberance*** (Latin.mentum=chin). At the two lateral ends of the triangle are the ***mental tubercles***. The mental protuberance and the mental tubercles form the chin. On either side on the external surface is the ***mental foramen*** that lies vertically below the second premolar tooth; the mental nerve and vessels emerge from this foramen. The faint ***external oblique line*** ascends posterosuperiorly from the mental tubercle, sweeps below the mental foramen and continues into the anterior border of the ramus. Below the incisor teeth, this surface shows a small ***incisive fossa***. The ***internal surface*** has an oblique mylohyoid line that runs down anteroinferiorly from behind the third molar tooth to the mental symphysis. It is sharp and clear posteriorly but faint anteriorly. A shallow ***submandibular fossa*** is seen below the line. A ***sublingual fossa*** is seen above the line anteriorly. On the posterior aspect of the symphysis

(and above the anterior ends of the mylohyoid lines), two tubercles, one above the other, are seen. These are the ***mental spines*** or genial tubercles (Greek.geneion=chin). A tiny lingual or genial foramen may be found above the mental spines. The ***mylohyoid groove*** that starts near the mandibular foramen on the internal aspect of the ramus runs down below the mylohyoid line. The internal surface may sometimes exhibit a rounded prominence medial to the molar teeth above the mylohyoid line. If present, the prominence is called the ***torus mandibularis*** (Latin. torus=bulge).

The upper border of the body bears sixteen alveoli for the roots of the upper teeth and so is called the ***alveolar process*** (Latin.alveus=cavity or trough). The lower border of the body is called the ***base***. On either side of the midline, the base has a digastric fossa. Behind the fossa, the base is thick and shows a gentle anteroposterior convexity. As it approaches the ramus, this convexity turns into an external concavity giving the base a sinuous shape.

Each ramus of the mandible has a posterior border, a sharp anterior border and an inferior border that is continuous with the base of the body. The posterior and inferior borders meet at the ***angle*** of the mandible.

The anterior border of the ramus is continued downward and forward on the lateral surface of the body as the ***oblique line***.

Arising from the upper part of the ramus are two processes. The anterior of these is the ***coronoid process*** (so named because it resembles a crow's beak or a bird's beak). It is flat (from side to side) and triangular in shape. The posterior or ***condylar process*** (or mandibular condyle; Greek.kondylodes=knuckle) is separated from the coronoid process by the ***mandibular notch***. The upper end of the condylar process is expanded to form the ***head*** of the mandible. The head is elongated transversely and is convex both transversely and anteroposteriorly. It bears a smooth articular surface that articulates with the mandibular fossa of the temporal bone to form the temporomandibular joint.

The part immediately below the head is constricted and forms the ***neck***. Its anterior surface has a rough depression called the ***pterygoid fovea***. A little above the centre of the medial surface of the ramus is the ***mandibular foramen***. It leads into the mandibular canal that runs forwards in the substance of the mandible. The medial margin of the foramen is formed by a projection called the ***lingula***. Beginning just behind the lingula and running downwards and forwards is the ***mylohyoid groove***. This groove runs below the mylohyoid line.

Attachments and Relations of the Mandible

Muscular attachments (Figs 2.14 and 2.15)

❑ ***Masseter*** is inserted into the lateral surface of the ramus and of the angle.

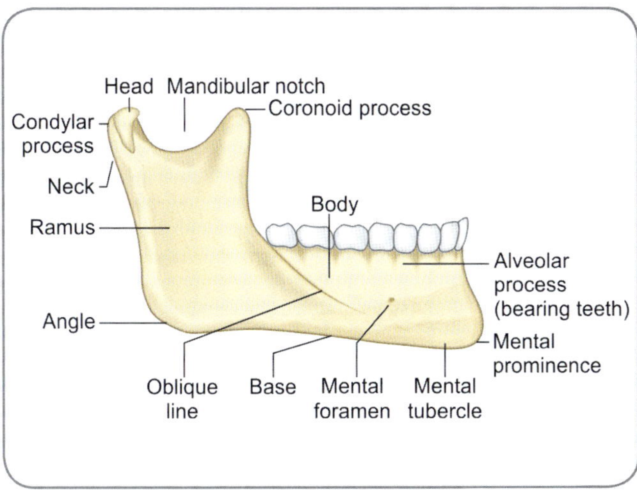

Fig. 2.12: Mandible seen from lateral side

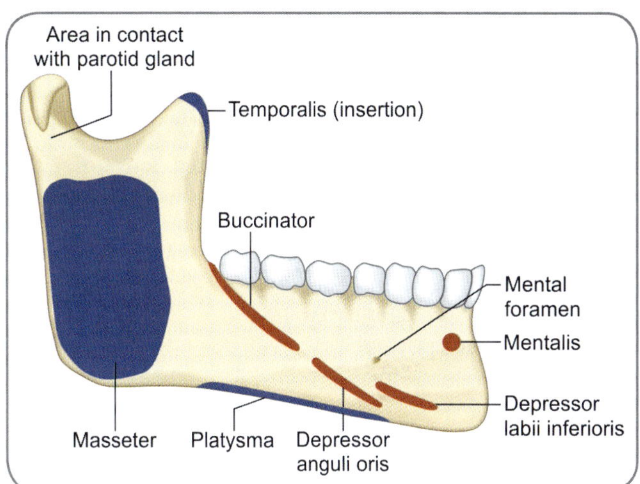

Fig. 2.14: Attachments on mandible seen from the lateral side

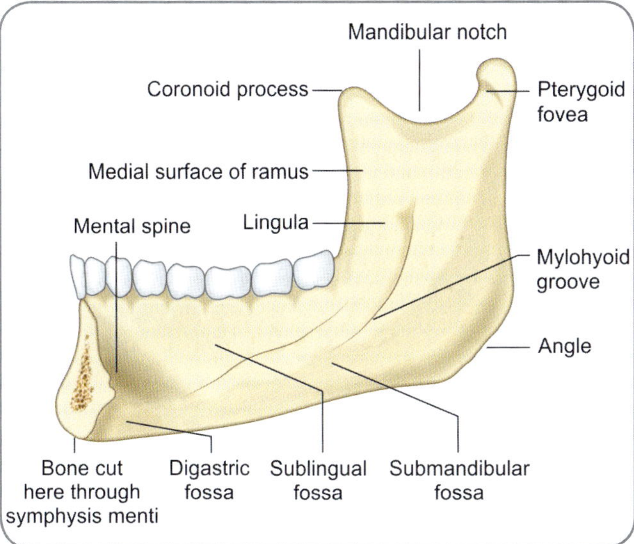

Fig. 2.13: Right half of mandible seen from the medial side

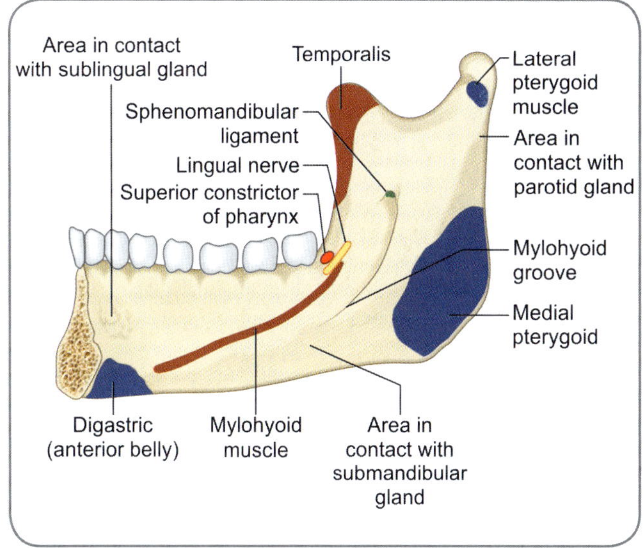

Fig. 2.15: Attachments on mandible as seen from the medial side

- ❑ *Temporalis* is inserted into the medial surface of the coronoid process including its apex, and its anterior and posterior borders.
- ❑ *Lateral pterygoid* is inserted into the fovea on the anterior aspect of the neck.
- ❑ *Medial pterygoid* is inserted into the medial surface of the angle and the adjoining part of the ramus.
- ❑ *Buccinator* arises from the outer surface of the body just below the molar teeth.
- ❑ *Anterior belly of the digastric* arises from the digastric fossa (on the anterior part of the base near the midline).
- ❑ *Mylohyoid* muscle arises from the mylohyoid line.

Other Closely Related Structures

- ❑ The *capsule of the temporomandibular joint* is attached around the margins of the articular surface.
- ❑ The inferior alveolar nerve and vessels enter the mandibular canal (that lies within the bone) through the mandibular foramen.

- ❑ The mylohyoid nerve and vessels run forwards in the mylohyoid groove.
- ❑ The facial artery is closely related to the mandible.
  - ○ Its initial part lies deep to the ramus, near the angle.
  - ○ The artery then runs downwards and forwards deep to the ramus.
  - ○ It reaches the lower border of the body of the mandible at the anteroinferior angle of the masseter.
  - ○ It then runs upwards and forwards superficial to the body of the mandible.
- ❑ The lingual nerve is closely related to the medial aspect of the body of the mandible just above the posterior end of the mylohyoid line.
- ❑ The sublingual gland lies over the sublingual fossa; and the submandibular gland over the submandibular fossa.
- ❑ The parotid gland is related to the upper part of the posterior border of the ramus.

Age Changes in Mandible

- ❑ *At birth:* The bone is in two halves united by the fibrous joint in the midline (symphysis menti). Body is fragile. Mandibular canal is near the lower border. Mental foramen opens below the first (deciduous) molar and directed forwards. The coronoid process projects above the condylar process. The angle between the ramus and the body is obtuse.
- ❑ *1 to 3 years of age:* Body elongates. Symphyseal fusion occurs. Chin develops—so mental foramen alters direction from facing forwards to backwards.
- ❑ *3 to 20 years of age:* Alveolar growth occurs due to eruption and growth of teeth; so the height of body increases. The posterior part of ramus grows while its anterior part resorbs to accommodate for newer permanent teeth. The differential growth of mandible causes the angle between ramus and body to reduce.
- ❑ *20 years to old age:* Features of adulthood persist. Alveolar and subalveolar bone are of equal height. Mental foramen midway between upper and lower borders. Mandibular canal almost parallel to mylohyoid line.
- ❑ *Old age:* Fall of teeth causes loss of alveolar bone. The sockets show healing changes. Alveolar border is closed because of loss of bone, resorption and healing. Mental foramen closer to upper border. Ramus-body angle obtuse. In extreme cases, mental foramen may disappear. Sometimes the mandibular canal may also be lost or exposed and the nerves lie just below the mucosa. Loss of maxillary and mandibular teeth decreases the vertical facial dimension and the mandible appears to close more. This is called *mandibular prognathism*.

Hyoid Bone

Other names: Os hyoideum, lingual bone, tongue bone.

The unpaired hyoid bone (Greek.hyoeides=shaped like upsilon; upsilon is a letter of the Greek alphabet and is written like a curved V; hyoid indicates anything that is U or V-shaped) is considered not a part of the skull by many authors. However, recent approaches to classification of bones have placed the hyoid in the skull bones group on grounds of attachments and development.

The hyoid bone is present in the front of the upper part of the neck. It is not attached to any other bone directly but is held in place by muscles and ligaments which are attached to it. It is, in fact, suspended from the styloid processes by the stylohyoid ligaments.

The bone consists of a central part called the *body and of two cornua* (greater and lesser) on either side (Figs 2.16 and 2.17).

The *body* is roughly quadrilateral. It has an anterior surface and a posterior surface. The convex anterior surface faces anterosuperiorly and the smooth, concave posterior surface faces posteroinferiorly.

The *greater cornua* (or greater horns or cornua majus; latin.cornu=horn) are attached to the lateral part of the

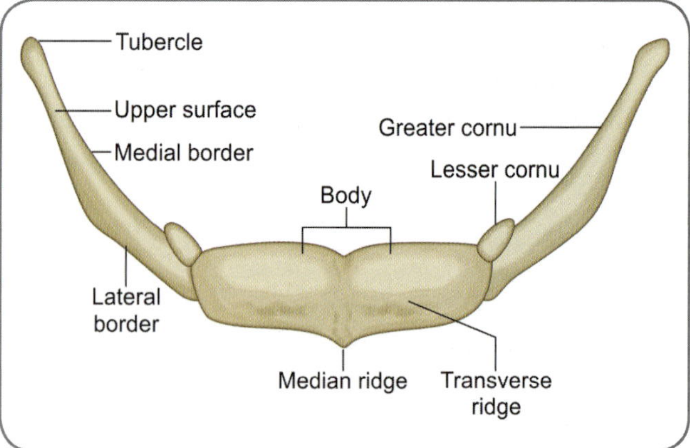

Fig. 2.16: Hyoid bone as seen from the front

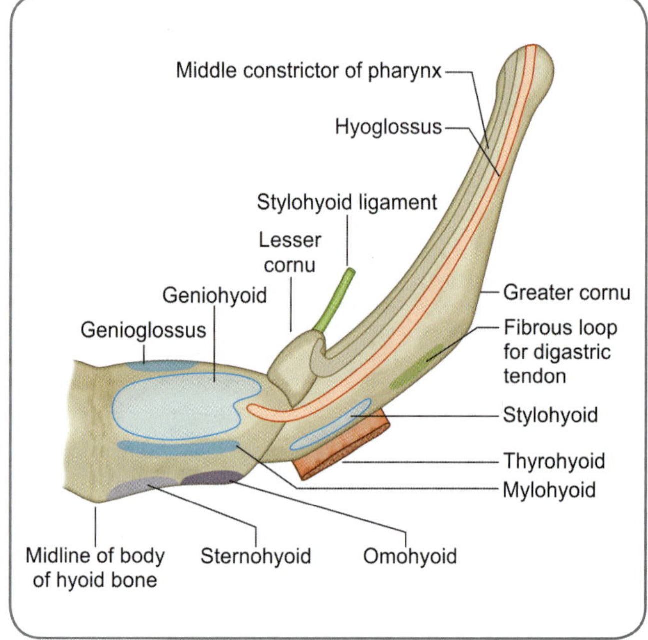

Fig. 2.17: Attachments of hyoid bone

body, from which they project backward and laterally. The connection to the body is cartilaginous in young age but ossifies by around midlife. Each cornu ends in a tubercle.

The *lesser cornua* (or lesser horns or cornua minus or styloid cornua) are small and conical. They project upwards and laterally from the junctions of the body and greater cornua.

Attachments to the Hyoid

Though a very small bone, the hyoid has several important muscles attached to it.

- ❑ The lowest fibres of the *genioglossus* muscle are inserted into the upper border of the body.
- ❑ The *geniohyoid* muscle is inserted to the anterior surface of the body.

❑ The *mylohyoid* muscle is inserted to the anterior surface of the body below the insertion of the geniohyoid.

❑ The *sternohyoid* muscle is inserted into the medial part of the inferior border of the body.

❑ The *superior belly of the omohyoid* muscle is attached to the lateral part of the inferior border of the body.

❑ The *stylohyoid* muscle is inserted into the upper surface of the greater cornu near its junction with the body.

❑ The *thyrohyoid* muscle is inserted into the anterior part of the lateral border of the greater cornu.

❑ The *hyoglossus* muscle arises from the upper surface of the greater cornu (lateral to the origin of the middle constrictor), and from the lateral part of the body.

Fracture of the Hyoid Bone

The hyoid bone can be fractured when the neck is forcibly pressed upon as in strangulation, or in hanging. Such a fracture is of considerable medicolegal significance.

Added Information

Cranial Curves and Buttresses

The cranium has bones which are not equally strong or dense. However, the bones are curved and have an external convexity in places where most strength is needed. This can typically be seen in the bones of the calvaria where they are curved (frontal, parietal, occipital and squamous temporal) to afford protection to the internally lying brain.

The cranium also has portions which are relatively thinner and weaker. Forces which are transmitted to the skull by way of muscular and ligamentous attachments should not compress these weaker zones. In order to circumvent the weaker zones and still obtain the best benefit of the forces, some of the cranial bones have developed 'buttresses'. Buttresses are thicker portions of bones which form 'bony bars' within the rest of the bones. Muscular and ligamentous forces travel through these bars, thus bypassing the weaker zones.

❑ *Frontonasal buttress:* This is bony condensation found in the maxilla and the frontal bone running from the region of the canine teeth up to the glabella in the gap between the openings of the nasal and orbital cavities.

❑ *Lateral orbital buttress:* This condensation runs up from the region of the molar teeth through the zygomatic bone and the zygomatic-frontal articulation to the lateral aspect of the superciliary arch.

❑ *Zygomatic arch buttress:* This is the zygomatic arch itself which relieves compressive and shearing forces from the temporal fossa.

❑ *Occipital buttresses:* Forces which reach the cranium from the vertebral column and the paravertebral muscles are diverted away from the foramen magnum. If these forces impact on the area around foramen magnum, they can cause irreparable damage to the spinomedullary junction and the contents of the posterior cranial fossa. A pair of occipital buttresses takes away the load on either side. One of the buttresses runs from near the occipital condyle to the parietal tuber; the other runs from the same region through the external occipital protuberance to the lambda.

contd...

Added Information *contd...*

❑ *Frontosagittal buttress:* This is a smaller bar of condensation from the area of frontal bone medial to the frontal tuber to the coronal suture.

❑ *Mandibular vertical buttress:* This is the dense condensation along the posterior border of the ramus of mandible.

❑ *Pterygomaxillary buttress:* This bar of strength runs along the posterior aspect of maxilla where the maxilla forms the anterior wall of the pterygomaxillary fissure.

❑ *Frontal bar:* This is a transverse bar along the superciliary arches.

❑ *Infraorbital buttress:* This is the thickening seen along the infraorbital margin.

❑ *Maxillary alveolar buttress:* This is the thickening seen along the alveolar margin of the maxilla and is reinforced by the presence of the hard palate.

Though forces from the paravertebral, temporal and mandibular regions impact on different parts of the cranium, the midface suffers regular onslaught during mastication. These masticatory forces tend to travel vertically upward. Therefore, the midface is surrounded by several vertical buttresses like the frontonasal, lateral orbital, mandibular vertical and pterygomaxillary bars. These vertical bars are reinforced by horizontal bars (maxillary alveolar, zygomatic, infraorbital and frontal bar) which interconnect them (like the cross bars of a scaffolding).

Special Regions of the Skull

Certain regions of the skull are specialised and perform protective and/or accommodative functions. Such regions include the orbital cavities, the nasal cavities, the paranasal sinuses and the fontanelles.

Orbit (Fig. 2.18)

Other names: Bony orbit, eye socket, orbita (*orbita* is singular and the plural form is *orbitae*).

❑ Each bony orbit (Latin.orbis=circle) is a cavity located in the upper part of the facial skeleton and is shaped like a pyramid with the base at the superficially placed orbital opening. The apex lies at the posterior end.

❑ The orbit has a roof, a floor, a medial wall and a lateral wall. None of the walls are sharply marked and each merges imperceptibly into the other.

❑ The *roof* or the superior wall is formed mainly by the orbital plate of the frontal bone. Posteriorly, a small part is formed by the lesser wing of the sphenoid. The anterolateral part of the roof has a depression called the *lacrimal fossa*. Close to the orbital margin, at the junction of the roof and the medial wall, there is a small depression called the *trochlear fossa*.

❑ The *floor or inferior wall* is formed mainly by the orbital surface of the maxilla. The anterolateral part is formed by the zygomatic bone.

❑ The *lateral wall* is formed in its anterior part by the zygomatic bone and in its posterior part by the greater wing of sphenoid.

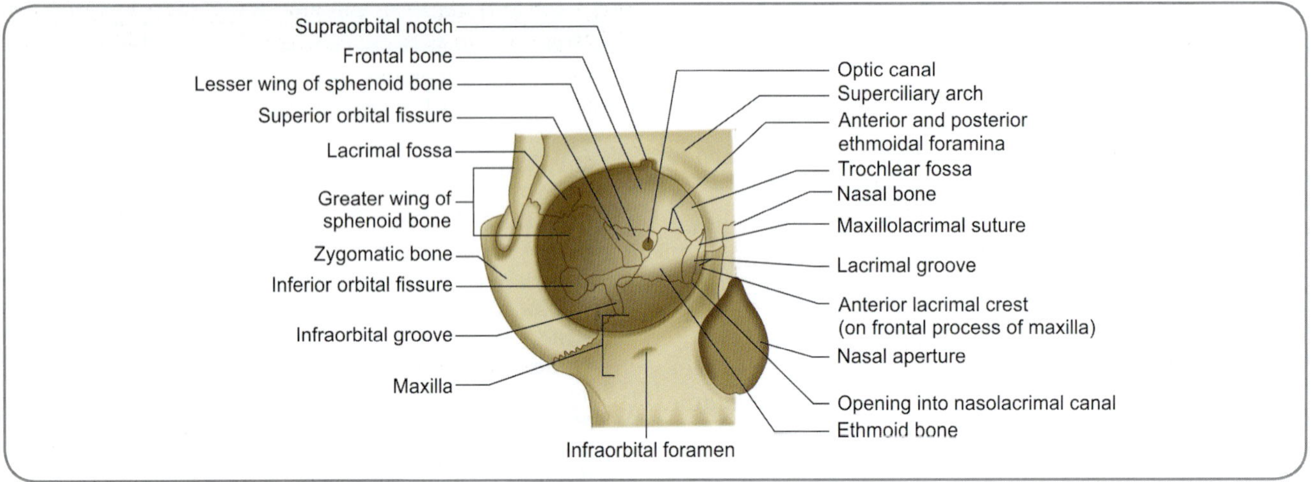

Fig. 2.18: Orbit and surrounding structures

❏ The ***medial wall*** is formed mainly by the orbital plate of the ethmoid. Anterior to the ethmoid, lacrimal bone and the frontal process of the maxilla contribute to this wall. The region of the medial wall formed by the lacrimal bone and the maxilla shows a deep ***lacrimal groove*** (for the lacrimal sac). The groove is continuous, inferiorly, with the ***nasolacrimal canal***, the lower end of which opens into the nasal cavity.

Apertures in the Orbit

❏ ***Superior orbital fissure:*** This is a prominent cleft that separates the posterior parts of the roof and the lateral wall. It is bounded above and medially by the lesser wing of sphenoid and below and laterally by the greater wing. It transmits the superior and inferior divisions of the oculomotor nerve, abducent nerve and the nasociliary branch of the ophthalmic nerve within the common tendinous ring; the trochlear nerve, the lacrimal and frontal branches of the ophthalmic nerve and the ophthalmic veins outside the tendinous ring.

❏ ***Inferior orbital fissure:*** This is a cleft between the posterior part of the floor and the lateral wall of the orbit. It is bounded above and laterally by the greater wing of the sphenoid; below and medially by the orbital surface of the maxilla. This fissure connects the orbit with the pterygopalatine and infratemporal fossae. It transmits the zygomatic and infraorbital branches of the maxillary nerve, their accompanying vessels, inferior ophthalmic vein or a connecting vein and the orbital rami of the pterygopalatine ganglion. The inferior orbital fissure is continuous anteriorly with the infraorbital groove on the maxilla.

❏ ***Optic canal:*** This opening is seen medial to the superior orbital fissure and at the apex of the orbit. It is the opening between the two roots of the lesser wing of sphenoid. It transmits the optic nerve and the ophthalmic artery.

Nasal Cavity

Other name: Cavum nasi

❏ The nasal cavity consists of right and left halves which are separated by a ***nasal septum*** (Figs 2.19 and 2.20). The cavity opens, anteriorly, on the front of the skull through the ***anterior nasal aperture***; and, posteriorly, on the base of the skull just above the posterior edge of the bony palate, through the right and left ***posterior nasal apertures***.

❏ Each half of the cavity has a lateral wall, a medial wall formed by the nasal septum, a floor formed by the upper surface of the palate, and a roof.

❏ The formation of the ***lateral wall*** is complicated. Several bones contribute to its formation and these bones are (Fig. 2.19):

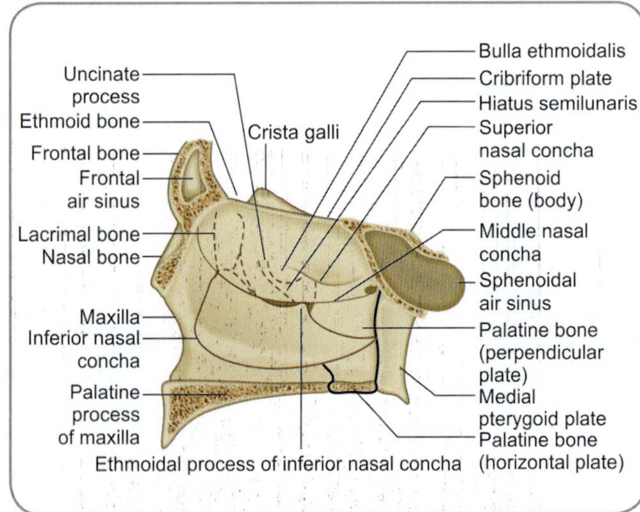

Fig. 2.19: Lateral wall of nasal cavity

- ○ The medial surface of maxilla;
- ○ The perpendicular plate of the palatine bone;
- ○ The lacrimal bone;
- ○ The inferior nasal concha and
- ○ The ethmoid bone.
- ❑ The *floor* of the nasal cavity is formed by the upper surface of the bony palate. Each half of the palate is formed anteriorly by the palatine process of the maxilla and posteriorly by the horizontal plate of the palatine bone.
- ❑ Several bones take part in forming the *roof* of the nasal cavity. From front to back these are parts of:
  - ○ The nasal bone,
  - ○ The frontal bone.
  - ○ The cribriform plate of ethmoid and
  - ○ The anterior surface of the body of sphenoid.
- ❑ The *medial wall* or *nasal septum* (Fig. 2.20) is formed as follows:
  - ○ Upper part by the perpendicular plate of the ethmoid bone,
  - ○ Lower part by the vomer,
  - ○ Anteriorly, a gap filled by cartilage and
  - ○ Around the edges of the septum, small contributions from the nasal, frontal, sphenoid, maxillary and palatine bones.

Apertures in the Nasal Cavity

In addition to the anterior and posterior nasal apertures and the openings of the paranasal sinuses, the nasal cavity also has the following apertures:

- ❑ Opening of the *nasolacrimal canal* into the inferior meatus—it connects the lacrimal system in the orbit to the nasal cavity and provides the drainage route for lacrimal secretions.
- ❑ Sphenopalatine foramen opens behind the superior meatus, just above the posterior end of the middle concha (Fig. 2.19)—it connects the nasal cavity with the pterygopalatine fossa.
- ❑ Apertures in the cribriform plate of ethmoid open in the roof of the nasal cavity—they connect the nasal cavity with the anterior cranial fossa allowing passage of the olfactory fibres.
- ❑ Openings of incisive canals in the anterior part of the floor of the nasal cavity—the canals open on the lower surface of the palate.

Paranasal Sinuses

Other name: Accessory nasal sinuses

- ❑ The paranasal sinuses are spaces present in bones around the nasal cavity and into which they open (Fig. 2.21). There are four of them—frontal, ethmoidal, sphenoidal and maxillary—located within bones of the same names.
- ❑ The right and left *frontal sinuses* are present in the frontal bone. Each sinus lies deep to a triangular area the angles of which lie at the nasion, at a point about 3 cm above the nasion and at a point on the supraorbital margin at the junction of the medial one-third with the lateral two-thirds. Each frontal sinus may extend for some distance into the orbital plate of the frontal bone between the roof of the orbit and the floor of the anterior cranial fossa. The sinus usually opens into the middle meatus through a funnel-like space, the *ethmoidal infundibulum* (Fig. 2.21) that is continuous with the upper end of the *hiatus semilunaris*.
- ❑ The *ethmoidal air sinuses* are located within the lateral part (or labyrinth) of the ethmoid bone. Each labyrinth (right or left) is bounded medially by the medial plate of the ethmoid and laterally by the orbital plate. The ethmoidal air sinuses lie between these plates. They can be divided into anterior, middle and posterior groups.

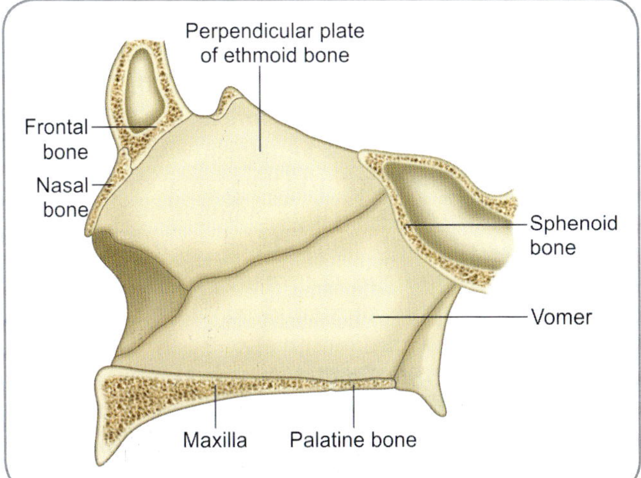

Fig. 2.20: Main bones taking part in the formation of nasal septum

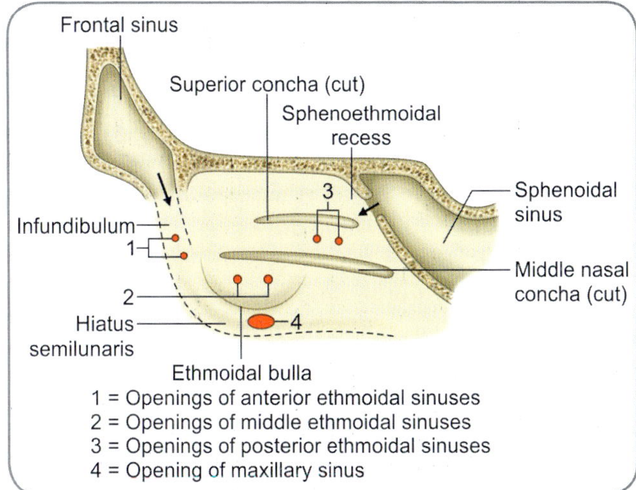

Fig. 2.21: Position of the openings of paranasal sinuses into the nasal cavity

- The anterior ethmoidal sinuses open into the ethmoidal infundibulum, or into the upper part of the hiatus semilunaris.
- The middle ethmoidal sinuses open on or near the bulla ethmoidalis.
- The posterior ethmoidal sinuses open into the superior meatus of the nasal cavity.
- The right and left *sphenoidal sinuses* are present in the body of the sphenoid bone. Each sinus opens into the corresponding half of the nasal cavity through an aperture on the anterior aspect of the body of the sphenoid. The part of the nasal cavity into which the sinus opens lies above the superior nasal concha and is called *sphenoethmoidal recess* (Fig. 2.21).
- The *maxillary sinus* lies within the maxilla of the same side. It opens into the middle meatus of the nasal cavity, via the middle part of the hiatus semilunaris.

Added Information

An air sinus may be trapped within the crista galli of the ethmoid and communicate with the frontal or anterior ethmoidal sinuses. It is called the *Palfyn's sinus*.

Pterygopalatine Fossa

The pterygopalatine fossa (or the pterygomaxillary fossa) is a small pyramidal space located below the apex of the orbit. It can be approached from the lateral aspect of the skull.

- *An anterior wall:* Formed by the superior part of the infratemporal surface of the maxilla;
- *A posterior wall:* Formed by the root of pterygoid process and adjoining greater wing of sphenoid bone;
- *A medial wall:* Formed by the perpendicular plate of palatine bone;
- *A lateral wall:* There is no lateral wall; the fissure replaces it, formed by the pterygomaxillary fissure.

The fossa communicates with other areas of the skull and receives various structures through various apertures. The foramina related to the pterygopalatine fossa are:

- *Sphenopalatine foramen:* On the medial aspect; provides communication to the nasal cavity; transmits nasopalatine nerve, posterior superior nasal nerves and sphenopalatine artery from the fossa.
- *Infraorbital fissure:* On the anterior aspect; provides communication to the orbit; transmits the zygomatic branch of maxillary nerve, the infraorbital nerve, orbital branches of the pterygopalatine ganglion and the infraorbital artery from the fossa.
- *Pterygomaxillary fissure:* On the lateral aspect; provides communication to the infratemporal fossa; transmits the maxillary artery to the fossa; transmits

the posterior superior alveolar artery onto the maxillary tuberosity.

- *Foramen rotundum:* On the posterior aspect; provides communication to the middle cranial fossa; transmits the maxillary nerve to the fossa.
- *Pterygoid canal:* On the posterior aspect; provides communication to the area of foramen lacerum; transmits the nerve of pterygoid canal to the fossa; transmits the artery of pterygoid canal to the nasopharynx and pharyngotympanic tube.
- *Greater palatine canal:* On the inferior aspect; provides communication to the palate; transmits the greater and lesser palatine nerves and artery.
- *Palatovaginal canal:* On the posterior aspect; provides communication to the nasopharynx; transmits the pharyngeal branch of the pterygopalatine ganglion and pharyngeal branch of the maxillary artery from the fossa.

Fontanelles

Checking the fontanelles in an infant gives the physician a lot of information.
- Status of growth of frontal and parietal bones—if growth impaired, fontanelles are depressed/asymmetrical/bulging/overriding;
- Degree of hydration of the infant—depressed fontanelle indicates dehydration;
- Status of intracranial pressure—bulging fontanelle indicates increased intracranial pressure.

Other names: Fonticuli, cranial fontanelles, fonticuli cranii. These are membranous intervals found at the angles of the cranial bones in the new born and infants. Six of them (two median and two lateral pairs) are usually present and are located in relation to the angles of the parietal bone.

- *Anterior fontanelle:* It is also called the *frontal fontanelle, bregmatic fontanelle* or *fonticulus anterior*. It is a median fontanelle and lies at the junction of the sagittal, coronal and frontal (metopic) sutures. It is the largest, measures about 4 cm anteroposteriorly and 2.5 cm transversely and closes by around 18 months of life.
- *Posterior fontanelle:* It is also called the *occipital fontanelle, lambdoid fontanelle* or *fonticulus posterior*. It is a median fontanelle with a triangular shape. It lies at the junction of the sagittal and lambdoid sutures and closes by the third month of life.
- *Anterolateral fontanelle:* It is also called the *sphenoidal fontanelle, fontanelle pterii* or *fonticulus anterolaterale*. It is present in relation to the pterion and closes by the third month of age.
- *Posterolateral fontanelle:* It is also called the *mastoidal fontanelle, fontanelle asterii* or *fonticulus posterolaterale*. It is present in relation to the asterion and closes by the end of first year of life.

Clinical Correlation

Congenital Malformations

- *Anencephaly* is a malformation in which the greater part of the vault of the skull is missing. It is caused by failure of the neural tube to close in the region where the brain is to be formed. The exposed neural tissue degenerates. The condition is fairly frequent and is not compatible with life.
- Establishment of the normal shape of skull depends on orderly closure of sutures.
  - Premature union of the sagittal suture gives rise to a boat-shaped skull (*scaphocephaly*).
  - Early union of the coronal suture results in a skull that is pointed upwards (*acrocephaly*). When the skull is not pointed but there is a generalised linear elongation, the condition is called *oxycephaly* or *turricephaly*.
  - Asymmetrical union of sutures (on the right and left sides) results in a twisted skull (*plagiocephaly*).
- *Congenital hydrocephalus* is a condition in which there is obstruction to the flow of cerebrospinal fluid. As a result, pressure in the ventricular system of the brain increases and leads to its dilatation. In turn, this leads to enlargement of the head and wide separation of the bones of the skull.
- *Primary craniosynostosis:* Premature closure of cranial sutures results in abnormal development of the bones and subnormal development of brain. Though the exact reason is unknown, abnormal development of cranial base is noted. This causes excess pressure on the developing dura leading to premature closure of the sutures. The condition is genetically determined and affects males more.
- Maxilla, mandible and zygomatic bone are derived from the first branchial arch. Occasionally, growth of this arch is defective so that the bones concerned remain underdeveloped and the face is deformed. The condition is called *mandibulofacial dysostosis*.
- Many bones of the skull are formed by intramembranous ossification. The clavicle is also formed in membrane. In the condition called *cleidocranial dysostosis*, formation of membrane bones is interfered with. Deformities of the skull are seen in association with absence of the clavicle.

Injuries and Fractures of Skull

General Features

- Fractures of skull are serious because of the likelihood of damage to the brain.
- The skull of a child is highly elastic and is seldom fractured. A localized blow on the skull of a child produces a depression on the area struck (pond fracture), the rest of the skull remaining unaffected. In contrast, a blow on the skull of an adult can shatter it, with cracks running in various directions from the area that is hit.
- Fractures run along lines which are weak.
  - Injury on the vault of skull can thus result in fractures of the base of skull (cranial fossae).
  - A blow over the parietal bone can lead to a fracture that extends into the squamous part of the temporal bone and into the middle cranial fossa.
  - In the cranial fossae, the fracture line often runs across foramina (which represent sites of weakness).

Clinical Correlation *contd...*

- Large bones of the skull are lined on both sides by compact bone. These are called the inner and outer tables. The inner table of the skull is more brittle than the outer table and so, a fracture may involve the inner table more extensively than the outer. Sometimes, injury to the head may fracture the inner table leaving the outside of the skull intact.
- Apart from its elasticity (in the young), other factors which tend to protect the skull from fractures are its rounded shape and the presence of muscles over the temporal fossa and the occipital region (where the skull wall is thin).

Injuries and Fractures of Specific Areas

Fractures of Middle Cranial Fossa

- The body and greater wing of sphenoid form part of the middle cranial fossa. A fracture involving the body of sphenoid can lead to leakage of blood and CSF into the sphenoidal air sinuses and through them into the nose and mouth. The 3rd, 4th and 6th cranial nerves lie in relation to the cavernous sinus (which lies against the body of the sphenoid bone). These nerves can be involved in fractures of the middle cranial fossa.
- Posteriorly, the middle cranial fossa is bounded by the petrous temporal bone. Involvement of this bone can lead to bleeding and discharge of CSF into the middle ear and external acoustic meatus. The 7th and 8th cranial nerves (which pass through the internal acoustic meatus) can also be injured in a fracture through the petrous temporal bone.

Fractures of Anterior Cranial Fossa

- This fossa is closely related to the nasal cavity and to the orbits. Fracture through the fossa can lead to bleeding or leakage of CSF through the nose (the blood or CSF flowing directly into nose through its roof or through the frontal air sinus).
- Bleeding into the orbit can push the eyeball forwards (exophthalmos). Blood in the orbit can seep into the eyelids resulting in a 'black eye'.

Fractures of Posterior Cranial Fossa

- Fractures through this fossa can lead to bleeding, the blood seeping into the muscles of the back of the neck. Blood often appears superficially over the mastoid process and the sternomastoid muscle.
- If the fracture passes through the jugular foramen, there can be injury to the 9th, 10th and 11th cranial nerves. The walls of the hypoglossal canal are strong and so the 12th cranial nerve usually escapes injury.

Fractures of Calvariae

The calvariae are convex and so the force of a blow gets distributed; the efeect tends to be minimal. Hard blows in thin areas cause *depressed fractures*—bone goes in and compresses the underlying brain. *Linear fractures* occur when blow is forceful—lines of fracture radiate in more directions. In *comminuted fractures*, bone is broken into many pieces. In *contrecoup fractures*, bone at the opposite side of cranium breaks.

Injuries on Face

- Injuries (e.g., blows) on the face can lead to fractures of the mandible, zygomatic bone, maxilla or nasal bones. The mandible is commonly fractured. Neck, body, angle or

contd…

ramus of the bone may be involved. The fracture can also be bilateral. A fracture through the body of the bone often takes place at the level of the canine socket (the deep socket making the bone weak here). Such a fracture can involve the inferior alveolar nerve.

- A fracture of the maxilla can deform the floor of the orbit causing ocular displacement. Involvement of the infraorbital nerve can produce anesthesia over the cheek and upper lip.
- Fractures of the maxilla and/or of the mandible, can cause malocclusion of teeth.

Craniometric Points

Measurement of dry skull and study of its topography is called *craniometry*. There are certain specific fixed points which serve as points of reference:

- **Nasion (nasal point):** Point on the middle of the nasofrontal suture;
- **Ophryon (supranasal or supraorbital point):** Point on the midline of the forehead just above the glabella;
- **Bregma:** Point at the junction of coronal and sagittal sutures;
- **Obelion:** Point on the sagittal suture between the parietal foramina;
- **Lambda:** Point at the junction of sagittal and lambdoid sutures;
- **Inion:** The most prominent point on the external occipital protuberance;
- **Pterion:** Region of the sphenoidal fontanelle;
- **Asterion:** Region of the mastoidal fontanelle;
- **Rhinion:** Lowest point of the internasal suture;
- **Dacryon:** Point at the junction of the frontomaxillary and the lacrimomaxillary sutures on the medial wall of orbit;
- **Acanthion:** Tip of anterior nasal spine;
- **Subnasal point** (apophysary point, Trousseau's point)—point on the midline immediately below the anterior nasal spine;
- **Jugale:** Point at the union of the temporal and frontal processes of the zygomatic bone;
- **Prosthion:** Most anterior point on the maxillary alveolar process in the midline;
- **Porion:** Central point on the upper margin of external acoustic meatus;
- **Pogonion (mental point):** Most anterior point of the mandible in the midline;
- **Gonion:** Most prominent point on the angle of mandible;
- **Gnathion:** Most inferior point of the mandible in the midine;
- **Metopion:** Point midway between the two frontal tuberosities;
- **Occipital point:** Most prominent posterior point on the occipital bone above the inion;
- **Sylvian point:** Nearest point on the surface of skull to the Sylvian fissure; marked 30 mm behind the zygomatic process of the frontal bone.

Foetal/Neonatal Skull

The skull at birth is not the same as that of the adult. The most important features of the neonatal skull are:

- Facial or masticatory portion is extremely small forming only one seventh of the cranial part.
- Ramus of mandible is in line with the bone's body (the angle is obtuse).

contd...

---

contd...

- Nasal cavities are small and the paranasal sinuses are rudimentary.
- Orbits are almost circular.
- There is no mastoid process. The stylomastoid foramen is exposed to the surface.
- The tympanic bone is only a ring and so the tympanic membrane is exposed.
- Frontal and parietal tuberosities are conical.
- Fontanelles are open.
- Additional fontanelles called the **median fontanelle** (in the interparietal suture) and the **metopic fontanelle** (in the frontal suture) may also be present.
- Bones are thin; there is no diploe; most of the bones are in variable number of pieces.

VERTEBRAL COLUMN OF THE NECK

The structure of a typical vertebra has already been considered in section on thorax. We shall now consider the special features of a typical cervical vertebra and also those of the atypical cervical vertebrae.

Typical Cervical Vertebra

- Each **transverse process** of a typical cervical vertebra is pierced by a **foramen transversarium** (Fig. 2.22). The part of the process in front of the foramen is called the **anterior root** and the part behind is called the **posterior root** (Fig. 2.23). The part lateral to the foramen is usually called the **costotransverse bar**, but it is more appropriate to call it the intertubercular bar. The anterior and posterior roots end in thickenings called the **anterior and posterior tubercles** respectively.
- When viewed from the lateral side the transverse process is seen to be grooved. The cervical spinal nerve (of the corresponding segment) lies in this groove after it passes out of the intervertebral foramen.
- The costal element (the element that forms the rib in the thoracic region) forms the anterior root, the

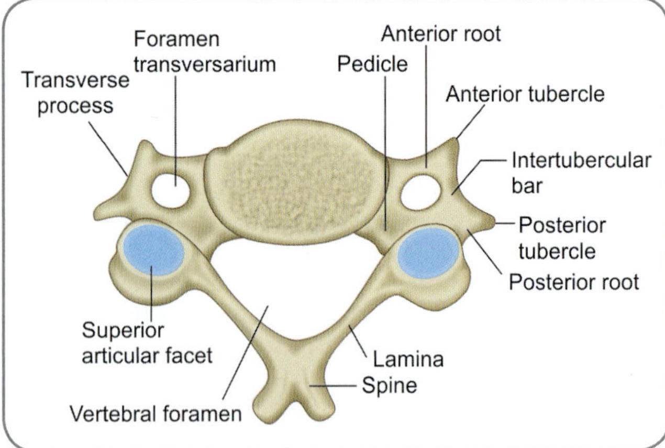

Fig. 2.22: Typical cervical vertebra as seen from above

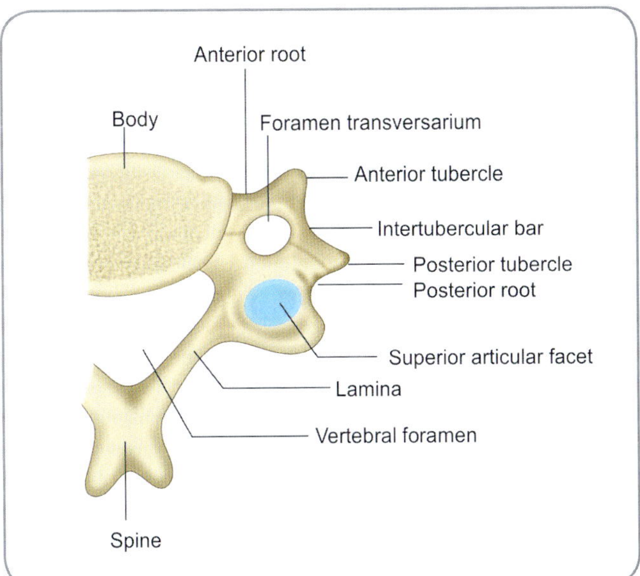

Fig. 2.23: Cervical transverse processes showing the parts derived from the costal elements

articular pillar that helps to transmit some weight from one vertebra to the next lower one.

Atypical Cervical Vertebrae

The first, second and seventh cervical vertebrae are atypical because they show some modifications.

❑ The *vertebral bodies* of the cervical vertebrae are small and appear oval in shape. The upper surface of the body is concave from side-to-side. The posterolateral parts of its edge are raised to form distinct lips. As a result of this, the superior vertebral notch is prominent.

❑ The *pedicles* are long and directed backwards and laterally. The *laminae* of cervical vertebrae are long (transversely) and narrow (vertically). The *spinous processes* are short and bifid.

❑ The *articular facets* are flat. The superior facets are directed equally backwards and upwards. The inferior facets are directed forwards and downwards. The superior and inferior articular processes form a solid articular pillar that helps to transmit some weight from one vertebra to the next lower one.

Atlas (First Cervical) Vertebra

❑ The first cervical vertebra is called the *Atlas*. It has been so named in the assumption that it supports the skull on itself similar to the way the Greek mythological titan supports the earth on his shoulders.

❑ It looks very different from a typical cervical vertebra as it has no body and no spine (Figs 2.24 and 2.25). It consists of two *lateral masses* joined, anteriorly, by a short *anterior arch*; and, posteriorly, by a much longer *posterior arch*. The arches give the atlas a ring-like appearance.

❑ A large transverse process, pierced by a foramen transversarium, projects laterally from the lateral mass. The superior aspect of each lateral mass shows an elongated concave facet that articulates with the

costotransverse bar and both the anterior and posterior tubercles in the cervical region.

❑ The *vertebral bodies* of the cervical vertebrae are small and appear oval in shape. The upper surface of the body is concave from side-to-side. The posterolateral parts of its edge are raised to form distinct lips. As a result of this, the superior vertebral notch is prominent.

❑ The *pedicles* are long and directed backwards and laterally. The *laminae* of cervical vertebrae are long (transversely) and narrow (vertically). The *spinous processes* are short and bifid.

❑ The *articular facets* are flat. The superior facets are directed equally backwards and upwards. The inferior facets are directed forwards and downwards. The superior and inferior articular processes form a solid

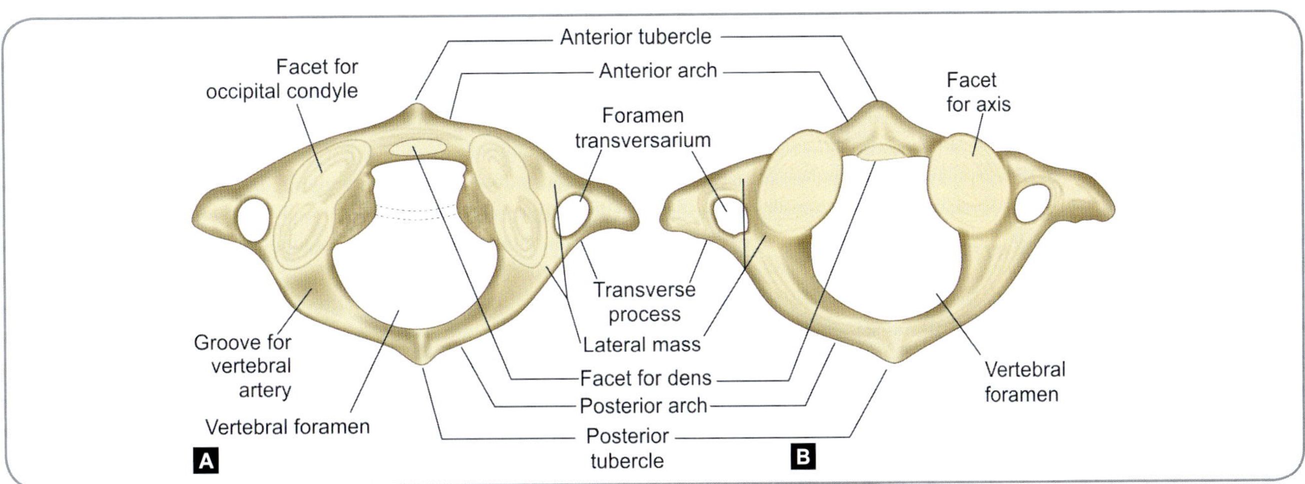

Fig. 2.24: Atlas vertebra seen from **A.** Above **B.** Below

corresponding condyle of the occipital bone (to form an ***atlanto-occipital joint***). Nodding and lateral movements of the head take place at the two (right and left) atlanto-occipital joints. The inferior aspect of each lateral mass (Fig. 2.24) shows a large oval facet for articulation with the corresponding superior articular facet of the axis (second cervical vertebra) to form a ***lateral atlanto-axial joint***.

❑ The medial side of the lateral mass shows a tubercle that gives attachment to the transverse ligament of the atlas (shown in dotted line in Fig. 2.24). This ligament divides the large foramen (bounded by the lateral masses and the arches) into anterior and posterior parts. The posterior part corresponds to the vertebral foramen of a typical vertebra. The spinal cord passes through it. The anterior part is occupied by the dens (which is an upward projection from the body of the axis).

❑ The dens articulates with the posterior aspect of the anterior arch, that bears a circular facet for it. The dens also articulates with the transverse ligament, these two articulations collectively forming the ***median atlanto-occipital joint***. In side-to-side movements of the head, the atlas moves with the skull around the pivot formed by the dens.

❑ The anterior arch bears a small midline projection called the ***anterior tubercle***. The posterior arch bears a similar projection, the ***posterior tubercle***, which may be regarded as a rudimentary spine. The upper surface of the posterior arch has a groove for the vertebral artery. The groove is continuous laterally with the foramen transversarium.

Relations of the Atlas

The atlas vertebra, by virtue of its position, has several important relations.

❑ The vertebral artery passes upwards through the foramen transversarium and then runs medially on the groove over the posterior arch.

❑ The first cervical spinal nerve crosses the posterior arch deep to the vertebral artery and divides here into the anterior and posterior primary rami.

❑ Structures passing through the vertebral canal include the spinal cord and its meninges, spinal part of the accessory nerve and the anterior and posterior spinal arteries.

Axis (Second Cervical) Vertebra (Fig. 2.25)

The most conspicuous feature of the second cervical vertebra, that distinguishes it from all other vertebrae, is the presence of a thick finger-like (tooth-like) projection arising from the upper part of the body. This projection is called the ***dens*** or ***odontoid process*** (Latin.dens=tooth, Greek.odous=tooth) (Fig. 2.23). The dens fits into the space between the anterior arch of atlas and its transverse ligament to form the median atlanto-occipital joint. The anterior aspect of the dens bears a convex oval facet for

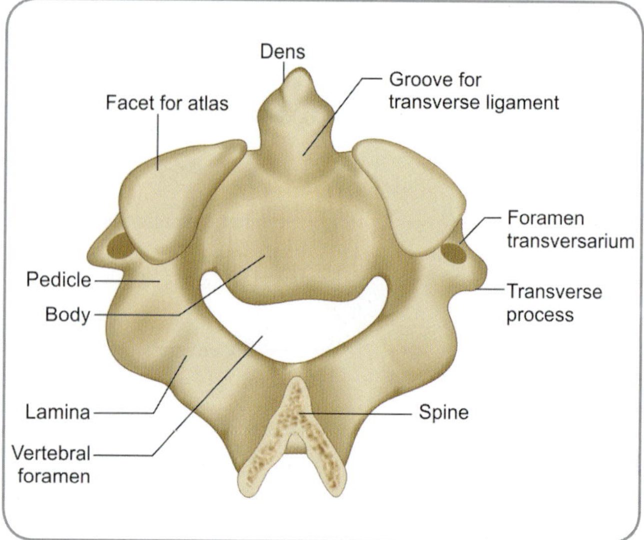

Fig. 2.25: The second cervical vertebra (axis) seen from the posterosuperior aspect

articulation with the anterior arch. Its posterior aspect shows a transverse groove for the transverse ligament.

On either side of the dens, the axis vertebra bears a large oval facet for articulation with the corresponding facet on the inferior aspect of the atlas. The transverse process of the axis lies lateral to this facet. It is small and ends in a single tubercle corresponding to the posterior tubercle of a typical cervical vertebra.

The transverse process is pierced by a foramen transversarium.

Seventh Cervical Vertebra

The seventh cervical vertebra differs from a typical cervical vertebra in having a long thick spinous process that ends in a single tubercle (Fig. 2.26). The tip of the process forms a prominent surface landmark. Because of this fact, this vertebra is often referred to as the ***vertebra prominens***.

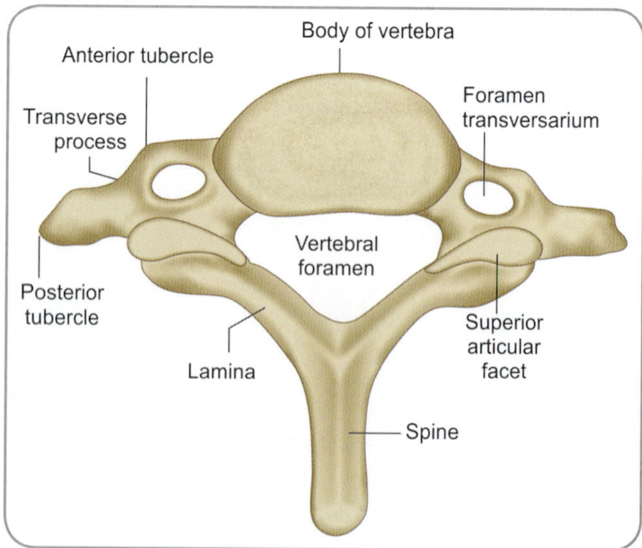

Fig. 2.26: Seventh cervical vertebra as seen from above

The transverse processes are also large and have prominent posterior tubercles.

The vertebral artery and vein ***do not*** traverse the foramen transversarium of this vertebra.

Clinical Correlation

- ***Congenital malformations***: The two halves of the neural arch may fail to fuse in the midline. This condition is called ***spina bifida***. If the gap between the neural arches is small, no obvious deformity may be apparent on the surface (***spina bifida occulta:*** occult = hidden). Two or more cervical vertebrae may be fused to one another resulting in a short neck (***Klippel-Feil syndrome***). The atlas vertebra may be fused to the occipital bone (***occipitalisation of atlas***). The dens may not be fused to the body of the axis vertebra.
- In a condition called ***infantile torticollis***, the mastoid process of one side is pulled towards the sternoclavicular joint of the same side (i.e, the two attachments of the sternocleidomastoid are pulled towards each other). As a result, the head is tilted to one side and rotated to the opposite side. The condition (previously regarded as congenital) is now believed to be a result of injury to the sternocleidomastoid muscle during birth and its gradual fibrosis. The altered position of the neck leads to deformity of cervical vertebrae which may become wedge shaped. The face can also become asymmetrical. The torticollis is often preceeded by a swelling (tumour) on the sternocleidomastoid.
- Degenerative changes taking place in the cervical spine, with age, often lead to stiffness and pain in the neck (***cervical spondylosis***). The intervertebral joints undergo inflammation that is associated with the formation of bony outgrowths (osteophytes). The outgrowths can encroach on intervertebral foramina narrowing them so that cervical nerves may be pressed upon. Such pressure (accentuated by the presence of oedema) can lead to pain that often radiates into the upper limb.
- Prolapse of an intervertebral disc can occur in the cervical region. The symptoms are similar to those of cervical spondylosis.
- Cervical vertebrae can be dislocated or fractured by fall on the head. This is because such falls cause acute flexion of the neck. In the thoracic and lumbar regions of the vertebral column adjoining vertebrae are maintained in position by close interlocking of the articular processes. Dislocation of the vertebral column can take place only after fracture of the articular processes. However, in the cervical region, the articular surfaces are flat and almost horizontal, so that dislocation is possible without fracture. Dislocation and fracture of the cervical vertebral column result in serious complications because of damage to the spinal cord.
- In death by hanging, the dens (of the axis) dislocates backwards (by tearing through the transverse ligament of the atlas), and crushes the lower medulla and the spinal cord. Sometimes there is fracture of the axis vertebra.
- A cervical vertebra (usually the atlas) may slip forwards over the next vertebra even in the absence of injury (***cervical spondylolisthesis***). This may be caused by failure of the dens to fuse with the body of the axis or by weakness of the transverse ligament of the atlas caused by inflammation.

JOINTS OF HEAD AND NECK

The joints of the head and neck region are
- Joints between the individual skull bones resulting in a compact skull;
- Joints between the cranial base and the mandible;
- Joints between the vertebral column and the skull;
- Joints between the vertebrae of the neck and
- Joints of the viscera.

JOINTS BETWEEN INDIVIDUAL BONES OF THE SKULL

Adjoining edges of bones of the skull (especially those of the skull cap) are united to each other by fibrous joints called ***sutures***. The frontoparietal (coronal), interparietal (sagittal) and the parieto-occipital (lambdoid) sutures are classic examples.

Some bones of the skull are united by ***synchondroses***. At such a joint, the two articulating surfaces are united by a plate of hyaline cartilage. As age advances, the cartilage is gradually invaded by bone and the union becomes bony. A typical example is the basisphenoid-basiocciput synchondrosis.

Temporomandibular Joints

These are the joints (one on either side) where the skull base articulates with the only mobile bone of the skull, the mandible. At this joint, the head of the mandible articulates with the articular fossa present on the temporal bone.

Atlanto-occipital Joints

These are the joints between the skull and the vertebral column.

Joints Between the Cervical Vertebrae

These are the joints between the individual components of the cervical vertebral column. Of these, the joints between the atlas and axis vertebrae are atypical. The remaining intervertebral joints are similar to typical intervertebral joints.

Joints of Viscera

Highly specialised joints are present between the ossicles of the middle ear and between the cartilages of the larynx.

OCCIPITO-ATLANTO-AXIAL JOINT SYSTEM

Though the occipito-atlantic and the atlanto-axial joints have been enumerated separately, they are closely aligned to each other functionally. Hence, they are considered together.

Atlanto-axial Joints

- The atlas and the axis vertebrae articulate with each other at three joints, one median, and two lateral (Fig. 2.27).

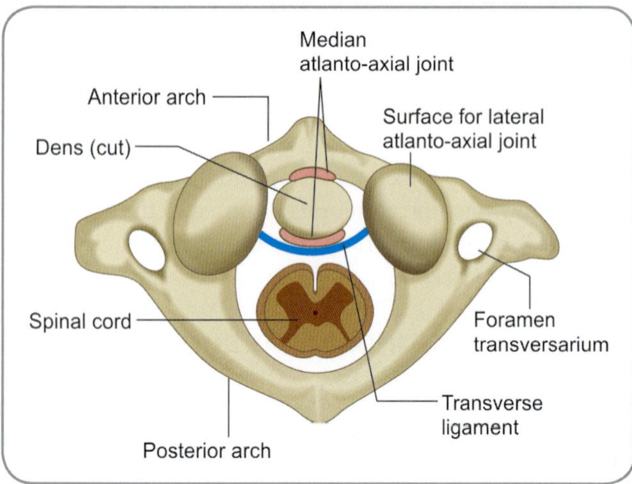

Fig. 2.27: Inferior aspect of atlas to show the atlantoaxial joints

- The ***median atlanto-axial joint*** is a synovial joint of the pivot variety.
- The dens of the axis (the pivot) is placed in the ring formed by the anterior arch of the atlas and its transverse ligament.
- In this situation, there are actually two synovial joints with independent capsules. There is one joint between the anterior surface of the dens and the posterior aspect of the anterior arch and another joint between the posterior surface of the dens and the transverse ligament. The transverse ligament is attached at each end to the medial surface of the lateral mass of the atlas.
- The ***lateral atlanto-axial joints*** are synovial joints of the plane variety.

Atlanto-occipital Joints

- On either side of the foramen magnum, there is a large convex occipital condyle (Fig. 2.15). The long axis of the condyle is directed forwards and medially. The condyle is convex both anteroposteriorly and from side to side. Each condyle articulates with a facet on the upper surface of the lateral mass of the atlas (Fig. 2.4). This facet is concave and corresponds in size and direction to the occipital condyle.
- These articular surfaces are enclosed in a capsule to form a synovial joint. From a functional point of view, the right and the left atlanto-occipital joints together form an ellipsoid joint.

LIGAMENTS UNITING THE ATLAS, AXIS AND OCCIPITAL BONE

Apart from the capsules of the atlanto-occipital joints, the atlas and axis are united to each other and to the occipital bone by a number of ligaments.

- ***Anterior longitudinal ligament*** (continued upwards from lower vertebrae) is attached to the front of the body of axis, to the anterior arch of atlas and to the basilar part of occipital bone (Fig. 2.28A).

Between the atlas and the occipital bone, the anterior longitudinal ligament is incorporated into the ***anterior atlanto-occipital membrane*** (Fig. 2.28A). This membrane is attached below to the upper border of the anterior arch of atlas and above to the anterior part of the margin of foramen magnum.

- ***Posterior atlanto-occipital membrane*** is attached above to the posterior margin of the foramen magnum

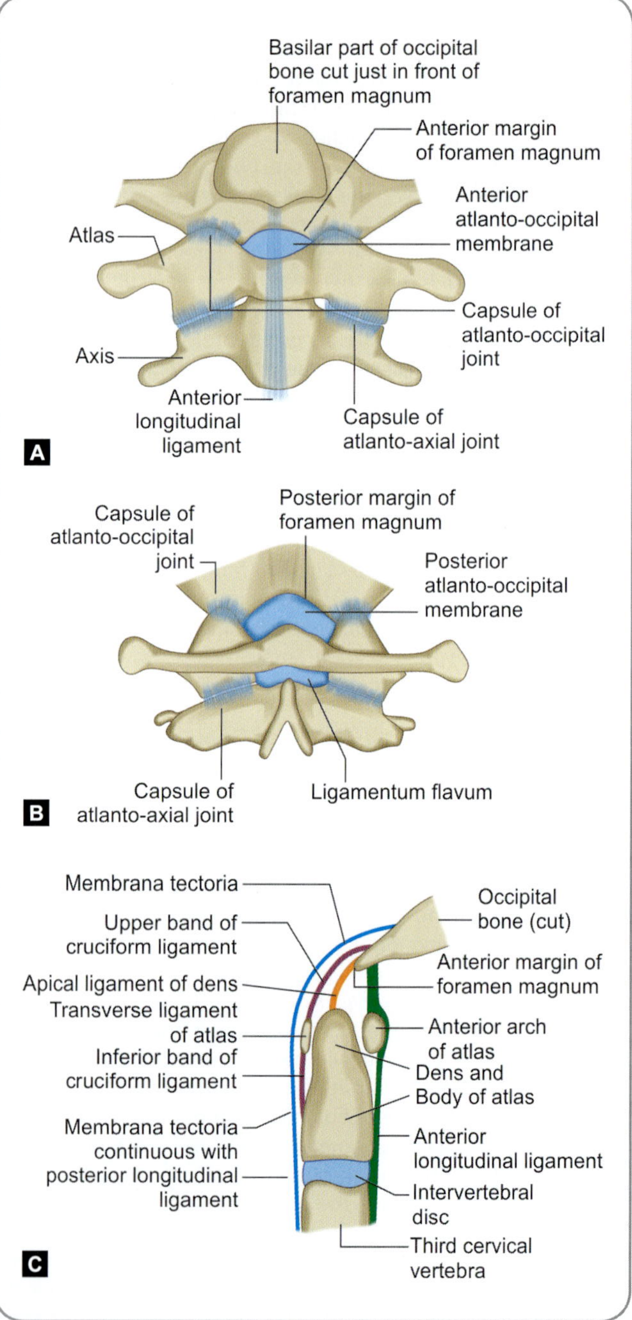

Figs 2.28A to C: Region of atlanto-occipital and atlanto-axial joints **A.** Seen from the front **B.** Seen from behind **C.** Median section

and below to the upper border of the posterior arch of atlas (Fig. 2.28B). (The ligament has a free inferolateral margin. The vertebral artery enters the vertebral canal by passing deep to this edge).

❑ The highest *ligamentum flavum* connects the posterior arch of atlas to the laminae of the axis vertebra (Fig. 2.28B).

❑ *Membrana tectoria* (Fig. 2.28C) is an upward continuation of the posterior longitudinal ligament (which connects the posterior surfaces of the bodies of adjacent vertebrae). It lies posterior to the transverse ligament of the atlas (Fig. 2.28C). Its lower end is attached to the posterior surface of the body of axis. Its upper end is attached to the occipital bone (basiocciput) above the attachment of the upper band of the cruciform ligament (Fig. 2.28C).

❑ The dens (of the axis) is connected to the occipital bone by the following:
  ○ *Apical ligament* passing upwards from the tip of the dens to the anterior margin of the foramen magnum (Fig. 2.28C);
  ○ Right and left *alar ligaments* which are attached below to the upper part of the dens lateral to the apical ligament and above to the occipital bone on the medial side of the condyle.

❑ *Transverse ligament of the atlas* stretches between the two lateral masses of the bone, behind the dens of the axis.

MOVEMENTS AT THE ATLANTO-OCCIPITAL AND ATLANTO-AXIAL JOINTS

❑ Being a pivot joint the median atlanto-axial joint allows the atlas (and with it the skull) to rotate around the axis provided by the dens (the second cervical vertebra is called the *axis* because of this role it plays). This rotation is accompanied by gliding movements at the lateral atlantoaxial joints. The extent of rotation at this joint is limited by the alar ligaments.

❑ From a functional point of view the two atlanto-occipital joints together form an ellipsoid joint. The main movements allowed by this ellipsoid are those of flexion and extension (of the head) as in nodding. Slight lateral movements are also allowed, but no rotation is possible.

Muscles responsible for these movements are as follows:

❑ *Side-to-side movements* (at the atlantoaxial joint) are produced by:
  ○ Sternocleidomastoid (of the opposite side),
  ○ Obliquus capitis inferior,
  ○ Rectus capitis posterior major and
  ○ Splenius capitis (of the same side).

❑ *Flexion* of the head, at the atlanto-occipital joints is produced by:
  ○ Longus capitis and
  ○ Rectus capitis anterior.

(Range of flexion is increased by movement at cervical intervertebral joints produced by the sternocleidomastoid, scaleni and longus cervicis.)

❑ *Extension* of the head, at the atlanto-occipital joint is produced by:
  ○ Rectus capitis posterior major and minor,
  ○ Obliquus capitis superior,
  ○ Erector spinae,
  ○ Splenius capitis,
  ○ Semispinalis capitis and
  ○ Upper fibres of the trapezius.

(Range of movement is increased by movements produced between cervical vertebrae by some of these muscles.)

❑ *Lateral flexion of the head at the atlanto-occipital joint is produced by:*
  ○ Splenius capitis,
  ○ Semispinalis capitis,
  ○ Upper fibres of trapezius and
  ○ Rectus capitis lateralis.

(The larger muscles also produce movements between cervical vertebrae. These are assisted by the sternocleidomastoid.)

Multiple Choice Questions

1. The nuchal lines extending from the external occipital crest are:
   a. The superior nuchal lines
   b. The inferior nuchal lines
   c. The hughest nuchal lines
   d. The supreme nuchal lines
2. One of the following does not form the floor of the temporal fossa. Which is it?
   a. Temporal bone
   b. Parietal bone
   c. Sphenoid bone
   d. Maxilla
3. Spheno-occipital synchondrosis:
   a. Is a cartilaginous joint between the greater wing of sphenoid and basiocciput
   b. Starts ossifying by around 6 to 8 years of age
   c. Contributes to the anteroposterior growth of skull
   d. Unifies the asterion
4. Anterior condylar canal is the other name for:
   a. Hypoglossal canal
   b. Jugular foramen
   c. Foramen lacerum
   d. Foramen magnum
5. The superior aspect of vomer articulates with:
   a. Sphenoid
   b. Ethmoid
   c. Palatine bone
   d. Occipital bone
6. Superolateral angles of dorsum sellae forms:
   a. Anterior clinoid processes
   b. Middle clinoid processes
   c. Posterior clinoid processes
   d. Dorsal processes
7. The bone forming the roof of the middle ear cavity is:
   a. Arcuate eminence
   b. Tegmen tympani
   c. Tympanic plate
   d. Tympanomastoid sheath
8. The reason for the viscerocranium being smaller than the neurocranium at birth is:
   a. Because the derivatives of pharyngeal arches are not formed
   b. Because the nasal cavities are not formed
   c. Because the dental alveolar processes have not yet developed
   d. Because vision is not fixed and so orbits are poorly developed
9. The optic canal is an opening between the roots of:
   a. Lesser wing of sphenoid
   b. Greater wing of sphenoid
   c. Pterygoid process
   d. Jugum sphenoidale
10. Sphenoidal fontanelle is related to:
    a. Pterion
    b. Asterion
    c. Bregma
    d. Lambda

ANSWERS

1. a   **2.** d   **3.** c   **4.** a   **5.** a   **6.** c   **7.** b   **8.** c   **9.** a   **10.** a

Clinical Problem-solving

Case Study 1: A 6-month-old child was brought to the paediatrician. The paediatrician examined for the 'soft spots' of the child's head.
- What are 'soft spots' in a child's head?
- What information can the paediatrician get by feeling for the soft spots?
- Can you associate the age old tradition of not using a comb or foreign object to groom a child's hair?

Case Study 2: A friend of yours had a road traffic accident and was badly injured over the head. As you saw him in a semi conscious condition in the intense care unit, you saw sticky fluid oozing from his nose. It was not blood.
- What do you think could be that fluid? If so, what would you call the condition?
- Which area of the skull should be involved to cause this condition?
- What would you call the condition if similar leaks happen in the ear?

(For solutions see Appendix).

Chapter 3

Scalp and Face (Including Lacrimal Apparatus and Parotid Region)

Frequently Asked Questions

- ❑ Write about the scalp in detail with regard to its composition, blood supply, nerve supply and functions.
- ❑ Write notes on: (a) Occipitofrontalis, (b) Tarsal plates, (c) Buccinator, (d) Lacrimal gland, (e) Danger area of face.
- ❑ Discuss the parotid gland with regard to its capsule, surfaces and parts, relations and structures within.
- ❑ Write briefly on: (a) Epicranius, (b) Lacrimal sac, (c) Orbicularis oculi, (d) Modiolus, (e) Orbicularis oris.

SCALP

The term 'scalp' (Skandinavian.skalpr=sheath) is applied to the soft tissues covering the vault of the skull. This region extends anteriorly up to the eyebrows (and, therefore, includes the forehead), posteriorly up to the superior nuchal lines and laterally up to the superior temporal lines.

LAYERS OF THE SCALP (FIG. 3.1)

The most superficial layer is the *skin*. Being hairy, it contains numerous sebaceous glands. However, except in the occipital region, it is generally thin. It has an abundant blood supply and also has many sweat glands. The skin is closely adherent to the underlying tissues.

The *superficial fascia* is represented by a dense connective tissue that is firmly united to the skin and to the underlying epicranial aponeurosis. Fibrous strands divide the superficial fascia into numerous small pockets. This layer also has a good blood and nerve supply.

The third layer of the scalp is called the *epicranius* which is partly muscular and partly fibrous. It corresponds to the deep fascia.

The greater part of this layer is formed by the *epicranial aponeurosis* (or galea aponeurotica).

The muscular part is formed by a muscle called the *occipitofrontalis.*

The three layers of the scalp described above are firmly united to one another. All the three layers move together over the fourth layer that is made up of *loose areolar tissue*. The extent of the layer of loose connective tissue corresponds to the extent of the scalp itself. Loose areolar tissue is traversed by emissary veins passing from the scalp to intracranial venous sinuses. This layer is sponge-like and has a number of potential spaces which can get enlarged in cases of fluid collection.

> The layers constituting the 'scalp' can be remembered by recollecting the letters of the term itself.
> **S** – skin, **C** – connective tissue (superficial fascia), **A** – aponeurosis, **L** – loose areolar tissue, **P** – pericranium.

The deepest layer of the scalp is the *pericranium* (which is the periosteum of the bones of the vault of the skull). It is important to note that at the edge of each bone of the skull, the pericranium is fixed because it is attached to the sutural ligaments of the sutural joints.

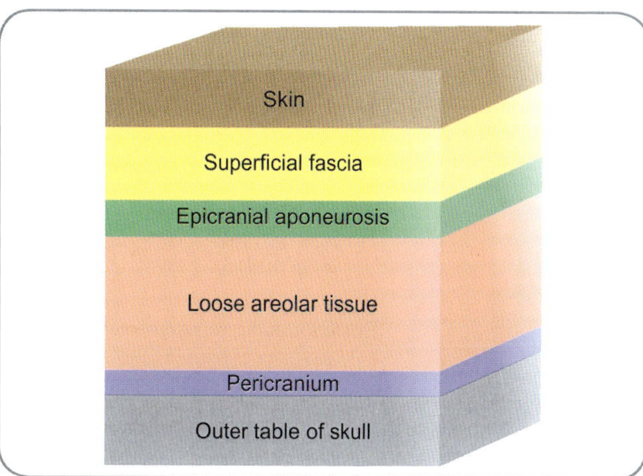

Fig. 3.1: Layers of scalp (schematic)

Dissection

Skin Incisions for scalp and face:
- ❏ A vertical incision from nasion to symphysis menti
- ❏ An oblique incision from angle of mouth to midpoint of tragus
- ❏ A curved coronal incision from one tragus to the other around the scalp
- ❏ A straight incision upwards from nasion to meet incision 3
- ❏ Small curved incisions encircling the eye, nose and mouth.

  After making the necessary incisions, slowly reflect the skin flaps in such a way that all the flaps are reflected laterally.

  With these incisions, you can now divert your attention towards the region that you want to study.

Epicranius

The epicranius, in the lower animals, is a single sheet of muscle. In human beings, most of the sheet has become aponeurotic; the anterior and posterior ends remain muscular and form the occipitofrontalis muscle.

Occipitofrontalis

The occipitofrontalis muscle (sometimes referred to as the fronto occipitalis) consists of a posterior *occipital part* (or *occipitalis*), and an anterior *frontal part* (or *frontalis*). Each of these is divided into a right half and a left half. These four parts are continuous with each other through the epicranial aponeurosis.

Parts and Attachments

Each *occipital part* arises from the occipital bone (lateral two-thirds of the highest nuchal line). Laterally, the line of origin extends onto the mastoid part of the temporal bone. The fibres of the occipitalis run upwards and forwards to end in the epicranial aponeurosis. The occipital parts of the two sides are separated from each other by a part of the epicranial aponeurosis that gains attachment to the external occipital protuberance and to the medial parts of the highest nuchal lines.

The *frontal parts* are attached posteriorly to the epicranial aponeurosis. Anteriorly, the fibres have no bony attachment.

The majority of the fibres of the frontalis merge with the upper edge of the orbicularis oculi. The medial-most fibres merge with the procerus, which is a small muscle overlying the upper part of the nose. More laterally, the deeper fibres merge with the corrugator supercilii, which is a small muscle underlying the medial part of the eyebrow.

Nerve Supply

The occipital part of the muscle is supplied by the posterior auricular branch of the facial nerve and the frontal part by the temporal branches of the same nerve.

Action

Acting alternatively, the frontal and occipital parts can move the scalp forwards and backwards over the vault of the skull.

In addition, the frontal parts raise the eyebrows (as in surprise). Acting from below, they produce transverse wrinkles of the forehead. Action causing wrinkling of forehead is a more powerful action when compared with that raising the eyebrows.

BLOOD VESSELS AND NERVES OF THE SCALP (FIG. 3.2)

The *arteries* supplying the scalp are:
- ❏ The supratrochlear and supraorbital branches of the ophthalmic artery in front;
- ❏ The anterior and posterior branches of the superficial temporal artery laterally;
- ❏ The occipital artery posteriorly.

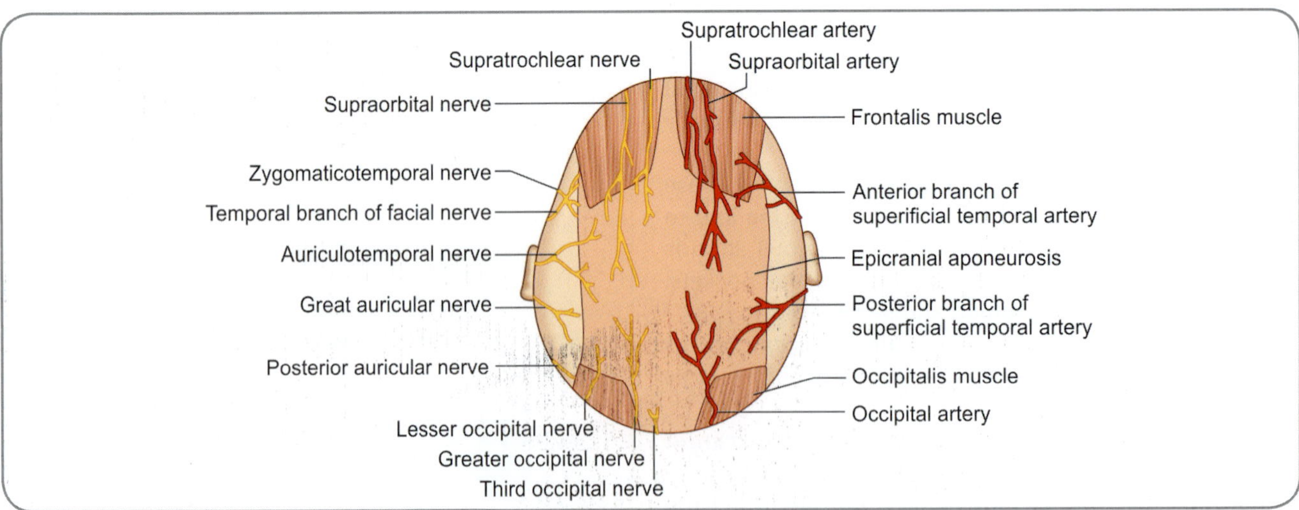

Fig. 3.2: Nerves and arteries of scalp

Dissection

If you want to study the scalp, extend the straight incision running up from the nasion to the coronal incision backwards and downwards till the external occipital protuberance. Reflect the skin flaps of the scalp in four quadrants: left anterior, left posterior, right anterior and right posterior. As you reflect the skin, notice that the subcutaneous tissue is tough. Look for the epicranial aponeurosis and the fibres of occipitofrontalis muscle. See the disposition of the muscle and the aponeurosis. Locate the various nerves and vessels in the scalp. In the anterior half, you can locate the supratrochlear nerve and vessels, supraorbital nerve and vessels, auriculotemporal nerve, zygomaticotemporal nerve, superficial temporal vessels, transverse facial artery and vein and temporal branches of facial nerve. In the posterior half of scalp, great auricular nerve, lesser and greater occipital nerves, occipital vessels, third occipital nerve and posterior auricular branch of the facial nerve can all be located.

Clean and define the frontal and occipital bellies of occipitofrontalis muscle. Shift your attention to the temporal region. Study the temporal fascia and the underlying temporalis muscle. Observe that the superficial three layers of the scalp move as a single unit over the underlying pericranium. Try if you can insinuate your finger or a blunt instrument under the aponeurosis and study the space occupied by loose areolar tissue.

The *veins* of the scalp accompany the corresponding arteries.

The *lymphatic drainage* of the scalp is into the occipital, retroauricular and superficial parotid nodes. Part of the forehead just above the root of the nose drains to the submandibular nodes.

The *nerves* of the scalp may be divided into motor nerves which supply the occipitofrontalis and sensory nerves which supply skin and other tissues of the scalp. The motor nerves are the temporal and posterior auricular branches of the facial nerve.

The sensory nerves are:

- In front—the supratrochlear and supraorbital nerves which are branches of the frontal nerve (which itself is derived from the ophthalmic division of the trigeminal nerve).
- Laterally—the zygomaticotemporal, the auriculo-temporal and great auricular nerves.
- Posteriorly—the greater occipital, lesser occipital and third occipital nerves.

Added Information

- Though the scalp is described to have five layers, the first three layers (skin, superficial fascia and epicranius) which are intimately attached to each other constitute the scalp proper.
- Occipitofrontalis is considered a single muscle by the clinicians. The frontal part is referred to as the frontal belly and the occipital part as the occipital belly.

Added Information *contd...*

- It is again customary for the clinicians to apply the name 'epicranius' to the collective group of occipitofrontalis, temporoparietalis and superior auricular muscles. The temporoparietalis can anatomically be considered a small but separate portion of the occipitofrontalis. The fibres arise from the temporal bone and ascend up to join the epicranial aponeurosis. The superior and anterior auricular muscles, which are separate muscles, take origin from the lower part of the epicranial aponeurosis. Because of their close proximity and all of them being supplied by the facial nerve, they are considered together.
- The sensory innervation of the scalp follows a sequential pattern. If a line is drawn at the level of the auricle of the external ear, the supply anterior to the line is by the three divisions of the trigeminal nerve and posterior to the line is by the spinal cutaneous nerves. The sequence of individual nerves from before backwards is as follows—supratrochlear, supraorbital (both cranial nerve V1), zygomaticotemporal (cranial nerve V2), auriculotemporal (cranial nerve V3), anterior rami of great auricular and lesser occipital (C2, C3) and posterior rami of greater occipital and third occipital (C2, C3).
- The arterial anastomosis of the scalp is between branches of the internal carotid system and the branches of the external carotid system.
- The arteries of the scalp supply very little blood to the neurocranium.

Clinical Correlation

Wounds of the Scalp

- The scalp is profusely supplied with blood, the arteries entering it from the sides, from the front and from behind and coursing in the superficial fascial layer. Because of rich blood supply, wounds of the scalp bleed profusely.
- Bleeding can be difficult to stop because:
  - The vessels anastomose freely with each other and across the midline with the contralateral vessels;
  - The walls of blood vessels are adherent to the dense subcutaneous fibrous tissue and do not constrict when cut; the cut openings tend to remain large and open.

  Direct pressure over the wound is needed to stop bleeding.
- The profuse blood supply also provides some advantages in dealing with scalp wounds.
  - Portions of scalp that are torn off (avulsed) retain adequate blood supply (even though the areas of attachment are narrow) and heal well when stitched back into position.
  - However, if a portion of scalp is completely detached, the surgeon finds it difficult to fill the gap. Hence, the surgeon makes it a point not to cut away parts of the scalp unless absolutely necessary.
- In relation to wounds of the scalp, it is necessary to remember that the skin and the epicranial aponeurosis are firmly united to each other by the layer of dense connective tissue. Also remember that the fibres of the aponeurosis run predominantly in an anteroposterior direction. A wound

contd...

contd... **45**

on the scalp does not gape (i.e., the edges do not separate) unless the epicranial aponeurosis is divided.

Even if the aponeurosis is divided in an anteroposterior cut, the wound will not gape because of the predominant direction of the fibres. It is only when the aponeurosis is cut transversely that the wound can gape widely.

❑ Injuries on the head can lead to bleeding into various layers of the scalp.
  ○ Bleeding into the superficial fascia is never extensive as the region is divided into small pockets by fibrous tissue;
  ○ Bleeding into the layer of loose areolar tissue spreads widely reaching the orbital margin anteriorly, the nuchal lines posteriorly, and the temporal lines laterally;
  ○ Bleeding deep to the pericranium (periosteum) does not extend beyond the margins of the bone as the pericranium is adherent to sutural ligaments which limit the spread of blood. The shape of the haematoma, therefore, corresponds to that of the underlying bone (cephalhaematoma).

❑ The arteries to the scalp arising from the sides of the head are protected by dense connective tissue. So, a partially detached scalp can be replaced with reasonable chance of healing.

❑ The arteries of the scalp supply very minimal blood to the calvarium (which is supplied by the middle meningeal arteries). As a result, loss of scalp does not cause necrosis of the skull bones.

Infections of the Scalp

An infection in the subcutaneous tissue is limited to a small area as the tissue is divided into small pockets by dense fibrous tissue. For the same reason, pressure on nerves in the area is high and leads to severe pain.

An infection in the layer of loose areolar tissue can spread to the limits of the space. As this layer is traversed by emissary veins, infection can pass through these veins to intracranial venous sinuses. Therefore, this layer of loose areolar tissue is called the ***dangerous area of the scalp***.

Osteomyelitis of the skull bones can also be caused by spread of infection through emissary veins.

Scalp infections cannot normally spread to the neck because the occipital bellies of the occipitofrontalis are attached to the occipital bone and mastoid processes. They also cannot spread to the lateral aspect of face because the epicranial aponeurosis and the temporal fascia are continuous and merge with the zygomatic arch.

Black Eyes

The occipitofrontalis muscle has no bony attachment in front. So, infection, fluid or pus collection can track into upper eyelids and root of nose. The skin of the eyelids is very thin (thinnest in the body). Subcutaneous tissue is also very loose. All these factors lead to easy accumulation of fluids even if the injury/inflammation are minimal. Black eyes or periorbital ecchymosis (Greek.ek=out, chymos=juice) occurs in scalp or forehead injuries. Blood extravasates into the subcutaneous tissue and skin of eyelids.

Swellings on the Scalp

Several kinds of swellings can occur in the scalp.

Sebaceous cysts: Because of the presence of numerous hair follicles and the sebaceous glands associated with them, the scalp is a common site for sebaceous cysts (pilar cysts; Latin. pilus=hair). The cysts are often multiple. A sebaceous cyst is caused by blockage to the discharge of secretion of a sebaceous gland. The secretion accumulates and leads to formation of the cyst. The interior of the cyst is lined with epithelial cells. A sebaceous cyst in the scalp will move with the scalp in relation to the underlying cranium.

Dermoid cysts: Dermoid cysts are formed by multiplication of epithelial cells which get buried under the skin surface. The cysts may be congenital or may be caused by pricks. A dermoid cyst can lead to erosion of underlying bone.

Neoplasms: Various types of tumours, benign and malignant, may occur in the scalp.

FACE

The face is the anterior portion of the head extending above downwards from the forehead to the chin and side to side from ear to ear. The identity for a human being is provided primarily by the face. Though features of the face are largely influenced by the underlying bones and their prominences and variations, the external features of the face are also important in that they serve as modalities of communication and interaction. The various parts of the face are worth noticing and have been anatomically described.

EXTERNAL NOSE

The prominence on the face that the layman refers to as the 'nose' is, strictly speaking, the external nose. The nasal cavities constitute the internal nose. Descriptive terms are applied to certain parts of the external nose and these are:

❑ The upper end (where the nose becomes continuous with the forehead) is the root or the ***angle***;
❑ The ridge-like free border passing down from the angle is the ***dorsum nasi*** (or the dorsum of the nose);
❑ The rounded prominence in which the dorsum nasi ends inferiorly is the tip or ***apex***;
❑ The sloping surfaces (on the left and right sides) which are continuous behind with the cheeks are the lateral surfaces;
❑ The lowest parts of the lateral surfaces which are rounded and mobile are the ***alae nasi*** (singular, ala nasi; Latin.alah=wing); and
❑ The upper parts of both lateral surfaces (just below the angle) together form the ***bridge*** (the place where one's spectacles would rest on the nose).

The shape of the nose is maintained by the presence of a skeleton made up partly of bone and partly of cartilage. The upper part of the lateral wall is formed by (1) the

contd...

nasal bone and (2) the frontal process of the maxilla. The anteroinferior part is reinforced by pieces of cartilage: an upper nasal cartilage; a lower nasal cartilage and some minor cartilages. The wall of the ala nasi is formed by fibro-fatty tissue. The ***external nares*** (or anterior nares; singular —naris or narium; Latin.naris=opening of nose) are the external openings of the nasal cavities. They are located on the inferior aspect of the nose and are bounded laterally by the alae nasi and medially by the lowest part of the nasal septum.

> As you reflect the skin flaps of the face, observe that there is practically no deep fascia in the face and muscle fibres are attached to the overlying skin and fascia. Trace as many nerve twigs and vessels as possible. Identify each one of them. From above downwards, locate the following sensory nerves or their twigs: palpebral branch of lacrimal nerve, infratrochlear nerve, infraorbital nerve, external nasal nerve, zygomaticofacial nerve, buccal nerve, great auricular nerve and mental nerve.
>
> As the next step, define the orbicularis oculi, orbicularis oris and the other facial muscles which lie between them. Identify and define the parotid gland. Clean the fascia over the gland. Look for the parotid duct and identify the branches of facial nerve which radiate from the anterior border of the gland. With utmost care, try to see the plexus formed by the branches of facial nerve and the various sensory nerves of the face.
>
> Identify the blood vessels of the region and define them. You can very easily locate the transverse facial artery running above the parotid duct, the tortuous facial artery coursing from base of mandible to angle of mouth and the buccal and infraorbital arteries along with their nerves. Facial vein can also be traced.

LIPS, CHEEKS AND CHIN

The ***lips***, upper and lower, are lined on the outside by skin and on the inside by mucous membrane. However, if you notice carefully, the texture of the skin changes from the prominent borders of the lips. From this border, usually called the ***vermillion border*** (because of its reddish hue), to the 'moist line' (usually the line where both lips meet in the normally closed position of the mouth) is the transition zone (also called the ***vermillion zone***). The moist line is the line of commencement of the intraoral mucous membrane; the vermillion border is the line of commencement of the extraoral skin. Though the transition zone has only skin covering it, the skin here is of a different quality; it is extremely thin and hairless; its keratinisation is also less. Because of these features, this skin is highly sensitive and is of a different colour (the underlying blood vessels are responsible for the reddish hue). The substance of the lip is formed by the orbicularis oris muscle and by numerous smaller muscles that blend with it. The points, on either side, where the upper and lower lips meet are the ***angles of the mouth***. The deep surface of each lip is connected

to the gum by a median fold of mucous membrane called the ***frenulum***. In most individuals, the median part of the upper lip has a small projection called the ***tubercle***. Superior to the tubercle and running up to the nasal septum is a shallow groove called the ***philtrum*** (Greek. phileo=love, philtron=love drug). Each lip is actually a fibromuscular fold. It can be seen that on the sides, the fibromuscular structure continues as parts of the cheeks. On the surface, the cheeks are separated from the lips by the nasolabial sulci (one on either side). The sulci gain differing degrees of prominence during smiling.

The ***cheeks*** are, like the lips, made up of an outer layer of skin, an inner layer of mucous membrane and an intervening layer of muscle, connective tissue and fat. The muscle layer is formed chiefly by the buccinator. The buccal pad of fat is especially prominent in infants and is responsible for the rounded appearance of the cheeks. Numerous glands are present in close proximity of the lips and cheeks. They open into the vestibule of the mouth.

The prominence of the ***chin*** is called the ***mental protuberance*** (Latin.mentum=chin). The horizontal depression between the lower lip and the chin is the mentolabial sulcus.

Fasciae of the Face

Before discussing the deeper structures of the face and related areas, it is better to make a study of the various fasciae. The fasciae of the face can be broadly grouped into three layers, which from superficial to deep are: (1) Subcutaneous fibrofatty layer, (2) Superficial musculoaponeurotic system, and (3) Parotidomasseteric fascia.

Subcutaneous fibrofatty layer: Also called the ***tela subcutanea***, this is a continuous layer immediately beneath the skin, but the amount of fat varies in different parts of the face. Over the lips and the zygomatic arches, the layer is more fibrous than fatty. One specific area that shows aggregation of fat is the cheek mass which contains the malar pad (Latin.malah=cheek) of fat that is inferolateral to the orbit.

Superficial musculoaponeurotic system (SMAS): Sometimes referred to as the ***superficial facial fascia***, this is a single layer that is deep to the subcutaneous fibrofatty layer but superficial to the muscles of face. It consists of muscle fibres in some places, fibrous tissue in some others and fibroaponeurotic in still other places. The muscle fibre or the fibrous tissue is not directly attached to the bone. Anteromedially this layer becomes continuous with some muscles but over the zygomatic arch, it becomes indistinct. Deep to the SMAS, a clear sub-SMAS plane is present. Only over the parotid gland, this plane is absent and the SMAS blends with the superficial layer of parotid fascia.

Added Information

- **Buccal pad of fat:** Otherwise called the **pad of Bichat**, this encapsulated mass of fat lies on the superficial surface of buccinator muscle. In the past, it was called the **suctorial pad** from its prominence in infants and association with sucking.
- A gross comparison drawn between the layers of scalp and layers of face and neck gives the following picture (the temporoparietal, temporal, parotidomasseteric and parotid fasciae are shown in their relative positions):

Scalp		Face		Neck	
Skin		Skin		Skin	
Dense connective tissue		Loose subcutaneous tissue		Thin superficial fascia with platysma	
Epicranius	Temporoparietal fascia	SMAS	Parotidomasseteric fascia		
		Muscles of face			
Loose connective tissue	Temporal fascia	Loose tissue and retaining ligaments	Parotid fascia and parotid gland	Investing layer of deep fascia with muscles enclosed within	
Pericranium	Temporalis muscle and deeper structures	Periosteum	Deeper structures		

Parotidomasseteric fascia: Also called the **deep facial fascia**, this layer is present only over the masseter muscle and the buccal pad of fat. It is a layer of thin areolar tissue and stretches over the branches of the facial nerve, parotid duct and the buccal pad of fat which is superficial to the buccinator (parotidomasseteric fascia should not be confused with the parotid fascia which is the superficial layer of parotid capsule).

Clinical Correlation

- Many parts of the face exhibit features of importance which can be used for diagnosis of internal conditions.
  - The lips are blue in cyanosis.
  - Inflammation at the angles of the mouth (angular stomatitis) occurs in vitamin deficiencies.
  - Conjunctiva becomes pale in anaemia, yellow in jaundice, reddish in conjunctivitis.
- Thyroid gland disorders show several changes on the face.
- Hypothyroidism in infants leads to cretinism. A child with cretinism has a puffed face with a protruding tongue, a bulky belly, and sometimes an umbilical hernia.
- Hypothyroidism in adults leads to deposition of mucopolysaccharides in subcutaneous tissue. The face is bloated. The lips are thick and protuberant and the expression is dull.
- In hyperthyroidism, the eyes become prominent (exophthalmos).
- Interruption of sympathetic supply to the head and neck results in Horner's syndrome. The features of this syndrome are as follows:
  - There is constriction of the pupil because of paralysis of the dilator pupillae. Unopposed action of the sphincter pupillae leads to constriction.
  - There is drooping of the upper eyelid (ptosis) because of paralysis of smooth muscle fibres present in the levator palpebrae superioris.
  - The eyeball is less prominent than normal (enophthalmos).

contd...

Clinical Correlation *contd...*

  - There is absence of sweating on the affected side of the face (secretomotor supply to sweat glands is through sympathetic nerves).
- Injuries (e.g., blows) on the face can lead to fracture of the mandible, the zygomatic bone, the maxilla or the nasal bones. The mandible is commonly fractured.
- The subcutaneous tissue of face is loose and there is no clear deep fascia. Because of these factors, lacerations of face gape widely. Fluid accumulation is faster and more. Considerable swelling can occur even with minimal inflammation or injury.
- Maxilla, mandible and zygomatic bone are derived from the first branchial arch. Occasionally, growth of this arch is defective so that the bones concerned remain underdeveloped, and the face is deformed. The condition is called **mandibulofacial dysostosis**.

EYELIDS AND CONJUNCTIVA

Though the eyeball is placed well within the orbital socket, part of it is clearly visible in the face. Along with the eyelids, it forms an integral part of the face and its location in the face made prominently visible by the eyebrow. The part of the eye seen on the face consists of a part that is white and an anterior circular area that looks dark. The so-called 'white of the eye' is formed by a layer of the eyeball called the **sclera**. The sclera is lined (on the outside) by a thin transparent membrane called the **ocular conjunctiva** (or the **bulbar conjunctiva**)/(Latin.conjunctos=binding together). The dark colour of the anterior circular area is due to the underlying **iris**. The colour of the iris is seen because the **cornea** which is in front of the iris and which is continuous with the sclera is transparent. At the centre of the iris, there is an aperture called the **pupil**. The pupil appears black (or dark) because the interior of the eye (which we are actually looking at when we are looking through the pupil) is dark.

When we view the 'eyes', we see only a small part of the eyeball in the interval between the upper and lower eyelids. This interval is called the **palpebral fissure** (or **rima palpebrae**). The upper and lower eyelids (or palpebrae; Latin.palpebra=eyelid) are lined on their internal surfaces by conjunctiva, which, on account of its location is called the **palpebral conjunctiva**. The eyelids protect the eyeball, especially the cornea, from injury in more than one way.

Firstly, they provide protection against mechanical injury by reflex closure when any object suddenly approaches the eye or when the cornea is touched.

Secondly, they help to keep the cornea moist; when the eyelids are closed (i.e., when the upper and lower eyelids meet) over the eyeball, a space is created between the eyelids anteriorly and the eyeball posteriorly. This capillary space that separates the posterior surfaces of the lids from the cornea and the sclera is the **conjunctival sac** (the space between the palpebral conjunctiva anteriorly and the ocular conjunctiva posteriorly). It contains a thin film of lacrimal fluid, which keeps the cornea and the conjunctiva moist. When the 'eyes' are open, the cornea has a tendency to dry up, but this is prevented by periodic, unconscious closure of the lids (blinking). Every time there is blinking of eye, the film of lacrimal fluid over the cornea is replenished. The conjunctival glands or Terson's glands (clusters of mucosal cells present in the conjunctival epithelium) also add a slightly viscous secretion to the lacrimal fluid. This viscosity prevents faster evaporation of the lacrimal fluid and helps in immobilisation of foreign particles.

Thirdly, the lids protect the eyes from sudden exposure to bright light by reflex closure. In bright light, partial closure of the lids may assist the pupils in regulating the light falling on the retina.

It is already noted that the space separating the upper and lower eyelids is the palpebral fissure. The medial and lateral ends of the fissure are called the **angles** (or **canthi**; singular, canthus; Greek.kanthos=corner of eye). The lateral canthus is in contact with the sclera. At the medial canthus, the upper and lower lids are separated by a triangular interval called the **lacus lacrimalis**. In the floor of this area is a rounded pink elevation called the **lacrimal caruncle** (Fig. 3.3). Lateral to the caruncle is a fold of conjunctiva called the **plica semilunaris**. It represents the third eyelid present in some lower animals (in whom it is called the **nictitating membrane**). The depressions above the upper eyelid and below the lower eyelid are the suprapalpebral and the infrapalpebral sulci respectively.

Each eyelid has a free edge to which eyelashes are attached. Lateral to the lacrimal caruncle, each lid margin has a slight elevation called the **lacrimal papilla**. On the summit of the papilla is a small aperture called the

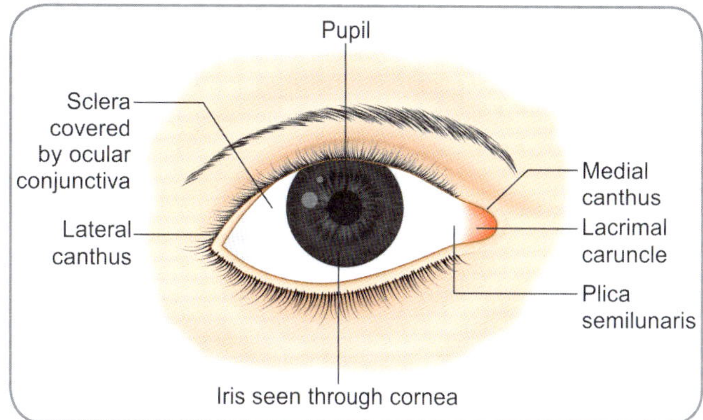

Fig. 3.3: Features of the eye as seen on the face—the eyelashes are omitted; the interval between the two eyelids is the palpebral fissure

lacrimal punctum. The punctum is normally in direct contact with the ocular conjunctiva.

Structure of the Eyelids and Related Structures

The eyelids are moveable fibromuscular folds which, when closed cover the eyeballs. As already noted, each eyelid is covered by the skin on the external aspect and the palpebral conjunctiva on the internal aspect. The palpebral conjunctiva becomes continuous with the corresponding bulbar conjunctiva at the deeper aspects of the sulci between the eyelids and the eyeball. These lines of reflection of the conjunctiva are called the **superior** and **inferior conjunctival fornices**.

The following features (from anterior to posterior) can be seen if a parasagittal section of an eyelid is taken (Fig. 3.4):

- ❏ Anteriorly, a layer of thin skin with very few hair and sweat glands;

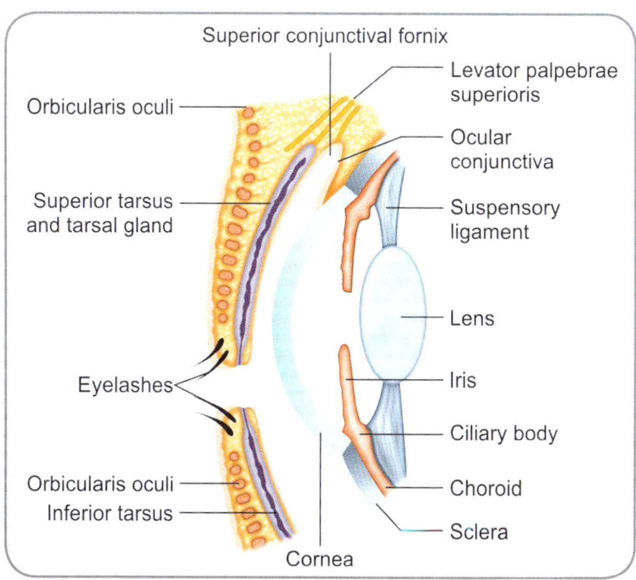

Fig. 3.4: Schematic sagittal section through the eyelids and anterior part of the eyeball

Development

The eyelids arise as ectodermal folds with inner core of mesenchyme at about 46 days of intrauterine life. The free edges grow rapidly and towards each other; the margins contact each other and unite by epithelial fusion. Fusion of the two lids cuts off a space from the exterior and this becomes the conjunctival sac later. The glands, hairs and other associated structures develop from the skin. The muscles and the tarsal plates develop from the underlying mesenchyme. The lids remain in contact until about the 7th month of intrauterine life. It should be remembered that the eyes remain closed until few days after birth in several other species of animals.

Dissection

Attempt dissecting the structures related to the lacrimal apparatus with help from faculty members and senior colleagues. Make a crescentic incision on the skin of the upper eyelid close to the orbital margin. Reflect the skin flaps up and down. With little prodding, locate the lacrimal gland at the upper lateral part of the orbit. Examine the gland. Study the tarsal plate by everting the eyelid. See the fibres of levator palpebrae superioris running to the tarsus, conjunctival fornix and skin of the upper eyelid. Look for the medial and lateral palpebral ligaments. Identify the lacrimal sac in the medial corner of the eye. Trace the fibres of orbicularis oculi in this region and see their relation to the lacrimal sac.

❏ Deep to the skin, a layer of delicate connective tissue, which normally does not contain fat;

❏ Fasciculi of the palpebral part of the orbicularis oculi muscle which considerably contribute to the bulk of the lid;

❏ A firm plate of fibrous tissue called the ***tarsus*** or the ***tarsal plate***;

❏ A series of vertical grooves on the deep surface of tarsi; these grooves lodge the ***tarsal glands*** (or ***Meibomian glands***); and

❏ A layer of palpebral conjunctiva.

Tarsal Plates (Fig. 3.5)

These are dense and firm plates of connective tissue which help maintain the shapes and strength of the eyelids. Each eyelid has a tarsal plate (Greek.tarsos=flat) and so there are the superior and inferior tarsal plates (or tarsi; singular, tarsus) related to each eye. The tarsal plates can be described to be the skeleton of the eyelids. The shape of each tarsus corresponds to that of the corresponding eyelid. They are usually semi-oval having a straight edge directed towards the palpebral margin and a convex edge directed towards the orbital margin. The 'height' of the tarsus is greatest midway between its medial and lateral

ends. It is about 10 mm in the upper lid and about 5 mm in the lower lid. The tarsi narrow down laterally and medially and become continuous with the ***lateral*** and ***medial palpebral ligaments*** through which they are attached to the walls of the orbit, just inside the orbital margin. The medial palpebral ligament is better developed than the lateral ligament. It gains attachment to the maxilla (lacrimal crest and adjoining part of frontal process). The lateral ligament is attached to the zygomatic bone. The upper and lower tarsi are attached to the corresponding parts of the orbital margin by a membrane called the ***orbital septum***.

On the deep surface of the tarsi and usually embedded in them are the tarsal glands (palpebral glands or Meibomian glands; named after the 17th century German anatomist Hendrik Meibom). Each gland has a duct that opens on the free margin of the lid. The tarsal glands are modified sebaceous glands and produce an oily (lipid) secretion. This secretion lubricates the edges of the eyelids and prevents them from sticking to each other when they close. It also spreads as a thin film over the lacrimal fluid (tears) and delays its evaporation.

The fibres of the palpebral part of orbicularis oculi are present anterior to the tarsal plates. In the upper eyelids, yet another muscle makes its appearance (Fig. 3.6.). The levator palpebrae superioris extends into the upper eyelid (of its corresponding side) anterior to the tarsal plate to gain insertion into skin of the eyelid. This muscle elevates the upper eyelid, while opening the palpebral fissure. Some fibres of the muscle, which are inserted into the superior conjunctival fornix, pull the fornix up during this movement. The fibres of the orbicularis oculi are disposed circularly around the eye and so contraction of this muscle closes the palpebral fissure (similar to sphincteric action).

The eyelashes are in the margins of the lids. Sweat glands adjacent to the hair follicles and small sebaceous glands opening into the hair follicles together constitute the ciliary glands (also called the ***glands of Moll***).

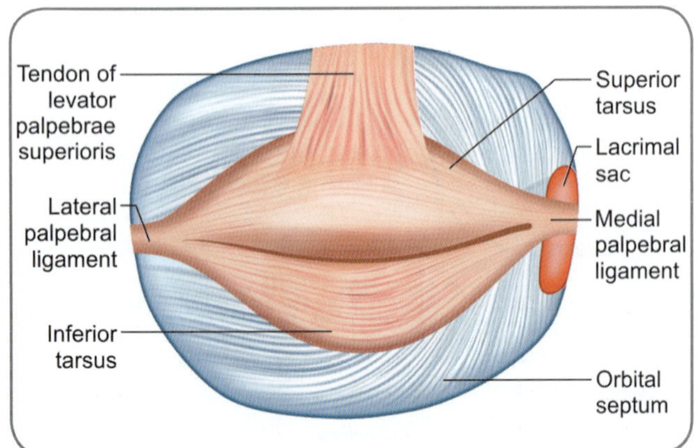

Fig. 3.5: Scheme to show the tarsal plate and the palpebral ligament as seen from the front

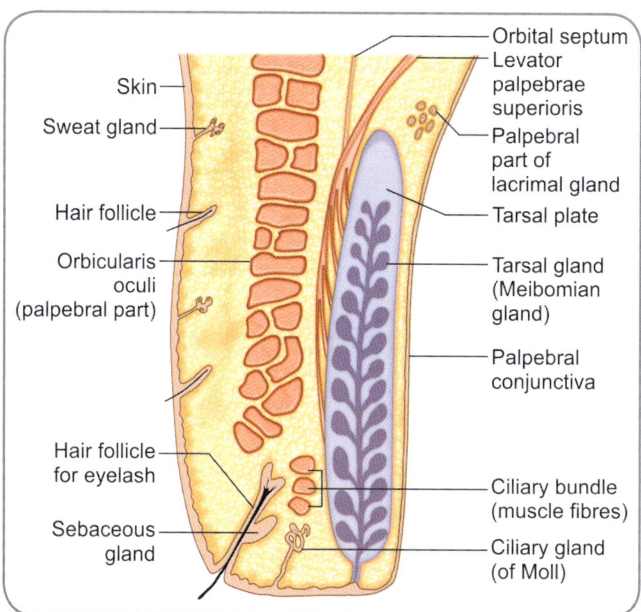

Fig. 3.6: Schematic sagittal section through the upper eyelid

Muscles of the Eyelids

The muscles of the eyelids can be classified into ***voluntary*** and ***involuntary groups***.

Voluntary group of muscles: Structurally, the palpebral part of the orbicularis oculi constitutes the muscular part of the eyelids. However, the orbital part of the muscle also plays an important role in closing the eyelids. Therefore, the entire orbicularis oculi can be considered to be a muscle of the eyelids.

For descriptive purposes, the voluntary group can be subclassified into three—(1) ***sphincteric part***, (2) ***dilator part*** and (3) ***accessory part***.

1. The ***sphincteric part*** comprises the orbicularis oculi muscle. The fibres of the palpebral part of the muscle arch from the medial to the lateral palpebral ligament. On contraction, they close the palpebral fissure. Though these are voluntary fibres and supplied by the facial nerve, they have an involuntary role as in involuntary closure of the eyelids in sleep and in blinking. The fibres of the orbital part are attached to the medial palpebral ligament. From here, they pass across the forehead and temple and arch back across the cheeks to reach the medial ligament again. Since there is no lateral attachment, on contraction, these fibres draw the eyelids medially.

2. The ***dilator part*** comprises the levator palpebrae superioris muscle. Thus, the dilator is present only in the upper eyelids. Contraction of the muscle causes elevation of the upper eyelids and, therefore, opens the palpebral fissure.

3. The ***accessory part*** comprises the pars lacrimalis portion (lacrimal part, also called the ***tensor tarsi***) of the orbicularis oculi. This is a small band of muscle fibres present in the deep part of the orbicularis oculi in the upper eyelids. These fibres pass behind the lacrimal sac and get attached to the posterior lacrimal crest. Contraction of these fibres has a dual effect. On one hand, the upper eyelids are kept closely approximated to the eyeball preventing lodging or accumulation of foreign particles; on the other, the lacrimal sac is kept dilated, creating a suction force to draw tears into it.

Involuntary group of muscles: These are two muscles, one in the upper and the other in the lower eyelids. Called the ***superior*** and the ***inferior tarsal muscles***, on contraction, they cause widening of the palpebral fissure.

Blood Vessels of the Eyelids

The eyelids are supplied by (a) the medial palpebral branches of the ophthalmic artery, and (b) the lateral palpebral branches of the lacrimal artery.

Nerves of the Eyelids

The upper eyelid receives branches from the following nerves (all are branches of the ophthalmic division of the trigeminal nerve)—(a) Lacrimal, (b) Supraorbital, (c) Supratrochlear and (d) Infratrochlear nerves.

The lower eyelid receives branches from the infraorbital and infratrochlear nerves.

Added Information

- The eyelashes are called ***cilia*** (Latin.silium=eyelash). The eyebrows are supercilia. The prominent arches seen immediately above the eyebrows are the superciliary arches.
- The lid margin of both upper and lower eyelids has a 'grey line'. This line corresponds to the mucocutaneous junction. The eyelashes are anterior to this line and the openings of the meibomian glands posterior.
- A small fold of skin covers the medial canthus of the eye in people of some races (eg., East Asians). This fold is called the ***epicanthal fold***.
- The orbital septum connects the tarsi to the orbital margins. It merges with the periosteum at the orbital margins. Spread of infection to and from orbit is prevented because of this merger.
- The orbital septum is derived from the fascia on the posterior aspect of the orbicularis oculi muscles.
- The lacrimal caruncle is developmentally a detached portion of the lower eyelid.
- The lacrimal puncta are directed backwards so as to be applied to the eyeball under normal circumstances; this will permit sucking of tears easily.
- The lateral five-sixths of the margins of the eyelids have eyelashes, are flat and constitute the ciliary part of the eyelid. The medial one-sixth which does not have any cilia is rounded and forms the lacrimal part.

contd...

Added Information *contd...*

- The lacrimal papilla and the puncta are at the junction of the ciliary and the lacrimal parts of the eyelid margins.
- The conjunctiva is adherent to the back of the eyelids and the cornea. Over the sclera and the fornices it is loose.
- The eyelids, by way of repeated blinking, keep a film of tears spread over the eyeball. In addition, the upper eyelids wipe away dust and foreign particles.
- The fibres of the orbital part of orbicularis oculi, on contraction, draw the eyelids medially. This action causes tears and dust particles to move to the medial angle of the eye, from where they can be removed easily in lacrimal circulation. These fibres cause the 'crow's feet' wrinkles when the eyes are tightly closed.
- Contraction of the orbital part of the orbicularis oculi causes vertical wrinkles above the root of nose, narrows palpebral fissure and bunches and protrudes the eyebrows. This results in the medial parts of the eyebrows and the eyelids project like a hood, thereby reducing the amount of light entering the eyes. This action is being supplemented by the procerus which produces horizontal wrinkling and drawing down of the medial part of eyebrows. Glare of bright sunlight is cut off by these actions.
- The older spelling of 'lachrymal' was based on Greek-Latin phonetics. The simplified 'lacrimal' has been widely and officially accepted for the sake of convenience.

Clinical Correlation

- Infections are common in the eyelids (**blepharitis**).
- If the duct of a ciliary gland is obstructed, a painful pus producing swelling occurs on the eyelid. This swelling is called a **stye** or **hordeolum**. Inflammation of a tarsal gland results in a localised swelling called **chalazion**. Obstructed ciliary glands project on the front of the eyelid. Obstructed tarsal glands project on the globe of the eyeball.
- Inflammation of the conjunctiva is called **conjunctivitis**. It may be caused either by infection or by allergy.

LACRIMAL APPARATUS (FIG. 3.7)

The lacrimal apparatus (Latin.lacryma=a tear) consists of the organs concerned with the secretion and drainage of the lacrimal fluid.

Lacrimal fluid is secreted by the **lacrimal gland** and is poured through the excretory ducts of the gland into the conjunctival sac at the superior conjunctival fornix. The fluid then passes downwards and medially to reach the lacus lacrimalis (Latin.lacus=lake, lacrimalis=of the lacrimal), which is the triangular space at the medial angle of the eye where tears collect. From here, the fluid passes through the lacrimal puncta into narrow tubes called the **lacrimal canaliculi**. These canaliculi open into the **lacrimal sac** (dilated superior part of the nasolacrimal duct). The **nasolacrimal duct** opens into the inferior nasal meatus.

Lacrimal gland: This is an almond-shaped structure lodged in the fossa for the lacrimal gland in the superolateral

Development

Most parts of the lacrimal apparatus arise by means of cord formation and subsequent canalisation. By the 6th week of intrauterine life, a number of solid epithelial cords arise from the superolateral part of the conjunctival sac. By the 3rd month, these cords branch repeatedly; canalisation occurs and an acinous gland is formed. Simultaneous with the changes occurring around the 3rd month of intrauterine life, solid epithelial buds arise near the medial ends of the eyelids. These will eventually canalise into the lacrimal canaliculi. In the lower lid, the portion medial to the canaliculus will differentiate into the caruncle and plica. The two canaliculi fuse medially and join another cord that develops vertically down between the maxillary and the frontonasal processes and that which canalises to form the lacrimal sac and the nasolacrimal duct.

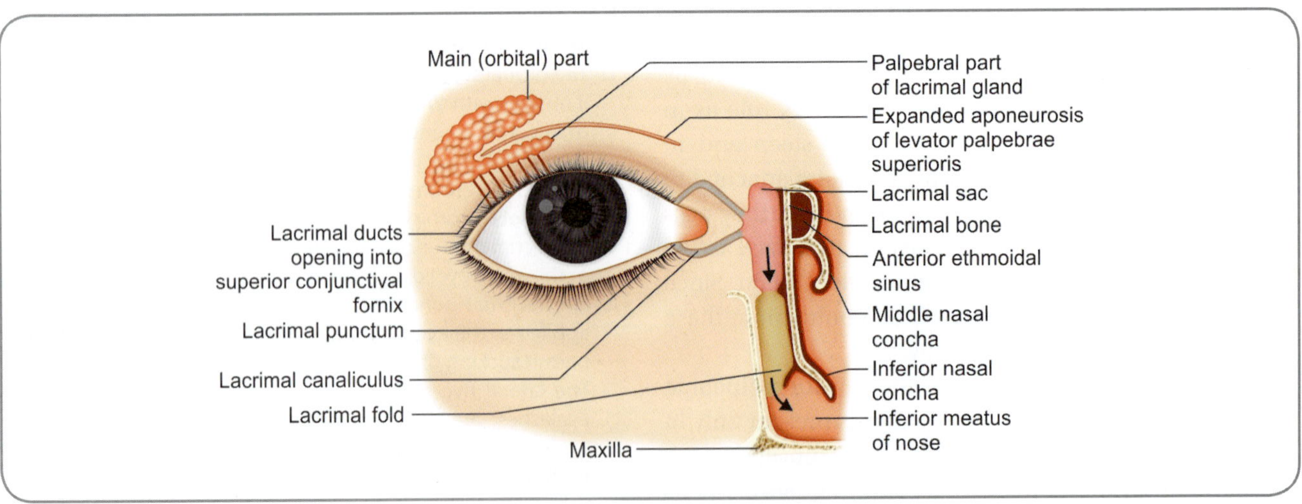

Fig. 3.7: Scheme to show the parts of lacrimal apparatus; arrows indicate the direction of flow of lacrimal fluid

part of the orbital cavity. A lateral expansion of the tendon of levator palpebrae superioris passes into the gland to divide it into two parts: the inferior palpebral part (also called *Galen's innominate gland*) and the superior orbital part. About 8 to 10 (lacrimal) excretory ducts open into the lateral aspect of the superior conjunctival fornix.

Lacrimal canaliculi: There are two lacrimal canaliculi; upper and lower, in relation to each eye. Each canaliculus is about 10 mm long. It is a narrow tube that starts at the lacrimal punctum and ends by joining the lacrimal sac. The canaliculi transmit lacrimal fluid from the lacus lacrimalis to the lacrimal sac.

Lacrimal sac: The lacrimal sac (Fig. 3.8) lies in the lacrimal groove on the medial wall of the orbit. The lacrimal sac is blind at its upper end. Inferiorly, it is continuous with the nasolacrimal duct. Laterally, it receives the lacrimal canaliculi near its upper end. The osseofibrous relations of the sac are important. Medially, a thin plate of bone separates it from the anterior ethmoidal sinus and the middle nasal meatus. Laterally the lacrimal fascia covers it. The medial palpebral ligament is anterior to the sac and the lacrimal fascia. The lacrimal part of the orbicularis oculi muscle, as it passes behind the lacrimal sac, gains some attachment to the lacrimal fascia. Every time the muscle contracts, there is pull on this fascia causing the lacrimal sac to expand and to suck lacrimal fluid into it.

Nasolacrimal duct: The nasolacrimal duct is a tube about 18 mm long. It may be regarded as the downward continuation of the lacrimal sac. It is closely related to the lateral wall of the nasal cavity and opens below into the inferior meatus of the nose. The wall of the nasolacrimal duct is made up of bone lined by mucous membrane. The lower end of the nasolacrimal duct is separated from the inferior meatus of the nose by a fold of mucous membrane called the ***lacrimal fold***.

Nerve supply of the lacrimal gland: The lacrimal gland has both sympathetic and parasympathetic innervation.

❑ ***Parasympathetic innervation:*** This is secretomotor. The preganglionic fibres arising from the lacrimatory nucleus go through the facial nerve, the greater petrosal nerve and the nerve of pterygoid canal to reach the pterygopalatine ganglion. They synapse with the cell bodies in the ganglion. The postganglionic fibres traverse through the maxillary nerve and its zygomatic branch to reach the lacrimal branch of the ophthalmic nerve and then enter the gland.

❑ ***Sympathetic innervation:*** This is vasoconstrictive. The postganglionic fibres arising from the superior cervical ganglion go through the internal carotid plexus, deep petrosal nerve, nerve of pterygoid canal to reach the pterygopalatine ganglion. They pass through the ganglion without any synapse to join the maxillary nerve. They follow the same path as the parasympathetic fibres, go through the zygomatic branch of the maxillary nerve, reach the lacrimal branch of the ophthalmic nerve and enter the gland.

Lacrimal circulation: Production of lacrimal fluid is stimulated by parasympathetic impulses. From the gland, the fluid flows through excretory ducts into the superior conjunctival fornix. From here, effect of gravity makes the lacrimal fluid move down. Whenever the eyes blink, the eyelids close sequentially from lateral to medial side. The lacrimal fluid is pushed over the cornea to the medial canthus. In this way, the cornea is cleansed of dust; foreign material and dust which could have settled on the cornea, sclera or conjunctiva are pushed into the medial area. The lacrimal fluid is sucked into the lacrimal

Histology

Histology of Lacrimal Gland

The lacrimal gland is a compound serous gland. The glandular substance is subdivided into several lobules by connective tissue septa. The lobules contain tubuloalveolar acini which are of different shapes and sizes. Each lobule may have about six to seven acini. Each acinus has a lumen surrounded by pyramidal or columnar secretory cells. The nuclei of the cells are rounded and occur at the bases of the cells. Secretory granules are present in the apical portions of the cells. Myoepithelial cells surround each acinus and are placed between the basement membrane and the secretory cells.

The secretions of an acinus are taken by its duct. Ducts from all the acini of the lobule join to form the intralobular excretory duct which has a small lumen surrounded by simple cuboidal cells. The intralobular ducts of many lobules join to form the interlobular duct which usually has two layers of cuboidal or columnar epithelium. Several such ducts carry the secretions to the conjunctival sac.

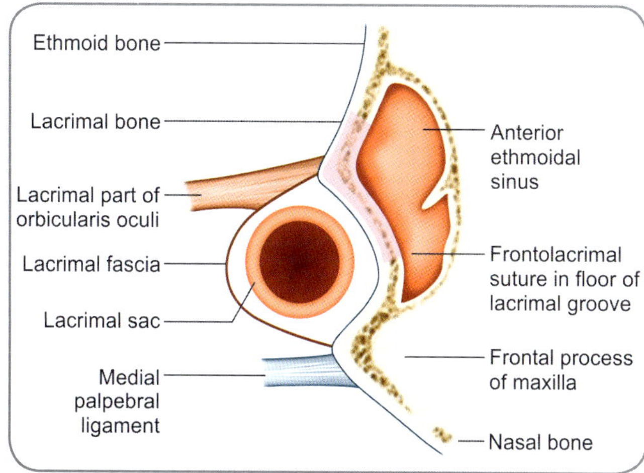

Fig. 3.8: Transverse section through lacrimal sac and surrounding structures

sac through the lacrimal puncta by capillary action. Dust and foreign material also are washed away. The fluid then passes through the nasolacrimal duct and reaches the inferior nasal meatus. It then drains posteriorly into the nasopharynx and subsequently swallowed. Apart from acting as a cleansing agent, the lacrimal fluid provides the cornea with oxygen and nutrients.

Accessory lacrimal glands: Small pieces of lacrimal glandular tissue may occur along the conjunctival fornices or in the central part of the upper eyelid. These secrete a watery fluid. These glands are variedly called the ***Krause's glands, Wolfring's glands*** and ***Ciaccio's glands***.

MUSCLES OF FACE (FIG. 3.9 AND TABLE 3.1)

The facial muscles are otherwise called the ***muscles of facial expression***. They are in the subcutaneous tissue of face and neck. They move the skin of the face and help express various moods. Most of these muscles have an attachment to the underlying bone or fascia and another to the skin. When they contract, they pull the skin to produce various expressions. All the muscles of the face are supplied by the facial nerve.

Dissection

While trying to dissect the deep muscles of the face, locate and identify the buccinator. Study the fascia over it. It is the buccopharyngeal fascia. Split the fascia and define the attachments of the muscle. Observe the orbicularis oris. Study the mingling of the fibres of the various muscles of the region.

Though it is customary to talk of all muscles of the face together, they can be classified according to their location and action. Thus, there are:

- Muscles of ocular opening and eyebrows
- Muscles of nose

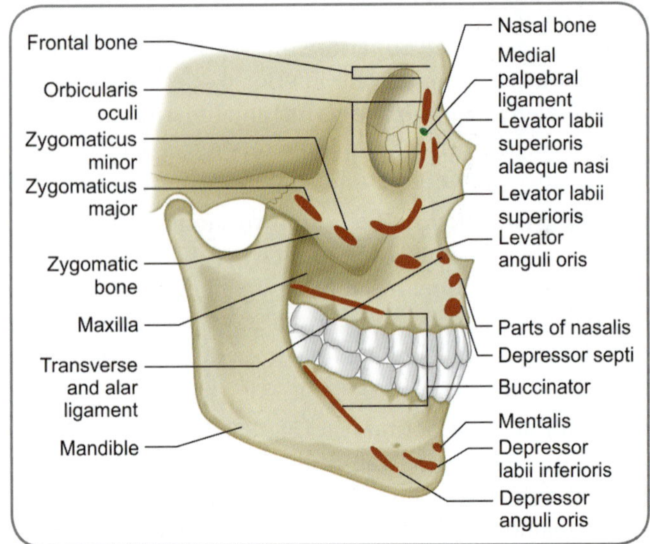

Fig. 3.9: Facial skeleton seen from the side to show muscular attachments

- Muscles of oral opening
- Muscles of cheeks and lips

In each group, there are two subgroups—(a) muscles which are primarily involved in actions of the area can be called the ***principal muscles of the area***; and (b) muscles which have a minor role can be called the ***secondary muscles of the area***. It should be noted that there is considerable overlap in the actions of various muscle groups and individual muscles, thus making the classification, one of descriptive convenience and not of exclusive functionality.

Muscles of Ocular Opening

- Principal muscle—Orbicularis oculi (Table 3.2).
- Secondary muscles—Procerus, corrugator supercilii, transverse part of nasalis.

Table 3.1: Muscles of face					
Name	**Location**	**Origin**	**Insertion**	**Action**	**Bony Attachment**
Corrugator supercilii (Latin. corruga=wrinkle)	Above eyebrow, deep to frontalis	Medial end of superciliary arch	Skin over the supraorbital margin	Produces vertical wrinkles on forehead and upper nose—***expression of worry***	
Depressor supercilii	Near eyebrow, merging with orbicularis oculi	Medial end of superciliary arch	Skin over forehead and root of nose	Depresses eyebrow	
Procerus (Latin. procero=stretched out)	Over bridge of nose	Aponeurosis covering nasal bone and lateral nasal cartilage	Skin between eyebrows	Produces transverse wrinkles over bridge of nose	
	The fibres of transverse part of nasalis remain close to and act along with procerus; they produce depression of medial end of eyebrow and wrinkling of skin over the dorsum nasi—***expression of dislike and irritation; expression of concentration.***				

contd...

contd...

Nasalis (compressor naris and dilator naris parts)	Over lower part of nose	Frontal process of maxilla (Inferomedial margin of orbit)	Skin of ala of nose and major alar cartilage	Constricts or dilates anterior nares	Maxilla
	The fibres of alar part (dilator part) of nasalis are close to and act along with levator labii superioris alaeque nasii. They depress the ala laterally when levator labii superioris alaeque nasi causes dilatation of the anterior naris. This causes *'flaring of nostrils'*—**expression of anger, exertion and anxiety**				
Depressor septi (this is considered a part of dilator naris)	Just below nasal septum	Fascia of upper lip and maxilla close to the anterior nasal spine	Lower end of nasal septum	Pulls nasal septum down and helps to dilate anterior nares	Maxilla
Levator labii superioris alaeque nasii	From upper part of nose • One slip to nose • One slip to upper lip	Frontal process of maxilla (Inferomedial margin of orbit)	Skin of upper lip and major alar cartilage	Raises ala nasi and upper lip	Maxilla, frontal process
Levator labii superioris	Above upper lip	Infraorbital margin of maxilla	Skin of upper lip	Raises upper lip	Maxilla and zygomatic bone
Levator anguli oris	Deep to levator labii superioris	Area of Canine fossa of maxilla	Modiolus (angle of mouth)	Raises angle of mouth and widens oral fissure—expression of grimace and grin	Maxilla below infraorbital foramen
Zygomaticus major	Above lateral angle of upper lip	Lateral aspect of Zygomatic bone	Modiolus (angle of mouth)	Pulls angle of mouth upwards and laterally as in laughing	Zygomatic bone
	• When the muscles of both sides act, the angles of mouth on both sides are pulled up—**expression of happiness and smile** • When the muscle of only one side acts, the angle of mouth of that side alone gets pulled up—**expression of disdain**				
Zygomaticus minor	Above lateral angle of upper lip	Anterior aspect of zygomatic bone	Skin of upper lip	Pulls up upper lip	Zygomatic bone
• Levator labii superioris and Zygomaticus minor usually act in unison. Together they produce elevation and eversion of upper lip and deepen the nasolabial furrow—**expression of sadness**					
Risorius (Latin.risor=laughter)	Merges with orbicularis oris at lateral angle of mouth	Parotid fascia and sometimes buccal skin	Modiolus (angle of mouth)	Retracts angle of mouth	
Depressor anguli oris	Below angle of mouth	Anterolateral basal area of mandible	Modiolus (angle of mouth)	Depresses angle of mouth	Mandible
• Depressor anguli oris and risorius act together. While acting bilaterally, they produce expression of frown					
Depressor labii inferioris	Merges with orbicularis oris from below	Platysma and anterolateral body of mandible	Skin of lower lip	Depresses lower lip; when acting bilaterally, eversion and pouting of lower lip occurs —**expression of sadness or loss**	Mandible
Mentalis (Latin.mentum=chin)	Over chin	Body of mandible below the incisor teeth	Skin of chin	Elevates and protrudes chin; Produces wrinkles on chin—**expression of doubt**	Mandible
Incisivus labii superioris	Above upper lip near midline	Maxilla near upper incisors	Skin of upper lip	Anchors upper lip to maxilla	Maxilla
Incisivus labii inferioris	Below lower lip near midline	Mandible near lower incisors	Skin of lower lip	Anchors lower lip to mandible	Mandible

Table 3.2: Large muscles of face

	Orbicularis oris	Orbicularis oculi	Buccinator

Arrangement of fibres	• The fibres surround the opening of the mouth
	• They form the muscular basis of the lips
	• Many facial muscles merge with it
	• The muscle is anchored to the maxilla by the incisivus superior and to the mandible by the incisivus inferior
		• Muscle fasciculi arranged concentrically around palpebral fissure
		• The innermost rings lie in the eyelids and form the palpebral part
		• Succeeding rings surround the margin of the orbit. They form the orbital part
		• Fibres closely related to the lacrimal sac form the lacrimal part
		• Fibres of orbital part are attached on the medial side to:
		  – Nasal part of frontal bone
		  – Medial palpebral ligament
		  – Frontal process of maxilla
		• The fibres of the palpebral part are attached medially to the medial palpebral ligament
		• They decussate on the lateral side to form the lateral palpebral raphe
			• The buccinator has a C-shaped line of origin from
			  – Maxilla just above molar teeth
			  – Pterygomandibular raphe
			  – Mandible below molar teeth.
			  (The pterygomandibular raphe is a fibrous raphe attached above to the ptergoid hamulus and below to the mandible)
			• Fibres run forward and become continuous with orbicularis oris

Nerve supply	Lower buccal and mandibular branches of facial nerve	Temporal and zygomatic branches of facial nerve	Lower buccal branches of facial nerve
Action	• Closes the lips • Performs complex movements required in speech, eating, drinking	• Closes palpebral fissure • The lacrimal part helps to suck lacrimal fluid into the lacrimal sac	• Aids mastication by pushing food between the teeth • Increases air pressure in the mouth as in blowing

Muscles of Nose

❏ Principal muscles—nasalis, levator labii superioris alaeque nasi.
❏ Secondary muscles—procerus, levator labii superioris.

Muscles of Oral Opening

❏ Principal muscles—orbicularis oris, buccinator (Table 3.2; Figs 3.10 and 3.11).
❏ Secondary muscles—All muscles which act on the upper and lower lips.

Muscles of Cheeks and Lips

❏ Principal muscles (of the upper lip) levator labii superioris, levator anguli oris. Depressor labii inferioris, depressor anguli oris (of the lower lip).

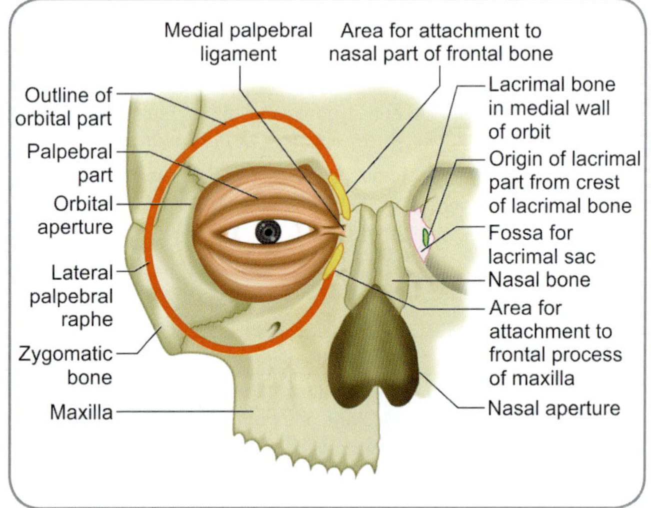

Fig. 3.11: Scheme to show the attachments of the orbicularis oculi muscle

❏ Secondary muscles (of the upper lip) zygomaticus major, zygomaticus minor, incisivus superior, risorius, mentalis, platysma, incisivus inferior (of the lower lip).

Additional Notes on Muscles of Oral Opening, Cheeks and Lips

❏ The orbicularis oris can be described as the first of the sphincters of the alimentary system. Encircling

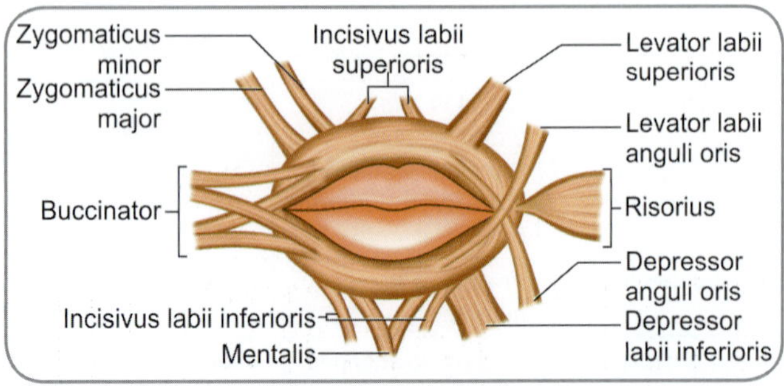

Fig. 3.10: Orbicularis oris and its relationship to various muscles attached to the lips

the mouth, this muscle controls entry into and exit from the oral fissure. It also plays an important role in articulation.

❏ The buccinator (Latin.buccinator=trumpeter) is a thin, flat, rectangular muscle occupying a deeper plane than the other muscles of the face. It passes deep to the mandible and, therefore, is more closely related to the buccal mucosa than to the skin of the face. The muscle is active while smiling. It also keeps the cheeks taut, thereby preventing the cheeks from getting caught between teeth during chewing and movements of the mouth (Fig. 3.12).

❏ The tone of buccinator keeps the cheeks and lips compressed against the teeth and gums. This gives a gentle but continued resistance to the natural tendency of the teeth to tilt outwards. Overactivity of retractors (retraction of upper lip is its elevation; retraction of lower lip is its depression) will neutralise this resistance which is required for normal positioning and occlusion of teeth.

❏ During chewing, food is kept between the occlusal surfaces of the teeth and is not pushed into the vestibule by the action of three muscles (or muscle groups). Each one of them acts from a different aspect. The orbicularis oris acts from the labial aspect, the tongue from the lingual aspect and the buccinator from the buccal aspect.

❏ The buccinator resists distension while blowing and also resists force produced during whistling and sucking.

❏ The buccinator is lined on its internal aspect by mucous membrane. On its external aspect, it is related to the parotid duct which pierces it opposite the third upper molar tooth.

❏ Several muscles help in the opening of the oral fissure and so are called the ***dilators of the mouth***. Levator labii superioris, levator anguli oris, zygomaticus major, zygomaticus minor, depressor labii inferioris, depressor anguli oris and Risorius are the dilators.

Modiolus

The modiolus or modiolus labii (Latin.modiolum=nave of a wheel) is a fibromuscular mass located just lateral to the buccal angle and formed by the interlacing of various muscle fibres. Atleast 7 to 9 muscles (the number depending on the classification) join each modiolus. It is defined as a conical structure with a truncated apex. The elliptical base, called the ***basis moduli***, lies adjacent to the mucosa and is in fact adherent to the latter. The truncated apex, called the ***apex moduli***, lies close to the dermis. The total thickness of the modiolus from the base to the apex is about 10 mm and is divided into three parts: the basal, central and apical parts. The central part usually has an oblique fibrous cleft for the facial artery to pass through; this arrangement partially protects the artery from being compressed by the buccolabial musculature. Two extensions from the basal part run towards the angle of the mouth. These extensions are the superior and the inferior cornua. The superior cornu extends into the soft tissues of the upper lip. The inferior cornu extends into the soft tissues of the lower lip but not to the extent the superior cornu does. With the cornua in position, the modiolar base acquires a reniform (kidney shape) disposition with the concavity (hilum) facing medially; the angle of the mouth projects into the hilum. The apex of the modiolus is adherent to panniculus carnosus.

The two modioli integrate the activities of the cheeks, lips and oral fissure and the jaws. This integration comes into play in all activities involving the above-mentioned structures; activities like biting, chewing, drinking, sucking, swallowing, speaking, singing, shouting, screaming, crying, grimacing, blowing and making facial signs call for modiolar integration. Modiolus itself can move to varying degrees during these movements. However, it becomes immobile when the mouth is wide open. It is shifted (pushed) superficially when the vestibule of the mouth is filled with contents or ballooned with air.

Retaining Ligaments of the Face

The retaining ligaments of the face are fascial bands which anchor the facial skin to the underlying bone. They originate in the fibrofascial tissue and periosteum of the bone, penetrate the muscle layer (SMAS) and insert into the dermis. At both ends, they are broad and have ramifications. Thus, they provide strong adhesion between the skin and the bone at specific locations.

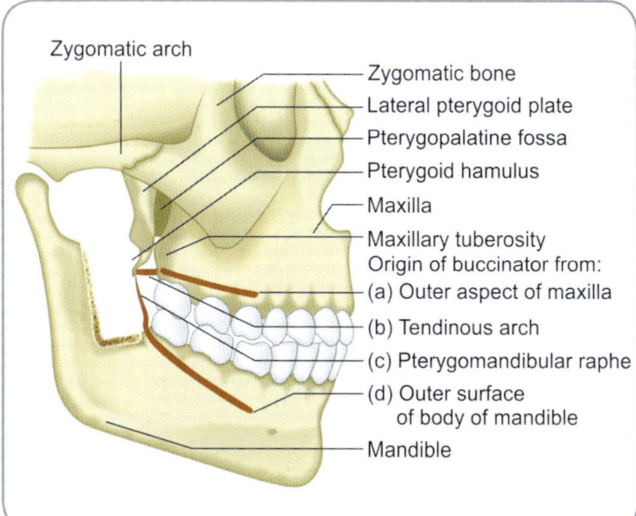

Fig. 3.12: Origin of the buccinator muscle

The following retaining ligaments have been identified:
- Parotid cutaneous ligament connects preauricular skin to the parotid fascia and the ramus of mandible;
- Zygomatic ligament (otherwise called the ***Mcgregor's patch***) connects the skin over the cheek prominences to the body and inferior border of zygoma;
- Masseteric ligament connects the skin over the anterior border of masseter to the ramus and body of mandible;
- Mandibular ligament connects skin over the body of mandible in the anterior third to the underlying bone;
- Orbital ligament connects infraorbital skin to the infraorbital rim.

The facial skin is more prone to gravitational sagging. This sagging is seen during ageing. In young age, muscle and subcutaneous tissue tone keep the skin fairly taut over the underlying tissues; as age advances, gradual decrease in tonus and the general cutaneous laxity of the region permit gravity to pull the skin down. In places where the retaining ligaments attach the skin to the underlying tissues, gravity is not able to cause sagging. The skin between the ligaments slides down due to gravity. This causes 'pouch formation' between the retaining ligaments; if the sagging pouches are prominent, the lines along the retaining ligaments (where there is adherence of tissues) appear as grooves or depressions.

The presence of mandibular ligaments keeps the skin over the chin closely applied to the underlying tissues and bone. Even when the rest of the face undergoes extreme sagging, the chin region alone remains unsagged. Even in early ageing, a bulge appears along the mandibular border, between the mandibular ligament and the masseteric ligament. This sagging bulge is called the ***jowl*** or ***jowl deformity*** (Old english.choll, ceol=throat; indicating any slack flesh of the cheek, jaw or throat region). A malar pouch develops between the orbital ligament and the zygomatic ligament. The malar pouch is seen in conditions of stress, reduced hydration and tiredness even when signs of ageing have not picked up in other parts of the face. The groove seen in the midcheek, aggravated in old age, overlies the zygomatic ligament.

The zygomatic ligament may cause a regular groove in many individuals even in young age. It serves as the major contributor to cheek dimpling during smiling and grinning.

PAROTID REGION

The parotid region is the posterolateral aspect of the face. Its boundaries are:
- Anteriorly—anterior border of Masseter muscle, posteriorly—external ear and anterior border of sternocleidomastoid,
- Superiorly—zygomatic arch, inferiorly—inferior border of mandible including the angle,
- Medially—ramus of mandible, laterally—skin of face.

The region includes the parotid gland and duct, the facial nerve embedded within the gland (this part is otherwise called the ***parotid plexus of facial nerve***), the retromandibular vein, the external carotid artery and the masseter muscle.

The most important content of the region is the parotid gland.

PAROTID GLAND (FIG. 3.13)

The parotid gland (Greek.para=near, otos=ear) is the largest of the three salivary glands. It lies on the lateral

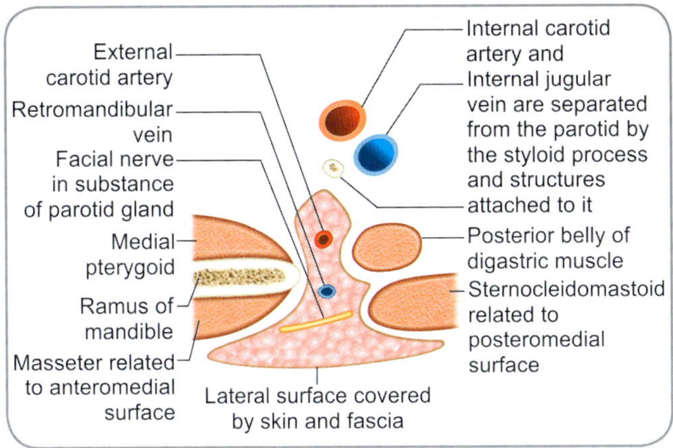

Fig. 3.13: Schematic transverse section through parotid gland to show its relations

side of the face in a depression below the external acoustic meatus, behind the mandible and in front of the sternocleidomastoid muscle.

Parotid Capsule

The gland is enclosed within a tough fascial sheath called the **parotid sheath** or **capsule**, which is derived from the investing layer of deep cervical fascia. At the lower end of the gland, the investing layer splits into two layers.

- One layer ascends superficial to the parotid gland to reach the zygomatic arch to which it is attached. This fascia is thick and strong and is called the **parotid fascia**.
- The other layer ascends deep to the gland. Part of this layer forms a thickened band extending from the posterior margin of the ramus of the mandible to the styloid process. This band is the **stylomandibular ligament**. It separates the parotid gland from the submandibular gland.

Dissection

After studying the various muscles of the face and locating the branches of the facial nerve, concentration can now be shifted on the parotid gland. Remove the fascia over the gland. Define the margins of the gland. Locate the parotid duct (this would have been done earlier). Look out for lymph nodes on the superficial aspect of the gland. With careful but piecemeal dissection, identify the various structures which run through the gland.

Surfaces of Parotid Gland

Though usually described to be triangular or conical, the gland actually has an irregular shape. The reason for the irregularity is that the space occupied by the gland, called the **parotid bed**, is irregular in shape. The gland is wedged between the ramus of mandible and the mastoid process. The wedging causes the medial aspect to be divided into

anteromedial and posteromedial aspects and so the gland appears triangular in cross-section.

Though irregular in outline, the superior aspect of the gland is broader than the inferior aspect; the inferior end which lies below and behind the angle of mandible is called the **apex**; the superior aspect which is related to the external acoustic meatus and the zygomatic arch is the base or the superior surface. Apart from the superior surface, three other surfaces of the gland can be described thus—(a) A (subcutaneous) lateral or superficial surface, which is almost flat, (b) An anteromedial surface and (c) A posteromedial surface.

Superior surface (or base): It is concave and fits under the external acoustic meatus. It is also in contact with the posterior surface of the temporomandibular joint. Close to this surface, the auriculotemporal nerve curves around the neck of mandible.

Apex: It is related to the posterior belly of digastric muscle.

Superficial surface: It extends from the apex to the base. Anteriorly, it is prolonged forward superficial to the masseter and, posteriorly, it overlaps the anterior margin of the sternocleidomastoid. This surface is covered by skin and fascia; the superficial fascia contains the facial branches of great auricular nerve, fibres of platysma and the lymph nodes of the superficial parotid group.

Anteromedial surface: This surface is in contact with the posterior border of the ramus of the mandible. As it slopes from before backwards, this surface is related to the masseter muscle, the ramus of mandible and the medial pterygoid muscle. Branches of facial nerve emerge on the face from the anterior border of this surface. The anteromedial surface is usually grooved by the posterior border of the ramus of mandible and subdivided into a larger superficial and a smaller deep part by the groove. The two parts are connected by an isthmus.

Posteromedial surface: This surface is in front of the mastoid process and the sternocleidomastoid. It is related, in its upper part, to the mastoid process and in the lower part, superficial and laterally, to the sternocleidomastoid and deep and medially, to the posterior belly of digastric. This surface is grooved by the external carotid artery before the latter enters the gland. It is also in close relation to the styloid process and the structures attached, which separate the gland from the internal carotid artery and the internal jugular vein.

Structures within the Gland (Fig. 3.14)

The parotid gland gains more significance because several nerves and vessels are related to it. The facial nerve, the external carotid artery and the retromandibular vein pass through it in such a way that they are virtually embedded in the glandular parenchyma.

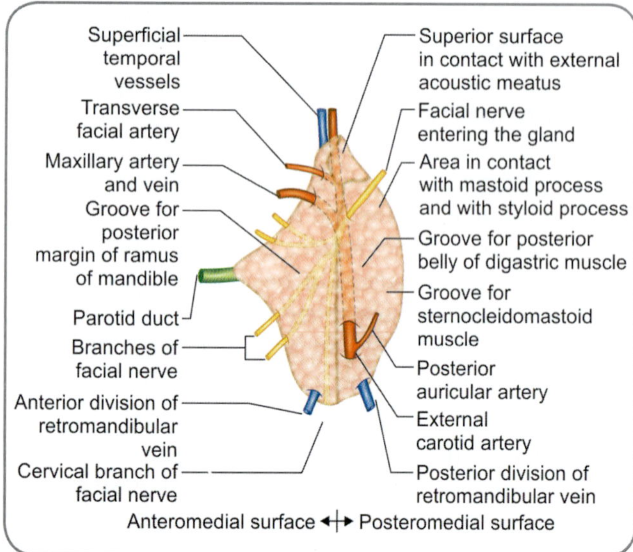

Fig. 3.14: Schematic view of parotid gland as seen from the medial side—structures entering or leaving the gland are shown

❑ The trunk of the ***facial nerve*** enters the upper part of the posteromedial surface. Within the gland, it divides into its terminal branches which emerge from the anteromedial surface near its anterior margin. In fact, the main trunk divides into many branches which form a plexus called the ***parotid plexus*** within the glandular tissue and the five terminal branches arise from this plexus. The five branches are (from above downwards) the temporal, the zygomatic, the buccal, the marginal mandibular and the cervical. The names indicate the areas supplied by the respective branches. The temporal branch, emerging from the superior surface of the gland, crosses the zygomatic branch to supply the auricularis superior, the auricularis anterior, the occipitofrontalis and the superior part of orbicularis oculi. The zygomatic branch, emerging from the anterosuperior angle, usually divides into two or three sub-branches which proceed anteroinferiorly to supply the lower part of orbicularis oculi and other infraorbital facial muscles. The buccal branch, emerges from the middle of the anterior aspect; running superficial to the buccinator muscle, it supplies the same muscle, muscles of upper lip and superior part of orbicularis oris. The marginal mandibular branch, emerging from the apical area of the gland, crosses the inferior border of mandible deep to the platysma to reach the lower jaw region. It supplies the risorius, muscles of the lower lip and muscles of the chin. The cervical branch, emerging from the inferior aspect of the gland, runs inferiorly to supply the platysma.

In many cases, before forming the parotid plexus, the main trunk of the nerve divides into two or more divisions which subsequently form the plexus. At the level of these divisions, the glandular tissue is tubular and narrow and is encircled by the divisions. This gives an appearance that the divisions of the nerve have separated the gland into two parts, namely the superficial and deep parts; the two parts are connected by the tubular masses of tissue.

❑ The ***external carotid artery*** enters the lower part of the posteromedial surface. Ascending within the substance of the gland, it divides into its terminal branches (superficial temporal and maxillary) which emerge on the anteromedial surface of the gland. The posterior auricular branch of the external carotid artery arises just before the latter enters the gland. Sometimes, it arises within the substance of the gland and emerges on the posteromedial surface. The transverse facial branch of the superficial temporal artery may arise within the substance of the gland.

❑ The ***retromandibular vein*** lies in the substance of the parotid gland, superficial to the external carotid artery and deep to the facial nerve. It runs downwards and emerges from the apex of the gland. At this level, it divides into an anterior branch that joins the facial vein and a posterior branch that joins the posterior auricular vein to form the external jugular vein.

❑ The ***parotid lymph nodes*** are found on the parotid sheath (on the lateral surface of the gland—superficial parotid nodes) and within the gland (deep parotid nodes).

Other Closely Related Structures

The ***auriculotemporal nerve*** passes laterally between the neck of the mandible and the superior surface of the gland. It gives branches to the gland (mainly to the parotid sheath).

The anterior (facial) branch of the great auricular nerve passes forwards over the superficial surface of the gland and supplies the overlying skin.

Histology

Histology of Parotid Gland

The parotid is a large serous gland of the compound tubuloacinar type. The surrounding capsule gives numerous connective tissue septa which run into the glandular substance and divides it into lobes and lobules.

The secretory unit is an acinus. Each lobule can have five to eight acini. Each serous acinus is a globular structure with a central luminous cavity lined by the acinar cells. The serous acinar cells are pyramidal in shape and are arranged around the lumen of the acinus. The apical parts of the pyramids face the lumen. The cells have spherical nuclei which are placed at the bases of the cells. The apical portions of the cells have small secretory granules (zymogen granules)

contd...

Histology *contd...*

which are eosinophilic. Cytoplasm at the basal portion of the cell is basophilic. The number and density of the secretory granules will vary according to the functional activity of the gland. Myoepithelial cells surround the acini and are located between the basement membrane and the serous cells.

The secretion of an acinus is poured into the intercalated duct that starts from the acinus. This duct has a lumen surrounded by small cuboidal epithelial cells (or rarely squamous cells) and may be surrounded externally by the myoepithelial cells. The secretion from the intercalated duct goes to the striated duct. This duct has striations on its basal aspect and hence the name. The lumen is slightly larger, the cells are simple columnar and the striations are due to infolding of the basal cell membrane. The secretion from the striated duct goes to the intralobular excretory duct (which indicates that the intercalated and striated ducts are intralobular structures). In other words, the striated ducts from all the acini of a lobule drain into the intralobular duct of the lobule. The intralobular duct has a still larger lumen surrounded by two rows of cuboidal cells. The duct as such is surrounded by connective tissue.

The intralobular ducts of several lobules join to form larger interlobular ducts which inturn unite to form the intralobar and interlobar ducts. The lumen gets larger and larger and the surrounding cells may proceed from columnar to stratified columnar. The terminal part of the interlobar duct continues as the duct of the gland and opens into the oral cavity. From the level of the interlobular duct which lies in the connective tissue septa, the ducts are surrounded by more and more of connective tissue.

Some lobules may have clusters of adipose (clear staining) cells between the acini (which appear dark-stained).

The connective tissue septa between the lobules and lobes contain blood vessels, nerves and the ducts of the gland.

To get familiarized with the technical names:
Gland has stroma and parenchyma. Stroma consists of connective tissue capsule and the septa sent in.

Parenchyma comprises the secretory part and the myoepithelial cells.

Secretory part consists of the serous acini. Acinar cells are the secretory cells.

Myoepithelial cells form the propulsive system of the gland. They help to push the secretions into the ducts.

Duct system of the gland is its excretory (and secretory) pathway.

PAROTID DUCT

Secretions of the gland are collected by a system of ducts which unite to form the ***parotid duct***. This duct emerges at the anterior margin of the gland and runs forwards across the masseter. At the anterior border of the masseter, it turns medially and pierces the buccinator. The terminal part of the duct runs forwards deep to the mucous membrane of the cheek. It opens into the vestibule of the mouth opposite the crown of the upper second molar tooth. The duct is about 5 cm long.

Development

Development of the Parotid Salivary Glands

An epithelial thickening appears on the deep aspect of the cheek by the 5th week of intrauterine life. It separates into a cord by the next few days and remains a solid cord until the third month of intrauterine life. The distal end of the solid cord starts branching repeatedly by the third month. By the sixth month, the solid cords develop lumina and get canalized. Small ducts of this developing gland, to start with, open into a long, narrow gutter. Closure of this gutter makes the duct of the gland appear longer. Such elongation of duct proceeds anteriorwards and the parotid duct opens into the oral cavity. Secretory activity commences close to birth; microstructural development occurs just before and after birth.

Chievitz's organ: An ectodermal thickening occurs in the cheek epithelium behind the developing parotid gland, somewhere between the eighth and twentieth weeks of intrauterine life. Though it later disappears, it is supposed to contribute to the microstructural development of the parotid gland. However, it has been described as a ***vestigeal salivary gland***. In some cases, the remnants of the Chievitz's organ may contribute to a teratoma.

Relations of Parotid Gland

- ❑ ***Anteriorly:*** Posterior border of ramus of mandible and the two muscles sandwiching the ramus, namely, the masseter and the medial pterygoid—the gland overflows on this aspect on to the masseter; the overflowing extension is the 'facial process' of the gland;
- ❑ ***Posteriorly:*** Mastoid process and anterior border of sternocleidomastoid;
- ❑ ***Superiorly:*** Capsule of the temperomandibular joint anteriorly and the external acoustic meatus posteriorly;
- ❑ ***Inferiorly:*** Stylohyoid and the Posterior belly of digastric – the gland overflows on this aspect approaching the carotid triangle; the overflowing extension is the 'carotid process' of the gland;
- ❑ ***Superficial:*** Skin, superficial fascia, parotid fascia (deep fascia) and a few lymph nodes resting on the parotid fascia;
- ❑ ***Deep:*** The fascia derived from the investing layer of deep cervical fascia and the styloid process with the muscles arising from it.

Blood Supply of Parotid Gland

The gland is supplied by small branches of the external carotid artery or its terminal divisions.

Veins drain into the retromandibular and external jugular veins.

Lymph vessels from the gland drain into the deep cervical nodes after passing through the superficial parotid nodes (lying on the lateral surface of the gland) and the deep parotid nodes (lying in the substance of the gland).

Nerve Supply of Parotid Gland

Secretomotor nerves reach the gland through branches of the auriculotemporal nerve. The preganglionic parasympathetic fibres arise from the inferior salivatory nucleus pass through the glossopharyngeal nerve, the tympanic branch of glossopharyngeal nerve, the tympanic plexus in the tympanic cavity and the lesser petrosal nerve to reach the otic ganglion and relay on the neurons there. The postganglionic fibres arising from the otic ganglion pass through the auriculotemporal nerve and its parotid branches to reach the gland. Parasympathetic stimulation produces thin and watery saliva.

Postganglionic sympathetic fibres are derived from the cervical ganglia and reach the gland through the external carotid nerve plexus around the external carotid artery. Sympathetic stimulation causes vasoconstriction resulting in reduced secretions from the gland.

Sensory fibres from the gland and fascia pass through the great auricular and auriculotemporal nerves.

Added Information

- The parotid bed is otherwise called the ***parotid mould***. It can be described as a vessel lined with fascia. The parotid gland fills this vessel and overflows in the anterior and superior aspects. The relations of the gland actually form the walls of the mould; the deep relations form the floor or bottom of the vessel.
- Because of the irregularities of the parotid bed and because of the irregular shape of the gland itself, several tongue like extensions or processes are given out at various parts of the gland. Two of such extensions are the facial process and the carotid process (which have been already noted). The largest of the processes is the facial process. Other processes are—(1) one that passes between the upper part of the mandibular ramus and the medial pterygoid; (2) another that passes between the temperomandibular joint and the external acoustic meatus and (3) another that passes anterior to the internal carotid artery to reach the superior constrictor.
- The facial process may be so large that it crosses over the parotid duct and appears above the duct as a small mass. This then is called an ***accessory parotid gland***. The part may also get detached from the main gland.
- The anteromedial and posteromedial surfaces meet at a margin called the ***medial border of the gland***. This border may be so sharp and projecting that it touches the lateral wall of the pharynx.
- Various structures emerging from the parotid gland or from near it cross other important structures of the region. These are—(a) The superficial temporal artery and vein, along with the auriculotemporal nerve cross the posterior root of zygoma; (b) The temporal branch of facial nerve crosses the zygomatic arch; (c) The zygomatic, buccal and mandibular branches of the facial nerve cross the masseter muscle; and (d) The posterior auricular artery and the posterior auricular nerve cross the mastoid.

Clinical Correlation

- The parotid gland (alone or along with the other salivary glands) is commonly infected with a virus that causes mumps. It causes swelling and pain. Other salivary glands may also be involved. Mumps also involves the parotid duct resulting in redness of the parotid papilla. This is the differentiating feature of early mumps infection from toothache which also can cause similar pain.
- The parotid gland can also be infected by spread of infection from the mouth. An abscess may form in the gland (pus formation leading to parotid abscess).
- The parotid fascia is very dense and allows very little expansion of the gland. Swellings of the parotid are therefore very painful. Any infection of the parotid (parotiditis) resulting in inflammation and swelling leads to intense pain. Pain is aggravated while chewing because the swollen gland is compressed against the mandibular ramus and mastoid process due to movements of the mouth. Referred pain from parotid may be felt over the auricle and external acoustic meatus since the auriculotemporal and great auricular nerves supply sensory fibres to the gland and its sheath and also supply the skin of temporal fossa, auricle and external meatus.
- Tumours may occur in salivary glands. They may be benign or malignant.
- Facial paralysis may occur by involvement of the facial nerve in a malignant growth of the parotid gland or by injury during removal of the gland. The surgeon tries to protect the nerve by removing the superficial and deep parts separately.
- Imaging of the parotid duct system can be done by injecting a radiopaque fluid into the parotid duct using a cannula. X-ray pictures can then be taken. Such technique is called ***sialography*** (Greek.sialos=saliva, grapho=writing) and the X-ray picture taken by this technique is called ***sialogram***. If there are dilatations or displacements in the duct system, they will be seen in the sialogram.
- The parotid duct may be blocked by calcified deposits. Such deposits are called ***sialoliths*** or ***calculus of the parotid duct*** (Latin.calculos=pebble). Pain that is aggravated on eating is the main symptom.
- Frey's syndrome: Otherwise called ***gustatory perspiration*** or ***auriculotemporal syndrome***, this condition occurs post parotid surgery. The autonomic nerve fibres supplying the parotid gland get damaged. During healing, the parasympathetic secretomotor fibres to the gland regenerate into the sheaths of the sympathetic fibres to the neighbouring sweat glands. Whenever there is salivary stimulation by taste or smell of food, the sweat glands secrete; sweating, warmth and redness of face occur.

Development

Development of face: Craniocaudal folding of the embryo has already resulted in a prominent head region and a notable chest (pericardial) bulge. The stomatodael pit and membrane occur between these prominences. Structures of face will have to develop around the stomatodael pit which in due course will develop into the mouth orifice. The

contd...

 **Development** *contd...*

development is contributed to mainly by the pharyngeal arches which develop on the sides of the stomatodael pit. Another factor of importance to be remembered is that the neural crest cells from the brain regions migrate to the future regions of face and pharynx. These cells variedly contribute to the initiation and formation of the several placodes (ectodermal thickenings which would form the optic, otic, lens and nasal placodes) of the face.

By the end of fourth week of intrauterine life, the facial prominences appear. These prominences primarily contain neural crest derived mesodermal tissue and occur as extensions of the first pharyngeal arch. The maxillary prominences, one on either side of the stomatodaeum and the mandibular prominences, one on either side but caudal to the stomatodaeum can be noted. Simultaneously, the midline frontonasal prominence makes its appearance cranial to the stomatodaeum (and ventral to the developing brain).

At about the same time, the nasal placodes make their appearance on the sides of the frontonasal prominence. The nasal placodes then dip in to form the nasal pits. This raises ridges on either side of each nasal pit; the ridges are called **nasal prominences**. Thus, the lateral nasal prominences are formed on the outer lips of the pits and the medial nasal prominences are formed on the inner lips of the pits.

Formation of lips and jaws: Growth continues at a rapid pace during the fifth and sixth weeks. The maxillary prominences continue to grow medially. The inferior edges of the lateral nasal prominences do not extend to the level as those of the medial prominences. The medial tip of the maxillary prominences is triangularly tapering. Because of these reasons, the medial ends of the maxillary processes come in contact with the medial nasal processes and gradually fuse with them. The rudiment of the upper lip, which is larger than the developed upper lip would be, is thus formed. The maxillary processes continue to push medially and thereby compress the medial nasal processes. As a result, the two medial nasal processes merge. The merged portion is called the **intermaxillary segment**. The most superficial portion of the intermaxillary segment (called the **labial component**) gives rise to the philtrum of the upper lip. The middle portion is the gingival component that gives rise to that part of the upper jaw which carries the four incisors. The deep portion develops an internal extension. This extension which is like a transverse triangle is called the **globular process** or **palatal process** and gives rise to the primary palate (also called the **premaxilla**). The intermaxillary segment is continuous cranially with the frontonasal process and contributes to the formation of the nose.

The mandibular prominences, which develop caudal to the stomatodaeum, also proceed to grow medially and fuse with each other. The lower lip and the lower jaw will form from the fused mandibular processes.

Formation of cheeks: Though the medial ends of the maxillary processes make contact directly with the medial nasal processes, the oblique upper borders of the maxillary processes lie parallel to the oblique lateral borders of the lateral nasal processes. The optic placodes are formed at the

contd...

 **Development** *contd...*

superolateral aspects of the frontal prominence. The junction between the lateral nasal and maxillary processes extend from the optic area to the lateral aspect of the developing nose. The deep furrow that separates the two processes is called the **nasolacrimal groove**. Ectodermal proliferation occurs in the floor of this groove and an epithelial cord is formed. The cord submerges inside and is separated off from the surface ectoderm. Around the sixth month of intrauterine life, canalisation of the cord occurs and the nasolacrimal duct is formed. The upper part of the duct dilates to form the lacrimal sac. Once the cord forming the nasolacrimal duct has separated off the surface, the maxillary process and the lateral nasal process merge. The maxillary process expands upwards and medially to form the cheeks. The nasolacrimal duct runs from the medial angle of eye to the nasal cavity.

Formation of external nose: Swellings appear in the inferior part of frontal process, the medial nasal processes and the lateral nasal processes. These swellings fuse to form the external nose. The bridge is derived from the frontal process, the crest and tip from the fused medial nasal processes and the alae and sides from the lateral nasal processes.

VESSELS OF FACE AND PAROTID REGION

ARTERIES

The arteries seen in the region of the face are the facial artery, transverse facial branch of the superficial temporal artery and facial branches of the maxillary and ophthalmic arteries.

Facial Artery

The facial artery is a branch of the external carotid artery and arises in the neck. It runs part of its course in the neck and the submandibular region. It then enters the face by passing around the lower border of the body of the mandible just in front of the masseter muscle. From here, it runs upwards and forwards across the body of the mandible and the buccinator to reach the angle of the mouth. Finally, it ascends along the side of the nose to reach the medial angle of the eye. From a very superficial position in the mandibular area, it becomes deeper to be covered by some of the facial muscles. The part of the facial artery beyond its terminal branch is called the **angular artery**.

The artery has a tortuous course; this factor is supposed to protect the artery from being damaged during jaw opening and various movements of the facial muscles.

The facial artery gives off the following named branches:

- ❑ **Premasseteric artery:** This is a 'not very constant' branch that runs along the anterior border of masseter muscle and supplies the adjacent areas.
- ❑ **Inferior labial artery:** Arising near the angle of the mouth, this branch runs upwards and forwards under

the depressor anguli oris muscle, penetrates the orbicularis oris muscle and runs along the margin of lower lip between the orbicularis muscle and the mucous membrane. It supplies the inferior labial glands, mucous membrane and muscles of the lower lip. It anastomoses freely with the fellow of the opposite side and the mental branch of inferior alveolar artery of the same side.

❑ ***Superior labial artery:*** Larger than the inferior labial branch, this artery has a similar course as that of the inferior labial artery. Apart from anastomosis with the fellow of the opposite side, it also gives a septal branch to the lower part of nasal septum and an alar branch to the ala of nose.

❑ ***Lateral nasal artery:*** Given at the side of the nose, this branch gives small branches to the dorsum and ala of the nose. It then anastomoses with the fellow of the opposite side.

Several unnamed branches arise from the facial artery throughout its course and especially from the angular artery and run to supply the neighbouring areas.

Superficial Temporal Artery

The superficial temporal artery arises from the external carotid artery within the parotid gland. It runs upwards behind the temporomandibular joint and ends by dividing into frontal and parietal branches which supply the scalp. It also gives off the ***transverse facial artery*** that runs forwards across the masseter muscle. The superficial temporal artery supplies the skin and muscles of the side of face, scalp, parotid gland and temporomandibular joint.

Its named branches are:

❑ ***Transverse facial artery:*** Arising within the parotid gland, it emerges out and crosses the masseter usually between the zygomatic arch and the parotid duct. It divides into numerous branches which supply the parotid gland, parotid duct, masseter muscle and neighbouring fasciae and skin.

❑ ***Auricular artery:*** Given out from the posterior aspect of the parent trunk, this artery gives branches which ramify on the external ear and external acoustic meatus.

❑ ***Zygomatico-orbital artery:*** Arising at the level of the palpebral fissure, this artery runs just above the zygomatic arch to the lateral angle of the eye. It supplies the orbicularis oculi muscle and anastomoses with the branches of the ophthalmic artery.

❑ ***Middle temporal artery:*** Arising just above the zygomatic arch, this artery pierces the temporal fascia to supply the temporalis muscle. It anastomoses with the deep temporal branches of the maxillary artery.

❑ ***Frontal artery:*** This is the smaller of the terminal branches of the superficial temporal artery given out

over the scalp. It passes anterosuperiorly towards the frontal tuberosity and supplies the skin, fasciae, muscles and pericranium of the region. It anastomoses with the branches of the ophthalmic artery and the fellow of the opposite side.

❑ ***Parietal artery:*** This is the larger of the two terminal branches of the superficial temporal artery. It curves posterosuperiorly but remains superficial to the temporal fascia. It anastomoses with the fellow of the opposite side and with other arteries of the neighbourhood.

Facial Branches of Maxillary Artery

Three named branches of the maxillary artery supply structures in the face. These arteries are the mental artery, buccal artery and the infraorbital artery.

❑ ***Mental artery:*** It arises as a terminal branch of the inferior alveolar artery (in turn a branch of the first part of maxillary artery). Emerging through the mental foramen on the face, it supplies the fasciae and muscles of the chin region.

❑ ***Buccal artery:*** Arising as a branch of the second part of maxillary artery, the buccal artery emerges from the infratemporal fossa. It crosses the buccinator and supplies the cheek region.

❑ ***Infraorbital artery:*** Arising from the third part of the maxillary artery, this artery emerges on the face through the infraorbital foramen and supplies the lower eyelid, lateral aspect of nose and the upper lip.

Dissection

Place back all the skin flaps of the face. Slowly, proceed to reflect them and trace all the vessels and nerves (which you have already done in different steps). Review the relationship of the vessels and nerves to other structures of the face.

Facial Branches of Ophthalmic Artery

Several branches of the ophthalmic artery supply structures in the face.

Supratrochlear artery: This artery emerges on the face through the frontal notch. It supplies the upper eyelid, forehead and scalp.

Supraorbital artery: Emerging on the face through the supraorbital foramen, this artery divides into superficial and deep branches, both of which supply the skin, fasciae and muscle of the upper eyelid, forehead and scalp. It also sends a branch to the diploe of the frontal bone and another to the mucoperiosteum of frontal sinus.

Lacrimal artery: From inside the orbit, this artery appears on the face at the superolateral corner of the orbit. It gives out a zygomatic artery within the orbit itself. This artery divides into the zygomaticofacial and zygomaticotemporal branches again within the orbit itself. The zygomaticofacial artery passes through the lateral wall of orbit and emerges on the face at the zygomaticofacial foramen. It supplies the region of the cheek. The zygomaticotemporal artery passes through the lateral wall of the orbit, emerges on the temple area through the zygomaticotemporal foramen and supplies the temple region. Appearing on the face, the lacrimal artery gives out the lateral palpebral arteries which supply the upper and lower eyelids and form a dense anastomosis with the medial palpebral arteries.

Medial palpebral arteries: Otherwise called the *internal palpebral arteries*, these are two in number—the superior medial palpebral artery and the inferior medial palpebral artery. Arising below the trochlea in the orbit, they descend behind the nasolacrimal duct and run into the respective eyelids. Each of them coursing between the orbicularis oculi and the tarsal plate forms an arterial arch in the respective lid. Thus the superior and inferior arches are formed. The arches establish dense anatomoses with adjacent arteries and with the corresponding lateral palpebral artery. The inferior arch anastomoses with the angular artery too.

Dorsal nasal artery: This artery, also called the *external nasal artery*, is terminal branch of the anterior ethmoidal branch of the ophthalmic artery. It supplies the skin of the nose.

VEINS

The veins seen in the face and parotid region are the superficial temporal vein, the retromandibular vein and the facial vein.

The *superficial temporal vein* forms in the scalp from a widespread network of veins as the anterior and posterior tributaries. The two tributaries unite to form the single superficial temporal vein. Accompanying its artery, the vein enters the parotid gland and joins with the maxillary vein to form the retromandibular vein. The tributaries received by the superficial temporal vein are the parotid veins, veins from the temporomandibular joint, anterior auricular vein and the transverse facial vein. One other major tributary is the middle temporal vein which joins the superficial temporal vein above the zygomatic arch. The middle temporal vein receives the orbital vein which is formed by the union of the superior and inferior lateral palpebral veins.

The *retromandibular vein* is formed within the upper part of the parotid gland by the union of the superficial temporal and maxillary veins. Its lower end divides, within the gland, into anterior and posterior divisions which emerge at the lower end of the gland. The posterior division, which is the main continuation of the retromandibular vein,

joins the posterior auricular vein to form the external jugular vein. The anterior division joins the facial vein to form common facial vein.

The *facial vein* is the main vein of the face. It receives the supratrochlear and supraorbital veins at its commencement and runs down the side of the nose, crosses obliquely and descends to the anterior border of the masseter. It then runs on the anterior surface of the muscle, crosses the body of mandible. In the neck, it is joined by the anterior division of the retromandibular vein to form the common facial vein that ends in the internal jugular vein. Its upper portion (the portion above the joining of the superior labial vein) is also called the *angular vein*. The tributaries of the facial vein in the face are the deep facial vein (that comes from the pterygoid venous plexus and joins below the level of the nose), inferior palpebral vein, superior labial vein, inferior labial vein, buccal vein, parotid vein and and masseteric vein. Tributaries in the neck join the common facial vein.

The most important connection of the facial vein is that with the cavernous sinus through the supraorbital and superior ophthalmic veins. Connnection with the pterygoid venous plexus through the deep facial vein ranks next. The facial vein does not have valves.

The infraorbital, buccal and mental veins drain the cheek and chin and enter into the pterygoid venous plexus.

contd...

> **⚕ Clinical Correlation** *contd...*
>
> vein with clot formation). It has been observed that such spread of infection is most likely to take place if the infection is over the upper lip or the lower part of the nose. This region is, therefore, called the **dangerous area of the face**. Anatomically, a triangular area from the root of nose to the upper lip usually drains into the facial vein. This triangle with the apex at the root of nose and the base at the upper lip is called the **danger triangle of the face**.

LYMPHATICS

The lymphatics of the face draining:

- The **forehead** above the root of nose pass to the submandibular nodes;
- The **lateral forehead, temporal region, auricular area, anterior wall of external acoustic meatus, lateral part of eyelids** and **zygomatic area** pass to the superficial parotid nodes;
- The **strip of scalp above the auricle** and **posterior wall of external acoustic meatus** pass to the retroauricular nodes.
- The **auricular lobule, floor of external acoustic meatus, skin of mandibular angle** and **lower parotid region** pass to the superficial cervical or upper deep cervical nodes.
- The **external nose, cheek, upper lip** and **lateral parts** of **lower lip** pass to the submandibular nodes.
- The **central part of the lower lip, buccal floor** and **tip of tongue** pass to the submental nodes.

Submandibular nodes lie deep to the deep cervical fascia in the submandibular triangle. One node lies at the anterior pole of the submandibular gland, two nodes flank the facial artery near the mandible and many others remain embedded in the gland. They receive afferents from the submental, buccal and lingual group of nodes; their efferents pass to the upper and lower deep cervical nodes.

Superficial parotid nodes lie anterior to the tragus either superficial to or deep to the parotid fascia. Their efferents pass to the upper deep cervical nodes.

Retroauricular nodes are superficial to the mastoid attachment of the sternocleidomastoid and their efferents pass to the upper deep cervical nodes.

Superficial cervical nodes lie along the external jugular vein; their efferents pass to the upper or lower deep cervical nodes.

Submental nodes lie on the mylohyoid muscle; their efferents pass to the submandibular and jugulo-omohyoid nodes.

NERVES OF FACE

The nerves of face can be classified as **motor** and **sensory**.

Motor Nerves

The motor nerves are the terminal branches of the **facial nerve** which emerge on the face at the borders of the parotid gland. These are:

- A **temporal** branch,
- A **zygomatic** branch,
- Upper and lower **buccal** branches,
- A **marginal mandibular** branch and
- A **cervical** branch.

These branches supply the various muscles of the face (muscles of facial expression).

Temporalis and masseter, two of the four muscles of mastication, are seen on the face. They are supplied by the mandibular division of the trigeminal nerve.

Sensory Nerves (Fig. 3.15)

The sensory nerves seen on the face are terminal ramifications of the **trigeminal nerve**.

Branches arising from the ophthalmic division are:

- **Supratrochlear,**
- **Supraorbital,**
- **Infratrochlear** and
- **External nasal** nerves.

Branches of the maxillary division of the trigeminal nerve are:

- **Infraorbital,**
- **Zygomaticofacial** and
- **Zygomaticotemporal** nerves.

Branches of the mandibular division are:

- **Auriculotemporal**
- **Buccal** and
- **Mental** nerves.

The areas of skin of the face supplied by the three divisions of the trigeminal nerve are large and have a specific

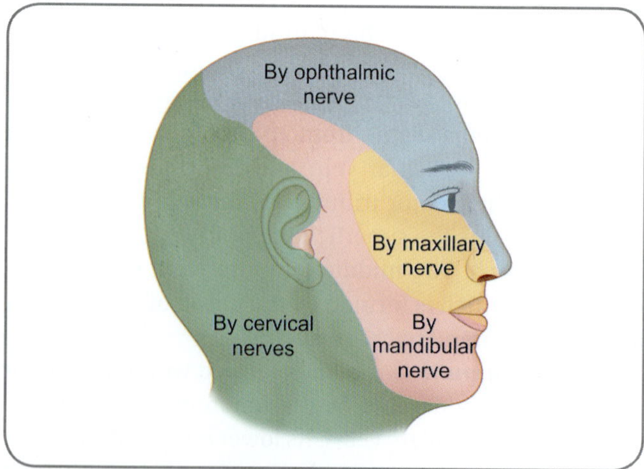

Fig. 3.15: Areas of skin supplied by the ophthalmic, maxillary and mandibular divisions of trigeminal nerve

distribution. These three areas curve upwards because the facial skin moves upwards in line with the growth of the brain and skull. Each division of the trigeminal nerve is related to a developmental process. The ophthalmic division to the frontonasal process, the maxillary nerve to the maxillary process and the mandibular nerve to the mandibular process.

A very small part of skin over the parotid gland is supplied by the great auricular nerve, a branch of the cervical plexus.

Clinical Correlation

- The ***trigeminal nerve*** is responsible for sensory supply of most of the face. When a lesion of the nerve is suspected, it is tested as follows:
  - Sensation of touch in the area of distribution of the nerve is tested by touching different areas of skin with a wisp of cotton-wool;
  - Sensation of pain is tested by gentle pressure with a pin;
  - Motor function is tested by asking the patient to clench his/her teeth firmly; contraction of the masseter can be felt by palpation when the teeth are clenched.
- The ***facial nerve*** supplies the muscles of the face including the muscles which close the eyelids and the mouth. The nerve is tested as follows:
  - The patient is asked to close eyes firmly. In cases of complete paralysis of the facial nerve, the patient will not be able to close the eye on the affected side. In partial paralysis, the closure is weak and the examiner can easily open the closed eye with his fingers (which is very difficult in a normal person).
  - The patient is asked to smile. In smiling, the normal mouth is more or less symmetrical, the two angles moving upwards and outwards. In facial paralysis, the angle fails to move on the paralysed side.
  - The patient is asked to fill his/her mouth with air. The examiner then presses both the cheeks of the patient and compares the resistance (by the buccinator muscle) on the two sides. The resistance is less on the paralysed side. On pressing the cheek, air may leak out of the mouth because the muscles closing the mouth are weak.
  - The sensation of taste should be tested on the anterior two-thirds of the tongue by applying substances of different tastes.

Facial changes in paralysis of facial nerve: The effects of paralysis are due to the failure of the muscles concerned to perform their normal actions. Some effects are as follows:

- The normal face is more or less symmetrical. When the facial nerve is paralysed on one side, the most noticeable feature is the loss of symmetry.
- Normal furrows on the forehead are lost because of paralysis of the occipitofrontalis.
- There is drooping of the eyelid and the palpebral fissure is wider on the paralysed side because of paralysis of the orbicularis oculi. The conjunctival reflex is lost for the same reason.
- There is marked asymmetry of the mouth because of paralysis of the orbicularis oris and of muscles inserted into the angle of the mouth. This is most obvious when a smile is attempted. As a result of asymmetry, the protruded tongue appears to deviate to one side, but is in fact in the midline.
- During mastication food tends to accumulate between the cheek and the teeth. (This is normally prevented by the buccinator.)

Defects in development of face in relation to neural crest cells and midline tissue: Failure of or improper fusion of the maxillary and lateral nasal processes result in oblique facial clefts. The nasolacrimal duct is not formed properly or is exposed to the surface.

Cleft lip deformities result from improper fusion of the contributing elements. It is already seen that the merged medial nasal tissue gives rise to the primary palate or the premaxilla. The junction between the primary palate and the secondary palate (secondary palate is that part of the palate which forms from the palatine shelves of the maxillary prominences) is indicated in the postnatal individual by the incisive foramen. Cleft deformities which are anterior to the incisive foramen are the anterior cleft defects and those posterior are the posterior cleft defects. Anterior cleft defects include lateral cleft lip, cleft upper jaw and cleft between primary and secondary palates.

- ***Lateral cleft lip:*** Improper fusion or non fusion of the maxillary and the medial nasal processes; can be unilateral or bilateral; the level of defect does not extend to the full thickness of the processes and only the superficial portion is involved.
- ***Cleft upper jaw:*** Improper fusion or non fusion of the maxillary and the medial nasal processes; can be bilateral or unilateral; the level of defect extends to the full thickness of the processes.
- ***Midline cleft lip:*** This condition results due to improper merging of the two medial nasal processes. Midline tissue formation of the craniofacial region involves a complicated developmental process. The merger of the medial nasal processes is parallel to developmental and expansive changes occurring in the interior. The upper portion of the frontal process undergoes differential changes to form the forehead and bridge of nose. The lower part of the frontal process and the merged medial nasal process undergo compaction and elevation to form the crest and tip of the external nose. When these external changes are happening, the midline of the brain is being established internally. Improper merger of the medial nasal processes is usually due to lack of adequate midline tissue which reflects internal defects too.

It should be remembered that the medial and lateral nasal processes are elevations formed in the original frontonasal processes because of the formation of the nasal pits. The (midline) tissue between the medial nasal processes is quite broad initially and then undergoes narrowing. Improper development in the initial stages results in inadequacy of midline tissue both externally (on the surface between the medial nasal processes) and internally (deeper aspect of the frontonasal process and in developing brain). This inadequacy results in improper fusion of medial nasal processes leading to midline cleft lip and to other external and internal defects.

✆ Clinical Correlation *contd...*

On the other side of the range, medial nasal processes do not form at all and the lateral nasal processes improperly fuse to create a single nasal opening. The midline descent of the nasal region does not take place. Lack of tissue in the frontal process causes the eyes to move more anteriorly (the eyes are laterally placed in the normal early development) and caudal to the nasal opening. At the extreme of this, the two eyes fuse to form a more or less midline synophthalmia (usually called the ***Cyclops eye or deformity***) and above the eye, is a stumpy proboscis with a single nasal opening. The head is narrow. Several internal changes occur. The most marked is the presence of a single (lateral) ventricle in the prosencephalon. Altered cholesterol mechanism, teratogen exposure and alcohol consumption in the first trimester of pregnancy have been implicated. A child with simple midline cleft lip is likely to have cognitive defects.

Neural crest cells migrate to the developing regions of face and contribute to the formation of the skeletal structures and the various placodes. The migratory paths of the neural crest cells appear to play a crucial role in providing axonal guidance for the various neural ganglia forming in the region of head and neck. However, for our consideration here, it is important to note that the skeleton of the face is derived from the neural crest cells which have migrated into the frontonasal, maxillary and mandibular prominences. It can be said that much of the craniofacial region develops from these cells. As a result, disorders involving crest cell development lead to craniofacial defects. Since neural crest cells also contribute to endocardial cushion development in the heart, infants with craniofacial deformities have associated cardiac abnormalities. Neural crest cells are prone to damage from alcohol and other chemical compounds.

Multiple Choice Questions

1. The pericranium is fixed at the edges of each skull bone because:
   a. It is continuous with the endocranium
   b. It is attached to the sutural ligaments
   c. It is pierced by emissary veins
   d. It is made up of areolar tissue
2. Vermillion border of lip is the:
   a. Line of commencement of oral mucosa
   b. Same as moist line
   c. Line of commencement of extra oral skin
   d. Line of mucocutaneous junction
3. Tarsal plates are attached to the walls of orbit through:
   a. Meibomian glands
   b. Palpebral ligaments
   c. Orbicularis oculi muscle fibres
   d. Conjunctiva of fornix
4. Risorius:
   a. Retracts the angle of mouth
   b. Raises upper lip
   c. Anchors upper lip
   d. Elevates chin
5. Apex of parotid gland overhangs:
   a. Masseter
   b. Medial pterygoid
   c. Posterior belly of digastric
   d. Temporalis

ANSWERS

1. b   2. c   3. b   4. a   5. c

Clinical Problem-solving

Case Study 1: A 52-year-old man presented with the following: sagging down of the left side of his face, left inferior eyelid everted, widened left palpebral fissure, difficulty in chewing and dribbling of saliva.
- ❑ What will be your provisional diagnosis?
- ❑ What are the anatomical reasons for the presenting features?
- ❑ What other symptoms/signs will you look for/elicit in this patient?

Case Study 2: A 22-year-old man presented with a swelling on the front of his right upper eyelid. It was a reddish and painful swelling.
- ❑ What is the probable diagnosis?
- ❑ How would you make out which gland is involved from the way the swelling projects?
- ❑ Explain the features of the various glands present in the region.

(For solutions see Appendix).

Chapter 4

Temporal and Infratemporal Regions (Including Temporomandibular Joint)

Frequently Asked Questions

- Discuss the infratemporal fossa in detail.
- Write notes on: (a) Temporal fascia, (b) Temporalis muscle, (c) Middle meningeal artery, (d) Pterygoid plexus of veins.
- Write notes on maxillary artery.
- Write briefly on: (a) Masseter, (b) Lateral pterygoid, (c) Nervus spinosus, (d) Auriculotemporal nerve, (e) Lingual nerve, (f) Movements of temporomandibular joint.
- Write in detail on the temporomandibular joint and its movements.

TEMPORAL REGION

The temporal region is that region of the head which overlies the temporal fossa present on the lateral aspect of the skull. It is bounded:

- Above by the superior temporal line
- Below by the zygomatic arch
- Anteriorly by the frontal and zygomatic bones
- Posteriorly by the superior temporal line.

The floor of the temporal region is formed by the *temporal fossa* which in turn is formed by parts of frontal bone, parietal bone, squamous part of temporal bone and greater wing of sphenoid bone. The anteroinferior boundary of the temporal fossa is the infratemporal crest. The temporal fossa communicates with the infratemporal fossa beneath the zygomatic arch.

The roof of the temporal region is formed by the temporal fascia.

CONTENTS

The temporal region contains (from superficial to deep):

- Temporoparietal fascia
- Temporoparietal fat pad

- Auriculotemporal nerve, superficial temporal vessels and temporal branch of facial nerve
- Parts of epicranial aponeurosis, subcutaneous muscles of the auricle and orbicularis oculi
- Temporal fascia
- Temporalis muscle
- Deep temporal vessels and nerves.

SUPERFICIAL STRUCTURES IN THE TEMPORAL REGION (SUPERFICIAL TO TEMPORAL FASCIA)

- *Temporoparietal fascia:* Otherwise called the superficial temporal fascia, this thin fascial layer is superficial to the more prominent temporal fascia. It is the same plane as that of the epicranial aponeurosis and blends with the latter in the superior aspect of the temporal region.
- *Temporoparietal fat:* Though called temporoparietal fat, this layer contains some amount of loose connective tissue. It is continuous with the subaponeurotic connective tissue of the scalp. The auriculotemporal nerve and its branches, the superficial temporal vessels and the temporal branch of the facial nerve are usually found in this layer.
  - Immediately in front of the external acoustic meatus, the *auriculotemporal nerve* which is a branch of the posterior division of mandibular nerve, emerges from under cover of the upper end of the parotid gland. The nerve ascends into the temporal region and scalp and divides into branches which supply them.
  - Just in front of the auriculotemporal nerve, the *superficial temporal artery,* a terminal branch of the external carotid artery is seen (Fig. 4.1). It is accompanied by the superficial temporal vein. The artery divides into branches and through them supplies the temporal region and scalp. The

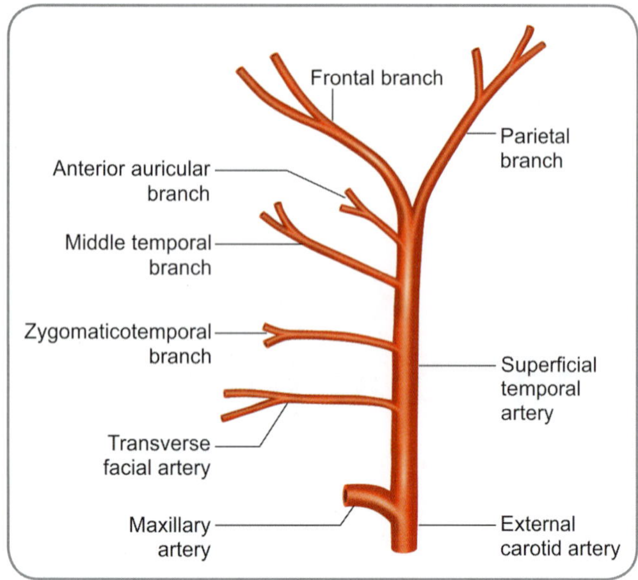

Fig. 4.1: Branches of superficial temporal artery

superficial temporal vein enters the parotid gland and forms the retromandibular vein after joining with the maxillary vein. Temporal branch of the facial nerve is usually seen coursing this space.

❑ *Temporal fascia:* Otherwise called the deep temporal fascia, this is a thick fibrous membrane described to be covering the temporal region and the temporalis muscle. The fascia is attached superiorly to the temporal lines, especially to the superior temporal line and blends with the periosteum. Inferiorly, it splits into two layers; these layers attach to the medial and lateral surfaces of the zygomatic arch. Some amount of fat is enclosed between these two layers; the zygomaticotemporal branch of the maxillary nerve and the zygomatico orbital branch of the superficial temporal artery are found in this space. The deep surface of temporal fascia gives attachment to the temporalis muscle which is also covered by the fascia. The superficial surface of the fascia is overlapped by the epicranial aponeurosis, the subcutaneous muscles of the auricle which arise from it, namely the *auricularis superior*, the *auricularis posterior* and the *auricularis anterior* and part of the orbicularis oculi muscle.

Posterior auricular branch of the facial nerve and the zygomaticofacial branch of the maxillary nerve may also be seen in this region.

DEEP STRUCTURES IN THE TEMPORAL REGION (DEEP TO THE TEMPORAL FASCIA)

The *temporalis muscle* is the main content of the temporal fossa and is situated deep to the temporal fascia. It is one of the muscles of mastication and is responsible for chewing movements. Deep to the temporalis muscle, the deep temporal nerves and vessels are present. The anterior, middle and posterior *deep temporal arteries* are branches of the maxillary artery; they run between the temporalis and the underlying pericranium, usually grooving the latter. They anastomose with branches of the superficial temporal arteries and supply the fasciae and muscles of the region. The *deep temporal nerves* (anterior and posterior) arise from the anterior division of the mandibular nerve, enter the deep surface of the temporalis and supply the muscle.

Dissection

Clean and define the masseter muscle. Detach it from its origin and reflect it downward. See the masseteric nerve and vessels entering the deep surface of the muscle. Remove the zygomatic arch using a saw. Using a bone forceps and chisel, remove the coronoid process of the mandible along with the insertion of the temporalis. Turn the temporalis upward. See and study the deep temporal nerve and vessels. After studying all the structures of the region, enter into the infratemporal region by removing a portion of the ramus of mandible.

Once in the infratemporal fossa, identify and clean the mandibular nerve and its branches. Define the pterygoid muscles. Identify and clean the maxillary artery and its branches. Trace the pterygoid plexus of veins.

INFRATEMPORAL FOSSA

The infratemporal region is an irregular space lying below the lateral part of the base of the skull and deep to the ramus of mandible. Due to its deep location, it is often called the infratemporal fossa. It presents the following boundaries:

❑ *Roof:* Formed mainly by the greater wing of the sphenoid bone, with a small contribution from the squamous temporal bone;
❑ *Medially:* Lateral pterygoid plate of the sphenoid bone;
❑ *Laterally:* Ramus of the mandible and its coronoid process;
❑ *Anteriorly:* Posterior surface of the maxilla;
❑ *Posteriorly:* Styloid process of temporal bone and carotid sheath;

Inferiorly the region is open and has no floor; it is continuous with the tissue spaces along the pharynx and oesophagus.

COMMUNICATIONS OF THE FOSSA

The infratemporal fossa is well connected to other regions of the head. It communicates:

❑ *In front:* With the orbit through the inferior orbital fissure;

❑ *Medially:* With the pterygopalatine fossa through the pterygomaxillary fissure;
❑ *Above and laterally:* With the temporal fossa through the gap between the zygomatic arch and the rest of the skull;
❑ *Above and medially:* With the middle cranial fossa through the foramen ovale and foramen spinosum.

CONTENTS

Several structures occupy the infratemporal fossa. These are:
❑ Lateral and medial pterygoid muscles;
❑ Lower part of temporalis;
❑ Maxillary artery and its branches;
❑ Pterygoid venous plexus;
❑ Mandibular nerve and its branches;
❑ Chorda tympani nerve;
❑ Otic ganglion.

The temporomandibular joint is also an anatomical content of the fossa.

Pterygoid Muscles

The pterygoids are two of the four muscles of mastication.

❑ The *lateral pterygoid muscle* is the key muscle of the infratemporal fossa. It has two heads, the upper and the lower heads and lies in the roof of the fossa. The fibres of the lower head run more or less horizontally and converge posteriorly to reach the neck of the mandible. Most of the contents of the infratemporal fossa have important relations to this muscle. The mandibular nerve and its branches and the deep head of the medial pterygoid muscle are deep to it. The buccal branch of the mandibular nerve runs between the two heads of the lateral pterygoid. The maxillary artery, most commonly, is superficial to the muscle; sometimes it is deep. The deep temporal vessels and nerves emerge at its superior border; the medial pterygoid muscle, the inferior alveolar and lingual nerves emerge at its inferior border. The pterygoid venous plexus lies within the substance of the muscle and around it.

❑ The *medial pterygoid muscle* is seen inferior to the lateral pterygoid muscle. The fibres of medial pterygoid run downwards and backwards towards the lower border of the mandible. Anteriorly, some fibres of this muscle lie superficial to the lower head of the lateral pterygoid; these fibres constitute the superficial head of the medial pterygoid.

Both the medial and lateral pterygoid muscles lie deep to the ramus of the mandible. The ramus separates the medial pterygoid from the masseter. On the anterior aspect, the medial pterygoid overlaps the posterior part of the buccinator muscle.

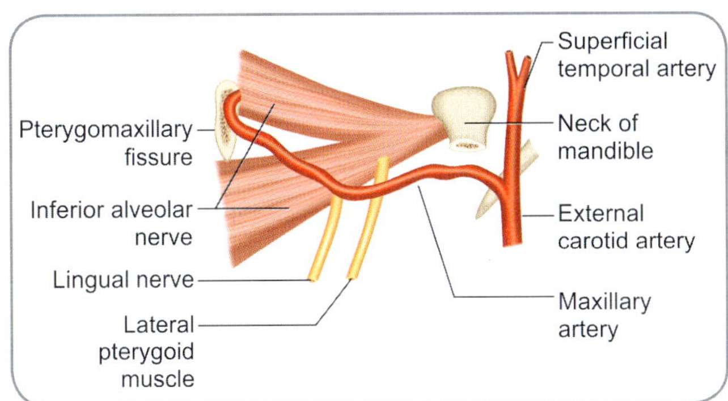

Fig. 4.2: Course of the maxillary artery

Tendon of Temporalis

The tendon of the temporalis muscle descends through the gap between the zygomatic arch and the lateral aspect of the skull. It, therefore, passes from the temporal fossa to the infratemporal fossa.

Maxillary Artery (Fig. 4.2)

The *maxillary artery* (also called the internal maxillary artery or the mandibulomaxillary artery) is the larger terminal branch of the external carotid artery and arises behind the neck of the mandible within the parotid gland. The course of the artery is divided into three parts by the lower head of lateral pterygoid muscle.

The *first part* (mandibular part) runs forwards and horizontally deep to the neck of the mandible to enter the infratemporal fossa and reaches the lower border of the lateral pterygoid muscle. In this part, it passes between the neck of mandible and the sphenomandibular ligament and lies parallel to and below the auriculotemporal nerve. Crossing the inferior alveolar nerve, it reaches the lower border of the lateral pterygoid. The *second part* (pterygoid part) passes upwards and forwards and then runs either superficial or deep to the lower head of lateral pterygoid muscle. Finally, it enters the interval between the two heads of the lateral pterygoid and disappears into the pterygomaxillary fissure to enter into the pterygopalatine fossa. The course of the artery within the pterygopalatine fossa is the *third part* (pterygopalatine part).

 **Development**

The maxillary artery is a remnant of the first pharyngeal arch artery; while most of the first pharyngeal arch artery disappears early in the foetus, a small portion of its middle part develops into the maxillary artery.

Branches of Maxillary Artery (Fig. 4.3)

The branches of the maxillary artery can be classified into three groups as according to the three parts of the artery itself.

71

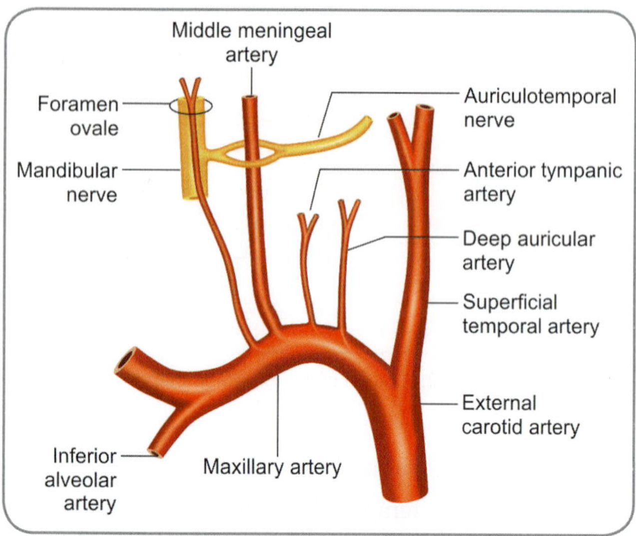

Fig. 4.3: Branches of first part of maxillary artery in the infratemporal region

Branches of the Mandibular Part

Five branches given out from the first or mandibular part. While still posterior to the mandible, the maxillary gives off the **anterior tympanic** and **deep auricular** arteries. The middle meningeal artery, the accessory meningeal artery and the inferior alveolar artery are given out in the infratemporal fossa.

❑ The **anterior tympanic artery** enters the tympanic cavity through the petrotympanic fissure and supplies the inner surface of the tympanic membrane. It is accompanied by the chorda tympani nerve and anterior ligament of malleus when it passes through the petrotympanic fissure.

❑ The **deep auricular artery** enters the external acoustic meatus and supplies the skin of the external acoustic meatus and part of the outer surface of tympanic membrane. It may also give a twig to the temporomandibular joint.

❑ The **middle meningeal artery** is the largest meningeal branch supplying the meninges and bones of the skull vault. It usually arises as a separate branch but may arise in common with the inferior alveolar branch. The artery ascends between the sphenomandibular ligament and the lateral pterygoid muscle, passes between the two roots of auriculotemporal nerve and leaves the infratemporal fossa through the foramen spinosum along with the meningeal branch of mandibular nerve to reach the middle cranial fossa. In the middle cranial fossa, it turns upwards and forwards on the greater wing of sphenoid and divides into anterior and posterior branches, both of which end by supplying the meninges. However, the main supply of this artery is to extracranial structures which include the medial and lateral pterygoid muscles, tensor veli palatini, pterygoid processes and greater wing of sphenoid bone, otic ganglion and mandibular nerve.

❑ The **inferior alveolar artery** (or the **inferior dental artery**) descends in the infratemporal fossa posterior to the inferior alveolar nerve between the mandible laterally and the sphenomandibular ligament medially. After giving a branch to mylohyoid, it enters the mandibular foramen along with the inferior alveolar nerve and ends by supplying the teeth of the lower jaw and the skin around the area of chin.

Added Information

The middle meningeal artery divides into anterior and posterior branches.

The anterior branch is the larger of the two. Most of its course is opposite the motor area of the cerebrum. It runs on the side of the cranial cavity to a point opposite the pterion where it often tunnels the bone. It then runs upward and backward towards the middle of the cranial vault.

The posterior branch runs upward and backward; its course corresponds to the lateral sulcus of the cerebrum. The artery grooves the squamous part of the temporal bone and the inferior part of the parietal bone.

The middle meningeal artery through its two branches supplies most of the supratentorial dura mater except for the floor of the anterior cranial fossa. Therefore, the artery is of great clinical importance. It may be fatally damaged in fractures of skull; it should be handled carefully in surgical procedures involving the skull base and the lateral aspect of the skull.

Surface marking of middle meningeal artery: The stem and the anterior and posterior divisions of the artery can be marked on the surface.

Stem: Point A is marked 1 mm in front of the preauricular point. Point B is marked 2 cm vertically above the midpoint of the zygomatic arch. The two points are joined by a line that curves a little forwards as it runs upward. The stem or trunk of the artery is marked on the surface by this line.

Anterior (frontal) branch: Mark Point B as noted above (it will be the upper end of the trunk of the artery where the artery divides into the anterior and posterior branches). Point C is marked on the centre of pterion (marking of pterion can be done using the frontozygomatic suture as landmark or using the Stile's method). Point D is marked midway between the root of nose (nasion) and external occipital protuberance (marked by inion which indicates the highest point of external occipital protuberance). Points B and C are connected by a line that runs upwards with a forward convexity. Point C and D are then connected by a wavy line. These lines indicate the anterior branch that runs up to the vault.

Posterior (parietal) branch: Point B is already marked. Point E is marked on the upper point of attachment of auricle. Point F is marked on the lambda or approximately 6 cm above the external occipital protuberance. Points B, E and F are joined by a line which is slightly convex upwards proximally and slightly convex downwards distally.

The artery is liable to be torn in blows to the side of the head and extensive bleeding occurs between the skull and the duramater. Surface marking is therefore surgically important.

Branches of the Pterygoid Part

Branches given out from the second or pterygoid part of the artery do not enter the bone and supply fasciae and muscles. These branches are the buccal artery, the pterygoid branches, the deep temporal arteries and the masseteric artery.

- The buccal artery runs anteriorly between the medial pterygoid and the temporalis to supply the mucosa and skin over the buccinator. It is accompanied by the lower portion of the buccal branch of the mandibular nerve. It may give a lingual branch that supplies the lingual nerve and floor structures of the mouth cavity.
- The pterygoid branches are many in number; they supply the two pterygoid muscles.
- The deep temporal branches, as already noted, are three in number—the anterior, middle and posterior deep temporal arteries. Passing deep to the temporalis, they supply the same muscle and the fasciae and muscles of the region. They anastomose with the branches of the superficial temporal artery.
- The masseteric artery is a small branch that runs posterior to the temporalis tendon and skirts the mandibular notch to enter the deep surface of and supply the masseter muscle. Its branches anastomose with the masseteric branches of the facial artery and the transverse facial branch of the superficial temporal artery.

Branches of the Pterygopalatine Part

Branches given out from the third or the pterygopalatine part of the maxillary artery accompany branches of the maxillary nerve. These branches are the posterior superior alveolar artery, infraorbital artery, greater palatine artery, pharyngeal artery, artery of pterygoid canal and the sphenopalatine artery, which is a continuation of the maxillary artery itself.

- The ***posterior superior alveolar artery*** runs through the pterygomaxillary fissure to reach the maxillary tuberosity where it ends by supplying the teeth of upper jaw and the maxillary air sinus. Some branches also supply the buccal mucosa.
- The ***infraorbital artery*** enters the orbit through the inferior orbital fissure. It runs on the floor of the orbit in the infraorbital groove and the infraorbital canal and comes out in the face through the infraorbital foramen. It supplies the lower eyelid, part of cheek, upper lip and nose. Within the infraorbital canal, it gives off the anterior superior alveolar artery which anastomoses with the posterior superior alveolar artery and supplies the teeth.
- The ***artery of the pterygoid canal*** (also called the ***Vidian artery***) passes through the pterygoid canal and anastomoses with the pharyngeal, ethmoidal and

sphenopalatine arteries in the pterygopalatine fossa and with the ascending pharyngeal, accessory meningeal, palatine arteies in the oropharynx and around the pharyngotympanic tube. It supplies the tympanic cavity, pharyngotympanic tube and the upper pharynx.

- The ***pharyngeal artery*** passes through the palatovaginal canal and supplies the nasal mucosa, nasopharynx, parts of the pharyngotympanic tube and the sphenoidal air sinus.
- The ***greater palatine artery*** (also called the ***descending palatine artery***) runs through the greater palatine canal and the greater palatine foramen to reach the hard palate. It supplies the inferior meatus of the nose, the hard palate and the palatal gingivae of the maxillary teeth. It gives a branch which runs through the incisive canal to anastomose with the sphenopalatine artery, thereby supplying the nasal septum. The lesser palatine arteries are branches of the greater palatine artery which are given out in the greater palatine canal and which emerge through the lesser palatine foramen to supply the soft palate.
- The ***sphenopalatine artery*** is the continuation of the maxillary artery. It passes through the sphenopalatine foramen to enter into the nasal cavity. It ends up by supplying the lateral wall of the nose through its posterior lateral nasal branches and nasal septum through the posterior septal branches. It anastomoses with the ethmoidal arteries and the nasal branches of the greater palatine artery to supply all the four paranasal air sinuses. A branch descends to the incisive canal to anastomose with the greater palatine artery and the septal branch of superior labial artery.

Pterygoid Venous Plexus

- This is a dense plexus of veins found between the temporalis and the lateral pterygoid and between the two pterygoids. The sphenopalatine vein, the inferior alveolar vein, the greater palatine vein, the masseteric veins, the pterygoid veins, the buccal veins, the deep temporal veins and the middle meningeal vein anastomose with each other to form the plexus. Some venous twigs from the inferior ophthalmic vein may also join the plexus. The plexus is connected to the facial vein through the deep facial vein; to the cavernous sinus through emissary veins passing through the foramen ovale, foramen lacerum and sphenoidal emissary foramen. The deep temporal veins are usually connected to the anterior diploic veins and thus to the middle meningeal veins.
- The pterygoid venous plexus ends as the maxillary vein which joins the superficial temporal vein at the level of the neck of mandible to form the retromandibular vein.

Mandibular Nerve

The mandibular nerve is the largest of the three divisions of the trigeminal nerve and is the nerve of the first pharyngeal arch. It is a mixed nerve and consists of a large sensory root and a small motor root.

- ❑ The large sensory root emerges from the lateral part of the trigeminal ganglion and exits the cranial cavity through the foramen ovale. The small motor root passes under the ganglion and through the foramen ovale to unite with the sensory root to form the trunk of mandibular nerve in the infratemporal fossa.

- ❑ The trunk lies between the tensor veli palatini medially and lateral pterygoid laterally. The otic ganglion is situated between the nerve trunk and the tensor veli palatini muscle. The middle meningeal artery lies behind the trunk. After a short course, the trunk divides into a thin anterior and a thick posterior division.

Branches of Mandibular Nerve (Fig. 4.4)

Branches are given out from the main trunk of the nerve and from the anterior and posterior divisions.

- ❑ *Branches from the trunk:* These are given out when the mandibular nerve passes between the tensor veli palatini and the medial pterygoid. They are two in number, the meningeal branch and the nerve to medial pterygoid.
  - ○ Otherwise called the nervus spinosus, the *meningeal nerve* enters into the cranial cavity through the foramen spinosum along with the middle meningeal artery. It divides into two branches, the anterior and the posterior, which accompany similar branches of the middle meningeal artery and supply the dura mater of the middle cranial fossa, parts of the anterior cranial fossa and the calvarium.

  - ○ The *nerve to the medial pterygoid* is a slender branch which ramifies on the deep aspect of medial pterygoid to supply the muscle. It also supplies the tensor tympani and tensor veli palatini muscles through branches which pass through the otic ganglion without interruption.

- ❑ *Branches from the anterior division:* The anterior division, through its branches supplies three of the four muscles of mastication. There are three motor branches and one sensory branch. The motor branches are the deep temporal nerves, the masseteric nerve and the nerve to lateral pterygoid. The sensory branch is the buccal nerve.
  - ○ Usually two in number (anterior and posterior), the *deep temporal nerves* enter the deep surface of the temporalis muscle and supply it. These two branches may arise in conjunction with any of the other branches.

  - ○ Often called the nerve to masseter, the *masseteric nerve* passes between the temporalis tendon and the temporomandbular joint to reach the mandibular notch. Passing through the notch along with the masseteric artery, it ramifies on the deep aspect of masseter muscle to supply it. The nerve also gives articular twigs to the temporomandibular joint.

  - ○ The *nerve to the lateral pterygoid* is a very small nerve. Arising from the anterior division, it immediately enters the deep surface of the lateral pterygoid muscle to supply it.

  - ○ The *buccal nerve* is the only sensory branch of anterior division. It emerges between the two heads of lateral pterygoid muscle and appears on the cheek beneath the anterior border of the masseter. It ends by supplying the skin and mucous membrane of the cheek. It does not supply the buccinator muscle.

- ❑ *Branches from the posterior division:* The posterior trunk is larger than the anterior trunk and is largely sensory. It gives off three branches, namely the auriculotemporal nerve, the lingual nerve and the inferior alveolar nerve.
  - ○ The *auriculotemporal nerve* arises by two roots which encircle the middle meningeal artery. It then runs deep to the lateral pterygoid and passes between the sphenomandibular ligament and the neck of the mandible. After coursing behind the temporomandibular joint, in relation to the upper part of the parotid gland, it emerges from behind the joint, runs up over the posterior root of zygoma and divides into superficial temporal branches. These are cutaneous branches which end up supplying the skin over the face and scalp, skin of the external auditory meatus and external surface of tympanic membrane. Articular twigs are given to the temporomandibular joint. The superficial temporal branches also give communicating branches to the facial nerve and otic ganglion. Postsynaptic

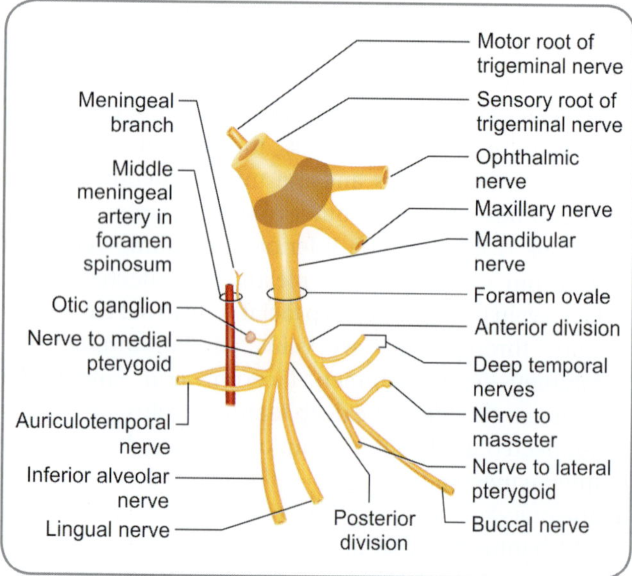

Fig. 4.4: Branches of mandibular nerve

parasympathetic secretomotor fibres from the otic ganglion pass through the auriculotemporal nerve to the facial nerve to be ultimately distributed to the parotid gland.

- The *lingual nerve* is sensory to the anterior two-thirds of the tongue, floor of the mouth and to the gingivae. After arising from the posterior division, it passes deep to the lateral pterygoid and superficial to the tensor veli palatini where it is joined posteriorly at an acute angle by the *chorda tympani* nerve which is a branch of the facial nerve. It then runs over the medial pterygoid and leaves the infratemporal region by passing beneath the superior constrictor and the pterygomandibular raphe, where it is closely applied to the periosteum of the third molar tooth. It then passes downwards and forwards over the deep surface of the mylohyoid, where it lies over the deep part of the submandibular gland. The nerve passes below the submandibular duct by which it is crossed from the medial to the lateral side. It then curves upward and forward to enter the tongue and ends by supplying the same. The lingual nerve is connected to the submandibular ganglion by two or three branches.

     Point A is marked on the centre of the zygomatic arch. Point B is marked below and behind the third molar tooth. Point C is marked immediately beneath the canine tooth. The three points are joined by a line that is curved with an upward concavity.

- The *inferior alveolar nerve* (also called the *inferior dental nerve*) emerges from under the lower border of the lateral pterygoid muscle and descends between the sphenomandibular ligament and the ramus of the mandible. It then enters the mandibular canal via the mandibular foramen and passes through it to supply the mandible and the lower teeth. It is accompanied by the inferior alveolar vessels within the canal. Just before it enters the mandibular canal, the inferior alveolar nerve gives off a branch, the *mylohyoid nerve* which pierces the sphenomandibular ligament and enters the mylohyoid groove on the medial surface of the mandible, descends into the submandibular region to reach the superficial surface of the mylohyoid muscle deep to the anterior belly of digastric muscle and ends by supplying both the muscles. Small cutaneous twigs are also given out by the mylohyoid nerve to supply the skin of the chin. Within the mandibular canal, the inferior alveolar nerve forms the *inferior dental plexus* through which it supplies the molar and the premolar teeth of the lower jaw. At the level of the first premolar, the inferior alveolar nerve ends up by dividing into its two terminal branches—the *incisive nerve* and the *mental nerve*. The incisive

nerve supplies the canine and incisors. The mental nerve emerges out through the mental foramen, gives out two branches which form an incisive plexus that supplies the gingivae and the periosteum of the lower jaw and ends in a third branch that supplies the skin over the chin and lower lip.

Surface marking of mandibular and inferior alveolar nerves: The inferior alveolar nerve, though technically only a branch of the posterior division, appears more to be a continuation of the mandibular nerve because of the same curvature. So the two nerves can be marked together. Point A is marked on the centre of the zygomatic arch. Point B is marked on the centre of the masseter muscle. This point will indicate the mandibular foramen. Point C is marked vertically below the interval between the two premolar teeth, but midway between the upper and lower borders of mandible. All the three points are joined by a broad line that curves with an upward concavity. The upper most part of the vertical segment represents the mandibular nerve and the rest of the line represents the inferior alveolar nerve.

Dissection

Identify the branches of the mandibular nerve. Trace the lingual nerve. Identify the chorda tympani that joins the lingual nerve. Trace the nervus spinosus. After defining the pterygoid muscles, cut across them to expose the deeper structures. Trace and define the third part of maxillary artery and its branches.

Note on the Chorda Tympani Nerve

The chorda tympani nerve carries taste sensation fibres from the anterior two-thirds of the tongue. It joins the lingual nerve in the infratemporal fossa. The chorda tympani also transmits parasympathetic secretomotor fibres destined for the submandibular and sublingual glands.

Note on the Otic Ganglion

This is a parasympathetic ganglion located in the infratemporal fossa. It is inferior to the foramen ovale, posterior to the medial pterygoid muscle and medial to the mandibular nerve. Presynaptic parasympathetic fibres from the glossopharyngeal nerve reach the otic ganglion and synapse. The postsynaptic secretomotor fibres start from the ganglion and go to the parotid gland through the auriculotemporal nerve and its communication with the facial nerve.

Added Information

Pterygospinous ligament: This is a fibrous cord stretching between the spine of sphenoid and the upper part of the posterior border of the lateral pterygoid. The ligament may be replaced by muscle fibres or ossified. If ossified, it creates a foramen through which the muscular branches of the anterior division of the mandibular nerve pass.

Clinical Correlation

- Mandibular nerve block is usually performed through the infratemporal fossa. The needle with the anaesthetic agent is passed through the mandibular notch of the ramus of the mandible into the fossa. The anaesthetic agent injected here also anaesthetises the inferior alveolar, auriculotemporal, lingual and buccal nerves.
- The inferior alveolar nerve can be blocked by injecting an anaesthetic agent near the mandibular foramen. The mandibular teeth till the median plane, the skin and mucosa of lower lip, gingivae of the region and the skin of the chin will be anaesthetised.

MUSCLES OF MASTICATION

The muscles of mastication are four in number and are very intimately connected with the movements of the temporomandibular joint. These are shown in Figure 4.5 and Table 4.1.

1. Temporalis
2. Masseter
3. Lateral pterygoid
4. Medial pterygoid

Additional Notes on Temporalis

The temporalis muscle (Fig. 4.6) is disposed like a fan. The anterior most fibres are almost vertical; the posterior most are more or less horizontal; the intervening fibres are oblique. The muscle closes the mouth by elevating the mandible. Contraction of the anterior fibres provides an upward pull and that of the posterior fibres provides a backward pull. Both pulls are required for the complete closure of mouth and also for approximation of teeth. Contraction of both temporalis muscles cause side-to-side grinding movements. Isolated contraction of the posterior fibres help a protracted mandible come back to normal position.

Additional Notes on Masseter

(Greek.masasthai=to chew). This muscle of mastication (Fig. 4.7) lies on the lateral aspect of the ramus of mandible. It can easily be palpated in a living individual and is made prominent by clenching the teeth. The muscle itself has three layers—the superficial, the middle and the deep layers. All the three layers merge anteriorly. The anterior part of the muscle overlaps the posterior part

Table 4.1: Muscles of mastication

Muscle	Origin	Insertion	Action	Nerve supply
Temporalis	Temporal fossa on the lateral side of skull upto the inferior temporal line, temporal fascia	Tendon passes deep to zygomatic arch and is inserted to the coronoid process of mandible and the anterior border of the ramus of the mandible	• Elevates mandible and closes the mouth • Retracts the mandible and also assists in side to side grinding movements	Deep temporal branches of anterior division of mandibular nerve
Masseter	• Superficial fibres from anterior 2/3 of lower border of zygomatic arch • Deep fibres from deep surface and posterior 1/3 of lower border of the arch	On the lateral surface of the ramus and the angle of mandible	• Elevates mandible and closes the mouth • Protraction of jaw	Branch from anterior division of mandibular nerve
Lateral pterygoid	• *Upper head:* Infratemporal surface and infratemporal crest of greater wing of sphenoid bone • *Lower head:* Lateral surface of the lateral pterygoid plate	• Fovea on the anterior surface of the neck of mandible • Intra-articular disc and capsule of temporomandibular joint	• Depresses the mandible and opens the mouth • Protraction of the mandible • With the medial pterygoid and muscles of the opposite side, causes side to side movements of jaw	Branch from the anterior division of mandibular nerve
Medial pterygoid	• Deep head from medial surface of lateral pterygoid plate • Superficial slip from pyramidal process of palatine bone and maxillary tuberosity	Medial surface of angle of mandible and adjoining part of the ramus	• Elevates the mandible and closes the mouth • Protraction of the mandible • With the lateral pterygoid and the muscles of the opposite side causes side to side movements of jaw	Branch from the trunk of the mandibular nerve

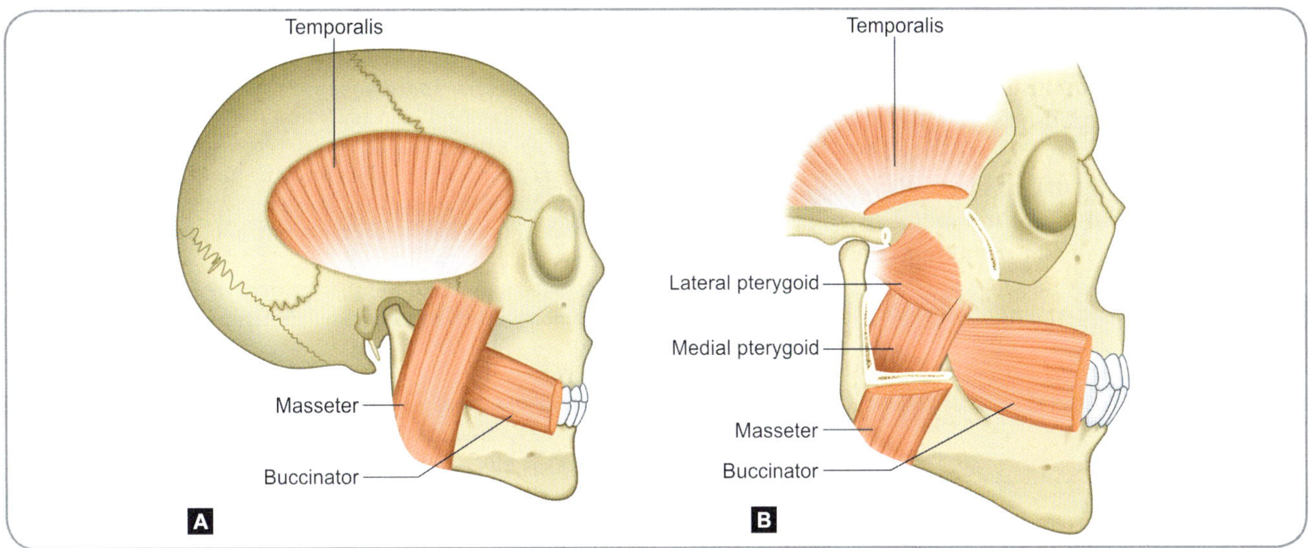

Figs 4.5A and B: Overall views of the muscles of mastication

of buccinator muscle. The masseter itself is overlapped by (from superficial to deep) skin and superficial muscles of the face, anterior part of the parotid gland, parotid duct, branches of the facial nerve and transverse facial vessels. Deep to the masseter are the ramus of the mandible, lower part of temporalis and the posterior part of buccinator.

Additional Notes on Pterygoid Muscles (Figs 4.8 and 4.9)

The medial and lateral pterygoids of both sides acting together protract the mandible. The medial and lateral pterygoids of one side, acting together, pull the mandibular condyle of that side forwards and medially. As a result,

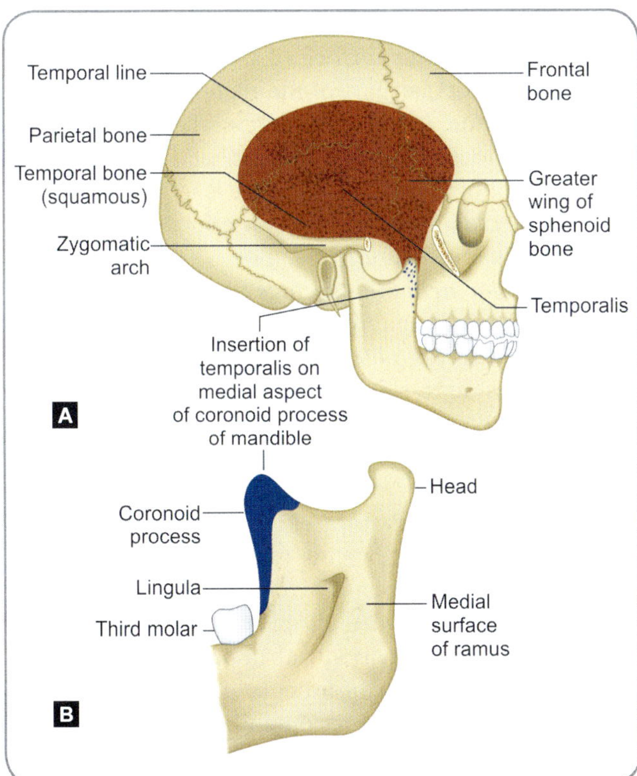

Figs 4.6A and B: A. Lateral view of the skull showing the origin of temporalis muscle from the temporal fossa **B.** Ramus of mandible seen from the medial side to show the insertion of temporalis

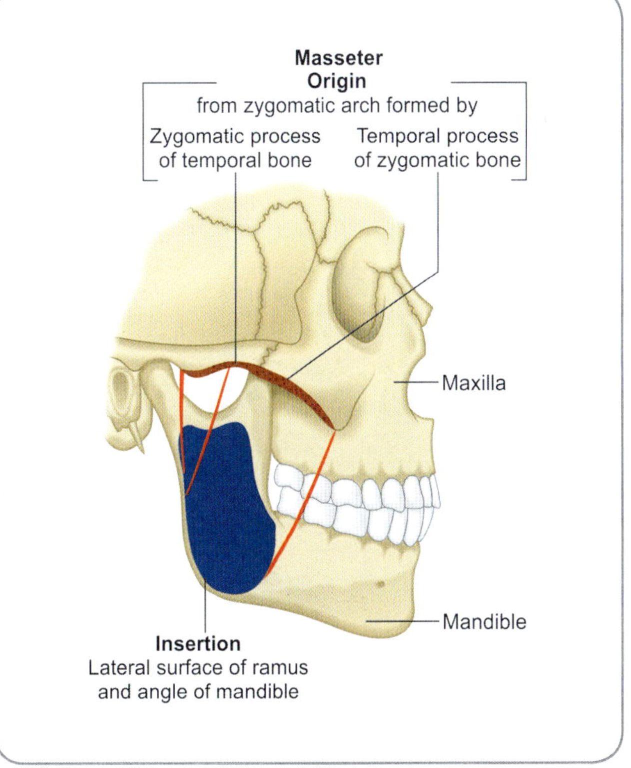

Fig. 4.7: Attachments of masseter muscle

the chin moves forwards and to the opposite side. This movement is facilitated by slight rotation of the head of mandible of the opposite side. Alternate action of the muscles of both the sides results in side-to-side chewing movements.

The two pterygoid muscles have opposite actions as far as opening and closing of the mouth is concerned. The lateral pterygoid helps in opening the mouth by pulling the head of the mandible forwards along with the intra-articular disc. The medial pterygoid elevates the mandible and thus helps in closure of the mouth.

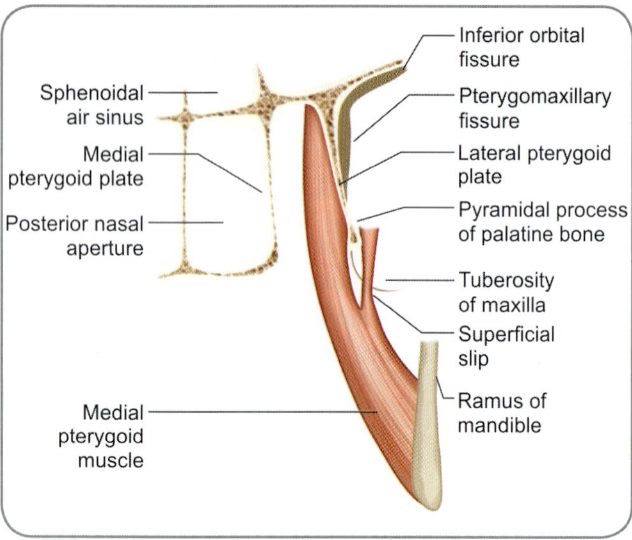

Fig. 4.8: Scheme to show the arrangement of medial pterygoid muscle

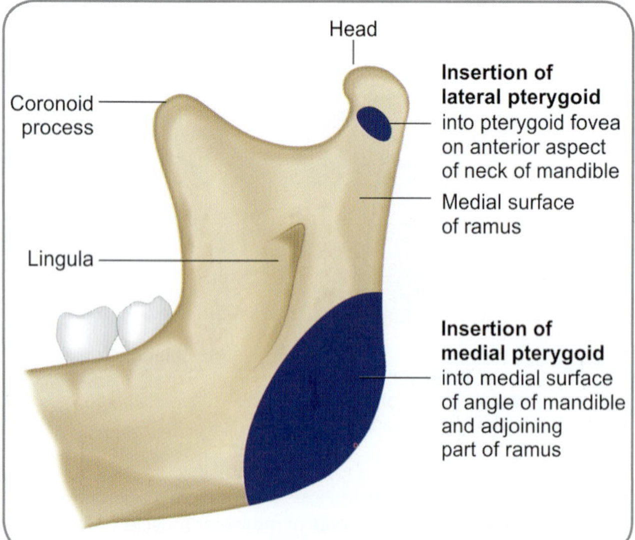

Fig. 4.9: Medial aspect of the ramus of mandible showing areas for insertion of lateral and medial pterygoid muscles

Clinical Correlation

- ❑ ***Submasseteric space infection:*** Infections from the lower third molar tooth can track into the submasseteric space. This space is between the masseter muscle and the ramus of the mandible. If pus tracks from the tooth and accumulates here, no visible swelling is caused; but the condition is painful, produces muscle spasm and restriction in the opening of the mouth.
- ❑ Fractures involving the middle third of the face inevitably damage structures of the infratemporal fossa. Muscles, nerves and blood vessels can be damaged. Injury to the maxillary or the mandibular nerves cause altered sensations in the oral cavity, jaws and parts of face. Involvement of chorda tympani causes impaired taste. Fractures can extend into the pterygopalatine fossa, orbit or the ear.
- ❑ Collection of pus or blood in the infratemporal fossa can have serious consequences since the collection can spread into the middle cranial fossa through the foramina in the skull base.

TEMPOROMANDIBULAR JOINT

The temporomandibular joint is a synovial joint of the condylar variety. The mandible articulates with the base of the skull at the mandibular fossa of the temporal bone. It is a complex joint as its cavity is divided into upper and lower parts by an intra-articular disc.

Articular Surfaces

The ***upper articular surface*** of the joint is formed by the anterior articular part of the mandibular fossa (also called the glenoid fossa) of the temporal bone. Anteriorly, the surface extends onto the articular tubercle. The anterior limit of the articular surface is a transverse ridge on the articular tubercle. The posterior limit is till the squamotympanic fissure.

The ***inferior articular surface*** of the joint is formed by the head of the mandible. The head of the mandible is markedly convex anteroposteriorly and more gently convex from side-to-side. The axes of both the mandibular heads are directed backwards and medially and lie in the arc of a circle passing through the anterior margin of the foramen magnum.

Both the articular surfaces are covered with white fibrocartilage and not hyaline cartilage as in most other synovial joints as the bones involved in the joint ossify in membrane. The ***articular disc*** is made of fibrocartilage and divides the joint cavity into an upper meniscotemporal compartment (between the articular disc and the temporal bone) and a lower meniscomandibular compartment (between the articular disc and the mandible).

Fibrous Capsule

The articular components of the joint are enclosed in a fibrous capsule which is compact and tight in the lower part but loose and flaccid in the upper part. The fibres of the deeper part of the capsule are in two sets—the upper set whose fibres are long and loose and attach between the mandibular fossa and the articular disc; the lower set whose fibres are short and tight and attach between the articular disc and the mandibular head. The superficial fibres are long and attach between the two bones. The upper attachment of the capsule (temporal attachment) is to the articular tubercle, squamotympanic fissure and the periphery of the articular fossa between the two. The lower attachment of the capsule (mandibular attachment) is around the neck of the mandible. Anteriorly, the capsule blends with the insertion of the lateral pterygoid muscle. The inside of the capsule is lined by synovial membrane except over the articular cartilages and the articular disc.

Ligaments

❑ ***Articular disc:*** Though the articular disc cannot be technically classified as a ligament of the joint, it can be called so on functional grounds since it accords stability to the joint. It is an oval plate of fibrocartilage and caps the head of the mandible. Morphologically, it represents the degenerated primitive insertion of the lateral pterygoid muscle. Upper surface of the disc is concavo-convex to fit the upper articular surface of the joint and the lower surface is concave into which fits the head of the mandible. The disc is attached to the inner aspect of the periphery of the fibrous capsule (both the upper and lower sets of deep fibres as explained above) and blends with the insertion of the lateral pterygoid anteriorly. Posteriorly, the disc is split into upper and lower lamellae by the presence of a venous plexus.

Through the two lamellae, the disc provides attachment; the upper lamella is fibroelastic and attaches to the squamotympanic fissure; the lower lamella is fibrous and attaches to the back of the mandibular condyle. The rest of the articular disc gives intra-articular support to the joint. Though the bilamellar portion of the disc has a venous plexus, the central part of the disc is avascular.

❑ ***Lateral temporomandibular ligament:*** Simply called the ***lateral ligament*** (Fig. 4.10)**,** this broad ligament strengthens the lateral part of the capsule by blending with the lateral part of the fibrous capsule. The upper end of the ligament is attached to the tubercle of the root of the zygoma and the lower end to the lateral aspect of the neck of the mandible. The fibres are directed inferolaterally.

In addition, the joint has two other ligaments. These two are customarily described as ***accessory ligaments*** since they are independent of the capsule and lie some distance away from it. They are the sphenomandibular ligament and the stylomandibular ligament. The ***sphenomandibular ligament*** (Fig. 4.11) is attached above to the spine of the sphenoid and below to the lingula of the mandibular foramen. Its extension is the anterior ligament of malleus which is attached to the anterior process of malleus. The ligament represents the unossified middle part of the Meckel's cartilage of the first pharyngeal arch. The ***stylomandibular ligament*** (Figs 4.10 and 4.11) is formed by the thickening of the deep cervical fascia and extends from the tip of the styloid process to the angle of the mandible.

Dissection

Define the lateral pterygoid muscle. See its attachment to the capsule of the temporomandibular joint. Clean the capsule. Identify and study the liagements of the joint. Carefully open the joint cavity and study the articular disc.

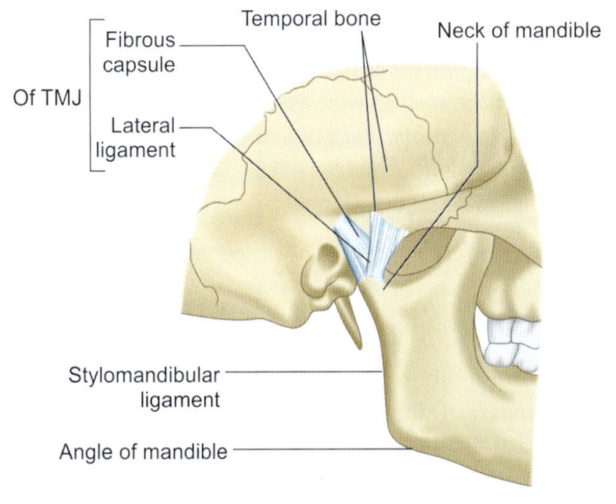

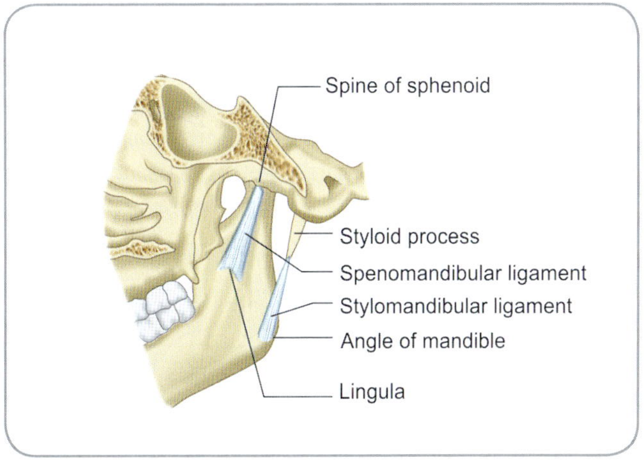

Fig. 4.10: Ligaments of the temporomandibular joint as seen from the lateral side (TMJ temporomandibular joint)

Fig. 4.11: Ligaments of temporomandibular joint as seen from the medial side

Relations

- *Anteriorly:* Lateral pterygoid, temporalis, masseteric vessels and nerve;
- *Posteriorly:* Parotid gland, superficial temporal vessels, auriculotemporal nerve and external acoustic meatus;
- *Medially:* From outside inwards-lateral pterygoid, roots of auriculotemporal nerve enclosing the middle meningeal artery, spine of sphenoid, sphenomandibular ligament and chorda tympani nerve;
- *Laterally:* Skin and fasciae;
- *Superiorly:* Floor of the middle cranial fossa.

Arterial Supply

The joint is supplied by the branches of superficial temporal artery and maxillary artery. Veins of the joint drain into the pterygoid venous plexus. Lymphatic drainage is into the upper cervical lymph nodes.

Nerve Supply

One branch each from the anterior and posterior divisions of mandibular nerve supply the joint; these are the *masseteric* and the *auriculotemporal* nerves respectively. The central part of the articular disc has no nerve supply. The joint capsule, the bilamellar part of the articular disc, the lateral ligament and the periarticular tissue have rich nerve supply which helps maintain mandibular posture and movements.

Movements

- Movements which occur at the temporomandibular joint are:
  - Protraction and retraction;
  - Depression and elevation of mandible;
  - Side-to-side chewing movements of the mandible.
- The movements at the joint can be divided into those between the upper articular surface and the articular disc (upper compartment) and those between the disc and the head of the mandible (lower compartment). Most movements occur simultaneously at the right and left temporomandibular joints. In forward movement or *protraction* of the mandible, the articular disc glides forwards over the upper articular surface with the head of the mandible moving along. The reversal of this movement is called *retraction*.
- In slight opening of the mouth (or *depression* of the mandible), the head of the mandible moves on the undersurface of the disc-like a hinge. In wide opening of the mouth, the hinge-like movement is followed by a forward gliding of the disc and the head of the mandible, as in protraction. At the end of this movement, the head comes to lie under the articular tubercle. These movements are reversed in closure of the mouth (or *elevation* of the mandible).

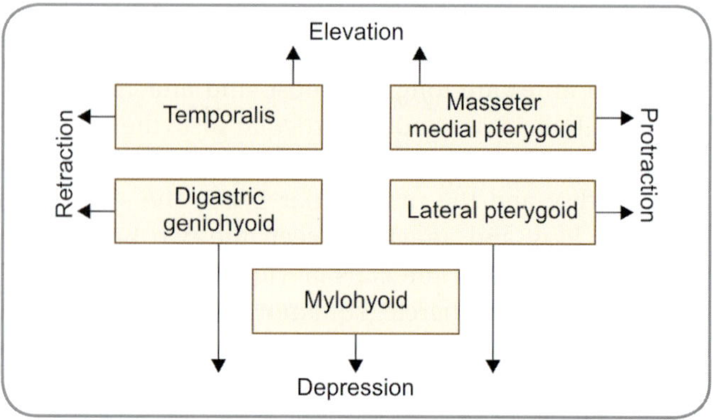

Fig. 4.12: Scheme to show the muscles responsible for movements at the temporomandibular joint

- *Chewing movements* involve side-to-side movements of the mandible. In these movements, the head of the mandible of one side glides forwards along with the articular disc as in protraction but the head of the opposite side merely rotates around a vertical axis. As a result of this, the chin moves forwards and to the side on which gliding has not occurred. Alternate movements of this kind on the two sides result in side-to-side movements of the lower jaw.

Muscles Producing Movements (Fig. 4.12)

- *Protraction:* Simultaneous action of lateral and medial pterygoid muscles of both sides.
- *Retraction:* Posterior fibres of temporalis muscle. Forceful retraction is assisted by the deep and middle fibres of masseter, digastric and geniohyoid muscle. Superficial fibres of masseter may play a role when retraction is more on one side.
- *Depression of mandible:* Lateral pterygoid; geniohyoid, mylohyoid and digastrics muscles.
- *Elevation of mandible:* Masseter, temporalis and medial pterygoid. Elevators are powerful antigravity muscles and therefore their sites of insertion make prominent markings in the mandible.
- *Side-to-side movements:* Contraction of lateral and medial pterygoid muscles of one side acting alternately with the other side.

Factors Maintaing Stability

- *Bones:* Forward displacement is prevented by the articular tubercles and backward displacement by postglenoid tubercles.
- *Ligaments:* Temporomandibular ligament prevents backward displacement.
- *Muscles:* Excessive protraction is resisted by the tension of temporalis and retraction by the tension of lateral pterygoid.
- *Position of mandible:* Occlusal position of mandible increases the stability of the joint.

Clinical Correlation

- The occlusal position (when the mouth is closed) is the stable position of the joint. The head of the mandible lies within the articular fossa in both the occlusal position and mouth in slightly open position. Hence, a blow on the chin, when the mouth is closed or slightly open causes fracture of the mandible rather than dislocation. Backward dislocation towards the external acoustic meatus is also prevented by the strong lateral temporomandibular ligaments.
- When the mouth is opened wide, the head of the mandible moves forwards and comes to lie just below the articular tubercle. In this position, the temporomandibular joint is highly unstable. A blow on the chin, or even sudden opening of the mouth as in yawning which involves sudden contraction of the lateral pterygoid muscle will cause the head of the mandible to slip forwards to the front of the articular tubercle. With the joint dislocated, the mouth cannot be closed. To reduce the dislocation, the surgeon inserts both his thumbs into the mouth and exerts downward pressure over the lower molar teeth. Simultaneously, the mandible is pressed backwards.
- The term derangement of the temporomandibular joint is applied to a condition in which part of the articular disc gets detached from the joint capsule which happens after injury. In this condition the movements of the jaw become painful, and clicking sounds may be produced while opening and closing the mouth.

Multiple Choice Questions

1. The temporoparietal fascia:
   a. Is the same as temporal fascia
   b. Is superficial to the temporal fascia
   c. Blends with temporal fascia
   d. Connects the temporal fascia to galea aponeurotica
2. The first part of maxillary artery:
   a. Is deep to the neck of mandible
   b. Enters the pterygopalatine fissure
   c. Enters the pterygoid canal
   d. Passes between the two heads of lateral pterygoid
3. Nerve to mylohyoid is a branch of:
   a. Inferior alveolar nerve
   b. Lingual nerve
   c. Buccal nerve
   d. Meningeal nerve
4. The masticatory muscle made prominent by clenching the teeth is:
   a. Lateral pterygoid
   b. Medial pterygoid
   c. Masseter
   d. Temporalis
5. What is untrue about the articular cartilage of the temporomandibular joint?
   a. Its posterior part is bilamellar
   b. Its lower surface is convex
   c. It is the degenerate portion of the lateral ptrygoid muscle
   d. It gives intra-articular support to the temporomandibular joint

ANSWERS

1. b   **2.** a   **3.** a   **4.** c   **5.** b

Clinical Problem-solving

Case Study 1: A 27-year-old man came to the hospital with an open mouth and complained that he could not close his mouth. Something appeared to block his closing the mouth. On examination, he said that the condition occurred after yawning.
- What condition is he probably suffering from?
- Why is he not able to close his mouth?
- In what way is yawning and this condition related?

Case Study 2: A 34-year-old woman presented with fever, sore throat and restriction in opening her mouth. There was no swelling. On close examination, she revealed that she had pain in the submandibular area.
- What condition do you think is the woman suffering from?
- Why is not there any visible swelling?
- Why is there a restriction in opening the mouth?

(For solutions see Appendix).

Chapter 5

Submandibular Region and Tongue

Frequently Asked Questions

- ❑ Discuss the submandibular salivary gland in detail.
- ❑ Write about the intrinsic and extrinsic muscles of the tongue.
- ❑ Write notes on: (a) Genioglossus, (b) Sublingual salivary gland, (c) Lymphatic drainage of the tongue, (d) Innervation of the tongue, (e) Suprahyoid muscles.
- ❑ Write briefly on: (a) Lingual artery, (b) Development of tongue, (c) Circumvallate papillae, (d) Lingual nerve, (e) Tongue tie.
- ❑ Correlate the innervation of tongue with its development.

SUBMANDIBULAR REGION

The submandibular region is that region between the mandible and the hyoid bone. Extending from the submandibular fossa of the body of the mandible to the hyoid bone, this region is sometimes called the suprahyoid region. It contains:

- ❑ Submandibular salivary gland;
- ❑ Sublingual salivary gland;
- ❑ Suprahyoid muscles—digastric, stylohyoid, mylohyoid and genioglossus;
- ❑ Submandibular ganglion;
- ❑ Lingual vessels and nerve;
- ❑ Glossopharyngeal and hypoglossal nerves.

Since the vessels and nerves related to the tongue are present in this region and the tongue itself is located close by, it is customary to study the tongue in relation to the submandibular region.

SUBMANDIBULAR GLAND

The submandibular gland is one of the three paired salivary glands. It is a mixed type of salivary gland with both serous and mucous components. It is located partly below and partly deep to the body of the mandible. The part below the mandible lies in the digastric triangle. The gland comprises a large superficial part and a small deep part. Both the parts are continuous around the posterior border of the mylohyoid muscle.

Superficial Part (Figs 5.1 and 5.2)

The superficial part of the salivary gland presents two ends—anterior and posterior and three surfaces—inferior, lateral and medial.

- ❑ The **anterior end** extends up to the anterior belly of the digastric muscle and the **posterior end** to the stylomandibular ligament. The stylomandibular ligament separates the submandibular salivary gland from the parotid gland. The **posterior end** also presents a groove for the lodgement of the ascending part of the cervical loop of the facial artery.

  This part of the gland is enclosed between two layers of the investing layer of deep cervical fascia. The fascia, extending up from the greater cornu of the hyoid bone, splits into two layers; the superficial layer gets attached

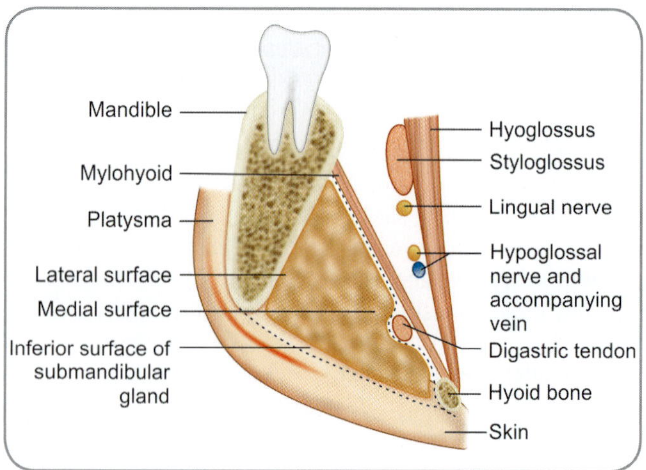

Fig. 5.1: Coronal section through the submandibular region to show the surfaces of the submandibular gland and their relations

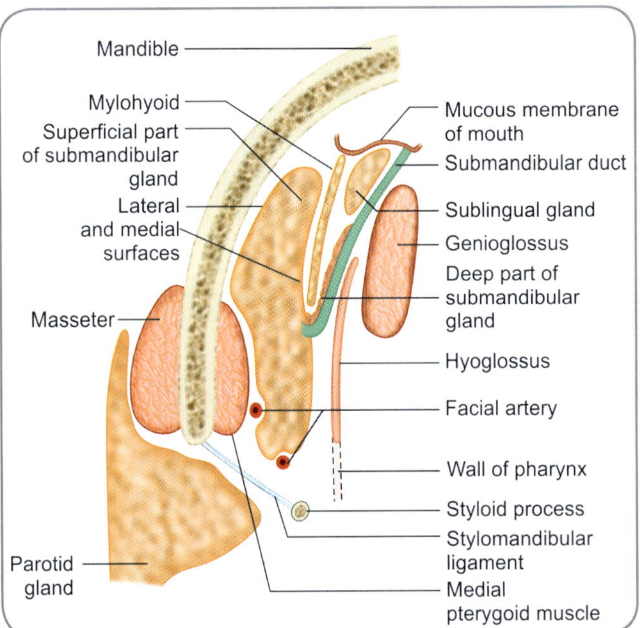

Fig. 5.2: Schematic transverse section through the body of mandible to show relations of the salivary glands

to the lower border of the body of mandible and covers the inferior surface of the gland; the deep layer gets attached to the mylohyoid line of the mandible and covers the medial surface.

- The ***inferior surface*** is superficial and is seen in the digastric triangle. It is directed downwards and somewhat laterally. It is covered by skin, superficial fascia, platysma and deep cervical fascia. This surface is related to the common facial vein and the cervical branch of the facial nerve beneath the platysma and to the submandibular lymph nodes deep to the deep fascia.
- The ***lateral surface*** is hidden from view by the body of the mandible. It lies in contact with the submandibular fossa of the body of the mandible, below the attachment

of the mylohyoid muscle. The posterior part of the lateral surface is separated from the mandible by the medial pterygoid muscle. The cervical loop of the facial artery descends down between the lateral surface of the gland and the ramus of the mandible, winds around the lower border of the mandible at the antero-inferior angle of masseter before ascending to the face.

- The ***medial surface*** rests on several structures. It can be subdivided into three parts, namely, anterior, intermediate and posterior parts.
  - The ***anterior part*** rests on the mylohyoid muscle, however separated from the muscle by the mylohyoid vessels and nerves and the submental branch of the facial artery.
  - The ***intermediate part*** rests on the hyoglossus muscle, however separated from the muscle by the lingual nerve, submandibular ganglion, hypoglossal nerve and intermediate tendon of digastric muscle.
  - The ***posterior part*** is related to the styloglossus, stylopharyngeus, glossopharyngeal nerve, posterior belly of digastric muscle, middle constrictor of pharynx, hypoglossal nerve and first part of the lingual artery.

Deep Part (Fig. 5.3)

The deep part of the submandibular gland extends in the interval between the mylohyoid (laterally) and hyoglossus (medially). It is related to the lingual nerve and the submandibular ganglion above and to the hypoglossal nerve and its accompanying veins below.

Blood Supply

The branches of the facial and lingual arteries supply the submandibular gland; the veins accompanying the arteries drain into the internal jugular vein.

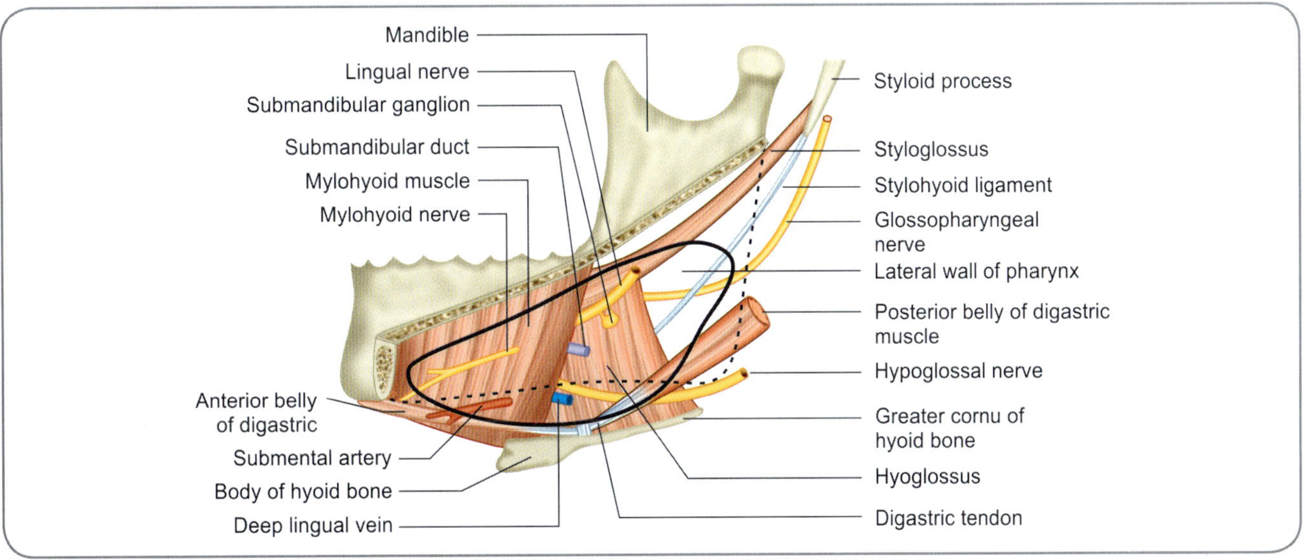

Fig. 5.3: Scheme to show structures deep to the submandibular gland – outline of the area covered by the gland is shown in bold line

Lymphatic Drainage

Lymphatics from the submandibular gland drain into the submandibular lymph nodes and through them into the deep cervical nodes, particularly the jugulo-omohyoid node.

Nerve Supply

The gland is supplied by both parasympathetic and sympathetic nerves. Both are secretomotor to the salivary glands. The parasympathetic innervation is responsible for a watery secretion and the sympathetic for a mucous secretion.

The *parasympathetic pathway* to the gland is as follows—preganglionic fibres arise from the superior salivatory nucleus which is located in the pons. The fibres pass through the facial nerve, its chorda tympani branch and the lingual nerve. Through the lingual nerve, they reach the submandibular ganglion and relay. The postganglionic fibres from the cells of the ganglion reach the submandibular gland and supply it.

The *sympathetic pathway* is as follows. Postganglionic sympathetic fibres from the superior cervical ganglion reach the submandibular ganglion through the sympathetic plexus around the facial artery. They do not synapse in the ganglion and run to the gland. They are both secretomotor to the gland and vasomotor to the blood vessels.

> **Surface Marking of Submandibular Gland**
>
> The outline of the gland is oval.
>
> Point A is marked on the angle of the mandible. Point B is marked on the middle of the base of the mandible (midway between the angle and the symphysis menti). Points A and B are joined by a line that is convex upwards (the convexity about 1.5 cm above the base of mandible). This line indicates the upper border of the gland.
>
> Point C is marked on the greater horn of hyoid. Points A and B are joined by a line that is convex downwards with the convexity reaching a little below the level of the greater horn of hyoid. This line indicates the lower border of the gland.
>
> The upper and lower borders are joined at their ends to complete the oval shape of the gland.

Submandibular Duct

The *submandibular duct* (otherwise called the Wharton's duct) is about 5 cm long and begins from the middle of the superficial part of the submandibular gland behind the posterior border of the mylohyoid. The duct passes upwards and backwards and then forwards through the deep part of the gland between the mylohyoid and the hyoglossus. It then runs anteriorly between the sublingual gland (laterally) and the genioglossus (medially) and finally opens into the oral cavity on the sublingual papilla located below the anterior part of the tongue on either side of the frenulum linguae. The lingual nerve is closely related to the duct. Initially it lies above the duct, then crosses its lateral side and finally ascends medially, winding around the lower border of the duct.

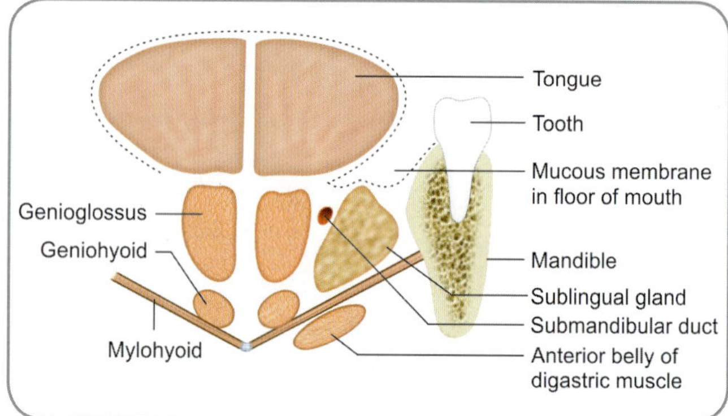

Fig. 5.4: Schematic coronal section through anterior part of the tongue and mouth to show relationship of the sublingual salivary gland and the oral diaphragm

SUBLINGUAL GLAND (FIG. 5.4)

The sublingual gland is the smallest of the three major salivary glands. It is located in the floor of the anterior part of the oral cavity deep to the oral mucous membrane and is responsible for raising the sublingual fold. Inferiorly, the gland rests on the mylohyoid muscle. The anterior end of the gland lies near the midline, close to the gland of the opposite side. Posteriorly, the gland is near the deep part of the submandibular gland. Laterally, it is in contact with the sublingual fossa of the mandible above the mylohyoid line. Medially, the sublingual gland rests on the genioglossus muscle separated by the submandibular duct. Secretions of the sublingual gland leave the gland mainly through a number of small ducts called the "***Ducts of Rivinus***" which open directly into the oral cavity on the sublingual fold. Small ducts from the anterior part of the gland join to form a sublingual duct called the 'Bartholin duct' which opens along with the submandibular duct.

Blood Supply

The sublingual gland receives its blood supply through the sublingual branch of the lingual artery and the submental branch of the facial artery. Venous drainage is into the internal jugular vein.

Lymphatic Drainage

The lymphatic drainage of the sublingual gland is into the submental lymph nodes.

Nerve Supply

The *secretomotor nerve supply* (parasympathetic) is similar to that of the submandibular gland for the pathway of the preganglionic fibres up to the submandibular ganglion. From the submandibular ganglion, some of the postganglionic fibres meant for sublingual gland re-enter the lingual nerve and reach it through the distal part of the lingual nerve to supply the gland, unlike the submandibular gland where the postganglionic fibres supply it directly.

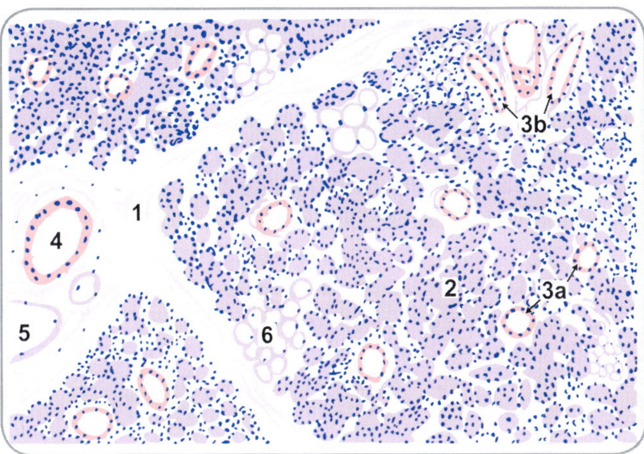

Fig. 5.5: Schematic drawing of a salivary gland

Key: 1. Interlobular connective tissue **2.** Serous acini
3. Intralobular duct **a.** Intercalated duct **b.** Striated duct
4. Interlobular duct **5.** Blood vessel **6.** Adipose tissue

Histology (Fig. 5.5)

Each salivary gland has numerous lobes, each of which is composed of smaller lobules separated by dense connective tissue which in turn is continuous with the capsule of the gland. Each lobule has a single duct, the branches of which terminate at the secretory end pieces, which are acinar (globular and bulb like) or tubular (cylindrical) in shape. Other ducts present are the intercalated ducts, striated ducts and excretory ducts. The secretory end pieces can be serous or mucous in nature. Mucous end pieces are often capped by serous cells at their closing ends and these appear as a crescent shaped structure called the serous demilune. Saliva secreted in the secretory end piece passes through the intercalated duct, the striated duct (which is the intralobular duct) and then the excretory duct (which is the interlobular collecting duct). The collecting ducts unite to form a main duct which discharges the secretion to the oral cavity. Saliva does not merely pass through the ducts but is modified in its electrolyte content and consistency. Cells of the intercalated ducts and the striated secrete immunoglobulin A, lysozyme and kallikrein into the saliva. These cells also actively transport various electrolytes and alter the viscosity.

In the submandibular gland, the secretory end pieces are predominantly serous acini with some mucous tubules and acini. Mucous tubules have serous demilunes. In the sublingual gland, mucous tubules and acini predominate, interspersed with few serous acini or demilunes.

Added Information

Apart from the three paired salivary glands, there are several minor salivary glands. These are the labial, buccal, palatal, palatoglossal and lingual glands. Specific mention should be made of the serous glands of Von Ebner, which are found near the circumvallate papillae. Their serous secretion helps in spreading taste stimuli over the taste buds and also in washing the food particles down the oesophagus.

Clinical Correlation

❏ The salivary glands are commonly infected with the virus that causes *mumps*. Swelling and pain occur in the affected gland. Mumps can involve more than one gland but the parotid is most commonly affected.

❏ The salivary glands can be infected by spread of infection from the mouth. An abscess may form in the gland.

❏ A calculus may form more commonly in the submandibular gland than the other salivary glands. Parotid gland has a predominantly serous secretion and hence calculi formation is rare. Though the sublingual gland has predominantly thick mucous secretions, the advantage of its location within the oral cavity assists in the drainage of secretions without stagnation. However, the drainage of the mixed secretions of the submandibular salivary gland is by the submandibular duct, passage through which is against gravity and also takes a longer course to reach the oral cavity. Hence, formation of calculus in the submandibular salivary gland is more common.

❏ In operations on the submandibular gland, the close relation of the facial artery and the marginal mandibular branch of facial nerve have to be kept in mind to avoid injury. Skin incision should be made atleast 2.5 cm inferior to the angle of mandible to avoid injury to the marginal mandibular branch. While handling the submandibular duct, caution should be exercised to avoid the lingual nerve which is directly inferior to the duct near the third molar tooth.

❏ Sialogram is a special type of radiograph, taken to visualise the secretory units of the salivary glands and the salivary ducts. Contrast medium is injected into the duct and the radiograph taken. Due to the exceedingly small size of the ducts of the sublingual gland, it is not possible to inject the contrast dye into them.

❏ The duct of any of the salivary glands can get blocked. As a result, the retained secretions accumulate inside and dilate the gland. The gland develops a retention cyst. The minor salivary glands are more affected than the major glands. When drainage of the sublingual gland is affected, the resultant retention cyst is found on the floor of mouth. Such a cyst is called a ranula. The swelling may push the tongue to one side.

Development

Development of the submandibular and sublingual salivary glands: An anteroposterior endodermal thickening appears between the developing tongue and the developing gums by the 5th week of intrauterine life. It separates into a cord by the next few days; subsequently, the cord develops into the submandibular salivary gland. The primordia of the greater sublingual glands appear immediately lateral to the developing submandibular glands by the 6th week. By the 7th week, primordial of five to ten lesser sublingual glands appear in the linguogingival groove.

Small ducts of these glands, to start with, open into long, narrow gutters. Closure of these gutters make the ducts appear longer. Such elongation of ducts proceeds in anteriorwards. Elongation of the submandibular duct causes

Development *contd...*

inclusion of the greater sublingual duct into the former, so that the two glands seem to share an opening. Connective tissue forms a common envelope around all the lesser sublingual glands so that in the adult they appear to be a single gland with several ducts.

The primordia of all the salivary glands remain as dense cords until about the third month of intrauterine life. The cords then branch repeatedly. By sixth month, the cords canalise. Secretory activity commences close to birth; microstructural development occurs just before and after birth.

SUPRAHYOID MUSCLES

The suprahyoid muscles are four in number—digastric, stylohyoid, mylohyoid and geniohyoid (Table 5.1).

The oral diaphragm is strengthened above by the geniohyoid muscle; and below by the anterior belly of the digastric muscle.

Additional Notes on the Suprahyoid Muscles

Two of the four suprahyoid muscles, namely the digastric and the stylohyoid, lie in the anterior triangle of the neck. The other two muscles, namely the mylohyoid and the geniohyoid, lie on the floor of the mouth. The posterior belly of the digastric is usually longer than the anterior belly. The intermediate tendon of the muscle perforates stylohyoid. The anterior borders of the two mylohyoids unite in a median raphe that extends from the hyoid to the symphysis menti. Together, the two mylohyoids form the *diaphragma oris*. The geniohyoids lie below the tongue with their medial borders in contact.

Dissection

Clean the digastric muscle. Trace its intermediate tendon and the attachment of this tendon to the hyoid bone. Trace and study the nerve supply of both the bellies—anterior belly from the mylohyoid nerve and posterior belly from the facial nerve. After cleaning and defining the stylohyoid muscle, detach the anterior belly of mylohyoid from its origin. Identify and define the mylohyoid muscle; secure its nerve supply. Trace the facial vein. You will see it joins the anterior branch of the retromandibular vein to form the common facial vein. You will be able to see that the common facial vein lies superficial to the submandibular salivary gland.

Once you have identified the submandibular gland, clean and define it. Notice that the gland has two parts—the superficial part lies on the mylohyoid and the deep part passes deep to the muscle by curving around its posterior border. Study the posterior aspect of the gland and identify the facial artery grooving this aspect. Trace the artery till the anteroinferior border of masseter; confirm your study by palpating the facial artery at this point in your own self or your colleague.

contd…

Table 5.1: Suprahyoid muscles				
Muscle	**Origin**	**Insertion**	**Action**	**Nerve supply**
Digastric	• Posterior belly from base of skull deep to mastoid process (mastoid notch of temporal bone) • Anterior belly from anterior part of base of mandible near midline (Fig. 5.6)	The two bellies join to form the intermediate tendon which is anchored to the hyoid bone through a fibrous pulley	• Elevates hyoid bone • Fixes the hyoid bone (along with other muscles)	• Anterior belly by mylohyoid branch of inferior alveolar nerve • Posterior belly by facial nerve
Stylohyoid	Posterior aspect of the middle of the styloid process (Fig. 5.7)	Tendon splits into two to enclose the intermediate tendon of the digastric muscle and is then attached to hyoid bone at the junction of body and greater cornu	Elevates and retracts hyoid bone	Facial nerve
Geniohyoid	Posterior aspect of symphysis menti of mandible from the inferior genial tubercle (below the origin of genioglossus)	Anterior aspect of the hyoid bone (Fig. 5.8)	• Pulls hyoid bone upwards and forwards • When hyoid bone is fixed, it can depress mandible	Fibres of C1 nerve travelling through hypoglossal nerve
Mylohyoid (Figs 5.9 and 5.10)	Mylohyoid line on medial surface of the body of the mandible	The fibres pass medially: • Most posterior of the fibres are attached to body of hyoid bone • Remaining fibres interdigitate with the opposite mylohyoid and are inserted into a median raphe extending from hyoid bone to mandible	The muscles of the two sides form the floor of the mouth called the *oral diaphragm*. They help in deglutition by raising the floor of the mouth	Mylohyoid branch of inferior alveolar nerve, which is its only motor branch. Inferior alveolar nerve is a branch of posterior division of mandibular nerve

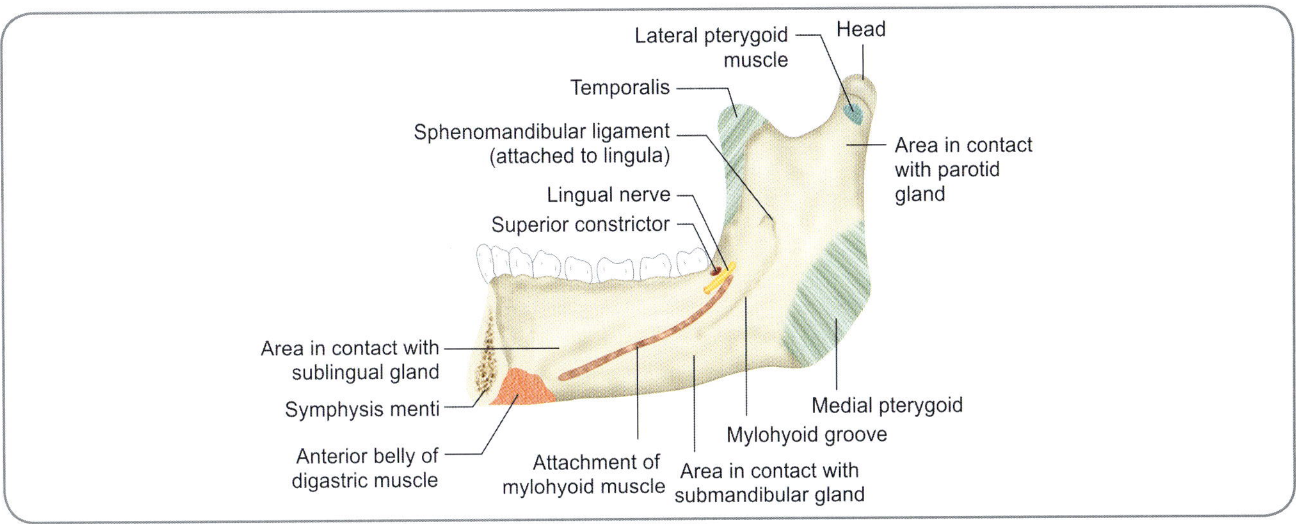

Fig. 5.6: Attachments on one-half of the mandible seen from the medial side

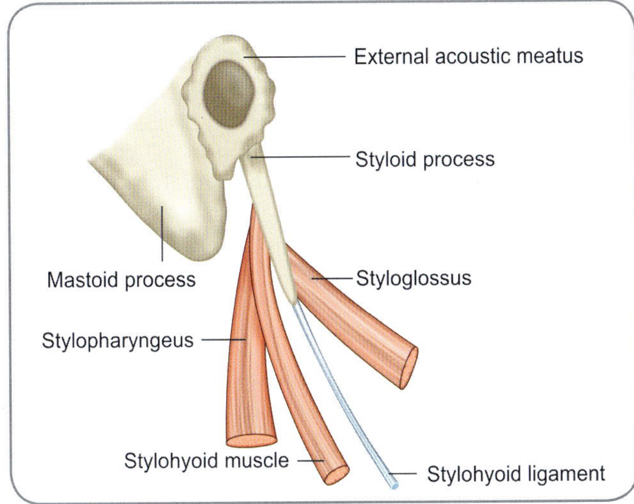

Fig. 5.7: Structures attached to the styloid process – the stylohyoid ligament is attached to the tip, the styloglossus muscle to the anterior aspect (lower part), the stylohyoid muscle to the posterior aspect (upper part) and the stylopharyngeus to the medial aspect

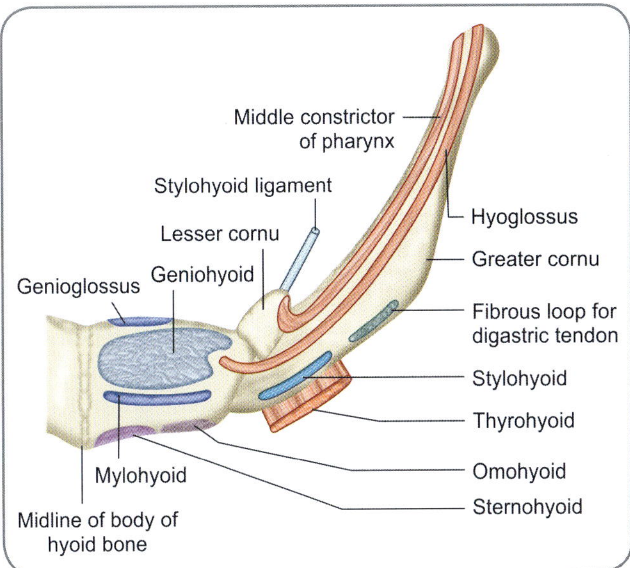

Fig. 5.8: Attachments on hyoid bone—anterosuperior aspect

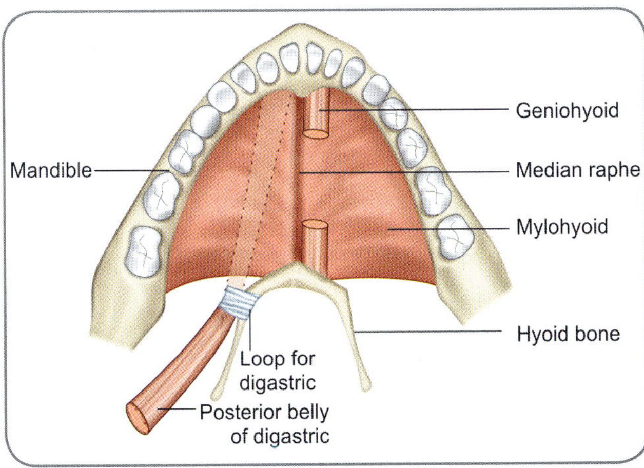

Fig. 5.9: Schematic diagram of floor of mouth seen from above – the layout of the mylohyoid and geniohyoid muscles shown

Dissection contd...

Slowly lift and displace the posterior aspect of the gland. Reflect the mylohyoid (if necessary, make a partial cut through its posterior border). The hyoglossus muscle and other deeper lying structures are exposed. Define the hyoglossus, genioglossus and the geniohyoid muscles. Study the adjacent structures and their relations to the muscles. As you define the lingual nerve, submandibular ganglion, submandibular duct and hypoglossal nerve (all of which lie on the hyoglossus), you will be able to make out that the deep part of the submandibular salivary gland is resting on the hyoglossus. Trace the lingual nerve and study its relationship (crossing and recrossing) to the submandibular duct.

Follow the lingual nerve proximally and identify the styloglossus muscle. Cut the hyoglossus near its origin and reflect up. The lingual artery is exposed. Study the artery and its branches.

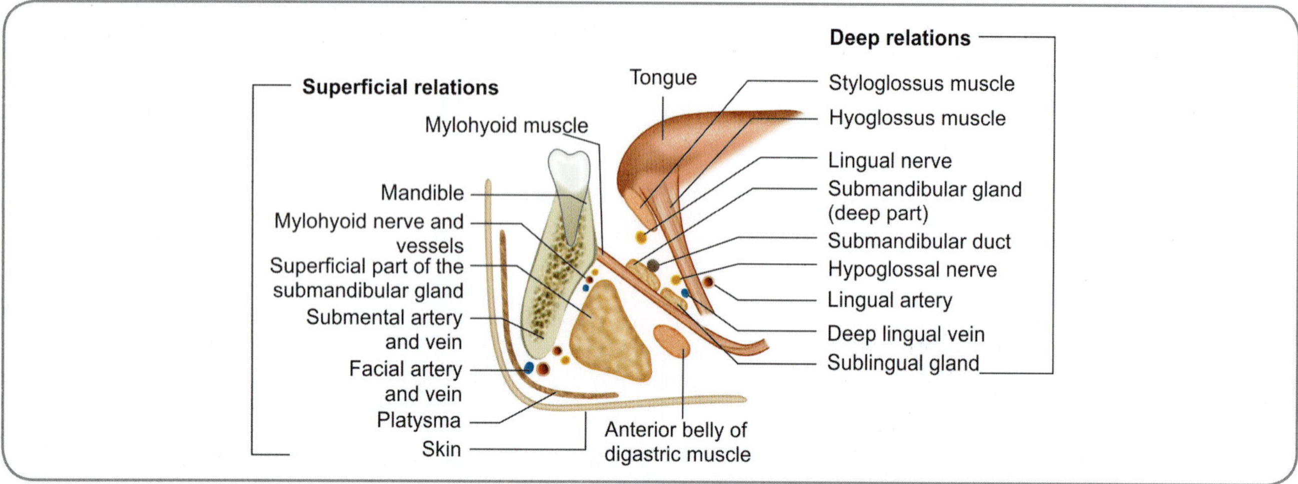

Fig. 5.10: Schematic coronal section of the submandibular region to show the relations of mylohyoid muscle

TONGUE

Tongue (Greek name.glossa, Latin name.lingua) is a solid and mobile muscular organ covered by mucous membrane. It lies partly in the oral cavity and partly in the pharynx and has the unique capacity to assume various shapes and positions. It contributes to the functions of mastication, deglutition, taste and articulation.

PARTS

The tongue presents a tip or apex, base, root and body.

- ❑ *Tip:* The anterior end of the tongue is the *tip* or *apex*; it rests against the incisor teeth and can be protruded out of the mouth.
- ❑ *Base:* The base is formed by the posterior one-third of the tongue and is directed backward towards the oropharynx. The base is connected to the epiglottis by a *median glossoepiglottic fold* in the centre and by a pair of *lateral glossoepiglottic folds*, one on either side. The space between the median and the lateral glossoepiglottic folds is called the *vallecula*; there are thus two (right and left) valleculae.
- ❑ *Root:* The part of the tongue that is attached below to the floor of the mouth is called the *root*. It extends from the symphysis menti to the hyoid bone.
- ❑ *Body:* All that portion of the tongue between the root and the apex is called the body; it includes the base.

Surfaces

The tongue has two surfaces—dorsal and inferior and two lateral margins.

- ❑ *Dorsal surface (Fig. 5.11):* The dorsal surface is more extensive and covers the superior and posterior aspects of the tongue. Otherwise called the dorsum of the tongue, it is convex on all sides and is covered

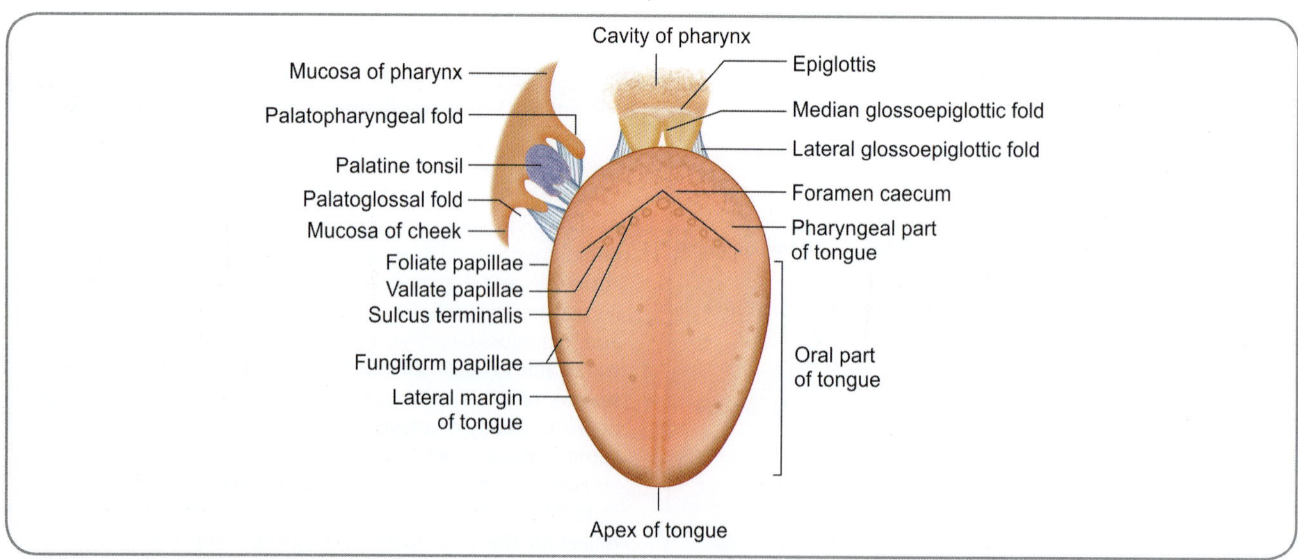

Fig. 5.11: Tongue and related structures as seen from above

with mucous membrane lined by non keratinised stratified squamous epithelium. Near its posterior end, the dorsum is marked by a V-shaped groove called the *sulcus terminalis*. The apex of the 'V' points backward and is marked by a depression called the *foramen caecum*, which indicates the commencement of the median thyroid rudiment as the thyroglossal duct. The limbs of the sulcus terminalis run forwards and laterally to the lateral margins of the tongue.

- The sulcus terminalis divides the dorsum into an anterior larger part (presulcal two-thirds) and a posterior smaller part (postsulcal one-third).
- The anterior part lies in the oral cavity and is, therefore, called the *oral part*. It faces upward and comes into contact with the palate. The posterior one-third faces backward and is called the *pharyngeal part*.
- The mucous membrane covering the *oral part* of the dorsum is rough because of the presence of numerous finger-like projections or *lingual papillae*. Each papilla is projection of the lamina propria covered by the overlying mucous membrane. There are different types of papillae.

Dissection

The tongue is best studied in a prosected specimen. Have a specimen of sagittal section of head and neck to aid in your study.

Study the dorsal surface of the tongue carefully. Look for the papillary texture of the mucous membrane, foliate papillae, the 'V' line of circumvallate papillae (and the sulcus terminalis), foramen caecum and the lymphoid nodules on the posterior part of the dorsum.

Observe the glossoepiglottic folds and the valleculae. Look for the frenulum linguae on the under surface. If an appropriate specimen is available, study the sublingual folds and the sublingual papillae.

With a senior colleague or a teacher by your side, make efforts to partially reflect the mucous membrane of the tongue over the dorsal and inferior aspects so as to expose the musculature. Study the interwoven muscular pattern of the substance of the tongue.

Histology

The mucous membrane of the tongue consists of stratified squamous epithelium resting on connective tissue stroma called the *lamina propria*. It covers the tip, dorsal surface, lateral margins and undersurface. The mucous membrane is adherent to the underlying structures except in the posterior part where there is a submucosa.

Lingual papillae are projections of lamina propria covered by overlying mucous membrane. There are different types of papillae (Fig. 5.12). They cover the presulcal part of the dorsum of tongue. They serve to increase the area of contact between the tongue and the contents of the mouth.

contd...

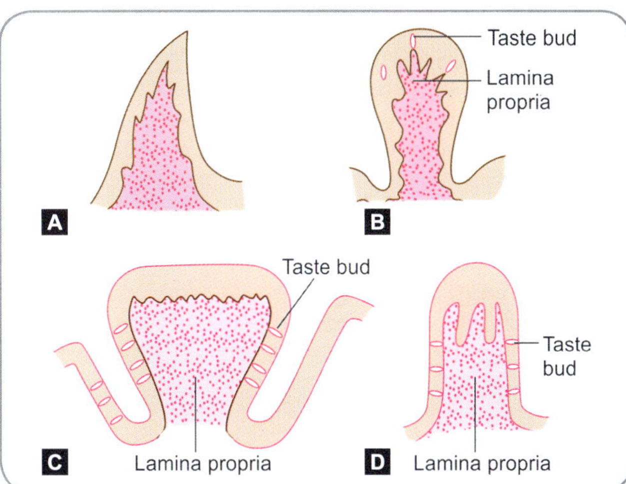

Figs 5.12A to D: Papillae (schematic diagram) **A.** Filiform **B.** Fungiform **C.** Circumvallate **D.** Foliate

Histology *contd...*

- Largest of these papillae are situated in a row just in front of the sulcus terminalis. These are the *circumvallate* papillae (Fig. 5.13). They are large, cylindrical, flat topped and surrounded by deep trenches. The trenches are studded with taste buds and the ducts of the serous glands (glands of Von Ebner) of the tongue open into the them. The outer wall of the trench is called the vallum and hence the name 'circumvallate'. Their epithelium is non keratinised stratified squamous epithelium.
- Most numerous of the papillae are the *filiform papillae*. These are small, conical in shape, pinkish grey, scaly and contain afferent nerve endings which are sensitive to touch. They are arranged in V-rows parallel to the sulcus terminalis except near the apex where they are in transverse rows. Their epithelium is keratinised. The

contd...

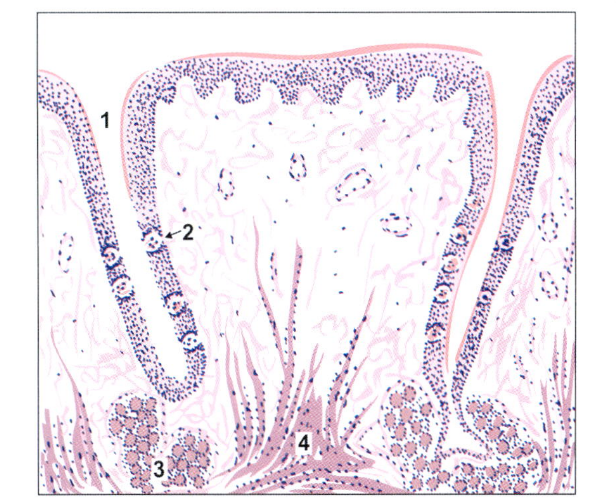

Fig. 5.13: Vallate papilla

Key: 1. Groove around papilla **2.** Taste bud **3.** Serous glands of von Ebner **4.** Muscle extending into papilla

contd...

Histology *contd...*

filiform papillae increase the friction between the food and the tongue thus facilitating movement of food within the oral cavity. They also have touch corpuscles. Due to the presence of keratin, their epithelium can become scaly; in animals like cats and cow, keratin causes thickening of the papillae making the tongue a rasp to grasp food.

- *Fungiform papillae* are pink and mushroom-shaped. Though they are scattered among the filiform papillae, they are more numerous at the apex and along the sides of the tongue. Their epithelium is nonkeratinised and the connective tissue core is highly vascular.
- Another type of papillae which deserve special mention are the *papillae simplex*. Unlike the other papillae, which can be seen by naked eye, these are microscopic and are quite different from the other papillae. They are not surface projections but are present at the junction of the lining epithelium with underlying tissues. They are the equivalent of dermal papillae seen in the skin.
- On the posterior parts of the lateral margins of the tongue, four or five vertical folds called the *foliate* papillae are present. These are poorly developed humans and are vestiges of larger papillae found in other mammals. However, they do bear taste buds.
- Taste buds are present in all the papillae except the filiform papillae.

Added Information

Taste buds are modified epithelial cells arranged as spherical masses within the epithelium of the tongue. They are numerous in the all types of papillae except the filiform papillae, but are also scattered over the dorsal and lateral aspects of the tongue. Each taste bud is a cluster of slender spindle-shaped cells; these cells converge on a gustatory pore, which is an opening on the mucosal surface. Some of these cells are gustatory in nature which perform the function of carrying the taste sensation and others are supporting cells which can become gustatory when need arises. The base of the bud is penetrated by the afferent gustatory nerve fibres. The spindle cells are in synaptic contact with these nerve endings. Since these cells are chemosensory in nature and share electrophysiological properties of neurons, they are often called the *paraneurons*. The taste buds are also present over the inferior surface of the soft palate, palatoglossal arches, posterior surface of the epiglottis, posterior surface of the oropharynx and aryepiglottic folds. Chemical substances in the food are dissolved in saliva; in the dissolved state these substances pass through the gustatory pore into the taste bud. They stimulate the spindle cells (or the taste receptor cells) and produce membrane depolarization. The taste impulse is then transmitted to the concerned gustatory nerve ending (facial or vagus or glossopharyngeal, depending on the area of the tongue) from where it travels towards the central nervous system (CNS).

○ The *pharyngeal part* faces backwards and forms part of the anterior wall of the oropharynx. Its

surface is not covered by papillae. However, it shows a number of rounded elevations which are produced by collections of lymphoid tissue lying deep to the mucosa. Each elevation is called a *lymphoid nodule*. Each nodule has a central crypt which receives ducts of underlying mucosal glands; the crypt opens on the surface. This lymphoid tissue is referred to, collectively, as the *lingual tonsil*. The posterior part of the tongue is connected to the palate on either side by a fold of mucous membrane called the *palatoglossal fold*. Immediately posterior to this fold, is seen the palatine tonsil. The mucous membrane lining the pharyngeal part of the tongue is continuous laterally with the mucosa covering the palatoglossal folds and with the mucosa covering the palatine tonsils.

Posteriorly, the tongue is closely related to the epiglottis (which is a part of the larynx).

- *Inferior surface (Fig. 5.14):* The inferior surface of the tongue is covered by a mucous membrane, which is devoid of papillae and is reflected on to the floor of the mouth. A delicate fold of mucosa in the midline passes from the tongue to the floor of the mouth; this is the *frenulum linguae*. On either side of the posterior end of the frenulum is a rounded projection called the *sublingual papilla*. The submandibular duct opens on the summit of the papilla. A *sublingual fold* produced by the underlying sublingual salivary gland is seen running laterally from the papilla.

A little lateral to the frenulum is seen a darkish line running towards the tip of the tongue. The deep lingual vein lies deep to this line. Further laterally, an irregular fold of mucosa called the *fimbriated fold* (or the *plica fimbriata*) is present.

- *Lateral margins:* The lateral margins of the tongue are lined by mucous membrane. The palatoglossal arch formed by the palatoglossus muscle meets the lateral margin at the junction of the anterior two-thirds and posterior one-third of the tongue.

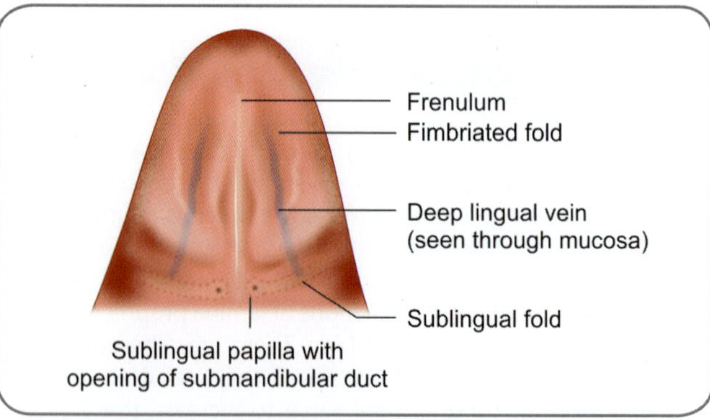

Fig. 5.14: Structures seen on the undersurface of the anterior part of tongue

MUSCLES

The substance of the tongue is mainly made up of muscles. The tongue is divided into two symmetrical halves by a median fibrous septum. Each half contains striated muscles which are grouped into two—*intrinsic muscles* which are confined to the tongue and *extrinsic muscles* which have origins from structures nearer to the tongue. The various extrinsic muscles enter the substance of the tongue through the root of the tongue. Through these muscles, the tongue is anchored to the hyoid bone and to the mandible. The intrinsic muscles *change the shape* of the tongue whereas the extrinsic muscles *alter the position* of the tongue.

Intrinsic Muscles

The intrinsic muscles comprise four pairs of muscles namely, the *superior longitudinal muscles*, the *inferior longitudinal muscles*, the *transversus linguae* and the *verticalis linguae.*

The median fibrous septum is posteriorly attached to the hyoid bone. Though it extends till the tip of the tongue, superiorly it does not extend till the dorsum.

- *Superior longitudinal muscles:* They lie beneath the mucous membrane of the dorsal surface of the tongue. The two muscles arise from the posterior aspect of the corresponding surface of the median fibrous septum, pass forwards and insert into the corresponding sides of the tongue.

  Action: The superior longitudinal muscles make the dorsal surface of the tongue concave and shorten the length of the tongue.

- *Inferior longitudinal muscles:* They lie beneath the mucous membrane of the inferior surface of the tongue. The two muscles arise from the corresponding sides of the tongue, pass forwards and insert into anterior aspect of the median fibrous septum.

  Action: The inferior longitudinal muscles make the dorsal surface of the tongue convex and shorten the length of the tongue.

- *Transversus linguae:* They arise from the median fibrous septum, pass laterally and insert into the sides of the tongue.

  Action: They reduce the width and increase the length of the tongue.

- *Verticalis linguae:* Each muscle arises from the lamina propria of the dorsal surface of the tongue and inserts into the lower parts of the sides of the tongue.

  Action: They increase the width of the tongue and make the dorsal surface concave from side-to-side.

The intrinsic muscles alter the shape of the tongue. These actions, in isolation or in combination, produce various combinations and give the tongue varied mobility. They play important roles in deglutition and in articulation. The transverse and vertical fibres decussate with the longitudinal fibres; this contributes to the special feature of the tongue; interwoven musculature. Pockets of fat are found between the muscle fibres in the posterior part of the tongue.

Extrinsic Muscles (Fig. 5.15)

The *extrinsic muscles* of the tongue are the styloglossus, the palatoglossus, the genioglossus and the hyoglossus (Table 5.2).

Additional Notes on the Extrinsic Muscles of the Tongue (Fig. 5.15)

- *Styloglossus* is the smallest and shortest of the three styloid muscles. At the side of the tongue, it divides into a longitudinal part that blends with the inferior

Table 5.2: Extrinsic muscles of tongue				
Muscle	**Origin**	**Insertion**	**Action**	**Nerve supply**
Styloglossus	Anterior and lateral aspect of the tip of the styloid process	Merges with side of tongue	Pulls tongue upwards and backwards	Hypoglossal nerve
Palatoglossus	Oral aspect of the palatine aponeurosis	Side of tongue	Muscles of both sides acting together bring the palatoglossal arches together closing the aperture from oral cavity to pharynx	Cranial part of accessory nerve
Genioglossus	Superior genial tubercles on the posterior surface of symphysis menti of mandible	Majority of fibres insert into the ventral aspect of tongue forming the bulk of the tongue muscle. Lower fibres insert into the hyoid bone	Protrudes tongue out of mouth by pulling posterior part forwards. Called the "*Safety muscle of the tongue*"	Hypoglossal nerve
Hyoglossus	Greater cornu and lateral part of body of hyoid	Side of tongue	Depresses tongue	Hypoglossal nerve

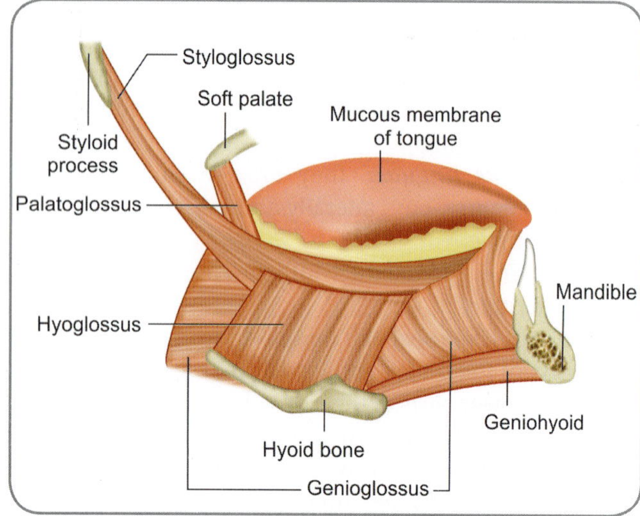

Fig. 5.15: Extrinsic muscles of tongue

Table 5.3: Relations of hyoglossus muscle (Fig. 5.16)	
Structures passing deep to posterior border	• Glossopharyngeal nerve • Stylohyoid ligament • Lingual artery
Structures superficial to muscle	• Lingual nerve • Submandibular ganglion (suspended from the lingual nerve) • Deep lingual vein • Submandibular gland (deep part) • Submandibular duct • Sublingual gland (part of) • Styloglossus muscle • Stylohyoid muscle • Hypoglossal nerve • Mylohyoid muscle • Digastric tendon and pulley
Structures deep to muscle	• Glossopharyngeal nerve • Stylohyoid ligament • Lingual artery and it dorsal lingual branches • Genioglossus muscle • Middle constrictor of pharynx • Inferior longitudinal muscle of tongue

longitudinal muscle and an oblique part that decussates with the hyoglossus.

❑ ***Palatoglossus*** along with its overlying mucosa forms the palatoglossal arch. When this muscle contracts, it pulls the palatoglossal arch up and approximates it to the contralateral arch. When the two palatoglossal arches come closer to each other, the oral cavity is shut off from the oropharynx. The palatoglossus, though described as a lingual muscle, functions along with the muscles of the palate and has a nerve supply same as that of the palatal muscles.

❑ ***Genioglossus*** is attached to the genial tubercle. This attachment keeps the tongue anterior and prevents it from sinking back into the oral cavity. This in turn prevents the tongue from 'falling back' and obstructing respiration. Genioglossus is, therefore, called the 'safety muscle'. Paralysis, fracture of the mandible or anaesthesia put the muscle out of action and the tongue falls back resulting in suffocation. If the genioglossus of one side is paralysed, protruded tongue points to the paralysed side. When both the genioglossi act, the central part of the tongue is depressed and made concave.

❑ ***Chondroglossus*** is another extrinsic lingual muscle. Sometimes described as a part of the hyoglossus, it is actually separated from the hyoglossus by the posterior fibres of genioglossus. Arising from the lesser cornu of the hyoid, the chondroglossus passes up to merge with the intrinsic lingual musculature between the hyoglossus and the genioglossus. Having the same blood and nerve supply as that of the hyoglossus, the chondroglossus also depresses the tongue.

❑ ***Hyoglossus*** is a quadrilateral muscle that runs up like a wall from the hyoid bone to the tongue. Several structures entering the tongue are closely related to this muscle. Apart from simply depressing the tongue, the hyoglossus pulls down the sides of the tongue and also shortens (retrudes) it.

The relations of hyoglossus muscle are shown in Figure 5.16 and discussed in Table 5.3.

BLOOD VESSELS (FIG. 5.17)

The chief artery of the tongue is the lingual artery. The tongue is also supplied by branches of the ascending palatine and tonsillar branches of the facial artery and ascending pharyngeal arteries (Fig. 5.17).

Lingual Artery (Fig. 5.18)

The lingual artery arises from the external carotid artery opposite the tip of the greater cornu of the hyoid bone. Its course is divided into three parts by the hyoglossus muscle. The first part lies before the hyoglossus, the second part deep to the muscle and the third part beyond it.

❑ The ***first part*** of the artery lies in the carotid triangle, superficial to the middle constrictor of the pharynx and extends from its origin to the posterior border of hyoglossus muscle. It forms a characteristic upward loop and is crossed superficially by the hypoglossal nerve.

❑ The ***second part*** of the artery lies deep to the hyoglossus muscle that separates the artery from the hypoglossal nerve and its accompanying venae commitantes. This part of the artery also lies on the middle constrictor.

❑ The ***third*** or ***deep part*** of the artery runs upwards along the anterior margin of the hyoglossus and then runs forwards beneath the mucous membrane of the tongue to reach the tip of the tongue. This part lies on the genioglossus muscle and is accompanied by the

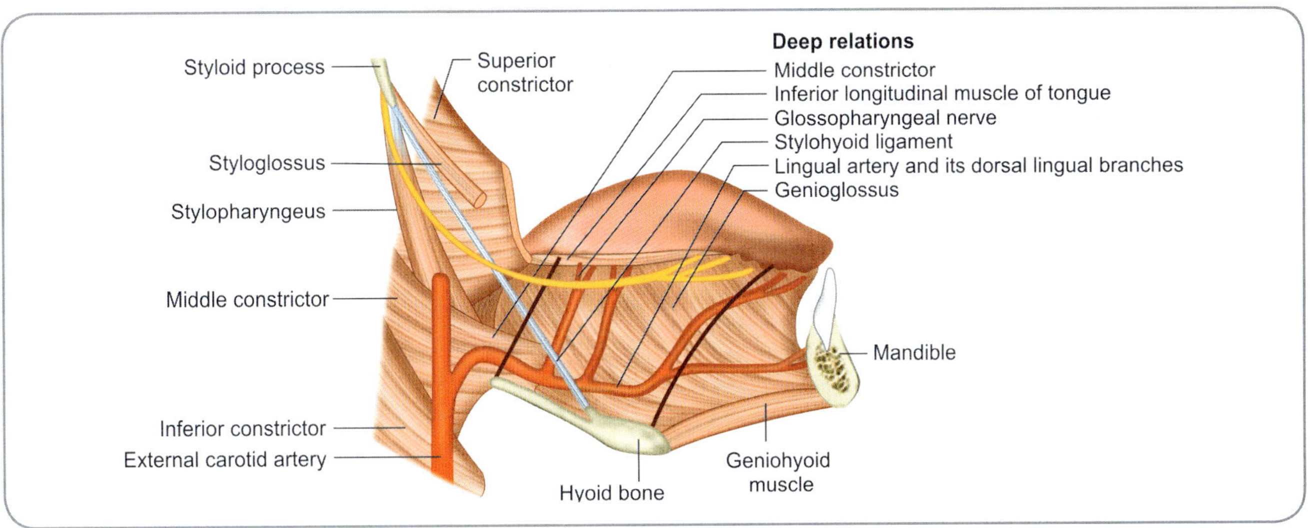

Fig. 5.16: Deep relations of hyoglossus muscle

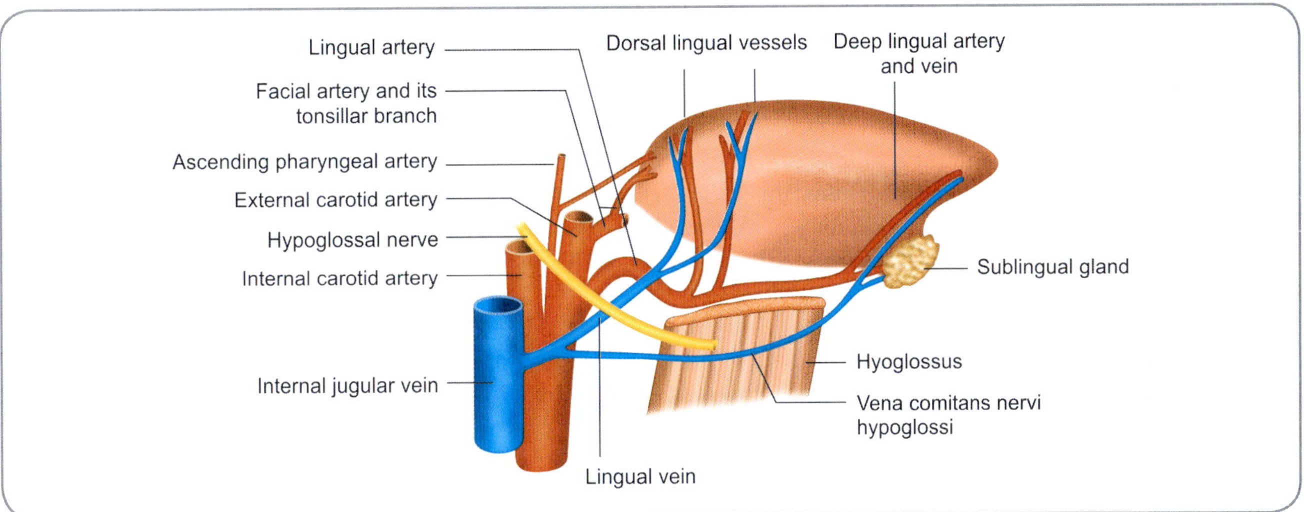

Fig. 5.17: Scheme to show the arteries and veins which supply the tongue

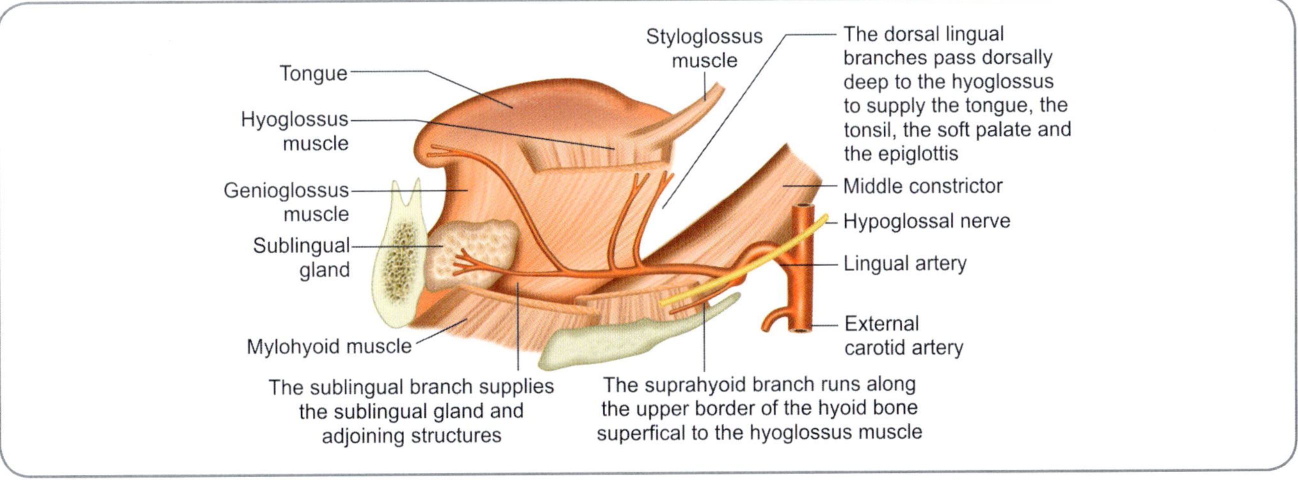

Fig. 5.18: Lingual artery and its branches

lingual nerve. At the tip of the tongue the lingual artery anastomoses with its fellow of the opposite side.

The branches of the lingual artery are:

- ***Suprahyoid artery from the first part:*** Usually a single branch that gives off muscular twigs to the suprahyoid muscles.
- ***Dorsal lingual arteries from the second part:*** Two or three small branches which arise medial to the hyoglossus and ascend to the posterior aspect of the dorsum of tongue; they supply the mucosa of tongue, palatoglossal arch, tonsil, soft palate and epiglottis.
- ***Sublingual artery from the third part:*** Arises at the anterior border of hyoglossus and passes forward on the floor of mouth to the sublingual gland; supplies the gland and the mucosa of sublingual, gingival and buccal areas.

The terminal continuation of the lingual artery is called the ***deep lingual artery***. It is found on the undersurface of the tongue and supplies the apical portion.

Venous Drainage

The veins of the tongue are arranged in two sets—superficial and deep. The superficial set, consisting of the dorsal lingual veins, drains the tip and the undersurface of the tongue, accompanies the hypoglossal nerve and drains into the internal jugular vein directly or through the lingual veins. The deep set consists of tiny veins which start near the tip; they unite to form a deep lingual vein that runs backward under the mucosa on the inferior surface. The deep lingual vein joins the sublingual vein (coming from the sublingual salivary gland) and forms the venae comitantes nervi hypoglossi which run along with the hypoglossal nerve to join the facial vein or the internal jugular vein or the lingual vein.

NERVES

The nerves supplying the tongue are of three functional types—nerves of general sensation, nerves of taste and nerves supplying muscles.

- ***Nerves of general sensation:*** Sensations like touch, pain and temperature are carried from the anterior two-thirds of the tongue by the lingual nerve (branch of the posterior division of the mandibular nerve) and from the posterior one-third by the glossopharyngeal nerve.
- ***Nerves of taste:*** Sensations of taste from the anterior two-thirds of the tongue except the circumvallate papillae are carried by the chorda tympani nerve through the peripheral processes of cells in the geniculate ganglion of the facial nerve. Sensations of taste from the posterior one-third and the circumvallate papillae are carried by the glossopharyngeal nerve. Taste fibres from the posterior most part of the tongue just in front of the epiglottis are carried by the superior laryngeal branch of the vagus nerve.
- ***Nerves supplying muscles:*** The musculature of the tongue except the palatoglossus is supplied by the ***hypoglossal nerve***. The palatoglossus is supplied by the cranial part of the accessory nerve.

Lingual Nerve (Fig. 5.19)

The lingual nerve is a branch of the posterior division of the mandibular nerve. At its origin in the infratemporal fossa, it runs downwards deep to the lateral pterygoid muscle

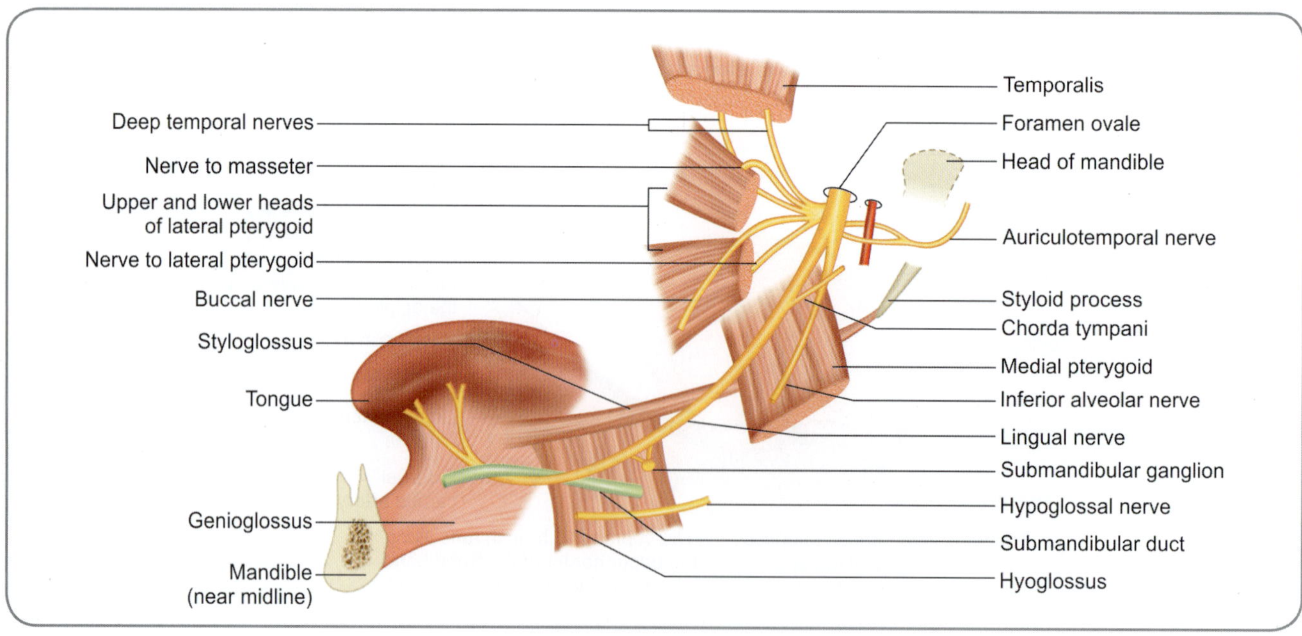

Fig. 5.19: Lingual nerve

and is joined by the ***chorda tympani*** nerve at an acute angle. A branch of the inferior alveolar nerve may also join here. Lower down, the lingual nerve runs downwards and forwards between the medial pterygoid and the ramus of the mandible. It then enters the mouth by passing deep to the mandibular attachment of the superior constrictor of the pharynx. Within the mouth, the nerve lies deep to the mucous membrane overlying the medial surface of the mandible just below the third molar tooth where it is very superficial and can be palpated. Leaving the gum, the nerve enters the side of the tongue, crosses the styloglossus and runs forwards across the lateral surface of hyoglossus between the hyoglossus and the mylohyoid muscles. At the anterior margin of hyoglossus, the nerve passes onto the genioglossus and divides into a number of branches which are distributed as described below. When the nerve is over the hyoglossus, the lingual nerve lies above the submandibular duct. Continuing forwards, the nerve crosses superficial to the duct and then hooks around it to reach its medial side. Lying on the hyoglossus, a little below the lingual nerve, is the submandibular ganglion. The nerve is connected to the ganglion by two or three branches.

The lingual nerve carries three types of fibres:

1. Most fibres are those of general sensation and carry the sensations of touch, pain and temperature from the anterior two-thirds of the tongue. They also carry sensations from the mucous membrane of the floor of the mouth and the gums related to the lower teeth.
2. The part of the lingual nerve distal to the joining of the chorda tympani carries ***fibres for taste*** (special sensation) from the part of the tongue in front of the sulcus terminalis excluding the circumvallate papillae. These fibres pass from the lingual nerve to the chorda tympani and then continue in the facial nerve.
3. ***Secretomotor fibres*** for the submandibular and sublingual glands reach the lingual nerve through the chorda tympani. They end in the submandibular ganglion. Postganglionic fibres reach the submandibular gland directly through branches to the gland from the ganglion. The fibres for the sublingual gland re-enter the lingual nerve and pass through its distal part to reach the gland.

Surface Marking

Point A is marked on the centre of the zygomatic arch. Point B is marked below and behind the third molar tooth. Point C is marked immediately beneath the canine tooth. The three points are joined by a line that is curved with an upward concavity.

Submandibular Ganglion

The submandibular ganglion lies on the hyoglossus muscle, suspended from the lingual nerve by two or more roots. It is one of the parasympathetic ganglia of the head and neck.

Functionally, the ganglion is related to the chorda tympani nerve and topographically to the lingual nerve.

The posterior root (or sometimes, roots) transmits parasympathetic secretomotor fibres and sensory fibres. The preganglionic parasympathetic fibres arise from the superior salivatory nucleus and travel through the facial nerve and its chorda tympani branch to reach the lingual nerve. From the lingual nerve, these fibres and the sensory fibres reach the ganglion. Postsynaptic sympathetic fibres from the superior cervical ganglion travel in the plexus around the facial artery and reach the ganglion.

The parasympathetic fibres relay in the ganglion. The sensory and sympathetic fibres have no relay in the ganglion and simply pass through it.

Postsynaptic parasympathetic fibres arising in the postsynaptic neurons of the submandibular ganglion accompany the arterial twigs and directly reach the submandibular salivary gland. Those postsynaptic parasympathetic fibres destined for the sublingual salivary gland pass through the anterior root to be distributed through the lingual nerve.

Development (Figs 5.20A to C)

The tongue develops from the floor of the mouth by the 4th to 5th week of intrauterine life. A median eminence called the tuberculum impar develops at the junction of the medial ends of the first pharyngeal arches. Soon, a pair of swellings arises at the medial ends of the first arches. These two swellings, called the lingual swellings, merge with the tuberculum impar and with each other. The anterior two-thirds of the tongue develop from this merged single eminence.

Another median swelling called the copula or the hypobranchial eminence develops at the junction of the medial ends of the second, third and part of the fourth arches. However, the third arch mesoderm overgrows and submerges the second arch mesoderm. The second arch mesoderm is thus pushed down and is separated from the surface of the adult tongue. The hypobranchial eminence grows into a dichotomous branching in such a way that the two branched extensions enclose the anterior portion like a 'v'. The line of junction between the third and the first arches is represented by the sulcus terminalis.

Yet another midline eminence develops at the posterior part of the medial ends of the fourth arches. This eminence develops into the epiglottis.

The pharyngeal endoderm gives rise to the mucous membrane of the tongue. The general connective tissue of the tongue (connective tissue that is underlying the mucous membrane in the adult tongue) develops from the pharyngeal mesoderm. Though some experimental evidence suggests that the musculature of the human tongue develops *in situ*, innervations evidence suggests that the lingual muscles develop from the occipital myotomes.

The lingual epithelium is initially single cell thick. By the 6th week, it becomes two or three cells thick. By the 7th week, it becomes stratified and papillae make their appearance.

contd…

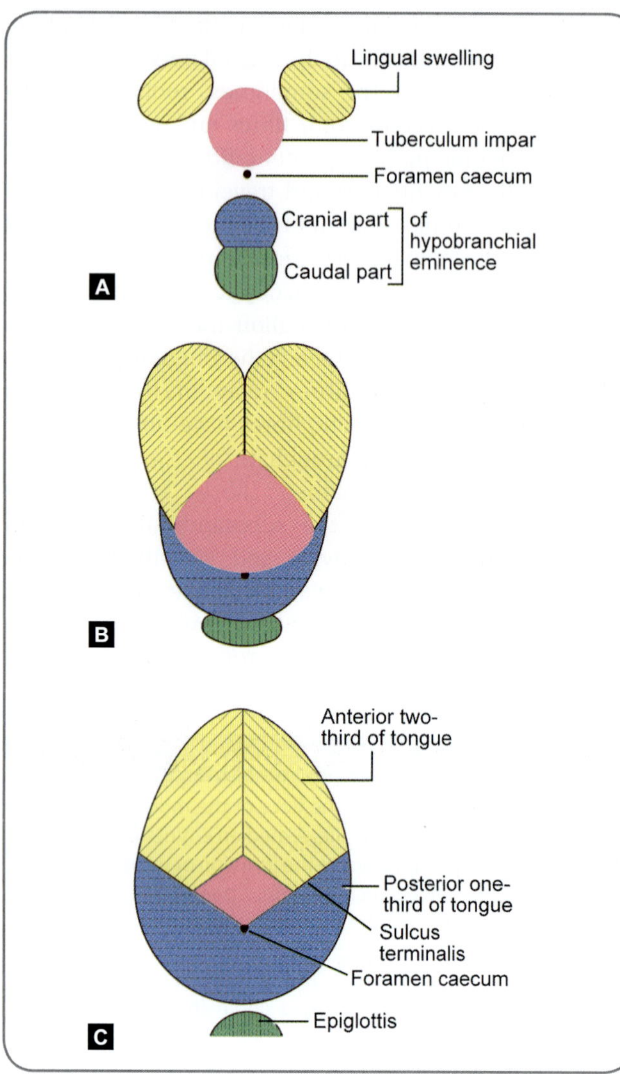

Figs 5.20A to C: Scheme to show the origin of different parts of the tongue

 **Development** *contd...*

The circumvallate and foliate papillae appear first; at the same time, the nerve endings of glossopharyngeal nerve grow into the posterior tongue. The nerve endings induce the development of the vallate and the foliate papillae. A little later, growth of facial nerve endings into the anterior tongue induces the development of the fungiform papillae. The filiform papillae appear by around the 5th week but have no nervous connection.

By the seventh month, reflex pathways between taste receptors and facial muscles seem to have developed. Appreciation of the sweet taste is also established by this time.

Glossopharyngeal Nerve (Fig. 5.21)

Glossopharyngeal nerve is the ninth cranial nerve. It leaves the cranial cavity by passing through the jugular foramen. It then descends in close relation to the internal carotid artery, passing deep to the styloid process and the structures attached to it. Reaching the posterior border of the stylopharyngeus muscle, it curves forwards along with the muscle and passes between the superior and the middle constrictor of pharynx. The nerve then reaches the side of the tongue where it lies deep to the hyoglossus muscle, superficial to the genioglossus and terminates by dividing into branches to the tongue. One set of branches of the glossopharyngeal nerve is that of the *lingual branches* which supply the part of the tongue behind the sulcus terminalis (other branches are studied elsewhere). They also supply the vallate papillae. These branches carry fibres for both general sensation and taste.

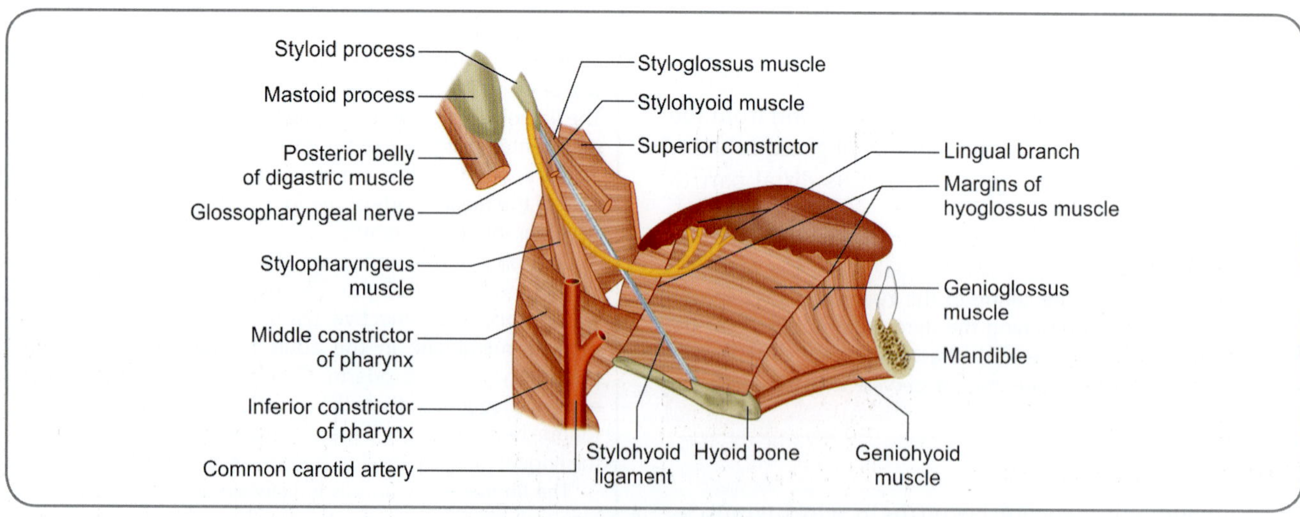

Fig. 5.21: Glossopharyngeal nerve

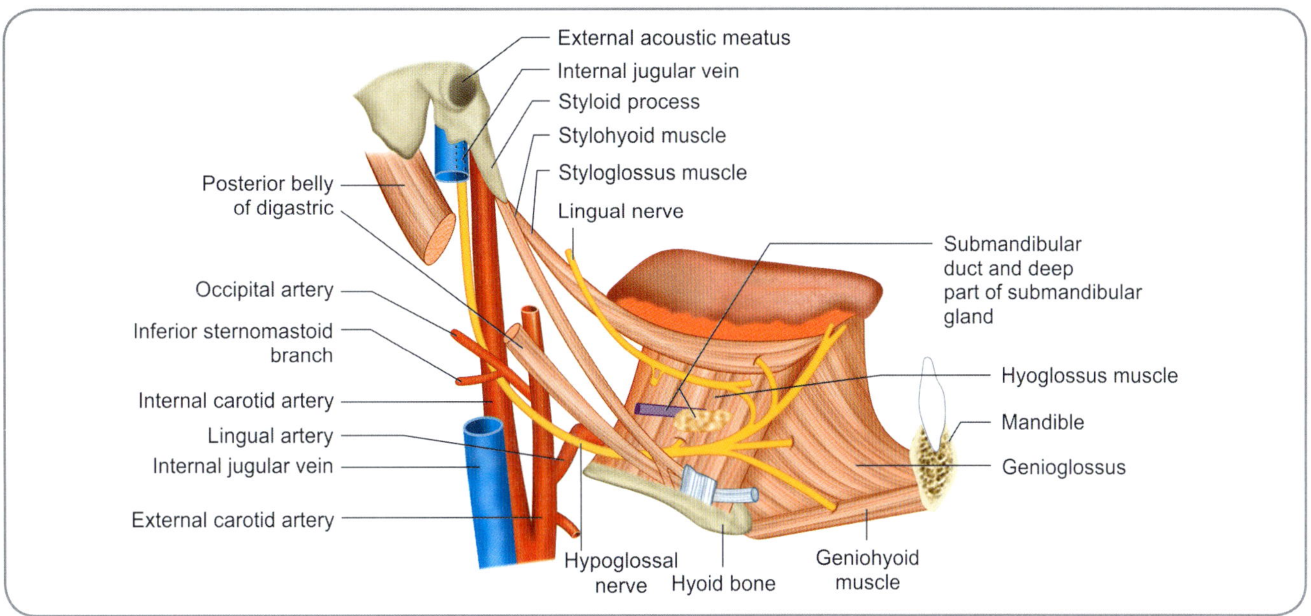

Fig. 5.22: Hypoglossal nerve

> **Surface Marking**
>
> Point A is marked on the lower part of the tragus on its anterior surface. Point B is marked just above the angle of the mandible. The two points are joined by a line that runs downwards and forwards with a downward convexity. The line then continues for a short distance just above the lower border of the mandible but along the latter. The entire line indicates the glossopharyngeal nerve on the surface.

Hypoglossal Nerve (Fig. 5.22)

Hypoglossal nerve is the twelfth cranial nerve. Its fibres are purely motor. They supply the muscles of the tongue. The hypoglossal nerve arises from the medulla and leaves the cranial cavity through the hypoglossal canal (or the anterior condylar canal). The nerve descends close to the internal jugular vein and internal carotid artery up to the level of the angle of the mandible. It then passes forwards crossing the internal and external carotid arteries and enters the submandibular region. In the submandibular region, the hypoglossal nerve, at first lies superficial to the hyoglossus muscle and then to the genioglossus. It ends by dividing into its terminal branches. These supply all the intrinsic and extrinsic muscles of the tongue except the palatoglossus which is supplied, along with other muscles of the palate, by the cranial accessory nerve.

> **Surface Marking**
>
> Point A is marked on the lower part of the tragus on its anterior surface. Point B is marked just above and behind the tip of the greater horn of hyoid bone. Point C is marked on the base of mandible midway between the angle and symphysis menti with the head tilted a little backwards. Points A and B are joined by a line that runs downwards; points B and C are joined by a line that runs upwards. At point B, there is a sharp upward curve. The entire line from point A to point C through point B indicates the hypoglossal nerve on the surface.

> ### Development
>
> *Innervation of tongue in relation to its development:* The tongue is supplied by several nerves. The differences in nerve supply are consequential to different sources contributing to the development.
>
> The anterior two-thirds are derived from the first arch mesoderm and adjacent endoderm. The general sensations are subserved by the lingual branch of the mandibular nerve (a derivative of the post-trematic nerve of the first arch) and the taste sensations by the chorda tympani nerve (pre-trematic nerve of the first arch).
>
> The posterior one-third is derived from the third arch mesoderm and adjacent endoderm. The general and taste sensations are subserved by the glossopharyngeal nerve which is the nerve of the third arch. This nerve also supplies the circumvallate papillae. The mucous membrane of the third arch is pulled anteriorly over the developing vallate papillae and hence this nerve supply.
>
> The posterior most portion and the epiglottis are subserved by the superior laryngeal branch of the vagus nerve, which is the nerve of the fourth arch.
>
> The lingual muscles are supplied by the hypoglossal nerve indicating their development from the occipital myotomes.

LYMPHATIC DRAINAGE (FIG. 5.23)

The tongue is a frequent site of carcinoma and hence knowledge of its lymphatic drainage is of special importance. The lymphatics of the tongue comprise of four sets—apical, marginal, central and dorsal.

1. *Apical set:* The lymph nodes of the apical set drain the tip and frenulum linguae; the efferents pass into the submental lymph nodes. A few efferents go to the submandibular and the jugulo-omohyoid lymph nodes.

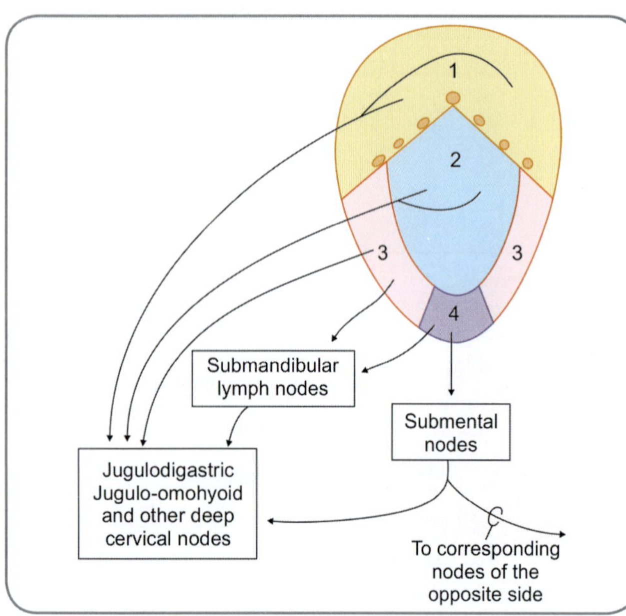

Fig. 5.23: Lymphatic drainage of tongue
Key: 1. Dorsal set **2.** Central set **3.** Marginal set **4.** Apical set

3. ***Central set:*** The nodes of the central set drain the dorsum of the tongue in front of the valate papillae and drain mainly into the jugulodigastric and jugulo-omohyoid lymph nodes. Few drain into the submandibular nodes.
4. ***Dorsal set:*** The nodes of the dorsal set drain the posterior one-third of the tongue including the valate papillae and terminate bilaterally into the jugulo-digastric nodes. Some also drain directly into jugulo-omohyoid nodes.

Lymph from areas of the tongue near the midline can pass to lymph nodes of either the right or left side.

2. ***Marginal set:*** The lymph nodes of the marginal set drain the sides of the tongue in front of the sulcus terminalis and terminate mainly into the submandibular nodes. Few drain into the jugulo-digastric and jugulo-omohyoid nodes.

Added Information

- Four types of taste sensations are perceived—sweet, salty, sour and bitter. Others are combinations of these types in various degrees. All areas of the tongue are responsive to all sensations.
- Perceptions of taste are influenced heavily by perceptions of smell.
- Postganglionic parasympathetic fibres from the sub-mandibular ganglion reach the lingual mucosa through the chorda tympani and supply the mucosal glands. Postganglionic sympathetic fibres arise in the carotid plexus and reach the tongue through the plexuses around the lingual artery. They innervate the glands and the vessels.

Clinical Correlation

Congenital malformations:
- The tongue may be too large (***macroglossia***) or too small (***microglossia***).
- Very rarely the tongue may be absent (***aglossia***).
- The tongue may be ***bifid***.
- The apical part of the tongue may be anchored to the floor of the mouth by a very short frenulum. The condition, which is called ***ankyloglossia*** or tongue tie, interferes with speech. Frenectomy (surgical cutting of the frenulum) may be required to free the tongue.
- Other rare anomalies include the presence of:
  - Fissures on the tongue
  - Lingual thyroid
  - Remnants of the thyroglossal duct in the form of cysts

Laceration of the tongue can occur if the tongue gets caught between the teeth. This is likely to occur while eating, during epileptic attacks and in injuries leading to fractures of the jaw. Because of the high vascularity of the tongue, lacerations can cause serious bleeding. Pressure is applied posterior to the area of laceration to stop bleeding as the lingual artery runs forwards. Pressure is applied either by holding the tongue between the thumb and index finger or by placing the thumb on the tongue and the index finger in the submental region.
- Muscles of the tongue are paralysed in lesions of the hypoglossal nerve. In unilateral paralysis, the tongue is protruded to the affected side as the unaffected genioglossus pushes the tongue out towards the affected side with no resistance from the affected side.
- Sensations of taste are lost from the anterior two-thirds of the tongue in facial nerve paralysis and from the posterior one-third in lesions of the glossopharyngeal nerve. In some cases, taste sensations over the anterior two-thirds of the tongue may be affected in lesions of the trigeminal nerve.
- The mucosa below the anterior part of the tongue is highly vascular. A drug placed here enters the blood stream faster than it does after an intramuscular injection. Such technique of administration of drugs is called sublingual application. This method is adopted for patients suffering from angina to give quick relief.
- If the anterior part of the tongue is touched, one does not feel uncomfortable. When the posterior part is touched, the individual gags. The ninth and tenth cranial nerves cause contraction of the pharyngeal muscles. This is the ***gag reflex***; afferents of the reflex pass through the glossopharyngeal nerve and the efferent limb is through the ninth and tenth nerves.
- In an unconscious patient, care has to be taken that the tongue does not fall backward towards the pharynx as this would obstruct respiration. The head is placed on one side and made to hang down.
- Carcinoma of the tongue (and other parts of the mouth) is common in India because of tobacco chewing.

Multiple Choice Questions

1. The preganglionic parasympathetic fibres for the submandibular salivary gland arise from:
   a. The superior salivatory nucleus
   b. The inferior salivatory nucleus
   c. The submandibular ganglion
   d. The otic ganglion
2. Stylohyoid is supplied by the:
   a. Inferior alveolar nerve
   b. Mylohyoid nerve
   c. Glossopharyngeal nerve
   d. Facial nerve
3. The submandibular duct opens on the:
   a. Sublingual fold
   b. Sublingual papilla
   c. Frenulum linguae
   d. Fimbriated fold
4. The safety muscle of the tongue is:
   a. Palatoglossus
   b. Styloglossus
   c. Genioglossus
   d. Chondroglossus
5. Taste sensation from the posteriormost part of the tongue is carried by:
   a. Chorda tympani nerve
   b. Lingual nerve
   c. Glossopharyngeal nerve
   d. Superior laryngeal nerve

ANSWERS

1. a   **2.** d   **3.** b   **4.** c   **5.** d

Clinical Problem-solving

Case Study 1: Your medical school colleague was trying to examine his own tongue. He tried touching one of the lymphoid tubercles behind the circumvallate papillary line. He suddenly gagged.
- Is this reaction normal in a person of this age group?
- What is this reaction called?
- Trace the pathways associated with this reaction.

Case Study 2: A 37-year-old man presented with a cystic swelling on the left side of the floor of his mouth. The tongue was slightly pushed to the right side. He also complained of pain on and off.
- What condition does this man have?
- Which gland is blocked to produce this kind of swelling?
- Which other gland(s) are likely to be involved in such a condition?

(For solutions see Appendix).

Chapter 6

Cranial Cavity

Cranium is the skeleton of the head; the cavity within the cranium is the cranial cavity. Though technically the cranium comprises two parts, namely the neurocranium and the viscerocranium, the term 'cranial cavity' refers only to the cavity of the neurocranium.

CONTENTS OF THE CRANIAL CAVITY

The most important content of the cranial cavity is the brain. The brain and the spinal cord (the latter being a content of the vertebral canal) are surrounded by three membranes called the **meninges** (Singular meninx; Greek. meninghs=membrane).

The meninges are:
❑ **Dura mater**
❑ **Arachnoid mater**
❑ **Pia mater**.

The dura mater is the outermost of the meninges. Beneath it is the arachnoid mater. The innermost layer is the pia mater. The dura mater contains a series of venous sinuses which drain the intracranial structures including the brain. Between the dura mater and the arachnoid is the subdural space. Between the arachnoid mater and` the pia mater is the **subarachnoid space** that contains **cerebrospinal fluid**.

In addition to the brain and meninges, the cranial cavity contains the proximal parts of **cranial nerves.** They travel from their attachments on the surface of brain to various foramina in the skull through which they leave the cranial cavity. The cranial cavity also contains **blood vessels** which supply the brain, the meninges and other intracranial structures.

Lying in close relationship to the brain are two endocrine glands of great importance and they are the hypophysis cerebri and the pineal gland.

MENINGES

As already noted, the three membranes of meninges envelop the brain and the spinal cord. That part of meninges enveloping the brain is referred to as cranial meninges and that part enveloping the spinal cord as spinal meninges. The layers of meninges are called 'mater' (Sanskrit.maathaa, Latin.mahta=mother) because they give protection and nourishment to the underlying structure. They also provide a supporting framework for the vessels. The subarachnoid space is an integral part of the meningeal system and is vital to the normal functioning of the brain. The names of the layers are indicative of their predominant qualities: dura mater is tough and firm, arachnoid is web like and pia is slender and clings to the brain tissue.

Dura Mater

Other names: Dura, pachymeninx (Greek.pachy=thick), ectomeninx, meninx fibrosa.

The dura mater (hard mother; Latin.durus=hard) is a thick fibrous membrane. The dura mater of the cranial cavity is called the cerebral or cranial dura (or dura mater encephali) while the dura mater covering the spinal cord is called the spinal dura (or dura mater spinalis).

The dura is composed of densely packed collagen bundles arranged in laminae. The bundles run in different directions in adjacent laminae thereby providing a lattice like arrangement. The cranial dura is in two layers: an inner meningeal layer and an outer periosteal layer. The periosteal layer is actually the periosteum (also referred to as the endosteum or endocranium) covering the internal surface of the cranial bones. The two layers are united except in places where they are separated by dural venous sinuses.

Histology

Histology of Dura Mater

The dura is a thick fibrous membrane with minimal cells. Fibroblasts occur scattered. Osteoblasts may be present in the periosteal layer.

The dura is firmly adherent (primarily through the periosteal layer and secondarily through the attachment of the meningeal layer to the periosteal layer) to the internal aspect of the cranial bones. Fibrous bands pass from the dura to the bones. Adherence of the dura to the bones is strongest at the sutures, around foramina and at the cranial base. However, when the sutures fuse in adults, the dura gets separated from them. On the other hand, as age advances, the dura becomes thicker, more firm and more adherent to the bones, including those of the calvaria. At the sutures and the foramina, the periosteal layer becomes continuous with the pericranium (periosteum on the external aspect of the cranial bones). At the margins of the superior orbital fissure, the periosteal layer is continuous with the orbital periosteum. When the cranial nerves pass through the various foramina, the meningeal layer forms a tubular covering for the nerves for a short distance; the tubular covering fuses with the epineurium of the nerve a short distance from the foramen. Where major vessels enter into the cranial cavity, there the dura (both layers) fuses with the adventitia of the vessels.

Dissection

The dura and its folds can be studied while removal of brain is being done and after the brain has been removed from the cranial cavity. The former gives an idea of the various relationships that the dural folds have with parts of brain. The latter gives a complete and comprehensive picture of the dura, its folds and their significance. Look for the grooves that these vessels make on the internal surfaces of the skull bones. Observe that the veins are superficial to the arteries.

Locate the various dural folds and study their shapes and attachments. Wherever a venous sinus is located within the layers of dura, try to open one of the dural layers to expose the sinus. Identify each of the sinuses and define their connections. Specifically identify and study the attachments of tentorium cerebelli, the triangle created by the crossing of its attachments, the nerves which appear in the concerned region and the confluence of sinuses.

Dural Partitions

In certain places, the meningeal layer separates from the periosteal layer and forms inwardly projecting folds by reduplicating on itself. These folds form partial septa which divide (ofcourse, partially) the cranial cavity into compartments. Four such septa are present.

1. Falx cerebri
2. Tentorium cerebelli
3. Falx cerebelli
4. Diaphragma sellae

Falx Cerebri (Fig. 6.1)

The first and largest of these septa is the falx cerebri (Latin. falx=sickle; indicating a sickle shaped structure). It is formed by the meningeal layer dipping in, reduplicating and reaching back the midsagittal line. The falx cerebri, therefore lies in the sagittal plane and occupies the great longitudinal fissure between the two cerebral hemispheres. It is sickle shaped. It has a convex upper edge that is attached to the vault of the skull in the midline (i.e., along the sagittal suture till the internal occipital protuberance). Its lower edge, which is free, is markedly concave downwards. The anterior end of the falx cerebri is narrow and pointed. It is attached to the crista galli. At its posterior end, the falx cerebri has a straight lower edge that is attached to the upper surface of the tentorium cerebelli (6.1). This edge slopes backward and downward.

At the upper attachment of the falx cerebri, the two layers of dura mater that form it diverge to enclose a triangular space. In other words, at this attachment, the meningeal layer dips down and folds back to form a double layered infolding which is the falx cerebri. The triangular space is in the gap where the meningeal layer dips in and reaches back. The third side of the triangle is formed by the endocranium. This space is occupied by the ***superior sagittal sinus*** (Figs 6.2 and 6.3). It is an example of a sinus walled partly by the meningeal layer of dura mater and partly by periosteal layer.

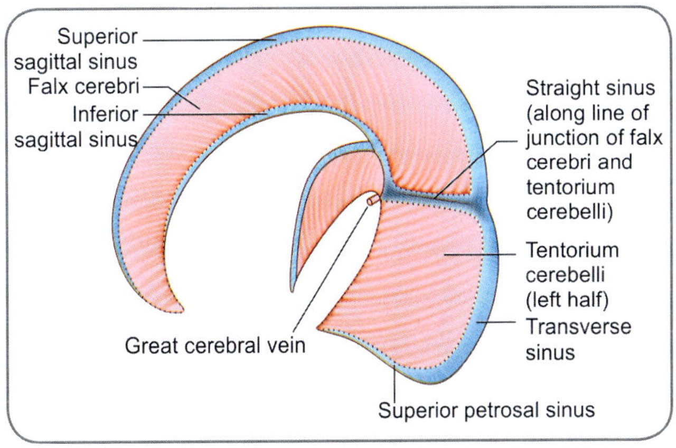

Fig. 6.1: Falx cerebri and tentorium cerebelli

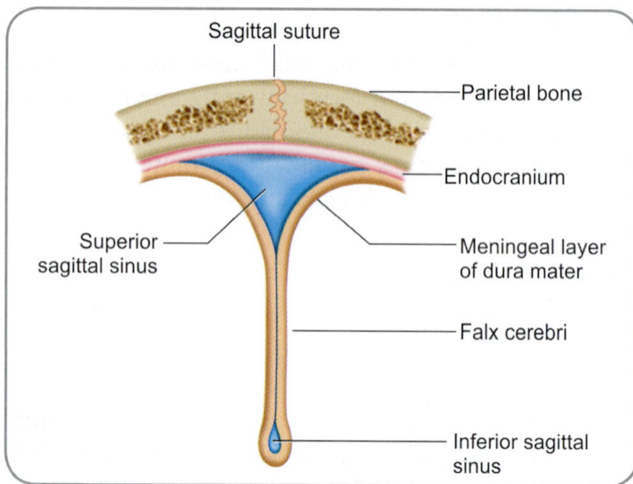

Fig. 6.2: Coronal section through falx cerebri midway between its anterior and posterior ends

At the lower end of the falx cerebri, the dura mater folds on itself to form the free lower edge. An oval space is left in the fold along the lower edge. This space is occupied by the *inferior sagittal sinus* (Fig. 6.2). The wall of this sinus is formed all around by the meningeal layer of dura mater (there being no participation of the endosteal layer).

At its posterior end, the falx cerebri joins the upper surface of the tentorium cerebelli. A triangular space is created between the diverging layers of falx and the upper layer tentorium. This space is occupied by the straight sinus which is also an example of a sinus between meningeal layers but of two different folds.

Tentorium Cerebelli (Figs 6.3 and 6.4)

The second largest of the dural partitions is the tentorium cerebelli (Latin.tendo=tent; a tent for the cerebellum). It appears like a tent that slopes down from a single pole (and hence its name). It is placed more or less transversely i.e., in a plane that is at right angles to that of the falx cerebri.

Its central part is higher than its right and left margins. It, therefore, forms a tent-like roof over the posterior cranial fossa in which the cerebellum lies.

The anterior part of the tentorium cerebelli is marked by a deep *tentorial notch or tentorial incisure*. The U-shaped edge of this notch is called the *free margin* of the tentorium cerebelli. Traced anteriorly, the free margin extends into the middle cranial fossa and gains attachment to the *anterior* clinoid process.

Posterolaterally, the tentorium cerebelli has a curved edge. Along this edge, the two layers of dura mater forming the tentorium separate and gain attachment to the lips of a broad groove (transverse sulcus) present over the internal surface of the occipital bone. The anterior part of this groove extends on to the internal aspect of the posteroinferior angle of the parietal bone. Along this attachment the two layers of dura mater forming the tentorium cerebelli separate to leave a triangular interval that forms the transverse sinus.

Anterolaterally, each half of the tentorium cerebelli is attached to the superior border of the petrous temporal bone. Medially, this edge is prolonged to reach the *posterior* clinoid process. In the gap created by the two layers of the tent getting attached to the petrous temporal bone, the superior petrosal sinus is lodged. Near the apex of the petrous temporal, the dura mater forming the lower layer of the tentorium cerebelli is prolonged forwards onto the anterior surface of the petrous temporal bone to form a pouch-like extension. This pouch is called the *trigeminal cave* (also called cavum trigeminale or Meckel's cave; after the 18th century German anatomist and obstetrician Johann Meckel Senior) because the trigeminal ganglion lies in it.

Anteriorly, the free margin of the tentorium cerebelli crosses above the attached margin, thus producing a small triangular depression of dura mater anterior to the crossing.

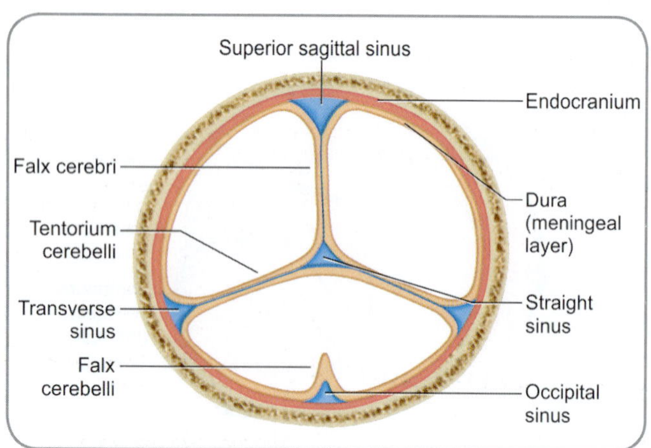

Fig. 6.3: Coronal section through posterior part of falx cerebri and tentorium cerebelli; the falx cerebelli is also shown

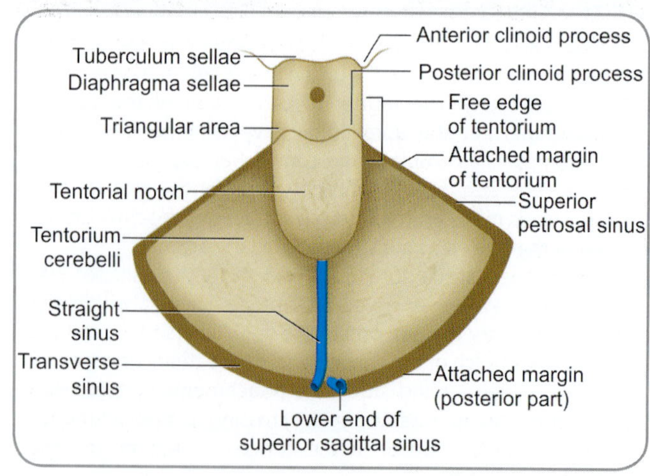

Fig. 6.4: Scheme to show the tentorium cerebelli and its attachments

The tentorium cerebelli divides the cranial cavity into an ***infratentorial compartment*** that is occupied by the cerebellum and brainstem and a ***supratentorial compartment*** that is occupied mainly by the cerebral hemispheres. The tentorial notch is the only communication between the two compartments. It is occupied by the midbrain and surrounding meninges. The narrow gap between the midbrain and the edge of the tentorial notch is a site at which flow of CSF through the subarachnoid space can be obstructed following inflammation.

Falx Cerebelli

In the median plane, over the floor of the posterior cranial fossa, is a short fold of dura mater called the falx cerebelli (also called the falcula) (Fig. 6.3).

The fold is crescentic in shape and is placed in the sagittal plane. It occupies the median groove (vallecula cerebelli) that separates the lower part of the right and left cerebellar hemispheres.

It has an anterior edge that is free and a posterior edge that is attached to the occipital bone in the middle line. At its upper end (or base), it is attached to the undersurface of the tentorium cerebelli. Its lower end (or apex) reaches the posterior edge of the foramen magnum. Here it may divide into two parts which pass forwards on either side of the foramen. The occipital sinus lies along the posterior attachment of the falx cerebelli.

Diaphragma Sellae

The smallest of the dural partitions is the diaphragma sellae. The body of the sphenoid bone occupies the median region of the middle cranial fossa. The hypophyseal fossa is a depression in this bone. The diaphragma sellae (Greek. diaphragma=mid partition, sella=saddle) is a horizontal fold of dura mater that roofs over the hypophyseal fossa. As the diaphragma sellae forms a transverse covering for the hypophysis cerebri, it is also called the tentorium of the hypophysis.

There is an aperture at the centre of the diaphragm which allows the infundibulum and the pituitary stalk to pass into the hypophyseal fossa. Anteriorly, the diaphragma is attached to the tuberculum sellae and posteriorly, to the dorsum sellae.

Diaphragma sellae and the cavernous sinus formation (Fig. 6.5)

We have already seen that the periosteal and the meningeal layers of the dura separate from each other in certain places providing sufficient space for the dural venous sinuses to be formed. One such separation is found in the middle cranial fossa and the resultant space is occupied by the cavernous sinus. However, it is important to know the

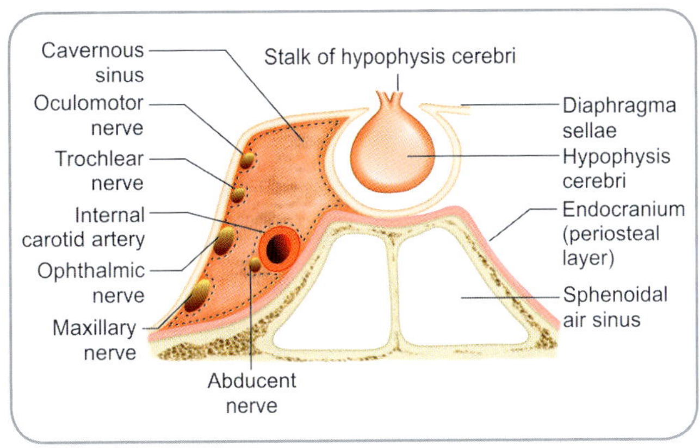

Fig. 6.5: Coronal section through hypophyseal fossa, cavernous sinus and diaphragma sellae

relationship that this sinus has with the hypophyseal fossa by virtue of the way the dura is disposed here.

On either side of the body of the sphenoid bone, the meningeal layer of the dura mater is widely separated from the periosteal layer to form a cuboidal space. This space is occupied by the ***cavernous venous sinus***. The dura mater forming the lateral wall of the sinus turns medially to form the roof of the sinus, and then continues medially over the hypophyseal fossa to form the upper layer of the diaphragma sellae. Reaching the central aperture in the diaphragma sellae, the dura mater curves on itself to form the lower layer of the diaphragma. It then descends forming the upper part of the medial wall of the cavernous sinus and passes medially to coat the hypophyseal fossa. In this way, the dura forms a sac within which the hypophysis cerebri lies. The stalk of the hypophysis passes through the aperture in the diaphragma sellae.

Blood Supply and Nerve Supply of Cerebral Dura Mater

A large number of meningeal arteries take part in supplying the cerebral dura mater. They also supply the endocranium and the skull bones (including the diploe). Although, they are called meningeal arteries, ***they do not supply*** the arachnoid mater and pia mater. They are actually periosteal arteries and lie embedded in the outer periosteal layer of dura; from here, they ramify into the dura, inner table of the bone concerned and diploe.

The largest meningeal artery is the middle meningeal branch of the maxillary artery. Assisted by a branch of the anterior ethmoidal artery, the middle meningeal artery supplies the ***supratentorial dura***. Small branches from the posterior ethmoidal additionally supply the dura of anterior cranial fossa; branches from the accessory meningeal and internal carotid arteries additionally supply the dura of middle cranial fossa. The ***infratentorial dura*** is supplied by branches from occipital artery, ascending pharyngeal artery and the vertebral artery. These branches to the infratentorial dura ascend into the distribution area through the various foramina of the skull base.

The veins draining the dura accompany the corresponding arteries. They communicate with the dural venous sinuses and with the diploic veins. Many of them join the middle meningeal veins which leave the cranial cavity either through the foramen spinosum or ovale and enter the pterygoid venous plexus.

The nerves to the dura mater are derived from various branches of the trigeminal nerve and from branches of the glossopharyngeal, vagus and upper three spinal nerves.

Added Information

- The anterior tip of the free margin of the tentorium cerebelli on either side is attached to the anterior clinoid process. As this margin is traced from the petrous temporal bone to the anterior clinoid process, it forms a sharp ridge which marks the junction of the roof and lateral wall of the cavernous sinus.
- The free and attached margins of the tentorium cross; due to this crossing, a small triangular depression is produced anterior to it. This depression is actually on the roof of the cavernous sinus and is pierced in front by the oculomotor nerve and behind by the trochlear nerve.
- At the opening in the diaphragma sellae, the dura, arachnoid and pia blend with each other and also with the capsule of the pituitary gland. Individual layers of the meninges cannot be distinguished in the sella turcica and the subarachnoid space is obliterated.
- The falces (plural term for falx) cerebri and cerebelli provide stability to the brain within the cranial cavity.
- The tentorium cerebelli is held in position and taut by the falces cerebri and cerebelli.
- The nerve supply of cerebral dura is clinically important and can be classified as supratentorial and infratentorial.
  Supratentorial nerve supply: Supply by the ophthalmic division of trigeminal—the ethmoidal branches of the nasociliary nerve supply the anterior cranial fossa and anterior part of falx cerebri; the tentorial nerve, a large recurrent branch of the ophthalmic division supplies the posterior part of falx cerebri, cavernous sinus and tentorium cerebelli.
  Supratentorial nerve supply: Supply by the maxillary and mandibular divisions—numerous branches of maxillary division (nervi meningeus medii), a recurrent branch from the mandibular division (nervus spinosus) supply the middle cranial fossa either directly or by joining the middle meningeal plexus. A twig from the glossopharyngeal nerve may also supply.
  Infratentorial nerve supply: Supply by other nerves—Several branches of the upper three cervical spinal nerves, vagus and hypoglossal nerves ascend through the foramen magnum and supply the infratentorial dura.
  The nerve fibres predominantly carry sensory impulses from the dura. Few sympathetic fibres have also been traced (probably vasomotor to the vessels). Sensory endings in the dura are more numerous in the region of the superior sagittal sinus and in the tentorium cerebelli.

Arachnoid Mater and Pia Mater

Other name: Meninges tenuis (Latin.tenuim=slender), endomeninx.

The arachnoid and the pia are together called the **leptomeninges** (Greek.lepto=slender). Though similar histologically, the arachnoid and the pia are different in their disposition.

Arachnoid Mater

Other names: Meninx serosa, arachnoidea

The ***arachnoid mater*** (Greek.arachne=spider, eidos=like; resembling cobweb; so named because of its appearance as a meshwork) is made up of many layers of cells which are very loosely packed. Because of this loose packing, the layer may vary in its thickness from a thin membrane on the superior aspect of brain to a fairly thick and opaque membrane on the basal aspect of brain. There are no blood vessels in it.

The arachnoid mater is loosely held against the inner surface of dura; very little force can separate the two. The space between the two is the ***subdural space***. Both the dura and the arachnoid are smooth at this dura-arachnoid interface; this facilitates slight (but smooth) movement between the dura and arachnoid.

The cerebral arachnoid invests the brain loosely. However, it does not enter the sulci of the brain, except for the great longitudinal fissure between the two cerebral hemispheres. In the great longitudinal fissure, it dips in and clothes the superior aspect of the corpus callosum. The arachnoid also dips into the pituitary fossa and coats it.

The space between the pia and arachnoid is the ***subarachnoid space***. Traversing the subarachnoid space there are numerous trabeculae that connect the pia and arachnoid, so that at many places the space resembles a sponge. The sponge like appearance is the reason for the name arachnoid (like the mesh or cobweb of a spider). The surface of the brain is not smooth, but has several ups and downs. The pia closely invests the brain and so, dips into all the grooves and gaps. The arachnoid does not do so. This causes gaps between the pia and the arachnoid in several places. Though many of these spaces remain small, in certain areas where the contour of the underlying brain is considerably irregular, the spaces are enlarged to form the subarachnoid cisterns. The subarachnoid spaces and cisterns are filled with cerebrospinal fluid.

Pia Mater

Other names: Meninx vasculosa, tenuis mater.

The ***pia mater*** (Latin.pius=tender) is a delicate vascular membrane which closely invests the surface of the brain. Blood vessels in it are important for supply of the underlying brain. It is difficult to separate the pia from the underlying brain tissue. However, at the microscopic level, a very thin subpial space is present between the two.

Despite being very slender, the pia has a regulatory role to play in the relationship between the subarachnoid space and the brain.

Histology

Histology of Arachnoid and Pia Mater

The arachnoid and pia mater are layers which have cells embedded in collagen. The cells of both the layers have similar properties. They are nucleated flat or cuboidal cells. In a discussion of histology of both layers, it is preferable to call these cells as '***leptomeningeal cells***' and not as cells of arachnoid or cells of pia separately, the reason for which will be evident as we see more details.

The arachnoid is made up of several layers of the leptomeningeal cells. There seems to be a natural stratification into outer, middle and inner zones. In the outer zone, which is the dura-arachnoid interface, there are four to six layers of flattened cells. The cells are united by tight junctions and desmososmes. A barrier is therefore created for the CSF to move from the subarachnoid space to the subdural space which is a requisite for normal healthy status. In the middle zone, 3–4 layers of cells are cuboidal, tightly packed and united by tight and gap junctions. In the inner zone, 3–4 layers of cells are loosely packed and are interspersed between collagen bundles. The inner zone cells have basememnt membrane but the cells of other two zones do not. Basement membrane is well defined where the cells are in contact with collagen.

The pia mater consists of one or two layers of flattened leptomeningeal cells which are connected by gap junctions and desmososmes. Collagen bundles and fibroblastic cells separate the pia from the underlying glia limitans. The space between the pia and the glia limitans is the subpial space which is traversed by tiny vessels.

The arachnoid and pia are connected by trabeculae. Trabeculae are sheets or finger like projections from the inner surface of arachnoid to the pia traversing the subarachnoid space. Each trabecula has a central core of collagen covered by the leptomeningeal cells. This covering is usually one or two cells thick. The covering layer is continuous with the inner zone of arachnoid at the arachnoid interface and with the cells of pia at the pial interface. Thus, it can be seen that the arachnoid and pia have cellular continuity. The vessels which traverse the subarachnoid space are also covered by the trabeculae. In other words vessels lie within the layers of trabeculae. Nerves which traverse the subarachnoid space are also covered by leptomeningeal cells.

The picture can be better understood if the development is thought of. The arachnoid and pia develop from a single stratum of endomeninx which initially has many layers of cells. Vacuolation occura and cells in the middle of the stratum are lost. It then appears as though the stratum is split into two, the outer arachnoid and the inner pia (much like the splitting of the somato and splanchnopleures).

contd...

Histology *contd...*

In certain places, vacuolation is incomplete and cells remain to form the trabeculae. Thus the cells of pia, trabeculae and arachnoid are in physical and functional continuity. This is the reason that it is better to call their cells as leptomeningeal cells and not as cells of individual layers.

Subarachnoid Space and Subarachnoid Cisterns

The space between the arachnoid and the pia is called the ***subarachnoid space*** (or the leptomeningeal space). Apart from the cerebrospinal fluid, this space also contains the arteries and veins of the brain. It has already been noted that the pia is in close approximation with the surface of the brain and follows the convexities and concavities of the latter. The arachnoid does not do so but bridges the concavities. As a result, the two layers of leptomeninges are in contact over the convexities of the brain surface but are separated from each other at the concavities. Due to this, a subarachnoid space of variable depth and dimension (due to variability in the size and shapes of the concavities) is formed. Wherever the space is expansive, it is called a ***subarachnoid cistern*** (Latin.cista=box; indicating a container or reservoir). Subarachnoid cistern can therefore be regarded as dilatations of specific portions of the general subarachnoid space.

The subarachnoid space is traversed by arachnoid trabeculae which run from the arachnoid to the underlying pia. It communicates with the fourth ventricle by way of three openings. The median foramen of magendie is in the roof of the fourth ventricle and opens into the cistern magna. The two lateral foramina of Luschka are in the lateral recesses of the fourth ventricle and open into the subarachnoid space at the cerebellopontine angle. Cerebrospinal fluid is formed in the ventricles of the brain, circulates through the ventricles, passes into the subarachnoid space through the foramina of Magendie and Luschka and is then reabsorbed at the arachnoid granulations.

Cisterna Magna

Otherwise called the ***cerebellomedullary cistern***, this cistern is formed where the arachnoid bridges the gap between the inferior aspect of cerebellum and the posterior surface of medulla. It is the largest of the subarachnoid cisterns and is continuous below with the subarachnoid space of spinal cord and above with the lumen of the fourth ventricle through the foramen of Magendie.

Pontine Cistern

This cistern is formed ventral to the pons (especially the lower part of the anterior convexity of pons and the pontomedullary junction) where the arachnoid is pushed well forward. It communicates below with the subarachnoid space of spinal cord, posteriorly with the cistern magna and above with the interpeduncular cistern. The basilar artery traverses through this cistern to reach the interpeduncular cistern.

Interpeduncular Cistern

The inferior surfaces of the temporal lobes of the cerebrum are a little inferiorly placed than the structures which occur between them. Thus, when the arachnoid bridges across the temporal lobes, space is created between it and the cerebral peduncles and other structures of the interpeduncular fossa. This space forms the interpeduncular cistern (also called the cisterna basalis or cisterna cruralis). It communicates below with the pontine cistern and anterosuperiorly extends till the optic chiasma. Though this cistern is not the largest in terms of volume, it is the widest. It contains the arterial circle of Willis (circulus arteriosus). The part of the interpeduncular cistern between the right and left cerebral peduncles is called the Tarin's space.

Cisterna Ambiens

(Latin.ambiens=going around) Otherwise called the **superior cistern** or the cistern of the great cerebral vein or **Bichat's canal**, this cistern is on the posterior aspect of the midbrain. On the posterior side, the arachnoid passes from the splenium of the corpus callosum to the posterosuperior aspect of the cerebellum. This causes a gap between these two structures which is closed anteriorly by the posterior aspect of the midbrain. This gap forms the cisterna ambiens. The great cerebral vein traverses the cistern. The pineal body projects from the midbrain and protrudes into this cistern.

Smaller cisterns also exist. These include the cistern of the lateral fossa (as the arachnoid bridges the lateral sulcus of brain across the opercula), the prechiasmatic and postchiasmatic cisterns (as the arachnoid covers the optic chiasma), the cistern of lamina terminalis (as the arachnoid passes ventral to it) and the supracallosal cistern (as the arachnoid tents over the corpus callosum). The cistern of lateral fossa contains the middle cerebral artery. The rest of them can be regarded as extensions of the interpeduncular cistern.

Dissection

In an intact but isolated specimen of brain with intact arachnoid, observe the way the arachnoid bridges across the sulci. Insert slender blunt probes under these bridges and locate the small subarachnoid spaces. Peel the arachnoid off in a small segment of the superolateral surface of the brain and see the underlying pia. Try to peel the pia in a small portion; see how adherent it is to the brain surface; also observe the change in the texture of the brain surface once the overlying pia is peeled off.

With the arachnoid intact on the inferior and posterior aspects, try to locate the various sub arachnoid cisterns.

MENINGEAL SPACES

Spaces which are formed between the different layers of meninges or spaces between the meninges on one side and other structures on the other are collectively grouped as meningeal spaces.

❑ **Spaces between different layers of meninges:** Interfaces occur at places where one layer of meninges comes in contact with another layer. Theoretically, a space should exist between the layers in contact at the particular interface. The spaces of this group are the subdural space (between the dura and arachnoid) and the subarachnoid space (between the arachnoid and pia). Of the two, the subdural space is only a 'potential' space.

1. **Subdural space:** This space is described to be present at the dura-arachnoid interface. It is not a natural space in the sense that it is not found or present under normal conditions of life. However, separation between the two layers occurs very easily on application of mild force. Veins from the sub arachnoid space pierce the arachnoid and meningeal layer of dura to reach the dural venous sinuses. As they pierce the two layers, there is a very small part of the vein that actually bridges the space between the two layers. Even mild force is sufficient to rupture such bridging veins. Blood oozing out separates the dura and the arachnoid thus making a space out of an original approximation. This feature can well be understood if it is remembered that there is no physical adhesion at the dura-arachnoid interface and the arachnoid is kept in approximation against the dura only because of the CSF pressure in the subarachnoid space. When blood oozes at the interface, hydrodynamic pressure on the dural side of the arachnoid increases than that in the sub arachnoid space. The force that attaches the arachnoid to the dura is lost; an additional force that pushes the arachnoid from the dura is created. Both these factors combine to separate the dura and the arachnoid. The sub dural space is evident only at times of pathology.

2. **Subarachnoid space:** This space is a natural normal space at the arachnoid-pia interface. The arachnoid and pia together form the leptomeninges. They develop from the parietal and the visceral layers of the mesenchyme surrounding the developing brain. The two layers are continuous with each other at their edges. The cavity of the original mesenchyme is the sub arachnoid space. Arachnoid trabeculae which run from the arachnoid to the pia traverse the space.

   ○ **Magendie's spaces** are those parts of sub arachnoid space which are trapped between the pia and the arachnoid at the sulci of the brain.

❑ ***Spaces between the meninges and other structures:*** Interfaces occur at places where any one layer of the meninges comes in contact with some other neighbouring or adjacent structure. The spaces of this group are the epidural space (between the dura and the cranial bones), subpial space (space between the pia and the brain tissue) and the perivascular spaces (spaces around the blood vessels of the brain).

○ ***Epidural space:*** Otherwise called the cranial epidural space or the extradural space, this space is described to be present between the external periosteal layer of the dura and the inner aspect of the cranial bone. This is not a natural space. The periosteal layer is closely and firmly adhered to the inner aspect or table of the cranial bone. There is no space under normal conditions of life. The space comes into existence at times of pathology, when blood from a ruptured vessel pushes the periosteal layer away from the bone.

○ ***Subpial space:*** This is a thin space that is seen between the piamater and the glia limitans of the underlying brain surface. Collagen bundles, fibroblast like cells and vessels occur in this space.

○ ***Perivascular spaces:*** Otherwise called the Virchow-Robin spaces or His's spaces, these are spaces seen around vessels in the brain. They can be considered as extensions of the subpial space.

Arachnoid Villi and Granulations (Fig. 6.6)

Arachnoid villi and granulations are extensions of the arachnoid mater and subarachnoid space through the walls of dural venous sinuses.

At several sites related to the intracranial dural venous sinuses, the arachnoid mater passes through minute apertures in the walls to project into the sinuses (Fig. 6.6). The arachnoid mater then is separated from the blood in the sinus only by endothelium. When these projections are small, they are referred to as ***arachnoid villi***. When they are large and visible to the naked eye they are called ***arachnoid granulations*** (Latin.granum=grain; indicating small masses). Arachnoid villi can be identified in all the dural venous sinuses. Arachnoid granulations are most numerous in relation to the superior sagittal sinus. They may also be seen in lateral extensions (lateral lacunae) present in relation to this sinus.

In children only arachnoid villi are present. Arachnoid granulations appear later in life and are most prominent in old persons in whom they may produce depressions on the skull bones.

The arachnoid villi and granulations are sites at which cerebrospinal fluid is absorbed into the bloodstream.

Added Information

❑ The arachnoid is merely held against the inner aspect of dura by the pressure of cerebrospinal fluid in the subarachnoid space. The pressure compresses the arachnoid against the dura and obliterates the subdural space. The subdural space is thereby, only potential. The subarachnoid space is actual.

❑ In places where major vessels enter the subarachnoid space, the arachnoid forms a sleeve around the vessel by getting reflected on to it and then becomes continuous with the pia.

❑ The arachnoid and pia develop from the same mesenchyme that surrounds the developing brain. The parietal part of the mesenchyme becomes the arachnoid and the visceral part becomes the pia (much like the development of the parietal and visceral layers of serous membranes like peritoneum and pericardium). This common developmental factor is reflected in the close approximation of the two layers in adult life too. In addition, traces of the common development are shown by the fact that the arachnoid is connected to the pia by means of trabeculae. Numerous trabeculae pass between the arachnoid and the pia and give the arachnoid its typical cobweb (network) appearance.

❑ The surface of the brain is marked by several grooves or ***sulci*** that are of varying depth. At such sites, the pia mater extends into the sulci lining them, but the arachnoid does not do so. In other words, the pia mater is closely adhered to the brain surface at all places, but the arachnoid jumps across the sulci. This means that the subarachnoid space extends into the sulci.

❑ Though thin and translucent, the arachnoid can be held and manipulated with instruments like the forceps.

❑ The pia is so adherent to the brain surface that it is difficult to see it separately. However, it gives the brain surface a shiny and smooth appearance.

❑ Francois Megendie, after whom the median opening of the fourth ventricle is named was an 18-19th century French physiologist and Hubert Luschka, after whom the paired lateral openings of the fourth ventricle have been named was a 19th century German anatomist.

❑ Pierre Tarin, after whom a part of the interpeduncular cistern has been named was an 18th century French anatomist.

❑ ***Histology of arachnoid granulation:*** Each granulation has a neck region, where the arachnoid projects out. The subarachnoid space is also dragged along. The projection extends out into the cavity of the venous sinus. As the

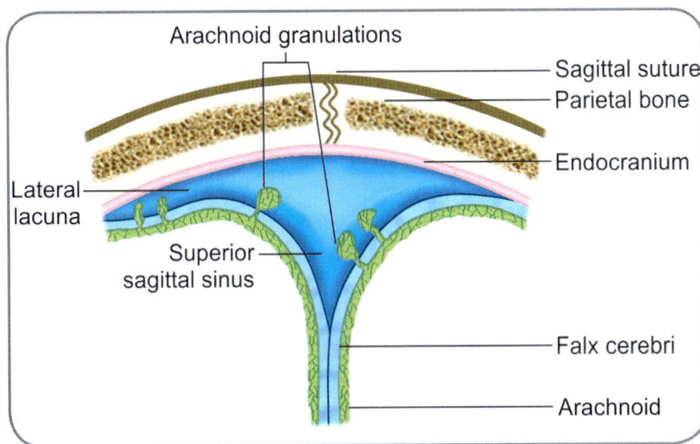

Fig. 6.6: Coronal section to show arachnoid granulations present in relation to the superior sagittal sinus

contd... **107**

Added Information *contd...*

projection extends out, the dura is also pulled along providing a covering. However, the dura fails to reach the apical portion of the extension which then is directly in contact with the endothelium of the venous sinus. The middle and inner zones of arachnoid form several layers of cells in the apical portion (it may not be suitable to call it the apical portion because it is expanded and does not conform to the shape of a pointed apex). The collagen of the trabeculae of the subarachnoid space forms a dense meshwork in the extension. Some collagen channels extend through the thick stratum of arachnoid cells at the apex. It can be seen that with all these arrangements, the surface area for exchange between the subarachnoid space and the venous sinus is increased multifold. In the apex area (also called the cap), the arachnoid cells and sinus endothelium are in direct contact.

Clinical Correlation

- Inflammation of pia and arachnoid is called meningitis. The condition is characterised by fever and severe headache. Spasm of extensor muscles leads to rigidity of the neck. The spasm is a result of irritation of cervical nerve roots as they pass through the subarachnoid space. CSF pressure and chemical composition show alterations. Samples of CSF may be obtained for examination by lumbar puncture. In some cases, lumbar puncture may fail to yield CSF. In such cases, CSF may be obtained by cisternal puncture. A needle is introduced from behind, through the interval between the atlas and axis vertebrae. The needle passes through the posterior atlanto-occipital membrane and enters the cisterna magna. Cisternal puncture is the method of choice in infants and very young children.
- *Hydrocephalus:* Accumulation of cerebrospinal fluid in the subarachnoid space and consequent enlargement of head is called hydrocephalus. The usual cause for such accumulation is some kind of obstruction to the flow of CSF. The condition is then called obstructive hydrocephalus. Common sites of obstruction are the cerebral aqueduct or one of the interventricular foramina. Common causes for obstruction are a nearby tumour pressing to reduce the passage, cellular debris resulting from intraventricular haemorrhages and infections of the central nervous system. Circulation of CSF is impaired; dilatation of ventricles superior to the point of obstruction occurs; as a result, there is pressure on the cerebrum and the cerebral cortex thins out. In the case of infants in whom the sutures have not yet fused and the fontanelles are still open, the bones of the calvaria separate. An artificial channel is produced so that CSF is drained and damage to brain is lessened.

  Absorption of CSF into the venous system may be impaired due to congenital absence of arachnoid villi or if the villi and arachnoid granulations are blocked due to blood cells when there is subarachnoid haemorrhage.
- *CSF leaks:* Fractures in the skull, especially in the basal part, may tear the meninges and cause CSF to leak. Leakage of CSF into

contd...

Clinical Correlation *contd...*

the nose (CSF rhinorrhoea) or ear (CSF otorrhoea) can occur. Such conditions also increase the risk of meningitis because infections can spread from the nose or ear.
- Bulging of diaphragma sellae can occur because of pituitary tumours. Tentorial herniation may occur when the pressure in the supratentorial compartment increases.
- The dura mater is sensitive to pain. The sensitivity is more in areas of venous sinuses and in relation to meningeal arteries. Pain from dura is usually referred to cutaneous and mucosal regions supplied by the same cervical nerve or division of trigeminal nerve. The headache after a lumbar puncture or spinal anesthesia is also of dural origin. When CSF is drawn out, there is a mild sagging of the brain with associated pull on the dura. This causes a head ache. Patients are often asked to keep their heads down after a lumbar puncture to minimise the dural pull.
- Inflammation of pia and arachnoid is called leptomeningitis. The common cause is a bacterial infection.

Meningeal Space Haemorrhages
- *Extradural haemorrhage:* This is usually of arterial origin. Blood collects in the extradural space between the endocranial layer of dura and the bone of calvaria. The extravasated blood separates the dura from the bone. Compression of brain occurs when blood mass increases. Immediate brief loss of consciousness is followed by a lucid interval. When the blood mass increases, the patient goes in for drowsiness and unconsciousness.
- *Subdural haematoma:* This is actually a dural border haematoma. The dural border cell layer of the arachnoid is separated from the dura and a new space created between the dura and the arachnoid. Hard blows to the head which jerk the brain are common causes for trauma which produce this kind of a haemorrhage. The haemorrhage is venous in origin. Accumulation of blood may split between the cells of the outer layer of arachnoid (dural border layer).
- *Subarachnoid haemorrhage:* This is bleeding into the subarachnoid space and is arterial. Aneurysms of arteries in the subarachnoid space rupture and result in bleeding. Cranial injuries may also cause bleeding.

Development

Development of Meninges

The meninges have been proposed to develop from the mesenchymatous tissue that surrounds the developing brain. However, it has been found that neural crest cells contribute to the leptomeninges and the ectomeninx is not formed exclusively from mesoderm but has an ectodermal component. The surrounding mesoderm therefore should have components from ectoderm and neural crest cells merging with it much early in development.

This surrounding tissue condenses around the developing brain. The portion close to the brain forms the endomeninx and the portion next to it forms the ectomeninx. The ectomeninx develops collagen bundles and attains a tough character. The mesoderm exterior to it is already developing into the cranial box around the developing brain. The mesodermal

contd...

Development *contd...*

component in the ectomeninx gives rise to fibroblastic and osteoblastic cells and the ectomeninx develops into the dura. The endomeninx closely adheres to the surfaces of the developing brain. However, as growth proceeds and sulci-gyri develop, the endomeninx splits into a parietal layer that becomes the arachnoid and a visceral layer that becomes the pia. Further growth of the underlying brain brings the pia and arachnoid close to each other in some places but they remain separate in others forming the subarachnoid cisterns.

NERVES AND ARTERIES IN THE CRANIAL CAVITY

Cranial nerves (Fig. 6.7)

The brain gives attachment to 12 pairs of *cranial nerves*. Parts of these nerves traverse the cranial cavity.

- In the anterior cranial fossa, is a median bony projection called the *crista galli*. The *cribriform plates* of the ethmoid bone bearing numerous foramina extend on either side of the crista galli. Bundles of olfactory nerve fibres enter the cranial cavity through these foramina and end in the *olfactory bulb*. Posteriorly, the olfactory bulb continues as the *olfactory tract* that is attached to the cerebral hemisphere.

- Medial to the anterior clinoid process, the *optic nerve* is seen. It reaches the cranial cavity from the orbit by passing through the optic canal. The terminal part of the internal carotid artery is posterolateral to the optic nerve.

- The *oculomotor nerve* emerges from the anterior aspect of the midbrain and passes forwards through the subarachnoid space. It penetrates the dura in the triangular interval between the free and attached margins of the tentorium cerebelli and enters the lateral wall of the cavernous sinus. It runs forwards and enters the orbit through the superior orbital fissure.

- The *trochlear nerve* lies a short distance posterior to the oculomotor nerve. It emerges from the *posterior aspect* of the midbrain and winds around its lateral side to reach the front of the midbrain. The nerve runs forward and penetrates the dura mater just below the free margin of the tentorium cerebelli, a little behind the posterior clinoid process. It then enters the lateral wall of the cavernous sinus and runs in it up to the superior orbital fissure. It enters the orbit through this fissure.

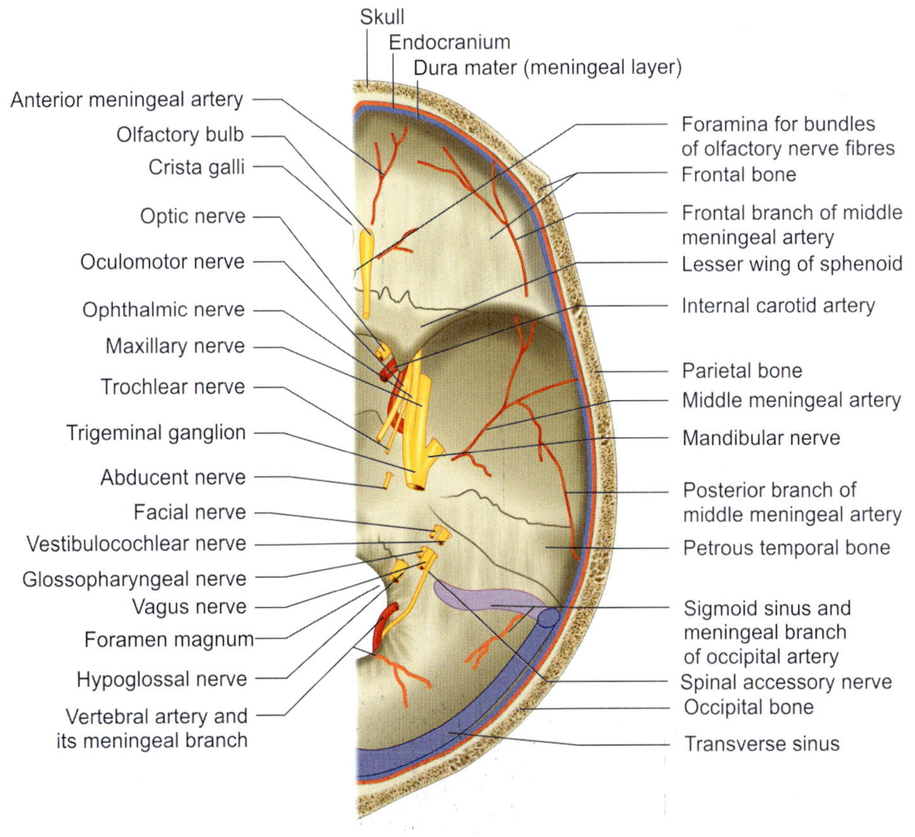

Fig. 6.7: Structures in the cranial fossae

- The **trigeminal ganglion** lies in a depression on the anterior aspect of the petrous temporal bone, near its apex. The ganglion lies within a recess of dura called the **trigeminal cave**. Posteriorly, the ganglion is continuous with the sensory root of the **trigeminal nerve**. Anteriorly, it is continuous with the ophthalmic, maxillary and mandibular divisions of the same nerve. The trigeminal ganglion (and cave) are closely related to the posterior end of the cavernous sinus.
  - The **ophthalmic nerve** runs forwards in the lateral wall of the cavernous sinus, below the trochlear nerve (Fig. 6.5). It divides into three branches (lacrimal, frontal, nasociliary) which enter the orbit by passing through the superior orbital fissure.
  - The **maxillary nerve** pierces the distal edge of the trigeminal cave and comes to lie in the lowest part of the lateral wall of the cavernous sinus (Fig. 6.7). It runs forward and enters the foramen rotundum through which it leaves the cranial cavity.
  - The **mandibular nerve** is formed by the union of two roots. The sensory root arises from the lateral part of the trigeminal ganglion and leaves the cranial cavity through the foramen ovale. The motor root passes forward deep to the trigeminal ganglion, passes through the foramen ovale and then joins the sensory root outside the skull.
- The **abducent nerve** runs upwards and laterally over the anterior wall of the floor of the posterior cranial fossa. It pierces the dura a little lateral to the dorsum sellae of the sphenoid bone. It then runs upwards to reach the upper border of the petrous temporal bone and bends around it to enter the middle cranial fossa. The nerve now comes to lie within the cavernous sinus where it is closely related to the internal carotid artery (Fig. 6.7). At the anterior end of the cavernous sinus, the nerve passes through the superior orbital fissure to enter the orbit.
- The **facial nerve** and the **vestibulocochlear nerve** enter the internal acoustic meatus located on the posterior aspect of the petrous temporal bone.
- A short distance below and medial to the internal acoustic meatus, the jugular foramen is seen. The **glossopharyngeal nerve**, the **vagus nerve** and the **accessory nerve** pass through it.* The spinal part of the accessory nerve is formed by union of a number of rootlets which emerge from the lateral aspect of the upper part of the spinal cord. The nerve ascends through the upper part of the vertebral canal, lateral to the spinal cord. It enters the cranial cavity through the foramen magnum and leaves it through the jugular foramen.

- The **hypoglossal nerve** passes through the hypoglossal canal (anterior condylar canal) a little above the lateral margin of the foramen magnum.
- Some branches of the cranial nerves are also seen in the cranial cavity. Such branches include the greater and the lesser petrosal nerves which course downwards and medially on the anterior surface of the petrous temporal bone and the deep petrosal nerve at the foramen lacerum. A number of small meningeal branches given out by various cranial nerves are present.

Arteries

Intracranial Part of Internal Carotid Artery

The internal carotid artery enters the skull through the carotid canal (a sinuous canal that starts at the base of skull and passes through the petrous temporal bone). After passing through the carotid canal, the artery reaches the foramen lacerum and crosses the latter to enter the cavernous sinus at the posterior end of the sinus. It then runs forward within the sinus (Fig. 6.7). Near the anterior end of the cavernous sinus the artery turns upwards, pierces the dura mater forming the roof of the cavernous sinus and comes into relationship with the cerebrum.

The artery, at this point, lies medial to the anterior clinoid process and lateral to the optic nerve and optic chiasma. It ends by dividing into the anterior and middle cerebral arteries which supply the brain. It also gives off the ophthalmic artery that runs forward into the orbit through the optic canal.

Vertebral Artery

The vertebral artery arises from the subclavian artery in the lower part of the neck, ascends through the foramina transversaria of the upper six cervical vertebrae, passes through the suboccipital region and then enters the cranial cavity through the foramen magnum lying anterolateral to the medulla.

After entering the cranial cavity, the vertebral arteries of both sides anastomose wih each other in the midline to form the basilar artery. The vertebral and basilar arteries (collectvely called the vertebrobasilar system) give branches which supply the brain and spinal cord.

Meningeal Arteries

The middle meningeal artery, a branch of the maxillary artery and several other smaller meningeal branches from various arteries course through different parts of the cranial cavity. They supply the dura mater.

---

* The cranial part of the accessory nerve runs along with the vagus nerve.

Congenital Malformations of Skull

- *Anencephaly* is a malformation of the skull in which the greater part of the vault is missing. It is caused by failure of the neural tube to close in the region where the brain is to be formed. Neural tissue, which is exposed to the surface degenerates. The condition is fairly frequent and is not compatible with life.
- Establishment of the normal shape of the skull depends on orderly closure of sutures.
  - Premature union of the sagittal suture gives rise to a boat shaped skull (*scaphocephaly*).
  - Early union of the coronal suture results in a skull that is pointed upwards (*acrocephaly*).
  - Asymmetrical union of sutures (on the right and left sides) results in a twisted skull (*plagiocephaly*).
- *Congenital hydrocephalus* is a condition in which there is obstruction to the flow of cerebrospinal fluid. As a result, pressure in the ventricular system of the brain increases and leads to its dilatation. In turn, this leads to enlargement of the head and wide separation of the bones of the skull.
- The maxilla, the mandible and the zygomatic bone are derived from the first pharyngeal arch. If growth of this arch is defective, the bones concerned remain underdeveloped and the face is deformed. The condition is called *mandibulofacial dysostosis*.
- Many bones of the skull are formed by intramembranous ossification. The clavicle is also formed in membrane. In a condition called *cleidocranial dysostosis,* formation of membrane bones is interfered with. Deformities of the skull are seen in association with absence of the clavicle.

Fractures of the Skull at the Cranial Fossae

Fractures Through the Anterior Cranial Fossa

- This fossa is closely related to the nasal cavity and to the orbit. Fracture through the fossa can lead to bleeding or leakage of CSF through the nose (the blood flowing directly into nose through its roof or through the frontal air sinus).
- Bleeding into the orbit can push the eyeball forward (exophthalmos). Blood in the orbit can seep into the eyelids resulting in 'black eye'.

Fractures Through the Middle Cranial Fossa

- The body and greater wing of the sphenoid are closely related to the middle cranial fossa. A fracture involving the body of the sphenoid bone can lead to leakage of blood and CSF into the sphenoidal air sinuses and through them into the nose and mouth.
- The 3rd, 4th and 6th cranial nerves lie in relation to the cavernous sinus (which lies against the body of the sphenoid bone). These nerves can be involved in fractures of the middle cranial fossa.
- Posteriorly, the middle cranial fossa is bounded by the petrous temporal bone. Involvement of this bone can lead to bleeding and discharge of CSF into the middle ear and external acoustic meatus.
- The 7th and 8th cranial nerves (which pass through the internal acoustic meatus) can also be injured in a fracture through the petrous temporal bone.

Fractures Through the Posterior Cranial Fossa

- Fractures through this fossa can lead to bleeding, the blood seeping into the muscles of the back of the neck. The blood often appears superficially over the mastoid process and the sternocleidomastoid muscle.
- If the fracture passes through the jugular foramen, there can be injury to the 9th, 10th and 11th cranial nerves. The walls of the hypoglossal canal are strong and so the 12th cranial nerve usually escapes injury.

Injuries to the Brain Following Injury to the Skull

- Fractures of the skull often lead to injury to the brain. An impact on the head can damage the brain even in the absence of a fracture. In this connection, it is to be remembered that the brain is a very delicate tissue. However, it is not damaged by normal bodily movements because:
  - It is surrounded by a cushion of CSF.
  - Its displacement is prevented by folds of dura mater like the falx cerebri, tentorium cerebelli, falx cerebelli.
  - Veins passing from the brain to the dural venous sinuses (especially the superior cerebral veins passing to the superior sagittal sinus) also help to keep the brain in position.
- The cushion of CSF prevents the brain from striking against the inner wall of the skull. However, a strong impact on the head can shake up the brain. When this shaking up leaves no apparent physical damage, the condition is called *cerebral concussion*. It is usually followed by a variable period of unconsciousness. On regaining consciousness, the patient may suffer from headaches and loss of memory.
- An injury in which there is superficial injury to brain tissue (but without any break on its surface) is called *cerebral contusion*; and when there is tearing of brain tissue it is *cerebral laceration*.
  - Severe damage to the brain, especially the brain stem, can lead to prolonged periods of deep unconsciousness (coma) and to death.
  - Patients who come out of coma may show various neurological symptoms depending on the part of the brain injured.
- Apart from direct injury to the brain, injury to the skull can affect the brain by causing haemorrhage. Such bleeding can be of various types.

contd...

Clinical Correlation *contd...*

- *Extradural haemorrhage:* Injury to meningeal vessels can cause bleeding into the potential space between the dura mater and the skull. The haematoma can press upon the brain surface and produce symptoms. Extradural haemorrhage is often caused by injury to the anterior division of the middle meningeal artery, resulting in pressure over the motor area of the brain.
- *Subdural haemorrhage:* Blood accumulates in the space between the dura and arachnoid. It is often caused by rupture of superior cerebral veins at their entry into the superior sagittal sinus. Such injury is most likely to be caused by a blow on the front or back of head (such an injury causes anteroposterior displacement of the brain stretching the veins).
- *Subarachnoid haemorrhage:* Blood flows into the subarachnoid space and mixes with CSF. Such haemorrhage can be caused by rupture of aneurysms which are not uncommon on arteries forming the circulus arteriosus.

Multiple Choice Questions

1. The venous sinus found at the posterior lower edge of the falx cerebri is:
    a. Superior sagittal sinus
    b. Inferior sagittal sinus
    c. Straight sinus
    d. Inferior petrosal sinus
2. The anterior attachment of the free margin of tentorium cerebelli is to:
    a. Anterior clinoid process
    b. Middle clinoid process
    c. Posterior clinoid process
    d. Petrous temporal border
3. What is true about the meningeal arteries?
    a. They are all branches of the maxillary artery
    b. They supply all the three layers of meninges
    c. They do not supply the leptomeninges
    d. Though they supply the endocranium, they do not supply the diploe
4. Apart from the great cerebral vein, another important structure found in cistern ambiens is:
    a. Basilar artery
    b. Optic chiasma
    c. Pineal body
    d. Branches of middle cerebral artery
5. Arachnoid villi:
    a. Occur in relation to all dural venous sinuses
    b. Occur in relation to the superior sagittal sinus only
    c. Occur only in old age
    d. Occur only during obstruction of cerebrospinal fluid

ANSWERS

1. c   **2.** a   **3.** c   **4.** c   **5.** a

Clinical Problem-solving

Case Study 1: A 27-year-old man presented with bacterial meningitis. He had fever, head ache and neck rigidity.
- Which one of these features provides the clinching diagnosis for meningitis?
- What is the reason for this particular feature to occur?
- Which tissue/layer(s) is involved in the inflammation?

(For solutions see Appendix).

Chapter 7

Vertebral Canal

Frequently Asked Questions

- Write notes on: (a) Spinal duramater, (b) Epidural space, (c) Internal vertebral venous plexus.
- Write briefly on: (a) Subarachnoid space and its clinical importance, (b) Ligamentum denticulatum.
- Write short notes on: (a) Vertebral venous plexuses, (b) Differences between spinal and cranial duramater.

The vertebral canal is a longitudinal passage formed collectively by the vertebral foramina placed one over the other. The vertebral foramina are not of the same shape and dimensions throughout the length of the canal. In the thoracic region, the canal is circular in shape in line with the cylindrical appearance of the spinal cord. In regions where the nerve plexuses spring from the spinal cord (brachial and lumbosacral plexuses) thereby increasing the total dimensions of the cord and its surrounds, the vertebral canal is larger and expanded.

In the cervical part, the vertebral canal is large. This is a matter of functional importance. The roominess of the vertebral canal prevents the medulla and upper spinal cord from being compressed or constricted during movements of head which involve the upper cervical vertebrae.

Clinical Correlation

For descriptive purposes, the vertebral canal is divided into three longitudinal zones: the middle (within the margins of the vertebral articular processes) and the right and left lateral (below the vertebral articular processes and extending into the intervertebral foramina). The foraminal extensions of the lateral parts form the radicular or root canals. Developmental anomalies, degenerative changes in the intervertebral discs and osteoarthritic changes in the small joints of the vertebrae can cause stenosis of any of the three zones of the vertebral canal leading to compression of the spinal cord or the spinal nerves. Compression of the associated vessels causes ischaemic changes in the neural tissue compounding the damage than what would be expected from mere compression.

MENINGES IN THE VERTEBRAL CANAL

Note: Basic details of the spinal meninges are dealt within the chapter on Spinal Cord in the section on Neuroanatomy. Intricate and additional details of the spinal meninges with emphasis on clinical anatomy are given here.

The spinal cord that lies within the vertebral canal is covered and protected by layers of meninges much like the brain being protected by meninges. The layers of meninges of the brain and those of the spinal cord are continuous with each other (Fig. 7.1). However, for the sake of convenience of description and keeping in mind the differences that exist in some features of the two meningeal systems, the meninges of the brain are called the cranial meninges and those of the spinal cord are called the spinal meninges.

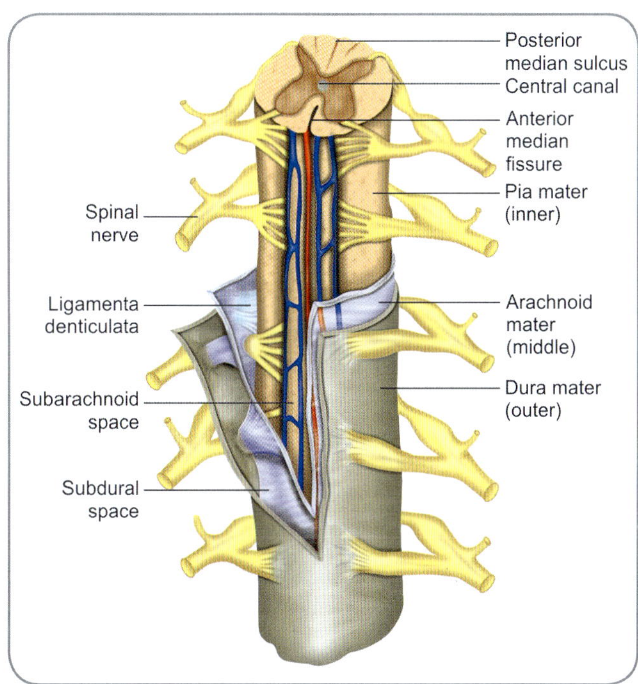

Fig. 7.1: Spinal meninges

The spinal meninges comprises the same three layers as that of the cranial meninges, namely the duramater, the arachnoid mater and the piamater. The characteristics of each of these layers are the same as those of the cranial meninges. The dura is tough and fibrous; the arachnoid is tender and cob webbed; the pia is slender and thin.

Spinal Dura Mater

The spinal dura mater surrounds the spinal cord throughout its length. It is composed of a single layer which is thought to be the spinal counterpart of both the layers of the cranial dura. In other words, the two layers of the cranial dura fuse with each other at the margin of foramen magnum and continue inferiorly as the 'true' spinal duramater. The vertebral canal has several tissues lining it, which include the periostea of the vertebrae, posterior longitudinal ligament and neighbouring connective tissue. These tissues are collectively called the lining tissues of the vertebral canal. The theory that the endocranial layer (outer periosteal layer) of the cranial dura continues as the periosteum of the vertebrae is not agreed upon for the simple reason that the lining tissues include structures other than the periostea (example being the posterior longitudinal ligament). It is also for the same reason that the spinal dura is sometimes described to be composed of a meningeal layer alone.

In whatever way it is described or called, the spinal duramater has a single layer that is tough and fibrous. It is adherent to the margin of the foramen magnum superiorly (where it is continuous with the cranial dura) and fixed to the coccyx caudally blending with the periosteum of the coccyx. It is predominantly a free tube within the vertebral canal. In certain places it has attachments to some of the surrounding structures. These attachments are to the atlanto-occipital membranes, to the posterior surfaces of the second and third vertebral bodies and to the posterior longitudinal ligament in the lower lumbar and caudal parts of the vertebral canal. The dural tube narrows at the lower border of the second sacral vertebra. It then invests the thin filum terminale, descends till the level of the coccyx and blends with the periosteum.

Histology

Histology of Spinal Dura

Much like the cranial dura, the spinal dura is also a predominantly fibrous structure. Unlike the cranial dura where the fibroblastic cells are scattered and sporadic, the cells are arranged in a sheet in the spinal dura. The spinal dura can be described to be in three strata: an outer stratum that has fibres loosely arranged; a middle stratum where the fibres are closely and densely arranged and an inner cellular stratum. Fibroblastic cells occur in the inner stratum and have interlacing extensions. There are no cell junctions.

Histology *contd...*

Whatever be the arrangement, the fibres of spinal dura are longitudinally disposed. Most of the fibres are collagenous; the collagen fibres run in densely packed bundles. The middle stratum is exclusively made up of these collagen bundles. Elastin fibres are seen in addition to the collagen bundles in the outer stratum. The longitudinal disposition of the fibres help in the absorption of the longitudinal shear and tension created in the vertebral column during various body movements. In the cervical portion where transverse shearing forces are likely to build up during movements of head, some of the fibre bundles of the spinal dura are also transversely disposed.

The spinal nerve roots and the nerves traverse the lateral portion of the vertebral canal to emerge out of the intervertebral foramina. Each nerve root as it comes out of the spinal segment drags a layer of pia closely applied to itself (Fig. 7.2). This pial coat extends till the point where the ventral and dorsal roots unite to form a spinal nerve. A sleeve of arachnoid is also drawn along each nerve root (which means, the subarachnoid space is also prolonged so). At the level of fusion of the roots and the commencement of the spinal nerve, the two layers of leptomeninges (the closely applied pia and the arachnoid that is separated from the pia by the subarachnoid space) merge with each other and with the perineurium of the nerve. The dura mater extends as a tubular prolongation (called the ***root sheath***, nerve sheath or the ***radicular dural tube***) around the nerve roots and nerves. However, it is a single prolongation around both the ventral and dorsal nerve roots which unite with each other within the dural

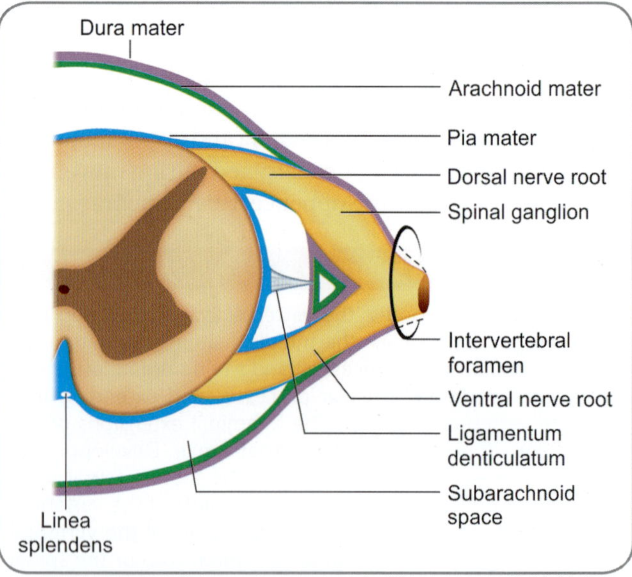

Fig. 7.2: Transverse section through spinal cord to show formation of meningeal sheaths over the roots of a spinal nerve

contd...

sheath immediately distal to the dorsal root ganglion. The dural sheath merges with the epineurium a little beyond the intervertebral foramen. Almost at this point the united spinal nerve gives out its ventral and dorsal rami. The dural prolongations of the upper spinal nerves are shorter; as the nerves become more and more oblique at lower levels, the dural prolongations lengthen accordingly.

Nerve Supply

The spinal dura is supplied by the recurrent meningeal branches of the spinal nerves. As each spinal nerve is formed by the union of its roots, two to four slender branches are given out from the mixed spinal nerve (the only branches from the mixed spinal nerve directly before they divide into ventral and dorsal rami), as the latter lies in the intervertebral foramen. These branches travel back into the vertebral canal through the intervertebral foramen (and hence the name, recurrent). Inside the vertebral canal they divide into ascending, transverse and descending branches to supply the lining tissues of the vertebral canal, ligamenta flava, spinal dura and intervertebral discs.

Added Information

- ***Dural ligaments***: The topographic attachments of the spinal dura to its surrounding attachments are called the dural ligaments. Many of these dural ligaments have been identified and given specific names. The fibrous bands connecting the spinal dura to the margin of foramen magnum (occipitodural ligament), to the adjacent parts of the atlanto-occipital membranes (atlantodural ligaments), to the posterior surfaces of C2 and C3 vertebral bodies (vertebrodural ligaments) are collectively called the ***cranial ligament of spinal dura mater.*** Attachments between the spinal dura and the posterior longitudinal ligament may be present in irregular pieces anywhere from the lower thoracic to the coccygeal end of the vertebral canal. However, these are often found only from the sacral level. The fibrous band connecting the spinal dura to the sacral part of posterior longitudinal ligament in the midline is called the anterior sacrodural ligament (or the ligament of Trolard). At every sacral segmental level, two fibrous connections, one on either side of the midline, between the spinal dura and the posterior longitudinal ligament can be made out. These are called the ***Hoffman's ligaments*** or the ***lateral sacrodural ligaments***. Rarely, Hoffman's ligaments may be found at the lumbar levels too. The anterior sacrodural and the Hoffman's ligaments are collectively called the meningeovertebral ligaments.
- The radicular dural tube is connected to the periosteum of the intervertebral foramen by fibrous tissue extending from its external surface to the bone. These irregular and intermittently placed fibrous connections are variedly called the transforaminal dural ligaments or the Opercula of Forestier. Thin fibrous bands may connect the radicular dural tube to the vertebral bodies. If present, these bands are called the Hoffman's lateral ligaments or the dorsolateral dural ligaments.

contd...

Added Information *contd...*

- These various ligaments, in the adult, act as lines of force transmission. Tensile and shearing forces are transmitted from the duramater spinalis to the duramater encephalii and periorbital structures. Balance of these forces is atleast partially responsible for maintenance of craniovertebral concordance and posture.
- The original role played by the various dural ligaments is embryologically and developmentally significant. The developing dural tube is held open and kept pressed against the surrounding structures (which eventually develop into the vertebral canal and tissues that maintain the continuity of the canal) by connective tissue masses in the epidural space, so that development of the underlying spinal cord is not constricted. As the foetus advances in age, this connective tissue mass reduces and is left out only in certain places forming the so called dural ligaments. The meningesvertebral ligaments gain more importance in growing children in whom the spinal cord is yet to reach its adult levels. The ligaments anchor and hold the dura caudally facilitating spinal cord growth and repositioning.

Spinal Arachnoid Mater

The arachnoid mater that surrounds the spinal cord is continuous above with the cranial arachnoid. An arachnoid sleeve gets prolonged along the nerves and vessels which traverse the subarachnoid space. The disposition of the leptomeninges around the spinal nerve root and spinal nerve and their merger with the perineurium is already explained (under spinal dura). The merging of the leptomeninges with the perineurium closes the subarachnoid space and prevents passage of any particulate matter into the nerve. As in the case of the cranial arachnoid, the spinal arachnoid is held against the spinal dura by the pressure of CSF in the subarachnoid space.

Spinal Pia Mater

The spinal pia closely invests the spinal cord as the cranial pia does so over the brain. It passes into the anterior median fissure and coats the surface of the spinal cord. It is coextensive with the spinal cord that ends at the level of the lower part of the first lumbar vertebra. At this level where the conus medullaris of the spinal cord is continuous with the filum terminale, the pia merges with the connective tissue of the filum and continues as a coating to it for some distance.

Histology

Histology of Spinal Pia Mater and Arachnoid

Both the pia and arachnoid are made up of leptomeningeal cells which are polygonal and nucleated. Pia is usually a leptomeningeal sheet that is two or three cells thick and arachnoid a sheet that is four or five cells thick. The intermediate arachnoid is thin and may have layers that are one or two cells thick. Collagen bundles are seen interspersed between the leptomeningeal cells on the neural surface of pia and in the parietal arachnoid.

Development

Development of Spinal Meninges

The process of development of the spinal meninges is the same as that of the cranial meninges. The ectodermal-mesodermal tissue surrounding the developing spinal cord condenses into an inner endomeninx and an outer ectomeninx. The endomeninx has neural crest cells contributing to its further specialisation and the ectomeninx has ectodermal contribution doing so. The endomeninx develops into the pia-arachnoid leptomeninges and the ectomeninx develops into the dura mater.

Ligamentum Denticulatum (Fig. 7.3)

Running longitudinally along each lateral margin of the spinal cord, there is a flat (anteroposteriorly flattened) sheet of pia mater that projects laterally. This sheet is the ligamentum denticulatum. Its medial border is connected to the spinal cord and its lateral border has a series of triangular projections (denticulate projections, which give the structure its name). The ligament denticulata anchor the spinal cord in position.

Linea Splendens

This term is applied to a narrow thickening of pia mater present over the anterior median line of the spinal cord.

Subpial Space and Ligamentum Denticulatum

The thin space between the pia mater and the neural tissue of the spinal cord is the subpial space. The spinal subpial collagenous layer is thicker than the glia limitans of the brain and so the subpial space is practically nonexistent. The subpial connective tissue and a double fold of pia are pulled out from each lateral margin of the spinal cord. This pulled out sheet is the ligamentum denticulatum. Therefore, the ligament has a central core of connective tissue covered by piamater. As already seen, the lateral edge of the ligament is thrown into a series of dentations or processes; the apices of these processes

contd...

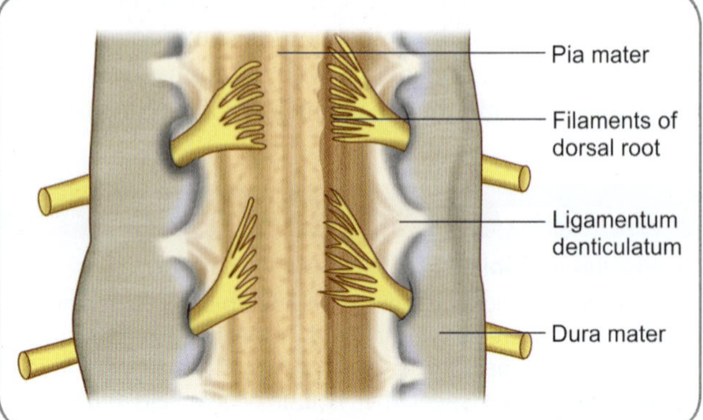

Fig. 7.3: Spinal pia mater seen from posterior aspect – Note the extension of pia mater as ligamenta denticulata attached to the dura mater

contd...

pass through the arachnoid and are attached at intervals to the duramater. Twenty one such dentations occur on each side. The first of these dentations is attached to the dura just above the rim of the foramen magnum. The last of these is between the 12th thoracic and first lumbar spinal nerves and passes obliquely down from the conus medullaris to the dura.

SPACES OF SPINAL MENINGES

- *Epidural space:* The spinal epidural space is the space between the spinal dura and the tissues lining the vertebral canal. It is a natural and normally existent space; it is occupied by loose connective tissue, epidural fat, internal vertebral venous plexus, few tiny arterial twigs, few lymphatics and the dural ligaments that connect the dura to the vertebral lining tissues. It extends from the foramen magnum to the coccyx and laterally ends at the intervertebral foramina. This epidural space is not in continuity with the cranial epidural space which actually is an unnatural space coming into existence only when there is pathology. The cranial epidural space is external to the periosteum while the spinal epidural space is internal to the periosteum. The epidural fat and epidural connective tissue permit displacement of the dural sac during movements and venous engorgement. This factor is of importance in providing protection to the spinal cord and preventing pressure on it.
- *Subdural space:* The spinal subdural space is only a potential space normally because the dura and arachnoid are in close approximation. Epidural injections may sometimes pierce into the subdural space resulting in compression of the spinal cord or toxic influence on it. Bleeding at the dura-arachnoid interface may create expansion of the potential space resulting in a subdural haematoma.
- *Subarachnoid space:* The spinal subarachnoid space is continuous with the cranial subarachnoid space and contains the cerebrospinal fluid. The spinal subarachnoid space is part of the cerebrospinal fluid circulatory system.

Dissection

With the help of senior colleagues and teachers, observe a dissected specimen of the vertebral canal in the cadaver. Try to locate the various structures. Take an isolated separate specimen of spinal cord with its coverings. Look out for individual layers and study their features.

The spinal dura mater forms a loose tubular covering for the spinal cord. It extends downwards upto the level of the lower border of the second sacral vertebra. The arachnoid mater also extends to the same level. The pia mater is coextensive with the spinal cord that ends at the level

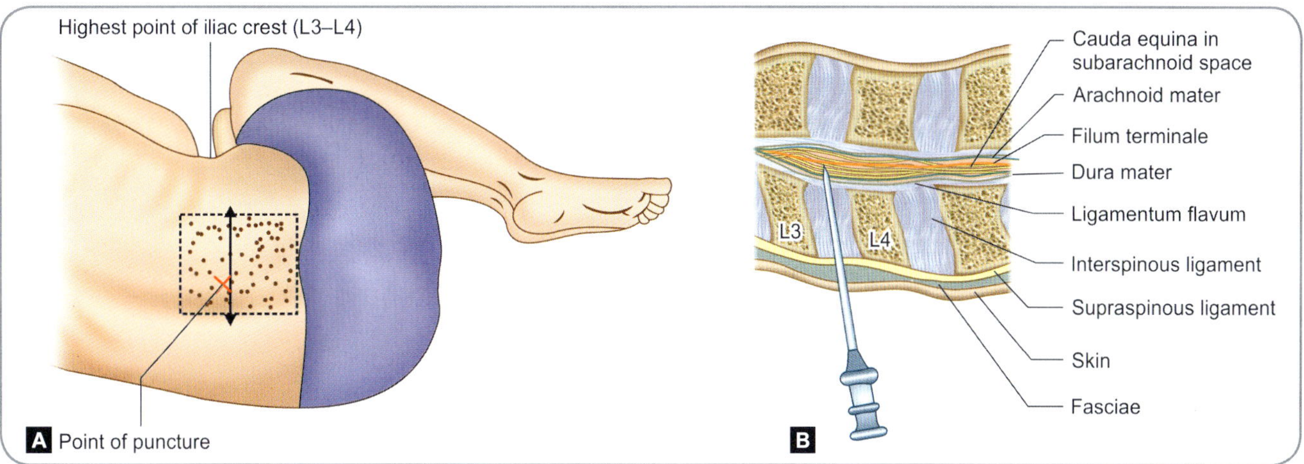

Figs 7.4A and B: A. The site of lumbar puncture **B.** Anatomical layers pierced to reach the subarachnoid space

of the lower part of the first lumbar vertebra. Therefore, opposite to the vertebrae L2 to S2, the vertebral canal contains cerebrospinal fluid (CSF) in the subarachnoid space, but not the spinal cord. A needle can be introduced into the subarachnoid space (to withdraw CSF or to inject substances) without danger of damage to the spinal cord. For this procedure, called the **lumbar puncture**, the needle is most often introduced through the interval between vertebrae L3 and L4 (Fig. 7.4). The part of the vertebral canal below the level of the spinal cord contains several roots of spinal nerves that collectively form the **cauda equina** (Fig. 7.5). These nerve roots are not injured during lumbar puncture as they are easily pushed aside by the needle.

Added Information

Specialised area of spinal arachnoid: The two layers of leptomeninges, pia and arachnoid, coat the spinal cord. In addition, there is a specialised part of arachnoid that is also found. This is the intermediate layer of arachnoid. It can be described as the spinal counterpart of the cranial arachnoid trabeculae. The intermediate arachnoid is in the form of a lacy and perforated network of leptomeningeal cell layers. In certain places it is more compact and condensed. On the dorsal aspect, these compactions are present in the three longitudinal lines in the subarachnoid space: one in the midline and two in the right and left dorsolateral lines. These compactions do not form continuous sheets and are, here and there, fenestrated and deficient. They connect the pia and the internal aspect of arachnoid. A discontinuous series of dorsal ligaments (in the midline) and two sets of dorsolateral ligaments are thus formed. On the ventral aspect, the intermediate arachnoid is less compact and looser. No ligament formation can be seen. On the lateral aspects of the spinal cord, there is no intermediate arachnoid. Histologically, the intermediate arachnoid is the same as the arachnoid trabeculae. A central core of collagen is clothed on either aspect by one (or rarely, two) layer of leptomeningeal cells. The presence of intermediate arachnoid and the histological similarity of pial and arachnoidal cells have given place to modified terms of nomenclature with respect to the spinal meninges.

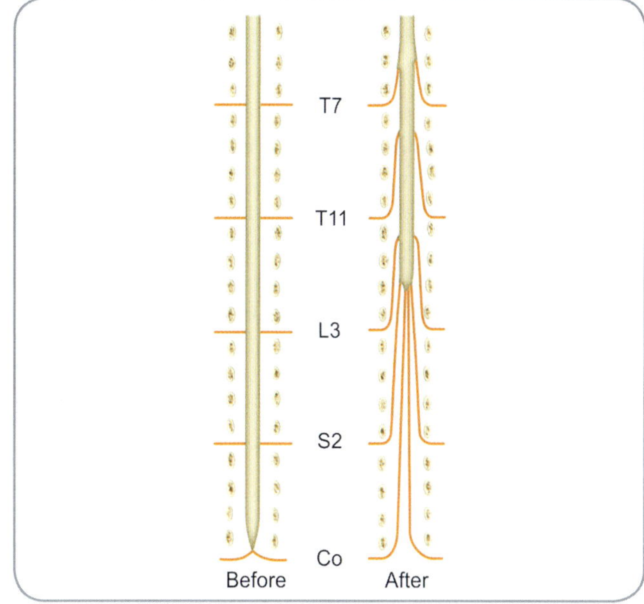

Fig. 7.5: Scheme to show the effect of recession of the spinal cord on course and length of the roots of spinal nerves

Added Information *contd...*

A comparative account of old and current nomenclature is as follows:

Old name	Anatomical entity	Explanatory usage	Current name
Pia mater	Pia mater	Inner pia	Visceral leptomeninx
Arachnoid	Arachnoid that lines the dura	Outer arachnoid	Parietal leptomeninx or parietal arachnoid
	Intermediate arachnoid (strands of lacy arachnoid in subarachnoid space)	Intermediate arachnoid	Intermediate leptomeninx

contd...

Clinical Correlation

Both the spinal subarachnoid and epidural spaces have been used for anaesthetic procedures. Epidural anesthesia is a preferred procedure for child birth.

VERTEBRAL VENOUS PLEXUSES (FIG. 7.6)

- Each vertebra is drained by basivertebral and intervertebral veins. Basivertebral veins emerge out of the vertebral bodies through the posterior foramina and reach the anterior part of epidural space. Intervertebral veins drain the spinal cord, spinal nerves and vertebrae and are found in the intervertebral foramina.
- These veins drain into the external and internal vertebral venous plexuses.

External Vertebral Venous Plexuses

These are two in number—the anterior and the posterior. The anterior venous plexus is a chain of interconnected veins lying anterior to the vertebral bodies. It communicates with the basivertebral and intervertebral veins and receives tributaries from the vertebral bodies. The posterior venous plexus is a chain of interconnected veins lying posterior to the laminae and articular processes of the vertebrae. They communicate with the anterior external and the internal venous plexuses.

Internal Vertebral Venous Plexuses

The internal plexuses are in four longitudinal chains, two anterior and two posterior. All four of them lie in the epidural space. The anterior chains lie on either side of the posterior longitudinal ligament and receive the basivertebral veins. The posterior chains lie in front of the vertebral arch.

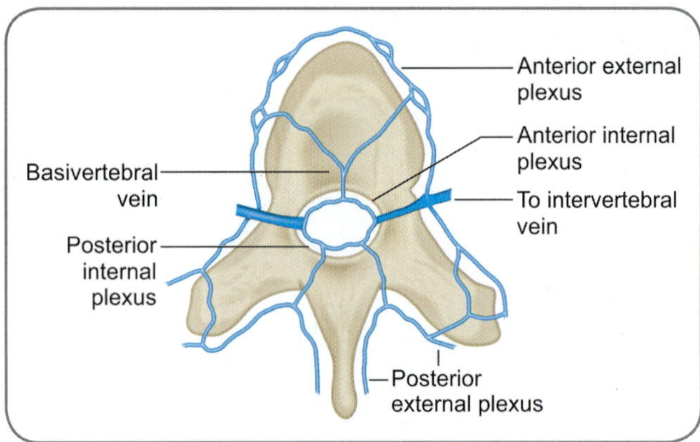

Fig. 7.6: Vertebral venous plexus

Both the venous plexuses and the surrounding veins anastomose freely. These veins and plexuses have no valves. They ultimately drain into the caval or the ascending lumbar veins/venous systems.

Clinical Correlation

- The absence of venous valves in the vertebral venous plexuses permits the veins of this system to engorge and form channels of collateral circulation. In patients with venous obstruction in the thorax and neck, these can form alternate routes of venous return.
- The absence of valves allows sepsis and malignancies to spread easily through these veins.
- The direction of blood flow in these veins can easily be reversed. Such reversal happens when there is increased intra-abdominal pressure or postural variations. When there is such a reversal, pelvic malignancies (commonest to do so being prostatic cancer) may spread to vertebral bodies and even to the skull through connections of the vertebral venous plexuses with the dural venous sinuses.

Multiple Choice Questions

1. The radicular dural sheath of a spinal nerve merges with:
   a. The epineurium of the nerve
   b. The perineurium of the nerve
   c. The arachnoid of the sheath
   d. The pia-arachnoid interface
2. The spinal epidural space:
   a. Is a potential space
   b. Is a pathological space
   c. Is filled with epidural fat
   d. Is devoid of connective tissue
3. Reversal of blood flow can happen in the vertebral venous plexuses because:
   a. They are connected to the dural venous sinuses
   b. They are broad and engorged
   c. They have no valves
   d. They are widely dilated by bony attachments
4. In a lumbar puncture, cerebrospinal fluid is attempted to be drawn from:
   a. The epidural space
   b. The subdural space
   c. The subarachnoid space
   d. The subpial space
5. A basivertebral vein emerges out:
   a. On the anterior aspect of the vertebral body
   b. On the posterior aspect of the vertebral body
   c. Into the intervertebral canal
   d. On the laminae of the vertebra

ANSWERS

1. a   **2.** c   **3.** c   **4.** c   **5.** b

Clinical Problem-solving

Case Study 1: An 84-year-old man had secondaries of malignancy in the vertebrae. He had been having difficulty in micturition and frequency of micturition for some years.
- What would be your opinion about the vertebral secondaries?
- Why should the vertebral bodies be involved when there is no direct connection between the primary site of malignancy and them?
- Explain the anatomical basis.

Case Study 2: A 53-year-old man had degenerative disease of his lumbar intervertebral discs.
- What possible consequences do you anticipate in this individual?
- From your anatomy knowledge, would you be able to say in which region would he have pain?

(For solutions see Appendix).

Muscles and Triangles of the Neck

NECK

Neck is that region of the body that extends from the base of cranium and inferior border of mandible to the superior thoracic aperture. It contains four important compartments, namely the midline visceral compartment, the midline musculoskeletal compartment and the right and the left neurovascular compartments.

The neck can be defined as a transition and conducting zone between the head and the rest of the body. As a result of the latter, the neck serves as a conduit for several important structures. In addition, it also houses some of the important viscera of the endocrine, respiratory and digestive systems. The neck necessarily remains slender so as to position the head to the best advantage. This factor is primarily responsible for the efficacy of sense organs like the eyes, ears and mouth. However, as much as the neck increases advantage to the head, it suffers its slender disposition. The structures occupying the neck are crowded; as there is no bony covering or cage, these structures also are prone to injury and damage.

FASCIAE OF NECK

Superficial Fascia

The superficial fascia of the neck is thin and has no specific features. On the anterolateral aspects on both sides, it contains the platysma. In many individuals, especially females, the superficial fascia houses considerable quantities of fat tissue. Cutaneous nerves, blood vessels and lymphatics traverse it as in other parts of the body.

Platysma (Fig. 8.1)

Other names: Musculus platysma myoides, musculus subcutaneous colli, musculus tetragonus.

Platysma (Greek.platysma=flat plate) is the most superficial muscle in the neck and lies in the superficial fascia like the muscles of the face. It is a remnant of an extensive sheet of subcutaneous muscle called the ***panniculosus carnosus*** seen in some animals. The muscle arises from the fascia that covers the pectoralis major and deltoid and crosses the clavicle to ascend medially in the neck. Its upper fibres cross

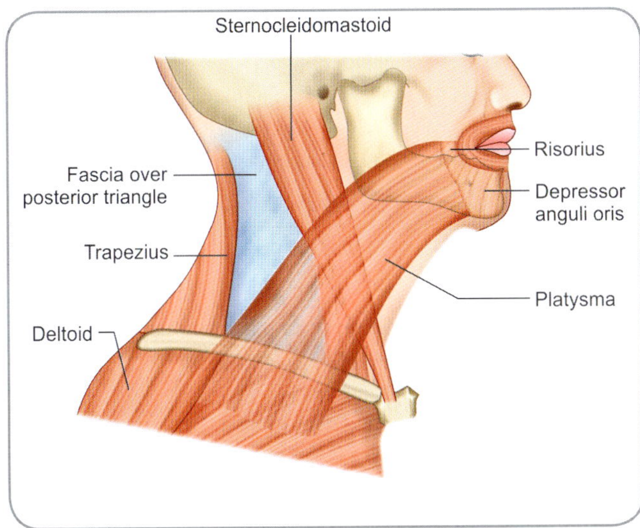

Fig. 8.1: Platysma muscle

the base of mandible and merge with the superficial fascia of the face. Its lower and anterior fibres merge with its fellow of the opposite side in the midline. Contraction of this muscle draws the lower jaw closer to the neck and reduces the angle between the two. The muscle can also pull the angle of the mouth down resulting in the expression of horror, tension and stress. Acting from below, it depresses the mandible and draws the corners of the mouth down; therefore, it is called the '*muscle of grimace*'.

Added Information

- Platysma develops superficial to all structures. The two first bones to ossify are the clavicle and the mandible. Platysma is superficial to these bones too and crosses the entire length of both bones.
- The muscle does not cover the lower part of the anterior triangle of neck but covers the lower part of the posterior triangle.
- Under normal circumstances, the side of the neck is concave. By its contraction, platysma obliterates the concavity and takes off the pressure on the underlying veins.

Clinical Correlation

If the platysma is paralysed, the skin of the neck falls in loose folds. Injury to the cervical branch of the facial nerve during surgeries is the common cause. Care should be taken not to damage the nerve. In addition, while suturing neck incisions or injuries, edges of platysma should be kept in apposition. If not, the healing wound would be pulled in different directions due to contraction of platysma fibres and a disfigured scar occurs.

Deep Cervical Fascia (Fig. 8.2)

As in other parts of the body, all muscles, blood vessels and nerves of the neck are sheathed by connective tissue, which is condensed to form recognisable sheets. These sheets are collectively referred to as deep cervical fascia or fascia colli. The deep fascia of the neck is conventionally divided into an investing layer, a pretracheal layer and a prevertebral layer. Natural cleavage planes occur between these three layers. Spread of abscesses from one zone to another is prevented by them. They also provide various structures of the neck with lubrication to move against each other.

Investing Layer

Other names: External layer of deep cervical fascia, external cervical fascia, lamina superficialis of cervical fascia.

The investing layer (investing=complete covering) of deep cervical fascia wraps around the neck as a collar beneath the superficial fascia; it is posteriorly attached to the ligamentum nuchae and the spine of the seventh cervical vertebra. It extends on either side of the neck and interlaces with the fascia of the opposite side in the midline. It splits to enclose various structures both in its horizontal and vertical dispositions.

- Horizontal disposition of the investing layer of deep cervical fascia: if the investing layer is traced in its horizontal disposition from its posterior attachment, it splits to enclose the trapezius, rejoins to form a single layer that roofs the posterior triangle, splits again to enclose the sternocleidomastoid muscle, then forms the roof of the anterior triangle and finally merges with the fellow of the opposite side in the midline. It also interlaces with the fibres of platysma and the fascia covering the strap muscles in the midline.
  - *Superior attachment:* The fascia is attached superiorly to the external occipital protuberance, periosteum of the superior nuchal line, mastoid process, entire base of the mandible, symphysis menti and the hyoid bone. Between the angle of the mandible and the anterior border of the sternocleidomastoid, it is particularly strong.
  - *Inferior attachment:* Inferiorly, it is attached to the spine of scapula, acromion process, clavicle and manubrium sterni. In its lower attachment it splits twice to enclose two spaces. Above the manubrium sterni, it splits into two; the anterior layer attaches to the anterior border of manubrium and the posterior layer to the posterior border of manubrium sterni and the interclavicular ligament. A space is enclosed between the anterior and posterior layers and is called the **suprasternal space of Burns** (named after the 18th century Scottish anatomist Allan Burns). The space contains the sternal heads of the sternocleidomastoid muscles, lower parts of the anterior jugular veins, jugular venous arch, some fat and a lymph node.

Above the clavicle, near the lower part of the roof of the posterior triangle, the investing layer splits into two;

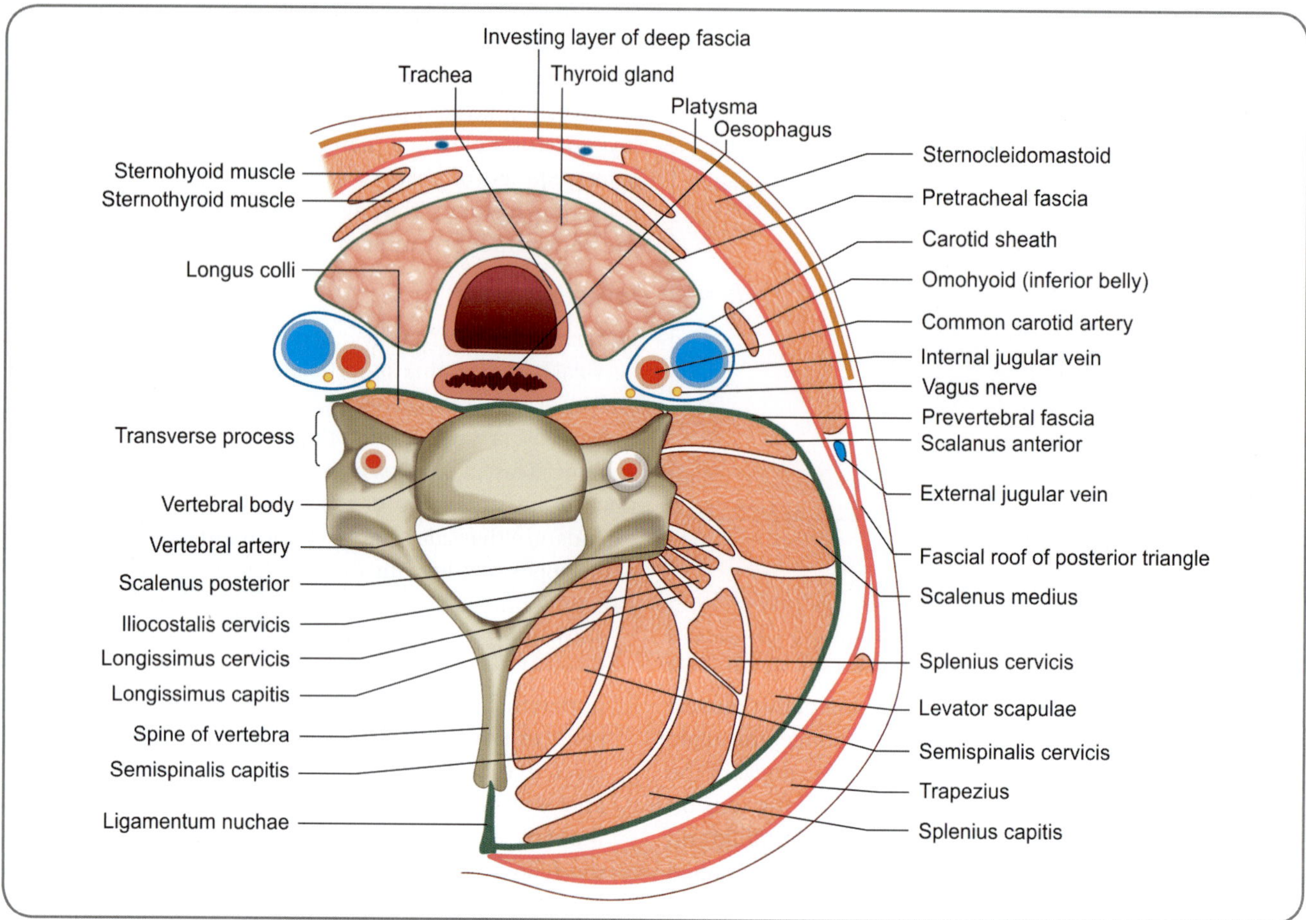

Fig. 8.2: Transverse section through lower part of the neck to show the various muscles and parts of the deep cervical fascia

the anterior (superficial) layer attaches to the anterior aspect of the clavicle; the posterior (deep) layer attaches to the posterior aspect of the clavicle; a space called the ***supraclavicular space*** is thereby enclosed between the two layers. This space contains a part of the external jugular vein which pierces the deep fascia to drain into the subclavian vein and the supraclavicular nerves.

❑ Vertical disposition of the investing layer of deep cervical fascia: If the investing layer is traced upwards in its vertical disposition, it covers the submandibular region, where it splits to enclose the submandibular gland. The superficial layer covering the submandibular gland is attached to the lower border of the mandible and the deep layer to the mylohyoid line of the mandible. Posterior to the submandibular gland, in the interval between the mandible and the mastoid process the investing layer again splits to enclose the parotid gland. The superficial layer covering the superficial surface of the parotid gland is attached above to the zygomatic arch, is thick and forms the ***parotid fascia*** (or the ***parotidomasseteric fascia)***. The deep layer that passes deep to the parotid gland is attached to the tympanic plate. The

thickened part of the deep cervical fascia between the angle of the mandible and the styloid process forms the ***stylomandibular ligament*** which intervenes between the parotid gland and the submandibular gland.

The accessory nerve is related closely to the investing layer in the posterior triangle.

Pretracheal Layer

Other names: Pretracheal fascia, middle cervical fascia, porter's fascia, lamina pretrachealis.

This is a thin layer of deep fascia limited to the anterior part of the neck. It is a coronal extension from the internal surface of the investing layer that encloses the thyroid gland and stretches in front of the trachea.

Superiorly, the pretracheal layer is attached to the hyoid bone. Attachments to the arch of cricoid cartilage and to the oblique line of thyroid cartilage have also been defined. Inferiorly, it passes along the trachea into the superior mediastinum of the thorax and merges with the fibrous pericardium of the heart. Laterally, it forms the anterior layer of the carotid sheath. Between the two lateral limits, the pretracheal layer encloses the infrahyoid muscles and the thyroid gland. The portion that encloses the muscles is called the *muscular part* and that, which encloses the gland is called the *visceral part*. The infrahyoid (strap) muscles overlap the thyroid gland in front. Therefore, the muscular part is anterior to the visceral part. The two parts are in continuity with each other.

The superior attachment of the pretracheal fascia is responsible for the movement of the thyroid gland with deglutition.

Posterosuperiorly, the pretracheal layer is continuous with the buccopharyngeal fascia of the pharynx.

Prevertebral Layer

Other names: Prevertebral fascia, lamina prevertebralis. A coronal extension from the internal aspect of the deep cervical fascia that runs anterior to the prevertebral muscles is called the prevertebral layer of deep cervical fascia. It lies behind the oesophagus and pharynx.

Traced laterally, it forms the posterior layer of the carotid sheath, then passes on the ventral aspects of the scalene muscles and levator scapulae and finally either merges with the layer of fascia that lies deep to the trapezius or passes around the deep muscles of the back to reach the ligamentum nuchae. The part of the fascia over the scalene muscles and levator scapulae forms the fascial carpet covering the floor of the posterior triangle. Superiorly, the prevertebral fascia reaches the base of the skull and inferiorly, it passes into the superior mediastinum of the thorax in front of the upper thoracic vertebrae and merges with the anterior longitudinal ligament at the level of T3/T4 vertebra.

The cervical nerves which emerge from the intervertebral foramina lie deep to the prevertebral fascia. The ***cervical and brachial plexuses*** formed by the union of the ventral rami of these nerves are also deep to this fascia. Though the muscular branches from these plexuses (namely nerve to rhomboids, nerve to serratus anterior, etc.) remain deep to the fascia, the cutaneous nerves pierce it to become superficial. As the brachial plexus and subclavian artery emerge from behind the scalenus anterior to go to the axilla through the cervicoaxillary canal, they drag a part of the prevertebral layer with them. This 'pulled away' extension forms a covering for the brachial plexus and the subclavian artery and is called the axillary sheath.

Anteriorly, the prevertebral layer is separated from the pharynx and the buccopharyngeal fascia by the retropharyngeal space. The cervical parts of the sympathetic trunks lie embedded in this layer.

Carotid Sheath

The condensation of the connective tissue around the great vessels of the neck forming a tubular sheath of fascia is called the *carotid sheath*. The sheath encloses the common carotid artery, internal carotid artery, internal jugular vein and the vagus nerve. The common carotid artery lies within the sheath up to the level of upper border of thyroid cartilage, where it bifurcates into internal and external carotid arteries. The internal carotid artery continues within the sheath in its entire extent, whereas the external carotid artery pierces the carotid sheath immediately and comes out. Similarly the tributaries of the internal jugular vein pierce the sheath to drain into it. The vagus nerve lies within the sheath in the interval between the common (or internal) carotid artery and the internal jugular vein and posterior to them.

The anterior layer of the carotid sheath is formed by the pretracheal layer and the posterior layer by the prevertebral layer of deep cervical fascia. Superiorly it extends to the base of the skull and inferiorly it merges with the fascia covering the arch of aorta. The sympathetic trunk is closely related to the posterior aspect of the sheath and the ansa cervicalis anterior to the carotid sheath.

> **Surface Marking of Carotid Sheath**
>
> Point A is marked on the sternoclavicular joint. Point B is marked midway between the angle of mandible and mastoid process. The broad line that joins the two points will indicate the carotid sheath.

TISSUE SPACES OF NECK

The disposition of the various layers of fascia result in the formation of potential tissue spaces in the neck. These spaces are not 'real' spaces, but are potential. In healthy conditions of life, these spaces are non-existent because they are either filled by connective tissue or their walls are in apposition. Only at times of pathology (like infection or inflammation) the spaces are filled with secretions and noted. The neck does not have any specific tissue barrier. Infections from one region can easily spread to other regions and the tissue spaces resulting in a number of complications.

The tissue spaces of the neck can be broadly classified into two groups—(1) suprahyoid spaces and (2) infrahyoid spaces. The suprahyoid spaces are in continuity with the regions of head (like the infratemporal fossa, submandibular fossa and tonsillar fossa) and infections from the latter can easily track down. It is not possible to confine the suprahyoid spaces either to the neck or to the head (they are discussed together because of this reason).

Suprahyoid Tissue Spaces

The spaces above the hyoid are in the submandibular region and the prevertebral region. They are the spaces around the lower jaw, pharyngeal spaces and the prevertebral space.

- **Spaces around the lower jaw:** Though the lower jaw anatomically belongs to the head, functional continuity assigns its position to be considered from the neck too. The potential spaces around the lower jaw are many; of these, those in direct contact with the regions of the neck are the submental space, submandibular space, sublingual space, peripharyngeal space and the peritonsillar space.

  - **Submental space:** This space is deep to the upper part of the investing layer of the deep cervical fascia and superficial to the mylohyoid muscles. It is in the midline and is bounded on either side by the anterior bellies of the digastric muscles.

  - **Submandibular space:** The submandibular spaces are paired. Each submandibular space is located deep to the concerned half of mandible between the anterior and posterior bellies of the digastric muscle of its side.

  - **Sublingual space:** The sublingual space lies deep to the mylohyoid muscles in the floor of the mouth. It is a paired space that is continuous with its fellow of the opposite side across the midline.

    The sublingual spaces are in contact with the submandibular spaces around the posterior free borders of the mylohyoid muscles. The submandibular spaces are in free communication with the submental space which is anatomically a cervical tissue space.

  - **Retropharyngeal space:** This is the most important of the tissue spaces of the neck. It is a potential space between the buccopharyngeal fascia of the pharynx (anteriorly) and the prevertebral layer (posteriorly). Laterally, on each side, the carotid sheath closes the space and superiorly the cranial base closes it. Normally, it contains some loose connective tissue. It permits smooth movements of the various viscera of the neck during deglutition and turning of head. It communicates with the peripharyngeal space. The retropharyngeal space is subdivided anteroposteriorly by a thin fascial layer called the alar fascia. This fascia extends from the midline of the buccopharyngeal fascia to the carotid sheath and from the skull base to the level of C7 vertebra. The alar fascia is very thin that during infections, it is not possible to discern its presence.

  - **Peripharyngeal space:** The peripharyngeal space (sometimes called the parapharyngeal space) is a paired space located one on either side of the posterolateral aspect of the pharynx. It is bounded medially by the superior constrictor of pharynx and laterally by the medial pterygoid. It communicates with the retropharyngeal space.

  - **Peritonsillar space:** The peritonsillar space is actually a part of the intrapharyngeal space. It is located around the palatine tonsils between the pillars of fauces. Bounded by the medial surface of the superior constrictor, this space is in continuity with the rest of the pharynx and the retropharyngeal space.

Infrahyoid Spaces

The infrahyoid spaces are the pretracheal space, retrovisceral space, prevertebral space and the carotid space.

- **Pretracheal space:** The pretracheal space (should actually be labelled the paratracheal space) is located behind the pretracheal fascia and the infrahyoid muscles and in front of the anterior wall of oesophagus. It surrounds the trachea. It is closed superiorly by the attachments of the infrahyoid muscles to the thyroid cartilage. Inferiorly the space is open into the superior mediastinum.

- **Retrovisceral space:** The retrovisceral space is located between the posterior wall of oesophagus and the prevertebral fascia. It is continuous superiorly with the retropharyngeal space and inferiorly extends into the superior mediastinum.

- **Prevertebral space:** The prevertebral space is located between the prevertebral fascia and the vertebral column. The prevertebral muscles are closely packed and so this space cannot extend laterally. *Superiorly*, it is closed by the base of skull. *Inferiorly*, it is continuous with the posterior mediastinum. *Anteriorly*, though the prevertebral fascia is present, it is weakest wall when compared with the rest of them especially at times of infection.

- **Carotid space:** The carotid space is the space around the contents of the carotid sheath demarcated by parts of the investing, pretracheal and prevertebral layers. Above and below it is closed because of the adhesion of the fascia to the adventitia of the vessels.

Added Information

- The pretracheal and prevertebral layers were previously described as coronal partitions which just had a single stratum each. However, studies have revealed that the pretracheal layer splits, encloses and rejoins around muscles and thyroid gland. Since the muscles and the gland are disposed anteroposteriorly, the split layers are not separate, but are attached to each other variedly. It can easily be said that the pretracheal layer forms a fasciomusculoglandular column with its enclosed structures.
- Similarly, the prevertebral layer also forms a fascio-osteomuscular column by surrounding the prevertebral muscles, paravertebral muscles and the vertebral column.

contd…

Added Information *contd...*

The fascia is not discernible as a single layer from where it passes under cover of trapezius. From here, it is seen as a mesh of loose areolar tissue. Its anterior part (the part where it passes in front of the prevertebral muscles) is well defined and so the name 'prevertebral' and the description of a single stratum partition.

❑ With the complete description of all the layers and parts of the deep cervical fascia, it can be understood that the investing layer forms a fascial column within which are located four other columns: the anterior fasciomusculoglandular column of the pretracheal fascia, the posterior fascio-osteomuscular column of the prevertebral fascia and two fascioneurovascular columns of the carotid sheaths.

❑ From the anterolateral aspects of the infrahyoid muscles, an extension of the pretracheal fascia passes over the intermediate tendon of the omohyoid. This extension helps tethering the muscle in position. Above the level of the hyoid, another extension of the pretracheal fascia passes up to form a pulley for the intermediate tendon of the digastric muscle.

Clinical Correlation

❑ The arrangement of fascial layers in the neck determines the direction of spread of infections in and around the neck. The investing layer tends to prevent spread of purulent infections from the superficial aspects of the neck to deep areas and thorax. Similarly, an infection in the region clothed by the muscular part of the pretracheal fascia does not usually spread to the thorax, but an infection in the region clothed by the visceral part spreads down easily.

❑ In tuberculosis of the cervical vertebrae, the pus formed collects between the vertebral column and the prevertebral fascia and produces a midline swelling in the posterior wall of the pharynx. The swelling is also referred to as a ***chronic retropharyngeal abscess*** because of the chronic nature of tubercular infection. The pus from such an abscess can further track down in front of the prevertebral muscles and can appear under the skin of the posterior triangle of the neck or track down to the axilla along the axillary sheath.

❑ The deep cervical lymph nodes are also often sites of tubercular infection. The pus formed may pierce through a small area of deep fascia and form a swelling under the skin. This results in two collections of pus namely, one deep to the deep fascia and another superficial to the fascia and the two are in communication through a narrow opening. Such an abscess is called ***collar stud abscess.***

❑ The retropharyngeal lymph nodes are located between the prevertebral fascia behind and the buccopharyngeal fascia in front. When these nodes get infected, an ***acute retropharyngeal abscess*** is formed. Unlike a chronic abscess that is situated in the midline, an acute abscess lies to one side of the midline, because the prevertebral fascia and buccopharyngeal fascia are adherent to each other along the midline. Pus can pass downwards behind the pharynx and oesophagus to reach the superior mediastinum or even the posterior mediastinum.

Clinical Correlation *contd...*

❑ An infection in the submandibular region is limited to a triangular area bounded by the two halves of the mandible and hyoid bone posteriorly because of the attachments of the investing layer of deep fascia to these bones. Abscess in this region produces a triangular swelling called ***Ludwig's angina*** which may also push the tongue upwards.

❑ The fascia covering the parotid gland (parotid fascia) is very dense. Therefore, in infections of the parotid gland, there is no space for expansion. This results in severe pain.

MUSCLES OF NECK (TABLE 8.1)

Muscles of the neck can be described in columns and layers within these columns.

❑ The outer or the superficial most column (called the subcutaneous column) is formed by the platysma which is a subcutaneous muscle. However, this muscle is present only in the anterior and lateral aspects of the neck and so, this outermost column does not completely surround the neck.

❑ The next column comprises the sternocleidomastoid and trapezius muscles which occupy the investing layer of deep fascia. This column is complete since it surrounds the neck completely; however, this is a fasciomuscular column (also called the ***deep fascial cylinder***) because the deep fascia forms a complete cylinder and the muscles are only a part of this cylinder. The sternocleidomastoid muscle is considered a key muscle because it divides the neck into anterior and lateral regions (anterior and posterior triangles respectively).

Within the subcutaneous and the fasciomuscular columns are located two other columns of muscles. These two columns do not occupy the entire diameter of the neck. They are of smaller diameters (than the diameter of the entire neck) and are present one behind the other within the deep fascial cylinder.

❑ The anterior column is of smaller dimensions (than the posterior column), lies in front of the aerodigestive viscera of the neck, is flattened and runs from the tongue to the sternum. Presence of the hyoid bone subdivides this column into a suprahyoid and an infrahyoid part. The infrahyoid part comprises the infrahyoid muscles which are actually part of the fasciomusculoglandular column of the pretracheal fascia.

❑ The posterior column is the musculoskeletal column (the fascio-osteomuscular column of the prevertebral fascia) and comprises muscles which are located around the vertebral column.
  - They are in three groups—(1) anterior vertebral group, (2) lateral vertebral group and the (3) posterior vertebral group.

contd...

Table 8.1: Groups of muscles of neck (Fig. 8.3)

Column	Position			Muscles
Outermost column/cylinder of neck	Subcutaneous			Platysma
Fasciomuscular column (within deep fascia)	Anterior			Sternocleidomastoid
	Posterior			Trapezius Levator scapulae
Anterior muscular column	Suprahyoid			Mylohyoid Geniohyoid Stylohyoid Digastric
	Infrahyoid			Sternohyoid Sternothyroid Thyrohyoid Omohyoid
Muscles of the musculoskeletal column (posterior column)	Prevertebral muscles	Anterior to the vertebral column—anterior vertebral		Rectus capitis anterior Rectus capitis lateralis Longus colli Longus capitis
		Lateral to the vertebral column—lateral vertebral		Scalenus anterior Scalenus medius Scalenus posterior Scalenus minimus
	Posterior to the vertebral column—posterior vertebral	Extrinsic muscles of back		Immigrant muscles—muscles of upper limb which have migrated to the back region; superficial respiratory muscles
		Intrinsic muscles of back	Superficial group	Splenius capitis Splenius cervicis Erector spinae
			Deep group	Transverso-spinalis Suboccipital
			Deepest group	Interspinal and Intertransverse muscles

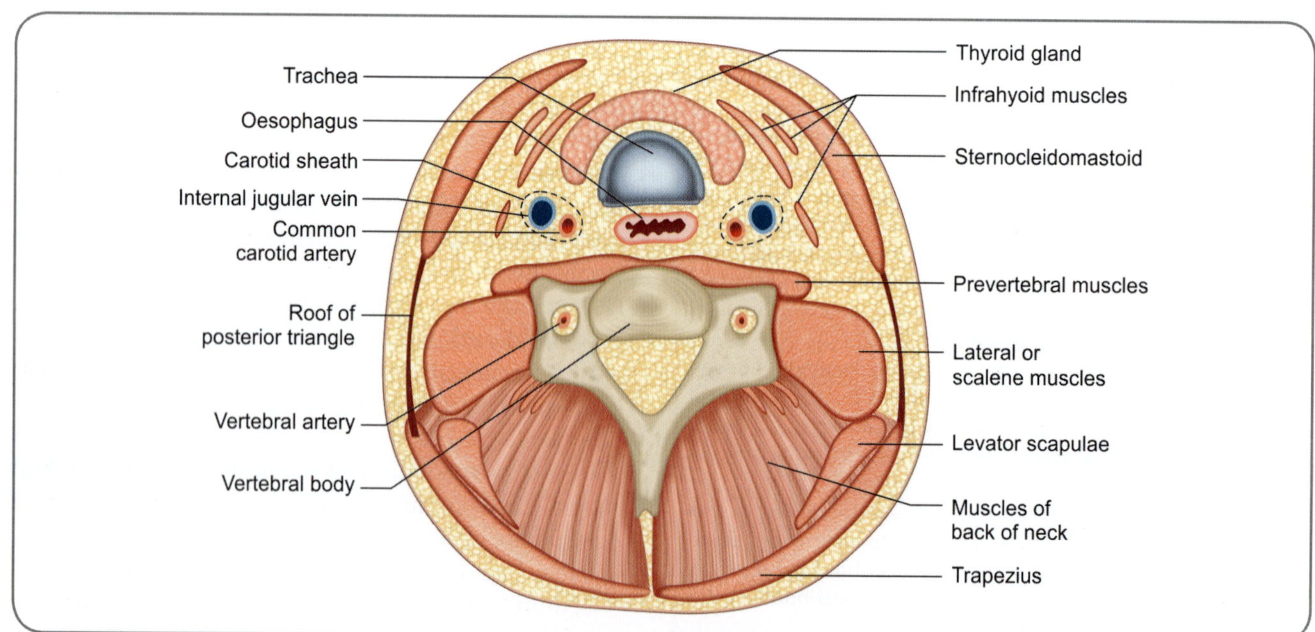

Fig. 8.3: Schematic transverse section through the lower part of the neck—relative position of various groups of muscles are shown

○ The posterior vertebral group is composed of true muscles of the back (called the intrinsic muscles of the back) and the muscles which have migrated from the upper limb to keep the upper limb attached to the trunk (called the extrinsic muscles of the back). The extrinsic muscles lie posterior and superficial to the intrinsic muscles and virtually clothe them. The intrinsic muscles of the back are further subdivided into superficial, deep and deepest groups. The superficial group consists of the splenius muscles in the neck and the erector spinae throughout the length of the trunk. The deep group consists of the suboccipital muscles in the head-neck junction and the transversospinal muscles in the cervical and thoracic regions. The deepest group consists of the interspinal and intertransverse muscles.

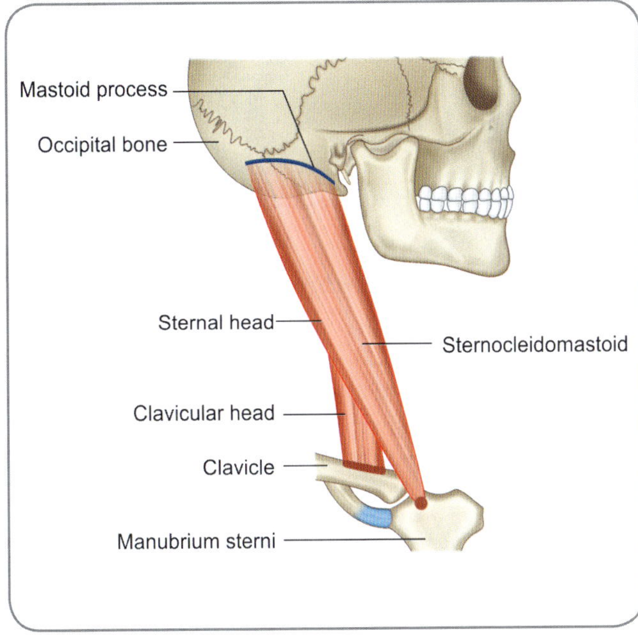

Fig. 8.4: Scheme to show the attachments of sternocleidomastoid muscle

Sternocleidomastoid and Trapezius

Sternocleidomastoid (Fig. 8.4) and trapezius form the muscles of the fasciomuscular column of the investing layer of deep cervical fascia. The investing layer of the deep cervical fascia splits and encloses these two muscles; sternocleidomastoid is placed anteriorly and trapezius posteriorly. Trapezius is a muscle of the upper limb (though part of it lies in the neck) and has been described in the section on upper limb. Sternocleidomastoid is described in Table 8.2.

Additional Notes on Sternocleidomastoid

As already noted, it is a key muscle of the neck dividing the latter into the anterior and posterior triangles. It arises by two heads; the **sternal head** is medial and rounded; the **clavicular head** is lateral and band like. At their attachments, the two heads are separate; the interval between the two can be marked on the surface by a shallow depression called the *lesser supraclavicular fossa* above the medial end of clavicle. The clavicular head spirals behind the sternal head near the middle of the neck and blends with its deep surface. The muscle is related to almost all the important structures in the neck. Superficially it is related to the parotid gland, superficial branches of the cervical plexus namely, the lesser occipital, great auricular, transverse cervical and medial supraclavicular nerves, external jugular vein and the platysma. A number of structures are present deep to the sternocleidomastoid. They can be grouped as muscles, nerves and vessels. The muscles which lie deep are the posterior belly of digastric, splenius capitis, levator scapulae, scalenus anterior, both superior and inferior bellies of omohyoid, sternohyoid and sternothyroid. The nerves which lie deep to the sternocleidomastoid are the spinal part of accessory nerve, roots of cervical and brachial plexuses, ansa cervicalis, phrenic nerve and vagus nerve. The arteries lying deep are the common carotid and its two divisions of external and internal carotid arteries, sternomastoid

Table 8.2: Platysma and sternocleidomastoid				
Muscle	**Origin**	**Insertion**	**Action**	**Nerve supply**
Platysma	Deep fascia over upper part of pectoralis major and anterior part of deltoid muscle	• Most anterior fibres interlace with those of the opposite side. • Other fibres cross the lower border of mandible and merge with muscles at the angle of mouth and continue as risorius.	• Produces wrinkles over skin of neck • Reduces concavity of the neck thereby releasing pressure from the vessels in the neck	Cervical branch of facial nerve
Sternocleido-mastoid	• Sternal head from anterior surface of manubrium sterni • Clavicular head from upper surface of medial part of clavicle	• Lateral half of superior nuchal line • Lateral surface of mastoid process from apex to upper border	• When muscle of one side contracts, the head is tilted to the same side and chin is rotated to the opposite side • When muscles of both sides contract the head and neck are flexed	• Spinal part of Accessory nerve • Ventral rami of spinal nerves C2, C3

branch of superior thyroid artery, transverse cervical, suprascapular and occipital arteries. The veins lying deep are the internal jugular, lingual and common facial veins. The sternoclavicular joint and the deep cervical lymph nodes also lie deep to the sternocleidomastoid muscle.

Making the Muscle a Surface Landmark

The head is rotated to one side and tilted forwards, the opposite muscle stands out prominently. The anterior border runs down from the mastoid process to the sternoclavicular joint. The posterior border runs down from the midpoint between inion and mastoid to the junction of medial and middle thirds of clavicle. These borders can be used as reference lines for various other surface marking.

Added Information

When the head and neck are fixed, bilateral contraction of sternocleidomastoid causes raising up of the clavicle and manubrium resulting in elevation of the upper thorax. This way, the muscle acts as an accessory muscle of respiration.

Clinical Correlation

- Testing the sternocleidomastoid muscle. The muscle is tested by turning the head to the opposite side against resistance. If the muscle is normal it can be palpated and seen.
- Branchial cysts usually present at the junction of the upper and middle thirds of the anterior border of the muscle. Branchial fistula, if present, appears as a small pit near the anterior border of the muscle in the lower third. These are conditions of anomalous development of the pharyngeal arches and pouches.

contd…

Clinical Correlation *contd…*

- ***Torticollis*** or ***wry neck***: This is a condition (Latin. tortus=twisted, collum=neck) where there is twisting of the neck and slanting of the head. Contraction or shortening of the neck muscles leads to the condition. A fibrous tumour of the sternocleidomastoid that develops just before or after birth causes unilateral fibrous contracture of the muscle leading to tilting of the face and head. Such condition is referred to as congenital wry neck or ***congenital torticollis***. In ***muscular torticollis***, the muscle is injured during child birth. The resultant haematoma gets converted into a fibrotic mass that entraps the accessory nerve and paralyses the muscle. Fibrosis and shortening of the muscle fibres may also occur. ***Intermittent torticollis*** occurs when there are spasmodic intermittent contractions of the muscle.

Infrahyoid Muscles

These are:
- Sternohyoid,
- Sternothyroid,
- Thyrohyoid and
- Omohyoid.

The details of these muscles are given in Table 8.3 and shown in Figure 8.5.

Additional Notes on Infrahyoid Muscles

The infrahyoid muscles are so organised that two of them, namely sternohyoid and omohyoid, lie superficial to the other two muscles. All of them are involved in movements of the hyoid bone and thyroid cartilage during swallowing, mastication and vocalisation. The two sternohyoids lie

Table 8.3: Infrahyoid muscles

Muscle	Origin	Insertion	Action	Nerve supply
Sternohyoid	• Posterior aspect of manubrium sterni • Posterior aspect of medial end of clavicle • Capsule of sternoclavicular joint	Body of hyoid bone	Depresses hyoid bone	Branch from ansa cervicalis
Sternothyroid	• Posterior aspect of manubrium sterni • Medial end of first costal cartilage	Oblique line on lamina of thyroid cartilage	Pulls larynx downwards	Branch from ansa cervicalis
Thyrohyoid	Oblique line on lamina of thyroid cartilage	Lower border of greater cornu of hyoid bone	• Depresses hyoid bone. • Raises larynx when hyoid bone is fixed	Fibres of C1 nerve travelling through hypoglossal nerve
Omohyoid (Greek.omos = shoulder)	• Inferior belly from upper border of scapula, near scapular notch • Superior belly from intermediate tendon	• Inferior belly passes upwards and medially across the floor of the posterior triangle. It joins the intermediate tendon deep to sternocleidomastoid • Superior belly passes upwards to be inserted into lower border of body of hyoid bone.	Depresses hyoid bone	• Superior belly from the superior ramus of ansa cervicalis • Inferior belly from ansa cervicalis itself

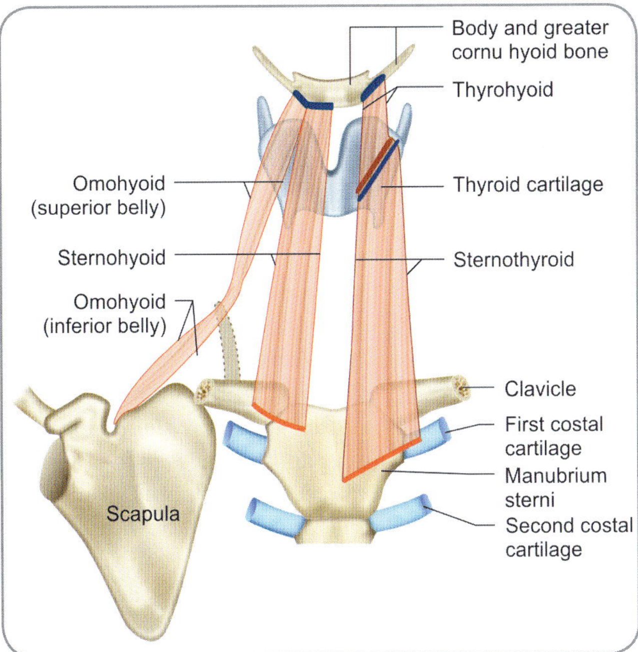

Fig. 8.5: Scheme to show the attachments of the infrahyoid muscles

almost contiguous. The omohyoid is a muscle that has two bellies interconnected by an intermediate tendon. The intermediate tendon of this muscle is held in position by a fibrous band from the deep cervical fascia. This anchorage is important for maintaining the angulated course of the muscle. The superior belly lies almost vertical next to the lateral border of sternohyoid. The omohyoid is supposed to be playing a crucial role during forced and prolonged inspiration. It tenses the lower part of the deep cervical fascia; this prevents the soft parts of the neck from being sucked into thorax during such efforts. The sternothyroid is the muscle that determines the upper extent of the thyroid gland by its attachment to the oblique line of thyroid lamina. Its action of pulling the larynx, especially after the latter had been elevated during deglutition or vocalisation, plays an imporatant role in low note singing. On the contrary, the 'pulling up' action of thyrohyoid muscle on the larynx helps in high note singing.

The infrahyoid muscles are usually referred to as the 'strap muscles' due to their ribbon like appearance. They belong to the same muscular sheet as that of the rectus abdominis and therefore, are called the *rectus cervicis* muscles by clinicians. The inferior belly of the omohyoid is a developmental extension of the superior belly. As the muscle keeps extending posteriorly, its middle portion undergoes fibrous changes due to friction (mechanical factor due to change of direction) and becomes the intermediate tendon. This is unlike the digastrics bellies, where the two bellies are of different origins and the fusion of their ends gives rise to the intermediate tendon.

Anterior Vertebral Muscles (Table 8.4 and Fig. 8.6)

These are:
- ❑ Rectus capitis anterior,
- ❑ Rectus capitis lateralis,
- ❑ Longus capitis and
- ❑ Longus colli.

Table 8.4: Anterior vertebral muscles

Muscle	Origin	Insertion	Action	Nerve supply
Rectus capitis anterior (anterior straight muscle of the head, rectus capitis anticus minor)	Lateral mass and transverse process of atlas vertebra	Inferior surface of the basilar part of occipital bone	Flexion of head	Ventral rami of C1, C2 spinal nerves
Rectus capitis lateralis (lateral straight muscle of the head)	Upper surface of transverse process of atlas vertebra	Jugular process of occipital bone	Flexes head laterally to the same side	Ventral rami of C1, C2 spinal nerves
Longus capitis (rectus capitis anticus major)	Anterior tubercles of transverse processes of C3-C6 vertebrae	Basilar part of occipital bone	Flexion of head	Ventral rami of C1, C2, C3 spinal nerves
Longus colli (upper oblique part)	Anterior tubercles of transverse processes of C3, C4, C5 vertebrae	Tubercle on the anterior arch of the atlas vertebra	• Flexion of neck • Lateral flexion especially by the oblique parts • Rotation of neck to the opposite side by the inferior oblique part	Ventral rami of C2 to C6 spinal nerves
Longus colli (vertical intermediate part)	Front of bodies of C5 to T3 vertebrae	Front of the bodies of C2, C3, C4 vertebrae		
Longus colli (inferior oblique part)	Front of bodies of upper 2-3 thoracic vertebrae	Anterior tubercles of transverse process of C5, C6 vertebrae		

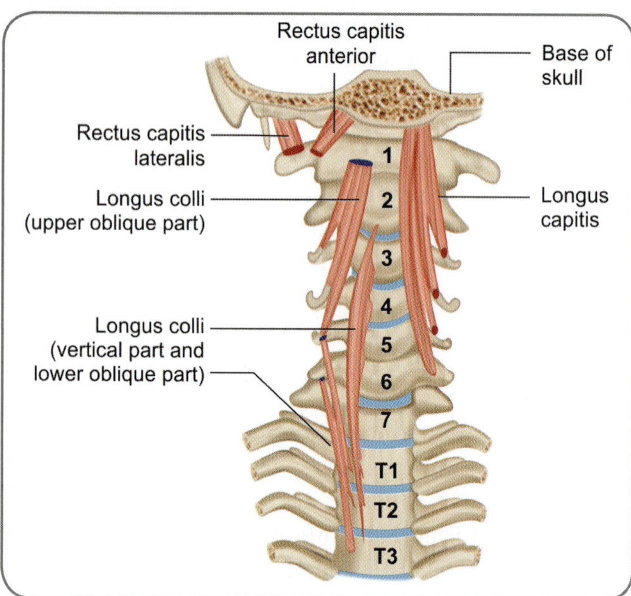

Fig. 8.6: The anterior vertebral (prevertebral) muscles

Additional Notes on the Anterior Vertebral Muscles

Together with the lateral vertebral muscles they form the prevertebral muscle group. They lie deep to the prevertebral fascia.

Lateral Vertebral Muscles (Table 8.5)

These are:
- ❑ Scalenus anterior (Fig. 8.7)
- ❑ Scalenus medius (Fig. 8.7)
- ❑ Scalenus posterior and (Fig. 8.8) and
- ❑ Scalenus minimus (Fig. 8.8)

Additional Notes on Lateral Vertebral Muscles

These muscles have been called the lateral vertebral muscles because they stretch between the upper two ribs and the cervical transverse processes. Scalenus Anterior forms an important landmark in the root of neck and is related to many important structures intimately. The phrenic nerve runs anterior to the muscle, the subclavian artery lies posterior to it and the brachial plexus lies at its lateral border. The proximity of this muscle to these structures can give rise to compression syndromes. Below its attachment to the sixth cervical vertebra, the medial border of the muscle is separated from longus colli by an angular interval sometimes called the *scaleno-vertebral triangle*. The vertebral artery and vein pass to the foramen transversarium of the sixth cervical vertebra through this space. The inferior thyroid artery and the sympathetic trunk with the middle and inferior cervical ganglia are also present in this triangle. In the left side, the thoracic duct crosses the triangle. All the three main scalene muscles are active during inspiration. They are particularly important during breathing in the erect posture.

Surface marking of scalenus anterior: This muscle serves as a guide for many internal structures and is often used as a reference structure. The lateral and medial borders of the muscle can be marked and the muscle band can be drawn on the surface.
Lateral border: Point A is marked 2.5 cm from the midline at the level of the upper border of thyroid cartilage. Point B is marked on the lower border of clavicle at the junction of the medial and middle thirds. The two points are joined by a straight line that runs inferolaterally. This line marks the lateral border of the muscle on the surface.

contd...

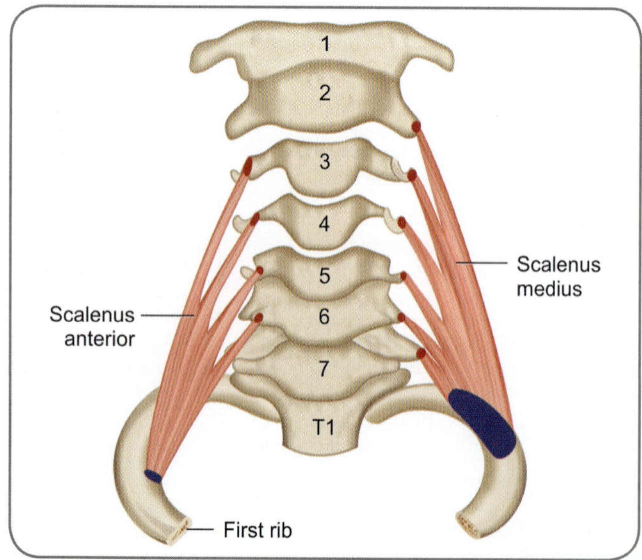

Fig. 8.7: Attachments of scalenus anterior and scalenus medius muscles

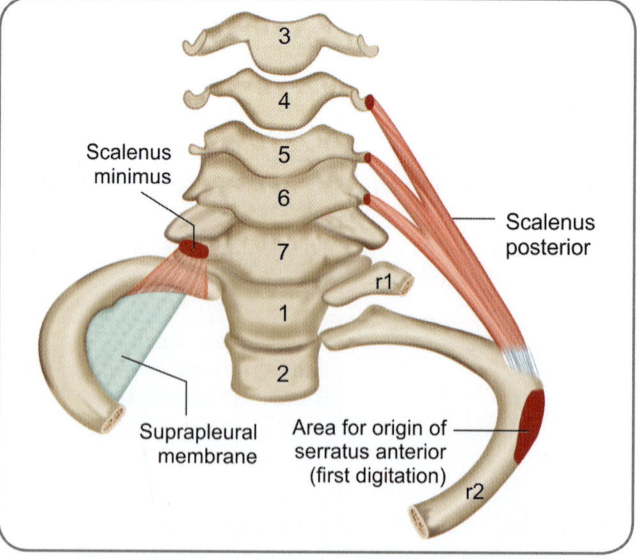

Fig. 8.8: Attachments of scalenus posterior and scalenus minimus muscles

Table 8.5: Lateral vertebral muscles

Muscle	Origin	Insertion	Action	Nerve supply
Scalenus anterior (anterior scalene muscle, scalenus anticus)	Anterior tubercle of transverse processes of C3 to C6 vertebrae	Inner border of the first rib on scalene tubercle	**Acting from below:** Bends neck forwards and laterally and turns face to opposite side **Acting from above:** Elevates first rib	Ventral rami of C4, 5, 6 spinal nerves
Scalenus medius (medial scalene muscle)	• Transverse process of C2 Vertebra • Posterior tubercle of transverse process of C3 to C7 vertebrae	Upper surface of first rib between tubercle of the rib and groove for subclavian artery	**Acting from below:** Bends cervical spine to the same side **Acting from above:** Raises the first rib	Ventral rami of C3 to C8 spinal nerves
Scalenus posterior (posterior scalene muscle, scalenus posticus)	Posterior tubercles of transverse process of C4 to C6 vertebrae	Outer surface of second rib behind the attachment of serratus anterior	• With the fixed second rib, it bends lower part of cervical spine to the same side • With the fixed upper attachment, it raises the second rib	Ventral rami of C6 to C8 spinal nerves
Scalenus minimus (smallest scalene muscle) (if present)	Transverse process of C7 Vertebra	• Inner border of the first rib behind the subclavian groove • Some fibres are continuous with suprapleural membrane	Not significant	C7 spinal nerve

contd…

> **Medial border:** Point C is marked 2.5 cm from the midline at the level of the arch of cricoid cartilage. Point D is marked 1 cm medial to the lower end of the lateral border. Points C and D are joined by a straight line that runs inferolaterally. This line marks the medial border of the muscle. The two borders together mark the extent of the scalenus anterior muscle.

DEEP MUSCLES OF THE BACK AND THEIR LAYOUT (FIG. 8.9 AND TABLE 8.6)

Posterior Vertebral Muscles

The musculature of the back is arranged in a series of layers, of which only the deeper muscles are true, intrinsic back muscles. The true back muscles are characterized by their position and by their innervation which is by the branches of the dorsal rami of spinal nerves. They lie deep to the posterior layer of thoracolumbar fascia.

The superficial muscles of the back are the *extrinsic muscles*. They are innervated by the ventral rami of spinal nerves indicating their morphological origin. They are in three layers. The superficial and the superficial intermediate layers (superficial two layers) form the immigrant muscles or the posterior axioappendicular muscles. These muscles attach the upper limb to the axial skeleton and are positioned in the back due to functional reasons. The superficial layer comprises the trapezius and latissimus dorsi and the superficial intermediate layer comprise the levator scapulae and rhomboid muscles. The posterior axioappendicular muscles are described in the section on upper limb. The third layer is the deep intermediate layer consisting of the serratus posterior muscles. These muscles form the superficial respiratory muscles; though they cannot strictly be labelled 'immigrant' as they connect the vertebrae with the ribs, they cannot be classified as intrinsic muscles of the back due to their innervation by the ventral rami of the thoracic spinal nerves. They are described in the section on thorax.

The intrinsic muscles (native muscles of the back) are also arranged in three layers (Fig. 8.9).

1. The *superficial layer* comprises the Splenius muscles in the neck and upper thorax and the erector spinae group in the trunk as a whole;

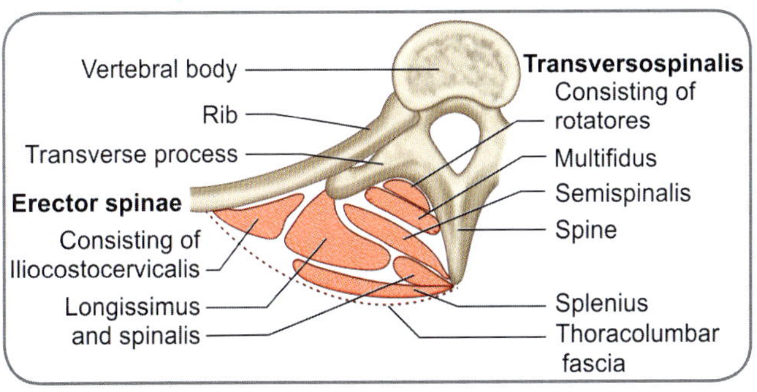

Fig. 8.9: Schematic transverse section to show the arrangement of deep muscles of the back

2. The ***deep layer*** comprises the transversospinal group (consisting of the semispinalis, multifidus and rotatores) and the suboccipital group of muscles.

3. The ***deepest layer*** constitutes the interspinal and intertransverse muscles.

The muscles are described in detail in the Table 8.6.

Table 8.6: Deep muscles of the back

Muscle	Origin	Insertion	Action	Nerve supply
Splenius capitis (splenius muscle of the head) (Fig. 8.10)	• Lower half of ligamentum nuchae • Vertebral spines C7 to T3	Mastoid process and area below lateral 1/3 of superior nuchal line	• Acting bilaterally, causes extension of head (capitis) neck(cervicis) • Acting unilaterally and synergistically with the opposite sternocleidomastoid, causes rotation of head(capitis) and neck(cervicis) to the same side	Dorsal rami of cervical nerves
Splenius cervicis (Splenius muscle of the neck, splenius colli)	Vertebral spines T3 to T6	Costo-transverse process of vertebrae C1, C2, C3		Dorsal rami of cervical nerves
Erector spinae (erector muscle of the spine, sacrospinalis)	Dorsal surface of sacrum through a U-shaped tendon	• The muscle divides into three main parts: a. Spinalis b. Longissimus c. Iliocostalis • The spinalis component is attached to the spines of the thoracic and upper lumbar vertebrae • The longissimus component attaches to the junction of the transverse and costal elements of the cervical, thoracic and lumbar levels • The Iliocostalis component attaches to the site homologous to ribs at all levels • Only three slips reach the neck – Iliocostalis cervicis reaches the transverse processes of lower cervical vertebrae – Longissimus cervicis reaches the transverse processes of upper cervical vertebrae – Longissimus capitis reaches the mastoid process of skull	• Acting bilaterally, it causes extension of the thoracic and lumbar spines • Acting unilaterally causes lateral flexion of vertebral column • Keeps the spine erect • Longissimus capitis restores head to neutral from rotated position	Dorsal rami of cervical nerves
Semispinalis capitis (Fig. 8.11) (Semispinal muscle of the head, Musculus complexus)	• Transverse processes of upper six thoracic vertebrae, and C7 vertebra • Articular processes of C4, C5, C6 vertebrae	Medial part of area between superior and inferior nuchal lines	• Extension of head • Slightly rotates face to opposite side	Dorsal rami of cervical spinal nerves
Semispinalis cervicis (semispinal muscle of the neck, semispinalis colli)	Posterior surfaces of upper five or six thoracic transverse process	Spinous processes of second to fifth cervical vertebrae.	Extension of vertebral column	
Semispinalis thoracis (semispinal muscle of thorax, semispinalis dorsi)	Transverse processes of sixth to tenth thoracic vertebrae	Spinous processes of lower two cervical and upper four thoracic vertebrae		

132

contd…

Contd...

Multifidus (multifidus spinae)	• Dorsal surface of sacrum between its spinous and transverse crests; • Mamillary processes in lumbar region; • Transverse process in thoracic region, articular process in cervical region	Crosses 3–5 vertebrae above and inserts into the vertebral spines	Rotates the trunk	Dorsal rami of spinal nerves
Rotatores (rotator muscles)	• Root of the transverse process of vertebra below	Spinous process of vertebra above		
Interspinales	Spine of vertebra above	Spine of vertebra below	Postural muscles	Dorsal rami of spinal nerves
Intertransversarii	Transverse process of vertebra above	Transverse process of vertebra below		

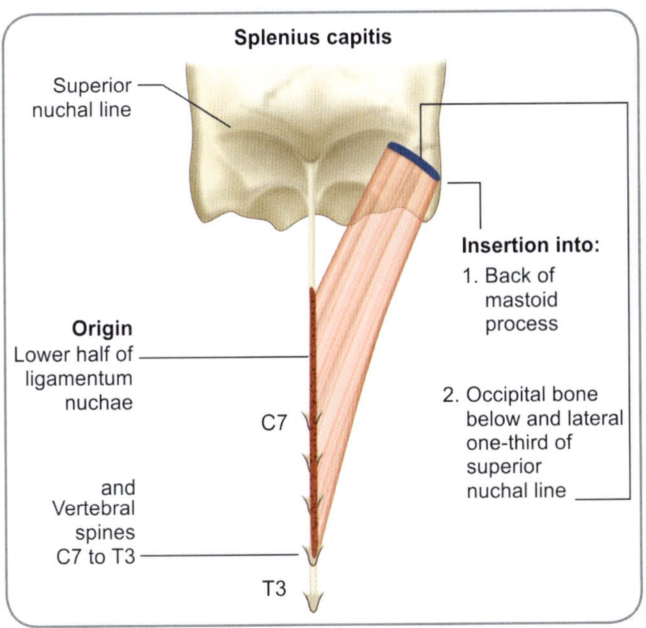

Fig. 8.10: Attachments of splenius capitis muscle

Additional Notes on the Deep Muscles of the Back

The splenii muscles have been named after their similarity to a bandage (Greek.splenion=bandage). Contraction of the splenei of both sides draws the head backwards. The splenei of one side turn the head and rotate the face to the same side. Therefore, the splenei are synergistic to contralateral sternocleidomastoid.

The erector spinae is not a single muscle but a large muscle complex. It is disposed in three columns, which from midline to lateral are the spinalis, longissimus and iliocostalis components. Each column has three parts: lumborum in the lumbar and lower thoracic region, thoracis in the thoracic region and cervicis in the cervical region. Erectores spinae extend the trunk. However, extension of the trunk is a delicate balance between the actions of erector spinae and

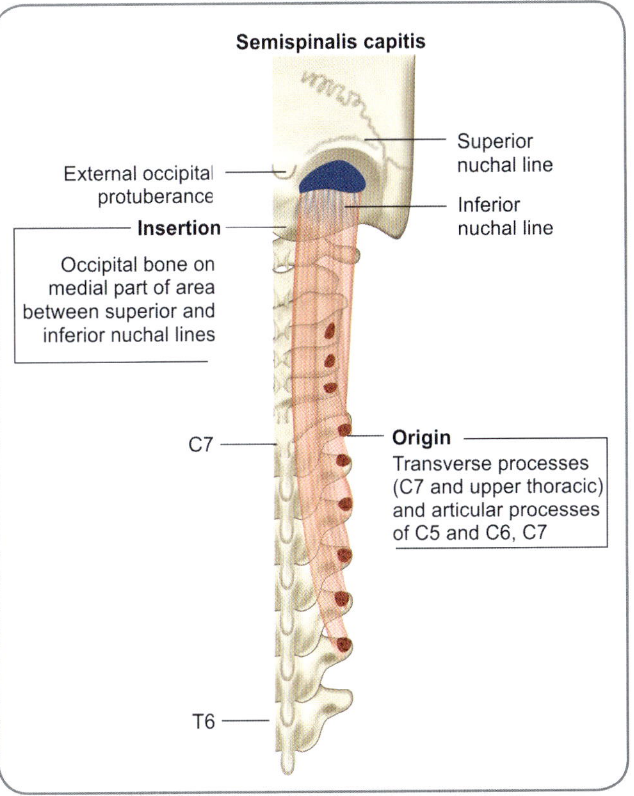

Fig. 8.11: Scheme to show the attachments of semispinalis capitis muscle

the abdominal muscles like the rectus abdominis and the oblique muscles. Flexion of the trunk is first initiated by the rectus abdominis; as the trunk flexes, centre of gravity of the body shifts forward. Erectores spinae then contract and establish extension. Paradoxically, the muscles are inactive in complete flexion of the trunk. The erector spinae work the maximum when an individual stoops forward to carry out some work at a slightly lower level in the standing position. The various components of the erector spinae are collectively

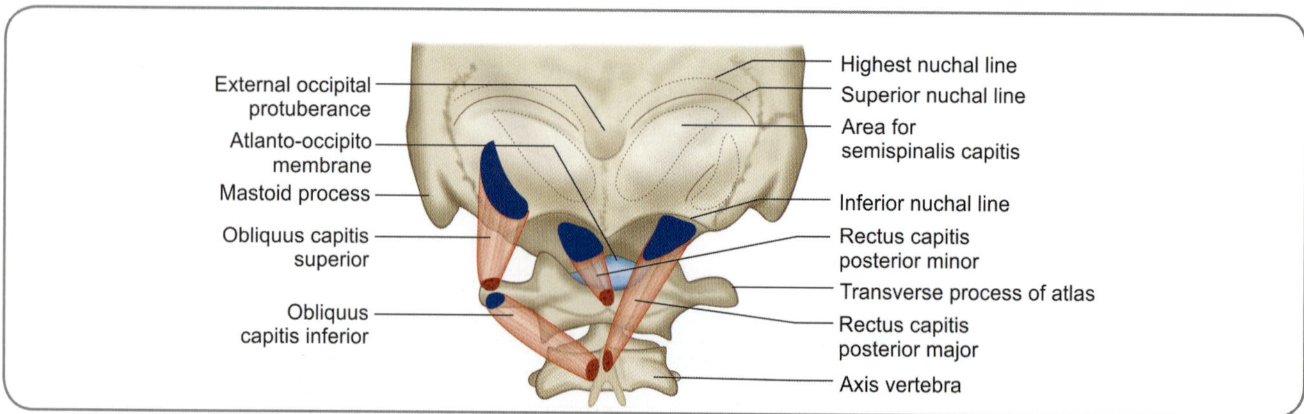

Fig. 8.12: Schematic diagram to show the attachments of the suboccipital muscles

Table 8.7: Suboccipital muscles				
Muscle	**Origin**	**Insertion**	**Action**	**Nerve supply**
Rectus capitis posterior minor (lesser posterior straight muscle of head)	Tubercle on the posterior arch of atlas	Medial part of area below inferior nuchal line	• Collectively, these muscles maintain position of the head • The two recti and the superior oblique extend the head at the atlanto-occipital joints • The rectus capitis posterior major along with obliquus capitis inferior rotates the head turning the face to its own side • Obliquus capitis superior flexes the head laterally to its own side	Dorsal ramus of first cervical nerve
Rectus capitis posterior major (greater posterior straight muscle of head)	Spine of axis vertebra	Lateral part of area below inferior nuchal line		
Obliquus capitis inferior	Spine of axis vertebra	Transverse process of atlas vertebra		
Obliquus capitis superior	Transverse process of atlas vertebra	Lateral part of area between superior and inferior nuchal lines		

referred to as the 'long muscles of the back'. The lateral border of each erector spinae is marked on the surface by a groove seen clearly on the posterior aspect of the trunk. This groove ascends up laterally and then vertically at the angles of the ribs; it then swerves medially and disappears under the scapula.

The transversospinalis group of muscles have been so named because they originate from the transverse processes of vertebrae and pass to the spinous processes of higher vertebrae. They occupy the gutter between the spines of vertebrae and the transverse processes. They are in three layers, which from superficial to deep are the semispinalis, multifidus and rotators. The longitudinal bulge seen on either side of the midline on the back of neck is produced by the semispinalis capitis. All components of the transversospinalis group except the semispinalis capitis are deep to the erector spinae muscles. The semispinalis capitis is comparatively superficial being deep only to the trapezius and splenius.

Rectus capitis anterior and rectus capitis posterior are modified intertransverse muscles. The scalene are modified intercostal muscles.

Suboccipital Muscles (Fig. 8.12)

The suboccipital muscles are a group of small muscles placed in the uppermost part of the back of neck, deep to the semispinalis capitis. They form the boundaries of the suboccipital triangle. The suboccipital muscles are described in Table 8.7.

Clinical Correlation

As seen above all the deep muscles of the back are supplied by dorsal primary rami of the spinal nerves of the levels concerned and hence the muscles can be cut transversely, without denervating parts above or below the incision.

TRIANGLES OF NECK (FIG. 8.13)

The neck appears as a quadrilateral area when viewed anterolaterally. It is limited **superiorly** by the base of mandible and an imaginary line between the angle of the mandible and mastoid process, **inferiorly** by the upper border of clavicle, **anteriorly** by the anterior median line

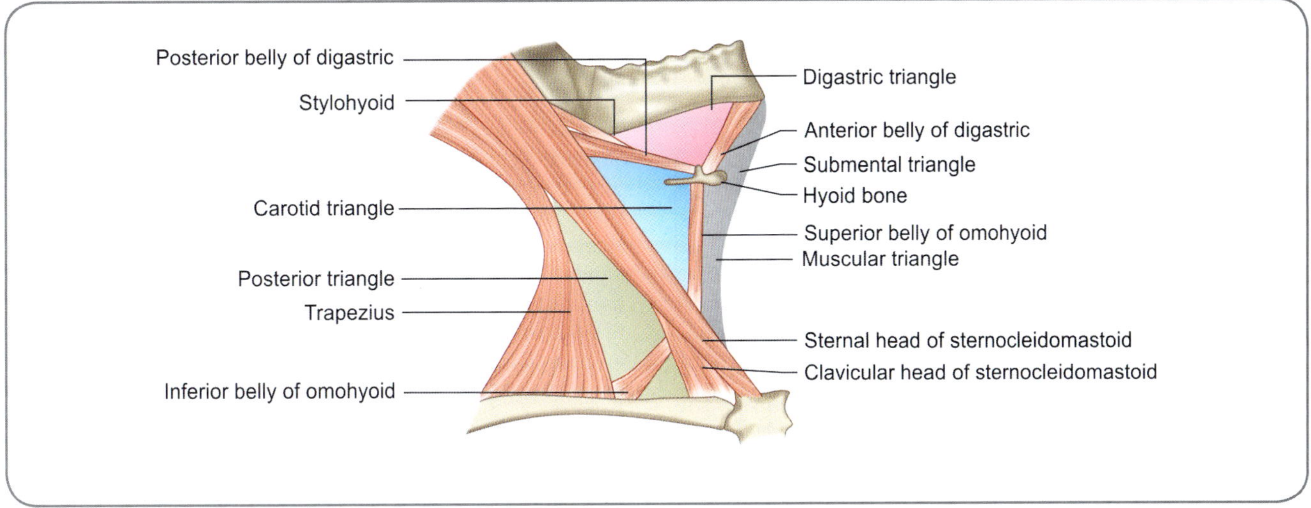

Fig. 8.13: Triangles of neck (lateral view)

and *posteriorly* by the anterior margin of trapezius. This quadrilateral area is further subdivided into anterior and posterior triangles by the sternocleidomastoid muscle.

> The boundaries of the **posterior triangle** are:
> *Anteriorly*, the posterior border of sternocleidomastoid.
> *Posteriorly*, the anterior border of trapezius.
> *Inferiorly* (the base of the triangle), the middle third of clavicle.
> The boundaries of the **anterior triangle** are:
> *Posteriorly*, anterior border of sternocleidomastoid.
> *Medially*, midline of the front of neck.
> *Superiorly (the base of the triangle)*, base of mandible and its projection to mastoid process.
> *Apex of the triangle*, manubrium sterni.
> The anterior triangle is further divided into four smaller triangles by the digastric and the omohyoid muscles into Digastric, Carotid, Muscular and Submental triangles. The space between the anterior bellies of the two digastric muscles is the submental triangle. It occupies the midline position and is common for both sides.
> In addition to the above, the **suboccipital triangle** is present deep in the upper part of the back of neck, and is bounded by the suboccipital muscles.

Posterior Triangle (Figs 8.14 and 8.15)

Other name: Lateral cervical region

Boundaries (Fig. 8.14)

- Anteriorly, posterior border of sternocleidomastoid.
- Posteriorly, anterior border of trapezius.
- Inferiorly (base), middle third of clavicle (between the sternocleidomastoid and trapezius).

Though often described a triangle, the space appears to be quadrilateral. The apex of the triangle is at the superior nuchal line of the occiput where the attachments of sternocleidomastoid and trapezius are placed close to

each other. This area is not pointed and so the 'triangle' gets a quadrangular appearance.

The lower part of the posterior triangle is crossed by the inferior belly of the omohyoid muscle which divides the posterior triangle into:

- ❑ An upper part called the ***occipital triangle and***
- ❑ A lower part called the ***supraclavicular or the subclavian triangle***

Floor

The floor of the posterior triangle is formed from above downwards by (Fig. 8.14):

- ❑ Semispinalis capitis,
- ❑ Splenius capitis,
- ❑ Levator scapulae,
- ❑ Scalenus posterior,
- ❑ Scalenus medius and
- ❑ Scalenus anterior.

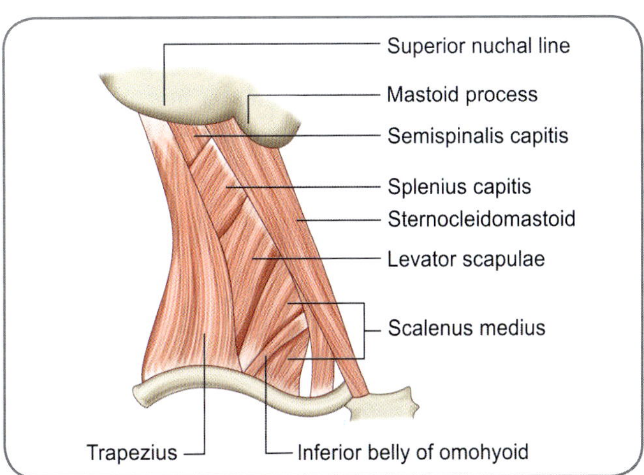

Fig. 8.14: Boundaries of posterior triangle of neck and muscles of its floor

135

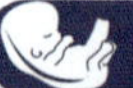

Development

The sternocleidomastoid and trapezius muscles are thought to have developed from the visceral arch mesoderm related to the sixth arch. The two muscles formed a single muscle mass in early embryonic life. Longitudinal cleavage occurs leading to the formation of the posterior triangle. The mastoid process of the temporal bone is drawn downwards, medially and anteriorly along the direction of pull of the sternocleidomastoid muscle.

The scalenus anterior muscle lies deeper to the sternocleidomastoid muscle and so, it is, often not included under the group of floor muscles of the posterior triangle. The muscles forming the floor are covered by the prevertebral layer of deep cervical fascia.

Roof

The roof of the posterior triangle is formed by the investing layer of deep cervical fascia. Superficial to the deep cervical fascia the posterior triangle is covered by superficial fascia and skin. Platysma forms a part of the roof within the superficial fascia in the lower part of the triangle. The cutaneous branches of the cervical plexus, namely, the supraclavicular, Lesser occipital, Great auricular and Transverse cervical nerves are seen in the roof. Also present in the roof are the external jugular vein and some of its tributaries.

External Jugular Vein

The external jugular vein is formed by the union of the posterior division of retromandibular vein and posterior auricular vein. The origin lies within the lower part of the parotid gland or just below it. The level corresponds to the angle of mandible. From its origin, the vein runs downwards and backwards and ends by joining the subclavian vein after piercing the deep fascia. The termination lies behind the middle of clavicle, near the lateral margin of scalenus anterior muscle.

As the greater part of the vein is superficial, being just covered by skin, superficial fascia and platysma, it can be clearly seen in the living. The vein crosses the sternocleidomastoid obliquely running downwards and backwards across it.

Apart from the formative tributaries, the external jugular vein receives:

- **Posterior external jugular vein** from the upper and posterior part of the neck,
- **Transverse cervical vein,**
- **Suprascapular veins** which accompany the corresponding arteries and
- **Anterior jugular vein.**

Dissection

The study of posterior triangle is essential for detailed study of the rest of the neck. Two skin incisions are first marked. The first incision is made from the mastoid process to the jugular notch along the anterior border of the sternocleidomastoid. The second incision is from the jugular notch to the acromion along the clavicle. Reflect the skin flap carefully paying attention to locate and define the platysma. Identify and study the external jugular vein, great auricular nerve, lesser occipital nerve, transverse cervical nerve and supraclavicular nerves. Identify the posterior triangle and define its boundaries. Locate the investing layer of deep cervical fascia. Identify the accessory nerve and trace it. Locate the prevertebral layer and look for the floor muscles. If need be, cut the clavicular head of sternocleido mastoid close to its lower attachment and reflect it. Locate the various structures in the triangle and define them.

Anterior Jugular Vein

The anterior jugular vein runs down the front of neck a short distance from the midline. It begins near the hyoid bone and extends downwards to a point a little above the sternoclavicular joint where it turns laterally deep to the sternocleidomastoid and superficial to the sternohyoid and sternothyroid muscles to end by joining the lower end of the external jugular vein. Occasionally, it may end in the subclavian vein. Just above the sternum, the right and left anterior jugular veins are united by a transverse vein called the ***jugular arch.***

Surface Marking of External Jugular Vein

Point A is marked immediately below the angle of mandible. Point B is marked on the clavicle immediately behind the posterior border of sternocleidomastoid. A line joining the two points marks the external jugular vein on the surface.

Contents of Posterior Triangle (Fig. 8.15)

- ***Cutaneous branches of the cervical plexus*** namely the supraclavicular, lesser occipital, great auricular and transverse cervical nerves enter the posterior triangle by piercing the fascia over its floor and run for some distance between the floor and the roof, before piercing the latter to become subcutaneous.
- ***Muscular branches*** arising from the cervical plexus to the levator scapulae and trapezius run deep to the fascia of the floor.
- The ***spinal accessory nerve***, before entering the posterior triangle, has already entered the substance of the sternocleidomastoid muscle and supplied it. The nerve then emerges a little above the middle of the posterior border of the muscle, runs downwards and laterally across the triangle (within or deep to the investing layer) over the levator scapulae but superficial to the prevertebral layer of deep cervical fascia. It passes

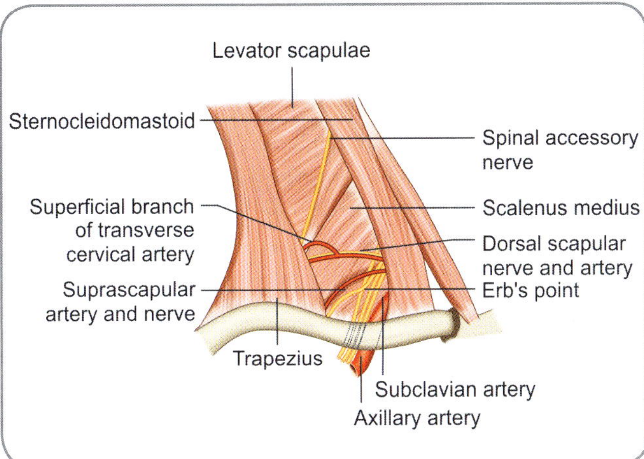

Fig. 8.15: Deeper contents of lower part of posterior triangle. The external jugular vein with its tributaries and inferior belly of omohyoid have been removed for better exposure of the contents. (Note the trunks of the brachial plexus and branches of subclavian artery)

deep to the anterior border of trapezius and enters the deep surface of the muscle.

❑ The ***occipital artery*** crosses the apex of the posterior triangle.

❑ The ***trunks of brachial plexus*** are seen in the lower part of the triangle. Several branches of the plexus are related to the triangle. These are:
  ○ The nerve to rhomboids which pierces the scalenus medius,
  ○ The nerve to serratus anterior,
  ○ The nerve to subclavius and
  ○ The suprascapular nerve.

❑ The ***subclavian artery*** (surrounded by the axillary sheath) runs across the lowest part of the posterior triangle.

❑ The ***suprascapular*** and the ***transverse cervical*** arteries are also seen in the lower part of the triangle. These are the lateral branches of the thyrocervical trunk. The suprascapular artery runs downwards and laterally across the scalenus anterior and then crosses the third part of subclavian artery. It then runs posteriorly to supply the muscles on the posterior aspect of scapula. The transverse cervical artery, also called the cervicodorsal trunk, divides into the superficial cervical and dorsal scapular arteries. Both of them run across the scalenus anterior muscle, pass between the trunks of brachial plexus and give out vasa nervorum to the brachial plexus. The superficial cervical artery passes deep to trapezius, supplies branches and joins the arterial anastomosis around the scapula. The dorsal scapular artery passes deep to levator scapulae and rhomboids, supplies the muscles and joins the arterial anastomosis around the scapula.

❑ Some lymph nodes are present—these nodes usually lie along the external jugular vein and superficial to the sternocleidomastoid. They are called the ***superficial cervical nodes***, receive afferents from the tissues around the area and send efferents to the deep cervical nodes.

Of the abovementioned contents, the trunks of brachial plexus, nerve to subclavius and subclavian artery with its branches are seen in the ***supraclavicular triangle*** and the rest are seen in the occipital triangle. However, the external jugular vein can be seen running behind the posterior border of sternocleidomastoid to reach the subclavian vein, in the roof of the supraclavicular triangle.

Added Information

❑ The posterior triangle is subdivided into an upper occipital and a lower supraclavicular triangle by the inferior belly of omohyoid.

❑ The boundaries of occipital triangle, which is the larger subdivision are anteriorly, sternocleidomastoid, posteriorly, trapezius and inferiorly, the inferior belly of omohyoid. The triangle is so called because of the occipital artery appearing at its apex. The most important structure of this triangle is the spinal accessory nerve crossing it.

❑ The boundaries of supraclavicular triangle are superiorly, the inferior belly of omohyoid, inferiorly, the middle 1/3rd of the clavicle and medially, posterior border of sternocleidomastoid. The *floor* is formed by the first rib, scalenus medius and the first slip of serratus anterior muscle. The supraclavicular triangle is also known as the omohyoid triangle and the subclavian triangle. It is called the subclavian triangle because of the third part of subclavian artery being located in it.

❑ In a living individual, the supraclavicular triangle corresponds to the prominent hollow called the greater supraclavicular fossa seen above the clavicle. It is possible to palpate the trunks of brachial plexus here if the neck is flexed to the opposite side and the examining fingers are run perpendicular to the trunks.

❑ The nerve point of the neck is situated in the anterior boundary of the posterior triangle. This is the point around the middle of the posterior border of sternocleidomastoid where the cutaneous branches of the cervical plexus emerge into the posterior triangle.

Surface Marking of Accessory Nerve in the Posterior Triangle

Point A is marked on the posterior border of the sternocleidomastoid muscle at the junction of the superior and middle thirds. Point B is marked on the anterior border of the trapezius muscle at the junction of the middle and lower thirds. A line joining the two points runs obliquely downwards and backwards and marks the course of the accessory nerve in the posterior triangle.

Surface Marking of Subclavian Artery in the Posterior Triangle

Point A is marked on the sternoclavicular joint. Point B is marked on the midpoint of clavicle. The two points are joined by a broad line that is curved with an upward convexity reaching about 2 cm above the clavicle. This line indicates the subclavian artery. The artery can be palpated by pressing immediately behind the junction of the medial and middle thirds of the clavicle.

Clinical Correlation

- The spinal accessory nerve is plastered to the roof of the posterior triangle. Inadvertent injury to the roof may result in lesions of the nerve. Paralysis of sternocleidomastoid causes weakness in turning the head to the opposite side. Atrophy and weakness of trapezius may also occur. The normal contour of the neck produced by the trapezius muscle is lost and there is drooping of the shoulder.
- The external jugular vein is clearly visible above the clavicle in most individuals. When the venous pressure of the individual is normal, only a small segment of the vein immediately above the clavicle is visible. When the venous pressure rises, the vein becomes prominent throughout its course. Observation of the vein should be routinely carried out during the physical examination of a patient so that information about internal conditions can be known. Because the external jugular vein is used for knowing internal information, it is dubbed the 'internal barometer' of the body.
- External Jugular vein present superficially may also be injured and may give rise to air embolism. As the external jugular vein pierces the deep fascia to drain into the subclavian vein, it is adherent to the rim of the fascial opening which prevents retraction of the vein. The lumen of the vein is held open and air is sucked in because of the negative air pressure in the thorax. Immediate compression of the injured site or a deeper cut incising the deep fascia of neck will cause retraction of the vein thereby preventing the occurrence of air embolism.
- For producing anaesthesia in the neck, anaesthetic block can be produced in the posterior triangle. The anaesthetic agent is injected at several points along, especially around the middle of the posterior border (nerve point of the neck) of the sternocleidomastoid. Anaesthesia of upper limb can also be achieved by injecting the anaesthetic agent in the posterior triangle superior to the midpoint of clavicle. This point will be around the supraclavicular part of the brachial plexus.
- The supraclavicular nerve may be damaged in the lower part of the posterior triangle. Fracture of the middle third of clavicle is the usual cause. Lateral rotation of the humerus is compromised; there is a compensatory medial rotation leading to a waiter's tip position.
- The pressure point for the subclavian artery is in the lower part of the posterior triangle. The artery can be palpated by pressing down immediately behind the junction of the medial and middle thirds of the clavicle. This is also the pressure point of the artery. The artery can be compressed against the first rib here; when haemorrhage occurs distally the artery can be compressed and bleeding stopped.

Anterior Triangle

Other name: Anterior cervical region.

As already noted, the anterior triangle is further subdivided into the unpaired submental and the paired digastric, carotid and muscular triangles.

Subdivisions of the Anterior Triangle

Submental Triangle

The submental triangle is an unpaired triangle below the chin, half of which lies on each side of the midline of neck. Its boundaries are: *Above and laterally*, on each side the anterior bellies of digastric muscle; *base* of the triangle, the hyoid bone and apex, the mandibular symphysis. The *floor* of the triangle is formed by the two mylohyoid muscles which meet in the midline.

Contents

A few submental lymph nodes and some small blood vessels, especially small veins which unite to form the anterior jugular vein.

Added Information

The mylohyoids and the anterior bellies of digastrics originally formed a single sheet of muscle. In some lower animals they still form a single layer. During human development, they separate to form two distinct layers. Their common development is indicated by their common nerve supply. In some individuals, they may be partially separated and partially attached.

Digastric Triangle

Other names: Submandibular triangle, trigonum submandibulare, submaxillary triangle.

Boundaries

- ***Above:*** Base of the mandible and an imaginary line extending from the angle of the mandible to the mastoid process.
- ***Below:*** Anteriorly the anterior belly of digastrics; posteriorly the posterior belly of digastric muscle and the stylohyoid.
- ***Floor:*** Mylohyoid, hyoglossus and anterior part of the middle constrictor of pharynx.
- ***Roof:*** Skin, superficial fascia containing the platysma, cervical branch of the facial nerve and some cutaneous nerves and the investing layer of deep fascia.

The digastric triangle is subdivided into anterior and posterior parts by the stylomandibular ligament. The posterior part is continuous with the parotid region above.

Contents

The main content of the *anterior part* of the digastric triangle is the superficial part of the **submandibular gland**. Other contents are—**facial vein** and **submandibular lymph nodes** which lies superficial to the submandibular gland, **facial artery** accompanied by the **marginal mandibular branch of facial nerve** close to the lower border of mandible, **mylohyoid vessels and nerve, submental vessels and nerve** and a part of the **hypoglossal nerve**.

The contents of the *posterior part* of the digastric triangle are the lower part of the **parotid gland**, **external carotid artery** and carotid sheath containing the internal carotid artery, internal jugular vein and vagus nerve.

> ### Clinical Correlation
>
> The lymph nodes in the digastric triangle are one of the clinically important groups which drain over a wide area of head and neck. One of the nodes lies close to the facial artery when it crosses the lower border of the mandible. Any infection of the head and neck region may lead to enlargement and suppuration of these nodes. An incision to open an abscess of these nodes especially near the lower border of the mandible may result in accidental injury of the marginal mandibular branch which lies superficially resulting in paralysis of muscles of lower lip.

Muscular Triangle

Other name: Inferior Carotid triangle, tracheal triangle, omotracheal triangle.

Boundaries

- **Posteroinferiorly:** Sternocleidomastoid muscle
- **Posterosuperiorly:** Superior belly of the omohyoid muscle
- **Anteriorly (or medially):** Anterior midline of the neck

Contents

It does not have any important content except the infrahyoid muscles. Since the thyroid and parathyroid glands lie in the midline of the neck, they are sometimes defined as contents of this triangle.

> ### Clinical Correlation
>
> Though the muscular triangle is technically defined to have no important structure, some structures which lie in this region are in a vulnerable position and surgical danger. These are the cross communication between the anterior jugular veins present in the suprasternal space, the brachiocephalic vein that may peep into the jugular notch, the inferior thyroid veins which descend to the brachiocephalic veins and the occasional thyroidea ima artery. All these structures are in front of the tracheal rings.

Carotid Triangle

Other names: Superior carotid triangle, trigonum caroticum, malgaigne's triangle

Boundaries (Fig. 8.16)

- **Posteriorly:** Anterior margin of sternocleidomastoid muscle.
- **Antero-superiorly:** Posterior belly of digastric muscle and stylohyoid.
- **Antero-inferiorly:** Superior belly of omohyoid muscle.

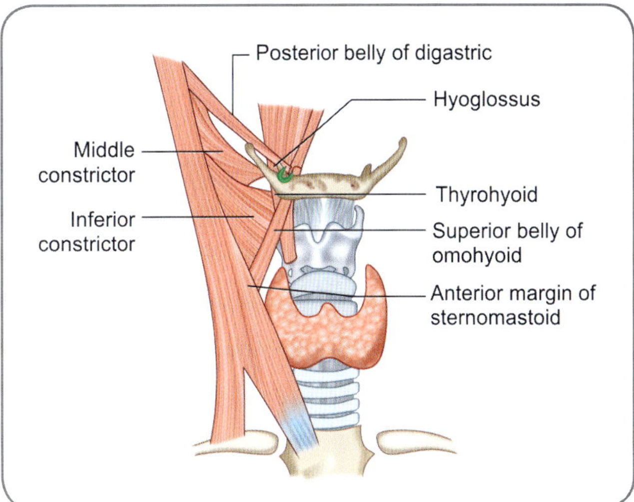

Fig. 8.16: Boundaries of carotid triangle (Note the muscles in the floor **1.** Middle constrictor **2.** Inferior constrictor **3.** Thyrohyoid **4.** Hyoglossus)

- **Roof:** Skin, superficial fascia (with platysma and cutaneous nerves) and deep cervical fascia.
- **Floor:** Thyrohyoid muscle, hyoglossus muscle, middle and inferior constrictors of pharynx.

Contents (Fig. 8.17)

The carotid triangle contains several important blood vessels and nerves. These are:

- Common carotid artery, along with carotid sinus and carotid body.
- Internal carotid artery.
- External carotid artery and the following branches arising from it:

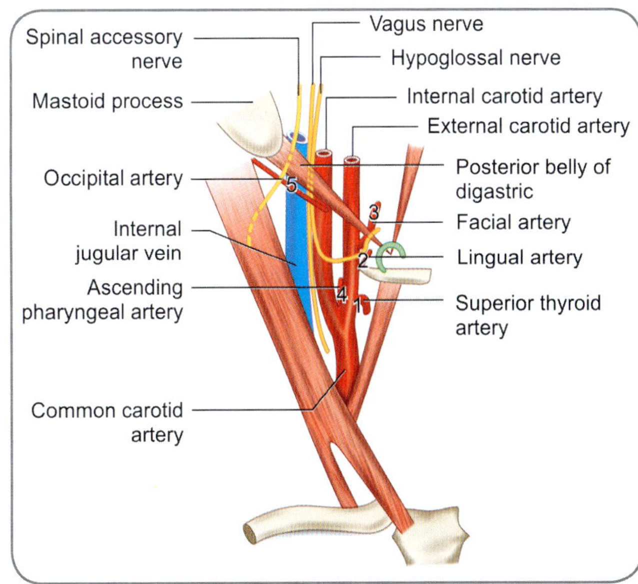

Fig. 8.17: Contents of carotid triangle (Note the bifurcation of common carotid artery and five branches of external carotid artery) **1.** Superior thyroid artery **2.** Lingual artery **3.** Facial artery **4.** Ascending pharyngeal artery **5.** Occipital artery

- ○ Superior thyroid artery
- ○ Lingual artery
- ○ Facial artery
- ○ Ascending pharyngeal artery
- ○ Occipital artery
- ❑ Internal jugular vein and some tributaries draining into it.
- ❑ Vagus nerve and its superior laryngeal branch dividing into external and internal laryngeal nerves. A cardiac branch that fuses with the cardiac branch of the sympathetic trunk is also seen here.
- ❑ Spinal accessory nerve.
- ❑ Hypoglossal nerve and upper root of ansa cervicalis.
- ❑ Sympathetic trunk.
- ❑ Lymph nodes.

The common carotid and internal carotid arteries, the internal jugular vein and the vagus nerve are surrounded by the carotid sheath. The ansa cervicalis is embedded in the anterior wall of the carotid sheath.

Added Information

It can be seen that the posterior belly of digastric muscle forms the dividing boundary between the digastric and the carotid triangles. The muscle can be labelled the key muscle of the submandibulocarotid region. The mastoid process conceals the posterior attachment of the muscle. The parotid and submandibular salivary glands overlap the posterior belly. However, it can easily be exposed because only three sets superficial structures cross it – (1) The tributaries of common facial vein, (2) Branches of great auricular nerve, (3) Cervical branch of facial nerve. On the contrary, several important structures pass deep to it, placing it in a 'key' position. The internal and external carotid arteries, the internal jugular vein, tenth, eleventh and twelfth cranial nerves and the sympathetic trunk pass under it. All structures in the carotid triangle which need to reach higher levels pass under the posterior belly of digastric muscle.

Common carotid arteries

The right common carotid artery arises from the brachiocephalic trunk and the left common carotid artery directly from the arch of aorta. The courses and relations of the cervical parts of the right and left common carotid arteries are similar. Starting behind the corresponding sternoclavicular joint, each artery runs upwards and somewhat laterally up to the level of the upper border of the thyroid cartilage, where it terminates by dividing into the internal and external carotid arteries. It normally gives no other branches.

The artery is enclosed in a fibrous ***carotid sheath*** that also encloses the internal jugular vein that is lateral to the artery and the vagus nerve that lies posterolateral in the interval between the artery and the vein.

Posterior relations

Behind the carotid sheath, the common carotid arteries are related to the sympathetic trunk, ascending cervical artery and the vertebral artery. The inferior thyroid artery runs transversely behind the lower part of the artery. In addition to these structures, on the right side, the artery is crossed posteriorly by the recurrent laryngeal nerve and on the left side, it is crossed by the thoracic duct.

Anterior relations

The artery is crossed anterolaterally by the intermediate tendon or the superficial belly of omohyoid at the level of cricoid cartilage. Below the omohyoid, the artery is situated deeply and is covered by the sternohyoid and sternothyroid muscles, sternocleidomastoid, deep cervical fascia, platysma and skin from within outwards. Above the omohyoid, it is more superficial and covered only by the skin, platysma, deep cervical fascia and the medial margin of sternocleidomastoid. It is also related anteriorly to the ansa cervicalis, superior thyroid vein, middle thyroid vein and the anterior jugular vein.

Medial relations

On the medial side, the common carotid artery is related to the recurrent laryngeal nerve between the trachea and the oesophagus, the corresponding lobe of the thyroid gland and a part of the inferior thyroid artery.

A fusiform dilatation involving the bifurcation of the common carotid artery and the beginning of the Internal carotid artery is called the ***carotid sinus***. The carotid sinus acts as a baroreceptor and controls the intracranial blood pressure. A small, oval neurovascular structure situated close to the bifurcation of the common carotid artery is called the carotid body. The carotid body acts as a chemoreceptor and monitors oxygen tension within the artery.

Surface Marking

Point A is marked on the sternoclavicular joint. Point B is marked on the anterior border of sternocleidomastoid at the level of the upper border of the thyroid cartilage. The two points are joined by a broad line. This line indicates the common carotid artery on the surface.

Clinical Correlation

- ❑ The carotid pulse can easily be palpated in the carotid triangle. It is palpated behind the anterior border of sternocleidomastoid at the upper border of the thyroid cartilage. The index and middle fingers are placed on the thyroid cartilage and pointed posterolaterally between the trachea and the sternocleidomastoid muscle. The artery can also be compressed against the transverse process of the sixth cervical vertebra(called the Chassaignac's tubercle)

contd…

and above this level it is easily felt. Absence of a carotid pulse usually indicates cardiac arrest. Palpation is done on the common carotid artery low in the neck so that pressure on the carotid sinus and a consequent drop in blood pressure and heart rate are avoided.

❏ Stimulation of the carotid sinus produces reflex fall in blood pressure and slowing of heart rate. In patients with *hypersensitive carotid sinus,* sudden attacks of fainting and slowing of heart rate occur when the individual suddenly rotates the head. Such symptoms constitute the *carotid sinus syndrome.* Pressure on the carotid artery also produces similar effects. Checking of carotid pulse is not preferred in such individuals.

❏ The carotid body acts as a chemoreceptor and its stimulation by anoxia (lack of oxygen) produces reflex rise in blood pressure and heart rate with changes in the depth and rate of respiration.

❏ The "*potato tumour*" of the neck is a rare condition caused due to enlargement of the carotid body. Though it is a disease of the carotid body, the symptoms produced are that of carotid sinus syndrome because of the pressure effect of the growing tumour on the sinus.

❏ Thrombotic blocks of the internal carotid artery result in ischaemic disorders of the brain. If a neurologic function is lost but is restored within 24 hours, the condition is called a transient ischaemic attack. Sometimes minor strokes may occur where the neurological deficit lasts for more than 24 hours but less than 3 weeks. Stenosis and block caused by thrombosis is relieved by a procedure called Carotid endarterectomy. The internal carotid artery is opened at its origin and stripped of the atherosclerotic / thrombotic plaque.

❏ The carotid triangle is a surgeon's delight because it provides easy access to the carotid vessels, internal jugular vein, vagus and hypoglossal nerves and the sympathetic trunks. However, the vagus or the recurrent laryngeal nerves may be damaged resulting in hoarseness of voice.

Internal Carotid Artery

Internal carotid artery in the neck

The internal carotid artery begins at the upper border of the thyroid cartilage and ascends to reach the base of skull where it enters the carotid canal. Each artery can be considered as the main upward continuation of the common carotid artery and occupies a similar position. Like the common carotid artery, it is surrounded along with the internal jugular vein and the vagus nerve, by a carotid sheath. It lies on the transverse processes of the upper cervical vertebrae being separated from them by the longus capitis and the superior cervical sympathetic ganglion and is crossed posteriorly by the superior laryngeal nerve. On the medial side, the artery is related to the pharynx. Pharyngeal veins separate it from the ascending pharyngeal artery and the superior laryngeal nerve.

At its upper end, the internal jugular vein lies posterior to the artery with the glossopharyngeal, vagus, accessory and hypoglossal nerves lying between them. Superficially, it is crossed by styloid process, stylohyoid muscle, posterior belly of digastric, stylopharyngeus muscle, glossopharyngeal nerve, hypoglossal nerve, superior root of ansa cervicalis, posterior auricular and occipital arteries, facial and lingual veins.

The internal carotid artery does not give any branch in the neck.

> **Surface Marking**
>
> Point A is marked on the anterior border of sternocleidomastoid at the level of the upper border of thyroid cartilage. Point B is marked immediately posterior to the condyle of mandible. The two points are united by a broad line that marks the internal carotid artery on the surface.

External carotid arteries

The external carotid arteries arise from the common carotid artery at the level of the upper border of the thyroid cartilage at the level of the disc between the third and fourth cervical vertebrae. The branches of the external carotid artery supply structures of the head and neck outside the cranial cavity.

From its origin, the artery runs upwards and terminates behind the neck of mandible within the parotid gland into its two terminal branches. The lower part of the artery is anterior and medial to the internal carotid. Its upper part is lateral to the internal carotid. Though the artery is medial to the internal carotid artery, it is named as external carotid because it supplies structures which are external to the cranial cavity, whereas the internal carotid artery supplies structures within the cranial cavity.

The lower part of the artery is located within the carotid triangle. Here, it is relatively superficial being covered by skin, superficial and deep fascia and by the anterior margin of the sternocleidomastoid muscle. Above the triangle, the artery lies deep to the hypoglossal nerve and its venae commitantes, posterior belly of the digastric muscle and the parotid gland with the facial nerve within it. Deep to the artery, the pharynx is located which is separated from the upper part of the artery by the styloid process, muscles attached to it and by the internal carotid artery. The glossopharyngeal nerve and the superior laryngeal nerve are also related posteriorly to the external carotid artery.

> **Surface Marking**
>
> Point A is marked on the anterior border of sternocleidomastoid at the level of the upper border of thyroid cartilage. Point B is marked midway between the angle of mandible and mastoid process. These two points are joined by a broad line that is slightly convex forwards in the lower half and convex backwards in the upper half. This sinuous line marks the external carotid artery on the surface.

Branches of external carotid artery

The branches of the external carotid artery and their levels of origin are as follows (in order of origin):

- ❑ ***Ascending pharyngeal*** artery arises from the deep aspect of the external carotid artery just above its lower end.
- ❑ ***Superior thyroid*** artery arises from the anterior aspect of the external carotid just below the level of the greater cornu of the hyoid bone.
- ❑ ***Lingual artery*** arises from the anterior aspect of the external carotid artery opposite the tip of the greater cornu of the hyoid bone.
- ❑ ***Facial artery*** arises from the anterior aspect of the external carotid a little above the origin of the lingual artery.
- ❑ ***Occipital artery*** arises from the posterior aspect of the external carotid opposite the origin of the facial artery.
- ❑ ***Posterior auricular artery*** arises from the posterior aspect of the external carotid just above the level at which the latter is crossed by the posterior belly of the digastric muscle.
- ❑ ***Superficial temporal artery*** and the ***maxillary artery*** are the terminal branches of the external carotid artery. They begin behind the neck of mandible, in the substance of the parotid gland.

Ascending pharyngeal artery

The ascending pharyngeal artery is the only medial branch of the external carotid artery and runs upwards to the base of the skull, lying between the pharynx and the internal carotid artery. It gives branches to the pharynx and adjacent structures like auditory tube, tonsil and soft palate. It also gives the inferior tympanic branch to the middle ear and meningeal branches which enter the skull though the jugular foramen and the hypoglossal canal.

Superior thyroid artery

The superior thyroid artery runs inferomedially accompanied by the external laryngeal nerve to reach the upper pole of the thyroid gland, where it divides into ***anterior*** and ***posterior thyroid branches.*** These branches ramify over the corresponding surfaces of the gland. The terminal part of the anterior branch runs across the upper part of the isthmus of the gland to anastomose with the artery of the opposite side. The posterior branch runs downwards along the posterior border of the thyroid to anastomose with the inferior thyroid artery.

Other branches of the superior thyroid artery are the ***infrahyoid branch***, the ***Superior laryngeal branch*** which accompanies the internal laryngeal nerve and pierces the thyrohyoid membrane to supply the larynx, the ***cricothyroid branch*** and the ***sternomastoid branch***.

Lingual artery

The lingual artery arises from the external carotid artery opposite the tip of the greater cornu of the hyoid bone. The hyoglossus muscle divides the lingual artery into three parts.

The ***first part*** of the artery lies in the carotid triangle, superficial to the middle constrictor of pharynx and forms a characteristic upward loop which is crossed by the hypoglossal nerve. The ***second part*** of the artery lies deep to the hyoglossus muscle that separates the artery from the hypoglossal nerve. This part of the artery also lies on the middle constrictor. The ***third or deep part*** of the artery runs upwards along the anterior margin of the hyoglossus and then forwards to the tip of the tongue. This part lies on the genioglossus muscle.

The branches of the lingual artery are the *suprahyoid artery* from the first part, *dorsal lingual arteries* from the second part and the *sublingual artery* from the third part. A few branches pierce the mylohyoid and anastomose with the submental artery.

Facial artery

The facial artery arises from the external carotid just above the greater cornu of the hyoid bone and terminates at the medial angle of the eye as the *angular artery* where it anastomoses with the dorsal nasal branch of the ophthalmic artery. The course of the artery can be divided into a cervical part and a facial part. The artery is remarkably tortuous which, in the neck allows for the expansion of pharynx and in the face for movements during mastication, smiling and other movements of the face.

Cervical part

The artery first runs upward deep to the posterior belly of digastric and the stylohyoid and then grooves the posterior border of the submandibular gland. It then runs downward and forward between the gland and the medial pterygoid muscle which is superficial to it and reaches the lower border of the mandible. Then, it pierces the deep cervical fascia and enters the face at the anterior edge of the masseter, where it is crossed by the marginal branch of mandibular nerve and the facial vein.

Facial part

Curving around the antero-inferior angle of masseter, the artery runs upward and forward across the superficial aspect of the body of mandible and across the buccinator muscle to reach the angle of the mouth. It then runs upward along the side of the nose to reach the medial angle of the palpebral fissure.

Branches of facial artery

The branches of the facial artery are as follows:

Branches of cervical part:

- *Ascending palatine* artery ascends on the lateral side of the pharynx. It supplies the pharynx, the palate, the tonsil and the auditory tube.
- *Tonsillar* branch reaches the tonsil by piercing the superior constrictor muscle.
- Glandular branches enter into the submandibular gland.
- *Submental* artery runs forward along the lower border of mandible over the mylohyoid muscle and supplies the muscles of the region including those of the chin and the lower lip.

Branches of the facial part:

- *Superior* and *inferior labial* branches supply the lips.
- *Lateral nasal* branch supplies the side of the nose and the *septal* branch supplies the nasal septum.
- The terminal part of the facial artery is called the *angular* artery which anastomoses with the dorsal nasal branch of the ophthalmic artery thereby establishing an anastomosis with the branch of Internal carotid artery.

Occipital Artery

The occipital artery arises from the posterior aspect of the external carotid opposite the origin of the facial artery. It runs backward along the lower border of the posterior belly of digastric muscle to reach the skull medial to the mastoid process. Here, it lies deep to the sternocleidomastoid, the digastric and the deep muscles of the back of neck. It crosses the apex of the posterior triangle of the neck before it pierces the trapezius to become superficial and ends by supplying the posterior part of the scalp. The artery is accompanied by the greater occipital nerve in the scalp.

The occipital artery gives off several branches. These are as follows:

- Two branches *to the sternocleidomastoid* which run backwards across the carotid sheath.
- *Auricular branch* supplies the pinna.
- *Mastoid branch* passes through the mastoid foramen.
- *Meningeal branches* enter the skull through the jugular foramen and the carotid canal.
- *Descending branch* runs down through the deep muscles of the back of the neck and divides into a *superficial branch* which anastomoses with the transverse cervical artery and a *deep branch* which anastomoses with branches of the vertebral and deep cervical arteries. These anastomoses establish collateral links between the external carotid artery and the subclavian artery.
- *Occipital branches* supply the posterior part of the scalp.

Posterior Auricular Artery

The posterior auricular artery arises from the external carotid just above the posterior belly of digastric and the stylohyoid muscle. It passes backward and upward deep to the parotid gland to reach the mastoid process. Apart from some *auricular* branches and *occipital* branches, the named branch it gives off is the *stylomastoid* artery which enters the stylomastoid foramen to supply the middle ear and related structures.

Superficial Temporal Artery

The superficial temporal artery is one of the terminal branches of the external carotid artery. It begins behind the neck of the mandible within the substance of the parotid gland. It then runs upwards behind the temporomandibular joint and ramifies in the scalp over the temporal region. The artery is accompanied by the auriculotemporal nerve. The branches of the artery are as follows:

- *Frontal branch:* Runs upward and forward over the temporal and frontal bones.
- *Parietal branch:* Runs backward over the temporal and parietal bones.
- *Anterior auricular branch:* Supplies part of the auricle and the external acoustic meatus.
- *Middle temporal artery:* Supplies the temporalis.
- *Zygomatico-orbital branch:* Runs forward along the upper border of the zygomatic arch up to the lateral angle of the eye.
- *Transverse facial branch:* Arises within the substance of the parotid gland and runs forward across the masseter above the parotid duct.

Clinical Correlation

- Ligature of superior thyroid artery in thyroidectomy should always be close to the gland to avoid injury to the external laryngeal nerve.
- In surgical removal of tongue, the first part of the lingual artery is ligated before any branch to the tongue or tonsil is given out.

Internal jugular veins

The internal jugular veins are the chief veins of the head and neck. They collect blood from the brain and superficial part of the face and neck. Each internal jugular vein begins as a continuation of the *sigmoid sinus* of that side at the base of skull in the posterior compartment of the jugular foramen. At its commencement, the vein presents a dilatation called the *superior bulb* which lodges in the jugular fossa and is related to the floor of middle ear. The vein passes downwards in the neck and behind the sternal end of clavicle; it joins the subclavian vein to form the corresponding brachiocephalic vein. Close to its termination another dilatation called the ***inferior bulb*** is present and is guarded by a pair of valves.

In the neck, it is enclosed, along with the internal carotid and common carotid arteries and vagus nerve, within the carotid sheath. Within the sheath, in the lower part, the vein lies lateral to the arteries with the vagus nerve lying posterior and between the two. In the upper part, the vein lies posterior to the internal carotid artery with the lower four cranial nerves between them.

Relations

- ***Superficial relations:*** Superficially the internal jugular vein is related to the sternocleidomastoid muscle in its entire length, posterior belly of digastric, superior belly of omohyoid, sternohyoid and sternothyroid in its lower part. Apart from the muscles, it is also related to accessory nerve, styloid process, posterior auricular and occipital branches of external carotid artery, descendens cervicalis nerve, anterior jugular vein and parotid gland in its upper part.
- ***Deep relations:*** The internal jugular vein is related posteriorly to rectus capitis lateralis muscle, transverse process of atlas, levator scapulae, scalenus medius and anterior muscles, phrenic nerve, vertebral vessels and first part of subclavian artery.

> **Surface Marking**
>
> Point A is marked on the upper part of the lobule of the ear. Point B is marked on the medial end of clavicle lateral to the sternoclavicular joint. The two points are joined by a broad line. The lower end of the line will be between the sternal and clavicular heads of the sternocleidomastoid muscle. Mark a dilatation in this part of the line to represent the inferior bulb of the vein.

Tributaries of the Internal Jugular Vein

The tributaries of the internal jugular vein are:

- ***Inferior petrosal sinus:*** It is usually the first tributary of the internal jugular vein. It connects the jugular vein with the cavernous venous sinus.
- ***Pharyngeal veins:*** These veins start from the pharyngeal venous plexus lying outside the pharynx. They drain the pharynx.
- ***Common facial vein:*** The common facial vein is formed by the union of the anterior division of the retromandibular vein and the facial vein in the upper part of the neck. The facial vein begins near the medial angle of the eye by the union of two superficial veins of the forehead, namely, the supratrochlear and the supraorbital veins which then run downwards and backwards across the face. The terminal part of the vein crosses the internal and external carotid arteries, the hypoglossal nerve and the loop formed by the lingual artery. A triangular interval is formed between the internal jugular vein, common facial vein and the posterior belly of digastric which is occupied by the jugulo-digastric group of lymph nodes.

- ***Lingual vein:*** Both the superficial venae commitantes accompanying the hypoglossal nerve and the deep lingual veins accompanying the lingual artery drain into the internal jugular vein as a common trunk or separately.
- ***Superior thyroid vein:*** The superior thyroid vein corresponds in its course and tributaries to those of the superior thyroid artery. After receiving the superior laryngeal and the cricothyroid veins, it drains into the internal jugular vein.
- ***Middle thyroid vein:*** The middle thyroid vein drains the lower part of the thyroid gland. It crosses the common carotid artery to enter the internal jugular vein. It is the first blood vessel to be encountered during a thyroid surgery and requires careful ligation.
- ***Occipital veins:*** They occasionally drain into the internal jugular vein.

  Thoracic duct on the left side and the ***right lymphatic duct*** on the right side. Each of these opens at the junction of the internal jugular vein and the subclavian vein of its own side.

Communications of the internal jugular vein:

- With the external jugular vein in the upper part of the neck through the oblique jugular vein;
- With cavernous sinus via the inferior petrosal sinus.

> ### Clinical Correlation
>
> - The internal jugular vein acts as a guide for surgeons during removal of deep cervical lymph nodes.
> - The internal jugular vein is used for IV access, in which case the vein is accessed through the lesser supraclavicular triangle(between the sternal and clavicular heads of the sternocleidomastoid muscle) to avoid injury to the pleural sac.
> - Queckenstedt's test: Under normal condition, the rate of flow of CSF is increased during lumbar puncture when both internal jugular veins are compressed due to the back flow. This is called Queckenstedt's test. If there is block in spinal subarachnoid space, the pressure will not rise when the internal jugular veins are compressed.

Suboccipital Triangle

The suboccipital triangle lies more or less in the horizontal plane when the head is in the normal anatomical position.

Boundaries

The suboccipital muscles form the boundaries of the suboccipital triangle as follows :

- ***Medially and above:*** Rectus capitis posterior major and minor muscles.
- ***Laterally and above:*** Obliquus capitis superior.

- ❑ ***Inferiorly:*** Obliquus capitis inferior.
- ❑ ***Roof:*** Trapezius, splenius capitis, semispinalis capitis muscles from superficial to deep. The semispinalis capitis muscle forms the immediate roof of the suboccipital triangle and longissimus capitis lies over the lateral part of the triangle. Deep to the semispinalis capitis, the triangle is covered by dense fascia.
- ❑ ***Floor:*** Posterior arch of atlas and the posterior atlantooccipital membrane.

Contents

- ❑ ***Vertebral artery:*** The third part of the vertebral artery enters the suboccipital region after emerging from the foramen transversarium of the atlas, winds backwards and medially behind the lateral mass of atlas and lodges in a groove on the upper surface of the posterior arch of atlas. It then enters the vertebral canal below the lower arched border of the posterior atlanto-occipital membrane to continue as the fourth part of vertebral artery. The third part of the vertebral artery gives muscular branches to supply the suboccipital muscles and meningeal branches to supply the posterior cranial fossa.
- ❑ ***Dorsal ramus of the first cervical nerve:*** It is also known as the ***suboccipital nerve.*** The first cervical nerve divides into ventral and dorsal rami behind the lateral mass of atlas. The dorsal ramus then runs backwards above the posterior arch of atlas, lying below the vertebral artery and gives branches to the suboccipital muscles and to the semispinalis capitis. The ventral ramus of C1 nerve winds forward around the posterior and lateral surfaces of the lateral mass of atlas, where it lies medial to the vertebral artery. It then descends in front of the root of the transverse process of atlas and supplies the rectus capitis anterior and lateralis muscles. It contributes to the formation of the cervical plexus and gives branches to superior belly of omohyoid, geniohyoid and thyrohyoid.
- ❑ ***Suboccipital plexus of veins:*** The suboccipital venous plexus lies around the suboccipital triangle, collects blood from the neighbouring muscles, occipital veins, transverse sinus through the emissary veins, internal vertebral plexus and finally drains into the vertebral veins after piercing the posterior atlanto occipital membrane.
- ❑ Other Structures related to the suboccipital triangle are:
  - ○ ***The greater occipital nerve (dorsal ramus of C2):*** It winds around the lower border of the obliquus capitis inferior and then runs upwards and slightly mediallyacross the suboccipital triangle. It leaves the triangle by piercing the semispinalis capitis. It is the thickest cutaneous nerve of the body.
  - ○ ***Third occipital nerve (dorsal ramus of C3 nerve):*** It pierces the semispinalis capitis and trapezius and ascends medial to the greater occipital nerve to supply the back of neck and scalp.
  - ○ ***Occipital artery:*** In the suboccipital region, the occipital artery gives the meningeal, mastoid, muscular and descending branches. The descending branch divides into superficial and deep branches which anastomose with the superficial branch of transverse cervical artery and deep cervical branch of costocervical trunk of subclavian artery respectively. This anastomosis helps to maintain collateral circulation in ligation of external carotid artery or subclavian artery.

> **Clinical Correlation**
>
> Cisternal puncture is usually done through the suboccipital triangle to collect CSF from the cisterna magna.

Triangle of Vertebral Artery

In addition to the various triangles of the neck, an additional arteriomuscular triangle is also defined. It is called the triangle of the vertebral artery or the scaleno-vertebral triangle and is present in the prevertebral region. The boundaries of the triangle are:

- ❑ ***Base:*** First part of subclavian artery;
- ❑ ***Medial line:*** Lateral border of longus colli muscle;
- ❑ ***Lateral line:*** Posteromedial border of scalenus anterior muscle;
- ❑ ***Apex:*** Anterior tubercle of the transverse process of the 6th cervical vertebra.

The posterior wall or the floor of the triangle is formed by (from above downwards) transverse process of C7 vertebra, ventral ramus of C8 spinal nerve, neck of first rib the cupola of pleura.

The anterior cover to the triangle is formed by the carotid sheath with its contents, phrenic nerve and in the lower part, the inferior thyroid artery and the thoracic duct.

Contents

- ❑ ***Vertebral artery:*** It runs from the base to the apex and enters the foramen transversarium of C6 vertebra.
- ❑ ***Vertebral vein:*** It runs down anterior to the vertebral artery, crosses the subclavian vein and enters the brachiocephalic vein.
- ❑ ***Sympathetic trunk with its ganglia and branches:*** It runs down close to the posterior wall.

The triangle of vertebral artery is important not because of its mere presence, but because its points to an important landmark that lies at its apex. The anterior tubercle of the transverse process of C6 vertebra is the landmark. The prominence of this landmark is accentuated because there is no such tubercle on the C7 vertebra.

Significant anatomical events occur at this level of C6 tubercle:

❑ The cricoid cartilage lies at this level; pharynx joins the oesophagus; larynx joins the trachea.

❑ Intermediate tendon of omohyoid lies at this level; So, this level will mark the boundary between the muscular and carotid triangles.

❑ The common carotid artery can be comprssed against the C6 anterior tubercle; the tubercle comes to be known as the carotid tubercle or Chassaignac' tubercle (named after the 19th century French surgeon Edouard Chassaignac).

Multiple Choice Questions

1. Which of the following is not a content of the suprasternal space of burns?
   a. Lower parts of the anterior jugular veins
   b. Clavicular heads of sternocleidomastoid muscles
   c. Jugular venous arch
   d. Adipose tissue
2. Which of the following is a part of the intrapharyngeal space?
   a. Parapharyngeal space
   b. Peripharyngeal space
   c. Retropharyngeal space
   d. Peritonsillar space
3. The nerve point of the neck is located on the:
   a. Anterior border of sternocleidomastoid
   b. Posterior border of sternocleidomastoid
   c. Anterior border of trapezius
   d. Posterior border of trapezius
4. Tonsillar branch of facial artery reaches the tonsils by piercing the:
   a. Medial pterygoid
   b. Buccinator
   c. Superior constrictor
   d. Middle constrictor
5. Suboccipital nerve is:
   a. The dorsal ramus of C1
   b. The ventral ramus of C1
   c. The greater occipital nerve
   d. The third occipital nerve

ANSWERS

1. b   **2.** d   **3.** b   **4.** c   **5.** a

Clinical Problem-solving

Case Study 1: A 12-year-old boy presented with pain in the neck, slanting and tilting of the head and neck. His mother said that this kind of slanting occurred when he was about 6 months of age. Before that, there was a swelling in the middle of his neck.
❑ What is the name given to the deformity where there is slanting of the neck and tilting of the head?
❑ What is the reason that this boy had it almost from birth?
❑ What are the other reasons for this condition?

Case Study 2: You are asked to elicit vcarotid pulsations in a patient.
❑ Where will you attempt to palpate the carotid artery?
❑ What is the bony point against which the carotid artery can be compressed in this region?
❑ Why should the carotid artery be palpated in the lower part of the neck and not in the higher part?

(For solutions see Appendix).

Oral and Nasal Regions

The oral region and the nasal region are so closely related, both anatomically and functionally, that it is easy to consider them together. The oral region consists of the oral cavity, teeth and gingivae, tongue, palate and palatine tonsils. The nose is the part of the respiratory tract which lies superior to the palate. Thus, the palate forms a boundary for both the nasal and the oral regions.

ORAL REGION

The oral region can be defined as the area of commencement of the digestive system and the digestive process. Food is ingested into the oral cavity and prepared for digestion. It is chewed with the help of teeth and the masticatory muscles; saliva from the salivary glands is admixed to form a manageable food bolus which can then be handled with ease by the rest of the digestive system. Deglutition is initiated in the oral cavity and the voluntary phase of deglutition takes place here.

ORAL CAVITY

The lay person uses the word 'mouth' loosely both for the external opening and for the cavity it leads to. Strictly speaking, the term 'mouth' should be applied only to the external opening which is also called the **oral fissure.** The cavity (containing the tongue and teeth) is the **mouth**

cavity or **oral cavity**. A look at the boundaries of the oral cavity will make us understand further the oral cavity is bounded:

- Laterally by the cheeks;
- Superiorly by the palate (which separates it from the nasal cavity); and
- Inferiorly by a floor to which the tongue is attached.

Projecting into the cavity from above and below, just medial to each cheek, are the alveolar processes of the upper and lower jaws which bear the teeth. When the mouth is closed bringing the upper and lower teeth into apposition, the oral cavity is seen to consist of:

- A part between the teeth of the two sides—the **oral cavity proper**; and
- A part between the alveolar processes and the cheeks —the **vestibule**.

In Figure 9.1, the vestibule is seen in two halves, right and left. When traced anteriorly, the two halves become continuous in the midline in front of the teeth. Here, the

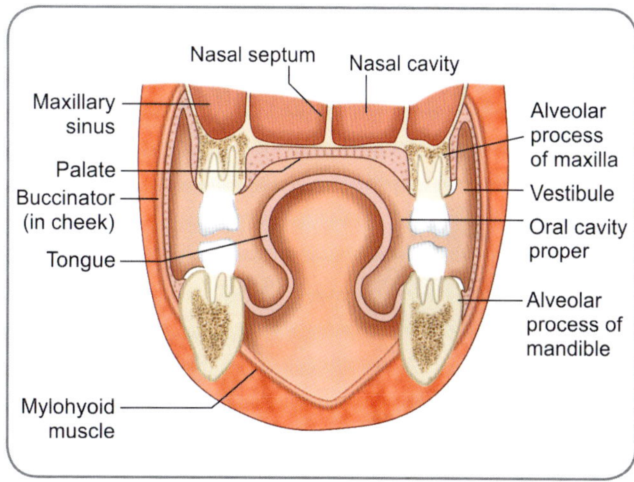

Fig. 9.1: Schematic coronal section through the oral cavity

vestibule communicates with the exterior through the oral fissure. Its external walls are formed by the upper and lower lips. When the teeth are in apposition, the vestibule communicates with the oral cavity proper through a space behind the last tooth. (This is a point of practical importance; it means that any liquid put into the vestibule will find its way into the mouth even if the jaws are kept closed). The size of the oral fissure is controlled by the perioral muscles; the sphincter of the fissure is the orbicularis oris which reduces the size and the dilators are the buccinators, risorius, elevators of lip and depressors of lip.

> The microstructure of the lips has four layers—cutaneous, muscular, glandular and mucous. Between the muscular and the glandular layers is an arterial circle formed by the upper and lower labial branches of the facial artery. If the lip is grasped tight between the thumb and a finger, the pulsations of the arteries can be felt.

With the exception of the teeth, all structures in the oral cavity are covered by mucous membrane. The mucous membrane over the alveolar processes of the jaws is firmly attached to the underlying bone and is referred to as the *gum*. The oral cavity proper communicates posteriorly with the oral part of the pharynx. The communication between the two is called the *oropharyngeal isthmus* (Fig. 9.2). When the mouth is closed in a position of rest, the oral cavity is completely occupied by the tongue.

The roof of the cavity is formed by the palate. The main structure in the floor is the tongue. The rest of the floor is formed by mucous membrane passing from the sides of the tongue to the gum. The anterior part of the tongue is not attached to the floor and can be protruded out of the mouth; the posterior part of the inferior surface is attached to the floor by a median fold of mucosa called the *frenulum linguae*.

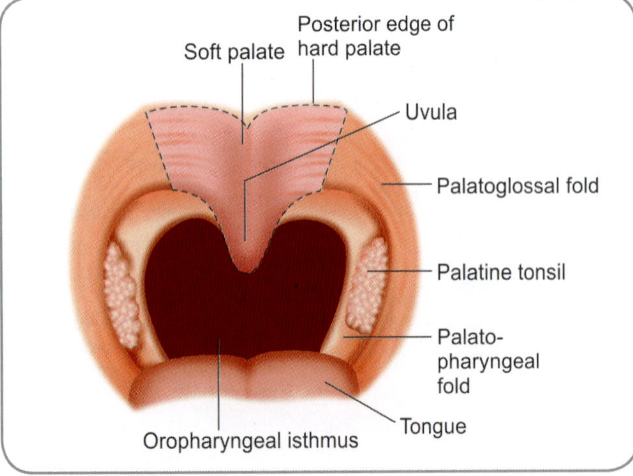

Fig. 9.2: Soft palate as seen through the mouth. The dotted line indicates its upper and lateral limits

Three pairs of salivary glands (parotid, submandibular and sublingual salivary glands) are present near the oral cavity and pour their secretions into it.

The secretions of the parotid glands are poured into the mouth through the right and left parotid ducts which open into the corresponding half of the vestibule, on the inner side of the cheek, opposite the crown of the second upper molar tooth. The duct for each submandibular gland opens on the *sublingual papilla* located just lateral to the frenulum linguae. The sublingual glands lie just below the mucosa on the floor of the mouth. Each gland raises a ridge of mucosa which starts at the sublingual papilla and runs laterally and backwards. This ridge is called the *sublingual fold*.

Lymphatics from the floor of the mouth, gums and teeth drain into submandibular nodes and submental nodes and from them into the deep cervical lymph nodes.

Clinical Correlation

The mouth is always examined as part of a general physical examination. Hence, a medical doctor must be aware of common conditions affecting the region. The lips are blue in cyanosis. Inflammation at the angles of the mouth (angular stomatitis) occurs in vitamin deficiencies.

PALATE

The palate separates the oral cavity from the nasal cavity. It is the arched roof of the oral cavity and the floor of the nasal cavity. It is divisible into an anterior, larger part, the *hard palate*, and a posterior, smaller part, the *soft palate*.

- The hard palate forms the anterior two-thirds of the total palate. It has a skeletal basis formed by the palatal processes of the right and left maxillae and the horizontal plates of the palatine bones. This bony basis is covered by periosteum. The lower surface of the hard palate is lined by mucous membrane of the mouth (oral mucosa packed with glands) and its upper surface by mucous membrane of the nasal cavity (nasal mucosa).

Histology

The oral mucosa of the hard palate is covered by keratinised stratified squamous epithelium. It is thick and appears brown because of the densely packed racemose mucous glands. The periosteum of the underlying bone is more adherent to the mucosa than the bone; therefore, the mucosa and the periosteum together constitute the mucoperiosteum. In the lateral regions of the palate, there is a submucosa that contains the neurovascular bundle of the palate.

- Depressions and foramina in the hard palate:
  - *Incisive fossa:* This is a midline depression immediately posterior to the central incisor teeth; the incisive canals open into it. The number of these

canals is variable; the nasopalatine nerves from inside the nasal cavity pass through these canals to reach the palate.

- ○ ***Greater palatine foramen:*** A large foramen is seen on the lateral border of the hard palate medial to the third molar tooth. The greater palatine nerve and vessels reach the palate through this foramen and run anteriorly.
- ○ ***Lesser palatine foramina:*** These foramina are seen posterolateral to the greater foramen and transmit the lesser palatine vessels and nerves to the soft palate.

☐ A thin ridge called the ***palatine raphe*** runs anteroposteriorly in the midline. An oval prominence called the ***incisive papilla*** lies at the anterior end of the raphe, covering the incisive fossa and the incisive canal. Thick transverse ridges called ***rugae*** run laterally from the palatine raphe. Each ridge has a dense connective tissue core. As already noted, several small mucous glands are present and these are the minor salivary glands. They open by small ducts which are dotted all over the hard palate. The rugae and the numerous small ducts give an orange peel appearance to the oral mucosa of the hard palate. Ducts of many glands in the posterior part of the hard palate join to form two large ducts, one on each side. These open in a pair of depressions called the ***palatine fovea*** seen on either of the palatine raphe close to the posterior border of the hard palate.

> The rugae are more numerous in most mammals, though in humans they are less in number and in importance. In other mammals, food bolus is triturated against the rugae by the tongue and this serves as an important masticatory aid.

☐ The soft palate is a mobile flap of musculoaponeurotic connective tissue suspended from the posterior aspect of the hard palate. It slopes down between the nasopharynx and oropharynx; it is for this reason that it is frequently described as part of the oropharynx. The soft palate is attached to the posterior margin of the hard palate. In the living, it appears red in colour with an yellowish tinge. In its normal relaxed position, it has one surface directed upwards and backwards (posterosuperior nasal surface) and another surface directed forwards and downwards (anteroinferior oral surface). In some individuals, the fovea palatini may be seen in the anterior part of the soft palate instead of the posterior edge of the hard palate. The median part of the soft palate is prolonged downwards as a conical projection called the ***uvula*** (Fig. 9.2). The lateral margins of the soft palate are continuous with two folds of mucous membrane. The anterior of these connects the palate to the lateral margin of the posterior part of the

tongue and is called the ***palatoglossal arch (or fold)***. The posterior fold connects the palate to the wall of the pharynx and is called the ***palatopharyngeal arch (or fold)***. The soft palate has no bony basis but consists of two layers of mucous membrane which are continuous with those lining the upper and lower surfaces of the hard palate. Between these layers of mucosa, there is a fibrous basis called the ***palatine aponeurosis*** which attaches to the posterior edge of the hard palate. The aponeurosis strengthens the anterior part of the soft palate which is consequentially called the ***aponeurotic part***. The aponeurosis, which by itself is the expanded tendon of tensor veli palatini, is thicker in front and thins posteriorly to blend with the muscles. The posterior part of the soft palate is muscular. During deglutition, the soft palate is first made taut; this allows the tongue to press against it and the food bolus is squeezed back to the oral cavity. At the next phase, the soft palate is elevated posterosuperiorly and pressed against the pharyngeal wall; food is prevented from entering the nasal cavity.

☐ If palpated carefully, a small bony prominence can be felt in the lateral area of the anterior part of the soft palate. This prominence is the pterygoid hamulus. A tendinous band called the ***pterygomandibular raphe*** runs down from the pterygoid hamulus to the posterior end of the mylohyoid line in the mandible. This band is between the buccinator and the superior pharyngeal constrictor serving as an anchor to both muscles and giving attachment to their fibres. It raises a prominent vertical mucosal fold when the mouth is opened; this fold can be seen as the posterior boundary of the cheek on the internal aspect.

Histology

The oral mucosa of the soft palate is stratified squamous epithelium. The posterior part of the upper surface of the soft palate comes in contact with the posterior pharyngeal wall during deglutition. This causes repeated friction and so this part of the mucosa is also stratified squamous epithelium. The rest of the upper surface is covered by ciliated epithelium of the respiratory type. A thick sheet of racemose mucous glands is present on the oral aspect of the soft palate.

☐ The fauces (Latin.faucium=throat) is the junctional area between the oral cavity and the pharynx, usually described as part of the oropharynx. It is closely related to the palate. The palatoglossal and palatopharyngeal arches form the anterior and posterior pillars of fauces. As the two arches of the same side diverge from the palate to the pharynx, a triangular fossa is formed between them. This is the tonsillar fossa in which lies the palatine tonsils.

Table 9.1: Muscles of the palate				
Name of muscle	**Origin**	**Insertion**	**Nerve supply**	**Actions**
Tensor veli palatini	Medial pterygoid plate of sphenoid, spine of sphenoid, cartilage of auditory tube	Palatine aponeurosis	Medial pterygoid nerve via the otic ganglion	Tenses the anterior part of the soft palate, opens the auditory tube while swallowing or yawning
Levator veli palatini	Cartilage of auditory tube and inferior surface of petrous part of temporal bone	Palatine aponeurosis	Pharyngeal branch of vagus (cranial part of accessory nerve) via the pharyngeal plexus	Elevates the posterior part of the soft palate during swallowing or yawning
Palatoglossus	Palatine aponeurosis	Side of tongue		Elevates posterior part of tongue and approximates the tongue and the palate
Palatopharyngeus	Palatine aponeurosis	Lateral wall of pharynx		Elevates the pharyngeal wall anterosuperiorly and medially during deglutition
Musculus uvulae	Palatine aponeurosis	Mucosa of uvula		Pulls the uvula up and shortens it

Muscles of the Soft Palate (Table 9.1)

The soft palate has five muscles and these are:

1. Tensor veli palatini
2. Levator veli palatini
3. Musculus uvulae
4. Palatoglossus
5. Palatopharyngeus

Additional Notes on Palatine Muscles

The palatine muscles are responsible for movements of the palate associated with deglutition and with speech.

❑ The ***tensor veli palatini*** (other names—tensor palati, dilator tubae, palatosalpingeus) arises from the base of the skull (Fig. 9.3). It descends and ends in a tendon which winds around the pterygoid hamulus and expands into a wide fibrous band called the ***palatine aponeurosis***. This aponeurosis forms the fibrous basis of the soft palate and gives attachment to other muscles of the palate. Acting together, they make the soft palate taut (especially the anterior part) and help in deglutition by pressing the food bolus between the palate and the tongue. Increased rigidity also aids in palatopharyngeal closure. However, the primary action of this muscle is to open the auditory tube during swallowing or yawning. It helps equalising air pressure between the nasopharynx and middle ear. The pterygoid hamulus acts as a trochlea and so the functional origin for this muscle. As a result, the tensor acts as a true tensor (pulling from the lateral aspect) and not as an elevator (not pulling from above). A bursa intervenes between the tendon and the pulley.

❑ The ***levator veli palatini*** (other names—levator palati, petrostaphylinus) also arises in relation to the base of the skull (Fig. 9.3). It is inserted into the upper surface of the palatine aponeurosis near the uvula. The fibres reach as far as the midline to unite and interlace with the fibres of the fellow of the opposite side. Thus, the two levators together form a sling. The two muscles contract during swallowing and close the pharyngeal isthmus by elevating the palate (especially the posterior part) and bringing it into contact with the posterior wall of the pharynx.

❑ The ***musculus uvulae*** (other names—azygos uvulae, palatouvularis) is attached to the posterior edge of the hard palate near the midline. Its fibres run backwards (on either side of the midline) through the palatine aponeurosis (sometimes enclosed within two sheets of the aponeurosis) (Fig. 9.3). They gain insertion into the mucous membrane of the uvula. This muscle retracts the uvula and thickens the soft palate. It acts in unison with the levator palatini in palatopharyngeal closure.

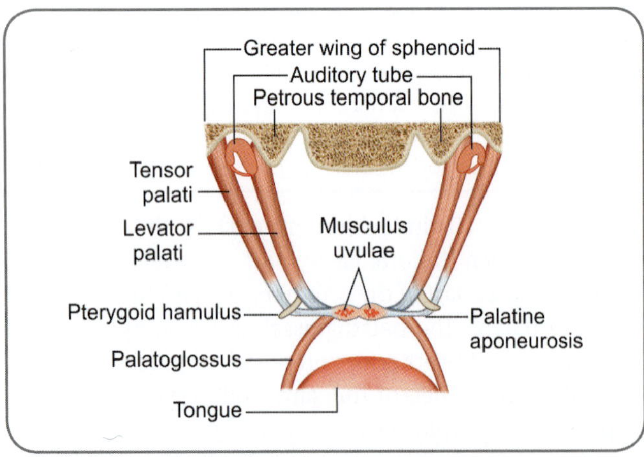

Fig. 9.3: Schematic coronal section through palate to show arrangement of muscles

The fibres of levator and musculus uvulae run at right angles to each other and close the nasopharynx.

- The **palatoglossus** (other name—glossopalatinus) arises from the inferior surface of palatine aponeurosis and is continuous with its fellow at the origin. It passes downwards, forwards and laterally in front of the palatine tonsils to reach the side of the tongue (Fig. 9.3). Along with its overlying mucosa, it forms the palatoglossal fold. Some of its fibres spread over the dorsum of the tongue; many of them pass into the substance and intermingle with the intrinsic fibres. The muscle elevates the posterior tongue and brings the palatoglossal arch closer to its fellow. Thus the oral cavity is shut off from the oropharynx.

- The **palatopharyngeus** (other name—pharyngo-palatinus) arises from the superior surface of the palatine aponeurosis by two fascicles which are separated by the insertion of the levator palatini. The fascicles unite at the posterior aspect of the soft palate and pass downwards behind the palatine tonsils. The muscle descends to the pharyngeal wall; some of its fibres which are in close contact with the salpingopharyngeus muscle get attached to the thyroid cartilage; many of them end in the lateral pharyngeal wall and merge with its fibrous tissue; the rest of them pass around posterorly and decussate with the opposite fellow. Thus the palatopharyngeus forms an incomplete longitudinal muscle coat of the pharynx.

- **Palatopharyngeal sphincter:** Otherwise called the **Passavant's muscle**, this muscle is described by many anatomists and clinicians. It is considered to be a separate palatine muscle that arises from the anterior part of the superior surface of the palatine aponeurosis, passes lateral to the levator palatini, blends with the upper border of the superior pharyngeal constrictor and encircles the pharynx as a sphincter. However, some consider this to be a part of the superior constrictor and palatopharyngeus. Whatever is its origin, when it contracts it produces a ridge called the **Passavant's ridge** (or **Passavant's bar** or **pad** or **cushion**; named after the 19th century German physician Phillias Passavant) on the posterior wall of the nasopharynx. Acting along with the palatopharyngeus and the levator palatini, the palatopharyngeal sphincter helps in closure of pharyngeal isthmus.

Nerve Supply, Blood Supply and Lymphatics of the Palate

The nerves supplying the palate are the greater palatine, lesser palatine and nasopalatine nerves.

- The **greater palatine nerve** (otherwise called the **anterior palatine nerve** or the **major palatine nerve**)

is a branch of the pterygopalatine ganglion. It descends through the greater palatine canal, emerges on the hard palate through the greater palatine foramen and runs forwards in a groove on the inferior surface of the hard palate till the incisive foramen. It supplies the gingivae and mucosa and glands of the hard palate. As it leaves the canal through the foramen, it gives off two or three branches which supply both surfaces of the soft palate.

Development

Before the palate makes its appearance in the embryo, the nasal and oral cavities were one and together. The side walls are formed from the maxillary process. The palate is also predominantly formed from the maxillary process. Thus, it comes to be supplied by the maxillary nerve, which is the nerve of the maxillary process.

- The **nasopalatine nerve** is also a branch of the pterygopalatine ganglion. Traversing the nasal cavity, it runs down the nasal septum and reaches the incisive foramen. It passes through the foramen to enter the oral cavity where it supplies the anterior part of the hard palate and communicates with the greater palatine nerve.

- The **lesser palatine nerve** (usually two in number and called the **middle** and **posterior palatine nerves**) are again branches of the pterygopalatine ganglion. They are smaller than the greater palatine nerve but descend in the same canal. In the lower part of the canal, they diverge and pass through the lesser palatine foramina to emerge on the soft palate. They supply the soft palate, uvula and the tonsils.

Dissection

It is preferable to study the oral cavity and the palate in prosected specimens. Locate the different regions of the oral cavity in a sagittal section of the head and neck. Identify the hard palate, soft palate and uvula. Mark the oropharyngeal and pharyngeal isthmuses. See the palatoglossal and palatopharyngeal arches. Try to see the various muscles.

Functional Components of Innervation

All the branches of the pterygopalatine ganglion are actually sensory branches of the maxillary nerve. The branches supplying the hard and the soft palates convey general sensations.

Taste sensations from the hard and soft palates are conveyed through the greater and lesser palatine nerves. The concerned fibres pass through the pterygopalatine ganglion without relay. From the ganglion, they go via the Vidian nerve (the nerve of pterygoid canal; named after the 16th century Italian anatomist Guido Guidi Vidius) and the greater petrosal nerve to reach the geniculate facial ganglion. Their cell bodies lie in this ganglion. The central processes start from here and pass through the nervus intermedius of Wrisberg; named after the 18th century German anatomist Heinrich Wrisberg) component of the facial nerve to the gustatory nucleus of the tractus solitarius.

Parasympathetic secretomotor fibres to the palatine mucous glands also travel in the palatine nerves. The preganglionic fibres arise in the superior salivatory or lacrimal nuclei (origin not very certain) and pass through the facial nerve, the greater petrosal nerve and the Vidian nerve to reach the pterygopalatine ganglion. After relay in the ganglion, the postganglionic fibres pass through the palatine nerves to supply the mucosal glands.

The hard palate is supplied by the greater palatine branch of the third part of maxillary artery; the soft palate is supplied by the ascending palatine branch of the facial artery and the palatine branch of the ascending pharyngeal artery.

❑ The greater palatine artery is given out in the pterygopalatine fossa and passes through the greater palatine canal along with its accompanying nerve to reach the hard palate. In the greater palatine canal, it gives off two or three lesser palatine arteries and some branches to supply the inferior nasal meatus. On emerging from the greater palatine foramen on the oral surface of the hard palate, it runs forward in a groove near the alveolar border to the incisive canal and supplies the hard palate, the gingivae and the palatine glands. A small twig passes up through the incisive canal, anastomoses with a branch of sphenopalatine artery and supplies the nasal septum. The lesser palatine arteries emerge on the soft palate through the lesser palatine foramina and supply the soft palate.

❑ The ascending palatine artery is given out of the facial artery in the carotid triangle near the hyoid bone. Ascending up between the styloglossus and the stylopharyngeus, it reaches the side of the oropharynx. As it continues to ascend up along the oropharyngeal wall to the skull base, it divides into two near the levator palatini. One branch winds around the upper border of superior constrictor and runs along the levator palatini to supply the soft palate and anastomose with the branches of greater palatine artery and its own fellow of the opposite side. The other branch pierces the superior constrictor to supply the palatine tonsils and the auditory tube.

❑ The palatine branch of the ascending pharyngeal artery is actually a ramus of the pharyngeal branch of the parent artery. It descends between the superior pharyngeal constrictor and the levator palatine to reach the soft palate and supply it.

The veins from the hard palate accompany their arterial counterparts and drain into the pterygoid venous plexus. Most of the veins of the soft palate also run along with their corresponding arteries and drain into the pterygoid plexus. Some of them may also end in the tonsillar venous plexus.

The lymph vessels drain into the deep cervical lymph nodes.

Added Information

Sucking in mammals—significance of palate—modification in muscles

The appearance and presence of a palate is peculiar to mammals. Without a palate, mammals would not be able to survive. The young ones of the mammals are fed from the mammary glands and therefore, should be able to suck. They grasp at the nipple with their lips. When they attempt to suck, the tongue develops a groove in the middle and recedes from the palate. The palate itself is made flat or convex upwards by the pull of its muscles. A vacuum is thereby created in the oral cavity because of the superior and inferior concavities produced and suction force created for sucking to be effective. At the same time, contraction of the two palatoglossi shuts of the oral cavity from the pharynx. Air in the nasal cavity is thus prevented from entering the oral cavity and the vacuum maintained for successful sucking. A child with a cleft palate cannot suck properly because air from the nasal cavity enters into the oral cavity through the cleft and vacuum formation is prevented.

It had become mandatory, as mammals evolved, for an extensive, firm and effective palate to develop so that the nasal cavities could be shut off from the oral cavity while sucking. However, one other difficulty arose. For an efficient functioning of the palate, some muscles were required. Nature being an effective resource manager, did not create new muscles but modified existing ones to suit the demands. The modification could be done to nearby muscles (native muscles) or distant muscles (immigrant muscles, because the modified muscle would have to migrate to the required area). When the palate developed, muscles from the superior constrictor group (or parts of superior constrictor itself) got modified as palatine muscles. Muscle fibres of the same plane as that of the superior constrictor but above the (present) superior border of the muscle developed into the levator palatini, musculus uvulae and palatoglossus. The inner lamina of the superior constrictor per se got detached and modified into the palatopharyngeus. The only other palatine muscle, tensor palatini, is an immigrant. It migrated behind from the anterior mandibular region. This developmental factor can be well seen in the innervation to these muscles. All except the tensor palatini are supplied by the cranial part of accessory nerve via the pharyngeal plexus indicating their relationship to superior constrictor which also has the same supply. The tensor veli palatini is supplied by the mandibular nerve through the nerve to medial pterygoid indicating the muscle's development from the first pharyngeal arch.

Clinical Correlation

Harelip and Cleft Palate

Embryologically, both the upper lip and the palate are derived from three elements. These are:

- The right and left maxillary processes
- The frontonasal process which is a median structure.

Harelip

Harelip is not an anomaly of the palate but of the lip. It is discussed here because both harelip and cleft palate have a common basis. The frontonasal process forms the median part of the upper lip. This part is called the **philtrum**. On each side, the frontonasal process fuses with the corresponding maxillary process. Abnormalities in fusion of these processes lead to clefts in the upper lip (called **harelip** because of the hare normally having an upper lip with a cleft). The defect may be unilateral or bilateral and is more common in males. When defect in fusion is minimal, only a small indentation may be seen in the margin of the lip.

When non-union is complete, the defect extends into the nostril and is usually continuous with a defect in the palate.

Cleft Palate

As stated above, the palate is derived from the frontonasal process and the right and left maxillary processes. The frontonasal process forms the part of the palate that bears the incisor teeth. This part of the palate is also called the **premaxilla**. The rest of the palate is formed by shelf-like projections of the right and left maxillary process. These processes grow towards the midline. Anteriorly, each maxillary process fuses with the corresponding edge of the premaxilla. Behind the level of the premaxilla, the two maxillary processes fuse with each other. From the manner of fusion it will be clear that the line of union of the three elements forming the palate is Y-shaped.

Defects in the process of union lead to the formation of different varieties of cleft palate as follows. Fusion of components of the palate starts anteriorly and proceeds posteriorly.

- Complete non-union gives rise to a Y-shaped cleft. Anteriorly the limbs of the 'Y' become continuous with clefts in the upper lip (i.e., bilateral harelip is also present).
- The premaxilla may fuse with the maxillary process on one side, but not on the other side. At the same time the two maxillary processes do not fuse with each other. This results in a defect that is oblique anteriorly and median posteriorly. It will be associated with a unilateral harelip.
- Both the maxillary processes fuse with the premaxilla but their fusion to each other is deficient. This can give rise to median defects of varying extent. The cleft may involve both the hard palate and the soft palate, may be confined to the soft palate, or may be represented only by a cleft in the uvula.

Cleft palate is more common in females. A harelip can exist without there being any cleft in the palate.

Greater palatine block: This anaesthetic procedure is done by injecting the anaesthetic agent into the greater palatine foramen. The nerve emerges between the 2nd and 3rd upper molar teeth. Greater palatine block anaesthetises the hard palate, its mucosa and the lingual gingivae behind the canine tooth.

contd...

Clinical Correlation *contd...*

Nasopalatine block: This anaesthetic procedure is done by injecting the anaesthetic agent into the incisive fossa in the hard palate. Both the right and the left nerves are anaesthetised by an injection given immediately behind the incisive papilla. The hard palate, palatal mucosa and lingual gingivae of the anterior six maxillary teeth are anaesthetised.

TEETH

As the teeth can be readily seen and felt, some facts about them are commonly known. We know that the newborn have no teeth. The first tooth appears when the infant is about six months old. The teeth in young children gradually fall off and are replaced by new ones, which can almost last throughout life. The teeth which appear in children and fall off with time are called **deciduous** (or milk) teeth. The teeth of the second set which gradually replace the deciduous teeth constitute the **permanent** teeth. The teeth, both deciduous and permanent, have varying shapes. Some have sharp cutting edges and are, therefore, called **incisors** and others which are sharp and pointed are called **canines** as they form the most prominent teeth in canine species (e.g., dogs). Still others have edges suitable for a grinding function—these are the **molars**. In the permanent set, there are also grinding teeth which are somewhat smaller than the molars and are called the **premolars** (as they lie in front of the molars).

- *A set of deciduous teeth consists of the following:* Beginning from the midline (in front) there is a central incisor, a lateral incisor (i.e., two incisors), one canine and two molars (distinguished from each other by being called the **first** and **second molars**). There are, thus, five teeth in each half of each jaw, i.e., twenty in all.

- *A set of permanent teeth consists of the following:* Beginning from the midline there is a central incisor, a lateral incisor, a canine, two premolars (first and second which replace the deciduous molars) and three molars (first, second and third). Thus, in each half of each jaw there are eight teeth, or thirty two in all.

The number of teeth in an individual is represented in a dental formula. The normal dental formula for full complement of permanent teeth is as follows:

$$L\,U\,3.\,2.\,1.\,2. \, — \, 2.\,1.\,2.\,3$$
$$L\,L\,3.\,2.\,1.\,2. \, — \, 2.\,1.\,2.\,3$$

There is considerable variation in the ages at which the various teeth erupt. The following scheme gives the approximate ages of appearance in a form easy to remember. At birth, the jaws are plain rigid bony bars. This is important because absence of teeth helps in grasping the nipple for sucking.

Deciduous Teeth

Central incisor	=	6 months
Lateral incisor	= (+ 2)	8 months
First molar	= (+ 4)	12 months
Canine	= (+ 4)	16 months
Second molar	= (+ 4)	20 months

Note: The first deciduous molar appears before the canine.

Permanent Teeth

First molar	=	6 years
Central incisor	= (+ 1)	7 years
Lateral incisor	= (+ 1)	8 years
Canine	= (+ 1)	9 years
Premolars	= (+ 1)	10 years
Second molar	= (+ 1)	11 years
Third molar	=	17 years +

Note: The first permanent tooth to appear is the first molar. Approximately, one tooth appears every year from the 6th to 11th years. The third molar teeth appear at the age of 17 years or later and are, therefore, called the **wisdom teeth**. Not infrequently one or more third molars may fail to erupt.

Structure of a Typical Tooth

A tooth consists of an upper part, the **crown**, which is seen in the mouth; and of one or more **roots** which are embedded in sockets in the jaw bone (mandible or maxilla). Vertical section through a typical tooth is shown in Fig. 9.4. The greater part of the tooth is formed by a bone-like material called **dentine**. In the region of the crown the dentine is covered by a much harder white material called the **enamel**. Over the root, the dentine is covered by a thin layer of **cement**. The cement is united to the wall of the bony socket in the jaw through a layer of vascular fibrous

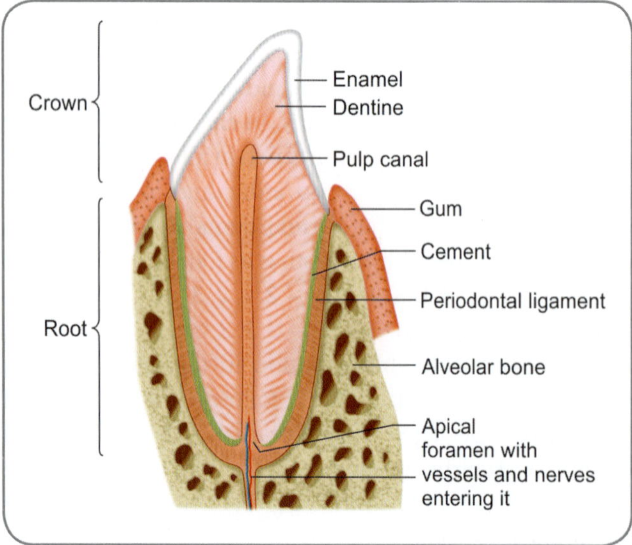

Fig. 9.4: Vertical section through a typical tooth

tissue called the **periodontal ligament or membrane**. The periodontal membrane is modified periosteum. The external surface of the alveolar process is covered by the gum which normally overlaps the lower edge of the crown. Within the dentine there is a **pulp canal** which contains a mass of cells, blood vessels and nerves which constitute the pulp. The pulp canal within the root is the root canal. The blood vessels and nerves enter the pulp canal at the apex of the root through an **apical foramen**.

Differences in Structure of Different Teeth

Details of the anatomy of individual teeth are beyond the scope of this book. However the following information is basic and essential. In describing teeth, certain specific terms are used. As the teeth are arranged in an arch, terms like anterior, posterior, medial or lateral are confusing. The surface of the tooth facing the lip or cheek is the **buccal** (or **labial**) surface. The surface towards the tongue is the **lingual** (or **palatal**) surface. The surface or aspect nearer to the midline is the **mesial** surface; this term corresponds to 'medial' in the case of the anterior teeth and to 'anterior' for the posterior teeth. The term 'proximal' may be used in place of 'mesial'. The surface or aspect opposite to the mesial surface is called the **distal** surface or aspect. The upper surfaces of the lower teeth and the lower surfaces of the upper teeth are referred to as the **occlusal (or masticatory)** surfaces (meaning surfaces which approximate with each other and occlude the gap in between, 'masticatory' because these surfaces help in chewing).

In the case of the incisors the occlusal surface forms a sharp cutting edge. In the canines, it is pointed. In the molars and premolars, the occlusal surface bears rounded elevations or **cusps**. Typically there are two cusps on each premolar and four or five on each molar. The incisors and canines have one root each. The premolars may have one or two roots. The molars have two or three roots, but the last molar may have only one. The meeting of the upper and lower teeth is called **occlusion**. In the relaxed mouth, the upper and lower teeth are a short distance apart. In proper occlusion there is minimal space between the upper and lower teeth, the cusps of one set fitting into the depressions on the occlusal surface of the other set. The upper incisors lie slightly in front of the lower.

Blood Supply and Nerve Supply of Teeth

The **lower teeth** and their periodontal membranes are supplied by branches from the inferior alveolar artery (branch of maxillary artery) and the inferior alveolar nerve (branch of mandibular nerve). The lingual part of the lower gum is supplied by the lingual nerve. The labial part of the gum related to the premolars and molars is supplied by the buccal nerve and that related to the canine and incisors is supplied by the mental nerve.

The ***upper teeth*** are supplied by the anterior and posterior superior alveolar branches of the maxillary artery and the anterior, middle and posterior superior alveolar nerves (branches of the maxillary nerve and its infraorbital continuation). The lingual part of the upper gum related to the premolars and molars is supplied by the greater palatine nerve and that part related to the canine and incisors is supplied by the nasopalatine nerve. The labial part of the upper gum related to the molars is supplied by the posterior superior alveolar nerve and that related to the premolars, canine and incisors is supplied by the infraorbital nerve.

⚕ Clinical Correlation

Teeth and Gums

The number of teeth present should be compared with the age of the paediatric patient. Delayed eruption may be a sign of malnutrition or some other growth disorder. In older people who have lost some teeth, the absence of sufficient teeth for mastication may result in digestive disorders. Peg-shaped upper central incisors (**Hutchinson's teeth**) may be an indication of syphilis. In this condition the teeth may show notches. Notching may also be present in rickets. In acromegaly, the lower jaw becomes relatively larger than the upper jaw so that the teeth go out of alignment.

Gums are often the site of chronic inflammation especially in persons with poor oral hygiene. Inflammation of gums is called **gingivitis**. The gums may be inflamed and bleed easily. Pockets may form and pus may be present in them (**pyorrhoea**). Infection in the gums can lead to digestive and respiratory problems. A blue line running along the edge of the gum may be a sign of lead poisoning. The gums are swollen and spongy and bleed easily in scurvy which is caused by a deficiency of vitamin C.

Dental caries is a very common disease of the teeth. Microorganisms produce small cavities that gradually enlarge. The patient is usually unaware of them until the cavity invades the dentine when the teeth become sensitive to hot and cold, or to sugar. If left untreated, the cavity ultimately reaches the pulp of the tooth resulting in severe pain. Dental caries can be prevented by teaching children to brush their teeth after meals. Prevention can also be ensured by regular check up by a dentist. If cavities are discovered early, they can be filled up and the tooth can be saved. Caries is common in milk teeth and is ignored on the assumption that these teeth are going to be replaced. However, caries can result in too early loss of milk teeth and this can result in abnormal eruption of permanent teeth.

Medicolegal Importance of Teeth

Being very hard, teeth are preserved for a very long time after death. They can be very useful in identifying a dead person especially if a dentist's record of the state of the teeth is available.

NASAL REGION

The nasal region includes the nasal cavities and the paranasal sinuses. The nasal cavity is divided by a median septum into the right and left halves. Each half of the nasal cavity opens on to the face through the anterior or ***external naris*** (naris is a latin term and indicates singular; the plural term is nares) and posteriorly it opens into the nasopharynx through the posterior nasal aperture or choana (Greek.choana=funnel; plural, choanae). A schematic coronal section through the nasal cavity shows that each half of the cavity is triangular. It has a vertical medial wall formed by the nasal septum, a sloping lateral wall, a relatively broad floor formed by the palate (which separates it from the oral cavity) and a narrow roof which lies at the junction of the medial and lateral walls. These walls have a skeletal basis that is made up predominantly of bone, but is cartilaginous at some places.

The skeletal basis is covered (over most of the nasal cavity) by mucous membrane. Typically the mucosa is moist and highly vascular. It serves to warm inspired air and also helps to remove dust (which sticks to the moist wall). For these reasons, the mucosa is referred to as ***respiratory***. The mucosa lining the uppermost part of the septum and the adjoining part of the lateral wall differs from that present elsewhere in the nasal cavity; it is characterised by the presence of receptor cells which are sensitive to smell. The mucosa in this region is, therefore, called the ***olfactory mucosa***. Olfactory nerves arise from the olfactory mucosa. A small area of the nasal cavity (near the anterior nares) is lined not by mucous membrane, but by skin. This skin bears hair that serve to trap dust present in the inspired air.

Each half of the cavity is divisible into three regions, namely the vestibule, chemosensitive olfactory area and the respiratory area. The vestibule is the anteroinferior portion that is immediately adjacent to the anterior naris. The olfactory area is the posterosuperior portion adjacent to the roof of the cavity. The rest of the area between the anterior and posterosuperior portions is the respiratory portion subserving the principal function of respiration. The nasal vestibule narrows in its posterior aspect (where it becomes continuous with the respiratory portion) to form the nasal valve. The nasal valve is the narrowest portion of the nasal airway.

The nasal cavity is in communication with some air filled spaces which are either present within the lateral nasal wall or within adjacent bones. These air filled spaces are the paranasal air sinuses.

EXTERNAL NOSE

The osseocartilaginous framework of the nasal cavity extends anteriorly to form the external nose. The skin over the external nose is thin and loosely attached to the underlying structures. It is thicker over the apex and the alae and is studded with numerous sebaceous glands. The openings of these glands are usually visible.

The lower part of the nasal septum and the skin of the alae are supplied by the lateral nasal and septal branches of the facial artery. The dorsum of the nose and the lateral aspects are supplied by the dorsal nasal branch of the ophthalmic artery and the infraorbital branch of the maxillary artery.

The veins of the dorsum and the lateral aspects drain into the facial vein. Those from the root of nose drain into the ophthalmic veins.

The lymphatic drainage from most of the external nose is into the submandibular group of nodes. Lymph from the root of nose drains into the superficial parotid nodes.

The skin of the external nose is supplied by the infratrochlear and external nasal branches of the nasociliary nerve and by the nasal twig of the infraorbital nerve. The muscles are supplied by the buccal branches of the facial nerve.

NASAL CAVITY

Medial Wall—Nasal Septum

The skeletal basis of the medial wall of the nasal cavity is formed by the osseocartilaginous components of the nasal septum. The bony components are:

- ❑ *The perpendicular plate of the ethmoid bone:* Forms the posterosuperior part of the bony septum,
- ❑ *The vomer:* Forms the posteroinferior part of the bony septum including the posterior border.

The wide anterior deficiency in the bony septum is filled by the septal cartilage.

Around the edges of the septum there are small contributions from the nasal, frontal, sphenoid, maxillary and palatine bones (Fig. 9.5). As a point of practical importance, it may be remembered that the septum is fairly often deflected to one side so that one-half of the nasal cavity may be larger than the other.

The nasal septum is relatively featureless. A few grooves related to the nasopalatine vessels and nerves may be found. The mucous membrane of the septum is adherent to the underlying bone forming a mucoperiosteum. In the posterosuperior part, the mucoperiosteum is thickened by vascular connective tissue.

Lateral Wall

The bones taking part are (Figs 9.6 and 9.7):

- ❑ The medial surface of maxilla—anteroinferiorly,
- ❑ The ethmoid bone—superiorly,
- ❑ The palatine bone—posteriorly,
- ❑ The inferior nasal concha—inferiorly and
- ❑ The lacrimal bone—anterosuperiorly.

The lateral wall of the external nose also forms the anterior part of the lateral wall of the nasal cavity. The predominant feature of the lateral wall is the presence of three projections of variable size. These projections are the nasal conchae.

Conchae

These are the three anteroposterior elevations on the lateral wall, called the *superior*, *middle* and *inferior nasal conchae* (or turbinates; singular, concha; Latin. concha=shell, turbinated=shaped like a top or scroll). Each concha has a core of bone covered by mucous membrane. The bony core of the superior and middle conchae is formed by parts of the ethmoid bone, while that of the inferior concha is an independent bone

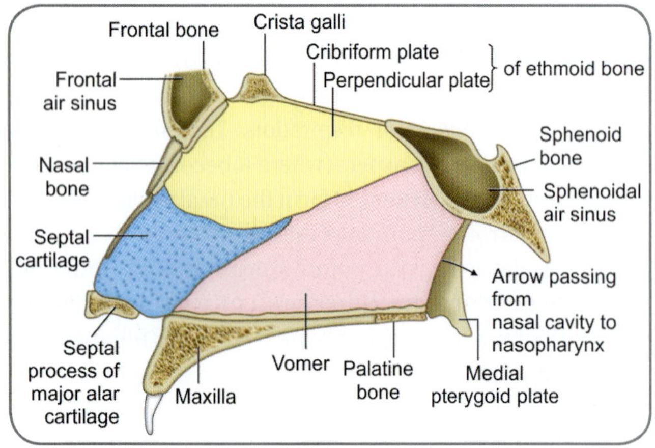

Fig. 9.5: Skeletal basis of the nasal septum

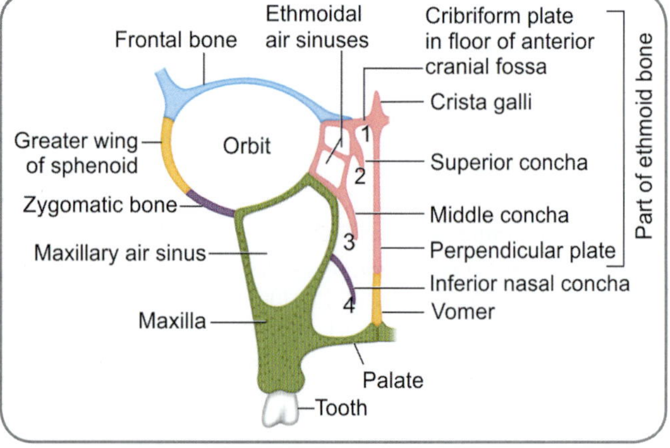

Fig. 9.6: Coronal section of the nasal cavity

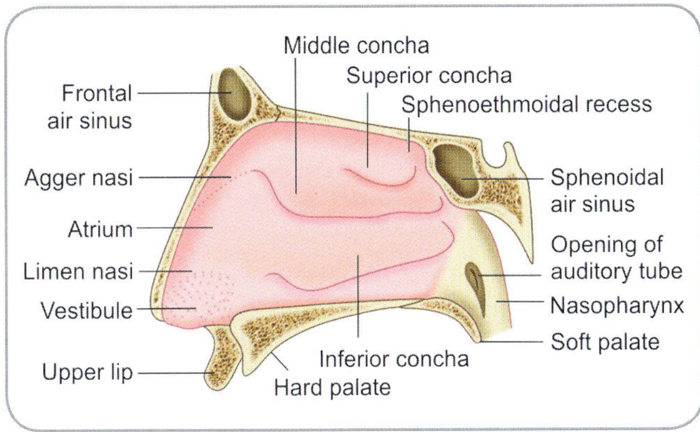

Fig. 9.7: Lateral wall of the nasal cavity with intact mucous membrane

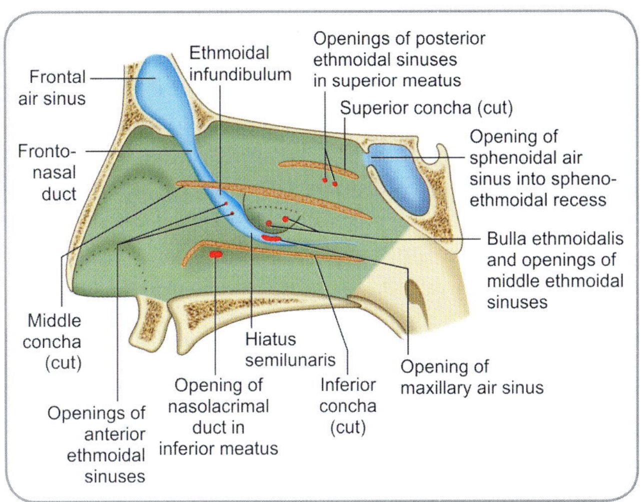

Fig. 9.8: Lateral wall of the nasal cavity after removing the conchae to reveal structures deep to them

(Fig. 9.6). Each concha has an upper border attached to the rest of the lateral wall and a free lower margin. The conchae generally curve inferomedially; each of them, therefore roofs a groove called the *meatus* that is open to the nasal cavity.

Meatuses

The grooves or spaces deep to the superior, middle and inferior conchae are called the *superior*, *middle* and *inferior meatuses* (Latin.meo=to go, meatus=passage) respectively (Fig. 9.8). Each meatus can be defined to lie inferior and lateral to its corresponding concha. Above the superior concha, is found a triangular space (Fig. 9.6). This is the *sphenoethmoidal recess* (Fig. 9.7). Occasionally, an additional concha (called the *highest nasal concha*) may be present on the lateral wall of the sphenoethmoidal recess.

Some structures in the meatuses of the lateral wall of the nose can be clearly seen only when the conchae are cut away (Fig. 9.8).

The superior meatus is a short oblique passage below and lateral to the superior concha. It extends only halfway along the upper border of the middle concha. The posterior ethmoidal air sinuses open into it through a variable number of openings.

The middle meatus is deeper in front than behind. It is continuous anteriorly with a shallow depression above the vestibule of the nose; this depression is the atrium (of the middle meatus). An ill defined curved ridge slopes downwards and forwards from the anterior free edge of the middle concha; this ridge is the agger nasi (Latin. agger = elevation or ridge, plural aggares) which is better developed in the newborn than adults. A rounded elevation seen in the middle meatus is the *bulla ethmoidalis*. Below and in front of the bulla is a curved groove called the *hiatus semilunaris*. The bulla itself is formed by the expansion of the underlying middle ethmoidal air sinuses and these sinuses open on it or just above it. The lower curved margin of the hiatus is sharp; it is produced by a process

of the ethmoid bone called the *uncinate process*. The anterior end of the hiatus is continuous with a depression called the *ethmoidal infundibulum,* into which open the anterior ethmoidal air sinuses. The upper end of the infundibulum is usually continuous with the *frontonasal duct* which connects the frontal sinus to the nasal cavity. The opening of the maxillary sinus into the nasal cavity is seen below the bulla, usually hidden by the flange of the uncinate process.

The inferior meatus is the largest and extends through the entire length of the lateral nasal wall. It reaches the nasal floor and is deepest at the junction of its anterior third with the middle third. This is the place where the nasolacrimal duct opens into it.

The presence of conchae in the lateral wall of the nasal cavity greatly increases the surface area of the nasal mucosa. This factor combined with the presence of a highly vascular and moist mucous membrane enables the nose to effectively warm inspired air and to make it moist.

The part of the nasal cavity just above the anterior nares is called the *vestibule*. The vestibule is lined by skin. At the upper limit of the vestibule (where skin meets mucous membrane), there is a curved elevation called the *limen nasi* (Fig. 9.7). Above the limen nasi is the *atrium*. The upper limit of the atrium, as already noted, is marked by another curved ridge called the *agger nasi*.

Roof

The roof of the nasal cavity lies at the junction of the medial and lateral walls (Fig. 9.6). When traced anteroposteriorly, is divisible into three parts:

1. The middle part is formed by the mucous membrane covering the ethmoid bone and is almost horizontal;
2. The anterior part slopes downwards and forwards and is formed by the mucous membrane covering the frontal bone above and the nasal bone below;

157

3. The posterior part slopes downwards and backwards and is formed by the mucous membrane covering the sphenoid bone and the body of the sphenoid.

The anterior slope formed by the nasal and the frontal bones actually contributes to the upper part of the external nose. The central horizontal portion is formed by the cribriform plate of the ethmoid bone which separates the nasal cavity from the anterior cranial fossa. An anterior foramen in the cribriform plate transmits the anterior ethmoidal nerve and vessels; there are numerous other foramina through which the olfactory nerves pass. The cribriform plate is so named because of the sieve like pattern produced by the foramina.

Floor

The floor of the nasal cavity is formed by the mucous membrane covering the upper surface of the hard palate. Each half of the hard palate is formed in its anterior three-fourths by the maxilla (palatine process) and in its posterior one-fourth by the palatine bone (horizontal plate). The floor itself is smooth and slopes from anterior to posterior. It is also concave transversely. Anteriorly and close to the septum, the floor exhibits a small opening that leads to the incisive canal which descends to the incisive fossa.

Blood Supply

Arterial Supply

The chief artery to the mucous membrane of the nose is the sphenopalatine branch of the maxillary artery. It gives off posterolateral nasal branches to the lateral wall of the nose (supplying the conchae and meatuses) and posterior septal branches to the septum (supplying the posteroinferior part of the nasal septum).

The sphenopalatine artery passes through the sphenopalatine foramen to reach the posterosuperior aspect of the nasal cavity. The posterolateral branches are given out here. These branches descend and ramify over the conchae and the meatuses; they also anastomose freely with branches of the ethmoidal arteries and the nasal branches of the greater palatine artery. Twigs from this anastomosis go to the frontal, maxillary, ethmoidal and sphenoidal air sinuses. Crossing on the sphenoid, the sphenopalatine artery then reaches the nasal septum and ends in a set of posterior septal branches which supply the septum and anastomose with the branches of the ethmoidal arteries. A slender branch of the sphenopalatine artery descends on the nasal septum to reach the incisive canal and anastomoses with the terminal part of the greater palatine artery and the septal branch of the superior labial artery.

The other arteries helping in the supply are:

- ❑ The anterior and posterior ethmoidal branches of the ophthalmic artery—they enter into the nasal cavity through the anterior and posterior ethmoidal foramina in the roof of the cavity; they supply the roof and the adjacent part of the septum; the anterior ethmoidal artery also supplies the ethmoidal air sinus, frontal air sinus and a small part of the lateral nasal wall; the posterior ethmoidal artery also supplies the posterior ethmoidal air sinuses.

- ❑ The greater palatine branch of the maxillary artery—as the greater palatine artery runs in the greater palatine canal, it gives off some branches (nasal branches) which supply the inferior meatus of the nose; after emerging out of the greater palatine foramen and running on the inferior surface of the hard palate, the terminal part of this artery ascends through the incisive canal to anastomose with the branches of the sphenopalatine artery, branches of anterior ethmoidal artery and septal branch of superior labial artery.

- ❑ The septal branch of the superior labial artery is given out near the inferior aspect of the septum in the upper lip. It ascends up the septum to anastomose with the branch of the sphenopalatine artery, terminal part of the greater palatine artery and the branches of the ethmoidal arteries.

Venous Drainage

The veins accompany the arteries. The smaller venules form a rich submucosal plexus; the plexus is particularly dense over the middle and inferior conchae and the posterior part of the septum. The plexuses over the conchae are cavernous and when engorged, increase much in size. The deep layer of the mucosa has numerous arteriovenous anastomoses. The venules collect into small sized veins. Veins from the anterior aspect of the nasal cavity drain into the anterior ethmoidal veins which in turn drain into the ophthalmic vein or the facial vein. Veins from the posterior aspect drain into the sphenopalatine vein and through that to the pterygoid venous plexus. A few veins near the roof may pass through the cribriform plate and drain into the veins on the orbital surface of the frontal lobe of brain. A vein may pass through the foramen caecum and drain into the superior sagittal sinus.

Lymphatic Drainage

Lymphatics from the anterior aspect run to the superficial aspect, join with the lymphatics of the external nasal skin and drain into the submandibular lymph nodes. Lymphatics from the posterior aspect join along with those of the paranasal sinuses and the nasopharynx and drain into the upper deep cervical nodes through the retropharyngeal nodes. The posterior nasal floor may drain into the parotid lymph nodes.

Nerve Supply

The nerve supply of the nasal cavity will fall into two categories—the olfactory (special sense) supply and the general sensory supply.

Dissection

Make a parasagittal section of the head and neck; study the nasal septum and its parts.

On the other part of the section, the lateral wall of the nose would be exposed. Study the nasal conchae. Identify each of the conchae. Lift the middle concha and study the features in the middle meatus. Identify the bulla ethmoidalis, hiatus semilunaris and the various apertures on them. Try to insert a slender probe into the frontonasal duct. Lift the inferior concha and see the inferior meatus. Insert a probe into the nasolacrimal duct.

Switch your attention to the superior meatus and the sphenoethmoidal recess. Locate the openings of the paranasal sinuses. Try to insert probes into them.

The olfactory supply is through the olfactory nerves.

The general sensory supply is associated with the branches of the ophthalmic and maxillary nerves. Touch, pain and temperature sensations are subserved by these nerves.

- The upper and anterior part of the cavity (part of roof, part of the lateral wall and part of the septum) is innervated by the anterior ethmoidal branch of the nasociliary nerve.
- The nasal vestibule is supplied by the infraorbital nerve.
- The lower and anterior part of the cavity (lower part of septum, anterior floor, anterior lateral wall till the maxillary sinus opening) is supplied by twigs from the anterior superior alveolar nerve.
- The upper and posterior part of the cavity (posterior three-fourths of the lateral wall, roof and floor) is supplied by the lateral posterior superior nasal branches of the pterygopalatine ganglion and the posterior inferior nasal branches of the greater palatine nerve.
- The lower and posterior part of the cavity (posteroinferior part of the septum) is supplied by the medial posterior superior nasal branches of the pterygopalatine ganglion and the nasopalatine nerve (which also is a branch of the pterygopalatine ganglion).
- The most upper part of the roof and the septum may be supplied by twigs from the Vidian nerve (nerve of pterygoid canal).

Autonomic Innervation

The postganglionic sympathetic fibres which are derived from the plexuses around the adjacent major arteries are distributed to the nasal blood vessels. The postganglionic parasympathetic fibres from the pterygopalatine ganglion provide the secretomotor supply to the nasal mucosal glands.

Added Information

- Minor contributions to the nasal septum are—antero-superior–nasal bone and nasal spine of frontal bone; posterosuperior–rostrum and crest of sphenoid; inferior–nasal crests of maxilla and palatine bone.
- A small depression may be seen in the lower edge of the septal cartilage above the incisive canal. This is a remnant of the foetal nasopalatine canal.
- A tiny blind pouch may be seen on either side of the septum close to its lower edge. This pouch will have the vomeronasal organ. This is an accessory olfactory organ similar to the olfactory epithelium present in lower animals like the reptiles. In humans, it is considered vestigial.
- The superior concha is the shortest and the shallowest of all the three conchae. In clinical parlance, it is usually referred to as Morgagni's concha.
- The middle concha is often called the **concha medialis**.
- The highest nasal concha, which may sometimes be present, is otherwise called the **supreme concha**, fourth concha or Santorini's concha.
- Sphenoidal conchae or Bertin's bones are small paired ossicles found in the roof of the nasal cavity. They project from the sphenoid.
- If a supreme concha is present, the passage immediately below it will be called the **supreme meatus**. An opening of the posterior ethmoidal sinus may be present in it.
- An air cell may occur within the middle concha itself, in which it is called a **concha bullosa**.

Clinical Correlation

Infection and inflammation of the nasal cavity is called **rhinitis**. Infection from the nasal cavity can spread to paranasal sinuses, to the middle ear, to the pharynx and larynx, and even to the anterior cranial fossa (through the roof of the cavity). Rhinitis may be caused by viruses (as in common cold), by bacteria and by allergy (allergic rhinitis). In allergic rhinitis, the nasal mucosa can undergo hypertrophy resulting in chronic blockage.

Children often insert foreign bodies into the nose and they can get impacted there.

The nasal septum is commonly deflected (i.e., it comes to lie to one side of the middle line). This can cause blocking of the nasal cavity on one side and may require surgical correction.

In cleft palate, the nasal cavity is in communication with the mouth. Because of leakage of air into the nose during speech the person has a nasal twang.

Variations in the structures of the lateral nasal wall may cause airway obstruction. Oversized bulla, internally curved middle or inferior conchae, presence of a concha bullosa are the common variations. Conchae may have to be excised to clear the airway.

In conditions of allergy or infection, the highly vascular mucoperiosteum of the lateral wall may be swollen. The venous plexus of the submucosa is cavernous and can get engorged so much that the air way could be obstructed. In some patients, a regular periodical cyclical swelling may cause impaired ventilation.

contd...

Paranasal Sinuses

Other names: Accessory air sinuses, accessory nasal sinuses.

These are spaces present in the substance of bones related to the nasal cavities. Each sinus opens into the lateral wall of the nasal cavity and is lined by mucous membrane continuous with that of the latter. Because of this communication, each sinus is normally filled with air.

Frontal Sinuses

Other names: Antrum frontale.

The right and left frontal sinuses are present in the part of the frontal bone deep to the superciliary arches. Each sinus lies deep to a triangular area, the angles of which lie (a) at the nasion, (b) at a point about 3 cm above the nasion and (c) at a point on the supraorbital margin at the junction of the medial one-third with the lateral two-thirds. The cavity of the frontal sinus extends for some distance into the orbital plate of the frontal bone between the roof of the orbit and the floor of the anterior cranial fossa. Each frontal sinus usually opens into the middle meatus through the frontonasal duct. This duct is usually continuous, below, with a funnel-like space the ethmoidal infundibulum (Fig. 9.8) that is continuous with the upper end of the hiatus semilunaris.

contd...

Sphenoidal Sinuses

Other name: Antrum sphenoidale.

The right and left sphenoidal sinuses are present in the body of the sphenoid bone. Each sinus opens into the corresponding half of the nasal cavity through an aperture on the anterior aspect of the body of the sphenoid. The part of the nasal cavity into which the sinus opens lies above the superior nasal concha and is called *sphenoethmoidal recess* (Fig. 9.9).

Maxillary Sinuses

Other names: Antrum maxillaris, antrum of Highmore.

The maxillary sinuses are the largest of the paranasal sinuses. Each maxillary sinus lies within the body of the maxilla of that side and is in the shape of a truncated pyramid. Every wall or boundary of the sinus has important relations. The truncated apex is lateral and may extend into the zygomatic bone. The base is medially placed and occupies most of the lateral nasal wall. Its posterior wall has canals which conduct the posterior superior alveolar vessels and nerves to the molar teeth. The floor of the sinus is formed by the alveolar process itself and lies below the level ofs the nasal floor; it is related to the roots of the lower premolar and molar teeth. The infraorbital canal forms a ridge on the roof and the bone here frequently has dehiscences. A thin canal from the infraorbital canal runs on the anterior wall of the sinus; this is the canalis sinuosus and conducts the anterior superior alveolar vessels

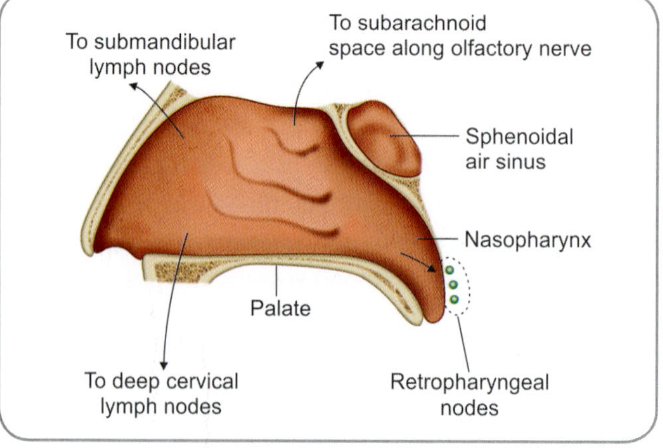

Fig. 9.9: Scheme to show the lymphatic drainage of nasal cavity

and nerve. The medial wall has a large posterosuperior deficiency and this forms the large maxillary hiatus. In the normal position, the large hiatus itself is partially closed by the inferior concha, uncinate process of the ethmoid bone, perpendicular plate of palatine bone, lacrimal bone and overlying mucosa. By such partial closure, only three smaller openings are seen. These are the ostium and the anterior and posterior fontanelles. All the three openings are close to the roof of the sinus (at the angle between the roof and the medial wall actually); the ostium opens into the middle part of the hiatus semilunaris of the middle meatus of the lateral nasal wall. The fontanelles open into the middle meatus as accessory openings.

> **Surface Marking**
>
> The surface outline of this sinus is irregular. A roughly triangular marking, more or less, marks the sinus. The superior border is marked parallel but just below the infraorbital margin. The medial border is marked immediately lateral to the lateral border of the nose but stops just above the alveolar process of the maxilla. The oblique lateral border is carried upwards and laterally from the lower limit of the medial border to the lateral limit of the superior border. All the angles of the triangle are curved and rounded.

Ethmoidal Air Sinuses

Other name: Antrum ethmoidale.

The ethmoidal air sinuses are located within the lateral part (or labyrinth) of the ethmoid bone. The ethmoid sinus of each side is not a single cavity like the other sinuses but consists of small multiple cavities. They can be divided into anterior, middle and posterior groups. Each group may have one large sinus or four to six small sinuses. Each labyrinth (right or left) is bounded medially by the medial plate and laterally by the orbital plate of the ethmoid bone. The orbital plate is usually paper thin and called the *lamina papyracea*. The ethmoidal air sinuses lie between these plates. Though these sinuses lie within the ethmoid bone, the walls of many of them are completed by parts of the frontal, maxillary, lacrimal, sphenoidal and palatine bones. The anterior ethmoidal sinuses open into the ethmoidal infundibulum or into the upper part of the hiatus semilunaris. The middle ethmoidal sinuses open on or near the bulla ethmoidalis. The posterior ethmoidal sinuses open into the superior meatus.

Palatine Sinuses

These are air spaces, one on each side in the orbital processes of palatine bones. They open into the sphenoidal or posterior ethmoidal sinus. They are considered extensions of posterior ethmoidal sinuses.

Other Openings in the Nasal Cavity

In addition to the openings of the paranasal sinuses, the lateral wall of the nose shows the opening of the *nasolacrimal duct.* This duct conveys lacrimal fluid from the conjunctival sac via the lacrimal sac. It opens into the anterior part of the inferior meatus (Fig. 9.8).

> ### Added Information
>
> - Mucous secreted within the paranasal sinuses is pushed into the nasal cavity by ciliary action.
> - Cilia are not present throughout the sinuses but are always present near their openings.
> - The frontal sinuses are rudimentary/not present at birth, well developed by the eighth year and reach full size by puberty. They are prominent in males. Their arterial supply is from the supraorbital and anterior ethmoidal arteries, venous drainage through the supraorbital and superior ophthalmic veins and lymphatic drainage to the submandibular lymph nodes. They are innervated by the branches of the supraorbital nerve, a branch of the ophthalmic nerve.
> - The sphenoidal sinuses are tiny cavities at birth and develop at puberty. The sinuses of both sides can overlap one another or intercommunicate. Each sinus may be subdivided by accessory septa and the air cavity may extend into other parts of the sphenoid bone. Each sinus is related above to the optic chiasma and the pituitary gland; and to the cavernous sinus and internal carotid artery laterally. Some of the bony walls may be dehiscent and in such cases, the mucosa of the sinus may be in contact with the cranial dura. The arterial supply is by the posterior ethmoidal branch of the ophthalmic artery and nasal branch of sphenopalatine artery, venous drainage through the posterior ethmoidal vein and lymphatic drainage to the retropharyngeal nodes. Their sensory innervation is through the posterior ethmoidal nerve, a branch of the nasociliary nerve, in turn a branch of the ophthalmic nerve.
> - The maxillary sinuses have very important relations. Any of their walls may be deficient resulting in communication of the mucosa with the structures of the cavity/fossa thus connected. They are rudimentary cavities at birth, enlarge during the eruption of the permanent teeth and develop fully at puberty. Their arterial supply is through the anterior, middle, posterior superior alveolar branches of the maxillary artery and the infraorbital and greater palatine arteries, venous drainage through corresponding veins which drain into the facial vein or pterygoid venous plexus and lymphatic drainage to the submandibular nodes. Their innervation is by the infraorbital and anterior, middle, posterior superior alveolar branches of the maxillary nerve.
> - Though three groups of ethmoidal sinuses have been described for long, clinicians prefer to group into two— the anterior and the posterior groups. The middle sinuses are incorporated into the anterior group. The anterior group comprises the anterior ethmoidal air cells which are also called the *peri-infundibular air cells* and the middle ethmoidal air cells which are otherwise called the *bullar sinuses*. The most anterior of the peri-infundibular cells are called the *agger nasi cells*. One or two of the agger nasi air cells may extend forwards to encroach on the frontal air sinus. Some of the air cells of this group may extend below the orbital floor and are then called the *Haller's cells*. The posterior group of air cells are very close to the optic nerve

and the optic canal. The ethmoidal sinuses are shallow diverticula attached to the nasal cavities at birth but are significant; they are susceptible to inflammation at that stage itself. They increase in size by 6-8 years and at puberty. Their arterial supply is by nasal branches of the sphenopalatine artery and the anterior and posterior ethmoidal branches of the ophthalmic artery and venous drainage through corresponding veins. The lymphatics from the anterior group drains to the submandibular nodes and the posterior group to the retropharyngeal nodes. Their innervation is from the anterior and posterior ethmoidal branches of the nasociliary nerve.

❑ ***Osteomeatal complex:*** This refers to the ostium of the maxillary sinus, ethmoid infundibulum, hiatus semilunaris and frontal recess. It is final pathway common for the frontal, maxillary, anterior and middle ethmoidal sinuses. Obstruction in the related areas can result in persistent sinusitis.

Sinusitis

❑ Paranasal sinuses are frequent sites of inflammation and infection (sinusitis). The infection usually reaches them from the nasal cavity. As the sinuses open into the nasal cavity through narrow openings, slight swelling of the mucosa or presence of thick secretions at the orifice, can block outflow of secretions that accumulate within the sinus. This is one reason why sinusitis so often becomes chronic. This is especially true in the case of the maxillary sinus because the level of the opening of the maxillary air sinus into the nose is placed at a higher level than the floor of the sinus, so that natural drainage is difficult; some residual secretions stagnate within the sinus.

❑ The secretions of the sinuses are pushed to the nasal cavity by the ***mucociliary escalator*** (or the mucociliary ladder) mechanism. The cilia beat towards the openings of the sinuses and push the mucous droplets and any other particle that may happen to get in. Though all sinuses have a mucociliary escalator mechanism, the natural drainage of the maxillary sinus depends only on its intactness and efficacy. The other sinuses have gravitational drainage and the maxillary sinus does not. The mucociliary escalator may be insufficient when the maxillary sinus is inflamed or when its secretions are more.

❑ To facilitate drainage of the maxillary sinus, it is sometimes necessary to make an artificial opening into the sinus through the inferior meatus of the nose (this opening will be closer to the floor of the sinus than its original opening). Such an opening is created in diseases of mucociliary clearance. The sinus can also be drained through the vestibule of the mouth near the canine tooth.

❑ A study of the anatomy of the lateral wall of the nose shows that secretions draining out of the frontal air sinus flow towards the opening into the maxillary sinus. For this reason frontal sinusitis often leads to maxillary sinusitis. Infections in the maxillary sinus can easily spread to the orbit.

❑ The prechambers of the sinuses are important anatomical zones in deciding the state of infection in a sinus. Prechambers are actually the areas where the sinuses exchange air with and drain their mucous into the nasal cavity. The middle meatus and the hiatus semilunaris are the prechambers for the anterior ethmoidal, maxillary and frontal air sinuses and for the osteomeatal complex. The sphenoethmoidal recess and the superior meatus are prechambers for the sphenoidal and posterior ethmoidal sinuses. Obstruction of the prechambers by inflamed mucosa and by polyps and tumours lead to stagnation of mucous secretions, overgrowth of bacteria and result in chronic infection.

❑ Tooth infections can cause maxillary sinusitis.

Diagnosis of Sinusitis

Infection in paranasal sinus leads to generalised headache. In addition, tenderness can often be elicited by pressure over the corresponding sinus. In frontal sinusitis, tenderness is felt by upward pressure over the medial part of the superior orbital margin. In maxillary sinusitis there may be tenderness over the cheek below the inferior orbital margin. In ethmoidal sinusitis, there will be tenderness over the medial wall of the orbit.

Pain originating in the paranasal sinuses may be referred to other sites.

❑ The frontal sinus is supplied by the supraorbital nerve which also supplies the skin of the forehead and anterior part of the scalp. Therefore, frontal headache is almost always present in frontal sinusitis.

❑ In maxillary sinusitis pain may be felt in the upper jaw and teeth.

Diagnosis of maxillary sinusitis can be confirmed by transillumination. In a darkened room, light (from a torch) is directed towards the wall of the sinus through the mouth or through the cheek. The presence of fluid, rather than air, reduces transmission of light.

Imaging of sinuses: Radiography (X-ray pictures) is useful in confirming presence of sinusitis. In normal X-rays, normal sinuses appear radiolucent. When diseases and inflamed, they show opacity.

❑ In posteroanterior skull X-rays, the frontal sinuses are clearly seen above the nasal cavity and the medial aspects of the orbits. The ethmoidal sinuses appear less clearly, superimposed on each other. The sphenoidal sinuses are not clearly made out because the ethmoidal sinuses obscure them. The maxillary sinuses are seen below the orbits and lateral to lower nasal cavities. They may extend inferiorly into the alveolar processes.

❑ In lateral views, the frontal sinuses are well seen, especially with regard to their posterosuperior extent into the frontal bone and extensions into the orbital roofs. The ethmoidal sinuses can be made out stretching from the frontal process of maxilla to the sphenoidal sinus. The sphenoidal sinuses can be made out below the hypophyseal fossa. However, the two sphenoidal sinuses superimpose each other. The maxillary sinus is also seen in this view.

❑ In occipitomental view X-rays, all the sinuses are well seen. The sphenoidal sinus is better viewed in an occipitomental

contd...

contd...

Clinical Correlation *contd...*

view (open mouth position). Occipitomental view is also called **Water's view**.

Computerised tomography images give high grade pictures of the sinuses.

In medicolegal cases, morphology of frontal sinus in radiological images can be used for identification of individuals.

The frontal sinus gives contour to the head. The obliquity of the forehead (sloping upwards) in males is due to the prominence of the frontal sinuses. On the contrary, the face profile in females and children is vertical or convex.

Extraction of upper molars may damage the floor of the maxillary sinus. Impacted molars may cause fractures in the walls of the sinus.

Tumours in the maxillary sinus:
- May displace the eyeball by pushing up the orbital floor;
- May cause nasal obstruction by projecting into the nasal cavity;
- May cause a swelling on the cheek;
- May press on the pterygoid muscles causing trismus by pushing back into the infratemporal fossa;
- May cause dental loosening and teeth problems by pushing down.

Development

Development of the nasal cavity, palate and paranasal sinuses: By the end of the 4th week of intrauterine life, developmental changes occur in the facial region of the embryo. An epithelial thickening called the **nasal** or **olfactory placode** occurs on either side of the front of head above the stomatodaeal pit. This enlarges, circumscribes and then sinks to form the **olfactory pit**. The pit deepens gradually. The sinking is enhanced by elevations which are produced by the proliferation of the underlying mesoderm on either side of the pit. The two elevations are the **medial** and **lateral nasal processes**. The two medial nasal processes (right and left) along with the intervening area between them form the **frontonasal process**. Thus, there is a frontonasal process in the middle and two lateral nasal processes, one on either side. There are also two **maxillary processes**, one on either side of the stomatodaeum. Each maxillary process grows medially, taps the lower edge of the lateral nasal process and then grows further medially across the lower limit of the olfactory pit and touches the lower edge of the frontonasal process. The surface limit of the olfactory pit now gets defined as the **anterior naris**. On the internal aspect, the olfactory pit deepens further. The medial growth of the maxillary process and its fusion with the lower edge of the frontonasal process also delimits a **primitive posterior naris**.

Meanwhile, the medially growing maxillary process splits the frontonasal process into a superficial and a deep layer. The deep layer grows backwards (towards the interior) to form the **primitive palate** which now lies inferior to the primitive posterior nares. The deepening olfactory pits now are called the **olfactory cavities** or **nasal cavities**. They remain separated from each other by the upper part of the posterior aspect of the frontonasal process which also grows backwards but as a thinner extension that forms the **primitive**

Development *contd...*

nasal septum. From the internal aspect of each maxillary process, a mesodermal extension called the **tectoseptal extension** goes upwards and medially. The tectoseptal extensions fuse with each other and are placed posterior to the primitive nasal septum growing from the frontonasal process. The fused extension fuses with the primitive nasal septum and together they form the **definitive nasal septum**.

From the internal aspect of each maxillary process, yet another medial extension starts but a little lower to the tectoseptal extension. This medial extension is quite thick and is the **palatal process**. The two palatal processes grow medially, at the same (horizontal) level with the primitive palate. The anterior parts of the free edges of the palatal processes fuse with the primitive palate on the corresponding side and then fuse medially with each other. On their superior surfaces, the lower edge of the nasal septum fuses with them. The **definitive palate** is thus formed and the nasal cavities are separated off from the oral cavity. This fusion also results in a slow migration of the posterior nares from their primitive anterior positions to their posterior definitive positions. In later development, membranous ossification occurs in the primitive palate; the ossified structure becomes the **premaxilla**. Membranous ossification extends into the palatal processes of the maxillary mesoderm giving rise to the palatine process of maxillae and the palate bones. The bony portion thus becomes the **hard palate**. No bone is formed in that part of the fused palatal processes that lies beyond the nasal septum; that portion becomes the **soft palate** and **uvula**. A small canal in the midline persists throughout foetal life between the primitive palate and the maxillary palatal processes; this is the **nasopalatine canal**. It closes eventually but is represented by the incisive canal in the adult.

As the palate develops, the nasal cavities have been delimited. The original olfactory placode area had sunk in and the surrounding mesoderm had proliferated and pushed up the ectoderm. We already know that the frontonasal process was thus formed. Further proliferation and remodelling result in the formation of the **external nose** and its cartilaginous and bony skeleton. The olfactory placode comes to occupy the roof area of the nasal cavities giving rise to the **olfactory (smell) epithelium**. Mesodermal proliferation leads to the appearance of elevations on the lateral walls of the nasal cavities and these elevations develop into the **nasal conchae**.

In the late foetal life, small diverticula from the lateral nasal walls make their appearance. These diverticula invade the adjacent developing bones. Those diverticula which invade into the maxillary bones and the sphenoid bone have done so just before birth. Therefore, the **maxillary sinuses** and the **sphenoidal sinuses** have made their appearances before birth and are rudimentary. Further invasion into the bones takes place in early childhood. The diverticula pertaining to the ethmoidal and frontal sinuses are mere dimples on the nasal walls at birth. Invasion takes place only in early childhood.

An epithelial canal is trapped in the line of fusion of the superficial aspects of the frontonasal process and the maxillary process on either side. This canal becomes the **nasolacrimal duct**. The duct grows actively down to open into the lateral nasal wall below the inferior nasal concha.

contd...

163

Multiple Choice Questions

1. The septal cartilage:
   a. Forms part of the vomer
   b. Fills the anterior deficiency in the nasal septum
   c. Occupies the posteroinferior part of the nasal septum
   d. Has no mucosal covering
2. The main opening of the maxillary air sinus opens:
   a. On the bulla ethmoidalis
   b. In the middle of hiatus semilunaris
   c. In the inferior meatus of nose
   d. Anterosuperior to the hiatus semilunaris
3. Palatine aponeurosis is the widened expansion of the:
   a. Tendon of tensor veli palatini
   b. Tendon of palatoglossus
   c. Tendon of musculus uvulae
   d. Tendon of Passavant's muscle
4. The feature increasing the surface area of nasal mucosa is:
   a. Adherence of mucosa to periosteum
   b. Presence of conchae
   c. Dense vascular mucosa
   d. Presence of meatuses
5. The primary artery to the nasal mucosa is:
   a. Greater palatine artery
   b. Nasopalatine artery
   c. Sphenopalatine artery
   d. Anterior ethmoidal artery

ANSWERS

1. b   **2.** b   **3.** a   **4.** b   **5.** c

Clinical Problem-solving

Case Study 1: A friend of yours, has had caries of his right upper second premolar for many years. Of late, he had been complaining of more pain, especially over his right cheek and right lateral part of his face. He also developed repeated upper respiratory tract infections.
- What is the possibility that any of the paranasal sinuses would be involved in this case?
- Which paranasal sinus is likely to be involved?
- Why would the involvement of this particular paranasal sinus result in chronic infection?

Case Study 2: A 72-year-old woman complained of epistaxis. On examination, no functional cause could be detected except that the woman had caused some digital trauma to her nose.
- Why should digital trauma lead to epistaxis?
- Which part of the nasal cavity is involved?
- Give the anatomical basis for this problem.

(For solutions see Appendix).

Chapter 10

Viscera of the Neck (Including Pharynx and Larynx)

Viscera of the neck predominantly consist of the organs of the upper digestive and upper respiratory tracts. The organs of the upper digestive tract are the pharynx and oesophagus. A lymphoid organ within the oropharynx is the palatine tonsils. The organs of the upper respiratory tract are the pharynx, the larynx and the trachea.

It can be seen that the pharynx belongs to both the digestive and respiratory pathways. The organs of upper digestive and the upper respiratory tracts are so closely related anatomically and functionally that they are usually referred to as the viscera of the aerodigestive tract.

PHARYNX

The pharynx is a median passage that is common to the alimentary and the respiratory systems (Fig. 10.1). It is a musculomembranous tube extending from the skull base to the lower level of the cricoid cartilage (coinciding with the level of the C6 vertebra) where it becomes continuous with the oesophagus. Because it forms a common passageway for both the food and air pathways, both the oral (food path) and the nasal (air path) cavities open into it and the food and the wind pipes start from it. The pharynx is made up of the pharyngeal muscles which provide the tone to keep it open.

RELATIONS

- ❑ **Anteriorly:** Posterior ends of the nasal and oral cavities, commencement of larynx.
- ❑ **Posteriorly:** Connective tissue in the prevertebral space, prevertebral fascia covering the paravertebral muscles in the neck and the cervical part of vertebral column.
- ❑ **Superiorly:** Basisphenoid and basiocciput.
- ❑ **Inferiorly:** Continues into the oesophagus.

The pharynx is divisible (from above downwards) into (Fig. 10.1):

- ❑ A **nasal part** (or **nasopharynx**) into which the nasal cavities open;
- ❑ An **oral part** (or **oropharynx**) which is continuous with the posterior end of the oral cavity and
- ❑ A **laryngeal part** (or **laryngopharynx**) which is continuous in front with the larynx and below with oesophagus.

The communication between the nasopharynx and the oropharynx is called the **pharyngeal isthmus**. This isthmus can be closed (e.g., during swallowing) by elevation of the soft palate. Closure is helped by contraction of muscle fibres in the pharyngeal wall which constitute the **palatopharyngeal sphincter**. Contraction of this sphincter produces a ridge (of Passavant) on the posterior pharyngeal wall with which the soft palate (situated in the anterior aspect of pharynx) comes in contact. The nasopharynx is therefore shut off from the digestive pathway.

The communication between the oral cavity and the pharynx is called the **oropharyngeal isthmus**. It is bounded above by the soft palate, below by the posterior part of the tongue and on either side by the palatoglossal

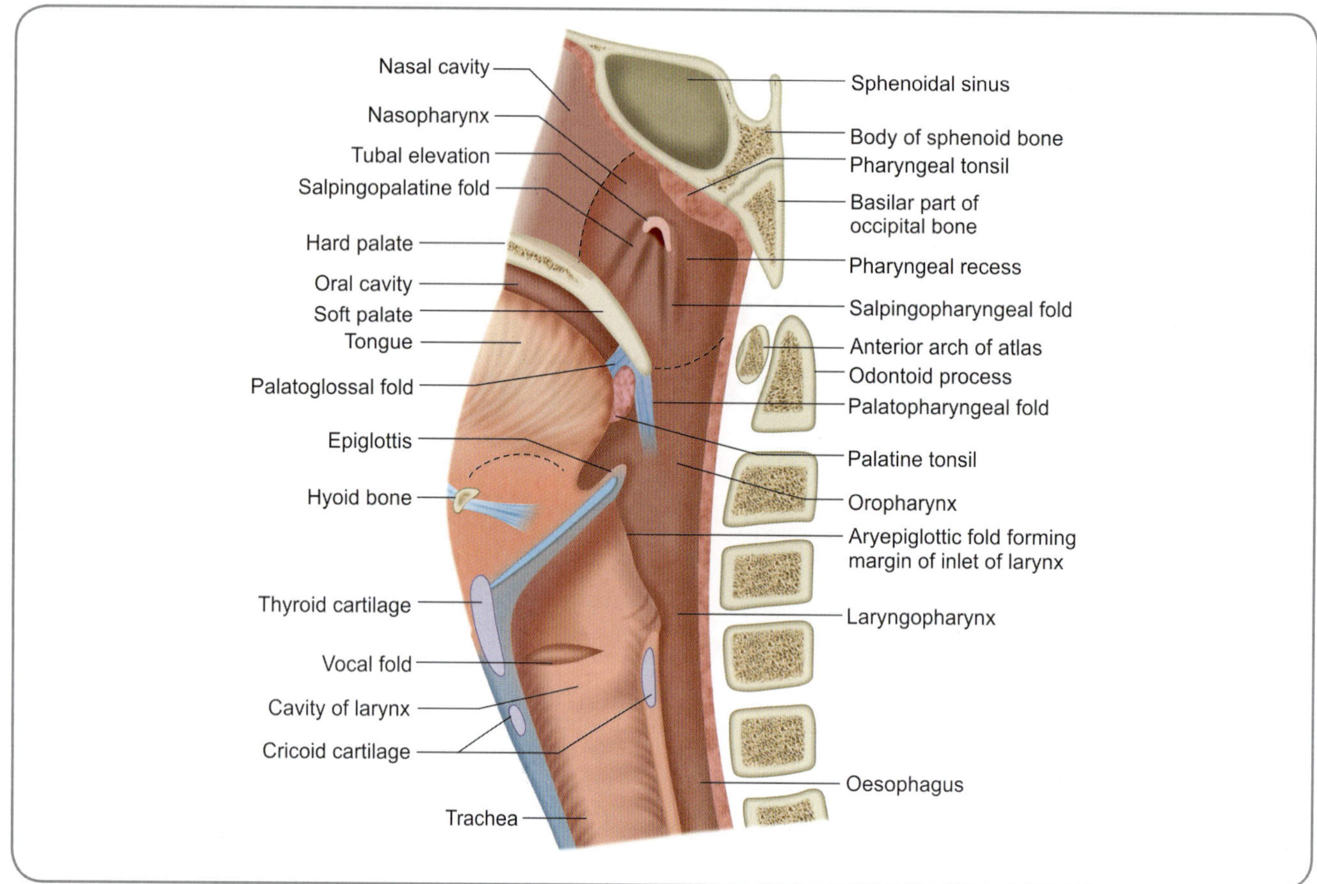

Fig. 10.1: Schematic median section through the pharynx (and neighbouring structures) to show its lateral wall. The limits of the subdivisions of the pharynx are indicated in dotted lines

arches. The oropharyngeal isthmus can be closed by contraction of the palatoglossi muscles. This closure also plays an important part in deglutition.

The laryngopharynx lies just behind the inlet and posterior wall of the larynx.

NASAL PART OF THE PHARYNX

The nasopharynx is the upper part of the pharynx into which the nasal cavities open through the posterior nasal choanae. It lies above the soft palate. The posterior choanae allow free passage of air between the nasal cavities and the nasopharynx. Superiorly, it is closed by the roof and inferiorly, it is continuous with the oropharynx. The walls of the nasopharynx are rigid (unlike the walls of the oro and laryngeal parts). This factor helps in keeping the nasopharynx open all the time, thereby helping in air passage.

The nasopharynx has a roof, a floor, a posterior wall and two lateral walls.

❑ The **roof** and the **posterior wall** form a continuous slope from the posterior limit of the nasal septum to the commencement of the oropharynx. They are actually formed by (from above downwards) the mucosa covering the basisphenoid and basiocciput

till the pharyngeal tubercle and the mucosa lining the pharyngobasilar fascia and upper fibres of superior constrictor muscle. Behind the pharyngobasilar fascia and the superior constrictor lies the anterior arch of the atlas vertebra. The mucosa of the median part of the roof shows a bulging produced by a mass of lymphoid tissue. This lymphoid tissue constitutes the **pharyngeal tonsil** (when enlarged, the pharyngeal tonsils are referred to as **adenoids**).

❑ The **floor** is formed by the upper surface of the soft palate.

❑ On each **lateral wall** of the nasopharynx, there is an opening which leads into the auditory tube (also called the pharyngotympanic or the Eustachian tube). This tube connects the nasopharynx to the middle ear. Above and behind the opening of the auditory tube, the wall of the nasopharynx shows a bulging called the **tubal elevation**. This elevation is produced by the projecting end of a cartilage which forms part of the wall of the auditory tube. A fold of mucous membrane starting at the tubal elevation passes down the pharyngeal wall behind the tubal opening. This is the **salpingopharyngeal fold** and is produced by a muscle called the **salpingopharyngeus**. Another mucosal fold passes from the tubal elevation to the soft palate in front of the tubal opening. This is the ***salpingopalatine fold***. A small mass of lymphoid tissue called the ***tubal tonsil*** (also called the ***Eustachian tonsil*** or ***Gerlach's tonsil***) lies in the mucosa immediately behind the tubal opening. Further behind the tubal elevation, the wall of the nasopharynx shows a depression called the ***pharyngeal recess*** (otherwise called the ***fossa of Rosenmuller***).

The nasopharynx communicates with the oropharynx through the pharyngeal isthmus. As already noted, this isthmus lies behind the posterior border of the soft palate and in front of the posterior pharyngeal wall. Any factor that causes the soft palate to get elevated and become horizontal (thereby approximating it to the posterior pharyngeal wall) and any factor that causes the posterior pharyngeal wall to get closer to the soft palate will close the isthmus. Both the actions of elevation of the soft palate and approximation of the posterior pharyngeal wall occur during deglutition by the action of the palatal muscles. The contraction of the fibres constituting the Passavant's palatopharyngeal sphincter contribute largely to such a closure by producing a sphincteric ridge at the level of the pharyngeal isthmus (the reason why these fibres have been called the palatopharyngeal sphincter).

Content

Though the pharynx is actually a passage way, some structures which lie on its walls deserve special mention and are called the contents. The content of the nasopharynx is the nasopharyngeal tonsil.

ORAL PART OF THE PHARYNX

The oropharynx is the middle part of the pharynx into which opens the posterior aspect of the oral cavity. It extends from the level of the soft palate to the upper border of the epiglottis and therefore, lies in front of the second and upper part of the third cervical vertebrae. Superiorly, it is continuous with the nasopharynx and inferiorly with the laryngopharynx.

The oropharynx has a posterior wall and two lateral walls. Its anterior wall can be described to be deficient.

- ❏ Anteriorly, the oropharynx opens into the oral cavity through the oropharyngeal isthmus and faces the posterior aspect of the tongue.
- ❏ The posterior wall is formed of the mucosa lining the pharyngeal muscles.
- ❏ The lateral walls show only two special features. The first is the presence of the ***palatopharyngeal folds*** (or arches). These stretch from the uvula to the lateral wall of the pharynx and enclose the palatopharyngeus muscle. The palatine tonsil lies between the palatoglossal and palatopharyngeal folds. The depression in which the palatine tonsil lies is called the ***tonsillar sinus***.

Content

The main content of the oropharynx is the palatine tonsil. However, many authors prefer to describe the soft palate as one because of the latter's location.

LARYNGEAL PART OF THE PHARYNX

The laryngopharynx (also called the **hypopharynx**) is the lower part of the pharynx from where the larynx begins. It extends from the upper border of the epiglottis to the level of the inferior border of the cricoid cartilage. The vertebral levels of its extent correspond to the front of the body of the third to the body of the sixth cervical vertebrae. Superiorly, it is continuous with the oropharynx and inferiorly with the oesophagus.

The laryngopharynx has an incomplete anterior wall, a posterior wall and two lateral walls.

- ❏ The upper part of its anterior wall (actually the deficient part of the anterior wall) is formed by the inlet of the larynx; and the lower part by the posterior surfaces of the arytenoid and cricoid cartilages. Lateral to the inlet of the larynx, on either side, the mucosa shows a depression called the ***piriform recess*** or fossa.
- ❏ The posterior wall is formed by the mucosa lining the pharyngeal muscles.
- ❏ The lateral walls are featureless and formed by mucosa lining the pharyngeal muscles.

> **Piriform Fossa**
>
> The piriform fossa is a small depression, one on either side of the laryngeal inlet, in the laryngopharynx. The boundaries of the fossa are anteromedially the aryepiglottic fold and anterolaterally the medial surfaces of the thyroid cartilage and the thyrohyoid membrane. Posteriorly, the fossa is open to the pharyngeal lumen. Branches of the internal laryngeal nerve and the recurrent laryngeal nerve lie deep to the mucous membrane of the piriform fossa. Foreign particles in the food can get lodged in the piriform fossa. Sharp articles can get hooked to the mucosa (e.g. Fish bone). While attempting to remove the particle, maneuvering with sharp instruments can damage the underlying nerves and result in paralysis of the laryngeal musculature. The piriform fossa lies deep to the superior pole of the thyroid gland. The fossa may be connected to the gland by a sinus tract which can become a potential site for recurrent infections. Developmental adherence of the thyroglossal duct to the laryngopharynx causes this condition.

WALLS

The walls of the pharynx are constituted mainly by muscles.

> **Microstructure of Pharyngeal Wall**
>
> The pharyngeal wall has four coats. From outer to inner, these are the (1) areolar coat, (2) muscular coat, (3) fibrous coat and (4) mucous coat.
>
> The ***areolar coat*** comprises the buccopharyngeal fascia. It contains the pharyngeal plexus of veins and nerves. The nervous

contd...

plexus is formed by the pharyngeal branches of vagus and glossopharyngeal nerves and the sympathetic nerves.

The *muscular coat* has an outer circular layer and an inner longitudinal layer.

The *fibrous coat* consists of the pharyngobasilar fascia.

The *mucous coat* has distinctive features in different parts of the pharynx and consists of lymphoid tissue at varied locations.

Two fascial layers provide support and strength to the muscles.

❑ The muscular layer is covered on the outer aspect by a layer of fascia called the *buccopharyngeal fascia*. This fascia extends forwards (from the superior constrictor) on the buccinator muscle (and hence its name). It is actually the epimysium on the outer aspect of the muscles and is thinner than the pharyngobasilar fascia.

❑ Between the mucous membrane and the muscular layer (that is the inner aspect of the muscular layer), there is a fascia. This fibrous layer of fascia supports the mucous membrane. Above the level of the superior constrictor, it is thickened to form the *pharyngobasilar* fascia. The pharyngobasilar fascia is attached above to the skull base and extends below into the pharynx (hence the name). Below the level of the superior constrictor, it is thinner. This layer is the epimysium of the muscles on their internal aspects.

A fibrous band descends in the midline from the pharyngeal tubercle of the occipital bone; this forms the *median pharyngeal raphe* that provides attachments to the pharyngeal constrictor muscles. It runs behind the fibrous layer that is continuous below from the pharyngobasilar fascia and provides support to the former.

MUSCLES (FIG. 10.2)

The muscular basis of the wall of the pharynx is formed by the circular and longitudinal muscles (Table 10.1 and Table 10.2). The circular muscles are also called the constrictors of the pharynx and the longitudinal muscles, the elevators.

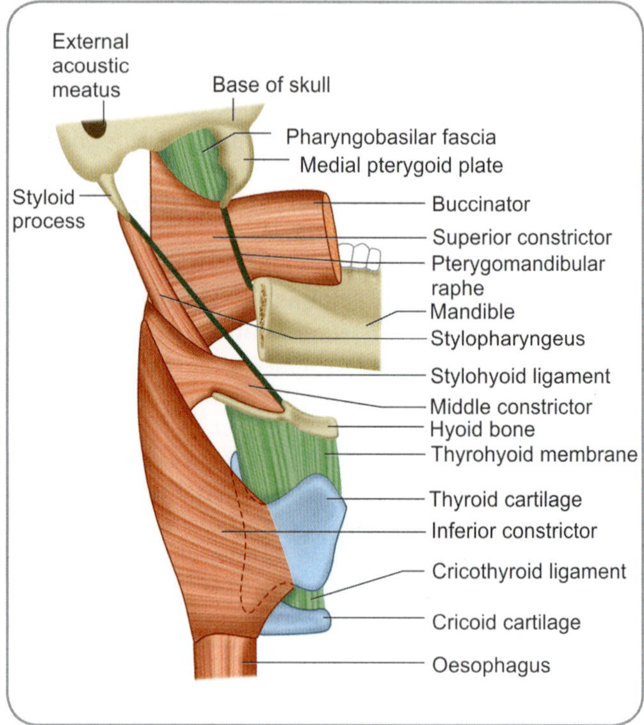

Fig. 10.2: Muscles of the pharynx seen from the lateral side

Constrictors of the Pharynx

There are three pairs of constrictors—(1) superior constrictor, (2) middle constrictor and (3) inferior constrictor.

The constrictors can be imagined to be three cones overlapping each other. Their origins are situated anteriorly and laterally in relation to the posterior openings of the nose, mouth and larynx (from above downwards). From here, the fibres pass into the lateral and posterior walls of the pharynx, the fibres of the two sides meeting posteriorly, in the midline in the median pharyngeal raphe. Therefore, the insertion of all the three muscles can be described to be the median pharyngeal raphe. The three constrictors are so arranged that the inferior overlaps the middle, which in turn, overlaps the superior.

Table 10.1: Circular muscles of the pharynx				
Name of the muscle	**Origin**	**Insertion**	**Nerve supply**	**Actions**
Superior pharyngeal constrictor	Pterygoid hamulus, pterygomandibular raphe and mylohyoid line of mandible	Median pharyngeal raphe	Cranial part of accessory nerve via the pharyngeal branch of vagus and pharyngeal plexus	Constricts the upper part of the pharynx
Middle pharyngeal constrictor	Lesser horn of hyoid, lower part of stylohyoid ligament and greater horn of hyoid	Median pharyngeal raphe	Cranial part of accessory nerve via the pharyngeal branch of vagus and pharyngeal plexus	Constricts the middle part of the pharynx
Inferior pharyngeal constrictor	Oblique line of thyroid lamina, small part of the inferior horn of thyroid cartilage and side of cricoid cartilage	Median pharyngeal raphe	Cranial part of accessory nerve via the pharyngeal branch of vagus and pharyngeal plexus	Constricts the lower part of the pharynx; the cricopharyngeus part acts as a sphincter between the laryngopharynx and the oesophagus

Additional Notes on Pharyngeal Constrictors (Table 10.1)

The ***superior constrictor***, on either side, is a quadrilateral sheet of muscle. It is thinner than the other two constrictors. Its upper fibres curve upward and backward to reach the median pharyngeal raphe at the latter's upper limit, which is at the pharyngeal tubercle of the occipital bone. This causes a gap, on the sides, between the base of the skull and the upper edge of the muscle. This gap is filled by the submucosal fibrous fascia, which thickens as the pharyngobasilar fascia in this location.

The ***middle constrictor*** is described to have two parts—the part arising from the lesser horn of hyoid and stylohyoid ligament, forming the ***chondropharyngeal*** part of the muscle and the part arising from the greater horn forming the ***ceratopharyngeal*** part. The insertion of the fibres of middle constrictor have a characteristic presentation: the upper fibres ascend and overlap the superior constrictor; the middle fibres pass transversely; the lower fibres descend deep to the inferior constrictor and reach the lower end of the pharynx. The muscle is more active during swallowing.

The ***inferior constrictor*** is the thickest of the constrictors. It is in two parts—the part arising from the thyroid lamina and the inferior horn of thyroid is the ***thyropharyngeus*** and that arising from the cricoid is the ***cricopharyngeus***. The fibres of thyropharyngeus ascend up to overlap the middle constrictor and the fibres of cricopharyngeus become continuous with the circular muscle of the oesophagus. The junction of the cricopharyngeus and the oesophagus forms the most narrow part of the pharynx.

Longitudinal Muscles (Table 10.2)

In addition to the constrictors, the pharynx has three muscles the fibres of which run longitudinally. These are—(1) stylopharyngeus, (2) palatopharyngeus and (3) salpingopharyngeus.

Important Relations of the Pharyngeal Muscles

With the three constrictors overlapping each other, three gaps result. With an additional gap above the upper border of the superior constrictor, a total of four gaps occur. Various structures pass through these gaps.

❑ The pharyngotympanic tube, the levator veli palatini muscle and the ascending palatine artery pass through the gap between the superior constrictor and the skull base.
❑ The stylopharyngeus muscle and the glossopharyngeal nerve spiralling around the muscle pass through the gap between the superior and the middle constrictors.
❑ The internal laryngeal nerve and the superior laryngeal vessels pass through the gap between the middle and the inferior constrictors.
❑ The recurrent laryngeal nerve and the inferior laryngeal artery pass through the gap between the inferior constrictor and the oesophagus.

> ### Development
>
> **Development of Pharynx**
>
> The pharynx develops from the cranial part of the foregut called the pharyngeal gut. The ***pharyngeal gut*** extends from the stomatodeal membrane to the point where the respiratory diverticulum is given out. The development of the tongue and the palate remodels the ventral wall of the pharyngeal gut and leads to the subdivision of the developing pharynx into the nasopharynx and the oropharynx. That part of the developing pharynx from where the respiratory diverticulum is given out becomes the laryngopharynx.

Actions of Muscles of the Pharynx

The muscles of the pharynx play an important part in deglutition. Food entering the oropharynx is carried downwards by successive contraction of the superior, middle and inferior constrictors. The stylopharyngeus, salpingopharyngeus and the palatopharyngeus help by pulling the pharynx upwards and by shortening it.

Table 10.2: Longitudinal muscles of the pharynx				
Name of the muscle	**Origin**	**Insertion**	**Nerve supply**	**Actions**
Stylopharyngeus	Base of styloid process	One part merges with the constrictors; one part merges with the palatopharyngeus and attaches to the thyroid cartilage	Glossopharyngeal nerve	Elevates pharynx and larynx
Palatopharyngeus	Palatine aponeurosis	Lateral wall of pharynx	Cranial part of accessory nerve via the pharyngeal branch of vagus and pharyngeal plexus	Draws the pharynx up and shortens it; draws the palatopharyngeal arches closer to each other
Salpingopharyngeus (Greek.salpinx= trumpet; meaning the pharyngotympanic tube here)	Inferior aspect of the cartilage of pharyngotympanic tube	Blends with palatopharyngeus	Cranial part of the accessory nerve via the pharyngeal branch of vagus and pharyngeal plexus	Elevates the pharynx and closes the tubal opening during swallowing

The thyropharyngeal part of the inferior constrictor is propulsive in contrast to the cricopharyngeal part which is 'sphincteric'. Incoordination in the action of the two parts may result in the formation of a diverticulum on the posterior wall of the pharynx at the junction of the two parts. The inner surface of the superior constrictor is lined by a band of muscle fibres arising from the sides of the palate. These fibres form the ***palatopharyngeal sphincter*** which produces a ridge (of Passavant) on the pharyngeal wall at the junction of nasopharynx with the oropharynx. Acting along with the soft palate, the palatopharyngeal sphincter closes the pharyngeal isthmus preventing food from entering the nasopharynx.

BLOOD VESSELS, LYMPHATICS AND NERVES

The pharynx receives numerous small branches which arise from the ascending pharyngeal, lingual, facial and maxillary arteries. The veins drain into a plexus which surrounds the pharynx and subsequently drains into the internal jugular and facial veins.

The ***lymph vessels*** of the pharynx drain into the deep cervical lymph nodes. Some of the lymph passes through the retropharyngeal nodes.

The nerve supply of the pharynx is through the ***pharyngeal plexus*** which is formed on the external surface of the pharynx by the pharyngeal branches of the glossopharyngeal and vagus nerves and fibres from the sympathetic trunk. The fibres of the cranial accessory nerve run through the vagus and constitute the main supply of the muscles of the pharynx. Sensory fibres travel through the glossopharyngeal and vagus nerves. Some sensory fibres also travel through the lesser palatine branch of the pterygopalatine ganglion.

Added Information

- The auditory tube (also called the tuba auditiva or tuba acoustica) is frequently referred to as '***Eustachian tube***' (after the 16th century Italian anatomist Bartolomeo Eustachi). Inflammation of the mucosa of the tube is referred to as eustachitis.
- The Eustachian opening in the nasoparynx lies about 10 mm behind and below the posterior end of the inferior nasal concha.
- The oropharyngeal isthmus and the pharyngeal isthmus are not one and same. The names are indicative. The isthmus (narrow passage between two larger portions; Greek. isthmos=passage) between two parts of the pharynx itself (naso and oro) is plainly 'pharyngeal' isthmus. The isthmus between the oral cavity and the pharynx is the oropharyngeal isthmus.
- The isthmus between the nasal cavity and the nasopharynx is the pharyngonasal isthmus. This isthmus is formed by the posterior choanae. The choanae are oblong rigid structures and cannot be shut off. Though the pharyngonasal isthmus is identified as a landmark, it is really not an isthmus and cannot be closed.

Added Information *contd...*

- The pharyngobasilar fascia is described as the submucosal fibrous tissue between the mucous and the muscular layers of the pharyngeal wall. In view of this description, the pharyngobasilar fascia is also called by the following names; tela submucosa pharyngis and aponeurosis pharyngea.
- At its upper end (close to the skull base), the pharynx is about 5-6 cm wide. But at its lower end (junction with oesophagus), it is the most narrow being only about 3 cm wide. This is the narrowest and the least dilatable part of the digestive tract. Any foreign body that passes through this point, is not likely to be arrested anyway farther down.
- ***Retropharyngeal space:*** This is a potential tissue space that exists between the pharynx anteriorly and the prevertebral fascia posteriorly. It extends upwards to the skull base and downwards to the retrovisceral space in the neck. It contains loose connective tissue.
- ***Parapharyngeal spaces:*** These are two potential spaces, one on either side of the pharynx. They are suprahyoid in location. Each space is bounded medially by the pharynx, laterally by the pterygoid muscles, superiorly by the skull base and inferiorly by sheath of submandibular gland. It can be seen that this space is partly located in the infratemporal fossa. Each parapharyngeal space surrounds that side of the pharynx and becomes continuous with the retropharyngeal space posteriorly.
- Along with the submandibular and submental spaces, the retropharyngeal and the parapharyngeal spaces are called the ***peripharyngeal spaces***.
- ***Intrapharyngeal space:*** This is a potential space that exists between the pharyngeal musculature and the pharyngeal mucosa. Infections in this space can spread to the peripharyngeal spaces.
- ***Peritonsillar space:*** This is the potential space around the palatine tonsils between the pillars of fauces.

Dissection

The pharynx can be studied the best in a sagittal section of the head and neck and in a prosected specimen.

Study the different parts of the pharynx in a sagittal section; locate the different features on the lateral walls of the different parts.

In a prosected specimen that has the pharynx and larynx, identify and study the laryngeal aditus, piriform fossae and the laryngopharynx. If the pharyngeal wall is intact, make a posterior midline cut, so that the right and the left halves of the wall can be reflected. See the cut ends of the wall and notice the muscle fibres. Reflect the mucosa in small loci at two or three places and study the muscular fibres.

Clinical Correlation

Hypopharyngeal Diverticula

The division of the inferior constrictor of the pharynx into a cricopharyngeal part and a thyropharyngeal part has been noted above. The cricopharyngeus is believed to act as a sphincter at the junction of the pharynx and oesophagus and prevent passage of air into the latter. The thyropharyngeus (along with other constrictors) help to push the bolus of food

contd...

contd...

down the pharynx. The posterior wall of the pharynx is weakest just above the cricopharyngeus. There may be a gap here between the thyropharyngeal and cricopharyngeal fibres. This gap is referred to as **Killian's dehiscence**. Pressure of food over this area can lead to the formation of a pouch (diverticulum). This is more likely to occur if there is incoordination of muscles (especially if the cricopharyngeus fails to relax when the thyropharyngeus is contracting). The pharyngeal pouch thus formed cannot expand posteriorly (because of the presence of the vertebral column). It, therefore, grows downwards and can press on the oesophagus resulting in dysphagia (difficulty in swallowing). The pouch may also trap part of food particles passing by; regurgitation of old food and aspiration pneumonitis may occur.

Pharyngeal Tone and Sleep Disorders

❑ The width of the pharyngeal passage depends on the tone of its musculature. During waking, the pharyngo-oesophageal junction remains closed due to the tone of the cricopharyngeal sphincter. In sleep, the muscle tone decreases.

❑ During inspiration, negative pressure is created in the lumen of the pharyngeal airway. This negative pressure is counteracted by the joint action of genioglossi, tensores veli palatine, geniohyoids and stylohyoids. These muscles keep the passage way dilated and are called the **dilators of the pharynx**. Sleep, alcohol, sedatives, hypothyroidism and some neurological disorders reduce the tone of these muscles. The walls of the pharynx fall flaccid and get approximated.

If the pharyngeal obstruction is intermittent, snoring results. If obstruction continues for long periods, apnoea and other sleep disturbances occur. Surgical intervention is done by methods like plicating the tonsillar pillars, reducing the bulk of soft palate, reducing the length of soft palate and tonsillectomy. Such methods are collectively called (when used for this purpose) uvulopalatopharyngoplasty.

Infection in the Pharyngeal Spaces

Infections from the oral cavity and tonsils can spread to the peripharyngeal and parapharyngeal spaces. Infections in the parapharyngeal spaces can produce trismus (by pressing on the pterygoid muscles) and pain. Swelling can press on the pharyngeal lumen leading to dysphagia. Spread into the retropharyngeal space will result in a midline swelling of the posterior pharyngeal wall; dyspnoea and nuchal rigidity can occur. Involvement of the prevertebral fascia can allow the infection to spread to the carotid sheath resulting in thrombosis of the internal jugular vein and compression of the cranial nerves. Infection can spread into the prevertebral space and can track down to the thorax leading to dyspnoea, chest pain and retrosternal discomfort.

TONSIL

NASOPHARYNGEAL TONSIL

Other names: Pharyngeal tonsil, third tonsil, Luschka's tonsil, Luschka's gland.

The nasopharyngeal tonsil is a median mass of lymphoid tissue located in the roof and posterior wall of the nasopharynx. Since it is associated with the mucosa, it forms part of such lymphoid tissue collectively called the **Mucosa associated lymphoid tissue (MALT)**. During childhood it is large in size and appears like a pyramid. The base of the pyramid is on the posterosuperior pharyngeal wall and the apex points to the nasal septum. The free surface exhibits a median blind recess projecting superolaterally. This recess is called the pharyngeal bursa and is the remnant of the cranial end of the embryological notochord. Several folds radiate from the bursa. It grows larger soon after birth and reaches its maximum size by 5 years of age. After 8-10 years, it starts involuting.

The nasopharyngeal tonsil is supplied by the ascending pharyngeal and ascending palatine arteries, pharyngeal branch of the maxillary artery, tonsillar branches of the facial artery and the artery of the pterygoid canal. It may also get a twig from the basisphenoid artery, a branch of the inferior hypophyseal artery. The veins draining the nasopharyngeal tonsil end in the pharyngeal venous plexus. They may connect with the paratonsillar veins, the facial vein, the internal jugular vein or the pterygoid venous plexus.

The nasopharyngeal tonsil forms part of the Waldeyer's ring and so, contributes to the defence mechanism of the upper aerodigestive tract.

When the nasopharyngeal tonsil is enlarged, it is called the **adenoids** (Gland like; Greek.adeno=gland, eidos=like). Inflammation of adenoids is **adenoiditis** (or **adenoid disease** or **Meyer's disease**). When compared with the relative size of the nasopharynx, the nasopharyngeal tonsil is the largest around 5 years of age. Adenoiditis at this stage can cause blockage of the posterior choanae. Passage of air from the nasal cavities to distal parts of the respiratory tract is difficult and the child switches to mouth breathing. A constantly open mouth can lead to deformities of the teeth and palate (as normal pressure of the tongue on the palate is not present). Infection from the adenoids can spread to the tubal tonsil resulting in tubal block and subsequent middle ear infection in chronic cases. Hearing impairment may occur. Adenoidectomy is done to relieve the nasal obstruction and other complications.

PALATINE TONSILS (FIG. 10.3)

Other name: Faucial tonsil.

The palatine tonsils (Latin.tonsilla = stake) are masses of lymphoid tissue. Each palatine tonsil (right or left) lies in the tonsillar sinus on the lateral wall of the oropharynx. The **tonsillar sinus** is bounded by the palatoglossal fold in front and the palatopharyngeal fold behind. Relative to the surface of the body, the palatine tonsil lies just in front of and above the angle of the mandible.

Histology

The luminal surface is covered with stratified squamous epithelium. The epithelium invaginates the tonsillar parenchyma in many places forming the tonsillar crypts. In sectional slides, the crypts are seen as linear spaces. The parenchyma is filled with lymphoid follicles and nodules of varying sizes. Follicles with germinal centres are also seen. The epithelial surface area is increased by the presence of crypts; this offers more area for antigen identification by the underlying tissue.

FEATURES

The medial surface of the palatine tonsil is covered by mucous membrane which is continuous with that of the palatoglossal folds and below, with the mucous membrane on the tongue. The mucosa shows a number of recesses called the **tonsillar crypts** which dip into the substance of the tonsils (Fig. 10.3). A deep **intratonsillar cleft** is seen in the upper part. It is semilunar in shape in conformity with the convex dorsal aspect of the tongue. The walls of all the crypts and the intratonsillar cleft contain lymphoid tissue. The lymphoid tissue in the upper wall of the intratonsillar cleft extends into the soft palate to form the pars palatina of the palatine tonsil.

Dissection

Study the oropharynx in a sagittal section of head and neck. Locate the pillars of the fauces. Identify the tonsillar fossa and the palatine tonsils. Review the details with your theory knowledge.

A triangular fold of mucosa called the plica triangularis stretches posteroinferiorly over the tonsil from the palatoglossal arch. It is usually seen only in a young adult and contains lymphoid tissue that represents the anteroinferior part of the tonsil.

Laterally, the lymphoid tissue of the tonsil extends upwards into the soft palate (pars palatina as seen above), downwards into the tongue and anteriorly, under the palatoglossal arch.

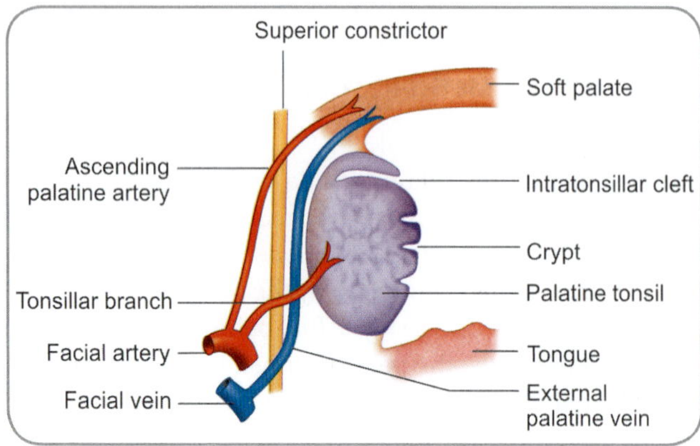

Fig. 10.3: Coronal section through the palatine tonsil

Tonsillar fossa or sinus is the shallow depression between the palatoglossal and palatopharyngeal folds. The palatine tonsils does not fill the fossa. The structures underlying the tonsils form the tonsillar bed. The **tonsillar bed** is formed by two areolar layers and two muscular layers. The muscular layers are between the areolar layers. These are from within outward are—pharyngobasilar fascia, palatopharyngeus muscle, superior constrictor muscle and buccopharyngeal fascia. The paratonsillar vein pierces the lower part of the bed from within outward to reach the pharyngeal plexus.

The lateral surface of the tonsil is covered by fascia which forms a **capsule** for it and separates it from the pharyngobasilar fascia and the superior constrictor of the pharynx. The capsule is categorically called a hemicapsule because it does not completely encircle the tonsil but covers only the lateral aspect. The hemicapsule blends with the palatoglossus and palatopharyngeus muscles and also with the musculature of the tongue. This blending provides mooring of the tonsil to the tongue so that tonsillar tissue is not scraped away and swallowed along with hard food substances which are likely to cause abrasion and separation of tissue.

The right and left palatine tonsils form the most conspicuous parts of a ring of lymphoid tissue (**Waldeyer's ring**) present near the oropharyngeal isthmus.

BLOOD SUPPLY OF TONSIL

The tonsil receives branches from several arteries in the region which are all branches of the external carotid artery. The chief supply is through the tonsillar branch of the facial artery. One other branch of the facial artery, the ascending palatine branch, also supplies it. The tonsillar branch of the facial artery (accompanied by its venae comitantes) reaches the tonsil by piercing the superior constrictor of the pharynx at the upper border of styloglossus and ramifies in its substance and in the posterior lingual musculature. Smaller tonsillar branches are derived from the ascending pharyngeal artery, dorsal lingual branches of the lingual artery and from the greater palatine artery.

Veins from the tonsil may join a vein descending from the soft palate across its lateral side (between the capsule and the fasciomuscular wall). This is the external palatine or paratonsillar vein. Some veins from the tonsil may pierce the superior constrictor muscle to end in the facial vein or in the pharyngeal plexus of veins.

Development

Development of Palatine Tonsils

The palatine tonsils are developed from the second pharyngeal pouch. By the sixth week of intrauterine life, the pharyngeal arches, clefts and pouches are formed. The epithelial lining of the second pharyngeal pouch proliferates and sends in buds into the adjacent mesoderm. The mesoderm invades the proliferated

contd...

Development *contd...*

tissue and forms the primordium of the palatine tonsils. Between the third and the fifth months of intrauterine life, lymphatic tissue infiltrates into the primordium. Part of the second pharyngeal pouch remains as a shallow depression which forms the palatine fossa.

NERVE SUPPLY

The nerve supply to the palatine tonsil is by the tonsillar branches of the maxillary and the glossopharyngeal nerves. These are sensory nerves. The branches of the maxillary nerve are actually given out as branches of the lesser palatine nerves. These branches and the tonsillar branches of the glossopharyngeal nerve form a plexus called the *circulus tonsillaris* around the tonsil. Twigs from this plexus ramify into the substance of the tonsil; some twigs also supply the soft palate and the oropharyngeal isthmus area.

LYMPHATICS

The tonsils do not receive afferents lymphatics. Instead, dense plexuses of slender lymphatic vessels surround each follicle. Efferents arise from these plexuses and run towards the hemicapsule. They pierce the superior constrictor and drain into the jugulodigastric nodes of the upper deep cervical group of lymph nodes.

Age changes in the tonsil: The size of the tonsil varies according to age. Both enlargement and involution of tonsillar tissue result in associated changes in the oropharyngeal region.
- ❑ First 5 to 6 years of age—enlarges rapidly in size;
- ❑ 12–14 years of age (around puberty)—maximum size;
- ❑ After 14 years of age—involution begins;
- ❑ Young adulthood—parts of tonsil persist prominently;
- ❑ Middle age—diminution in size.

The pars palatina which extends into the soft palate regresses usually at the age of five.

SURFACE MARKING

A small oval drawn just in front of, and above, the angle of the mandible indicates the palatine tonsils on the surface. The marking will lie over the masseter muscle.

Added Information

- ❑ The intratonsillar cleft (also called the recessus palatinus) is sometimes erroneously referred to as the supratonsillar fossa. It is not above or outside the tonsillar tissue. It is within the substance of the tonsils.
- ❑ Between the hemicapsule and the pharyngobasilar fascia, under normal healthy conditions, a layer of loose areolar tissue exists. This layer permits the muscles which are on the lateral aspect (palatopharyngeus and superior constrictor) to contract without any impediment from the tonsil. In tonsillitis, this layer of tissue may be obliterated or adhered to other structures.

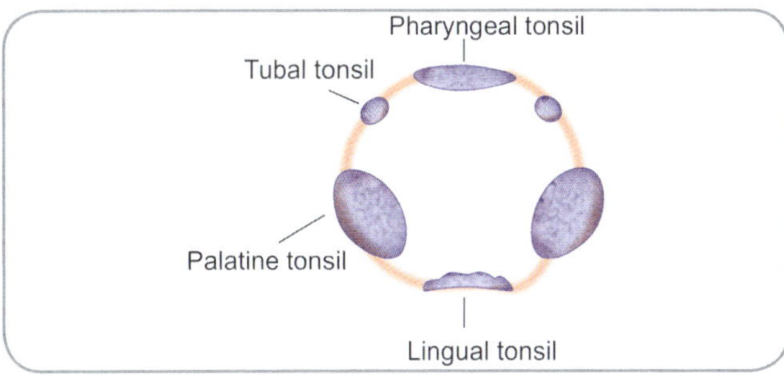

Fig. 10.4: Waldeyer's ring of lymphoid tissue

Clinical Correlation

The term 'tonsils' generally refers to the palatine tonsil. Infection of the tonsils is common and is referred to as *tonsillitis*. Infection can spread from them to peritonsillar tissues leading to a *peritonsillar abscess* (also called *quinsy*). Infected tonsils can be responsible for spread of infection to the nasal cavities, the ears and the respiratory passages.

The tympanic branch of the glossopharyngeal nerve supplies the middle ear. Branches from glossopharyngeal nerve supply the tonsils. Infections and any cause of inflammation of the tonsils and the tonsillar fossa can produce referred pain in the ear.

Surgical removal of the tonsil is called *tonsillectomy*.
- ❑ The *external palatine* or *paratonsillar vein*, descends from the soft palate lateral to the hemicapsule. This vein may be damaged during tonsillectomy and be the cause for extensive bleed from the upper angle of the tonsillar fossa.
- ❑ The *ascending palatine artery* and sometimes even the facial artery is separated from the tonsillar fossa only by the thin superior constrictor muscle. These arteries may be injured in a crudely performed tonsillectomy.
- ❑ The *tonsillar artery* and its venae comitantes may frequently lie in the palatoglossal fold. These vessels would bleed if the fold is damaged during surgery.

WALDEYER'S RING (FIG. 10.4)

Waldeyer's ring (named after the 19th-20th century German anatomist Heinrich Waldeyer) is the circumpharyngeal ring of mucosa associated lymphoid tissue. The ring surrounds the openings of the aerodigestive tract. Anteroinferiorly, the ring consists of the lingual tonsil (tonsillar tissue in the tongue, close to the foramen caecum and in the posterior part); laterally, the palatine and the tubal tonsils; posterosuperiorly, the nasopharyngeal tonsil. The ring is completed by smaller masses of lymphoid tissue which are interspersed in the intertonsillar intervals.

LARYNX

The larynx is a space that communicates above with the laryngeal part of the pharynx and below with the trachea. Apart from being a respiratory passage, the larynx is the organ where voice is produced.

173

Near the middle of the larynx are the ***vocal folds*** (one right and one left) which project into the laryngeal cavity. Between these folds is an interval called the ***rima glottidis***. The rima is fairly wide in ordinary breathing. When we wish to speak, the two vocal folds come together narrowing the rima glottidis. Expired air passing through the narrow gap causes the vocal folds to vibrate resulting in the production of sound. Variations in the loudness of sound are produced by the force with which air is expelled through the rima glottidis. Variations in pitch are achieved by the stretching of the vocal folds to different degrees. The differences in the voice of a man and that of a woman (or of a child) are due to the fact that the vocal folds are considerably longer in the male adult. The larynx moves with deglutition. The structure of the larynx has to be studied keeping these facts in view.

The larynx has a rigid framework made up of cartilages. The cartilages are joined to one another by ligaments. A number of muscles are attached to the cartilages. They produce movements of the vocal folds which are necessary for speech. The cartilages, ligaments and muscles are covered on the inner aspect by mucous membrane that is continuous above with that of the laryngeal part of the pharynx and below with that of the trachea.

Development

Development of Larynx

During the fourth week of intrauterine life, the respiratory diverticulum is given out from the ventral wall of the developing pharynx. It grows ventrally and caudally into the surrounding mesoderm. As it extends caudally, the originally broad communication between the pharynx and the respiratory bud is narrowed by two coronal ridges called the tracheo-oesophageal ridges. These ridges eventually fuse and form a tracheo-oesophageal septum that separates the ventrally placed respiratory diverticulum and the dorsally placed oesophagus. However, the septation does not extend to the cranial part of the connection. The respiratory diverticulum maintains its communication with the pharynx through the laryngeal aditus.

The mesoderm of the developing arches is closeby. The endoderm of the upper part of the respiratory diverticulum develops into the laryngeal epithelium. The mesoderm of the fourth and the sixth arches surrounds the developing larynx and contributes to the cartilages and the muscles of larynx.

When the cartilages and the muscles are forming, the epithelium proliferates very rapidly and as a result, the laryngeal lumen is closed. Canalisation occurs later. Remodelling leads to the adult shape and pattern. Since the cartilages and muscles are derived from the fourth and the sixth arches, the superior laryngeal and the recurrent laryngeal nerves supply them.

CARTILAGES

These are seen from the front in Fig. 10.5, from the lateral side in Fig. 10.6, an from above in Fig. 10.7.

The cartilages of the larynx are of two categories—The unpaired and the paired.
1. There are three unpaired cartilages—thyroid cartilage, cricoid cartilage and epiglottis cartilage.
2. There are three paired cartilages—arytenoid cartilages, corniculate cartilages and cuneiform cartilages.

A fourth pair, tritiate cartilages, has also been described.

Unpaired Cartilages

Thyroid Cartilage

The thyroid cartilage (Greek. Thyreos=shield; thyroid= shield like) consists of right and left ***laminae.*** Each lamina is roughly quadrilateral. The laminae are placed obliquely relative to the midline; their posterior borders are far apart but the anterior borders approach each other at an angle of about 90° in the male and about 120° in the female. The lower parts of the anterior borders of the right and left laminae fuse and form a median projection called the ***laryngeal prominence***. The upper parts of the anterior borders (of the laminae) do not meet, they are separated by a ***notch*** (superior thyroid notch or incisure). The posterior margin of each lamina is prolonged upwards to form a projection called the ***superior cornu*** and downwards to form a smaller projection called the ***inferior cornu***. The medial side of each inferior cornu articulates with the corresponding lateral aspect of the cricoid cartilage. The lateral surface of each lamina is marked by an ***oblique line*** that runs downwards and forwards. At its upper and lower ends the oblique line ends in projections called the ***superior*** and ***inferior tubercles*** respectively.

Cricoid Cartilage

The cricoid cartilage (Greek.Krikos=ring; cricoid=ring shaped) is shaped like a ring. This fact is best appreciated when the cartilage is viewed from above. The posterior part of the ring is enlarged to form a roughly quadrilateral ***lamina*** (Fig. 10.8A). The rest of the cartilage is called the ***arch***. The vertical diameter of the arch is greatest where it joins the lamina and least near the midline in front. The anterior part of the cricoid cartilage lies below the thyroid cartilage. The posterior part of the cricoid cartilage extends upwards into the interval between the laminae of the thyroid cartilage (Fig. 10.8A). The inferior cornua of the thyroid cartilage articulate with the lateral sides of the arch of the cricoid cartilage. On each side, the superolateral aspect of the lamina of the cricoid cartilage bears a facet for articulation with the arytenoid cartilage.

Epiglottis Cartilage

The cartilage of the epiglottis is tongue-shaped, having a broad upper part and a narrow lower end. The upper broad part lies just behind the tongue. The lower end is attached to the angle formed by the anterior margins of the laminae of the thyroid cartilage, just below the notch. The lower part of

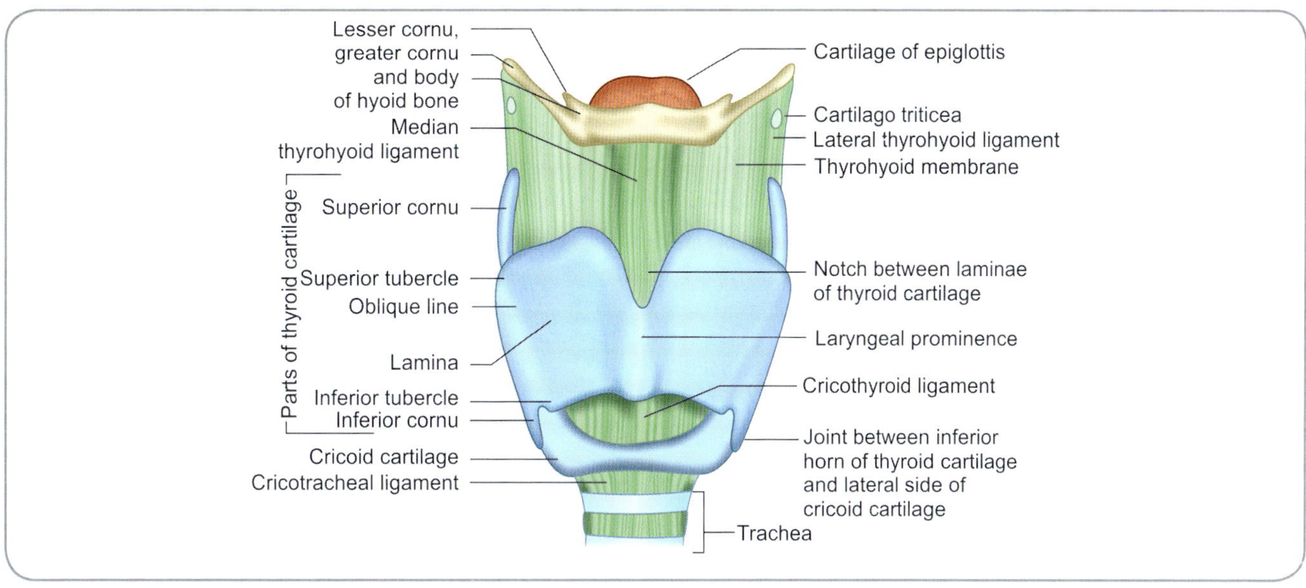

Fig. 10.5: Cartilages of the larynx as seen from the front

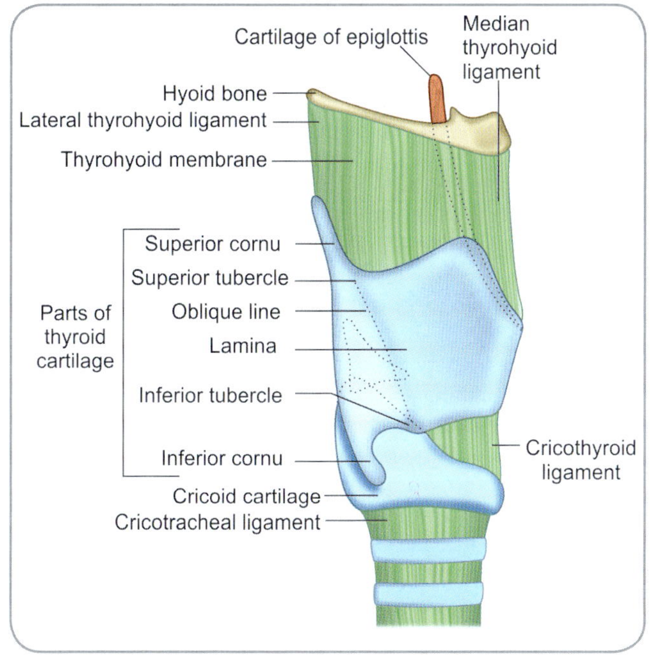

Fig. 10.6: Cartilages of the larynx seen from the lateral side

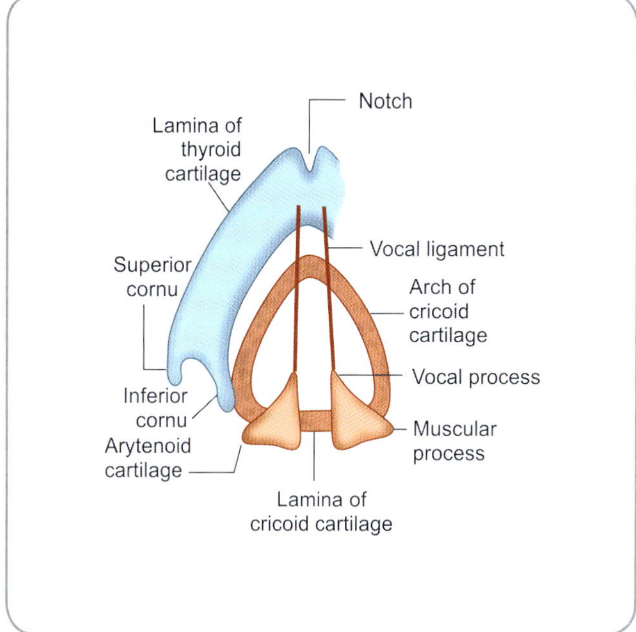

Fig. 10.7: Cartilages of the larynx seen from above

the epiglottis shows a backward projection in the midline. This projection is called the ***tubercle*** of the epiglottis.

Paired Cartilages

Arytenoid Cartilage

Each arytenoid cartilage (right or left) (Greek. Arytain=ladle; arytenoid=ladle shaped) is pyramidal (Fig. 10.8B). It has a ***base*** (below) which articulates with the cricoid cartilage, an ***apex*** which is directed upwards and to which the corniculate cartilage is attached and three *surfaces—(1) medial, (2) posterior* and *(3) anterolateral.*

The anteroinferior angle of the cartilage is prolonged forwards to form the ***vocal process.*** The inferolateral angle is enlarged to form the ***muscular process.***

Dissection

Study the larynx in a prosected specimen in which the posterior wall of the pharynx is opened by a vertical midline incision. Identify the various cartilages and membranes. Locate the vestibular and vocal folds. Try to insert a slender probe into the laryngeal vestibule. Peel the mucosa in small pockets at different places and study the internal architecture.

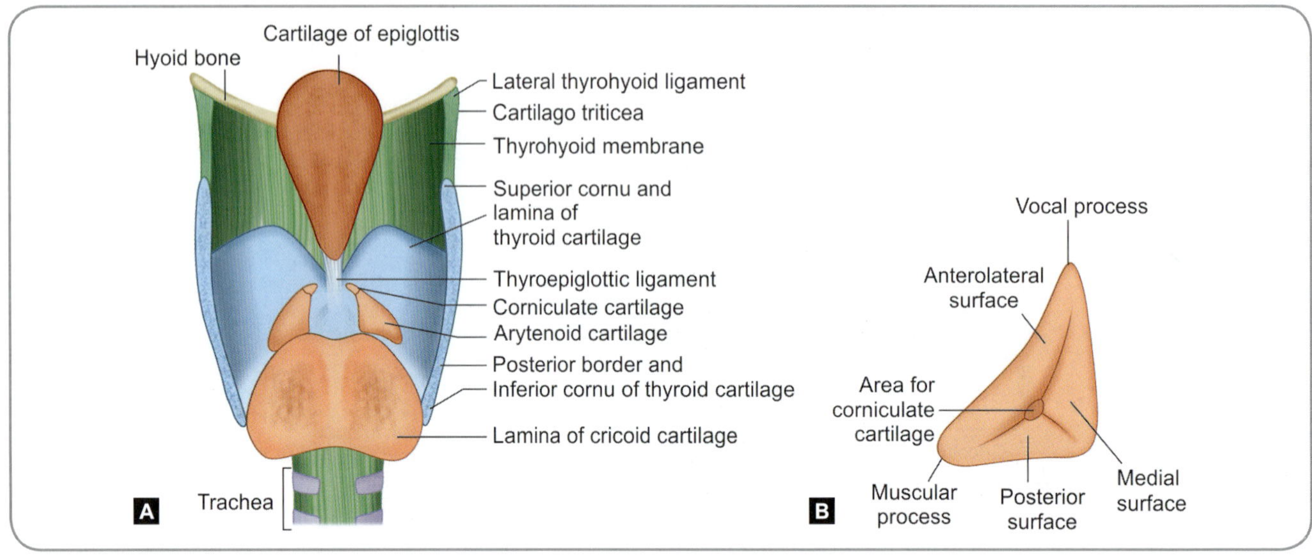

Figs 10.8A and B: A. Cartilages of the larynx seen from behind **B.** Arytenoid cartilage of the left side seen from above

Corniculate Cartilage

Each corniculate cartilage (Latin.corniculatus=horned) is a conical nodule articulating with the apex of the corresponding arytenoids and may sometimes be fused with the latter.

Cuneiform Cartilage

Each cuneiform cartilage (Latin.cuneus=wedge; cuneiform=wedge shaped) is club shaped and is placed posterosuperior to the corniculate cartilage in the aryepiglottic fold.

Tritiate Cartilage

Otherwise called the cartilago triticea, each tritiate cartilage (Latin.triticum=grain of wheat; triticeate=like a grain of wheat) is situated above the superior cornu of the thyroid cartilage within the posterior edge of the thyrohyoid membrane.

Added Information

- The epiglottic, corniculate, cuneiform, tritiate cartilages and the apices of the arytenoid cartilages are made up of elastic cartilage.
- The thyroid and cricoid cartilages and most of the arytenoid cartilages are made up of hyaline cartilages.
- Those laryngeal cartilages which are made up of hyaline cartilage have a tendency to calcify. Calcification starts between 18 and 25 years of age; more predisposed in males. Calcification usually starts at the posteroinferior parts of thyroid cartilage and proceeds to other parts.
- Though it has become customary to describe the 'tritiate' cartilage, the correct term should be 'triticeate' (meaning wheat grain). 'Tritiate' would mean 'having atom of tritium'.
- ***Relationship between the tongue and the epiglottis:*** The epiglottis, whose highest edge forms the upper boundary

Added Information *contd...*

of the larynx, is in close relation with the tongue. The mucosa covering its free anterior surface is reflected on to the posterior aspect of the tongue and the lateral walls of the pharynx as the median glossoepiglottic and the lateral glossoepiglottic (right and left) folds respectively. The depression on either side of the median fold (and bounded laterally by its corresponding lateral fold) is the vallecula. The lower part of its anterior surface is separated from the thyrohyoid membrane by a small space containing thin adipose tissue; this is the preepiglottic space. The posterior surface of the epiglottis is lined with ciliated respiratory mucosa. Small perforations in the epiglottis connect the posterior space of epiglottis (the laryngeal space proper) with the preepiglottic space.

- ***Role of epiglottis during swallowing:*** The epiglottis derives its name from its lid like action on the laryngeal inlet (loosely called the glottis and eventually leads to the glottis proper). It can be seen that a posterior bending of the epiglottis will close the inlet. During deglutition, the hyoid bone moves upward. Tongue is already undergoing changes in its shape. As a result, the base of tongue presses on the epiglottis; the aryepiglottic muscles also contract. The epiglottis bends backward over the laryngeal inlet and permits the food bolus to slip on its anterior surface. The food bolus splits and passes to the piriform fossae which form the lateral food passages at this level. However, studies indicate that the epiglottis is not essential for swallowing, respiration or phonation.
- As already noted, the internal angle of the laryngeal prominence is about 90 degrees in men and 120 degrees in women. Because of the acute angle in men, the prominence itself is larger and the vocal cords longer.
- The functional role of the cartilage triticea is unknown. The cartilaginous nodules are supposed to strengthen the lateral thyrohyoid ligaments in which they actually lie.

contd...   *contd...*

Added Information *contd...*

❑ The corniculate cartilage is a separated portion of the arytenoid cartilage and the cuneiform is a separated portion of the epiglottis.
❑ The vertebral levels of the various laryngeal cartilages are important:
  ○ Upper border of thyroid cartilage—bifurcation of common carotid artery—C3-C4 junction
  ○ Thyroid cartilage—C4-C5 junction
  ○ Cricoid cartilage—C6
❑ The larynx, at rest lies opposite the C3 to C6 vertebrae in the adult male; in females and children, it is a little higher.
❑ The male and the female larynges are of the same size until puberty. After that, the male larynx enlarges much in size. All cartilages become larger and the thyroid cartilage projects forwards in the midline (forming the Adam's apple projection).

INTERIOR

Almost midway between the upper and lower ends of the larynx, two pairs of mucosal folds project into its cavity. The upper folds are the right and left ***vestibular folds*** (or ***false vocal cords***). The lower folds are the right and left ***vocal folds*** (or ***true vocal cords***). The part of the laryngeal cavity lying above the vestibular folds is called the ***vestibule***. The narrow recess between the levels of the vestibular and vocal folds (on either side) is called the ***sinus*** or ***ventricle*** of the larynx. A pouch like diverticulum arises from the anterior part of each sinus. This pouch is called the ***saccule*** of the larynx.

The upper aperture of the larynx is called its ***inlet*** or ***aditus.*** The aperture is directed backwards and a little upwards (so that its anterior margin lies distinctly higher than the posterior margin). It is bounded on either side by the ***aryepiglottic folds***. These folds are made of mucous membrane that extend from the sides of the epiglottis (in front) to the arytenoid cartilages (behind).

Each vestibular fold encloses a bundle of fibres that constitutes the ***vestibular ligament*** (or ***ventricular ligament***). The ligament is attached, in front, to the angle of the thyroid cartilage; and, behind, to the anterolateral surface of the arytenoid cartilage. The fissure separating the right and left vestibular folds is called the ***rima vestibuli*** (Fig. 10.9) (or ***false glottis*** or ***glottis spuria)*** (Latin. Rima=slit; indicates a narrow opening).

Each vocal fold contains a bundle of elastic fibres that constitutes the ***vocal ligament***. The ligament is attached in front to the angle of the thyroid cartilage (below the attachment of the vestibular ligament) and behind to the vocal process of the arytenoid cartilage (Fig. 10.7). The function of voice production (by vibration) demands that the vocal folds be firm and of uniform thickness. This aim is achieved by close adherence of the lining epithelium to

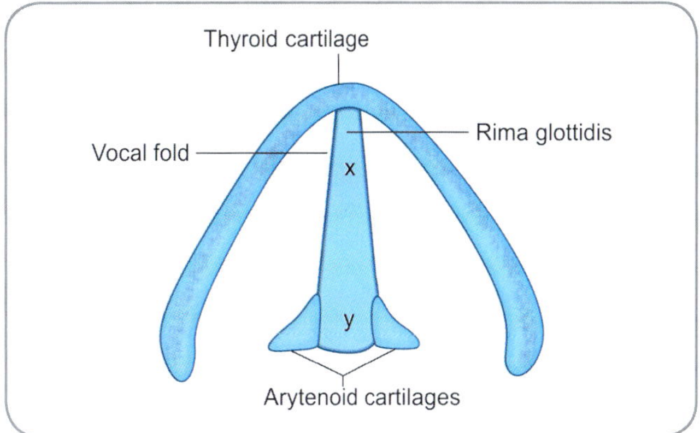

Fig. 10.9: Scheme to show the intermembranous (x) and intercartilaginous (y) parts of the rima glottidis

the vocal ligaments and by the absence of blood vessels. As a result, the vocal folds withstand the stress of repeated and intense vibration. The right and left vocal folds are separated by the anterior or ***intermembranous*** part of a fissure called the ***rima glottidis (or true glottis*** or ***glottis vera)***. The posterior part of the fissure lies between the two arytenoid cartilages and is, therefore, called the ***intercartilaginous*** part (Fig. 10.10). The shape of the rima varies in different phases of respiration and of phonation.

Added Information

❑ The gap between the vestibular folds and vocal folds on either side opens into a recess called the laryngeal ventricle. The ventricle extends upwards lateral to the vestibular folds but medial to the thyroid cartilage. Its upper portion is continuous with another pouch called the ***saccule*** that also extends upwards medial to the thyroid cartilage. The saccule opens into the vestibule guarded by a thin mucosal fold called the ***ventriculosaccular fold***. The saccule is significant by the presence of numerous mucosal glands within itself.

contd…

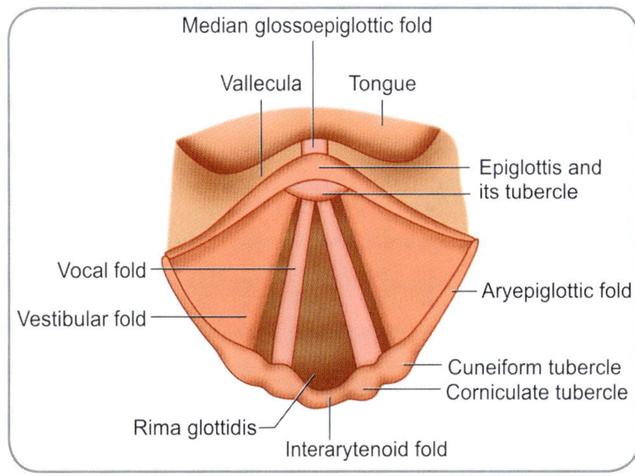

Fig. 10.10: Features of the larynx as seen through a laryngoscope (i.e., from above); the gap between the two vestibular folds is the rima vestibuli

Added Information *contd...*

The glands open on to the luminal surface of the saccule. The vestibular ligament and the thyroepiglottic muscle which are in close proximity to the saccule apply a compression force on the saccule so that the mucosal secretions are poured on to the vocal cords thereby lubricating and moisturising the latter. Hoarseness of voice experienced during excessive speaking, shouting or at times of dehydration is due to inadequacy of the saccular mucosal secretions.

❑ The mucosa over the vocal ligaments is thin and is directly adherent to it without the intervention of any submucous tissue. The vocal cords appear pearly white in colour because of this. Immediately above the vocal ligaments, there is a small space between the mucosa and the underlying tissue. This potential space is called the **Reinke's space** and is a space prone to have fluid accumulation.

❑ The two vocal ligaments meet anteriorly and the point of meeting is the anterior commisure. The fibres of the vocal ligaments from either side pass through the thyroid cartilage at this point of the anterior commisure and blend with its perichondrium. This connection of the fibres with the perichondrium is called the **Broyle's ligament**.

JOINTS

Some of the laryngeal cartilages connect with each other through joints. The joints, thus formed are:

❑ *Cricothyroid joint:* Synovial joint between the inferior cornu of the thyroid cartilage and the side of the cricoid cartilage on either side; a capsule connects the articulating cartilages; a posterior ligament reinforces the capsule.

❑ *Cricoarytenoid joint:* Synovial joint between the lateral part of the cricoid lamina and the arytenoids base on either side; a capsule connects the two cartilages; a posterior cricoarytenoid ligament reinforces the capsule.

❑ *Arytenocorniculate joint:* A synovial or cartilaginous joint connecting the two concerned cartilages on either side.

The joints permit various movements of the cartilages in relation to each other, thereby, permitting various positions of the vocal cords. The length and width of rima glottidis can, therefore, be altered and modified.

The laryngeal inlet or 'aditus laryngis' is the opening between the larynx and the pharynx. It is through this opening that the larynx is entered into from the superior aspect. It is bounded anteriorly by the superior rim of the epiglottis, laterally by the aryepiglottic folds and posteriorly by the interarytenoid fold.

The laryngeal introitus is a clinical term. It indicates the space between the laryngeal inlet and the vestibular folds. It is bounded anteriorly by the posterior surface of the epiglottis, laterally by the medial surfaces of the aryepiglottic folds and posteriorly by the interarytenoid fold.

FIBROUS STRUCTURES

The laryngeal cartilages are connected by fibrous structures which can be bands or sheets. The bands form the ligaments and the sheets form the membranes. These fibrous structures can be classified into the extrinsic and intrinsic structures.

❑ *Extrinsic fibrous structures:* They are attached on one hand to a part of the larynx and on the other to some other neighbouring structure. The most important of these is the thyrohyoid membrane.

  ○ *Thyrohyoid membrane:* It is attached above to the superior border of the body and greater horn of hyoid and below to the superior border of the thyroid laminae. The membrane from its lower attachment ascends behind the hyoid from which it is separated by a bursa. The thickened central part is called the *median thyrohyoid ligament*. The posterior free borders (which connect the superior cornua of thyroid to the posterior ends of the greater horn of hyoid) form the lateral thyrohyoid ligaments.

  ○ *Hyoepiglottic ligament:* This ligament connects the internal aspect of the middle of the body of hyoid to the stalk of the epiglottis.

  ○ *Cricotracheal ligament:* This ligament unites the lower border of the cricoid cartilage to the upper border of first tracheal cartilage.

❑ *Intrinsic fibrous structures:* The intrinsic fibrous structures, true to their classification, connect different parts of the laryngeal cartilages and provide support to them. These structures are actually specialised or condensed portions of the fibroelastic membrane of the larynx.

The fibroelastic membrane lies interior to the cartilaginous skeleton and exterior to the mucosa. The horizontal cleft between the vestibular and the vocal folds divides the membrane into two parts—the upper and the lower.

Histology

Histology of Laryngeal Mucosa

The laryngeal mucosa lines the inner aspect of the entire larynx including the saccule and the ventricle. It is thickened over the vestibular folds. Over the vocal folds it is thin and is attached to the underlying vocal ligaments. On the inner aspects of most of the larynx, the laryngeal epithelium is of the ciliated pseudostratified type. It also has a ciliary clearance mechanism like the rest of the air passage. Over other areas, the epithelium changes in type. It is of the nonkeratinised stratified squamous type over the vocal folds, anterior surface of epiglottis and the exterior surfaces which merge with the pharyngeal walls. The vocal cords are subject to considerable mechanical stress during phonation and voice production. The other areas (where stratified squamous

contd...

Histology *contd...*

epithelium is present) are subject to similar stress, but during deglutition due to the abrasion caused by swallowed food. Hence these areas have stratified squamous epithelium.

The laryngeal mucosa has the laryngeal mucosal glands. These are more numerous over the epiglottis (forming the epiglottic glands) and in the aryepiglottic folds (forming the arytenoids glands). Large mucosal glands are present in the saccule and secrete mucosa over the vocal cords which themselves are devoid of glands. The vocal cords require moisture and lubrication due to their constant efforts; they tend to become dry because of the stratified squamous epithelium.

Taste buds may be present on the posterior epiglottic surface and aryepiglottic folds.

The ***upper part*** of the fibroelastic membrane forms the quadrangular membrane. It extends from the side of the epiglottis to the arytenoids of the same side. Its free posterosuperior border is the aryepiglottic ligament and free inferior border is the vestibular ligament.

The ***lower part*** of the fibroelastic membrane forms the cricothyroid membrane, median cricothyroid ligament and the cricovocal membrane. The cricothyroid membrane runs up from the upper border of cricoid cartilage to the lower border of the thyroid cartilage. Its anterior midline portion is thickened as the median cricothyroid ligament. The cricovocal membrane (otherwise called the conus elasticus) starts from the inner surface of the cricoid cartilage near the lower margin. It passes up deep to the thyroid cartilage. Its upper edge gets attached to the angle of the thyroid cartilage anteriorly and to the vocal process of the arytenoids posteriorly. Between these two attachments, the rest of the upper border is free and forms the vocal ligament.

MUSCLES

The muscles of the larynx are extrinsic and intrinsic. The ***extrinsic muscles*** are those in which one end of the muscle is attached to a cartilage of the larynx and the other end is attached to neighbouring structures. These muscles raise or lower the larynx as a whole during deglutition and phonation. The extrinsic muscles are the infrahyoid strap muscles (sternothyroid, sternohyoid and thyrohyoid), the inferior constrictor of the pharynx and the stylopharyngeus and palatopharyngeus. The role played by the extrinsic muscles during respiration is minimal. However, they alter the pitch and tone of the voice by raising or lowering the larynx.

The ***intrinsic muscles*** are confined to the larynx. They may be classified in accordance with their actions as follows:

❏ ***Muscles which increase or decrease the tension of the vocal folds***: cricothyroid, thyroarytenoid and vocalis.

❏ ***Muscles which open or close the glottis***: Posterior cricoarytenoid, lateral cricoarytenoid and transverse arytenoid.

❏ ***Muscles which open or close the inlet of the larynx***: Oblique arytenoids, aryepiglottic and thyroepiglottic muscles.

Additional Notes on the Laryngeal Muscles

The cricothyroid is the only intrinsic muscle of the larynx that is seen on its outer aspect. It passes from the cricoid cartilage to the thyroid cartilage. The cricothyroid lengthens the vocal folds and makes them tense.

Contraction of vocalis muscle can cause different lengths of the vocal ligament to become tense, thus varying the pitch of the voice. The muscle is therefore called the ***tuning fork of the larynx***.

The posterior cricoarytenoid muscles abduct the vocal folds; these are the only laryngeal muscles which open the glottis. The rima glottidis becomes actually triangular (due to lateral rotation of the arytenoid cartilages) when the posterior cricoarytenoid muscles contract. When all other intrinsic muscles have a sphincteric action, these muscles have an 'opening' action on the glottis. Paralysis of both posterior cricoarytenoid can lead to closure of rima glottidis and in turn to suffocation (interfering with air passage). For this reason, these are called the "***safety muscles of the larynx***" (Table 10.3).

'Glottis' is the name given to the vocal apparatus of the larynx. It consists of the vocal folds investing the vocal ligaments and the vocales muscles and rima glottidis, the aperture between the vocal folds. Very often the term is loosely applied to the rima or to the vocal cords.

The vocal fold can be labeled as the organ that controls sound production. Each vocal fold consist of the vocal ligament and the vocalis muscle. The free edge of the vocal fold is the vocal cord. The vocal folds are the source of laryngeal sounds. They vibrate and produce audible sounds when their free margins come close to each other and air is forced out intermittently.

Alterations in Shape of Rima Glottidis (Figs 10.11A to D)

The rima glottidis (or plainly glottis) has an anterior (intermembranous) part placed between the two vocal folds and a posterior (intercartilaginous) part placed between the medial surfaces of the two arytenoid cartilages (Fig. 10.11A). The size and shape of the glottis undergoes changes during different phases of respiration and of speech. During quiet respiration (and after death) the glottis is moderately wide, the anterior part being triangular and the posterior part rectangular (Fig. 10.11A). In this position, the laryngeal muscles are relaxed. The glottis is narrowest during speech (Fig. 10.11B), both the two vocal folds and the two arytenoid cartilages being close together. This position is produced mainly by the contraction of the transverse arytenoid which

Table 10.3: Muscles of larynx

Name of the muscle	Origin	Insertion	Nerve supply	Actions
Cricothyroid (paired muscle)	External surface of the arch of cricoid cartilage	Lower border of thyroid lamina and anterior border of inferior horn of thyroid cartilage	External branch of superior laryngeal nerve	Stretches the vocal ligament by tilting the thyroid cartilage forwards and downwards thus increasing the distance between the thyroid and the vocal processes of the arytenoids
Thyroarytenoid (paired muscle)	Angle of thyroid cartilage and cricothyroid ligament	Anterolateral surface of the arytenoid cartilage	Recurrent laryngeal nerve; may also get a branch from the external laryngeal nerve	Draws the arytenoid to the thyroid cartilage thus shortening and relaxing the vocal ligaments
Vocalis (deeper fibres of thyroarytenoid) (paired muscle)	Angle of thyroid cartilage and cricothyroid ligament	Vocal process of arytenoid	Recurrent laryngeal nerve	Rotates the arytenoid medially– approximation of vocal cords and closure of rima glottidis
Posterior cricoarytenoid (paired muscle)	Posterior surface of cricoid lamina	Muscular process of arytenoid	Recurrent laryngeal nerve	Rotates the arytenoid laterally, thus separates the vocal process and vocal cord; opens the glottis
Lateral cricoarytenoid (paired muscle)	Upper border of cricoid arch	Muscular process of arytenoid	Recurrent laryngeal nerve	Rotates the arytenoid anteromedially and closes the glottis
Transverse arytenoid (unpaired muscle)	The muscle extends from the back of the muscular process of one arytenoid to the back of the muscular process of the other arytenoid		Recurrent laryngeal nerve; may also receive twigs from the internal laryngeal nerve	Pulls the two arytenoid cartilages together and thus closes the posterior part of the rima glottidis
Oblique arytenoid	Back of the muscular process of arytenoid	Apex of the opposite arytenoid cartilage	Recurrent laryngeal nerve	Adducts the aryepiglottic folds and closes the laryngeal inlet
Aryepiglotticus (fibres of oblique arytenoid muscle which extend into the aryepiglottic fold)	Apex of the arytenoid cartilage	Aryepiglottic fold	Recurrent laryngeal nerve	Acts, along with the oblique arytenoid, as a sphincter of laryngeal inlet

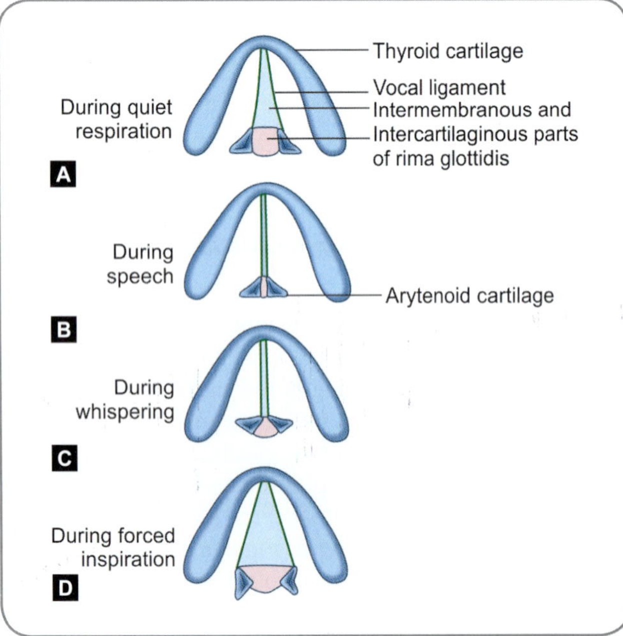

Figs 10.11A to D: Alterations in shape of the rima glottidis during respiration and during speech

pulls the cartilages towards each other. In whispering (Fig. 10.11C), the glottis is closed in its anterior part, but the posterior part is wide open. The glottis is widely opened in forced respiration, both the anterior and posterior parts becoming triangular (Fig. 10.11D).

VESSELS AND NERVES

The *arteries* supplying the larynx are the superior laryngeal branch of the superior thyroid artery and the inferior laryngeal branch of the inferior thyroid artery. That part of the larynx from the epiglottis to the vocal cords, most of its tissues and musculature are supplied by the superior laryngeal artery. The lower part of the larynx is supplied by the inferior laryngeal artery. There is a rich anastomosis between the ipsilateral arteries and the corresponding contralateral arteries.

❑ The *veins* accompany the arteries.
❑ The *lymphatic drainage* (Fig. 10.12) can be described in two parts. The vocal cords which have a strongly adherent mucosa and no lymphatics serve as the border for the two parts—the supraglottic part (above the vocal

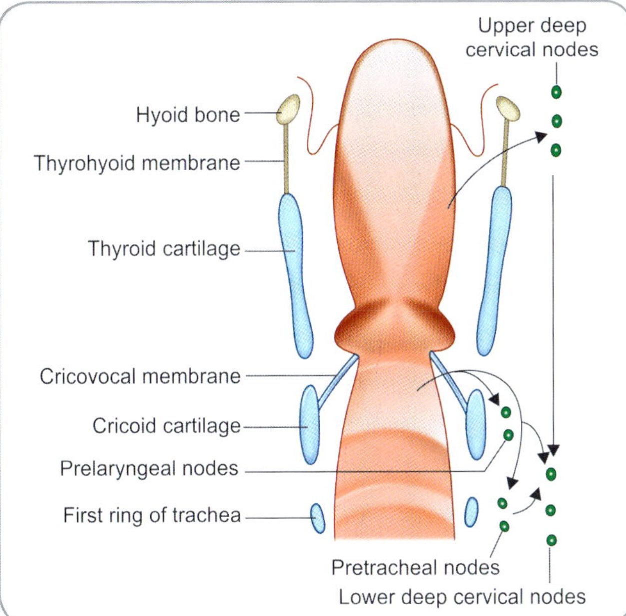

Fig. 10.12: Scheme to show the lymphatic drainage of the larynx

cords) and the infraglottic part (below the vocal cords). The supraglottic lymphatics accompany the superior laryngeal artery, pierce the thyrohyoid membrane and end in the upper deep cervical lymph nodes. The infraglottic lymphatics pass through the cricovocal membrane and end in the prelaryngeal or pretracheal nodes. Some of them may reach the lower deep cervical nodes.

The sensory *innervation* of the part of the larynx above the vocal folds is by the internal laryngeal nerve. Sensations of taste from the epiglottis also travel through this nerve. The part of the larynx below the vocal folds receives its sensory innervation through branches of the recurrent laryngeal nerve.

Most of the intrinsic muscles of the larynx are supplied by the recurrent laryngeal nerve. The only exception is the cricothyroid that is supplied by the external laryngeal nerve.

> ***Gender differences in the larynx:*** No gender differences are seen until puberty. Testosterone causes strengthening of the laryngeal walls and enlargement of all laryngeal structures. Laryngeal prominence becomes very conspicuous. Vocal folds lengthen and thicken with the anteroposterior diameter of rima glottidis also increasing. The longer vocal cords take a longer time for vibration and so pitch decreases. Voice becomes deeper.

Clinical Correlation

❑ Entry of any foreign matter (including water or food) into the larynx can result in suffocation and death. The intense cough that is set up on the entry of even a small particle is a protective mechanism against suffocation.

contd…

Clinical Correlation *contd...*

❑ Inflammation of the mucous membrane of the larynx (***laryngitis***) can cause ***hoarseness of voice***, or even complete loss of voice, because of oedema above the level of the vocal folds. The laryngeal mucous membrane is firmly adherent to the vocal folds and that is why fluid accumulates above this level. In the presence of severe inflammation or irritation, the oedema may be of such a degree as to lead to suffocation. In such cases it may become necessary to create an artificial opening in the trachea (***tracheostomy***) or in the larynx itself below the vocal cords through the cricothyroid ligament (laryngotomy) to save the life of the person.

❑ Paralysis of one or more muscles of the larynx also leads to hoarseness of voice. The hoarseness is temporary in case of injury to the external laryngeal nerve (as the function of the paralysed cricothyroid is gradually taken up by the muscle of the normal side). When the recurrent laryngeal nerve is injured, hoarseness is permanent. In this connection, it may be noted that in some cases only some groups of muscles may be affected. When all the intrinsic laryngeal muscles are paralysed, the vocal folds are immobile and lie midway between abduction and adduction. This is the position occupied by the vocal folds after death. Paralysis of abductors alone leads to adduction of the vocal folds (by unopposed action of the adductors); this leads to closure of the glottis with consequent suffocation. Hence, paralysis of abductors alone is much more dangerous than paralysis of both abductors and adductors of the vocal folds.

❑ Improper muscle balance during phonation and voice production can cause the vocal cords to rub against each other. The point of contact is at the junction of the anterior third with the posterior two-thirds of the vocal ligament. When people speak with force, abuse the voice level or sing with improper techniques, repeated rubbing occurs resulting in trauma to the epithelium, subepithelial haemorrhages and bruising. Chronic occurrence of such damage can lead to scarring and nodule formation. These nodules are referred to as *'singer's nodules'* or 'clergyman's nodes'.

❑ Throughout the larynx, its mucous membrane is loosely attached except for certain specific areas like the anterior surface of epiglottis, vocal folds and over the smaller cartilages. In acute infections, excessive fluid secretion occurs and leads to accumulation of the secreted fluid. Mucosal and submucosal swelling thus caused may compromise the airway and result in respiratory complications. Due to the adherence of the mucosa over the vocal cords, fluid cannot pass from the upper to the lower compartment or vice versa. In cases of swelling above the vocal cords, the Reinke's space gets swollen. This condition is called Reinke's oedema. The vocal cords also become flabby due to swelling in the space immediately above. Smoking and vocal abuse are the common causes of ***Reinke's oedema***.

❑ The laryngeal ventricle and saccule may get infected. If the ventricular aditus (opening of the ventricle to the main laryngeal cavity) is closed or partially closed due to inflammation and consequent swelling, fluid accumulation occurs inside. The condition is complicated by the presence of mucous within the saccule. A mucous filled cyst may occur and is then called a ***laryngocoele***. Airway obstruction, chronic and repeated respiratory infections can occur because of the cyst pressing on neighbouring structures or getting infected further.

contd…

TRACHEA

The trachea is a wide tube lying on the front of the neck more or less in the midline. The upper end of the trachea is continuous with the lower end of the larynx. The laryngotracheal junction lies opposite to the lower part of the body of the sixth cervical vertebra. At the root of the neck, the trachea passes into the superior mediastinum of the thorax.

Relations of the Trachea in the Neck

The trachea is related to a large number of structures in the neck as follows:

- ***Posteriorly:*** Related to the oesophagus which runs vertically behind it, and separates it from the bodies of vertebrae C6 and C7 (Fig. 10.13).
- ***Anteriorly:*** Covered by skin, superficial and deep fascia over its entire length. The deep fascia is represented by a superficial part, the investing layer and a deeper part, the pretracheal fascia. The right and left sternohyoid and sternothyroid muscles overlap that part of the trachea near the inlet of the thorax. Near its upper end (over the 2nd, 3rd and 4th rings) the trachea is covered anteriorly by the isthmus of the thyroid gland.
- ***Right and left sides:*** Overlapped on the corresponding sides by the right and left lobes of the thyroid gland. The

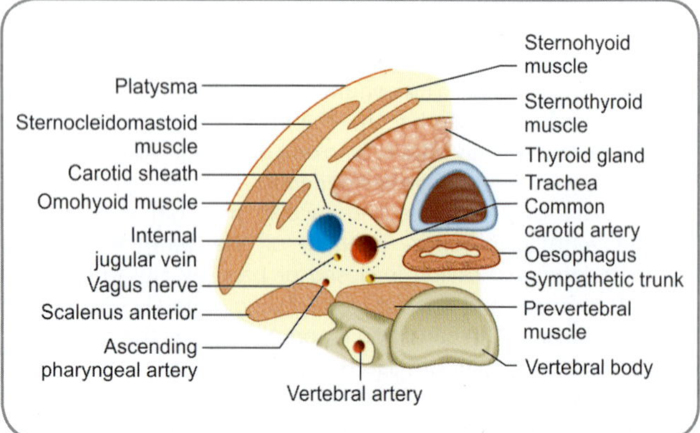

Fig. 10.13: Transverse section through the upper part of the trachea to show some of its relations in the neck

right and left common carotid arteries ascend along the corresponding side of the cervical part of the trachea.
- On either side the inferior thyroid artery is closely related to the trachea and the recurrent laryngeal nerve lies in the groove between the trachea and the oesophagus (on the left side the nerve is related to both the thoracic and cervical parts of trachea, but on the right side it is related to the cervical part only).
- ***Smaller structures lying in front of the trachea are:***
  - An anastomosis between branches of the right and left superior thyroid arteries lying along the upper margin of the isthmus of the thyroid gland;
  - Inferior thyroid veins passing from the lower border of the isthmus of the thyroid gland to their termination in the left brachiocephalic vein;
  - A small artery, the thyroidea ima, may sometimes be present (It arises from the brachiocephalic trunk or from the arch of the aorta and runs upwards in front of the trachea to reach the isthmus of the thyroid gland);
  - The jugular arch connecting the right and left anterior jugular veins and running across the trachea just above the manubrium sterni.

Lymphatic vessels of the trachea drain into pretracheal and paratracheal nodes and through them (or directly) to the deep cervical nodes.

OESOPHAGUS

The oesophagus is a tubular structure which starts at the lower end of the oropharynx (i.e., in front of the sixth cervical vertebra). It descends through the lower part of the neck and enters the thorax through its inlet. After passing through the thorax, the oesophagus enters the abdomen and ends by joining the cardiac end of the stomach. The cervical part of the oesophagus receives branches from the inferior thyroid artery. Veins drain into the inferior thyroid vein. Lymphatics from the cervical part of the oesophagus drain to the deep cervical nodes. Some lymph passes through the paratracheal nodes.

Relations of Oesophagus in the Neck

Posteriorly, the oesophagus is related to the sixth and seventh cervical vertebrae (covered by prevertebral muscles and fascia). Anteriorly, it is related to the trachea. Laterally, the oesophagus is related to the corresponding common carotid artery. The upper part of the oesophagus is overlapped, laterally, by the corresponding lobe of the thyroid gland. The thoracic duct ascends along the left side of the oesophagus in the lower part of the neck. The right and left recurrent laryngeal nerves lie anterolateral to the oesophagus in the corresponding grooves between it and the trachea.

Multiple Choice Questions

1. The term 'Hypopharynx' indicates the:
   a. Nasopharynx
   b. Oropharynx
   c. Laryngopharynx
   d. Larynx
2. Thyropharyngeus muscle is:
   a. A part of superior constrictor
   b. A part of middle constrictor
   c. A part of inferior constrictor
   d. A separate muscle
3. The plexus of circulus tonsillaris consists of:
   a. Branches of glossopharyngeal and lesser palatine nerves
   b. Branches of tonsillar and ascending palatine arteries
   c. Branches of glossopharyngeal and chorda tympani nerves
   d. Venules joining the paratonsillar vein
4. The vocal process is a part of:
   a. Thyroid cartilage
   b. Cricoid cartilage
   c. Arytenoid cartilage
   d. Epiglottis
5. The anterior part of glottis is closed but the posterior part open during:
   a. Normal Speaking
   b. Quiet respiration
   c. Forced respiration
   d. Whispering

ANSWERS

1. c   **2.** c   **3.** a   **4.** c   **5.** d

Clinical Problem-solving

Case Study 1: A 48-year-old musician developed singer's nodules.
- What is the area of problem and what does the musician suffer from?
- What is the primary cause for this condition?
- What would be your advice to avoid such a condition?

Case Study 2: A 12-year-old boy complained of left ear pain for some time. One day he presented with the left side of his face swollen, respiratory infection and mild difficulty in breathing while speaking. On examination, his palatine tonsils were inflamed.
- In what way would you associate the boy's respiratory infection and tonsillitis?
- Was the ear pain a cause of or consequence of the tonsillitis? Reason out your answer.
- What is quinsy?

(For solutions see Appendix).

Chapter 11

Orbit and Eye

Frequently Asked Questions

- Discuss briefly the boundaries and contents of the orbit.
- Write notes on: (a) Superior orbital fissure, (b) Recti muscle, (c) Superior oblique muscle, (d) Levator palpebrae superioris.
- Write a detailed account of the extraocular muscles and their actions.
- Write notes on: (a) Ciliary ganglion, (b) Nasociliary nerve, (c) Oculomotor nerve, (d) Venae vorticosae.
- Discuss the retina and the photoreceptive pathway.
- Write briefly on: (a) Iris, (b) Ciliary body, (c) Cornea, (d) Aqueous humour.

ORBIT

The orbits are bilateral pyramidal bony cavities in the face; each of these cavities contains the eyeball of its side and the accessory visual structures.

Before we proceed to study the various structures related to the eyes, it is essential to learn the definition of certain terms.

- ***Orbit:*** The bony cavity in the upper part of the facial skeleton that contains the eyeball and accessory visual structures; other names are orbita (a singular term; the plural form is 'orbitae'), orbital cavity and eye socket.
- ***Eye:*** The organ of vision that comprises the eyeball and the optic nerve.
- ***Orbital region:*** The area of the face that overlies the orbital cavity and includes the eyelids and the lacrimal apparatus.
- ***Accessory visual structures:*** Otherwise called the adnexa oculi, these include the eyelids, conjunctiva, lacrimal apparatus, related nerves and vessels, extraocular muscles and orbital fascia.

BOUNDARIES

Each orbit has a base, an apex, medial wall, lateral wall, superior wall and an inferior wall. The base is on the surface of the facial skeleton and is directed anterolaterally. The apex is directed posteromedially. The medial wall is more or less straight and is parallel to the fellow of the opposite side but is separated from it by the ethmoidal sinuses. The lateral wall runs obliquely backward; the two lateral walls are at right angles to each other.

The ***base*** is marked by the ***orbital margin*** (or the ***orbital rim***) which in turn surrounds the ***orbital opening*** (or the ***orbital aditus***). The upper part of the anteromedial rim, the superior rim and the upper part of the anterolateral rim are formed by the frontal bone. The lower part of the anteromedial rim and the medial half of the inferior rim are formed by the maxilla. The lower part of the anterolateral rim and the lateral half of the inferior rim are formed by the zygomatic bone. The bony orbital margin is not circular but spiral. The lower part of the anteromedial rim, if traced medially, becomes the anterior lacrimal crest of the maxilla. The upper part of the anteromedial rim, if traced medially, becomes continuous with the posterior lacrimal crest of the lacrimal bone. The two lacrimal crests bound the lacrimal fossa in front and behind. Thus, the bony orbital margin is not circular but somewhat spiral. The orbital margin is reinforced by the membranous orbital septum, to which it also gives attachment.

> The orbital margin is somewhat like a rounded square. It has four angles. Each angle has a notable feature behind it. The features are—***superomedially, trochlea for superior oblique; superolaterally, the fossa for lacrimal gland; inferomedially, nasolacrimal canal and inferolaterally, the lateral end of inferior orbital fissure.***

The *apex* of the orbit is marked by an opening called the optic canal. The optic canal is present in the lesser wing of sphenoid, medial to the superior orbital fissure and forms part of the middle cranial fossa.

The *medial wall* is formed predominantly by the orbital plate of ethmoid bone. The frontal process of maxilla, the lacrimal and the sphenoid bones also contribute smaller parts of the medial wall. In the anterior aspect of this wall are found the lacrimal groove and the lacrimal fossa for the lacrimal sac. Superiorly, the trochlea for the tendon of superior oblique muscle is seen. The medial wall is extremely thin.

The *lateral wall*, which is the thickest and the strongest, is formed by the frontal process of the zygomatic bone and the greater wing of sphenoid. The posterior part of this wall separates the orbit from the temporal and middle cranial fossae. The lateral wall is the most exposed and is prone to injuries.

> Perforations made in the walls of the orbit will connect to:
> ❏ Anterior most part of roof—frontal air sinus
> ❏ Rest of the roof—anterior cranial fossa
> ❏ Floor—maxillary air sinus
> ❏ Most anterior part of medial wall—atrium of nasal cavity
> ❏ Middle part of medial wall—ethmoidal air sinus
> ❏ Posterior part of medial wall—sphenoidal air sinus
> ❏ Anterior part of lateral wall—temporal fossa
> ❏ Posterior part of lateral wall—middle cranial fossa

The *superior wall*, also called the roof of the orbit, is formed primarily by the orbital part of the frontal bone. Though the entire roof appears to be more or less horizontal, the inferior surface of the roof is gently concave. The lesser wing of sphenoid contributes a small portion of the roof near the apex. In the anterolateral aspect of this wall is a shallow depression called the lacrimal fossa that lodges the lacrimal gland.

The *inferior wall*, otherwise called the floor of the orbit or the orbital floor, is formed by the maxilla, zygomatic and palatine bones. This wall slopes down from the apex to the base and separates the orbit from the maxillary sinus. The wall is thin. The inferior orbital fissure demarcates the inferior wall from the lateral wall. An infraorbital groove runs forwards on the floor and sinks deep to form the infraorbital canal.

❏ *Orbital fissures and the optic canal:* Two orbital fissures, namely the superior and the inferior orbital fissures open into the orbit and transmit important structures. The optic canal opens at the apex.
  ○ The *superior orbital fissure* connects the cranial cavity with the orbit. It is bounded medially by the body of sphenoid, superiorly by the lesser wing of sphenoid and inferiorly by the medial border of the orbital surface of sphenoid. Its apex, where the superior and inferior boundaries meet, is bounded by the frontal bone. The fissure, which is wide medially and narrows laterally, transmits the oculomotor, trochlear and abducent nerves, branches of the ophthalmic nerve and the ophthalmic veins.
  ○ The *inferior orbital fissure* connects the orbit with the pterygopalatine and infratemporal fossae. It is bounded superiorly by the greater wing of sphenoid, inferiorly by the maxilla and the orbital process of palatine bone and laterally by the zygomatic bone. The fissure transmits the infraorbital and zygomatic branches of the maxillary nerve, orbital rami of the pterygopalatine ganglion, vessels accompanying the branches of the maxillary nerve and connecting veins between the inferior ophthalmic veins and the pterygoid venous plexus.
❏ The optic canal is an opening found at the apex of the orbit and connects the orbit with the middle cranial fossa. The lesser wing of sphenoid is attached to the body of the bone by two roots. The foramen that lies between the two roots is the optic canal.

> The optic canal is the grand royal entrance to the orbit; the superior orbital fissure is the general entrance and the inferior orbital fissure is the accessory side entrance. The nasolacrimal duct, infraorbital groove, anterior ethmoidal and posterior ethmoidal foramina, zygomatic foramen and supraorbital foramen are exits.

CONTENTS

The contents of the orbit can be classified into two groups—the main contents and the accessory contents. The accessory contents are as equally important as the main contents and have applied significance.

The main contents are as follows:
❏ The eyeball;
❏ Muscles of the orbit including the extraocular muscles;
❏ Nerves of the orbit:
  ○ Optic nerve,
  ○ Oculomotor nerve,
  ○ Trochlear nerve,
  ○ Abducent nerve,
  ○ Ophthalmic division of trigeminal nerve and its branches and
  ○ Part of maxillary division of trigeminal nerve;
❏ Ciliary ganglion;
❏ Arteries of the orbit:
  ○ Ophthalmic branch of internal carotid artery,
  ○ Infraorbital branch of the maxillary artery;
❏ Veins of the orbit:
  ○ Superior and inferior ophthalmic veins;
❏ Lacrimal gland.

The accessory contents are as follows:
- Orbital septum;
- Orbital fat;
- Fascial sheath of the eyeball;
- Suspensory ligament of eye;
- Suspensory ligament of lacrimal gland and
- Palpebral ligaments.

Main Contents

Eyeball

The eyeball is the most important main content of the orbit. All other contents are, in some way or the other, helpful in the maintenance, protection and functioning of the eyeball (the eyeball is discussed in detail, elsewhere in the same chapter).

Muscles of the Orbit

The muscles of the orbit can be broadly classified into the skeletal muscle group and the smooth muscle group. The *extraocular muscles,* which are the main muscles of the orbit, constitute the skeletal muscle group. The muscles of this group are:
- The four recti (superior, inferior, medial and lateral) (singular, rectus);
- Two oblique muscles (superior and inferior) and
- The levator palpebrae superioris.

Muscles made up of smooth muscle fibres and supplied by autonomic nerves constitute the second group and are the *superior tarsal, inferior tarsal* and *orbitalis* muscles. The *sphincter pupillae, dilator pupillae* and *ciliaris* muscles lie within the eyeball and will be considered when the eyeball is discussed. The orbicularis oculi muscle and its parts (especially the Horner's muscle and the Riolan's muscle) are part of the face and will be dealt with when facial muscles are discussed.

Extraocular Muscles (Table 11.1 and Fig. 11.1)

The extraocular or extrinsic muscles are seven in number. Of these, one muscle, namely the levator palpebrae superioris acts on the upper eyelid. The other six act on the eyeball proper and produce various movements of the eye.

Table 11.1: Extraocular muscles

Muscle	Origin	Insertion	Action	Nerve supply
Superior rectus *Other names:* Attollens oculi, superior straight muscle	All recti arise from posterior part of orbit through a common tendinous ring which surrounds the optic canal, and encloses part of the superior orbital fissure. The superior, inferior, medial and lateral recti arise from the corresponding parts of the ring.	The recti run forward to reach the corresponding side of the sclera about 6 mm behind the junction of the sclera and cornea. Insertions are in front of the equator of the eyeball.	• Upward movement of eyeball. • Inward movement of eyeball. • Intorsion of eyeball.	Oculomotor nerve
Inferior rectus *Other name:* Inferior straight muscle			• Downward movement of eyeball. • Inward movement of eyeball. • Extorsion of eyeball.	Oculomotor nerve
Medial rectus *Other names:* Rectus internus, medial straight muscle			Inward movement of eyeball.	Oculomotor nerve
Lateral rectus *Other names:* Rectus externus, abducens oculi, lateral straight muscle			Outward movement of eyeball.	Abducent nerve
Superior oblique	Body of sphenoid bone just above and medial to optic canal.	• Runs forward in upper part of orbit. • Near orbital margin the muscle ends in a tendon that passes through a tendinous pulley. • Tendon then runs backwards and laterally to be inserted into the upper lateral quadrant of eyeball behind the equator.	• Outward movement of eyeball. • Intorsion of eyeball.	Trochlear nerve
Inferior oblique	Anterior and medial part of floor of orbit (maxilla).	Muscle winds around eyeball to reach the lateral part of the sclera behind equator of eyeball.	• Outward movement of eyeball. • Extorsion of eyeball.	Oculomotor nerve
Levator palpebrae superioris	Posterior part of orbit (lesser wing of sphenoid) above optic canal.	Upper eyelid (tarsus and superior conjunctival fornix).	Elevates eyelid and keeps palpebral fissure open.	Oculomotor nerve

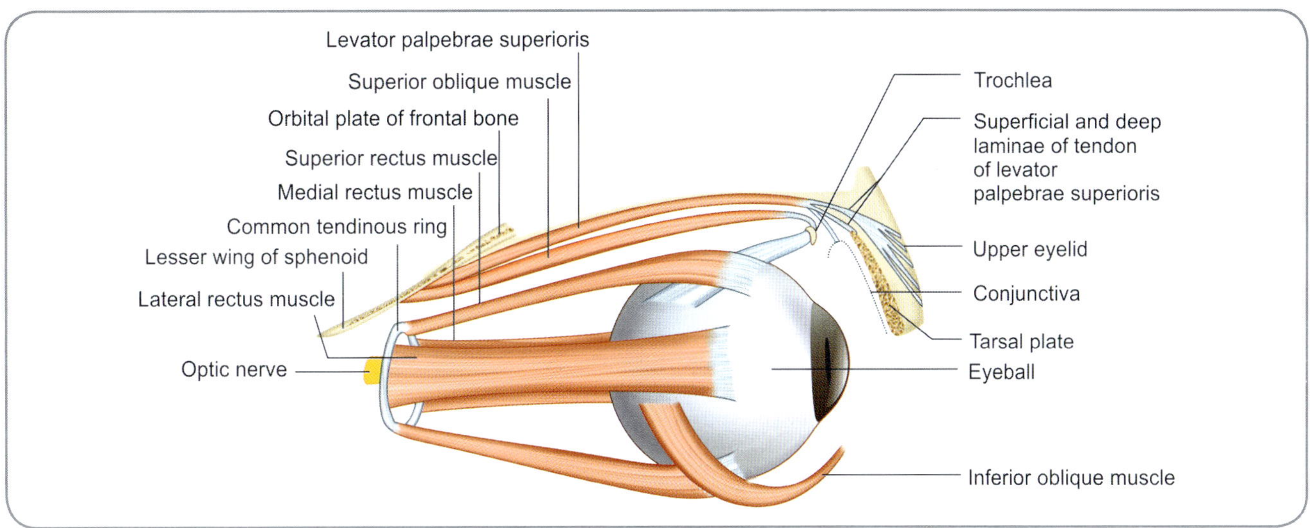

Fig. 11.1: Scheme to show extraocular muscles as seen from the lateral side

Common Tendinous Ring (Fig. 11.2)

The four recti arise from a common fibrous ring called the common tendinous ring or the common annular ocular tendon or the ligament of Zinn. This ring surrounds the optic canal and the medial part of the superior orbital fissure. It is attached inferiorly to the small tubercle on the inferior border of the superior orbital fissure and superiorly to the inferior surface of the lesser wing of sphenoid. The four recti muscles arise from the superior, inferior, medial and lateral aspects of this ring and pass forward to form a muscular cone. Structures which pass through the optic canal and that part of the superior orbital fissure which is enclosed within the ring, lie within the cone of recti. These structures are the optic canal and the ophthalmic artery

(transmitted through the optic canal), the superior and inferior divisions of the oculomotor nerve, the nasociliary branch of the ophthalmic nerve and the abducent nerve (all transmitted through the superior orbital fissure). The structures passing through the lateral part of the superior orbital fissure and those passing through the inferior orbital fissure lie outside the common tendinous ring.

Additional Notes on Levator Palpabrae Superioris

Levator palpabrae superioris is a thin triangular muscle arising from the inferior surface of lesser wing of sphenoid, immediately superior to the common tendinous ring and the superior rectus. Though its origin is narrow, the muscle ends in a wide aponeurosis at its insertion. Some fibres of this aponeurosis pass directly into the upper eyelid and attach to the anterior surface of the tarsal plate; this is described as the deep stratum of the bilaminar aponeurosis. Majority of the fibres radiate, pierce the orbicularis oculi muscle and get attached to the skin of the upper eyelid; this is described as the superficial stratum of the bilaminar aponeurosis. As the lateralmost aponeurotic fibres are traced, they pass between the orbital and palpebral parts of the lacrimal gland to get attached to a small tubercle called the **Whitnall's tubercle** on the zygomatic bone. The medial aponeurotic fibres become thin and merge with the medial palpebral ligament. The connective tissue coverings of the adjoining surfaces of the levator palpebrae and the superior rectus are fused to each other. When the two muscles diverge away to their respective insertions, the connective tissue forms a thickened mass to which the superior conjunctival fornix is attached. This provides an additional attachment to the levator palpebrae and is helpful during movements of the upper eyelid. The place where the muscle fibres convert into aponeurosis,

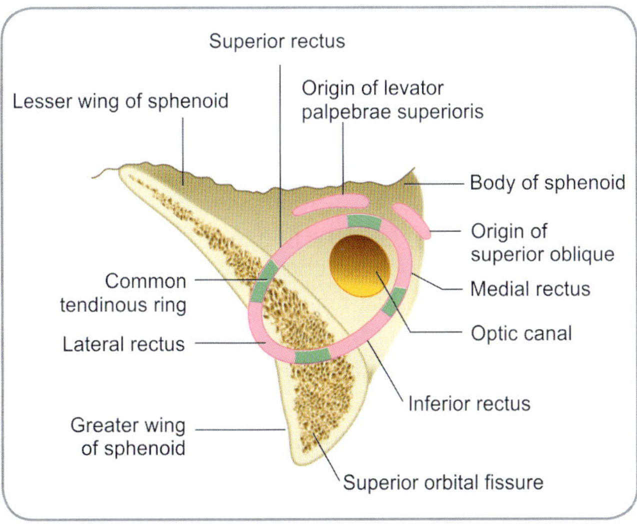

Fig. 11.2: Diagram showing the common tendinous ring for origin of the rectus muscles of the orbit. Note its relationship to the optic canal and to the superior orbital fissure

the covering fasciae of the muscle fibres unite to form a continuous band. This band appears like a ligament and is called the **Whitnall's ligament**. The levator palpebrae muscle is thus described to pierce the Whitnall's ligament and become an aponeurosis. A sympathetic component of this muscle is usually described and constitutes the superior tarsal muscle. The muscle is supplied by a branch of the superior division of the oculomotor nerve which enters the inferior surface of the muscle. The sympathetic component is supplied by sympathetic fibres derived from the plexus around the internal carotid artery. Levator palpabrae superioris can be described as the 'dilator' of the palpebral fissure and an antagonist to the orbicularis oculi which functions as the sphincter constrictor of the palpebral fissure.

Additional Notes on Recti Muscles

The four recti (Latin.rectos=straight) muscles, arising from the common tendinous ring, run anteriorly to be attached to the eyeball. They are named for their individual positions relative to the eyeball. All of them attach to the sclera of the eyeball anterior to its equator (and posterior to the margin of cornea). Their contractions, therefore, would bring the anterior part of the eyeball closer to their respective origins. The recti are strap like and form thin tendons at their ocular attachments (attachments to the eyeball). Though commonly said to be attached to the sclera about 6 mm from the corneal margin or the limbus, the individual ocular attachments are as follows—medial rectus: 5.5 mm from the limbus, inferior rectus: 6.6 mm from the limbus, lateral rectus: 7 mm from the limbus and superior rectus: 7.7 mm from the limbus. If the insertions starting from the medial rectus to superior rectus are connected by a clockwise line, it would make a spiral. This is the spiral of Tillaux (named after the 19th century French physicia Paul Tillaux). Each rectus is enclosed in an areolar tissue sheath. The margins of adjacent sheaths are attached to each other; a complete musculo areolar cone is established and this is referred to as the **muscle cone of recti**. The superior rectus is the largest of the recti. The medial rectus is the shortest but the strongest. The lateral rectus has an additional muscular slip arising from the orbital surface of the **greater wing of sphenoid lateral** to the common tendinous ring.

Additional Notes on Oblique Muscles (Fig. 11.3)

The two oblique muscles of the eyeball, true to their names, have an oblique disposition within the orbital cavity and in relation to the eyeball. Both of them are attached to the sclera of the eyeball, posterior to the equator. The superior oblique, the longer of the two, arises from the body of the sphenoid superomedial to the optic canal and the common tendinous attachment of the superior rectus passing forwards, the muscle ends in a rounded tendon. The

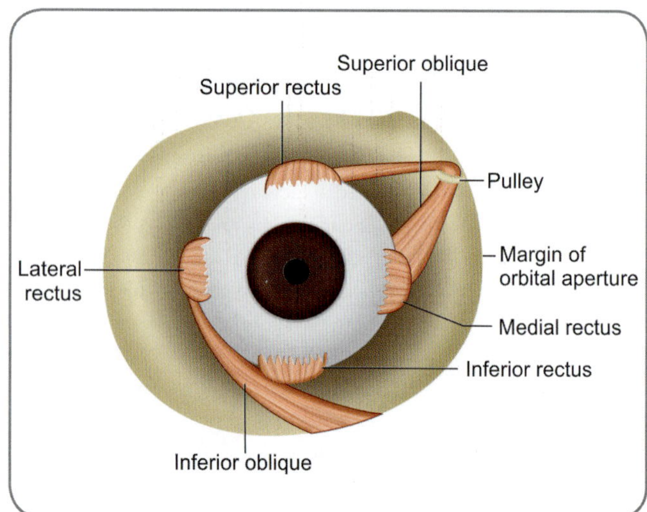

Fig. 11.3: Schematic anterior view of the orbit to show the attachments of extraocular muscles to the eyeball. Particularly note the arrangement of the superior and inferior oblique muscles

tendon passes through a fibrocartilaginous loop called the trochlea (Greek.trochileia=pulley, trecho=run) found in the trochlear fossa of the frontal bone on the superior aspect of the medial wall of the orbit. Having passed through the trochlea, the tendon turns posteroinferolaterally beneath the superior rectus and gets inserted into the superolateral posterior quadrant of the eyeball. The fourth cranial nerve supplies this muscle by entering its superior aspect; this nerve is called the trochlear nerve because of the muscle tendon passing through the trochlea. The inferior oblique muscle arises from the anterior part of the orbital surface of maxilla (that is the orbital floor) and is the only extraocular muscle to arise from the anterior part of the orbit; it ascends posterolaterally, first between the orbital floor and the inferior rectus and then between the eyeball and the lateral rectus. It is inserted into the inferolateral posterior quadrant of the eyeball.

Movements of the Eyeball

Axes of orbit and eyeball: The four recti and the two obliques are primarily responsible for the various movements of the eyeball. The lines of pull and the actions of these muscles will be comprehensible only if the physical and mechanical features of the eyeball and the muscles are properly understood.

❑ **Orbital axis:** The two orbital cavities are two pyramids placed in such a way that their medial walls are parallel to each other. Their lateral walls are at right angles to each other. The **axis of the orbit** (also called the **orbital axis** or the **long axis of the orbital cavity**) is a line passing from the apex of the orbit to the centre of its base, i.e., from the optic canal to the centre of the orbital opening. It is directed anterolaterally and is the axis around which the recti muscles are arranged.

❑ ***Optical axis:*** The eyeball occupies the anterior part of the orbital cavity. The line joining the anterior and posterior poles of the eyeball is the optical axis (also called the anteroposterior axis or the sagittal axis of the eye). The optical axis is the axis of gaze or the line of sight and runs directly anteriorly (resulting in looking straight ahead) in the normal anatomical position when the eyeball is also in its primary position. The right and the left optical axes are parallel to each other and to the medial walls of the right and the left orbital cavities.

❑ ***Visual axis:*** The rays of light come to a focus on the retina at its posterior pole; this point of retina is the yellow spot or the macula lutea. A straight line from the object seen (which, for purposes of vision, is the point of origin of the light rays) through the centre of the pupil to the macula lutea is then the line of vision. This also is the visual axis. The optical axis diverges from the visual axis by about 5 or more degrees.

Direction of movements and their axes: As a convention, movements of the eyeball are described with reference to its anterior end (or more simply, the cornea). There are three axes around which the movements occur. These are the sagittal, horizontal and the vertical axes. The equator of the eyeball encircles it midway between the anterior and posterior poles. The sagittal or the anteroposterior axis, as already noted, is the line joining the anterior and posterior poles. The horizontal or the transverse axis passes through the right and left points on the equator. The vertical axis passes through the upper and lower points on the equator.

❑ The cornea can move upwards or downwards, the movement occurring around the horizontal axis. Upward movement can be produced by pulling the anterior part of the eyeball upwards or by pulling the posterior part downwards. This action is also called ***elevation***. Similarly, downward movement can be produced by pulling the anterior part downwards or by pulling the posterior part upwards. This action is also called ***depression***.

❑ The cornea can move medially or laterally around the vertical axis. Medial movement can be produced by pulling the anterior part of the eyeball medially or by pulling the posterior part laterally. The cornea appears to move closer to the medial canthus and therefore this action is otherwise called ***adduction***. Lateral movement can be produced by pulling the anterior part of the eyeball laterally or by pulling the posterior part medially. The cornea appears to move away from the medial canthus and therefore this action is called ***abduction***.

❑ The cornea can move obliquely by combination of the movements described above, e.g., upwards and medially, downwards and laterally and so on.

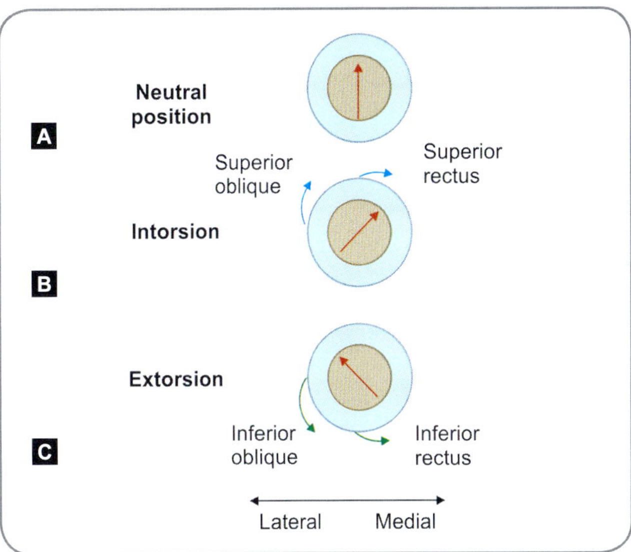

Figs 11.4A to C: Schematic diagram of the eye to explain the movements of intorsion and extorsion; the inner circle represents the cornea, and the outer circle represents the sclera

❑ The eyeball undergoes torsional movements around the anteroposterior or optical axis. To understand these movements, imagine a vertical line drawn through the middle of the cornea dividing it into medial and lateral halves (Fig. 11.4A). Torsion is described with reference to the ***upper*** end of this line (indicated by the arrow). When the eyeball rotates so that the upper end of the line moves medially, the movement made is ***intorsion*** (Fig. 11.4B) and when it moves laterally the movement made is ***extorsion*** (Fig. 11.4C).

Factors guiding the movements: Varied movements of the eyeball occur (Fig. 11.5). These movements, however, are not simple but are the result of interplay of several factors.

❑ The medial and the lateral recti act only on one axis. Each of the other muscles acts on multiple axes.

❑ In the primary position, the orbital axis runs anterolaterally. The optical axis runs straight forwards and cuts the orbital axis.

❑ The recti muscles are arranged around and are parallel to the orbital axis. Since they are attached to the eyeball anterior to the equator, their contraction would move the anterior part of the eyeball.

❑ The oblique muscles are attached to the eyeball posterior to the equator and their contraction would move the posterior part of the eyeball. Both muscles have their effective points of fixation on the anterior aspect of the orbit. The superior oblique, though arising from the body of sphenoid, passes through the trochlea in the anteromedial aspect of the orbit. This effectively changes the line of pull of the muscle from posteromedial to 'anteromedial'. The inferior oblique arises on the

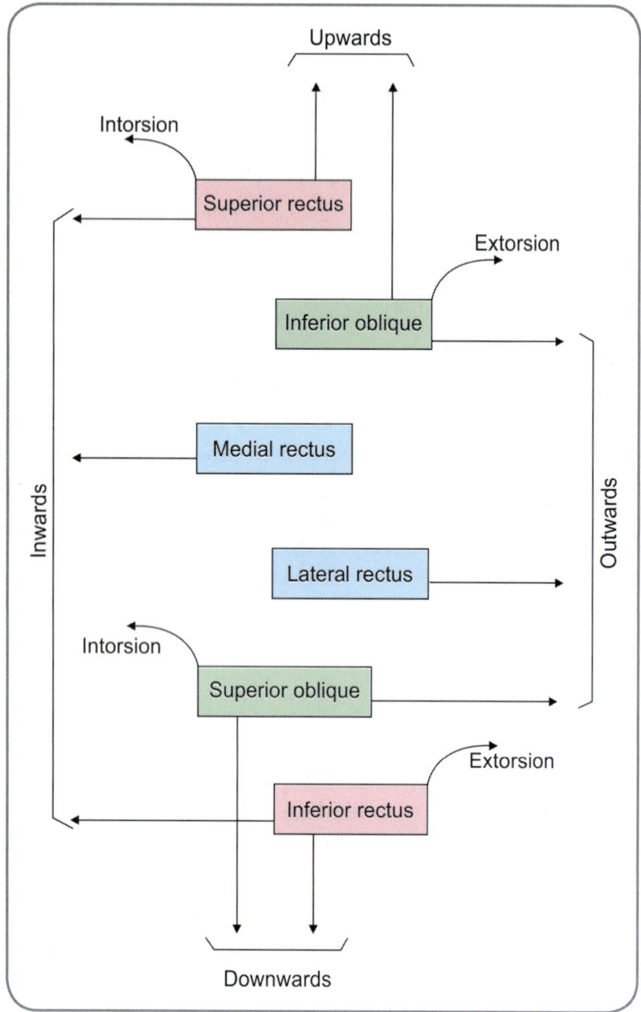

Fig. 11.5: Scheme showing the movements of the eyeball produced by individual extraocular muscles, and the muscles responsible for each movement

anterior orbital floor and passes posterolaterally. Hence, its line of pull is also anteromedial.

Actions of the extraocular muscles: The actions of each of the six muscles can be deduced by working out the mechanics of the muscle (Table 11.2).

❑ ***Medial rectus:*** It pulls the anterior part of the eyeball medially; hence is an adductor. Effective movement—***adduction***.

❑ ***Lateral rectus:*** It pulls the anterior part of the eyeball laterally; hence is an abductor. The sixth cranial nerve that supplies the lateral rectus is called the '***abducent***' nerve because of its role in abduction of the eyeball. Effective movement—***abduction***.

❑ ***Superior rectus:*** It pulls the anterior part of the eyeball upward; hence is an elevator. In addition, by virtue of it being parallel to the orbital axis which cuts the optical axis medial to lateral from posterior to anterior, its pull will move the anterior eyeball medially; hence is an adductor. The muscle is attached to the upper part of the eyeball and pulls medially; this will cause the 12 o'clock position of the cornea to move internally and produce intorsion. Effective movements—***elevation, adduction and intorsion***.

❑ ***Inferior rectus:*** It pulls the anterior part of the eyeball downward; hence is a depressor. In addition, by virtue of it being parallel to the orbital axis which cuts the optical axis medial to lateral from posterior to anterior, its pull will move the anterior eyeball medially; hence is an adductor. The muscle is attached to the lower part of the eyeball and pulls medially; this will cause the 12 o'clock position of the cornea to move externally and produce extorsion. Effective movements—***depression, adduction and extorsion***.

❑ ***Superior oblique:*** It pulls the posterior part of the eyeball upward and so the anterior eyeball downward; hence is a depressor. In addition, by virtue of its anteromedial line of pull, the posterior eyeball is moved medially producing a lateral movement of the anterior eyeball; hence is an abductor. The muscle is attached to the superior posterolateral quadrant of the eyeball and pulls anteromedially; since the line of pull is superior to the anteroposterior axis, the 12 o'clock position of the cornea will move internally and result in intorsion. Effective movements—***depression, abduction and intorsion***.

Table 11.2: Individual movements of the eyeball and the muscles responsible for them	
Elevation: Upward movement	By pulling anterior eyeball upward—superior rectus By pulling posterior eyeball downward—inferior oblique
Depression: Downward movement	By pulling anterior eyeball downward—inferior rectus By pulling posterior eyeball upward—superior oblique
Adduction: Medial movement	By pulling anterior eyeball medially—medial rectus, superior rectus and inferior rectus (No muscle pulls the posterior eyeball laterally)
Abduction: Lateral movement	By pulling anterior eyeball laterally—lateral rectus By pulling posterior eyeball medially—superior oblique and inferior oblique
Intorsion: 12 o'clock internal	By medial pull superior to anteroposterior axis—superior rectus and superior oblique
Extorsion: 12 o'clock external	By medial pull inferior to anteroposterior axis—inferior rectus and inferior oblique

❑ *Inferior oblique:* It pulls the posterior part of the eyeball downward and so the anterior eyeball upward; hence is a elevator. In addition, by virtue of its anteromedial line of pull, the posterior eyeball is moved medially producing a lateral movement of the anterior eyeball; hence is an abductor. The muscle is attached to the inferior posterolateral quadrant of the eyeball and pulls anteromedially; since the line of pull is inferior to the anteroposterior axis, the 12 o'clock position of the cornea will move externally and result in extorsion. Effective movements—*elevation, abduction and extorsion.*

Smooth Muscles of the Orbit

❑ *Tarsal muscles:* The tarsal muscles are two small bundles of smooth muscle fibres. A thin, well-defined sheet of smooth muscle extends from the deep stratum of the levator palpebrae superioris (the stratum that gets attached to the superior tarsal plate) to the upper border of the superior tarsal plate. This sheet constitutes the superior tarsal muscle (or Muller's muscle). Similarly, a thin sheet of poorly developed smooth muscle extends forward from the inferior rectus muscle to the lower border of the inferior tarsal plate. This constitutes the inferior tarsal muscle.

These smooth muscles help in elevation and depression of the upper and lower eyelids respectively. They act according to the mood of the individual and to factors which cause involuntary muscle action. The size of the palpebral fissure alters accordingly.

❑ *Orbitalis:* Otherwise called the orbital muscle of Muller, this smooth muscle lies in the posterior part of the orbital cavity. It bridges across the infraorbital groove. When this muscle contracts, the eyeball is protruded slightly forwards.

All the three smooth muscles receive sympathetic innervation from the superior cervical ganglion through the internal carotid plexus.

Nerves of the Orbit

Several nerves are found in the orbit. They can be classified as the somatic motor, somatic sensory and the autonomic motor nerves.

❑ *Somatic motor nerves:* These are the oculomotor, trochlear and abducent nerves; they supply the extraocular muscles.
❑ *Somatic sensory nerves:* These are the optic nerve and branches of the ophthalmic and maxillary nerves. The optic nerve is concerned with visual impulses. Many branches of the ophthalmic and maxillary nerves simply pass through the orbit on their way to face and scalp.
❑ *Autonomic motor nerves:* These are the sympathetic and parasympathetic fibres. The sympathetic fibres supply the dilator papillae muscle and the arteries. The parasympathetic fibres supply the sphincter papillae and the ciliaris muscles (from the oculomotor nerve through the ciliary ganglion) and supply the lacrimal gland and the choroid (from the facial nerve through the pterygopalatine ganglion). Parasympathetic fibres also supply the arteries.

Optic Nerve

The optic nerve forms an important part of the visual pathway. It is a purely sensory nerve and transmits visual impulses. By convention, it is described as the second cranial nerve but is an anterior extension of the forebrain. Therefore, it actually is a central nervous system fibre tract. It lies partly in the orbit and partly in the cranial cavity. Its anterior end is attached to the posterior pole of the eyeball. The attachment lies a little medial to the anteroposterior axis of the eyeball. From here, the nerve passes backwards and medially first through the orbit, next through the optic canal and finally through part of the cranial cavity. The nerve ends by joining the nerve of the opposite side to form the *optic chiasma*. The total length of the nerve is about 40 mm. Of this 25 mm is in the orbit, 5 mm in the optic canal and 10 mm in the cranial cavity.

The intraorbital part of the nerve is surrounded by the superior, inferior, medial and lateral rectus muscles. The ciliary ganglion is placed between the optic nerve and the lateral rectus muscle. The ophthalmic artery is inferolateral to the nerve in the optic canal and in the posterior most part of the orbit. The artery then crosses above the nerve from lateral to medial side. Apart from the ophthalmic artery, the optic nerve is crossed from lateral to medial side by the nasociliary nerve which crosses above the optic nerve and the branch from the oculomotor nerve to the medial rectus muscle which crosses below the nerve.

Throughout its intraorbital course, the optic nerve is enclosed within extensions of the cranial meninges and subarachnoid space. A thin layer of cerebrospinal fluid occupies the subarachnoid space. Of the three layers of the extended meninges, the pia closely adheres to and covers the surface of the nerve. The intraorbital extensions of the arachnoid and the dura constitute the *optic nerve sheath* which becomes continuous anteriorly with the fascial sheath of the eyeball and the sclera.

The central artery of the retina first lies below the optic nerve. Piercing the dural sheath of the nerve it runs forwards for a short distance between these two. About 12 mm behind the eyeball the artery enters the substance of the nerve and runs forwards in its substance to reach the eyeball.

Dissection

The bony orbit should first be studied in bony specimens of the cranium and cranial fossae. See the relationship of the orbit to the anterior and middle cranial fossa and how closely structures in the orbit would be located.

contd...

Dissection *contd...*

In the cadaver, carefully remove the orbital plate of the frontal with chisel and hammer, under the supervision of a guide/teacher. The orbital periosteum is first exposed. Study it. Carefully incise and reflect the periosteum without injuring the underlying structures. Locate the lacrimal, trochlear and frontal nerves. With the help of the trochlear nerve, locate the superior oblique muscle. Levator palpebrae superioris and superior rectus muscles will clearly be seen. Clean and define the muscles. Divide the levator palpebrae and superior rectus to expose the deeper contents; when you do this step, spare the frontal nerve without cutting and leave it in place. You will now be able to see structures passing through the common tendinous ring. Look out for the nasociliary nerve and its branches. See the optic nerve and the ophthalmic artery.

At the next step, locate the medial and the lateral recti muscles. Once you identify the lateral rectus, trace the abducent nerve. Carefully, by a slow dissection and with help from senior staff, try to locate the ciliary ganglion. The short ciliary nerves may guide you.

Identify the oculomotor nerve and its divisions and branches. Define the recti muscles. Cut the optic nerve at some distance from the eyeball and lift and reflect the eyeball forwards. You will be able to see the inferior rectus muscle. Slowly, by careful turning of the structures, locate the common tendinous ring and see the origin of the recti from this ring. Observe the origin and insertion of all the extraocular muscles. Secure all their nerve supply.

Review all the contents once again with your theory knowledge.

Oculomotor Nerve

The oculomotor is the third cranial nerve. It arises from the midbrain and runs part of its course in the cavernous sinus. Here, it divides into the superior and the inferior rami. Both the rami enter the orbit by passing through that part of the superior orbital fissure (Fig. 15.28), which lies within the common tendinous ring and are separated by the nasociliary branch of the ophthalmic nerve within the ring. Within the orbit, the superior ramus supplies the superior rectus and the levator palpebrae superioris; it first enters the superior rectus on the muscle's inferior surface, supplies and pierces through to enter the levator palpebrae on its inferior surface. The inferior division supplies the medial rectus, the inferior rectus and the inferior oblique (Fig. 15.27). On entry into the orbit, the inferior division divides into three branches—***medial, central and lateral branches***. The medial branch passes medially to enter the lateral surface (also the ocular surface, because this surface faces the eyeball) of the medial rectus. The central branch passes downward and forward to enter the superior surface (also the ocular surface) of inferior rectus. The lateral branch runs anterolaterally to enter the inferior surface (also the orbital surface) of inferior oblique. The lateral branch communicates with the ciliary ganglion.

The oculomotor nerve also carries secremotor fibres. Preganglionic parasympathetic fibres arising from the Edinger-Westphal nucleus travel through the trunk of the nerve and then through its inferior division. They then pass into the lateral branch of the inferior division (nerve to the inferior oblique muscle). A short twig arising from the lateral branch conveys the fibres to the ciliary ganglion. This branch to the ciliary ganglion from the nerve to the inferior oblique constitutes the **motor root** of the ganglion. As the preganglionic fibres relay to the cells of the ganglion, postganglionic fibres arising from these cells pass through the short ciliary nerves to reach the sphincter pupillae and the ciliaris muscles.

When near vision is required, as in reading, the oculomotor nerve comes into play. It causes all the actions required for near vision. It causes convergence (by acting on the adductors medial, superior and inferior recti), accommodation of the lens (through the parasympathetic component acting on the ciliaris) and pupillary constriction (by acting on the sphincter papillae). Peripheral light is thus shut off and near vision effected.

Trochlear Nerve

The trochlear nerve is the fourth cranial nerve. It enters the orbit through the upper part of the superior orbital fissure. In the fissure, it lies above the common tendinous ring and medial to the frontal and lacrimal nerves. Having entered the orbit, the nerve runs forwards and medially above all the orbital muscles and enters the superior surface of the superior oblique muscle. It supplies only the superior oblique muscle and is named '***trochlear nerve***' because of being the nerve to the muscle that passes through the trochlea.

Abducent Nerve

This is the sixth cranial nerve. The nerve passes through the superior orbital fissure within the tendinous ring to enter the orbit. It lies, first, below and then between the two divisions of the oculomotor nerve. It then runs forward to enter the ocular surface of the lateral rectus muscle and supply it. The nerve is named 'abducent' because of being the nerve to the muscle that produces abduction of the eyeball.

Ophthalmic Nerve

The fibres of the ophthalmic nerve are purely sensory. However, some sympathetic fibres for the eyeball travel part of their course through this nerve and some of its branches. Arising from the trigeminal ganglion in the middle cranial fossa, it passes along the lateral wall of the cavernous sinus. Just before reaching the superior orbital fissure, it divides into its three main branches, namely the ***lacrimal nerve, the frontal nerve and the nasociliary*** *nerve*. These branches enter the orbit by passing through the superior orbital fissure.

The *lacrimal nerve* enters the orbit through the lateral part of the superior orbital fissure above the common tendinous ring and lateral to the frontal and trochlear nerves. It runs along the lateral wall of the orbit (along the upper border of the lateral rectus). It ends in the lacrimal gland (and hence its name). Some of its branches pass through the gland to supply the conjunctiva and the skin of the upper eyelid. The lacrimal nerve is joined by a twig from the zygomaticotemporal branch of the maxillary nerve; the postganglionic parasympathetic secretomotor fibres from the pterygopalatine ganglion reach the lacrimal gland through this connection.

The *frontal nerve*, the largest of the branches of the ophthalmic division, enters the orbit through the superior orbital fissure above the common tendinous ring. It lies lateral to the trochlear nerve and medial to the lacrimal nerve. It runs forward between the levator palpebrae superioris and the periosteum lining the roof of the orbit. Half way through the orbit, it divides into the supraorbital and supratrochlear branches. The *supraorbital nerve* continues in the line of the frontal nerve (Fig. 15.33), lying between the levator palpebrae superioris and the roof of the orbit. Reaching the orbital margin it passes through the supraorbital notch (on the medial part of the upper margin of the orbit) and curves upwards into the forehead. It supplies the mucosa of the frontal air sinus, skin and conjunctiva of the upper eyelid and the skin of the forehead and scalp. The postganglionic sympathetic fibres for the sweat glands of the supraorbital region reach the ophthalmic nerve through a communication from the abducent nerve in the cavernous sinus; they then travel through the frontal and supraorbital nerves. The *supratrochlear nerve* runs forwards and medially above the orbital muscles, and medial to the supraorbital nerve. It passes above the trochlea (for the tendon of the superior oblique muscle) and hence the name. Reaching the upper margin of the orbital aperture, near its medial end, the nerve turns upwards into the forehead giving branches to the skin over its lower and medial part. Other branches given off by the supratrochlear nerve are a descending branch which joins the infratrochlear branch of the nasociliary nerve (Fig. 15.33) and branches to the conjunctiva and skin of the upper eyelid.

On entering the orbit, the *nasociliary nerve* lies between the optic nerve and the lateral rectus. It then runs medially crossing above the optic nerve (Fig. 15.33). As it does so it lies below the superior rectus muscle. Reaching the medial wall of the orbit, the nerve ends by dividing into the *anterior ethmoidal* and *infratrochlear* nerves. The communications and branches of the nasociliary nerve are (Fig. 15.33):

❑ Just after entering the orbit, the nasociliary nerve gives off the *sensory root of the ciliary ganglion.*

❑ The *long ciliary nerves* (two or three) arise from the nasociliary nerve as it crosses the optic nerve. They run forwards to the eyeball where they pierce the sclera; and then run between the sclera and the choroid. They supply sensory fibres to the ciliary body, the iris and the cornea. They also carry postganglionic sympathetic fibres meant for the dilator pupillae. These sympathetic fibres begin in the superior cervical sympathetic ganglion (Fig. 15.34) and travel through a plexus surrounding the internal carotid artery. They pass to the nasociliary nerve while the latter lies in the wall of the cavernous sinus.

❑ The *posterior ethmoidal branch* enters the posterior ethmoidal foramen (on the medial wall of the orbit) and supplies the ethmoidal and sphenoidal air sinuses.

❑ The *anterior ethmoidal nerve* has a complicated course through the orbit, the anterior cranial fossa, and the nasal cavity. It first passes through the anterior ethmoidal foramen into the cranial cavity; then runs on the upper surface of the cribriform plate. Entering the nasal cavity through a slit, it runs down on the internal surface of the nasal bone. After giving out the two internal nasal branches, it continues and emerges as the external nasal nerve at the lower border of the nasal bone. The external nasal nerve supplies the skin of the ala of the nose, the apex and the vestibule. The medial internal nasal and the lateral internal nasal nerves supply the nasal septum and the lateral nasal wall respectively.

❑ The *infratrochlear nerve* (Fig. 15.33) runs forwards on the medial wall of the orbit and ends by supplying part of the skin of the upper and lower eyelids and skin over the upper part of the nose. The nerve also gives branches to the conjunctiva, the lacrimal sac and the lacrimal caruncle.

Branches of Maxillary Nerve in the Orbit

The maxillary nerve is the second division of the trigeminal nerve. It gives out its branches in the pterygopalatine fossa. The zygomatic and the infraorbital branches pass into the orbit through the inferior orbital fissure. The zygomatic nerve, as it passes on the lateral orbital wall close to the floor, divides into the zygomaticotemporal and zygomaticofacial nerves. Both the branches pass through canals in the zygomatic bone. The zygomaticotemporal nerve emerges into the temporal fossa through a foramen in the medial surface of the zygomatic bone and pierces the temporal fascia to supply the skin over the temporal fossa. It gives a communicative branch to the lacrimal nerve. The postganglionic parasympathetic secretomotor fibres from the pterygopalatine ganglion and destined for the lacrimal gland, run through the maxillary nerve, the zygomatic nerve, the zygomaticotemporal branch, the communication to the lacrimal nerve and the lacrimal nerve to reach the lacrimal gland. The zygomaticofacial nerve emerges out on the lateral surface of the zygomatic bone and supplies the skin of the cheek. The infraorbital nerve lies in the infraorbital groove on the orbital floor. Close to the anterior end of the orbit, it enters into the infraorbital canal and passes through the canal to emerge on the face at the infraorbital foramen. It supplies the skin and conjunctiva of the lower eyelid and the skin of the upper jaw. The infraorbital nerve also gives out the middle and the anterior superior alveolar nerves.

Orbital Branches of the Pterygopalatine Ganglion

Small twigs called the **rami orbitales** arise from the pterygopalatine ganglion and pass directly into the orbit through the inferior orbital fissure. They transmit postganglionic parasympathetic fibres to the lacrimal gland, ophthalmic artery and the choroid of the eyeball.

Ciliary Ganglion

The ciliary ganglion is one of the peripheral parasympathetic ganglia of the head and neck. It is a small reddish coloured swelling, about 1 to 2 mm in diameter, located near the apex of the orbit in the loose fat of the orbital cavity. It lies lateral to the optic nerve, between it and the lateral rectus muscle. Most of its neurons are large in size.

Posteriorly, the ganglion has three roots: Motor or parasympathetic, sympathetic and sensory roots. The **motor or parasympathetic root** is derived from the branch of the oculomotor nerve to the inferior oblique and carries fibres for the sphincter pupillae and ciliaris. The preganglionic parasympathetic fibres start from the Edinger-Westphal nucleus and travel through the oculomotor nerve and its branch. Reaching the ciliary ganglion, they relay in the neurons of the ganglion. The postganglionic fibres arising from the ganglion pass through the short ciliary nerves to the sphincter papillae and ciliaris muscles. The **sympathetic root** arises from the sympathetic plexus on the internal carotid artery. It carries postganglionic sympathetic fibres that begin in the superior cervical sympathetic ganglion. These fibres pass through the ciliary ganglion without relay and enter the short ciliary nerves through which they reach the blood vessels of the eyeball. The fibres for the dilator pupillae normally follow a separate route (sympathetic plexus around the internal carotid artery–ophthalmic nerve–nasociliary nerve–long ciliary branches–dilator papillae muscle) but sometimes they may pass through the ciliary ganglion and short ciliary nerves. The **sensory root** of the ciliary ganglion is a branch of the nasociliary nerve. It carries sensory fibres which begin in the cornea, the iris and the choroid; pass through the short ciliary nerves and then through the ciliary ganglion; and finally enter the sensory root through which they reach the nasociliary nerve.

Eight to ten slender nervous twigs called the **short** ciliary nerves emerge from the anterior aspect of the **ciliary ganglion**. They are usually in two or three bundles. They run forwards above and below the optic nerve along with the ciliary arteries; divide into fifteen to twenty branches and pierce the posterior aspect of the sclera around the optic nerve. They then run in tiny grooves on the internal surface of the sclera. They carry sympathetic or parasympathetic or sensory fibres to the internal structures of the eyeball.

Arteries of the Orbit

The main vessel supplying the orbital structures is the ophthalmic artery (internal carotid stream). The infraorbital branch of the maxillary artery (external carotid stream) also supplies some of the orbital structures.

Ophthalmic Artery

The ophthalmic artery, a branch of the internal carotid artery given out medial to the anterior clinoid process, passes forwards to enter the cavity of the orbit through the optic canal. In this canal, it lies inferolateral to the optic nerve. Having entered the orbit, the artery is at first lateral to the optic nerve (Fig. 14.8). It then crosses above the nerve to reach the medial wall of the orbit and runs forward along this wall. Near the medial end of the upper eyelid, it divides into its two terminal branches, namely, the supratrochlear and dorsal nasal arteries. Most of the branches accompany correspondingly named nerves.

Branches of the ophthalmic artery

- The **central artery of the retina** is the first branch of the ophthalmic artery. It arises from the ophthalmic artery when the latter is still within the optic canal. It first lies below the optic nerve. It pierces the dural sheath of the nerve and runs forwards for a short distance within the sheath. It then enters the substance of the nerve and runs forwards in its centre to reach the optic disc. Here, it divides into branches that supply the retina (Fig. 14.9).
- The largest branch of the ophthalmic artery is the **lacrimal artery** (accompanies the lacrimal nerve); it is given out near the optic canal and runs forward along the lateral wall of the orbit, traverses and supplies the lacrimal gland, ends in the upper and lower eyelids as the upper and lower lateral palpebral arteries. Just near its origin from the ophthalmic artery, the lacrimal artery gives off a **recurrent meningeal** branch that runs backward to enter the middle cranial fossa through the superior orbital fissure and anastomoses with the middle meningeal artery. The lacrimal artery gives off two **zygomatic** branches that enter canals in

the zygomatic bone. One branch appears on the face through the zygomaticofacial foramen, supplies the region and anastomoses with the other arteries of the area. The other branch appears on the temporal surface of the bone through the zygomaticotemporal foramen, supplies the region and anastomoses with the deep temporal arteries.

- The ***posterior ciliary arteries (long and short)*** arise as the ophthalmic artery crosses the optic nerve. They run forward around the optic nerve and supply the eyeball. Of the six or more posterior ciliary arteries, two are long and called the ***long posterior ciliary arteries***. After piercing the sclera, they run forward between the sclera and the choroid (one on either side) to anastomose with the anterior ciliary arteries.

- The ***supraorbital*** branch (accompanies the supraorbital nerve) is also given out where the ophthalmic artery crosses the optic nerve; it supplies the skin of the forehead.

- The ***anterior*** and ***posterior ethmoidal*** branches (accompanies the anterior and posterior ethmoidal nerves) enter the anterior and posterior ethmoidal foramina in the medial wall of the orbit to supply the ethmoidal (and frontal) air sinuses, then enter the anterior cranial fossa and subsequently terminate in the nasal cavity supplying structures there.

- The ***anterior ciliary arteries*** arise from the ophthalmic artery (mainly from the muscular branches to recti) near the anterior part of the eyeball; they pierce the sclera just behind the sclerocorneal junction, anastomose with the long posterior ciliary artery of the same side and supply the ciliary body and iris. Before piercing the sclera, the anterior ciliary arteries give out twigs to the deep conjunctival plexus around the corneal margin.

- The ***medial palpebral*** branches supply the eyelids.

- The ***supratrochlear artery*** (accompanies the supra-trochlear nerve) is one of the terminal branches of the ophthalmic artery. It supplies the skin of the forehead (along with the supraorbital artery).

- The ***dorsal nasal branch*** (accompanies the infra-trochlear nerve) supplies the upper part of the nose.

- In addition to these branches, the ophthalmic artery and its branches give off several ***muscular twigs*** to the muscles of the orbit.

Infraorbital Branch of the Maxillary Artery

The infraorbital branch of the maxillary artery enters the orbit through the inferior orbital fissure. It passes through the infraorbital groove on the orbital floor and enters into the infraorbital canal to reach the face. While traversing the infraorbital groove, it gives branches to the inferior rectus and inferior oblique muscles and the nasolacrimal sac.

Veins of the Orbit

The veins draining the structures of the orbit are the ***superior and inferior ophthalmic veins***. The superior ophthalmic vein forms from tributaries joining with each other posterior to the upper eyelid. Running posterior along the ophthalmic artery, it receives the corresponding tributaries (those veins which accompany the arterial branches), two superior vortex veins of the eyeball and the central vein of retina. The superior ophthalmic vein then traverses the superior orbital fissure above the common tendinous ring and ends in the cavernous sinus. The inferior ophthalmic vein starts in a venous network at the anterior part of the orbital floor. It receives tributaries from the muscles of the orbit and the two inferior vortex veins. Passing through the inferior orbital fissure, it either directly drains into the cavernous sinus or through the superior ophthalmic vein. The inferior orbital vein is seen on the floor of the orbit; drains structures of the orbital floor; passes backward through the inferior orbital fissure and drains into the pterygoid venous plexus.

Lacrimal Gland

The lacrimal gland lies in relation to the upper part of the lateral wall of the orbit (formed here by the zygomatic process of the frontal bone) (Fig. 3.7). It is a serous gland and has two parts—the main or orbital part and the palpebral part. The two parts are continuous with each other around the lateral side of the aponeurosis of the levator palpebrae superioris. The orbital part is lodged in the lacrimal fossa on the lateral wall of the orbit. It lies above the levator palpebrae superioris and the lateral rectus muscles.

Its lower surface is attached to the connective sheath of the levator palpebrae and the upper surface is connected to the orbital periosteum. The palpebral part extends below the levator palpebrae into the lateral part of the upper eyelid. The lacrimal gland drains into the superior conjunctival fornix through about twelve ducts. The ducts of the orbital part also pass through the palpebral part. Accessory lacrimal glands may be present in relation to the superior conjunctival fornix or less commonly in relation to the inferior fornix. The lacrimal gland is supplied by twigs from the lacrimal branch of the ophthalmic artery. The secretomotor fibres to the gland follow a complicated course; preganglionic parasympathetic fibres start from the lacrimatory nucleus and pass through the facial nerve. They further travel through the greater petrosal branch of the facial nerve and the nerve of pterygoid canal to reach the pterygopalatine ganglion. After relay in the pterygopalatine ganglion, the postganglionic fibres either pass through the zygomatic–zygomaticotemporal–lacrimal nerves or through the rami orbitale to reach the lacrimal gland.

Histology

Histology of lacrimal gland: The lacrimal gland is of the compound tubuloacinar type. The secretory units are the acini and their cells. The secretory cells, which lie on a basement membrane, are short columnar cells with number of microvilli on their luminal surfaces. Three categories of secretory cells have been identified by electron microscopy depending upon their electron densities—light, medium and dark secretory cells. Though it had been believed for long that the lacrimal gland is a serous gland and is common knowledge that tears are patently serous, mucous cells have also been identified. Myoepithelial cells and lymphocytes are seen here and there between the secretory cells and the basement membrane. Myoepithelial cells are associated with tear production and secretion. Each acinus ends in a small duct called the **terminal duct**. Terminal ducts of adjacent acini join together and form the branching duct. The branching ducts of three or more such acinar bunches join to form a duct. Each lacrimal gland has about six to ten ducts.

The secretion of the lacrimal glands (tears or lacrimal fluid) is alkaline and contains lysozyme which is a bactericidal enzyme. Immunoglobulin A, lactoferrin, tear specific prealbumin are some other substances found in tears.

Accessory Contents of the Orbit

Periorbita: The bones forming the orbit are covered with periosteum. This periosteum forms a lining for the orbital cavity and is called the **periorbita** or the **orbital fascia**. The periorbita is continuous with the periostea of adjacent areas in the following way:

- ❑ With the periosteal layer of dura mater at the optic canal and the superior orbital fissure;
- ❑ With the orbital septa at the orbital margin of the orbital opening;
- ❑ With the periosteum of the various bones at the inferior orbital fissure and the orbital margin; and
- ❑ With the fascial sheaths of the extraocular muscles and that of the eyeball.

The periorbita sends expansions into the eyelids. These extensions form the **orbital septum**. The lacrimal fascia and the trochlea for the superior oblique tendon can be regarded as special regions of the periorbita.

Orbital Septum

This is a thin membranous sheath attached all around the orbital margin. At its bony attachment, it is continuous with the periosteum of the concerned bones. From the upper attachment, the upper part of the septum passes into the upper eyelid and blends with the tarsal plate. From the lower attachment, the lower part of the septum passes into the lower eyelid and blends with the tarsal plate. The upper part is thickest laterally, where it is superficial to the lateral palpebral ligament. On the medial aspect, it passes deep to the medial palpebral ligament and the

nasolacrimal sac. The upper part is pierced by the levator palpebrae superioris and the lower part by the inferior rectus. The lacrimal, supratrochlear, infratrochlear and supraorbital nerves and vessels which go to the scalp and face pass through the septum.

Orbital Fat

As already noted, several structures are located within the orbital cavity. The spaces between these main structures are occupied by orbital fat. This fat helps in stabilising the various structures, especially the eyeball, in position within the cavity. It also provides a 'soft socket' for the eyeball to rotate smoothly and freely. Conditions like hyperthyroidism wherein the fat content is increased, cause the eyeball to protrude forwards.

Fascial Sheath of the Eyeball (Fig. 11.6)

Otherwise called the **vagina bulbi** (Latin.vagina=sheath, bulbi=related to eyeball) or the fascia bulbi or the capsula bulbi or the **eye capsule** or **Tenon's capsule** (named after the 18th century French anatomist and oculist Jacques Tenon), this is a thin fascial sheath that surrounds the eyeball from the optic nerve to the sclerocorneal junction. It separates the eyeball from the orbital fat and forms a complete socket. Posteriorly it fuses with the sclera and the sheath of the optic nerve; it again fuses with the sclera immediately posterior to the sclerocorneal junction. In the rest of the places, its internal surface is loosely attached to the sclera by thin strands of episcleral connective tissue. The space between the fascial sheath and the sclera is the **episcleral space**.

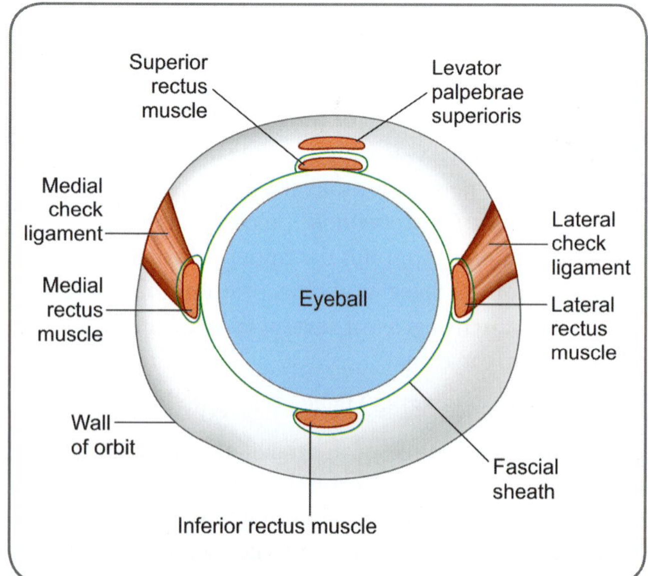

Fig. 11.6: Scheme to show some parts of the fascial sheath of the eyeball. Note the 'hammock' formed by the lower part of the sheath and the check ligaments

The extraocular muscles, as they reach their ocular attachments, perforate the sheath and receive extensions from it. These extensions form tubular sheaths (called the muscular fascia or the areolar tissue sheath) around the muscles. The sheath of the superior oblique extends till the trochlea. The sheaths of the four recti fuse with the sheaths of the adjacent muscles at their margins thus forming a muscular areolar cone. This forms the muscular cone of recti.

Fascial extensions from the muscular fascial sheaths reach and attach to neighbouring bones. Two such extensions are important. These are from the fascial sheaths of the medial and the lateral recti; are triangular and strong; are attached to the lacrimal and the zygomatic bones respectively. They limit the actions of the two recti (place a check) and so are called the ***medial and lateral check ligaments*** (together called the Mauchart's ligaments; named after the 18th century German anatomist Burkhard Mauchart). Extensions from the sheaths of the other muscles are not strong enough to cause restriction of movements.

Suspensory Ligament of the Eye

Though described separately, the suspensory ligament (or the ***Lockwood's ligament*** or ***ligament of Lockwood***; named after the 19th century English surgeon Charles Lockwood) is actually a part of the fascial sheath of the eye. We have already seen that the extraocular muscles receive fascial extensions from the fascial sheath. The sheaths of the inferior oblique and the inferior rectus thicken and blend with each other on their adjoining surfaces (inferior surface of inferior rectus and superior surface of inferior oblique). However, at their medial and lateral margins, the sheaths of these two muscles are continuous with the sheaths of the medial and lateral rectus muscles respectively. The sheaths of the medial and lateral recti extend to the medial and lateral walls of the orbit as the medial and lateral check ligaments. It can now be seen that from the medial wall to the lateral wall, a continuous fascial layer exists below the eyeball (lacrimal bone–medial check ligament–sheath of medial rectus–blended sheaths of inferior rectus and inferior oblique – sheath of lateral rectus–lateral check ligament–zygomatic bone). This provides a kind of a hammock for the eyeball; the eyeball rests on this hammock and is so well supported that even when the maxilla (forming the orbital floor) is removed, the eye is retained in position. Since the hammock provides support to the eyeball, it is called the ***suspensory ligament*** of the eye.

Suspensory Ligament of Lacrimal Gland

The suspensory ligament of the lacrimal gland (otherwise called the ***superior transverse ligament*** of the eye, the

Whitnall's ligament, named after the 19–20th century English anatomist Samuel Whitnall; the Soemmering ligament, named after the 18–19th century German anatomist Samuel von Soemmering) is actually a modification of the muscular fascia of levator palpebrae superioris. The muscle, arising from the posterior aspect of the orbit runs forward to the orbital margin and the upper eyelid. Just before the orbital margin, it converts into an aponeurosis. The place where the muscle fibres convert into aponeurosis, the covering fasciae of the muscle fibres unite to form a continuous band. This band appears like a ligament and is called the ***Whitnall's ligament***. The levator palpebrae muscle is thus described to pierce the Whitnall's ligament and become an aponeurosis. The Whitnall's ligament attaches to the muscular fascial sheath of the underlying superior rectus; it extends medially to the trochlea of the superior oblique and laterally to the lateral orbital wall close to the orbital part of the lacrimal gland (about 1 cm above the Whitnall's tubercle). It accords fixation to the levator palpebrae and the superior oblique. Inadvertent damage to the ligament during eye surgery can cause ptosis. The ligament also supports the upper eyelid and the lacrimal gland. The latter action (physically supporting the orbital part of the lacrimal gland) gives the ligament its name. The ligament is the identification landmark for the superior border of the levator aponeurosis.

Palpebral Ligaments

Though the palpebral ligaments (or canthal ligaments) are located in the eyelids (which are technically outside the orbital cavity), their attachments to the orbital margin qualify them as accessory contents of the orbit. The palpebral ligaments connect the tarsal plates of both eyelids to the orbital margin. The medial palpebral ligament connects the medial ends of the upper and lower tarsal plates to the frontal process of maxilla. The lateral palpebral ligament connects the lateral ends of the upper and lower tarsal plates to the tubercle (Whitnall's tubercle) on the zygomatic bone.

Accessory Modifications of the Orbital Fasciae

❑ ***Superior and inferior tarsal expansions:*** The muscular fascial sheaths of the superior rectus and the levator palpebrae superioris thicken and blend with each other on their adjoining surfaces. An anterior fascial extension passes from this thickening to the upper eyelid. It is reinforced by the fibres of the superior tarsal muscle, attaches to the superior tarsal plate and to the superior conjunctival fornix and is called the ***superior tarsal fascial expansion***. Whenever the superior rectus contracts to produce an upward gaze, the upper eyelid and the superior conjunctival fornix are also drawn up and elevated to help in clear, unobstructed vision. This drawing up of the upper eyelid and the conjunctival fornix are made more effective by the superior tarsal fascial extension than the drawing up produced by the action of levator palpebrae muscle alone. The muscular fascial sheaths of the inferior rectus and the inferior oblique thicken and blend (as already seen in the discussion on the suspensory ligament). An anterior fascial thickening passes from this thickening to the lower eyelid. It is reinforced by the fibres of the inferior tarsal muscle, attaches to the inferior tarsal plate and is called the *inferior tarsal fascial expansion*. Whenever the inferior rectus contracts to produce a downward gaze, the lower eyelid is also drawn down to help in clear, unobstructed vision. This drawing down is due to the inferior fascial tarsal expansion.

❑ *Transverse ligament of the eye:* This is a collective name applied to the medial and lateral extensions of the vagina bulbi to the medial and lateral orbital walls. Though its importance is questionable, it may play an important role as a fulcrum for levator movements of the eyeball.

❑ *Radial fascial septae:* These are thin fascial sheaths which extend from the vagina bulbi to the orbital walls. Since they extend peripherally from the globe of the eyeball, they are disposed radially. They provide compartments for orbital fat. The orbital septum is considered to be the most anterior of these radial septae. Many of these septae may have smooth muscle cells.

Added Information

❑ Microscopic muscle fibres are found in the fascia surrounding the eyeball. Collectively these fibres are called the **Landstrom's muscle** (named after the 19–20th century Swedish surgeon John Landstrom). They are attached anteriorly to the eyelids and the orbital septum. Their action is to pull the eyeball forward and the lids backward. They are supposed to resist the action of the recti muscles.

❑ The name of Muller is used for more than one muscle—the superior tarsal muscle, the orbitalis muscle and the circular fibres of the ciliaris muscle. Heinrich Muller was 19th century German anatomist.

❑ Though the extraocular muscles and their actions are described separately, the actual role played by them is a result of the play of multiple factors. Except for the medial and lateral recti, the other muscles act in all three planes. However, their actions are classified as primary, secondary and tertiary according to the power of action. For superior rectus, primary action is elevation; secondary, adduction and tertiary, intorsion. For inferior rectus, primary action is depression; secondary, adduction and tertiary, extorsion. For superior oblique, primary action is abduction; secondary, depression and tertiary, intorsion. For inferior oblique, primary action is abduction; secondary, elevation and tertiary, extorsion.

contd...

Added Information *contd...*

❑ In the primary position of the eyeball, the line of vision is directed to the horizon.

❑ When the eye is abducted (that is, the visual axis brought parallel and in line with the orbital axis), the superior and the inferior recti act as pure elevator and depressor respectively.

❑ When the eye is adducted, the superior and inferior oblique muscles act as pure depressor and elevator respectively.

❑ When both the superior and inferior recti act together and their pulling powers are equal, their elevation-depression and intorsion-extorsion actions neutralise each other and they adduct (rotate medially) the eye. They assist the medial recti in converging the visual axes.

❑ When both superior and inferior oblique muscles act together and equally, their depression-elevation and intorsion-extorsion actions neutralise and they abduct (rotate laterally) the eye. They assist the lateral recti in diverging the visual axes.

❑ When the gaze of both the eyes is directed medially (seeing near objects or reading–convergence), both the superior oblique muscles act together causing depression. When the gaze is directed up (as in looking up the page during reading), the two inferior oblique muscles act and cause elevation.

❑ Every movement of the eyeball requires the action of several muscles in the same eye and both eyes. The muscles act as synergists when they assist each other and as antagonists when they oppose each other. Muscles which are synergistic for one action may be antagonistic for another. In the primary position, both superior rectus and inferior oblique elevate the eyeball, thus being synergists. But in the same action, they are antagonists because the adduction caused by superior rectus is neutralised by the abduction caused by inferior oblique. Only then, elevation occurs. Similarly, in the primary position, superior oblique and inferior rectus synergistically produce depression while their other actions neutralise each other.

❑ For binocular vision, paired action of opposed muscles takes place. For gaze to the left side, the left lateral rectus and right medial rectus will have to act together. Such functionally paired contralateral muscles are called **yoke muscles**.

❑ The spiral of Tillaux serves as an external indication of the ora serrata.

Clinical Correlation

❑ *Fractures and injuries of orbit:*
   ○ The orbital margin may be fractured at the sutures of the constituent bones if blows are direct and powerful.
   ○ A blow to the eye may fracture the medial and inferior orbital walls (because these walls are thin), even when the orbital margin is intact.
   ○ Indirect injury may displace the orbital walls; such an injury is called a *'blow out' fracture*.
   ○ Fractures of the medial wall may involve the ethmoidal and sphenoidal air sinuses. Orbital floor fractures involve the maxillary air sinus.

contd...

EYEBALL

INTRODUCTION

The greater part of the eyeball (posterior five-sixths) is shaped like a sphere and has a diameter of about 24 mm. The anterior one-sixth is much more convex than the posterior part. It represents part of a sphere having a diameter of about 15 mm. The posterior five-sixths of the eyeball are covered by a thick white opaque membrane called the *sclera*. The wall of the anterior one-sixth of the eyeball is transparent and is called the *cornea*. When the 'eye' is viewed from in front in the living person, only a small part of the eyeball is seen in the interval between the upper and lower eyelids (i.e., in the palpebral fissure). The 'white' of the eye is formed by the sclera. The dark central part is formed by the cornea. The cornea itself is transparent. The dark appearance is because of the presence of a pigmented diaphragm, the *iris*, deep to the cornea. In the centre of the iris, there is an aperture called the *pupil*. The pupil appears black in colour as the interior of the eye is dark.

A horizontal section across an eyeball is shown in Fig. 11.7. The following features:

The wall of the eyeball is made up of three main layers.

1. The outermost layer is called the *fibrous coat*. It is formed posteriorly by the sclera and anteriorly by the cornea.
2. The next (inner) layer is the *vascular coat*. It has the following subdivisions—the part lining the inner surface of most of the sclera is thin and is called the *choroid*; near the junction of the sclera with the cornea the vascular coat is thick and forms the *ciliary body*; the ciliary body is continuous with the *iris* that lies a short distance behind the cornea.

   The space between the iris and the cornea is called the *anterior chamber*. The space between the iris and the front of the lens is called the *posterior chamber*.
3. The innermost layer of the wall of the eyeball is called the *retina.* Light falling on the retina has to pass through a number of *refracting media* before reaching the retina to form an image on it. These are—*cornea, aqueous humour*–the fluid that fills the anterior and posterior chambers, *crystalline lens* and *vitreous body*—a jelly like substance that fills the eyeball posterior to the lens.

The centre of the cornea is called the *anterior pole* of the eyeball. The opposite end of the eyeball is called the *posterior pole*. The *visual axis* of the eye passes from the anterior pole to the posterior pole. In (Fig. 11.7) note that the optic nerve is attached to the back of the eyeball a short distance medial to the posterior pole. An imaginary line passing round the eyeball midway between the anterior and posterior poles is called the *equator* of the eyeball. Any line passing through both the poles (i.e., at right angles to the equator) is called a *meridian*.

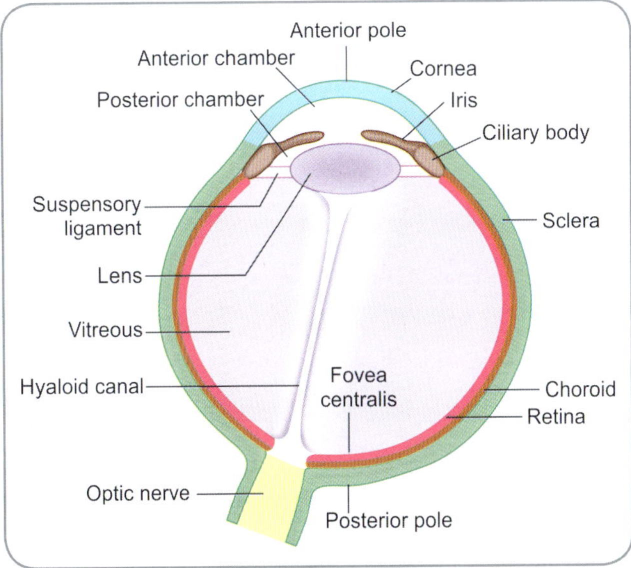

Fig. 11.7: Horizontal section across the eyeball to show the main features of its structure

Fibrous Coat

Other names: Ocular fibrous tunic, tunica fibrosa bulbi, tunica externa oculi

This is the outer most coat of the eyeball. It consists of the ***sclera*** and the ***cornea***.

Sclera (Fig. 11.8)

The sclera (or the sclerotica or the sclerotic coat or white of the eye) is a dense, hard layer made up mainly of fibrous tissue (Greek.skleros=hard). Its external surface gives attachment to the extrinsic muscles of the eyeball. Posteriorly, in the region of the attachment of the optic nerve, the sclera is perforated like a sieve. This area is called the ***lamina cribrosa*** (Fig. 11.8). Bundles of nerve fibres arising in the retina pass through these perforations to form the optic nerve. At the centre of the lamina cribrosa, there is a larger aperture for passage of the central artery and vein of the retina. Just around the lamina cribrosa, the sclera becomes continuous with the dural sheath of the optic nerve. A short distance from the lamina cribrosa the sclera is perforated by the short ciliary nerves and arteries, and further away by the long ciliary nerves and arteries. A little behind the equator of the eyeball, the sclera has four or five apertures for veins which are called the ***venae vorticosae*** (Fig. 11.8).

Anteriorly, the sclera becomes continuous with the cornea at the ***sclerocorneal junction***. As this junction is circular, it is also called the ***limbus***. A circular channel called the ***sinus venosus sclerae*** (or ***canal of Schlemm;*** named after the 18–19th century German anatomist Friedrich Schlemm) is located in the sclera just behind the sclerocorneal junction (Fig. 11.12). A triangular mass of scleral tissue projects into the cornea just medial to the sinus. This projection is called the ***scleral spur***. The sclera over the anterior part of the eyeball is covered by the ocular conjunctiva. The rest of the sclera is covered by ***fascial sheath (vagina bulbi or Tenon's capsule)*** that surrounds the eyeball. Between the fascial sheath and the external surface of the sclera is the episcleral space that contains loose fibrous tissue with few blood vessels. The deep surface of the sclera is separated from the choroid by the ***perichoroidal space***. Delicate pigmented connective tissue present in this space constitutes the ***suprachoroid lamina*** (or ***lamina fusca sclera;*** Latin.fuscus=dark or dusky, indicating brown or dark colour). The blood supply and the nerve supply to sclera are both, sparse.

The contents of the eyeball are under positive pressure. This pressure is resisted by the sclera, which is not extensible. These two factors together maintain the shape of the eyeball. The sclera provides a smooth external surface that facilitates movements of the eyeball. This surface also provides attachment to the extrinsic muscles of the eyeball.

Cornea (Fig. 11.9)

The cornea (Latin.coneus=horny) is the anterior, projecting transparent part of the external coat of the eyeball. It is not exactly circular. Its horizontal diameter (usually about 11.7 mm in an average adult) is somewhat greater than its vertical diameter (usually less than 11 mm). As the cornea is more convex than the sclera the junction of the two is marked, on the exterior of the eyeball, by a groove called the ***sulcus sclerae***. Microscopically, the cornea is made up

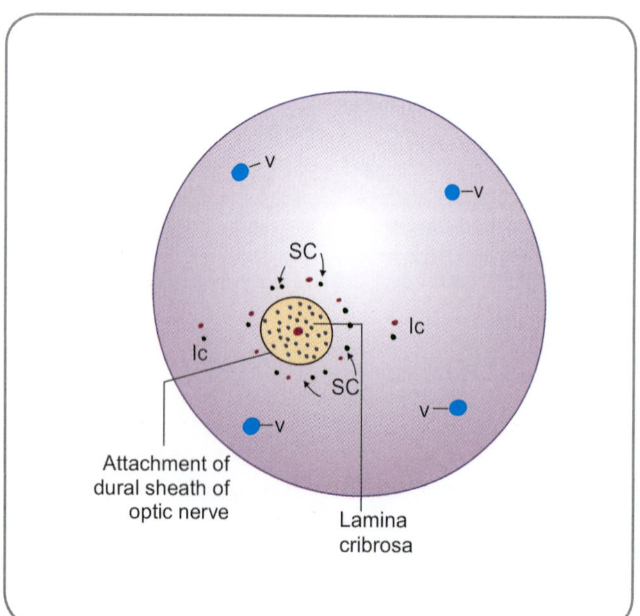

Fig. 11.8: Eyeball seen from behind to show the openings in the sclera. SC—foramina for short ciliary nerves and arteries Ic—foramina for long ciliary nerves and arteries v—foramina for venae vorticosae

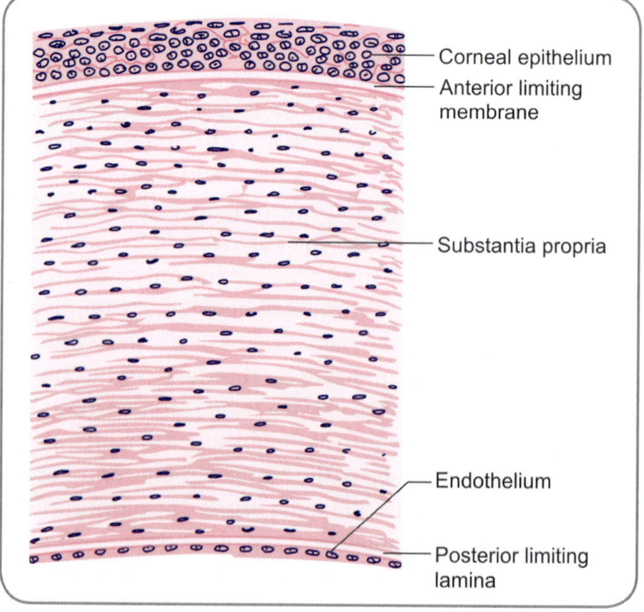

Fig. 11.9: Diagram of a section through the cornea to show its layers

of five layers (Fig. 11.9). The cornea is avascular containing neither blood vessels nor lymphatics. The capillaries of the sclera and the conjunctiva end in loops near the periphery of the cornea.

Histology

Histology of sclera: The sclera has dense collagenous tissue interspersed with few elastic fibres and flat fibroblasts. The collagen fibres are arranged in different patterns in different parts of the sclera. Around the optic disc and around apertures in the lamina cribrosa, they are arranged circumferentially. In other parts, they are reticular. In the scleral spur, collagen fibres are circularly disposed; further, more elastic fibres are seen.

Histology of cornea: Histologically, the cornea has five layers. From before backwards, these layers are corneal epithelium, anterior limiting lamina, substantia propria, posterior limiting lamina and endothelium. The corneal epithelium that covers the anterior surface has five layers:

❑ Deepest layer has columnar cells with flat bases, rounded apices and large nuclei.

❑ Second layer has polyhedral cells.

❑ Three superficial layers have progressively flatter cells with flat nuclei and smooth surface.

At the sclerocorneal junction, the corneal epithelium becomes continuous with the conjunctival epithelium. Cornea has no epithelial stem cells. Cell replacement is done by centripetal migration of limbal stem cells from the edges of the cornea. The anterior limiting lamina is about 12 microns thick and has a dense mass of collagen fibrils set in a matrix. This layer has no fibroblasts. The substantia propria forms the major bulk of cornea. It is a transparent layer consisting of about 200 to 250 lamellae. Each of these lamellae is made up of slender, parallel collagen fibrils. Interconnecting fibroblasts form a meshwork between the lamellae. Alternate lamellae are oriented at different angles to each other. The fibrils in the anterior lamellae are smaller than those in the posterior lamellae. The fibrils are smaller than the wavelength of light. This feature and the regularity of the spacing of the fibrils are the principal reasons for corneal transparency. The posterior limiting lamina is thin and is the basement membrane of the endothelium. At the sclerocorneal junction, this lamina gets merged with the trabecular tissue. The endothelium is a layer of polygonal cells which interdigitate with each other.

Vascular Coat

Other names: Uvea, uveal tract, uveaformis (Latin. uvaea=grape; named after the mulberry like appearance caused by network of blood vessels), ocular vascular tunic, tunica vasculosa bulbi, haller's tunica vasculosa.

This is the middle coat of the eyeball and consists of the ***choroid, ciliary body*** and ***iris***.

Choroid

The choroid (Greek.chorioeides=like a membrane), which is a thin and a highly vascular layer, consists of a network of blood vessels supported by connective tissue containing many pigmented cells that give it a dark brown colour. It is the dark colour of the choroid that darkens the interior of the eyeball. It also prevents reflection of light within the eyeball. Both these factors are necessary for formation of sharp images on the retina.

Within the choroid, there is an ***outer vascular layer*** made up of small arteries and veins interspersed with loose connective tissue and pigment cells, a middle ***capillary layer and an inner structureless basal layer***. The short ciliary arteries which pierce the posterior part of the sclera run forward in the vascular layer. However, most of the vascular layer is filled by veins that converge on four or five ***venae vorticosae*** (or principal vorticose veins; Latin.vorticosus=whorl; arranged in whorls). These veins pierce the sclera a short distance behind the equator of the eyeball. The capillary layer, which is separated from the retina only by the basal layer, consists of a dense network of large capillaries and so provides nutrition to the retina. The basal layer (otherwise called the membrane of Bruch, Bruch membrane, lamina basalis choroideae, lamina vitrea) appears glassy and homogenous but has a large elastic fibre mesh. It may be associated with the passage of fluid and solutes from the capillaries to the retina.

The choroid is separated from the sclera by the ***suprachoroid lamina.***

Ciliary Body

The ciliary body (or corpus ciliare; ciliary=relating to cilia) represents an anterior continuation of the choroid. Anteromedially, it becomes continuous with the iris. It is made up of vascular tissue, muscle and connective tissue.

When the eyeball is cut coronally some distance behind the lens and is viewed from behind, the following features can be made out:

❑ The retina proper ends anteriorly (a short distance behind the sclerocorneal junction) in a wavy line called the ***ora serrata*** (Latin.ora=margin, serrata=serrated).

❑ This line also represents the junction of the choroid with the ciliary body.

❑ The ciliary body extends forwards (and medially) from the ora serrata to end near the periphery of the lens.

❑ At the ora serrata, the retina is reduced to two layers of epithelial cells; this double layer of simple epithelium continues forward lining the inner surface of the ciliary body as pars ciliaris retinae and further forward on the posterior surface of the iris as pars iridica retinae.

The ciliary body can be divided into a posterior part called the ***ciliary ring*** and an anterior part made of the ***ciliary processes***. It appears brown because of the presence of melanin in the deeper layer of its epithelium.

❑ The ***ciliary ring*** (***other names:*** corpus orbiculus ciliaris, annular orbiculus ciliaris, pars plana) is about

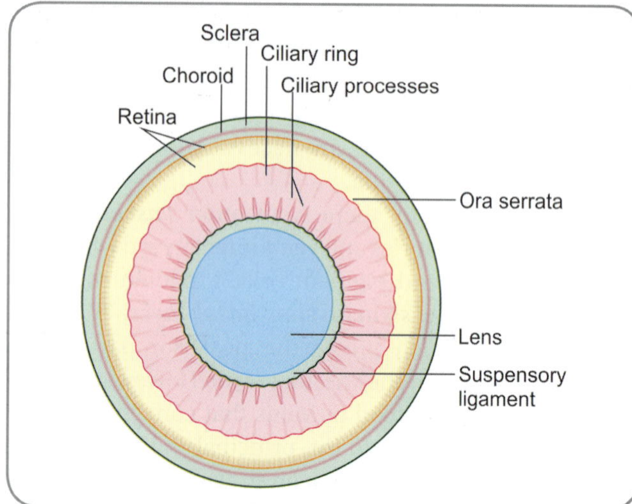

Fig. 11.10: Anterior part of the eyeball viewed from behind after cutting the eyeball in the coronal plane

3.5 mm wide and forms more than half the meridional width of the ciliary body. It has a relatively flat surface on which there are shallow radial grooves (Fig. 11.10). The peripheral rim of the ciliary ring is the ora serrata. The ora serrata is scalloped (serrated, on a gross appearance). Thin dentate processes from the retina flank the ciliary bays at this scallop.

❑ The *ciliary processes* (*other names:* corpus corona ciliaris, corona ciliaris, pars plicata) lie between the ciliary ring and the base of the iris (Fig. 11.10). The ciliary processes secrete the aqueous humour. They may also produce some components of the vitreous body. There are about 70 processes arranged radially (meridionally). Each process may be regarded as an infolding of the surface of the ciliary body and having a very irregular surface because of the presence of secondary folds. The substance of each process is made up of a plexus of blood vessels covered by the double layer of retinal epithelium mentioned above. Minor ridges called *ciliary plica* are found in many of the valleys between the processes. The crests of the processes are less pigmented and appear whitish. When the posterior aspect of the anterior half of the eyeball is viewed, the ciliary processes appear as white or light coloured striae and simulate the appearance of a circle of projections or lashes. The name 'ciliary' (related to lashes) is derived from this appearance.

The ciliary processes and the furrows between them give attachment to numerous delicate fibres that pass to them from the periphery of the lens. These are the fibres of the zonule of the suspensory ligament of the lens. Many of these fibres run into the ciliary ring to be attached to its epithelium. Their sites of attachment extend as thin lines from the valleys of the ciliary processes to the retinal dentate processes of ora serrata.

In a meridional section (i.e., a section along a line joining the anterior and posterior poles of the eyeball), the ciliary body appears to be triangular. Lateral to the ciliary processes (and medial to the anterior part of the sclera), is seen the triangular mass of the *ciliary muscle* (musculus ciliaris or Bowman's muscle). The fibres of the ciliary muscle take origin from the sclera spur and are arranged in two main groups. The *circular fibres* (fibrae circulars or Muller's muscle or Rouget's muscle) lie in the anterior and inner part of the ciliary body. They swerve from the sclera spur to the periphery of the lens and form a ring of muscle, contraction of which relaxes the fibres of the suspensory ligament. Release of tension on the lens makes it more convex. The *meridional fibres* (fibrae meridionales or Brucke's muscle or Crampton's muscle) pass backwards and get attached to the ciliary processes and the ciliary ring. Some fibres extend into the choroid. When these fibres contract the ciliary ring is drawn forwards (i.e., towards the lens). As a result, the suspensory ligament is again relaxed. A third set of fibres is also described. This is the set of radial oblique fibres which interconnect the circular and meridional fibres in a lattice. The circular and meriodonal fibres act together (and the radial fibres too) to make the lens more convex; and enable it to focus images of near objects on the retina. This phenomenon is called *accommodation*. The ciliary muscle (or simply, the ciliaris) is supplied by parasympathetic nerve fibres which arise in the Edinger-Westphal nucleus and travel in the oculomotor nerve, ciliary ganglion and short ciliary nerves. Parasympathetic stimulation contracts the ciliaris. The muscle also has a very sparse sympathetic innervation which is likely to relax it.

Iris (Fig. 11.11)

The iris (Greek.iris=rainbow) is the most anterior part of the vascular coat of the eyeball. It forms a diaphragm

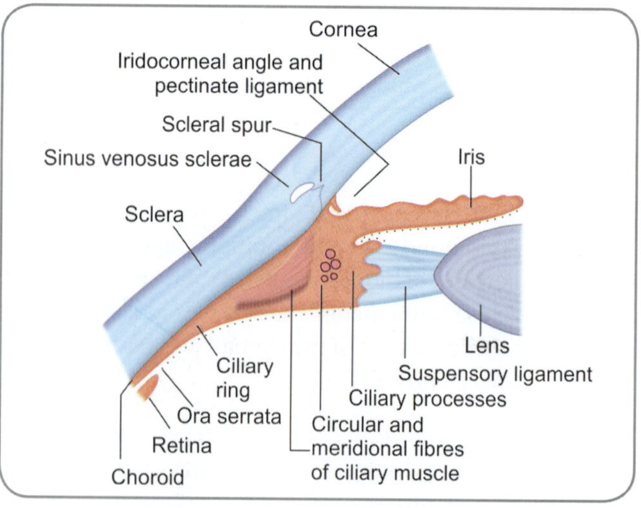

Fig. 11.11: Meridional section through the ciliary body and iris

placed immediately in front of the lens. At its periphery it is continuous with the ciliary body. In its centre is an aperture called the ***pupil***. The iris is composed of a stroma of connective tissue that contains numerous pigment cells and in which are embedded blood vessels and smooth muscle.

Histology

Histology of iris: The iris has anterior cellular layer, a posterior epithelium and a stroma. The stroma has fibroblasts and melanocytes embedded in a loose collagenous matrix. The intercellular spaces of the stroma communicate with the anterior chamber and exchange of fluids takes place freely. Aggregation of smooth muscle cells in the stroma forms the pupillary sphincter. The anterior layer is not a separate epithelium but the condensation of the superficial part of the stroma; condensation occurs because of the presence of more fibroblasts and melanocytes. The posterior epithelium covers the posterior aspect of the iris. It is bilaminar, is a continuation of the bilaminar epithelium of the ciliary body and curves as a thin border called the ***pigment ruff*** on to the anterior surface through the pupillary aperture. Of the two layers of the epithelium, the deeper (from posterior aspect) is called the ***anterior epithelium***. It has pigmented cells and gives rise to the dilator papillae. The superficial or posterior layer forms the posterior epithelium and its cells have more pigment. The posterior surface of the iris is thrown into radial folds by variations in the thickness of stroma; these folds help in pushing the aqueous fluid from the posterior to the anterior chamber.

The pupil regulates the amount of light passing into the eye. In bright light the pupil contracts and in dim light it dilates. In this way, the optimum amount of light required for proper vision reaches the retina within a considerable range in intensity of illumination. Changes in the size of the pupil are produced by the smooth muscle of the iris that consists of two parts—the ***sphincter*** and the ***dilator papillae***.

The iris is not exactly a flat diaphragm. The lens, which lies posterior to the iris, makes the latter bulge a little in the centre. The iris should properly be described as a shallow cone that is truncated by the papillary aperture lens. The iris is interposed between the lens and the cornea. It therefore divides the aqueous segment into an anterior chamber and a posterior chamber.

The iris is named after the rainbow. However, its colour range is from blue to brown only. Pigment cells (melanocytes) are present in the stroma of the iris. Varying concentrations of these pigment cells and differing quantities of pigment are responsible for producing varying colours of the iris. Pigment is largely absent at birth and the colour of the iris is light blue. The colour can vary between the two eyes of the same individual.

Pupillary membrane: This is a foetal structure. It is seen as a thin, vascular membrane closing the pupil in the foetus. Its blood vessels, which come from the irideal angle and the lens capsule, pass towards the centre and form loops. The central part of the membrane, by itself, is avascular. By around the 6th week of intrauterine life, absorption of the membrane begins at the centre and proceeds to the periphery. At birth, the membrane is fully absorbed except from scattered fragments which may be seen at the pupillary rim. Persistence of the membrane in rare cases will interfere with vision.

Muscles of the iris: Sphincter pupillae is a ring of circularly arranged muscle situated just around the pupil; its contraction narrows the pupil. ***Dilator pupillae*** is in the form of muscle fibres which are arranged radially in the iris; these fibres are intimately related to the anterior of the two layers of epithelium lining the posterior surface of the iris; near the pupil, these radial fibres merge with the sphincter pupillae; their contraction dilates the pupillary opening.

The sphincter pupillae has a parasympathetic nerve supply (similar to that of the ciliary muscle). Preganglionic neurons located in the Edinger-Westphal nucleus (in the upper part of the midbrain) give off axons which pass through the oculomotor nerve and its branches to reach the ciliary ganglion. After relay in the ganglion, postganglionic fibres pass through the short ciliary nerves to reach the muscle. The dilator pupillae is supplied by sympathetic nerves. The preganglionic neurons concerned are located in segment T1 of the spinal cord. The postganglionic neurons lie in the superior cervical sympathetic ganglion. The postganglionic fibres pass through the sympathetic plexus around the internal carotid artery, enter the ophthalmic nerve and its nasociliary branch and then run through the long ciliary nerves.

Iridocorneal Angle (Fig. 11.12)

The angle between the peripheral margins of the iris and the cornea is a region of importance. From meridional and coronal sections of the eyeball, it will be seen that the most anterior part of the ciliary body appears in this angle. Laterally, the angle is related to the sinus venosus sclerae. The lateral wall of the sinus is formed by a groove in the sclera and the medial wall is related to the scleral spur and the pectinate ligament. The ***pectinate ligament*** is formed by numerous interlacing filaments which are continuations of the posterior limiting lamina of the cornea into the iris. These filaments are separated by spaces called the ***spaces of the iridocorneal angle*** or ***trabecular spaces***. These spaces communicate medially with the anterior chamber and laterally with the sinus venosus sclerae.

Aqueous humour secreted by the ciliary processes, first passes into the posterior chamber of the eye (i.e., the

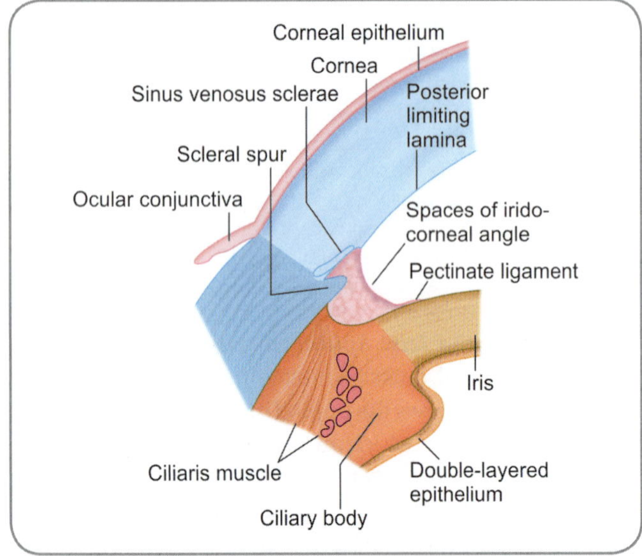

Fig. 11.12: Some features of the iridocorneal angle. Blood vessels in the ciliary body and iris are omitted

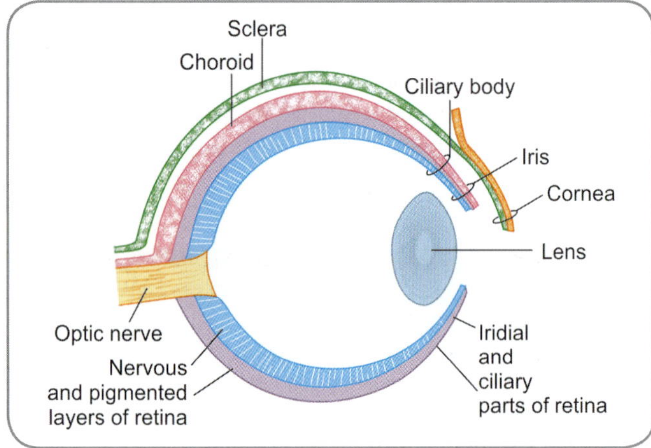

Fig. 11.13: Scheme to show some features of the developing eye

space between the posterior surface of the iris and the lens). It then passes through the pupil into the anterior chamber (space between the iris and the cornea). From here, it filters through the spaces of the iridocorneal angle to enter the sinus venosus sclerae through which it is drained into the veins of the region. The clinician often refers to the iridocorneal angle as the **angle of filtration**. With advancing age, the trabecular spaces may get blocked resulting in increased tension within the eyeball. This disease, called **glaucoma,** may have serious consequences.

Sensory Coat

Other names: Tunica interna bulbi, nervous coat of the eyeball, optomeninx, tunica nervea.
This is the inner coat of the eyeball and consists of the retina.

Retina

The retina (Latin.rete=network) is the sensory layer of the eyeball. It has a complex structure and can be considered as a special extension of the brain. To understand the structure of the retina, brief reference to its development is necessary. The retina develops as an outgrowth from the diencephalon part of the brain. The proximal part of the diverticulum remains narrow and is called the **optic stalk**. It later becomes the optic nerve. The distal part of the diverticulum forms a rounded hollow structure called the **optic vesicle**. This vesicle is invaginated by the developing lens (and other surrounding tissues) so that it gets converted into a two-layered **optic cup**. At first, each layer of the cup is made up of a single layer of cells. The outer layer persists as a single layered epithelium that

becomes pigmented and forms the **pigment cell layer** of the retina (Fig. 11.13). Over the greater part of the optic cup, the cells of the inner layer multiply to form several layers of cells which become the **nervous layer** of the retina. The anterior limit of the region where this transformation takes place is later marked by the ora serrata. The part of the retina behind the ora serrata now consists of an outer, pigmented layer, and an inner nervous layer. The two are separated by a cleft representing the original cavity of the optic cup. It is only this part of the retina that is light sensitive and is referred to as the **optic part** of the retina (or the retina proper, pars optica retinae, pars cerebrale retinae, neural part of retina). Anterior to the ora serrata, both layers of the optic cup remain single layered. These two layers of epithelium (one being the pigmented layer and the other being a columnar layer) line the inner surface of the ciliary body forming the **ciliary part** of the retina (pars ciliaris retinae), and the posterior surface of the iris forming the **iridial part** of the retina (pars iridica retinae).

The layer of pigment cells has greater adherence to the choroid than to the nervous layer of the retina. In some cases, part of the retina may get detached from its normal position. In such **detachment of the retina**, it is only the nervous layer that gets detached, the pigment cells remaining attached to the choroid. The essential features of the structure of the retina are shown in (Fig. 11.14).

The optic part of retina extends from the optic disc to the ora serrata. It contains photoreceptors which convert the stimulus of light into nervous impulses. These receptors are of two kinds—**rods** and **cones**. There are about seven million cones in each retina. The rods are far more numerous; they number more than 100 million. The cones respond best to bright light. They are responsible for sharp vision and for discrimination of colour.

Opposite the posterior pole of the eyeball, the retina shows a **central region** about 6 mm in diameter. This region is responsible for sharp vision. In the centre of this

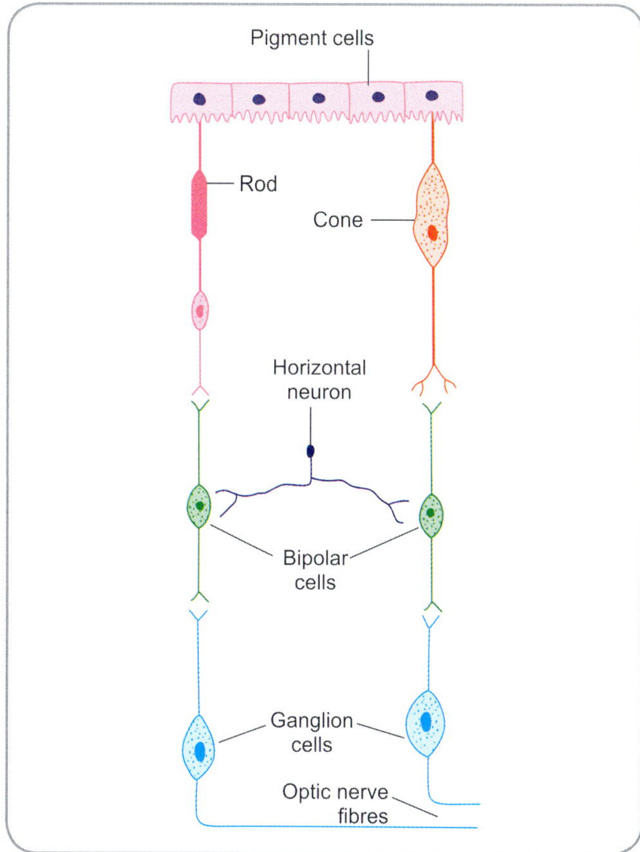

Fig. 11.14: Simplified scheme to show the main elements of the retina

region, an area about 2 mm in diameter appears yellow due to xanthophyll derivatives and is called the ***macula lutea.*** The macula has a central depression of about 0.4 mm diameter called the ***fovea centralis (or the foveola).*** Cones are numerous in the central region; the fovea centralis is believed to contain cones only and visual resolution is highest here. Rods can respond to poor light and especially to movement across the field of vision. They predominate in the peripheral parts of the retina.

Each rod or cone can be regarded as a modified neuron. It consists of a cell body, a peripheral process and a central process. The cell body contains the nucleus. The peripheral process is rod-shaped in the case of rods and cone-shaped in the case of cones (hence the names, rods and cones). The ends of the peripheral processes are separated from one another by processes of pigment cells. The central processes of rods and cones are like those of neurons; they end by synapsing with other neurons within the retina.

The basic neuronal arrangement within the retina follows a simple but relay pattern (Fig. 11.14). The peripheral processes of the rods and cones are the photoreceptive ends. The central processes of rods and cones synapse with the peripheral processes of ***bipolar cells.*** The central processes of bipolar cells synapse with

dendrites of ***ganglion cells.*** Axons arising from ganglion cells form the ***fibres of the optic nerve.***

The various elements mentioned above form a series of layers within the retina. By convention, the retina is described to have ten layers. The layers are named after the primary components in them and their position in the entire thickness of retina. Those structures/layers closer to vitreous are designated 'inner' and those closer to choroid are 'outer'.

1. The outermost layer (layer close to the choroid) called the ***retinal pigment epithelium*** is formed by the pigment cells; this layer has a simple low cuboid epithelium that forms the boundary with the choroid; it is separated from the choroid by a thick basal lamina.

2. ***Layer of rods and cones:*** This layer contains the outer segments and the outer part of the inner segments of the rod and cone cells.

3. ***External limiting membrane:*** This layer that appears as a distinct line under light microscope contains the intercellular junctions between the glial cells and photoreceptor cell processes.

4. ***Outer nuclear layer:*** This layer consists of the cell bodies of the rods and cones and their nuclei.

5. ***Outer plexiform layer:*** The central processes of the rods and cones synapse with the peripheral processes of the bipolar cells and the dendrites of the horizontal cells in this layer. The synaptic network gives name to this layer.

6. ***Inner nuclear layer:*** This layer has three strata of nuclei; horizontal cell bodies and nuclei lie outermost; bipolar cell bodies and nuclei along with the cell bodies and nuclei of glial cells lie next; cell bodies and nuclei of amacrine cells lie innermost.

7. ***Inner plexiform layer:*** This again is a layer of synapses. The central processes of the bipolar cells and those of the amacrine cells synapse with the dendrites of the ganglion cells.

8. ***Ganglion cell layer:*** As the name indicates, this layer has the cell bodies and nuclei of the ganglion cells.

9. ***Nerve fibre layer:*** The axons of the ganglion cells pass as a cluster through this layer.

10. ***Internal limiting membrane:*** This layer is the glial boundary between the retina and the vitreous.

The layer of nerve fibres is opposed to the vitreous. It can be understood that light rays entering the eyeball through the cornea will have to go through the aqueous, pupil, lens and vitreous before reaching the retina. Light also has to pass through several layers of the retina to reach the rods and cones. At the fovea centralis, the other layers are 'swept aside' to allow direct access of light to the cones here.

The optic nerve is attached to the eyeball a short distance medial to the posterior pole. The nerve fibres

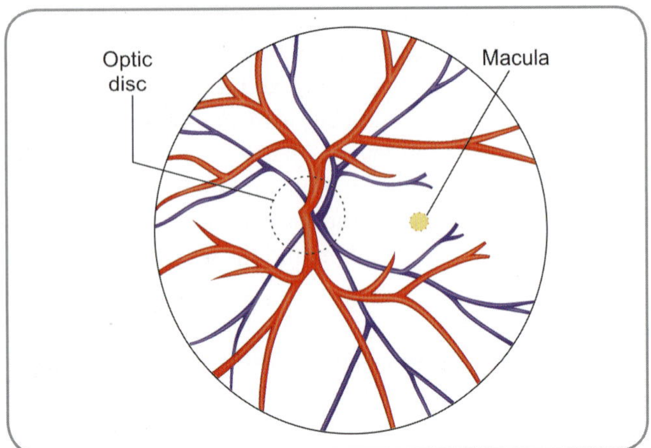

Fig. 11.15: Blood vessels of the retina as seen through an ophthalmoscope

arising from the ganglion cells all over the retina converge to this region, where they pass through the lamina cribrosa to form the optic nerve. When viewed from the retinal side this region is seen as a circular area called the **optic disc** (Fig. 11.15). There are no photoreceptors here. The optic disc is, therefore, insensitive to light and is called the **blind spot**. The optic disc is pierced, near its centre, by the central artery and vein of retina.

OCULAR REFRACTIVE MEDIA

The components of the eye which transmit light rays can be collectively called the **refractive media of the eye**. They do not belong to a single coat or layer. They occupy different parts of the eyeball and have varying physical and anatomical properties. The refractive media are (from before backwards) cornea, aqueous humour, lens and vitreous body.

Cornea

The cornea has already been discussed under the ocular fibrous coat. The corneal epithelial cells have flat nuclei and present an optically perfect, smooth surface. Of the total refractive power of the eye, nearly 66% is provided by the cornea.

Aqueous Humour

Aqueous humour (Greek.aqua = water, aqeuo = watery) can be defined as the fluid occupying the anterior segment of the eyeball. It is part of the eye's own circulatory apparatus. To satisfy the demands of vision and to provide a congenial environment for the same, the eye develops its own circulatory apparatus and attempts to maintain appropriate fluid media. The aqueous humour is the main component and result of this effort. It is secreted into the posterior chamber by the non-pigmented epithelium of the ciliary processes. It then passes into the anterior chamber through the pupil and drains at the iridocorneal angle through the trabecular spaces into the sinus venosus sclerae. From here, it joins the venous stream. Itself being a refractive media (light entering the eyeball has to pass through the aqueous), it also provides nutrition and maintains the metabolism of other avascular refractive media like the lens, vitreous and cornea. The aqueous maintains the intraocular pressure by regulating the balance between its production and drainage. Interference with its drainage increases the intraocular pressure causing a condition called **glaucoma**.

Lens

The lens lies in front of the vitreous body and behind the iris. It is transparent and is enclosed in a capsule. It has convex anterior and posterior surfaces (therefore, biconvex) and a peripheral margin (or equator) to which the suspensory ligament is attached. The anterior surface is less convex than the posterior surface. It comes into contact with the iris near the margin of the pupil but further away the gap between the iris and the lens gives rise to the posterior chamber. The posterior surface of the lens lies in a depression in the vitreous body called the **hyaloid fossa**.

The central points of the anterior and posterior surfaces are the anterior and posterior poles of the lens; a line connecting the two poles is the axis of the lens. In adults, the lens is avascular, colourless, transparent and soft. In old age, the anterior surface becomes more convex. In cataract, the lens becomes opaque.

Suspensory Ligament of Lens

The suspensory ligament of the lens is also called the **zonule** (Fig. 11.11). It is made up of fibres passing from the equator of the lens to the ciliary processes and to the recesses between them. Changes in the tension on the suspensory ligament, produced by contraction of the ciliary muscle, cause alterations in the convexity of the lens and enable it to focus objects at varying distances from the eye.

Vitreous Body

Other names: Corpus vitreum, hyaloid body, vitreum.
The vitreous body (Latin.vitrum=glass, vitreus=glassy) fills the vitreous chamber and occupies about four-fifths of the eyeball. Anteriorly it has a deep concavity called the **hyaloid fossa** (Greek.hyalos=glass, oeidos=like; glass like) on which rests the lens. It is colourless and consists of more than 99% water. It is gel like in the peripheral areas but liquid in its central zone. Loose networks of collagen fibrils and scattered mononuclear type of cells called the **hyalocytes** have been described to be present in the gel portion of its substance. The gel portion is called the **cortex of vitreous**.

Blood Vessels and Nerves of the Eyeball

The arteries supplying the eyeball are all branches of the ophthalmic artery. The ***short (posterior) ciliary arteries*** are several in number. They pierce the sclera near the lamina cribrosa, end in numerous branches that form a plexus in the choroid and supply it (Fig. 11.16). The capillary plexus in the choroid is also responsible for providing nutrition (by diffusion) to the outer part of the retina (i.e., the part nearer the choroid).

The ***long (posterior) ciliary arteries*** are two in number (one medial and one lateral). They pierce the sclera a short distance to the corresponding side of the lamina cribrosa. These arteries then run forward between the sclera and the choroid to reach the region of the ciliary body. Here, each artery divides into one upper and one lower branch (Fig 11.17). The upper branches of the medial and lateral

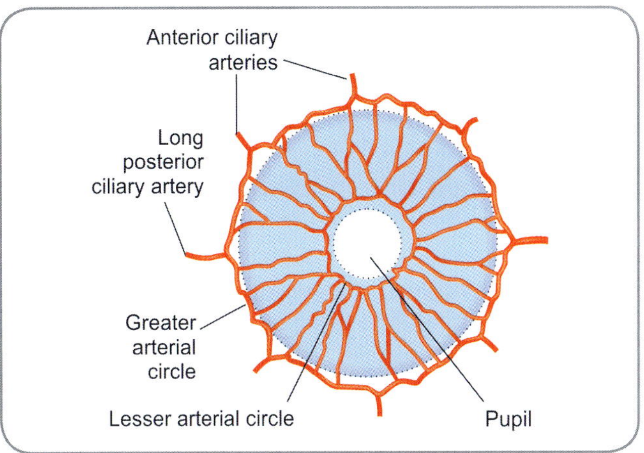

Fig. 11.17: Arteries of the iris

sides anastomose with each other. Similarly, the lower branches of the two sides anastomose. In effect, the upper and lower anastomoses are connected to each other to form the ***greater arterial circle*** at the periphery of the iris. This circle is joined by small ***anterior ciliary arteries*** which reach the region through the attachments of the rectus muscles to the eyeball. Branches pass radially into the iris from the greater circle and join each other just around the pupil to form the ***lesser arterial circle***.

The veins draining the iris, the ciliary body and the choroid form a dense plexus deep to the sclera. The veins of this plexus converge on four or five ***venae vorticosae***. These veins pierce the sclera a little behind the equator of the eyeball to end in the ophthalmic veins. The ophthalmic veins pass through the superior orbital fissure and drain into the cavernous sinus.

The main blood supply to the retina reaches it through the central artery of the retina. This artery runs forwards through the distal part of the optic nerve to enter the retina through the optic disc. It divides into upper and lower branches, each of which divides into medial (or nasal) and lateral (or temporal) branches. Further ramifications supply the entire retina. These arteries can be seen in the living subject by looking into the eye through the pupil using an instrument called the ophthalmoscope (Fig. 11.15). Branches of the central artery of the retina are end arteries, i.e., they do not anastomose with each other. Blockage of any branch results in death of the part of the retina supplied by it and to consequent loss of the part of the field of vision concerned. No large vessels are to be seen in the central region of the retina.

The blood from the retina is drained by tributaries that correspond to the branches of the central artery but do not accompany them closely. These tributaries end in two veins, superior and inferior, which pierce the lamina cribrosa and join each other to form the ***central vein of the retina***. This vein usually drains directly into the cavernous sinus but may join one of the ophthalmic veins.

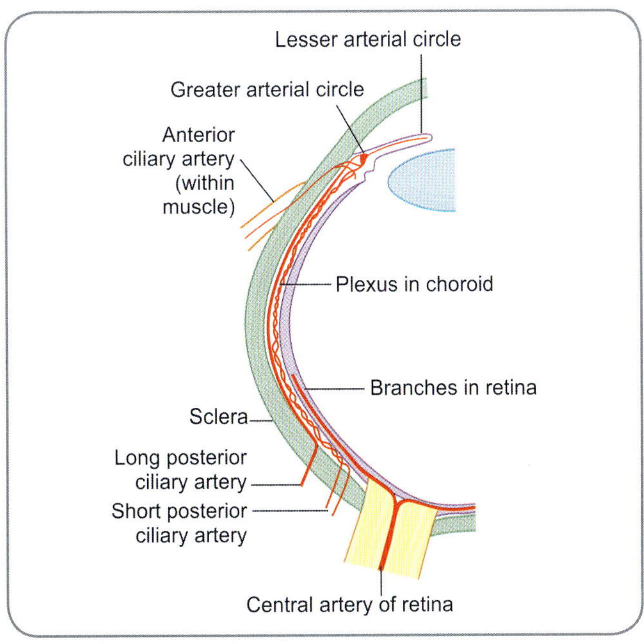

Fig. 11.16: Simplified scheme to show the arteries supplying the eyeball

The nerves (other than the optic nerve) which supply the eyeball are the ***long and short ciliary nerves***. The long ciliary nerves are branches of the nasociliary nerve, while the short ciliary nerves arise from the ciliary ganglion. They enter the eyeball along with the corresponding arteries. The short ciliary nerves carry parasympathetic postganglionic fibres from the ciliary ganglion to the ciliary muscle and to the sphincter pupillae. They also carry sympathetic postganglionic fibres to blood vessels of the eyeball. Sympathetic postganglionic fibres meant for the dilator pupillae normally travel through the long ciliary nerves. Occasionally, they may pass through the short ciliary nerves. In addition to autonomic efferents, the long and short ciliary nerves carry sensory fibres from various structures in the eyeball including the ciliary body, the iris and the cornea.

Clinical Correlation

Eyes and Related Structures

- ***Inflammation and infections:*** Infections are common in the eyelids (***blepharitis***). A pus-filled swelling near the edge of the eyelid is a ***stye (hordeolum)***. It is caused by infection in the large sebaceous glands (ciliary glands) of the eyelids and obstruction of their ducts. Inflammation of a tarsal gland results in a localised swelling called (tarsal) ***chalazion***; this swelling protrudes toward the eyeball from the eyelid; everytime the eyelids blink it rubs against the eyeball. Cysts of the sebaceous glands may also produce chalazia. The conjunctiva may become hyperaemic due to irritation or dust. Inflammation of the conjunctiva is called ***conjunctivitis***. Other structures in the eye which may be infected are the iris (***iriditis***), the ciliary body (***cyclitis***) and a combination of both of these (***iridocyclitis***). Inflammation of the cornea is called ***keratitis***. It can lead to the formation of a corneal ulcer. Corneal ulcers can also be caused by injury or by foreign bodies that enter the eye. A superficial infection of the sclera is ***episcleritis*** and a deeper infection is ***scleritis***. Inflammation of the vascular tunic is ***uveitis***. Inflammation of the retina is ***retinitis***. Inflammation of the optic nerve is ***optic neuritis***.
- ***Subconjunctival haemorrhages:*** Rupture of small, slender subconjunctival capillaries causes subconjunctival haemorrhages; these may be caused by irritation, dust, local injury, blow to any part of the eye, hard blowing of the nose and sudden, violent coughing or sneezing.
- ***Cornea:*** The cornea may be injured and damaged by foreign particles; corneal abrasions and lacerations may be caused. There will be sudden stabbing pain and copious flow of tears. Such injuries can result in ***corneal opacities*** which can lead to blindness. Damage to the sensory innervation of the cornea (trigeminal nerve–ophthalmic division) makes the cornea prone to injuries and subsequent opacities. Blindness due to corneal opacity can be cured by corneal grafting. What are being talked of as 'eye transplants' are actually corneal transplants. A whole eye cannot be transplanted. Alterations in the curvature of the cornea can lead to errors of refraction.

Clinical Correlation *contd...*

- ***Coloboma of Iris:*** Absence of a portion of the iris is called ***coloboma*** (Greek.koloboo=mutilated, damaged). It is usually a congenital defect. It may occur also because of penetrating injuries or surgical iridectomy.
- ***Glaucoma:*** Aqueous humour is produced by ciliary body and drained by the sinus venosus sclerae. The rate of production and drainage should be equal. If the outflow tract is blocked, then aqueous gets stagnated in the anterior and posterior chambers. This causes increase in pressure inside the eyeball leading to the condition called glaucoma. If normal intraocular pressure is not maintained, increased pressure causes compression of the internal structures of the eyeball, especially the retina, resulting in blindness.
- ***Hyphaema:*** This is a condition where there is bleeding into the anterior chamber and accumulation of blood. Blunt injury to the eyeball is the usual cause.
- ***Lens:*** Opacity in the lens of the eyeball is called ***cataract***. It occurs in oldage and is a common cause of blindness. Vision can be restored by removing the opacified lens and this is one the most common surgical operations done on the eyeball. Focussing power is restored by use of appropriate spectacles or by implanting an artificial lens into the eye (***intraocular lens transplantation***). In an extracapsular cataract extraction, the lens is removed after leaving the capsule intact; a synthetic intraocular lens is placed in the capsule. In an intracapsular cataract extraction, the lens is removed along with the capsule and a synthetic implant is placed in the anterior chamber.
- ***Errors of refraction:*** Defects in the shape of the eyeball or in the refractive media of the eye can lead to errors of refraction in which images are not focussed on the retina. Common errors are as follows:
  - ***Myopia (near sightedness):*** The image comes into focus in front of the plane of the retina. Objects can be seen clearly only when placed close to the eyes. Myopia is corrected with convex (plus) lenses.
  - ***Hypermetropia:*** The image comes to a focus behind the plane of the retina. Only distant objects can be seen. It is corrected with concave (minus) lenses.
  - ***Presbyopia:*** This is a condition caused by decreased elasticity of the lens in persons over 40 years old (Greek. presbyos=old). Focussing power of the lens is lost. Presbyopia can be corrected by convex lenses.

  In ***astigmatism,*** the image is correctly focussed in some planes but not in others. This is due to abnormalities in curvature of the cornea or of the lens. Correction requires the use of cylindrical lenses. The degree of refractive error is determined by a procedure called ***retinoscopy***.
- ***Fundus Examination:*** The retina can be examined through an ***ophthalmoscope (also called a fundoscope)***. The procedure is commonly referred to as fundus examination. Apart from diagnosis of diseases of the retina itself, such examination also helps in assessing the status of patients in diseases like hypertension and diabetes. A pale and oval optic disc on the medial side, arteries and veins radiating from the centre of the disc, pulsation of retinal arteries and a darker looking macula in the centre are some of the normal observations in such an examination.

contd...

contd...

Clinical Correlation *contd...*

Papilloedema is a condition that may be observed in an ophthalmoscopic examination. There is increase in CSF pressure which in turn causes reduced venous return in the retina resulting in oedema of the retina. Under normal conditions, the optic disc appears flat. But in the case of increased intracranial pressure, the disc appears elevated and like a papilla.

- ❏ ***Tumours:*** Tumours may occur in the eye. A ***melanoma*** may arise from the iris or the choroid. Another tumour seen in the eye is a ***retinoblastoma***.
- ❏ ***Retinal Detachment:*** The retina consists of a nervous layer (the retina proper) and an outer layer of pigment cells. In the embryo, the two layers are separated by a cavity called the ***intraretinal space***. The space gets obliterated in early foetal life. The pigment layer becomes firmly fixed to the choroid but its attachment to the neural retina is not firm. The nervous layer can get detached from the pigment layer leading to the condition called ***retinal detachment***. It causes serious impairment of vision. Detachment usually results from trauma to the eye; seepage of fluid occurs between the two layers as a result of the injury.
- ❏ ***Injury to the nerves of the orbit and eyeball:*** Lesion of oculomotor nerve results in ptosis–drop of upper eyelid (paralysis of levator palpebrae superioris). Lesion of facial nerve results in incomplete closure of eyelids (paralysis of orbicularis oculi) and loss of normal frequent protective blinking. Orbicularis oculi loses its tone. The lower eyelid is also affected and gets everted. The cornea is exposed and becomes dry. It is left unprotected from dust and foreign particles. If the secretomotor fibres are involved, lacrimation is also affected.

Important Procedures of Examination

Pupillary Reflex

Testing for pupillary light reflex is a simple but important procedure in the neurological examination of a patient. The pupils constrict in response to light. The afferent limb of the reflex is through the optic nerve and the efferent limb is the oculomotor nerve. Even when light enters one eye, both the pupils constrict. If the oculomotor nerve is compressed, there is ipsilateral slowing of the pupillary response.

Corneal Reflex

This again is a simple procedure performed during a neurological examination. The examiner touches the cornea of the patient with a wisp of cotton. The afferent limb of the reflex is through the ophthalmic division of the trigeminal nerve and the efferent limb is through the facial nerve (motor supply to orbicularis oculi muscle). In a normal positive response, the patients blinks. Lesion in the afferent or efferent limb can abolish the corneal reflex. If the patient is wearing contact lens, a negative response may result.

Development

The different parts of the eyeball develop from varied sources.

In the early embryo, bilateral thickenings appear on the anterior end of the neural tube. As developmental changes occur in the neural tube, these also develop into the optic vesicles. The optic vesicles are lateral diverticula from the prosencephalon, projecting into the surrounding mesoderm. To start with, the cavity of the optic vesicle is continuous with that of the prosencephalon. Differential growth causes the proximal part of the optic vesicle to remain constricted and this portion becomes the optic stalk. The distal portion, as it proceeds to grow longer, gets closely related to the overlying ectoderm which in turn is undergoing appropriate changes. The ectoderm becomes thickened to form the lens placode. The formation of lens placode is induced by the optic vesicle which releases a chemical substance called the ***evocator***. The lens placode is soon depressed from the surface level; it then is converted into a lens vesicle. By the fifth week, the lens vesicle is separated off from the surface ectoderm. The formation of lens vesicle causes changes in the optic vesicle. The latter is invaginated and so gets transformed into a double-layered optic cup. Invagination also occurs in the lower part of the distal segment of the optic stalk, resulting in the formation of the choroid fissure. The two layers of the optic cup are separated by a thin lumen called the ***intraretinal space***.

The outer layer of the optic cup shows pigmented cells. This develops into the pigment epithelium of the retina. The inner layer undergoes considerable modifications. Posterior four fifths of this layer develops into the pars optica retinae and the anterior one-fifth into the pars caeca retinae. In the pars optica, cells bordering the intraretinal space develop into the photoreceptive rods and cones. It should be remembered that the optic vesicle (and so the optic cup) is an extension of the prosencephalon. Therefore, further development is on the lines of similar development in the CNS. The rods and cones layer can be likened to the neuroepithelial layer of CNS development. Outer to this layer will be the mantle layer which in CNS gives rise to major neurons. The neuroepithelial cells of the developing retina give rise to cells of the mantle layer which develop into the ganglion cells, bipolar cells, horizontal cells and other cells of the retina. Similar to the CNS, a marginal layer also develops. This is a layer on the surface with fibrous tissue and contains axons of the deeper lying nerve cells. The axonal fibres of the ganglion cells which develop to form the optic nerve are derived from this layer. The pars caeca retinae remains one cell thick. It is in close contact with the pigment cell layer. Together they form the pigment epithelium of the ciliary body (pars ciliaris retinae) and the iris (pars iridica retinae).

Meanwhile, mesoderm fills in the space between the optic cup and the lens vesicle. As this mesodermal tissue condenses outer to the pigment layer of the pars ciliaris and pars iridica, blood vessels and connective tissue develop from it. The pigment layer of the iris portion contributes to the formation of the muscles of the iris, namely the ***sphincter*** and ***dilator papillae***. In the ciliary portion, the ciliaris muscle develops from the mesodermal tissue. The internal aspect of this portion is close to the developing lens and contributes to the suspensory ligament of the lens.

contd...

 **Development** *contd...*

When invagination of the optic cup and choroidal fissure formation are taking place, the lens vesicle separates off completely from the surface ectoderm. It is then spherical, hollow and has a wall that has a single layer of columnar cells. As the size of the lens increases, the cells of the superficial portion become cuboidal. They form the anterior lens cells of the **anterior lens capsule**. The cells of the deep portion become elongated and lose their nuclei. As these elongated cells project into the lens cavity, the cavity is gradually obliterated. The elongated cells, after losing their nuclei, get converted into fibres. The fibres formed from the deep cells become the primary lenticular fibres and occupy the central zone of the lens. As growth proceeds, the cells of the equatorial region also undergo similar modifications. Their fibres are added circumferentially forming the secondary lenticular fibres. The mesoderm lying immediately adjacent to the lens condenses to form the lens capsule. The lens capsule receives its blood supply posteriorly from the hyaloid artery and anteriorly from the annular artery. Vascularisation of the lens regresses before birth. The entire lens except the anterior epithelium is therefore, formed from the deep and equatorial cells.

Initially there is no mesoderm between the developing lens and the internal aspect of the optic cup. As growth proceeds, a jelly like substance is seen surrounding the blood vessels which have entered the optic cup through the **choroidal fissure**. This substance is the primordium of the vitreous and is called the hyaloid substance. It is thought to be of both ectodermal (neurectoderm of the optic cup) and mesodermal (mesoderm that insinuates through the mouth of the optic cup) in origin. A thin hyaloid membrane surrounds the developing vitreous; in the central part of the vitreous runs a hyaloid artery that reaches the posterior surface of the lens.

By the end of the fifth week of intrauterine life, the eye primordium (the optic cup and the lens vesicle) is surrounded by mesodermal tissue. A parallel may be drawn to the basic pattern of development of the brain. Brain tissue develops from the embryonic neural tissue; Optic cup and optic stalk develop from the embryonic neural tissue. Mesoderm condenses around the developing brain and forms two basic layers: the endomeninx and the ectomeninx. Endomeninx further forms the pia and the arachnoid. Ectomeninx forms the dura. Similar pattern is seen in the development of the layers of the eyeball. The mesoderm close to the eye primordium is the endomeninx, though there is no separation of pia and arachnoid here. The endomeninx forms the vascular choroid coat around the retina. As the endomeninx is closely applied to the neural tissue, it initially covers (or lines) the internal invagination of the cup and forms the inner vascular layer. Various changes occur in the inner vasculature and the central artery and vein are seen entering into the optic stalk through the choroid fissure. During later development, inner vasculature undergoes necessary regressive changes. The vascular choroid outer to the retina becomes the definitive choroid. Close to the rim of the cup, the choroid forms the ciliary body. The part of the endomeninx in front of the lens vesicle splits into two (as the endomenix splits into the pia

 **Development** *contd...*

and the arachnoid in the brain); the pia portion (close to neural tissue) forms the iridopupillary membrane in front of the lens.

Mesoderm outer to the endomeninx condenses to form the tough and firm ectomeninx (dura). It develops into the sclera in the posterior part; in the anterior part, it contributes to the substantia propria of the cornea. However, additional sources contribute to the cornea. The arachnoid portion of the split endomeninx adheres to the posterior aspect of the substantia propria. The surface ectoderm adheres to the anterior surface of the substantia propria. Thus the cornea is formed by (from before backwards) epithelium from surface ectoderm, substantia propria or stroma of ectomeninx and epithelial layer from the endomeninx. The anterior chamber is formed between the iridopupillary membrane and the cornea (in a way the subarachnoid space of the eye; similar to the subarachnoid space of the brain that contains cerebrospinal fluid, the anterior chamber contains fluid). The ciliary body is similar to the choroid plexus and secretes the aqueous humour.

The iridopupillary membrane disappears completely and the posterior chamber is clearly established.

By the end of sixth week, the ectoderm overlying the eye primordium shows two transverse ridges and soon develops the eyelid primordia. The eye primordium sinks more into the underlying mesoderm; the ridges demarking the eyelids deepen further and the upper and the lower eyelids overlap and cover the eye primordium. Each eyelid is an ectodermal fold with a central core of mesoderm. They grow rapidly towards each other, come in contact and unite by epithelial fusion. This fusion cuts off a space from the exterior. This space becomes the conjunctival sac. Differential changes in the ectoderm give rise to a serous conjunctiva. The eyelids remain fused until the seventh month of intrauterine life. The tarsal plates and the lid muscles develop in the enclosed mesoderm.

 **Histology**

- **Cells of retina:** Though histology of retina is primarily discussed in terms of its layers, it is essential to know details about its cellular components, for even the layers are described and designated by the types of cells found in them.
- **Pigment cells:** These are cuboidal cells with abundant melanin granules in the cytoplasm. They form a continuous single layer from the optic disc to the ora serrata and continue into the pigment epithelium of the ciliary body. Being pentagonal or hexagonal on a surface view, they number about 5 to 6 million in the human retina. On the apical aspect (that aspect towards the rods and cones), they bear numerous microvilli which project between the rods and cones. The tips of the outer segments of rods and cones project between the microvilli but there are no cell junctions connecting them. This is the line of separation in retinal detachment. The pigment cells have a multifold function—(a) they phagocytose and degrade the shed parts of the rods and cones thereby contributing to the

contd...

contd...

Histology *contd...*

turnover of the photoreceptive components; the end products of this degradation are the lipofuscin granules found in the pigment cells; (b) the epithelium is an anti-reflection device and cuts light from bouncing back onto the photoreceptors; (c) the tight junctions between the pigment cells make the pigment layer act as a blood-retina barrier; the ionic environment of the retina is thus protected and entry of leukocytes into retina that may cause immunological problems is prevented.

❑ *Rods and cones:* These are the photoreceptor cells of the retina. They are long cells and the photoreceptive portion is close to the pigment epithelium. The other end of the cell is the synaptic portion. From the outer (choroidal/pigment epithelial) end inwards, the parts of each of these cells are outer segment, inner segment, soma with nucleus and either the rod spherule or the cone pedicle. The outer and inner segments together constitute the rod or the cone process. The cone process is wider but has a cone shape; the rod process has a cylindrical rod shape; hence, their names. The outer segments are embedded in the pigment epithelium cone cells (total number around 65 million) are responsible for spatial resolution and colour vision (photopic vision). They are of three types—red, green and blue-depending upon their colour sensitivities. Though they are found throughout the retina, they are densely concentrated in the fovea, leading to the visual acuity of the area. Rods (total number around 120 million) are responsible for monochromatic sensitivity in low illumination too (scotopic vision). They are not found

Histology *contd...*

in the fovea. The photosensitivity of the rods and cones is due to the photoreceptive proteins. The rods have the protein rhodopsin (visual purple); cones have different proteins with different properties. The photoreceptive proteins are found in the flattened membranous discs called the lamellae in the outer segments of the cells. These lamellae are infoldings of the plasma membrane. New lamellae are formed at the proximal end of the outer segment and older lamellae are budded off free and shed in the distal end (the end embedded in the pigment epithelium). The shed off pieces are phagocytosed in the pigment epithelium. This kind of turnover is rapid in the rods but much less in the cones.

❑ *Amacrine cells:* These cells do not have axons and their dendrites act as axons too. They make synaptic contacts with axons of bipolar cells, dendrites of ganglion cells and dendrites of other amacrine cells. They modulate photoreceptive signals and modify colour and luminosity inputs depending on the light conditions. They are also responsible for directional movement detection.

❑ *Ganglion cells:* These are the final pathway multipolar neurons of the retina. They synapse with the bipolar and amacrine neurons through their dendrites. Their axons form the fibres of the optic nerve. The fibres first form a layer of nerve fibres on the inner surface of retina. They then turn to the optic disc through which they leave the eyeball as optic nerve.

❑ *Retinal glial cells:* These are of three types, namely the radial cells of Muller, the astrocyes and the microglial cells.

contd...

Multiple Choice Questions

1. The shape of the lens of the eyeball is altered by the action of:
   a. The pupillary muscles
   b. The ciliaris and pupillary muscles
   c. The superior oblique and ciliaris muscles
   d. The ciliaris muscle
2. Which of these muscles acts only in one axis?
   a. Superior rectus
   b. Superior oblique
   c. Medial rectus
   d. Inferior rectus
3. The frontal nerve:
   a. Is the smallest of the branches of the ophthalmic nerve
   b. Enters the orbit within the common tendinous ring
   c. Divides into supraorbital and supratrochlear nerves
   d. Has both motor and sensory fibres
4. The angle of filtration is:
   a. Iridocorneal angle
   b. Sclerocorneal junction
   c. Angle of ora serrata
   d. Hyaloid fossa
5. The central processes of the rods and cones synapse with:
   a. The central processes of bipolar cells
   b. The central processes of ganglion cells
   c. The peripheral processes of bipolar cells
   d. The peripheral processes of ganglion cells

ANSWERS

1. d   **2.** c   **3.** c   **4.** a   **5.** c

Clinical Problem-solving

Case Study 1: One of your friends wants you to explain about cataract since his grandfather has had cataract surgery.
- How would you explain 'cataract'?
- What is the principle behind cataract surgery?
- How is vision restored in a person who has undergone cataract surgery?

Case Study 2: A 62-year-old man has suffered retinal detachment. Your professor insists that you take a detailed history from the patient.
- What is retinal detachment?
- Correlate development of retina with the condition.
- In what way is the history important?

(For solutions see Appendix).

Chapter 12

Ear and the Organs of Hearing and Balance

Frequently Asked Questions

- ❏ Discuss the middle ear cavity with reference to its walls, ossicles and muscles.
- ❏ Write notes on: (a) Ear ossicles, (b) Mastoid antrum, (c) Bony cochlea, (d) Tensor tympani muscle.
- ❏ Write briefly on: (a) External acoustic meatus, (b) Auditory tube, (c) Vestibule, (d) Stapes.

EAR

Anatomically speaking, the ear is made up of three main parts (Fig. 12.1):

- ❏ *External ear,*
- ❏ *Middle ear* and
- ❏ *Internal ear.*

The external ear and the middle ear are concerned exclusively with hearing. The internal ear has a *cochlear part* concerned with hearing and a *vestibular part* which provides information to the brain regarding the position and movements of the head.

The part of the ear that is seen on the surface of the body (i.e., the part that the lay person calls the ear) is anatomically speaking the *auricle* or *pinna*. Leading inwards from the auricle, is a tube called the *external acoustic meatus*. The auricle and the external acoustic meatus together form the *external ear*. The inner end of the external acoustic meatus is closed by a thin membranous diaphragm called the *tympanic membrane* or *tympanum* (Greek.tympanon=drum; from which we get the adjective 'tympanic' applied to structures connected with the ear drum or middle ear). This membrane separates the external acoustic meatus from the middle ear.

The *middle ear* is a small space placed deep within the petrous part of the temporal bone. Medially, the middle ear is closely related to parts of the internal ear. The cavity of the middle ear is continuous with that of the nasopharynx through a passage called the *auditory tube*. Within the cavity of the middle ear, are three small bones which are collectively called the *ossicles* of the ear (Fig. 12.1). The ossicles are:

- ❏ *Malleus,*
- ❏ *Incus* and
- ❏ *Stapes*.

The three ossicles form a chain that is attached on one side to the tympanic membrane and at the other to a part of the internal ear.

The *internal ear* is in the form of a cavity within the petrous temporal bone having a very complex shape. This bony cavity (or *bony labyrinth*) has a central part called the *vestibule*. Continuous with the front of the vestibule, there is a spiral-shaped cavity, the bony *cochlea*. Posteriorly, the vestibule is continuous with three *semicircular canals*.

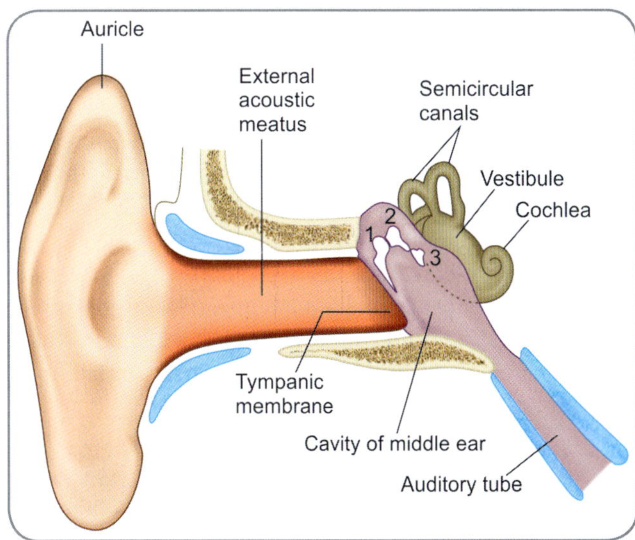

Fig. 12.1: Scheme to show the parts of the ear **1.** Malleus **2.** Incus **3.** Stapes

Sound waves travelling through air reach the ears. Waves striking the tympanic membrane produce vibrations in it. These vibrations are transmitted through the chain of ossicles present in the middle ear to reach the internal ear. Specialised end organs in the cochlea act as transducers which convert the mechanical vibration into nervous impulses. These impulses travel through the cochlear part of the vestibulocochlear nerve to reach the brain. Actual perception of sound takes place in the auditory (or acoustic) areas in the cerebral cortex.

AURICLE

Other names: Pinna (Latin.pinna=feather), ala auris.
The auricle (Latin.auris=ear), which collects sound waves, is made up of a skeleton of elastic cartilage and fibrous tissue, which is covered on both sides by a layer of thin skin. The cartilage of the auricle is continuous with that of the external acoustic meatus. The auricle has an external surface facing laterally and an inner or cranial surface that lies against the side of the head. The skin over the external surface is continuous with the skin lining the external acoustic meatus. Part of it passes forward to become continuous with the skin over the parotid gland. The skin over the cranial surface is continuous with the skin covering the head behind the auricle [These facts help in understanding why the nerves and blood vessels that supply the cranial surface of the auricle reach it (mainly) from the back, and those supplying the external surface reach it from the front].

The cartilage of the auricle is curved on itself in a complicated manner so that a number of elevations and depressions are produced. Elevations on the external surface correspond to depressions on the cranial surface and vice versa. These elevations and depressions are given names which are shown in figure 12.2. The lowest part of the auricle is soft. It does not contain cartilage and is composed only of a fold of skin with enclosed connective tissue. This part is called the **lobule**. At the centre of the external surface of the auricle, is a large depression called the **concha** (Latin.conchum=shell; named after the shape). Its anterior part is continued into the external acoustic meatus. On the cranial surface, the position of the concha is marked by an elevation called the **eminentia conchae**. The concha is partially divided by the anteroinferior tip of the outer rim. The upper subdivision is the cymba concha and indicates on the surface the suprameatal triangle of the temporal bone.

Auricular Muscles

Extrinsic auricular muscles pass from the skull to the auricle. These are the auricularis anterior, auricularis superior and auricularis posterior. The smallest is the auricularis anterior and the largest is the auricularis

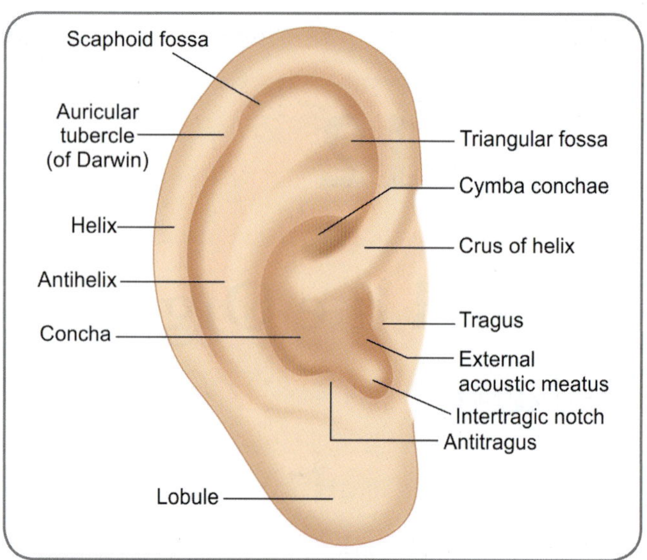

Fig. 12.2: Scheme to show elevations on the external surface corresponding to depressions on the cranial surface of the auricle and vice versa

superior. They tend to draw the auricle towards their respective directions. Intrinsic muscles are confined to the auricle and connect different parts of the same. These are the helicis major, helicis minor, tragicus, antitragicus, transversus auriculae and obliquus auriculae. They tend to modify the shape of the auricle. In man, all the auricular muscles are to be regarded as vestigial structures. They have no important role to play in moving the auricle or modifying the shape of it. All of them receive nerve supply through branches of the facial nerve.

Blood Vessels, Lymphatics and Nerves of the Auricle

The auricle is supplied by the posterior auricular branch of the external carotid artery, the anterior auricular branch

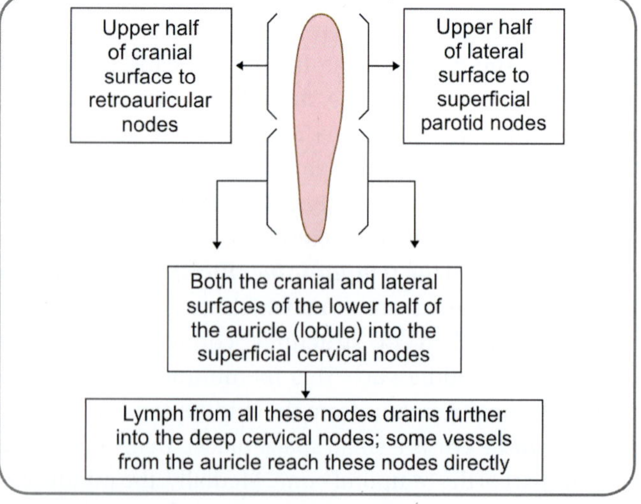

Fig. 12.3: Scheme to show the lymphatic drainage of the auricle

of the superficial temporal artery and branches of the occipital artery.

The veins accompany the arteries.

The lymphatics drain into the parotid group of nodes (especially to the node in front of the tragus), upper deep cervical group of nodes and mastoid lymph nodes (Fig. 12.3).

The sensory nerves supplying the auricle are the auriculotemporal branch of the mandibular nerve, the great auricular nerve and the auricular branch of the vagus nerve.

EXTERNAL ACOUSTIC MEATUS

Other names: External auditory meatus, Antrum auris.

The external acoustic meatus is a tube passing medially from the bottom of the concha of the auricle. It is closed medially by the tympanic membrane. The total length of the tube is approximately 24 mm.

Of this, the walls of the outer 8 mm are cartilaginous, while those of the inner 16 mm are bony. The cartilage or bone forming the walls of the meatus is covered by a layer of thin skin that is continuous with that over the concha. Laterally (that is, externally) the cartilage of the meatal walls is continuous with the cartilage of the auricle. It does not form a complete tube, but is deficient in its posterosuperior part. Medially (that is, internally) it is firmly attached to the rough edge of the bony part of the tube.

The wall of the bony part of the meatus is formed mainly by the tympanic plate of the temporal bone. The posterosuperior region of the wall is formed by the squamous part of the temporal bone. The medial end of the meatus is closed by the tympanic membrane. Here, the bony wall is marked by a groove called the ***tympanic sulcus***, into which is attached the perimeter of the tympanic membrane. The membrane is placed obliquely both in the anteroposterior and vertical planes. As a result, the floor and anterior wall are longer than the roof and posterior wall.

The external acoustic meatus is not straight, but follows an S-shaped course. This is so because the cartilaginous part is not in line with the bony part and is also bent on itself. The cartilaginous part first passes medially, forward and upward. It then passes medially, backward and upward. The bony part runs medially, forward and downward.

In clinical examination of the meatus and through it of the tympanic membrane, the auricle is pulled upward, backward and somewhat laterally. This renders the meatus more or less straight. It then has a uniform medial, forward and downward direction.

The meatus shows a narrowing at the junction of the cartilaginous and bony parts. It shows another narrowing called the ***isthmus*** about 4 mm from the tympanic membrane (i.e. 20 mm from the floor of the concha). The floor of the meatus shows a depression immediately lateral

to the tympanic membrane. Foreign bodies entering the meatus can get stuck in this depression. The skin lining the external acoustic meatus contains numerous ***ceruminous glands***. These are modified sweat glands which produce the wax of the ear or ***cerumen*** (Latin.cera=wax).

Development

The external acoustic meatus develops from the first pharyngeal cleft. The depressed cleft is closed by a meatal plug, which is formed by a proliferation of epithelial cells from the bottom of the cleft. The plug that forms in the third month of intrauterine life dissolves in the seventh month and the definitive external acoustic meatus is established.

The first pharyngeal cleft and the first pharyngeal pouch are separated by a thin strip of mesodermal tissue. Three layers are present at this area and develop into the three layers of the tympanic membrane. The ectodermal epithelial layer at the bottom of the cleft, the endodermal epithelial layer of the pouch and the intervening mesoderm are the three layers and they give rise to the outer cuticular layer, inner mucosal layer and the intermediate fibrous layer of the tympanic membrane respectively.

The first pharyngeal cleft is bounded by the first pharyngeal arch above and the second arch below. By early sixth week, mesodermal proliferations are seen as swellings around the cleft. A total of six swellings appear—three on the first arch and three on the second arch. These swellings are the auricular hillocks. Starting from the seventh week, the hillocks fuse with each other in a complicated and differential manner to form the auricle of the external ear. The ear opening is transversely disposed in the early embryo when the mandible is rudimentary. With growth of mandible, the ear opening and the auricle are repositioned to the adult disposition. Anomalies in the fusion of the auricular hillocks is common resulting in developmental abnormalities of the auricle. Failure of dissolution of the meatal plug will cause congenital deafness.

Relations of External Acoustic Meatus

The external acoustic meatus is related in front to the temporomandibular joint and behind to the mastoid process in the substance of which are the mastoid air cells. Inferiorly, it is related to the parotid gland and superiorly (in its deeper part) to the middle cranial fossa.

Blood Vessels, Lymphatics and Nerves of External Acoustic Meatus

The external acoustic meatus is supplied by the posterior auricular branch of the external carotid artery, the auricular branches of the superficial temporal artery and the deep auricular branch of the maxillary artery.

The veins of the meatus drain into the external jugular vein, the maxillary vein and the veins of the pterygoid plexus.

The lymphatics follow those of the auricle and drain into the parotid and upper deep cervical group of nodes (Fig. 12.4).

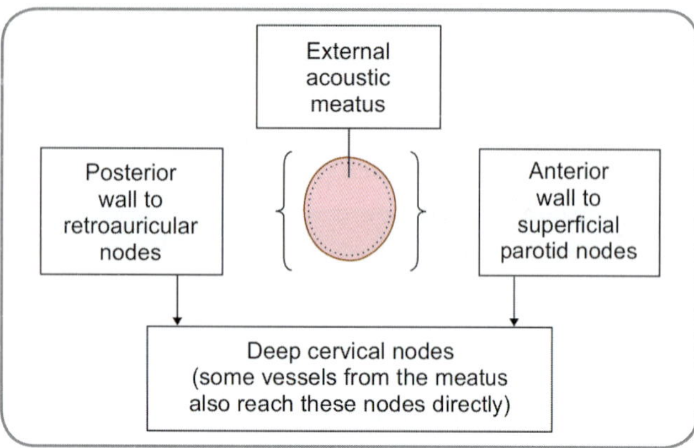

Fig. 12.4: Scheme to show the lymphatic drainage of the external acoustic meatus

The sensory nerve supply to the anterior wall and the roof of the meatus is derived from the auriculotemporal nerve; that to the posterior wall and floor is derived from the auricular branch of the vagus nerve.

Added Information

- The lobule of the pinna is non-cartilaginous and consists of fibrous tissue, fat and blood vessels.
- The tragus is a projection that overlaps the opening of the external acoustic meatus; it projects backwards from the cheek side of the face. Its name is intriguing (Greek. tragos=goat). This name is given because hairs tend to grow from this formation resembling the goat's beard.

Clinical Correlation

- ***Boxer's ear:*** This is a condition that results from trauma, collection of blood between the auricular cartilages, induration and distortion of the cartilages. It is also called the ***cauliflower ear***. A haematoma may form as a result of trauma and in due course may compromise the blood supply to cartilages. The resultant fibrosis causes extensive distortion. It is usually seen in boxers and wrestlers.
- ***Blainville's ears:*** This is a condition where there is asymmetry in the shapes and sizes of the pinnae of both sides.
- ***Aztec's or Cagot's ears:*** This is a condition where the lobule is absent.
- ***Scroll ears:*** This is a condition where the pinnae are rolled forwards.
- ***Mozart's ears:*** This is a condition where the upper part of the pinna is distorted causing a bulging appearance. Composer Mozart is supposed to have had this kind of a pinna.
- ***Otoscopic examination:*** The insertion of an otoscope into the external acoustic meatus has to be done carefully. The first step is to straighten the meatus. In adults, this is done by grasping the pinna and pulling it posterosuperiorly (upward, outward and backward). The meatus is very short in the infants. Insertion of an otoscope or even plain examination may cause damage to the tympanic membrane. Extra care should be exercised. The meatus is straightened in infants by pulling the pinna inferoposteriorly.

contd...

Clinical Correlation *contd...*

- ***Otitis externa:*** This is a condition where there is infection and inflammation of the skin of the external meatus. Pulling the pinna and pressure on tragus cause pain. Swimmers, in whom the external meatuses can easily be dampened or clogged with water, suffer more from this condition.

MIDDLE EAR

Other names: Tympanic cavity, Cavum tympani, Cavity of middle ear, Tympanum.

The middle ear is an irregular space in the petrous temporal bone. It is separated from the external acoustic meatus by the tympanic membrane. Normally, it is lined with mucous membrane and filled with air. The air reaches the middle ear from the nasopharynx through the auditory tube. It also contains the ear ossicles which form a chain connecting the lateral and medial walls of the cavity.

The middle ear is bounded laterally by the tympanic membrane and medially by the lateral wall of the inner ear (Fig. 12.5). It has a roof, a floor and medial, lateral, anterior and posterior walls. Vertically and anteroposteriorly, it is almost of the same dimension, measuring about 15 mm both ways. However, mediolaterally (side to side) it is compressed; the transverse diameter is 6 mm near the roof, 2 mm in the middle and 4 mm near the floor (Fig. 12.15). The middle ear communicates anteriorly with the cavity of the nasopharynx through the ***auditory tube***

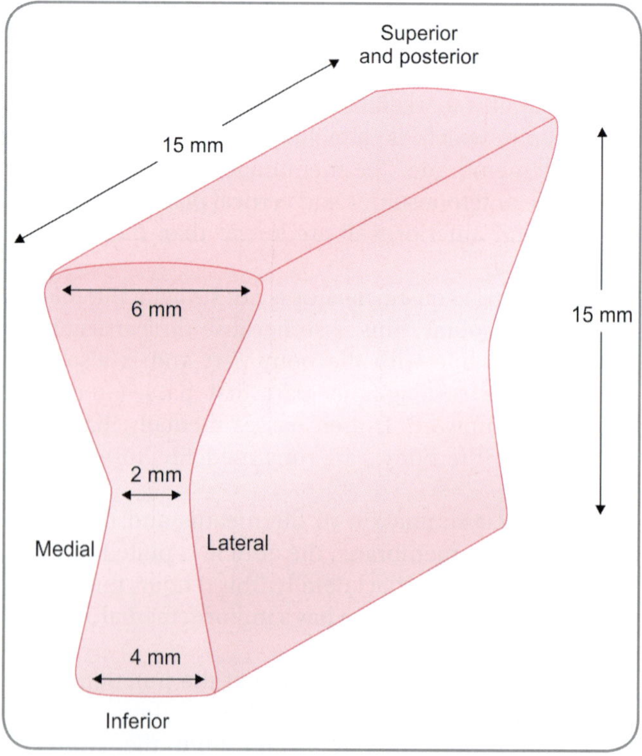

Fig. 12.5: Dimensions of the tympanic cavity

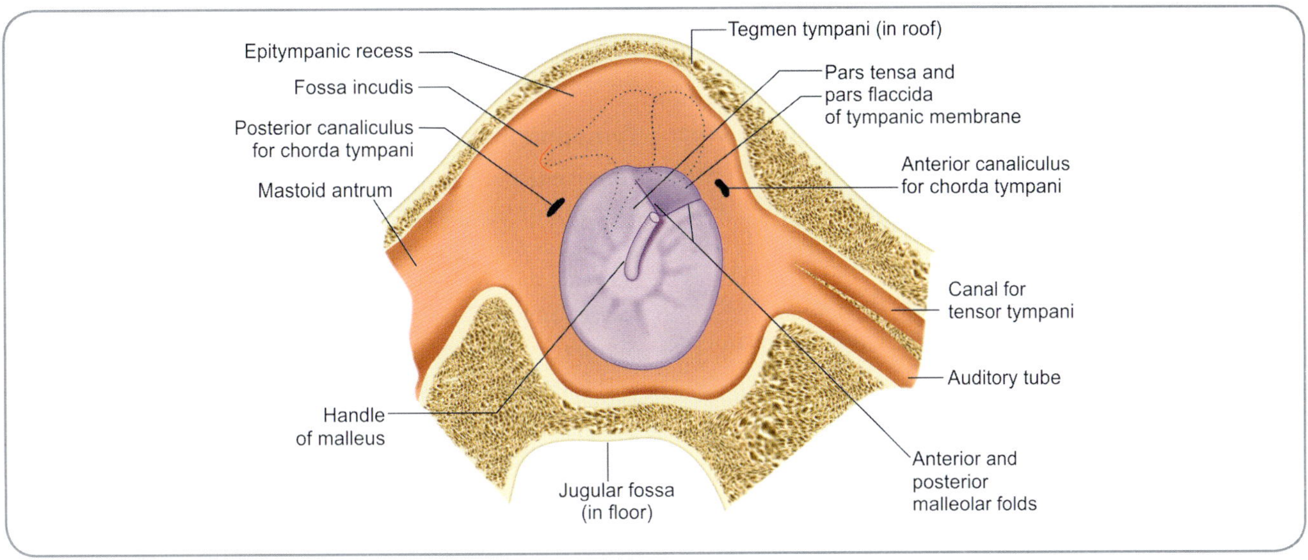

Fig. 12.6: Lateral wall of tympanic cavity. Parts of roof, floor, anterior and posterior walls (adjoining the lateral wall) are also seen; the positions of the upper part of the malleus and of the incus are shown in dotted line

and posteriorly with a large space in the petrous part of the temporal bone called the ***mastoid antrum*** and with smaller spaces within the mastoid process called the ***mastoid air cells***. These spaces, the tympanic cavity itself and the auditory tube are all lined by mucous membrane. Because of their communication with the nasopharynx, these spaces are filled with air.

It can be seen that a part of the tympanic cavity (Fig. 12.6) lies opposite the tympanic membrane and a part lies above the level of the tympanic membrane; the part opposite is called the ***tympanic cavity proper*** and the part above is called the ***epitympanic recess***.

Ossicles of Middle Ear

Other names: Ossicula auditus, Otostea, Ear bones.
The three ear ossicles are present in the middle ear cavity (Fig. 12.7) and connect the tympanic membrane to the medial wall of the middle ear. They transfer sound waves which strike the tympanic membrane to the fenestra vestibuli. They form articulations with each other and form an osseous chain.

Malleus

The malleus is so called because it resembles a hammer (Latin.malleus=hammer). It is the largest of the ear ossicles and has an upper rounded part called the ***head*** to which is attached a relatively long ***handle*** (or ***manubrium***). At the junction of the head with the handle, there is a slight constriction called the ***neck***. Just below the neck, the bone gives off two ***processes: anterior and lateral.*** The head is the upper end of the bone and lies in the epitympanic recess. It articulates posteriorly with the incus. The anterior process is connected to the petrotympanic fissure by slender

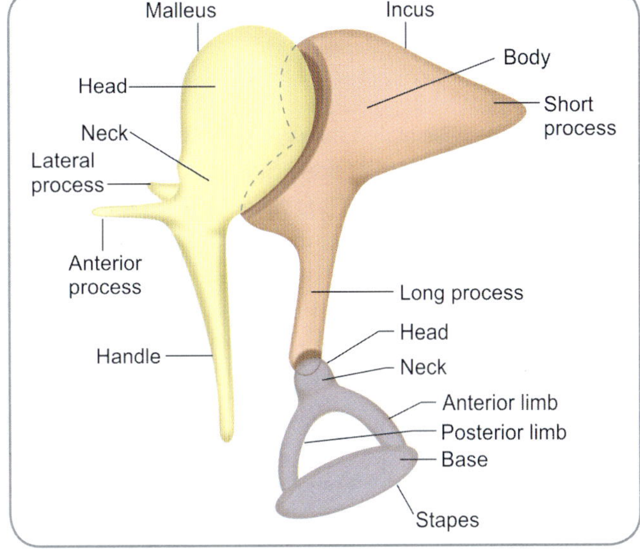

Fig. 12.7: Ossicles of the ear as seen from the medial side

fibres. The lateral process is attached to the upper part of the tympanic membrane at the apex of pars flaccida.

Incus

The incus (Latin.ingkus=anvil) has a main part or ***body*** and ***two processes: long and short***. The long process is directed downwards parallel to the handle of the malleus. The short process is directed backwards. On the anterior surface of the body is the articular facet for malleus. At the lower end of the long process is a rounded projection called the ***lenticular process*** which articulates with the head of stapes. The short process is attached by fibres to the fossa incudis.

217

Stapes

The stapes is shaped like a stirrup (Latin.stapez=stirrup). It has a **head**, that is connected by two **limbs** or **crura** (anterior and posterior), to an oval plate called the **base**. The constricted part adjoining the head is called the **neck**. The head is directed laterally and has an articular facet for the lenticular process of incus. The tendon of stapedius muscle is attached to the posterior aspect of the neck. The base is otherwise called the **footplate**. The footplate sits on the fenestra vestibuli and is attached along the perimeter by the annular ligament.

Roof of Middle Ear

The roof (otherwise called the **tegmental wall**) of middle ear cavity is formed by a plate of bone called the **tegmen tympani** (Latin.tego=cover; this plate forms the intracranial surface of the petrous temporal bone that is seen in the floor of the middle cranial fossa). The same plate of bone extends forwards to form the roof of the canal for tensor tympani muscle and backwards to form the roof of the mastoid antrum.

Floor of Middle Ear

The floor (otherwise called the **jugular wall**) of middle ear is formed by a thin plate of bone that separates it from the superior bulb of the internal jugular vein. This plate of bone may sometimes be deficient, in which case only mucous membrane and connective tissue intervene between the middle ear and the vein.

Lateral Wall of Middle Ear

The lateral wall of the middle ear is otherwise called the **membranous wall**. The greater part of this wall is formed by the tympanic membrane (Fig. 12.6). The part of the cavity lying above the level of the tympanic membrane is the epitympanic recess and the lateral wall of the epitympanic recess is formed by part of the temporal bone. Three apertures of importance are seen in the upper part of this wall. These apertures, from behind forwards, are the posterior canaliculus, petrotympanic fissure and the anterior canaliculus. At the upper part of the junction of the posterior and lateral walls, the posterior canaliculus for chorda tympani opens into the middle ear cavity. This opening transmits the chorda tympani branch of the facial nerve and the tympanic branch of the stylomastoid artery. Just above the tympanic membrane on the lateral wall is the petrotympanic fissure. Just about 2 mm in length, this fissure contains the anterior process of the malleus bone and the anterior ligament of malleus. It transmits the anterior tympanic branch of the maxillary artery. At the anteromedial end of the petrotympanic fissure is another tiny opening. It leads to the anterior canaliculus of chorda tympani and transmits the chorda tympani nerve out of the middle ear.

Tympanic Membrane

Other names: Ear drum, Drum membrane, Membrane of the tympanum, Drum head.

This is an oval membrane that separates the external acoustic meatus from the middle ear. It is about 9 mm in its longest diameter and about 8 mm in its shortest diameter (Fig. 12.8). The long diameter passes downwards and forwards. The membrane is placed obliquely both in the vertical and the anteroposterior planes. In the vertical plane its upper end is distinctly lateral to its lower end and the membrane forms an acute angle of about 55° with the floor of the external acoustic meatus (Fig. 12.9).

With the exception of a small area in its anterosuperior part, the circumference of the tympanic membrane is thickened because of the presence of fibrocartilage. This ring of fibrocartilage fits into a groove, the **tympanic sulcus**, present at the medial end of the external acoustic meatus. This tympanic sulcus is deficient superiorly and so a notch is seen.

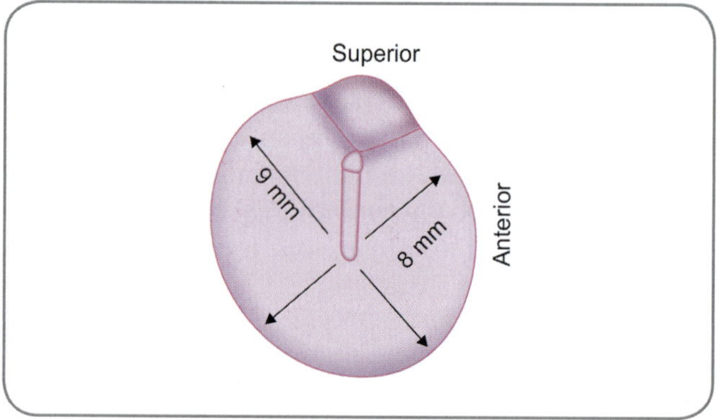

Fig. 12.8: Dimensions of tympanic membrane

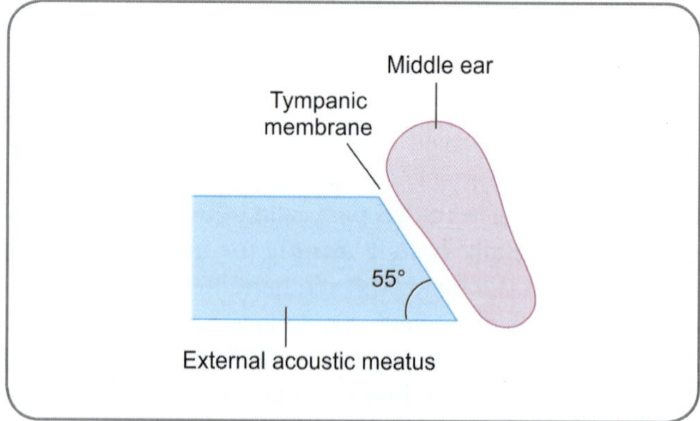

Fig. 12.9: Angle between tympanic membrane and floor of external acoustic meatus

Histology

Histology of tympanic membrane: Histologically the tympanic membrane has three layers—the outer cuticular layer, the intermediate fibrous layer and the inner mucous layer. The cuticular layer is continuous with the skin of the external acoustic meatus. It has keratinized squamous epithelium but has no hairs. The fibrous layer has an external coat of radial fibres and an inner coat of circular fibres. The radial fibres radiate from the handle of the malleus and the circular fibres are dense peripherally. In the pars flaccida the fibrous layer is replaced by loose connective tissue. The mucous layer is continuous with the mucosa of the tympanic cavity and is thick in the upper portion. It consists of flat squamous cells with several cell junctions.

The handle of malleus is closely attached to the medial side of the tympanic membrane and its lower end lies approximately at the centre of the membrane. From here the handle passes upwards and forwards ending a little short of the circumference of the membrane. The lateral process of the malleus projects towards the membrane at this point. The anterior and posterior malleolar folds pass from the anterior and posterior ends of the above mentioned notch of the tympanic sulcus to the lateral process of malleus.

The small triangular area of the tympanic membrane lying between the two malleolar folds is not stretched like the rest of membrane, is lax and thin. It is, therefore, called the ***pars flaccida***. In contrast, the rest of the membrane is called the ***pars tensa***.

On the whole, the tympanic membrane is convex medially. The point of greatest convexity corresponds to the lower end of the handle of the malleus and is called the ***umbo*** (Latin.umbonis=knob). However, in passing from the umbo to the tympanic sulcus the layers of the membrane show a slight convexity outwards.

Structurally, the tympanic membrane has three layers. The outer layer is continuous with the skin lining the external acoustic meatus. The inner layer is formed by the mucous membrane of the tympanic cavity. Between these two there is a layer of fibrous tissue. Some of the fibres in this tissue are arranged radially and some circularly. Though the tympanic membrane, as a whole is convex internally, its radiating fibres are concave internally.

Tympanic Membrane as seen through the External Acoustic Meatus

Many of the features of the tympanic membrane described in the preceding paragraphs can be seen in the living through the external acoustic meatus. The pars flaccida can be seen in the anterosuperior corner of the membrane and the anterior and posterior malleolar folds can also be distinguished. The lateral process of the malleus can

be seen as a white dot where these folds meet. Running downwards and backwards from this dot to the centre of the membrane is the handle of the malleus. A little behind and parallel to the upper part of the handle of the malleus the long process of the incus may be visible as a faint white streak. The anteroinferior part of the membrane (between the lower end of the handle of the malleus and the circumference of the membrane) reflects light more than the rest of the membrane and is referred to as the ***cone of light***.

Blood Vessels and Nerves of Tympanic Membrane

The blood vessels and nerves to the external surface of the tympanic membrane are derived from those supplying the external acoustic meatus. Those to its internal surface are derived from vessels and nerves which supply the middle ear.

The ***external surface*** is supplied by the deep auricular branch of the maxillary artery and drains into the external jugular vein. It is supplied by the auriculotemporal nerve and the auricular branch of the vagus.

The ***internal surface*** is supplied by the tympanic branches of the maxillary artery and the stylomastoid branch which arises either from the posterior auricular artery or from the occipital artery. The nerves to the internal surface are derived from the tympanic branch of the glossopharyngeal nerve.

Relationship of Chorda Tympani to the Tympanic Membrane

The chorda tympani nerve has an intimate relationship to the tympanic membrane.

❑ Just posterior to the upper part of the tympanic membrane is the tiny opening of the ***posterior canaliculus*** for the chorda tympani.

❑ Just above the upper part of the tympanic membrane is a transverse slit of the ***petrotympanic fissure***. At the medial end of this fissure, is the opening of the ***anterior canaliculus*** for the chorda tympani.

❑ The anterior ligament of the malleus passes into the petrotympanic fissure.

❑ The handle of the malleus is actually embedded within the tympanic membrane and lies between its fibrous and mucosal layers.

The chorda tympani, a branch of the facial nerve, arises from the parent nerve within the substance of the temporal bone, 6 mm above the stylomastoid foramen; it ascends upwards and forwards in the posterior canaliculus which itself is anterior to the facial canal. The nerve enters the middle ear through the opening of the posterior canaliculus. It then passes forwards through the substance of the upper part of the tympanic membrane, lying between the fibrous and mucosal layers. As it does so it crosses medial to the handle of the malleus, near the upper end of the latter. The chorda tympani leaves the middle ear by passing through the anterior canaliculus to reach the infratemporal fossa.

> ### Clinical Correlation
>
> ***Normal tympanic membrane:*** The bright cone of light that radiates anteroinferiorly (as a reflection of the otoscope illuminator) is seen only in a healthy ear.
>
> It is sometimes necessary to incise the tympanic membrane to let out pus from the middle ear. Such an incision is always made in the lower part to avoid damage to the chorda tympani. Another advantage of such an incision is that the lower part of the membrane is less vascular.
>
> Perforation of tympanic membrane may occur due to multiple causes. Trauma, foreign bodies in the external meatus, excess pressure (such as in conditions like scuba diving) and middle ear infections are the usual causes. Minor ruptures heal spontaneously. Larger tears require surgical intervention.

Anterior Wall of Middle Ear

The medial and lateral walls of the middle ear are fairly close to each other. It follows that the anterior and posterior walls are narrow. In the upper part of the anterior wall (otherwise called the ***carotid wall***) are two openings. The upper opening leads into a canal in which the tensor tympani muscle lies. The lower opening is that of the auditory tube. The two openings are separated by a thin bony septum. This septum extends onto the medial wall of the middle ear where it curves on itself to form a pulley (***processus trochleariformis***). The tendon of the tensor tympani winds around this pulley (to change from a backward direction to a lateral one). The part of the anterior wall lying below the opening of the auditory tube is formed by a plate of bone that separates the middle ear from the carotid canal (through which the internal carotid artery passes). Minute apertures in this plate give passage to branches from the internal carotid artery to the middle ear and also to ***caroticotympanic nerves*** which arise from the sympathetic plexus surrounding the artery to reach the middle ear.

Posterior Wall of Middle Ear

The posterior wall (otherwise called the ***mastoid wall***) is wider above than below. The upper part (which actually is the upper part of the posterior wall of the epitympanic recess) shows a large round aperture through which the middle ear (the epitympanic recess portion) communicates with the mastoid antrum. This aperture is called the ***aditus to the mastoid antrum***. On the medial wall of the aditus, is a bulging produced by the lateral semicircular canal (which is within the internal ear). Inferolateral to the aditus is the lower part of the posterior wall of the

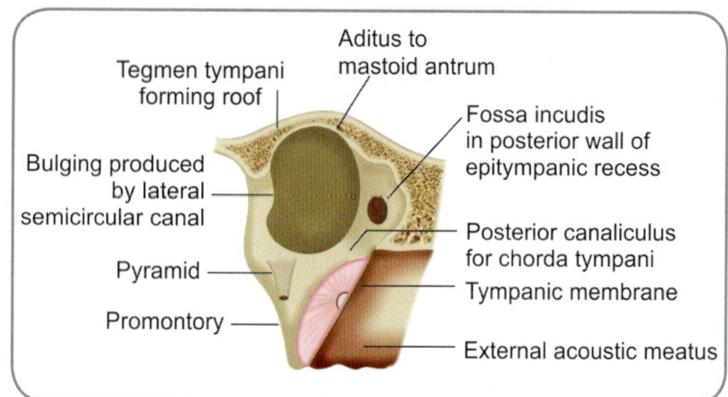

Fig. 12.10: Features on the posterior wall of the middle ear

epitympanic recess. On it is seen a depression called the ***fossa incudis (Fig. 12.10)***. The tip of the short process of the incus extends into this depression and is attached to the fossa by fibres of the posterior ligament of the incus. Inferior to the aditus, the medial end of the posterior wall of the middle ear bears a conical elevation called the ***pyramid***. The tip of the pyramid projects forwards and has an opening that leads into a canal in which the stapedius muscle is lodged. The tendon of the stapedius emerges from the opening at the tip of the pyramid and runs forwards to be inserted into the posterior surface of the neck of the stapes. The facial nerve is closely related to the internal ear and to the medial and posterior walls of the middle ear. Part of it runs vertically downwards in a bony canal placed along the junction of the medial and posterior walls. The canal containing the stapedius muscle is in front of the vertical part of the facial canal; the two canals are connected by a tiny opening through which the stapedial branch of the facial nerve passes to the muscle.

Medial Wall of Middle Ear

The medial wall (otherwise called the ***labyrinthine wall***) of the middle ear is also the lateral wall of the internal ear. It can be properly understood only after examining some features of the internal ear (Fig. 12.11).

The bony internal ear consists of a central part called the ***vestibule*** which is connected anteriorly to the cochlea and posteriorly to three semicircular canals. The cochlea is in the form of a spiral and that its 'turns' are narrow in the centre and become wider as they proceed outwards. The lowest (or basal) turn of the cochlea is large and is continuous posteriorly with the vestibule. The interior or cavity of the cochlea consists of two parts, upper and lower. The upper part, which is called the ***scala vestibuli*** (Latin.scala=stairway; indicates the spiral winding), becomes continuous with the cavity of the vestibule. The vestibule communicates with the middle ear through a large oval aperture called the ***fenestra vestibuli*** (Latin.

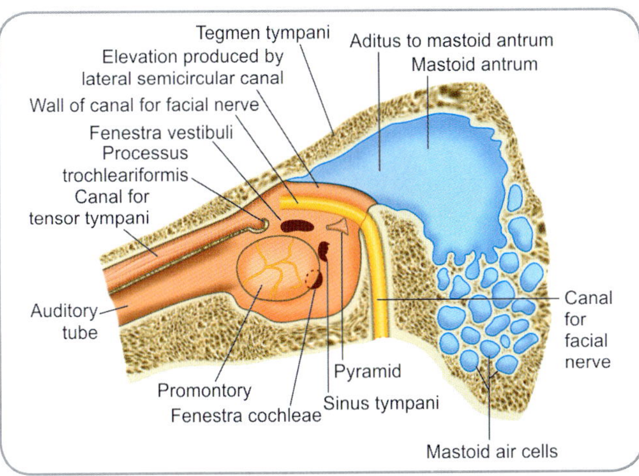

Fig. 12.11: Medial wall of middle ear—parts of the anterior and posterior walls, and of the mastoid antrum and mastoid air cells are also seen

Fenestrum=opening). Thus, the scala vestibuli is in indirect communication with the middle ear. The lower part of the cavity of the cochlea called the ***scala tympani***, opens independently into the middle ear through a round opening called the ***fenestra cochleae***. This fenestra is situated inferior to the vestibule. The three semicircular canals are named anterior, posterior and lateral. The lateral canal bulges in a lateral direction and is, therefore, closest to the middle ear.

With these details, the features on the medial wall of the middle ear can now be studied. Superiorly the medial wall meets the roof, formed by the tegmen tympani; and inferiorly it meets the floor. The most prominent feature to be seen on the medial wall is the ***promontory***. This is a large circular bulging produced by the basal turn of the cochlea. Its surface bears a number of grooves in which lie the nerves of the tympanic plexus. The apex of the cochlea comes in contact with the medial wall of the middle ear just in front of the promontory.

Posterosuperior to the promontory, is the fenestra vestibuli (or ***fenestra ovalis*** or ***oval window***). It is actually a kidney-shaped opening with its long diameter being horizontal and convex border directed upwards. The base of the stapes fits into this opening and is attached to its margins by the annular ligament.

Posteroinferior to the promontory, is a round aperture called the ***fenestra cochleae*** (or ***fenestra rotunda*** or ***round window***). This aperture is only partially seen (or not seen at all) because it is overlapped by the lower edge of the promontory which overhangs it. It is continuous with the scala tympani part of the cochlear cavity and closed by the ***secondary tympanic membrane***.

Posterior to the promontory, is a depression called the ***sinus tympani***, which lies between the fenestra vestibuli and the fenestra cochleae. A part of the posterior

semicircular canal (ampulla) lies deep to the sinus tympani. The part of the medial wall above the promontory and the fenestra vestibuli, is marked by two rounded ridges that run anteroposteriorly. The upper of these is produced by the lateral semicircular canal (and this ridge continues on the medial wall of the aditus to mastoid). The lower ridge is the wall of a canal through which the facial nerve runs backwards. Only a thin layer of bone separates the Facial nerve from the cavity of the middle ear. Occasionally, this bone may be missing and the nerve may lie just under the mucosa. Running backwards across the medial wall of the middle ear, the facial canal reaches the aditus to the mastoid antrum. Here it bends downwards and runs through bone just behind the angle between the medial and posterior walls of the middle ear to reach the stylomastoid foramen.

Muscles of the Middle Ear (Fig. 12.12)

Two muscles are present in the middle ear and connect to the ossicles. These are the tensor tympani and the stapedius.

The ***tensor tympani*** lies in a canal that opens into the anterior wall of the middle ear.

At its other end, this canal opens on the base of the skull. Muscle fibres arise from the wall of this canal. Some fibres arise from the cartilaginous part of the auditory tube and some from the base of the skull formed here by the greater wing of the sphenoid. The muscle ends in a tendon that reaches the middle ear cavity near its medial wall. Here, it bends sharply to the lateral side by passing around the processus trochleariformis. It is inserted into the upper end of the handle of the malleus. The muscle is supplied by a branch of the nerve to medial pterygoid which in turn is a branch of the mandibular nerve.

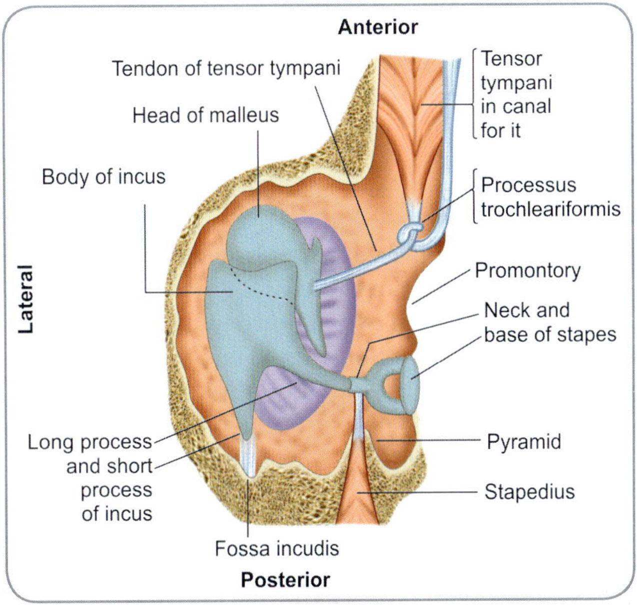

Fig. 12.12: Scheme to show the muscles of the middle ear—they are viewed from above after removing the roof of the middle ear

The ***stapedius*** is a small muscle lying in a bony canal that is related to the posterior wall of the middle ear. Posteriorly and below, this canal is continuous with the vertical part of the canal for the facial nerve. Anteriorly, the canal opens into the middle ear at the apex of the pyramid. The fibres of the stapedius arise from the walls of this canal. They end in a tendon that enters the middle ear through the pyramid and runs forwards to be inserted into the posterior surface of the neck of the stapes. The stapedius muscle is supplied by a branch from the facial nerve.

Both the tensor tympani and the stapedius protect the ear against very loud sounds by restricting the vibrations of the tympanic membrane and the ossicles. Paralysis of the muscles (especially of the stapedius) gives rise to a condition called ***hyperacusis*** in which even normal sounds appear too loud.

Mastoid Antrum

The mastoid antrum (also called the ***tympanic antrum***) is of considerable importance as it is a frequent site of infection, which may be difficult to eradicate. Furthermore, infection may spread from it to neighbouring structures with serious consequences. Although, it is called the ***mastoid antrum***, this space actually lies in the ***petrous*** part of the temporal bone. Anteriorly, the antrum opens, through its aditus, into the epitympanic recess. The medial side of the aditus is related to the lateral semicircular canal.

Superiorly, the roof of the antrum is formed by the tegmen tympani that separates it from the middle cranial fossa and from the temporal lobe of the cerebral hemisphere. Inferiorly, the mastoid antrum is continuous with the mastoid air cells. Anteriorly, below the aditus, the antrum is related to the facial nerve as it descends within its bony canal. Posteriorly, the antrum is close to the posterior surface of the temporal bone (i.e., to the posterior cranial fossa) and here, it may be separated only by a thin plate of bone from the sigmoid sinus. Medially, behind the aditus, the antrum is related to the posterior semicircular canal.

The lateral wall of the mastoid antrum is related to the ***suprameatal triangle***. This triangle is bounded above by the supramastoid crest, anteroinferiorly by the posterosuperior margin of the (bony) external acoustic meatus and posteriorly by a vertical line drawn as a tangent to the posterior margin of the meatus. The thickness of bone separating the mastoid antrum from the surface of the skull is only about 2 mm at birth, but it increases by about 1 mm for every year of age until it is about 13 to 14 mm thick (in other words, it attains its full thickness by puberty) (Fig. 12.13).

Mastoid Air Cells

These are a series of intercommunicating spaces of variable sizes present within the mastoid process. They

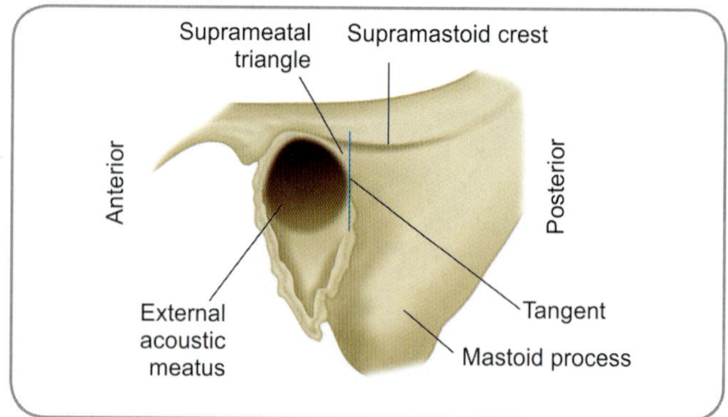

Fig. 12.13: Boundaries of the suprameatal triangle

communicate above with the mastoid antrum. Their number varies considerably. Sometimes they are just a few and confined to the upper part of the mastoid process. In other cases, they may extend throughout the process. Occasionally, they may extend beyond the mastoid process into the squamous or petrous part of the temporal bone. Some of the cells may occasionally lie very close to the sigmoid sinus, in the roof of the external acoustic meatus, in the floor of the tympanic cavity close to the jugular bulb and in the medial part of the petrous temporal bone in relation to the internal ear, the carotid canal, the auditory tube and the abducent nerve. Infection can reach the mastoid air cells though the tympanic cavity and the mastoid antrum and can spread to any of the structures related to them.

Auditory Tube

Other names: Pharyngotympanic tube, eustachian tube. The auditory tube provides a communication between the nasopharynx and the middle ear. Because of this communication, air passes into the tympanic cavity (and into the mastoid antrum and air cells). As a result, air pressure on both sides of the tympanic membrane is the same. This is important for proper vibration of the tympanic membrane. However, the auditory tube is not patent all the time. It opens during deglutition or even during the swallowing of saliva. When we suddenly ascend to a higher altitude (as in going up a hill in a car) the air pressure on the outside of the tympanic membrane (on the external acoustic meatus side) falls, but that on its inner side (middle ear side) remains the same as before. This inequality in pressure gives rise to a change in the quality of sound perceived. However, on swallowing of saliva, and the consequent equalisation of pressure, the sound suddenly returns to normal. The same phenomenon takes place much more acutely during the take off of an aircraft, and can give rise to distress in the ear; more so in persons who have a mild infection (the sweets handed out just before take-off are meant to keep passengers swallowing so that discomfort is avoided or lessened).

Clinical Correlation

The communication between the pharynx and the middle ear is a path along which infection frequently reaches the middle ear. This occurs more commonly in children in whom the auditory tube is shorter and wider than in the adult. In the presence of infection, the tube is easily blocked. When this happens, air within the tympanic cavity is gradually absorbed and pressure on the outside of the tympanic membrane becomes greater than on the inside. The tympanic membrane is retracted inside and its movements are restricted. This can give rise to discomfort that can be relieved by introducing air into the auditory tube through a catheter. If obstruction to the auditory tube is prolonged, pus can accumulate in middle ear, resulting in severe pain. The pus may burst through the tympanic membrane leading to discharge from the ear, and to the formation of a perforation in the membrane. It is for these reasons that the anatomy of the auditory tube is of much practical importance.

Blood Vessels, Lymphatics and Nerves of Middle Ear

The middle ear receives several small branches which arise from arteries of the neighbourhood. Branches are received from anterior tympanic branch of maxillary artery, middle meningeal artery, artery of pterygoid canal, stylomastoid branch of posterior auricular artery (or of occipital artery), ascending pharyngeal artery and directly from internal carotid artery.

The veins of the middle ear drain downward (along the auditory tube) towards the infratemporal fossa where they end in the pterygoid plexus. Some veins drain through apertures in the petrous temporal bone to end in the superior petrosal sinus.

The lymphatics from the middle ear and the mastoid air cells end in the parotid lymph nodes while those from the auditory tube reach the deep cervical nodes (Fig. 12.14).

Development

The dorsolateral end of the first pharyngeal pouch receives contributions from the second and (sometimes) the third pouches and extends laterally. This extension of the pharyngeal pouch is called the ***tubotympanic recess***. The middle ear and the auditory tube are developed from this tubotympanic recess.

The dorsal part of the first pharyngeal arch cartilage lies cranial to the tubotympanic recess and that of the second pharyngeal arch lies caudal. From the dorsal end of the first arch cartilage, a small condensation becomes separated and develops into the incus (first to appear) and the malleus. Similarly, a condensation from the second arch cartilage becomes separated and is pierced by the second arch artery. The artery soon disappears. But the condensation develops two crura and becomes the stapes eventually. The developing stapes soon extends towards the lateral wall of the developing inner ear and engages there.

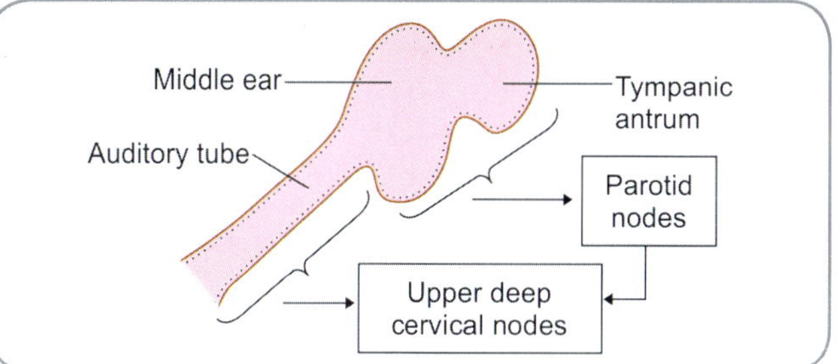

Fig. 12.14: Scheme to show the lymphatic drainage of the middle ear

Development *contd...*

Close to the ossicles thus developed, the tensor tympani and the stapedius muscles develop from the first and second arch mesoderm.

The lateral part of the tubotympanic recess (where all these changes are taking place) expands. This expansion encloses the ossicles, muscles, ligaments and the developing nerve fibres. This expansion becomes the middle ear cavity with the ear ossicles, their muscles, their ligaments and the chorda tympani nerve as its contents. The middle ear cavity expands further posteriorly and develops into the mastoid antrum and mastoid air cells.

The unexpanded medial portion of the tubotympanic recess develops into the pharyngotympanic or auditory tube.

The nerves supplying the mucous membrane of the middle ear, the mastoid antrum and air cells, and the auditory tube are derived from the tympanic plexus that lies over the promontory. The tympanic plexus is formed mainly by branches from the tympanic branch of the glossopharyngeal nerve. It also receives some fibres from the sympathetic plexus around the internal carotid artery (caroticotympanic nerves). The tympanic plexus gives off the lesser petrosal nerve, which ends in the otic ganglion.

Added Information

❑ The middle ear can be likened to a narrow long box that has concave sides. It has six walls. All the walls separate the middle ear cavity from important structures.
  ○ Roof—from dura mater of the floor of middle cranial fossa;
  ○ Floor—from the superior bulb of internal jugular vein;
  ○ Medial wall—from internal ear;
  ○ Lateral wall—from external ear;
  ○ Posterior wall—from mastoid antrum and air cells and facial nerve;
  ○ Anterior wall—from internal carotid artery.
❑ Under normal circumstances, the walls of the cartilaginous portion of the auditory tube are in apposition. When active equalisation of air pressure between its two sides is sought (as in trying to do so in plane journeys or at high altitudes), the tube must be actively opened. Contraction of levator veli palatini and tensor veli palatini causes expansion of the tubal girth and opens it. Because these muscles are muscles of the soft palate, equalizing pressure requires opening the mouth, swallowing or yawning.

contd...

contd...

Added Information *contd...*

- The secondary tympanic membrane is otherwise called the **Scarpa's membrane**. It is concave towards the tympanic cavity and convex to the cochlea. It has three layers— **external layer derived from the tympanic mucosa, intermediate fibrous layer and an inner layer derived from the cochlear lining membrane**.
- The auditory ossicles are connected to the walls of the middle ear by ligaments. There are three ligaments for malleus and two for incus and one for stapes. The anterior ligament of malleus connects the bone to the anterior wall; the lateral ligament of malleus connects the bone to the posterior aspect of the tympanic sulcus; the superior ligament of malleus connects the bone to the roof of middle ear. The posterior ligament of incus connects the bone's short process to the fossa incudis. The superior ligament of incus connects the bone to the roof of the cavity. The annular ligament of stapes connects the foot plate to the rim of the oval window.
- Both the incudomalleolar and the incudostapedial joints are synovial. The former is a saddle and the latter ball and socket.
- **Movements of the auditory ossicles:** The auditory ossicles serve the purpose of transmitting sound waves to the fenestra vestibuli and thereby to the scala vestibuli which contains the perilymph. When sound waves hit the tympanic membrane, it moves inwards along with the handle of malleus. The incus also moves in the same direction and pushes the stapes inwards at the oval window. This causes the perilymph in the labyrinth to move. As a result, a compensatory bulge is produced at the fenestra cochleae. The footplate of stapes does not move like a piston at the oval window but swings like a door. It has a rocking movement around a vertical axis. However, this door-like movement occurs only at moderate intensities of sound. At both low and loud sounds, the stapes moves around a horizontal axis. Change in the direction of the axis protects the footplate and prevents excessive displacement of perilymph.
- The auditory ossicles are the first set of bones to be fully ossified during foetal development. They are fully formed at birth. These ossicles are covered by the mucous membrane of the middle ear cavity but do not have a covering periosteum.
- Contraction of tensor tympani tightens the tympanic membrane and dampens vibrations. It also pushes the stapes into the oval window. Contraction of stapedius pulls the stapes out thereby again dampening vibrations.

Clinical Correlation

- **Otitis media:** Infection of middle ear is otitis media. Severe ear ache and reddish tympanic membrane are the clinical manifestations. The condition usually occurs secondary to respiratory infections. Swelling and inflammation of the middle ear cavity leads to blockage of the auditory tube. Chronic or untreated otitis media may result in scarring of the auditory ossicles and hearing impairment.

contd...

Clinical Correlation *contd...*

- **Mastoiditis:** This is a condition where there is infection and inflammation of the mastoid antrum and air cells. Infections from middle ear can easily spread to the mastoid. Infections from here can spread to the middle cranial fossa through the tegmen tympani.
- Stapedius muscle is paralysed in facial nerve lesions. Uninhibited movements of stapes occur. The dampening effect of the muscle is lost and even very small sounds (low in intensity and frequency) are 'heard'. This excessive acuteness of hearing is called **hyperacusis**.

INTERNAL EAR

The internal ear, which contains the organs of hearing and balance, is in the form of a complex system of cavities within the petrous temporal bone. Because of the complex shape of these intercommunicating cavities, it is referred to as the **labyrinth**. The outer part of the labyrinth is made up of bone and so is called the **bony labyrinth**. It is made up of dense bone and its inner surface is lined by periosteum. Lying within the bony labyrinth, is a system of ducts which constitute the **membranous labyrinth** (Fig. 12.15).

The spaces within the membranous labyrinth are filled with a fluid called the **endolymph**. The space between the membranous labyrinth and the bony labyrinth is filled with another fluid called the **perilymph**.

The internal acoustic meatus is a passage that leads to the internal ear from the posterior cranial fossa. Though very closely associated with the internal ear, the meatus is not usually described as part of the internal ear as in the case of the external acoustic meatus with relation to the external ear.

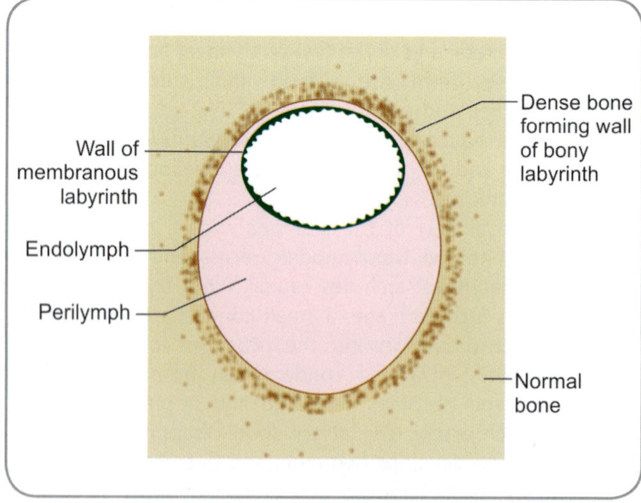

Fig. 12.15: Basic structure of the internal ear as seen in a section through a semicircular canal

Internal Acoustic Meatus

Other names: Internal auditory meatus, internal auditory canal.

The cranial opening of the internal acoustic meatus is seen in the middle of the posterior surface of petrous temporal bone which also forms the anterior wall of the posterior cranial fossa. The meatus runs a short distance of about 1 cm and is closed at its lateral end by a vertical plate of bone. This vertical plate is also the medial wall of the internal ear.

A transverse crest (also called the ***horizontal crest*** or the ***horizontal bar***) subdivides the meatal side of the wall into a smaller upper and a larger lower portion. The anterosuperior part opens into the facial canal; the facial nerve runs through the opening and the canal. The posterosuperior part is called the ***superior vestibular area*** and shows numerous openings for the nerves to the utricle, anterior and lateral semicircular nerves (all branches of the superior division of the vestibular nerve and so collectively called the ***superior vestibular nerve***) to pass through. The facial part and the superior vestibular area are separated by a vertical bar of bone called the ***Bill's bar***. Below the horizontal crest, an anterior cochlear area shows a spiral of small foramina. This spiral set of foramina goes by the term ***tractus spiralis foraminosus*** and transmits branches of the cochlear nerve to the cochlea. The posterior area below the horizontal crest is the inferior vestibular area and transmits through numerous small foramina nerve branches to the saccule. Behind the inferior vestibular area, a single large foramen is seen. This is the foramen singulare that transmits the nerve to the posterior semicircular canal.

Dissection

Before attempting to dissect parts of ear, observe the external ear in a live individual and try to see the various parts in a model of the ear.

In a cadaver, incise and reflect a part of the skin of the auricle and study. With appropriate guidance, chip through the roofs of the internal acoustic meatus, middle ear, internal ear and mastoid air cells. Identify the different parts. Review with a model.

BONY LABYRINTH (FIG. 12.16)

The parts of the bony labyrinth are the ***vestibule***, ***semicircular canals*** and the ***cochlea***.

Vestibule

The vestibule (Latin.vistibulum=antechamber or entrance area) is the central part of the bony labyrinth; it is anterior to the semicircular canals and posterior to the cochlea. It is more or less oval in shape and is about 5 mm in diameter. It has anterior, posterior, medial and lateral walls. The

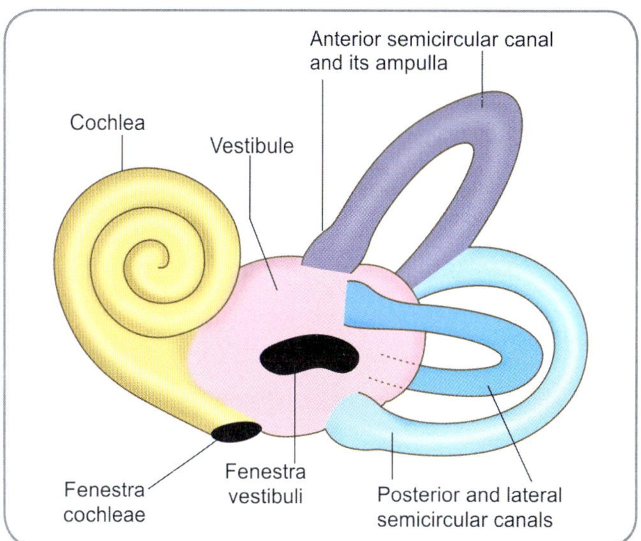

Fig. 12.16: Bony labyrinth seen from the lateral side

anterior and posterior walls are smaller than the medial and lateral walls. The lower half of the anterior wall has a large opening through which the vestibule communicates with the scala vestibuli of the cochlea. The posterior wall has five openings for the semicircular canals. The lateral wall is also the medial wall of the middle ear. There is a large aperture called the ***fenestra vestibuli*** in the middle of the lateral wall; the vestibule communicates with the middle ear cavity through the fenestra vestibuli which is closed by the footplate of the stapes bone. The medial wall of the vestibule is the lateral wall of the internal acoustic meatus. At the middle of the medial wall, is a vertical crescentic crest of bone called the ***vestibular crest***; in front of the crest is a spherical recess which lodges the saccule; the wall of this recess is perforated by several holes and is called the ***macula cribrosa media***. The upper part of the crest forms the vestibular pyramid. Posterior to the crest is an elliptical recess that lodges the utricle (saccule and utricle are parts of the membranous labyrinth). The vestibular pyramid and the wall of the elliptical recess are perforated by several holes; this perforated plate of the medial wall is called the ***macula cribrosa superior***. An opening is seen below the elliptical recess; this is the opening of the vestibular aqueduct. Inferior to the vestibular crest is another small shallow depression called the ***cochlear recess*** (Fig. 12.17).

Semicircular Canals

There are three semicircular canals, ***anterior*** (or superior), ***posterior*** and ***lateral*** (or horizontal); they are located posterior to the vestibule. One end of each canal is dilated and the dilatation is called an ***ampulla***. The non-ampullated ends of the anterior and posterior canals join to form a common channel, the ***crus commune***. As a result,

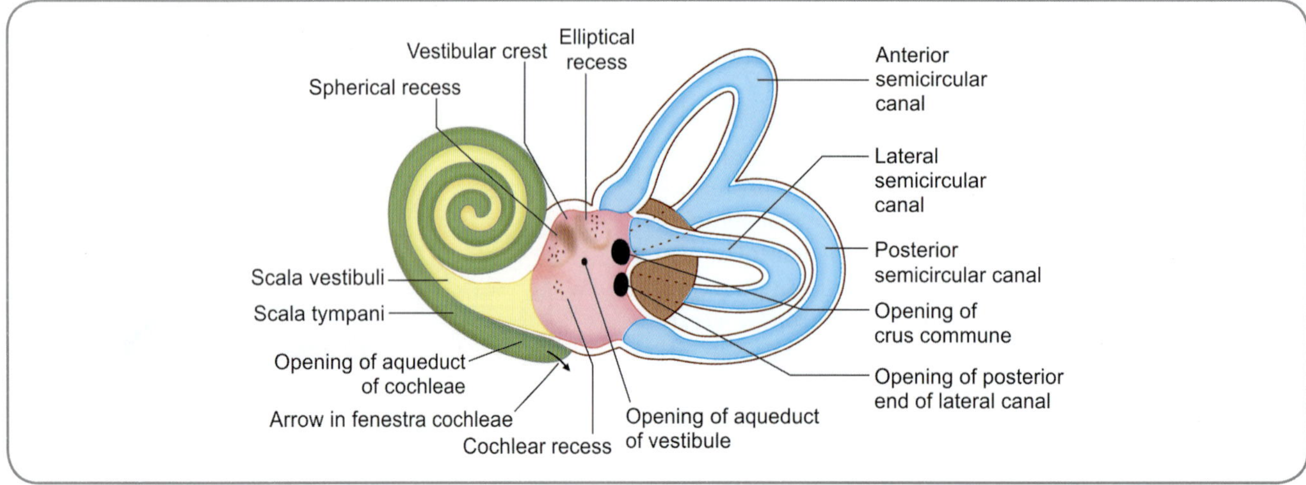

Fig. 12.17: Interior of the bony labyrinth as seen from the lateral side

the semicircular canals open into the vestibule through five (and not six) openings. The most important feature of the semicircular canals concerns their orientation. The three canals lie in planes at right angles to one another. The anterior and posterior canals are both vertical, while the lateral canal is horizontal. The plane of the posterior canal is parallel to the long axis of the petrous temporal bone. The plane of the anterior canal is at right angles to this axis. The plane of the lateral canal is transverse. The right and left lateral semicircular canals lie in the same plane. The anterior canal of one side lies in the same plane as the posterior canal of the opposite side (Fig. 12.18).

Bony Cochlea

The cochlea (Greek.coklos=snail) is the most anterior part of the bony labyrinth and is continuous posteriorly with the vestibule. The cochlea is basically a tube that is coiled on itself for two and three quarter turns. The diameter of the tube is greatest at its junction with the vestibule and this part is called the **basal turn** of the cochlea. The tube becomes progressively narrower towards the centre or **apex** (or cupula) of the cochlea. The basal turn of the cochlea produces an elevation, the promontory, on the medial wall of the middle ear. The apex of the cochlea is related to this wall of the middle ear just in front of the promontory. The cochlea has a central bony pillar called the **modiolus** (Latin.modiolum=nave of a wheel). A spiral canal runs around it. A thin plate of bone called the **spiral lamina** projects from the modiolus into the bony spiral canal like the thread of a screw and partially subdivides into two. Two membranes, one below the other, run from the free edge of the spiral lamina to the opposite wall of the canal. The upper membrane is the Reissner's membrane and the lower is the basilar membrane (Fig. 12.19). Three channels are, therefore, formed:

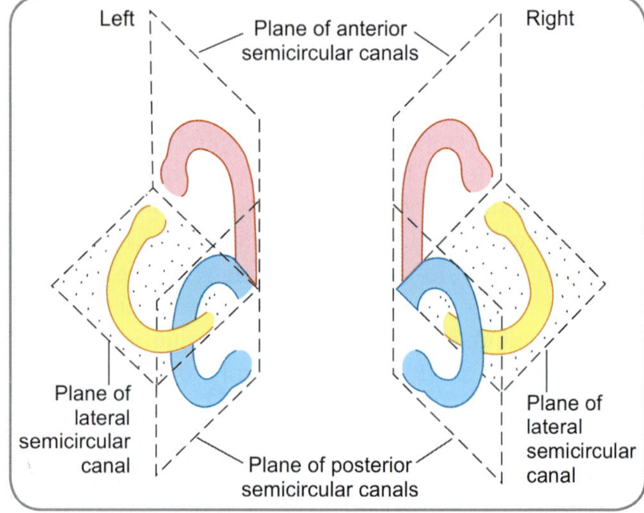

Fig. 12.18: Planes in which the semicircular canals lie as seen from behind

- ❑ The middle channel is between the Reissner's membrane and the basilar membrane and is called the **scala media** (or **cochlear duct**); it is a blind channel and ends at the apex of cochlea.
- ❑ Between the spiral lamina and Reissner's membrane on one side and the outer wall of the cochlea on the other is the scala vestibuli, the upper channel.
- ❑ Between the spiral lamina and the basilar membrane on one hand and the outer wall of the cochlea on the other is the scala tympani, the lower channel.

The scala vestibuli and the scala tympani are continuous with each other at the modiolar apex through a thin slit called the **helicotrema**. The scala vestibuli is continuous with the vestibule and the scala tympani opens into the middle ear through the fenestra cochleae, which is closed by the **secondary tympanic membrane**.

Added Information

- The entire internal ear is within the petrous temporal bone. Both the middle and the posterior cranial fossae are, therefore, in close proximity.
- The vestibule communicates anteriorly with cochlea, posteriorly with semicircular canals, laterally with tympanic cavity through the oval window and footplate of stapes and medially with posterior cranial fossa through the vestibular aqueduct that opens on its lateral aspect into the vestibule and on its medial aspect on the posterior surface of petrous temporal.
- The perilymph is similar in composition to extracellular fluid and the endolymph to intracellular fluid.
- The membranous labyrinth technically houses the organs of hearing and balance. It, therefore, has to have adequate and appropriate nerve supply. Branches of the vestibular nerve reach the saccule through the macula cribrosa media which is the perforated wall of the spherical recess. Nerve branches to the utricle and the semicircular canals pass through the macula cribrosa superior which is the perforated wall of the elliptical recess. Macula cribrosa superior overlies the superior vestibular area. Nerves to the cochlear duct pass through the perforations in the wall of the cochlear recess.
- Modiolus is otherwise called the ***Columella cochleae***.
- Reissner's membrane (named after the 19th century German anatomist Ernst Reissner), which is otherwise called the ***vestibular membrane***, separates the scala media from the scala vestibuli.
- The basilar membrane separates the scala media from the scala tympani.
- The bony spiral lamina and the basilar membrane together are called the ***spiral membrane***.
- The broad base of the modiolus lies against the medial wall of the internal ear. On the meatal side of this wall (lateral wall of the internal meatus), the tractus spiralis foraminosus overlies the modiolar base. Fascicles of the cochlear nerve destined to the turns of cochlea run through the foramina of tractus spiralis foraminosus. Those for the first 1¾ turn run in the smaller foramina of the spiral and those for the apical turn (the distal 1 turn) run through the large foramen in the centre of the smaller foramina. Each small foramen continues into a small canal. All these small canals run into the modiolus and open in a spiral sequence along the base of the osseous spiral lamina. At this point (along the base of the osseous spiral lamina), all small canals are connected by a common canal called the ***Rosenthal's canal***. The Rosenthal's canal, as should be expected, is spiral, runs along the base of the osseous spiral lamina and contains the spiral ganglion. The central fascicle continues through the centre of the modiolus to the apex.
- From the Rosenthal's canal, several tiny canals run radially to the rim of the osseous spiral lamina. These are the *habenula perforata* and carry nerve fascicles to the Organ of Corti.
- It has been noted that the osseous spiral lamina projects into the cavity and the basilar membrane continues from its free edge to the outer wall and completes the partition. This, however, is a much simplified statement. Another smaller osseous projection called the ***secondary spiral lamina*** projects from the outer wall opposite the osseous spiral lamina. The basilar membrane only connects the two spiral laminae. The gap between the two laminae gradually increases towards the apex of the cochlea; thus the basilar membrane is broader at the apex than at the base.
- A small canal called the ***cochlear canaliculus*** or the ***cochlear aqueduct*** runs from the base of the cochlea to the posterior cranial fossa close to the inferior petrosal sinus. A tiny vein usually runs in this aqueduct and drains into the inferior petrosal sinus. The aqueduct which was initially presumed to be connecting the scala tympani to the subarachnoid space of the posterior cranial fossa is now realized to be closed by a membrane at the petrosal end.

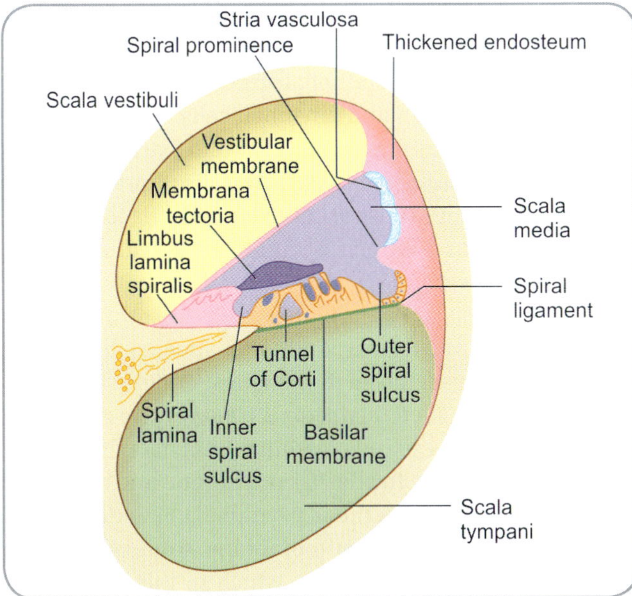

Fig. 12.19: Transverse section through one turn of cochlea

MEMBRANOUS LABYRINTH

The parts of the membranous labyrinth are the vestibular apparatus and the cochlear duct. The bony labyrinth has a cavity and this cavity is lined by periosteum (the endosteum, actually). Within the cavity of the bony labyrinth is a bag of membranes. Since the wall of the bag is made up of membranes, it is called the ***membranous labyrinth***. Though it is a single bag, it has different shapes at different places and, therefore, can be subdivided into the ***vestibular apparatus*** and the ***cochlear duct***. The membranous labyrinth is separated from the endosteum of the bony labyrinth by a space that contains the perilymph and a network of blood vessels.

Vestibular Apparatus

The membranous labyrinth within the vestibule and the semicircular canals of the bony labyrinth comprises the vestibular apparatus. It consists of the two vestibular sacs

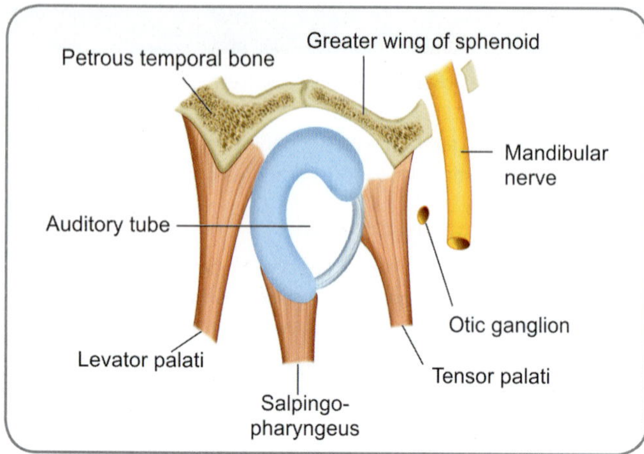

Fig. 12.20: Some relations of the cartilaginous part of the auditory tube

called the *utricle* and the *saccule* (Fig. 12.20) and the three semicircular ducts. The utricle (Latin.uter=leather bag) and the saccule (Latin.saccus=sac or pouch) are membranous sacs present within the bony cavity of the vestibule. The three semicircular ducts are the membranous tubules present within the corresponding bony cavity of the semicircular canal.

The utricle (*other names:* utriculus, sacculus communis, vestibular organ) is the larger of the two vestibular sacs and occupies the posterosuperior portion of the vestibule. The fundal portion (the closed and highest part of the bag) of the utricle is in contact with the elliptical recess. The ampullae of the anterior and lateral semicircular ducts open to this highest part; the non-ampullated ends of the lateral duct and the crus commune of the anterior and posterior ducts open into its central part; the ampulla of the posterior duct opens into its lower portion. The utricle is connected to the saccule by the utriculosaccular duct which leaves the utricle in its inferomedial portion. Due to irregularity in shape, the highest portion comes to have a small floor that is continuous with its curved anterior wall. This anterior wall and floor are thickened to form the macula of the utricle.

The saccule (*other names:* sacculus, sacculus proprius, sacculus vestibuli) occupies the spherical recess of the bony vestibular cavity, somewhat globular in shape and is anterior to the utricle. The utriculosaccular duct joins it on its posterosuperior part. The medial wall of the utricle is thickened to form the macule of the saccule. The saccule is connected to the cochlea by the ductus reuniens that starts from the lower part of the saccule and curves up to join the base of the cochlear duct. A slender endolymphatic duct starts from the lower part of the utriculosaccular duct; it runs into the vestibular aqueduct and ends under the dura on the posterior aspect of petrous temporal bone as the endolymphatic duct.

The semicircular ducts occupy the semicircular canals (it is important to distinguish between the terms semicircular canal and semicircular duct). Each duct is much smaller than its bony canal; it is attached to the endosteum of the bony canal by one part of its circumference while the rest of its circumference bulges into the bony cavity. Each duct has a dilated end called the *ampulla*; the ampulla of the duct fills the corresponding bony ampulla. The floor of each ampulla is thickened and forms a transverse ridge called the *crista* (or *ampullary crest*).

Cochlear Duct

Other names: Scala media, lowenberg's scala, lowenberg's canal, membranous cochlea.

The cochlear duct is the membranous labyrinth housed inside the bony cochlea. As already seen, it is the spirally arranged canal lying between the Reissner's membrane and the basilar membrane. Its basal end is blind, lies at the cochlear recess of the medial wall of the bony vestibule and is called the *caecum vestibulare*. Its apical or upper end is also blind, lies at the cupula of the bony cochlea and is called the *caecum cupulare* or *caecum lagena*. The helicotrema is a slit between the caecum cupula and the bony wall. The cochlear duct is connected to the saccule of the vestibular apparatus by the ductus reuniens. The floor of the cochlear duct contains the organ of hearing called the *Organ of Corti*.

Specialised End Organs in the Membranous Labyrinth

Organ of Corti

Other names: Spiral organ, corti's organ, acoustic papilla

The end organ for hearing is the *organ of Corti* (named after the 19th century Italian anatomist Alfonso Corti). It lies in the cochlear duct, just above the basilar membrane and consists of specialised epithelial structures. The outer and inner pillar cells, outer and inner hair cells, and supporting cells like the Deiter's phalangeal cells and external limiting cells make up these epithelial specialisations. The hair cells are the actual sensory receptor cells. Each hair cell is elongated, has a bunch of modified apical microvilli (called the *stereocilia*) and synaptic contacts with cochlear axons at its base. All these cells are arranged in a specific pattern and form a continuous sheet on the basilar membrane from the basal to the apical end of cochlea. A gelatinous layer called the *tectorial membrane* covers the apical aspects of these cells. Within the cochlear duct is the endolymph. Beneath the basilar membrane and above the Reissner's membrane is the perilymph. Due to diffusional characteristics of the basilar membrane, some parts of the organ of Corti have perilymph, which in this situation is called the *cortilymph*.

Anatomy of sound reception: Sound waves travelling through air produce vibrations in the tympanic membrane. These are transmitted through the malleus and incus to the stapes. The footplate of the stapes (that fits into the fenestra vestibuli) transmits these vibrations to the perilymph of the vestibule. From here, the vibrations pass into the scala vestibuli. Each time the base of the stapes moves inwards into the vestibule, it creates a pressure wave that extends along the perilymph filling the entire length of the scala vestibuli. Reaching the helicotrema, the pressure wave passes into the perilymph filling the scala tympani. Traversing the entire length of the scala tympani, it reaches the secondary tympanic membrane (which closes the fenestra cochleae) causing it to bulge into the middle ear.

Development

Development of Inner Ear

By the late third week, thickening of the surface ectoderm occurs on either side of the developing head region. This region overlies the developing rhombencephalon. The thickenings are called the **otic placode** (Greek.otikos=ear). The otic placodes enlarge rapidly and sink into the underlying mesoderm. The sunken pits deepen further and are separated off the surface ectoderm to form the otic vesicles. By the mid fifth week, the otic vesicles show division into a ventral and a dorsal component.

By the sixth week, the ventral component expands and gives out a tubular evagination at its lower end. The evagination penetrates the surrounding mesoderm in a spiral fashion till the eighth week. By then it completes two and three-quarter turns and becomes the cochlear duct. The upper part of the ventral portion becomes the saccule and the connection between the cochlear duct and the saccule develops into the ductus reuniens. The mesoderm surrounding the cochlear duct undergoed condensation and differentiates into cartilage. This cartilage undergoes vacuolation by the tenth week and thereby two cavities are formed above and below the cochlear duct resulting in **scala vestibuli** and scala tympani. The wall of the cochlear duct next to scala vestibuli becomes the vestibular membrane and that which is next to scala tympani becomes the basilar membrane. The lateral wall of the cochlear duct remains in close contact with the cartilage; the tapering medial portion is connected to a plate of cartilage that would eventually develop into the modiolus. Differentiaton of the cells of the cochlear duct gives rise to the hair cells and the tectorial membrane which together constitute the organ of Corti.

By the sixth week, the dorsal component of the otic vesicle develops three flattened evaginations. The walls of the central portions of these evaginations come into contact with each other and fuse. They eventually break down resulting in the formation of the three semicircular canals. The canals develop the ampullae by differential growth. The remaining original portion of the dorsal component develops into the utricle. A slender evagination from its dorsal end gives rise to the endolymphatic duct.

contd...

Development *contd...*

Differentiation in the epithelial cells of the utricle and saccule begins by the eleventh week and the maculae acousticae are formed. Simultaneously, differentiation in the ampullary cells result in the formation of cristae.

Much early in development (by the fourth week), some cells of the otic vesicle break away and form a statoacoustic ganglion. The ganglion which remains close to the developing otic vesicle also receives neural crest cells. The ganglion then splits into the cochlear and vestibular portions and provides the sensory cells of the acoustic and vestibular systems.

These changes take place instantaneously. In this way, vibrations are set up in the perilymph. These in turn produce vibrations of the basilar membrane and of the organ of Corti. The basilar membrane, though a continuous sheet, varies in its width, mass (including the exact number of cells and their junctional properties) and stiffness at different locations from the base to apex of cochlea. Each part of the basilar membrane is specific for intensity and frequency of sound. When a sound wave arrives, a continuous chain of vibrations is set from the tympanic membrane to the perilymph. As a result, the basilar membrane also vibrates. Vibrations of the basilar membrane actually cause a mechanical wave on it; this wave is called the ***travelling wave***. As this wave travels along the basilar membrane, whichever part of the membrane that would respond the maximum to that particular intensity and frequency of sound, vibrates the maximum; the wave dies away. The hair cells also vibrate and the different patterns of vibration produce different kinds of forces on their apices. Vibrations of the basilar membrane also produce movements in the endolymph. Local movements of the endolymph cause stimulation of the stereocilia of the concerned hair cells. Ionic exchange occurs between the endolymph and the hair cells resulting in depolarization of the hair cells and neurotransmitter release at the base. The cochlear afferents at the base of the hair cells are stimulated. These afferent fibres are actually the peripheral processes of neurons located in the spiral ganglia. These processes reach the hair cells through canals in the spiral lamina. Impulses generated in the hair cell bases travel through the peripheral processes and reach their cell bodies in the spiral ganglia. Central processes arising from the ganglionic cell bodies constitute the cochlear nerve.

Maculae

Information about changes in the position of the head is provided by end organs called ***maculae*** (singular = macula) present in the utricle and saccule (collectively called the ***maculae acousticae***). Each macula can be described to have three layers: a layer of sensory epithelium that has specialised epithelial cells; a layer of extracellular gelatinous material and a layer of otoliths. Each macula is actually a thickening of a specific portion of the membrane of the membranous labyrinth.

The macula of the utricle (or the utricular macula or the utricular spot) is present on the floor and anterior wall of its highest part. It lies horizontally; its basal layer is the epithelial layer which is continuous with the epithelium of the rest of the membranous wall of the utricle; at the area of the macula, this epithelium shows specialised cells. The specialised cells are the mechanosensitive hair cells. These hair cells have on their apical surfaces a bunch of large microvilli called the *stereocilia* and a single cilium called the *kinocilium*. The hair cells are interspersed among non-sensory supporting cells. At their bases, the hair cells make synaptic contact with the vestibular nerve fibres. A layer of extragelatinous material lies over the epithelial layer. The 'hairs' (stereocilia and kinocilia) of the hair cells are embedded in the gelatinous material. Over the gelatinous material is another layer of crystals of calcium carbonate. These crystals are called *otoliths* (Greek.otikos=ear, lithos=stone) or statoliths (Greek.statos=standing; so named because, in some way, they are associated with the position of head and body). The gelatinous and the otolith layers together are called the *otolithic membrane*. When the head is in its normal position, the macula of each utricle is horizontal. Changes in position of the head in the horizontal plane cause corresponding changes in all inner parts. However, the otolithic membrane lags behind due to its inertia when the epithelial layer has moved. As a result, the otoliths and the gelatine material cause deflection of the hairs of the hair cells. Impulses thus generated are carried by the vestibular nerve endings.

The macula of the saccule (or the saccular macula or the saccular spot) is present on the medial wall of the saccule. It lies vertically. The details of its layers and cells are the same as those of the utricle. The only and the major difference between the utricular and saccular maculae is that while the utricular macula is horizontal when the

head is in its normal position, the saccular macula is vertical. The mechanism of action is also the same except that the saccular macula is sensitive to changes in position in the vertical plane. It is the primary gravitational sensor.

Cristae Ampullae

Information about angular movements (acceleration) of the head is provided by end organs called the *ampullary crests* (or *cristae ampullae*) one of which is present in the ampulla of each semicircular duct. The floor of each ampulla has a transverse thickening called the *transverse septum*. The central portion of this septum is much more elevated than the sides and is called the *crista*. The crista consists of hair cells and supporting cells. The features of the hair cells are the same as those of the hair cells of the maculae. The bases of these cells have contact with nerve fibres. The apical parts of the cells are surmounted by a gelatinous covering that forms almost a pillar (cupula or cupola) projecting into the cavity of the ampulla. Movements of the head produce currents in the endolymph within the semicircular ducts. These cause the cupoolae to move resulting in deformation of hair cells and production of nerve impulses.

The nerve fibres innervating the cristae of the semicircular ducts and the maculae of the utricle and saccule are peripheral processes of neurons located in the vestibular ganglion. This ganglion lies in the internal acoustic meatus. The central processes of cells of the ganglion form the vestibular nerve.

Blood Vessels of the Internal Ear

The internal ear is supplied by the labyrinthine artery which usually arises from the anterior inferior cerebellar artery. It sometimes arises from the basilar artery. The internal ear also receives some twigs from the stylomastoid artery that supplies the middle ear.

The veins of the vestibule and the semicircular canals join with the cochlear veins to form the labyrinthine vein that ends in the superior petrosal sinus or in the transverse sinus.

Added Information

- ❑ The ductus reuniens is referred to as the Hensen's duct or uniting duct by the clinicians.
- ❑ The membranous labyrinth does not float in/on the perilymph. It is suspended into the cavity of the bony labyrinth by thin filaments of connective tissue. In the case of the cochlear duct, the secondary spiral lamina (usually called the *spiral ligament*) keeps the former anchored to the bony labyrinth. The filamentous network that suspends the membranous labyrinth resembles the arachnoid mater of the meninges.

Clinical Correlation

- ❑ *Hearing loss:* Hearing loss is usually of two types—conduction loss and sensorineural loss. Conduction loss occurs when something intervenes with conduction of sound waves through the external and middle ears. The ear ossicles may be fixed and unable to move. Any factor that intervenes with the movements of the oval and round windows also causes deafness of this category. Surgical intervention and hearing aid implants are used in treatment. Sensorineural loss result from defects in the pathway from cochlea to brain. Cochlear implants have been attempted in treatment.
- ❑ *Motion sickness:* During movements, the maculae and cristae are stimulated variedly. During travel, it is possible that the head undergoes tilting movements. When there is discordance between the visual stimulation and the vestibular stimulation produced by the movements, motion sickness results.
- ❑ When different parts of the internal ear are injured, different symptoms may occur. Vertigo (dizziness) occurs when the semicircular canals suffer damage. When the cochlear duct alone is injured, tinnitus (ringing in the ears) occurs.
- ❑ *Otic barotrauma:* When there is imbalance in pressure between the surrounding air and the air in middle ear cavity, otic barotraumas occurs. Various parts of the ear are damaged.
- ❑ *Meniere's disease:* Excessive production of endolymph results in this disease. Tinnitus, vertigo, hearing loss are the symptoms. Sensitivity to noises may also be seen.

Multiple Choice Questions

1. Epitympanic recess is:
   a. The part of middle ear above the level of tympanic membrane
   b. The part of middle ear opposite the tympanic membrane
   c. The part of middle ear that lodges the anterior process of the malleus
   d. The part of middle ear that lodges the lenticular process of incus
2. The point of greatest convexity of tympanic membrane corresponds to:
   a. The cone of light
   b. The tympanic sulcus
   c. The lower end of malleal handle
   d. The apex of pars flaccid
3. Suprameatal triangle corresponds to:
   a. Fossa incudis
   b. Fenestra cochleae
   c. Tegmen tympani
   d. Mastoid antrum
4. The membranous labyrinth consists of:
   a. Vestibular apparatus and cochlear duct
   b. Utricle and cochlear duct
   c. Modiolus, cochlear duct and utricle
   d. Saccule, utricle and cochlear duct
5. Intrinsic muscles of the auricle:
   a. Include the auricularis anterior
   b. Connect the skull and the auricle
   c. Are supplied by branches of facial nerve
   d. Move the auricle in different directions

ANSWERS

1. a   **2.** c   **3.** d   **4.** a   **5.** c

Clinical Problem-solving

Case Study 1: A 52-year-old man was diagnosed to be suffering from sensorineural hearing loss. Having read about stapedial surgeries, he wanted to have it done for him.
- ❑ Do you think stapedial surgery would offer a solution for this problem?
- ❑ If not, substantiate your answer.
- ❑ What is the anatomical basis of a sensorineural defect?

Case Study 2: A 27-year-old man felt that his ears were blocked after plane travel.
- ❑ What is the reason for his ear block?
- ❑ What is the anatomic basis for this condition?
- ❑ What measures would you advocate to avoid such blocks?

(For solutions see Appendix).

Chapter 13

Endocrine Glands of Head and Neck, Carotid Sinus and Carotid Body

Frequently Asked Questions

- ❑ Discuss the hypophysis cerebri, its important relations and secretions. Add a note on its development.
- ❑ Write notes on: (a) Adenohypophysis, (b) Pineal gland, (c) Thyroid follicles, (d) Corpora arenacea.
- ❑ Discuss the thyroid gland with reference to its location, parts, surfaces and borders, relations and blood supply.
- ❑ Write briefly on: (a) Parathyroid glands, (b) Parafollicular cells, (c) Thyroglossal duct, (d) Carotid body.

ENDOCRINE GLANDS

The endocrine glands located in the head and neck region are:
- ❑ The hypophysis cerebri,
- ❑ The pineal gland,
- ❑ The thyroid gland, and
- ❑ The parathyroid glands.

Some endocrine functions are also ascribed to the carotid body.

HYPOPHYSIS CEREBRI

The hypophysis cerebri is also called the ***pituitary*** (Latin. pituita=thick secretion) gland. It is placed in the cranial cavity, on the floor of the middle cranial fossa. It lies in a depression on the superior surface of the body of the sphenoid bone called ***hypophyseal fossa*** or ***sella turcica*** (Latin.sella=saddle, turcica=Turkish) and is suspended from the floor of the third ventricle of the brain by a narrow funnel shaped stalk called ***infundibulum*** (Latin. infundibulum=funnel). It is called the ***hypophysis*** (Greek. hypo=under, physia=growth) because of its location below the brain where it appears to be an undergrowth

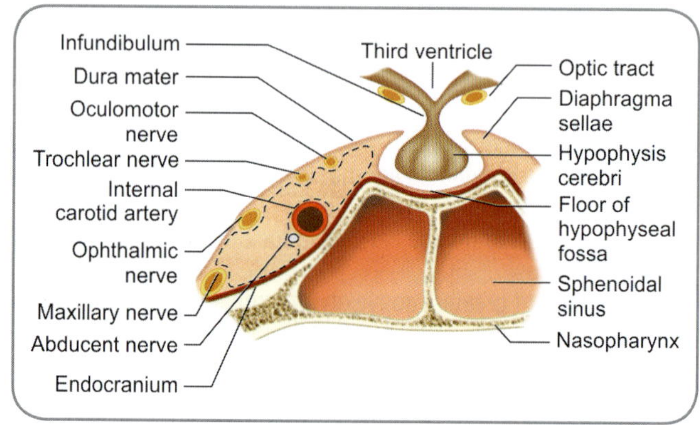

Fig. 13.1: Coronal section through hypophysis cerebri to show its relations

of the latter. The gland appears as a small ovoid structure measuring about 13 mm from side to side, about 10 mm front to back and about 8 mm in vertical diameter (Fig. 13.1).

The hypophyseal fossa is lined by dura mater. Superior to the hypophysis, the dura mater is folded on itself to form the ***diaphragma sellae***. The infundibulum passes up through an aperture in the diaphragma to join the inferior wall of the third ventricle. The optic chiasma lies antero-superior to the hypophysis cerebri being separated from it by the anterior part of the diaphragma. It lies anterior to the infundibulum. Inferiorly, the hypophysis cerebri is related to the sphenoidal air sinuses and beyond them to the nasopharynx. On the right and left sides, the hypophysis cerebri is related to the corresponding cavernous sinus (and to structures in its wall). The cavernous sinuses are connected across the midline by anterior and posterior intercavernous sinuses, which are located within the layers of the diaphragm sellae (Fig. 13.2).

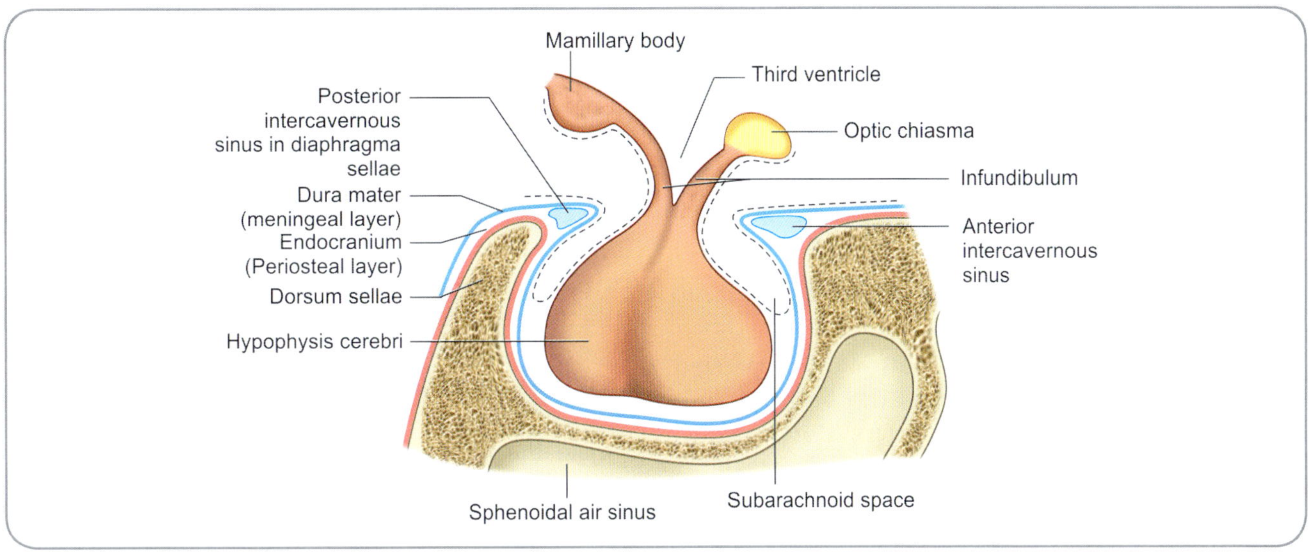

Fig. 13.2: Sagittal section through the hypophysis cerebri to show its relations

Clinical Correlation

The relations of the hypophysis cerebri help us to understand the effects of pressure by a tumour of the organ. Pressure on the optic chiasma leads to loss of both temporal halves of the fields of vision (bitemporal hemianopia). Lateral pressure may lead to paralysis of muscles supplied by the oculomotor nerve. The trochlear and abducent nerves are rarely affected (as they are too low down). Downward pressure of an enlarging hypophysis leads to 'ballooning' of the hypophyseal fossa; and backward pressure can cause erosion of the dorsum sellae. These features can be recognised on a skiagram.

Subdivisions

Different types of classification have been used to subdivide the parts of the hypophysis cerebri. Old classification described an anterior and a posterior lobe. Subsequent classifications mentioned about different parts like an anterior part—the **pars anterior**, an intermediate part — the **pars intermedia** and a posterior part—the **pars posterior** (or **pars nervosa**) (Fig. 13.3).

Development

Developmental factor: A distinction between the adenohypophysis and the neurohypophysis can also be made on the basis of their development. The neurohypophysis is formed as a down growth from the floor of the third ventricle. In contrast, the adenohypophysis is derived from **Rathke's pouch** that arises from the ectoderm lining the roof of the primitive mouth (stomatodaeum). This pouch has a cavity. The pars anterior is formed in the anterior wall of the pouch and the pars intermedia in its posterior wall. The original cavity of Rathke's pouch may persist as a cleft separating the pars anterior and the intermedia.

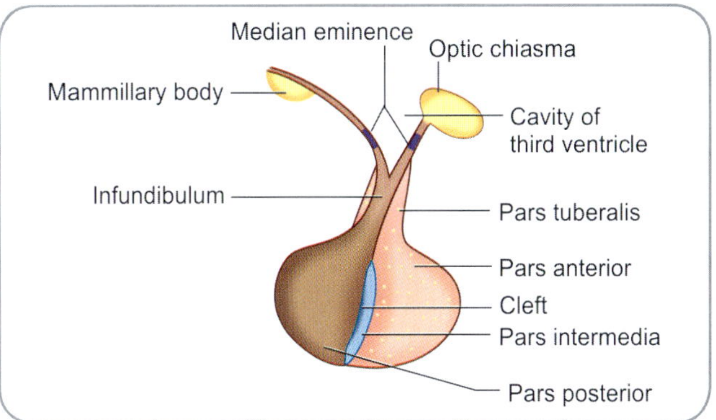

Fig. 13.3: Sagittal section to show the subdivisions of the hypophysis cerebri

It is preferable to classify the pituitary into an adenohypophysis and a neurohypophysis. Both these parts differ in their origin, structure and function.

The **adenohypophysis** (Greek.adeno=gland; indicating the glandular portion that is responsible for several secretions) comprises the pars anterior, pars intermedia and pars tuberalis. The **pars anterior** (also called the **pars distalis** or **pars glandularis**) and the **pars intermedia** are both made up of cells having a direct secretory function. An extension of the pars anterior surrounds the central nervous core of the infundibulum. Because of the tubular shape of this extension, it is called the **pars tuberalis**.

The **neurohypophysis** (indicating that portion which is connected to the neural tissue) comprises the pars posterior, infundibulum and median eminence. The **pars posterior** contains numerous nerve fibres. It is directly continuous with the central core of the **infundibulum**, which is made up of nervous tissue. The area of the floor of the third ventricle (tuber cinereum) immediately adjoining the attachment to it of the infundibulum is referred to as the ***median eminence***.

Structure and Hormones

Adenohypophysis

This is the highly vascular portion of the pituitary consisting of epithelial cells and vascular sinusoids supported by a mesh of connective tissue.

Pars Anterior

The pars anterior is made up of cords of cells separated by sinusoids. Several types of cells can be recognised on the basis of their staining characters. More recently, the cells responsible for production of individual hormones have been distinguished by a technique called ***immunofluorescence***. Using routine staining procedures, the cells of the adenohypophysis can be divided into ***chromophil*** cells (Greek.chromo=colour, phileo=loving; colour loving cells; readily staining cells) which have brightly staining granules in their cytoplasm and ***chromophobe*** cells (Greek.phobos=fear; colour fearing cells; cells not staining) in which the granules are not present.

Chromophil cells

Chromophil cells are further classified as ***acidophil***s which stain with acid dyes (like eosin or orange G) and ***basophils*** which stain with basic dyes (like haematoxylin). The acidophil cells are often called ***alpha cells*** and the basophils are called ***beta cells***. Basophils are more numerous in the central part of the gland. Both acidophils and basophils can be divided into subtypes on the basis of structural details and on the basis of the hormones produced by them.

Types of acidophil cells

- ❑ ***Somatotrophs (or somatotropes):*** These cells produce the somatotropic hormone [also called ***somatotropin, somatotropic hormone (STH)*** or ***growth hormone***]. which controls body growth, especially before puberty.
- ❑ ***Mammotrophs*** (or ***mammotropes***): These cells produce the ***mammotropic hormone*** [also called ***mammotropin, prolactin, lactogenic hormone*** or ***luteotropic hormone (LTH)***] which stimulates the growth and activity of the female mammary gland, during pregnancy and lactation.

Types of basophil cells

- ❑ ***Corticotrophs*** (or ***corticotropes***): These cells produce the ***corticotropic hormone*** [also called ***adrenocorti-cotropin*** or ***adrenocorticotropic hormone (ACTH)***] which stimulates the secretion of some hormones of the adrenal cortex [The staining characters of these cells are intermediate between acidophils and basophils; and they are sometimes classified amongst acidophils].
- ❑ ***Thyrotrophs*** (or ***thyrotropes***)***:*** These cells produce the ***thyrotropic hormone*** (***thyrotropin*** or ***thyroid stimulating hormone***) which stimulates the activity of the thyroid gland.
- ❑ ***Gonadotrophs*** (***gonadotropes*** or ***delta basophils***): These cells produce two types of hormones; each type has separate actions in the male and female. In the female, the first of these hormones stimulates the growth of ovarian follicles. It is, therefore, called the ***follicle stimulating hormone (FSH)***. It also stimulates the secretion of oestrogens by the ovaries. In the male, the same hormone stimulates spermatogenesis. In the female, the second hormone stimulates the maturation of the corpus luteum and the secretion by it of progesterone. It is called the ***luteinising hormone (LH)***. In the male, the same hormone stimulates the production of androgens by interstitial cells of the testes and is called the ***interstitial cell stimulating hormone (ICSH)***.

The secretion of hormones by the cells of the adenohypophysis is under the control of the hypothalamus.

Chromophobe cells

The chromophobe cells probably represent alpha or beta cells which have been depleted of their granules. Some of them are stem cells which give rise to new chromophil cells.

Pars tuberalis

The pars tuberalis consists of numerous blood vessels amidst which are found clusters of undifferentiated cells and gonadotrophs.

Pars intermedia

The pars intermedia contains follicles of chromophobe cells.

Neurohypophysis

The neurohypophysis is that part of the pituitary which is in close association with hypothalamus.

Pars Posterior

The pars posterior consists of numerous unmyelinated nerve fibres and cells called ***pituicytes***. The nerve fibres are the axons of neurons located in the hypothalamus. Situated between these axons are the pituicytes. These cells have long dendritic processes many of which lie parallel to the nerve fibres. The axons descending into the pars posterior from the hypothalamus end in terminals closely related to capillaries. The pars posterior of the hypophysis is associated with the release into the blood of two hormones. One of these is ***vasopressin***

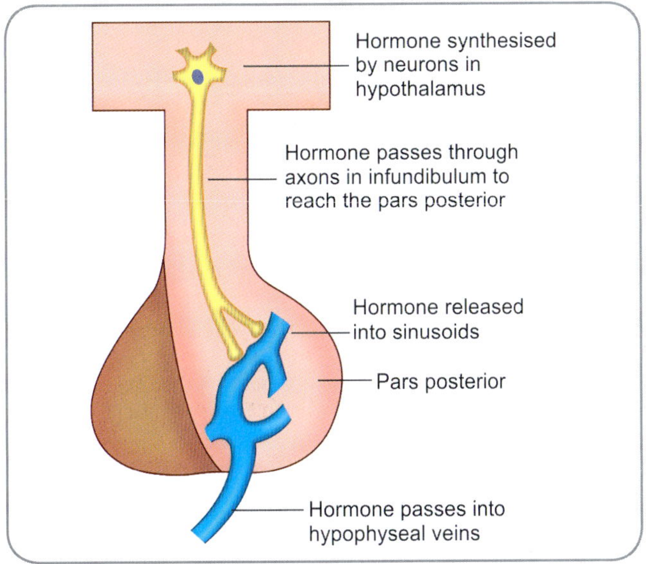

Fig. 13.4: Scheme to show the relationship of the hypothalamus and the pars posterior of the hypophysis cerebri

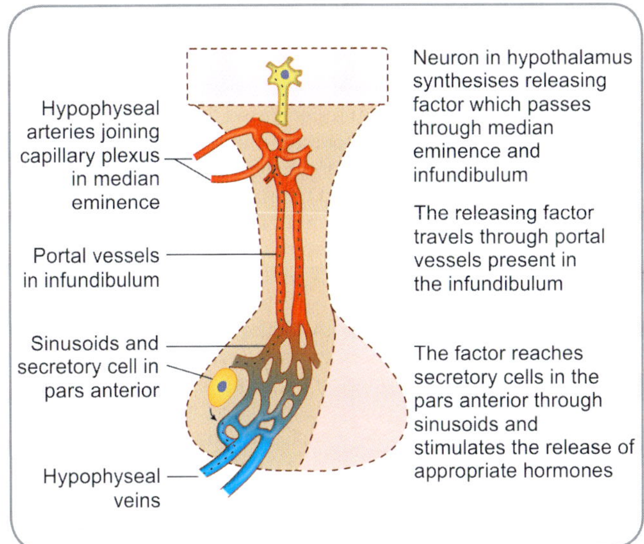

Fig. 13.5: Scheme to show the hypothalamo-hypophyseal portal circulation and the control of secretions of the adenohypophysis by the hypothalamus

(also called the ***antidiuretic hormone*** or ***ADH***). This hormone controls the reabsorption of water by kidney tubules. The second hormone is ***oxytocin***. This hormone controls the contraction of smooth muscle of the uterus and also of the mammary gland. It is now known that these two hormones are not produced in the hypophysis at all. They are synthesised in neurons located mainly in the supraoptic and paraventricular nuclei of the hypothalamus. Vasopressin is produced in the supraoptic nucleus and oxytocin in the paraventricular nucleus. These secretions (which are bound with other proteins) pass down the axons of the neurons concerned, through the infundibulum into the pars posterior. Here they are released into the capillaries of the region and enter the general circulation (Fig. 13.4).

Dissection

Locate the sella turcica in the middle cranial fossa. Locate the diaphragm sella. See the opening in the diaphragm for the pituitary stalk. Remove the diaphragm and study the pituitary gland.

Blood Supply

The hypophysis cerebri is supplied by superior and inferior hypophyseal branches arising from the internal carotid arteries. Some branches also arise from the anterior and posterior cerebral arteries. The inferior hypophyseal arteries are distributed mainly to the pars posterior.

Branches from the superior set of arteries supply the median eminence and infundibulum. Here they end in capillary plexuses from which ***portal*** vessels arise. These portal vessels descend through the infundibulum and end in the sinusoids of the pars anterior. The sinusoids are drained by veins which end in neighbouring venous sinuses. It will be noticed that the above arrangement is unusual in that two sets of capillaries intervene between the arteries and veins. One of these sets is in the median eminence and the upper infundibulum. The second set of capillaries is represented by the sinusoids of the pars anterior.

This arrangement is referred to as the ***hypothalamo-hypophyseal portal system*** (Fig. 13.5). The functional significance of this system is associated with the secretion of hormones by the adenohypophysis.

Control of Secretion of Hormones by Adenohypophysis

The secretion of hormones by the adenohypophysis takes place under higher control by neurons in the hypothalamus, notably those in the median eminence and in the infundibular nucleus. The axons of these neurons end in relation to capillaries in the median eminence and in the upper part of the infundibulum. Different neurons produce specific ***releasing factors*** (or releasing hormones) for each hormone of the adenohypophysis. These factors are released into the capillaries mentioned above. Portal vessels arising from the capillaries carry these factors to the pars anterior of the hypophysis. Here, they stimulate the release of appropriate hormones. Some factors inhibit the release of hormones.

🔔 Clinical Correlation

- Various types of tumours may arise in the hypophysis cerebri. The effects of the tumour (adenoma) may be caused by pressure on surrounding structures or by increased or decreased production of hormones.
  An adenoma arising from chromophobe cells can become quite large and can produce pressure effects as follows:
    - Pressure on the walls of the hypophyseal fossa (sella turcica) leads to its enlargement and this enlargement can be seen in a skiagram. The enlarging tumour presses on and destroys other cells (acidophils, basophils) of the pars anterior and gives rise to deficiency of the hormones produced by them.
    - Pressure on the pars posterior can lead to diabetes insipidus.
    - Pressure on the optic chiasma can lead to loss of vision in the temporal halves of vision in both eyes (bitemporal hemianopia). Stretching of the optic nerves can lead to optic atrophy.
    - Pressure on the hypothalamus can interfere with various visceral functions. It can lead to **Frohlich's syndrome** that is characterised by obesity, poor development of sex organs including the gonads and altered secondary sex characters.
    - Pressure on the third ventricle can result in obstruction to flow of CSF and raised intracranial tension.
- An adenoma can be removed surgically. The approach can be through the roof of the pharynx and through the sphenoidal sinuses. In extensive tumours, the cranial cavity has to be opened so that the tumour can be seen directly.
    - In an eosinophil adenoma, pressure effects are negligible. The main effects arise from excessive production of growth hormone. In a young individual (before epiphyseal fusion), the condition results in excessive growth (**gigantism**). If the adenoma is formed after the epiphyses have fused, overgrowth mainly affects the head, hands and feet (**acromegaly**). The scalp, lips, tongue and face become thick because of increased amount of subcutaneous tissue, and the same happens to the hands and feet. The paranasal sinuses enlarge making the facial region prominent. There is overgrowth of hair and the man's appearance tends to resemble that of an ape.
    - In the case of a basophil adenoma, there is excessive secretion of adrenotropic hormones. It leads to **Cushings's syndrome** that can also be caused by a tumour of the adrenal cortex. The syndrome is seen mostly in females. Abnormal deposition of fat takes place over the face, neck and trunk. The limbs remain thin and weak.
- The posterior lobe of the hypophysiscerebri produces antidiuretic hormone (ADH) that is responsible for reabsorption of water from renal tubules. Destruction of the posterior lobe (e.g., by pressure from an adenoma), can result in **diabetes insipidus**. In this condition, large amounts of urine are passed (polyuria). The urine is of very low specific gravity and contains no sugar or albumin. The resultant dehydration leads to excessive thirst (polydipsia) and to dryness of skin. Apart from pressure by a tumour, diabetes insipidus can be caused by trauma to the region. In many cases, there is no obvious cause for the condition.

PINEAL BODY

The pineal body or the pineal gland (Latin.pineus=relating to pine; here, indicating the pine shape) is a small piriform structure present in relation to the posterior wall of the third ventricle of the brain. It is also called the **epiphysis cerebri** (Greek.epi=above, physea=growth) because of the reason that it appears like a extra growth from the dorsal aspect of the brain. It is about 8 mm in length and about 4 cm in width and thickness. It is situated in the median plane just below the splenium of the corpus callosum and just above the superior colliculi of the midbrain (Fig. 13.6).

The pineal body is made up mainly of cells called **pinealocytes**. In the adult, sections of the pineal gland show aggregations of salts containing calcium. These are referred to as **corpora arenacea** or 'brain sand'. They are by products of secretory activity.

The pineal body has for long been regarded as a vestigial organ of no functional importance. Recent investigations have shown that the pineal body is an endocrine gland of great importance. It produces hormones which seem to have an important regulatory influence on many other endocrine organs (including the adenohypophysis, the neurohypophysis, the thyroid, the parathyroids, the adrenal cortex and medulla, endocrine pancreas and the gonads). The hormones of the pineal body reach the target organs both through the CSF and through blood.

The best known hormone of the pineal body is the amino acid **melatonin** (so called because it causes changes in skin colour in amphibia). The pineal body may act as a kind of biological clock that may produce circadian rhythms (variations following a 24-hour cycle) in various parameters.

The pineal body is supplied by the pineal arteries which are branches of the medial posterior choroidal arteries which in turn are branches of the posterior cerebral artery.

Added Information

- The pineal body is attached to the midbrain by a peduncle called the **pineal stalk**. The peduncle splits into the superior and inferior laminae which are separated by an extension of the third ventricle called the **pineal recess**.
- Postganglionic adrenergic sympathetic fibres derived from the superior cervical ganglion enter the dorsal aspect of the pineal body from near the tentorium cerebella. This bunch of fibres (sometimes paired bunch) is called the **nervus conarii**. The pineal body itself is also called **conarium** (Greek. konarion, konos=pine).
- The effects of the pineal on other glands like the adenohypophysis, thyroid, parathyroids, adrenal and endocrine pancreas are inhibitory in nature.
- Histologically, the pineal body has clusters of pinealocytes amidst neuroglial cells. The pineal stalk has only glial cells. Pinealocytes are specialised neurones with several

contd...

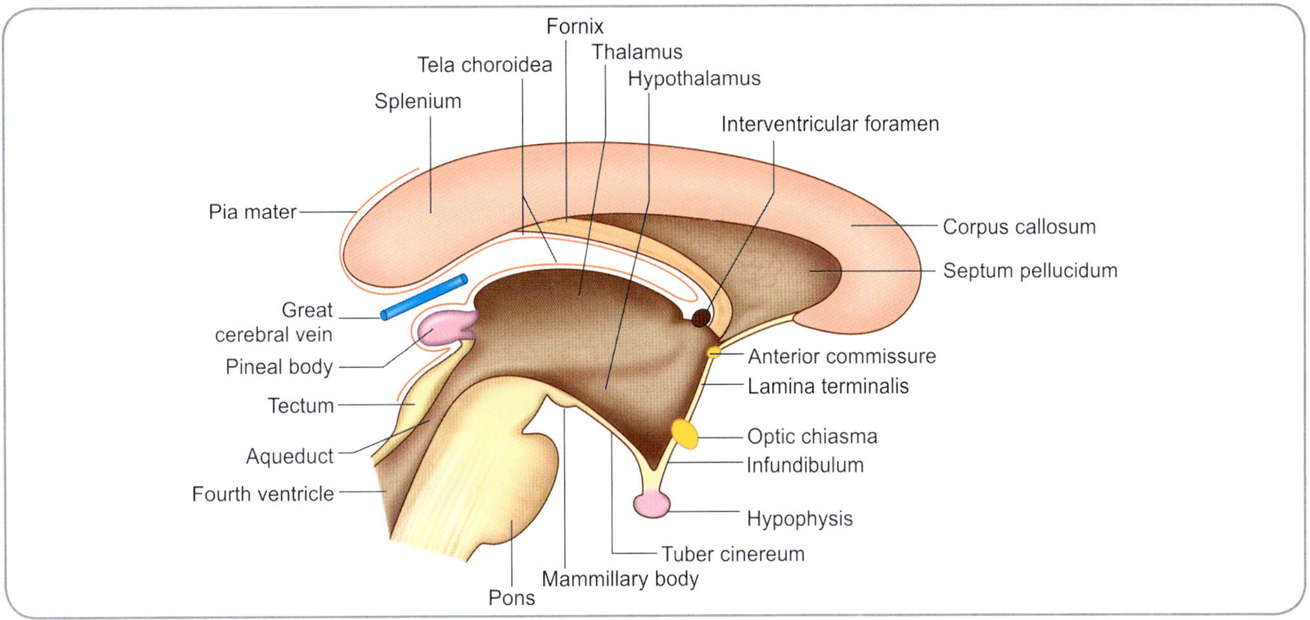

Fig. 13.6: Diagram showing the position of the hypophysis cerebri and of the pineal body relative to the third ventricle of the brain

synapses and many processes. The processes extend and end in terminal bulbs near the capillaries or ependyma of pineal recess. The terminal bulbs show dense vesicles which store melatonin. From 25 years of age, calcareous deposits accumulate in the extracellular matrix of the body. These are the corpora arenacea (Latin.arena=sand).

Development

Developmentally, the structures which develop in relation to the lateral wall of the diencephalon are given names which recapitulate this relation. Those structures which develop in the superior part of the lateral wall become the epithalamus, those which develop in the central part of the lateral wall become the thalamus and those in the inferior part become the subthalamus and hypothalamus.

The epithalamus in the adult consists of the paraventricular nuclei, habenular nuclei, stria medullaris thalami and the pineal gland.

Clinical Correlation

Tumours of the pineal gland can press on the tectum of the midbrain. This can damage the oculomotor nucleus and can thus lead to paralysis of the oculomotor nerve. Pressure of the tumour may obstruct the aqueduct and cause hydrocephalus.

THYROID GLAND (FIG. 13.7A)

The thyroid gland (Greek.thyreos=oblong shield), which forms the endocrine stratum of the neck viscera, lies in the front of neck, in front of the lower part of the larynx and the upper part of the trachea (Fig. 13.7B).

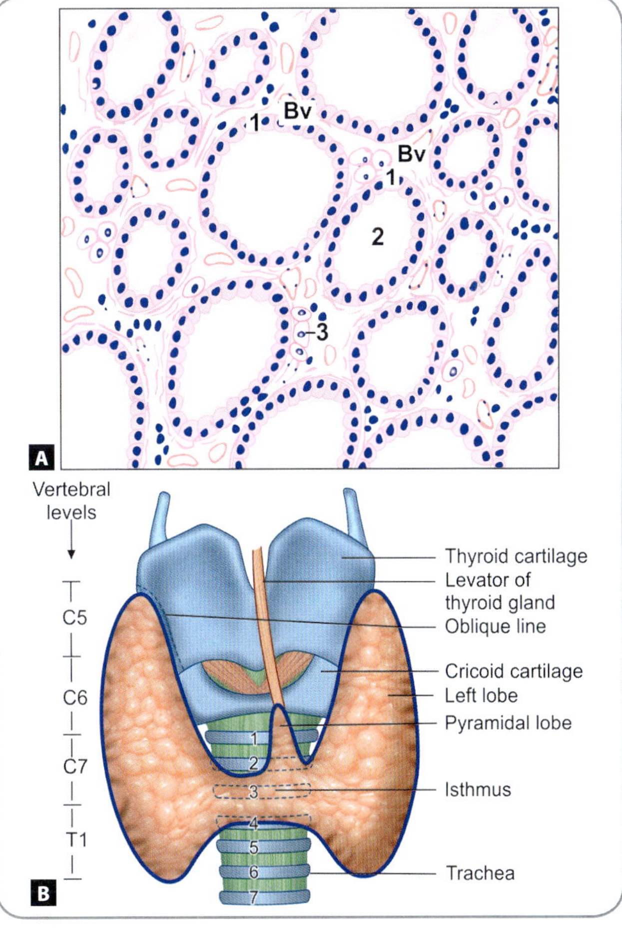

Figs 13.7A and B: A. Thyroid gland **B.** Outline of the thyroid gland as seen from the front, and its relationship to the larynx and trachea

Key: 1. Follicles lined by cuboidal epithelium **2.** Pink stained colloidal material **3.** Parafollicular cells **Bv.** Blood vessels

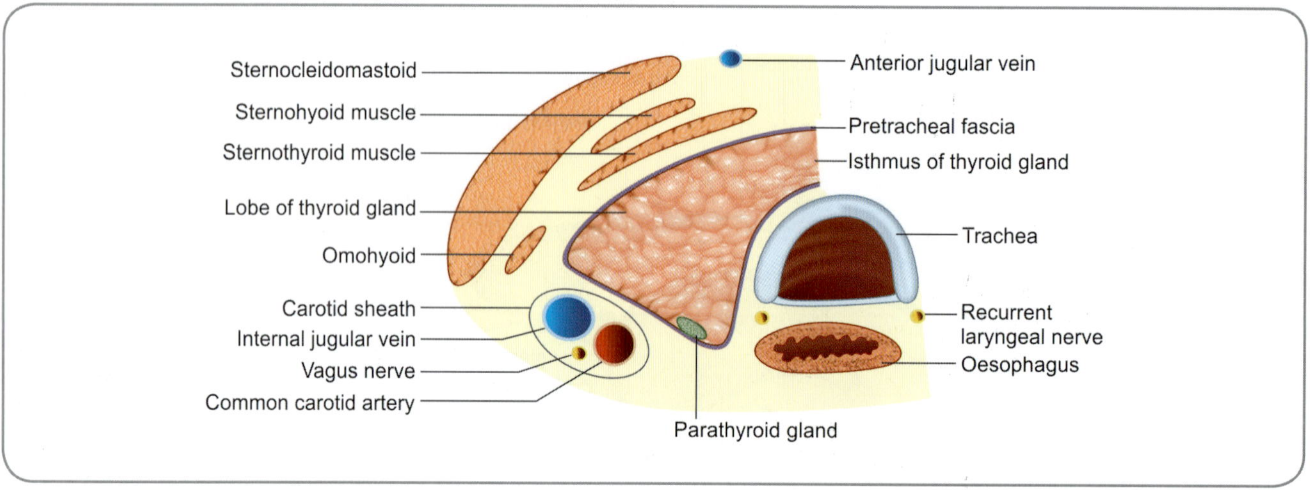

Fig. 13.8: Transverse section across the thyroid gland and related structures

It consists of **right and left lobes** which are joined across the midline by an **isthmus** (Fig. 13.7B).

The vertical diameter of each lobe is about 5 cm (2 inches) and that of the isthmus is about 1 cm (half inch). The anteroposterior diameter of each lobe is about 2 cm and the transverse diameter of the entire thyroid is about 8 cm. The gland is enclosed by the deep cervical fascia and weighs about 25 gms. It is heavier in females and enlarges during menstruation and pregnancy.

Surfaces and Part

Each lobe of the thyroid has three surfaces: lateral (or superficial), medial and posterior; two borders: anterior and posterior.

The **lateral surface** is convex and directed forwards and laterally (Fig. 13.8). It is covered by (from inner to outer aspects) the sternothyroid muscle, sternohyoid and superior belly of omohyoid, anterior part of the sternocleidomastoid, skin and fasciae.

The **medial surface** lies over the thyroid and cricoid cartilages of the larynx, uppermost parts of the trachea and oesophagus. Other structures deep to it are:

❏ **Parts of two muscles**—inferior constrictor of the pharynx and cricothyroid;

❏ **Two important nerves**—recurrent laryngeal nerve and external laryngeal nerve.

The two muscles actually separate the gland from the thyroid lamina and the cricoid cartilage.

The recurrent laryngeal nerve is deep to the thyroid as it ascends in the groove between the trachea and oesophagus while the external laryngeal nerve lies deep to the thyroid as it descends to reach the cricothyroid (Fig. 13.9).

The **posterior surface** of the lobe is directed posterolaterally. It is in contact with the carotid sheath and its contents.

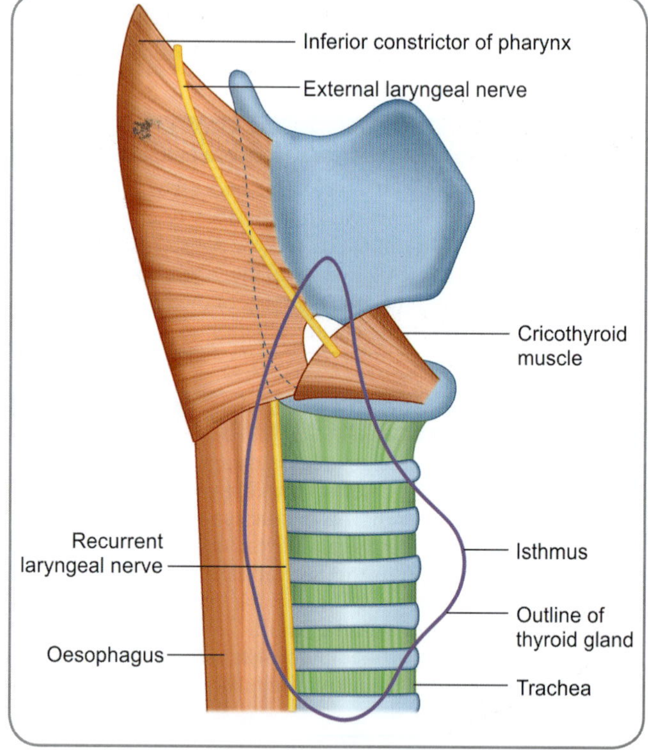

Fig. 13.9: Medial relations of the thyroid gland. The outline of the gland is shown in blue line

The lateral and medial surfaces of the lobe are separated by a sharp **anterior border**. A branch of the superior thyroid artery descends along this border.

The posterior and medial surfaces of the lobe are separated by the **posterior border** which is rounded. It is related to inferior thyroid artery, parathyroid glands and on the left side, to the thoracic duct (Fig. 13.10).

The **upper end** of each lobe extends up to the oblique line of the thyroid cartilage (Fig. 13.7B) [It is prevented

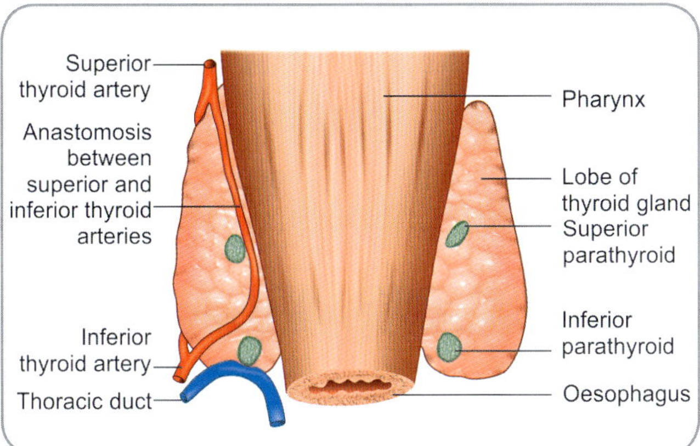

Fig. 13.10: Thyroid and parathyroid glands seen from behind

from extending further upwards by the insertion of the sternothyroid muscle to the oblique line]. The upper end lies opposite vertebra C5.

Dissection

The thyroid gland may be dissected if the anterior triangle of neck had already been studied. After examining the infrahyoid mucles, detach the sternothyroid from its origin and reflect it upward. The thyroid gland is now exposed. Study the capsule of the gland. Remove the capsule and observe the lobes of the gland. Locate the isthmus. Study the extent of the lobes and the isthmus. Observe the various vessels and nerves related to the gland. Divide the isthmus and try to reflect one of the lobes so that the posterior surface may be seen. Locate and study the parathyroid glands.

The ***lower end*** of the lobe lies at the level of the fifth or sixth tracheal ring (corresponding to vertebra T1). The level of the lower end is variable. An enlarging thyroid usually extends downwards.

The isthmus of the thyroid gland lies in front of the second, third and fourth rings of the trachea. It is covered in front by skin and fascia, sternothyroid and sternohyoid muscles and anterior jugular veins. The superior thyroid arteries of both sides anastomose along the upper border of the isthmus. The inferior thyroid veins leave the gland along the inferior border. A finger like projection of thyroid tissue frequently arises from the upper border of the isthmus. This is called the ***pyramidal lobe*** (Fig. 13.7B). Its upper end is attached to the hyoid bone by a cord of fibrous tissue or a band of muscular tissue. If a clear muscular band is present, it is then called the ***levator of the thyroid gland (levator glandulae thyroideae)***.

Capsule of Thyroid Gland

The thyroid gland is surrounded by a thin capsule made up of loose connective tissue. The capsule sends in septa into the gland. This is the ***true capsule*** of the gland. Outside this capsule, the thyroid has another sheath (called the ***false capsule***) formed by the pretracheal fascia. On each side, this fascia is thickened posteromedially to form a band connecting the lobe of the thyroid gland to the side of the cricoid cartilage. This band is called the ***lateral thyroid ligament*** or the ligament of Berry.

Blood Supply and lymphatic Drainage

Arterial Supply

The thyroid has a rich blood supply. The arteries supplying it are the superior thyroid branch of the external carotid artery, the inferior thyroid branch of the thyrocervical trunk, a small artery called the ***arteria thyroidea ima*** arising from the brachiocephalic trunk and accessory thyroid arteries derived from those supplying the trachea and oesophagus.

The superior thyroid artery, usually arising as the first branch of the external carotid artery, descends to the superior pole of the gland, pierces the false capsule and divides the anterior and posterior branches. The anterior branch supplies the anterior surface and the posterior branch supplies the lateral and medial surfaces.

The inferior thyroid artery, a branch of the thyrocervical trunk of the subclavian trunk, reaches the posterior surface of the gland; it usually divides into an ascending and an inferior branch which pierce the false capsule to enter into the gland substance. The ascending branch supplies the posterior surface and the parathyroid gland. The inferior branch supplies the inferior aspect of the gland.

The anterior branch of the superior thyroid artery, after running down the anterior border of the lobe and supplying the glandular substance, continues along the upper border of the isthmus, to anastomose with the corresponding artery of the opposite side. An anastomotic branch joining the superior and inferior thyroid arteries (of the same side) runs along the posterior border of the lobe.

Venous Drainage

The gland is drained by three veins—superior, middle and inferior. The superior thyroid vein drains the upper part of the gland and enters into the internal jugular vein. The middle thyroid vein drains the lower part of the gland and enters into the internal jugular vein. The inferior thyroid vein forms a plexus with the fellow of the opposite side; the plexus lies in front of the upper trachea; two veins are reformed from the plexus; the left drains into the left brachiocephalic vein and the right into the right brachiocephalic vein.

Development

The Thyroid gland arises as a median outgrowth of pharynx. When the pharyngeal arches and pouches are formed, a median diverticulum is given out from the floor of the primitive pharynx between the anterior and posterior rudiments of tongue. This diverticulum grows into the adjacent mesoderm and elongates further. Since it is a tube from the level of the tongue, it is called the thyroglossal duct. and lies in the midline of the rows downward ventral to the hyoid (second arch cartilage), thyroid and cricoid cartilages (fourth and sixth arch cartilages) and trachea. On reaching the level of the upper trachea, the tip of the diverticulum divides into two parts which spread out on the sides of the developing trachea and form the lobes of the thyroid gland (one lobe on either side). Due to rotational and regressional changes which happen during development, the thyroglossal duct passes ventral to the hyoid, winds around the bone's inferior border, goes up the posterior aspect and then descends ventral to the laryngeal cartilages. In adult life, the site of origin of the diverticulum is indicated by the foramen caecum. The pyramidal lobe indicates the original dichotomous branching of the distal end of the diverticulum.

Lymphatics (Fig. 13.11)

The lymphatics of the gland run in the interlobular connective tissue and form a capsular network of lymphatics. The vessels which arise from the network pass to the prelaryngeal, pretracheal and paratracheal nodes. Efferents from the pretracheal nodes pass to the upper deep cervical nodes and from the pretracheal and paratracheal nodes to the lower deep cervical nodes. Lymphatics from the lateral part of the gland may pass directly to the lower deep cervical nodes and from the inferior part to the brachiocephalic nodes. Some lymphatics, without an intervening node, may directly drain into the thoracic duct.

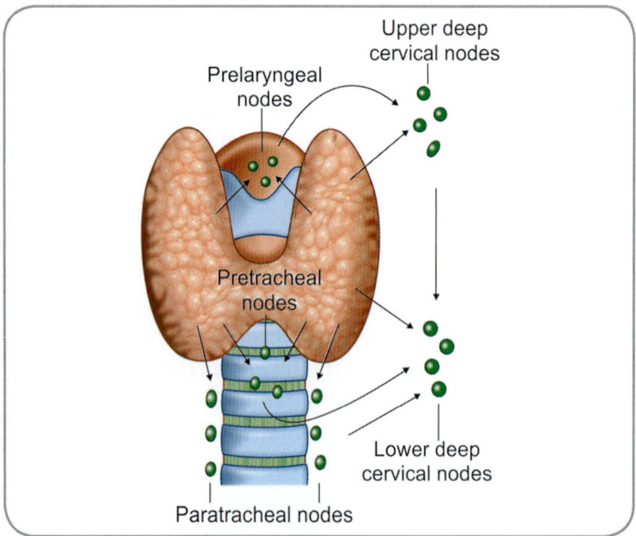

Fig. 13.11: Scheme to show the lymphatic drainage of the thyroid gland

Nerve Supply

Postganglionic sympathetic nerve fibres from the superior, middle and inferior cervical sympathetic ganglia supply the gland. They reach the gland through the cardiac plexus and the peri arterial plexuses surrounding the superior and inferior thyroid arteries.

Histology

Surface Marking

The isthmus and the lobes of the thyroid gland can be marked on the surface.

Isthmus: A 1.5 cm line is drawn across the trachea 1 cm below the arch of cricoids. This line marks the upper border of isthmus. Another line drawn parallel to this but 2cm below marks the lower border.

Lateral lobe: Point A is marked 1 cm below the lateral end of the lower border of the isthmus. Point B is marked 2.5 cm below and lateral to the lateral end of the lower borer of isthmus. Points A and B are at the same horizontal level. Point C is marked immediately in front of the anterior border of sternocleidomastoid at the level of the laryngeal prominence. Points A and B indicate the lower border of the lateral lobe and Point C marks the upper pole. Point C is joined to points A and B by two separate lines which are slightly convex outwards. Points A and B are joined by a transverse line. The outline so obtained marks the lateral lobe.

The thyroid gland is covered by a fibrous **capsule**. Septa extending into the gland from the capsule divide it into irregularly **lobules** (Fig. 13.7). Each lobule is made up of an aggregation of **follicles**. The functional unit of the gland is the follicle. Each follicle is more or less spherical and is lined by a layer of **follicular cells**. Typical follicular cells are cubical and rest on a basement membrane. The follicle has a cavity that is filled by a homogeneous material called **colloid** (which appears pink in haematoxylin and eosin stained preparations). The colloid contains the inactive stored form of thyroid hormones; this inactive form is the iodinated glycoprotein called the **iodothyroglobulin**. Follicular epithelial cells secrete this glycoprotein. Each follicle is surrounded by thin connective tissue stroma which contains capillaries, lymphatics and sympathetic nerve fibres.

The size and type of the follicular cells vary according to their level of activity. During moderate activity of the thyroid gland, the follicular cells are cubical, and a moderate amount of colloid is present in the follicles. When the gland is inactive, the follicles get distended with stored colloid and the cells are flattened. In highly active states, there is little colloid in the follicles and the cells become columnar.

The follicular cells secrete thyroglobulin and push it to the follicular lumen by exocytosis. When the circulating level of TSH increases, endocytosis of colloid occurs at the luminal aspect. Apical microvilli, which are sort in a resting follicular cell, become elongated, branch and surround pieces of colloid. They then engulf the colloid piece And bring it into the cell. Lysosomes move towards the engulfed colloid and fuse with it to form what is called a **secondary lysosome**. The lysosomal enzymes degrade the thyroglobulin and release the thyroid hormones. The hormones are pushed out of the cell through the basal aspect of the cell. They are taken up by the blood vessels in the stroma. To keep in tune with these steps, stromal vascularity increases when colloid endocytosis starts.

contd...

Histology *contd...*

The thyroid hormones influence the rate of metabolism. Iodine is an essential constituent of these hormones. One hormone containing three atoms of iodine in each molecule is called *triiodothyronine* or T3. Another hormone containing four atoms of iodine is called *tetraiodothyronine* or T4.

At some places, cells of a different type intervene between the follicular cells and the basement membrane. These are called the *parafollicular cells* and are completely different in function from the follicular cells. They produce a hormone called *calcitonin*. The parafollicular cells are also called the *'C' cells* meaning clear cells from their pale staining property. Sometimes, clus ters of parafollicullar cells may be found in the interfollicular connective tissue stroma. This last fact had initiated the name *parafollicular*.

Added Information

❑ The thyroid gland is the largest endocrine gland of the body having its effect and influence on all areas and parts of the body except itself, spleen, testis and uterus.

❑ The sympathetic nerve fibres which supply the gland are not secretomotor but are only vasomotor.

❑ Secretion of the thyroid gland is hormonally controlled by the pituitary.

❑ The gland clings to the four visceral tubes of the neck – pharynx, oesophagus, larynx and trachea. Together with these four tubes, the thyroid gland forms the 'viscera of the neck'.

Clinical Correlation

Congenital Anomalies

❑ Parts of the thyroid (isthmus or one lobe) may be absent.

❑ A pyramidal lobe may pass upwards from the isthmus or from one of the lobes. It may be just a short stump or may reach right up to the hyoid bone.

❑ Abnormal thyroid tissue may develop anywhere along the course of the thyroglossal duct. Thyroid tissue present under the mucosa of the dorsum of the tongue is called a *lingual thyroid*. Thyroid tissue may be embedded within the lingual musculature, may lie in the midline of the neck above or below the hyoid bone and may even be found in the thorax.

A fact of considerable surgical importance is that before removing an abnormally situated mass of thyroid tissue (*called ectopic thyroid*), the surgeon must make sure that a thyroid gland is present at the normal site. Sometimes, it may not be present, in which case the ectopic thyroid may be the only gland.

❑ Apart from these masses of thyroid tissue present along the path of the thyroglossal duct, ectopic thyroid tissue may be present in other locations. These include larynx, trachea, oesophagus, pleura or pericardium and ovaries.

❑ Remnants of the thyroglossal duct may persist and may form *thyroglossal cysts* which may be seen anywhere along the course of the duct. When a cyst opens on to the surface of the neck, a fistula is created. In treating these conditions, the surgeon has to remove the entire track of the thyroglossal duct. The duct forms a loop deep to the hyoid bone and, therefore, a part of this bone may have to be excised for complete removal of the duct.

❑ Remnants of the thyroglossal duct may give rise to a carcinoma.

Hypothyroidism and Hyperthyroidism

❑ In examining a patient in whom an enlargement of the thyroid is suspected, it is useful to remember that the thyroid moves up and down during swallowing.

❑ Deficient intake of iodine (common in areas where drinking water does not contain iodine) can lead to benign enlargement of the thyroid gland. The enlarged thyroid is referred to as *goitre*. The symptoms are those of hypothyroidism. Hypothyroidism in infants leads to *cretinism*. A child with cretinism has a puffed face with a protruding tongue, a bulky belly and sometimes an umbilical hernia. Hypothyroidism in adults is manifested by symptoms including a slow pulse, cold intolerance, mental and physical lethargy and a hoarse voice. In advanced cases, the condition is called *myxoedema*. There is deposition of mucopolysaccharides in the subcutaneous tissue at various sites resulting in a non-pitting oedema. The face is bloated, the lips are thick and protuberant and the expression is dull. Hypothyroidism can be caused by underdevelopment of the thyroid gland and this, in turn can be caused by maternal and foetal iodine deficiency. It can also occur as a result of destruction of thyroid tissue because of carcinoma, thyroiditis, surgical removal of the thyroid gland and prolonged use of antithyroid drugs.

❑ Hyperthyroidism is also referred to as *thyrotoxicosis* or *toxic goitre*. The condition is much more common in women than in men. It is marked by nervousness, loss of weight, tachycardia and palpitation, excitability, tremors of the outstretched hands and exophthalmos. The most important causes are *Graves' disease* (or *diffuse toxic goitre* in which no nodules are felt on palpation of the thyroid), *multinodular goitre* and *toxic adenoma*.

Tumours of Thyroid Gland

Tumours of the thyroid may be benign or malignant. A tumour can press upon or involve the trachea or carotid sheath. Involvement of the recurrent laryngeal nerve may occur. Upward expansion of a tumour of the thyroid is limited by the fact that the sternothyroid muscles, which cover the thyroid gland in front, are attached above to the thyroid cartilage. The tumours, therefore, tend to grow downwards and can even enter the thorax (retrosternal goitre).

contd...

Clinical Correlation *contd...*

An operation for removal of the thyroid gland is called *thyroidectomy*. This may be required in some cases of hyperthyroidism. Normally, the posterior parts of the gland (and so, the parathyroids) are left behind. It is of interest to note that the part left behind receives an adequate blood supply through branches from the tracheal and oesophageal arteries (even after the main thyroid arteries have been ligated). In thyroidectomy, the main arteries have to be cut and tied. The superior thyroid artery is intimately related to the external laryngeal artery at a high level, but they separate near the upper pole of the gland. The surgeon, therefore, always cuts this artery as near the gland as possible so that the nerve is not cut. In contrast, the inferior thyroid artery is closely related to the recurrent laryngeal nerve near the gland and has to be ligated as far away as possible from the gland. The veins of the gland form a plexus deep to the capsule of the gland. In order to avoid injury to them, the surgeon usually removes the thyroid along with its true capsule. The possibility of the presence of a thyroidea ima artery should be remembered because the artery is a potential source of bleeding.

Inflammation of Thyroid

Inflammation of the thyroid gland is called *thyroiditis*. It can be caused by infection or by various other causes. The most important form of thyroiditis is caused by an autoimmune process (*autoimmune or lymphatic thyroiditis*). It is also called *Hashimoto's disease*. The thyroid gland is enlarged and is infiltrated with lymphocytes.

Imaging of Thyroid

The thyroid gland can be imaged using radioactive iodine. Different patterns of take up of iodine help in diagnosis of disorders. The parathyroid glands can also be seen. Ultrasound examination of the thyroid is also a useful diagnostic technique.

PARATHYROID GLANDS

The parathyroid glands are so called because they lie in close relationship to the thyroid gland. Normally, there are two glands, one superior and one inferior, on either side, there being four glands in all. However, the number can be more or less. Each gland is roughly oval and weighs about 50 mg.

On each side, the ***superior parathyroid gland*** lies near the middle of the posterior border of the thyroid gland. It is relatively constant in position. The ***inferior parathyroid gland*** lies near the lower end of the posterior border of the thyroid gland. Its position is variable. It may lie above or below the inferior thyroid artery. It may lie outside the false capsule of the thyroid, between the false and true capsules or deep to the true capsule within the substance of the thyroid. Occasionally, the inferior parathyroid gland may lie below the level of the thyroid gland and may even descend into the superior or posterior mediastinum.

The superior parathyroid gland receives a branch from the anastomotic channel connecting the superior and inferior thyroid arteries. The inferior gland receives a branch from the inferior thyroid artery.

Development

Epithelium of the third pharyngeal pouch differentiates into the (inferior) parathyroid gland in the 5th week of intrauterine life. The ventral region of the same pouch develops into the thymus. As both these primordial lose their connections to the pharyngeal wall, the thymus migrates caudally. It pulls the parathyroid primordium also. At about the same time, the epithelium of the fourth pharyngeal pouch develops into the (superior) parathyroid. It loses contact with the pharyngeal wall and attaches to the posterior surface of the thyroid gland that is developing nearby. Due to the pull of the thymus, the third pouch parathyroid reaches more caudal to the fourth pouch parathyroid (thus becoming the inferior parathyroid) and attaches to the posterior aspect of thyroid gland. If migration is incomplete, the inferior parathyroid may occupy a higher level or may be lodged at abnormal sites.

Histology

Histology and Function

Each parathyroid gland has a very thin connective tissue capsule that sends in intraglandular septa; however, there are no distinct lobules. Until puberty, the parenchyma consists essentially of columns of chief cells separated by sinusoids.

At around puberty and after, two changes occur:

- Adipose cells accumulate in the stroma;
- A second cell type makes its appearance and increases in number with age. These are the oxyphil cells.

The ***chief cells*** or ***principal cells*** can be in the active or inactive state. Active cells have large Golgi complexes and less number of granules. Inactive cells have abundant granules. In the normal adult gland, inactive cells outnumber the active cells. The active and inactive states of these cells depend on the levels of serum calcium. The chief cells secrete parathormone.

The oxyphil cells or eosinophil cells are so named due to the presence of more cytoplasm which stains deep with eosin. They have smaller nuclei and abundant mitochondria. Their function is unknown.

contd...

Histology *contd...*

The parathyroid glands produce a hormone called the **parathyroid hormone** (PTH) or **parathormone**. This hormone helps to maintain a suitable level of calcium ions in blood. When there is a tendency for serum calcium levels to fall, calcium is removed from the bone stores in bringing serum levels back to normal. Simultaneously, the excretion of calcium by the kidney is decreased and calcium absorption by the intestines is increased.

Calcitonin secreted by the parafollicular cells of the thyroid gland has effects opposite to those of the parathyroid hormone. A decrease in serum calcium level stimulates the secretion of parathyroid hormone while an increase stimulates the secretion of calcitonin.

Clinical Correlation

The variations in position of parathyroid glands are of considerable importance to a surgeon trying to locate the glands. The parathyroid glands can be seen when the thyroid is imaged using radioactive iodine. The areas where radioactive material is located can be recorded on a gamma camera. Computer separation of images reveals the location of the parathyroids.

Hyperparathyroidism: Excessive amounts of circulating parathormone can be present in tumours of the parathyroid gland (parathyroid adenoma). As a result, calcium is depleted from bones that become weak (and can fracture). Increased urinary excretion of calcium may lead to formation of urinary calculi.

Hypoparathyroidism: Calcium levels in the blood decrease leading to muscular irritability and convulsions. The condition may be spontaneous or may occur following accidental removal of parathyroid glands during thyroidectomy.

CAROTID SINUS

The term '***carotid sinus***' is applied to a dilated segment of the common carotid body located at its bifurcation. The dilatation usually extends on to the initial part of the internal carotid artery and may be confined to it. In the region of the dilatation, the tunica media in the arterial wall is thin, but the adventitia is thick. Numerous nerve endings are seen in the adventitia.

The region of the carotid sinus is surrounded by a nerve plexus. The main contribution to this plexus is by the carotid branch of the glossopharyngeal nerve. The plexus also receives fibres from the superior cervical sympathetic ganglion and from the vagus nerve.

The afferent nerve terminals present over the carotid sinus are stimulated by alterations in blood pressure. Afferent impulses arising from the sinus play an important role in reflex control of blood pressure.

Mesoderm condenses around the third pharyngeal arch artery. This is the primordium of the carotid body. Nerve fibres from the 3rd arch nerve (glossopharyngeal) start supplying the condensation. Gradually neuroblasts migrate along the nerve fibres into the condensation and differentiate into chemoreceptor cells. Similar condensations appear in relation to fourth arch arteries (aortic bodies) and are supplied by the vagus nerves. The rest of the developmental process is the same.

CAROTID BODIES AND PARAGANGLIA

These are small oval structures, present one on each side of the neck, at the bifurcation of the common carotid artery (i.e., near the carotid sinus). The main function of the carotid bodies is that they act as chemoreceptors which monitor the oxygen and carbon dioxide levels in blood. They exercise reflex control on the rate and depth of respiration through respiratory centres located in the brainstem. In addition to this function the carotid bodies are also believed to have an endocrine function.

The most conspicuous cells of the carotid body are called the ***glomus cells*** (or type I cells). These are large cells which have several similarities to neurons. Apart from the glomus cells, various other cells are also present in the carotid bodies. They include a few sympathetic and parasympathetic postganglionic neurons.

The carotid bodies are generally included under the description of ***paraganglia***. This term is used to describe small collections of neuroendocrine cells present in association with autonomic nerves. Structures similar to the carotid bodies present in relation to the inferior aspect of the arch of the aorta are also included in paraganglia. They are called the ***aortic bodies*** or ***aortico-pulmonary paraganglia.***

Multiple Choice Questions

1. The basophilic cells of adenohypophysis which are sometimes classified as acidophils due to their intermediary nature are:
   a. Mammotrophs
   b. Corticotrphs
   c. Gonadotrophs
   d. Thyrotrophs
2. One of the following is not part of the neurohypophysis. Which is it?
   a. Pars posterior
   b. Median eminence
   c. Infundibulum
   d. Pars glandularis
3. Nervus conarii contains:
   a. Preganglionic sympathetic fibres
   b. Preganglionic parasympathetic fibres
   c. Postganglionic sympathetic fibres
   d. Postganglionic parasympathetic fibres
4. Thyroid follicular cells when active:
   a. Are cuboidal
   b. Leave the basement membrane
   c. Adhere to parafollicular cells
   d. Are columnar
5. Oxyphil cells found in the parathyroid gland:
   a. Secrete parathormone
   b. Are eosinophilic
   c. Induce thyrocalcitonin
   d. Have no mitochondria

ANSWERS

1. b   **2.** d   **3.** c   **4.** d   **5.** b

Clinical Problem-solving

Case Study 1: A 46-year-old man presents with thick lips, oversized hands and feet, thickened scalp and enlarged and angulated face. Though he is not able to narrate properly, he says the physician whom he saw some weeks ago told him that he has a tumour somewhere near his brain.
- What do you think is the condition that the man has?
- What is the primary cause for these symptoms?
- If the same disorder had occurred in earlier life, what would the man have suffered from?

Case Study 2: A 54-year-old woman presents with hoarseness of voice. She says that she had not been able to withstand cold climates and has been feeling extremely dull, lethargic and 'down'. You notice that her face is bloated. There is oedema especially over her shin.
- What condition does the woman have?
- What is the reason for her face to get bloated?
- In what way does examination of the shin give an idea of the diagnosis?

(For solutions see Appendix).

Blood Vessels of Head and Neck

The main arterial supply to structures of the head and neck region is from the common carotid and subclavian arteries.

COMMON CAROTID ARTERIES

The right common carotid artery arises from the brachiocephalic trunk and the left common carotid artery directly from the arch of aorta. It follows that the left common carotid artery runs parts of its course in the thorax. The courses and relations of the cervical parts of the right and left common carotid arteries are similar.

Course of the Common Carotid Artery in the Neck

Starting behind the corresponding sternoclavicular joint, each artery runs upwards and somewhat laterally up to the level of the upper border of the thyroid cartilage, where it terminates by dividing into the internal and external carotid arteries (Fig. 14.1). It normally gives no other branches except a few tiny twigs to the carotid sheath and the carotid body.

The artery is enclosed in a fibrous *carotid sheath* that also encloses the internal jugular vein that is lateral to the artery and the vagus nerve that lies posterolateral in the interval between the artery and the vein.

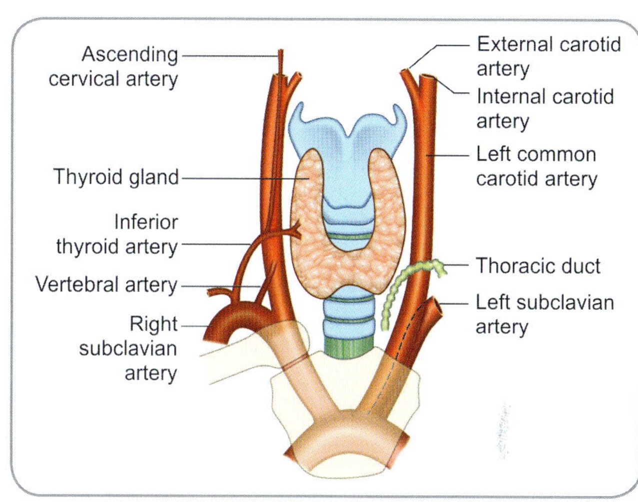

Fig. 14.1: Relationship of common carotid artery to the larynx, trachea and thyroid. Some structures deep to the artery are also shown

Posterior Relations

Behind the carotid sheath, the common carotid arteries are related to the sympathetic trunk, ascending cervical artery and the vertebral artery. The inferior thyroid artery runs transversely behind the lower part of the artery. In addition to these structures, on the right side, the artery is crossed posteriorly by the recurrent laryngeal nerve and on the left side, it is crossed by the thoracic duct.

Still posterior to the sympathetic trunk and the prevertebral fascia are the cervical transverse processes and the prevertebral muscles. As the artery runs up, it rests on the transverse process of the C6 vertebra and can easily be compressed against the anterior tubercle of the process. The tubercle is named the carotid tubercle for the same reason and is at the level of the cricoid cartilage.

Anterior Relations

The artery is crossed anterolaterally by the intermediate tendon or the superficial belly of omohyoid at the level of cricoid cartilage. Below the omohyoid, the artery is situated deeply and is covered by the sternohyoid and sternothyroid muscles, sternocleidomastoid, deep cervical fascia, platysma and skin from within outwards (Fig. 14.2). Above the omohyoid, it is more superficial and covered only by the skin, platysma, deep cervical fascia and the medial margin of sternocleidomastoid. It is also related anteriorly to the ansa cervicalis, superior thyroid vein, middle thyroid vein and the anterior jugular vein.

Medial Relations

On the medial side, the common carotid artery is related to the recurrent laryngeal nerve between the trachea and the oesophagus, the corresponding lobe of the thyroid gland and a part of the inferior thyroid artery.

A fusiform dilatation involving the bifurcation of the common carotid artery and the beginning of the Internal carotid artery is called the **carotid sinus**. The carotid sinus acts as a baroreceptor and controls the intracranial blood pressure. A small, oval neurovascular structure situated close to the bifurcation of the common carotid artery is called the **carotid body**. The carotid body acts as a chemoreceptor and monitors oxygen tension within the artery.

> **Surface Marking**
>
> Point A is marked on the sternoclavicular joint. Point B is marked on the anterior border of sternocleidomastoid at the level of the upper border of the thyroid cartilage. The two points are joined by a broad line. This line indicates the common carotid artery on the surface.

Internal Carotid Artery

The internal carotid artery arises from the common carotid artery, ascends to the base of skull, enters the carotid canal,

passes into the cranial cavity and supplies the brain. The course of the artery can be divided into two parts—**course in the neck** and **course in the cranial cavity**.

Internal Carotid Artery in the Neck

The internal carotid artery begins at the upper border of the thyroid cartilage and ascends to reach the base of skull where it enters the carotid canal. Each artery can be considered as the main upward continuation of the common carotid artery and occupies a similar position. Like the common carotid artery, it is surrounded along with the internal jugular vein and the vagus nerve, by a carotid sheath (Fig. 14.3). It lies on the transverse processes of the upper cervical vertebrae being separated from them by the longus capitis and the superior cervical sympathetic ganglion and is crossed posteriorly by the superior laryngeal nerve. On the medial side, the artery is related to the pharynx. Pharyngeal veins separate it from the ascending pharyngeal artery and the superior laryngeal nerve (Fig. 14.4).

At its upper end, the internal jugular vein lies posterior to the artery with the glossopharyngeal, vagus, accessory and hypoglossal nerves lying between them. Superficially, it is crossed by styloid process, stylohyoid muscle, posterior belly of digastric, stylopharyngeus muscle, glossopharyngeal nerve, hypoglossal nerve, superior root of ansa cervicalis, posterior auricular and occipital arteries, facial and lingual veins.

The Internal Carotid artery does not give any branch in the neck.

> **Surface Marking**
>
> Point A is marked on the anterior border of sternocleidomastoid at the level of the upper border of thyroid cartilage. Point B is marked immediately posterior to the condyle of mandible. The two points are united by a broad line that marks the cervical part of the internal carotid artery on the surface.

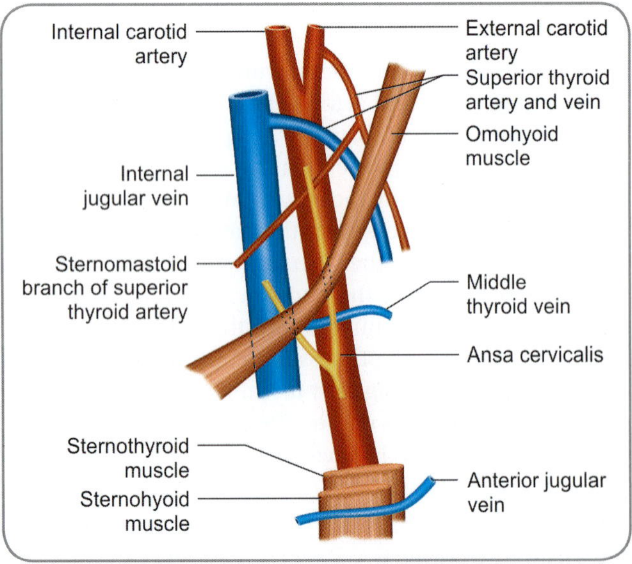

Fig. 14.2: Right lateral view showing structures crossing superficial to the common carotid artery

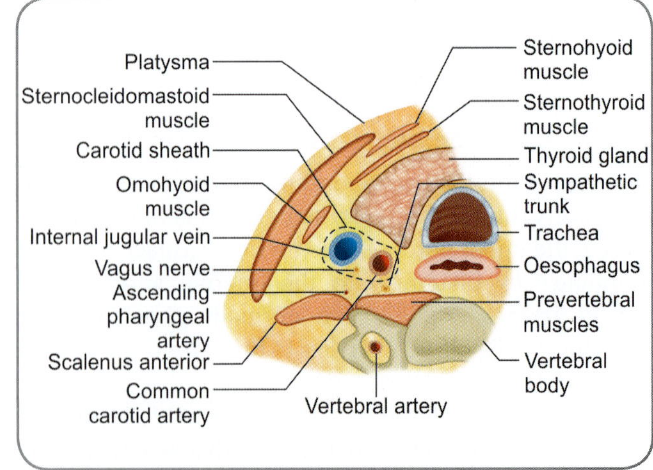

Fig. 14.3: Transverse section showing structures crossing superficial to the common carotid artery

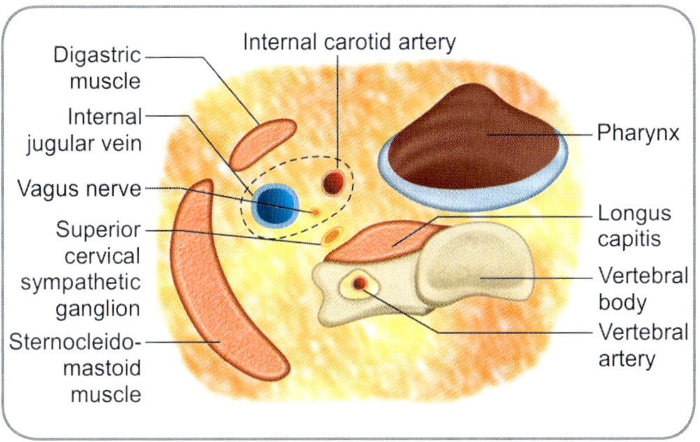

Fig. 14.4: Transverse section of neck to show relations of the internal carotid artery

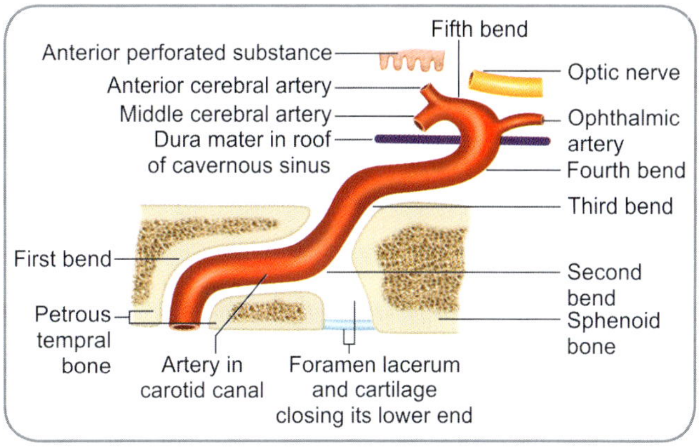

Fig. 14.5: Scheme to show the intracranial course of the internal carotid artery

---

Structures Crossing Internal Carotid Artery

Superficial to Artery

- Styloid process
- Stylohyoid muscle
- Stylopharyngeus muscle
- Digastric, posterior belly
- Glossopharyngeal nerve
- Pharyngeal branch of vagus nerve
- Hypoglossal nerve
- Ansa cervicalis, superior root
- Posterior auricular artery
- Occipital artery
- Sternomastoid branch of occipital artery
- Facial vein
- Lingual vein

Deep to Artery

- Superior laryngeal branch of vagus nerve.

Internal Carotid Artery in the Cranial Cavity

The cranial part of the internal carotid artery has a complicated course. Successive parts of the artery run vertically and horizontally (Fig. 14.5).

On reaching the base of the skull, the artery enters the petrous part of the temporal bone through the external opening of the carotid canal. It then bends sharply (Fig. 14.5, first bend) to run horizontally forward and medially through the carotid canal to reach the foramen lacerum. It now undergoes a second bend to run vertically through the upper part of this foramen (above the fibrocartilage which fills the foramen in life) to enter the cranial cavity. It pierces the periosteum and enters the cavernous sinus. Here it undergoes a third bend to run forward on the side of the body of the sphenoid bone. Near the anterior end of the body of the sphenoid bone, it again bends upward (fourth bend) on the medial side of the anterior clinoid process. Here it pierces through the dura mater forming the roof of the cavernous sinus and comes into relationship with the cerebrum. The artery now turns backwards (fifth bend) to

reach the anterior perforated substance of the brain. The artery terminates here by dividing into the anterior and middle cerebral arteries.

The intracranial course of the internal carotid artery can, therefore, be divided into a ***petrous part***, a ***cavernous part*** and a ***cerebral part***. The petrous part is where the artery runs in the carotid canal which is within the petrous temporal bone. The cavernous part is where the artery runs within the cavernous sinus. The cerebral part is where it is in relation to the brain and divides into its terminal branches.

Relations

Throughout its course the artery is surrounded by a plexus of sympathetic nerves (called the ***internal carotid plexus***) derived from the superior cervical sympathetic ganglion and by a plexus of veins that connects the intracranial veins to those outside the skull.

As it lies in the carotid canal, the artery is closely related to the middle ear, the auditory tube and the cochlea.

As the artery passes through the cavernous sinus (in the inferomedial part of the sinus), it is related to several cranial nerves which are embedded in the lateral wall of the sinus (Fig. 14.6). From above downward, these are:

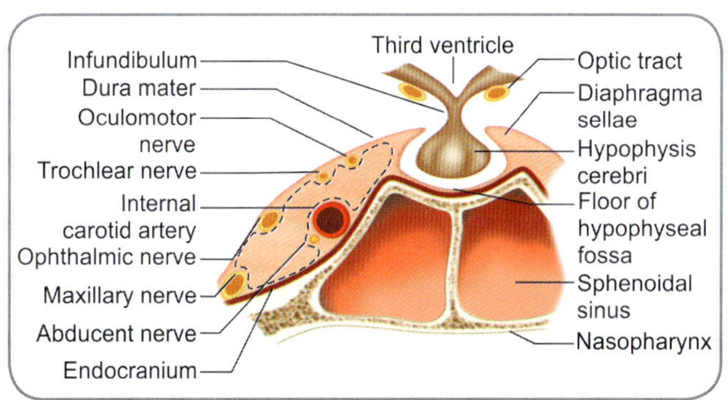

Fig. 14.6: Coronal section through the cavernous sinus showing the internal carotid artery and related structures

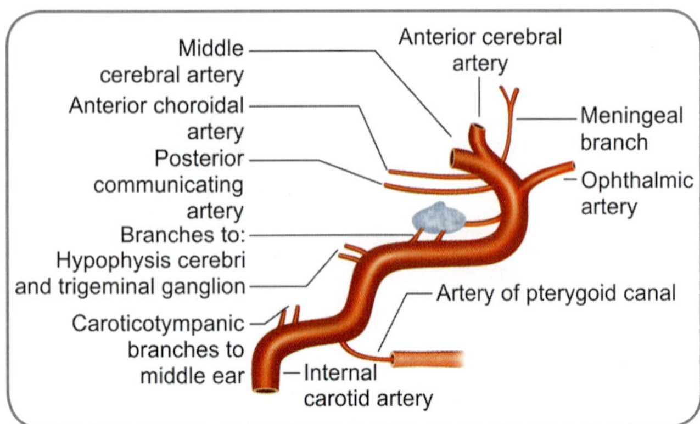

Fig. 14.7: Scheme to show the branches given off by the internal carotid artery

- The oculomotor, trochlear and ophthalmic divisions of the trigeminal nerve;
- The maxillary division of the trigeminal nerve;
- The abducent nerve—(the artery has an intimate relationship with this nerve that runs in close contact with the inferolateral side of the artery);

After piercing the dura mater, the artery has the optic nerve above it and the oculomotor nerve below it.

Branches of Internal Carotid Artery

The internal carotid artery gives off three large branches. These are (Fig. 14.7):
- The **ophthalmic artery** to the orbit and
- The **anterior** and **middle cerebral arteries** to the brain.
  Apart from these major branches, the internal carotid artery also gives smaller branches. These are:
- Caroticotympanic branch that is given out in the petrous temporal bone and that which perforates the carotid canal to enter the tympanum;
- Twigs to the trigeminal ganglion, cranial nerves and duramater which are given out in the cranial cavity;

- Branches to hypophysis given out as the main artery turns up medial to the anterior clinoid process;
- Posterior communicating artery that is given out just proximal to the termination of the internal carotid artery and
- Anterior choroid artery that is given out near the termination of the internal carotid artery.

Ophthalmic Artery

The ophthalmic artery, a branch of the internal carotid artery given out medial to the anterior clinoid process, passes forwards to enter the cavity of the orbit through the optic canal. In this canal, it lies inferolateral to the optic nerve. Having entered the orbit, the artery is at first lateral to the optic nerve (Fig. 14.8). It then crosses above the nerve to reach the medial wall of the orbit and runs forward along this wall. Near the medial end of the upper eyelid, it divides into its two terminal branches, namely, the supratrochlear and dorsal nasal arteries. Most of the branches accompany correspondingly named nerves.

Branches

- The central artery of the retina is the first branch of the ophthalmic artery. It arises from the ophthalmic artery when the latter is still within the optic canal. It first lies below the optic nerve. It pierces the dural sheath of the nerve and runs forwards for a short distance within the sheath. It then enters the substance of the nerve and runs forwards in its centre to reach the optic disc. Here, it divides into branches that supply the retina (Fig. 14.9).
- The largest branch of the ophthalmic artery is the lacrimal artery (accompanies the lacrimal nerve); it is given out near the optic canal and runs forward along the lateral wall of the orbit, traverses and supplies the lacrimal gland, ends in the upper and lower eyelids as the upper and lower lateral palpebral arteries.

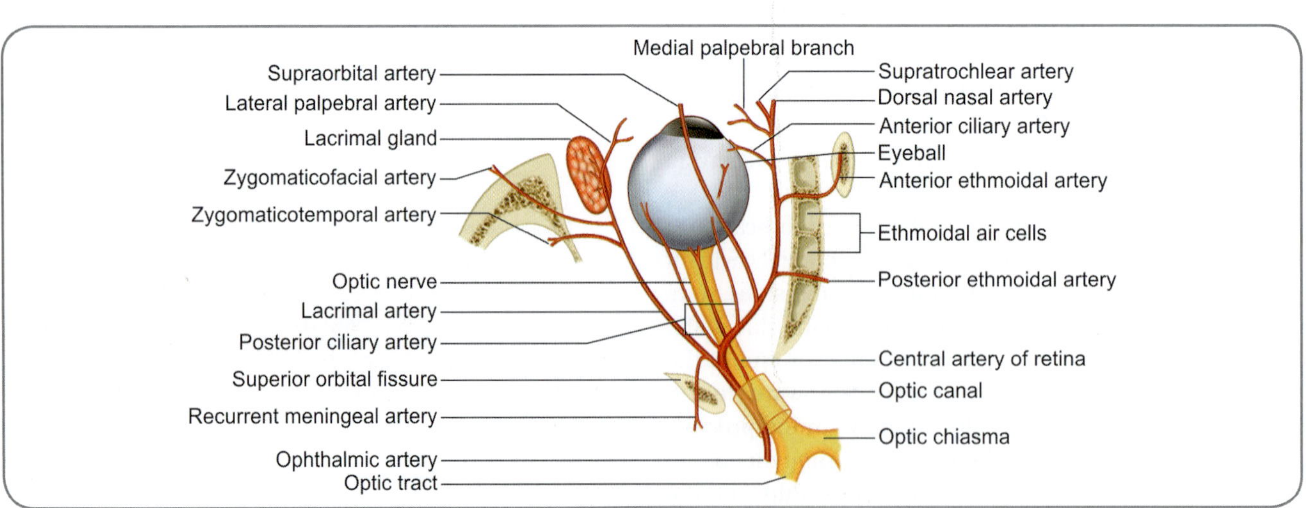

Fig. 14.8: Scheme to show branches of the ophthalmic artery

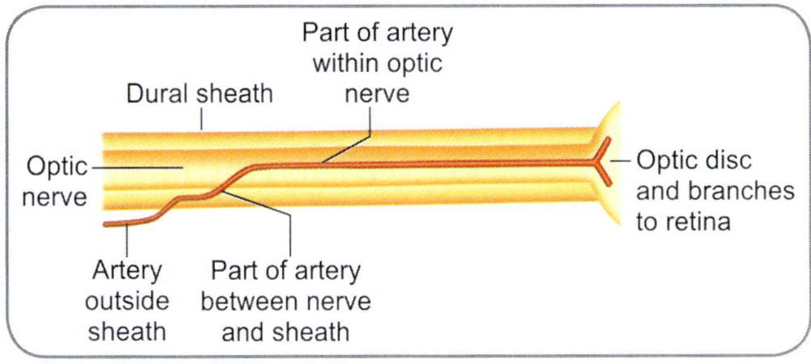

Fig. 14.9: Course of central artery of retina

Just near its origin from the ophthalmic artery, the lacrimal artery gives off a recurrent meningeal branch that runs backward to enter the middle cranial fossa through the superior orbital fissure and anastomoses with the middle meningeal artery. The lacrimal artery gives off two zygomatic branches that enter canals in the zygomatic bone. One branch appears on the face through the zygomaticofacial foramen, supplies the region and anastomoses with the other arteries of the area. The other branch appears on the temporal surface of the bone through the zygomaticotemporal foramen, supplies the region and anastomoses with the deep temporal arteries.

❑ The posterior ciliary arteries (long and short) arise as the ophthalmic artery crosses the optic nerve. They run forward around the optic nerve and supply the eyeball. Of the six or more posterior ciliary arteries, two are long and called the ***long posterior ciliary arteries***. After piercing the sclera, they run forward between the sclera and the choroid (one on either side) to anastomose with the anterior ciliary arteries.

❑ The supraorbital branch (accompanies the supraorbital nerve) is also given out where the ophthalmic artery crosses the optic nerve; it supplies the skin of the forehead.

❑ The anterior and posterior ethmoidal branches (accompanies the anterior and posterior ethmoidal nerves) enter the anterior and posterior ethmoidal foramina in the medial wall of the orbit to supply the ethmoidal (and frontal) air sinuses, then enter the anterior cranial fossa and subsequently terminate in the nasal cavity supplying structures there.

❑ The anterior ciliary arteries arise from the ophthalmic artery (mainly from the muscular branches to recti) near the anterior part of the eyeball; they pierce the sclera just behind the sclerocorneal junction, anastomose with the long posterior ciliary artery of the same side and supply the ciliary body and iris. Before piercing the sclera, the anterior ciliary arteries give out twigs to the deep conjunctival plexus around the corneal margin.

❑ The medial palpebral branches supply the eyelids.

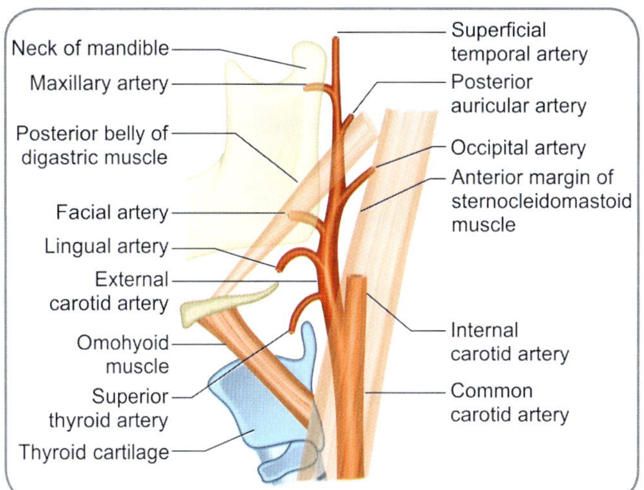

Fig. 14.10: Scheme to show the landmarks to which the external carotid artery and its branches are related—the boundaries of the carotid triangle are shown

❑ The supratrochlear artery (accompanies the supra-trochlear nerve) is one of the terminal branches of the ophthalmic artery. It supplies the skin of the forehead (along with the supraorbital artery).

❑ The dorsal nasal branch (accompanies the infratrochlear nerve) supplies the upper part of the nose.

❑ In addition to these branches, the ophthalmic artery and its branches give off several muscular twigs to the muscles of the orbit.

Details of arterial supply to the brain have been considered in the section on ***Brain***.

External Carotid Arteries

The external carotid arteries arise from the common carotid artery at the level of the upper border of the thyroid cartilage and at the level of the disc between the third and fourth cervical vertebrae. The branches of the external carotid artery supply structures of the head and neck outside the cranial cavity (Fig. 14.10).

Course and Relations

The external carotid artery, which is the narrower of the two terminal branches of the common carotid artery, extends from the upper border of the thyroid cartilage to the back of the neck of mandible.

From its origin, the artery runs upwards and terminates behind the neck of mandible within the parotid gland into its two terminal branches. The lower part of the artery is anterior and medial to the internal carotid. Its upper part is lateral to the internal carotid. Though the artery is medial to the internal carotid artery, it is named as external carotid because it supplies structures which are external to the cranial cavity, whereas the internal carotid artery supplies structures within the cranial cavity.

249

The lower part of the artery is located within the carotid triangle. Here it is relatively superficial being covered by skin, superficial and deep fascia and by the anterior margin of the sternocleidomastoid muscle. Above the triangle, the artery lies deep to the hypoglossal nerve and its venae commitantes, posterior belly of the digastric muscle and the parotid gland with the facial nerve within it. Deep to the artery, the pharynx is located which is separated from the upper part of the artery by the styloid process, muscles attached to it and by the internal carotid artery. The glossopharyngeal nerve and the superior laryngeal nerve are also related posteriorly to the external carotid artery.

> **Surface Marking**
>
> Point A is marked on the anterior border of sternocleidomastoid at the level of the upper border of thyroid cartilage. Point B is marked midway between the angle of mandible and mastoid process. These two points are joined by a broad line that is slightly convex forwards in the lower half and convex backwards in the upper half. This sinuous line marks the external carotid artery on the surface.

Branches

The branches of the external carotid artery and their levels of origin are as follows (in order of origin) (Fig. 14.10):

- *Ascending pharyngeal artery* arises from the deep aspect of the external carotid artery just above its lower end.
- *Superior thyroid artery arises* from the anterior aspect of the external carotid just below the level of the greater cornu of the hyoid bone.
- *Lingual artery* arises from the anterior aspect of the external carotid artery opposite the tip of the greater cornu of the hyoid bone.

- *Facial artery* arises from the anterior aspect of the external carotid a little above the origin of the lingual artery.
- *Occipital artery* arises from the posterior aspect of the external carotid opposite the origin of the facial artery.
- *Posterior auricular artery* arises from the posterior aspect of the external carotid just above the level at which the latter is crossed by the posterior belly of the digastric muscle.
- *Superficial temporal artery* and the maxillary artery are the terminal branches of the external carotid artery. They begin behind the neck of mandible, in the substance of the parotid gland.

Ascending Pharyngeal Artery

The ascending pharyngeal artery is the only medial branch of the external carotid artery and runs upwards to the base of the skull, lying between the pharynx and the internal carotid artery (Fig. 14.11). It gives branches to the pharynx and adjacent structures like auditory tube, tonsil and soft palate. It also gives the inferior tympanic branch to the middle ear and meningeal branches which enter the skull though the jugular foramen and the hypoglossal canal.

Superior Thyroid Artery

The superior thyroid artery runs inferomedially accompanied by the external laryngeal nerve to reach the upper pole of the thyroid gland, where it divides into *anterior* and *posterior thyroid branches.* These branches ramify over the corresponding surfaces of the gland. The terminal part of the anterior branch runs across the upper part of the isthmus of the gland to anastomose with the artery of the opposite side (Fig. 14.12). The posterior branch runs downwards along the posterior border of the thyroid to anastomose with the inferior thyroid artery.

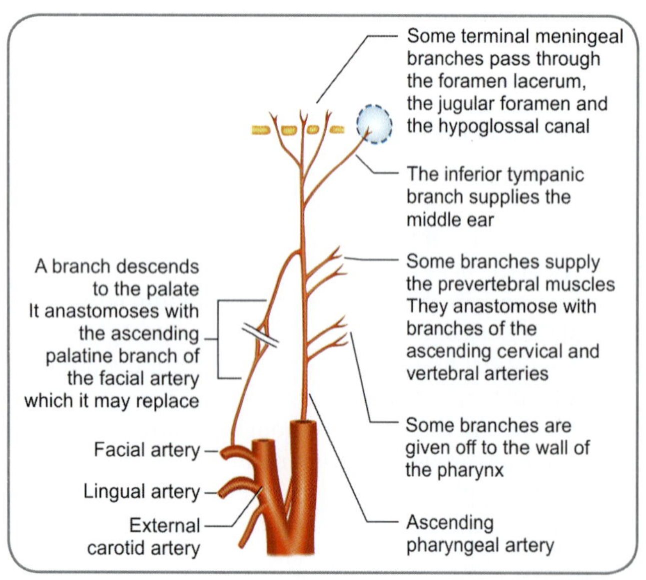

Fig. 14.11: Scheme to show the distribution of the ascending pharyngeal artery

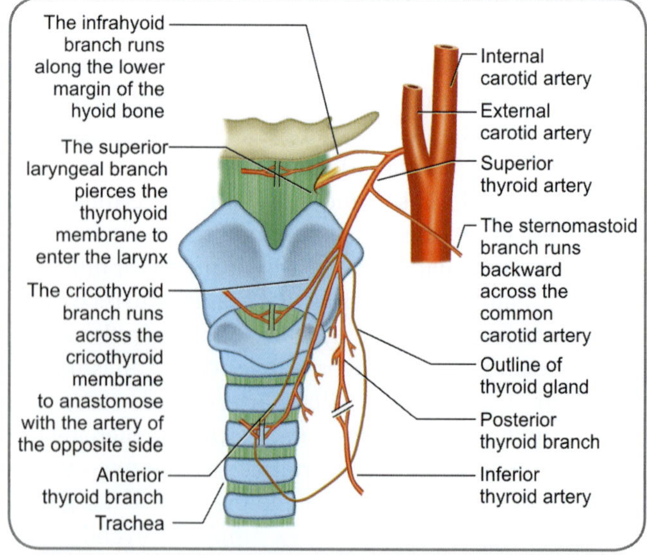

Fig. 14.12: Scheme to show the distribution of the superior thyroid artery

Other branches of the superior thyroid artery are the *infrahyoid branch*, the *superior laryngeal branch* which accompanies the internal laryngeal nerve and pierces the thyrohyoid membrane to supply the larynx, the *cricothyroid* branch and the *sternomastoid* branch.

Structures Crossing External Carotid Artery

Superficial to artery
Stylohyoid muscle
Posterior belly digastric muscle
Facial nerve (in parotid gland)
Hypoglossal nerve
Common facial vein
Superior thyroid vein (sometimes)
Veins along hypoglossal nerve

Deep to Artery

Styloid process
Stylopharyngeus
Glossopharyngeal nerve
Pharyngeal branch of vagus nerve
Superior laryngeal branch of vagus nerve

Lingual Artery

The lingual artery arises from the external carotid artery opposite the tip of the greater cornu of the hyoid bone. The hyoglossus muscle divides the lingual artery into three parts.

The *first part* of the artery lies in the carotid triangle, superficial to the middle constrictor of pharynx and forms a characteristic upward loop which is crossed by the hypoglossal nerve. The *second part* of the artery lies deep to the hyoglossus muscle that separates the artery from the hypoglossal nerve. This part of the artery also lies on the middle constrictor. The *third* or *deep part* of the artery runs upward along the anterior margin of the hyoglossus and then forwards to the tip of the tongue. This part lies on the genioglossus muscle.

The branches of the lingual artery are the *suprahyoid artery* from the first part, *dorsal lingual arteries* from the second part and the *sublingual artery* from the third part. A few branches pierce the mylohyoid and anastomose with the submental artery (Fig. 14.13).

Facial Artery

The facial artery (also called the *external maxillary artery*) arises from the external carotid just above the greater cornu of the hyoid bone and terminates at the medial angle of the eye as the *angular artery* where it anastomoses with the dorsal nasal branch of the ophthalmic artery (Fig. 14.14). The course of the artery can be divided into a cervical part and a facial part. The artery is remarkably tortuous which, in the neck allows for the expansion of pharynx and in the face for movements during mastication, smiling and other movements of the face.

Cervical part

The artery first runs upward deep to the posterior belly of digastric and the stylohyoid and then grooves the posterior border of the submandibular gland. It then runs downward and forward between the gland and the medial pterygoid muscle which is superficial to it and reaches the lower border of the mandible. Then it pierces the deep cervical fascia and enters the face at the anterior edge of the masseter, where it is crossed by the marginal branch of mandibular nerve and the facial vein.

Facial part

Curving around the anteroinferior angle of masseter, the artery runs upward and forward across the superficial aspect of the body of mandible and across the buccinator muscle to reach the angle of the mouth. It then runs upward along the side of the nose to reach the medial angle of the palpebral fissure.

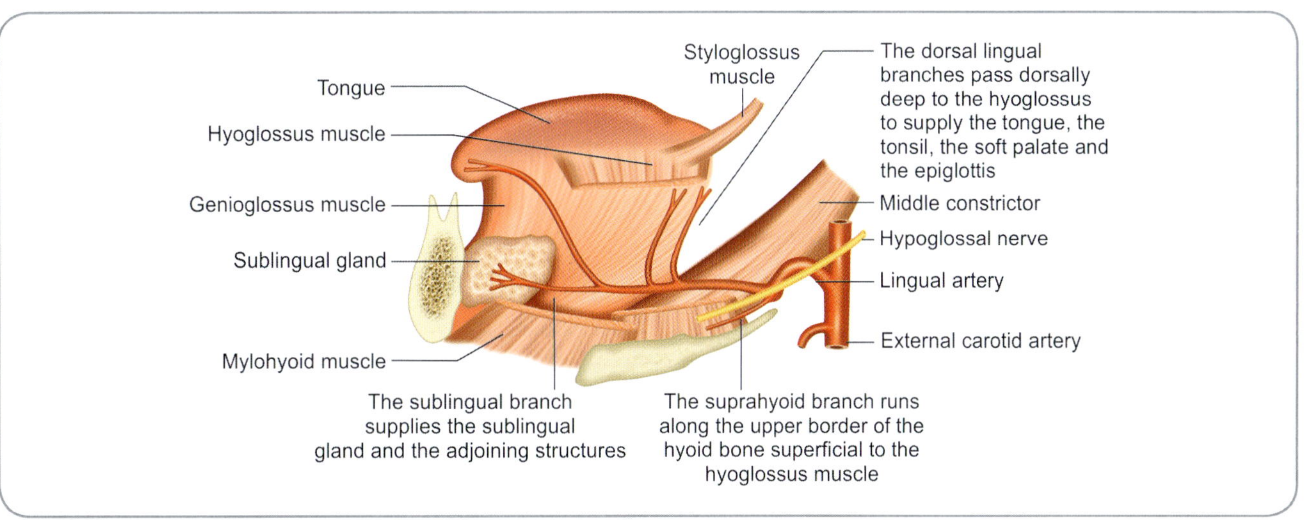

Fig. 14.13: Scheme to show the branches of the lingual artery

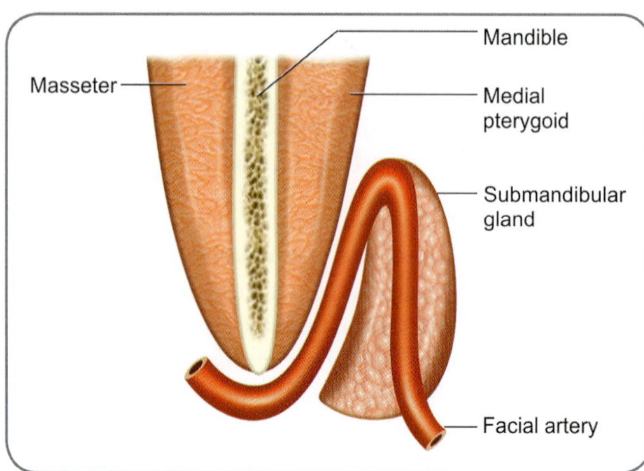

Fig. 14.14: Scheme to show the relationship of the facial artery to the submandibular gland

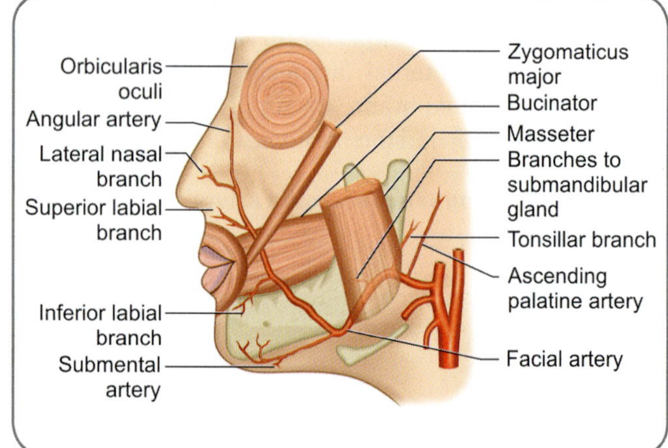

Fig. 14.15: Scheme to show branches of facial artery

Branches

The branches of the facial artery are as follows:

Branches of cervical part:
- Ascending palatine artery ascends on the lateral side of the pharynx. It supplies the pharynx, the palate, the tonsil and the auditory tube.
- Tonsillar branch reaches the tonsil by piercing the superior constrictor muscle.
- Glandular branches enter into the submandibular gland.
- Submental artery runs forward along the lower border of mandible over the mylohyoid muscle and supplies the muscles of the region including those of the chin and the lower lip.

Branches of facial part (Fig. 14.15):
- Superior and inferior labial branches supply the lips.
- Lateral nasal branch supplies the side of the nose and the septal branch supplies the nasal septum.
- The terminal part of the facial artery is called the ***angular artery*** which anastomoses with the dorsal nasal branch of the ophthalmic artery thereby establishing an anastomosis with the branch of Internal carotid artery.

Occipital Artery

The occipital artery arises from the posterior aspect of the external carotid opposite the origin of the facial artery. It runs backward along the lower border of the posterior belly of digastric muscle to reach the skull medial to the mastoid process. Here, it lies deep to the sternocleidomastoid, the digastric and the deep muscles of the back of neck. It crosses the apex of the posterior triangle of the neck before it pierces the trapezius to become superficial and ends by supplying the posterior part of the scalp. The artery is accompanied by the greater occipital nerve in the scalp.

The occipital artery gives off several branches. These are as follows:
- Two branches to the sternocleidomastoid which run backwards across the carotid sheath.

- Auricular branch supplies the pinna.
- Mastoid branch passes through the mastoid foramen.
- Meningeal branches enter the skull through the jugular foramen and the carotid canal.
- Descending branch runs down through the deep muscles of the back of the neck and divides into a superficial branch which anastomoses with the transverse cervical artery and a deep branch which anastomoses with branches of the vertebral and deep cervical arteries. These anastomoses establish collateral links between the external carotid artery and the subclavian artery.
- Occipital branches supply the posterior part of the scalp.

Posterior Auricular Artery

The posterior auricular artery arises from the external carotid just above the posterior belly of digastric and the stylohyoid muscle (Fig. 14.16). It passes backward and upward deep

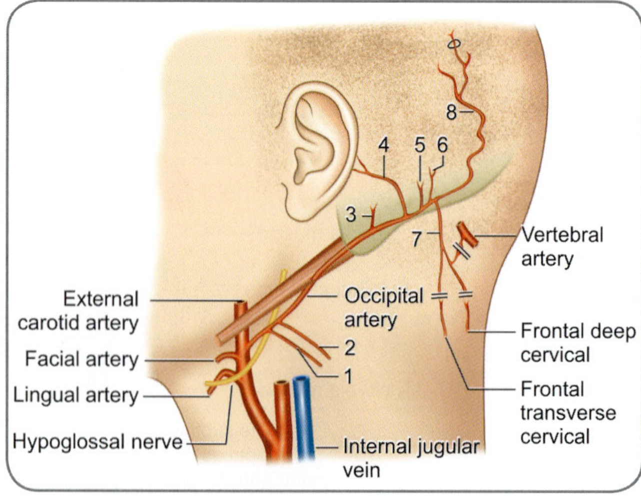

Fig. 14.16: Distribution of posterior auricular artery

Key: 1, 2. To sternocleidomastoid **3.** Stylomastoid branch
4. Auricular branch **5.** Mastoid branch **6.** Meningeal branches
7. Descending branch **8.** Occipital

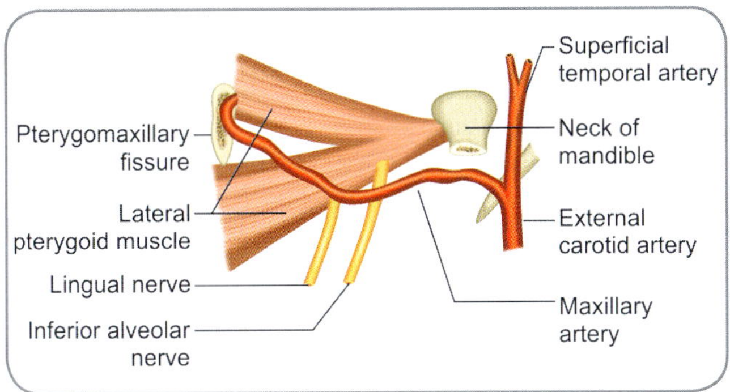

Fig. 14.17: Diagram to show the course of maxillary artery

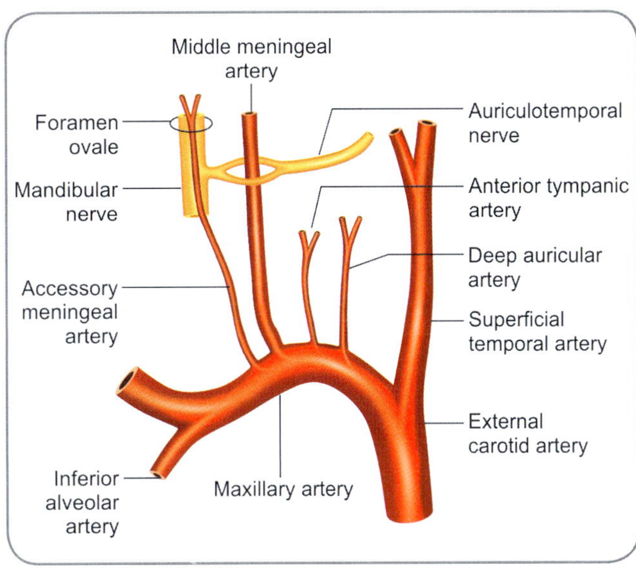

Fig. 14.18: Branches of the first part of maxillary artery

to the parotid gland to reach the mastoid process. Apart from some *auricular* branches and *occipital* branches, the named branch it gives off is the **stylomastoid** artery which enters the stylomastoid foramen to supply the middle ear and related structures.

Maxillary Artery

The **maxillary artery** (also called the **internal maxillary artery** or the **mandibulomaxillary artery**) is the larger terminal branch of the external carotid artery and arises behind the neck of the mandible within the parotid gland. The course of the artery is divided into three parts by the lower head of lateral pterygoid muscle (Fig. 14.17).

The **first part** (mandibular part) runs forwards and horizontally deep to the neck of the mandible to enter the infratemporal fossa and reaches the lower border of the lateral pterygoid muscle (Fig. 14.18). In this part, it passes between the neck of mandible and the sphenomandibular ligament and lies parallel to and below the auriculotemporal nerve. Crossing the inferior

alveolar nerve, it reaches the lower border of the lateral pterygoid. The **second part** (pterygoid part) passes upwards and forwards and then runs either superficial or deep to the lower head of lateral pterygoid muscle. Finally, it enters the interval between the two heads of the lateral pterygoid and disappears into the pterygomaxillary fissure to enter into the pterygo palatine fossa. The course of the artery within the pterygopalatine fossa is the **third part** (pterygopalatine part).

Branches

The branches of the maxillary artery can be classified into three groups as according to the three parts of the artery itself (Fig. 14.19).

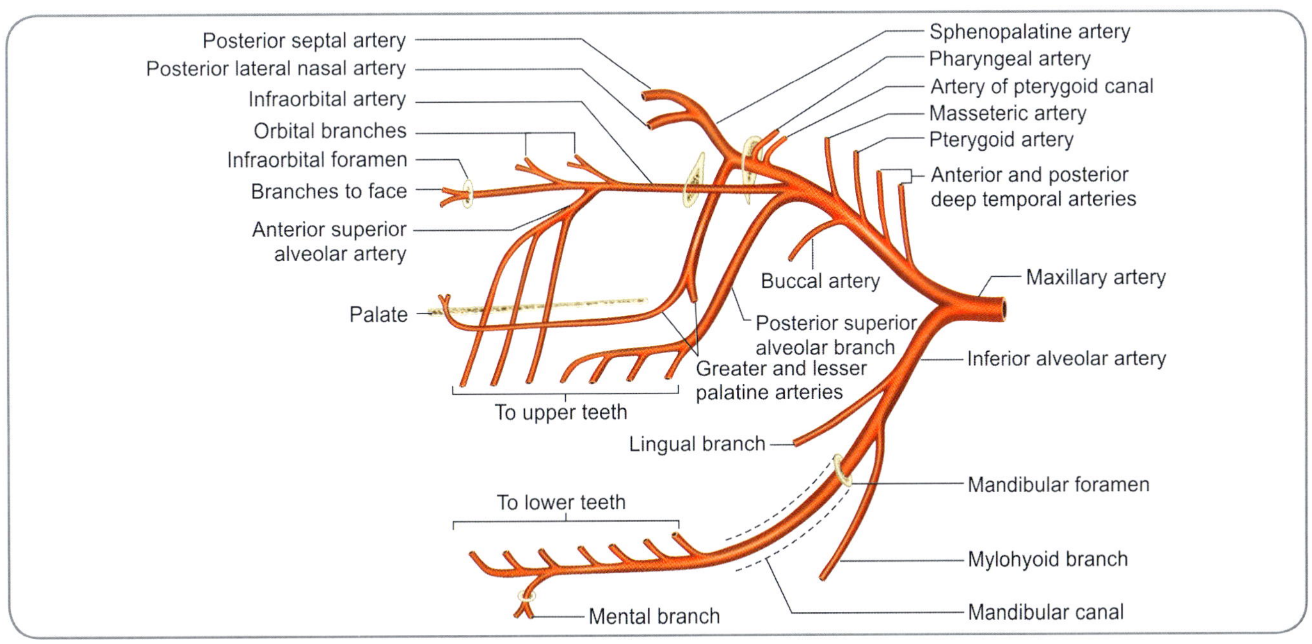

Fig. 14.19: Branches of maxillary artery

A. Branches of the mandibular part: Five branches given out from the first or mandibular part. While still posterior to the mandible, the maxillary gives off the anterior tympanic and deep auricular arteries. The middle meningeal artery, the accessory meningeal artery and the inferior alveolar artery are given out in the infratemporal fossa.

The **anterior tympanic artery** enters the tympanic cavity through the petro tympanic fissure and supplies the inner surface of the tympanic membrane. It is accompanied by the chorda tympani nerve and anterior ligament of malleus when it passes through the petro tympanic fissure.

The **deep auricular artery** enters the external acoustic meatus and supplies the skin of the external acoustic meatus and part of the outer surface of tympanic membrane. It may also give a twig to the temporomandibular joint.

The **middle meningeal artery** is the largest meningeal branch supplying the meninges and bones of the skull vault. It usually arises as a separate branch but may arise in common with the inferior alveolar branch. The artery ascends between the sphenomandibular ligament and the lateral pterygoid muscle, passes between the two roots of auriculo temporal nerve and leaves the infratemporal fossa through the foramen spinosum along with the meningeal branch of mandibular nerve to reach the middle cranial fossa. In the middle cranial fossa, it turns upwards and forwards on the greater wing of sphenoid and divides into anterior and posterior branches, both of which end by supplying the meninges. However, the main supply of this artery is to extracranial structures which include the medial and lateral pterygoid muscles, tensor veli palatini, pterygoid processes and greater wing of sphenoid bone, otic ganglion and mandibular nerve (Fig. 14.20).

The middle meningeal artery divides into anterior and posterior branches. The anterior branch is the larger of the two. Most of its course is opposite the motor area of the cerebrum. It runs on the side of the cranial cavity to a point opposite the pterion where it often tunnels the bone.

It then runs upward and backward towards the middle of the cranial vault. The posterior branch runs upward and backward; its course corresponds to the lateral sulcus of the cerebrum. The artery grooves the squamous part of the temporal bone and the inferior part of the parietal bone. The midlle meningeal artery through its two branches supplies most of the supratentorial duramater except for the floor of the anterior cranial fossa. Therefore, the artery is of great clinical importance. It may be fatally damaged in fractures of skull; it should be handled carefully in surgical procedures involving the skull base and the lateral aspect of the skull.

The **inferior alveolar artery** (or the **inferior dental artery**) descends in the infra temporal fossa posterior to the inferior alveolar nerve between the mandible laterally and the sphenomandibular ligament medially. After giving a branch to mylohoid, it enters the mandibular foramen along with the inferior alveolar nerve and ends by supplying the teeth of the lower jaw and the skin around the area of chin.

B. Branches of the pterygoid part: Branches given out from the second or pterygoid part of the artery do not enter the bone and supply fasciae and muscles. These branches are the **buccal artery**, the **pterygoid branches**, the **deep temporal arteries** and the **Masseteric artery**.

The buccal artery runs anteriorly between the medial pterygoid and the temporalis to supply the mucosa and skin over the buccinator. It is accompanied by the lower portion of the buccal branch of the mandibular nerve. It may give a lingual branch that supplies the lingual nerve and floor structures of the mouth cavity.

The pterygoid branches are many in number; they supply the two pterygoid muscles.

The deep temporal branches, as already noted, are three in number—the anterior, middle and posterior deep temporal arteries. Passing deep to the temporalis, they supply the same muscle and the fasciae and muscles of the region. They anastomose with the branches of the superficial temporal artery.

The masseteric artery is a small branch that runs posterior to the temporalis tendon and skirts the mandibular notch to enter the deep surface of and supply the masseter muscle. Its branches anastomose with the masseteric branches of the facial artery and the transverse facial branch of the superficial temporal artery.

C. Branches of the pterygopalatine part: Branches given out from the third or the pterygopalatine part of the maxillary artery accompany branches of the maxillary nerve. These branches are the posterior superior alveolar artery, infraorbital artery, greater palatine artery, pharyngeal artery, artery of pterygoid canal and the sphenopalatine artery, which is a continuation of the maxillary artery itself.

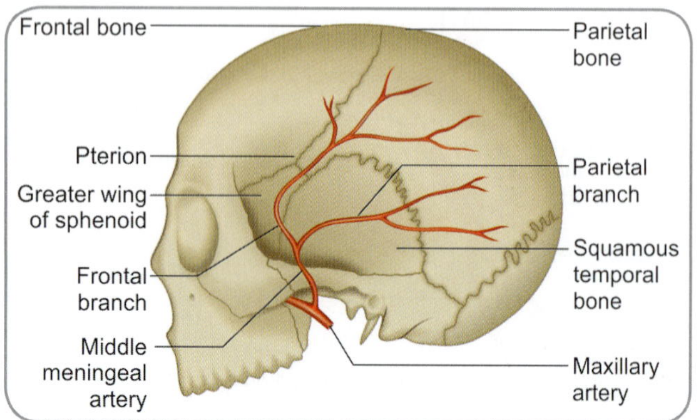

Fig. 14.20: Schematic diagram showing the course of the middle meningeal artery as projected onto the lateral aspect of the skull

The ***posterior superior alveolar artery*** runs through the pterygomaxillary fissure to reach the maxillary tuberosity where it ends by supplying the teeth of upper jaw and the maxillary air sinus. Some branches also supply the buccal mucosa.

The ***infraorbital artery*** enters the orbit through the inferior orbital fissure. It runs on the floor of the orbit in the infra orbital groove and the infra orbital canal and comes out in the face through the infraorbital foramen. It supplies the lower eyelid, part of cheek, upper lip and nose. Within the infra orbital canal, it gives off the anterior superior alveolar artery which anastomoses with the posterior superior alveolar artery and supplies the teeth.

The ***artery of the pterygoid canal*** (also called the ***Vidian artery***) passes through the pterygoid canal and anastomoses with the pharyngeal, ethmoidal and sphenopalatine arteries in the pterygopalatine fossa and with the ascending pharyngeal, accessory meningeal, palatine arteries in the oropharynx and around the pharyngotympanic tube. It supplies the tympanic cavity, pharyngotympanic tube and the upper pharynx.

The ***pharyngeal artery*** passes through the palatinovaginal canal and supplies the nasal mucosa, nasopharynx, parts of the pharyngotympanic tube and the sphenoidal air sinus.

The ***greater palatine artery*** (also called the ***descending palatine artery***) runs through the greater palatine canal and the greater palatine foramen to reach the hard palate. It supplies the inferior meatus of the nose, the hard palate and the palatal gingivae of the maxillary teeth. It gives a branch which runs through the incisive canal to anastomose with the sphenopalatine artery, thereby supplying the nasal septum. The lesser palatine arteries are branches of the greater palatine artery which are given out in the greater palatine canal and which emerge through the lesser palatine foramen to supply the soft palate.

The ***sphenopalatine artery*** is the continuation of the maxillary artery. It passes through the sphenopalatine foramen to enter into the nasal cavity. It ends up by supplying the lateral wall of the nose through its posterior lateral nasal branches and nasal septum through the posterior septal branches. It anastomoses with the ethmoidal arteries and the nasal branches of the greater palatine artery to supply all the four paranasal air sinuses. A branch descends to the incisive canal to anastomose with the greater palatine artery and the septal branch of superior labial artery.

Superficial Temporal Artery

The superficial temporal artery is one of the terminal branches of the external carotid artery. It begins behind the neck of the mandible within the substance of the parotid gland. It then runs upwards behind the

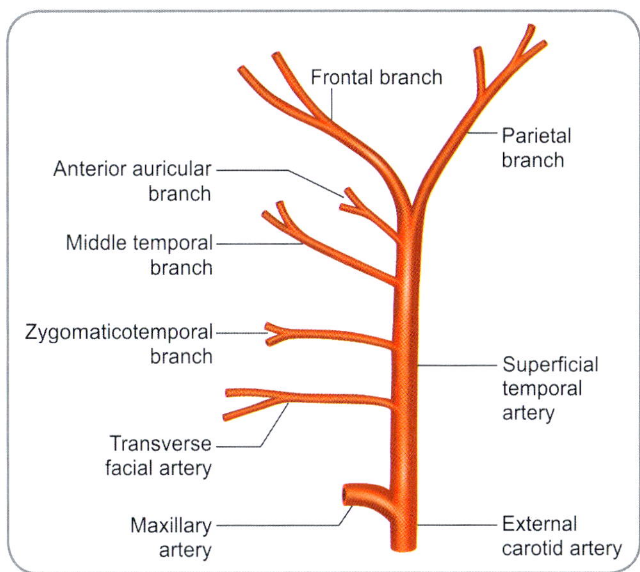

Fig. 14.21: Scheme to show the branches of superficial temporal artery

temporomandibular joint and ramifies in the scalp over the temporal region. The artery is accompanied by the auriculotemporal nerve. The branches of the artery are as follows (Fig. 14.21):

❏ Frontal branch—runs upward and forward over the temporal and frontal bones.

❏ Parietal branch—runs backward over the temporal and parietal bones.

❏ Anterior auricular branch—supplies part of the auricle and the external acoustic meatus.

❏ Middle temporal artery—supplies the temporalis.

❏ Zygomatico-orbital branch—runs forward along the upper border of the zygomatic arch up to the lateral angle of the eye.

❏ Transverse facial branch—arises within the substance of the parotid gland and runs forward across the masseter above the parotid duct.

SUBCLAVIAN ARTERIES

The right subclavian artery is a branch of the brachiocephalic trunk and begins behind the right sternoclavicular joint. The left subclavian artery is a direct branch from the arch of aorta. The right subclavian artery has a thoracic part which ends behind the left sternoclavicular joint. Thereafter, the course and relations of the right and left subclavian arteries are almost similar.

Each subclavian artery is the initial part of a long channel that supplies the upper limb. It enters the neck behind the corresponding sternoclavicular joint and further loops upwards into the neck. It then leaves the neck by passing into the axilla, where it becomes the axillary artery.

The whole of the right subclavian artery and the cervical part of left extend from the sternoclavicular joint to the outer border of the first rib.

Relations of Subclavian Arteries (Table 14.1)

Anterior Relations

The subclavian artery is crossed anteriorly by the lower part of the **scalenus anterior** muscle on its way to the insertion into the first rib just in front of the artery (Fig. 14.22). The crossing of the scalenus anterior muscle divides the artery into three parts namely:

- **First part**—medial to the muscle,
- **Second part**—deep to the muscle, and
- **Third part**—lateral to the muscle.

Other structures which are anterior to the subclavian artery are the common carotid artery, which crosses over the medial most part of the artery. Immediately lateral to the common carotid artery, the internal jugular vein runs vertically across the subclavian artery to join the subclavian vein. The subclavian vein lies below and in front of the artery separated from it by the scalenus anterior muscle. The artery is also crossed vertically by the vagus nerve which lies behind the internal jugular vein. In other words, the medial part of the subclavian artery is crossed by all structures enclosed by the carotid sheath. The following structures also lie anterior to the artery:

- The ansa subclavia which is a nerve cord that descends from the middle cervical sympathetic ganglion to the front of the first part of the artery where it loops around

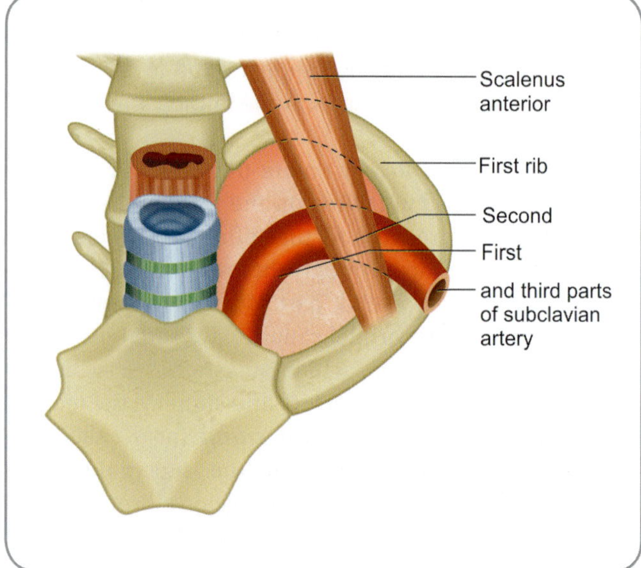

Fig. 14.22: Relationship of the subclavian artery to scalenus anterior

it and ascends behind it to reach the inferior cervical (cervicothoracic) sympathetic ganglion;

- The vertebral vein which descends across the first part of the subclavian artery to end in the brachiocephalic vein;
- The terminal part of the thoracic duct which descends in front of the left subclavian artery near the medial border of the scalenus anterior to terminate by joining the junction of the left internal jugular and subclavian veins;

Table 14.1: Summary of relations of subclavian artery		
Common to all three parts	**Anterior Relations**	**Posterior Relations**
	Skin Platysma and superficial fascia Deep fascia	Lung Cervical pleura Suprapleural membrane
First Part	Sternothyroid muscle Sternohyoid muscle Anterior jugular vein Sternocleidomastoid muscle Common carotid artery Internal jugular vein Vagus nerve Phrenic nerve Thoracic duct (on the left side only) Vertebral artery Ansa subclavia (both anterior and posterior relation to first part as it loops over the artery)	
Second Part	Sternocleidomastoid Anterior jugular vein Scalenus anterior	Lower trunk of brachial plexus
Third Part	External jugular vein Transverse cervical vein Suprascapular vein Anterior jugular vein Clavicle	Lower trunk of brachial plexus Above and lateral relation: Upper and middle trunk of brachial plexus

❑ The external jugular vein which descends across the third part of the subclavian artery to end in the subclavian vein. In front of the artery, the external jugular vein is joined by the transverse cervical, suprascapular and anterior jugular veins.

Apart from the aforementioned structures, certain structures have differing relationship on the right and left sides. They are the recurrent laryngeal branch of vagus nerve and the phrenic nerve.

The right vagus nerve gives off the right recurrent laryngeal branch just as it reaches the lower margin of the subclavian artery which then curves around the inferior and posterior aspects of the subclavian artery and runs medially to reach the trachea-oesophageal groove (Fig. 14.23). However, the left recurrent laryngeal nerve arises from the vagus below the arch of the aorta, winds around the ligamentum arteriosum and then ascends in the trachea-oesophageal groove. Hence, the left recurrent laryngeal nerve is not closely related to the left subclavian artery in the neck.

With respect to the phrenic nerves, both of them descend across the corresponding scalenus anterior muscle (Fig. 14.23). On the left side, the phrenic nerve crosses the medial border of the scalenus anterior and then runs onto the front of the first part of the subclavian artery. But on the right side, the phrenic nerve usually crosses the medial border of the scalenus anterior lower down, so that the nerve does not come into direct contact with the first part of the subclavian artery. It is separated from the second part by the scalenus anterior. Occasionally, however, the relationship may be the same as on the left side.

Posterior Relations

The subclavian artery crosses the apex of the lung ,the cervical pleura and the suprapleural membrane as it arches across the lower part of the neck.

Relation with brachial plexus deserves a special mention. The first part of subclavian artery lies below the level of the trunks of brachial plexus. The lower trunk of brachial plexus is present posterior to the second and third parts and the upper and middle trunks are present above and lateral to the third part of subclavian artery.

Branches of Subclavian Artery

The branches of the subclavian artery can be classified into three groups—*branches from first part, branches from second part* and *branches from third part.*

❑ *Branches from first part (Fig. 14.24):*
  ○ *Vertebral artery* which runs upwards to enter the foramen transversarium of the sixth cervical vertebra.
  ○ *Internal thoracic artery* which runs downward into the thorax.
  ○ *Thyrocervical trunk* which shortly divides into the inferior thyroid, suprascapular and transverse cervical arteries.
  ○ *Costocervical trunk* on the left side only and runs backwards to reach the neck of the first rib where it divides into the deep cervical and superior intercostal arteries.
❑ *Branches from second part:* Costocervical trunk on the right side only
❑ *Branches from third part:* Dorsal scapular artery which when present, replaces the deep branch of the transverse cervical artery.

Vertebral Artery

The vertebral artery arises from the first part of the subclavian artery. It ascends to enter the foramen transversarium of the sixth cervical vertebra and then continues upwards through the foramina of higher

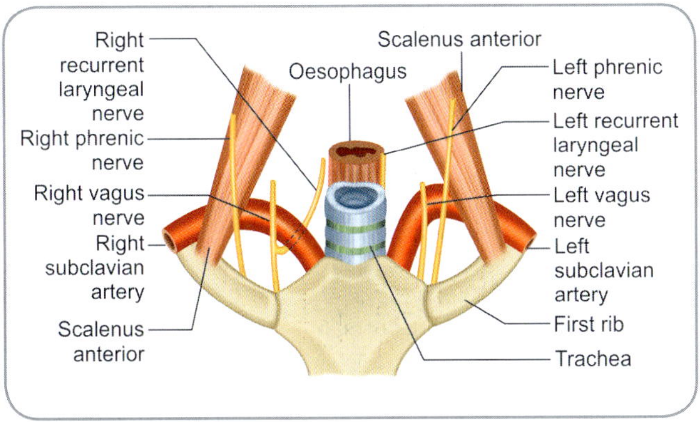

Fig. 14.23: Relationship of the subclavian artery to vagus nerve – the phrenic nerve is also shown

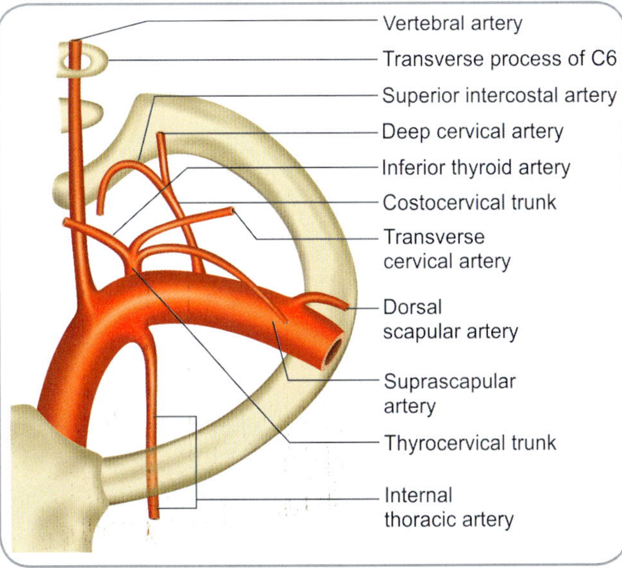

Fig. 14.24: Scheme to show the branches of the subclavian artery

vertebrae. After the artery emerges out of the foramen transversarium of atlas, it winds around the lateral mass of atlas, where it lies in the groove on the upper surface of the posterior arch of atlas. Finally, it passes forward into the vertebral canal and passes upward through the foramen magnum to enter the cranial cavity. Here it lies lateral to the lower end of medulla oblongata. It gradually passes forwards and medially over the medulla and ends at the lower border of the pons by anastomosing with the opposite vertebral artery to form the **basilar artery**.

Course and Relations

The course of the vertebral artery can be divided into four parts.

- ❑ The part of the artery between its origin from the subclavian artery and its entry into the foramen transversarium of C6 vertebra constitutes its first part (Fig. 14.25). It lies deep in the lower part of the neck in a triangular interval bounded laterally by the scalenus anterior, medially by the longus colli and inferiorly by the subclavian artery. This part of the artery lies deep to the common carotid artery (Fig. 14.26).
- ❑ The part of the artery passing through the foramina transversaria of the upper six cervical vertebrae constitutes the second part. It is surrounded by a plexus of veins that unite to form the vertebral vein and by a plexus of sympathetic nerves derived from the cervicothoracic (or stellate) ganglion.
- ❑ The part of the vertebral artery winding around the lateral mass of atlas is the third part. It lies in the suboccipital triangle. Its most important relation is the first cervical nerve (Fig. 14.27). The dorsal ramus of the nerve runs backwards between the artery and the posterior arch of atlas and the ventral ramus runs

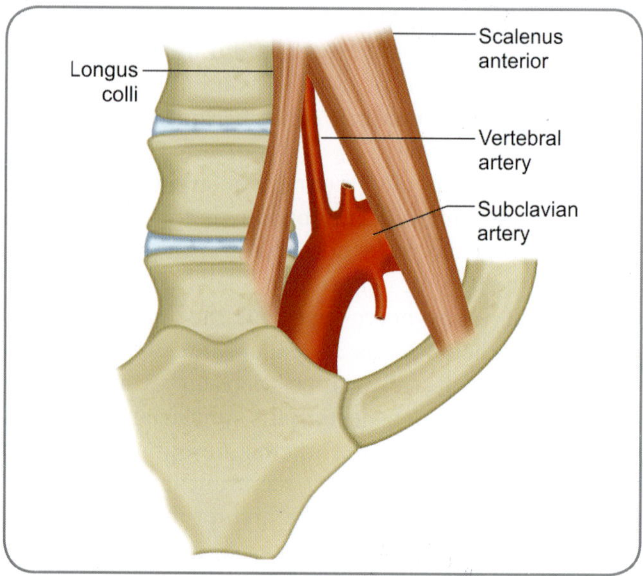

Fig. 14.25: Position of the first part of the vertebral artery

forwards around the lateral mass of atlas, where it lies medial to the artery.

- ❑ The vertebral artery then enters the vertebral canal below the inferior border of the posterior atlanto-occipital membrane and continues as fourth part .

Branches (Fig. 14.28)

- ❑ **Spinal branches:** These pass through the intervertebral foramina to reach the vertebral canal. They supply the upper five or six segments of the spinal cord and the cervical vertebrae.
- ❑ **Muscular branches:** These branches arise in the suboccipital triangle and supply the muscles in that region.

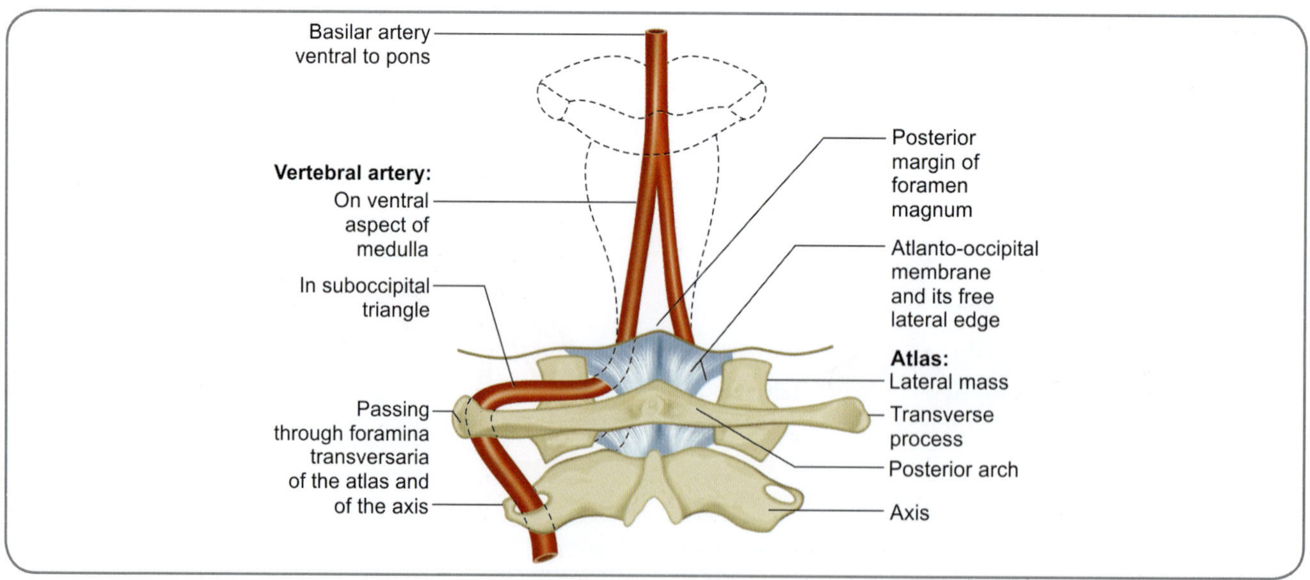

Fig. 14.26: Scheme to show the course of the third and fourth parts of vertebral artery

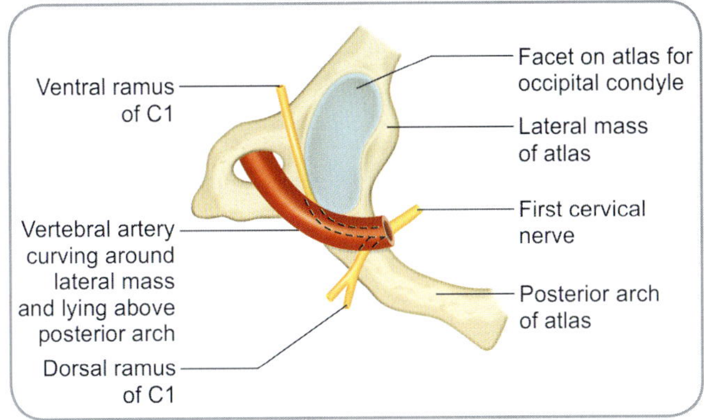

Fig. 14.27: Relationship of the vertebral artery to atlas vertebra

❑ *Anterior spinal artery:* This artery is given from the terminal part of the vertebral artery. The arteries of both sides join to form a single trunk in the middle of the medulla, descend down through the foramen magnum on the anterior aspect of the spinal cord in the midline to supply the spinal cord. It also supplies the medulla oblongata.

❑ *Posterior spinal artery:* The posterior spinal artery either arises from the fourth part or from the posterior inferior cerebellar artery and descends through the foramen magnum along the posterolateral aspect of the spinal cord. Each artery divides into two branches, one lying in front of the dorsal nerve roots and the other lying behind them.

Both the anterior and posterior spinal arteries are joined by spinal branches arising from the vertebral artery itself and from various other arteries like the *ascending cervical*, *posterior intercostal* and *first lumbar arteries* to supply the spinal cord.

❑ *Posterior inferior cerebellar artery:* This is the largest branch of vertebral artery and arises from the fourth part. Sometimes the posterior spinal artery arises from it instead of vertebral artery. It takes part in supplying the brain.

Internal Thoracic Artery

The internal thoracic artery (or the internal mammary artery), though a branch of the first part of subclavian artery, descends down, enters the thorax and ends up by forming an anastomosis with the inferior epigastric artery through one of its terminal branches called the *superior thoracic artery*. It supplies the structures in the mediastinum, sternum, intercostal spaces, pericardium, phrenic nerve and mammary gland in females.

Thyrocervical Trunk

The thyrocervical trunk is a short wide artery that arises from the first part of the subclavian artery near the medial border of the scalenus anterior muscle. After a short course, the trunk divides into three branches, namely:

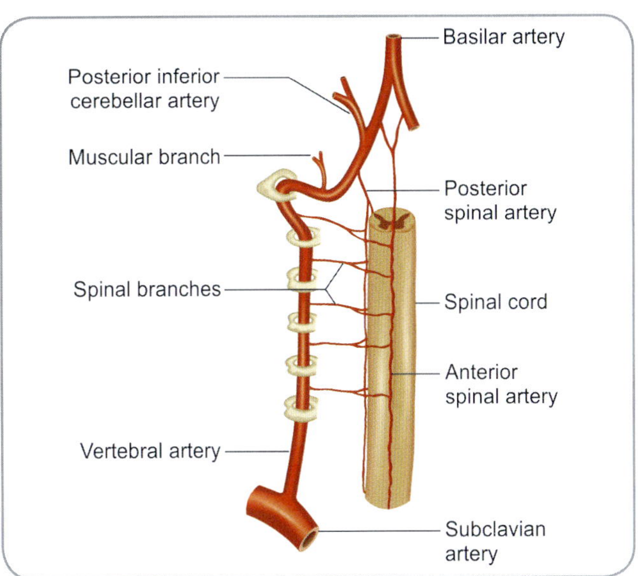

Fig. 14.28: Scheme to show the branches of vertebral artery

❑ Inferior thyroid artery
❑ Suprascapular artery
❑ Superficial cervical artery.

In one third of individuals, the superficial cervical artery arises by a common trunk (from the first part of subclavian) along with the dorsal scapular artery in which case it is called the *transverse cervical artery*.

Inferior Thyroid Artery

The inferior thyroid artery is a branch of the thyrocervical trunk. It runs upwards anterior to the medial border of scalenus anterior, turns medially at the level of transverse process of C6 cervical vertebra, then descends on the longus colli to reach the lower pole of the thyroid gland. In its course, it lies between the vertebral artery behind and carotid sheath and its contents in front. The sympathetic trunk and the middle cervical sympathetic ganglion lie in front of the inferior thyroid artery. The artery is distributed mainly to the thyroid gland through an ascending and a descending *glandular branch*. Close to the lower pole of thyroid gland, the recurrent laryngeal nerve is closely related to the inferior thyroid artery. It may either cross in front of the terminal branches of the artery or pass behind them or may have some branches both in front and back of the nerve. In the left side, the thoracic duct crosses laterally in front of the inferior thyroid artery to reach its termination.

Branches (Fig. 14.29)

❑ Glandular branches to thyroid;
❑ Ascending cervical artery that supplies the cervical vertebrae and spinal cord;
❑ Inferior laryngeal artery that accompanies the recurrent laryngeal nerve and supplies the laryngeal muscles and the mucous membrane of larynx below the level of vocal cords;

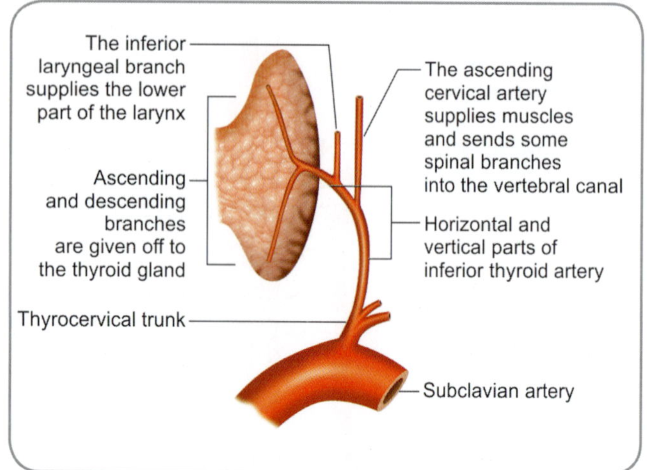

Fig. 14.29: Scheme to show the branches of inferior thyroid artery

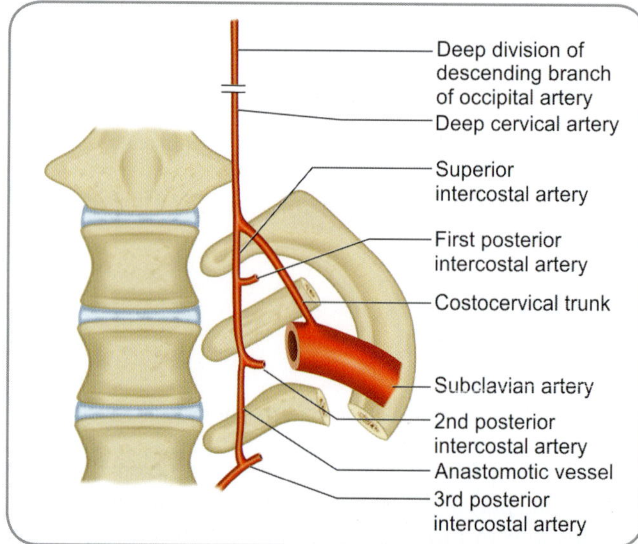

Fig. 14.30: Scheme to show the branches of costocervical trunk

❑ Branches to the pharynx, trachea, oesophagus and the infrahyoid muscles.

Suprascapular Artery

The suprascapular artery (sometimes called the ***transverse scapular artery***) lies below the superficial cervical artery and passes across the scalenus anterior, phrenic nerve, third part of subclavian artery, brachial plexus and behind the clavicle, subclavius and inferior belly of omohyoid to reach the supraspinous fossa where it takes part in the scapular anastomosis.

Superficial Cervical Artery

It passes laterally and upward anterior to scalenus anterior and phrenic nerve and appears in the posterior triangle in front of the trunks of brachial plexus and levator scapulae. The artery then ascends beneath the trapezius and anastomoses with the descending branch of the occipital artery.

Costocervical Trunk

The costocervical trunk takes origin from the posterior aspect of the first part of subclavian artery on the left side and from the second part on the right side. It runs backwards with an upward convexity following the curve of the cervical pleura to reach the neck of the first rib. Here, it divides into the deep cervical and superior intercostal arteries (Fig. 14.30).

❑ The ***deep cervical artery*** (or the profunda cervicalis artery) passes backwards above the neck of the first rib to reach the back of neck, where it ascends through the deep muscles and finally anastomoses with the deep division of the descending branch of the occipital artery. It supplies the back muscles.

❑ The ***superior intercostal artery*** (or the highest intercostal artery) descends across the neck of the first rib to reach the first intercostal space where it gives

off the posterior intercostal arteries for the first two intercostal spaces.

Dorsal Scapular Artery

The dorsal scapular artery usually arises from the third part of subclavian artery and passes laterally between the trunks of brachial plexus. It passes deep to levator scapulae and descends along the medial border of scapula deep to the rhomboid muscles along with the dorsal scapular nerve and ends up by taking part in scapular anastomosis.It also supplies the rhomboid muscles. In one third of individuals, the dorsal scapular artery arises from the first part by a common trunk with the superficial cervical artery,in which case, the common trunk is called the ***transverse cervical artery*** (or the ***transverse colli artery***). Sometimes, it arises from the second part too.

Clinical Correlation

❑ In obstruction of subclavian artery proximal to the origin of the vertebral artery, blood is diverted to the affected arm through the vertebral artery which has communication with the opposite vertebral artery. This condition in which blood is diverted from the brain to the arm is called **subclavian steal syndrome** (blood is stolen from one part and given to another). It may result in ischaemic neurological symptoms due to defective blood supply to brain.

❑ The relationship of the inferior thyroid artery to the recurrent laryngeal nerve is particularly important during thyroidectomies. During a thyroidectomy, ligation of the artery is done some distance away from the lower pole of the gland to avoid injury to the recurrent laryngeal nerve. (In contrast the superior thyroid artery is closely related to the external laryngeal nerve some distance away from the gland and is safely ligated close to the gland).

❑ The close relation of the ascending cervical artery (branch of inferior thyroid artery) to the phrenic nerve serves as a guide for location of phrenic nerve in neck surgeries.

contd...

VEINS

The veins of head and neck comprise the following:
- Internal jugular vein and its tributaries,
- Subclavian vein and its tributary,
- Veins of scalp,
- Veins of orbit, nose and infratemporal region, and
- Intracranial venous sinuses and veins of cranium.

Veins of the brain, which are also part of the veins of head and neck, are considered in the section on brain for the sake of better understanding and convenience.

Internal Jugular Veins

The internal jugular veins are the chief veins of the head and neck. They collect blood from the brain and superficial part of the face and neck. Each internal jugular vein begins as a continuation of the *sigmoid sinus* of that side at the base of skull in the posterior compartment of the jugular foramen. At its commencement, the vein presents a dilatation called the *superior bulb* which lodges in the jugular fossa and is related to the floor of middle ear. The vein passes downwards in the neck and behind the sternal end of clavicle; it joins the subclavian vein to form the corresponding brachiocephalic vein. Close to its termination another dilatation called the *inferior bulb* is present and is guarded by a pair of valves.

In the neck, it is enclosed, along with the internal carotid and common carotid arteries and vagus nerve, within the carotid sheath. Within the sheath, in the lower part, the vein lies lateral to the arteries with the vagus nerve lying posterior and between the two. In the upper part, the vein lies posterior to the internal carotid artery with the lower four cranial nerves between them.

Relations (Table 14.2)

- ***Superficial relations:*** Superficially the internal jugular vein is related to the sternocleidomastoid muscle in its entire length, posterior belly of digastric, superior belly of omohyoid, sternohyoid and sternothyroid in

Table 14.2: Structures superficial and deep to internal jugular vein	
Superficial	**Deep**
Muscles 1. Sternocleidomastoid (in entire length of vein) 2. Posterior belly, digastric 3. Superior belly, omohyoid, cross the vein 4. Sternohyoid 5. Sternothyroid over lower part	From above downwards 1. Rectus capitis lateralis muscle 2. Transverse process of atlas 3. Levator scapulae muscle 4. Scalenus medius muscle 5. Scalenus anterior muscle, with phrenic nerve over it 6. Vertebral artery and vein 7. Subclavian arery (first part)
Other than muscles	
1. Accessory nerve 2. Styloid process 3. Posterior auricular 4. Occipital branches of external carotid artery 5. Descendens cervicalis nerve 6. Anterior jugular vein (Superficial to sternohyoid and sternothyroid) 7. Parotid gland (in upper part)	

its lower part. Apart from the muscles, it is also related to accessory nerve, styloid process, posterior auricular and occipital branches of external carotid artery, descendens cervicalis nerve, anterior jugular vein and Parotid gland in its upper part.
- ***Deep relations:*** The internal jugular vein is related posteriorly to rectus capitis lateralis muscle, transverse process of atlas, levator scapulae, scalenus medius and anterior muscles, phrenic nerve, vertebral vessels and first part of subclavian artery.

Surface Marking

Point A is marked on the upper part of the lobule of the ear. Point B is marked on the medial end of clavicle lateral to the sternoclavicular joint. The two points are joined by a broad line. The lower end of the line will be between the sternal and clavicular heads of the sternocleidomastoid muscle. Mark a dilatation in this part of the line to represent the inferior bulb of the vein.

Tributaries

The tributaries of the internal jugular vein are:
- ***Inferior petrosal sinus:*** It is usually the first tributary of the internal jugular vein. It connects the jugular vein with the cavernous venous sinus.
- ***Pharyngeal veins:*** These veins start from the pharyngeal venous plexus lying outside the pharynx. They drain the pharynx.
- ***Common facial vein:*** The common facial vein is formed by the union of the anterior division of the retromandibular vein and the facial vein in the upper part of the neck. The facial vein (also called the ***anterior***

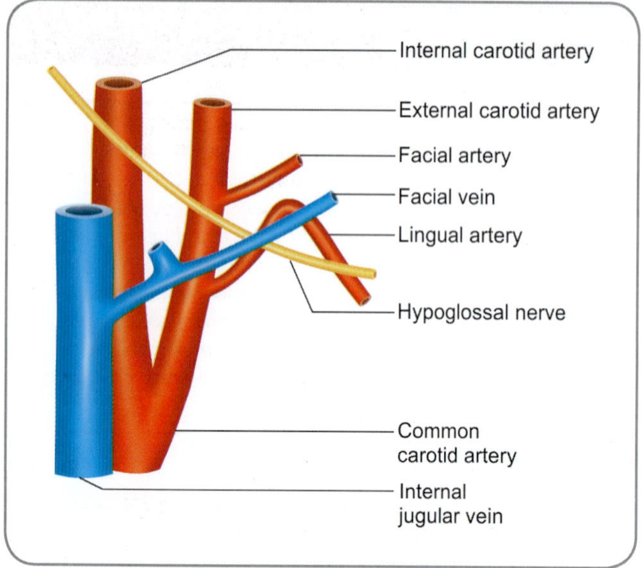

Fig. 14.31: Relations of the terminal part of the facial vein

facial vein) begins near the medial angle of the eye by the union of two superficial veins of the forehead, namely, the supratrochlear and the supraorbital veins which then run downwards and backwards across the face. A little below the angle of the jaw, it is joined by the posterior division of the retromandibular vein (This division being called the *posterior facial vein*) and forms the common facial vein. The terminal part of the common facial vein crosses the internal and external carotid arteries, the hypoglossal nerve and the loop formed by the lingual artery to drain into the inernal jugular vein (Fig. 14.31). A triangular interval is formed between the internal jugular vein, common facial vein and the posterior belly of digastric which is occupied by the jugulo-digastric group of lymph nodes. The facial vein receives tributaries corresponding to all branches of its counterpart artery (except for the ascending palatine and tonsillar branches which have no accompanying veins) and communicates with the cavernous sinus through the superior ophthalmic vein and through the pterygoid venous plexus. An anastomotic channel called the *deep facial vein runs* deep to masseter on the maxilla from the pterygoid venous plexus to join the facial vein.(Fig. 14.32).

❑ *Lingual vein:* Both the superficial venae commitantes accompanying the hypoglossal nerve and the deep lingual veins accompanying the lingual artery drain into the internal jugular vein as a common trunk or separately.

❑ *Superior thyroid vein:* The superior thyroid vein corresponds in its course and tributaries to those of the superior thyroid artery. After receiving the superior laryngeal and the cricothyroid veins, it drains into the internal jugular vein.

❑ *Middle thyroid vein:* The middle thyroid vein drains the lower part of the thyroid gland (Fig. 14.33). It crosses the common carotid artery to enter the internal jugular vein. It is the first blood vessel to be encountered during a thyroid surgery and requires careful ligation.

❑ *Occipital veins:* They occasionally drain into the internal jugular vein.

Thoracic duct on the left side and the *right lymphatic duct* on the right side—each of these opens at the junction of the internal jugular vein and the subclavian vein of its own side.

Communications of Internal Jugular Vein

❑ With the external jugular vein in the upper part of the neck through the oblique jugular vein;

❑ With cavernous sinus via the inferior petrosal sinus (Fig. 14.34).

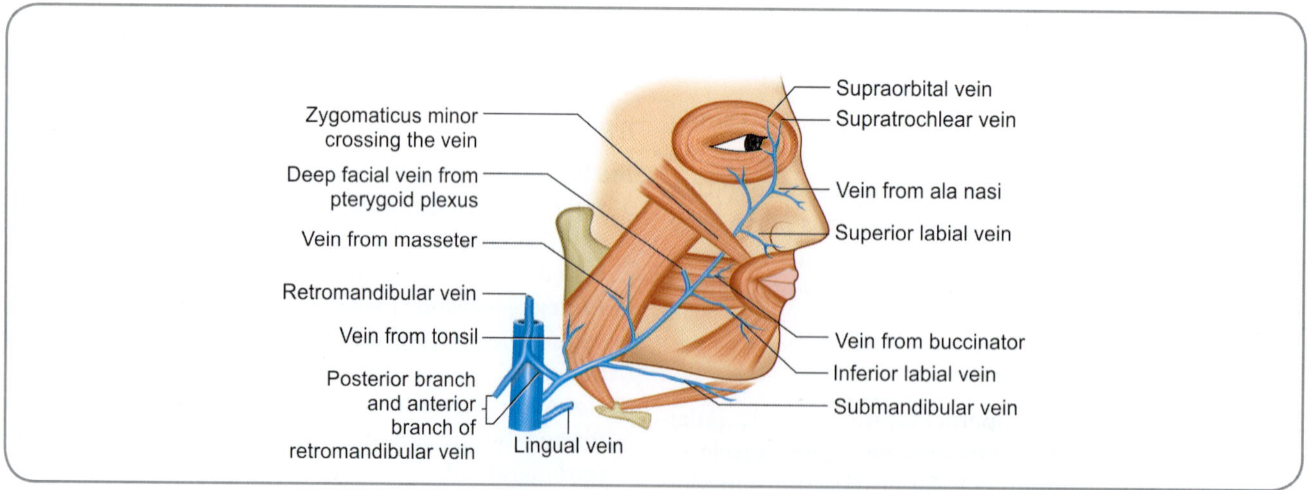

Fig. 14.32: Scheme to show the course and tributaries of the facial vein

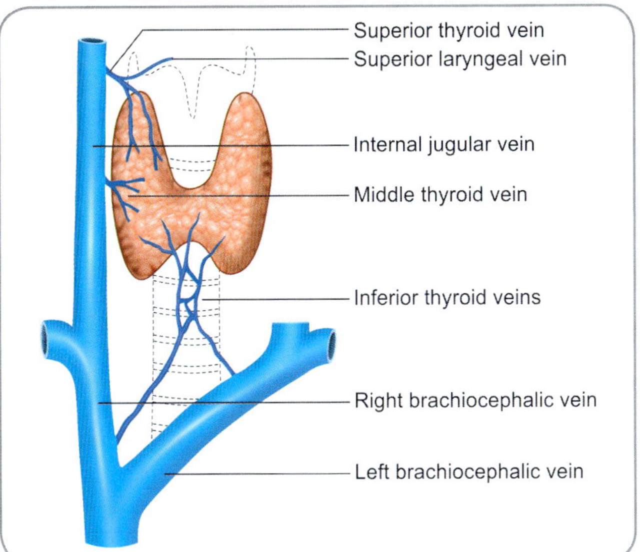

Fig. 14.33: Scheme to show the veins draining the thyroid gland

medially parallel to the subclavian artery, but lies anterior and inferior to the artery. The two vessels are separated by the scalenus anterior. The subclavian vein ends at the medial margin of the scalenus anterior muscle by joining the internal jugular vein (Fig. 14.35).

Anteriorly, the subclavian vein is related to the clavicle. Below, the vein rests on the first rib and on the cervical pleura.

The ***external jugular vein*** is the main tributary of the subclavian vein. Other minor or occasional tributaries are the dorsal scapular vein and the anterior jugular vein (Fig. 14.36). At its junction with the internal jugular vein, the left subclavian vein receives the thoracic duct and the right subclavian vein receives the right lymphatic duct.

Clinical Correlation

Subclavian vein is often used as a route for central venous access in patients on critical care.

Clinical Correlation

- The internal jugular vein acts as a guide for surgeons during removal of deep cervical lymph nodes.
- The internal jugular vein is used for IV access, in which case the vein is accessed through the lesser supraclavicular triangle (between the sternal and clavicular heads of the sternocleidomastoid muscle) to avoid injury to the pleural sac.
- ***Queckenstedt's test:*** Under normal condition, the rate of flow of CSF is increased during lumbar puncture when both internal jugular veins are compressed due to the back flow. This is called ***Queckenstedt's test***. If there is block in spinal subarachnoid space, the pressure will not rise when the internal jugular veins are compressed.

Subclavian Veins

Each subclavian vein begins at the outer border of the first rib as a continuation of the axillary vein. It then runs

External Jugular Vein

The external jugular vein is formed by the union of the posterior division of retromandibular vein and posterior auricular vein. The origin lies within the lower part of the parotid gland or just below it. The level corresponds to the angle of mandible. From its origin, the vein runs downwards and backwards and ends by joining the subclavian vein after piercing the deep fascia. The termination lies behind the middle of clavicle, near the lateral margin of scalenus anterior muscle.

As the greater part of the vein is superficial, being just covered by skin, superficial fascia and platysma, it can be clearly seen in the living. The vein crosses the sternocleidomastoid obliquely running downwards and backwards across it.

Apart from the formative tributaries, the external jugular vein receives:

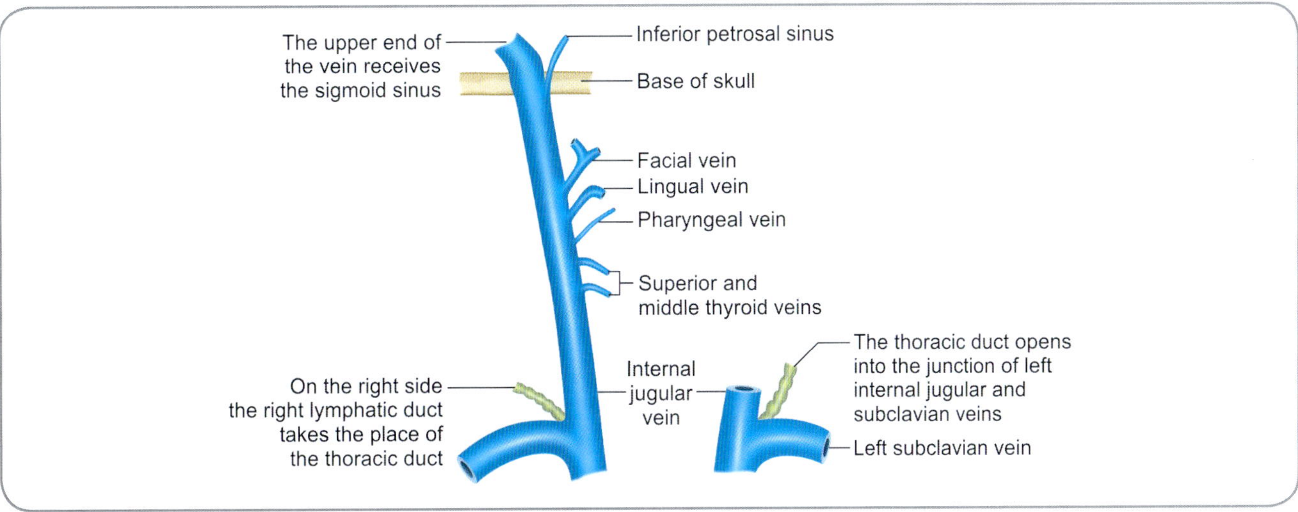

Fig. 14.34: Scheme to show the tributaries of internal jugular vein

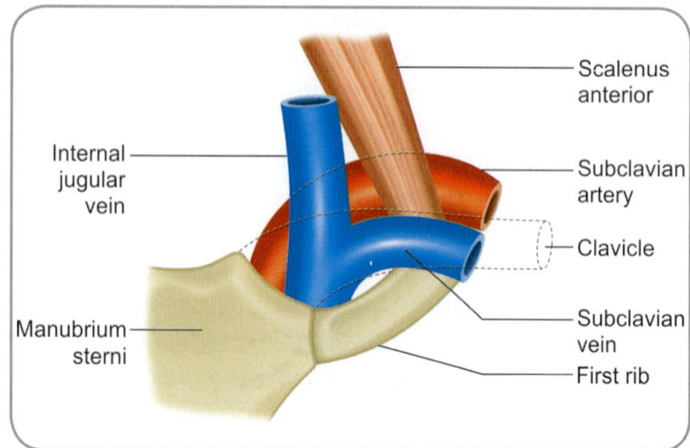

Fig. 14.35: Some relations of subclavian vein

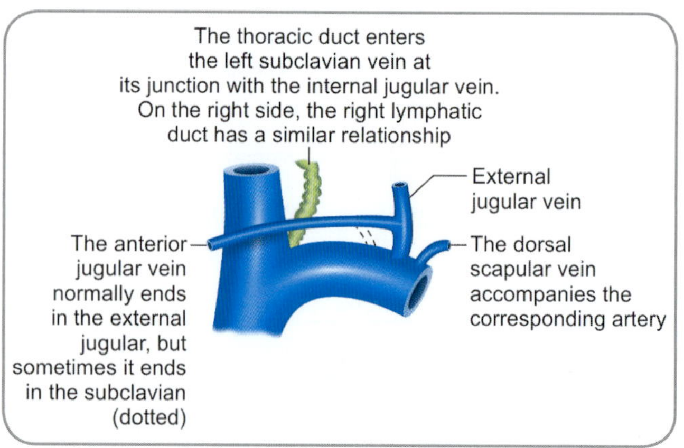

Fig. 14.36: Tributaries of subclavian vein

- ***Posterior external jugular vein*** from the upper and posterior part of the neck,
- ***Transverse cervical vein,***
- ***Suprascapular veins*** which accompany the corresponding arteries and
- ***Anterior jugular vein.***

> **Surface Marking**
>
> Point A is marked on the angle of the jaw. Point B is marked on the clavicle at the junction of the medial and the middle thirds. The two points are joined bya line that marks the external jugular vein on the surface.

Veins of Scalp

The veins of scalp can be grouped into those draining the superficial portion of scalp and those draining the deep portion of scalp.

Veins draining superficial portion of scalp:

- Supratrochlear and supraorbital veins drain the anterior part of scalp and unite to form the facial vein. The supraorbital vein communicates with the superior ophthalmic vein by a twig that receives the frontal diploic vein as it passes through the superior orbital fissure.
- Superficial temporal vein is formed by union of veins accompanying branches of the superficial temporal artery. The vein formed so, descends to the zygomatic arch, receives the middle temporal vein, crosses the arch and enters the substance of the parotid gland. Within the parotid gland, it unites with the transverse facial vein and with the maxillary vein to form the retromandibular vein.
- Posterior auricular vein begins by union of tributaries present in the posterior part of the scalp. It passes downwards and forwards behind the auricle and receives veins from its cranial surface. Finally, it ends by joining the posterior division of the retromandibular vein to form the external jugular vein.

- Occipital vein begins by union of some veins draining the posterior part of the scalp (Fig. 14.43). It descends in the scalp a few centimetres behind the auricle. Reaching the attachment of the trapezius to the superior nuchal line, it pierces the muscular attachment and becomes deep. It then reaches the suboccipital triangle where it ends in a plexus from which the deep cervical and vertebral veins begin.

Veins draining deep portion of scalp: The veins draining the deep portion of scalp are the ***deep temporal veins*** which drain into the pterygoid venous plexus.

Veins of orbit, nose and infratemporal region: The veins of the orbit, nose and nasal cavities and the infratemporal region are closely associated with each other.

Veins of Orbit

The veins of orbit, except the supratrochlear and supraorbital veins, correspond to the branches of the ophthalmic artery. They pass backwards and converge to form two main venous channels, namely the superior and inferior ophthalmic veins.

The ***superior ophthalmic vein*** accompanies the ophthalmic artery. Anteriorly, it communicates with the facial vein and posteriorly, it passes through the superior orbital fissure and ends in the cavernous sinus. The superior ophthalmic vein receives a supraorbital tributary from the supraorbital vein which is a tributary of the facial vein. Thus, the superior ophthalmic vein acts as a communication between the facial vein and the cavernous sinus. Infections on the face can spread to the cavernous sinus through this route.

The ***inferior ophthalmic vein*** lies below the eyeball. It terminates in the cavernous sinus either directly or by joining the superior ophthalmic vein. It very often communicates with the pterygoid plexus.

The ***central vein of the retina*** accompanies the artery of the same name. It ends in the cavernous sinus directly or through the superior ophthalmic vein.

Veins of Nose

Many of the ***internal nasal veins*** join to form the ***sphenopalatine vein*** that ends in the pterygoid venous plexus. Some of the internal nasal veins end in the ethmoidal tributaries of the superior ophthalmic vein. Some more of them end in the septal tributaries of the superior labial vein and/or in the external nasal veins. The superior labial and the ***external nasal veins*** pass to the facial vein.

Veins of Infratemporal Region

The veins of the infratemporal region form the ***pterygoid venous plexus.***The plexus covers the lateral surface of the medial pterygoid muscle and surrounds the lateral pterygoid. It receives tributaries from veins corresponding to the branches of the maxillary artery. The plexus communicates with the cavernous sinus by an emissary vein through the foramen ovale, with the inferior ophthalmic vein usually by a small unnamed vein and with the facial vein by a deep facial branch. The plexus drains into the maxillary vein.

- The ***maxillary vein*** is a short vein that accompanies the first part of the maxillary artery. It unites with the superficial temporal vein to form the retromandibular vein.
- The ***retromandibular vein*** is formed by the union of the maxillary vein and the superficial temporal vein within the substance of the parotid gland. Within the gland, this vei is superficial to the external carotid artery and deep to the facial nerve. It bifurcates within the gland. The anterior division joins the facial vein. The posterior division goes to form the external jugular vein.

Intracranial Venous Sinuses

The intracranial venous sinuses are otherwise called the ***dural venous sinuses*** and form a complex network of venous channels which drain blood from the brain and the cranial bones. They occupy the space between the periosteal and meningeal layers of the duramater, except for the inferior sagittal and the straight sinuses. These two sinuses are present within the infolding of the meningeal layer of duramater. In addition to the brain, meninges and diploe of skull, they also drain the internal ear and the orbit.

The dural venous sinuses possess the following unique characteristics unlike other veins:

- They are lined by endothelium only and lack the muscular coat.
- They are valveless.
- They absorb CSF through the arachnoid granulations.
- They receive valveless emissary veins which help in the maintenance of the equilibrium of venous pressure both within and outside the cranial cavity.

The dural venous sinuses are classified into ***paired sinuses*** and ***unpaired sinuses*** (Fig. 14.37).

Unpaired Sinuses

The unpaired dural venous sinuses are:
- Superior sagittal sinus,
- Inferior sagittal sinus,
- Straight sinus,
- Occipital sinus,
- Anterior intercavernous sinus and posterior intercavernous sinus

Superior Sagittal Sinus

Other name: Superior longitudinal sinus.

The ***superior sagittal sinus*** runs in the attached convex margin of the falx cerebri (Fig. 14.38). It begins anteriorly in front of crista galli and runs backwards deeply grooving the internal surfaces of the frontal bone in the midline, of the adjacent margins of the two parietal bones (where they join at the sagittal suture) and of the occipital bone again in the midline. The sinus enters the confluence of sinuses at the internal occipital protuberance. At the confluence, the sinus shows a dilatation and usually deviates to the right and becomes continuous with the right transverse sinus. Occasionally, it may be continuous with the left transverse sinus.The superior sagittal sinus becomes broader as it runs backwards and is triangular in cross section.The arachnoid granulations project into the interior of the sinus for absorption of CSF.

The superior sagittal sinus receives:
- The superior cerebral veins which drain the superolateral and medial surfaces of the cerebral hemisphere,
- Cortical veins from the frontal lobe and
- The diploic and meningeal veins through three venous lacunae. The venous lacunae (also called the ***lacunae***

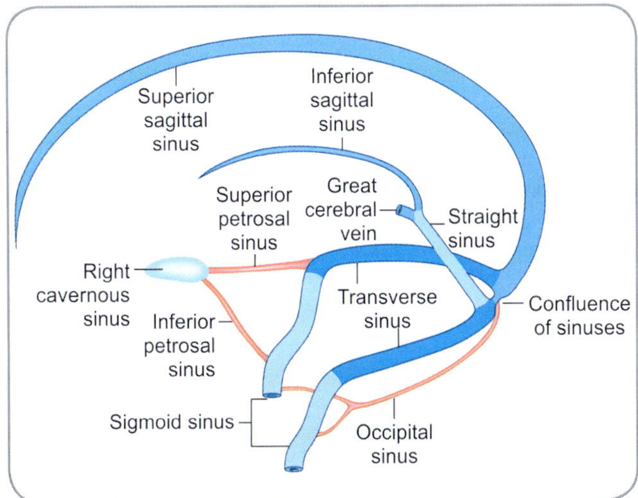

Fig. 14.37: Scheme to show the intracranial venous sinuses—the cavernous and petrosal sinuses are paired, but are shown only on one side for the sake of clarity

laterales) are irregular venous spaces present within the dural layers and are frontal, parietal and occipital in positions.

The superior sagittal sinus communicates with the veins of the scalp through the parietal emissary vein and the cavernous sinus through the superior anastomotic vein (vein of Trolard) and the superficial middle cerebral vein.

Surface Marking

Point A is marked on the midline on glabella. Point B is marked on the inion. The two points are joined by a line around the median plane of the vertex; the line should be narrow in front and widen to about 1.2 cm at the inion.

The **confluence of sinuses** (also called the **confluens sinuum** or **Torcular Herophili**) is the meeting place of the posterior ends of the superior sagittal, straight, occipital and the right and left transerse sinuses. The right transverse sinus is connected to the confluence directly but left transverse sinus communicates through a narrow channel. The posterior end of the superior sagittal sinus is dilated as it joins the confluence; the dilatation and the confluence are of the same size and merge imperceptibly; some authors tend to call the dilated posterior end of the superior sagittal sinus as the confluence because of this merger.

Inferior Sagittal Sinus

Other name: Inferior longitudinal sinus.

The **inferior sagittal sinus** lies within the posterior half of the lower free margin of the falx cerebri (Fig. 14.38). It increases in size posteriorly and ends by joining the straight sinus. It receives veins from the falx cerebri and sometimes from the medial surfaces of the cerebral hemispheres.

Straight Sinus

Other names: Sinus rectus, tentorial sinus.

The **straight sinus** lies in the triangular interval where the lower edge of the posterior part of falx cerebri joins the tentorium cerebelli. Anteriorly, it begins as a continuation of the inferior sagittal sinus, runs backward and ends in the

transverse sinus of the side opposite to that, with which the superior sagittal sinus is continuous, i.e. usually the left side. Its terminal part is also connected to the confluence of sinuses (Fig. 14.39).

The straight sinus receives:

- ❑ The inferior sagittal sinus and
- ❑ The great cerebral vein (of Galen) at its commencement (the straight sinus is sometimes described as formed by the union of the inferior sagittal sinus and the great cerebral vein).
- ❑ Some cerebellar veins drain into the straight sinus.

Occipital Sinus

The **occipital sinus** lies in the midline in relation to the floor of the posterior cranial fossa within the attached margin of the falx cerebelli. The anterior end of the occipital sinus bifurcates into two channels which pass around either side of the foramen magnum to join the corresponding sigmoid sinus. Posteriorly, it ends in the confluence of sinuses. Occasionally, the sinus may be paired wherein the sinus of each side will run from the transverse sinus of one side to the inferior end of the sigmoid sinus of the same side through the falx cerebella and the margin of foramen magnum.

Intercavernous Sinuses

The right and left cavernous sinuses are connected by anterior and posterior intercavernous sinuses which are located in the anterior and posterior attached margins of the diaphragma sellae respectively and thus form a complete circular venous sinus (also called the **Ridley's circle**) along with the cavernous sinuses (Fig. 14.40). They receive small irregular veins from the pituitary gland. A few tiny intercavernous sinuses run between the two cavernous sinuses across the floor of the hypophyseal fossa.

Basilar Venous Plexus

Other names: Basilar sinus, plexus basilaris.

It is a plexiform venous network that lies on the clivus of the skull. It connects laterally with the inferior petrosal

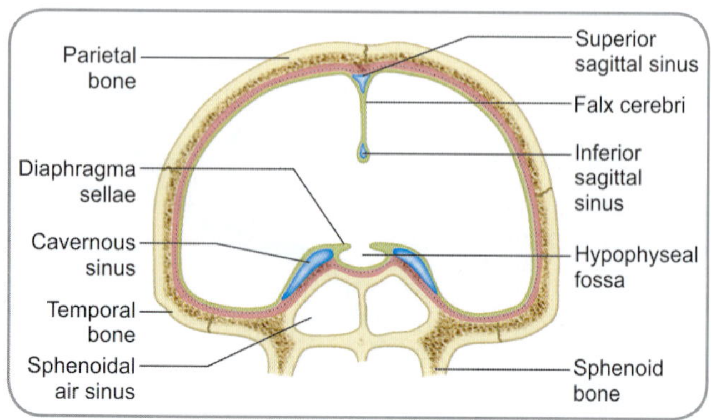

Fig. 14.38: Coronal section through middle cranial fossa to show the position of intracranial venous sinuses

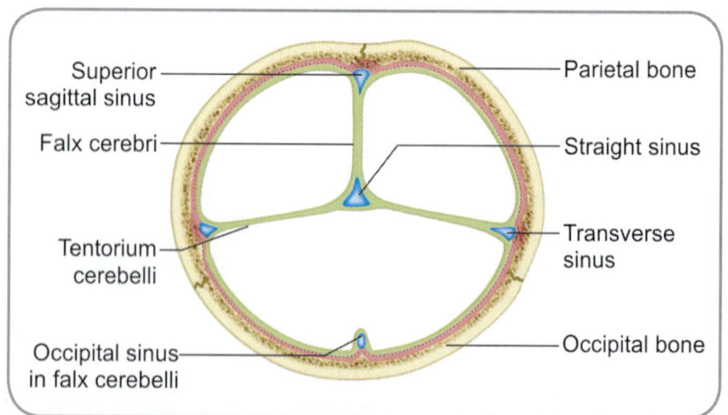

Fig. 14.39: Coronal section through the posterior cranial fossa (behind the foramen magnum) to show the position of intracranial venous sinuses

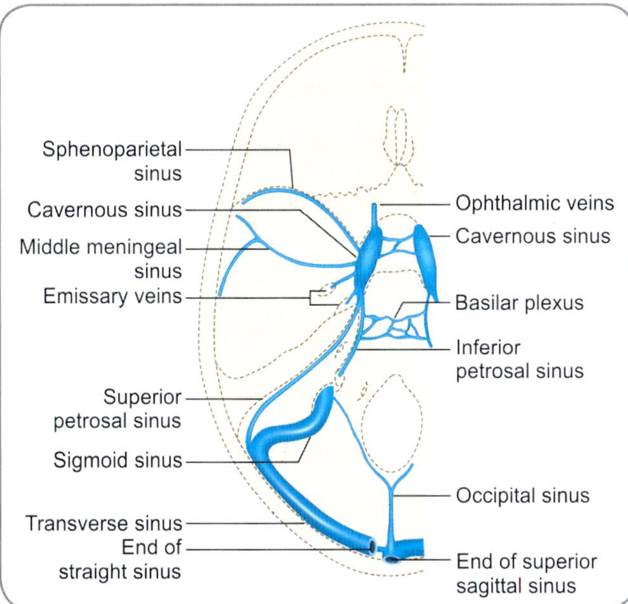

Fig. 14.40: Relationship of the cranial venous sinuses to the floor of the cranial cavity

sinuses of both sides and communicates below through the foramen magnum with the internal vertebral venous plexus. It is considered an unpaired intracranial venous sinus by some authors because though in the form of a plexus, it lies between the two layers of duramater.

Paired Sinuses

The paired dural venous sinuses are:

- ❑ Transverse sinuses,
- ❑ Sigmoid sinuses,
- ❑ Cavernous sinuses,
- ❑ Superior petrosal sinuses,
- ❑ Inferior petrosal sinuses,
- ❑ Sphenoparietal sinuses,
- ❑ Petrosquamous sinuses and
- ❑ Middle meningeal sinuses

Of these, the transverse, sigmoid and cavernous sinuses are large and the others are much smaller.

Transverse Sinuses

Other name: Lateral sinuses.

The right and left ***transverse sinuses*** lie horizontally as indicated by their names (Fig. 14.40). Each sinus begins posteriorly at the internal occipital protuberance. The right sinus is usually a continuation of the superior sagittal sinus and the left a continuation of the straight sinus, but this arrangement may sometimes be reversed. Each sagittal sinus runs in a curve, proceeding at first laterally and then forwards, along the line of attachment of the tentorium cerebelli. The sinus produces a transverse groove on the inner surface of the occipital bone and on the posteroinferior angle of the parietal bone. Finally, it

reaches the posterolateral part of petrous temporal bone, where it becomes continuous with the sigmoid sinus.

The transverse sinus receives the inferior cerebral, inferior cerebellar, diploic and inferior anastomotic veins. It is also joined by the superior petrosal sinus at the point of continuation as sigmoid sinus. The ***inferior anastomotic vein (vein of Labbe)*** connects the transverse sinus with the superficial middle cerebral vein.

> **Surface Marking**
>
> Point A is marked just above the inion. Point B is marked on the upper part of the back of the root of auricle. The two points are joined by a line that is slightly convex upwards. This line marks the ipsilateral transverse sinus on the surface.

Sigmoid Sinuses

The right and left ***sigmoid sinuses*** are continuations of the corresponding transverse sinuses. As indicated by the name, each sigmoid sinus is S-shaped. It first runs downward and medially in a deep groove on the internal aspect of the mastoid part of the temporal bone and then across the jugular process of the occipital bone. Finally, it runs forward to reach the posterior part of the jugular foramen where it ends by becoming continuous with the superior bulb of the internal jugular vein. Anteriorly, a thin plate of bone separates it from the mastoid antrum and air cells.

The sigmoid sinus receives the mastoid and condylar emissary veins, cerebellar veins and labyrinthine veins.

> **Surface Marking**
>
> Point A is marked on the upper part of the back of the root of auricle. Point B is marked on the curve of the root of the auricle at the level of the lower margin of the external acoustic meatus. Point C is marked on the lower margin of the external acoustic meatus. Points A and B are joined by a broad line running along the back of the root of the auricle. Points B and C are then joined by extending this broad line forwards. The entire curved line marks the sigmoid sinus on the surface.

Cavernous Sinuses

The right and left ***cavernous sinuses*** are so called because their cavities are traversed by delicate strands of tissue which appear to subdivide each sinus into a number of smaller spaces (or caverns). The cavernous sinuses are placed anteroposteriorly on either side of the body of the sphenoid bone in the middle cranial fossa. Each sinus is located between the periosteal and meningeal layers of the duramater. The ***roof and lateral wall*** are formed by the meningeal layer and the ***floor and medial wall*** by the periosteal layer (details of formation are dealt with in the chapter on cranial cavity).

Extent

Anteriorly, each sinus reaches the medial end of the superior orbital fissure. Posteriorly, it reaches the apex of the petrous part of the temporal bone.

Structures passing through the cavernous sinus

Several important structures pass through the cavernous sinus.

- The internal carotid artery surrounded by the sympathetic plexus of nerves (called the ***internal carotid plexus***) passes anteriorly through the inferomedial portion of the sinus and leaves the sinus by piercing the roof of the sinus. The artery is accompanied by the abducent nerve that lies below and lateral to the artery.
- Four cranial nerves are embedded in the lateral wall of the sinus. From above downwards, these are the oculomotor nerve, the trochlear nerve, the ophthalmic division of the trigeminal nerve and the maxillary division of the trigeminal nerve. The oculomotor nerve and the trochlear nerve enter the sinus by piercing the roof of the sinus and leave it through the superior orbital fissure. The ophthalmic and the maxillary divisions of the trigeminal nerve enter by piercing the cavum trigeminale and the lateral wall of the sinus.

External relations of cavernous sinus (Fig. 14.41)

- ***Medially***, the sinus is related to the hypophysis cerebri, and the sphenoidal air sinus from which it is separated by a plate of bone.
- ***Laterally***, it is related to cavum trigeminale, which is a pouch like extension of dura mater containing the trigeminal ganglion and to the uncus of the cerebral temporal lobe.
- ***Superiorly***, it is related to the optic chiasma and a part of the internal carotid artery.
- ***Inferiorly***, it lies on the bone at the junction of the body and greater wing of sphenoid.

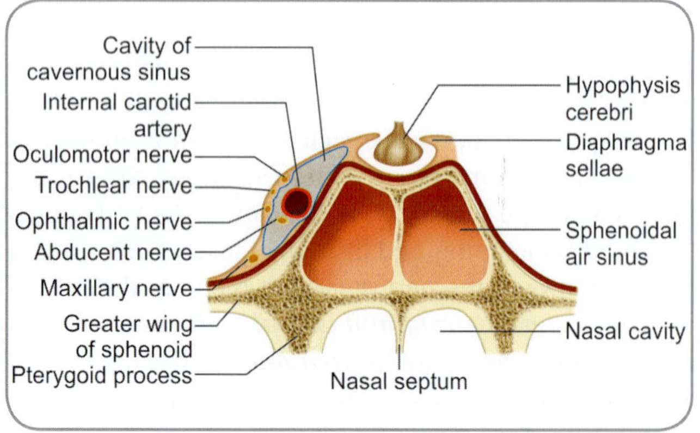

Fig. 14.41: Coronal section through cavernous sinus to show its relations

Tributaries

Being one of the large venous sinuses and situated in the middle cranial fossa, the cavernous sinus receives many tributaries. These tributaries include:

- Superior Ophthalmic vein,
- A branch from the inferior ophthalmic vein,
- Occasionally, central vein of retina,
- Superficial middle cerebral vein,
- Few inferior cerebral veins,
- Sphenoparietal sinus and
- Middle meningeal vein (anterior trunk).

Communications

The cavernous sinus communicates

- With the transverse sinus—via superior petrosal sinus,
- With the tnternal jugular vein—through inferior petrosal sinus,
- With the pterygoid venous plexus—through emissary veins,
- With the facial vein—through both superior ophthalmic and angular veins; and pterygoid venous plexus and deep facial vein,
- With the opposite cavernous sinus—through the intercavernous sinuses and
- With the superior sagittal sinus—through the superficial middle cerebral vein and superior anastomotic vein.

Superior Petrosal Sinuses

Each ***superior petrosal sinus*** begins at the posterior end of the cavernous sinus. It runs backwards and laterally along the sharp upper margin of the petrous temporal bone (i.e. along the attached margin of the tentorium cerebelli). It terminates by joining the junction of the sigmoid sinus and transverse sinus. It receives the cerebellar, inferior cerebral and tympanic veins and connects with the inferior petrosal sinus and basilar plexus.

Inferior Petrosal Sinuses

Other name: Englisch's sinuses.

Each ***inferior petrosal sinus*** begins at the posterior end of the cavernous sinus. It runs downwards and somewhat laterally in the groove between the petrous temporal bone and the basilar part of the occipital bone. It passes through the anterior part of the jugular foramen and terminates by joining the upper end of the internal jugular vein. The inferior petrosal sinuses of the right and left sides are connected by a ***basilar plexus of veins*** lying on the basal parts of the sphenoid and occipital bones. The sinus receives the labyrinthine veins and tributaries from medulla, pons and inferior cerebellar surface.

Sphenoparietal Sinuses

Other names: Sinuses alae parvae, Breschet's sinuses.

Each ***sphenoparietal sinus*** runs medially along the undersurface of the posterior border of the lesser wing of sphenoid in the floor of the anterior cranial fossa. The sinus ends by joining the anterior end of the cavernous sinus. Each sinus receives some temporal diploic veins and occasionally the anterior trunk of the middle meningeal vein.

Petrosquamous Sinuses

Other name: Luschka's sinuses.

These sinuses when present are lodged in the petro squamous fissures of both sides and end by draining into the respective transverse sinus.

Middle Meningeal Sinuses

The vein accompanying the middle meningeal artery is called the **middle meningeal sinus.** The sinus has frontal (anterior) and parietal (posterior) trunks corresponding to the branches of the artery. Both the trunks communicate superiorly with the superior sagittal sinus through the venous lacunae. In addition, the anterior trunk drains into either the cavernous sinus or the spheno parietal sinus or into the pterygoid venous plexus by passing through the foramen ovale. The posterior trunk passes through the foramen spinosum and ends in the pterygoid venous plexus.

> ### Clinical Correlation
>
> - Thrombosis of the superior sagittal sinus may occur due to spread of infection from the nose, scalp and diploic tissue. The clinical manifestation will be that of increased intracranial tension due to defective absorption of CSF.
> - As the upper part of the sigmoid sinus is separated only by a thin plate of bone from the mastoid antrum, infection from the mastoid antrum or the middle ear can spread to the sigmoid sinus resulting in thrombosis of sigmoid sinus. Backward extension of the thrombosis to the superior sagittal sinus (through the transverse sinus) can lead to hydrocephalus due to defective absorption of CSF.
> - Septic thrombosis of the cavernous sinus can occur due to its numerous communications with the danger areas of the face, orbit and pharynx. The clinical manifestations correspond to the affliction of the structures passing through the sinus. They are:
>   - Severe pain in the eye and forehead along the distribution of the ophthalmic nerve,
>   - Ophthalmoplegia due to involvement of 3rd, 4th and 6th cranial nerves, and
>   - Marked oedema of eyelids with exopthalmos due to congestion of orbital veins.
>
> Unilateral clinical manifestations may become bilateral due to the extension of the infection to the opposite cavernous sinus through the intercavernous sinuses.
> - A fracture of the base of skull causing rupture of internal carotid artery may sometimes result in the establishment of an arterio-venous communication between the artery and the cavernous sinus. It will also show features mentioned above but the exophthalmos will be pulsatile in nature
>
> contd…

> ### Clinical Correlation   *contd…*
>
> (because of the arterial pressure). In addition, a loud systolic murmur will be heard over the temporal region.
> - Thrombosis of inferior petrosal sinus can occur from infection of middle and internal ears with ipsilateral involvement of trigeminal and abducent nerves. This syndrome is called **Gradenigo's syndrome**.
> - The middle meningeal veins are closer to the bones than the arteries and the grooves in the parietal bone are formed by the veins. Hence, fracture of skull will lead to tearing of the middle meningeal veins rather than the arteries.

Other Veins of Cranial Cavity

Several small venous channels are seen in the cranial cavity and cranial walls. These are the diploic veins and the meningeal veins.

Diploic Veins

The diploic veins are anastomosing spaces lined with endothelium located within the diploe of the skull bones. Efferent vessels emerge from these spaces and are called the **definitive diploic veins.** They drain into another vein or sinus. There are four such definitive diploic veins on each side.

- **Frontal diploic vein:** Passes through the supraorbital notch and ends in the supraorbital vein.
- **Anterior temporal diploic vein:** Emerges from the greater wing of sphenoid and ends in the sphenoparietal sinus or the deep temporal veins.
- **Posterior temporal diploic vein:** Runs out of the parietal bone at its posteroinferior angle and ends in the transverse sinus.
- **Occipital diploic vein:** Being the largest of the diploic veins, runs down the occipital bone to join the transverse sinus or the occipital emissary vein.

Smaller **unnamed diploic veins** pierce the inner table of the skull and end in the venous lacunae located along the superior sagittal sinus.

Meningeal Veins

A slender but dense venous plexus exists in the duramater. Two sets of meningeal veins emege from this plexus. The **inner plexus of meningeal veins** drains into the cranial venous sinuses. The **outer plexus of meningeal veins** accompanies the meningeal arteries. These veins lie outer to the corresponding arteries in the grooves in the bones, leave the skull through the foramina ovale or spinosum and end in the pterygoid venous plexus. The outer plexus is liable for injury since it lies close to the bones and outer to the arteries. The veins may be torn in fractures.

Emissary Veins

These are veins which connect the intracranial dural venous sinuses with veins outside the cranial cavity. Blood can flow in either direction in these veins. The incidence and size of an emissary vein is variable. However, a few of them are constantly noted.

- *Frontal emissary vein:* Present in a child; unpaired; passes through the foramen caecum; connects the superior sagittal sinus with veins of nasal cavity.
- *Parietal emissary vein:* One on each side; passes through the parietal foramen; connects the superior sagittal sinus with the occipital veins.
- *Mastoid emissary vein:* One on each side; passes through the mastoid foramen; connects the sigmoid sinus with the occipital or posterior auricular vein.
- *Condylar emissary vein:* Inconstant; passes through the condylar canal if the latter is present; connects the sigmoid sinus with the suboccipital plexus of veins.
- *Mandibular emissary plexus:* A plexus of veins surrounding the mandibular nerve as the latter passes through the foramen ovale; connects the cavernous sinus with the pterygoid plexus.
- *Carotid emissary veins:* Small veins along the course of the cranial part of the internal carotid artery; connect the cavernous sinus with the pharyngeal plexus of veins or rarely, with the internal jugular vein.
- *Unnamed emissary veins:* Pass through the foramen lacerum; connect the cavernous sinus with the pterygoid plexus or pharyngeal veins.

Other Veins of Head and Neck Region (Fig. 14.42)

Deep Cervical Vein

The deep cervical vein begins in the venous plexus present in the suboccipital region. It accompanies the corresponding artery through the deep muscles of the back of the neck and ends by joining the lower part of the vertebral vein (Fig. 14.43).

Vertebral Vein

The vertebral vein begins in the suboccipital venous plexus.It then enters the foramen transversarium of the atlas and runs downwards in the form of a dense plexus around the vertebral artery. At the foramen transversarium of the sixth cervical vertebra, the plexus takes the form of a single vessel and then runs downwards behind the internal jugular vein and ends in the upper part of the corresponding brachiocephalic vein (Fig. 14.43).

Clinical Correlation

Arteries

Congenital anomalies: The large arteries of the head and neck develop from a series of aortic arch arteries which lie in the embryonic pharyngeal arches. Disorders in development can result in various kinds of anomalies ranging from malpositioning of the vessels to complete absence.
- Normally, the left common carotid artery is a direct branch of the arch of the aorta. Sometimes, it may arise as a branch of the brachiocephalic trunk (the brachiocephalic trunk

contd…

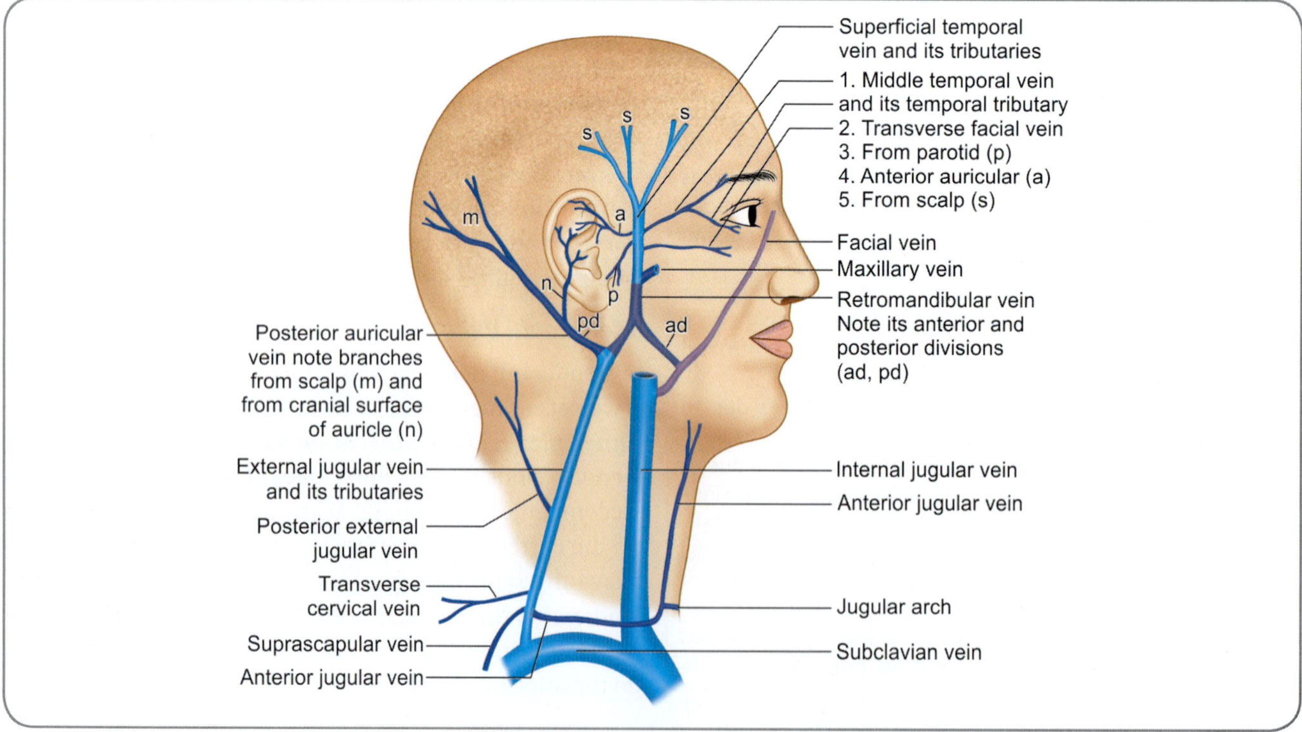

Fig. 14.42: Scheme to show some veins of the head and neck

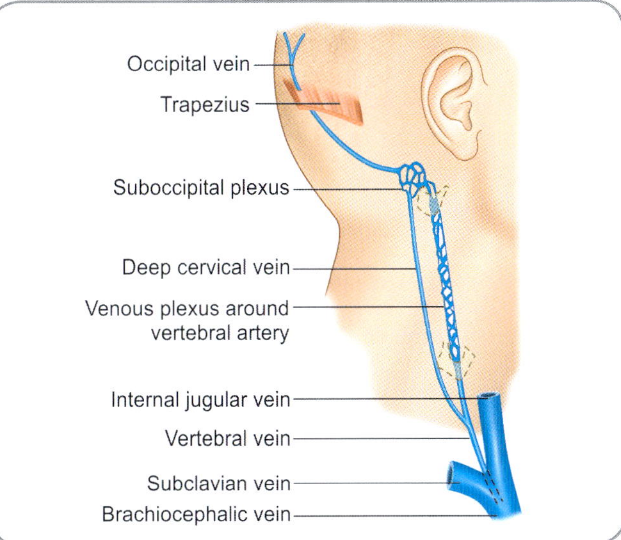

Fig. 14.43: Scheme to show the occipital, vertebral and deep cervical veins

---

Clinical Correlation *contd...*

then divides into three branches, viz. right subclavian, right common carotid, and left common carotid). Sometimes, the left common carotid and left subclavian arteries may arise from the arch of the aorta by a common stem (which is then called the ***left brachiocephalic trunk***).

- The right subclavian artery may arise as the last branch of the arch of the aorta (or of the descending thoracic aorta). To reach the neck, the artery has to run upwards behind the oesophagus. In such cases, along with the arch of the aorta, the artery forms an arterial ring enclosing the trachea and oesophagus. The ring may obstruct the oesophagus leading to dysphagia. This is called ***dysphagia lusoria***. The ring can also be a cause of obstruction of the trachea.
- The left vertebral artery may arise as a direct branch of the arch of the aorta (instead of arising from the subclavian artery).
- ***Pulsations of the arteries:*** Pulse of the superficial temporal artery can be felt near the zygomatic arch; that of the facial artery as it crosses the lower border of the mandible (near the anterior border of the masseter). Pulsations of the common carotid artery (carotid pulse) can be felt at the level of the superior border of the thyroid cartilage, beneath the anterior border of the sternocleidomastoid muscle.
- ***Aneurysm:*** Weakening of the wall of any artery can lead to dilatation of the artery called ***aneurysm*** at that site. Aneurysm of any intracranial artery can burst and can be a cause of bleeding into the subarachnoid space. An aneurysm of the third part of the subclavian artery can press on the brachial plexus leading to motor and sensory symptoms of the upper limb.
- ***Atherosclerotic blocks:*** Any artery of the body can undergo degenerative changes associated with deposition of fatty substances in the arterial wall. Such deposits are called

contd...

Clinical Correlation *contd...*

atheroma. The vessel that gets blocked by fatty deposits becomes rigid and narrow. The condition is called ***atherosclerosis*** (Greek.athere=gruel, skleroses=hardness). Blood flow in an atherosclerosed vessel is reduced. The deposit may partially or completely block the vessel.

- Blockage of the common carotid or internal carotid arteries in this way can interfere with blood supply of the brain and of the eyeball.
- In suspected cases of blockage, the carotid system of arteries is investigated by ***carotid angiography***. A radioopaque dye is injected into the common carotid artery either directly or through a catheter (passed up through the femoral artery to reach the arch of aorta). Skiagrams taken after the injection reveal the site of blockage. In such skiagrams the petrous, cavernous and cerebral parts of the internal carotid artery cast a typical S-shaped shadow to which the term ***carotid syphon*** is applied.
- Atheroma may affect the vertebral arteries, especially their first and fourth parts. These arteries may also be pressed upon by osteophytes in cervical spondylosis. Inadequacy in blood flow through the vertebral and basilar arteries can give rise to attacks of transient ischaemia in which the patient complains of dizziness. Obstruction of one vertebral artery usually does not produce symptoms because of blood flow through the contralateral vertebral artery. Collateral anastomoses also exist through the ascending cervical, thyrocervical and occipital arteries.
- ***Cervical rib:*** One cause of possible obstruction to flow of blood in the subclavian artery is the presence of a cervical rib. The artery has to loop over the cervical rib to enter the axilla and is, therefore, pressed upon.
- ***Pressure on arteries:*** Compression of arteries can be used to stop haemorrhage in the area of distribution.
- The common carotid artery can be compressed against the carotid tubercle (which is the anterior tubercle of transverse process of the sixth cervical vertebra).
- It should be remembered that pressure on the carotid sinus causes slowing of heart rate and fall of blood pressure.
- When the neck is extended, the carotid sheath is retracted. It is because of this fact that in suicidal attempts, slitting of the neck often fails to reach the carotid arteries.
- Bleeding caused by severe injuries in the proximal part of the upper extremity can be controlled by pressure over the subclavian artery. Pressure is directed downward and backward (and somewhat medially). The artery gets compressed against the first rib.

   Injury to the middle meningeal artery is an important cause of an extradural haemorrhage. The anterior branch is commonly involved. Extradural haemorrhage from an injured posterior division of the middle meningeal artery is less common. The blood can press upon the superior temporal gyrus and lead to deafness in the ear of the opposite side.

contd...

> ### ♪ Clinical Correlation *contd...*
>
> ❑ *Veins*
>   - *Variability:* As in other parts of the body, the arrangement of veins, especially that of superficial veins, is highly variable. The anterior jugular vein (which is normally paired) may be represented by a single median vein, or may be absent on one side.
>   - *Injury to veins:* Superficial veins are liable to be involved in injuries to the neck.
>   - The external jugular vein is particularly vulnerable.
>   - Normally, when a vein is cut, its walls retract thus limiting the amount of bleeding. However, near its lower end the external jugular vein pierces the deep fascia. Here the wall of the vein is adherent to fascia and this prevents the wall from retracting if the vein is injured. During inspiration, pressure in the vein can be negative and air can be sucked into the vein. This can have very serious consequences if it leads to air embolism.
>
> ❑ *Dangerous Area of Face:* Near the medial angle of the eye the supraorbital vein, which is a tributary of the facial vein, communicates with the superior ophthalmic vein (lying in the orbit). The superior ophthalmic vein drains into the cavernous sinus. In this way the facial vein is brought into communication with the cavernous sinus. The facial vein also communicates with the cavernous sinus through the deep facial vein and the pterygoid plexus. Because of these communications, an infection in the face can spread to the cavernous sinus leading to cavernous sinus thrombosis. It has been observed that such spread of infection is most likely to take place if the infection is over the upper lip or the lower part of the nose. As a result, this region is called the *dangerous area of the face*.
>
> ❑ *Venous Pressure:* Clinicians examining the cardiovascular system use the level of blood in the external jugular vein as an indication of venous pressure.
>   - When the patient lies flat, the vein is at the same level as the right atrium and the entire vein is full of blood.
>   - When the head rests on a pillow, the level of blood in the vein lies at junction of the lower and middle-thirds of the neck.
>   - When the patient sits up, the vein becomes empty.
>
> Venous pressure can be raised if there is right heart failure, raised intrathoracic pressure from any cause or obstruction to the superior vena cava. Increased venous pressure leads to dilatation of veins in the neck. The internal jugular vein can be markedly dilated.
>
> ❑ *Involvement in Malignancy:* Lymph nodes from all parts of the head and neck ultimately drain into the deep cervical lymph nodes which are, therefore, frequently involved in malignancy and in infections such as tuberculosis. In the treatment of malignancy, the surgeon often has to remove these nodes. Enlarged nodes become adherent to the internal jugular vein and it is sometimes necessary for the surgeon to remove a segment of the vein along with the lymph nodes.
>
> ❑ *Intracranial Venous Sinuses:* Intracranial venous sinuses are of importance because they receive blood from the brain. Infection in these sinuses can spread to brain tissue with serious consequences. Infections in the scalp can spread into the venous sinuses through emissary veins which pass through the foramina in the skull. Emissary veins traverse the layer of loose areolar tissue of the scalp that is, therefore, called the *dangerous area of the scalp.*
>   - *Pulsating exophthalmos:* Because of the peculiar relationship of the cavernous sinus to the internal carotid artery a communication between the two may occur as a result of injury. When this happens, arterial pressure is communicated through the sinus to veins of the orbit (which open into the sinus). As a result, the eyeball becomes prominent and pulsates with each heart beat. The condition is called *pulsating exophthalmos*.
>   - *Thrombosis in the cavernous sinus:* The cavernous sinus can be infected by spread of infection from the dangerous area of the face. Infection can also reach it from the nose and paranasal sinuses. Symptoms can be produced by blockage to blood flow or involvement of cranial nerves. As veins of the orbit drain into the cavernous sinus they become congested. Accumulation of fluid in the orbit pushes the eyeball forwards (resulting in exophthalmos). The eyelids and the root of the nose show swelling. Involvement of the ophthalmic nerve leads to severe pain in the region of distribution of the nerve (eye and over the forehead). Involvement of the oculomotor, trochlear and abducent nerves can lead to paralysis of extraocular muscles.
>   - *Thrombosis in the superior sagittal sinus:* Infection can spread to this sinus from the scalp or from the nasal cavities. An infected sigmoid or transverse sinus can also be a source. The arachnoid granulations in the superior sagittal sinus are the main sites at which CSF is absorbed into the blood stream. When the sinus is thrombosed, this absorption is interfered with, leading to increased pressure in CSF. This leads to various signs of increased intracranial tension. In a child it can lead to hydrocephalus. Venous blood from the cerebral hemisphere is drained predominantly by superior cerebral veins into the superior sagittal sinus. Infection from the sinus can extend along these veins to the cerebral hemisphere.
>   - *Thrombosis in the sigmoid sinus:* Middle ear and the mastoid air cells are very frequent sites of infection. The close relationship of the sigmoid sinus to the middle ear (lying in the petrous temporal bone) and the mastoid process often leads to spread of infection to this sinus. Thrombosis in the sinus can extend downwards into the internal jugular vein. It can spread upwards along the transverse sinus into the superior sagittal sinus.

Multiple Choice Questions

1. The cranial nerve passing through the cavity of the cavernous sinus is:
   a. Oculomotor nerve
   b. Trochlear nerve
   c. Abducent nerve
   d. Maxillary nerve
2. The venous sinus receinving the Great Cerebral vein of Galen is:
   a. Superior sagittal sinus
   b. Occipital sinus
   c. Straight sinus
   d. Cavernous sinus
3. The vein acting as a communication between the facial vein and the cavernous sinus is:
   a. Superior ophthalmic vein
   b. Inferior ophthalmic vein
   c. Supratrochlear vein
   d. Posterior auricular vein
4. One of the following is not a branch of the vertebral artery. Which is it?
   a. Posterior spinal artery
   b. Anterior spinal artery
   c. Posterior inferior cerebellar artery
   d. Posterior communicating artery
5. The internal carotid artery:
   a. Is the main continuation of the common carotid artery
   b. Enters the cranial cavity through the foramen lacerum
   c. Gives out the ophthalmic branch while traversing the petrous temporal bone
   d. Gives out a pharyngeal branch in the neck

ANSWERS

1. c   **2.** c   **3.** a   **4.** d   **5.** a

Clinical Problem-solving

Case Study 1: A 20-year-old young woman acnes over her upper lip and cheeks.
- What complication do you expect in case of the acnes getting infected?
- What is the anatomical basis for such a complication?
- What other areas of the face can predispose to similar complications?

Case Study 2: When you went on a rounds through the wards, you saw a patient with pulsating exophthalmos.
- Where is the lesion causing such a condition?
- Why is the eye pulsating in this condition?
- Why does cavernous sinus infection cause exophthalmos?

(For solutions see Appendix).

Nerves of Head and Neck

Chapter 15

Frequently Asked Questions

- Write notes on: (a) Suboccipital nerve, (b) Supraclavicular nerves, (c) Phrenic nerve, (d) Ansa cervicalis, (e) Greater occipital nerve.
- Describe the course and branches of the hypoglossal nerve.
- Trace the course of the facial nerve. Add a note on its functional components.
- Discuss the oculomotor nerve in detail.
- Discuss the glossopharyngeal nerve in detail.

CERVICAL NERVES

In the thoracic, lumbar and sacral regions the number of spinal nerves corresponds to that of the vertebrae, each nerve lying **below** the numerically corresponding vertebra. However, in the neck, there are seven cervical vertebrae, and eight cervical nerves. The reason for this is that the first spinal nerve (first cervical spinal nerve also) emerges out from the vertebral canal above the first cervical vertebra (Fig. 15.1). Therefore, the **upper seven cervical nerves lie above the numerically corresponding vertebrae**. The eighth cervical nerve lies below vertebra C7. The spinal nerves on each side, after being formed by the union of dorsal root and ventral root, exit the vertebral canal via the intervertebral foramen and divide again to form dorsal ramus and ventral ramus.

Dorsal Rami of Cervical Nerves

The dorsal ramus of a typical spinal nerve is smaller than the ventral ramus, as its area of supply is less when compared to the ventral ramus. It passes backwards and divides into medial and lateral branches which supply the deep muscles and skin of the back (Fig. 15.2). The dorsal rami of the upper three cervical nerves have some atypical features and are briefly described herewith.

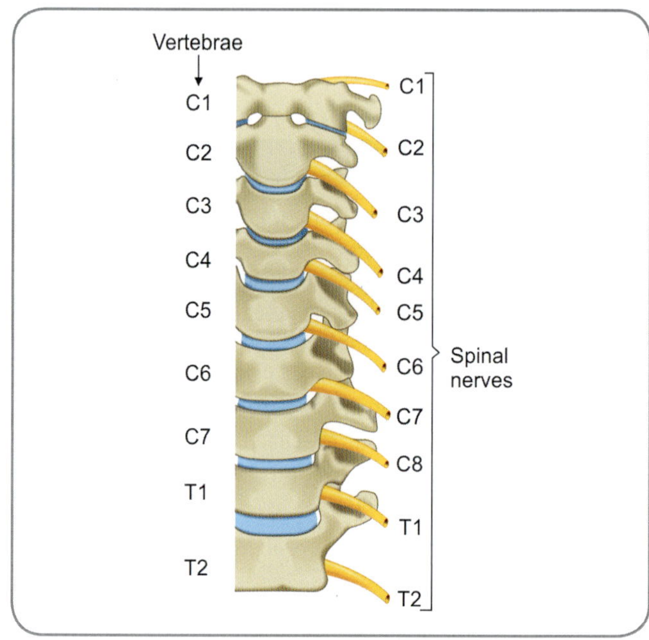

Fig. 15.1: Scheme to show the relationship of cervical and upper thoracic nerves to vertebrae

Dorsal Ramus of the First Cervical Nerve (Suboccipital Nerve)

The dorsal ramus of the first cervical nerve is larger than the ventral ramus. It is seen in the suboccipital triangle which it reaches by passing above the posterior arch of atlas (Fig. 15.3). Here the nerve lies between the arch and the vertebral artery. The dorsal ramus divides into branches which supply the rectus capitis posterior major and minor, the superior and inferior oblique muscles and the semispinalis capitis (Fig. 15.4A). Some cutaneous branches may reach the skin of the scalp and connect with the greater and lesser occipital nerves.

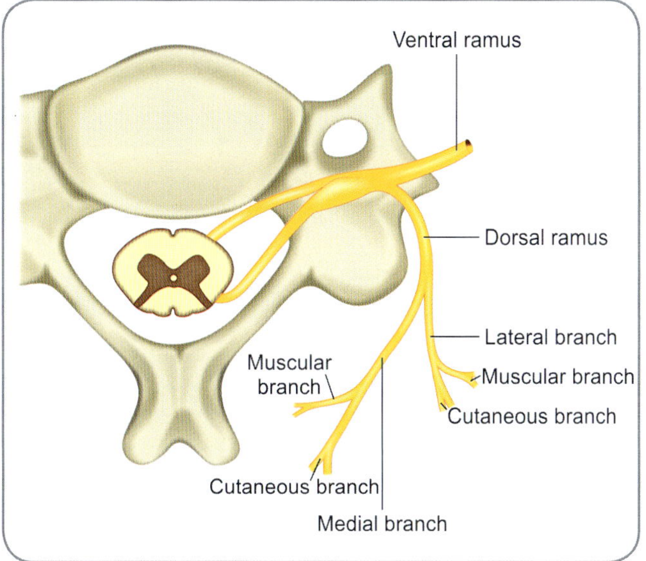

Fig. 15.2: Dorsal ramus of a typical spinal nerve

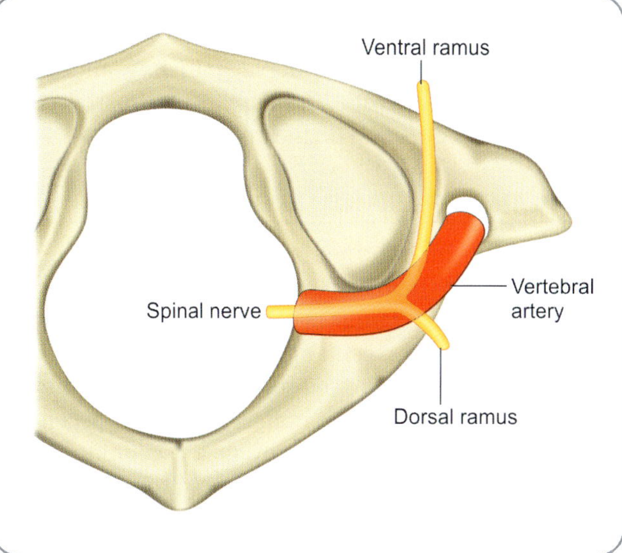

Fig. 15.3: Dorsal ramus of the first cervical nerve

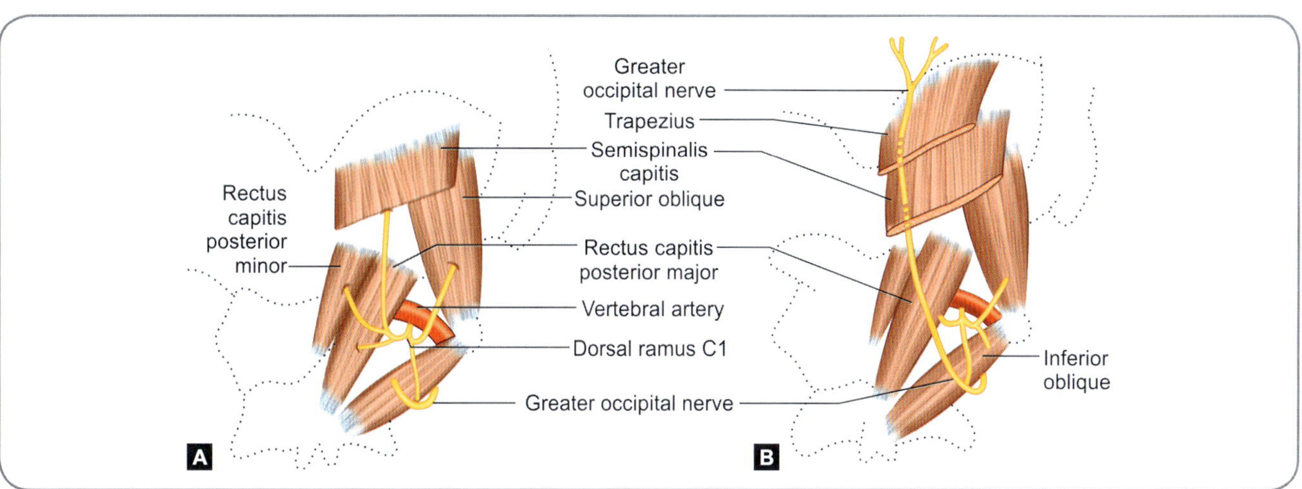

Figs 15.4A and B: A. Distribution of the dorsal ramus of the first cervical nerve **B.** Course of the greater occipital nerve

Dorsal Ramus of the Second Cervical Nerve

The dorsal ramus of the second cervical nerve is large. It reaches the suboccipital region by passing below the posterior arch of atlas, and below the inferior oblique muscle. It divides into medial and lateral branches. The medial branch that is much more prominent is called the ***greater occipital nerve*** (Fig. 15.4B). Winding around the lower border of the inferior oblique muscle, this nerve passes upwards and medially across the suboccipital triangle, lying deep to the semispinalis capitis. It becomes superficial by piercing first the semispinalis capitis and then the trapezius. Finally, it divides into branches which ramify in the scalp supplying its posterior part, till the coronal suture. It also gives a branch to the semispinalis capitis muscle. It gives communicating branches to the third occipital and lesser occipital nerves.

Clinical Correlation

Greater occipital neuralgia—there is pain and paraesthesia in the area of supply of greater occipital nerve in the scalp, because of involvement of the dorsal root ganglia of C2.

Dorsal Ramus of the Third Cervical Nerve

The dorsal ramus of the third cervical nerve behaves like a typical dorsal ramus. The only special feature about it is that it also gives a small branch to the skin of the occipital region. This branch is called the ***third occipital nerve.*** This too gives communicating branches to greater and lesser occipital nerves.

Ventral Rami of Cervical Nerves

The ventral rami of the first, second, third and fourth cervical nerves unite with each other to form the ***cervical*** *plexus*. The ventral rami of the fifth, sixth, seventh and eighth cervical nerves, and the greater part of the ventral ramus of the first thoracic nerve, join one another to form the brachial plexus. The brachial plexus has been considered in the section on the upper limb.

CERVICAL PLEXUS AND ITS BRANCHES

The cervical plexus is formed by the ventral rami of the first, second, third and fourth cervical nerves (Fig. 15.5). With the exception of the ramus of the first cervical nerve, each of them divides into ascending and descending branches. The ascending branch of the second nerve joins the first nerve; and its descending branch joins the ascending branch of the third nerve. Similarly, the descending branch of the third nerve joins the ascending branch of the fourth nerve. The descending branch of the fourth nerve is small and joins the fifth cervical nerve.

The cervical plexus gives off a large number of branches. The *cutaneous branches* are shown in Figure 15.6 and are as follows:

- The *lesser occipital* nerve arises from the second cervical nerve.
- The *great auricular* nerve and the *transverse cutaneous nerve of neck* arise from the second and third nerves.
- The *supraclavicular* nerves arise from the third and fourth nerves.

The *muscular branches* of the cervical plexus are numerous. For ease of visualisation they may be subdivided as follows (Fig. 15.7):

- *Branches to muscles forming boundaries of the posterior triangle:* Sternocleidomastoid receives a branch from C2 (and sometimes from C3). Levator scapulae, scalenus medius and trapezius receive branches from C3 and C4.

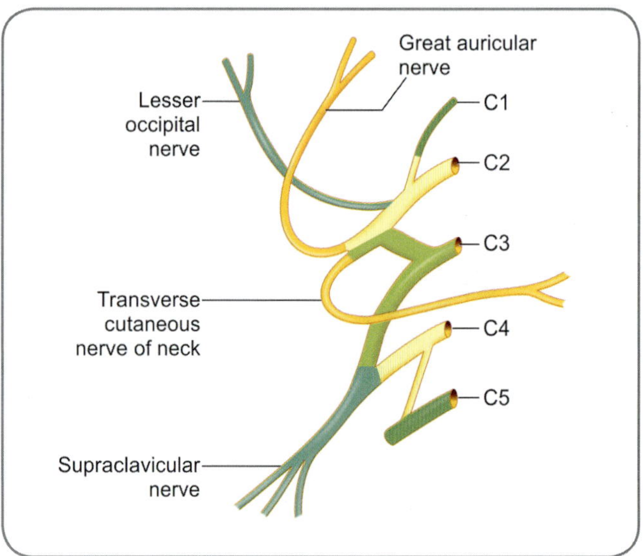

Fig. 15.6: Cervical plexus and its cutaneous branches

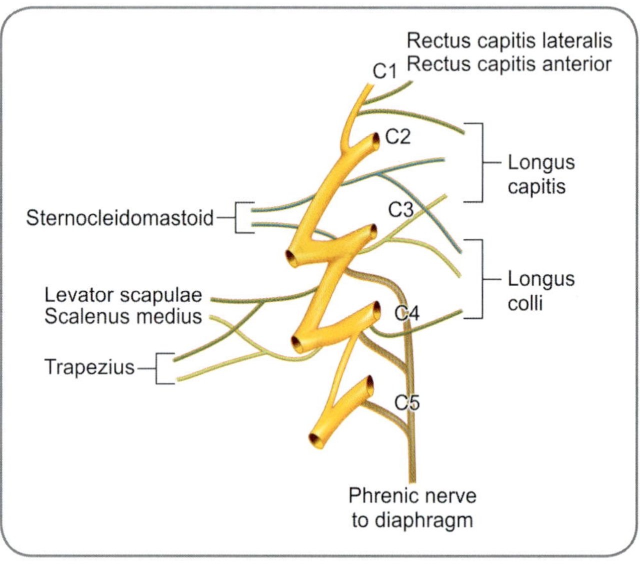

Fig. 15.7: Scheme to show the muscular branches of cervical plexus

- *Phrenic nerve* (which supplies the diaphragm) arises by separate roots from C3, C4 and C5.
- *Branches to infrahyoid muscles* reach them through the hypoglossal nerve and through the ansa cervicalis.
- *Branches to prevertebral muscles:* Rectus capitis lateralis and rectus capitis anterior receive branches from C1. Longus capitis receives branches from C1, C2 and C3. Longus colli receives branches from C2, C3 and C4.

Cutaneous Nerves Arising from Cervical Plexus

Lesser Occipital Nerve

The lesser occipital nerve arises from the ventral ramus of the second cervical nerve. The origin lies deep to the sternocleidomastoid muscle (Fig. 15.8A). The nerve forms

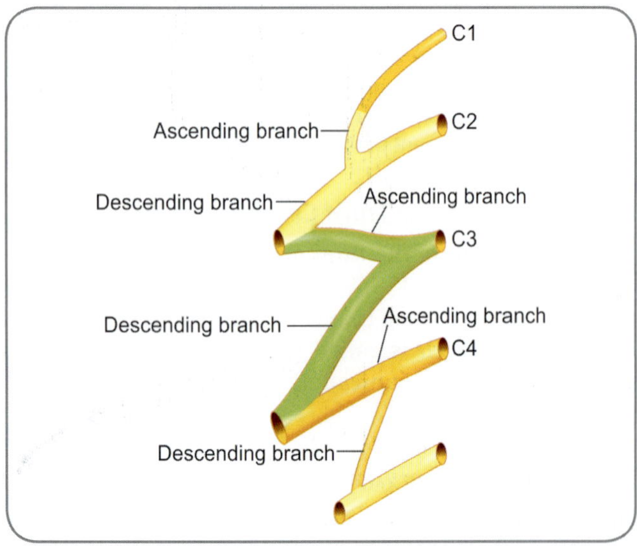

Fig. 15.5: Plan of cervical plexus

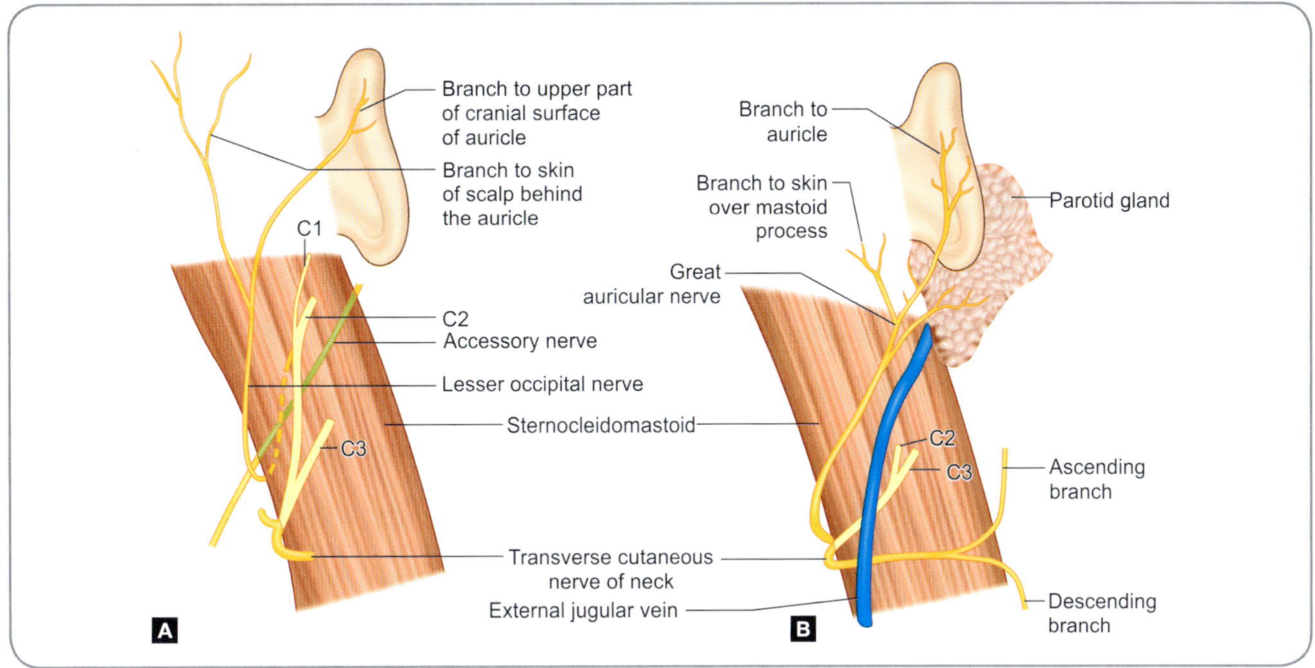

Figs 15.8A and B: Course and distribution of lesser occipital nerve, great auricular and transverse cutaneous nerves

a loop round the accessory nerve. It then runs upwards and backwards for some distance along the posterior border of the sternocleidomastoid. Higher up, it lies superficial to the muscle (between the muscle and the deep fascia). It becomes subcutaneous behind the auricle and divides into branches that supply the skin of this region. It also gives off an ***auricular branch*** that supplies the upper part of the cranial surface of the auricle (Fig. 15.6).

Great Auricular Nerve

The great auricular nerve arises from the ventral rami of the second and third cervical nerve. Its origin lies deep to the sternocleidomastoid (Fig. 15.8B). Winding around the posterior border of the sternocleidomastoid muscle it reaches the superficial surface of the muscle. It pierces the deep fascia and runs upwards and somewhat forwards over the surface of the sternocleidomastoid muscle. Here it is accompanied by the external jugular vein (Fig. 15.9A). A little below the auricle it divides into anterior and posterior branches. The anterior branch supplies the skin over the parotid gland. The posterior branch supplies most of the cranial surface of the auricle (except the part supplied by the lesser occipital (Fig. 15.9B)), the posteroinferior part of the lateral surface of the auricle including the lobule and concha and the skin over the mastoid process (Figs 15.6 and 15.8B to 15.10).

Transverse Cutaneous Nerve of Neck

The transverse cutaneous nerve of neck is also called the anterior cutaneous nerve. It arises from the ventral rami of the second and third cervical nerves. It first runs laterally deep to the sternocleidomastoid and at its posterior border it curves, running forwards across the muscle. The

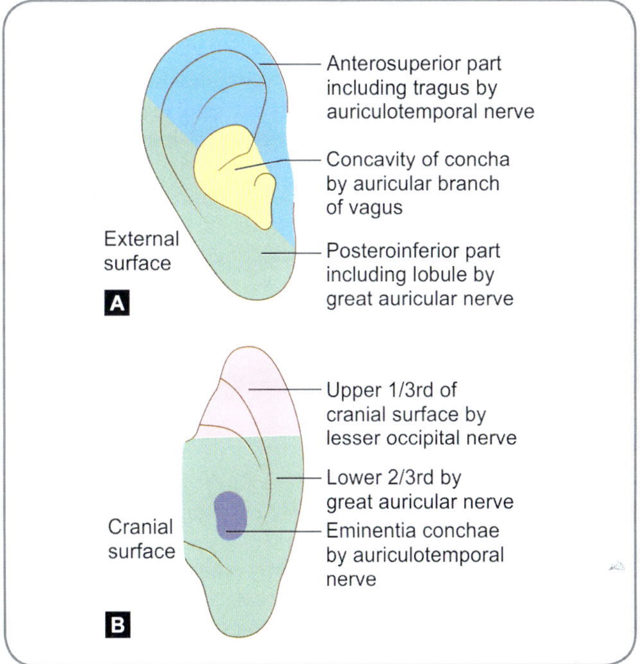

Figs 15.9A and B: Nerve supply of the auricle **A.** External surface **B.** Cranial surface

nerve becomes superficial and divides into ascending and descending branches which supply the skin on the front of the neck (Figs 15.6 and 15.8B).

Supraclavicular Nerves

The supraclavicular nerves arise (from a single common trunk) from the third and fourth cervical nerves (Fig. 15.6). The origin lies deep to the sternocleidomastoid. The nerve

277

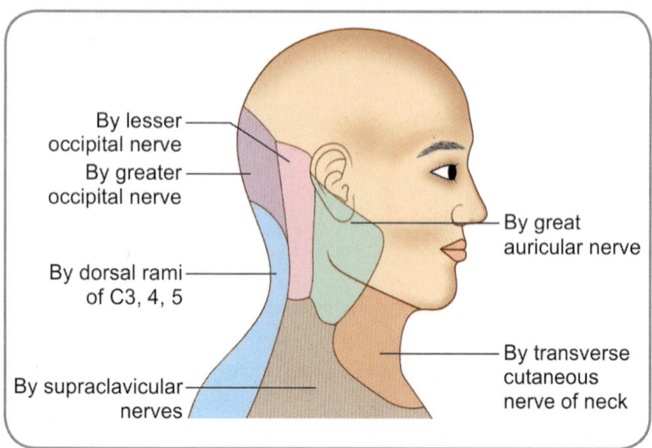

Fig. 15.10: Areas of skin of neck supplied by various cutaneous nerves

trunk runs downwards and backwards deep to the muscle and appears at its posterior border. Here the trunk divides into three branches called the ***medial, intermediate*** and ***lateral*** supraclavicular nerves. These branches descend over the posterior triangle of the neck giving some branches to the skin here. They pierce the deep fascia a little above the clavicle and then run downwards across this bone to reach the pectoral region.

The union of the consecutive descending and ascending parts of the cervical plexus gives the appearance of loop formation. The plexus comes to have two loops—first loop (C2 and 3) and second loop (C3 and 4). A general pattern is seen in the distribution of branches. Cutaneous and muscular branches to parts of head and neck arise from the first loop. Cutaneous and muscular branches to the shoulder and chest arise from the second loop.

In accordance with this scheme, a simple pattern is followed in the classification of the branches of the cervical plexus. The branches are superficial and deep. The superficial branches are those which pierce the deep fascia to supply the skin (cutaneous branches). The deep branches are those which supply muscles (muscular branches).

The superficial branches once again are classified as ascending and descending—lesser occipital, great auricular and transverse cervical cutaneous nerves are ascending; supraclavicular nerves are descending.

The deep branches are either medial or lateral. Branches to prevertebral muscles, ansa cervicalis and phrenic nerve form the medial set. The branches to the muscles of posterior triangle of neck form the lateral set.

Muscular Branches Arising from Cervical Plexus

Among the muscular branches of the cervical plexus is the inferior root of the ansa cervicalis.

Ansa Cervicalis

The ***ansa cervicalis*** is also called the ***ansa hypoglossi***. It is a nerve loop lying in front of the common carotid artery (Latin.ansa=loop) (Fig. 15.11). Branches arising from the ansa cervicalis innervate the sternohyoid, the sternothyroid and the omohyoid muscles (viz., all infrahyoid muscles other than the thyrohyoid). The ansa cervicalis is formed by union of two roots (Fig. 15.11). The superior root appears to arise from the hypoglossal nerve (hence the term ansa hypoglossi given to the loop). However, the root is really made up of fibres derived from the first cervical nerve. These fibres reach the hypoglossal nerve through a communicating branch from the first cervical nerve. Apart from forming the superior root of the ansa cervicalis, these fibres from the first cervical nerve (travelling along the hypoglossal nerve) innervate the thyrohyoid and geniohyoid muscles (see Fig. 15.11). The inferior root of the ansa cervicalis arises from the second and third cervical nerves. It descends at first lateral to the internal jugular vein and then superficial to it to join the superior root superficial to the common carotid artery.

The inferior root which arises from the cervical plexus is called the ***nervus descendens cervicalis*** and the superior root which appears to arise from the hypoglossal nerve is called the ***nervus descendens hypoglossi***.

Phrenic Nerve

The phrenic nerve is one of the most important nerves in the body as it is the only motor supply to the diaphragm. This nerve arises from the (ventral rami of) spinal nerves

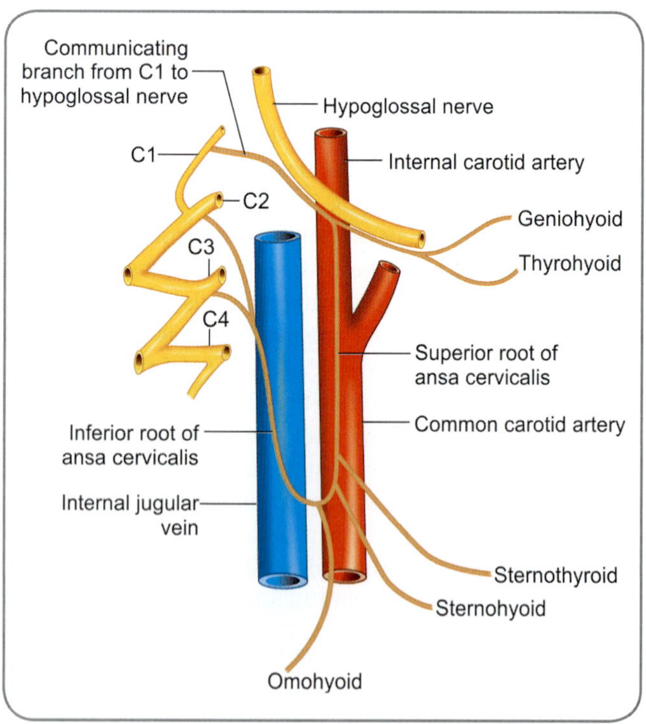

Fig. 15.11: Scheme to show the mode of innervation of the infrahyoid muscles from the cervical plexus

C3, C4 and C5, with the contribution from C4 being the greatest. The nerve descends vertically through the lower part of the neck and then through the thorax to reach the diaphragm. Some terminal branches enter the abdomen.

Course and Relations of Phrenic Nerve in the Neck

It is formed at the lateral border of the scalenus anterior muscle and descends vertically over the anterior surface of the muscle (Fig. 15.12). After crossing the medial (or lower) border of the muscle, it crosses in front of the first part of the subclavian artery (see Fig. 14.23). On the right side, however, the nerve is usually separated from the artery by a part of the scalenus anterior.

Throughout its course in the neck the nerve lies deep to the sternocleidomastoid muscle. It can be viewed by lifting up the lateral border of this muscle. While over the scalenus anterior the nerve lies deep to the prevertebral fascia. It is crossed superficially by the superior belly of the omohyoid, the transverse cervical artery and the suprascapular artery. On the left side it is also crossed by the thoracic duct (Fig. 15.12).

Lower down the nerve lies behind the lower end of the internal jugular vein. Still lower down the nerve passes behind the medial end of the subclavian vein. The nerve then continues into the thorax.

> ### Added Information
>
> ❑ ***Accessory phrenic nerve:*** The root of the phrenic nerve from C5 may sometimes follow a complicated course. Instead of arising from C5 itself, it may arise from the nerve to subclavius.
>
> *contd...*

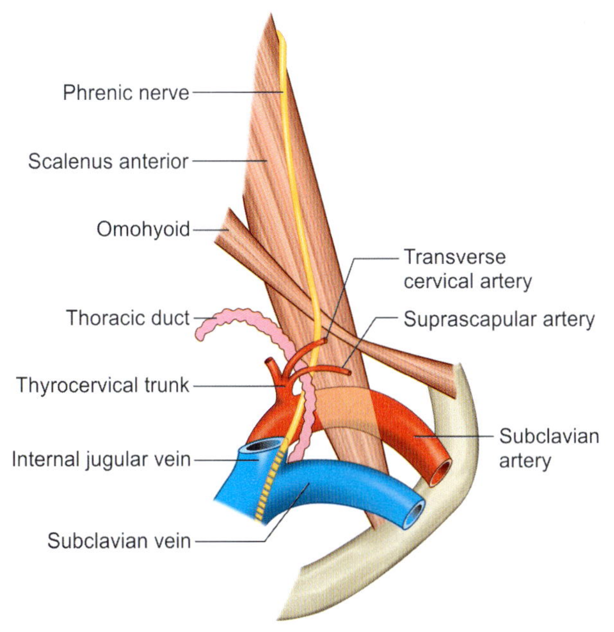

Phrenic nerve

Scalenus anterior

Omohyoid

Transverse cervical artery

Thoracic duct

Suprascapular artery

Thyrocervical trunk

Subclavian artery

Internal jugular vein

Subclavian vein

Fig. 15.12: Relations of phrenic nerve

> ### Added Information *contd...*
>
> From here the root descends through the neck lateral to the main phrenic nerve and joins it in the upper part of the thorax. Such a root from C5 constitutes the accessory phrenic nerve.
>
> ❑ Communicating branches from the cervical plexus go to vagus nerve, hypoglossal nerve (which include the superior root of ansa cervicalis) and sympathetic trunk.

CRANIAL NERVES

There are twelve pairs of cranial nerves which emerge from the surface of the brain. They are identified by number (in craniocaudal sequence) and also bear names as follows:

1. The ***first*** cranial nerve is called the ***olfactory*** nerve. It is the nerve of smell (olfaction = smell).
2. The ***second*** cranial nerve is called the ***optic*** nerve. It is the nerve of sight (optics = science of formation of images).
3. The ***third*** cranial nerve is called the ***oculomotor*** nerve as it supplies several muscles that move the eyeball (ocular = pertaining to the eye).
4. The ***fourth*** cranial nerve is called the ***trochlear*** nerve. It is so called because it supplies a muscle (superior oblique) that passes through a pulley (trochlea = pulley).
5. The ***fifth*** cranial nerve is called the ***trigeminal*** nerve because it has three major divisions. These are:
   i. The ***ophthalmic*** division to the orbit,
   ii. The ***maxillary*** division to the upper jaw and
   iii. The ***mandibular*** division to the lower jaw.
6. The ***sixth*** cranial nerve is called the ***abducent*** nerve because it supplies a muscle (lateral rectus) that 'abducts' the eyeball.
7. The ***seventh*** cranial nerve is the ***facial*** nerve because it supplies the muscles of the face.
8. The ***eighth*** cranial nerve is called the ***vestibulocochlear*** nerve because it supplies structures in the vestibular and cochlear parts of the internal ear. It is sometimes called:
   ○ Auditory nerve (auditory = pertaining to hearing) or
   ○ Statoacoustic nerve (stato = pertaining to equilibrium; acoustic = pertaining to sound or hearing).
9. The ***ninth*** cranial nerve is called the ***glossopharyngeal*** nerve as it is distributed to the pharynx and to part of the tongue (glossal = pertaining to the tongue).
10. The ***tenth*** cranial nerve is called the ***vagus***. It has an extensive course through the neck, the thorax and the abdomen and thus roams about like a vagabond.
11. The ***eleventh*** cranial nerve is called the ***accessory*** nerve because it appears to be a part of the vagus nerve (or 'accessory' to the vagus).

12. The ***twelfth*** cranial nerve is called the ***hypoglossal*** nerve (because it runs part of its course below the tongue before supplying the muscles in it (hypo = below; glossal = pertaining to tongue).

Types of Fibres in Nerves

Efferent and Afferent Fibres

Both spinal and cranial nerves contain fibres that can be classified into several types on the basis of their function. It is necessary to consider these details here as cranial nerves can contain some functional types of fibres not encountered in spinal nerves. Fibres in a peripheral nerve are basically of two types. Some fibres carry impulses from the spinal cord or brain to a peripheral organ like muscle. Impulses passing along such nerves cause the muscle to contract and thus result in movement. Such nerve fibres are called ***motor fibres***. They are also called ***efferent fibres*** (efferent=go away from; Latin.effero=bring out). Some efferent fibres end in glands and their stimulation produces secretion. They are, therefore, called ***secretomotor fibres***.

Other nerve fibres carry impulses in the opposite direction, i.e. towards the spinal cord or brain. They are called ***afferent fibres*** (afferent=come towards; Latin. Affero=to bring to). Fibres bringing the sensations of touch from the skin, of sight from the eye, or of hearing from the ear are examples of afferent fibres. It is through these nerve fibres that we become conscious of such sensations. They are, therefore, called ***sensory fibres.*** Some afferent fibres carry impulses (e.g., from muscles or joints) of which we may not become conscious, but which are nevertheless very important for maintenance of posture and for proper control of movements. Such impulses are referred to as ***proprioceptive*** (as they have their origin within the body 'proper' and not from outside the body). Similar impulses arising from viscera also reach the brain. We are conscious of some of them (e.g., distension of the urinary bladder). Many of them do not reach our consciousness, but help to regulate the functions of the viscera.

Somatic and Visceral Fibres

Both afferent and efferent fibres can further be classified on the basis of the tissues supplied by them. The tissues and organs that make up the body can be broadly divided into two major parts, somatic and visceral.

- ***Somatic*** structures are those present in relation to the body wall or soma (Greek.soma=body); they include the tissues of the limbs (which represent a modified part of the body wall). Thus the skin, bones, joints and skeletal muscles of the limbs and body wall are classified as somatic.
- In contrast, the tissues which make up the internal organs like the heart, lungs or stomach are classified as ***visceral*** (Latin.visco=soft and internal). These include the lining epithelia of hollow viscera, and smooth muscles (including the smooth muscle in the walls of blood vessels).

A distinction between somatic and visceral structures may also be made on embryological considerations as follows:

- Structures derived from ectoderm, including those developing from specialised areas of ectoderm, e.g., the retina or the membranous labyrinth, are classified as somatic.
- Structures derived from endoderm (example, the epithelium of the tongue which is of endodermal origin) is classified as visceral.

Somatic and Visceral Muscles

It is essential to know the somatic and visceral divisions of skeletal muscles. Skeletal muscles may be derived embryologically from three distinct sources. These sources are:

1. The somites developing in the paraxial mesoderm,
2. The somatopleuric mesoderm of the body wall and
3. The mesoderm of the branchial arches.

The musculature of the limbs and body wall develops partly from somites and partly *in situ* from the mesoderm (somatopleuric mesoderm) of the body wall. The nerves supplying this musculature are classified as somatic. The muscles that move the eyeball and the muscles of the tongue are also derived from somites and the nerves supplying them are, therefore, classified as somatic.

However, skeletal muscle that develops in the mesoderm of the branchial arches is classified as visceral (or branchial).

Functional Components of Nerve Fibres

Keeping in view the distinction between afferent and efferent fibres on one hand, and somatic and visceral structures on the other, fibres in the peripheral nerves are grouped into four broad categories as follows:

1. Somatic efferent
2. Visceral efferent
3. Somatic afferent
4. Visceral afferent.

With the exception of somatic efferent fibres, each of the other categories is subdivided into a ***general*** and a ***special*** group. Thus, there are a total of seven functional components as follows:

1. ***Somatic efferent*** fibres supply skeletal muscle of the limbs and body wall; they also supply the extrinsic muscles of the eyeball and the muscles of the tongue.
2. ***General visceral efferent*** fibres supply smooth muscle and glands; the nerves to glands are called secretomotor nerves.
3. ***Special visceral efferent*** fibres supply skeletal muscles developing in branchial arch mesoderm; they are frequently called the ***branchial efferent*** fibres;

the muscles supplied include muscles of mastication, muscles of facial expression, muscles of pharynx and larynx.

4. ***General somatic afferent*** fibres are those that carry:
   - ○ Sensations of touch, pain and temperature from the skin (exteroceptive sensations) and
   - ○ Proprioceptive impulses arising in muscles, joints and tendons; these convey information regarding movement and position of joints.
5. ***Special somatic afferent*** fibres carry impulses of (a) vision, (b) hearing and (c) equilibrium (from the vestibular apparatus).
6. ***General visceral afferent*** fibres carry sensations, e.g., pain from viscera (visceroceptive sensations).
7. ***Special visceral afferent*** fibres carry the sensations of taste.

A typical spinal nerve contains fibres of the four general categories. The special categories are present in cranial nerves only.

Types of Neurons

Neurons Supplying Typical Skeletal Muscles

In the spinal cord, the cell bodies of neurons supplying skeletal muscles derived from somites lie in the ventral grey column. The axons of these neurons leave the spinal cord through the ventral nerve root (Fig. 15.13). They then travel through the spinal nerve (and its branches) to reach a muscle and supply it. The axon may divide into several branches along its course, each of which ends by supplying one muscle fibre (Fig. 15.13). In the case of cranial nerves, the cell bodies of neurons supplying most skeletal muscles are situated in masses of grey matter present in the brainstem.

The neurons supplying skeletal muscles (other than the muscles derived from branchial arches) are called ***somatic efferent neurons***. In the case of cranial nerves, the neurons are located in ***somatic efferent nuclei***. Some skeletal muscles are derived from branchial (pharyngeal) arches. These include the muscles of mastication and the

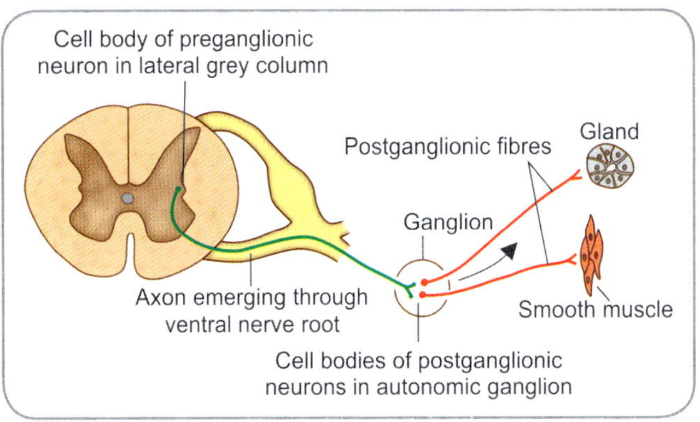

Fig. 15.14: Scheme to show the arrangement of neurons supplying smooth muscle or gland (general visceral efferent neurons)

musculature of the face, pharynx and larynx. Such muscles are innervated by neurons which are present in the special visceral efferent nuclei of the brainstem and are referred to as ***special visceral efferent neurons*** or ***branchial efferent neurons***.

Neurons Supplying Smooth Muscles and Glands

Nerve fibres innervating smooth muscles and glands are present in both spinal and cranial nerves. The pathway for supply of a smooth muscle or a gland always consists of two neurons which synapse in a ganglion (Fig. 15.14). The first neuron carries impulses from the brain or spinal cord to the ganglion and is, therefore, called the ***preganglionic neuron.*** The second neuron carries impulses from the ganglion to the smooth muscle or the gland and is, therefore, called the ***postganglionic neuron.***

Neurons innervating smooth muscle or glands are ***general visceral efferent neurons***. They constitute the greater part of the autonomic nervous system.

Afferent Neurons

The cell bodies of neurons which give rise to afferent fibres are located outside the central nervous system. In the case of spinal nerves, they lie in the dorsal nerve root ganglia (Fig. 15.15). In the case of cranial nerves, they lie

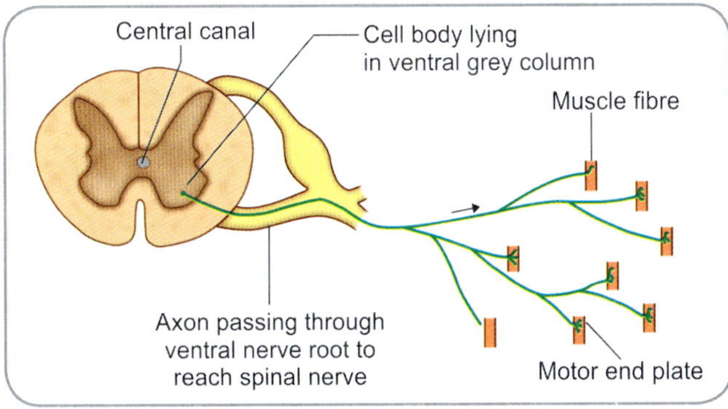

Fig. 15.13: Scheme to show the typical arrangement of a neuron supplying skeletal muscle (somatic efferent neuron)

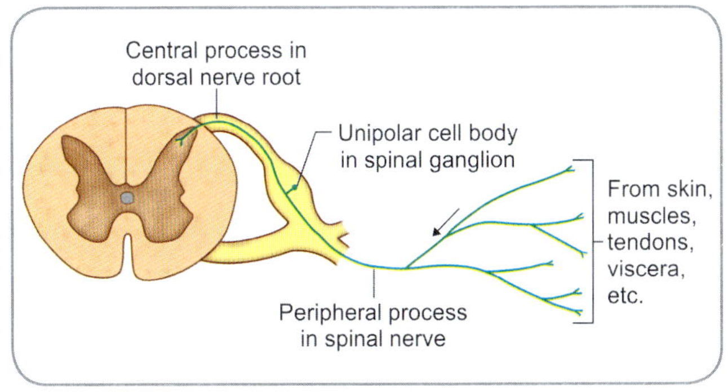

Fig. 15.15: Scheme to show the arrangement of an afferent neuron

in sensory ganglia (e.g., trigeminal ganglion) associated with the nerves. The arrangement can be explained with reference to a spinal nerve (Fig. 15.15).

The cells of dorsal nerve root ganglia are of the (pseudo) unipolar type. The cell body gives off a single process that divides into a *peripheral process* (which is really a dendrite) and a *central process* (which is the axon). The peripheral process extends into the spinal nerve and courses through its branches to reach the tissue or organ supplied. It may branch repeatedly during its course. *Such peripheral processes constitute the afferent fibres of peripheral nerves.* The sensory impulses brought by these processes from various organs of the body are conveyed to the spinal cord by the central processes. The central processes pass through the dorsal roots of spinal nerves to enter the spinal cord. In the case of cranial nerves they pass into the brainstem through the sensory roots of the cranial nerves.

As already noted, afferent nerve fibres can be divided into four categories viz., general somatic afferent, special somatic afferent, general visceral afferent and special visceral afferent. The basic arrangement of neurons that give origin to all four categories of afferent fibres is similar and the description given above applies to all of them.

Cranial Nerve Nuclei

Cranial nerves begin or end in groups of neurons, or *nuclei*, present in the brain. The olfactory and optic nerves are present in relation to the cerebral hemispheres and are actually extensions of the brain tissue. So, no separate nuclei are seen.

The nuclei of the remaining cranial nerves are located in the brainstem. These cranial nerve nuclei are arranged in seven groups that correspond to the seven functional types of nerve fibres described above. In the early embryo, the nuclei related to the various types of fibres present in cranial nerves are arranged in vertical rows (or columns). With further development, parts of these columns disappear, so that each column no longer extends through

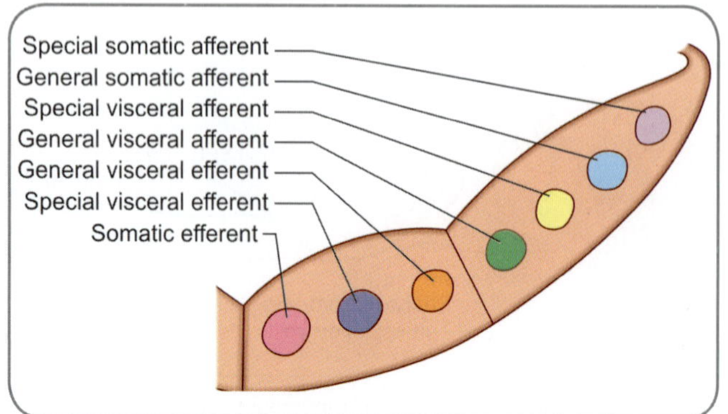

Fig. 15.16: Arrangement of the columns of cranial nerve nuclei as seen in the embryo

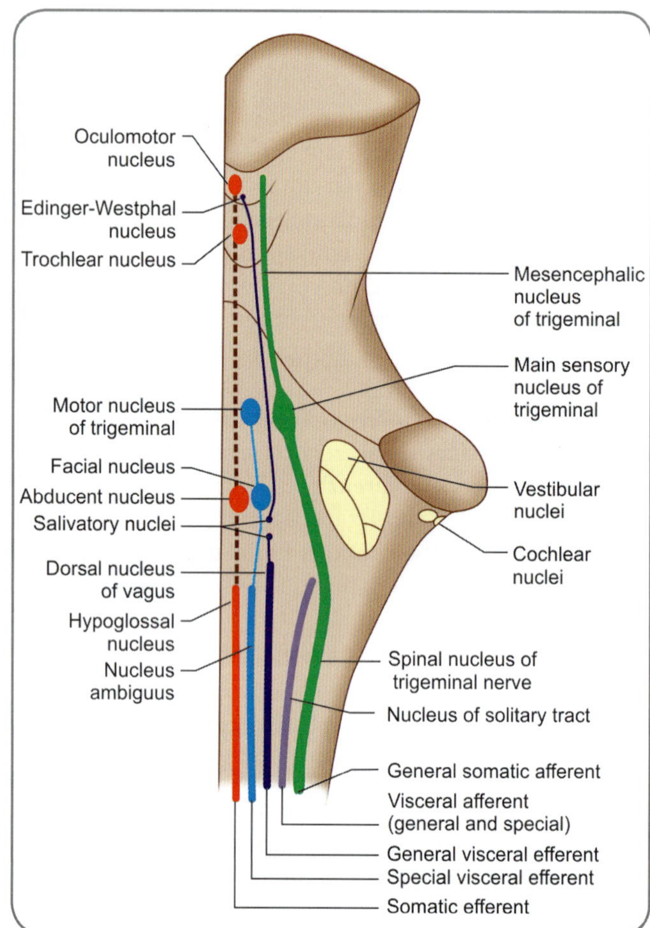

Fig. 15.17: Schematic view of cranial nerve nuclei projected onto the posterior aspect of the brainstem

the whole length of the brainstem, but is represented by one or more discrete nuclei or in parts. These are shown in Figure 15.17.

Somatic Efferent Nuclei

Somatic efferent nuclei supply skeletal muscles which are derived [embryologically (Fig. 15.16)] from somites. They are as follows:

- ❑ *Oculomotor nerve nucleus* is situated in the midbrain (upper part, at the level of the superior colliculus); the nuclei of the two sides form a single complex that lies in the central grey matter, ventral to the aqueduct (Fig. 15.18F).
- ❑ *Trochlear nerve nucleus* is situated in the midbrain (lower part, at the level of the inferior colliculus); the nucleus lies ventral to the aqueduct in the central grey matter (Fig. 15.18E).
- ❑ *Abducent nerve nucleus* is situated in the lower part of the pons; it lies in the grey matter lining the floor of the fourth ventricle, near the midline (Fig. 15.18C).
- ❑ *Hypoglossal nerve nucleus* is situated in the medulla; it is an elongated column extending into both the open and closed parts of the medulla; its upper part lies deep to the hypoglossal triangle in the floor of the fourth

ventricle; when traced downwards, it lies next to the midline in the central grey matter ventral to the central canal (Figs 15.18A and B).

Special Visceral Efferent Nuclei

These nuclei supply skeletal muscles derived from the branchial arch mesoderm.

- ❏ *Motor nucleus of the trigeminal nerve* lies in the upper part of the pons, in its dorsal aspect (Fig. 15.18D).
- ❏ *Nucleus of the facial nerve* lies in the lower part of the pons (Fig. 15.18C).
- ❏ *Nucleus ambiguus* lies in the medulla; it forms an elongated column that extends through both the open and closed parts of the medulla (Figs 15.18A and B); it is a composite nucleus and contributes fibres to the glossopharyngeal, vagus and accessory nerves.

General Visceral Efferent Nuclei

The nuclei in this column give origin to preganglionic fibres which end in peripheral ganglia. Postganglionic fibres arising in these ganglia supply smooth muscles or glands. These neurons, preganglionic and postganglionic, form part of the parasympathetic nervous system. The nuclei of this group are as follows:

- ❏ *Edinger-Westphal nucleus* is situated in the midbrain (Fig. 15.18F); it is closely related to the oculomotor complex. Fibres arising in this nucleus pass through the oculomotor nerve. They relay in the *ciliary ganglion* to supply the sphincter pupillae and the ciliaris muscles.
- ❏ *Salivatory nuclei* (superior and inferior) are situated in the pons (dorsal part) just above its junction with the medulla (Fig. 15.17). The superior nucleus sends fibres into the facial nerve. These fibres relay in the *submandibular ganglion* to supply the submandibular and sublingual salivary glands. The inferior nucleus sends fibres into the glossopharyngeal nerve. These fibres relay in the *otic ganglion* to supply the parotid gland. Other neurons located near the salivatory nuclei send out fibres which supply the lacrimal gland, after relaying in the *pterygopalatine ganglion*. These fibres travel through the facial nerve. These nuclei, are sometimes collectively called the lacrimatory nuclei.
- ❏ *Dorsal (motor) nucleus of the vagus* is situated in the medulla. It is a long nucleus lying vertically. Its upper end lies deep to the vagal triangle in the floor of the fourth ventricle. When traced downwards it extends into the closed part of the medulla (Fig. 15.18A). Fibres

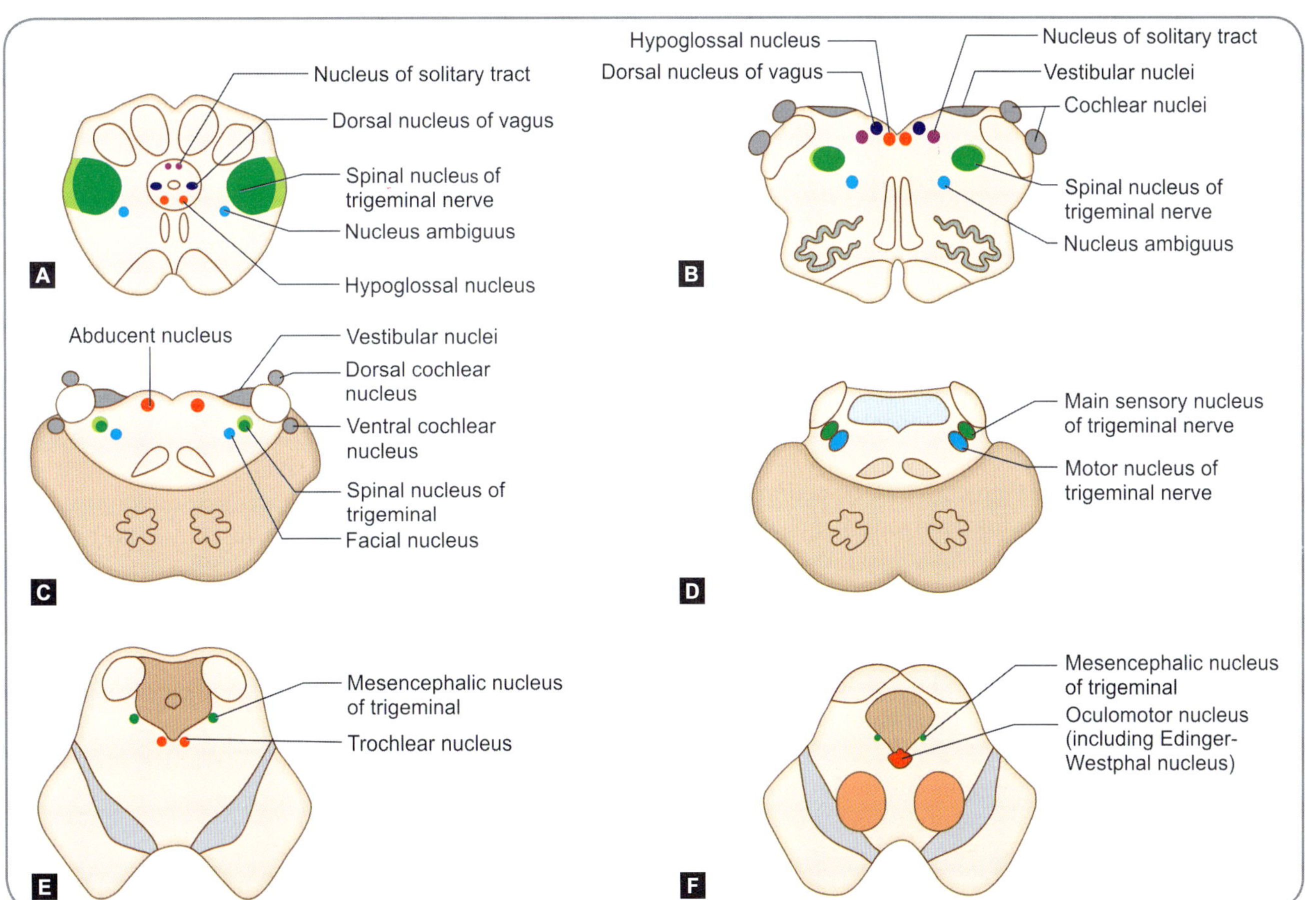

Figs 15.18A to F: Transverse sections through the brainstem to show the position of cranial nerve nuclei **A.** and **B.** Lower and upper parts of medulla oblongata **C.** and **D.** Lower and upper parts of pons **E.** and **F.** Lower and upper parts of midbrain

arising in this nucleus supply several thoracic and abdominal viscera. They end in ganglia closely related to these viscera. Postganglionic fibres arising in these ganglia run a short course to supply smooth muscle and glands in these viscera.

General and Special Visceral Afferent Nuclei

Both these columns are represented by the *nucleus of the solitary tract*, present in the medulla (Figs 15.17, 15.18A and B). Like other cranial nerve nuclei in the medulla, the cells of this nucleus form an elongated column extending into both the open and closed parts of the medulla. The nucleus of the solitary tract receives fibres carrying general visceral sensations through the vagus and glossopharyngeal nerves. Fibres of taste (special visceral afferent) carried by the facial, glossopharyngeal and vagus nerves end in the upper part of the nucleus. This part is sometimes called the *gustatory nucleus*.

General Somatic Afferent Nuclei

The general somatic afferent column is represented by the sensory nuclei of the trigeminal nerve. These are as follows:
- *Main sensory nucleus* lies in the upper part of the pons, lateral to the motor nucleus of the nerve (Figs 15.17, 15.18D).
- *Spinal nucleus* extends from the main nucleus down into the medulla (Figs 15.17, 15.18A and B). It also extends into the upper two segments of the spinal cord. In addition to the fibres of the trigeminal nerve, it also receives general somatic sensations carried by the facial, glossopharyngeal and vagus nerves.
- *Mesencephalic nucleus* of the trigeminal nerve extends from the upper end of the main sensory nucleus into the midbrain (Figs 15.17, 15.18E and F). The peripheral processes of these cells are believed to carry proprioceptive impulses from muscles of mastication and possibly also from muscles of the eyeball, face and tongue. The central processes of the neurons in the nucleus probably end in the main sensory nucleus of the trigeminal nerve.

Special Somatic Afferent Nuclei

These are the cochlear and vestibular nuclei.
- *Cochlear nuclei* are two in number—dorsal and ventral. They are placed dorsal and ventral to the inferior cerebellar peduncle (Figs 15.18D and E) at the level of the junction of the pons and the medulla. These nuclei receive fibres from end organs in the cochlea that are concerned with hearing.
- *Vestibular nuclei* lie in the grey matter underlying the lateral part of the floor of the fourth ventricle (Figs 15.17, 15.18B and C). They lie partly in the medulla and partly

in the pons. They receive fibres from end organs in the vestibular part of the internal ear. They are functionally related to the maintenance of equilibrium.

OLFACTORY NERVES

Other names: First cranial nerve, nerve of smell, nervi olfactorii.

The olfactory nerves are purely sensory and are concerned with smell (it is customary to talk of this nerve in plural, because it is a collective term for numerous olfactory filaments). The peripheral end organ for smell is the olfactory mucosa that lines the upper and posterior part of the nasal cavity (both on the lateral wall and on the septum). Nerve fibres arising in this mucosa collect to form about twenty bundles that together constitute an olfactory nerve. The fibres of the olfactory nerves are processes of olfactory receptor cells located in the olfactory epithelium. Each cell gives off a short peripheral process directed towards the lumen of the nasal cavity, and a larger central process that passes into the mucosa forming one fibre of the olfactory nerve. These fibres collect to form about twenty bundles that collectively constitute one olfactory nerve.

The bundles pass through foramina in the cribriform plate of the ethmoid bone to enter the cranial cavity (anterior cranial fossa) where they terminate in the *olfactory bulb* (Fig. 15.19). Olfactory impulses carried by these fibres pass to other neurons located in the olfactory bulb. From the bulb they pass into the olfactory tract and ultimately end in several small areas located on the inferior surface of the cerebral hemisphere.

The most important of these is the *entorhinal area* which includes the *uncus* and the anterior part of the *parahippocampal gyrus* (Fig. 15.20). Smell is perceived in this region of the brain. As the olfactory mucosa is derived from ectoderm, the functional component of olfactory nerves is classified as *special somatic afferent* (along with fibres for vision and hearing). However, smell is often regarded as a visceral sensation because of its close association to eating. Some authorities, therefore, classify the fibres of the olfactory nerves as special visceral afferent (like those of taste).

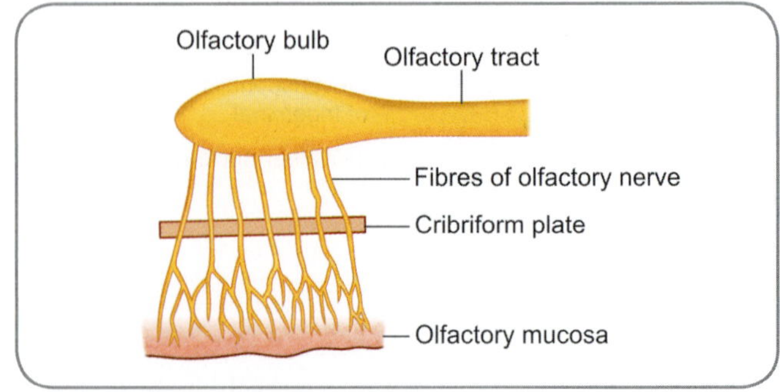

Fig. 15.19: Scheme to show the course of olfactory nerve

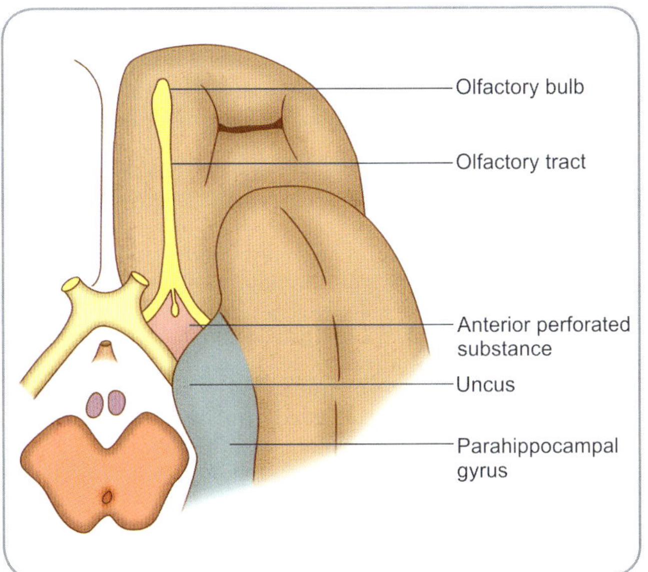

Fig. 15.20: Diagram showing the entorhinal area made up of uncus and anterior part of the parahpipocampal gyrus

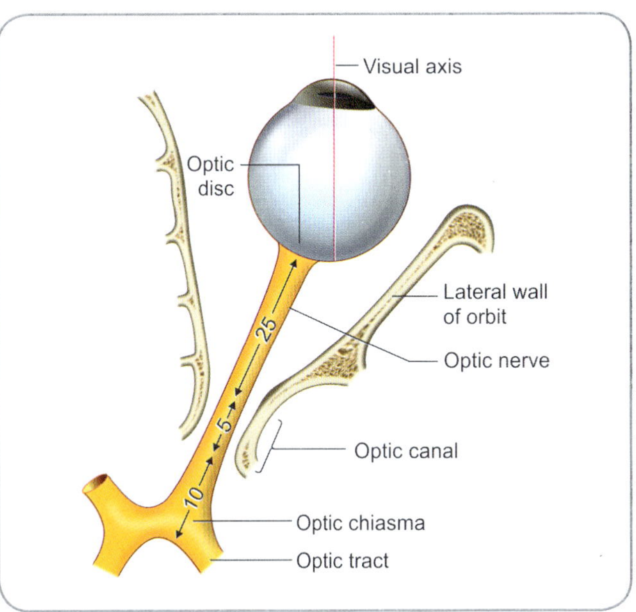

Fig. 15.21: Scheme to show course of the optic nerve

The bipolar neurons of the olfactory epithelium are continuously replaced by cell division even during adult life. The second order neurons from the olfactory bulb reach their ipsilateral cortical centres without relay in thalamus, which is the central relay station for all other senses.

Clinical Correlation

The olfactory nerve is tested by asking the patient to recognise various odours. The right and left nerves can be tested separately by closing one nostril and putting the substance near the open nostril.

Optic Nerve

Other names: Second cranial nerve, nervus opticus.
The optic nerve forms an important part of the visual pathway. It lies partly in the orbit and partly in the cranial cavity. Its anterior end is attached to the posterior pole of the eyeball. The attachment lies a little medial to the anteroposterior axis of the eyeball. From here the nerve passes backwards and medially through the orbit, the optic canal and through part of the cranial cavity. The nerve ends by joining the nerve of the opposite side to form the ***optic chiasma*** (Fig. 15.21). The total length of the nerve is about 40 mm. Of this 25 mm is in the orbit, 5 mm is in the optic canal and 10 mm is in the cranial cavity.

The intraorbital part of the nerve is surrounded by the superior, inferior, medial and lateral rectus muscles. The ciliary ganglion is placed between the optic nerve and the lateral rectus muscle. The ophthalmic artery is inferolateral to the nerve in the optic canal and in the posterior most part of the orbit. The artery then crosses above the nerve from lateral to medial side. It is also crossed from medial to lateral side by the nasociliary nerve above and the branch

from the oculomotor nerve to the medial rectus muscle, below. The central artery of the retina first lies below the optic nerve, piercing the dural sheath of the nerve. Then it runs forwards for a short distance between these two. About 12 mm behind the eyeball the artery enters the substance of the nerve and runs forwards in its substance to reach the eyeball.

The part of the optic nerve within the optic canal is related to the ophthalmic artery that lies below and lateral to it. On the medial side, the nerve is separated only by a thin plate of bone from the sphenoidal air sinus.

The intracranial part of the optic nerve is related to the internal carotid artery that is on its lateral side; and to the anterior cerebral artery that crosses above it (Fig. 15.22).

Visual Pathway

The peripheral receptors for light are situated in the retina. Nerve fibres arising in the retina converge upon an area on the posteromedial part of the eyeball called the optic disc. Here the fibres pass through the thickness of the retina, the choroid and the sclera. In this location, the sclera has

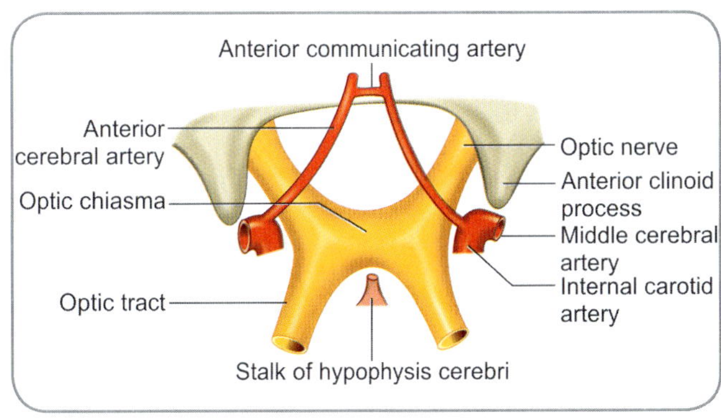

Fig. 15.22: Relations of optic nerve in its intracranial part

numerous perforations and is, therefore, called the ***lamina cribrosa*** (crib = sieve). The optic nerve is formed by the aggregation of fibres passing out through the lamina cribrosa.

The two optic nerves join to form the optic chiasma in which many of their fibres cross to the opposite side. The uncrossed fibres of the optic nerve, along with the fibres that have crossed over from the opposite side form the ***optic tract.*** The optic tract terminates predominantly in the ***lateral geniculate body***. Fresh fibres arising in the lateral geniculate body form the ***geniculocalcarine tract*** (or optic radiation) that ends in the ***visual areas*** of the cerebral cortex. Vision is actually perceived in the cerebral cortex.

Development

Though the optic nerve is seen as a nerve, it is developmentally a prolongation of white matter of brain. The optic nerves are surrounded by sheaths formed by prolongation onto them of the dura mater, the arachnoid mater and the pia mater. Hence, they are susceptible to the changes in intracranial pressure, as they have a surrounding subarachmoid space.

The fibres of the optic nerve arising in the medial (or nasal) half of each retina enter the optic tract of the opposite side after crossing in the optic chiasma. Fibres of the lateral (or temporal) half of each retina enter the optic tract of the same side. Thus the right optic tract contains fibres from the right halves of both retinae; and the left tract from the left halves. The optic tract carries these fibres to the lateral geniculate body of the corresponding side. From here they are relayed to the corresponding cerebral hemisphere.

The ***functional component of the optic nerve belongs to the special somatic afferent column*** (because the retina is derived from ectoderm). The optic nerve contains about 1,200,000 fibres.

Fibres responsible for sharp vision arise from an area of the retina called the ***macula***. The fibres arising from the macula form the ***papillomacular bundle.*** Close to the eyeball the macular fibres occupy the lateral part of the optic nerve, but by the time they reach the chiasma they lie in the central part of the nerve.

Clinical Correlation

❏ ***Testing the optic nerve:*** To test the optic nerve, first ask the patient if his/her vision is normal. Acuity (sharpness) of vision can be tested by making the patient read letters of various sizes printed on a chart from a fixed distance. It must be remembered that loss of acuity of vision can be caused by errors of refraction or by the presence of opacities in the cornea or the lens (the latter condition being called cataract). As part of a normal clinical examination the field of vision can be tested as follows:
  ○ Ask the patient to sit opposite you (about half a meter away) and look straight forwards at you;

contd...

Clinical Correlation *contd...*

  ○ As one eye should be tested at a time, ask the patient to place a hand on one eye so that he/she can see only with the other eye;
  ○ Stretch out one of your arms laterally so that your hand is about equal distance from your face and that of the patient;
  ○ Keep moving your index finger and gradually bring the hand towards yourself until you can just see the movements of the finger; this gives you an idea of the extent of your own visual field in that direction;
  ○ By asking the patient to tell you as soon as he/she can see the moving finger (making sure that he/she does not turn his/her head) you can get an idea of the patient's field of vision in the direction of your hand;
  ○ By repeating the test placing your hand in different directions, a good idea of the field of vision of the patient can be obtained;
  ○ If an abnormality is suspected, detailed testing can be done using a procedure called ***perimetry***.

If there is any doubt about the integrity of optic nerve, the retina is examined using an ophthalmoscope. With this instrument we can see the interior of the eye through the pupil of the eye. The optic disc and blood vessels radiating from it can also be seen.

❏ ***Effects of injury to visual pathway:*** Injuries to different parts of the visual pathway can produce various kinds of defects. Loss of vision in one half (right or left) of the visual field is called ***hemianopia***. If the same half of the visual field is lost in both eyes the defect is said to be ***homonymous*** and if different halves are lost the defect is said to be ***heteronymous***. Note that the hemianopia is named in relation to the visual field and not to the retina. Injury to the optic nerve will produce total blindness in the eye concerned. Damage to the central part of the optic chiasma (e.g., by pressure from an enlarged hypophysis cerebri) interrupts the crossing fibres derived from the nasal halves of the two retinae resulting in ***bitemporal heteronymous hemianopia***. It has been claimed that macular fibres are more susceptible to damage by pressure than peripheral fibres and are affected first. When the lateral part of the chiasma is affected, ***nasal hemianopia*** results. This may be unilateral or bilateral. Complete destruction of the optic tract, the lateral geniculate body, the optic radiation or the visual cortex of one side, results in loss of the opposite half of the field of vision. A lesion on the right side leads to ***left homonymous hemianopia***. Lesions anterior to the lateral geniculate body also interrupt fibres responsible for the pupillary light reflex.

❏ ***Pupillary light reflex:*** Constriction of pupils in response to light is called pupillary light reflex. The sensory input passes from ganglionic cells of retina to the pretectal nucleus on midbrain. From here, they are relayed to the Edinger-Westpal nucleus and thence to the ciliary ganglion. Postganglionic fibres pass through short ciliary nerves, which supply the sphincter pupillae. The optic nerve forms the afferent limb and the oculomotor nerve forms the efferent limb.

❏ ***Accomodation reflex:*** This reflex, mediated by the cerebral cortex, is observed when the person focuses the eye from a distant object to a nearby object. This is seen as convergence, pupillary constriction and lens thickening due to ciliaris contraction.

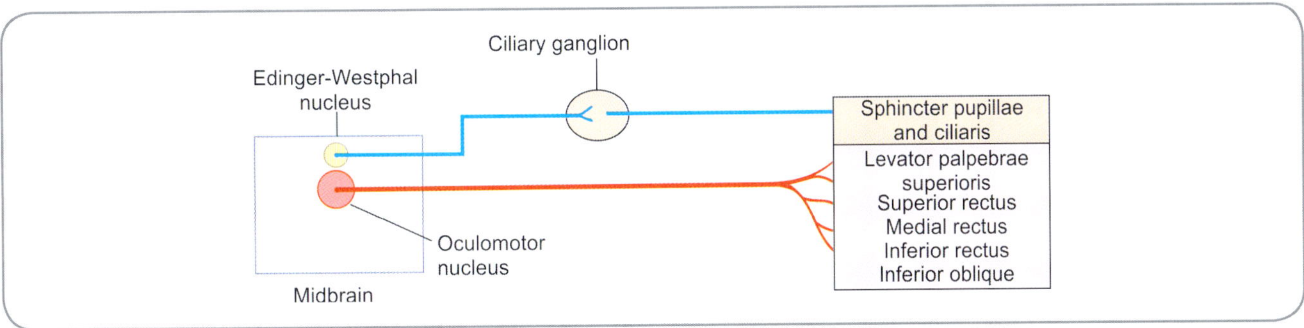

Fig. 15.23: Scheme to show the distribution of oculomotor nerve

Oculomotor Nerve

Other name: Nervus oculomotoricus.

The oculomotor is the third cranial nerve. The distribution of this nerve is shown in Figure 15.23. Most of the fibres of this nerve arise from the oculomotor nuclear complex that is situated in the upper part of the midbrain (ventral to the aqueduct. The right and left nuclei are fused to form a midline complex. Fibres arising in the complex supply all extrinsic muscles of the eyeball except the lateral rectus and the superior oblique. These fibres also supply the levator palpebrae superioris.

Some fibres of the oculomotor nerve arise from the **Edinger-Westphal** nucleus that forms part of the oculomotor complex. Fibres arising in this nucleus relay in the ciliary ganglion. Postganglionic fibres arising in this ganglion supply the sphincter pupillae and the ciliaris muscle. Arising from these nuclei the fibres of the oculomotor nerve pass forwards through the substance of the midbrain and emerge from the cerebral peduncle (Figs 15.24 and 15.25). Just in front of the midbrain, in the subarachnoid space the nerve passes between the superior cerebellar artery (which lies below it) and the posterior cerebral artery (which lies above it). The nerve passes forwards and laterally and pierces the arachnoid and the meningeal layer of dura mater in the triangular interval bounded by the free and attached margins of the tentorium cerebelli (Fig. 15.24). The nerve now comes to lie in the lateral wall of the cavernous sinus (Fig. 15.26).

In the anterior part of this wall, the nerve divides into superior and inferior divisions. The superior and inferior rami of the oculomotor nerve enter the orbit by passing through the superior orbital fissure (Fig. 15.28). They pass through the part of the fissure that lies within the common tendinous ring. Within the orbit, the superior division supplies superior rectus and levator palpebrae superioris muscles. The inferior division supplies medial rectus, inferior rectus and inferior oblique muscles (Fig. 15.27).

The *functional component of the oculomotor nerve* that supplies *the skeletal muscles (extraocular muscles) belongs to somatic efferent column*, while that which supplies *the smooth muscles* (sphincter pupillae, ciliaris) belongs to *general visceral efferent column*.

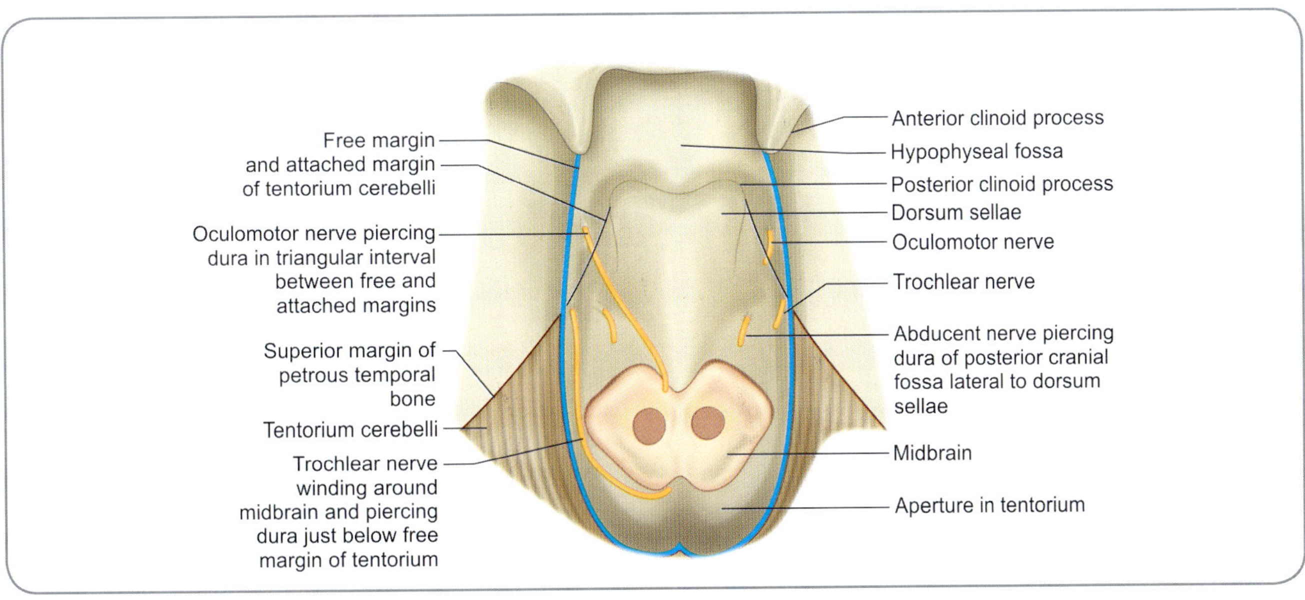

Fig. 15.24: Diagram to show the sites of penetration of dura mater by oculomotor, trochlear and abducent nerves

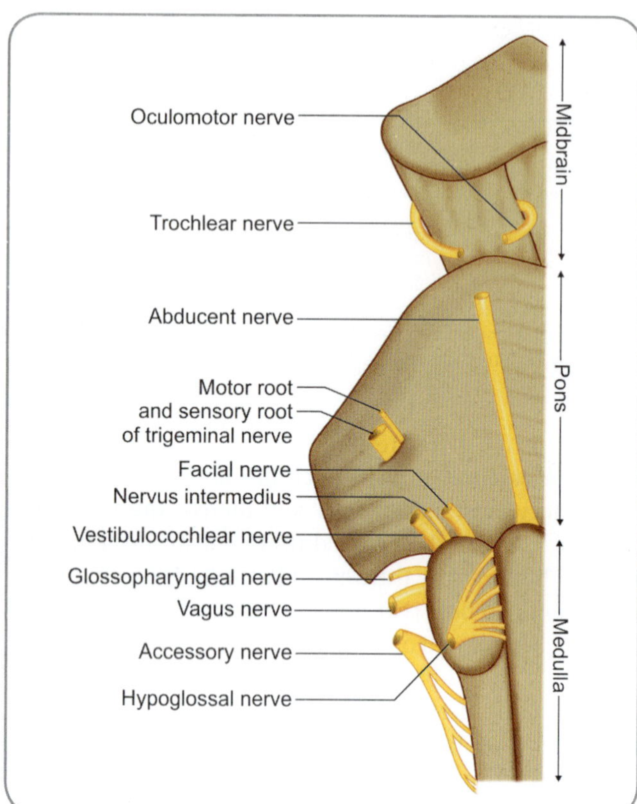

Fig. 15.25: Attachment of cranial nerves to the surface of brainstem

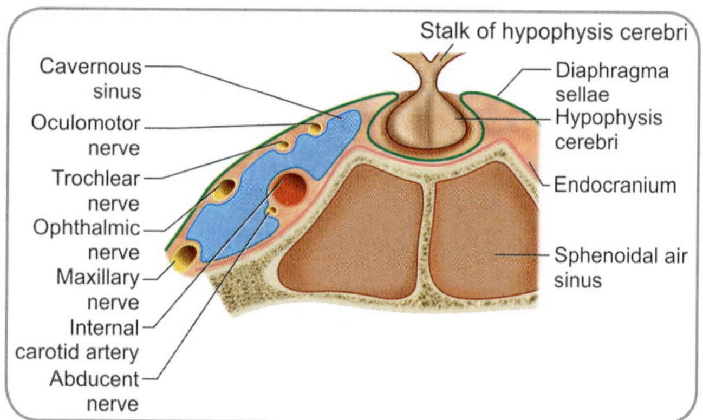

Fig. 15.26: Relationship of cranial nerves to cavernous sinus

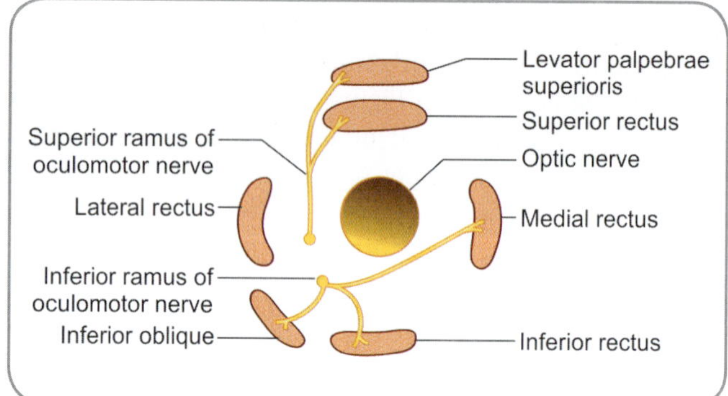

Fig. 15.27: Distribution of oculomotor nerve

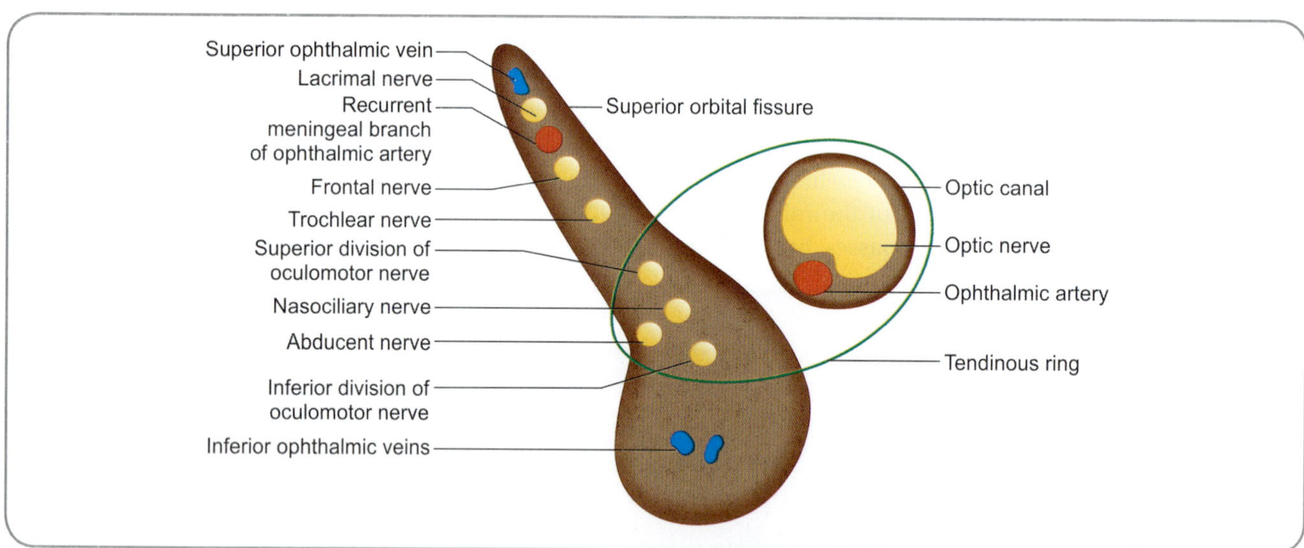

Fig. 15.28: Position of cranial nerves in the superior orbital fissure

Ciliary Ganglion

The ciliary ganglion is one of the four peripheral parasympathetic ganglia lying in the head and neck region. It is topographically and functionally related to the oculomotor nerve.

It lies lateral to the optic nerve and medial to the lateral rectus muscle. Posteriorly, it receives three roots—motor or parasympathetic, sympathetic and sensory.

❑ The motor or *parasympathetic* root is derived from the inferior division of the oculomotor nerve. It consists of the preganglionic fibres arising from the Edinger-Westphal nucleus. These fibres travel through the trunk of the oculomotor nerve and then through its inferior division. They then pass into the branch to the inferior oblique muscle. A short branch arising from this nerve conveys the fibres to the ciliary ganglion. This branch to the ciliary ganglion from the nerve to the inferior oblique constitutes the ***motor root*** of the ganglion. The cells of the ganglion give origin to postganglionic fibres which pass through the short ciliary nerves to reach the sphincter pupillae and the ciliaris muscles.

❑ The ***sympathetic root*** arises from the sympathetic plexus on the internal carotid artery. It carries postganglionic sympathetic fibres that begin in the superior cervical sympathetic ganglion. Sympathetic fibres pass through the ciliary ganglion, without relay and enter the short ciliary nerves through which they reach the blood vessels of the eyeball. The fibres for the dilator pupillae normally follow a separate route, but sometimes they may pass through the ciliary ganglion.

❑ The ***sensory root*** of the ciliary ganglion is a branch of the nasociliary nerve. It carries sensory fibres that begin in the cornea, the iris and the choroid; pass through the short ciliary nerves and then through the ciliary ganglion; and finally enter the sensory root through which they reach the nasociliary nerve.

It should be noted that only the parasympathetic fibres relay in the ciliary ganglion. The sympathetic and sensory fibres do not relay.

Trochlear Nerve

Other names: Fourth cranial nerve, pathetic nerve, nervus trochlearis.

The trochlear nerve is made up of fibres which arise in the trochlear nucleus located in the lower part of the midbrain (ventral to the aqueduct). Fibres arising in this nucleus pass ***backwards*** winding around the central grey matter of the midbrain to reach the upper part of the anterior (or superior) medullary velum (Fig. 15.29). Here the fibres of the right and left sides ***cross.*** After this decussation the fibres emerge on the posterior surface of the brainstem just below the inferior colliculus. Having emerged from the midbrain, the nerve winds around the cerebral peduncle to reach the front of the brainstem (Fig. 15.25). While winding around the cerebral peduncle, the nerve lies between the posterior cerebral artery (above it) and the superior cerebellar artery (below it). The nerve now runs forwards and pierces the dura mater just below

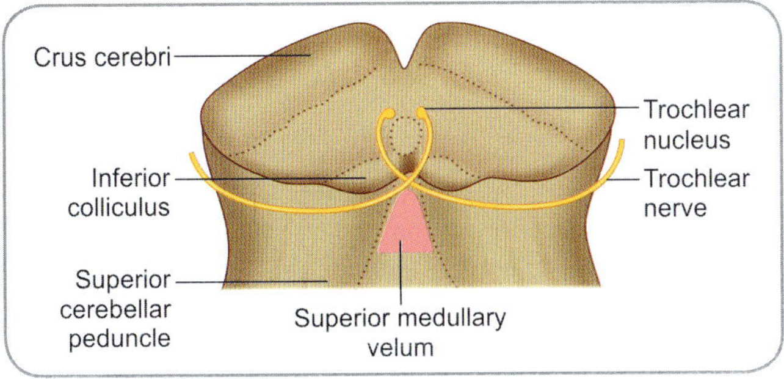

Fig. 15.29: Schematic posterior view of the course of trochlear nerve

the free margin of the tentorium cerebelli (Fig. 15.24). The point of penetration of the dura mater lies a little behind the posterior clinoid process.

After piercing the dura mater, the nerve comes to lie in the lateral wall of the cavernous sinus, below the oculomotor nerve, but above the ophthalmic division of the trigeminal nerve (Fig. 15.26). Continuing forwards, it crosses lateral to the oculomotor nerve to enter the upper part of the superior orbital fissure (Fig. 15.28). In the fissure, it lies above the tendinous ring for origin of the recti of the eyeball (Fig. 15.28). After entering the orbit, the nerve runs forward, passes medially above the origin of levator palpebrae superioris and ends in the superior oblique muscle.

Added Information

The trochlear nerve is the only cranial nerve that emerges on the posterior (or dorsal) aspect of the brainstem. It is also the only nerve that undergoes decussation. As a result of the decussation the left trochlear nerve is formed by fibres arising from the right trochlear nucleus and vice versa. The functional component of the trochlear nerve belongs to the somatic efferent column.

The next cranial nerve, in order of numerical sequence, is the fifth or trigeminal nerve. However, the abducent nerve (sixth) is considered first as its course and distribution are similar to those of the oculomotor and trochlear nerves.

Abducent Nerve

Other names: Sixth cranial nerve, nervus abducens, nervus abducentis.

This nerve is made up of fibres which arise in the abducent nucleus and supply the lateral rectus muscle of the eyeball. The abducent nucleus (Figs 15.17, 15.18C) is located in the lower part of the pons, in relation to the upper part of the floor of the fourth ventricle. The position of the nucleus in relation to the floor of the fourth ventricle is indicated by an elevation called the ***facial colliculus.*** Arising from this nucleus, the fibres of the nerve pass through the substance of the pons and emerge on the surface of the brainstem at the lower border of the pons cranial to the pyramid (Fig. 15.25).

The nerve then runs upwards, forwards and laterally. It pierces the dura mater lateral to the dorsum sellae of the sphenoid bone (Fig. 15.24). It then runs upwards to reach the upper border of the petrous temporal bone and bends around it to enter the middle cranial fossa. The nerve now comes to lie within the cavernous sinus, where it is closely related to the internal carotid artery. The nerve is first lateral to the artery and then inferolateral to it. At the anterior end of the cavernous sinus, it passes through the superior orbital fissure to enter the orbit. It ends by supplying the lateral rectus muscle.

The ***functional component of the abducent nerve belongs to the somatic efferent column*** (like all nerve fibres supplying extrinsic muscles of the eyeball).

Clinical Correlation

Oculomotor, Trochlear and Abducent Nerves

These three nerves are responsible for movements of the eyeball. In a routine clinical examination, the movements are tested by asking the patient to keep his/her head fixed and to move his/her eyes in various directions, i.e. upwards, downwards, inwards and outwards. An easy way is to ask the patient to keep his/her head fixed, and to follow the movements of the examiner's finger with his/her eyes. Such an examination can detect a gross abnormality in the movements of the eyes. Sometimes, one of the ocular muscles may not be completely paralysed but may be weak.

Diplopia

Diplopia means double vision (Greek.diplo=double, ops=eye) or that the person sees two images of a single object. To understand this phenomenon, it is essential to remember that objects lying in different parts of the visual field produce images over different spots on the retina. The brain judges the position of an object by the position at which its image is formed on the retina. Normally the movements of the right and left eyes are in perfect alignment, and an object casts images on corresponding spots on the two retinae so that only one image is perceived by the brain. When a muscle of the eyeball is weak, and a movement involving that muscle is performed, the movement of the defective eye lags behind that of the normal eye. As a result, images of the object on the two retinae are not formed at corresponding points but over two points near each other. The brain therefore 'sees' two images, one from each retina.

Squint (or Strabismus)

Squint is the condition (Greek.strabos=distorted) when the two eyes look in different directions. The squint becomes obvious when the eye movement involves a muscle that is paralysed or weak, because the weak muscle cannot keep up with the muscle of the normal side. Squint will usually be accompanied by diplopia. However, the patient compensates for lack of movement of the eyeball by turning the head in the direction of the object and on doing so, the diplopia disappears. If the normal eye is closed, the patient is unable to judge the position of objects in the field of vision correctly (because the image of the object does not fall on the part of the retina that corresponds to the true position of the object). These are features of *paralytic squint*.

Concomitant squint is another type which is congenital, and manifests itself in early childhood. This type of squint is present in all positions of the eyeball. There is no muscular weakness and movements are normal in all directions and there is no diplopia.

Oculomotor Nerve Palsy

In this condition, all movements of the eyeball are lost in the affected eye. When the patient is asked to look directly forward, the affected eye is directed laterally (by the lateral rectus) and downwards (by the superior oblique). There is lateral squint (*external strabismus*) and diplopia. As the levator palpebrae superioris is paralysed, there is drooping of the upper eyelid (*ptosis*). As parasympathetic fibres to the sphincter pupillae

contd...

Clinical Correlation   *contd...*

pass through the oculomotor nerve, the sphincter pupillae muscle is also paralysed. Unopposed action of sympathetic nerves produces a fixed and dilated pupil. Normally the pupil contracts when exposed to light (light reflex). It also contracts when the relaxed eye is made to concentrate on a near object (accommodation reflex). Both these reflexes are lost. The power of accommodation is lost because of paralysis of the ciliaris muscle.

Trochlear Nerve Palsy

The superior oblique muscle (supplied by the trochlear nerve) moves the eyeball downwards and laterally. The inferior rectus (supplied by the oculomotor nerve) moves it downwards and medially. For direct downward movement, a synchronised action of both muscles is required. When the superior oblique muscle is paralysed, the eyeball deviates medially on trying to look downwards.

Abducent Nerve Palsy

Abducent nerve supplies the lateral rectus muscle that moves the eyeball laterally. In looking forwards the lateral pull of the lateral rectus is counteracted by the medial pull of the medial rectus and so the eye is maintained in the centre. When the lateral rectus is paralysed, the affected eye deviates medially (*medial squint or internal strabismus*).

Trigeminal Nerve

Other names: Fifth cranial nerve, trifacial nerve, nervus trigeminus.

The trigeminal nerve is so called because it consists of three main divisions—ophthalmic nerve, maxillary nerve and mandibular nerve.

These nerves arise from a large trigeminal ganglion. The ganglion is connected to the brainstem (pons) by a thick sensory root. The trigeminal nerve also has a motor root (Fig. 15.25) which emerges from the pons medial to the

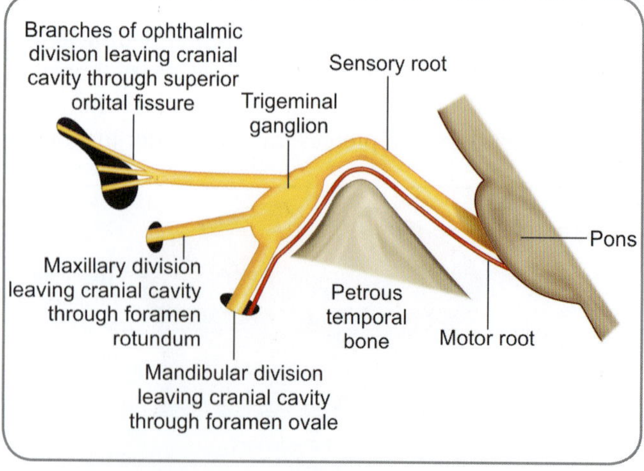

Fig. 15.30: Roots and divisions of trigeminal nerve

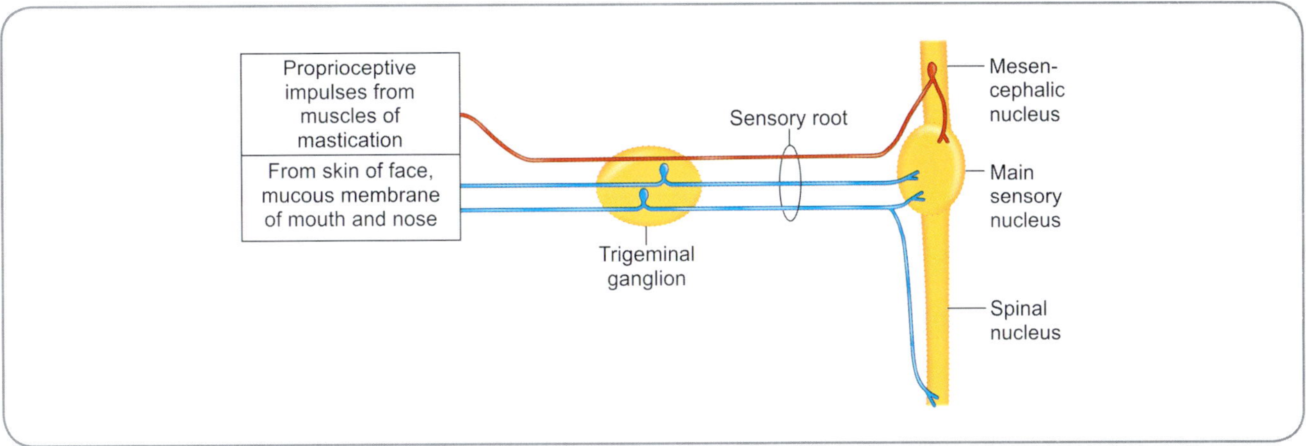

Fig. 15.31: Scheme to show the arrangement of afferent neurons of trigeminal nerve—fibres that descend along the spinal nucleus constitute the spinal tract of the nerve

sensory root. The motor root passes deep to the trigeminal ganglion and joins the mandibular nerve (Fig. 15.30). The trigeminal nerve contains both afferent and efferent fibres. Afferent fibres are peripheral processes of (pseudo) unipolar neurons located in the trigeminal ganglion. They are distributed through all the three divisions of the nerve. They carry exteroceptive sensations from the skin of the face, the mucous membrane of the mouth and the mucous membrane of the nose.

The central processes of the neurons in the trigeminal ganglion form the sensory root. After entering the pons, these processes terminate in relation to neurons in the main sensory nucleus and in the spinal nucleus of the trigeminal nerve (Fig. 15.31). Another group of afferent fibres carry proprioceptive impulses from the muscles of mastication (and possibly from the ocular, facial and lingual muscles). These fibres are believed to be peripheral processes of unipolar neurons present in the mesencephalic nucleus of the trigeminal nerve, within the brainstem. The central processes of these neurons end in relation to the main sensory nucleus.

The muscles of mastication (and some other muscles) are supplied through the mandibular division of the trigeminal nerve. The cell bodies of the neurons giving origin to these fibres are located in the motor nucleus of the trigeminal nerve. Axons arising from these neurons collect to form the motor root of the nerve. The motor root joins the mandibular nerve and is distributed through it. The muscles supplied by the motor fibres are as follows:

- ❑ ***Muscles of mastication***: Masseter, temporalis, medial and lateral pterygoids.
- ❑ ***Other muscles***: Mylohyoid, anterior belly of digastric, tensor palatini, tensor tympani.

Functional components of trigeminal nerve: The afferent fibres of the nerve (from skin and mucosa and proprioceptive from muscles) belong to the ***general somatic afferent*** column. The efferent fibres belong to the ***special visceral efferent*** column as the muscles supplied are derived (during development) from the mesoderm of the first branchial arch.

Trigeminal Ganglion

Other names: Semilunar ganglion, flat sensory ganglion, gasserian ganglion (named after the 18th century Vienna anatomist Johann Gasserio).

The trigeminal ganglion is shaped like a crescent (Fig. 15.32), whose convex border faces anterolaterally and concave border faces posteromedially. The convex border is continuous with the ophthalmic, maxillary and mandibular nerves; while the concave posterior border is continuous with the sensory root. The ganglion is placed in a depression (called the ***trigeminal impression***) on the anterior aspect of the petrous temporal bone (near its apex). The ganglion is enclosed within a pouch-like recess of dura mater called the ***trigeminal cave (or cavum trigeminale or Meckel's cave)***.

Ophthalmic Nerve

The fibres of the ophthalmic nerve are purely sensory. However, some sympathetic fibres for the eyeball travel

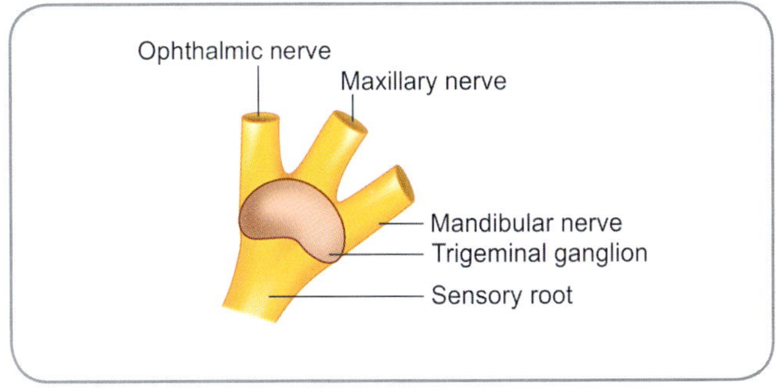

Fig. 15.32: Trigeminal ganglion seen from above

for part of their course through the nerve and some of its branches. The ophthalmic nerve arises from the trigeminal ganglion (Fig. 15.33). It pierces the dura mater of the trigeminal cave and comes to lie in the lateral wall of the cavernous sinus, below the trochlear nerve (Fig. 15.26). It runs forwards and divides into three branches, namely lacrimal, frontal and nasociliary nerves. These branches enter the orbit by passing through the superior orbital fissure. On entering the orbit, the lacrimal and frontal branches lie above the extraocular muscles, while the nasociliary nerve lies between them, lateral to the optic nerve.

The *lacrimal nerve* runs along the lateral wall of the orbit (along the upper border of the lateral rectus) (Fig. 15.33) and ends in the lacrimal gland (and hence its name). Some branches pass through the gland to supply the conjunctiva and the skin of the upper eyelid. The lacrimal nerve is joined by a twig from the zygomaticotemporal branch of the maxillary nerve.

The *frontal nerve* runs forward between the levator palpebrae superioris and the periosteum lining the roof of the orbit. It ends by dividing into supraorbital and supratrochlear branches. The *supraorbital nerve* continues in the line of the frontal nerve (Fig. 15.33), lying between the levator palpebrae superioris and the roof of the orbit. On reaching the orbital margin it passes through the supraorbital notch (on the medial part of the upper margin of the orbit) and curves upwards into the forehead. Here it divides into medial and lateral branches which supply the scalp as far back as the lambdoid suture, the

conjunctiva, the skin of the upper eyelid and the mucous membrane of the frontal air sinus.

The *supratrochlear nerve* runs forwards and medially above the orbital muscles and medial to the supraorbital nerve. It passes above the trochlea and reaching the upper margin of the orbital aperture, near its medial end, the nerve turns upwards into the forehead giving branches to the skin over its lower and medial part. Other branches given off by the supratrochlear nerve are, a descending branch which joins the infratrochlear branch of the nasociliary nerve (Fig. 15.33) and branches to the conjunctiva and skin of the upper eyelid.

The *nasociliary nerve,* on entering the orbit lies between the optic nerve and the lateral rectus. The nerve then runs medially crossing above the optic nerve (Fig. 15.33). As it does so, it lies below the superior rectus muscle. On reaching the medial wall of the orbit, the nerve ends by dividing into the *anterior ethmoidal* and *infratrochlear* nerves. The branches of the nasociliary nerve are as follows (Fig. 15.33):

- Just after entering the orbit, the nasociliary nerve receives the *sensory root of the ciliary ganglion.*
- The *long ciliary nerves* (two or three) arise from the nasociliary nerve as it crosses the optic nerve. They run

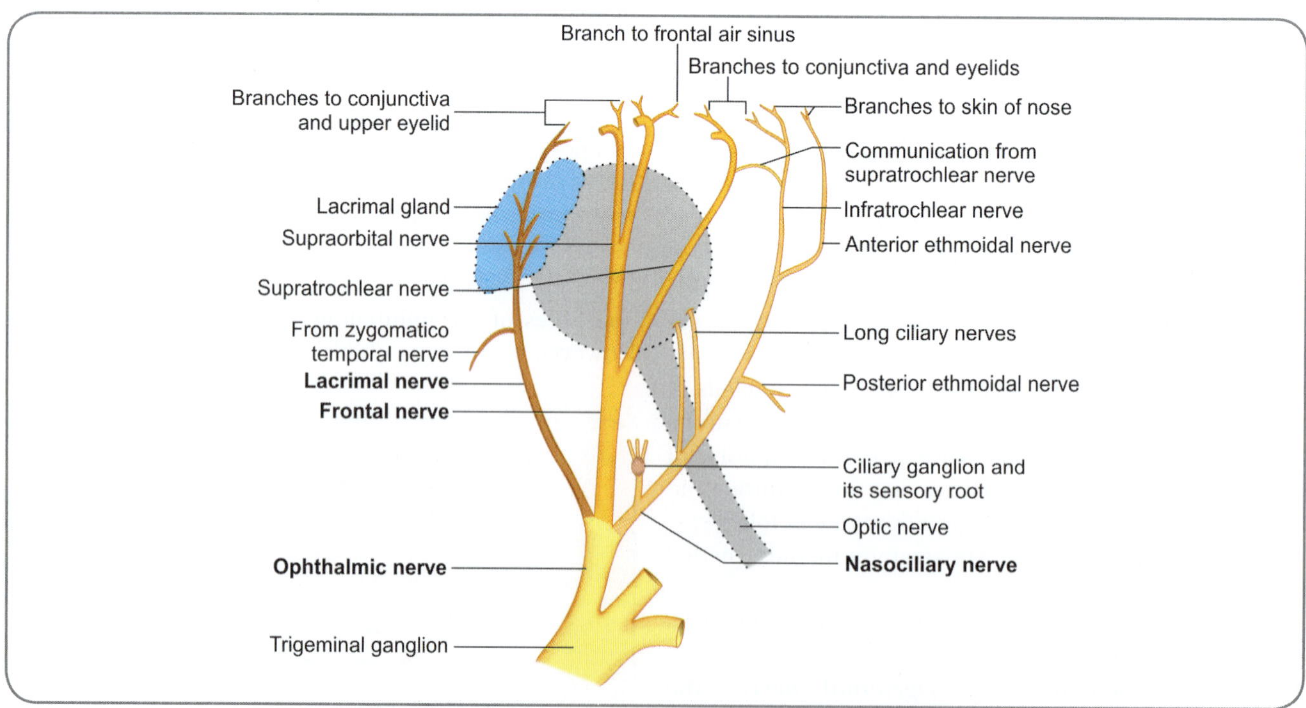

Fig. 15.33: Scheme to show the distribution of ophthalmic nerve

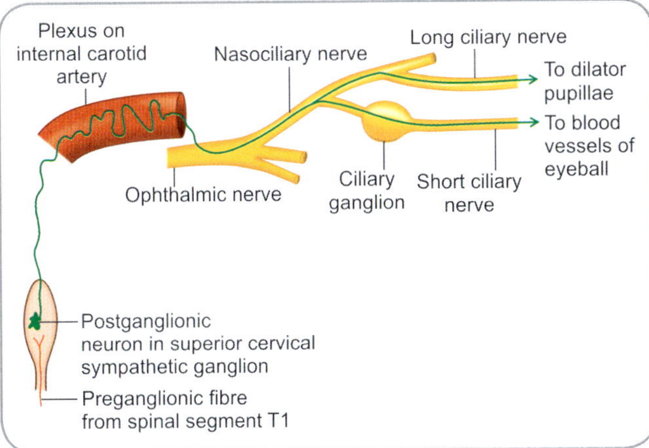

Fig. 15.34: Two pathways followed by sympathetic fibres for the eyeball—occasionally, the fibres for the dilator pupillae may pass through the ciliary ganglion

forwards to the eyeball where they pierce the sclera; and then run between the sclera and the choroid. They supply sensory fibres to the ciliary body, the iris and the cornea. They also carry postganglionic sympathetic fibres meant for the dilator pupillae, which are derived from the superior cervical sympathetic ganglion (Fig. 15.34) and travel through a plexus surrounding the internal carotid artery.

- The **posterior ethmoidal branch** enters the posterior ethmoidal foramen (on the medial wall of the orbit) and supplies the ethmoidal and sphenoidal air sinuses.
- The **anterior ethmoidal nerve** has a complicated course through the orbit, the anterior cranial fossa, and the nasal cavity. Its terminal part is seen on the face as the external nasal nerve. It gives **internal nasal branches** to the nasal septum and to the lateral wall of the nasal cavity. At the lower border of the nasal bone, the nerve leaves the nasal cavity, becomes superficial and supplies the skin over the lower part of the nose. This part of the nerve is called the **external nasal nerve**.
- The **infratrochlear nerve** (Fig. 15.33) runs forwards on the medial wall of the orbit and ends by supplying part of the skin of the upper and lower eyelids and over the upper part of the nose. The nerve also gives branches to the conjunctiva, the lacrimal sac and the lacrimal caruncle. The infratrochlear and supratrochlear nerves are joined to each other by a communicating twig.

Maxillary Nerve

The maxillary nerve is purely sensory. It arises from the middle of the distal edge of the trigeminal ganglion (Fig. 15.30). It pierces the dura mater and in the distal edge of the trigeminal cave, it comes to lie in the lower part of the lateral wall of the cavernous sinus. The nerve leaves the middle cranial fossa through the foramen rotundum to reach the pterygopalatine fossa. The nerve crosses the short distance between the anterior and posterior walls of the pterygopalatine fossa and leaves it by passing into the orbit through the inferior orbital fissure. The part of the maxillary nerve (Fig. 15.35) distal to the inferior orbital fissure is called the **infraorbital nerve.** This nerve lies first in the infraorbital groove and thereafter in the infraorbital canal. It appears on the face through the infraorbital foramen and ends here by dividing into a number of terminal branches. Several branches are also given off by the maxillary and infraorbital nerves along their courses as follows.

Branches Arising in the Pterygopalatine Fossa

Apart from branches arising from the pterygopalatine ganglion, the following branches arise directly from the maxillary nerve while the latter is in the pterygopalatine fossa (Fig. 15.36).

- The **zygomatic nerve** enters the orbit through the inferior orbital fissure and runs forwards along its lateral wall. It divides into two branches, the zygomaticotemporal and the zygomaticofacial nerves. Both these branches enter foramina present on the orbital surface of the zygomatic bone. Travelling through the zygomatic bone the **zygomaticotemporal nerve** emerges from the temporal surface of the bone. The nerve ends by supplying the skin over the temple. The **zygomaticofacial nerve** also passes through the

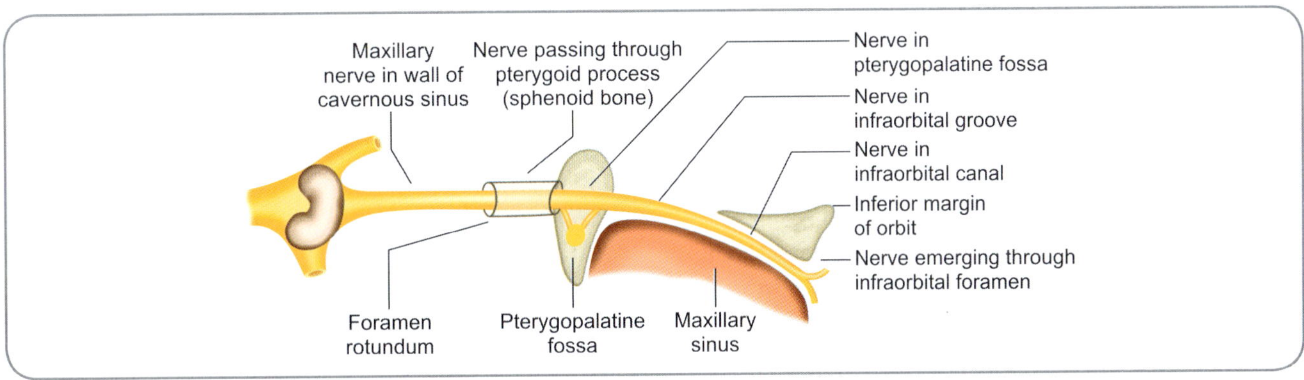

Fig. 15.35: Scheme to show the course of the maxillary nerve

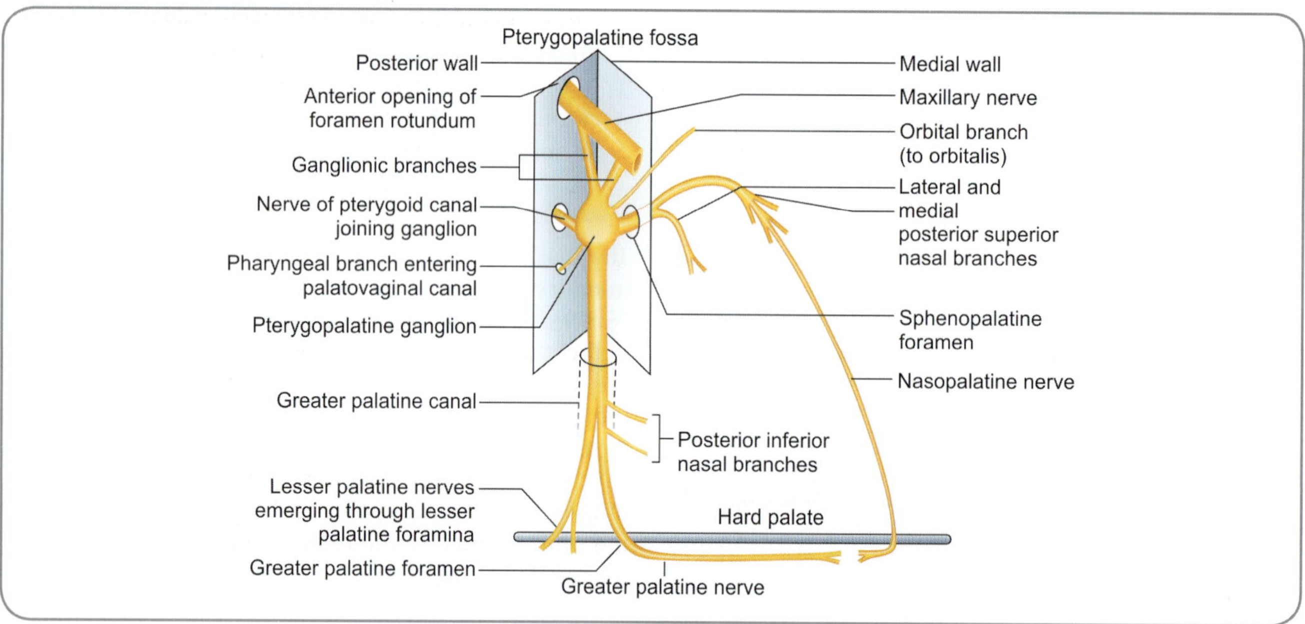

Fig. 15.36: Scheme to show branches of pterygopalatine ganglion

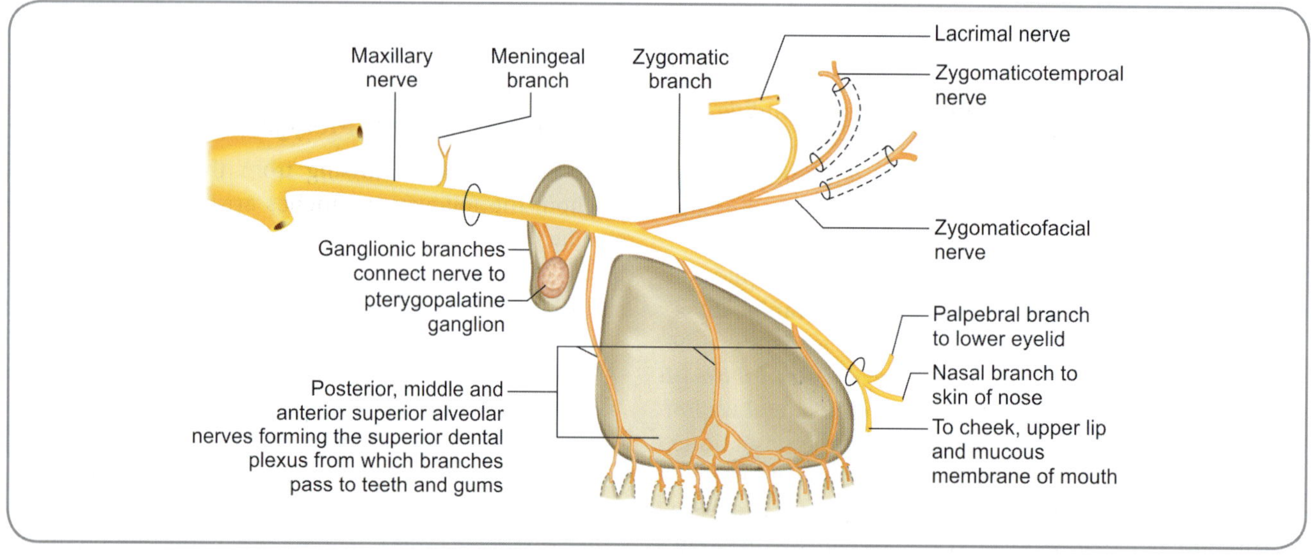

Fig. 15.37: Scheme to show direct branches arising from the maxillary nerve (including its infraorbital continuation)

substance of the zygomatic bone. It emerges from the bone through the zygomaticofacial foramen present on the lateral surface of the bone and supplies the skin of the cheek.

- The *posterior superior alveolar nerve* arises from the maxillary nerve (Fig. 15.37) in the pterygopalatine fossa. It runs down on the posterior surface of the maxilla (which forms the anterior wall of the pterygopalatine fossa) and then pierces it. The nerve now comes to lie in the wall of the maxillary sinus, which it supplies. It then divides into branches that form a plexus over the roots of the molar teeth. Branches from the plexus supply these teeth and the related gums.

Branches Arising in the Infraorbital Groove and Canal

The *middle superior alveolar nerve* arises from the infraorbital nerve as the latter lies in the infraorbital groove. Entering a foramen in the floor of the groove, it reaches the maxillary sinus and joins the posterior superior alveolar nerve in forming a plexus above the roots of the upper teeth. The fibres of this nerve supply the premolar teeth through the plexus.

The *anterior superior alveolar nerve* arises from the infraorbital nerve as the latter lies in the infraorbital canal. It then enters a bony canal that follows a complicated course through the maxillary bone (hence called the canalis sinuosus) to reach the incisor and canine teeth which it supplies. Branches from the posterior, middle and anterior superior alveolar nerves form a continuous plexus (***superior dental plexus***) in relation to the roots of the teeth borne on the maxilla. The anterior superior alveolar nerve also gives minute branches to the nasal cavity.

Branches of Infraorbital Nerve in the Face

The infraorbital nerve divides into the following branches after emerging from the infraorbital foramen:
- ***Palpebral branches*** supply the lower eyelid.
- ***Nasal branches*** supply the skin on the lateral side of the nose.
- ***Superior labial*** branches supply the skin of the upper lip and part of the cheek. Some branches reach the mucous membrane of the mouth.

Mandibular Nerve

The mandibular nerve is formed by the union of two roots. The ***sensory root*** arises from the lateral part of the trigeminal ganglion, and leaves the skull through the foramen ovale. The ***motor root*** also passes through the foramen ovale and unites with the sensory root just below the foramen. After it emerges from the foramen ovale, the nerve enters the infratemporal fossa. After a short downward course, the trunk of the mandibular nerve divides into a smaller ***anterior division*** and a larger ***posterior division*** (Fig. 15.38).

The trunk and both divisions give off a number of branches.

Branches from the Trunk (or Main Stem)

- ***Meningeal branch*** is given off from the nerve just after the union of the motor and sensory roots. It then re-enters the cranial cavity along with the middle meningeal artery through the foramen spinosum. Its branches are distributed to the dura mater (mainly) of the middle cranial fossa.
- ***Nerve to the medial pterygoid*** supplies the same muscle and also gives a branch to the otic ganglion. The fibres in this branch pass through the ganglion without relay and supply the tensor tympani and the tensor palatini muscles (Fig. 15.39).

Branches from the Anterior Division

Buccal nerve is sensory. It runs downwards and forwards through the muscles of the infratemporal fossa to reach the surface of the buccinator muscle. Here it supplies the skin superficial to the muscle and the mucous membrane lining its deep surface (note that this buccal nerve should not be confused with the buccal branch of the facial nerve).

The remaining branches of the anterior division of the mandibular nerve are motor. They supply the masseter, the lateral pterygoid and the temporalis as follows (Fig. 15.39):

The ***nerve to masseter*** passes laterally in front of the neck of the mandible (i.e., in the posterior part of the mandibular notch) to reach the masseter. The ***nerve to lateral pterygoid*** may be independent or may arise from the buccal nerve. The temporalis is supplied through the ***anterior, middle and posterior deep*** temporal nerves. These nerves pass upwards (above the lateral pterygoid) to reach the deep surface of the temporalis (Fig. 15.39).

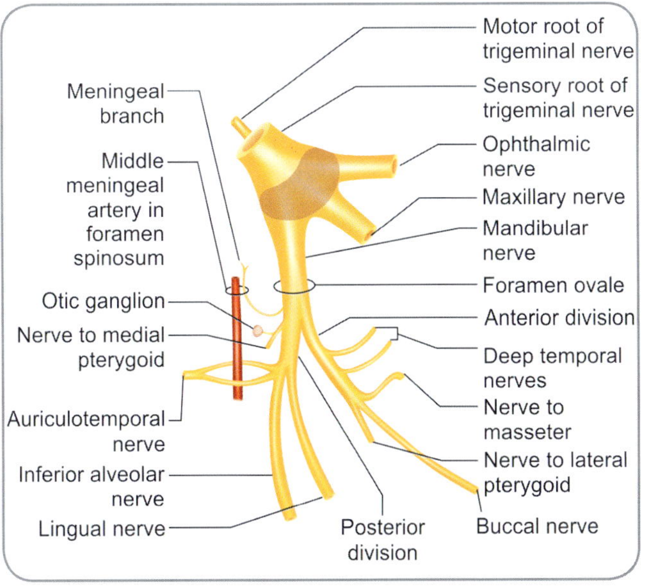

Fig. 15.38: Scheme to show the branches of mandibular nerve

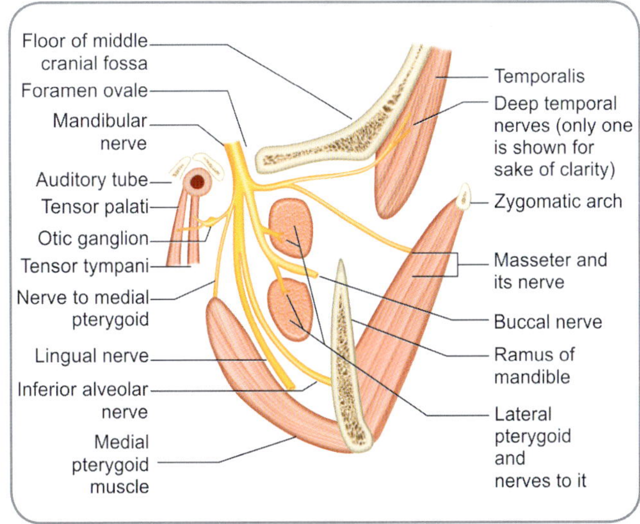

Fig. 15.39: Schematic coronal section through the infratemporal fossa to show the relationship of some branches of mandibular nerve to the muscles of mastication

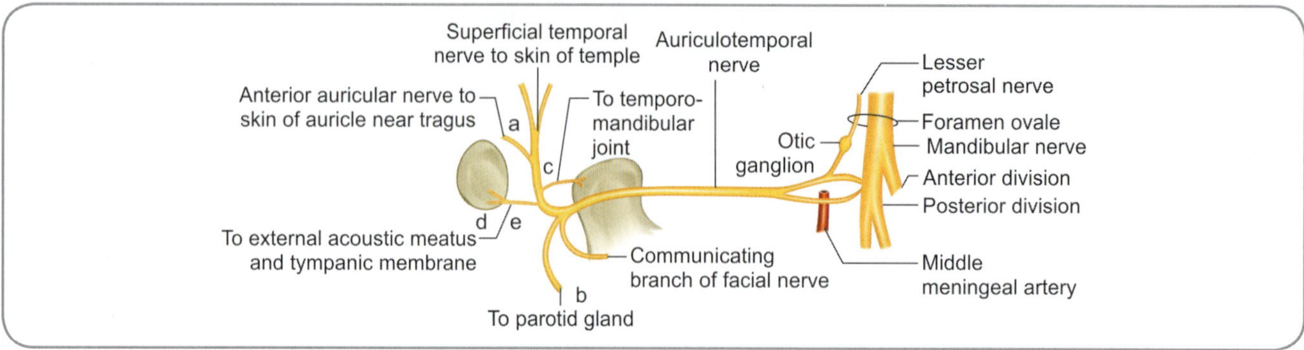

Fig. 15.40: Branches of auriculotemporal nerve

Branches from the Posterior Division

The posterior division of the mandibular nerve gives rise to three important nerves. These are the auriculotemporal, the lingual and the inferior alveolar nerves.

Auriculotemporal Nerve

The auriculotemporal nerve arises by two roots that form a ring through which the middle meningeal artery passes upwards (Figs 15.38 and 15.40). The nerve runs backwards deep in the infratemporal fossa and crosses deep to the neck of mandible. It then turns laterally behind the temporomandibular joint. In this part of its course, it is closely related to the upper part of the parotid gland. The nerve finally turns upwards into the scalp and ends by dividing into branches which supply the skin over the temple. The auriculotemporal nerve serves as a pathway for secretomotor fibres to the parotid gland. Preganglionic fibres travelling through the lesser petrosal nerve relay in the otic ganglion. Postganglionic fibres starting in this ganglion reach the roots of the auriculotemporal nerve through communicating twigs. They travel through this nerve and through its branches to the parotid gland (Fig. 15.40).

> The auriculotemporal nerve supplies:
> ('a' in Fig. 15.40)—**Skin** (of the temple);
> ('b' in Fig. 15.40)—A **gland** (parotid);
> ('c' in Fig. 15.40)—A **joint** (temporomandibular);
> ('d' in Fig. 15.40)—A **tube** (external acoustic meatus); and
> ('e' in Fig. 15.40)—A **membrane** (tympanic).

Lingual Nerve

The lingual nerve arises from the posterior division of the mandibular nerve (Fig. 15.41). Its upper part runs downward deep to the lateral pterygoid muscle (Fig. 15.39). It is joined here by the **chorda tympani** nerve (a branch of the facial nerve). The lingual nerve runs downwards and

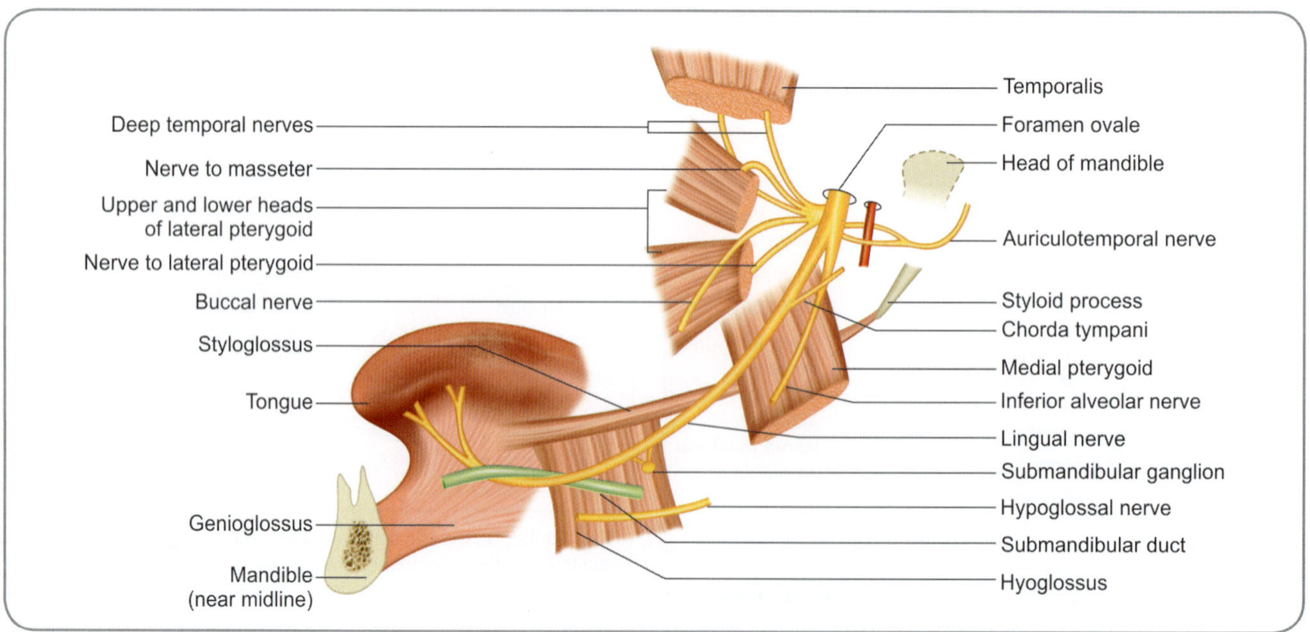

Fig. 15.41: Course and relations of lingual nerve and of other branches of mandibular nerve

forwards between the medial pterygoid (deep to it) and the ramus of the mandible (superficial to it). It then enters the mouth (by passing deep to the mandibular attachment of the superior constrictor of the pharynx). Within the mouth, the nerve lies deep to the mucous membrane overlying the medial surface of the mandible just below the third molar tooth. It can be felt here. Then the nerve enters the side of the tongue.

The nerve then crosses the styloglossus and runs forwards across the lateral surface of the hyoglossus (i.e. between the hyoglossus and the mylohyoid). While running across the hyoglossus, the lingual nerve lies above the submandibular duct, crosses superficial to the duct and then hooks around it to reach its medial side (Fig. 15.41). This is called the ***triple relation*** between the submandibular duct and the lingual nerve. The lingual nerve is connected to the submandibular ganglion by two or three branches. Still lower down the hypoglossal nerve runs across the hyoglossus. At the anterior margin of the hyoglossus, the nerve passes onto the genioglossus and divides into a number of branches.

The lingual nerve carries three types of fibres which are distributed as follows:

❑ Most of the fibres of the lingual nerve are those of ordinary sensation. They carry the sensations of touch, pain and temperature from the anterior two-thirds of the tongue. They also supply the mucous membrane of the floor of the mouth and the gums related to the lower teeth.

❑ The part of lingual nerve distal to the attachment of chorda tympani carries ***fibres for taste*** from the part of

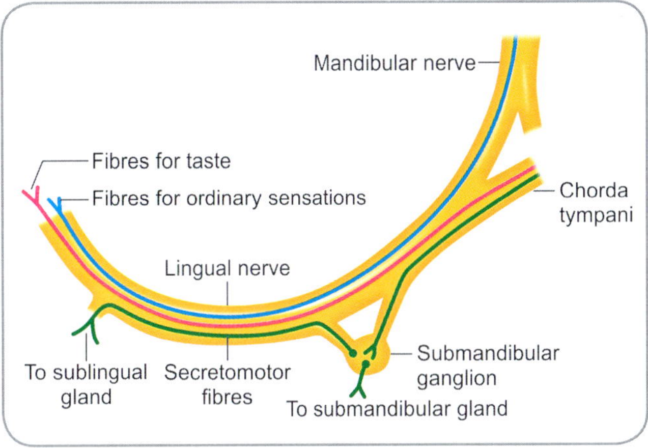

Fig. 15.43: Course of three types of fibres carried by the lingual nerve

tongue in front of the sulcus terminalis, but excluding the vallate papillae (Fig. 15.42).

❑ ***Secretomotor fibres*** for the submandibular and sublingual glands reach the lingual nerve through the chorda tympani. They relay in the submandibular ganglion. Postganglionic fibres reach the submandibular gland through branches to it from the ganglion. The fibres for the sublingual gland re-enter the lingual nerve and pass through its distal part to reach the gland (Fig. 15.43).

Inferior Alveolar Nerve

The ***inferior alveolar nerve*** (or the inferior dental nerve) is a branch of the posterior division of the mandibular nerve. At its upper end the nerve lies deep to the lateral pterygoid (Fig. 15.39). It emerges at the lower border of this muscle and then runs downwards and forwards deep to the ramus of the mandible (Fig. 15.41). Here, it is separated from the medial pterygoid by the sphenomandibular ligament. On reaching the mandibular foramen, it passes through it

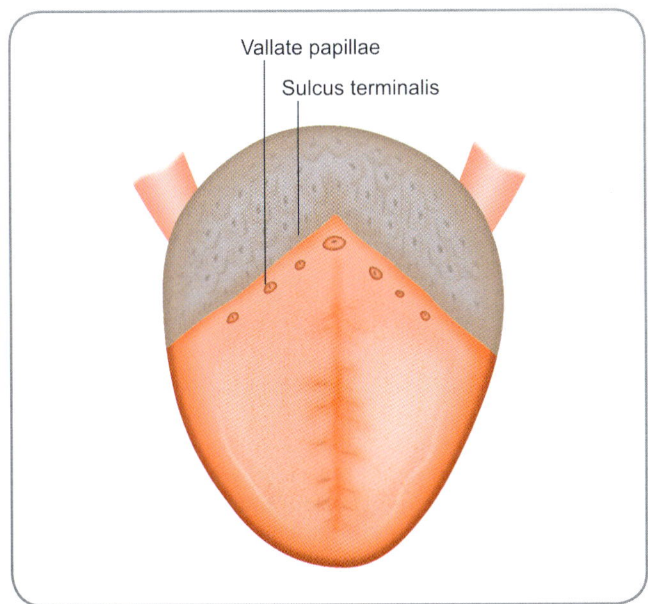

Fig. 15.42: Area of tongue supplied by the lingual nerve (shaded pink)

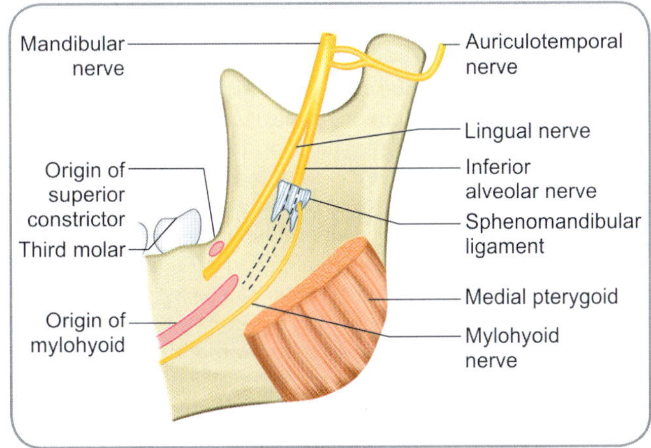

Fig. 15.44: Lingual, inferior alveolar and auriculotemporal nerves viewed from the medial side

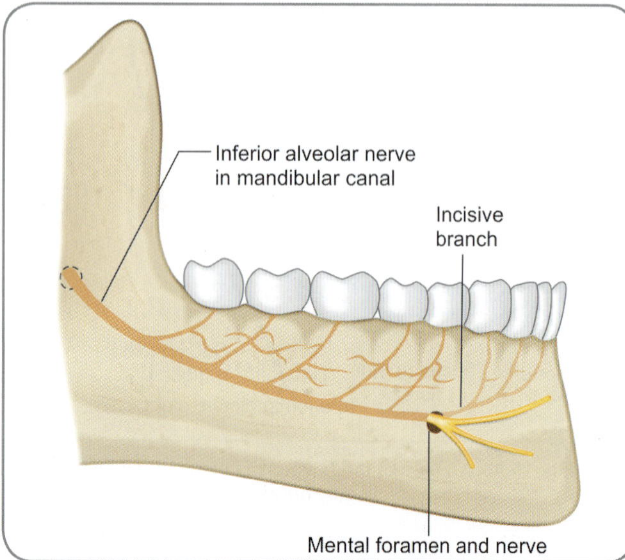

Fig. 15.45: Course of inferior alveolar nerve through the mandibular canal as projected on the lateral surface of the bone

into the mandibular canal. It runs forwards in this canal just below the teeth, and ends at the mental foramen by dividing into incisive and mental branches (Fig. 15.45). The nerve is distributed as follows:

❑ Within the mandibular canal, the nerve gives branches which supply the molar and premolar teeth.

❑ The ***incisive branch*** continues in the part of the mandibular canal anterior to the mental foramen. It supplies the canine and incisor teeth.

❑ The ***mental branch*** emerges from the mental foramen. It divides into branches that supply the skin over the chin, and that over the lower lip.

❑ The ***mylohyoid nerve*** arises from the inferior alveolar nerve just before the latter enters the mandibular foramen (Fig. 15.44). It runs in a groove on the medial side of the mandible, below the mylohyoid muscle and supplies the mylohyoid, and the anterior belly of the digastric muscle.

Clinical Correlation

❑ ***Testing of trigeminal nerve:*** The trigeminal nerve has a wide sensory distribution. It also supplies the muscles of mastication. The sensation of touch in the area of distribution of the nerve can be tested by touching different areas of skin with a wisp of cotton wool. The sensation of pain can be tested by gentle pressure with a pin. Motor function is tested by asking the patient to clench the teeth firmly. Contraction of the masseter can be felt by palpation when the teeth are clenched.

❑ ***Effects of injury or disease:*** Injury to the trigeminal nerve causes paralysis of the muscles supplied and loss of sensations in the area of supply. The muscles of mastication are responsible for all movements of the mandible. Contraction of these muscles on one side moves the chin

Clinical Correlation *contd...*

to the opposite side. Normally the chin is maintained in the midline by the balanced tone of the muscles of the right and left sides. In paralysis of the pterygoid muscles of one side, the chin is ***pushed to the paralysed side*** by muscles of the opposite side.

❑ Loss of sensation in the ophthalmic division (specially the nasociliary nerve) is of great importance. Normally the eyelids close as soon as the cornea is touched (corneal reflex). Loss of sensation in the cornea abolishes this reflex leaving the cornea unprotected. This can lead to the formation of ulcers on the cornea that can in turn lead to blindness.

❑ Pain arising in a structure supplied by one branch of the nerve may be felt in an area of skin supplied by another branch. This is called referred pain. Some examples are as follows:

  ○ Caries of a tooth in the lower jaw (supplied by the inferior alveolar nerve) may cause pain in the ear (auriculotemporal).

  ○ If there is an ulcer or cancer on the tongue (lingual nerve), pain may be felt over the ear and temple (auriculotemporal).

  ○ In frontal sinusitis (sinus supplied by a branch from the supraorbital nerve), pain is referred to the forehead (skin supplied by supraorbital nerve).

In fact, headache is a common symptom when any structure supplied by the trigeminal nerve is involved (e.g., eyes, ears, teeth).

❑ A source of irritation in the distribution of the nerve may cause severe persistent pain (***trigeminal neuralgia***). In some cases, no cause can be found. In such cases, pain can be relieved by injection of alcohol into the trigeminal ganglion, into one of the divisions of the nerve or into its sensory root. In some cases, it may be necessary to cut the fibres of the sensory root. In this connection, it is important to know that the fibres for the maxillary and mandibular divisions can be cut without destroying those for the ophthalmic division. This is possible as the fibres for the ophthalmic division lie separately in the upper medial part of the sensory root. Trigeminal pain can also be relieved by cutting the spinal tract of the trigeminal nerve; this procedure is useful especially for relieving pain in the distribution of the ophthalmic division as pain can be abolished without loss of the sense of touch and, therefore, without the abolition of the corneal reflex.

❑ ***Mandibular nerve block:*** This is used for anaesthesia of the lower jaw (for extraction of teeth). The anterior margin of the ramus of the mandible is palpated and just medial to it the pterygomandibular raphe (ligament) can be felt. The needle is inserted in the interval between the ramus and the raphe. The tip of the needle is now very near the inferior alveolar nerve, just before it enters the mandibular canal. Anaesthetic drug injected here blocks the nerve.

❑ The lingual nerve lies very close to the medial side of the third molar tooth, just deep to the mucosa. The nerve can be injured in careless extraction of a third molar. In cases of cancer of the tongue where there is intractable pain, the lingual nerve can be cut at this site to relieve pain.

contd...

Facial Nerve

Other names: Seventh cranial nerve, motor nerve of the face, nervus facialis.

The facial nerve emerges from the junction between pons and medulla. In its course, it first traverses through the pons, then the petrous temporal bone and then it runs within the parotid gland.

Intrapontine Course of Facial Nerve

The ***facial nucleus*** lies deep in the reticular formation of the pons, medial to the spinal nucleus of the trigeminal nerve (Figs 15.17 and 15.18C). The fibres arising in the nucleus follow an unusual course. They first run backwards and medially to reach the lower pole of the abducent nucleus. They then ascend on the medial side of that nucleus. Finally, the fibres turn forwards and laterally passing above the upper pole of the abducent nucleus. The abducent nucleus and the facial nerve fibres looping around it together form a surface elevation, the ***facial colliculus*** (Latin.collis=hill; also called the facial hillock, eminentia abducentis or facial eminence) in the floor of the fourth ventricle.

It is attached to the brainstem by two roots—a large ***motor root*** and a smaller ***sensory root.*** These roots are attached in the lateral part of the groove between the lower border of the pons and the upper border of the medulla, above the olive. The motor root is medial to the sensory root. The sensory root is attached midway between the motor root (medially) and the vestibulocochlear nerve (laterally). It is, therefore, called the ***nervus intermedius*** (also called the ***portio intermedia*** or the ***nerve of Wrisberg;*** named after the 18th century German anatomist Heinrich Wrisberg). From this attachment the motor and sensory roots pass forwards and laterally and leave the posterior cranial fossa by entering the internal acoustic meatus (on the posterior aspect of the petrous temporal bone).

Intrapetrous Course of Facial Nerve

After entering the internal acoustic meatus, the facial nerve enters a bony canal. Within this canal, the nerve first passes laterally above the vestibule ('1' in Fig. 15.46) to reach the anterosuperior angle of the medial wall of the middle ear. At this point ('2' in Fig. 15.46) the nerve bends sharply backwards. This bend is thickened by the presence of the ***genicular ganglion*** (or the geniculate ganglion) and is called external genu of the nerve.The nerve then runs horizontally backwards in a canal projecting into the medial wall of the middle ear ('3' in Fig. 15.46). Reaching the junction of the medial and posterior walls of the middle ear ('4' in Fig. 15.46) the nerve turns downward. Continuing downwards along the junction of the medial and posterior wall of the middle ear ('5' in Fig. 15.46), the nerve reaches the stylomastoid foramen through which it leaves the skull. It immediately enters the substance of the parotid gland and runs forwards within it and ends behind the neck of the mandible by dividing into several branches.

Added Information

'Genu' indicates a bend (Latin.genu=knee; anything that resembles the curve of the knee is named 'genu'). The facial nerve has two bends. The first bend or curve is where the fibres take a course around the abducent nerve nucleus in the pons. This is called the internal genu of facial nerve (since it is inside the cranial cavity) or plainly, genu nervi facialis. The second bend is outside the cranial cavity and that is where the nerve curves to change its direction within the ear. This is called the external genu of facial nerve.

The genicular ganglion is actually a ganglion of the nervus intermedius. Therefore, it is sometimes referred to as the intermedia ganglioformis.

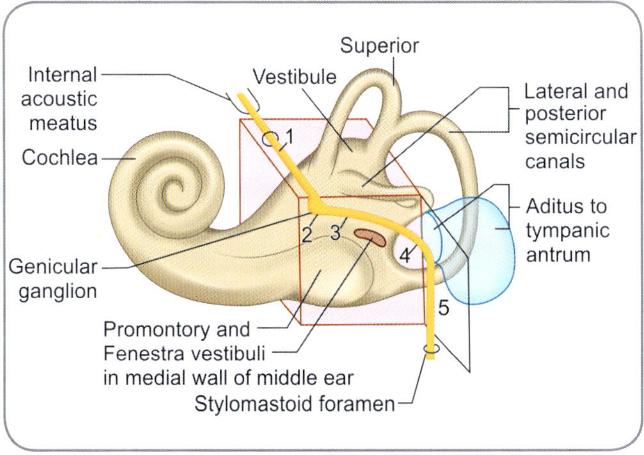

Fig. 15.46: Scheme to show the course of facial nerve within the petrous temporal bone

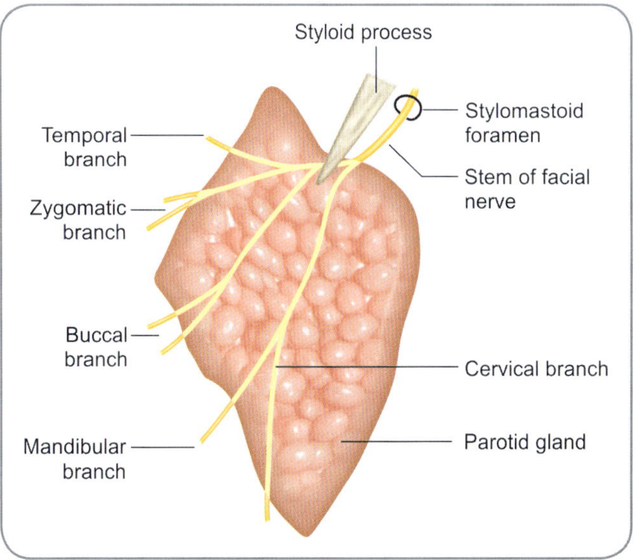

Fig. 15.47: Course of facial nerve through parotid gland

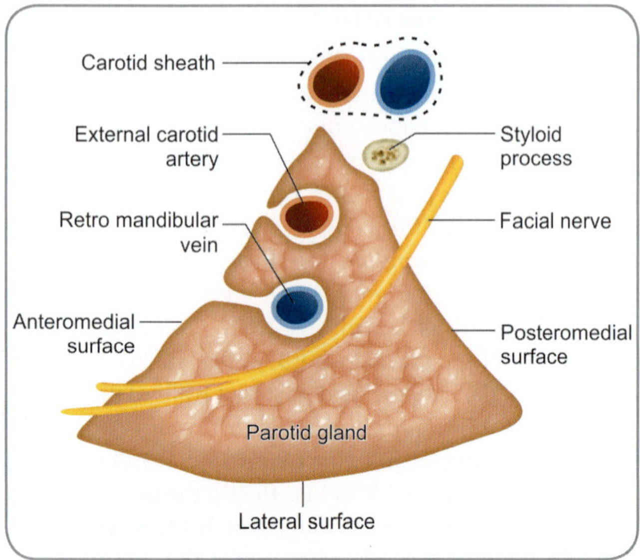

Fig. 15.48: Transverse section through parotid gland to show relationship to facial nerve

Intraparotid Course of Facial Nerve

As the facial nerve runs forward through the parotid gland, it crosses the styloid process, the retromandibular vein and the external carotid artery (Fig. 15.48). It divides into several branches while still within the gland. These branches emerge from the anteromedial surface of the gland and come into view along the anterior margin of the gland (Figs 15.47 and 15.48).

Branches of Facial Nerve

The branches of facial nerve (Fig. 15.49) which arise before entering into the parotid gland are as follows:

- ❑ *Greater petrosal nerve* that arises from the genicular ganglion;
- ❑ *Nerve to stapedius* that arises from the facial nerve as the latter turns downwards along the junction of the medial and posterior walls of the middle ear; this runs forwards through a short canal in the petrous temporal bone to reach the stapedius;
- ❑ *Chorda tympani* that arises from the intrapetrous part of the facial nerve about 6 mm above the stylomastoid foramen;
- ❑ *Posterior auricular nerve* that is given off just after the facial nerve emerges from the stylomastoid foramen; this nerve runs upwards into the scalp passing behind the external acoustic meatus; it divides into an *auricular branch*, which supplies some muscles of the auricle and an *occipital branch* which supplies the occipital belly of occipitofrontalis;
- ❑ *Nerve to posterior belly of digastric muscle* and the *nerve to stylohyoid* which arise near the stylomastoid foramen; both these nerves end by supplying the muscles concerned;

The remaining branches of the facial nerve arise within the parotid gland. These are:

- ❑ *Temporal branches* which enter the scalp in the temporal region–these nerves supply the frontal belly of the occipitofrontalis, corrugator supercilii and some

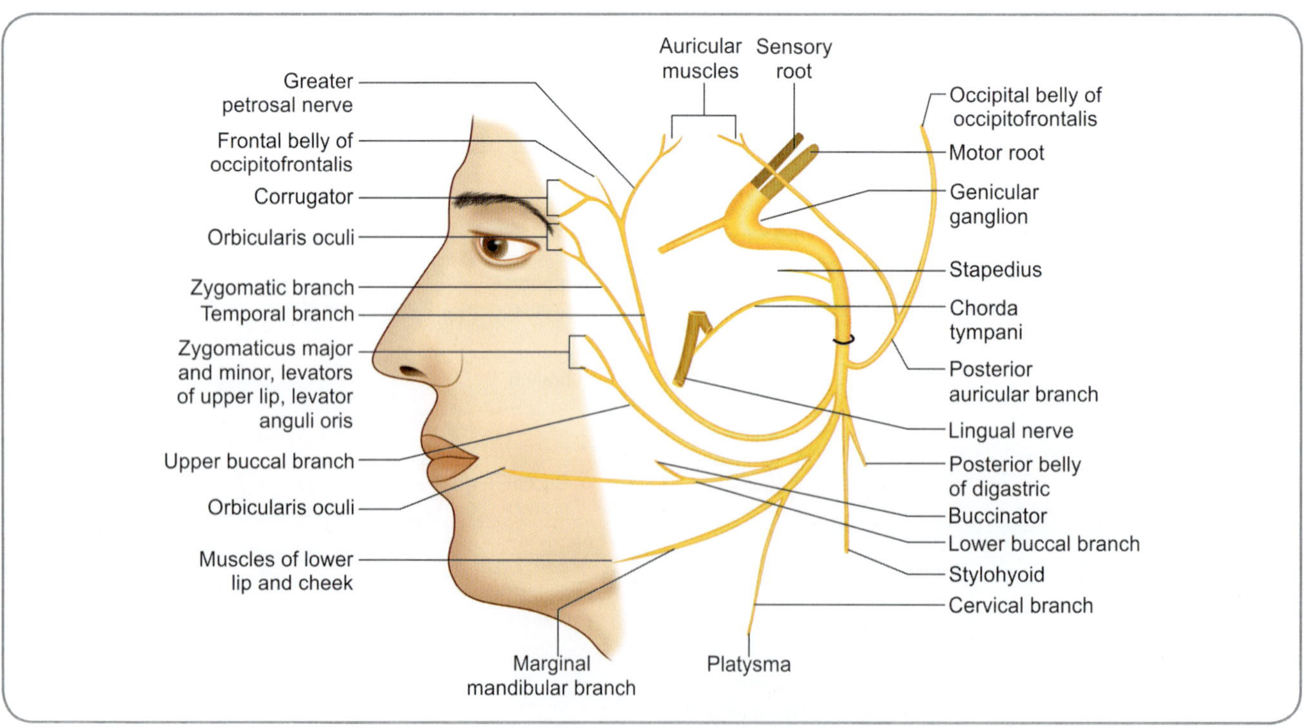

Fig. 15.49: Scheme to show the branches of the facial nerve

muscles of the auricle; some twigs are also given to the orbicularis oculi;

❏ ***Zygomatic branches*** which supply the orbicularis oculi;

❏ ***Buccal branches*** which are in two sets—upper and lower; the upper branches (sometimes called the lower zygomatic branches) supply zygomaticus major and minor, levator labii superioris, levator anguli oris, levator labii superioris alaeque nasi and small muscles related to the nose; the lower buccal branches supply buccinator and orbicularis oris;

❏ ***Marginal mandibular branch*** that is related to the lower border of the mandible—this nerve supplies the muscles of the lower lip and chin;

❏ ***Cervical branch*** that emerges from the parotid gland near its lower end—this nerve enters the neck and supplies the platysma.

Greater Petrosal Nerve

Other names: Greater superficial petrosal nerve, major petrosal nerve.

The ***greater petrosal nerve*** arises from the genicular ganglion. It passes through a canal in the petrous temporal bone and reaches its anterior surface by passing through an aperture called the hiatus (aperture) for the greater petrosal nerve. The nerve then runs forwards and medially in a groove on the anterior surface of the bone, passing deep to the trigeminal ganglion. On reaching the foramen lacerum, it ends by joining the deep petrosal nerve to form the nerve of the pterygoid canal. The ***deep petrosal nerve*** consists of sympathetic fibres derived from the plexus around the internal carotid artery. The ***nerve of the pterygoid canal*** (Vidian nerve) passes forwards in the pterygoid canal to enter the pterygopalatine fossa. It ends by joining the pterygopalatine ganglion. The greater petrosal nerve and the nerve of the pterygoid canal serve as pathways for secretomotor fibres to the lacrimal gland and to the glands of the nasal and palatine mucosa.

Chorda Tympani Nerve

The chorda tympani nerve is so called because it has an intimate relationship with the middle ear (tympanum). It arises from the facial nerve as the latter descends towards the stylomastoid foramen. The chorda tympani is of considerable importance as it carries fibres of taste from the anterior two-thirds of the tongue and secretomotor fibres to the submandibular and sublingual glands.

The origin of the nerve is about 6 mm above the foramen. The nerve enters the middle ear through the posterior canaliculus for the chorda tympani, which opens on the posterior wall of the middle ear, close to the posterior part of the margin of the tympanic membrane (Fig. 15.50). The nerve now passes forwards through the substance of the tympanic membrane (lying between its fibrous layer and the mucous membrane lining its internal surface) and crosses the handle of the malleus (which is embedded in the membrane). On reaching the anterior margin of the tympanic membrane, the chorda tympani enters the anterior canaliculus (in the anterior wall of the middle ear). It passes through this canaliculus and emerges on the base of the skull through the medial end of the petrotympanic fissure. It then passes medially, forwards and downwards, crosses medial to the spine of the sphenoid. (Note: the auriculotemporal nerve passes lateral to the spine, Fig. 15.40). The chorda tympani ends by joining the lingual nerve from behind. The junction of the chorda tympani with the lingual nerve lies deep to the lateral pterygoid muscle.

Added Information

Functional Components of the Facial Nerve

Special visceral efferent fibres: Most of the branches of the facial nerve are motor. Many of them supply the muscles of facial expression. Motor fibres also supply occipitofrontalis, muscles of auricle, stapedius, platysma, stylohyoid and posterior belly of digastric. All these muscles are derived from the mesoderm of the second branchial arch. The nerve fibres concerned in their innervation arise in the facial nucleus which lies in the lower part of the pons, and belong to the ***special visceral efferent*** column (Figs 15.17 and 15.18D). The axons of neurons located in this nucleus collect to form the motor root of the facial nerve.

General visceral efferent fibres: These are the parasympathetic secretomotor fibres in three groups.

❏ Preganglionic secretomotor fibres for the submandibular and sublingual glands (Fig. 15.51) arise from neurons located in the ***superior salivatory nucleus*** (this nucleus lies in the lower part of the pons). Fibres leave the pons through the nervus intermedius and run for some distance in the intrapetrous part of the facial nerve. They then enter the chorda tympani to reach the lingual nerve. They leave the lingual nerve through branches to the submandibular ganglion and relay in the ganglion. Postganglionic neurons located in this ganglion give out the postganglionic fibres which supply the submandibular and sublingual glands.

❏ Preganglionic secretomotor fibres for the lacrimal gland (Fig. 15.52) arise in the lacrimatory nucleus, which lies near the salivatory nuclei. They leave the pons through the nervus intermedius, pass into the greater petrosal nerve and through it into the nerve of the pterygoid canal to end in the pterygopalatine ganglion. Postganglionic neurons are located in this ganglion. Fibres arising from them pass successively through the pterygopalatine ganglion to the maxillary nerve, its zygomatic branch, then the zygomaticotemporal branch of the zygomatic nerve and then the communication to lacrimal nerve and finally through the lacrimal nerve to reach the lacrimal gland.

❏ Preganglionic secretomotor fibres for glands in the nasal and palatine mucosa arise in neurons the location of which

contd...

Added Information *contd...*

is uncertain. They probably lie near the salivary nuclei. The preganglionic fibres follow the same path as for the lacrimal gland. Postganglionic fibres arising in the pterygopalatine ganglion pass through its greater and lesser palatine branches to reach glands in the palate; and through its nasal branches to reach glands in the nasal mucosa.

Special visceral afferent fibres: The facial nerve contains ***special visceral afferent*** fibres which carry sensations of taste from the part of the tongue in front of the sulcus terminalis; and from the soft palate. These fibres are processes of unipolar neurons located in the genicular ganglion. Peripheral processes reach the tongue by passing through the chorda tympani and the lingual nerve (Fig. 15.51).

Fibres for the soft palate pass through the greater petrosal nerve, the nerve of the pterygoid canal, the pterygopalatine ganglion (without relay) and the lesser palatine nerves (Fig. 15.52). The central processes leaving the genicular ganglion pass through the nervus intermedius to reach the brainstem. Here they terminate in relation to the upper part of the nucleus of the solitary tract (Figs 15.17 and 15.18B).

General somatic afferent fibres: These fibres supply skin. Although the facial nerve gives no direct branches to skin there is evidence that some of its fibres reach part of the skin of the auricle. They probably pass through communications between the facial nerve and the auricular branch of the vagus. These fibres are ***general somatic afferent*** and are peripheral processes of unipolar cells located in the genicular ganglion. Central processes arising in these neurons pass through the nervus intermedius and end in the nucleus of the spinal tract of the trigeminal nerve. When the genicular ganglion is affected by an infection called ***herpes zoster,*** vesicles appear on the skin over part of the auricle.

Pterygopalatine ganglion

This ganglion is located in the pterygopalatine fossa and is suspended from the maxillary nerve by two ganglionic branches (Fig. 15.36). Functionally, the ganglion is autonomic and is a peripheral ganglion of the cranial parasympathetic outflow. Its motor (or parasympathetic) root is formed by the nerve of the pterygoid canal that conveys pre-ganglionic secretomotor fibres for the supply of the lacrimal gland, and for the glands of the nasal and palatine mucosa. Neurons located in this ganglion give off postganglionic fibres that innervate these glands. The pathways concerned have already been described (Fig. 15.52).

The ganglion also receives some sympathetic fibres. These fibres, which are postganglionic, arise in the superior cervical sympathetic ganglion, run along the internal carotid plexus, pass into the deep petrosal nerve and then into the nerve of the pterygoid canal to reach the ganglion. They pass through the ganglion, without relay, and enter its orbital branches to supply the orbitalis muscle.

Numerous sensory fibres from the palate (greater and lesser palatine nerves), the nose (nasal branches) and

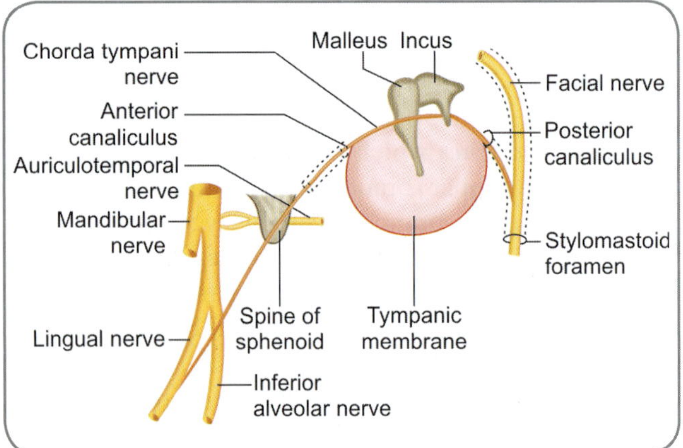

Fig. 15.50: Course of chorda tympani as seen from the medial aspect

the pharynx (pharyngeal branch) reach the ganglion, and pass through it without relay to enter the maxillary nerve through its ganglionic branches.

Fibres of taste from the soft palate reach the ganglion through the lesser palatine nerves (Fig. 15.52). They pass through the nerve of the pterygoid canal and the greater petrosal nerve to reach the genicular ganglion.

Submandibular ganglion

This ganglion lies over the hyoglossus muscle, suspended from the lingual nerve by two or more roots (Fig. 15.41). It is an autonomic ganglion connected with the cranial parasympathetic outflow. Functionally the ganglion is concerned with the secretomotor innervation of the submandibular and sublingual salivary glands. These fibres are parasympathetic. The pathway concerned is shown in Figure 15.51 and has already been described.

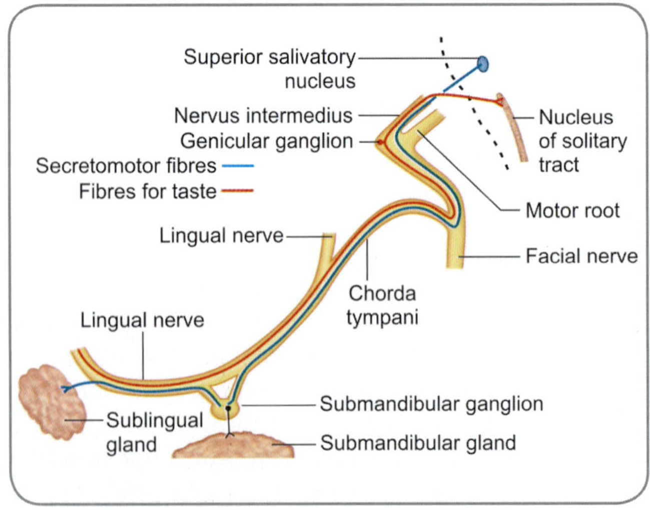

Fig. 15.51: Scheme to show the secretomotor pathway for the submandibular and sublingual glands and the pathway for taste from the anterior two-thirds of the tongue

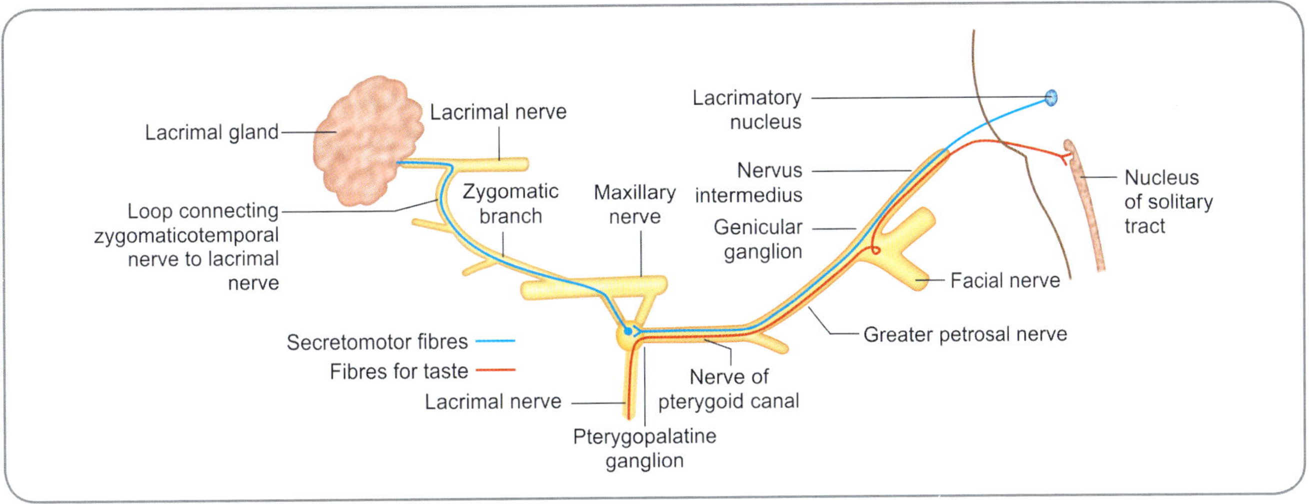

Fig. 15.52: Scheme to show the secretomotor pathway for the lacrimal gland and the pathway for fibres of taste from the soft palate

The ganglion receives sympathetic fibres from the plexus on the facial artery. These are postganglionic fibres arising from the superior cervical sympathetic ganglion. They pass through the ganglion without relay and supply the blood vessels of the submandibular and sublingual glands.

> ### Clinical Correlation
>
> **Testing the Facial Nerve**
>
> - Ask the patient to close his eyes firmly. In complete paralysis of the facial nerve, the patient will not be able to close the eye on the affected side. In partial paralysis, closure is weak and the examiner can easily open the closed eye with fingers (which is very difficult in a normal person).
> - Ask the person to smile. In smiling, the normal mouth is more or less symmetrical, the two angles moving upwards and outwards. In facial paralysis, the angle fails to move on the paralysed side.
> - Ask the patient to fill his mouth with air. Press the cheek with fingers and compare the resistance (by the buccinator muscle) on the two sides. The resistance is less on the paralysed side. On pressing the cheek, air may leak out of the mouth because the muscles closing the mouth are weak.
> - Sensation of taste should be tested on the anterior two-thirds of the tongue (as described under glossopharyngeal nerve).

Paralysis of facial nerve

Paralysis of the facial nerve is fairly common. It can occur due to injury or disease of the facial nucleus (nuclear paralysis) or of the nerve anywhere along its course (infranuclear paralysis). In the most common type of infranuclear paralysis called ***Bell's palsy,*** the nerve is affected near the stylomastoid foramen. Facial nerve can also be paralysed by interruption of corticonuclear fibres running from the motor cortex to the facial nucleus; this is referred to as ***supranuclear paralysis.***

- The effects of paralysis are due to the failure of the muscles concerned to perform their normal actions. Some effects are as follows:
  - The normal face is more or less symmetrical. When the facial nerve is paralysed on one side, the most noticeable feature is loss of symmetry.
  - Normal furrows on the forehead are lost because of paralysis of occipitofrontalis.
  - The palpebral fissure is wider on the paralysed side because of paralysis of orbicularis oculi.
  - The conjunctival reflex is lost for the same reason.
  - There is marked asymmetry of the mouth because of paralysis of orbicularis oris and of muscles inserted into the angle of the mouth. This is most obvious when a smile is attempted.
  - As a result of a symmetry, the protruded tongue appears to deviate to one side, but is in fact in the midline.
  - During mastication food tends to accumulate between the cheek and the teeth (this is normally prevented by the buccinator).
- Additional effects are observed in injuries to the facial nerve at levels higher than the stylomastoid foramen, as follows:
  - If the injury is proximal to the origin of the chorda tympani, there is loss of sensation of taste on the anterior two-thirds of the tongue.
  - The intensity of loud sounds reaching the internal ear is normally decreased by contraction of the stapedius muscle. When a lesion of the facial nerve is located proximal to the origin of the branch to stapedius this muscle is paralysed. As a result even normal sounds appear too loud (***hyperacusis***).

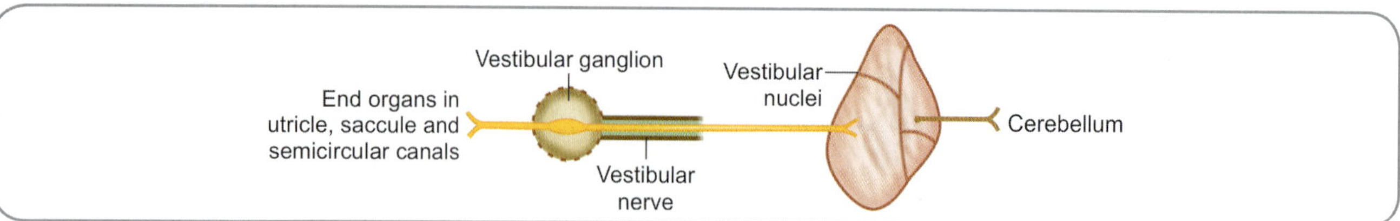

Fig. 15.53: Basic plan of vestibular nerve

- ○ In fractures of the temporal bone or in lesions near the exit of the nerve from the brain, the vestibulocochlear nerve may also be affected (leading to deafness).
- ○ In nuclear lesions (within the brainstem), other neighbouring nuclei may be affected leading to lesions of the abducent or trigeminal nerves.
- ❑ Supranuclear lesions can be distinguished from nuclear or infranuclear lesions because these are usually accompanied by hemiplegia. Movements of the lower part of the face are more affected than those of the upper part. The explanation for this is that the corticonuclear fibres concerned with movements of the upper part of the face are bilateral, whereas those concerned with movements of the lower part of the face are unilateral. Another difference is that while voluntary movements are affected, emotional expressions appear to be normal. It has been suggested that there are separate pathways from the cerebral cortex to the facial nucleus for voluntary and emotional movements, and that usually only the former are involved.

Vestibulocochlear Nerve

Other names: Eighth cranial nerve, statoacoustic nerve, acoustic nerve, octave nerve, nervus vestibulocochlearis. The vestibulocochlear nerve consists of two distinct parts—vestibular and cochlear. Both of them are purely sensory. The vestibular nerve carries impulses necessary for the maintenance of equilibrium from the vestibular part of the internal ear. The cochlear nerve carries impulses of hearing from the cochlear part of the internal ear. The vestibulocochlear nerve is attached to the surface of the brainstem at the lower border of the pons, posterolateral to the attachment of the facial nerve (Fig. 15.25). From here, the nerve passes forwards and enters the internal acoustic meatus along with the motor and sensory roots of the facial nerve. Here the nerve divides into vestibular and cochlear parts.

Vestibular Nerve

Other names: Radix superior, static nerve.
A plan of the vestibular nerve is shown in Figure 15.53. The fibres of the vestibular nerve are processes of cells in the vestibular ganglion (Scarpa's ganglion) that is located within the internal acoustic meatus. The cells of this ganglion are bipolar (not pseudounipolar as in typical sensory ganglia). Peripheral processes arising from neurons in the ganglion supply end organs in the vestibular part of the membranous labyrinth. The central processes of these neurons in the vestibular ganglion form the vestibular nerve. Entering the brainstem they end in relation to the vestibular nuclei (Figs 15.17, 15.18B and C). Fibres arising in the vestibular nuclei carry impulses for equilibrium to the cerebellum. Some fibres from the vestibular ganglion pass directly to the cerebellum.

Cochlear Nerve

Other names: Radix inferior, acoustic nerve, auditory nerve.

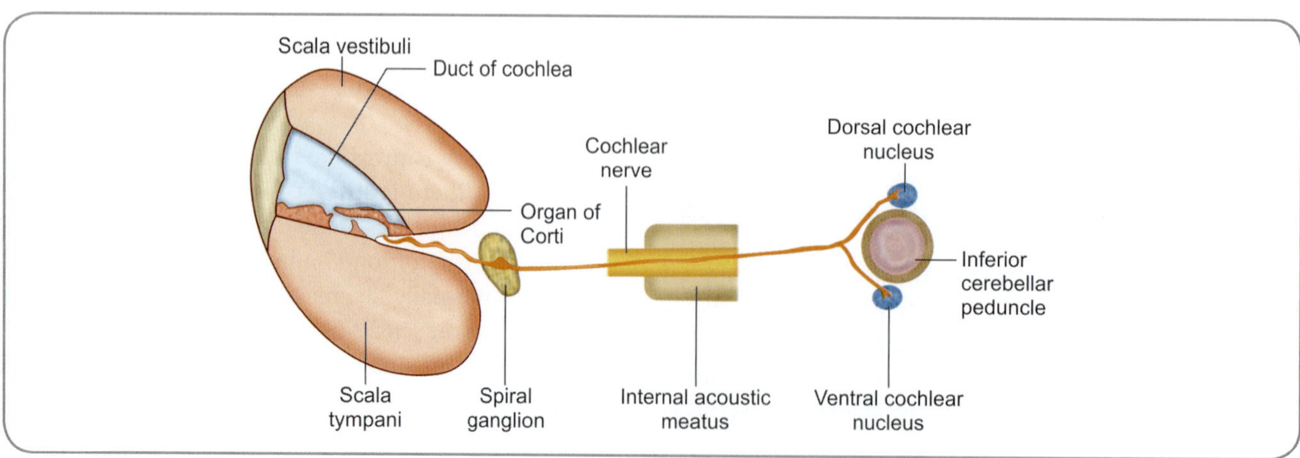

Fig. 15.54: Simplified scheme to show the course of cochlear nerve

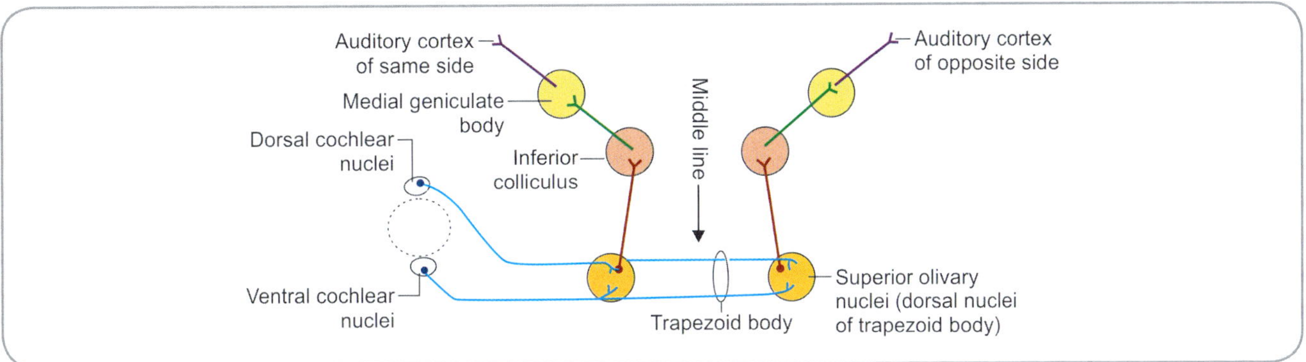

Fig. 15.55: Scheme to show the pathway from cochlear nuclei to the auditory cortex

The cochlear nerve (Fig. 15.54) is made up of the central processes of neurons located in the spiral ganglion (Corti's ganglion). The spiral ganglion is so called because it has a spiral configuration parallel to the turns of the cochlea. The neurons in this ganglion are bipolar (like those of the vestibular ganglion). Peripheral processes of the neurons of the spiral ganglion reach the organ of Corti, which is the peripheral receptor for sound. The fibres of the cochlear nerve enter the internal acoustic meatus. On reaching the brainstem they terminate in the dorsal and ventral cochlear nuclei located respectively on the dorsal and ventral sides of the (Fig. 15.55) inferior cerebellar peduncle. Sound is ultimately perceived in the auditory area of the cerebral cortex.

The fibres in both the ***vestibular and cochlear nerves*** belong to the ***special somatic afferent column***, as the epithelium of the membranous labyrinth is of ectodermal origin.

Clinical Correlation

Testing the Vestibulocochlear Nerve

This nerve is responsible for hearing (cochlear part) and for equilibrium (vestibular part). Normally, the cochlear part is tested.

Hearing of the patient can be tested by using a watch.

- ❑ First place the watch near one ear so that the patient knows what he is expected to hear.
- ❑ Next ask him to close his eyes and say so when he hears the ticking of the watch.
- ❑ The watch should be held away from the ear and then gradually brought towards it.
- ❑ The distance at which the sounds are first heard should be compared with the other ear.

In doing this test, it must be remembered that loss of hearing can occur from various causes such as the presence of wax in the ear, or middle ear disease.

Nerve deafness can be distinguished from deafness due to a conduction defect (as in middle ear disease) by noting the following:

contd...

Clinical Correlation *contd...*

- ❑ Sounds can be transmitted to the internal ear through air (normal way), and can also be transmitted through bone. Normally conduction through air is better than that through bone, but in defects of conduction, the sound is better heard through bone.
- ❑ Air conduction and bone conduction can be compared by using a tuning fork. Strike the tuning fork against an object so that it begins to vibrate producing sound. Place the tuning fork near the patient's ear and then immediately put the base of the tuning fork on the mastoid process. Ask the patient where he hears the sound better (this is called ***Rinne's test***).
- ❑ In another test, the base of a vibrating tuning fork is placed on the forehead. The sound is heard in both ears but is clearer in the ear with a conduction defect (this is **Weber's test**).

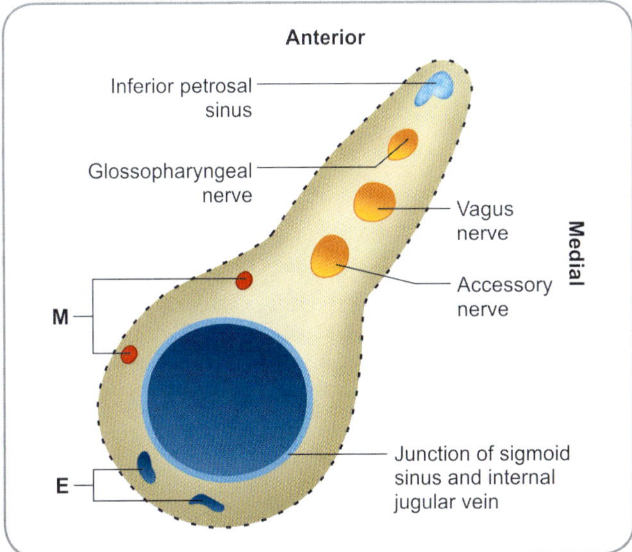

Fig. 15.56: Relative position of structures passing through the jugular foramen M: meningeal arteries E: emissary veins

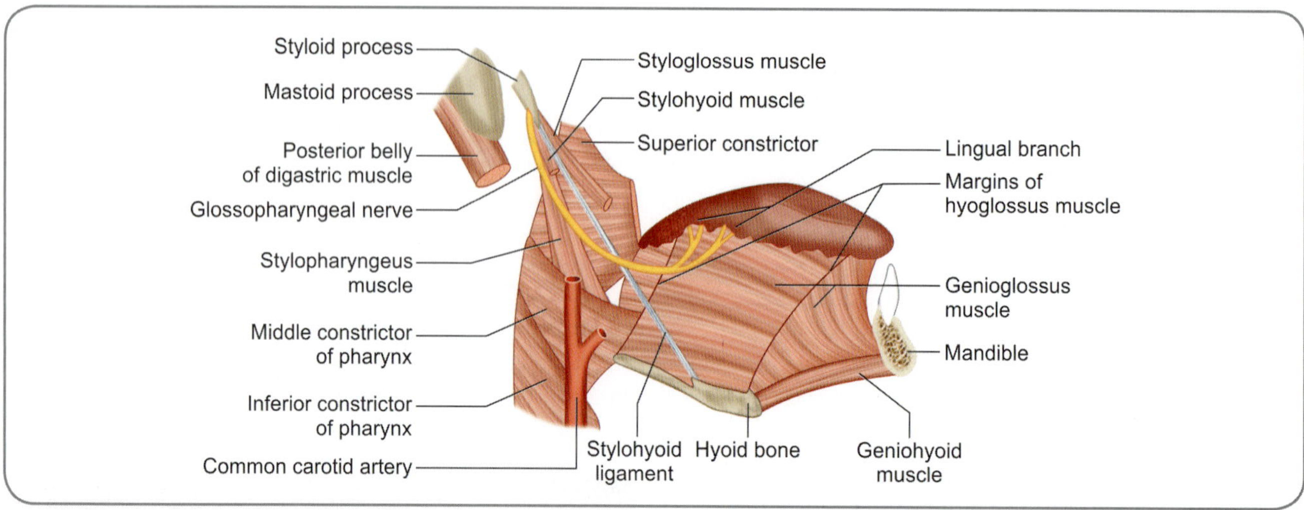

Fig. 15.57: Relations of glossopharyngeal nerve—structures deep to the hyoglossus are shown as if the muscle was transparent

Glossopharyngeal Nerve

Other names: Ninth cranial nerve, pharyngolingual nerve. This nerve is attached to the lateral side of the upper part of the medulla (between the olive and the inferior cerebellar peduncle) by three or four roots (Fig. 15.25). It runs forwards and laterally and leaves the cranial cavity by passing through the jugular foramen (Fig. 15.56). After emerging at the base of skull, the nerve passes forwards and laterally between the internal jugular vein and the internal carotid artery. It then descends in front of the internal carotid artery, passing deep to the styloid process and the structures attached to it. On reaching the posterior border of the stylopharyngeus muscle, it curves forwards passing lateral to the muscle (Fig. 15.57). The nerve then enters the pharynx by passing through the interval between the lower border of the superior constrictor of pharynx and the upper border of the middle constrictor and reaches

the side of tongue. Here it passes deep to the hyoglossus muscle, superficial to the genioglossus and terminates by dividing into branches to the tongue.

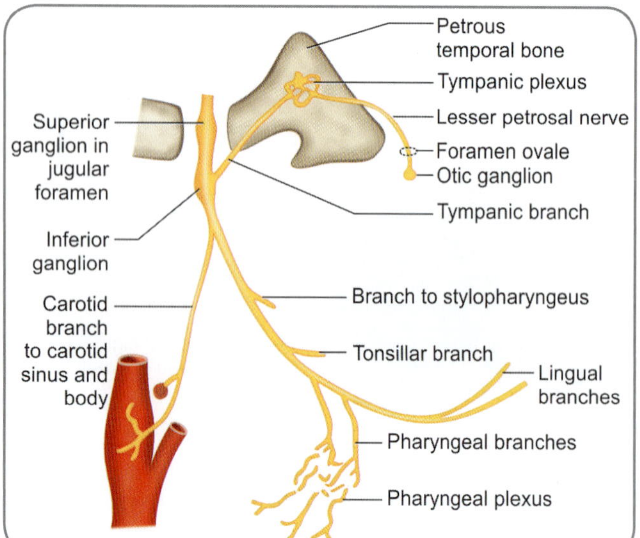

The proximal part of the glossopharyngeal nerve bears two ganglia, superior and inferior. The ***superior ganglion*** is small and lies within the jugular foramen. The ***inferior ganglion*** is larger and lies just below the foramen (Fig. 15.58).

Branches of Glossopharyngeal Nerve

- ***Tympanic branch*** arises from the inferior ganglion. It enters a canal (called the inferior tympanic canaliculus) within the petrous temporal bone and passes through the canaliculus to reach the tympanic cavity; it forms a plexus (***tympanic plexus***) over the promontory. Branches arising from this plexus supply the mucous membrane of the tympanic cavity, auditory tube and the mastoid air cells. The plexus also gives rise to the lesser petrosal nerve.
- ***Lesser petrosal nerve*** (lesser superficial petrosal nerve or the minor petrosal nerve) leaves the tympanic cavity through a canal that opens on the anterior surface of the petrous temporal bone (lateral to the greater petrosal nerve). It leaves the cranial cavity by passing through the foramen ovale (or through a small foramen, the canaliculus innominatus present lateral to the foramen ovale). The nerve ends by joining the otic ganglion.
- ***Carotid branch*** arises soon after the glossopharyngeal nerve emerges on the base of the skull. It supplies the carotid sinus and carotid body (forming a plexus with sympathetic fibres from the superior cervical ganglion, and from the vagus).
- ***Pharyngeal branches*** are given off to the mucous membrane of the pharynx. Their fibres pass through the pharyngeal plexus (which is also joined by sympathetic fibres and by fibres from the vagus).

Fig. 15.58: Branches of glossopharyngeal nerve

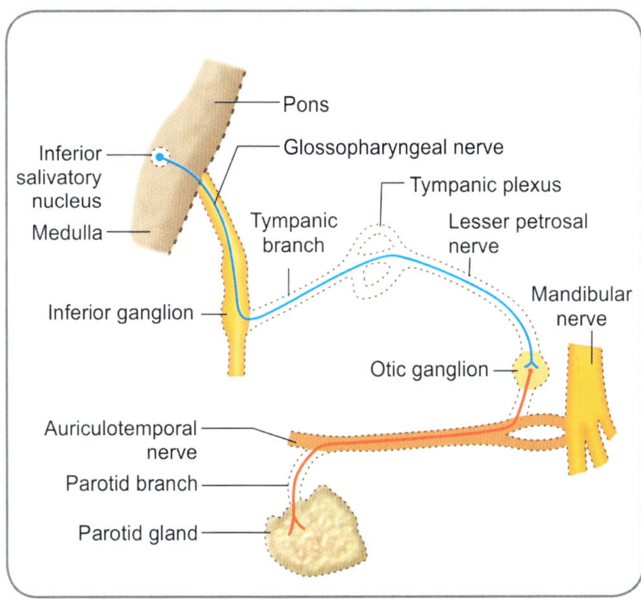

Fig. 15.59: Secretomotor pathway for the parotid gland

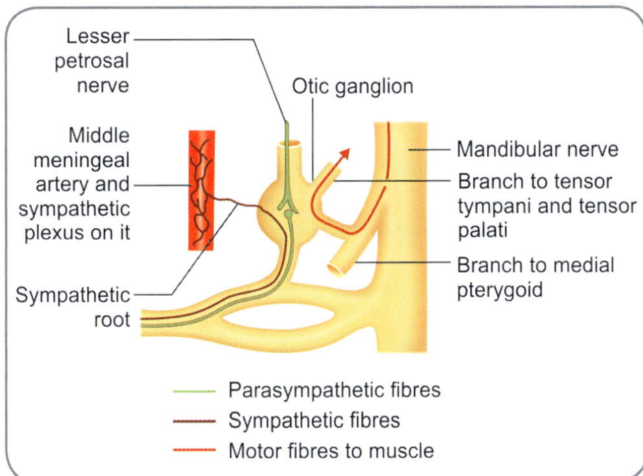

Fig. 15.60: Connections of otic ganglion

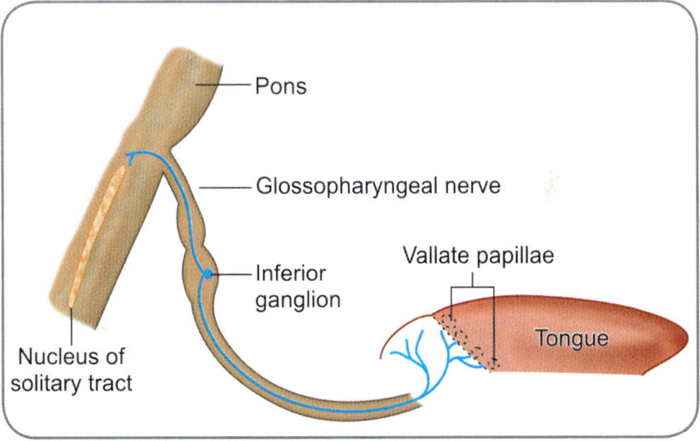

Fig. 15.61: Pathway for taste from the posterior one-third of tongue

❑ As the glossopharyngeal nerve winds around the ***stylopharyngeus*** it supplies this muscle. (This is the only motor branch of the nerve).

❑ ***Tonsillar branches*** supply the palatine tonsil and the soft palate.

❑ ***Lingual branches*** supply the part of the tongue (mucous membrane) behind the sulcus terminalis. They also supply the circumvallate papillae (Fig. 15.61). These branches carry fibres for both general sensation and taste.

Added Information

Functional Components of the Glossopharyngeal Nerve

Special visceral efferent fibres arising in the nucleus ambiguus (Figs 15.17, 15.18A and B) supply the stylopharyngeus muscle which develops from the mesoderm of the third branchial arch. The glossopharyngeal is the nerve of this arch.

General visceral efferent fibres supply the parotid gland (Fig. 15.59). The preganglionic neurons concerned are located in the inferior salivatory nucleus (Fig. 15.17) which lies at the junction of the pons and medulla just below the superior salivatory nucleus. These pass successively through, the proximal part of the glossopharyngeal nerve, then its tympanic branch, tympanic plexus and lesser petrosal nerve to end in the otic ganglion. Postganglionic fibres arising from neurons located in the otic ganglion (Fig. 15.60) pass through a nerve connecting the otic ganglion to the auriculotemporal nerve, and then through the auriculotemporal nerve itself. They leave the latter through its parotid branch to reach the parotid gland.

General visceral afferent fibres carry general visceral sensations. They are the peripheral processes of unipolar cells in the inferior ganglion of the nerve. They pass through the pharyngeal, tonsillar and lingual branches to supply the mucous membrane of the pharynx, posterior part of tongue, tonsil and soft palate. Central processes of the neurons concerned pass through the proximal part of the glossopharyngeal nerve into the brainstem. Here they end in the nucleus of the solitary tract (Figs 15.17, 15.18A and B).

Special visceral afferent fibres carry the sensation of taste from the part of the tongue behind the sulcus terminalis, and from the vallate papillae. The fibres are peripheral processes of pseudounipolar cells in the inferior ganglion of the glossopharyngeal nerve. They pass through the glossopharyngeal nerve and its lingual branches to reach the tongue. Central processes of the neurons concerned pass through the proximal part of the glossopharyngeal nerve and enter the medulla where they end in the upper part of the nucleus of the solitary tract.

According to some authorities, the glossopharyngeal nerve also contains some general somatic afferent fibres. These include some fibres from the skin of the auricle and proprioceptive fibres from the stylopharyngeus. They end in the spinal nucleus of the trigeminal nerve.

> ### ✂ Clinical Correlation
>
> **Testing the Glossopharyngeal Nerve**
>
> Testing of this nerve is based on the fact that it carries fibres of taste from the posterior one-third of the tongue and that it provides sensory innervation to the pharynx. Sensations of taste can be tested by applying substances which are salty (salt), sweet (sugar), sour (lemon) or bitter (quinine) to the posterior one-third of the tongue. The mouth should be rinsed and the tongue dried before the next substance is applied. Touching the pharyngeal mucosa causes reflex constriction of pharyngeal muscles. The glossopharyngeal nerve provides the afferent part of the pathway for this reflex.

Vagus Nerve

Other names: Tenth cranial nerve, wanderer's nerve, wandering nerve, pneumogastric nerve.

The vagus nerve is the longest cranial nerve and carries the main parasympathetic outflow to most of the viscera in the thorax and the abdomen. It arises from the lateral side of the medulla, by about ten rootlets which are attached to the groove between the olive and the inferior cerebellar peduncle, below the rootlets of the glossopharyngeal nerve (Fig. 15.25). Thus formed, the nerve leaves the skull through the jugular foramen. The part of the nerve within the jugular foramen shows an enlargement called the ***superior ganglion***. Just below the foramen the nerve has a much larger enlargement called the ***inferior ganglion***. The vagus nerve descends vertically in the neck and is enclosed within the carotid sheath. Here it lies in the interval between the posterior part of the internal or common carotid artery and the internal jugular vein. In the lower part of the neck the nerve crosses anterior to the first part of the subclavian artery and enters the thorax. It passes through the thorax and enters the abdomen, via the oesophageal opening of the diaphragm, where it has a wide distribution.

Branches of the Vagus Nerve in the Neck (Fig. 15.62)

The vagus nerve gives off numerous branches in the neck, in the thorax and in the abdomen. Branches arising in the thorax and the abdomen are considered in the appropriate sections. Branches which arise in the neck are considered here.

- ❑ ***Meningeal branch*** arises near the upper end of the nerve and supplies the dura mater in the region.
- ❑ ***Auricular branch*** arises from the superior ganglion. Soon after its origin the nerve enters the mastoid canaliculus (which opens on the lateral wall of the jugular fossa) and passes laterally through it within the temporal bone. The nerve emerges from the bone through the tympanomastoid fissure and divides into two branches. One of these supplies part of the skin of the auricle and the other branch supplies the skin lining

the posterior wall and floor of the external acoustic meatus, and the posteroinferior part of the outer layer of the tympanic membrane (The remaining parts of the meatus and of the membrane are supplied by the auriculotemporal nerve).

- ❑ ***Pharyngeal branch*** arises from the inferior ganglion and passes forwards on the side-wall of the pharynx crossing superficial to the internal carotid artery and deep to the external carotid artery. It divides into numerous branches which form the ***pharyngeal plexus***. Fibres continuing through the plexus supply the muscles of the pharynx and of the soft palate (except the tensor palati that is supplied by the mandibular nerve through the otic ganglion). The pharyngeal plexus is joined by branches from the glossopharyngeal nerve (sensory fibres) and by branches from the sympathetic trunk.
- ❑ One or more branches are given off to the carotid body. They may arise from the inferior ganglion or from the pharyngeal branch
- ❑ ***Superior laryngeal nerve*** arises from the inferior ganglion and descends on the side-wall of the pharynx posterior to the internal carotid artery. It then curves forwards passing deep to the artery and ends by dividing into the internal and external laryngeal nerves.
  - ○ The ***internal laryngeal nerve*** is sensory. It enters the larynx by piercing the thyrohyoid membrane.

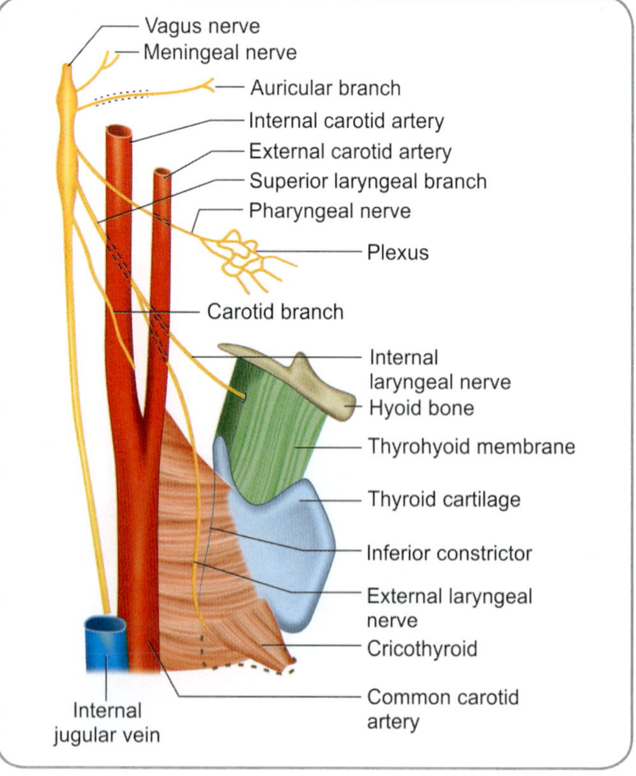

Fig. 15.62: Scheme to show branches of vagus nerve in the neck

It divides into branches which supply the mucous membrane of the upper half of larynx (up to the vocal folds), part of pharynx, epiglottis, vallecula, and posteriormost part of tongue.

- The *external laryngeal nerve* descends over the inferior constrictor muscle and supplies the cricothyroid muscle.

- *Recurrent laryngeal nerve:* The course of this nerve is different on the right and left sides.

  - The *right recurrent laryngeal nerve* arises from the right vagus as the latter passes in front of the subclavian artery. It passes backwards below the artery and then upwards behind the artery forming a loop. The nerve then runs upwards and medially deep to the common carotid artery to reach the side of the trachea.

  - The *left recurrent laryngeal nerve* arises from the left vagus in the thorax, as the latter crosses lateral to the arch of the aorta. The nerve winds below the arch, immediately behind the ligamentum arteriosum and then passes upwards and medially to reach the side of the trachea.

Having reached the trachea *both the right and left nerves* ascend in the groove between it and the oesophagus, deep to the medial surface of the thyroid gland. At the upper end of the trachea and oesophagus, the nerve passes deep to the lower border of the inferior constrictor muscle and enters the 7larynx. The nerve provides motor supply to all intrinsic muscles of larynx (except the cricothyroid supplied by the external laryngeal nerve) and sensory supply to the mucous membrane of the lower half of larynx, i.e. the part below the level of the vocal folds. It gives sensory branches to the trachea, oesophagus and inferior constrictor muscle. It also gives branches to the cardiac plexus. At the lower pole of the thyroid gland, the recurrent laryngeal nerve is intimately related to the terminal branches of the inferior thyroid artery. Variations in the relationship are of surgical importance.

- *Cardiac branches:* Each vagus nerve gives one (or more) superior cervical cardiac branch in the upper part of the neck and an inferior cervical cardiac branch in its lower part. Additional cardiac branches arise from the nerve in the superior mediastinum and also from the recurrent laryngeal branches. These branches end in the superficial and deep cardiac plexuses.

Added Information

Functional Components of the Vagus Nerve (Fig. 15.63)

The vagus nerve is composed predominantly of parasympathetic fibres, that is, *general visceral efferent.* These fibres are very widely distributed. The vagi are responsible for parasympathetic innervation of the thoracic viscera including the heart and bronchi and of the greater part of the gastrointestinal tract. Because of this fact, the term 'vagal' is synonymous with 'parasympathetic' while referring to these organs. The fibres in the vagus nerves are preganglionic. They arise from the dorsal nucleus of the vagus (Figs 15.17, 15.18E and F) and pass through the nerve and its ramifications to reach the viscera supplied. As a rule postganglionic neurons are located in

contd...

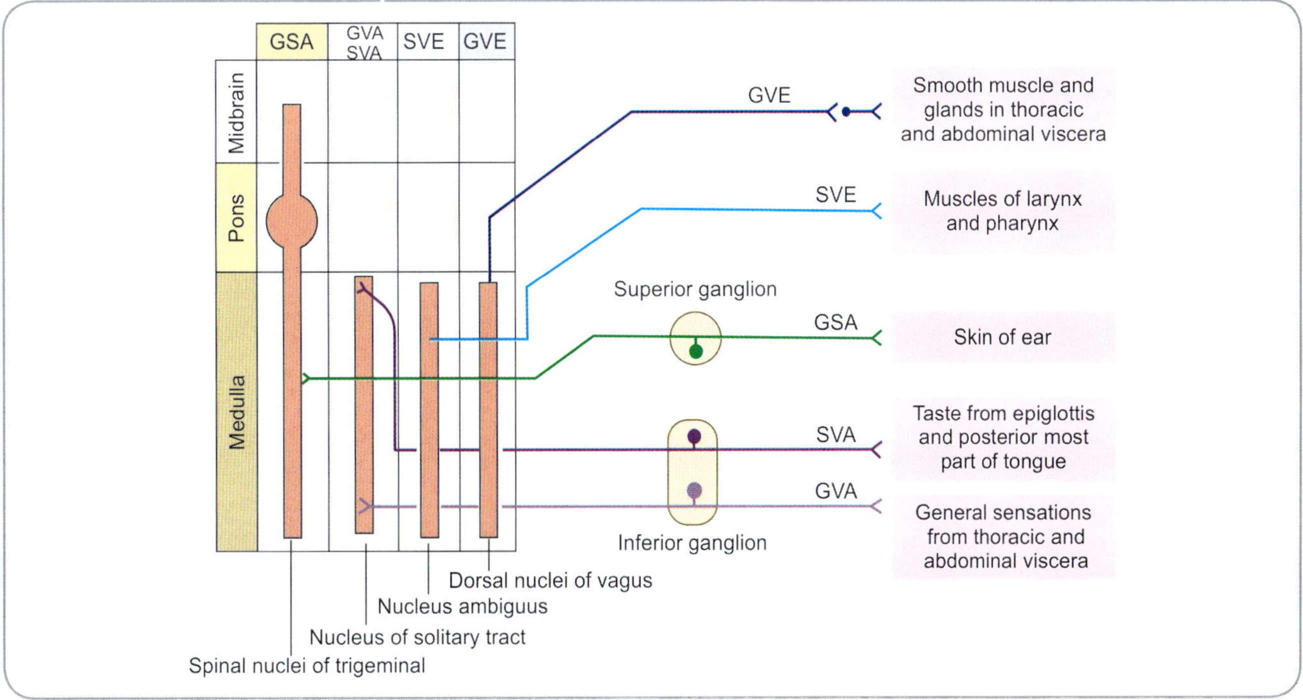

Fig. 15.63: Scheme to show the functional components of vagus nerve

Added Information *contd...*

plexuses situated close to the viscera, or in the walls of the viscera themselves. They innervate the smooth muscle and the glands present in the walls of the viscera.

Special visceral efferent fibres arise from the nucleus ambiguus (Figs 15.17, 15.18E and F). They pass through branches of vagus to innervate some of the musculature derived from the branchial arches. The superior laryngeal branch is the nerve of the fourth arch and the recurrent laryngeal branch that of the sixth arch. Through these branches (and through the pharyngeal branches) the vagus supplies muscles of the pharynx, soft palate and larynx.

The vagus contains numerous **general visceral afferent** fibres. The neurons concerned are pseudounipolar and are located in the inferior ganglion. The peripheral processes pass through the vagus and its branches to reach the pharynx, larynx, trachea, and oesophagus; and the thoracic and abdominal viscera. Central processes of these neurons end in the nucleus of the solitary tract.

The vagus carries the sensation of taste from the posteriormost part of the tongue and from the epiglottis. These fibres are **special visceral afferent.** The neurons concerned lie in the inferior ganglion of the nerve. Their peripheral processes pass through the superior laryngeal nerve to reach the tongue and epiglottis. The central processes end in the nucleus of the solitary tract (Figs 15.17, 15.18E and F).

Finally the vagus also contains **general somatic afferent** fibres which supply skin. The neurons concerned lie in the superior ganglion of the nerve. Peripheral processes pass through the auricular branch to reach the skin of the auricle.

Clinical Correlation

Vagus Nerve (and Cranial Part of Accessory Nerve)

This nerve has an extensive distribution but testing is based on its motor supply to the soft palate and to the larynx.

- ❑ Ask the patient to open the mouth wide and say 'aah'. Observe the movement of the soft palate. In a normal person the soft palate is elevated. When one vagus nerve is paralysed, the palate is pulled towards the normal side. When the nerve is paralysed on both sides, the soft palate does not move at all.
- ❑ In injury to the superior laryngeal nerve, the voice is weak due to paralysis of the cricothyroid muscle. At first there is hoarseness but after some time the opposite cricothyroid muscle compensates for the deficit and hoarseness disappears.
- ❑ Injury to the recurrent laryngeal nerve also leads to hoarseness, but this hoarseness is permanent.
  - ○ On examining the larynx through a laryngoscope it is seen that on the affected side the vocal fold does not move. It is fixed in a position midway between adduction and abduction.
  - ○ In cases where the recurrent laryngeal nerve is pressed upon by a tumour it is observed that nerve fibres which supply the abductors are lost first.
- ❑ In paralysis of both recurrent laryngeal nerves, voice is lost as both vocal folds are immobile.
- ❑ It may be remembered that the left recurrent laryngeal nerve runs part of its course in the thorax. It can be involved in bronchial or oesophageal carcinoma, or in secondary growths in mediastinal lymph nodes.

Accessory Nerve

Other names: Eleventh cranial nerve, spinal accessory nerve, accessorius willissii, nervus accessorius.

This nerve consists of two distinct parts, cranial and spinal. Both parts consist predominantly of efferent fibres. The fibres of the ***cranial part*** arise from the nucleus ambiguus. These fibres join the vagus nerve and are distributed through its pharyngeal and laryngeal branches to muscles of the pharynx, soft palate and larynx. The fibres of the spinal part arise from the lateral part of the ventral grey column of the upper five or six cervical segments of the spinal cord. They supply the sternocleidomastoid and trapezius muscles.

Cranial Part of Accessory Nerve

The cranial part of the nerve is attached, by four or five rootlets, to the side of the medulla in the groove between the olive and the inferior cerebellar peduncle. The rootlets are attached in line with, but below, those of the vagus nerve (Figs 15.25 and 15.64). From here the nerve runs laterally to reach the jugular foramen where it is joined by the spinal root. After passing through the jugular foramen the cranial root again separates from the spinal root and merges with the inferior ganglion of the vagus. The fibres of the cranial root of the accessory nerve pass into the pharyngeal and recurrent laryngeal branches of the vagus. They contribute to the innervation of the muscles of the pharynx and larynx. It is believed that fibres of the accessory nerve supply all the muscles of the soft palate (except the tensor palatini).

Spinal Part of Accessory Nerve

The spinal part of the accessory nerve is formed by the union of a number of rootlets which emerge from the

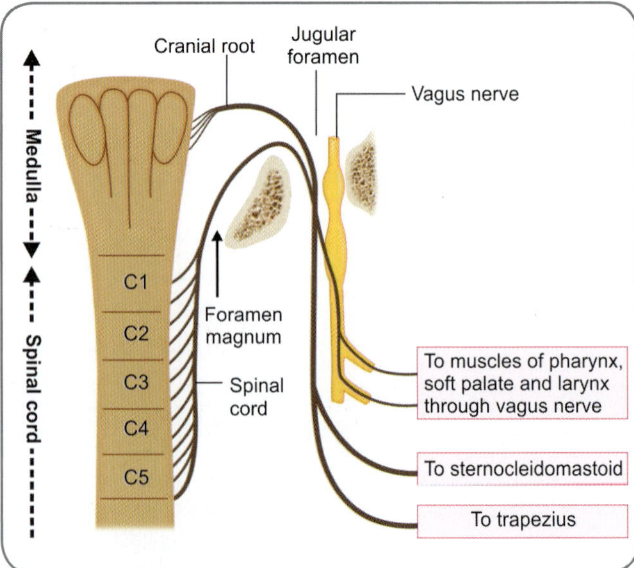

Fig. 15.64: Scheme to show the formation and distribution of accessory nerve

upper five or six cervical segments of the spinal cord. The rootlets emerge along a vertical line midway between the line of attachment of the ventral and dorsal roots of the spinal nerves and ascend lateral to the spinal cord. Forming a single cord, they enter the skull through the foramen magnum. Here, the cord (that is, the spinal root) lies behind the vertebral artery. The spinal root then runs upwards and laterally to reach the jugular foramen. The spinal root joins the cranial root within the foramen, but leaves it again on emerging from the foramen.

Course of Spinal Part in the Neck

In the neck the spinal accessory nerve first runs backwards and laterally to reach the transverse process of the atlas. It then runs downwards and backwards across the lateral side of the neck (Fig. 15.65).

In this part of its course, the nerve passes through the sternocleidomastoid. It enters the deep surface of the muscle and passing through it, emerges at its posterior border (near the middle). The nerve now runs downwards and backwards across the posterior triangle to reach the anterior margin of the trapezius about 5 cm above the clavicle. The terminal part of the nerve runs down the back deep to the trapezius. The spinal part of the accessory nerve supplies the sternocleidomastoid (as it passes through it) and the trapezius (by its terminal branches). These muscles also receive branches from the cervical plexus, but they are generally regarded as having only proprioceptive fibres.

Relations in the Neck

Between the jugular foramen and the transverse process of the atlas the nerve usually passes posterior to the internal jugular vein. In this part of its course, the nerve lies deep to the styloid process and the posterior belly of the digastric muscle. Over the transverse process of the atlas the nerve is crossed by the occipital artery (Fig. 15.65).While crossing the posterior triangle of the neck, the nerve lies on the levator scapulae (Fig. 15.65).

The ***functional component of the accessory nerve*** belongs to the ***special visceral efferent*** column as the muscles supplied are derived from branchial arches.

> ### Clinical Correlation
>
> The accessory nerve is tested as follows:
> - Put your hands on the right and left shoulders of the patient and ask him/her to elevate (shrug) the shoulders. In paralysis, this movement will be weak on the paralysed side (due to paralysis of the trapezius).
> - Ask the patient to turn his/her face to the opposite side (against resistance offered by your hand). In paralysis, the movement is weak on the affected side (due to paralysis of the sternocleidomastoid muscle).

Hypoglossal Nerve

Other names: Twelfth cranial nerve, nervus hypoglossi. The hypoglossal nerve is purely motor and supplies the muscles of the tongue. The neurons which give origin to the fibres of this nerve are located in the hypoglossal nucleus. The hypoglossal nerve emerges from the medulla by ten to fifteen rootlets which are attached in the vertical groove separating the pyramid from the olive (Fig. 15.25). (The rootlets are in line with those of the ventral root of the first cervical nerve). The hypoglossal nerve leaves the cranial cavity through the hypoglossal canal (or anterior condylar) canal. The distribution of the hypoglossal nerve is shown in Fig. 15.66.

In the initial part of its course, the nerve passes laterally behind the internal carotid artery, the glossopharyngeal nerve and the vagus. The nerve then winds around the

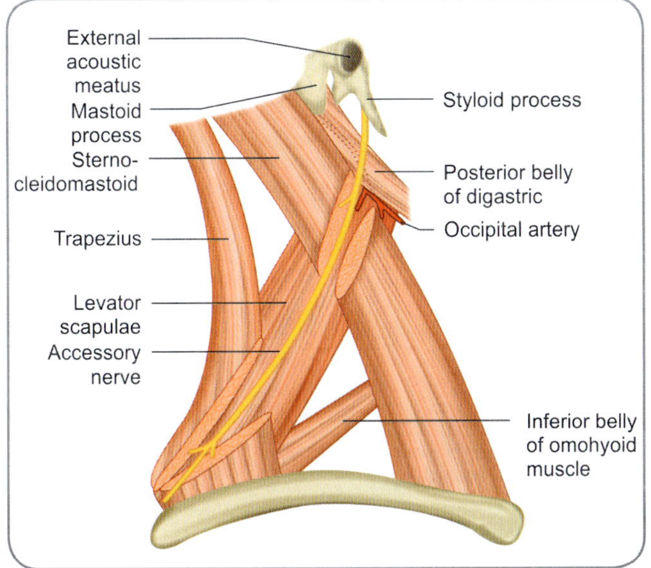

Fig. 15.65: Course and relations of accessory nerve

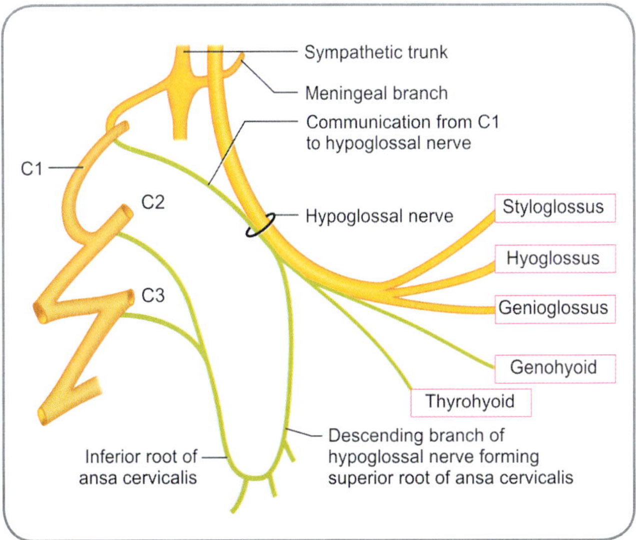

Fig. 15.66: Scheme to show the distribution of the hypoglossal nerve

lateral side of the inferior ganglion of the vagus to reach the front of the latter nerve. Just before the nerve turns forwards (near the angle of the mandible) it lies deep to the posterior belly of the digastric muscle. After emerging from under this muscle the nerve loops around the inferior sternocleidomastoid branch of the occipital artery (Fig. 15.67). As the nerve runs forward in the neck it crosses the internal carotid artery, external carotid artery and the loop formed by the lingual artery. The loop of the lingual artery is crossed just above the tip of the greater cornu of the hyoid bone. As the nerve runs forwards above the greater cornu of the hyoid bone, it is crossed by the digastric tendon and the stylohyoid. As it crosses the hyoglossus, lingual nerve, submandibular duct and deep part of the submandibular gland lie above it (Fig. 15.67). This part of the nerve is deep to the mylohyoid muscle.

On emerging at the base of skull, the nerve lies deep (medial) to the internal jugular vein and internal carotid artery. It passes downwards to reach the interval between these vessels, and then runs vertically between them, up to the level of the angle of the mandible (Fig. 15.67). Here the nerve passes forwards crossing the internal and external carotid arteries, and enters the submandibular region. In the submandibular region, the hypoglossal nerve at first lies superficial to the hyoglossus muscle and then to the genioglossus. It ends by dividing into its terminal branches. These supply all the intrinsic and extrinsic muscles of the tongue (except the palatoglossus that is supplied, along with other muscles of the palate, by the cranial accessory nerve).

Branches of Hypoglossal Nerve

The branches of the hypoglossal nerve may be divided into:
- Branches of the nerve proper and
- Branches that represent fibres which reach it from the first cervical nerve.

Branches of Hypoglossal Nerve Proper

The hypoglossal nerve itself supplies the muscles of the tongue (styloglossus, hyoglossus, genioglossus, and intrinsic muscles)—muscular branches.

Branches Carrying Fibres of Cervical Nerves

A **meningeal branch** arises from the nerve as it passes through the hypoglossal canal. It supplies the dura mater of the posterior cranial fossa. The fibres of this branch are probably derived from the upper cervical nerves and from the superior cervical sympathetic ganglion.

The hypoglossal nerve gives a **descending branch** that forms the superior root of the ansa cervicalis. Its fibres are derived from the first cervical nerve (Figs 15.11 and 15.66).

Branches from the hypoglossal nerve also supply the thyrohyoid and geniohyoid muscles (Fig. 15.66). Like the fibres of the descending branch, the fibres of these branches are also derived from the first cervical nerve.

Added Information

The fibres of the hypoglossal nerve belong to **somatic efferent column** because the muscles of the tongue develop from somites (occipital somites).

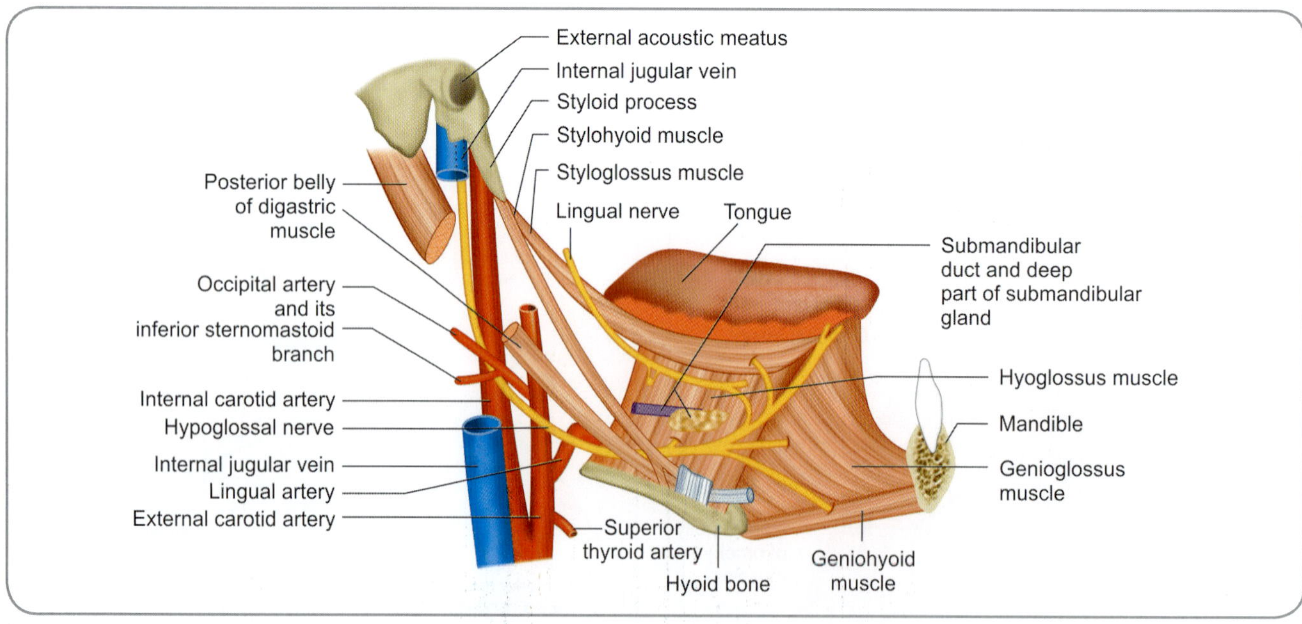

Fig. 15.67: Relations of hypoglossal nerve

Clinical Correlation

This nerve supplies muscles of the tongue. To test the nerve, ask the patient to protrude the tongue. In a normal person the protruded tongue lies in the midline. If the hypoglossal nerve is paralysed, the tongue deviates to the paralysed side. Protrusion of the tongue is produced by the pull of the right and left genioglossus muscles. The origins of the right and left genioglossus muscles lie anteriorly (on the hyoid bone), and the insertions lies posteriorly (onto the posterior part of the tongue). Each muscle draws the posterior part of the tongue forwards and medially. Normally, the medial pull of the two muscles cancel out, but when one muscle is paralysed, it is this medial pull of the intact muscle that causes the tongue to deviate to the opposite side.

Deviation of the tongue should be assessed with reference to the incisor teeth, and not to the lips. In facial paralysis the tongue may protrude normally, but may appear to deviate to one side because of the asymmetry of the mouth.

CERVICAL PART OF SYMPATHETIC TRUNK

The cervical part of the sympathetic trunk bears three ganglia—superior, middle and cervicothoracic.

- The **superior ganglion** represents fused ganglia C1 to C4, and lies in front of the transverse processes of vertebrae C2 and C3.
- The **middle ganglion** represents ganglia C5 and C6 and lies in front of C6.
- The **cervicothoracic ganglion** represents ganglia C7, C8 and T1 and lies between the transverse process of C7 and the neck of the first rib. This ganglion has numerous branches that give it a star-like appearance and so it is also called the **stellate ganglion**.

Branches of Superior Cervical Ganglion

The branches of the superior cervical sympathetic ganglion are as follows (Fig. 15.69).

- The **internal carotid nerve** arises from the upper pole of the ganglion. It is composed mainly of postganglionic fibres arising from the superior cervical ganglion. The nerve ascends along the internal carotid artery and divides into branches that form a plexus over it. This plexus has numerous connections the more important of which are as follows (Fig. 15.68).
  - In the carotid canal, the plexus gives off the **caroticotympanic nerves** which enter the middle ear and join the tympanic plexus.
  - In the foramen lacerum, the internal carotid plexus gives off the **deep petrosal nerve** which joins the greater petrosal nerve to form the nerve of the pterygoid canal through which the fibres reach the pterygopalatine ganglion. They pass through this ganglion, without relay, and travel through its orbital branches to supply the orbitalis muscle.
  - While the internal carotid nerve lies in the cavernous sinus, it communicates with the ophthalmic division of the trigeminal nerve. These fibres pass into its nasociliary branch, and then into the long ciliary nerves to reach the eyeball where they supply the dilator pupillae muscle, and the blood vessels of the eyeball. Some fibres pass from the nasociliary nerve, or directly from the plexus around the internal carotid artery, to the ciliary ganglion. These fibres pass through this ganglion without relay and then pass into the short ciliary nerves to supply blood vessels of the eyeball. The fibres to the dilator

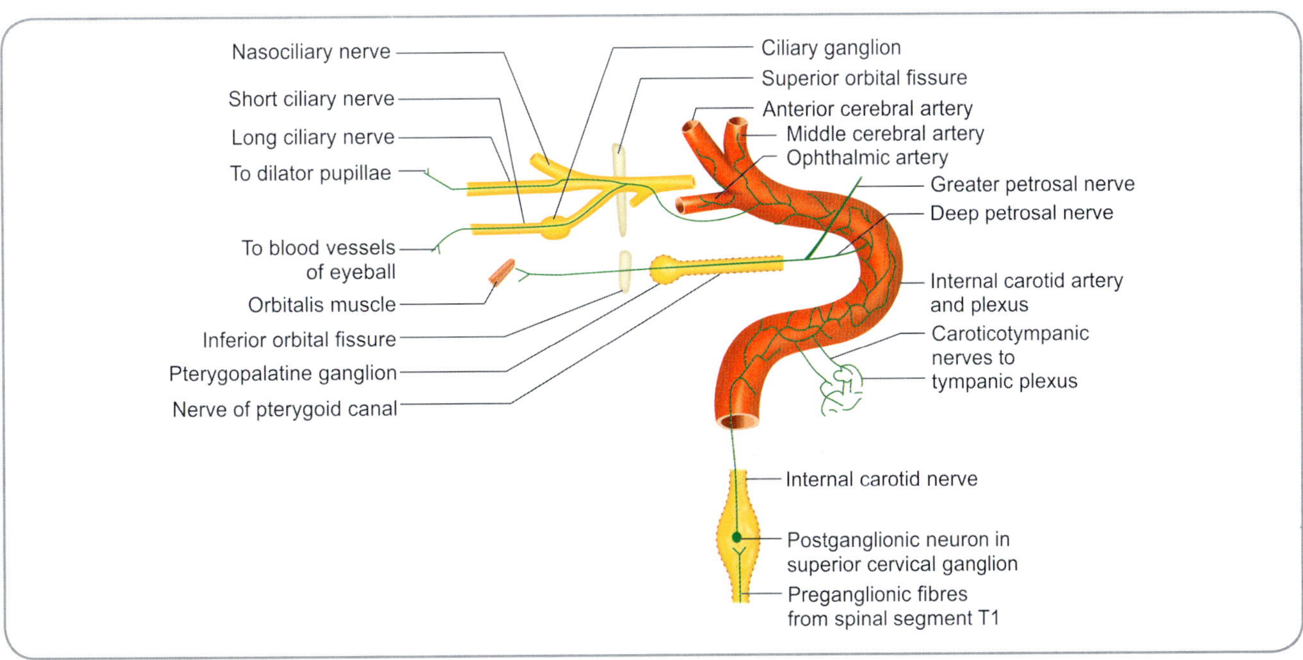

Fig. 15.68: Distribution of internal carotid nerve

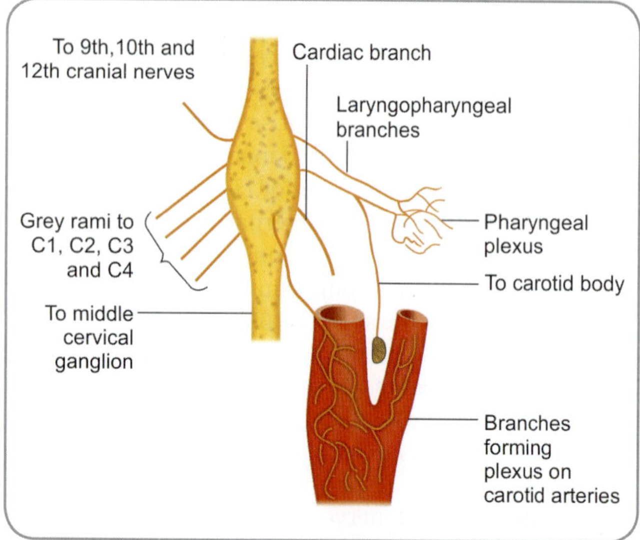

Fig. 15.69: Branches of superior cervical sympathetic ganglion

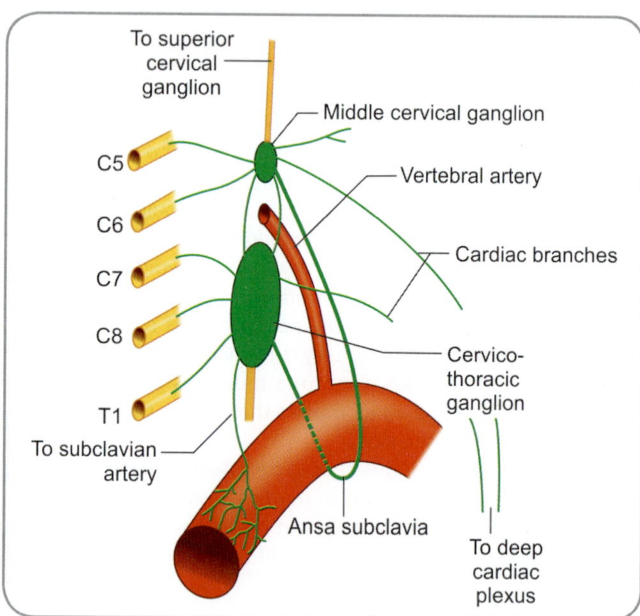

Fig. 15.70: Branches of middle cervical sympathetic and cervicothoracic ganglia

pupillae sometimes follow the second route. It may be noted here that preganglionic sympathetic fibres for the eyeball begin in segment T1 of the spinal cord and ascend in the sympathetic trunk to the superior cervical ganglion in which the postganglionic neurons lie.

❏ The superior cervical sympathetic ganglion gives grey rami to spinal nerves C1 to C4. It also communicates with the glossopharyngeal, vagus and hypoglossal nerves.

❏ *Laryngopharyngeal branches* arising from the ganglion join the pharyngeal plexus. They also supply the carotid body.

❏ The *cardiac branch* descends into the thorax along the common carotid artery. On the left side the nerve runs across the lateral side of the arch of the aorta and ends in the superficial cardiac plexus. On the right side it joins the deep cardiac plexus.

❏ Branches are also given off from the ganglion to the common carotid and external carotid arteries and form plexuses over them. These plexuses are seen on the external carotid artery, on the middle meningeal artery (gives a branch to the otic ganglion) and on the facial artery (to the submandibular ganglion and sweat glands of the face).

Branches of Middle Cervical Ganglion

The branches of the middle cervical ganglion and its connections are (Fig. 15.70):

❏ The ganglion is connected superiorly to the superior cervical ganglion.

❏ It is connected inferiorly to the cervicothoracic ganglion by an anterior and two posterior cords, which enclose the vertebral artery. The anterior cord passes down across the subclavian artery, forms a loop below it and then passes up behind the artery to join the cervicothoracic ganglion. The loop formed is called the *ansa subclavia*.

❏ The middle cervical ganglion gives grey rami to nerves C5 and C6.

❏ A cardiac branch descends into the thorax and ends in the deep cardiac plexus.

❏ Branches are also given off to the thyroid and parathyroid glands. They pass along the inferior thyroid artery.

Branches of Cervicothoracic Ganglion

The branches of the cervicothoracic ganglion (or stellate ganglion) are (Fig. 15.70):

❏ There are connections of the ganglion to the middle cervical ganglion, including that through the ansa subclavia.

❏ The ganglion gives grey rami to nerves C7, C8 and T1.

❏ A cardiac branch arising from this ganglion ends in the deep cardiac plexus.

❏ The cervicothoracic ganglion gives branches that form a plexus on the subclavian artery. This plexus extends onto the axillary artery and the branches of the subclavian artery including the vertebral and inferior thyroid arteries. The plexus on the vertebral artery extends along the artery into the cranial cavity and communicates with the internal carotid plexus. The branches from the vertebral plexus also pass to the cervical nerves.

Sympathetic Innervation of Upper Limb

The cervicothoracic ganglion is intimately concerned with the sympathetic innervation of the upper limb. The preganglionic neurons concerned lie in spinal segments T2 to T6 and emerge through the corresponding spinal nerves. These fibres ascend in the sympathetic trunk to reach the cervicothoracic ganglion in which they relay. Postganglionic fibres starting in the ganglion reach the blood vessels and sweat glands of the upper limb through the brachial plexus and its branches (mainly through nerves C8 and T1). The plexus on the subclavian artery was at one time considered to be the main pathway for sympathetic innervation of the upper limb, but it is now known that the fibres in the plexus do not extend much beyond the first part of the axillary artery.

Clinical Correlation

Horner's Syndrome

Interruption of sympathetic supply to the head and neck results in ***Horner's syndrome.*** The features of this syndrome are as follows:

- There is ***constriction of the pupil*** because of paralysis of the dilator pupillae. Unopposed action of the sphincter pupillae leads to ***constriction of the pupil***.
- There is drooping of the upper eyelid (***ptosis***) because of paralysis of smooth muscle fibres present in the levator palpebrae superioris.
- The eyeball is less prominent than normal (***enophthalmos***).
- There is *absence of sweating* on the affected side of the face (secretomotor supply to sweat glands is through sympathetic nerves).

Multiple Choice Questions

1. Which of the following is a ganglion of the craniosacral outflow tract?
   a. Gasserian ganglion
   b. Geniculate ganglion
   c. Ciliary ganglion
   d. Stellate ganglion
2. Right optic tract contains fibres from:
   a. Right halves of both retinae
   b. Left halves of both retinae
   c. Right half of right retina and left half of left retina
   d. Right half of left retina and left half of right retina
3. Fibres arising from Edinger-Westphal nucleus relay in the:
   a. Ciliary ganglion
   b. Coeliac ganglion
   c. Pterygopalatine ganglion
   d. Superior cervical ganglion
4. The lingual nerve has triple relations with the:
   a. Lingual artery
   b. Chorda tympani nerve
   c. Submandibular duct
   d. Submandibular ganglion
5. What is false about vagus?
   a. Its superior ganglion is located in the jugular foramen
   b. Its auricular branch arises from the inferior ganglion
   c. Its pharyngeal branch arises from the inferior ganglion
   d. It enters the abdomen through the oesophageal foramen

ANSWERS

1. c   **2.** a   **3.** a   **4.** c   **5.** b

Clinical Problem-solving

Case Study 1: A 37-year-old man presents with the following: wide palpebral fissure on the left side, smoothening of the forehead (no wrinkles) and a generalized asymmetry of the face. He complains that while eating he has difficulty because his food seems to be getting caught between his cheek and teeth. On examination, it is noticed that his conjunctival reflex is lost.
- What is your probable diagnosis about the man's condition?
- What other features would you test for in this man, to ascertain the level of lesion?
- Can you substantiate for the various features you see?

Case Study 2: A 42-year-old man presents with the following: left lateral squint, ptosis of left upper eyelid, fixed and dilated pupil on the left side and loss of light and accommodation reflexes on the left side.
- Can you state as to which nerve is affected?
- Analyse each of the presenting features and give reason for the same.

(For solutions see Appendix).

Chapter 16

Cross-Sectional, Radiological and Surface Anatomy of Head and Neck

Frequently Asked Questions

- Write notes on: (a) Reid's base line, (b) Lambda, (c) Suprameatal triangle, (d) Pterion.
- Draw a neat labelled diagram of the cross-section of neck at the level of C7 vertebra.
- Give details of the surface anatomy of the following structures: (a) Common carotid artery, (b) Parotid duct, (c) Thyroid gland, (d) Internal jugular vein, (e) Superior sagittal sinus, (f) Glossopharyngeal nerve.

CROSS-SECTIONAL ANATOMY OF HEAD

Transverse sections of head predominantly show sections of brain. The orbital and nasal cavities and the hypophyseal fossa are well seen in these sections.

TRANSVERSE SECTION AT THE LEVEL OF THE INTERNAL CRANIAL BASE (FIG. 16.1)

This section shows the features of the cranial fossae. The anterior, middle and posterior cranial fossae are well demarcated. The frontal air sinus is seen within the frontal bone in the anterior rim of the section.

The following are seen from before backwards in the median region:

- Crista galli
- Cribriform plate
- Optic canal and its contents
- Hypophyseal fossa
- Dorsum sellae
- Clivus
- Foramen magnum
- Internal occipital crest.

The orbital plates of the frontal bone, the lesser wings of sphenoid (alae minores), the greater wings of sphenoid

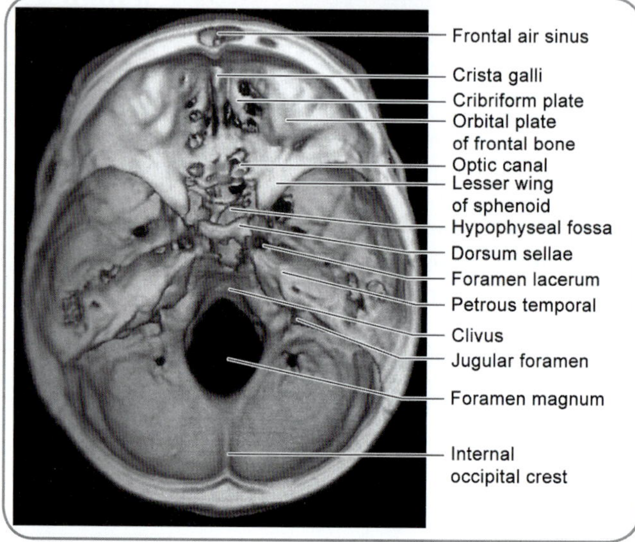

Fig. 16.1: Transverse section at the level of the internal cranial base

(alae majores), the foraminal crescents of middle cranial fossae, the petrous temporal, jugular foramina and the cerebellar fossae are seen bilaterally.

TRANSVERSE SECTION 2.5 CM ABOVE THE REID'S BASE LINE (FIG. 16.2)

Major part of this section is filled with the cerebrum. The eyeballs are seen anteriorly. The nasal septum is seen in the midline and the ethmoidal air cells on either side. Immediately behind and spanning across the right and left sets of ethmoidal cells is the sphenoidal air sinus. The zygomatic arch and the temporal fossa are seen posterolateral to the eyeball.

The cranial cavity is filled with the temporal lobes of cerebral hemispheres. Immediately behind the sphenoidal air sinus, the internal carotid arteries, infundibulum

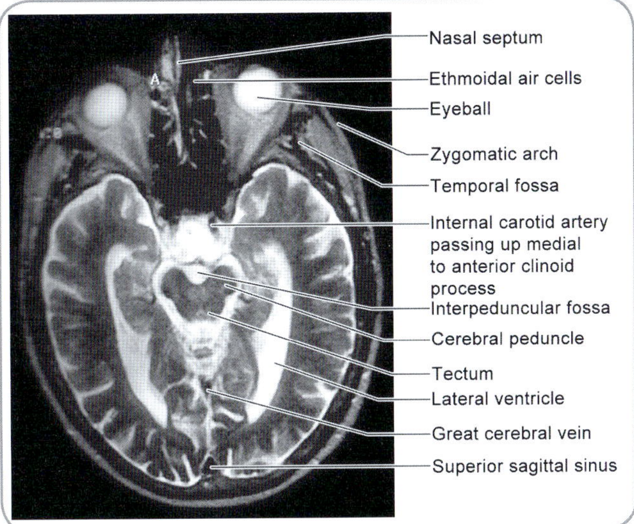

Fig. 16.2: Transverse section 2.5 cm above the Reid's Base Line

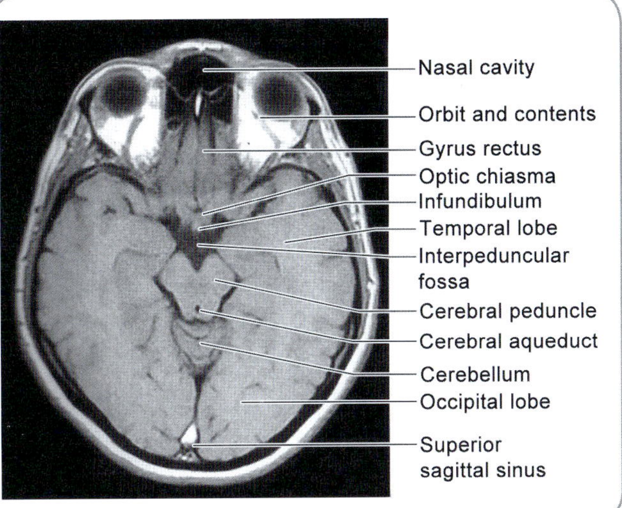

Fig. 16.3: Transverse section at the supraorbital-TEA (top of ear auricle) line

and dorsum sellae can be made out. The two cerebral peduncles and the interpeduncular area between the two peduncles are clearly seen. The temporal horns of the lateral ventricles and the cistern of great cerebral vein are also noted. The posterior curve of the superior sagittal sinus is usually seen in the midline between the posterior ends of the cerebral hemispheres.

TRANSVERSE SECTION AT THE SUPRAORBITAL-TEA (TOP OF EAR AURICLE) LINE (FIG. 16.3)

Since this section scrapes below the orbital plates of the frontal bone, the eyeballs and orbital cavities are seen. A small part of the nasal cavity is seen anteriorly. Gyri recti of the frontal lobe are made out. Optic chiasma, infundibulum, interpeduncular fossa, cerebral peduncles, cerebral aqueduct and cerebellum are seen one behind the other in the median region. The temporal and occipital lobes of the cerebrum fill most of the cranial cavity.

TRANSVERSE SECTION AT MIDLEVEL OF FOREHEAD (AND PARALLEL TO THE SUPRAORBITAL-TEA LINE) (FIG. 16.4)

The cranial cavity is entirely filled with cerebral hemispheres. The frontal, parietal and occipital lobes are seen. The longitudinal cerebral fissure and the falx cerebri are noted. Both the lateral ventricles are seen on either side of the midline and are bordered by the genu of the corpus callosum anteriorly. The septum pellucidum lies in the midline between the ventricles.

TRANSVERSE SECTION AT THE LEVEL OF UPPER FOREHEAD (FIG. 16.5)

This section cuts through the upper part of the cranial cavity and is filled with the cerebral hemispheres

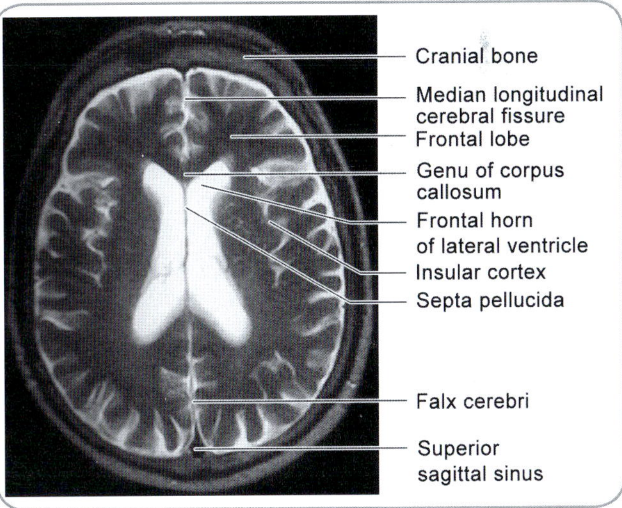

Fig. 16.4: Transverse section at midlevel of forehead (and parallel to the supraorbital-TEA line)

completely. Sulci and gyri and the longitudinal cerebral fissure are noted.

Axial tomographic sections of the head and the brain are analysed as part of diagnosis and treatment in patients. Computed tomographic images are taken with reference to some of the transverse planes and lines. The commonly used reference lines are the orbitomeatal line and the Reid's base line. When the orbital cavities and the interpeduncular area need to be visualized, sections are taken at planes parallel to these lines. When only the cranial cavity needs to be seen, the supraorbital-TEA line is used and sections are taken parallel to this line. When the posterior cranial fossa needs to be seen, sectional images below and parallel to the Reid's line and orbitomeatal line are taken.

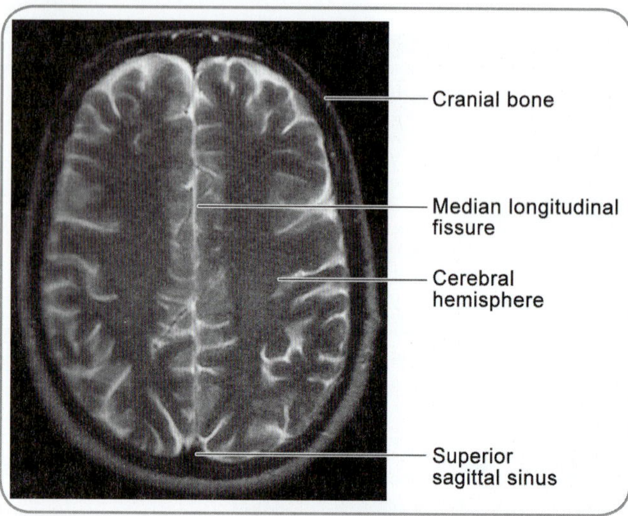

Fig. 16.5: Transverse section at the level of upper forehead

TRANSVERSE SECTIONS AT THE INFERIOR LEVELS

- Cerebellar hemispheres (posteriorly)
- Medulla or pons (depending on the level of section) in the median region just anterior to the cerebellar hemispheres
- Fourth ventricle between the cerebellum and the medulla/pons
- Parts of temporal lobe of cerebrum, and
- Parts of pituitary.

The jugular foramen with the nerves passing through, the internal carotid artery, the basilar artery with its branches, the foramina of fourth ventricle, the nerves emerging at the pontomedullary area can all be seen.

TRANSVERSE SECTIONS AT THE SUPERIOR LEVELS (SUPRATENTORIAL LEVELS)

- Lobes of cerebrum and cerebral cortical areas
- Midbrain in the median area
- Parts and branches of the internal carotid, middle and posterior cerebral arteries
- Parts of optic nerves and optic tracts
- Parts of the orbital contents and extraocular muscles
- Nasal cavities and nasal septum.

Depending upon the level of the section, cranial nerves, paranasal sinuses and sulci-gyri of the cerebrum can also be seen.

CROSS-SECTIONAL ANATOMY OF NECK (FIG. 16.6)

Transverse sections of the neck show the important viscera, vertebral column and muscles.

TRANSVERSE SECTION AT THE LEVEL OF C7 VERTEBRA

The bulk of paravertebral muscles is seen at this level. The section shows trachea and oesophagus one behind the other in the anterior portion. The thyroid gland is seen surrounding the trachea in front and sides and extending at the sides to overlap the tracheo-oesophageal groove and the oesophagus. The cross sections of the strap muscles

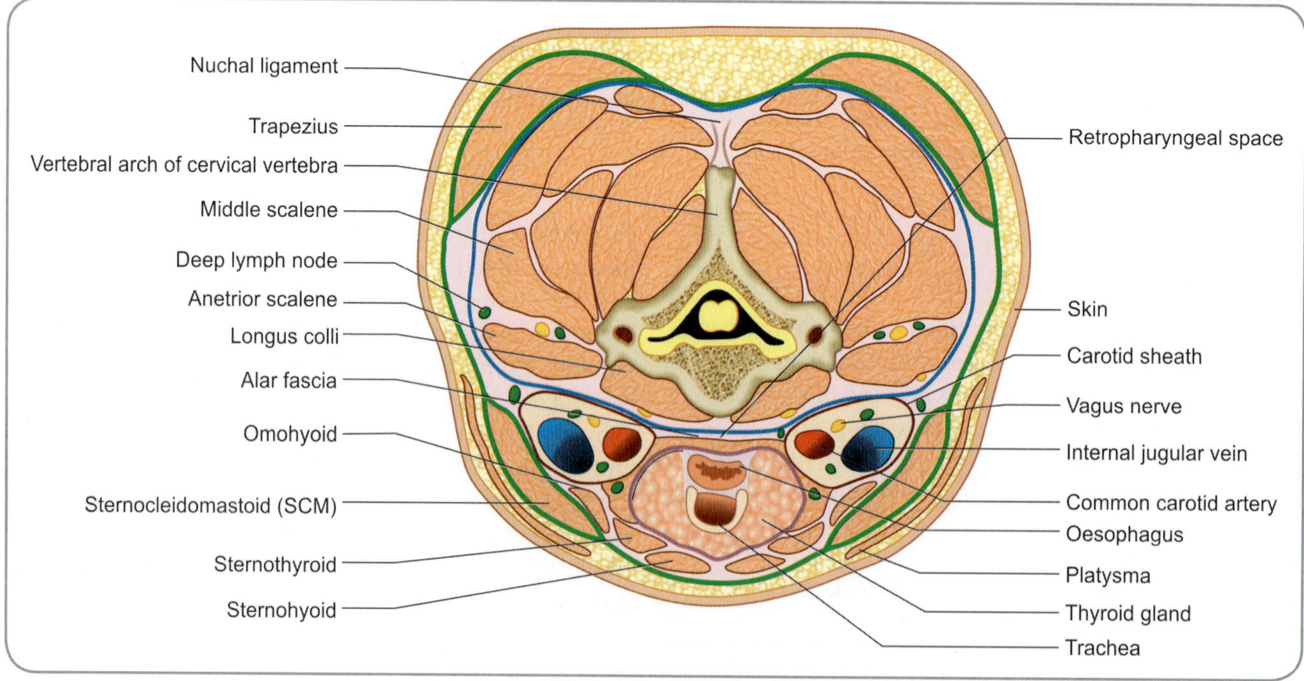

Fig. 16.6: Cross-sectional anatomy of neck

are seen on the anterior and anterolateral aspects of the thyroid gland. Posterolateral to the thyroid gland, on either side, are the carotid sheaths with their contents. The common carotid artery and the internal jugular vein with the vagus nerve posteriorly in the groove between them can be clearly noted within the carotid sheath. The seventh cervical vertebra (at which level this section is taken) occupies the central part of the section prominently. The posterior vertebral and paravertebral muscles are seen around the vertebra and its processes. The nuchal ligament can be seen posterior to the spine of the vertebra in the midline. All the structures of the neck can be seen to be surrounded by the investing layer of deep cervical fascia. The pretracheal and prevertebral fasciae with their fascial ramifications can be made out.

TRANSVERSE SECTION AT THE LEVEL OF THYROID CARTILAGE

The vertebra and the muscles around it form the main bulk of this section. The laryngeal vestibule is seen in the anterior median area. The carotid sheath and its neurovascular structures are seen posterolateral to the larynx. Laryngopharynx is seen posterior to the larynx. Facial vein, external jugular vein and internal jugular vein can be made out.

RADIOLOGICAL ANATOMY OF HEAD AND NECK

Radiographs of the head and neck region are commonly studied in lateral and anteroposterior views. Lateral view is generally preferred.

LATERAL VIEW OF SKULL (FIG. 16.7)

The general contour of the neuro and viscerocrania can be well made out. The cranial cavity is seen. The floor of the anterior cranial fossa is seen as a dense white line sloping down backwards from the anterior aspect. The frontal sinus can be located in front of the anterior end of this line and is seen as a radiolucent triangle. The hypophyseal fossa and the dorsum sellae can be clearly seen around the centre of the radiograph. The orbital cavity, maxillary antrum, external acoustic meatus, mastoid air cells and hard palate can all be seen. The internal acoustic meatus can be noted by the trained eye or with extra care. Some of the suture lines may be seen as wavy and corrugated markings.

While examining an X-ray of the skull, the general proportion of structures within the cranial cavity should be noted and areas of abnormal radiolucency or radiopacity should be identified. Erosion of bone, flattening of surfaces which usually are rounded or corrugated and bone rarefaction should be searched for.

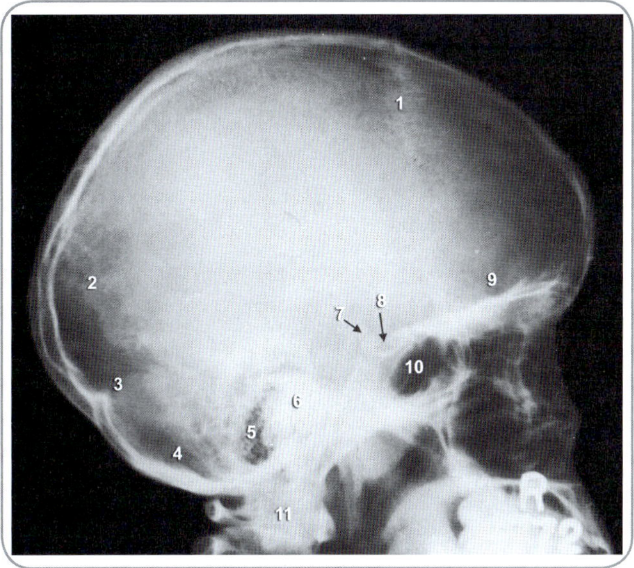

Fig. 16.7: Radiograph of the skull—lateral view 1. Coronal suture 2. Lambdoid suture 3. Internal occipital protuberance 4. Floor of posterior cranial fossa 5. Mastoid air cells 6. Petrous temporal bone 7. Dorsum sellae 8. Hypophyseal fossa 9. Floor of anterior cranial fossa 10. Sphenoidal sinus 11. Cervical vertebrae

It is possible to mark the location of some intracranial structures with the structures seen in a lateral view. The centrally located hypophyseal fossa should be studied. Its upper margin indicates the lower border of the telencephalon and the diencephalon. The hypothalamus is located 1 cm above the sellar floor. The pineal body is located 1 cm posterior and 3 cm superior to the external acoustic meatus.

ANTEROPOSTERIOR VIEW OF SKULL (FIG. 16.8)

This view gives less information than the lateral view. The frontal sinuses, orbital cavities, nasal cavity, nasal septum, maxillary sinuses and alveolar margins can be made out clearly. Symmetry of structures on both sides can be studied. The greater and lesser wings of the sphenoid can also be noted.

Different views of anteroposterior radiograph of the skull are taken and studied depending on the structure that needs to be specifically visualized. Towne's view is the commonly taken anteroposterior radiograph of the skull; the foramen magnum area with the middle-posterior cranial fossa border structures (dorsum sellae, posterior clinoid processes, petrous temporal) superimposed within the foramen magnum are studied. Caldwell's view (occipitofrontal view) is taken for visualizing the frontal sinuses better; opacities occur and frontal scalloping is lost in frontal sinusitis; all the paranasal air sinuses can be seen in this view; upper part of the face and anterior cranial fossa structures are seen well. Water's view

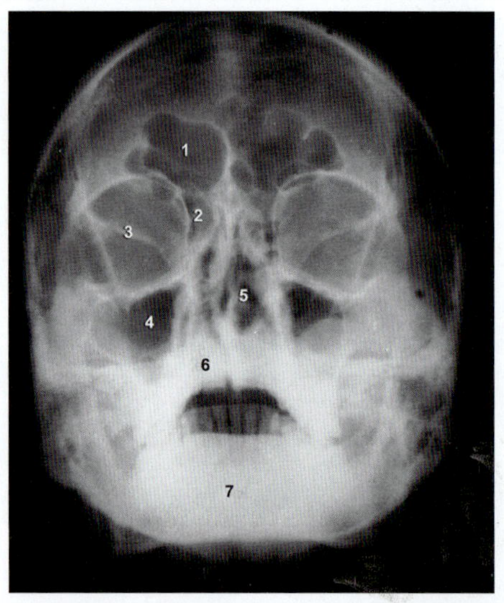

Fig. 16.8: Radiograph of the skull to show paranasal sinuses (PNS—Water's view) **1.** Frontal sinus **2.** Ethmoidal sinuses **3.** Orbit **4.** Maxillary sinus **5.** Nasal cavity **6.** Maxilla **7.** Mandible

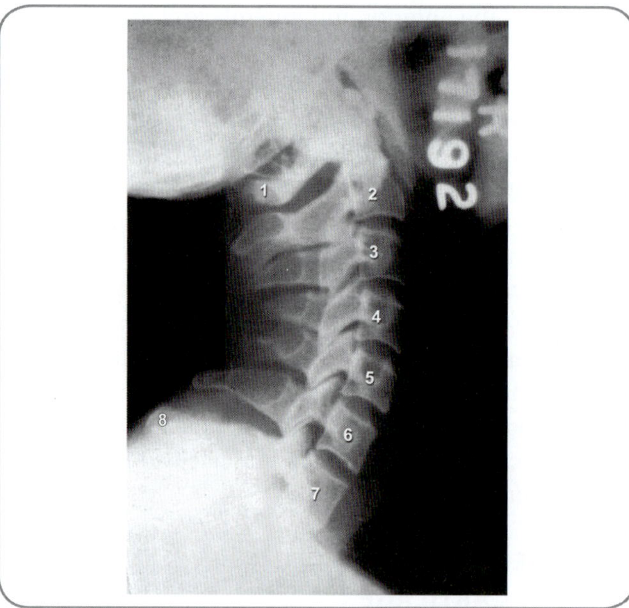

Fig. 16.9: Radiograph to show the cervical spine (lateral view) **1.** Posterior arch of atlas **2.** to **7.** Bodies of second to seventh cervical vertebrae **8.** Tip of spine of C7

(occipitomental view) is taken for visualizing the maxillary sinus in particular; Water's view taken with the mouth open gives better visualization of the sphenoidal sinuses.

RADIOGRAPHS OF NECK

The neck is radiographed for visualising the cervical vertebral column. Lateral or anteroposterior views can be taken. Since the cervical vertebral column is prominently seen, a neck X-ray is usually referred to as 'X-ray of cervical spine'. Fractures, straightening and degenerative changes of the cervical spine (spondylosis), joint dislocations, tumours or cysts of bone, thinning of bone (osteoporosis) and abnormal growth of bone tissue (bone spurs) can be made out.

Lateral View (Figs 16.9 and 16.10)

A lateral view radiograph of the cervical spine shows the bodies and spines of cervical vertebrae, intervertebral discs and the joints associated with the spine. The shape and height of the cervical vertebral bodies, height of the intervertebral discs and curvature of the cervical spine can be well studied in this view. Soft tissue structures like the larynx, trachea and oesophagus can also be seen. Air inside these structures makes them appear radiolucent. The anterior line of vertebral bodies, posterior line of vertebral bodies, line of vertebral spines should be studied. The prevertebral tissue should be specifically checked. Normally, the prevertebral tissue reduces in thickness till C4 level and then widens.

Radiograph of this view is of special and specific importance in the diagnosis of cervical spondylosis. Loss of cervical curvature, osteophytic growths and spurs,

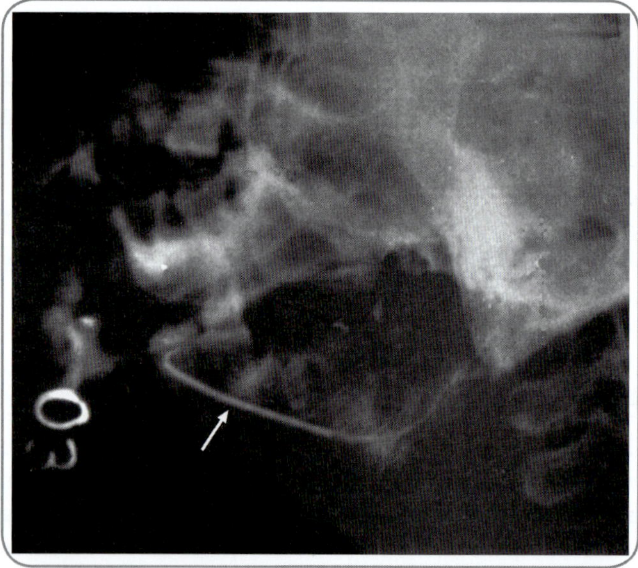

Fig. 16.10: Radiograph after injecting contrast medium into the submandibular duct (sialography)—the arrow points to the duct

reduction in the height of the cervical intervertebral discs and disc herniation indicate spondylotic changes.

Anteroposterior View

Width of the cervical vertebrae, spacing of spinous processes and normal/deviated course of trachea are some of the information that should be sought in this X-ray.

Other Views

An odontoid peg view (cervical spine AP view with open mouth) is taken to view intricate details of the atlantoaxial area. A swimmer's view (lateral view with the shoulders drooped) is taken to view the cervicothoracic junction.

SPECIAL RADIOGRAPHS OF HEAD AND NECK

Angiograms of internal carotid arteries, Venograms of the dural venous sinuses, Sialograms of salivary glands and ducts and Ventriculograms of the ventricles of brain are taken for specific purposes. Computed tomography pictures of brain and magnetic resonance imaging pictures of head and neck are also taken and analysed depending upon diseases and injuries which affect them.

IMPORTANT LINES AND AREAS

- ❑ *Base line:* A line drawn from the most inferior point on the infraorbital margin to the midpoint of the occiput cutting the external acoustic meatus. This line corresponds to the base of skull.
- ❑ *Orbitomeatal line (OML):* Otherwise called the *linea basale orbitomeatale, canthomeatal line* or *radiographic base line*. A line drawn from the outer canthus of the eye to the auricular point (centre of the opening of the external acoustic meatus) and extended backwards. Traditionally, this line had been used as the imaging reference line in computed tomographic imaging.
- ❑ *Reid's base line:* Otherwise called the *inferior orbitomeatal line (IOML)*. A line drawn from the most inferior point on the infraorbital margin to the auricular point (centre of the opening of the external acoustic meatus) and further extended backwards. This line is also used as a referral zero base line in computed tomography of head and brain.

  A difference of 7 degrees exists between the orbitomeatal line and the Reid's base line. Frankfort line is also sometimes referred to. However, it is used more in anthropological and morphometric studies than in radiological and treatment modalities.

- ❑ *Frankfurt plane:* Otherwise called the *Anthropological base line, linea basale anthropologica* or *Frankfurt-Virchow line*. It is a transverse plane that runs to connect the inferior margins of the both the orbital apertures and the superior margins of the external acoustic meati of both sides. This plane represents the base of skull. The Reid's base line deviates downwards from this plane in the posterior aspect.

  Some overlap occurs in the nomenclature of these lines. It is essential to understand that there are three different lines and the term 'orbitomeatal line or plane' should be used only to mean the canthomeatal line.

- ❑ *Camper's line:* Otherwise called the *ala-tragus line*. A line from the inferior border of ala of the nose to the superior border of the tragus of the ear. This line is parallel to the line of occlusion of upper and lower teeth.

- ❑ *Acanthomeatal line:* A line connecting the acanthion to the superior margin of the external acoustic meatus.
- ❑ *Mentomeatal line:* A line connecting the mental point to the superior margin of the external acoustic meatus.
- ❑ *Poirier's line:* A line from the nasion to a little above lambda.
- ❑ *Topinard's line:* A line from the glabella to the mental point. This line helps in the estimation of chin protrusion in studies of cranial evolution.
- ❑ *Frontal triangle:* A triangle bounded above by the maximum frontal diameter and laterally by lines joining the tips of this diameter to the glabella.
- ❑ *Inferior occipital triangle:* A triangle bounded inferiorly by a line joining the two mastoid processes and superolaterally by lines joining the tips of this line to the external occipital protuberance.
- ❑ *Suprameatal triangle:* Otherwise called the *suprameatal pit, mastoid fossa, foveola suprameatica* or *Macewen's triangle*. A triangular area bounded superiorly by the posterior root of the zygomatic process, posteriorly by the posterior border of external acoustic meatus and anteroinferiorly by the superior border of the external acoustic meatus. This triangle serves an important landmark while performing mastoidectomy and any instrument pushed through the triangle enters the mastoid antrum. The triangle lies under the cymba concha of the auricle.

SUPERNUMERARY BONES

The commonest variety of supernumerary bones in the cranium is the *Wormian group of bones*. These are the sutural bones and are found in the sutures of the skull. They may represent the detached portions of the adjacent bones or may represent the appearance of separate abnormal ossification centres. Wormian bones are found commonly in healthy individuals though their incidence has been related to congenital bony disorders. They are usually seen in the lambdoid and squamous sutures.

Ossa incae: Otherwise called the *incarial bones, interparietal bone, inca bone* or *intercalary bone*. This is a separate bone found at the lambda; it is actually the detached portion of the squama of the occipital bone that lies between the posterior aspects of the two parietal bones. The suture that separates the inca bone from the rest of the occipital bone is called the *sutura mendosa*.

SURFACE MARKING OF IMPORTANT BONY POINTS AND STRUCTURES OF HEAD AND NECK

BONY POINTS AND PLANES OF HEAD AND CRANIUM

- ❑ *Frontal tubers:* These prominences can easily be identified and felt on the lateral aspects of the forehead.

- ❑ *Mental protuberance:* This bony projection can be felt at the chin.
- ❑ *External occipital protuberance:* This prominence is felt at the upper end of the nuchal furrow at the back of neck.
- ❑ *Supraorbital notch:* This slight depression can be felt at the junction of the medial one third and lateral two thirds of the supraorbital margin.
- ❑ *Infraorbital foramen:* This opening is marked in the same vertical line as that of the supraorbital notch 1 cm below the infraorbital margin.
- ❑ *Angle of mandible:* The outer angle of the mandible (between the ramus and the body) can be felt anteroinferior to the ear lobule.
- ❑ *Condyloid process of mandible:* This can be felt immediately anterior to the lower part of tragus. If the individual is asked to alternately open and close the mouth, this prominence can be felt to move.
- ❑ *Frontomaxillary suture:* This is the junction of the frontal and maxillary bones through their processes on the lateral orbital margin. It can be felt as a small depression in the upper part of the lateral orbital margin.
- ❑ *Mastoid process:* This rounded prominence can be felt behind the auricle of the ear.
- ❑ *Preauricular point:* This point is located immediately in front of the tragus of the auricle of the ear.

Marking the Craniometric Points

Several craniometric points which have already been studied (in the chapter on *bones*) can be marked on the surface. These points help in locating the positions of internal structures.

- ❑ *Nasion (nasal point):* Point on the middle of the nasofrontal suture; it is marked on the midline at the clear depression felt at the root of nose.
- ❑ *Ophryon (supranasal or supraorbital point):* Point on the midline of the forehead just above the glabella.
- ❑ *Bregma:* Point at the junction of coronal and sagittal sutures; it is marked at the middle of a line drawn from one external acoustic meatus to the other through the vertex of the skull, when the head is held in its normal erect position.
- ❑ *Inion:* The most prominent point on the external occipital protuberance; it is marked on the external occipital protuberance.
- ❑ *Maximum occipital point:* The maximum posterior convexity felt above the external occipital protuberance.
- ❑ *Lambda:* Point at the junction of sagittal and lambdoid sutures; it is marked at the centre of an irregular shallow depression above the maximum occipital point made out on palpation. This point should be approximately 7 cm above the inion.

- ❑ *Pterion:* Region of the sphenoidal fontanelle; it is marked as a small area of 2 cm diameter, 3.5 cm behind and 1.5 cm above the frontozygomatic suture.

  An easier and simpler way of marking the pterion is to use the Stile's method. The right thumb of the examiner is invertedly placed posterior to the frontal process of zygoma. The left index and middle fingers of the examiner are placed above the zygomatic arch but parallel to it. The angle made by the thumb and the upper horizontal finger indicates the petrion.
- ❑ *Asterion:* Region of the mastoidal fontanelle; it is marked as a small area of 1.5 cm diameter, about 2 cm behind and 1.25 cm above the posterior border of mastoid process.
- ❑ *Rhinion:* Lowest point of the internasal suture; the lower limits of the nasal bones can easily be made out as the external nose is palpated. This point can be marked in the midline.
- ❑ *Subnasal point (apophysary point, Trousseau's point):* Point on the midline immediately below the anterior nasal spine; this can approximately be marked immediately below the nasal septum.
- ❑ *Porion:* Central point on the upper margin of external acoustic meatus.
- ❑ *Pogonion (mental point):* Most anterior point of the mandible in the midline; it can be marked by palpating at the chin.
- ❑ *Gonion:* Most prominent point on the angle of mandible; it can be marked by palpating the angle and identifying its maximal posteroinferior projection.
- ❑ *Gnathion:* Most inferior point of the mandible in the midine; it can be marked by running a finger down from the pogonion and feeling for the posterior slope of the chin.
- ❑ *Metopion:* Point midway between the two frontal tuberosities; it can be marked in the centre of a line that joins the centres of the two frontal tubers.
- ❑ *Occipital point:* Most prominent posterior point on the occipital bone above the inion; it is marked as the maximum occipital point.
- ❑ *Sylvian point:* Nearest point on the surface of skull to the Sylvian fissure; marked 30 mm behind the zygomatic process of the frontal bone. It will be within the area marked as pterion.

BONY POINTS AND PLANES OF NECK

- ❑ *C7 vertebra:* It can be palpated as a pointedly round prominence at the lower end of the nuchal furrow.
- ❑ *Carotid tubercle:* It is marked 3 cm from the anterior midline at the level of the arch of cricoids cartilage.
- ❑ *Transverse process of atlas:* The tip of the transverse process of atlas can be marked midway between the mastoid process and angle of mandible.

❑ *Midline structures:* The structures lying in the median area of the neck can be palpated by running a finger down from the symphysis menti. These are (from above downwards): body of hyoid, Thyroid cartilage, Cricoid cartilage, ring of trachea, thyroid gland and its isthmus and suprasternal notch.

Marking of these structures can be coincided with vertebral levels:

❑ Body of hyoid – C3
❑ Laryngeal cartilage – C4, C5
❑ Cricoid cartilage – C6

VESSELS OF HEAD AND NECK

Middle Meningeal Artery

The stem and the anterior and posterior divisions of the middle meningeal artery artery can be marked on the surface.

Stem: Point A is marked 1 mm in front of the preauricular point. Point B is marked 2 cm vertically above the midpoint of the zygomatic arch. The two points are joined by a line that curves a little forwards as it runs upward. The stem or trunk of the artery is marked on the surface by this line.

Anterior (frontal) branch: Mark point B as noted above (it will be the upper end of the trunk of the artery where the artery divides into the anterior and posterior branches). Point C is marked on the centre of pterion (marking of pterion can be done using the frontozygomatic suture as landmark or using the Stile's method). Point D is marked midway between the root of nose (nasion) and external occipital protuberance (marked by inion which indicates the highest point of external occipital protuberance). Points B and C are connected by a line that runs upwards with a forward convexity. Points C and D are then connected by a wavy line. These lines indicate the anterior branch that runs up to the vault.

Posterior (parietal) branch: Point B is already marked. Point E is marked on the upper point of attachment of auricle. Point F is marked on the lambda or approximately 6 cm above the external occipital protuberance. Points B, E and F are joined by a line which is slightly convex upwards proximally and slightly convex downwards distally.

Subclavian Artery in the Posterior Triangle

Point A is marked on the sternoclavicular joint. Point B is marked on the midpoint of clavicle. The two points are joined by a broad line that is curved with an upward convexity reaching about 2 cm above the clavicle. This line indicates the subclavian artery. The artery can be palpated by pressing immediately behind the junction of the medial and middle thirds of the clavicle.

Facial Artery

Point A is marked on the anteroinferior angle of masseter by palpating for the artery and feeling its pulsations. Point B is marked 1.2 cm lateral to the angle of the mouth. Point C is marked at the medial angle of the eye. Points A and B are connected by a line that is curved with a slight convexity anteriorwards; points B and C are connected by a line that is also curved with the convexity of the curvature reaching the ala of nose and then proceeding up straight along the lateral aspect of the nose. The whole stretch indicates the facial artery on the surface.

Common Carotid Artery

Point A is marked on the sternoclavicular joint. Point B is marked on the anterior border of sternocleidomastoid at the level of the upper border of the thyroid cartilage. The two points are joined by a broad line. This line indicates the common carotid artery on the surface.

Internal Carotid Artery

Point A is marked on the anterior border of sternocleidomastoid at the level of the upper border of thyroid cartilage. Point B is marked immediately posterior to the condyle of mandible. The two points are united by a broad line that marks the internal carotid artery on the surface.

External Carotid Artery

Point A is marked on the anterior border of sternocleidomastoid at the level of the upper border of thyroid cartilage. Point B is marked midway between the angle of mandible and mastoid process. These two points are joined by a broad line that is slightly convex forwards in the lower half and convex backwards in the upper half. This sinuous line marks the external carotid artery on the surface.

Internal Jugular Vein

Point A is marked on the upper part of the lobule of the ear. Point B is marked on the medial end of clavicle lateral to the sternoclavicular joint. The two points are joined by a broad line. The lower end of the line will be between the sternal and clavicular heads of the sternocleidomastoid muscle. Mark a dilatation in this part of the line to represent the inferior bulb of the vein.

External Jugular Vein

Point A is marked on the angle or immediately below the angle of mandible. Point B is marked on the clavicle immediately behind the posterior border of sternocleidomastoid. A line joining the two points marks the external jugular vein on the surface. An alternate way is to mark point B on the clavicle at the junction of the medial and the middle thirds.

Facial Vein

The facial vein can be marked on the surface just behind the marking of the facial artery with the same reference points.

Subclavian Vein

Point A is marked on the clavicle a little medial to its midpoint. Point B is marked on the medial end of the clavicle. A broad line joining the two points indicates the subclavian vein. The vein lies at a lower level than the subclavian artery.

INTRACRANIAL VENOUS SINUSES

Superior Sagittal Sinus

Point A is marked on the midline on glabella. Point B is marked on the inion. The two points are joined by a line around the median plane of the vertex; the line should be narrow in front and widen to about 1.2 cm at the inion.

Transverse Sinus

Point A is marked just above the inion. Point B is marked on the upper part of the back of the root of auricle. The two points are joined by a line that is slightly convex upwards. This line marks the ipsilateral transverse sinus on the surface.

Sigmoid Sinus

Point A is marked on the upper part of the back of the root of auricle. Point B is marked on the curve of the root of the auricle at the level of the lower margin of the external acoustic meatus. Point C is marked on the lower margin of the external acoustic meatus. Points A and B are joined by a broad line running along the back of the root of the auricle. Points B and C are then joined by extending this broad line forwards. The entire curved line marks the sigmoid sinus on the surface.

NERVES OF HEAD AND NECK

Mandibular and Inferior Alveolar Nerves

The inferior alveolar nerve, though technically only a branch of the posterior division, appears more to be a continuation of the mandibular nerve because of the same curvature. So the two nerves can be marked together. Point A is marked on the centre of the zygomatic arch. Point B is marked on the centre of the masseter muscle. This point will indicate the mandibular foramen. Point C is marked vertically below the interval between the two premolar teeth, but midway between the upper and lower borders of mandible. All the three points are joined by a broad line

that curves with an upward concavity. The upper most part of the vertical segment represents the mandibular nerve and the rest of the line represents the inferior alveolar nerve.

Lingual Nerve

Point A is marked on the centre of the zygomatic arch. Point B is marked below and behind the third molar tooth. Point C is marked immediately beneath the canine tooth. The three points are joined by a line that is curved with an upward concavity. The entire line indicates the lingual nerve on the surface.

Auriculotemporal Nerve

Point A is marked on the centre of the zygomatic arch. Point B is marked on the centre of the masseter muscle. These two points are joined by a line. From the middle of this line, another line that runs backwards across the neck of the mandible is marked. The line is then turned upwards passing immediately in front of the tragus.

Maxillary Nerve

Point A is marked on the jugal point. This is between the temporal border of the zygomatic bone and the upper border of the zygomatic arch. Point B is marked at the infraorbital foramen. The two points are joined by a line that is slightly curved with an upward concavity.

Facial Nerve

The extracranial part of the nerve before it enters the parotid gland and a general course of the nerve inside the gland can be marked.

Extracranial course: Point A is marked at the middle of the anterior border of mastoid process. A small horizontal line drawn from this point to just in front of the lobule of auricle marks the extracranial preparotid course of the nerve.

Intraparotid course: The nerve divides into its branches inside the gland. However, a general course and the initial part of the buccal branch can be marked by drawing a line parallel to but below the parotid duct for about 5–6 cm from the lobule of the auricle.

Glossopharyngeal Nerve

Point A is marked on the lower part of the tragus on its anterior surface. Point B is marked just above the angle of the mandible. The two points are joined by a line that runs downwards and forwards with a downward convexity. The line then continues for a short distance just above the lower border of the mandible but along the latter. The entire line indicates the glossopharyngeal nerve on the surface.

Vagus Nerve

Point A is marked on the anterior part of tragus. Point B is marked on the medial end of clavicle. The two points are joined by a straight line that indicates the vagus nerve in the neck.

Spinal Accessory Nerve

Point A is marked on the lower part of the anterior surface of tragus. Point B is marked midway between the mastoid process and the angle of mandible. Point C is marked on the posterior border of the sternocleidomastoid muscle at the junction of the superior and middle thirds. The three points are joined by a line with a mild anterior convexity.

Accessory Nerve in the Posterior Triangle

Point C (as seen above) is marked on the posterior border of the sternocleidomastoid muscle at the junction of the superior and middle thirds. Point D is marked on the anterior border of the trapezius muscle at the junction of the middle and lower thirds. A line joining points C and D runs obliquely downwards and backwards, marks the course of the accessory nerve in the posterior triangle.

Hypoglossal Nerve

Point A is marked on the lower part of the tragus on its anterior surface. Point B is marked just above and behind the tip of the greater horn of hyoid bone. Point C is marked on the base of mandible midway between the angle and symphysis menti with the head tilted a little backwards. Points A and B are joined by a line that runs downwards; points B and C are joined by a line that runs upwards. At point B, there is a sharp upward curve. The entire line from point A to point C through point B indicates the hypoglossal nerve on the surface.

Phrenic Nerve

Point A is marked 3.5 cm from the midline at the level of the upper border of thyroid cartilage. Point B is marked on the sternocleidomastoid midway between its anterior and posterior borders at the level of cricoid cartilage. Point C is marked on the sternal end of clavicle. The three points are joined by a straight line.

Sympathetic Chain in the Neck

Point A is marked just behind the condyle of the mandible. Point B is marked on the sternoclavicular joint. The two points are joined by a line that runs down the neck. The three cervical sympathetic ganglia can be marked on this line. The superior ganglion is marked on the line as an ovoid that extends from the level of the transverse process of atlas vertebra to the level of greater cornu of hyoid. The middle ganglion is marked on the line at the level of the

cricoids cartilage. The inferior ganglion is marked on the line 3 cm above the sternoclavicular joint.

Cutaneous Branches of the Cervical Plexus

Point X is marked on the midpoint of the posterior border of sternocleidomastoid. This is the zero point and indicates the nerve point of the neck.

Point A is marked just above point X. Point B is marked below the lobule of the ear. Points A and B are joined by a line that indicates the great auricular nerve. Point C is marked on the posterior aspect of mastoid. Points A and C are joined by a line that runs along the posterior border of sternocleidomastoid and indicates the lesser occipital nerve. Point D is marked on the anterior border of the muscle just opposite point X. Points D and X are joined by a line that indicates the transverse cutaneous nerve.

Point Y is marked immediately below point X. Three lines are drawn from this point, each towards the lateral, middle and medial thirds of clavicle and represent the three supraclavicular nerves.

GLANDS OF HEAD NECK

Pituitary Gland

This gland can be marked on the surface as a small sphere that lies about 6–7 cm posterior to nasion on a line joining the nasion to the inion.

Thyroid Gland

The isthmus and the lobes of the thyroid gland can be marked on the surface.

- ❑ *Isthmus:* A 1.5 cm line is drawn across the trachea 1 cm below the arch of cricoids. This line marks the upper border of isthmus. Another line drawn parallel to this but 2 cm below marks the lower border.
- ❑ *Lateral lobe:* Point A is marked 1 cm below the lateral end of the lower border of the isthmus. Point B is marked 2.5 cm below and lateral to the lateral end of the lower borer of isthmus. Points A and B are at the same horizontal level. Point C is marked immediately in front of the anterior border of sternocleidomastoid at the level of the laryngeal prominence. Points A and B indicate the lower border of the lateral lobe and Point C marks the upper pole. Point C is joined to points A and B by two separate lines which are slightly convex outwards. Points A and B are joined by a transverse line. The outline so obtained marks the lateral lobe.

Parotid Salivary Gland and Parotid Duct

Parotid Gland

Three borders of the parotid gland can be separately marked and then joined to get the marking of the gland.

Anterior border: Point A is marked on the upper border of the head of mandible. Point B is marked at the centre of the masseter muscle. Point C is marked 2 cm posteroinferior to the angle of mandible. Points A and B are joined by a line that runs downwards and forwards; points B and C are joined by a line that runs downwards and backwards. The entire stretch from point A to point C marks the anterior border.

Posterior border: Point C (lower end of the anterior border) is already marked. Point D is marked on the apex of the mastoid process. Points C and D are joined by a line that runs upwards and backwards. This line marks the posterior border.

Superior border: This is marked by a curved line joining points D and A. This curved line passes below the external acoustic meatus with a slight concavity upwards and should actually be marked by lifting the lobule of the auricle upwards.

Parotid Duct

Point A is marked on the lower border of the tragus of auricle. Point B is marked midway between the ala of nose and the vermilion border of the upper lip. Points A and B are connected by a line. This line is divided into three parts and the middle one-third of the line indicates the parotid duct on the surface.

Submandibular Salivary Gland

The outline of the gland is oval.

Point A is marked on the angle of the mandible. Point B is marked on the middle of the base of the mandible (midway between the angle and the symphysis menti). Points A and B are joined by a line that is convex upwards (the convexity about 1.5 cm above the base of mandible). This line indicates the upper border of the gland.

Point C is marked on the greater horn of hyoid. Points A and B are joined by a line that is convex downwards with the convexity reaching a little below the level of the greater horn of hyoid. This line indicates the lower border of the gland.

The upper and lower borders are joined at their ends to complete the oval shape of the gland.

PARANASAL SINUSES

Frontal Air Sinus

The surface projection of this sinus is triangular and lies above the medial part of the orbit.

Point A is marked on nasion (that is on the midline just above the depression between the forehead and the upper end of the nose—the point where the internasal and frontonasal sutures meet). Point B is marked on the midline 2.5 cm above point A. Point C is marked on the superior orbital margin at the junction of the medial and middle thirds. Points A and B are connected by a straight line and points A

and C by another line that runs slightly upwards. The triangle is completed by connecting the points B and C.

Maxillary Air Sinus

The surface outline of this sinus is irregular. A roughly triangular marking, more or less, marks the sinus. The superior border is marked parallel but just below the infraorbital margin. The medial border is marked immediately lateral to the lateral border of the nose but stops just above the alveolar process of the maxilla. The oblique lateral border is carried upwards and laterally from the lower limit of the medial border to the lateral limit of the superior border. All the angles of the triangle are curved and rounded.

OTHER STRUCTURES OF HEAD NECK

Palatine Tonsils

A small oval drawn just in front of, and above, the angle of the mandible indicates the palatine tonsils on the surface. The marking will lie over the masseter muscle.

Carotid sheath

Point A is marked on the sternocleidomastoid joint. Point B is marked midway between the angle of mandible and mastoid process. The broad line that joins the two points will indicate the carotid sheath.

Sternocleidomastoid

The head is rotated to one side and tilted forwards, the opposite muscle stands out prominently. The anterior border runs down from the mastoid process to the sternoclavicular joint. The posterior border runs down from the midpoint between inion and mastoid to the junction of medial and middle thirds of clavicle. These borders can be used as reference lines for various other surface marking.

❑ ***Scalenus anterior:*** This muscle serves as a guide for many internal structures and is often used as a reference structure. The lateral and medial borders of the muscle can be marked and the muscle band can be drawn on the surface.

❑ ***Lateral border:*** Point A is marked 2.5 cm from the midline at the level of the upper border of thyroid cartilage. Point B is marked on the lower border of clavicle at the junction of the medial and middle thirds. The two points are joined by a straight line that runs inferolaterally. This line marks the lateral border of the muscle on the surface.

❑ ***Medial border:*** Point C is marked 2.5 cm from the midline at the level of the arch of cricoid cartilage. Point D is marked 1 cm medial to the lower end of the lateral border. Points C and D are joined by a straight line that runs inferolaterally. This line marks the medial border of the muscle. The two borders together mark the extent of the scalenus anterior muscle.

PLACES WHERE PULSATIONS CAN BE FELT

- ❑ *Carotid pulse:* The index and middle fingers can be placed on the thyroid cartilage and gently pushed backwards and laterally; the pulse can be palpated medial to sternocleidomastoid.
- ❑ *Facial pulse:* The masseter muscle is made prominent by clenching the teeth. The facial pulse can be palpated by pressing against the inferior border of mandible just in front of the anterior border of masseter.

- ❑ *Temporal pulse:* The superficial temporal pulse can be palpated just anterior to the auricle and immediately above the zygomatic arch.

MAKING MUSCLES AND/OR TENDONS PALPABLE

- ❑ *Sternocleidomastoid:* The muscle can be made palpable by turning the head to the opposite side against resistance.
- ❑ *Trapezius:* The superior border of trapezius can be palpated by shrugging the shoulder against resistance.
- ❑ *Masseter:* The muscle is made prominent by clenching the teeth.
- ❑ *Temporalis:* The muscle and its tendon are made prominent by clenching the teeth.

Multiple Choice Questions

1. In a lateral X-ray of the skull, the upper border of the sella turcica corresponds to the:
   a. Lower border of hypothalamus
   b. Upper border of thalamus
   c. Lower limit of telencephalon
   d. Middle of uncus
2. Frankfort plane corresponds to the:
   a. Radiological base line
   b. Anthropological base line
   c. Line of occlusion of teeth
   d. Ala-tragus line
3. Inca bone is a detached part of the:
   a. Squamous occipital bone
   b. Squamous temporal bone
   c. Parietal bone
   d. Mastoid temporal bone
4. Sylvian point corresponds to the:
   a. Asterion
   b. Glabella
   c. Pterion
   d. Prognathion
5. The surface projection of the palatine tonsils will lie on the:
   a. Medial pterygoid muscle
   b. Lateral pterygoid muscle
   c. Masseter muscle
   d. Temporalis muscle

ANSWERS

1. c   **2.** b   **3.** a   **4.** c   **5.** c

Clinical Problem-solving

Case Study 1: A 45-year-old has been presenting with symptoms of paranasal sinusitis. You want to have radiologic confirmation.
- ❑ What X ray would you advise for this patient?
- ❑ If you want to visualize the maxillary air sinus specifically, would the same radiograph be sufficient?
- ❑ If you would want another radiograph, what would be your choice?

Case Study 2: One of your colleagues in the medical school is an overtly enthusiastic person. He was, the other day, trying to elicit his own bilateral carotid pulses. As he was doing so, he felt dizzy.
- ❑ Can you figure out any reason for such a symptom?
- ❑ What procedure or precaution should be adopted to avoid the symptom?
- ❑ What are the other arterial pulses which could be palpated in the facial region?

(For solutions see Appendix).

Neuroanatomy

Chapter 17

Introduction to Central Nervous System

- Discuss a typical multipolar neuron.
- Write notes on (a) Neuron, (b) Grey and white matter, (c) Ganglia, (d) Internuncial neuron.

The **central nervous system** is made up of the brain and spinal cord. In contrast, the **peripheral nervous system** consists of the cranial nerves and the spinal nerves (Fig. 17.1).

The brain consists of the (Fig. 17.2):

- Cerebrum which is made up of two large cerebral hemispheres,
- Cerebellum and
- Brainstem.

The brainstem is made up of:

- Midbrain,
- Pons and
- Medulla oblongata.

The medulla is continuous below with the spinal cord.

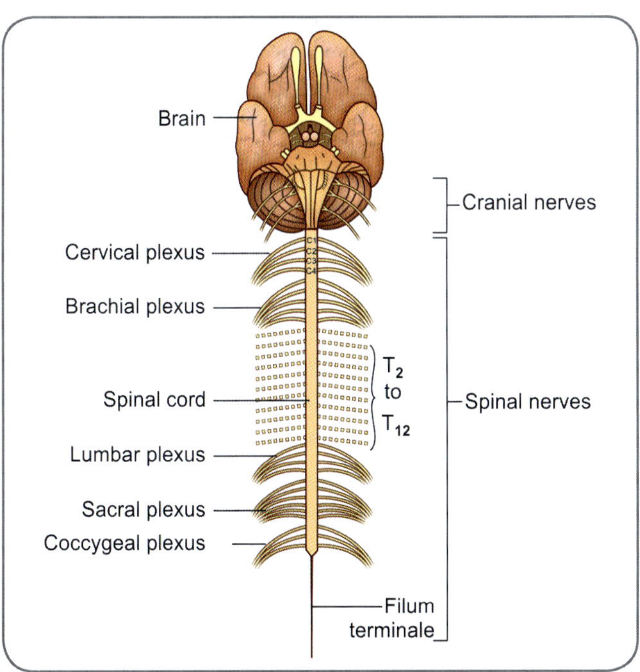

Fig. 17.1: Anatomical divisions of the nervous system – the central nervous system consists of the brain and spinal cord – the peripheral nervous system consists of cranial nerves and spinal nerves

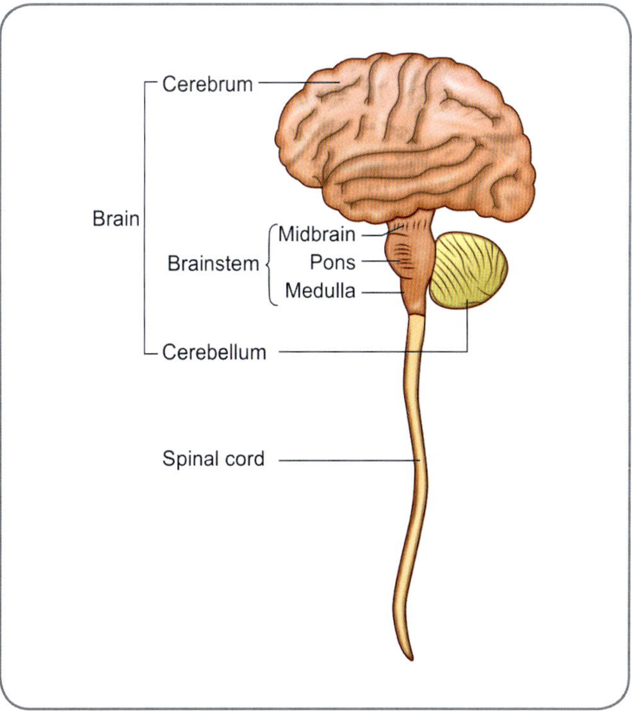

Fig. 17.2: Diagram showing parts of the central nervous system

ELEMENTARY FACTS ABOUT NEURONS

The nervous system is made up, predominantly, of tissue that has the special property of being able to conduct impulses rapidly from one part of the body to another.

The specialised cells that constitute the functional units of the nervous system are called **neurons**.

Within the brain and spinal cord, neurons are supported by a special kind of connective tissue that is called **neuroglia.**

Nervous tissue, composed of neurons and neuroglia, is richly supplied with blood, but lymph vessels are not present.

NEURON

A neuron consists of a **cell body** that gives off a variable number of **processes** (Figs 17.3 and 17.4).

The cell body is also called the **soma** or **perikaryon**.

The processes arising from the cell body of a neuron are called **neurites**. These are of two kinds. Most neurons give off a number of short branching processes called **dendrites** and one longer process called an **axon**.

There are various structural differences between dendrites and axons. However, the most important difference is its functional activity. In a dendrite, the nerve impulse travels **towards the cell body** whereas in an axon the impulse travels **away from the cell body**.

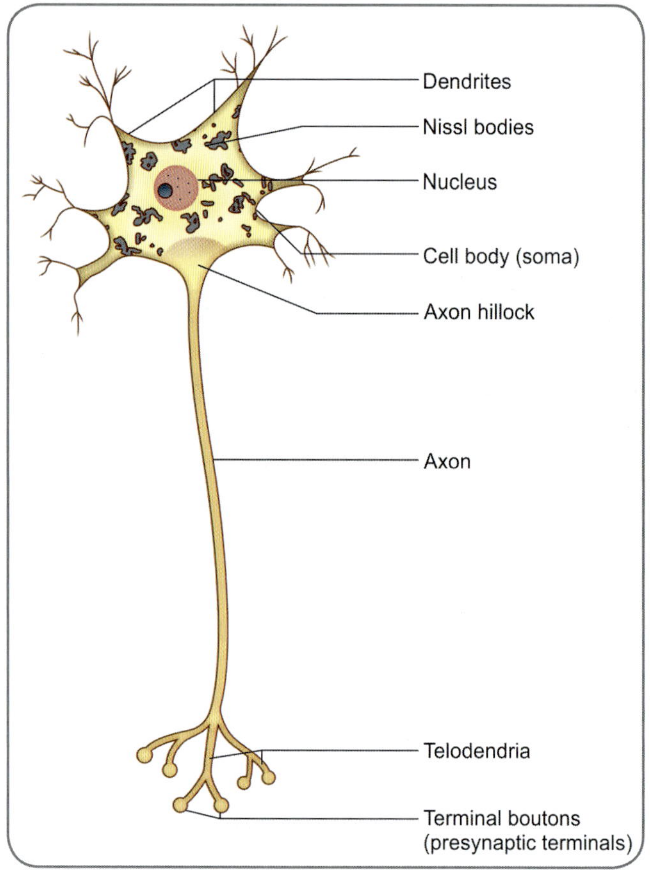

Fig. 17.3: Diagram showing the main parts of a typical neuron

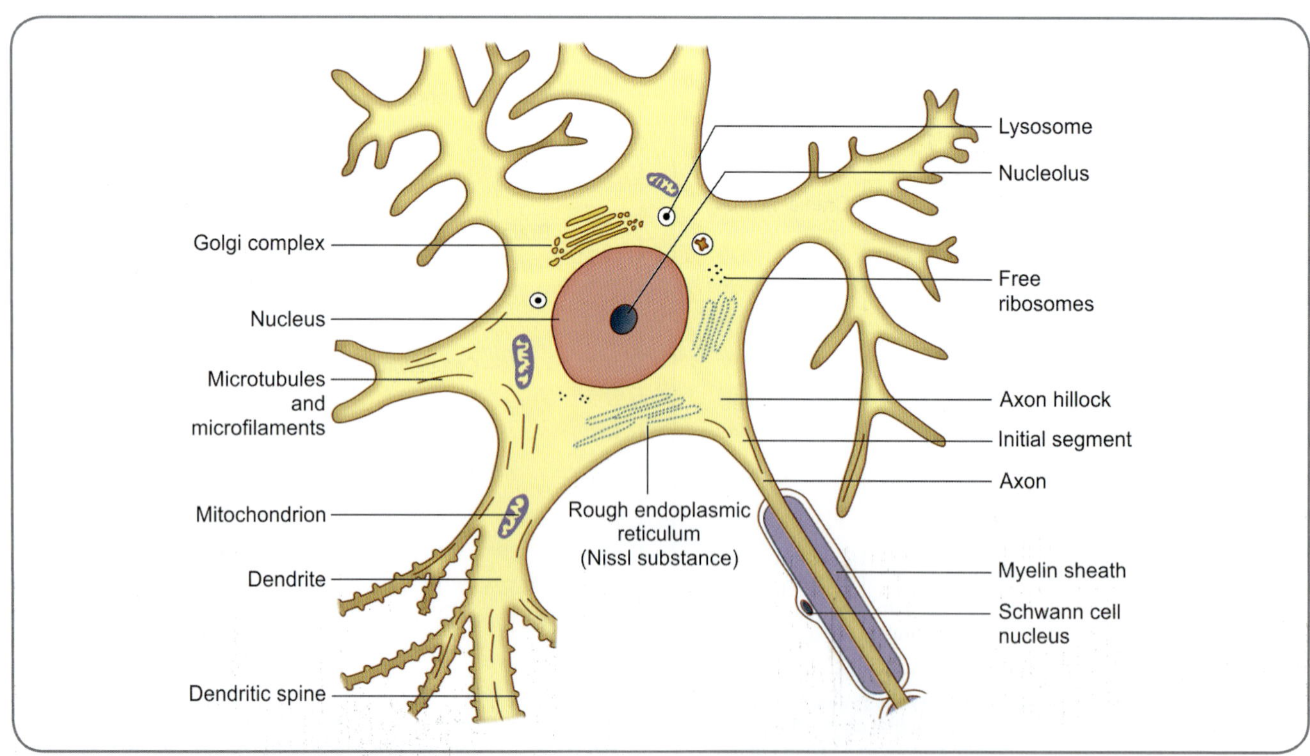

Fig. 17.4: Schematic representation of some structural features of neuron as seen under electron microscope

Peripheral nerves are made up of aggregations of axons (and in some cases of dendrites).

During its formation, each axon (and some dendrites) comes to be associated with certain cells that provide a sheath for it.

The cells providing this sheath for axons lying outside the central nervous system is called **Schwann cells**.

Axons lying within the central nervous system are provided a similar covering by a kind of neuroglial cell called an **oligodendrocyte**.

All nerve fibres in peripheral nerves have an outer sheath called the **neurilemma**. This forms the outer covering of the nerve fibre.

Deep to the neurilemma, most nerve fibres also have a second sheath that is rich in lipids. This is the **myelin sheath**.

Axons having a myelin sheath are called **myelinated axons**. Those without this sheath are **unmyelinated axons**.

An axon (or its branches) can terminate in two ways:

1. Within the central nervous system, it always terminates by coming in intimate relationship with another neuron, the junction between the two neurons being called a **synapse**.
2. Outside the central nervous system, the axon may end in relation to an effector organ (e.g., muscle or gland), or may end by synapsing with neurons in a peripheral ganglion.

Neurons vary considerably in the size and shape of their cell bodies (somata) and in the length and manner of branching of their processes.

The shape of the cell body is dependent on the number of processes arising from it (Figs 17.5 and 17.6).

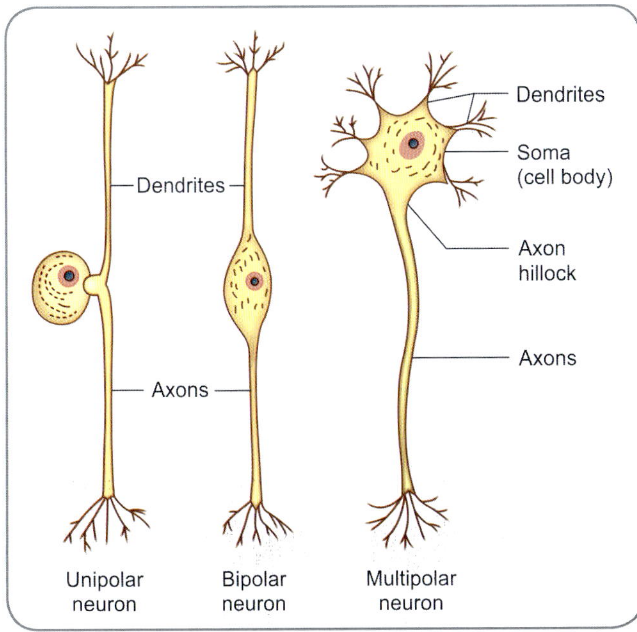

Fig. 17.5: Unipolar, bipolar and multipolar neurons

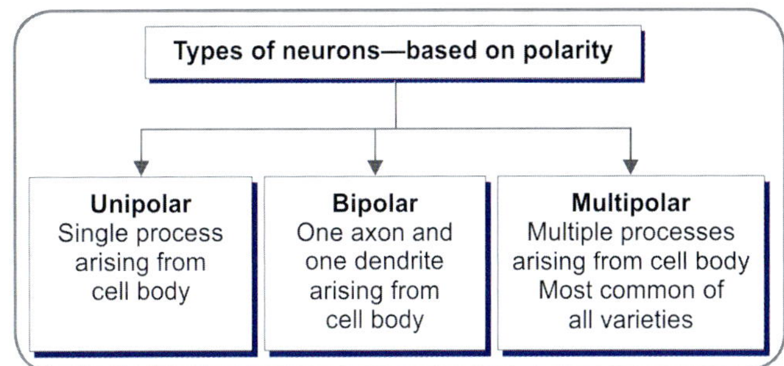

Fig. 17.6: Types of neurons—anatomical classification

The most common type of neuron gives off several processes and the cell body is, therefore, **multipolar**.

Some neurons have only one axon and one dendrite and are **bipolar**.

Another type of neuron has a single process and is therefore described as **unipolar**. However, this process almost immediately divides into two. One of the divisions represents the axon; the other is functionally a dendrite, but its structure is indistinguishable from that of an axon.

Depending on the shapes of their cell bodies some neurons are referred to as **stellate** (star-shaped) or **pyramidal**.

Apart from structural differences neurons can also be classified on the basis of their functions.

A classification of types of nerve fibres to be seen in peripheral nerves; and the arrangement of neurons giving origin to various functional types of nerve fibres has been discussed later.

These descriptions will be essential in understanding some aspects of the structure of brain.

GREY AND WHITE MATTER

Sections through the spinal cord or through any part of the brain show certain regions that appear whitish, and others that have a darker greyish colour. These constitute the **white** and **grey matter** respectively.

Microscopic examination shows that the grey matter contains:

❑ Cell bodies of neurons
❑ Dendrites and axons starting from or ending on the cell bodies
❑ Most of the fibres within the grey matter are unmyelinated.

On the other hand, the white matter consists predominantly of myelinated fibres. It is the reflection of light by myelin that gives this region its whitish appearance.

Neuroglia and blood vessels are present in both grey and white matter.

The arrangement of the grey and white matter differs at different situations in the brain and spinal cord.

In the spinal cord and brainstem the white matter is on the outside whereas the grey matter forms one or more masses embedded within the white matter.

In the cerebrum and cerebellum there is an extensive, but thin layer of grey matter on the surface. This layer is called the ***cortex***.

Deep to the cortex there is white matter, but within the latter several isolated masses of grey matter are present. Such isolated masses of grey matter, present anywhere in the central nervous system, are referred to as ***nuclei***.

As grey matter is made of cell bodies of neurons (and the processes arising from or terminating on them) nuclei can be defined as groups of cell bodies of neurons.

Aggregations of the cell bodies of neurons may also be found outside the central nervous system. Such aggregations are referred to as ***ganglia***.

Ganglia are of two distinct functional types.

1. ***Sensory ganglia*** contain neurons that give off processes that form the afferent fibres of peripheral nerves. Examples are the dorsal nerve root ganglia of spinal nerves and the trigeminal ganglion.
2. ***Autonomic ganglia*** includes ***sympathetic ganglia*** located on the sympathetic chain, and ***parasympathetic ganglia***. Examples for the latter are the ciliary ganglion and the submandibular ganglion.

Some autonomic neurons are located in ***nerve plexuses*** (often called ***ganglia***) present in close relationship to some viscera.

The axons arising in one mass of grey matter terminate very frequently by synapsing with neurons in other masses of grey matter.

The axons connecting two (or more) masses of grey matter are frequently numerous enough to form recognisable bundles.

Such aggregations of fibres are called ***tracts***.

Larger collections of fibres are also referred to as ***funiculi, fasciculi*** or ***lemnisci***.

Large bundles of fibres connecting the cerebral or cerebellar hemispheres to the brainstem are called ***peduncles***.

ARRANGEMENT OF NEURONS WITHIN THE CENTRAL NERVOUS SYSTEM

We have seen that neurons giving rise to fibres of peripheral nerves are basically of two types which are as follows:

1. Neurons having cell bodies that lie within the brain and spinal cord, send out ***efferent processes*** that leave the CNS to form the motor fibres of peripheral nerves.
2. Neurons having cell bodies located in ganglia outside the CNS, give origin to peripheral processes that form ***afferent nerve fibres*** of peripheral nerves. Their central processes form the sensory roots of the nerves in question. They enter the brain or spinal cord and synapse with neurons within them.

The bulk of the CNS is, however, made up of neurons that lie entirely within it.

As explained earlier, the cell bodies of these neurons are invariably located in masses of grey matter.

The axons may be short, ending in close-relation to the cell body, or may be long and may travel to other masses of grey matter lying at considerable distances from the grey matter of origin.

The neurons within the CNS are interconnected in an extremely intricate manner. The description that follows illustrates some of the basic arrangements is encountered.

The simplest pathways are those concerned with reflex activities, such as the contraction of a muscle in response to an external stimulus. For example, if the skin of the sole of a sleeping person is scratched, the leg is reflexly drawn up. Let us see how this happens.

The simplest possible arrangement is shown in Fig. 17.7.

The stimulus applied to skin gives rise to a nerve impulse that is carried by the peripheral process of a unipolar neuron to the dorsal nerve root ganglion.

From here the impulse passes into the central process that terminates by directly synapsing with an anterior grey column cell supplying the muscle which draws the leg up.

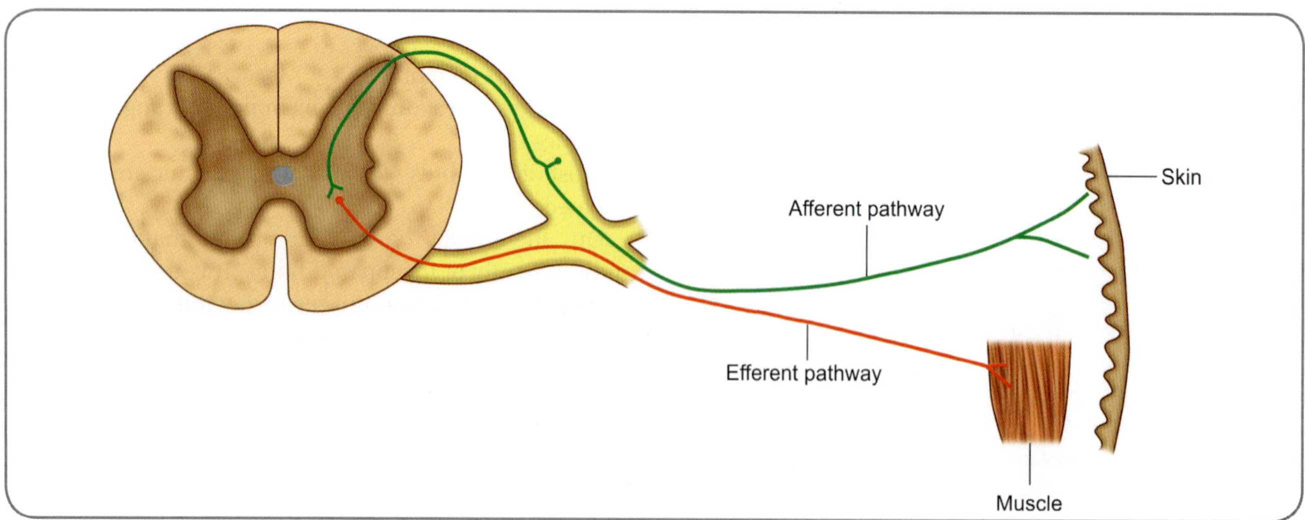

Fig. 17.7: A monosynaptic spinal reflex arc composed of two neurons

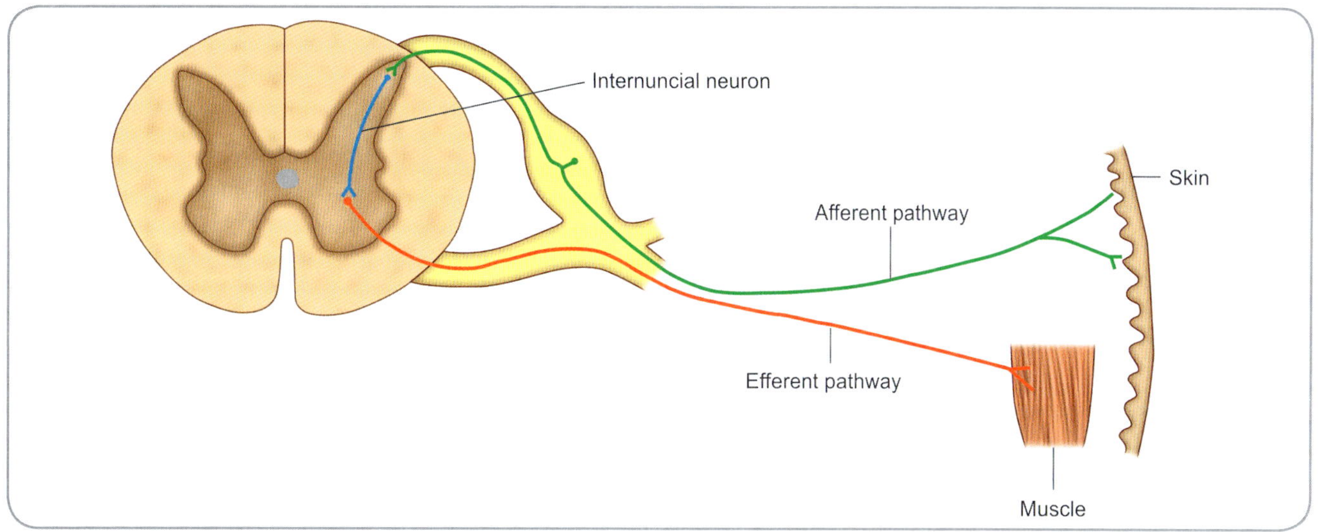

Fig. 17.8: A polysynaptic spinal reflex arc composed of three neurons

The complete pathway constitutes a ***reflex arc*** and in the above example it consists of two neurons—(1) afferent neuron and the (2) efferent neuron.

In actual practice, however, the reflex arc is generally made up of three neurons as shown in Fig. 17.8.

The central process of the dorsal nerve root ganglion cell ends by synapsing with a neuron lying in the posterior grey column.

This neuron has a short axon that ends by synapsing with a neuron in the anterior grey column, thus completing the reflex arc.

The third neuron interposed between the afferent and efferent neurons is called an ***internuncial neuron,*** or simply an ***interneuron***.

The purpose served by an interneuron may be basically of three types:

- Firstly, the axon arising from an interneuron may divide into a number of branches and may synapse with a number of different efferent neurons. As a result an impulse coming along a single afferent neuron may result in an effector response by a large number of efferent neurons

- Secondly, afferent impulses brought by a number of afferent neurons may converge on a single efferent neuron through the agency of interneurons

  Some of these impulses tend to induce activity in the efferent neuron (i.e. they are ***facilitatory)*** while others tend to suppress activity (i.e. they are ***inhibitory)***

- Thirdly, through interneurons an afferent neuron may establish contact with efferent neurons in the opposite half of the spinal cord, or in a higher or lower segment of the cord.

Everytime a stimulus reaches a neuron it does not mean that it must become active and must produce an impulse.

A neuron receives inputs from many neurons (in some cases from hundreds of them).

Some of these inputs are facilitatory and others are inhibitory.

Activity in the neuron (in the form of initiation of an impulse) depends on the sum total of these inputs.

Thus, each neuron may be regarded as a decision-making centre.

The greater the number of neurons involved in any pathway, the greater the possibility of such interactions.

Viewed in this light it will become clear that interneurons interposed in a pathway increase the number of levels at which 'decisions' can be taken.

It will also be appreciated that most of the neurons within the nervous system are, in this sense, interneurons that are involved in numerous highly complex interactions on which the working of the nervous system depends.

From what has been said above it will be seen that some activities occur due to reflex action, and may involve only neurons within the spinal cord.

However, most activities of the spinal cord are subjected to influence from higher centres.

In the more complicated types of activity, several higher centres may be involved and the pathways may be extremely complicated.

Afferent impulses reaching these higher centres (e.g. the cerebral cortex) would appear to be somehow stored and this stored information (of which we may or may not

Added Information

The singular terms for 'funiculi, fasciculi and lemnisci' are funiculus (small rope like; Latin.funis=rope), fasciculus (bundle; Latin.fascis=bundle) and lemniscus (band or ribbon; Greek. lemniskos=fillet) respectively. The singular term for ganglia is ganglion (cyst like or swelling; Greek.gangliose=swelling).

be conscious) is used to guide responses to similar stimuli received in future. This accounts for memory and for learning processes.

PROJECTION, ASSOCIATION AND COMMISSURAL FIBRES

When a considerable number of fibres pass from a mass of grey matter in one part of the brain to another mass of grey matter in another part of the brain or spinal cord, these are referred to as *projection fibres*.

Fibres interconnecting different areas of the cerebral cortex or of the cerebellar cortex are called *association fibres*.

Fibres connecting identical areas of the two halves of the brain are called *commissural fibres*.

When fibres originating in a mass of grey matter in one-half of the CNS end in some other mass of grey matter in the opposite half they are referred to as *decussating fibres,* and the sites where such crossings take place are referred to as *decussations*.

Multiple Choice Questions

1. A neuron interposed between an afferent and an efferent neuron is:
   a. A bipolar neuron
   b. An interneuron
   c. A multipolar neuron
   d. A reflex neuron
2. A nucleus with relation to the nervous system is:
   a. An isolated mass of grey matter
   b. An aggregation of fibres
   c. A junction between two neurons
   d. The cell body of a neuron
3. The processes arising from the perikaryon of a neuron are:
   a. Neurites
   b. Neuroglia
   c. Soma
   d. Neurilemma
4. Fibres interconnecting different areas of the cerebellar cortex are:
   a. Projection fibres
   b. Decussating fibres
   c. Commissural fibres
   d. Association fibres
5. A monosynaptic spinal reflex arc has:
   a. Two participating neurons
   b. Three participating neurons
   c. A single participating neuron
   d. Many participating neurons

ANSWERS

1. b   **2.** a   **3.** a   **4.** d   **5.** a

Clinical Problem-solving

Case Study 1: A mother noticed some dirt on the plantar surface of her son's right foot. She tried removing it. In the process, she scratched the boy's sole.
- What consequence do you expect to this?
- What is the neurological basis for such a reaction?
- Where do you expect the afferent neuron to be present?

(For solutions see Appendix).

Chapter 18

Spinal Cord—Gross Anatomy and Internal Structure

Frequently Asked Questions

❑ Write notes on (a) Spinal meninges, (b) Cauda equina, (c) Epidural space, (d) Modifications of spinal pia mater, (e) Spinal nerve.
❑ Discuss the internal structure of the spinal cord in detail.
❑ Write in detail about the blood supply to the spinal cord.
❑ Describe the lateral grey column of the spinal cord.

GROSS ANATOMY

The spinal cord is the lower elongated part of the central nervous system (CNS) and is the most important content of the vertebral canal.

It begins as a downward extension of medulla oblongata at the level of the upper border of the first cervical vertebra (C1) and extends down to the level of the lower border of the first lumbar vertebra (L1) (Fig. 18.1A). Thus, it occupies the upper two-thirds of the vertebral column. The level is, however, variable and the cord may terminate one vertebra higher or lower than this level. The level also varies with flexion or extension of the spine.

The lowest part of the spinal cord is conical and is called the ***conus medullaris***. The conus is continuous, below, with a fibrous cord called the ***filum terminale*** (Fig. 18.1B), which is a prolongation of pia mater and is attached to the posterior surface of the coccyx.

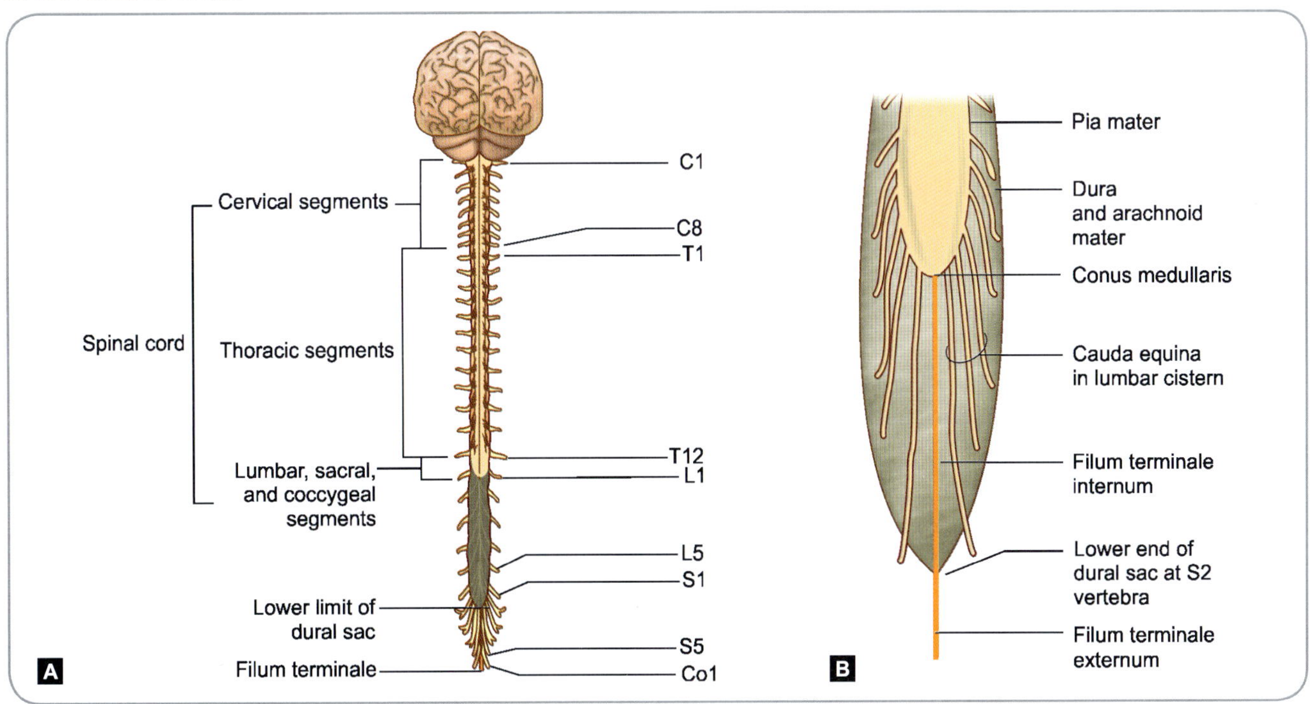

Figs 18.1A and B: A. The spinal cord **B.** The lower end of spinal cord magnified showing the conus medullaris, cauda equina and filum terminale

DIMENSIONS OF THE CORD

The length of the cord varies from 42 to 45 cm. The spinal cord is not of uniform thickness. The spinal segments that contribute to the nerves of the upper limbs are enlarged to form the ***cervical enlargement*** of the cord. Similarly, the segments innervating the lower limbs form the ***lumbar enlargement***.

Functions of Spinal Cord

The spinal cord has three major functions:

1. It acts as a conduit for motor information, which travels down the spinal cord.
2. It serves as a conduit for sensory information in the reverse direction.
3. It is a centre for coordinating simple reflexes.

Development

Age-wise Changes

In early foetal life (third month), the spinal cord is as long as the vertebral canal and each spinal nerve arises from the cord at the level of the corresponding intervertebral foramen. In subsequent development, the spinal cord does not grow as much as the vertebral column, and its lower end, therefore, gradually ascends to reach the level of the third lumbar vertebra at the time of birth and to the lower border of the first lumbar vertebra in the adult.

As a result of this upward migration of the cord, the roots of spinal nerves have to follow an oblique downward course to reach the appropriate intervertebral foramen (Fig. 18.2). This also makes the roots longer. The obliquity and length of the roots is most marked in the lower nerves and many of these roots occupy the vertebral canal below the level of the spinal cord. These roots constitute the ***cauda equina*** (Figs 18.1 and 18.2).

The brain and the spinal cord develop from a neural tube. To start with, that part of the neural tube which will form the future spinal cord has two thick lateral walls and thin roof and floor plates. The lumen is narrow. Each lateral wall can be subdivided into three zones—an ependymal zone close to the lumen, a marginal zone at the outer periphery and a mantle zone between the two. The ependymal zone has germinal cells or cells which give rise to future neuroblasts and neurons and is, therefore, called the ***germinal layer***. As a germinal cell develops into a neuroblast, it passes into the mantle zone and further processes of development and differentiation take place in this zone. Hence this zone is also called the ***nuclear layer***. The marginal zone is devoid of neuronal cells but has neuroglial cells and processes of the neuronal cells of the mantle layer.

In the developing spinal cord, the mantle zone in the ventral half of each lateral wall starts proliferating (because of neuronal differentiation going on within it) and becomes thick; as a result, the ventral part of the lumen is compressed. As the ventral part grows larger, it is separated from the thinner dorsal part by a limiting groove called the ***sulcus limitans***. Eventual enlargement, forward projection and approximation of the ventral parts cause the floor plate to be

contd...

Development *contd...*

'dipped in', leading to the formation of the anterior median sulcus. The ventral parts are now called **basal laminae** and will give rise to the motor components of the spinal cord. The thinner dorsal parts also will develop subsequently to become the alar laminae and contribute to the sensory components. The expansion of the alar laminae is not as much as the basal laminae; the posterior median fissure is formed by the shallow gap created by a very slight dipping of the roof plate. Both the basal and alar laminae keep increasing in size, leading to a reduction in the lumen; the posterior part of the lumen is almost completely obliterated and the fusion of its walls forms the posterior median septum. The remaining much reduced lumen forms the central canal. The motor cells differentiate into the ventral and lateral horn cells. The alar laminae cells differentiate into the sensory receptor cells of the posterior horns. The motor cells are comparatively older than the sensory cells which develop later. The sulcus limitans forms the boundary between the primarily motor and primarily sensory components.

As the neuronal cells grow and differentiate through and through the length of the spinal cord, their processes become the several ascending and descending fibres. As these fibres traverse up and down in the marginal zone, the said zone is thickened to form the white matter of the spinal cord. The growth and differentiation of the 'grey' matter (formed by the neuronal mantle layer) into the ventral, lateral and posterior horns, subdivides the white matter into ventral, lateral and dorsal funiculi. The differentiation of the various columns can be made out in 14 to 16 weeks of intrauterine development.

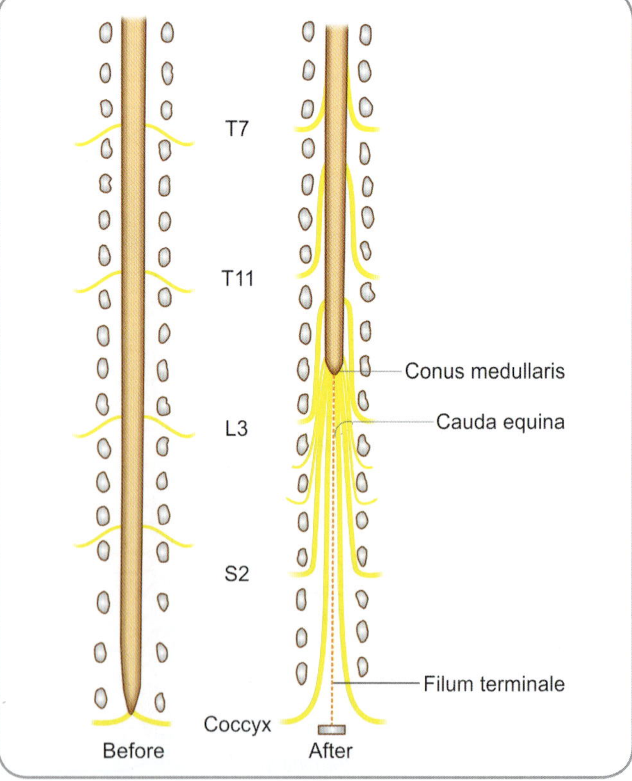

Fig. 18.2: Scheme to show the effect of recession of the spinal cord (during development) on the course of the roots of spinal nerves

<table>
<tr><td>

Dissection

Study the spinal cord, preferably *in situ*. Try to have a close look, when possible, at the meningeal layers and the roots of the spinal nerves.

In an isolated spinal cord specimen, see and study the following:
- ❑ The cervical and the lumbar enlargements
- ❑ Conus medullaris
- ❑ Filum terminale
- ❑ Coccygeal ligament

If the meninges are intact, the denticulate ligaments may be looked for.

</td></tr>
</table>

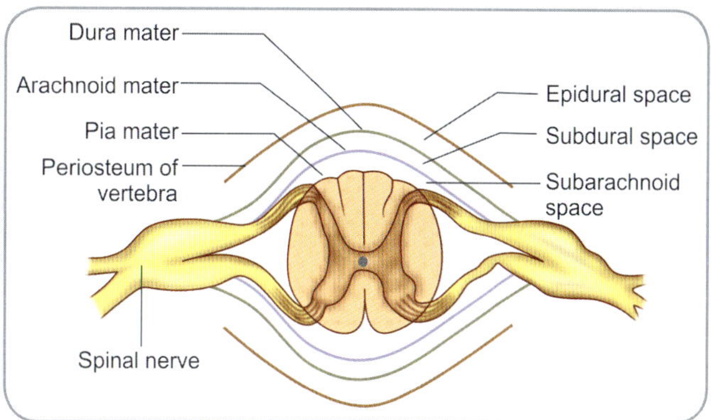

Fig. 18.3: Transverse section through spinal cord showing the spinal meninges

SPINAL MENINGES

The spinal cord is surrounded by the **meninges**. From outside to inside, these are the **dura mater**, the **arachnoid mater** and the **pia mater** (Fig. 18.3).

Dura Mater

The dura mater (Latin.durus=hard, mater=mother) is a thick fibrous membrane extending from foramen magnum to the lower border of second sacral vertebra (S2). The spinal dura mater forms a loose tubular covering for the spinal cord. The space between the dura mater and the wall of the vertebral canal is called the **extradural** (or **epidural**) **space** (Fig. 18.3). The spinal epidural space is filled with loose areolar tissue, fat, and the internal vertebral venous plexus.

The dorsal and ventral roots of spinal nerves pass through the spinal dura mater separately. Sheaths derived from dura extend over the nerve roots. These dural sheaths reach up to the intervertebral foramina and are attached to the margins of these foramina. The dorsal and ventral nerve roots unite in the intervertebral foramina to form the trunks of spinal nerves. The pia mater and arachnoid mater also extend on to the roots of spinal nerves as sheaths. These sheaths reach up to the site, where the nerve roots pass through the dura mater.

<table>
<tr><td>

♪ **Clinical Correlation**

Epidural Anesthesia

Epidural anesthesia is given in the epidural space to numb the spinal nerves that traverse the epidural space. This approach is used by the obstetricians in labour ward for painless delivery, postoperative pain relief after major surgeries, and to relieve the pain in cancer patients.

</td></tr>
</table>

Arachnoid Mater

This is a transparent (Greek.arachne=cobweb or spider; eidos=resembling) and avascular membrane, which is continuous above with the cranial subarachnoid mater and below, it extends upto the lower border of the second sacral vertebrae (S2).

The space between the dura and the arachnoid is called the **subdural space** (Fig. 18.3).

The arachnoid and pia mater are separated by the **subarachnoid space**, which contains the cerebrospinal fluid (Fig. 18.3). The spinal cord (and overlying pia mater) extends downwards only up to the lower border of the L1 vertebra. The dura and arachnoid (along with the subarachnoid space containing cerebrospinal fluid), however, extend up to (S2).

Between these two levels, the subarachnoid space around the filum terminale, becomes roomy and contains cauda equina in a pool of cerebrospinal fluid called **lumbar cistern**.

In this region, a needle can be introduced into the subarachnoid space, without a danger of injury to the spinal cord. This procedure is called **lumbar puncture**.

<table>
<tr><td>

♪ **Clinical Correlation**

Lumbar Puncture

Lumbar puncture is performed to obtain samples of cerebrospinal fluid for various diagnostic and therapeutic purposes. In this procedure, a needle is introduced into the subarachnoid space through the interval between the third and fourth lumbar vertebrae.

With the patient lying on his or her side or in the upright sitting position, with the vertebral column well flexed, the space between adjoining laminae in the lumbar region is increased to a maximum. Taking full aseptic precautions, the lumbar puncture needle is inserted into the vertebral canal above or below the fourth lumbar spine. An imaginary line joining the highest points on the iliac crests passes over the fourth lumbar spine, this is taken as a landmark to insert the spinal needle.

Structures through which the needle passes during a lumbar puncture are:
- ❑ Skin
- ❑ Superficial fascia
- ❑ Supraspinous ligament
- ❑ Interspinous ligament
- ❑ Ligamentum flavum
- ❑ Areolar tissue containing the internal vertebral venous plexus
- ❑ Dura mater
- ❑ Arachnoid mater

</td></tr>
</table>

contd...

Clinical Correlation *contd...*

Purpose of Lumbar Puncture

- ❑ The pressure of cerebrospinal fluid can be estimated roughly by counting the rate at which drops of cerebrospinal fluid flow out of the needle or more accurately, by connecting the needle to a manometer
- ❑ Samples of cerebrospinal fluid can be collected for examination. The important points to note about cerebrospinal fluid are its colour, cellular content, and chemical composition (specially the protein and sugar content)
- ❑ Lumbar puncture may be used for introducing air or radioopaque dyes into the subarachnoid space for certain investigative procedures. Drugs may also be injected for treatment
- ❑ Anesthetic drugs injected into the subarachnoid space act on the lower spinal nerve roots and render the lower part of the body insensitive to pain. This procedure called **spinal anesthesia** is frequently used for operations on the lower abdomen or on the lower extremities.

Pia Mater

It is a vascular (Latin.pia=tender) membrane that closely invests the surface of the spinal cord.

The pia mater is modified in some places as follows:

- ❑ The **filum terminale** is a slender filament of pia mater beyond the conus medullaris. It is about 20 cm long. It pierces the blind end of the dural sac and is subdivided into filum terminale internum, inside the dural sac (15 cm) and filum terminale externum, outside the dural sac (5 cm). It leaves the sacral canal through the sacral hiatus and is fused with the dorsal surface of the first piece of coccyx.
- ❑ **Linea splendens** is a thickened band of pia mater along the anterior median fissure of the spinal cord.
- ❑ The **ligamenta denticulata** are lateral extensions of the pia mater between the attachments of ventral and dorsal nerve roots. There are 21 teeth-like projections, which pass through the subarachnoid space to gain attachment to the inner surface of the dura mater. The first of such pair of ligamentum denticulatum lie at the level of foramen magnum, while the last lie between T12 and L1 spinal nerves. Its identification helps the surgeon in locating the first lumbar nerve during operation.

SURFACE FEATURES OF THE SPINAL CORD

The anterior surface of the spinal cord is marked by a deep anterior median fissure, which contains the anterior spinal artery (Fig. 18.4A).

The posterior surface is marked by a shallow posterior median sulcus (Fig. 18.4B).

The anterior median fissure and posterior median sulcus divide the surface of the cord into two symmetrical halves. Each half of the cord is further subdivided into

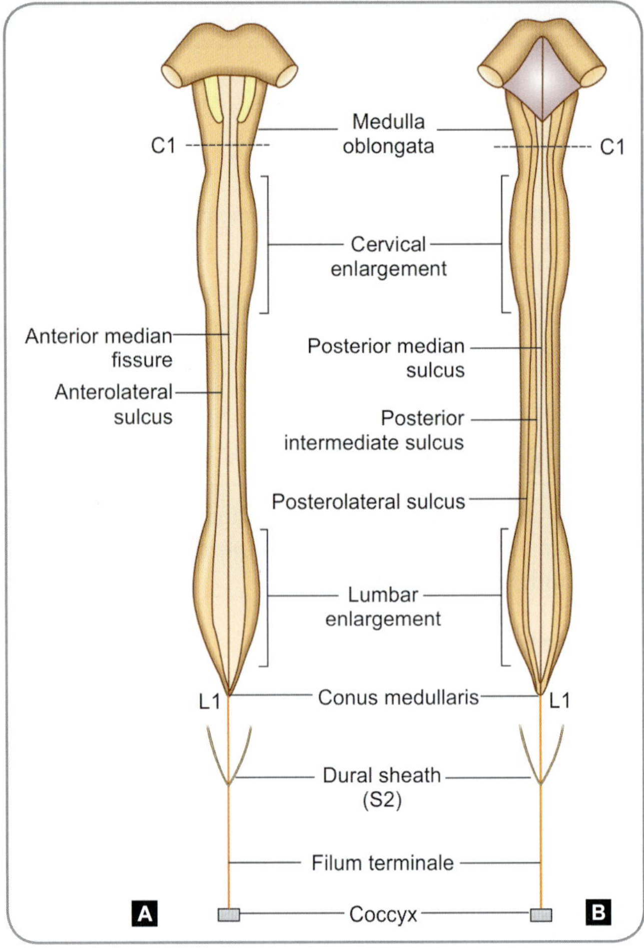

Figs 18.4A and B: External features of the spinal cord
A. Anterior aspect **B.** Posterior aspect

posterior, lateral, and anterior regions by anterolateral and posterolateral sulci (Figs 18.4A and B).

The rootlets of the dorsal or sensory roots of spinal nerves enter the cord at the posterolateral sulcus on either side.

The rootlets of the ventral or motor roots of spinal nerves emerge through the anterolateral sulcus on either side.

SPINAL NERVES

The spinal cord gives attachment on either side to 31 pairs of spinal nerves. Of these, 8 are cervical, 12 thoracic, 5 lumbar, 5 sacral, and 1 is coccygeal.

Each spinal nerve arises by two roots, (1) anterior motor root and (2) posterior sensory root (Fig. 18.5).

Each root is formed by aggregation of a number of rootlets that arise from the cord over a certain length (Fig. 18.6).

The rootlets that make up the dorsal nerve roots are attached to the surface of the spinal cord along a vertical groove (called the **posterolateral sulcus**) opposite the tip of the posterior grey column. The rootlets of the ventral nerve roots are attached to the anterolateral aspect of the cord opposite the anterior grey column.

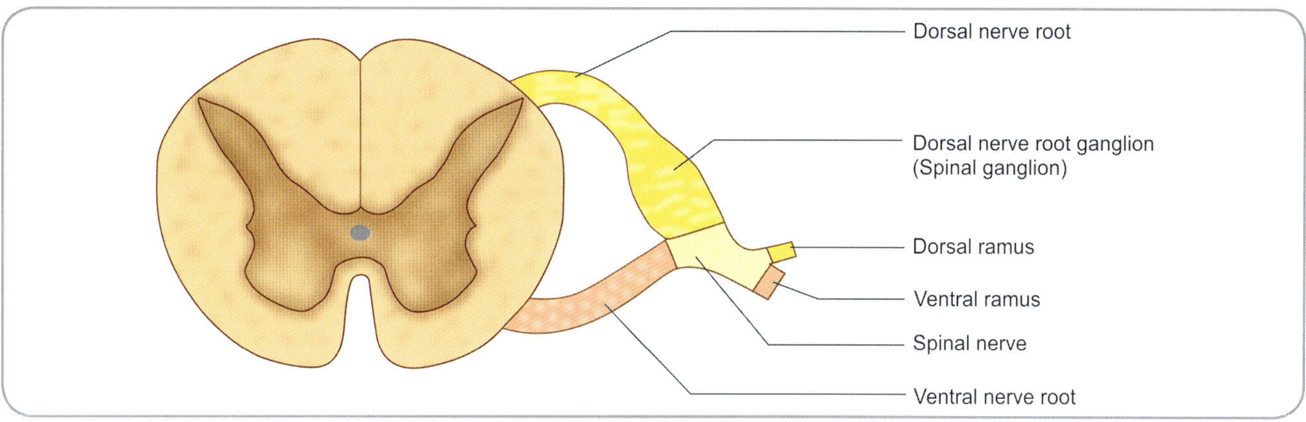

Fig. 18.5: Relationship of a spinal nerve to the spinal cord

Both the roots of spinal nerve receive a tubular prolongation from the spinal meninges and enter the corresponding intervertebral foramen.

In the intervertebral foramen, anterior and posterior spinal nerve roots unite to form the mixed spinal nerve trunk. Thus, a spinal nerve is made up of a mixture of motor and sensory fibres.

Just proximal to the junction of the two roots, the dorsal root is marked by a swelling called the ***dorsal nerve root ganglion*** or ***spinal ganglion*** (Fig. 18.5).

After emerging from the intervertebral foramen, each spinal nerve is divided into a dorsal and a ventral ramus (Fig. 18.5).

The dorsal ramus passes posteriorly around the vertebral column to supply the muscles and skin of the back. The ventral ramus continues anteriorly to supply the muscles and skin over the anterolateral body wall and all the muscles and skin of the limbs.

Spinal Segments

As discussed each spinal nerve arises from the spinal cord by two roots—(1) anterior (or ventral) and (2) posterior (or dorsal). Each nerve root is formed by the aggregation of a number of rootlets that arise from the cord over a certain length. The length of the spinal cord giving origin to the rootlets for one spinal nerve constitutes one spinal segment (Fig. 18.6). However, this definition applies only to the superficial attachment of nerve roots. The neurons associated with one spinal nerve extend well beyond the confinement of a spinal segment. So, the spinal cord is made up of 31 such segments; 8 cervical, 12 thoracic, 5 lumbar, 5 sacral and 1 coccygeal.

Clinical Correlation

The dorsal nerve root ganglia (and the sensory ganglia of cranial nerves) can be infected with a virus. This leads to the condition called ***herpes zoster***. Vesicles appear on the skin over the area of distribution of the nerve. The condition is highly painful.

Note: In the cervical and coccygeal regions, the number of spinal segments and spinal nerves, does not correspond to the number of vertebrae.

Vertebral Levels of Spinal Segments

Since the length of spinal cord (45 cm) is smaller than that of vertebral column (65 cm), the spinal segments are short and crowded, more so in the lower part of the cord. Thus, the spinal and vertebral segments (spines) do not lie at the same level. ***The spinal segments as a rule always lie above their numerically corresponding vertebral spines***. As a rough guide, it may be stated that in the cervical region, there is a difference of one segment (for example, the fifth cervical spine overlies the sixth cervical segment); in the upper thoracic region, there is a difference of two segments (for example, the fourth thoracic spine overlies the sixth thoracic segment); and in the lower thoracic region, there is a difference of three segments (for example, the ninth thoracic spine lies opposite the twelfth thoracic segment). Approximate spinal segments

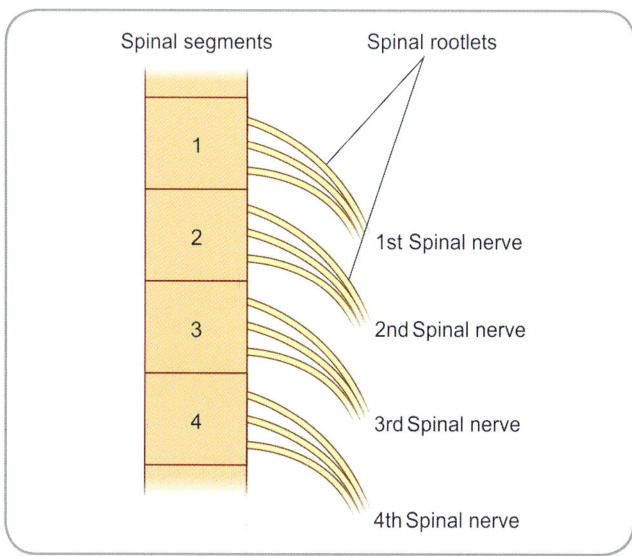

Fig. 18.6: Scheme to show the concept of spinal segments

Table 18.1: Approximate vertebral levels of the spinal cord segments

Region	General rule	Spinal segments	Vertebral level
Upper cervical	Same level	C2	C2
Lower cervical	Difference of one vertebra	C6	C5
Upper thoracic	Difference of two vertebrae	T5	T3
Lower thoracic	Difference of three vertebrae	T10	T7
Lumbar	Difference of three to five vertebrae	L1–L5	T10–T11
Sacral and coccygeal	Difference of six to ten vertebrae	S1–S5 and coccygeal segment	T12–L1

and the corresponding vertebral levels are shown in Table 18.1.

This is clinically important for assessing the level of cord compression following an injury or disease of the surrounding vertebrae.

Exit of Spinal Nerves

Each spinal nerve emerges through the intervertebral foramen. The cervical nerves leave the vertebral canal above the corresponding vertebrae with the exception of eighth, which emerges between seventh cervical and first thoracic vertebrae. Rest of the spinal nerves emerge below the corresponding vertebrae.

INTERNAL STRUCTURE

The cross-section of the spinal cord shows that it consists of an innercore of grey matter and a peripheral zone of white matter.

When seen in transverse section, the grey matter of the spinal cord forms an H-shaped mass (Fig. 18.7).

In each half of the cord, the grey matter is divisible into a larger ventral mass, the **anterior (or ventral) grey column** and a narrow elongated **posterior (or dorsal) grey column**. In some parts of the spinal cord, a small lateral projection of the grey matter is seen between the ventral and dorsal grey columns. This is the **lateral grey column**. The grey matter of the right and left halves of the spinal cord is connected across the middle line by the **grey commissure,** which is traversed by the **central canal** (Fig. 18.7). The lower end of the central canal expands to form the **terminal ventricle,** which lies in the conus medullaris. The cavity within the spinal cord continues for a short distance into the filum terminale. The central canal of the spinal cord contains cerebrospinal fluid. The canal is lined by the ependymal cells.

The white matter of the spinal cord is divided into right and left halves, in front by a deep **anterior median fissure** and behind by the **posterior median septum.** In each half of the cord, the white matter medial to the dorsal grey column forms the **posterior funiculus** (or posterior white column). The white matter medial and ventral to the anterior grey column forms the **anterior funiculus** (or anterior white column), while the white matter lateral to the anterior and posterior grey columns forms the **lateral**

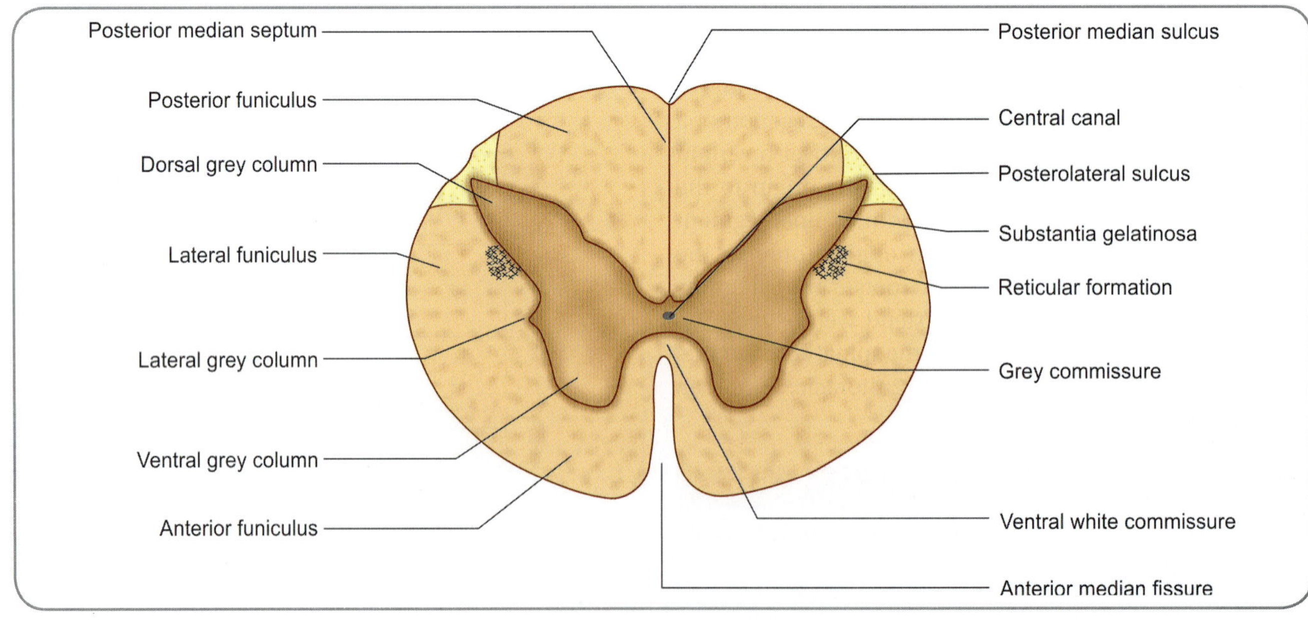

Fig. 18.7: Main features to be seen in a transverse section through the spinal cord

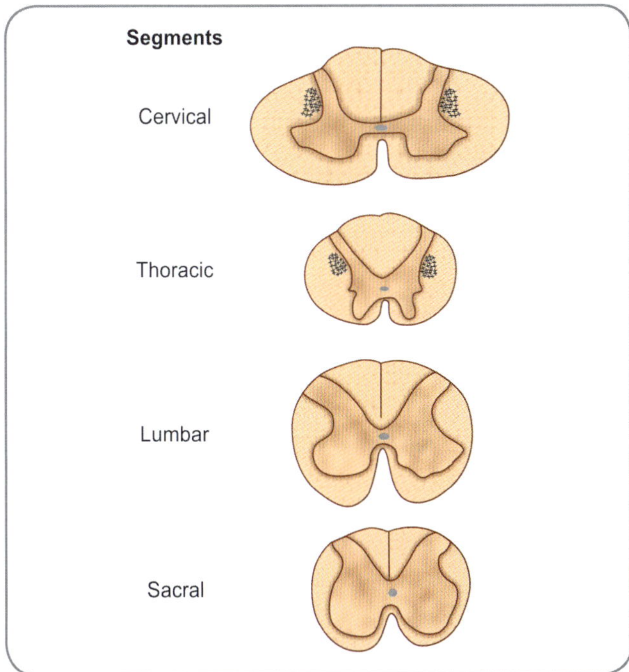

Fig. 18.8: Diagrams to show differences in the appearance of transverse sections through various levels of the spinal cord

funiculus. The anterior and lateral funiculi are collectively referred to as the ***anterolateral funiculus (Fig. 18.8)***.

The white matter of the right and left halves of the spinal cord is continuous across the middle line through the ***ventral white commissure***, which lies anterior to the ***grey commissure***. There is no clear cut dorsal white commissure. However, some myelinated fibres running transversely in the grey commissure, posterior to the central canal are referred to as the ***dorsal white commissure***.

VARIATIONS IN THE INTERNAL STRUCTURE OF SPINAL CORD AT DIFFERENT LEVELS

The relative amount of grey and white matter and the shape and size of the grey columns, vary at different levels of the spinal cord (Fig. 18.8). The amount of white matter undergoes progressive increase as we proceed up the spinal cord. This results from the fact that:

- Progressively, more and more ascending fibres are added in the upper cord, and
- The number of descending fibres decreases in the lower cord, as some of them terminate in each segment.

The amount of the grey matter, seen at a particular level, is well correlated with the mass of tissue it supplies. It is, therefore, maximum in the regions of cervical and lumbar enlargements, which supply the limbs and their associated girdles. The horns of grey are, thus, largest in the regions of cervical and lumbar enlargements.

GREY MATTER OF SPINAL CORD

Subdivisions of Grey Matter

The grey matter of the spinal cord may be subdivided in more than one manner.

Traditionally, the ventral grey column has been divided into a ventral part, the ***head***, and a dorsal part, the ***base***.

Similarly, the dorsal grey column has been subdivided (from anterior to posterior side) into a ***base***, a ***neck***, and a ***head***. These subdivisions have, however, been found to have little importance.

An attempt has been made to show the discrete collections of neurons (or nuclei) in various regions of the spinal grey matter. These are illustrated in the left half of spinal cord in the diagram given below and will not be described in detail. However, identify (Fig. 18.9) the following:

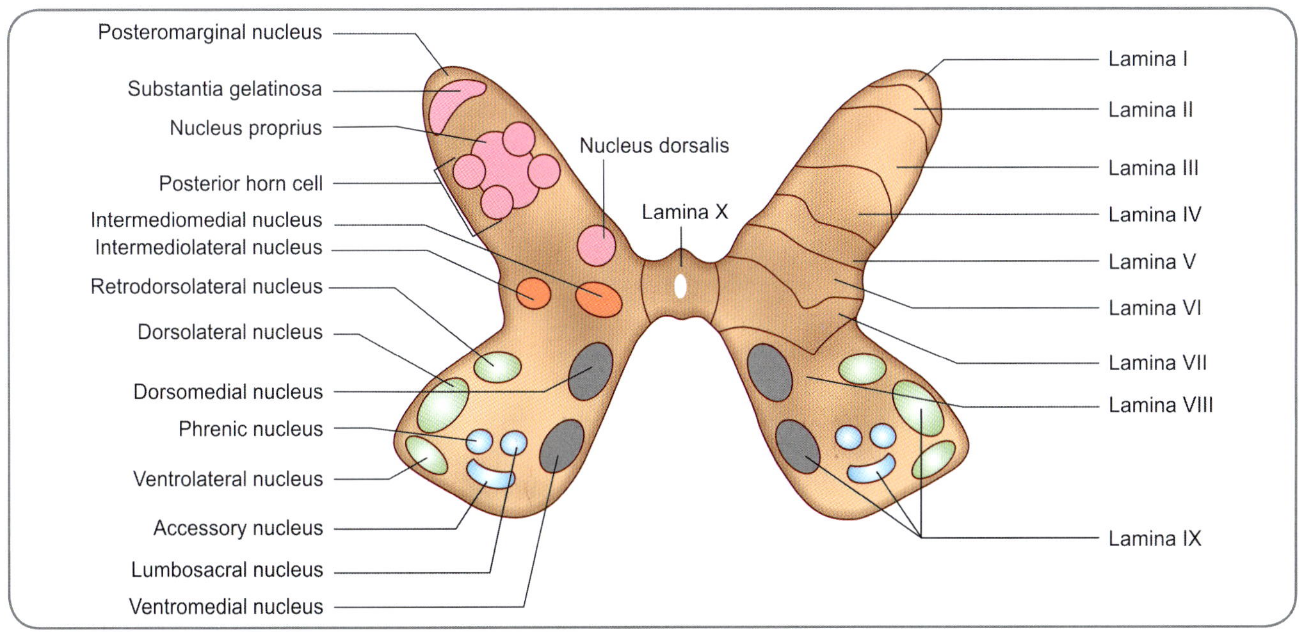

Fig. 18.9: Subdivisions of the grey matter of the spinal cord—the left half of the figure shows the cell groups and the right half shows the laminae

Table 18.2: Rexed laminae and nuclear groups (Fig. 18.9)	
Laminae	**Corresponding grey column nuclei**
I	Posteromarginal nucleus (lamina marginalis)
II	Substantia gelatinosa
III and IV	Nucleus propius
V and VI	Neck and base of dorsal grey column
VII	Intermediate region of grey matter including nucleus dorsalis (Clarke's column), interomediomedial, and interomediolateral nuclei of lateral horn
VIII and IX	Medial and lateral groups of ventral horn nuclei
X	Grey matter around the central canal consisting mainly of neuroglial cells

Substantia gelatinosa near the apex of the dorsal column.

Dorsal nucleus (also called the **thoracic nucleus** or **Clark's column**) lying on the medial side of the base.

Between the ventral and dorsal grey columns an intermediate zone is sometimes described. This contains the ***intermediolateral*** and ***intermediomedial*** nuclei.

Division of Spinal Grey Matter into Laminae

From the point of view of neuronal connections, the grey matter of the spinal cord may be divided into 10 areas or laminae, mapped out in the 1950s by Bror Rexed, a swedish Neuroscientist. This mapping is based on cellular structure. The laminae are shown in the right half of Fig. 18.9 and Table 18.2.

Lamina I corresponds to the posteromarginal nucleus (lamina marginalis; fibres conveying pain and temperature relay here), ***lamina II*** to the substantia gelatinosa, and ***laminae III*** and ***IV*** to the nucleus proprius (laminae I to IV correspond to the head of the dorsal grey column; some workers include lamina III in the substantia gelatinosa). Afferent fibres carrying sensations from the skin end predominantly, in laminae I to IV. ***Lamina V*** corresponds to the neck of the dorsal grey column, and ***lamina VI*** to the base. The lateral part of lamina V corresponds to the formatio reticularis. Proprioceptive impulses reach laminae V and VI. These laminae also receive numerous fibres from the cerebral cortex. ***Lamina VII*** corresponds to the intermediate grey matter and includes the intermediomedial, intermediolateral, and dorsal nuclei. Lamina VII gives off fibres that reach the midbrain and cerebellum (through spinocerebellar, spinotectal, and spinoreticular tracts). It receives fibres from these regions through tectospinal, reticulospinal, and rubrospinal tracts. At the level of the cervical and lumbar enlargements, this lamina extends into the lateral part of the ventral horn. Renshaw cells are located in a forward extension of lamina VII (into the interval between laminae VIII and IX). ***Lamina VIII*** occupies most of the ventral horn in the thoracic segments, but at the level of the limb enlargements, it is confined to the medial part of the ventral horn. Lamina VIII is made up of mainly interneurons that receive fibres from various sources. They give efferents to γ-neurons and, thus, influence muscle spindles.

Lamina IX is made up of the various discrete groups of ventral horn cells. Lamina IX contains the α- and γ-neurons that give off efferent fibres to skeletal muscle. Motor neurons in different parts of the ventral grey column show remarkable differences in the orientation and extent of their dendritic fields. Many of the dendrites of neurons in the ventromedial, central, and ventrolateral columns run longitudinally in the form of bundles. These neurons, therefore, come into intimate contact with each other. This arrangement is seen in relation to neurons that supply postural muscles. In contrast, the dendrites of neurons in the dorsolateral column have very little contact with those of neighboring neurons. This column supplies muscles in distal parts of the limbs and the discrete nature of the neurons may be associated with fine control necessary for movements produced by these muscles. ***Lamina X*** forms the grey matter around the central canal.

From the functional and neuronal organizational point of view, it has been proposed that (instead of division into ventral and dorsal grey columns) the spinal grey matter should be divided into a **central core** where the organization of neurons is diffuse and non-discriminative and into **dorsal and ventral appendages**. Laminae VII and VIII have been (tentatively) assigned to the central core, laminae I to VI to the dorsal appendage, and laminae IX to the ventral appendage.

Types of Neurons in the Spinal Grey Column

Anterior Grey Column

The ***ventral horn cells*** of the spinal cord may be functionally divided as follows:

- ***Alpha (α) neurons:*** The most prominent neurons with large cell bodies and prominent Nissl substance are designated as α-*neurons*. These are somatic efferent neurons. Their axons (α-***efferents***) leave the spinal cord through the ventral nerve roots of the spinal nerves and innervate skeletal muscle. They occupy lamina IX of the ventral grey column.

- *Gamma (γ) neurons:* Some smaller neurons designated as *γ-neurons* and are also located in lamina IX. They supply intrafusal fibres of muscle spindles. Sensory impulses arising in the spindle travel to the spinal cord and reach the α-neurons. The γ-neurons, thus, influence the activity of α-neurons indirectly through muscle spindles.
- *Interneurons:* A considerable number of smaller neurons in the ventral grey column are internuncial neurons. They are most abundant in lamina VIII. Some ramifications of the central processes of cells in the dorsal nerve root ganglia (bringing afferent impulses from the periphery) and axons descending from higher centres terminate in relation to these internuncial neurons. The axons of internuncial neurons convey these impulses to α- and γ-neurons.
- *Renshaw cell:* Another variety of neuron that is believed (on physiological grounds) to exist in the ventral grey column is the so called *Renshaw cell*. These cells receive the terminations of collaterals arising from the axons of α-neurons. The axons of Renshaw cells carry the impulses back to the cell bodies of the same α-neurons and, thus, help regulate their activity.

Posterior Grey Column

The *neurons of the dorsal grey column* may be subdivided as follows:

- Some of these are internuncial neurons similar to those in the ventral grey column.
- Many dorsal column neurons receive afferent impulses through the central processes of neurons in dorsal nerve root ganglia. These dorsal column neurons give off axons that enter the white matter of the spinal cord, either on the same or opposite sides. These axons may behave in one of the following ways:
  - They may ascend or descend for some segments before terminating in relation to neurons at other levels of the spinal cord. Such axons constitute *intersegmental tracts*.
  - A considerable number of axons arising from dorsal column neurons run upwards in the spinal cord and constitute *ascending tracts*, which terminate in various masses of grey matter in the brain. These tracts form a considerable part of the white matter of the spinal cord.

Lateral Grey Column

The *neurons of the intermediolateral group* (lateral grey column) are visceral efferent neurons. They are present at two levels of the spinal cord.

- One group is present in the thoracic and upper two or three lumbar segments. These are preganglionic neurons of the sympathetic nervous system. Their axons terminate in relation to postganglionic neurons in sympathetic ganglia (and occasionally, in some other situations). Axons of these postganglionic neurons are distributed to various organs and to blood vessels.
- The second group of visceral efferent neurons is found in the second to fourth sacral segments of the spinal cord. These are preganglionic neurons of the parasympathetic nervous system. Their axons leave the spinal cord through the ventral nerve roots to reach spinal nerves. They leave the spinal nerves as the *pelvic splanchnic nerves*, which are distributed to some viscera in the pelvis and abdomen. They end by synapsing with ganglion cells located in the wall of the viscera concerned. The postganglionic fibres arising in these ganglia are short and supply smooth muscles and glands within these viscera.

WHITE MATTER OF SPINAL CORD

The anterior, lateral and posterior funiculi of the spinal cord are made up of nerve fibres running up or down the cord. These constitute the ascending and descending tracts which are described in chapter 20.

BLOOD SUPPLY OF SPINAL CORD

Arterial Supply

The arterial supply of the cord is derived from the following arteries (Fig. 18.10):

- Anterior spinal artery,
- Two posterior spinal arteries and
- The radicular arteries.

In addition to these channels, the pia mater covering the spinal cord has an arterial plexus (called the *arterial vasocorona*), which also sends branches into the substance of the cord.

Anterior Spinal Artery

The anterior spinal artery is formed in the posterior cranial fossa by the union of the right and left anterior spinal arteries (which are the branches of the fourth part of the vertebral artery). The anterior spinal artery descends through the foramen magnum and runs down in the anterior median fissure of the spinal cord.

Posterior Spinal Arteries

The right and left posterior spinal arteries are the branches of the fourth part of the vertebral arteries. Each posterior spinal artery descends through the foramen magnum as two branches, which pass one in front and the other behind the dorsal roots of the spinal nerves.

Radicular Arteries

The main source of blood to the spinal arteries is from the vertebral arteries (from which the anterior and posterior

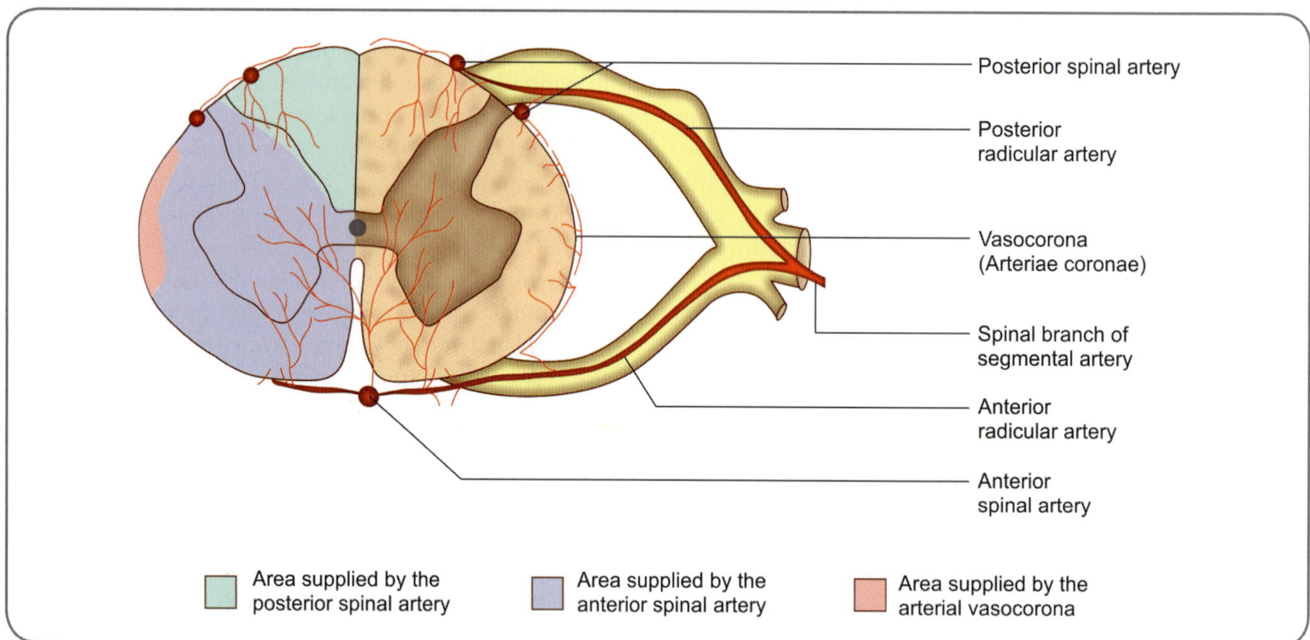

Fig. 18.10: Intrinsic arterial supply of the spinal cord—In the left half of the figure, the area showing green shading is supplied by the posterior spinal artery. The part showing pink shading is supplied by the arterial vasocorona and the area with purple shading is supplied by the anterior spinal artery

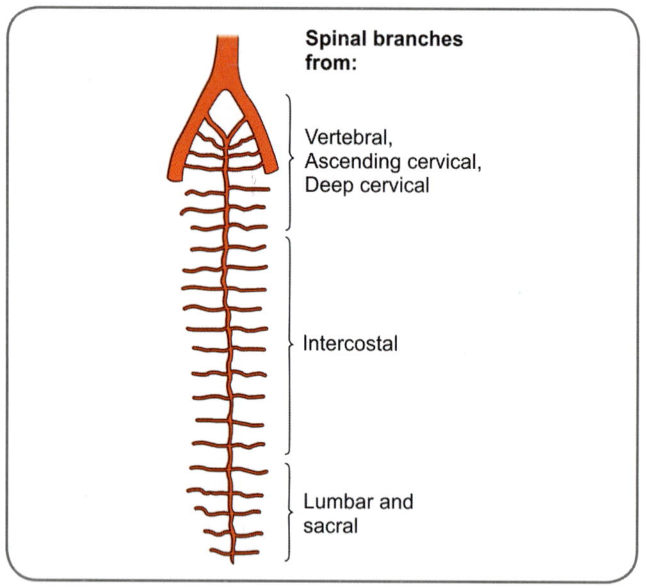

Fig. 18.11: Radicular arteries that contribute blood to the spinal arteries

spinal arteries take origin). However, the blood from the vertebral arteries reaches only up to the cervical segments of the cord. Lower down, the spinal arteries receive blood through radicular arteries that reach the cord along the roots of spinal nerves. These radicular arteries arise from spinal branches of the vertebral, ascending cervical, deep cervical, intercostal, lumbar, and sacral arteries (Fig. 18.11).

Arteria Radicularis Magna

Many of these radicular arteries are small and end by supplying the nerve roots. A few of them, which are larger, join the spinal arteries and contribute blood to them. Frequently, one of the anterior radicular branches is very large and is called the **arteria radicularis magna**. Its position is variable. This artery may be responsible for supplying blood to as much as the lower two-thirds of the spinal cord.

Intrinsic Blood Supply

The greater part of the cross-sectional area of the spinal cord, roughly the anterior two-third, is supplied by branches of the anterior spinal artery (Fig. 18.10). These branches enter the anterior median fissure (or sulcus) and are, therefore, called **sulcal branches**. Alternate sulcal branches pass to the right and left sides. They supply the anterior and lateral grey columns and the central grey matter. They also supply the anterior and lateral funiculi. The rest (posterior one-third) of the spinal cord is supplied by the posterior spinal arteries (Fig. 18.10). As already mentioned, branches from the arterial vasocorona also supply the cord.

Clinical Correlation

Anterior Spinal Artery Syndrome

Thrombosis in the anterior spinal artery produces a characteristic syndrome. The territory of supply includes the corticospinal tracts. This leads to an upper motor neuron paralysis below the level of lesion. The spinothalamic tracts are also involved. This leads to loss of sensations of pain and temperature below the level of lesion. Touch and conscious proprioceptive sensations are not affected as the posterior column tracts are not involved. The extent of damage varies depending on the efficiency of anastomoses in the region.

Venous Drainage

The veins draining the spinal cord are arranged in the form of six longitudinal channels (Fig. 18.12). These are:

- *Two median longitudinal channels*, one in the anterior median fissure called the *anteromedian channel*, and the other in the posteromedian sulcus called the *posteromedian channel*.
- *The paired anterolateral channels*, one on either side, posterior to the anterior nerve roots.
- *The paired posterolateral channels*, one on either side posterior to the posterior nerve roots.

These channels are interconnected by a plexus of veins that form a *venous vasocorona*. The blood from these veins is drained into radicular veins that open into a venous plexus lying between the dura mater and the bony vertebral canal (*epidural* or *internal vertebral venous plexus*) and through it into various segmental veins.

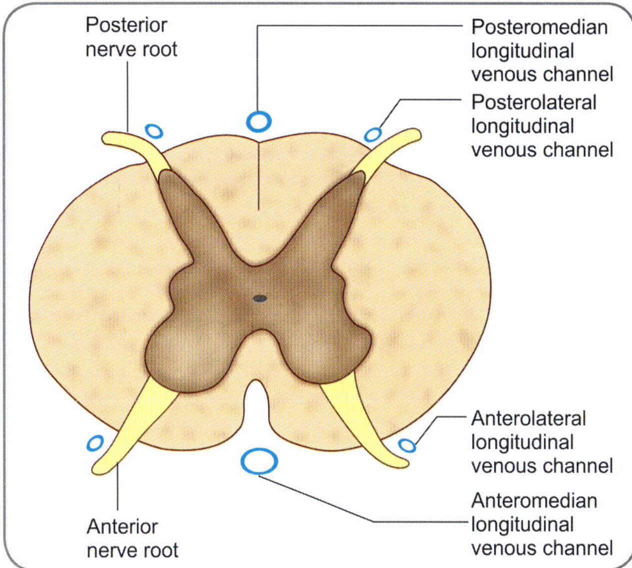

Fig. 18.12: Venous drainage of the spinal cord

Multiple Choice Questions

1. Extending up to the level of S2 vertebra:
   a. Is the dura mater only
   b. Is the arachnoid mater only
   c. Are the dura, arachnoid, and pia maters
   d. Are the dura and arachnoid maters
2. Substantia gelatinosa occupies the:
   a. Dorsal column of the spinal cord
   b. Lateral column of the spinal cord
   c. Medial part of the ventral column of the spinal cord
   d. Lateral part of the ventral column of the spinal cord
3. Venous vasocorona:
   a. Is the venous plexus connecting the longitudinal venous channels of spinal cord
   b. Is the other name for internal vertebral venous plexus

   c. Is the arteriovenous channel between the radicular arteries and their corresponding veins
   d. Is the epidural venous plexus
4. The preganglionic neurons of the sympathetic nervous system are located in the:
   a. Lower two thoracic and upper three sacral spinal segments
   b. Thoracic and lumbar spinal segments
   c. Lumbosacral spinal segments
   d. The sympathetic ganglia
5. Alpha neurons are present in:
   a. Lamina V
   b. Lamina VI
   c. Lamina IX
   d. Lamina X

ANSWERS

1. d   **2.** a   **3.** a   **4.** c   **5.** c

Clinical Problem-solving

Case Study 1: A 37-year-old man complained of severe pain on his back. On examination, the pain could be located on the lower left lumbar segments. Two days later, vesicles appeared on the skin over the same segments.
- What do you think is the condition the man is suffering from?
- Which neurological part/structure is specifically infected?
- Will the muscles of the region be affected? If yes, why? If no, why?

Case Study 2: A 45-year-old patient in the neurological ward of the hospital was subjected to lumbar puncture for examination of cerebrospinal fluid. You are a member of the medical team. Your senior colleague wants you to evaluate the speed with which CSF flows out.
- Is there any scientific basis for his directive?
- If so, how can such evaluation be done?
- What other features of the cerebrospinal fluid will help in establishing a diagnosis?

(For solutions see Appendix).

Chapter 19

Brainstem—Gross Anatomy and Internal Structure

INTRODUCTION

The brainstem consists of the midbrain, pons, and medulla from above downwards (Fig. 19.1):

Superiorly, the brainstem (midbrain) is continuous with the structures forming the forebrain—thalamus, hypothalamus, and cerebral hemispheres (Fig. 19.2). Inferiorly, it is continuous with the spinal cord (Fig. 19.1).

Posteriorly, the pons and medulla are separated from the cerebellum by the fourth ventricle (Fig. 19.2). The ventricle is continuous below with the central canal, which traverses the lower part of the medulla and

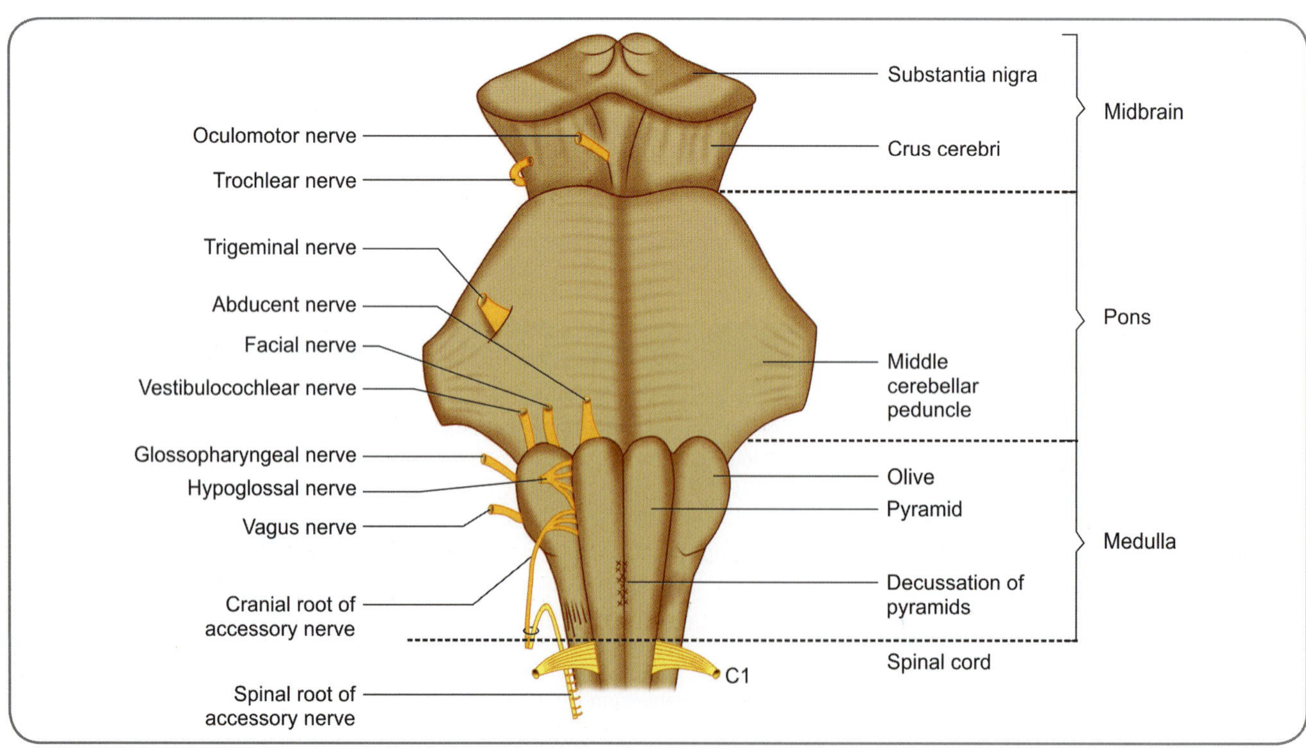

Fig. 19.1: Ventral aspect of the brainstem

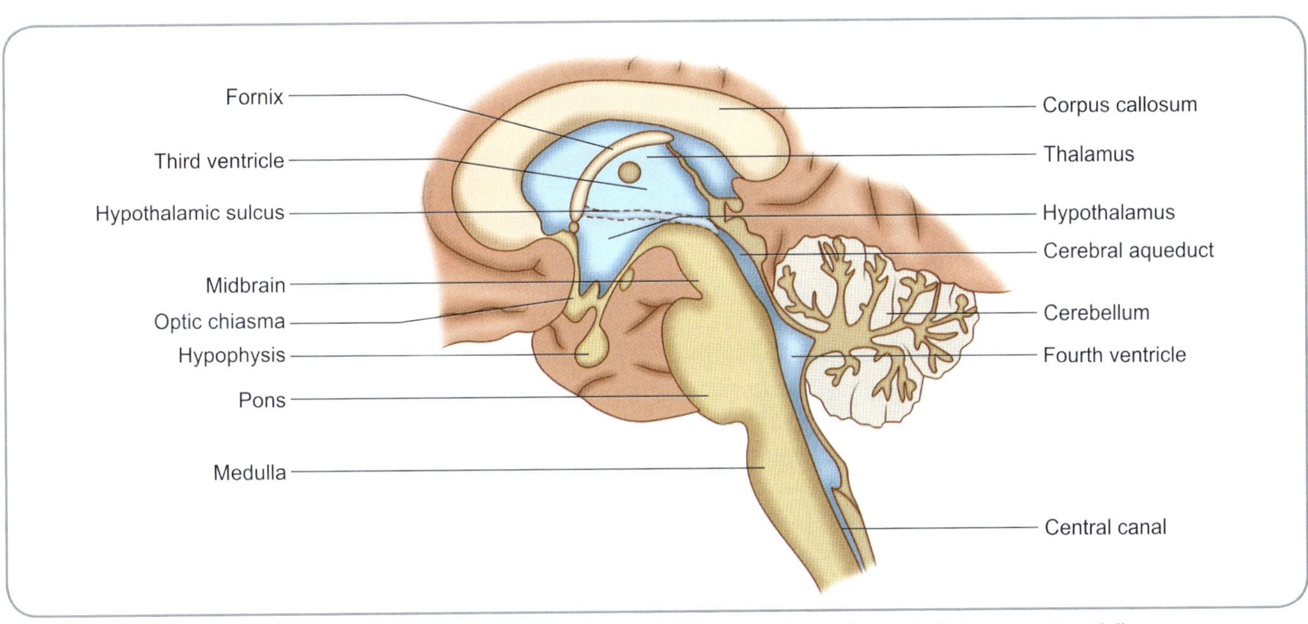

Fig. 19.2: Midsagittal section of the brain showing midbrain, pons and medulla – note that pons and medulla are separated from the cerebellum by the fourth ventricle

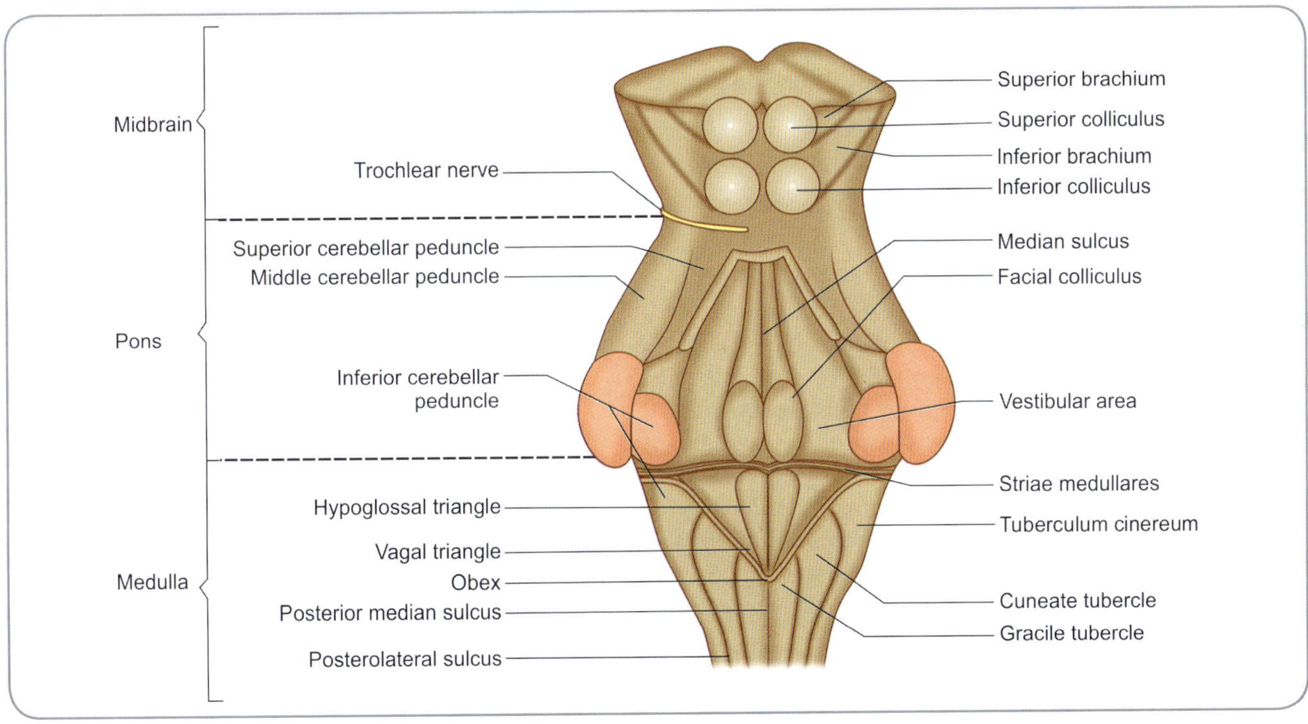

Fig. 19.3: Dorsal aspect of the brainstem

becomes continuous with the central canal of the spinal cord. Cranially, the fourth ventricle is continuous with the aqueduct, which passes through the midbrain (Fig. 19.2).

The midbrain, pons, and medulla are connected to the cerebellum posteriorly by the superior, middle, and inferior cerebellar peduncles, respectively (Fig. 19.3).

A number of cranial nerves are attached to the brainstem. The third and fourth nerves emerge from the surface of the midbrain and the fifth, from the pons. The sixth, seventh, and eighth nerves emerge at the junction of the pons and medulla. The ninth, tenth, eleventh, and twelfth cranial nerves emerge from the surface of the medulla (Fig. 19.1).

MEDULLA

Gross Anatomy

The medulla (or medulla oblongata; latin.medis=middle, oblongus=long) is broad above where it joins the pons and narrows down below where it becomes continuous with the spinal cord. Its length is about 3 cm, and its width is about 2 cm at its upper end. The junction of the medulla and cord lies at the level of the upper border of the atlas vertebra. The transition is, in fact, not abrupt but occurs over a certain distance.

The medulla is divided into a lower closed part, which surrounds the central canal and an upper open part, which is related to the lower part of the fourth ventricle.

External Features

The surface of the medulla is marked by a series of fissures or sulci that divide it into a number of regions. The *anterior median fissure* and the *posterior median sulcus* are upward continuations of the corresponding features seen in the spinal cord. Each half of the medulla is marked by two sulci, the anterolateral and posterolateral sulci which are continuations of the corresponding sulci of the cord. The *anterolateral sulcus* lies in line with the ventral roots of spinal nerves. The rootlets of the hypoglossal nerve emerge from this sulcus. The *posterolateral sulcus* is in line with the dorsal nerve roots of spinal nerves and gives attachment to rootlets of the glossopharyngeal, vagus, and accessory nerves. The anterolateral and posterolateral sulci with nerve roots divide the surface of each half of the medulla oblongata into anterior, posterior, and lateral regions like that in the spinal cord.

Anterior (Ventral) Aspect

- **Pyramid:** The region between the anterior median sulcus and the anterolateral sulcus is occupied (on either side of the midline) by an elevation called the *pyramid* (Fig. 19.1). The elevation is caused by a large bundle of fibres that descend from the cerebral cortex to the spinal cord. Some of these fibres cross from one side to the other in the lower part of the medulla, obliterating the anterior median fissure. These crossing fibres constitute the *decussation of the pyramids*.

  Some other fibres emerge from the anterior median fissure, above the decussation and wind laterally over the surface of the medulla. These are the *anterior external arcuate fibres*.
- **Olive:** In the upper part of the medulla, the region between the anterolateral and posterolateral sulci shows a prominent, elongated, oval swelling, named the *olive*. This swelling is about half an inch long. It is produced by a large mass of grey matter called the *inferior olivary nucleus* (Fig. 19.1).
- **Rootlets of the hypoglossal nerve:** These emerge from the anterolateral sulcus between the pyramid and the olive (Fig. 19.1).
- ***Inferior cerebellar peduncles*** of the left and right side attach the medulla with the cerebellum. They also form the inferolateral boundaries of the lower half of fourth ventricle on the posterior aspect of the open part of medulla.

- ***Rootlets of the ninth, tenth, and twelfth (cranial part) cranial nerves:*** These emerge through the posterolateral sulcus separating the olive from the inferior cerebellar peduncle.

Posterior (Dorsal) Aspect

The posterior surface of the lower part (closed part) medulla, between the posterior median sulcus and the posterolateral sulcus, contains tracts that enter from the posterior funiculus of the spinal cord. These are the *fasciculus gracilis,* next to the midline, and the *fasciculus cuneatus,* placed laterally. These fasciculi end in rounded elevations called the *gracile and cuneate tubercles*. These tubercles are produced by masses of grey matter called the *nucleus gracilis* and the *nucleus cuneatus,* respectively.

The lower part of the medulla, immediately lateral to the fasciculus cuneatus, is marked by another longitudinal elevation called the *tuberculum cinereum*. This elevation is produced by an underlying collection of grey matter of the *spinal nucleus of the trigeminal nerve*. The grey matter of this nucleus is covered by a layer of nerve fibres that form the *spinal tract of the trigeminal nerve*.

The posterior surface of the upper medulla (open part) forms the lower part of the floor of the fourth ventricle. This fossa is bounded on either side by the inferior cerebellar peduncles.

Internal Structure

The arrangement of grey and white matter in the lowermost part of medulla is similar to that of spinal cord. However, above this, its internal structure changes gradually. The change in the arrangement of grey and white matter in the upper part of the medulla is mainly due to the presence of the fourth ventricle.

The internal structure of medulla is generally studied at the following levels (Fig. 19.4):
1. At the level of the pyramidal decussation
2. At the level of the sensory decussation
3. At the level of the olivary nucleus.

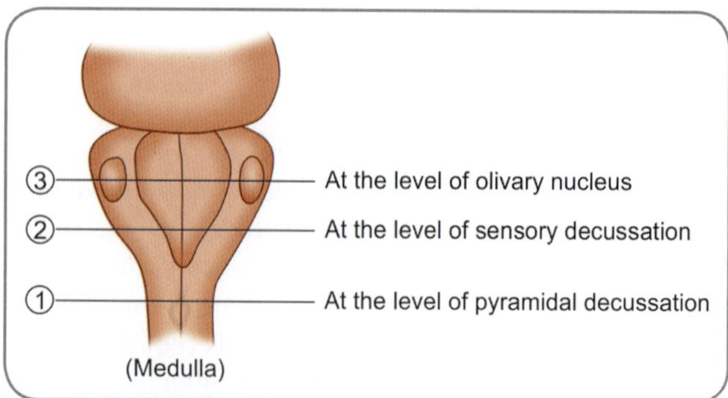

Fig. 19.4: Illustration shows the three levels at which transverse sections of medulla are studied

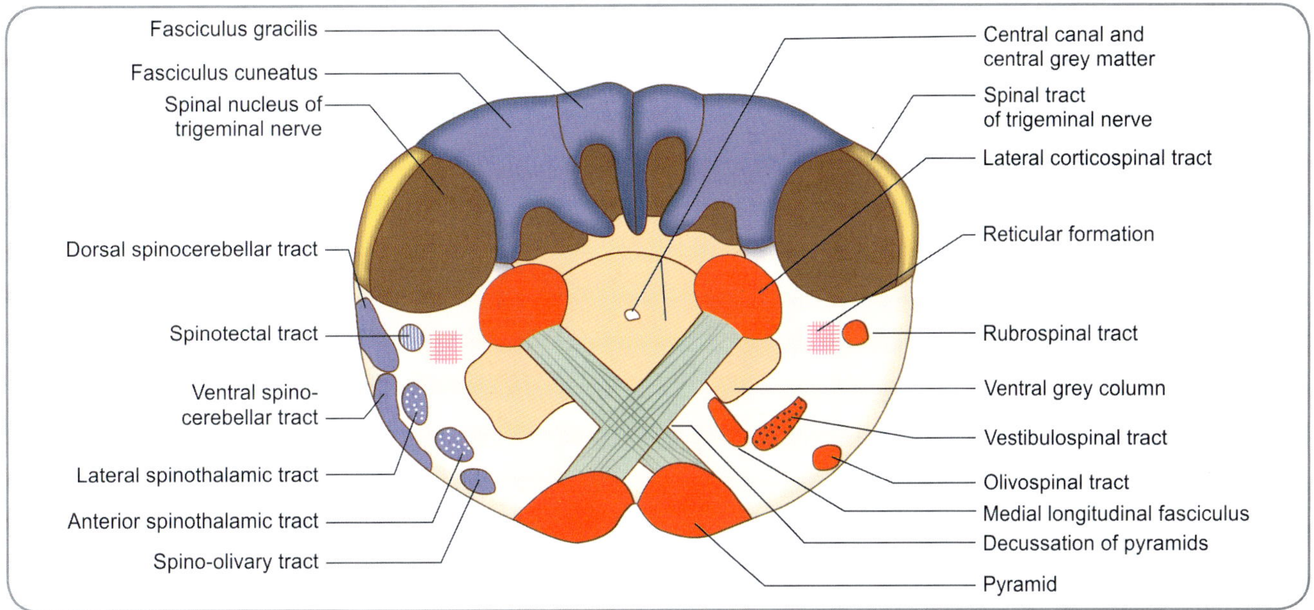

Fig. 19.5: Transverse section through the medulla at the level of the pyramidal decussation

A. Section through the Medulla at the Level of the Pyramidal Decussation

A section at the level of the pyramidal decussation (Fig. 19.5) shows some similarity to sections through the spinal cord.

The **central canal** is surrounded by **central grey matter**. The **ventral grey columns** are present but are separated from the central grey matter by the **decussating pyramidal fibres**.

The neurons in the ventral grey column give origin to the uppermost rootlets of the first cervical nerve and to some fibres in the spinal root of the accessory nerve. The area between the ventral grey column and the spinal nucleus of the trigeminal nerve is occupied by a network of fibres and scattered nerve cells called **reticular formation**.

The region behind the central grey matter is occupied by the **fasciculus gracilis** medially and by the **fasciculus cuneatus** laterally. Closely related to these fasciculi, there are two tongue-shaped extensions of the central grey matter. The medial of these extensions is the **nucleus gracilis** and the lateral is the **nucleus cuneatus**. More laterally, there is the **spinal nucleus of the trigeminal nerve**. When traced inferiorly, this nucleus reaches the second cervical segment of the spinal cord, where it becomes continuous with the substantia gelatinosa. Above, the nucleus extends as far as the upper part of the pons. The spinal nucleus of the trigeminal nerve is related superficially to the **spinal tract** of the nerve.

The main descending fibres seen at this level are the corticospinal fibres, on their way from the cerebral cortex to the spinal cord. At this level in the medulla, many of these fibres run backwards and medially to cross in the midline. These crossing fibres constitute the **decussation of the pyramids.** After crossing the middle line, these fibres turn downwards in the region lateral to the central grey matter to form the **lateral corticospinal tract**.

Those fibres of the pyramids that do not cross, descend into the ventral funiculus of the spinal cord to form the **ventral corticospinal tract**.

Other descending tracts seen at this level (in the anterolateral part of the medulla, Fig. 19.5, right half) are

❑ Rubrospinal tract,
❑ Vestibulospinal tract,
❑ Olivospinal tract and
❑ Tectospinal tract.

The tectospinal tract is incorporated within the medial longitudinal fasciculus.

Among the descending tracts may also be included the spinal tract of the trigeminal nerve, which forms a layer of fibres superficial to the spinal nucleus of this nerve.

The **ascending tracts** seen at this level include (Fig. 19.5, left half),

❑ Fasciculus gracilis and fasciculus cuneatus (which occupy the areas behind the corresponding nuclei),
❑ Spinothalamic tract,
❑ Spinocerebellar tract,
❑ Spinotectal tract and
❑ Spino-olivary tract.

B. Section through the Medulla at the Level of the Sensory Decussation (Lemniscal Decussation)

This level lies a little above the level of the pyramidal decussation (Fig. 19.6).

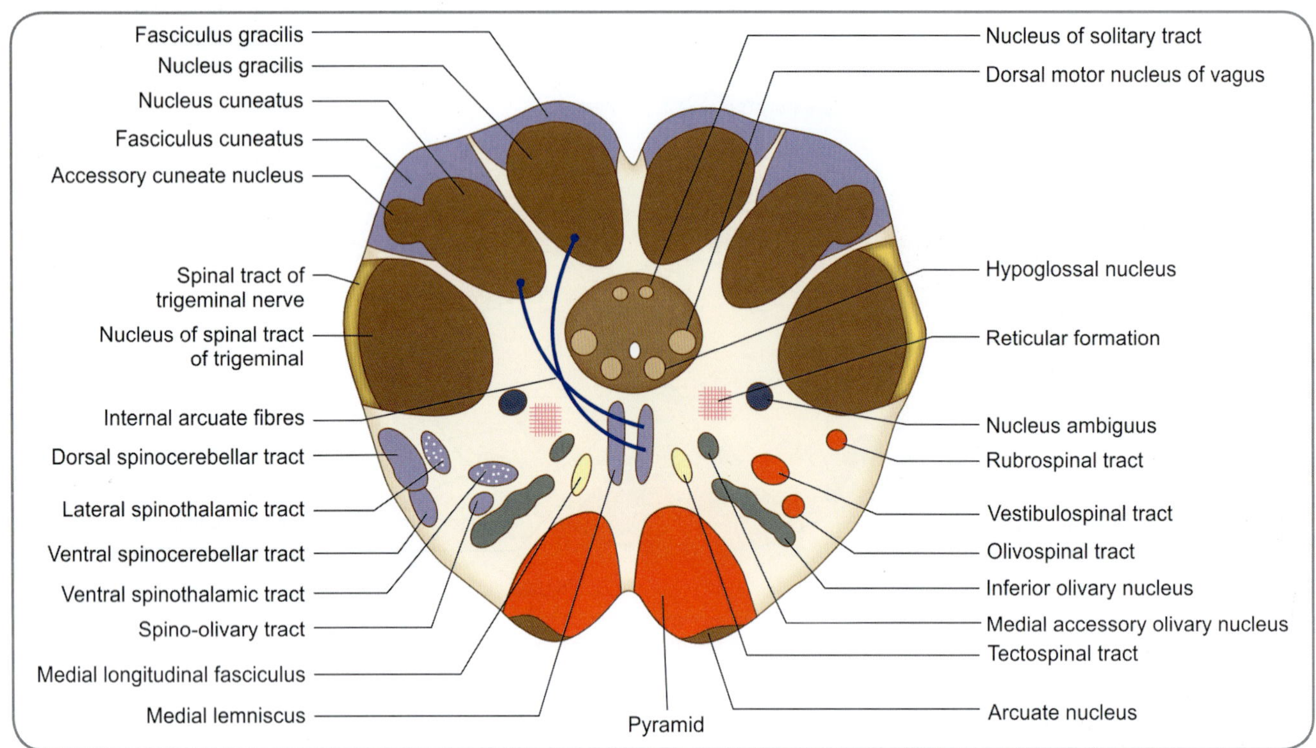

Fig. 19.6: Transverse section through the medulla at the level of sensory decussation

The central canal is surrounded by the central grey matter. The nucleus gracilis, the nucleus cuneatus, the spinal nucleus of the trigeminal nerve, and the pyramids occupy the same positions as at lower levels.

The nucleus gracilis and the nucleus cuneatus are, however, much larger and are no longer continuous with the central grey matter. Internal arcuate fibres arising in these nuclei arch ventrally and medially around the central grey matter to cross the middle line. These crossing fibres constitute the *lemniscal (or sensory) decussation*. After crossing the midline, these fibres turn cranially to constitute the *medial lemniscus*. As the fibres from the nucleus gracilis and the nucleus cuneatus pass ventrally, they cross each other so that the fibres from the nucleus gracilis come to lie ventral to those from the nucleus cuneatus. The most medial fibres (from the legs) lie most anteriorly in the medial lemniscus. These are followed by fibres from the trunk and upper limb in that order. At higher levels in the brainstem, the medial lemniscus changes its orientation; its long axis (as seen in crosssection) becomes transverse (Fig. 19.10). The most anterior fibres become lateral and the posterior fibres become medial. In its course through the medulla, the medial lemniscus is joined by the anterior spinothalamic tract.

Fibres in the medial lemniscus are arranged in layers corresponding to spinal segments; those from segment C1 are most medial and those from S4 are most lateral.

The *accessory cuneate nucleus* is placed dorsolateral to the cuneate nucleus. It receives proprioceptive impulses from the upper limb through fibres arising in spinal grey matter of cervical segments of the cord. Efferents of the accessory cuneate nucleus constitute the *posterior external arcuate fibres*. They reach the cerebellum through the inferior cerebellar peduncle of the same side.

A number of cranial nerve nuclei can be identified at this level. Several of these are present in relation to the central grey matter. The hypoglossal nucleus is located ventral to the central canal just lateral to the middle line. The dorsal vagal nucleus lies dorsolateral to the hypoglossal nucleus. The nucleus of the solitary tract is seen dorsal to the central canal near the middle line. The lower ends of these nuclei become continuous with each other to form the commissural nucleus of the vagus. The nucleus ambiguus lies in the reticular formation medial to the spinal nucleus of the trigeminal nerve.

Other masses of grey matter that may be recognised at this level are:

- The lowest part of the *inferior olivary nucleus,*
- The *medial accessory olivary nucleus,* which lies dorsal to the medial part of the inferior olivary nucleus,
- The *lateral reticular nucleus* lying in the lateral part of the reticular formation and
- *Arcuate nuclei* lying on the anterior aspect of the pyramids.

The region lateral to the medial lemniscus contains scattered neurons mixed with nerve fibres. This region is

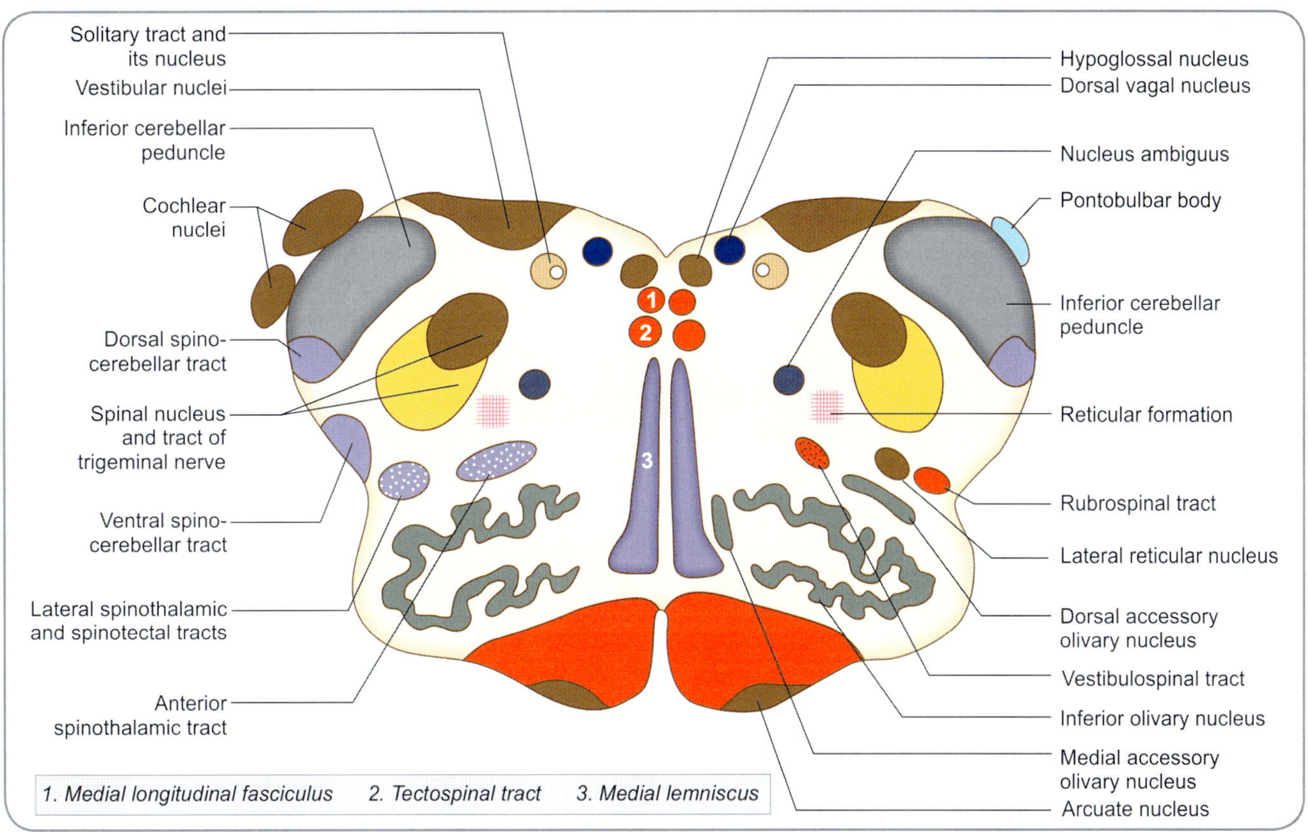

Fig. 19.7: Transverse section through the medulla at the level of the olive

the *reticular formation*. More laterally, there is a mass of white matter containing various tracts.

The ascending tracts present at this level are (Fig. 19.6, left half)

- ❑ The gracile and cuneate fasciculi—these are much smaller than at lower levels, as the fibres of these tracts progressively terminate in the gracile and cuneate nuclei,
- ❑ Spinothalamic tract,
- ❑ Spinocerebellar tract,
- ❑ Spinotectal tract and
- ❑ Spino-olivary tracts all of which lie in the anterolateral region.

The descending tracts present are (Fig. 19.6, right half)

- ❑ Pyramids,
- ❑ Rubrospinal tract,
- ❑ Vestibulospinal tract,
- ❑ Olivospinal tract and
- ❑ Medial longitudinal fasciculus which includes the tectospinal tract.

C. Section through the Medulla at the Level of the Olive

A section through the medulla at the level of the olive is shown in Figure 19.7.

The pyramids, medial lemniscus, spinal nucleus and tract of the trigeminal nerve, and reticular formation are present in the same relative position as at lower levels. The medial lemniscus is, however, much more prominent and is somewhat expanded anteriorly. Lateral to the spinal nucleus (and tract) of the trigeminal nerve, a large compact bundle of fibres is seen. This is the ***inferior cerebellar peduncle,*** which connects the medulla to the cerebellum. Posteriorly, the medulla forms the floor of the fourth ventricle.

Several cranial nerve nuclei can be recognized in relation to the floor of the fourth ventricle (Fig. 19.7). From medial to lateral side, these are the hypoglossal nucleus, the dorsal vagal nucleus, and the vestibular nuclei. The solitary tract and its nucleus lie ventrolateral to the dorsal vagal nucleus. The nucleus ambiguus is located much more ventrally within the reticular formation.

The dorsal and ventral cochlear nuclei are seen in relation to the inferior cerebellar peduncle. They are shown schematically in Figure 19.7. They are prominent at higher levels of the medulla, near its junction with the pons.

Other masses of grey matter present are the medial and dorsal accessory olivary nuclei (lying medial and dorsal, respectively, to the inferior olivary nucleus), the lateral reticular nucleus and arcuate nuclei, which occupy the same relative positions as at lower levels. The pontobulbar body lies on the dorsolateral aspect of the inferior cerebellar peduncle (Fig. 19.7, right half).

The descending tracts seen at this level (Fig. 19.7, right half)

- Pyramids,
- Tectospinal tract,
- Vestibulospinal tract,
- Rubrospinal tract and
- Spinal tract of trigeminal nerve.

The ascending tracts seen at this level are (Fig. 19.7, left half)

- Medial lemniscus, which forms an anteroposterior L-shaped band next to the middle line,
- Spinothalamic tract,
- Spinocerebellar tract and
- Spinotectal tract.

At this level, the dorsal spinocerebellar tract lies within the inferior cerebellar peduncle. The ventral spinocerebellar tract lies more anteriorly near the surface of the medulla. The spinothalamic tracts lie dorsolateral to the inferior olivary nucleus. The medial longitudinal fasciculus lies dorsal to the medial lemniscus.

Arterial Supply of Medulla

The medulla is supplied by the following arteries:

- Vertebral arteries (two in number),
- Anterior and posterior spinal arteries,
- Anterior and posterior inferior cerebellar arteries and
- Basilar artery.

Clinical Correlation

Injury to Medulla

Injury to the medulla is usually fatal because vital centres controlling the heart and respiration are located here. Paralysis due to a lesion in the medulla is called **bulbar palsy**. In this condition, the ninth, tenth, eleventh and twelfth cranial nerves are affected. The tracts are closely packed as they pass through the brainstem; hence, lesions produce widespread effect. This may result in paralysis of the muscles on the opposite side (due to damage to the corticospinal tract) and loss of sensation of the opposite side (due to damage to ascending sensory tracts).

PONS

Gross Anatomy

Pons is a part of the brainstem, situated between the medulla, below and midbrain, above (Figs 19.1 to 19.4).

It lies in the posterior cranial fossa on the clivus, anterior to the cerebellum.

Pons, in a literal sense, means *'the bridge'.* It is so named because it acts as a conduit for the passage of fibres from one side of the cerebellum to the other by its transverse fibres constituting the middle cerebellar peduncle as well as vertically between the medulla below, and midbrain, above. Pons is important physiologically, as the centre of

respiration is present in it. Nuclei of the cranial nerves, V (trigeminal), VI (abducent), VII (facial), and VIII (vestibulocochlear) originate from pons.

External Features

Pons shows a convex anterior surface, marked by prominent transversely running fibres. Laterally, these fibres collect to form a bundle, the **middle cerebellar peduncle**.

The **trigeminal nerve** emerges from the anterior surface and the point of its emergence is taken as a landmark to define the plane of junction between the pons and the middle cerebellar peduncle.

The anterior surface of the pons is marked, in the midline, by a shallow groove, the **sulcus basilaris**, which lodges the basilar artery.

The line of junction between the pons and the medulla is marked by a groove through which a number of cranial nerves emerge. The abducent nerve emerges from just above the pyramid and runs upwards in close relation to the anterior surface of the pons. The facial and vestibulocochlear nerves emerge from the interval between the olive and the pons.

The posterior aspect of the pons forms the upper part of the floor of the fourth ventricle.

On either side of the lower part of the pons, there is a region called the **cerebellopontine angle**. This region lies near the lateral aperture of the fourth ventricle. The facial, vestibulocochlear, and glossopharyngeal nerves, the nervus intermedius, and sometimes, the labyrinthine arteries lie in this region.

Divisions of Pons

The pons is divisible into a ***ventral part*** (basilar) and a ***dorsal part*** (tegmentum) (Fig. 19.8).

Structure of the Basilar Part of Pons

The ventral (or basilar) part contains numerous transverse and vertical fibres. Amongst the fibres are the groups of cells that constitute the **pontine nuclei** (Fig. 19.8).

- ***The pontine nuclei:*** The pontine nuclei (or ***nuclei pontis***) are small masses of grey matter scattered between longitudinal and transversally arranged fibres. They are a relay station in the corticopontocerebellar pathway, i.e. between the cerebral cortex and contralateral cerebellar hemisphere.

  They receive corticopontine fibres from the frontal, temporal, parietal, and occipital lobes of the cerebrum. Their efferents form the transverse fibres of the pons known as pontocerebellar fibres because they terminate at the cerebellum. Most of these fibres cross to the opposite side, but some may end ipsilaterally.

  The pontine nuclei also receive fibres from various other sources, including the tectum (superior colliculus),

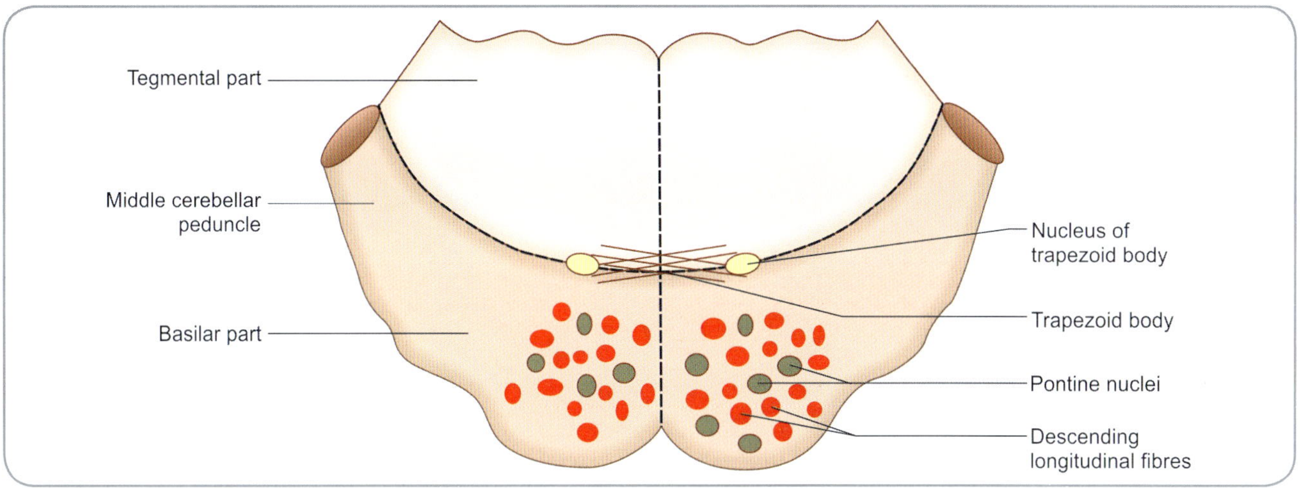

Fig. 19.8: Diagramatic representation of lower part of the pons showing the basilar part of pons lying ventral to the trapezoid body and the tegmental part of the pons lying dorsal to it

the mamillary body, the lateral geniculate body, the nuclei gracilis and cuneatus, trigeminal nuclei, hypothalamus, cerebellar nuclei, and reticular formation.

It has been estimated that there are about 20 million neurons in pontine nuclei. Most of them are glutaminergic. About 5% are gamma-aminobutyric acid (GABA) ergic and are inhibitory.

❏ ***Descending longitudinal fibres:*** The descending longitudinal fibres (Fig. 19.8) consist of—

○ ***Corticospinal fibres*** as they traverse the pons and converge again to form pyramid in medulla

○ ***Corticonuclear fibres*** as they descend along with the corticospinal fibres to form pyramids in medulla. However, most of them terminate in the contralateral (and to some extent ipsilateral) motor nuclei of the cranial nerves.

○ ***Corticopontine fibres*** as discussed, these fibres arise from frontal, temporal, parietal, and occipital cortices and terminate on pontine nuclei of same side.

❏ ***Transverse pontine fibres: Transverse fibres*** arise in the pontine nuclei and cross to the opposite side to form the middle cerebellar peduncle. These are ***pontocerebellar fibres***.

Structure of the Tegmental Part of Pons

The dorsal part (or tegmentum) of the pons may be regarded as an upward continuation of the part of the medulla behind the pyramids. Superiorly, it is continuous with the tegmentum of the midbrain. It is bounded posteriorly by the fourth ventricle. Laterally, it is related to the inferior cerebellar peduncles in its lower part (Fig. 19.10) and to the ***superior cerebellar peduncles*** in its upper part (Fig. 19.11). The spinal nucleus and tract of the trigeminal nerve lie just medial to these peduncles.

Internal Structure

The region adjoining the ventral part (of the pons) is occupied by important ascending tracts. The medial lemniscus occupies a transversely elongated oval area next to the midline. Lateral to this are the trigeminal and the spinal lemniscus (lateral spinothalamic tract). The fibres of the spinotectal tract run along with the spinal lemniscus, while those of the ventral spinothalamic tract lie within the medial lemniscus. Still more laterally, there is the lateral lemniscus.

Ventral to these lemnisci, there are conspicuous transversely running fibres that form the ***trapezoid body***.

The ventral spinocerebellar tract lies ventromedial to the inferior cerebellar peduncle in the lower part of the pons (Fig. 19.10). In the upper part of the pons, it is seen within the superior cerebellar peduncle.

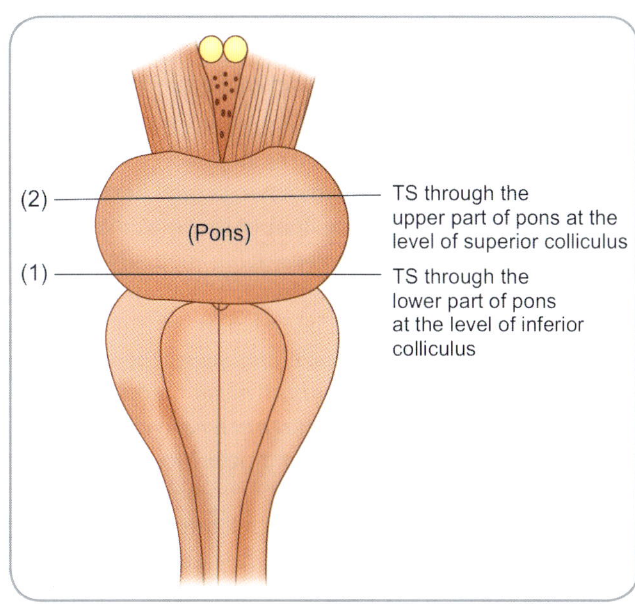

Fig. 19.9: Illustration showing levels of transverse section (TS) of pons

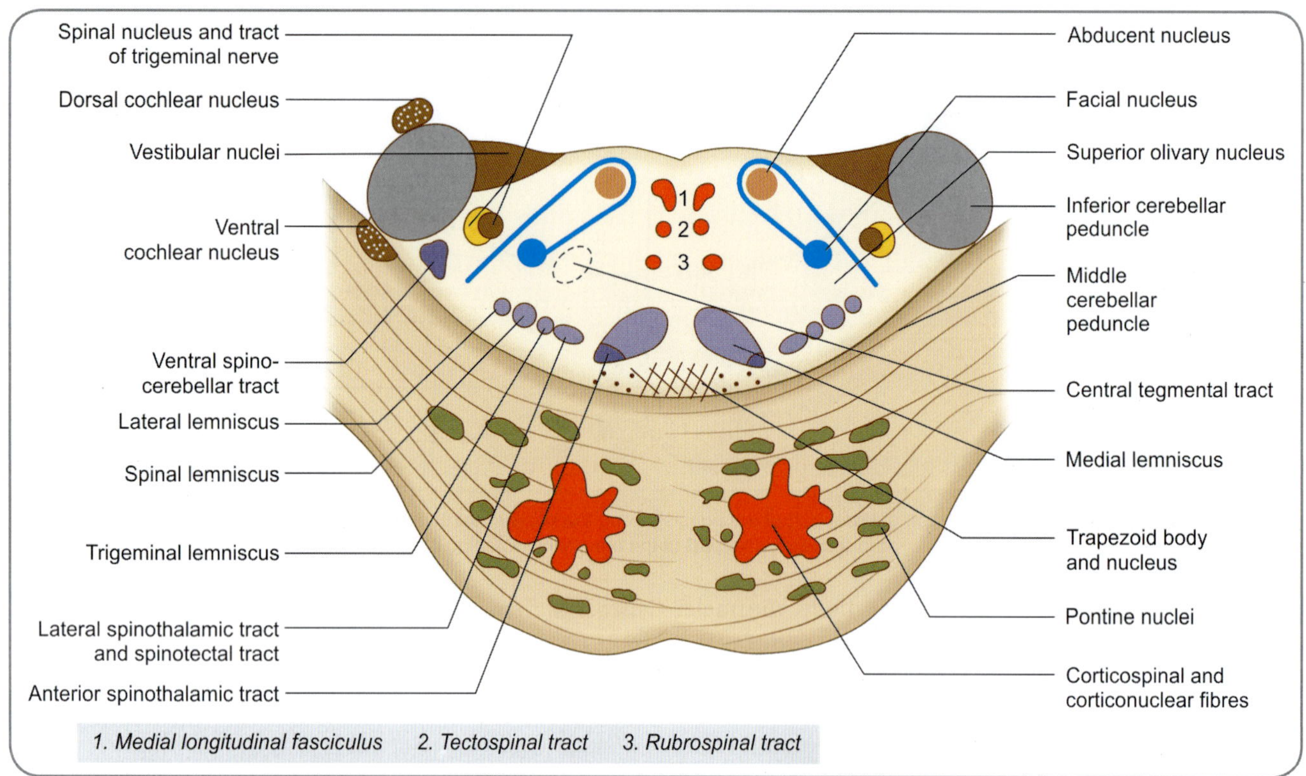

Fig. 19.10: Transverse section through the lower part of pons

Descending tracts passing through the dorsal part of the pons are the tectospinal tract and the rubrospinal tract. The medial longitudinal fasciculus lies dorsally near the midline.

However, the structure of the tegmentum is different in the upper and lower parts of pons. Hence, it is customary to study the internal structure of pons at two different levels—lower part, transverse section passing through the facial colliculus and upper part, transverse section passing through the trigeminal nuclei (Fig. 19.9).

Section through the Lower Part of Pons

The transverse section through the lower part of pons corresponds to the level of facial colliculus.

This section (Fig. 19.10) shows two cranial nerve nuclei that are closely related to the floor of the fourth ventricle. These are the **abducent nucleus**, lying medially and the **vestibular nuclei** that lie laterally.

At a deeper level in the lateral part of the reticular formation, two additional nuclei are seen. These are the **spinal nucleus of the trigeminal nerve** (along with its tract), lying laterally and the **facial nucleus,** lying medially. The dorsal and ventral cochlear nuclei lie dorsally and ventrally, respectively to the inferior cerebellar peduncle.

The fibres arising from the facial nucleus follow an unusual course (Fig. 19.11). They first run dorsally and medially to reach the lower pole of the abducent nucleus. They then ascend on the medial side of that nucleus. Here,

the fibres are closely related to the medial longitudinal fasciculus. Finally, the fibres of the facial nerve turn forwards and laterally passing above the upper pole of the abducent nucleus. As they pass ventrally, the fibres lie between the facial nucleus medially and the spinal nucleus of the trigeminal nerve laterally. The abducent nucleus and the facial nerve fibres looping around it, together form a surface elevation, the **facial colliculus,** in the floor of the fourth ventricle.

The vestibular nuclei occupy the vestibular area in the lateral part of the floor of the fourth ventricle. These nuclei are to be seen in the lower part of the pons and in the upper part of the medulla (Figs 19.6 and 19.10).

Other masses of grey matter to be seen in the lower part of the pons are the **superior olivary complex** (made up of several nuclei), which lies dorsomedial to the lateral lemniscus and the nuclei of the trapezoid body, which consist of scattered cells lying within this body.

Section through the Upper Part of the Pons

The transverse section through the upper part of pons passes through the motor and principal sensory nuclei of the trigeminal nerve (Fig. 19.9).

At this level (Fig. 19.11), the dorsal part is bounded laterally by the superior cerebellar peduncles. Medial to each peduncle, there is the main sensory nucleus of the trigeminal nerve and further medially, there is the motor nucleus of the same nerve. The superior olivary nucleus extends to this level

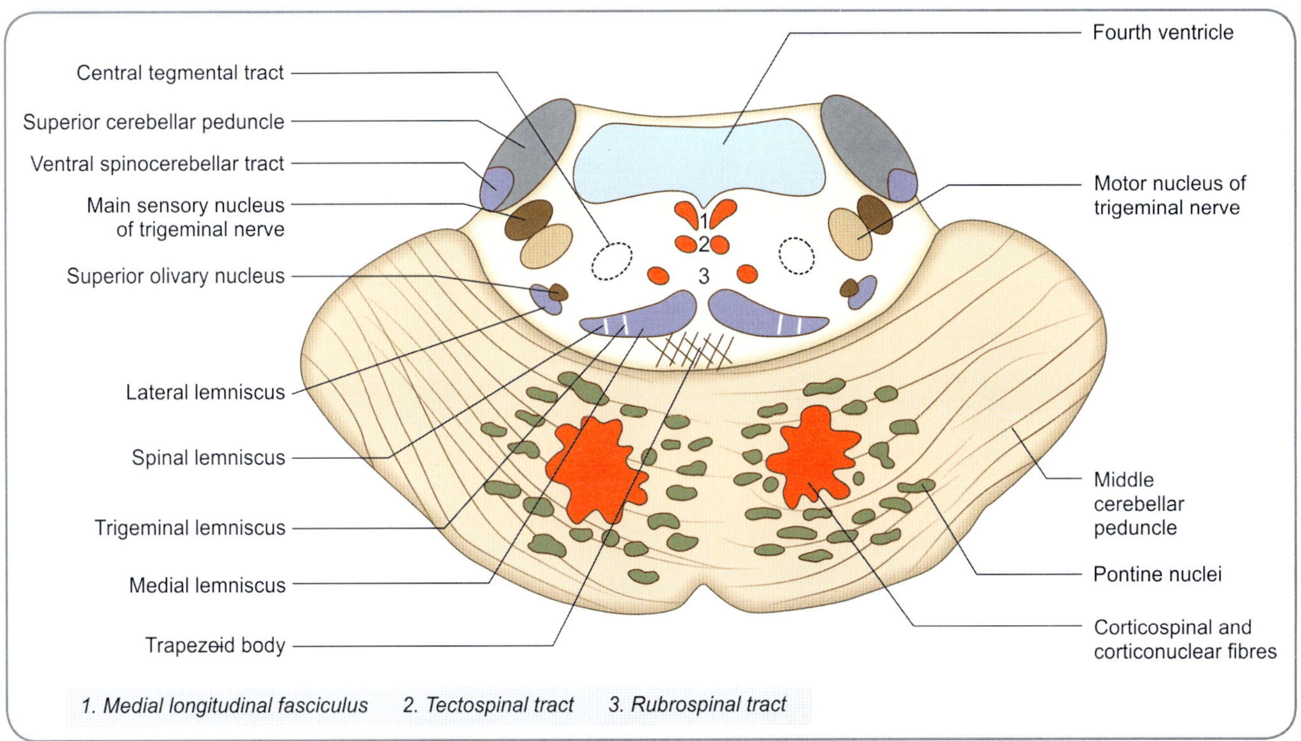

Fig. 19.11: Transverse section through the upper part of pons

but is less prominent, while the lateral lemniscus forms a more conspicuous bundle. Some fibres of the trapezoid body can be seen ventral to the medial lemniscus.

Arterial Supply of Pons

The pons is supplied by the following arteries:
- Pontine branches from the basilar artery and
- Anterior inferior cerebellar artery.

MIDBRAIN

Gross Anatomy

The midbrain is the uppermost part of the brainstem, connecting the pons and cerebellum with the forebrain. It is about 2 cm in length. Its cavity, the cerebral aqueduct, connects the third ventricle to the fourth ventricle (Fig. 19.2).

The midbrain contains nuclei of origin for cranial nerves III (oculomotor) and IV (trochlear). Apart from the cranial nerve nuclei, the midbrain also has nuclei that coordinate the movement of eyeball in response to visual stimuli located in superior colliculi (Fig. 19.3). Nuclei, which coordinate movements of head and trunk in response to auditory stimuli are located in inferior colliculi (Fig. 19.3).

External Features
Anterior Aspect

When the midbrain is viewed from the ***anterior aspect***, we see two large bundles of fibres, one on each side of the middle line. These are the ***crura*** of the midbrain. The crura are separated by a deep fissure. Near the pons, the fissure is narrow, but broadens as the crura diverge to enter the corresponding cerebral hemispheres. The parts of the crura just below the cerebrum, form the posterior boundary of a space called the ***interpeduncular fossa***. The oculomotor nerve emerges from the medial aspect of the crus (singular of crura) of the same side.

Posterior Aspect

The posterior aspect of the midbrain is marked by four rounded swellings (Fig. 19.3). These are the ***colliculi***, one superior and one inferior on each side. Each colliculus is related laterally to a ridge called the ***brachium***. The ***superior brachium*** (also called the superior quadrigeminal brachium or brachium of superior colliculus) connects the superior colliculus to the lateral geniculate body. Similarly, the ***inferior brachium*** (also called the inferior quadrigeminal brachium or brachium of inferior colliculus) connects the inferior colliculus to the medial geniculate body. Just below the colliculi, there is the uppermost part of a membrane, the ***superior medullary velum***, which stretches between the two superior cerebellar peduncles and helps form the roof of the fourth ventricle. The trochlear nerve emerges from the velum and then winds around the side of the midbrain to reach its ventral aspect.

Note: The trochlear nerve is the only cranial nerve that takes origin from the dorsal aspect of the brainstem.

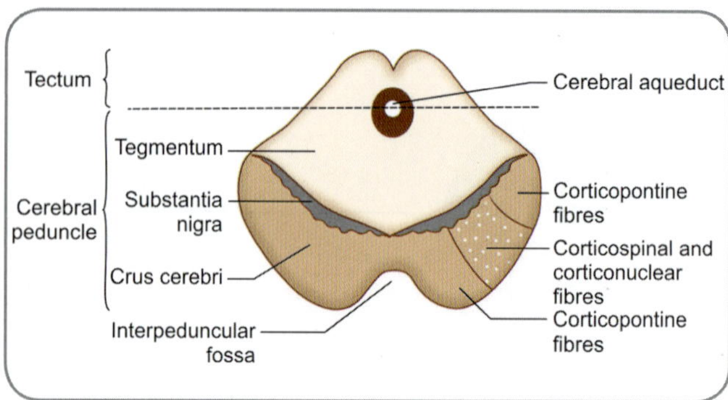

Fig. 19.12: Transverse section of the midbrain showing its main subdivision

Internal Structure

For convenience of description, the internal structure midbrain may be divided as follows (Fig. 19.12):

- The part lying behind a transverse line drawn through the cerebral aqueduct is called the *tectum*. It consists of the *superior and inferior colliculi* of the two sides.
- The part lying in front of the transverse line is made up of right and left halves called the *cerebral peduncles*. Each peduncle consists of three parts. From anterior to posterior side, these are the *crus cerebri* (or *basis pedunculi*), the *substantia nigra,* and the *tegmentum*.

Crus Cerebri (Basis Pedunculi)

The crus cerebri consists of a large mass of vertically running fibres, which descend from the cerebral cortex. The fibres in the crus cerebri consist of the following:

- Corticopontine fibres,
- Corticospinal fibres and
- Corticonuclear fibres.

Its medial one-sixth is occupied by corticopontine fibres descending from the frontal lobe and the lateral one-sixth is occupied by similar fibres from the temporal, occipital, and parietal lobes. The intermediate two-thirds of the crus cerebri are occupied by corticospinal and corticonuclear fibres (Fig. 19.12). The fibres for the leg are most lateral and those for the head are most medial.

Substantia Nigra

The *substantia nigra* lies immediately behind and medial to the crus cerebri (Fig. 19.12). It appears dark in unstained sections, as neurons within it contain a pigment (*neuromelanin*).

The substantia nigra is divisible into a dorsal part, the *pars compacta* and a ventral part, the *pars reticularis*. The pars compacta contains dopaminergic and cholinergic neurons. Most of the neurons in the pars reticularis are GABAergic. Superiorly, the pars reticularis becomes continuous with the globus pallidus. The substantia nigra is closely connected, functionally, with the corpus striatum.

Connections of Substantia Nigra

The main connections (both afferent and efferent) of substantia nigra are with the striatum (i.e., caudate nucleus and putamen). Dopamine produced by neurons in the substantia nigra (pars compacta) passes along their axons to the striatum (*mesostriatal dopamine system*).

> **Clinical Correlation**
>
> Dopamine is much reduced in patients with a disease called *Parkinsonism,* in which there is a degeneration of the striatum.

Along with other groups of dopaminergic neurons present in the ventral part of the tegmentum, the substantia nigra is believed to be a neural centre for "adaptive behavior". Efferents of this system are widely distributed (discussed in Authors' Textbook of Neuroanatomy).

The midbrain is traversed by the cerebral aqueduct, which is surrounded by central grey matter. Ventrally, the central grey matter is related to cranial nerve nuclei (oculomotor and trochlear) (Figs 9.14 and 9.15). The region between the substantia nigra and the central grey matter is occupied by the reticular formation.

Tegmentum

The tegmentum is the region of midbrain that lies between substantia nigra and tectum (Fig. 19.12).

The tegmentum of the two sides is continuous across the midline. It contains important masses of grey matter as well as fibre bundles. The largest of the nuclei is the *red nucleus* present in the upper half of the midbrain. The tegmentum also contains the *reticular formation,* which is continuous below with that of the pons and medulla.

The internal structure of tegmentum and tectum varies at different levels of midbrain; hence, the internal structure of midbrain is studied by transverse sections at two different levels—lower part, transverse section passing through the inferior colliculus and upper part, transverse section passing through the superior colliculus (Fig. 19.13).

Section through Midbrain at Level of Inferior Colliculus

The *inferior colliculus* is a large mass of grey matter lying in the tectum. It forms a relay station in the auditory pathway and is probably concerned with reflexes involving the auditory stimuli.

> **Clinical Correlation**
>
> Lesions of the inferior colliculus produce defects in appreciation of tones, localization of sound, and reflex movements in response to sound.

A section through the midbrain at the level of the inferior colliculus shows the following features (Fig. 19.14).

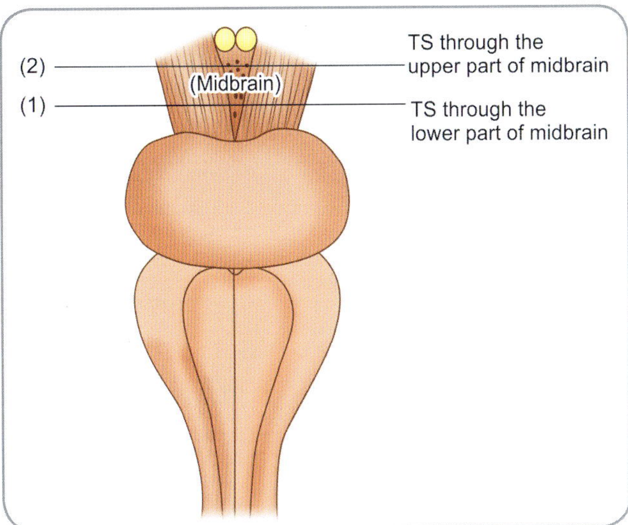

Fig. 19.13: Illustration shows the two levels at which transverse sections (TS) of midbrain are studied

❑ The ***trochlear nucleus*** lies in the ventral part of the central grey matter. Fibres arising in this nucleus follow an unusual course. They run dorsally and decussate (in the superior medullary velum) before emerging on the dorsal aspect of the brainstem.

❑ The ***mesencephalic nucleus of the trigeminal nerve*** lies in the lateral part of the central grey matter.

❑ A compact bundle of fibres lies in the tegmentum dorsomedial to the substantia nigra. It consists of the medial lemniscus (lies just behind the substantial nigra lateral to the red nucleus), the trigeminal lemniscus, and the spinal lemniscus in that order from medial to lateral side.

❑ The medial lemniscus includes fibres of the ventral spinothalamic tract, while the spinal lemniscus (made up mainly of the lateral spinothalamic tract) includes fibres of the spinotectal tract. More dorsally, the lateral lemniscus forms a bundle ventrolateral to the inferior colliculus (in which most of its fibres end).

❑ Important fibre bundles are also located near the middle line of the tegmentum. The medial longitudinal fasciculus lies ventral to the trochlear nucleus, and ventral to the fasciculus, there is the tectospinal tract.

❑ The region ventral to the tectospinal tracts is occupied by decussating fibres of the superior cerebellar peduncle. These fibres have their origin in the dentate nucleus of the cerebellum. They cross the middle line in the lower part of the tegmentum. Some of these fibres end in the red nucleus while others ascend to the thalamus.

❑ The part of the tegmentum ventral to the decussation of the superior cerebellar peduncle is occupied by the rubrospinal tracts.

Section through Midbrain at Level of Superior Colliculus

A section through the upper part of the midbrain (Fig. 19.15) shows two large masses of grey matter not seen

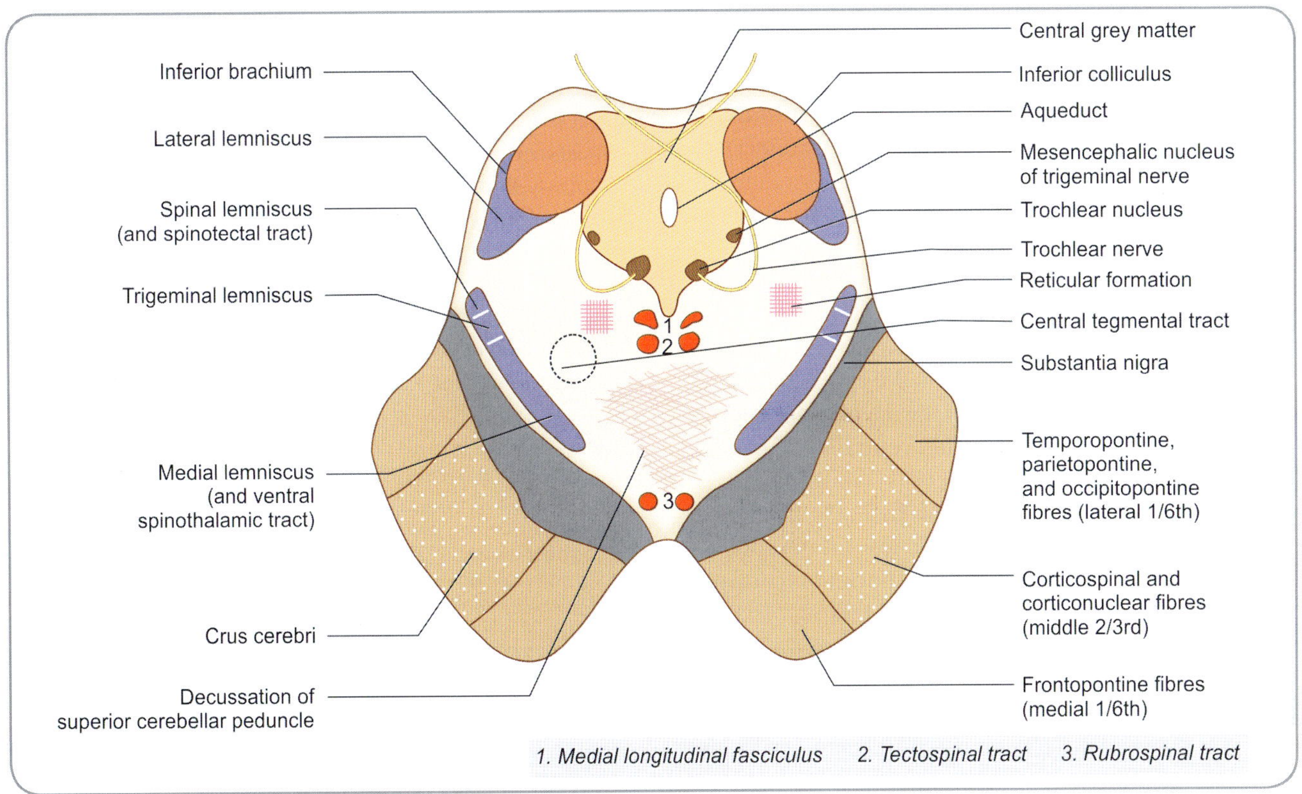

1. Medial longitudinal fasciculus   2. Tectospinal tract   3. Rubrospinal tract

Fig. 19.14: Transverse section through the lower part of the midbrain at the level of inferior colliculus

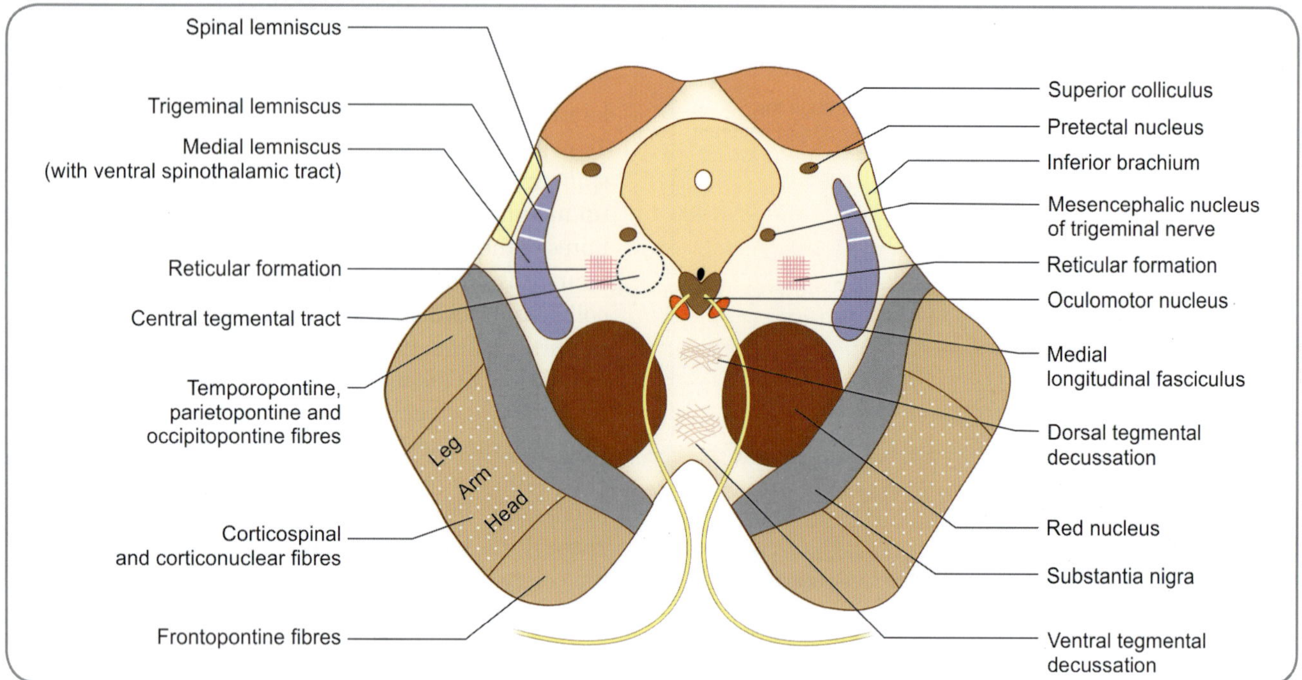

Fig. 19.15: Transverse section through the upper part of midbrain

at lower levels. These are the **superior colliculus** in the tectum and the **red nucleus** in the tegmentum.

The superior colliculus is a centre concerned with visual reflexes.

The **red nucleus** lies in the anterior part of the tegmentum dorsomedial to the substantia nigra. It is so called, because of a reddish colour in fresh material. The colour is produced by the presence of an iron pigment in its neurons. The red nucleus consists of a cranial **parvocellular part** and a caudal **magnocellular part**. The magnocellular part is prominent in lower species, but in man, it is much reduced, and is distinctly smaller than the parvocellular part. It is an important motor nucleus of the extrapyramidal system.

The **oculomotor nucleus** lies in relation to the ventral part of the central grey matter. The nuclei of the two sides lie close together forming a single complex. The **Edinger-Westphal nucleus** (which supplies the sphincter pupillae and ciliaris muscle) forms part of the oculomotor complex. The oculomotor complex is related ventrally to the medial longitudinal fasciculus.

Closely related to the cranial part of the superior colliculus, there is a small collection of neurons that constitute the **pretectal nucleus**. This nucleus is concerned with the pathway for the pupillary light reflex.

The pretectal nucleus extends cranially to the junction of the midbrain with the diencephalon. It receives retinal fibres through the optic tract. It also receives some fibres from the superior colliculus and from the visual cortex. The main efferents of the nucleus reach the oculomotor nuclei (of both sides). Some efferents reach the superior colliculus and the pulvinar.

The bundle of ascending fibres consisting of the medial lemniscus, the trigeminal lemniscus and the spinal lemniscus lies more dorsally than at lower levels (because of the presence of the red nucleus). The lateral lemniscus is not seen at this level as its fibres end in the inferior colliculus. However, the **inferior brachium** that conveys auditory fibres to the medial geniculate body can be seen near the surface of the tegmentum. The region of the tegmentum near the middle line shows two groups of decussating fibres. The **dorsal tegmental decussation** consists of fibres that have their origin in the superior colliculus and cross to the opposite side to descend as the tectospinal tract. The **ventral tegmental decussation** consists of fibres that originate in the red nucleus and decussate to form the rubrospinal tracts.

Added Information

Connections of the Red Nucleus (Fig. 19.16)
Afferent Fibres

The red nucleus receives its main afferents from:
- The cerebral cortex (from motor area—area 4 and 6 of frontal cortex of same side),
- The cerebellum (dentate, emboliform, and globose nuclei of opposite side),
- Globus pallidus of the same side,
- Subthalamic nucleus, hypothalamus, substantia nigra, and tectum.

Efferent Fibres

The efferent fibres from the red nucleus cross in the ventral tegmental decussation and then go to the spinal cord (rubrospinal tract), cranial nerve motor nuclei III, IV, V, VI, and VII (rubronuclear), inferior olivary nucleus (central tegmental fasciculus), reticular formation (rubroreticular), substantia nigra, cerebral cortex, and thalamus.

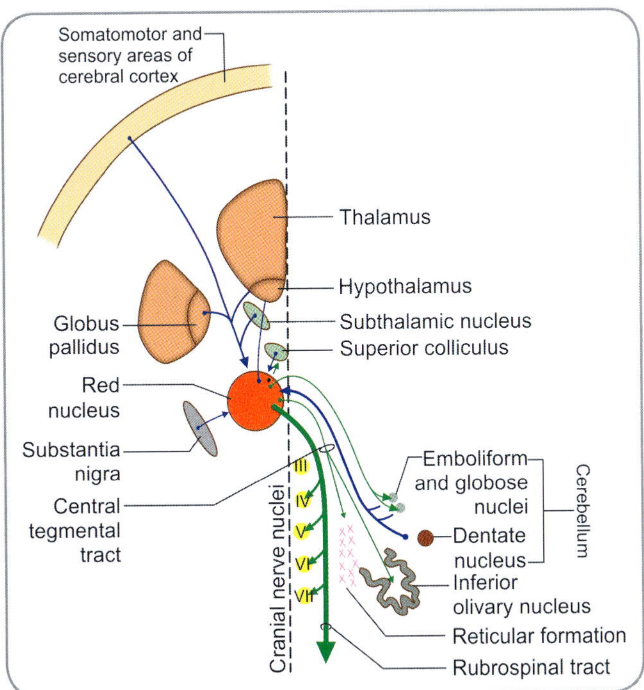

Fig. 19.16: Scheme to show the connections of the red nucleus

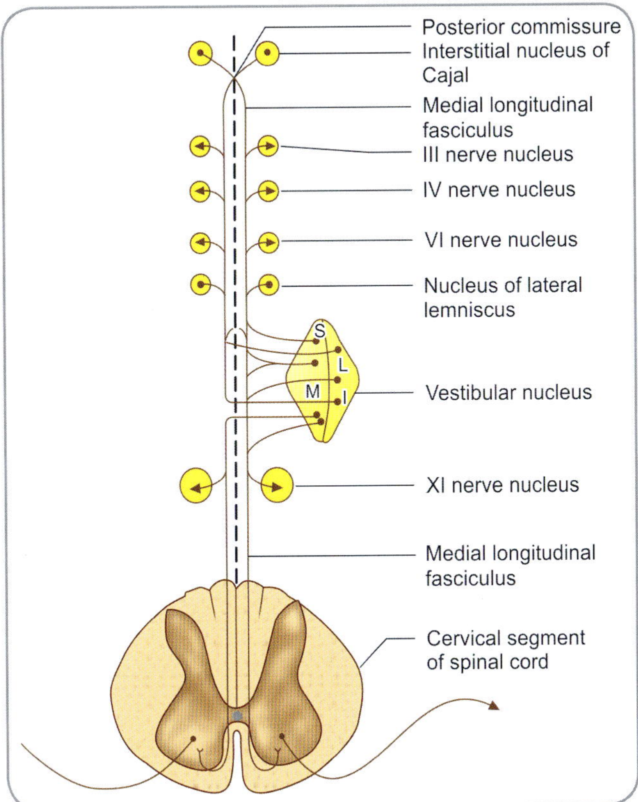

Fig. 19.17: Fibres in medial longitudinal fasciculus and vestibular nucleus

Arterial Supply of Midbrain

The midbrain is supplied by the following arteries:

- Basilar artery through its posterior cerebral and superior cerebellar arteries. Basilar artery also supplies midbrain through direct branches.
- Branches of posterior communicating and anterior choroidal arteries.

SOME FIBRE BUNDLES SEEN IN THE BRAINSTEM

In addition to the various ascending and descending tracts described in Chapter 20, there are a number of fibre bundles seen in the brainstem. These include:

- Medial longitudinal fasciculus (or bundle),
- Central tegmental tract and
- Dorsal longitudinal fasciculus.

Parts of the medial forebrain bundle and of the mamillary peduncle are also seen.

Medial Longitudinal Fasciculus

The medial longitudinal fasciculus consists of fibres that lie near the midline of the brainstem. Above, it reaches up to the level of the third ventricle. (The ascending fibres end in the interstitial nucleus of Cajal, the nucleus of the posterior commissure, and the nucleus of Darkschewitsch).

Below, the medial longitudinal fasciculus becomes continuous with the anterior intersegmental tract of the spinal cord.

The medial longitudinal fasciculus consists of fibres arising mainly from right and left medial vestibular nuclei in the medulla (however, some fibres also from nucleus of lateral leminiscus and interstitial nucleus of cajal) (Fig. 19.17).

The fasciculus is closely related to the nuclei of the third, fourth, sixth, and twelfth cranial nerves (all of the somatic efferent column and lying next to the midline). It is also related to the fibres of the seventh nerve (as they wind round the abducent nucleus) and some fibres arising from the cochlear nuclei. In the spinal cord, it establishes connections with ventral horn cells that innervate the muscles of the neck.

Functions of Medial Longitudinal Fasciculus

- Fibres arising in the vestibular nuclei of the same side, as well as those of the opposite side ascend or descend in the fasciculus to reach nuclei supplying the muscles of the eyeball and neck. These connections ensure harmonious movements of the eyes and head in response to vestibular stimulation.
- Some fibres of the fasciculus are connected to some nuclei of the auditory pathway. These are the nucleus of the trapezoid body and the nucleus of the lateral

lemniscus. Some fibres arising in the dorsal cochlear nucleus are closely related to the fasciculus. Through these connections, movements of the head and of the eyes can take place in response to auditory stimuli.

❑ The medial longitudinal fasciculus affords a pathway for fibres interconnecting the nuclei related to it.

Connections between the facial and hypoglossal nuclei may facilitate simultaneous movements of the lips and tongue as in speech.

❑ Fibres from the medial (and other) vestibular nuclei descend in the fasciculus to the spinal cord. These fibres form the ***medial vestibulospinal tract.***

Multiple Choice Questions

1. With regard to the pyramid, what is false?
   a. It is an elevation on the anterior aspect of the medulla
   b. It is caused by the corticospinal fibres
   c. It has the rootlets of hypoglossal nerve emerging lateral to it
   d. It merges imperceptibly with the olive on the external aspect
2. The superior brachium connects:
   a. The medial geniculate body to the superior colliculus
   b. The lateral id brain geniculate body to the superior colliculus
   c. The medial geniculate body to the inferior colliculus
   d. The lateral geniculate body to the inferior colliculus
3. The posterior external arcuate fibres arise from the:
   a. Nucleus gracilis
   b. Nucleus arcuatus
   c. Accessory cuneate nucleus
   d. Nucleus ambiguus
4. Most of the neurons in the pontine nuclei are:
   a. Gamma amino butyric
   b. Cholinergic
   c. Adrenergic
   d. Glutaminergic
5. With regard to the red nucleus:
   a. It is an important component of the pyramidal system
   b. It has a cranial magnocellular part
   c. Its colour is due to iron pigment in its neurons
   d. It lies in the tectum part of the midbrain

ANSWERS

1. d   **2.** b   **3.** c   **4.** d   **5.** c

Clinical Problem-solving

Case Study 1: Your friend informs you that his grandfather suffers from parkinsonism.
❑ What is the neurotransmitter involved in this disease?
❑ Which structure in the hindbrain produces this neurotransmitter?
❑ What structures show degeneration in this disease?

Case Study 2: A 55-year-old patient suffers a lesion of the medulla at the level of the pyramidal decussation.
❑ What other structures/tracts will also be affected apart from the pyramidal tract?
❑ What are the nuclei which are likely to be affected?
❑ What are the cranial nerve nuclei present at this level?

(For solutions see Appendix).

Chapter 20

Tracts of the Spinal Cord and Brainstem

Frequently Asked Questions

- ❏ Enumerate the ascending tracts of the spinal cord. Give a brief account of the spinothalamic tracts.
- ❏ Write notes on (a) Anterior corticospinal tract, (b) Spinocerebellar tracts, (c) Dorsal tegmental decussation.
- ❏ Trace the great motor pathway through the corticospinal tracts.

A collection of nerve fibres within the central nervous system (CNS) that connects two masses of grey matter is called a tract. A tract may be defined as a collection of nerve fibres having the same origin, course, and termination. Tracts may be ascending or descending. They are usually named after the masses of grey matter connected by them. Thus, a tract beginning in the cerebral cortex and descending to the spinal cord is called the corticospinal tract, while a tract ascending from the spinal cord to the thalamus is called the spinothalamic tract. Tracts are sometimes referred to as fasciculi (= bundles) or lemnisci (= ribbons). The major tracts passing through the spinal cord and brainstem are shown schematically in Figure 20.1 and Table 20.1. The position of the tracts in a transverse section of the spinal cord is shown in Figure 20.2.

Table 20.1: Classification of the tracts of the spinal cord	
Descending (motor) tracts	**Ascending (sensory) tracts**
Pyramidal tracts • Corticonuclear tract • Corticospinal tract	• Fasciculus gracilis • Fasciculus cuneatus • Anterior spinocerebellar tract
Extrapyramidal tracts • Rubrospinal tract • Tectospinal tract • Vestibulospinal tract • Olivospinal tract • Reticulospinal tract	• Posterior spinocerebellar tract • Anterior spinothalamic tract • Lateral spinothalamic tract • Spinotectal tract

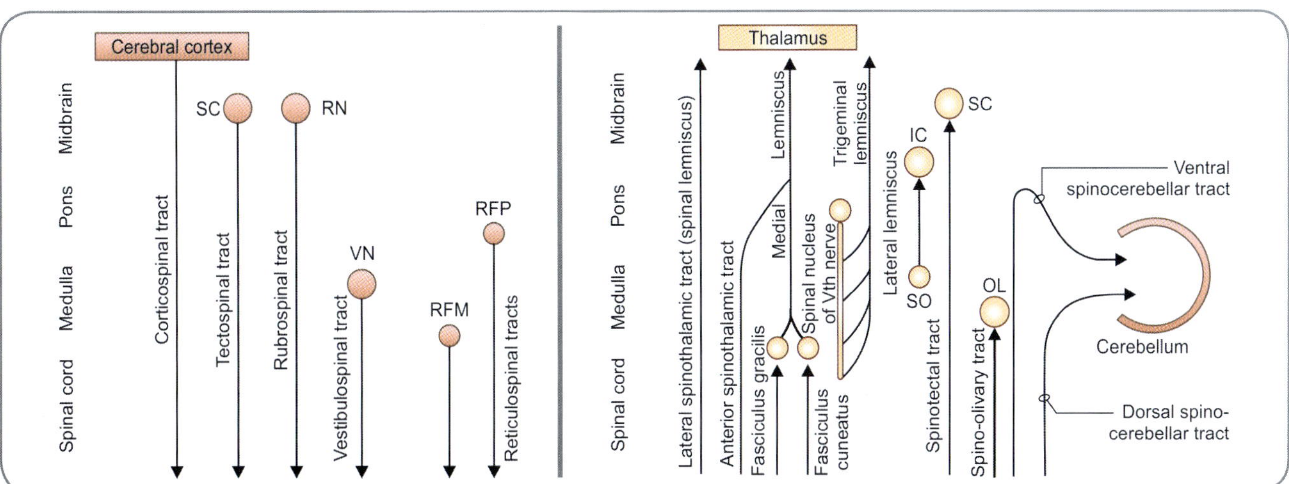

Fig. 20.1: Scheme to show the major tracts passing through the brainstem. SC = superior colliculus. RN = red nucleus. VN = vestibular nuclei. RFP = reticular formation of pons. RFM = reticular formation of medulla. IC = inferior colliculus. SO = superior olivary nucleus. OL = accessory olivary nucleus

Note—the tracts are named after the names of masses of grey matter connected by them. The name usually consists of two components (or terms)—the first term denotes the origin and second the termination of the tract.

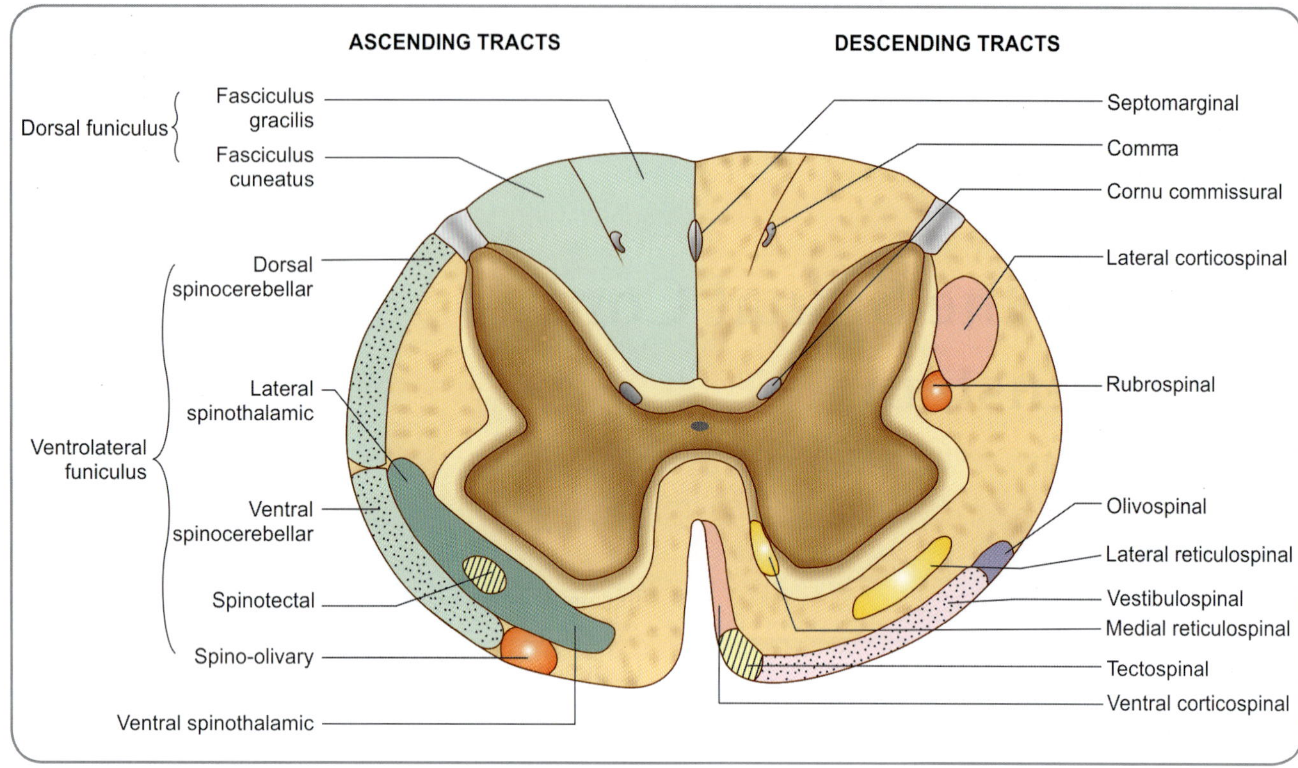

Fig. 20.2: Simplified scheme to show the location of the main ascending and descending tracts in the spinal cord – note that the positions of the tracts vary at different levels of the cord and that the area occupied by the fibres of different tracts overlap considerably

DESCENDING TRACTS ENDING IN THE SPINAL CORD

The various descending tracts described below that end in relation to ventral horn cells, influence their activity and, thereby, have an effect on contraction and tone of skeletal muscle. Although a small number of the fibres of these tracts may synapse directly with ventral horn cells, most of them influence these cells through intervening internuncial/interneurons. This influence is exerted on both α-neurons and γ-neurons.

The γ-neurons indirectly influence the activity of α-neurons via muscle spindles. Hence, all these influences ultimately reach the α-neurons. Such influences may be either facilitatory or inhibitory. The corticospinal and rubrospinal tracts form lateral descending pathway, which facilitates the flexors and inhibits the extensors. The vestibulospinal tract has the opposite effect. The medial reticulospinal tract is generally facilitatory and the lateral tract is inhibitory. All descending tracts other than the corticospinal and rubrospinal tracts form the medial descending pathway. The motor control is far more complex than such simple statements may suggest.

The corticospinal tracts are often referred to as *pyramidal tracts*. Traditionally, all other descending tracts projecting on motor neurons have been collectively referred to as *extrapyramidal tracts*.

The autonomic nerve fibres arise from neurons located in the general visceral efferent nuclei of the brainstem and from the intermediolateral region of the spinal grey matter. These regions are under control of higher centres through descending autonomic pathways.

CORTICOSPINAL TRACT (PYRAMIDAL TRACT)

Origin

The corticospinal tracts are predominantly made up of axons of the neurons lying in the motor area of the cerebral cortex (area 4) (Fig. 20.3). Some fibres also arise from the premotor area (area 6) and some from the somatosensory area (areas 3, 2, and 1). A few fibres arise in the parietal cortex (area 5).

Course

From the origin, fibres pass through the corona radiata to enter the internal capsule, where they lie in the posterior limb. After passing through the internal capsule, the fibres enter the crus cerebri (of the midbrain); they occupy the middle two-thirds of the crus. The fibres then descend through the ventral part of the pons to enter the pyramids in the upper part of the medulla.

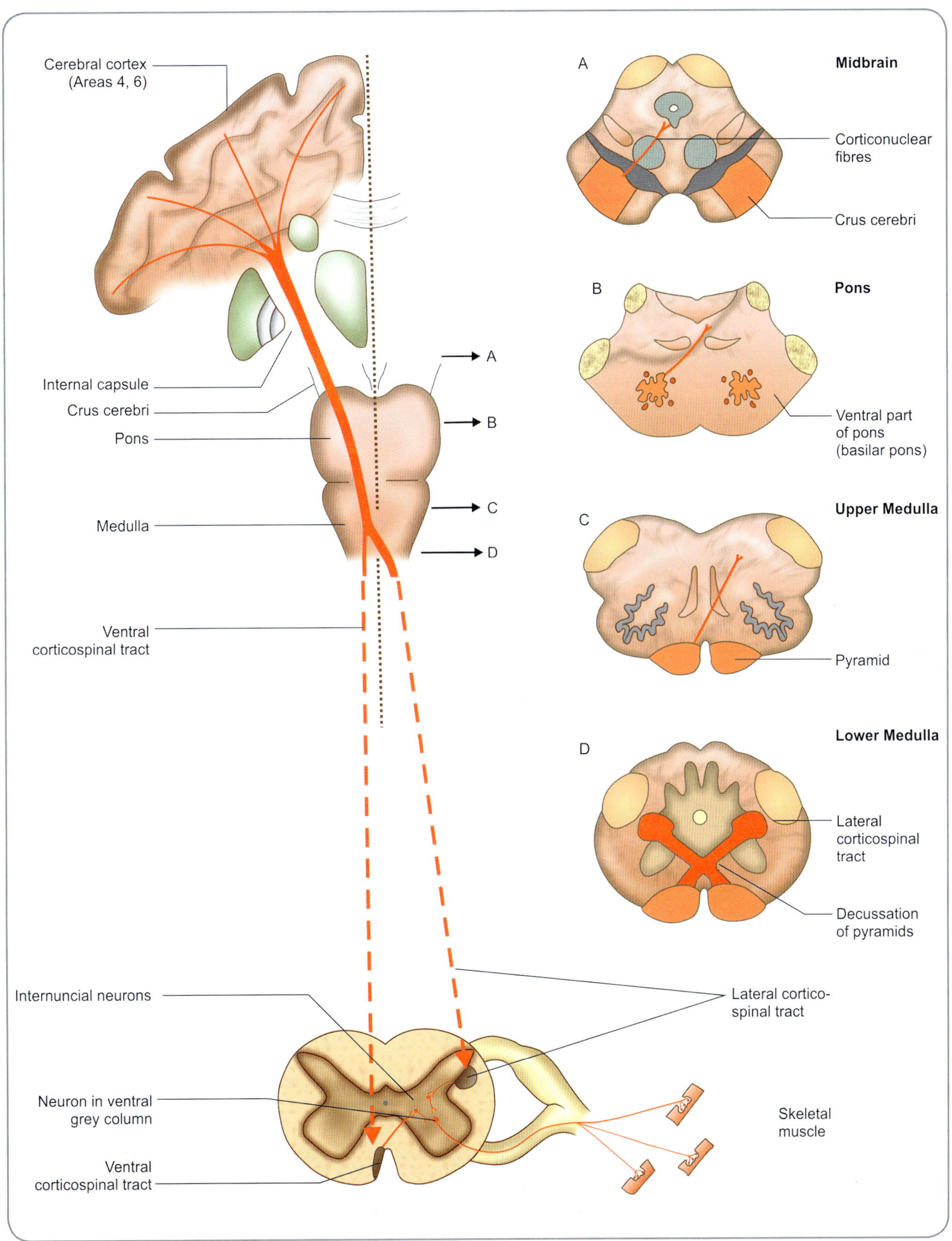

Fig. 20.3: Scheme to show the course of the corticospinal tracts – note the position of the corticospinal tracts (marked in dark red colour) at various levels of the brainstem

Near the lower end of the medulla, about 80% of the fibres cross to the opposite side. The crossing fibres of the two sides constitute the decussation of the pyramids.

The fibres that have crossed in the medulla enter the lateral funiculus of the spinal cord and descend as the *lateral corticospinal tract* (Fig. 20.3).

The corticospinal fibres that do not cross in the pyramidal decussation enter the anterior funiculus of the spinal cord to form the ***anterior corticospinal tract*** (also called the ***ventral corticospinal tract***) (Figs 20.2 and 20.3). On reaching the appropriate level of the spinal cord, the fibres of this tract cross the midline (through the anterior white commissure) to reach the grey matter on the opposite side of the cord. Their manner of termination is similar to that of fibres of the lateral corticospinal tract. In this way, the corticospinal fibres of both the lateral and anterior tracts ultimately connect the cerebral cortex of one side with ventral horn cells in the opposite half of the spinal cord.

Termination

The fibres of both lateral and anterior corticospinal tracts terminate in the grey matter at various levels of the spinal cord. Most of them end by synapsing with internuncial neurons at the base of the dorsal and ventral grey columns (laminae IV to VII). The internuncial neurons carry the impulses brought by fibres of the tract to ventral horn cells. Some fibres of the tract terminate directly on the ventral horn cells (lamina IX, dorsolateral, central, and ventrolateral groups).

Note: The corticospinal tract is also named as the pyramidal tract, because it traverses the medullary pyramids. Conventionally the term pyramidal tract refers specifically to corticospinal tract but clinically, it includes both corticospinal and corticonuclear fibres.

The fibres of the pyramidal tract which synapse with the motor neurons situated in the ventral horn of the spinal cord constitute the ***corticospinal tract***, while the fibres of the pyramidal tract which synapse with the motor neurons of cranial nerve nuclei in the brainsteam, constitute the ***corticonuclear tract***.

Function

The cerebral cortex controls voluntary movements through the corticospinal tract. The influence of this tract is said to be facilitatory for flexors and inhibitory for extensors.

Apart from their motor functions, corticospinal fibres influence conduction in ascending tracts.

Interruption of the tract anywhere in its course leads to paralysis of the muscles concerned.

Note: The neurons that give origin to the fibres of the corticospinal tracts are often referred to as ***upper motor neurons*** in distinction to the ventral horn cells and their processes which constitute the ***lower motor neurons***. Interruption of either of these neurons leads to paralysis, but the nature of the paralysis is distinctive in each case.

Clinical Correlation

Lesions of Corticospinal Tract

- ❑ The lesion of the corticospinal tract is the damage to the upper motor neurons of the pyramidal system. This leads to paralysis of the skeletal muscles. The part of the body paralysed will depend upon the site of lesion.
- ❑ The fibres of upper motor neurons meant for the limbs cross the middle in the lower part of the medulla (in the decussation of the pyramids) and those for the cranial nerves cross just above the level of their termination. A lesion above the level of crossing, produces a paralysis in the opposite half of the body and a lesion below this level produces paralysis on the same side.
- ❑ In case of upper motor neuron lesion, the muscles are not actually paralysed but the control of upper motor neurons on the lower motor neurons supplying the muscles is lost. As a result, lower motor neurons become hyperactive and the tone of muscles is increased leading to **spastic paralysis**, increased knee jerk reflex, loss of superficial reflexes, and appearance of Babinski's sign (refer to Author's Textbook of Neuroanatomy).

EXTRAPYRAMIDAL SPINAL TRACTS

The extrapyramidal system consists of all the motor tracts of the brain and spinal cord, which do not pass through the medullary pyramids.

The major descending tracts of the spinal cord (Fig. 20.4) included in the extrapyramidal system are:

- ❑ Rubrospinal,
- ❑ Tectospinal,
- ❑ Vestibulospinal,
- ❑ Olivospinal and
- ❑ Reticulospinal tracts.

The extrapyramidal system is responsible for regulating muscle tone, as well as posture and equilibrium of the body.

The extrapyramidal system works hand in hand with the pyramidal system to perform voluntary movements (Figs 20.5 and 20.6).

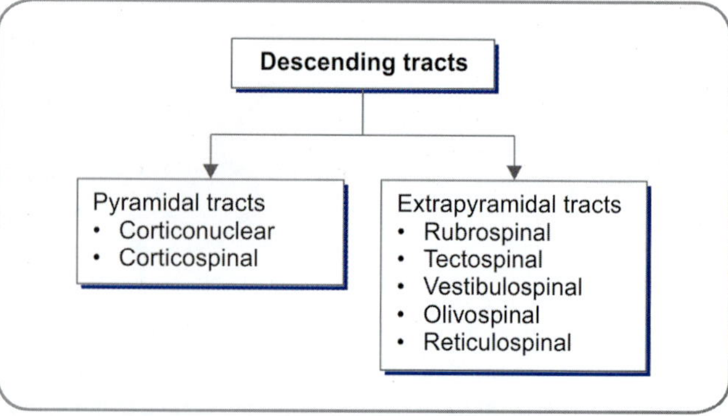

Fig. 20.4: Classification of descending tracts

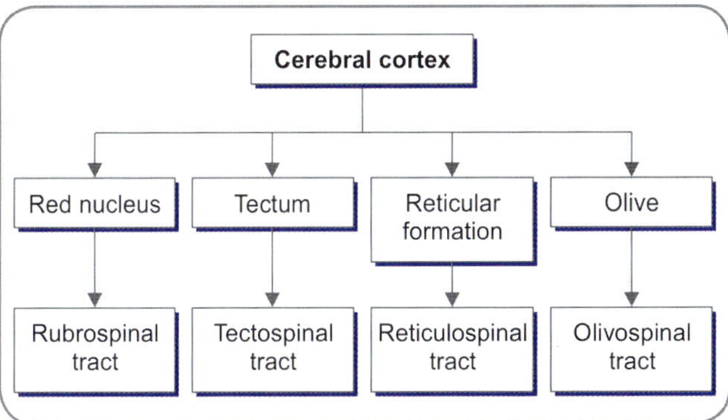

Fig. 20.5: Indirect pathways through which the cerebral cortex may influence the spinal cord – these are often described as extrapyramidal tracts

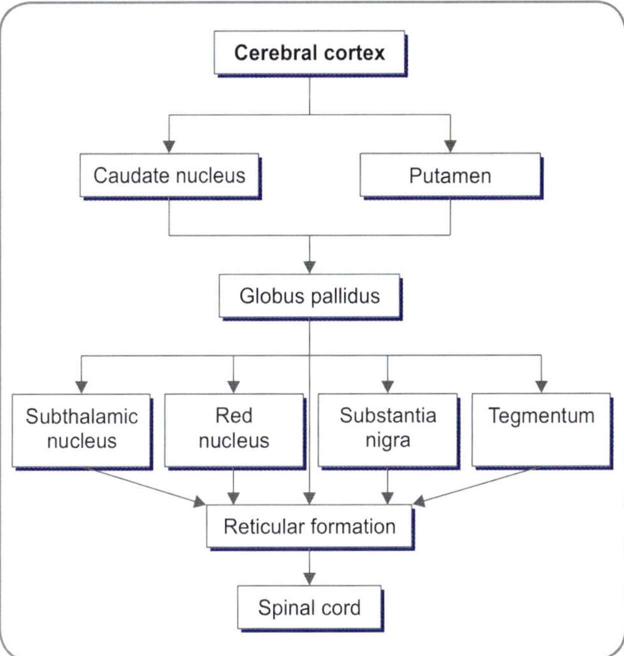

Fig. 20.6: Indirect pathways through which the corpus striatum may influence the spinal cord

The cerebral cortex exercises control on various masses of grey matter (other than cranial nerve nuclei and pontine nuclei) of the brainstem. Fibres from the cortex end in the red nucleus, tectum, substantia nigra, inferior olivary nucleus, interstitial nucleus, and reticular formation. These centres can also be influenced by the cortex indirectly through the striatum (caudate nucleus and putamen of lentiform nucleus) and the globus pallidus. In turn, these centres influence the spinal cord directly through tracts descending from them (rubrospinal, tectospinal, olivospinal, and reticulospinal) and indirectly, through the reticular formation and the reticulospinal tracts. These connections are often described as part of the extrapyramidal pathways referred above.

Rubrospinal Tract

This tract is made up of axons of neurons lying in the red nucleus (which lies in the upper part of the midbrain) (Fig. 20.7). The fibres of the tract cross to the opposite side in the lower part of the tegmentum of the midbrain. The crossing fibres constitute the ***ventral tegmental***

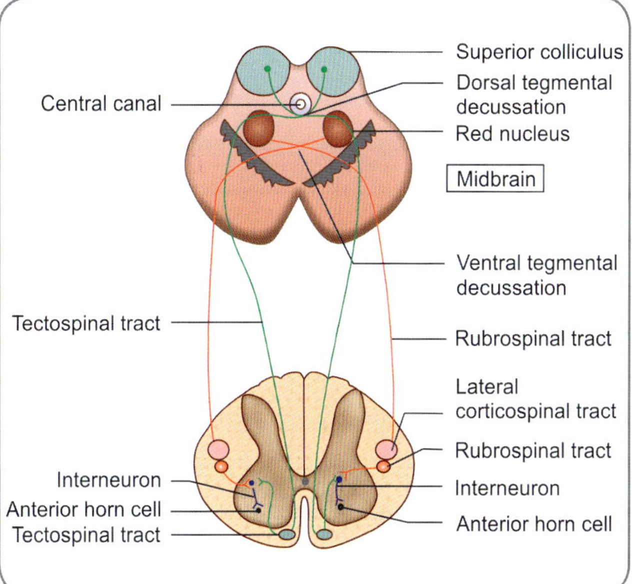

Fig. 20.7: Scheme to show the main features of tectospinal and rubrospinal tracts – note that the fibres of both the tracts cross to the opposite side in midbrain (in dorsal and ventral tegmental decussation)

decussation (Fig. 20.7). The tract descends through the pons and medulla to enter the lateral funiculus of the spinal cord (Fig. 20.2). Here, the tract lies just in front of the lateral corticospinal tract. The fibres of the rubrospinal tract end by synapsing with ventral horn cells through internuncial neurons located in laminae V–VII of the spinal grey matter.

The tract forms a part of extrapyramidal system and maintains the equilibrium and posture of the body and limbs. It is facilitatory to flexors and inhibitory to extensors.

Note: The rubrospinal tract is much better developed in some other species than in man. In man, the tract reaches only the upper three cervical segments of the spinal cord.

Tectospinal Tract

The fibres of this tract arise from neurons in the superior colliculus (midbrain). The fibres cross to the opposite side in the upper part of the tegmentum of the midbrain. The crossing fibres form ***the dorsal tegmental decussation*** (Fig. 20.7). The tract descends through the pons and medulla into the anterior funiculus of the spinal cord. The fibres terminate by synapsing with ventral horn cells in cervical segments of the cord, through internuncial neurons located in laminae VI–VIII of the spinal grey matter. The tectospinal tract is concerned with spinovisual reflexes, i.e. conveys impulses for reflex postural movements in response to visual stimuli.

Vestibulospinal Tracts

There are two vestibulospinal tracts, lateral and medial (Fig. 20.8).

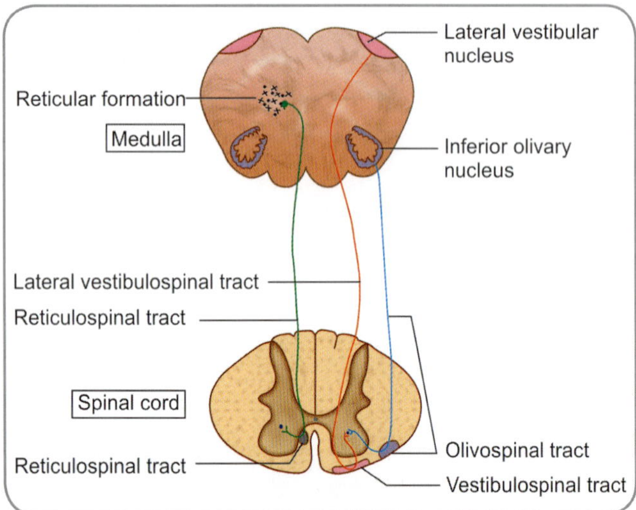

Fig. 20.8: Scheme to show the main features of vestibulospinal, olivospinal and reticulospinal tracts

Lateral Vestibulospinal Tract

The neurons of origin of the lateral vestibulospinal tract lie in the lateral vestibular nucleus. This tract is uncrossed and lies in the anterior funiculus of the spinal cord (Fig. 20.2) (shifting medially as it descends). Its fibres end in relation to neurons in the ventral grey column (laminae VII and VIII). This tract is an important efferent path for equilibrium. The lateral vestibulospinal tract is facilitatory to motor neurons supplying extensor muscles (of the neck, back, and limbs) and is inhibitory to flexor muscles.

Note: The fibres of the lateral vestibulospinal tract are somatotropically arranged. Fibres to cervical segments arise from the cranioventral part of the lateral vestibular nucleus and those to the thoracic segments from the central part and those to the lumbosacral segments from the dorsocaudal parts of the nucleus.

Medial Vestibulospinal Tract

The medial vestibulospinal tract arises mainly from the medial vestibular nucleus (with some fibres from the inferior and lateral nuclei). The tract descends through the anterior funiculus (within the sulcomarginal fasciculus). The fibres are partly crossed and partly, uncrossed. They end in cervical segments of the cord in laminae VII and VIII.

The medial tract is inhibitory to muscles of the neck and back.

Olivospinal Tract

This tract arises from the inferior olivary nucleus (medulla) and terminates in relation to ventral horn cells of the spinal cord (Fig. 20.8). However, recent research suggests that fibres do not descend from the inferior olive to the spinal cord.

Reticulospinal Tracts

The reticular formation is connected to spinal grey matter through the medial and lateral reticulospinal tracts (Fig. 20.8).

Medial Reticulospinal Tract

Fibres arise from the medial part of the reticular formation of both the pons and the medulla (mainly from the nucleus gigantocellularis reticularis of the medulla and the oral and caudal reticular nuclei of the pons). The fibres, which are crossed and uncrossed, descend in the anterior funiculus (near the anterior median fissure). The fibres reach all levels of the spinal cord. They end directly or through interneurons on α- and γ-motor neurons. The tract is facilitatory to muscles of the trunk and limbs, but some fibres are inhibitory to neck muscles. The tract is concerned with postural adjustments of the head, trunk, and limbs.

Lateral Reticulospinal Tract

This tract is constituted by fibres arising in the ventrolateral part of the reticular formation of the pons (caudal and oral pontine reticular nuclei). The fibres cross to the opposite side in the medulla and run down in the lateral funiculus.

Apart from the control of motor function, the reticulospinal tracts may influence transmission of pain through ascending tracts.

Note: Reticulospinal fibres terminate widely in spinal grey matter, but the exact laminae in which they end are controversial. They possibly terminate in all laminae other than II and III, with the majority ending in laminae VI–VIII. Some fibres reach the lateral cervical nucleus.

Clinical Correlation

Effects of Lesions of Extrapyramidal Tracts
- *Paralysis* with little or no muscular atrophy (except that due to disuse)
- *Spasticity or hypertonicity* of the muscles
- *Exaggerated deep reflexes, viz.* knee jerk, ankle jerk, etc.
- *Clasp-knife rigidity*.

Rigidity

When muscle tone is excessive, the body becomes rigid. Distinctive types of rigidity are recognised. If we try to flex a rigid limb, and there is sudden loss of resistance, the rigidity is described as *clasp-knife rigidity*. When the resistance is uniform over the range of movement, the condition is called *lead pipe rigidity*. When resistance is intermittent, the rigidity is said to be of the *cogwheel type*.

Note: After studying the connections and physiological roles of both pyramidal and extrapyramidal tracts, it is important to compare the tracts of these two systems. Table 20.2 compares the differences between pyramidal and extrapyramidal descending tracts.

contd...

Table 20.2: Differences between pyramidal tract and extrapyramidal tracts

Pyramidal tracts	Extrapyramidal tracts
It consists of corticospinal and corticonuclear tracts	These consists of rubrospinal, recticulospinal, tectospinal, vestibulospinal, and olivospinal tracts
It is phylogenetically recent and present only in mammals	Phylogenetically older than pyramidal system
Cortical fibres of the tract arise predominantly from the cerebral cortex (motor cortex)	Cortical fibres of these tracts arise predominantly from subcortical areas like red nucleus, reticular formation, tectum, vestibular nuclei, and olivary nucleus
It is connected directly to the lower motor neurons	They are connected indirectly following a polysynaptic pathway to lower motor neurons
It is responsible for non-postural, precise movements of small muscles involved in skillful activity	They are responsible for gross postural (stereotyped) movements involving large groups of muscles
Pyramidal tract lesions leads to an increase in muscle tone (spastic paralysis) due to loss of control of the upper motor neurons on the lower motor neurons	Extrapyramidal tract lesions leads to increased muscle tone (spasticity), clasp knife rigidity, and exaggerated deep reflexes

Clinical Correlation *contd...*

Naturally occurring lesions in man rarely, if ever, involve pyramidal pathway without simultaneous involvement of extrapyramidal pathways; therefore, the division of motor pathways into pyramidal and extrapyramidal systems is of little or no clinical relevance.

DESCENDING TRACTS ENDING IN THE BRAINSTEM

CORTICONUCLEAR TRACTS

The nuclei of cranial nerves that supply skeletal muscle (i.e. somatic efferent and special visceral efferent nuclei) are functionally equivalent to ventral horn cells of the spinal cord. They are under cortical control through fibres that are closely related in their origin and course to corticospinal fibres. At various levels of the brainstem, these fibres cross to the opposite side to end by synapsing with cells in cranial nerve nuclei, either directly or through interneurons.

CORTICOPONTOCEREBELLAR PATHWAY

Fibres arising in the cerebral cortex of the frontal, temporal, parietal, and occipital lobes descend through the corona radiata and internal capsule to reach the crus cerebri. The frontopontine fibres occupy the medial one-sixth of the crus and the temporopontine fibres (along with occipitopontine and parietopontine fibres) occupy the lateral one-sixth of the crus. These fibres enter the ventral part of the pons to end in pontine nuclei of the same side.

Axons of neurons in the pontine nuclei form the transverse fibres of the pons. These fibres cross the midline and pass into the middle cerebellar peduncle of the opposite side. The fibres of this peduncle reach the cerebellar cortex.

The corticopontocerebellar pathway forms the anatomical basis for control of cerebellar activity by the cerebral cortex.

Clinical Correlation

Disorders of Motor Function

Inability to move a part of the body is referred to as *paralysis*. This can be produced by interruption of motor pathways anywhere between the motor area of the cerebral cortex and the muscles themselves. We have seen that the pathway from cortex to muscle involves at least two neurons. The first of these is located in the cerebral cortex. Its axon terminates in the spinal cord or in motor cranial nerve nuclei in the brainstem referred to as the *upper motor neuron*. The second neuron located in the anterior grey column of the spinal cord (or in motor nuclei of the brainstem), which sends out an axon that travels through a peripheral nerve to innervate muscle, is referred to as the *lower motor neuron*.

When lower motor neurons are destroyed or their continuity interrupted, the muscles supplied by them lose their tone (i.e. they become flaccid) and in due course of time, the muscles undergo atrophy. Changes in electrical responses of the muscles also take place. These alterations constitute the *reaction of degeneration*. In addition, because of interruption of the efferent part of reflex pathways tendon reflexes are abolished.

Destruction or interruption of the upper motor neuron is not followed by any of these changes. On the other hand, it is usually accompanied by an increase in muscle tone and exaggeration of tendon reflexes. It is, therefore, possible to distinguish between an upper motor neuron paralysis (often called *spastic paralysis*) and a lower motor neuron (or *flaccid*) paralysis.

Paralysis may be confined to one limb (*monoplegia*) or to both limbs on one side of the body (*hemiplegia*). Paralysis of both lower limbs is called *paraplegia* and that of all four limbs is called *quadriplegia* (Table 20.3). Paralysis of muscles supplied by one or more cranial nerves may occur in isolation or in combination with hemiplegia. Destruction of a particular region often destroys lower motor neurons situated at that level, resulting in a localised flaccid paralysis of muscles supplied from

contd...

<table>
<tr><td colspan="2">

🦠 **Clinical Correlation** *contd...*
</td></tr>
</table>

that level. This lesion may at the same time interrupt descending tracts (representing upper motor neurons), resulting in a spastic paralysis below the level of the lesion. ***The presence of a localised flaccid paralysis can, thus, serve as a pointer to the level of lesion***.

It is also important to remember that the fibres of upper motor neurons meant for the limbs cross the midline in the lower part of the medulla (in the decussation of the pyramids) and those for the cranial nerves cross just above the level of their termination. A lesion above the level of crossing produces a paralysis in the opposite half of the body and a lesion below this level produces paralysis on the same side.

Effects of lesions of the motor pathways at various levels

- ❑ ***Lesion in cerebral cortex:*** Because of the large extent of the ***motor areas of the cerebral cortex,*** lesions here produce a relatively localised paralysis for example, a monoplegia.

- ❑ ***Lesion in internal capsule:*** A lesion in the ***internal capsule*** is capable of producing widespread paralysis on the opposite half of the body (hemiplegia), which may also involve the lower part of the face and the tongue. The cranial nerves having a bilateral corticonuclear supply are spared. A lesion in the internal capsule is most likely to result from thrombosis or rupture of one of the arteries supplying the capsule. The artery most often involved is Charcot's artery of cerebral haemorrhage.

- ❑ ***Lesion in brainstem:*** Lesions of corticospinal fibres at various levels in the ***brainstem,*** above the level of the pyramidal decussation, can produce contralateral hemiplegia. If the lesion crosses the midline symptoms can be bilateral. Involvement of motor cranial nerve nuclei (or of fibres arising from them) may result in various combinations. For example, a lesion in the upper part of the midbrain can produce a paralysis of muscles supplied by the oculomotor nerve on the side of lesion, along with a hemiplegia on the opposite side (***Weber's syndrome***). A similar lesion in the pons results in a paralysis of the lateral rectus muscle (abducent nerve) on the side of lesion with hemiplegia on the opposite side (***Raymond's syndrome***). Alternatively, facial paralysis of one side can be combined with contralateral hemiplegia (***Millard Gubler syndrome***). Various such combinations may result depending on the level of lesion.

- ❑ ***Lesion in spinal cord:*** Lesions affecting the lateral corticospinal tract in the spinal cord produce an upper motor neuron paralysis of muscles on the same side of the body. Lesions above the fifth cervical segment result in paralysis of both upper and lower extremities (quadriplegia), while lesions below the first thoracic segment affect only the lower limbs (paraplegia). As in the brainstem, lesions in the spinal cord may be bilateral. Involvement of lower motor neurons at the level of lesion produces a flaccid paralysis of muscles supplied from that level, along with spastic paralysis below the level of injury. A knowledge of muscles supplied by individual spinal segments can, thus, help in locating the level of a lesion in the spinal cord.

Note: In the above description, we have (for the sake of simplicity) considered involvement of motor neurons in isolation. However, disease at any level can involve other structures resulting in sensory and other disturbances.

Table 20.3: Terminology used in relation to motor dysfunction

Terminology	Meaning
Paralysis	Inability to move a part of the body
Paresis	Slight or partial paralysis
Monoplegia	Paralysis of one limb only
Hemiplegia	Paralysis of one-half of the body involving both upper and lower limbs of that half (right side hemiplegia will involve right upper limb and lower limb; left side hemiplegia will involve left upper limb and lower limb)
Paraplegia	Paralysis of either both upper limbs or both lower limbs
Quadriplegia	Paralysis of all the four limbs

ASCENDING TRACTS ENDING IN THE SPINAL CORD

The ascending tracts of the spinal cord and brainstem represent one stage of multineuron pathways by which afferent impulses arising in various parts of the body are conveyed to different parts of the brain.

The ***first order neurons*** of these pathways are usually located in spinal (dorsal nerve root) ganglia (Fig. 20.9). The neurons in these ganglia are unipolar (pseudounipolar). Each neuron gives off a peripheral process and a central process. The peripheral processes of the neurons form the afferent fibres of peripheral nerves. They end in relation to sensory end organs (receptors) situated in various tissues. The central processes of these neurons enter the spinal cord through the dorsal nerve roots. Having entered the cord, the central processes, as a rule, terminate by synapsing with cells in spinal grey matter. Some of them may run upwards in the white matter of the cord to form ascending tracts.

The majority of ascending tracts are, however, formed by axons of cells in spinal grey matter. These are ***second order*** sensory neurons. In the case of pathways that convey sensory information to the cerebral cortex, the second order neurons end by synapsing with neurons in the thalamus.

Third order sensory neurons located in the thalamus carry the sensations to the cerebral cortex (Fig. 20.9).

PATHWAYS CONNECTING THE SPINAL CORD TO THE CEREBRAL CORTEX

The Posterior Column— Medial Lemniscus Pathway

There are two tracts in the posterior column. The medially placed tract is called fasciculus gracilis (tract of Goll) and laterally placed tract is called fasciculus cuneatus (tract of Burdach) (Fig. 20.10). They carry the sensations

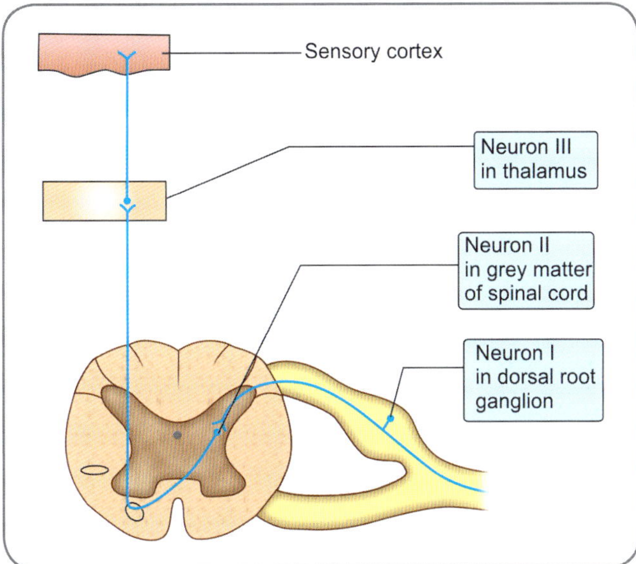

Fig. 20.9: Scheme to show the levels of neurons in ascending tract

of discriminative or fine touch, pressure, vibration, conscious sense of position and movements (conscious proprioception), and stereognosis.

The fibres derived from the lowest ganglia are situated most medially, while those from the highest ganglia are most lateral. The fasciculus gracilis, which lies medially is, therefore, composed of fibres from the coccygeal, sacral, lumbar, and lower thoracic ganglia, while the fasciculus cuneatus, which lies laterally consists of fibres from upper thoracic and cervical ganglia.

❑ The first order neurons in the spinal ganglia receive the sensations. The central processes of these neurons enter the spinal cord via the dorsal rootlets and form the fasciculus gracilis and fasciculus cuneatus.

Note: Fasciculus gracilis (tract of Goll) and fasciculus cuneatus (tract of Burdach) are unique in the way that they are formed predominantly by central processes of neurons located in dorsal nerve root ganglia, i.e. by first order sensory neurons.

❑ The fasciculus gracilis and fasciculus cuneatus terminate in the gracile and cuneate nuclei, respectively in medulla oblongata. Their axons run forwards and medially (as ***internal arcuate fibres***) to cross the middle line. The crossing fibres of the two sides constitute the ***sensory decussation (or lemniscal decussation)***. Having crossed the middle line, the fibres turn upwards to form a prominent bundle called the ***medial lemniscus*** (Fig. 20.10). The medial lemniscus runs upwards through the medulla, pons, and midbrain to end in the thalamus (ventral posterolateral nucleus).

❑ Third order sensory neurons located in the thalamus give off axons that pass through the internal capsule and the corona radiata to reach the somatosensory areas of the cerebral cortex.

This pathway carries:

❑ Some components of the sense of touch. These include deep touch and pressure, the ability to localise exactly the part touched (tactile localisation), the ability to recognise as separate two points on the skin that are touched simultaneously (tactile discrimination), and the ability to recognise the shape of an object held in the hand (stereognosis).

❑ Proprioceptive impulses that convey the sense of position and of movement of different parts of the body.

❑ The sense of vibration.

Clinical Correlation

❑ The ***lesions of posterior white columns*** (i.e. fasciculus gracilis and cuneatus) cause loss of sensations of proprioception (sense of relative position of parts of body), two-point discrimination, and vibration on the same side of body below the level of lesion.

❑ The ***lesions of medial lemniscus*** produce contralateral loss of sensations of proprioception (sense of relative position of parts of body), two-point discrimination, and vibration.

Tabes Dorsalis

This condition caused by syphilis affects the posterior column tracts (Fig. 20.11). There is numbness, paraesthesia, and severe pain in some parts of the body; while there is hypersensitivity in others. Maintenence of posture is affected. The patient is not able to tell the position or movements of his/her lower limbs, unless (s)he sees them. Therefore, when he is asked to stand up with his eyes closed and feet together, he staggers and falls because he cannot maintain his position due to lack of proprioceptive information (***Romberg's sign positive***).

SPINOTHALAMIC PATHWAYS

Anterior and Lateral Spinothalamic Tracts

❑ The first order neurons of this pathway are located in spinal ganglia. The central processes of these neurons enter the spinal cord and terminate in relation to spinal grey matter (Fig. 20.12). They may ascend in the dorsolateral tract (situated near the tip of the dorsal grey column, for one or more segments, before ending in grey matter).

❑ The second order neurons of this pathway are located mainly in laminae IV, V, VI, and VII. The axons of these neurons constitute the anterior and lateral spino-thalamic tracts. They cross to the opposite side of the spinal cord in the white commissure (but some fibres may remain on the same side). The crossing of the fibres is oblique. The fibres for the lateral spinothalamic tract cross within the same segment of the cord, while those of the anterior spinothalamic tract may ascend for one or more segments before they cross to the opposite side. The tracts also carry some uncrossed fibres (about 10%).

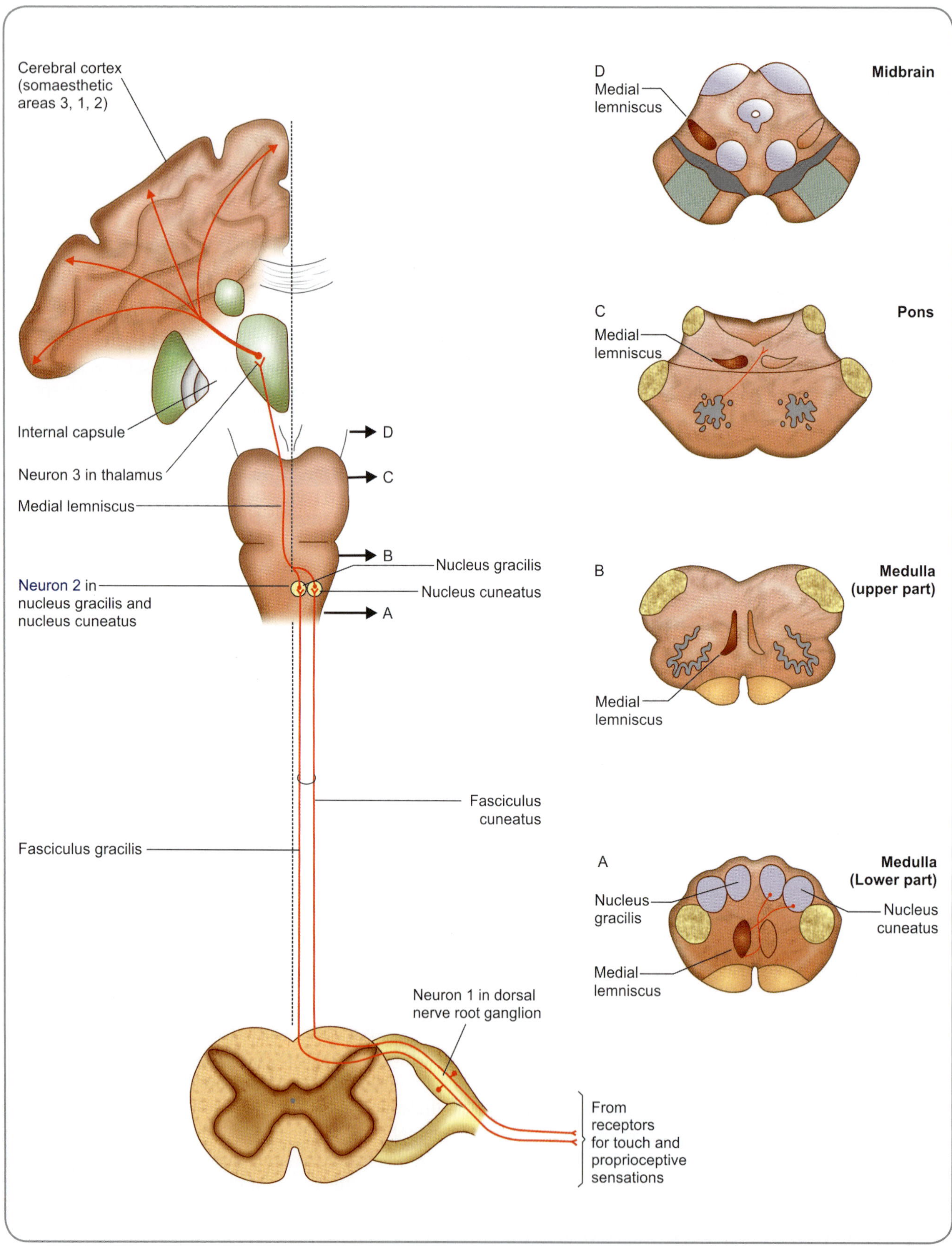

Fig. 20.10: Scheme to show the main features of the posterior column medial lemniscus pathway – note the position of the medial lemniscus at various levels of the brainstem

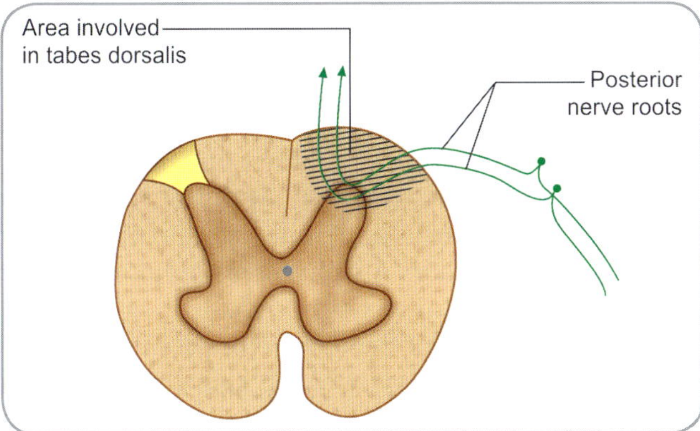

Fig. 20.11: Site of lesion in tabes dorsalis

The fibres for the anterior spinothalamic tract enter the anterior funiculus (Figs 20.2 and 20.12), where they lie medial to emerging fibres of ventral nerve roots. The fibres for the lateral spinothalamic tract enter the lateral funiculus. The two tracts form one continuous band that runs up the spinal cord. In the medulla, the spinothalamic tracts run through the brainstem as a separate bundle called the ***spinal lemniscus,*** which ends in the thalamus. All spinothalamic fibres end in the ventral posterolateral nucleus of the thalamus.

Traditionally, it has been said that the anterior spinothalamic tract carries sensations of crude touch and pressure, while the lateral tract carries sensations of pain and temperature. Although different fibres within the spinothalamic tracts carry different types of sensations, the anterior and lateral spinothalamic tracts constitute a single functional unit.

Clinical Correlation

Lesion of spinothalamic tract: A unilateral damage to the spinothalamic tract leads to loss of pain and temperature (hot and cold) sensations from the skin of the opposite side of the body, below the lesion.

Cordotomy

Sometimes, a patient may be in severe pain that cannot be controlled by drugs. As an extreme measure, pain may be relieved by cutting the spinothalamic tracts. The operation is called cordotomy.

The ligamentum denticulatum serves as a guide to the surgeon. For relief of pain, the incision is placed anterior to this ligament (***anterolateral cordotomy***).

Pain can also be relieved by cutting the posterior nerve roots in the region. This operation is called ***posterior rhizotomy***.

Syringomyelia

In a disease called ***syringomyelia,*** the region of the spinal cord near the central canal undergoes degeneration with formation of cavities (Fig. 20.13). Fibres of spinothalamic tracts crossing in this region are interrupted. Sensations of pain and temperature are lost over the part of the skin from which fibres are interrupted, but touch is retained, as it also passes through the posterior column tracts. This phenomenon is called ***dissociated anesthesia***.

Dorsolateral Spinothalamic Tract

In addition to the anterior and lateral spinothalamic tracts, a separate dorsolateral spinothalamic tract has been described. The fibres of this tract arise from neurons in lamina I, cross to the opposite side and ascend in the dorsolateral fasciculus to reach the ventral posterolateral nucleus of the thalamus. The tract carries impulses arising

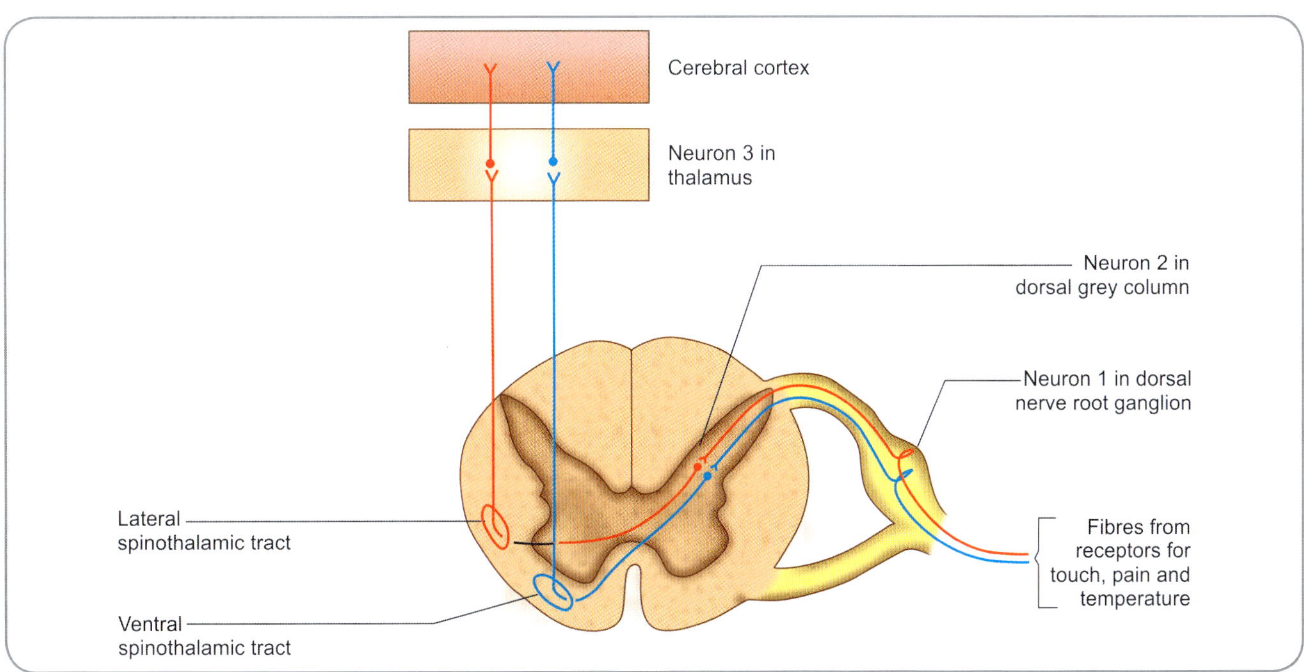

Fig. 20.12: Scheme to illustrate the main features of the spinothalamic pathways

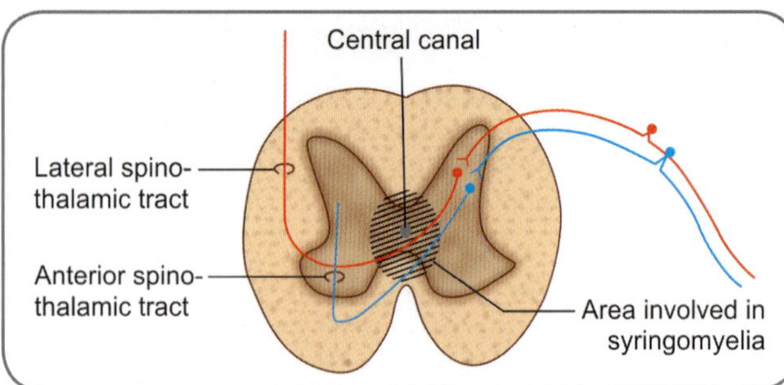

Fig. 20.13: Site of lesion in syringomyelia

in skin (mainly, pain and temperature). Relief of pain observed after dorsolateral cordotomy may be a result of the cutting of these fibres.

Spinocervicothalamic Pathway

This is yet another pathway through which cutaneous sensations (touch, pressure, pain, and temperature) reach the thalamus. Axons arising from neurons located in laminae III and V of spinal grey matter collect to form a *spinocervical tract* that ascends through the dorsolateral fasciculus. They end in the *lateral cervical nucleus* (which is a small collection of neurons lying amongst the fibres of the lateral funiculus in spinal segments C1 and C2). New fibres arising here project to the ventral posterolateral nucleus of the thalamus.

ASCENDING PATHWAYS ENDING IN THE BRAINSTEM

A number of tracts arising in spinal grey matter, and ending in masses of grey matter in the brainstem are described. Some of them are as follows.

SPINORETICULAR TRACTS

Spinoreticular fibres begin from spinal neurons mainly in lamina VII (also V and VIII). The fibres are partly crossed and partly uncrossed. The fibres ascend in the ventrolateral part of the spinal cord, intermingling with spinothalamic tracts. They end in the reticular formation of the medulla and pons. The tract probably carries pain.

SPINO-OLIVARY TRACT

The *spino-olivary tract* is also a crossed tract. It lies at the junction of the anterior and lateral funiculi of the spinal cord. The fibres of the tract end in accessory olivary nuclei.

SPINOMESENCEPHALIC TRACTS

A number of tracts travel from spinal cord to different areas in the midbrain. They are collectively referred to as spinomesencephalic tracts.

The *spinotectal tract* connects the spinal grey matter to the superior colliculus. It is a crossed tract. It carries impulses that regulate reflex movements of the head and eyes, in response to stimulation of some parts of the body. According to some authorities, the tract may carry sensations of pain and temperature. Other spinal fibres reach the pretectal nuclei and some nuclei in the reticular formation of the midbrain. Spinal cells projecting to the midbrain are located mainly in lamina I. They also lie in laminae IV–VII. Most of the fibres cross the midline and ascend in the anterior part of the spinal cord. Painful stimuli may pass through these fibres.

Clinical Correlation

Sensory Disorders

Interruption of ascending pathways carrying various sensations results in loss of sensory perception (**anesthesia**) over parts of the body concerned. In case of peripheral nerves, the area of anesthesia following injury is often much less than the area of distribution of the nerve. This is so because of considerable overlap in the areas supplied by different nerves. The area of skin supplied from one spinal segment is called a **dermatome**. Dermatomes for adjoining segments overlap a given area of skin being innervated by two or more segments.

- ❑ *In the case of spinal cord lesions*, the level of disease can be inferred from the level of sensory loss. In this connection, it must be remembered that the finer modalities of touch are carried by the posterior column tracts, which are uncrossed. Crude touch, pain, and temperature are carried by the spinothalamic tracts which are crossed. Thus, a unilateral lesion in the spinal cord can result in loss of the power of tactile localisation, tactile discrimination, and of stereognosis, on the side of lesion, with loss of crude touch, pain and temperature on the opposite side. Because of a double pathway for touch, loss of sensations of pain, and temperature is often more obvious than interference with touch.
- ❑ While crossing the midline, fibres of the spinothalamic tracts do not run horizontally. They ascend as they cross, so that their path is oblique. The degree of obliquity is greater in the case of fibres carrying touch as compared to those carrying pain and temperature. Because of this, there can be a difference of a few segments in the level at which (or below which) these sensations are lost when the crossing fibres are interrupted by a lesion.
- ❑ Sensory disturbances can also result from *lesions in the brainstem* because of damage to the medial lemniscus or to the spinal and trigeminal lemnisci.
- ❑ Lesions of the thalamus can produce bizarre sensory disturbances.
- ❑ Lesions in the internal capsule can cause sensory loss in the entire opposite half of the body as thalamocortical fibres pass through this region.
- ❑ Pressure on sensory areas of the cerebral cortex can result in various abnormal sensations or in anesthesia over certain regions (Table 20.4). Damage to pathways carrying special sensations of smell, vision, and hearing can result in various defects.

contd...

Table 20.4: Terms used to describe sensory disturbances

Terminology	Meaning
Anaesthesia	Loss of sensory perception
Hypoesthesia	Reduced perception for touch
Hypoalgesia	Reduced perception for pain
Hyperesthesia	Increased perception for touch
Hyperalgesia	Increased perception for pain
Paresthesia	Abnormal sensations

Clinical Correlation *contd...*

Referred Pain

It sometimes happens that when one of the viscera is diseased, pain is not felt in the region of the organ itself, but is felt in some part of the skin and body wall. This phenomenon is called *referred pain*. This pain is usually (but not always) referred to areas of skin supplied by the same spinal segments which innervate the viscus. Some classical examples of referred pain are shown in the Table 20.5.

Thalamic syndrome

❑ Threshold for appreciation of touch, pain, or temperature is lowered.
❑ Sensations that are normal may appear to be exaggerated or unpleasant.
❑ There may be spontaneous pain.
❑ Emotions may be abnormal.

Table 20.5: Referred pain

Site of origin of pain	Site of referred pain
Pain arising in the diaphragm	Shoulder (C4)
Pain arising in the heart	Lower cervical and upper thoracic segments (it is felt in the chest wall and along the medial side of the left arm. It may also be referred to the neck or jaw)
Pain arising in the stomach	Epigastrium (T6 to T9)
Pain arising in the intestine	Epigastrium and around the umbilicus (T7 to T10)
Pain from the gallbladder	Epigastrium (it may also be referred to the back just below the inferior angle of the right scapula)

SPINOCEREBELLAR PATHWAYS

These pathways carry proprioceptive impulses arising in muscle spindles, Golgi tendon organs, and other receptors to the cerebellum (Fig. 20.14). They constitute the afferent component of reflex arcs involving the cerebellum, for control of posture. Recent investigations have shown that some exteroceptive sensations (for example, touch) may reach the cerebellum through these pathways.

❑ The first order neurons of these pathways are located in dorsal nerve root ganglia. Their peripheral processes

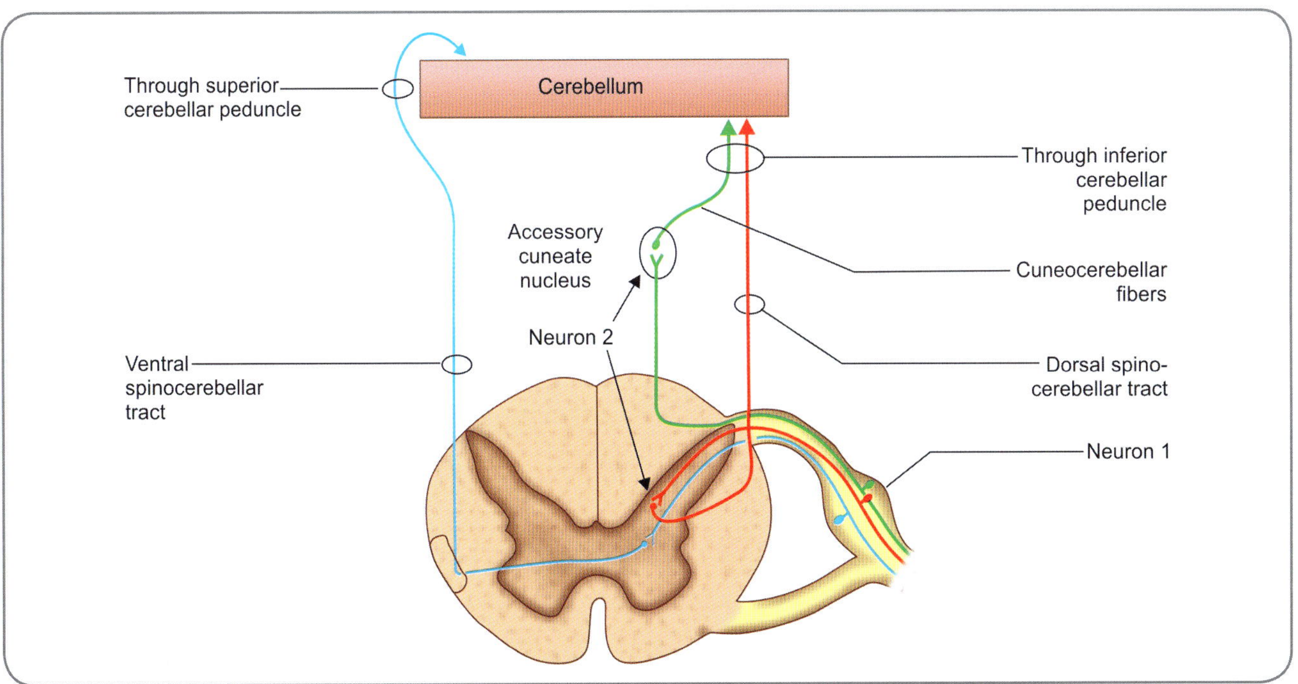

Fig. 20.14: Scheme to show the main features of the spinocerebellar pathways

end in relation to muscle spindles, Golgi tendon organs and other proprioceptive receptors. Some fibres are related to end organs, concerned with exteroceptive sensations (touch, pressure). The central processes of the neurons concerned ascend in the posterior funiculi for varying distances before ending in spinal grey matter. Some of them ascend all the way to the medulla and end in the accessory cuneate nucleus.

❑ The second order neurons of the pathway are arranged in a number of groups:
  ○ Neurons located in the ***nucleus dorsalis – Clarke's column*** (situated on the medial side of the base of the dorsal grey column in segments C8 to L3 of the spinal cord, Figure 18.9); give origin to fibres of the ***dorsal (posterior) spinocerebellar tract***. This is an uncrossed tract lying in the lateral funiculus (Fig. 20.2). It begins in the lumbar segments of the spinal cord and ascends to the medulla where its fibres become incorporated in the inferior cerebellar peduncle and pass through it to reach the vermis of the cerebellum (Fig. 20.14).
  ○ The neurons giving origin to the ***ventral (anterior) spinocerebellar tract*** are probably located in the junctional area between the ventral and dorsal grey columns (laminae V, VI, and VII) in the lumbar and sacral segments of the cord. Some of the neurons concerned may lie in the ventral grey column (spinal border cells). The fibres of this tract are predominantly crossed. They ascend in the lateral funiculus, anterior to the fibres of the dorsal spinocerebellar tract (Fig. 20.2) and pass through the medulla and pons. At the upper end of the pons, the fibres turn downwards to enter the superior cerebellar peduncle through which they reach the vermis of the cerebellum (Fig. 20.14).

From a functional point of view, both the ventral and dorsal spinocerebellar tracts are concerned mainly with the lower limbs and trunk. The dorsal tract carries impulses concerned with fine coordination of muscles controlling posture and with movements of individual muscles. On the other hand, the ventral tract is concerned with movements of the limb as a whole.

Spinocerebellar fibres reaching the cerebellum end as mossy fibres. Representation in the cerebellar cortex is somatotropic (i.e. fibres carrying sensations from different parts of the body end in definite areas of cortex). Conduction in the spinocerebellar tracts can be inhibited or facilitated under the influence of descending tracts (corticospinal, reticulospinal, and rubrospinal).

○ The central processes of some first order neurons (related to cervical segments) reach the accessory cuneate nucleus (also called the lateral or external cuneate nucleus) in the medulla. Second order neurons lying in this nucleus give origin to ***posterior external arcuate fibres*** which enter the inferior cerebellar peduncle (of the same side) to reach the cerebellum. This ***cuneocerebellar tract*** carries impulses from the upper limb. It may be regarded as the forelimb equivalent of the dorsal spinocerebellar tract.

Clinical Correlation

Disorders of Equilibrium

The maintenance of equilibrium and of correct posture is dependent on reflex arcs involving various centres including the spinal cord, cerebellum, and vestibular nuclei. Afferent impulses for these reflexes are carried by the posterior column tracts (fasciculus gracilis and fasciculus cuneatus), the spinocerebellar tracts, and others. Efferents reach neurons of the ventral grey column (anterior horn cells) through rubrospinal, vestibulospinal, and other 'extrapyramidal' tracts. Interruption of any of these pathways or lesions in the cerebellum, the vestibular nuclei and other centres concerned can result in various abnormalities, involving maintenance of posture and coordination of movements.

Inability to maintain the equilibrium of the body while standing or while walking is referred to as ***ataxia***. This may occur as a result of the interruption of afferent proprioceptive pathways (***sensory ataxia***). Disease of the cerebellum itself or of efferent pathways results in more severe disability. Lack of proprioceptive information can be compensated to a considerable extent by information received through the eyes. The defects are, therefore, much more pronounced with the eyes closed (***Romberg's sign***).

Multiple Choice Questions

1. Fibres arising from the premotor area 6 of cerebral cortex and ending in the ventral horn cells of the spinal cord are part of:
   a. The extrapyramidal tract
   b. The pyramidal tract
   c. The medial descending pathway
   d. The prepyramidal tract
2. The lateral vestibulospinal tract is:
   a. Facilitatory to flexor muscles
   b. Inhibitory to flexor muscles
   c. Facilitatory to extensor muscles
   d. Inhibitory to extensor muscles
3. The axons arising from the neurons of nucleus gracilis are:
   a. First order neurons
   b. Second order neurons
   c. Third order neurons
   d. A mix of second and third order neurons

4. The fibres of spinocervical tract end in:
   a. The neurons located in the lateral funiculus of upper cervical spinal segments
   b. The neurons located in the lower parietal cortex
   c. The neurons located in the ventral funiculus of the upper cervical spinal segments
   d. The neurons located in the lateral funiculus of lower thoracic spinal segments
5. The forelimb equivalent of the dorsal spinocerebellar tract is:
   a. The ventral spinocerebellar tract
   b. The cuneocerebellar tract
   c. The spino-olivary tract
   d. The spinotectal tract

ANSWERS

1. b   **2.** c   **3.** b   **4.** a   **5.** b

Clinical Problem-solving

Case Study 1: A 57-year-old man suffered a neurological stroke. He has widespread paralysis of the muscles on the right side of his body. The lower half on the right side of his face is also affected.
- Which part of the motor pathway is likely to be affected in this patient?
- Give reasons to say that it is not likely to be a lesion in the cerebral cortex or brainstem.
- Which vessel is the most common to be involved in such lesions?

Case Study 2: A 76-year-old man exhibited the following symptoms—unpleasant tickling even if touched lightly, abnormal bouts of crying, sudden spontaneous pain in different parts of the body and very limited tolerance for even warm objects.
- What condition, do you think, is this man suffering from?
- Why are sensations altered in this condition?
- Why is the motor pathway spared?

(For solutions see Appendix).

Chapter 21

Cranial Nerves—
Nuclei and Functional Aspects

Frequently Asked Questions

- ❏ Write notes on: (a) Somatic efferent column, (b) Edinger-Westphal nucleus, (c) Salivatory nuclei, (d) Functional components of the facial nerve.
- ❏ Discuss the functional components of the trigeminal nerve and the various nuclei related to it.
- ❏ Describe the supranuclear and infranuclear types of facial paralyses.

INTRODUCTION

There are 12 pairs of cranial nerves:

I	–	Olfactory
II	–	Optic
III	–	Oculomotor
IV	–	Trochlear
V	–	Trigeminal
VI	–	Abducent
VII	–	Facial
VIII	–	Vestibulocochlear
IX	–	Glossopharyngeal
X	–	Vagus
XI	–	Accessory
XII	–	Hypoglossal

The seven functional components, to which the fibres of a cranial nerve may belong, are as follows:

- ❏ Somatic efferent
- ❏ General visceral efferent
- ❏ Special visceral efferent
- ❏ General somatic afferent
- ❏ Special somatic afferent
- ❏ General visceral afferent
- ❏ Special visceral afferent.

Each functional component has its own nuclei of origin (in the case of efferent fibres) or termination (in the case of afferent fibres).

DEVELOPMENTAL ASPECTS

In the embryo, the nuclei related to the various components are arranged in vertical rows (or columns) in a definite sequence in the grey matter related to the floor of the fourth ventricle (Fig. 21.1). The sequence is easily remembered, if the following facts are kept in mind:

- ❏ Each half of the floor of the ventricle is divided into a medial part and a lateral part by the **sulcus limitans**. **Efferent nuclei** lie in the medial part (called the **basal lamina**) and **afferent nuclei** in the lateral part (called the **alar lamina**) (Table 21.1).
- ❏ In each part (medial or lateral), **visceral nuclei** lie nearer the sulcus limitans than **somatic nuclei**.
- ❏ Within each category (for example, visceral efferents, somatic afferents, etc.), the **general nucleus** lies nearer the sulcus limitans than the **special nucleus**.

Thus, in proceeding laterally from the midline, the sequence of nuclear columns is as follows:

- ❏ **Somatic efferent:** This column is not subdivided into **general** and **special parts**;
- ❏ Special visceral (or branchial) efferent;
- ❏ General visceral efferent;
- ❏ General visceral afferent;
- ❏ Special visceral afferent;
- ❏ General somatic afferent; and
- ❏ Special somatic afferent.

Table 21.1 : Nuclear columns in basal and alar laminae	
Nuclear columns of basal lamina	**Nuclear columns of alar lamina**
Somatic efferent	General visceral afferent
Special visceral efferent	Special visceral afferent
General visceral efferent	General somatic afferent
	Special somatic afferent

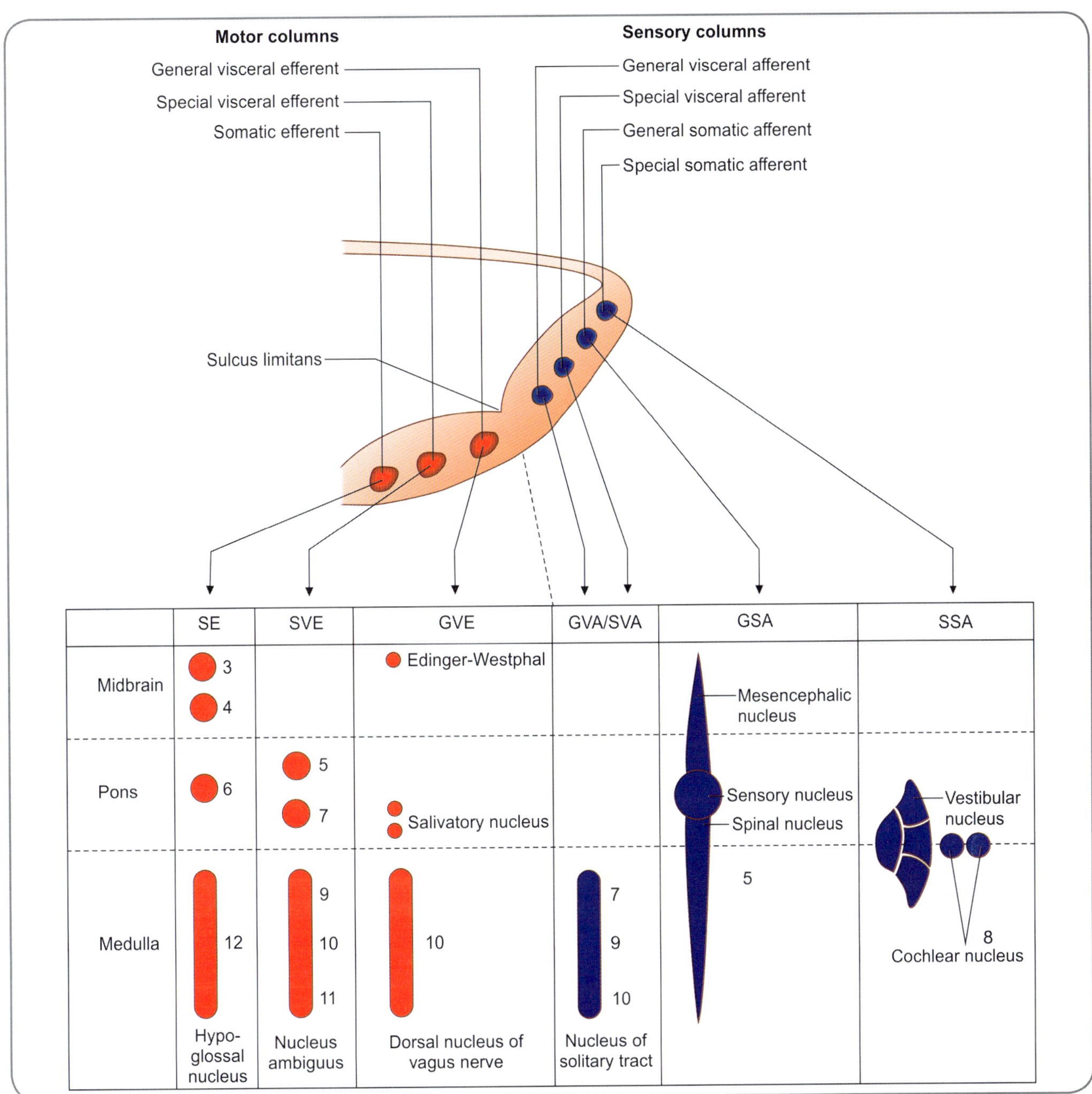

Fig. 21.1: Functional classification of cranial nerve nuclei. The upper figure shows the arrangement of nuclear columns in the brainstem of the embryo. The lower figure shows the nuclei derived from each column. Numbers indicate the cranial nerves connected to the nuclei

Note that basal lamina consists of motor cell columns: Somatic efferent (SE), special visceral efferent (SVE), general visceral efferent (GVE), and alar lamina consists of sensory columns: General visceral afferent (GVA), special visceral afferent (SVA), general somatic afferent (GSA) and special somatic afferent (SSA).

As development proceeds, parts of these columns disappear, so that each of them no longer extend to the whole length of the brainstem, but is represented by one or more discrete nuclei. These nuclei are shown schematically in the lower half of Fig. 21.1. Some nuclei retain their original positions in relation to the floor of the fourth ventricle, but some others migrate deeper into the brainstem. The position of the nuclei relative to the posterior surface of the brainstem is illustrated in Fig. 21.2.

In the description that follows, the nuclei of the third to twelfth cranial nerves are considered as they are located in the brainstem. The cranial nerves III and IV belong to the midbrain, V, VI, VII and part of VIII to the pons; and the remaining to the medulla.

From the functional point of view, the first two cranial nerves are pure sensory nerves. The remaining cranial nerves are mixed nerves with both sensory and motor types of fibres present within them.

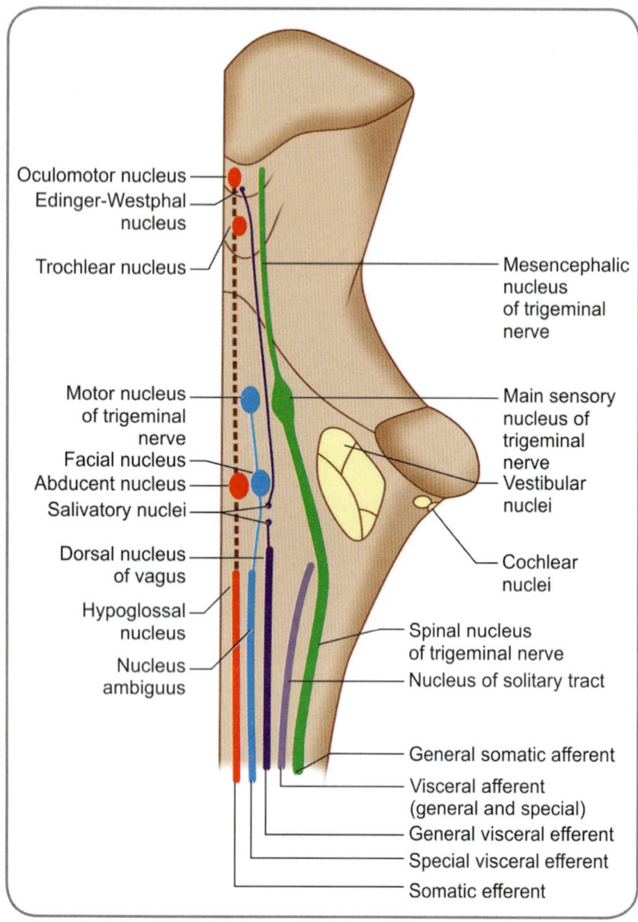

Fig. 21.2: Cranial nerve and nuclei as projected onto the dorsal aspect of the brainstem

The functional components of cranial nerve nuclei are shown in Table 21.2. It should be noted that some of these nuclei contribute fibres for more than one nerve. Similarly, some of the cranial nerves consist of fibres arising from more than one nucleus.

SOMATIC EFFERENT NUCLEI

The somatic efferent column consists of the following nuclei that supply striated (skeletal) muscles of somatic origin:

- The ***oculomotor nucleus*** is situated in the upper part of the midbrain at the level of the superior colliculus. The nuclei of the two sides form a single complex that lies in the central grey matter, ventral to the aqueduct.
- The ***trochlear nucleus*** is situated in the lower part of the midbrain at the level of the inferior colliculus (Figs 21.1 and 21.2). The nucleus lies ventral to the aqueduct in the central grey matter.
- The ***abducent nucleus*** is situated in the lower part of the pons. It lies in the grey matter lining the floor of the fourth ventricle near the midline (Figs 21.1 and 21.2).
- The ***hypoglossal nucleus*** lies in the medulla. It is an elongated column extending into both the open and closed parts of the medulla. Its upper part lies deep to the hypoglossal triangle in the floor of the fourth ventricle. When traced caudally, it lies next to the midline in the central grey matter, ventral to the central canal (Figs 21.1, 21.2 and 21.3).

Table 21.2: The functional nuclear columns and cranial nerve nuclei of brainstem

	Sensory (Afferent)			Motor (Efferent)		
	GVA/SVA	**GSA**	**SSA**	**GSE**	**SVE**	**GVE**
Midbrain		(V) Mesencephalic nucleus of trigeminal nerve		(III) Oculomotor (IV) Trochlear		(III) Edinger-Westphal nucleus
Pons		(V) Sensory nucleus of trigeminal nerve		(VI) Abducent	(V) Motor nucleus of trigeminal nerve (VII) Facial	Salivatory nuclei (VII, IX)
			(VIII) Vestibulo-cochlear nuclei			
Medulla	(VII, IX, X) Nucleus of the solitary tract	(V) Nucleus of the spinal tract of the trigeminal nerve	(VIII) Vestibulo-cochlear nucleus	(XII) Hypoglossal nucleus	(IX, X, XI) Nucleus ambiguus	(X) Dorsal nucleus of vagus

GVA, general visceral afferent; SVA, special visceral afferent; GSA, general somatic afferent; SSA, special somatic afferent; GSE, general somatic efferent; SVE, special visceral efferent; GVE, general visceral efferent.

SPECIAL VISCERAL EFFERENT NUCLEI

These nuclei are also called **branchial efferent** or **branchiomotor nuclei**. They supply striated (skeletal) muscles derived from the branchial arches.

- The **motor nucleus of the trigeminal nerve** lies in the upper part of the pons, in its dorsal part (Figs 21.1 and 21.2). It is situated in the lateral part of the reticular formation, medial to the main sensory nucleus of the trigeminal nerve.
- The **nucleus of the facial nerve** lies in the lower part of the pons and occupies a position similar to that of the motor nucleus of the trigeminal nerve. The spinal nucleus and tract of the trigeminal nerve lie lateral to it (Figs 21.1 and 21.2).
- The **nucleus ambiguus** lies in the medulla. It forms an elongated column lying deep in the reticular formation, both in the open and closed parts of the medulla (Figs 21.1 and 21.2). Inferiorly, it is continuous with the spinal accessory nucleus. It is a composite nucleus and contributes fibres to the glossopharyngeal, vagus and accessory nerves.

GENERAL VISCERAL EFFERENT NUCLEI

The nuclei of this column give origin to preganglionic fibres that constitute the cranial parasympathetic outflow. These fibres end in the peripheral ganglia. Post-ganglionic fibres arising in these ganglia supply smooth muscle or glands. The nuclei are as follows:

- The **Edinger-Westphal nucleus** (or **accessory oculomotor nucleus**) lies in the midbrain (Figs 21.1 and 21.2). It is closely related to the oculomotor complex. Fibres arising in this nucleus pass through the oculomotor nerve. They relay in the ciliary ganglion to supply the sphincter pupillae and the ciliaris muscle.
- The **salivary** (or **salivatory**) **nuclei** (superior and inferior) lie in the dorsal part of the pons, just above its junction with the medulla (Figs 21.1 and 21.2). They are located just above the upper end of the dorsal nucleus of the vagus nerve. The superior nucleus sends fibres into the facial nerve. These fibres relay in the submandibular ganglion to supply the submandibular and sublingual salivary glands. The inferior nucleus sends fibres into the glossopharyngeal nerve. These fibres relay in the otic ganglion to supply the parotid gland. The parotid gland may also receive some fibres from the superior salivary nucleus through the submandibular ganglion.

    Other neurons probably located near the salivary nuclei send out fibres that supply the lacrimal gland, after relaying in the pterygopalatine ganglion. These fibres travel through the facial nerve.

- The **dorsal (motor) nucleus of the vagus** (or **dorsal vagal nucleus**) lies in the medulla. It is a long nucleus lying vertically. Its upper end lies deep to the vagal triangle in the floor of the fourth ventricle. When traced downwards, it extends into the closed part of the medulla where it lies in the lateral part of the central grey matter. Fibres arising in this nucleus supply the heart, lungs, bronchi, oesophagus, stomach, small intestine and large intestine up to the right two-thirds of the transverse colon. They end in ganglia (or nerve plexuses) closely related to these organs. Post-ganglionic fibres arise in these ganglia and run a short course to supply smooth muscle and glands in these organs.

GENERAL AND SPECIAL VISCERAL AFFERENT NUCLEI

Both these columns are represented by the **nucleus of the solitary tract** present in the medulla (Figs 21.1 and 21.2). Like other cranial nerve nuclei of the medulla, the cells of this nucleus form an elongated column lying deep in the reticular formation. Its upper part lies ventrolateral to the dorsal nucleus of the vagus. When traced downwards, it extends into the closed part of the medulla; here, it lies dorsomedial to the vagal nucleus. The lower ends of the nuclei of the two sides fuse to form the **commissural nucleus** of the vagus.

Fibres of taste (special visceral afferent) carried by the facial, glossopharyngeal and vagus nerves end in the upper part of the nucleus of the solitary tract, which is sometimes called the **gustatory nucleus**. The lower portion of the nucleus receives the general visceral sensations from pharynx (glossopharyngeal and vagus) and from oesophagus and abdominal part of alimentary canal up to right two-thirds of the transverse colon (vagus).

GENERAL SOMATIC AFFERENT NUCLEI

The general somatic afferent column is represented by the three sensory nuclei of the trigeminal nerve. These are as follows:

- The **main** (or **superior**) **sensory nucleus** of trigeminal nerve lies in the upper part of the pons in the lateral part of the reticular formation. It lies lateral to the motor nucleus of the trigeminal (Figs 21.1 and 21.2). It is mainly concerned in mediation of proprioceptive impulses, touch and pressure (from the region to which the trigeminal nerve is distributed).
- The **spinal nucleus** of trigeminal nerve extends from the main nucleus down into the medulla (Figs 21.1 and 21.2) and into the upper two segments of the spinal cord. Its lower end is continuous with the substantia gelatinosa of the cord. In addition to fibres of the trigeminal nerve, the nucleus also receives general somatic sensations carried by the facial, glossopharyngeal and vagus nerves.

The spinal nucleus is concerned mainly with the mediation of pain and thermal sensibility. Different parts of the nucleus correspond to different areas innervated.

❏ The ***mesencephalic nucleus*** of the trigeminal nerve extends cranially from the upper end of the main sensory nucleus into the mid-brain. Here it lies in the central grey matter lateral to the aqueduct. Functionally, this nucleus appears to be similar to sensory ganglia of cranial nerves and to the spinal ganglia rather than to afferent nuclei. The neurons in it are unipolar. The peripheral processes of these neurons carry proprioceptive impulses from muscles of mastication and, possibly, also from muscles of the eyeballs, face, tongue and teeth. The central processes of the neurons in the nucleus end in the main sensory or motor nucleus of the trigeminal nerve.

The mesencephalic nucleus is the centre for the jaw jerk.

The main afferents of the nuclei of the general somatic afferent column (made up of the main and spinal nuclei of the trigeminal nerve) are central processes of neurons in the trigeminal ganglion. After entering the pons, many of these processes divide into ascending and descending branches, while others ascend or descend. The ascending fibres end in the main sensory nucleus. Those that descend form a large bundle of fibres that constitutes the ***spinal tract of the trigeminal nerve*** (Fig. 21.3). The tract is closely related to the dorsolateral aspect of the spinal nucleus of the trigeminal nerve. The tract descends from the pons into the medulla and then into the upper part of the spinal cord. Apart from trigeminal fibres, the tract also carries general somatic afferent fibres from the facial, glossopharyngeal and vagus nerves.

SPECIAL SOMATIC AFFERENT NUCLEI

These are the ***cochlear*** and ***vestibular nuclei.***

Cochlear Nuclei

The cochlear nuclei are two in number, dorsal and ventral. They are placed dorsal and ventral, respectively to the inferior cerebellar peduncle at the level of the junction of the pons and medulla. The two nuclei are continuous, being separated only by a layer of nerve fibres.

Vestibular Nuclei

The vestibular nuclei lie in the grey matter underlying the lateral part of the floor of the fourth ventricle. They lie partly in the medulla and partly in the pons. Four distinct nuclei are recognised. These are medial, lateral, inferior and superior. The lateral nucleus is also called ***Dieter's nucleus***.

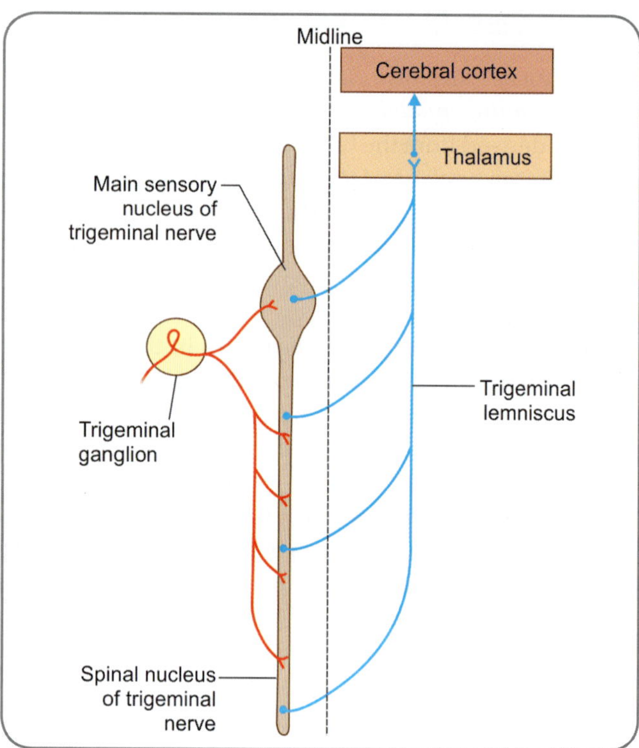

Fig. 21.3: Connections of the sensory nuclei of the trigeminal nerve

Connections of Vestibular Nuclei

The ***vestibular nuclei receive the following afferents (Fig. 21.4):***

❏ The main afferents are central processes of bipolar neurons of the vestibular ganglion. These fibres constitute the vestibular part of the vestibulocochlear nerve. They convey impulses from end organs in the

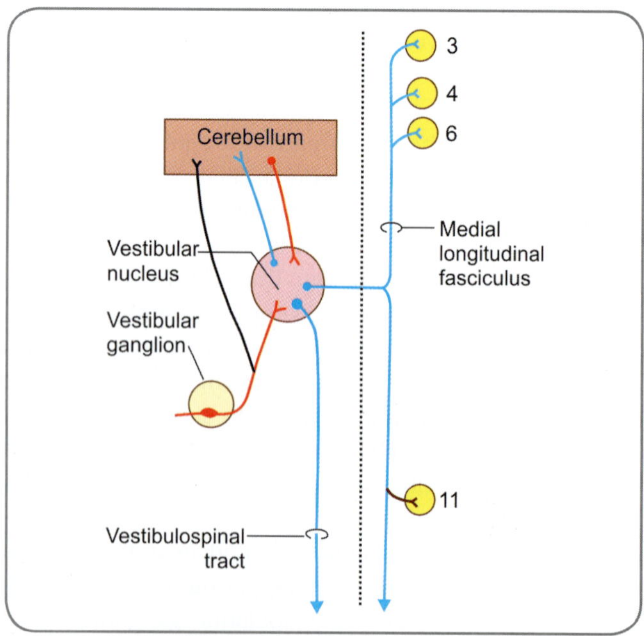

Fig. 21.4: Connections of the vestibular nucleus

semicircular ducts, utricle and saccule. These are necessary for the maintenance of equilibrium.

❑ The vestibular nuclei also receive fibres from some parts of the cerebellum.

FUNCTIONAL COMPONENTS OF INDIVIDUAL CRANIAL NERVES

Having considered the cranial nerve nuclei and their connections, it is now possible to work out the functional components of each cranial nerve.

I. Olfactory Nerve

This is the nerve of smell. As the olfactory mucosa is derived from ectoderm (of the nasal placodes), this nerve is classified as *special somatic afferent* (along with vision and hearing). However, in view of the close relationship between the sensations of smell and taste, some authorities classify this nerve as special visceral afferent.

> ### ✆ Clinical Correlation
>
> The olfactory nerve is tested by asking the patient to recognise various odours. The right and left nerves can be tested separately by closing one nostril and putting the substance near the open nostril.

II. Optic Nerve

This is the nerve of vision. Its fibres are regarded as *special somatic afferent*. From the point of view of its structure and development, this nerve is to be regarded as a tract of the brain rather than as a peripheral nerve.

III. Oculomotor Nerve

Functional Components

This nerve has the following components (Fig. 21.5):

❑ *Somatic efferent* fibres arising in the oculomotor nucleus supply all extrinsic muscles of the eyeball except the lateral rectus and the superior oblique.

❑ *General visceral efferent* fibres (preganglionic) arise in the Edinger-Westphal nucleus and terminate in the ciliary ganglion. Postganglionic fibres arising in this ganglion supply the sphincter pupillae and the ciliaris muscle.

The general visceral efferent component of the oculomotor nerve is involved in accommodation of lens (for the near vision) and constriction of pupil. The accommodation of lens is due to contraction of ciliary muscles and the constriction of pupils is due to contraction of sphincter pupillae muscle of iris.

IV. Trochlear Nerve

Functional Components

This nerve is made up of *general somatic efferent* fibres arising in the trochlear nucleus and supplying the superior oblique muscle of the eyeball.

V. Abducent Nerve

Functional Components

This nerve consists of *somatic efferent* fibres that arise from the abducent nucleus and supply the lateral rectus muscle of the eyeball.

> ### ✆ Clinical Correlation
>
> **Oculomotor, Trochlear and Abducent Nerves**
>
> These three nerves are responsible for movements of the eyeball. In a routine clinical examination, the movements are tested by asking the patient to keep his head fixed and to move his eyes in various directions, i.e. upwards, downwards, inwards and outwards. An easy way is to ask the patient to keep his head fixed and to follow the movements of your finger with his eyes. Such an examination can detect a gross abnormality in movement of the eyes.
>
> Sometimes one of the ocular muscles may not be completely paralysed but may be weak. Two indications of such weakness are as follows:
>
> **Diplopia**
>
> This term means that objects are seen double. To understand this phenomenon, remember that objects lying in different parts of the visual field produce images over different spots on the retina. The brain judges the position of an object by the
>
> *contd...*

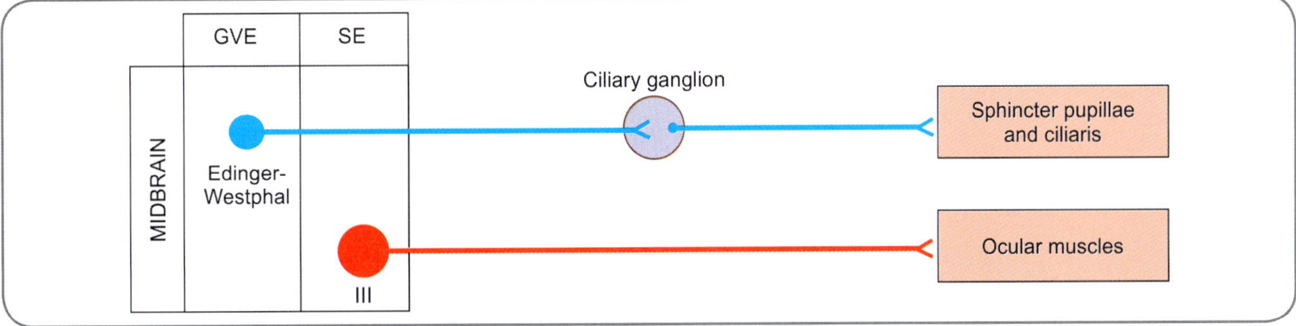

Fig. 21.5: Scheme to show the functional components of oculomotor nerve

---

Clinical Correlation *contd...*

position at which its image is formed on the retina. Normally, the movements of the right and left eyes are in perfect alignment and an object casts an image on corresponding spots on the two retinae so that only one image is perceived by the brain. When a muscle of the eyeball is weak and a movement involving that muscle is performed, the movement of the defective eye is slightly less than that of the normal eye. As a result, images of the object on the two retinae are not formed at corresponding points but over two points near each other. The brain, therefore, 'sees' two images, one from each retina.

To understand the causation of diplopia, you can do a little experiment on yourself. Fix your gaze on any object. Place a finger below one eyeball and gently push it upwards. In addition to the normal bright image of the object, you will see a second fainter image above the normal image. This illustrates that diplopia will be produced by any factor that distorts the normal alignment of the two eyes relative to each other.

Squint (Strabismus)

This is a condition in which the two eyes do not look in the same direction. The squint becomes obvious when the eye movement involves a muscle that is paralysed or weak, because the weak muscle cannot keep up with the muscle of the normal side.

As explained above, squint will be accompanied by diplopia. However, the patient compensates for lack of movement of the eyeball by turning the head in the direction of the object, and in doing so, the diplopia disappears.

If the normal eye is closed, the patient is unable to judge the position of objects in the field of vision correctly (because the image of the object does not fall on the part of the retina that corresponds to the true position of the object). All the features described above are those of paralytic squint.

There is another type of squint called **concomitant squint**. This condition is congenital and manifests itself in early childhood. Squint is present in all positions of the eyeball. There is no muscular weakness and movements are normal in all directions. There is no diplopia.

Paralysis of Oculomotor nerve

All movements of the eyeball are lost in the affected eye. When the patient is asked to look directly forward, the affected eye is directed laterally (by the lateral rectus) and downwards (by the superior oblique). There is lateral squint (external strabismus) and diplopia.

As the levator palpebrae superioris is paralysed, there is drooping of the upper eyelid (ptosis). As parasympathetic fibres to the sphincter pupillae pass through the oculomotor nerve, the sphincter pupillae is paralysed. Unopposed action of sympathetic nerves produces a fixed and dilated pupil.

Normally, the pupil contracts when exposed to light (light reflex). It also contracts when the relaxed eye is made to concentrate on a near object (accommodation reflex). Both of these reflexes are lost. The power of accommodation is lost because of paralysis of the ciliaris muscle.

Paralysis of Trochlear Nerve

The superior oblique muscle (supplied by the trochlear nerve) moves the eyeball downwards and laterally and the inferior rectus (supplied by the oculomotor nerve) moves it downwards and medially. For direct downward movement, synchronised

---

Clinical Correlation *contd...*

action of both muscles is required. When the superior oblique muscle is paralysed, the eyeball deviates medially on trying to look downwards.

Paralysis of Abducent Nerve

This nerve supplies the lateral rectus muscle, which moves the eyeball laterally. In looking forward, the lateral pull of the lateral rectus is counteracted by the medial pull of the medial rectus, and so the eye is maintained in the centre. When the lateral rectus is paralysed, the affected eye deviates medially (medial squint or internal strabismus).

VI. Trigeminal Nerve

Functional Components

This nerve contains the following components (Fig. 21.6).

- *Special visceral efferent* fibres arise from the motor nucleus of the nerve and supply the muscles of mastication.
- *General somatic afferent* fibres of the nerve are peripheral processes of unipolar neurons in the trigeminal ganglion. They carry exteroceptive sensations from the skin of the face and the mucous membrane of the mouth and nose. The central processes of the neurons in the ganglion constitute the sensory root of the nerve. They terminate in the main sensory nucleus and in the spinal nucleus of the nerve.

Another group of general somatic afferent neurons carry proprioceptive impulses from the muscles of mastication (and possibly from ocular, facial and lingual muscles). These fibres are believed to be peripheral processes of unipolar neurons located in the mesencephalic nucleus of this nerve.

The trigeminal nerve is attached to the ventrolateral surface of the pons by two roots, a very large lateral sensory root and a small medial motor root. Close to the attachment on pons, the sensory root contains a ganglion (trigeminal ganglion). The trigeminal ganglion divides into three branches: ophthalmic, maxillary, and mandibular (Fig. 21.7). The ophthalmic and maxillary nerves are sensory nerves, while the mandibular nerve has both motor and sensory fibres.

Clinical Correlation

The trigeminal nerve has a wide sensory distribution. It also supplies the muscles of mastication.

The sensation of touch in the area of distribution of the nerve can be tested by touching different areas of skin with a wisp of cotton wool. The sensation of pain can be tested by gentle pressure with a pin.

Motor function is tested by asking the patient to clench his teeth firmly. Contraction of the masseter can be felt by palpation when the teeth are clenched.

---

contd...

contd...

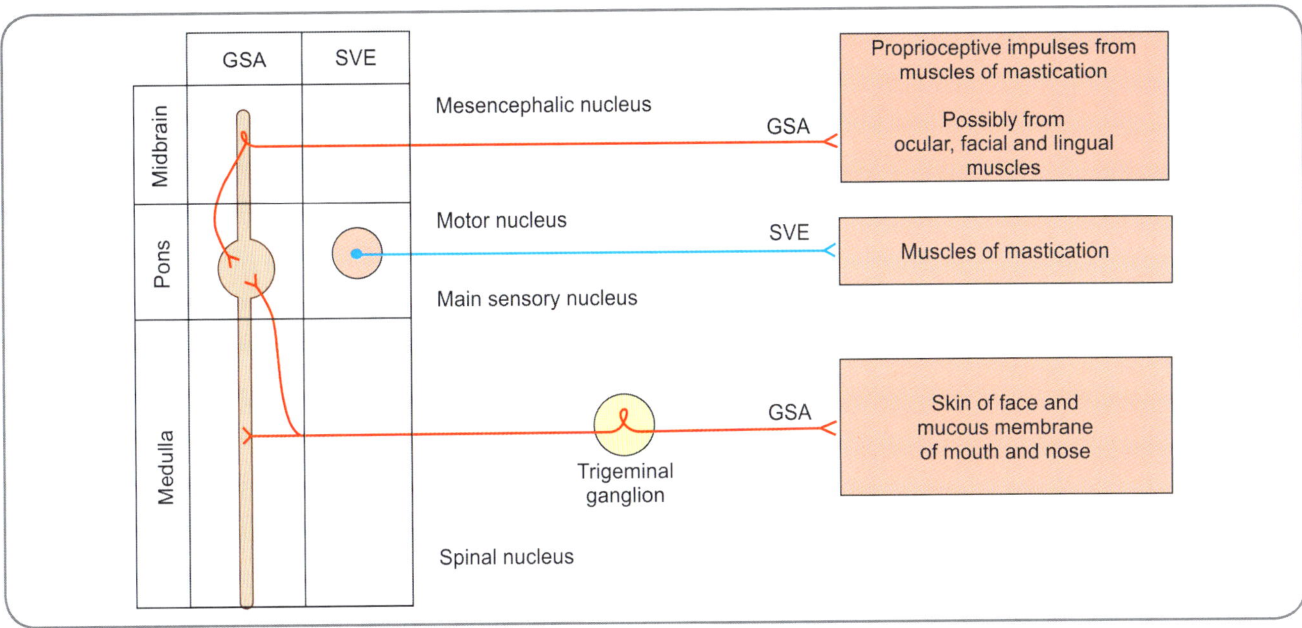

Fig. 21.6: Scheme to show the functional components of the trigeminal nerve—GSA, general somatic afferent; SVE, special visceral efferent

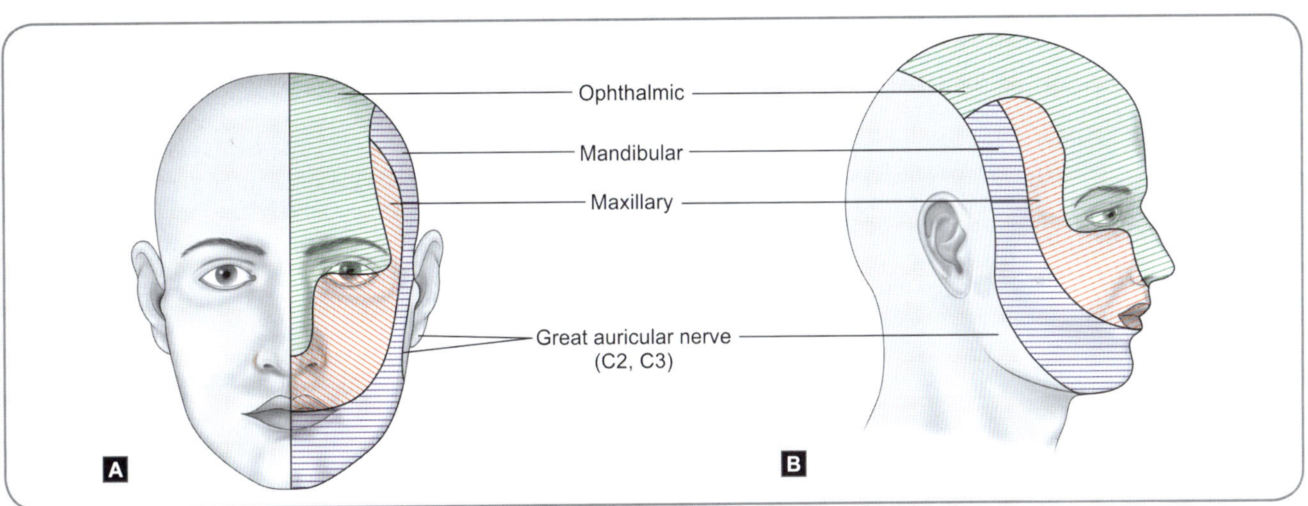

Fig. 21.7: Scheme to show the cutaneous territories of three divisions of the trigeminal nerve **A.** Frontal view and **B.** Lateral view

Clinical Correlation *contd...*

Effects of Injury or Disease

Injury to the trigeminal nerve causes paralysis of the muscles supplied and loss of sensations in the area of supply. Some features of special importance are as follows:

- Apart from their role in opening and closing the mouth, the muscles of mastication are responsible for side-to-side movements of the mandible. Contraction of these muscles on one side moves the chin to the opposite side. Normally, the chin is maintained in the midline by the balanced tone of the muscles of the right and left sides. In paralysis of the pterygoid muscles of one side, the chin is pushed to the paralysed side by muscles of the opposite side.
- Loss of sensation in the ophthalmic division (specially, the nasociliary nerve) is of great importance. Normally, the eyelids close as soon as the cornea is touched (corneal reflex). Loss of sensation in the cornea abolishes this reflex leaving the cornea unprotected. This can lead to the formation of ulcers on the cornea, which can in turn lead to blindness.
- Pain arising in a structure supplied by one branch of the nerve may be felt in an area of skin supplied by another branch. This is called *referred pain*. Some examples are as follows:
  - Caries of a tooth in the lower jaw (supplied by the inferior alveolar nerve) may cause pain in the ear (auriculotemporal).
  - If there is an ulcer or cancer on the tongue (lingual nerve), the pain may again be felt over the ear and temple (auriculotemporal).
  - In frontal sinusitis (sinus supplied by a branch from the supraorbital nerve), the pain is referred to the forehead (skin

contd... **385**

> ## ✆ Clinical Correlation *contd...*
>
> supplied by supraorbital nerve). In fact, headache is a common symptom when any structure supplied by the trigeminal nerve is involved (for example, eyes, ears, and teeth).
>
> ❑ A source of irritation in the distribution of the nerve may cause severe persistent pain (*trigeminal neuralgia*). Removal of the cause can cure the pain. However, in some cases, no cause can be found. In such cases, pain can be relieved by injection of alcohol into the trigeminal ganglion, one of the divisions of the nerve, or into its sensory root. In some cases, it may be necessary to cut fibres of the sensory root. In this connection, it is important to know that the fibres for the maxillary and mandibular divisions can be cut without destroying those for the ophthalmic division. This is possible as the fibres for the ophthalmic division lie separately in the upper medial part of the sensory root. Finally, it may be noted that trigeminal pain can also be relieved by cutting the spinal tract of the trigeminal nerve. This procedure is useful, specially for relieving pain in the distribution of the ophthalmic division as pain can be abolished without loss of the sense of touch and, therefore, without the abolition of the corneal reflex.
>
> ❑ *Mandibular nerve block:* This is used for anesthesia of the lower jaw (for extraction of teeth). Palpate the anterior margin of the ramus of the mandible. Just medial to it,   you will feel the pterygomandibular raphe (ligament). The needle is inserted in the interval between the ramus and the raphe. The tip of the needle is now very near the inferior alveolar nerve, just before it enters the mandibular canal. Anesthetic injected here blocks the nerve.
>
> ❑ The lingual nerve lies very close to the medial side of the third molar tooth, just deep to the mucosa. The nerve can be injured in careless extraction of a third molar. In cases of cancer of the tongue, having intractable pain, the lingual nerve can be cut at this site to relieve pain.

Reflexes Mediated by Trigeminal Nerve

The trigeminal nerve is involved in a number of reflexes summarized in Table 21.3.

VII. Facial Nerve

Functional Components

The components of this nerve are as follows (Fig. 21.8):

❑ *Special visceral efferent* fibres begin from the motor nucleus and supply the various muscles to which the nerve is distributed.

❑ *General visceral efferent* fibres (preganglionic) arise in the superior salivary nucleus. They relay in the submandibular ganglion from which post-ganglionic fibres arise to supply the submandibular and sublingual salivary glands.

The facial nerve also carries general visceral efferent fibres for the lacrimal gland. The preganglionic neurons concerned are said to be located near the salivary nuclei. Their axons terminate in the pterygopalatine ganglion, from which post-ganglionic fibres arise to supply the gland.

❑ *Special visceral afferent* fibres are peripheral processes of cells in the geniculate ganglion of the nerve. They supply taste buds in the anterior two-thirds of the tongue (and some in the soft palate). The central processes of the ganglion cells carry these sensations to the upper part of the nucleus of the solitary tract.

❑ *General somatic afferent* fibres are also peripheral processes of some cells of the geniculate ganglion. They innervate a part of the skin of the external ear. The central processes of these cells end in the spinal nucleus of the trigeminal nerve.

> ## ✆ Clinical Correlation
>
> The facial nerve supplies the muscles of the face, including the muscles that close the eyelids and the mouth. The nerve is tested as follows:
>
> ❑ Ask the patient to close his eyes firmly. In complete paralysis of the facial nerve, the patient will not be able to close the eye on the affected side. In partial paralysis, the closure is weak and the examiner can easily open the closed eye with his fingers (which is very difficult in a normal person).
>
> ❑ Ask the person to smile. In smiling, the normal mouth is more or less symmetrical, the two angles moving upwards and outwards. In facial paralysis, the angle fails to move on the paralysed side.
>
> ❑ Ask the patient to fill his mouth with air. Press the cheek with your finger and compare the resistance (by the buccinator muscle) on the two sides. The resistance is less on the paralysed side. On pressing the cheek, air may leak out of the mouth because the muscles closing the mouth are weak.
>
> ❑ The sensation of taste should be tested on the anterior two-thirds of the tongue (as described under glossopharyngeal nerve).

Table 21.3: Important reflexes mediated by the trigeminal nerve		
Reflex	**Afferent Limb**	**Efferent Limb**
Corneal reflex	Ophthalmic nerve	Facial nerve
Conjunctival reflex	Ophthalmic nerve	Facial nerve
Lacrimation reflex	Ophthalmic nerve	Facial nerve
Sneezing reflex	Maxillary nerve	Vagus nerve
Jaw-jerk (masseteric) reflex	Mandibular nerve	Mandibular nerve

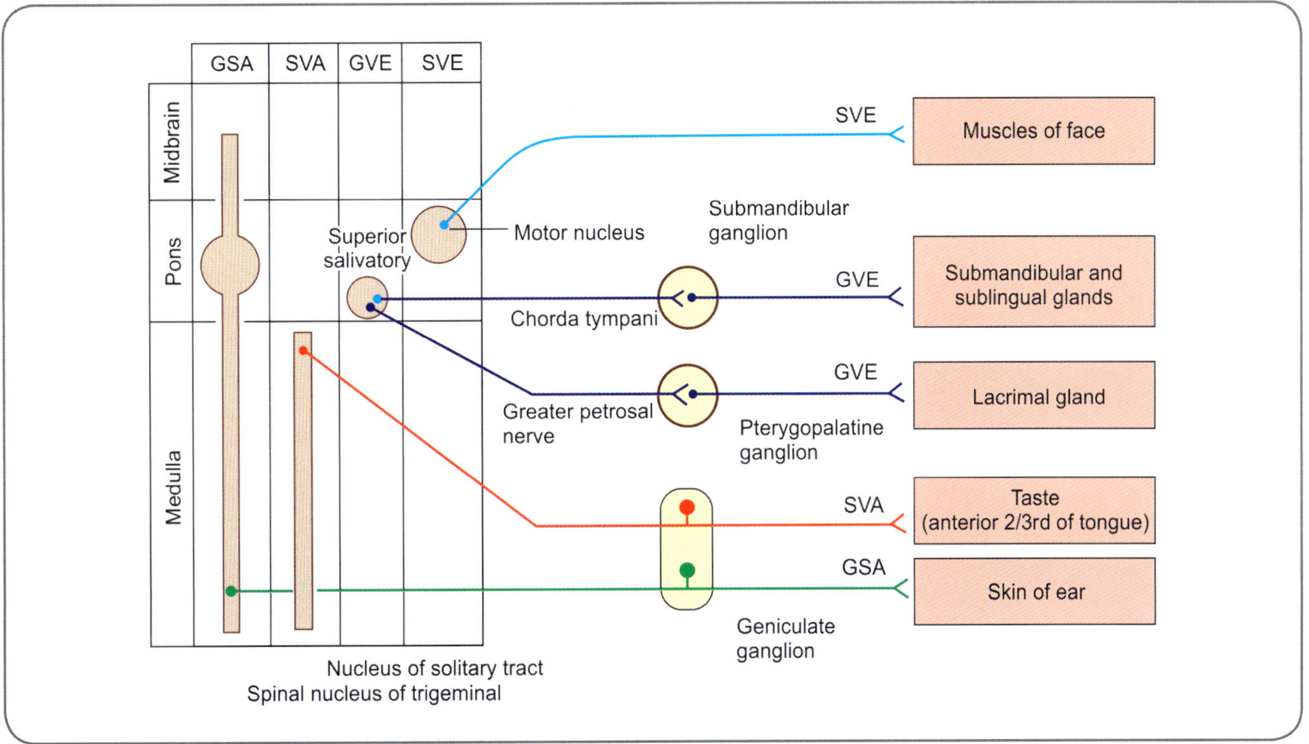

Fig. 21.8: Scheme to show the functional components of the facial nerve—GSA, general somatic afferent; SVA, special visceral afferent; GVE, general visceral efferent; SVE, special visceral efferent

Concept of Nuclear, Supranuclear and Infranuclear Facial Nerve Paralysis

The motor nucleus of the facial nerve is located in pons. Facial nerve paralysis occurring due to injury or disease of the facial nucleus is termed as ***nuclear paralysis*** (lower motor neuron-type paralysis).

The motor nucleus is innervated by corticonuclear fibres (supranuclear fibres or upper motor neuron fibres) arising from the contralateral cerebral cortex. A lesion anywhere in the course of upper motor neuron fibres is ***supranuclear facial paralysis*** (upper motor neuron type paralysis).

Note: Although the motor nucleus of the facial nerve receives supranuclear fibres arising from the contralateral cerebral cortex; however, a part of the motor nerve nucleus of the facial nerve is innervated by the cortical fibres from the ipsilateral side also. This means a part of the nucleus is bilaterally innervated.

The efferent fibres arising from the facial nerve nucleus are referred to as ***infranuclear fibres*** or ***lower motor neuron fibres***. These fibres innervate all facial muscles supplied by the facial nerve.

Facial nerve paralysis occurring due to injury or disease of the infranuclear fibres is termed as ***infranuclear facial nerve paralysis*** (lower motor neuron type paralysis).

Clinical Correlation

The part of motor nucleus of facial nerve supplying the muscles of the lower part of the face receives the corticonuclear fibres from the opposite cerebral hemisphere, while the part of motor nucleus of facial nerve, which supplies the muscles of the upper part of the face, receives corticonuclear fibres from both cerebral hemispheres. As a result, in supranuclear lesions (i.e. lesions involving the upper motor neurons) of the facial nerve, the upper half of the face on both sides is spared and the lower half of the face is affected on the opposite side; on the other hand, in nuclear and infranuclear lesions, i.e. lower motor neuron lesions, whole of the face is affected on the side of lesion (Fig. 21.9).

Paralysis of Facial Nerve

Paralysis of the facial nerve is fairly common. It may occur due to lesion anywhere in the course of upper motor or lower motor neurons.

Supranuclear Lesions (Upper Motor Neuron Lesion)

Supranuclear lesions can be distinguished from nuclear or infranuclear lesions because these are usually accompanied by hemiplegia. Only movements of the lower part of the face are affected and not those of the upper part: the explanation for this is that the corticonuclear fibres concerned with movements of the upper part of the face are bilateral, whereas those for movements of the lower part of the face are unilateral. Another difference is that while voluntary movements are affected,

contd…

emotional expressions appear to be normal. It has been suggested that there are separate pathways from the cerebral cortex to the facial nucleus for voluntary and emotional movements, and usually, only the former are involved.

Nuclear Lesion (Lower Motor Neuron Lesion)

The lesion of facial nerve nucleus is very rare. It may occur in poliomyelitis or progressive bulbar paralysis. In this type of lesion, all the facial muscles on the side of lesion are paralysed. The other components of facial nerve, i.e. special visceral afferent component for taste sensation and general visceral efferent for lacrimation and salivation, are not affected as these nuclei are not located very close to the motor nucleus of facial nerve. In nuclear lesions (within the brainstem), other neighbouring nuclei may be affected leading to simultaneous lesion of the abducent or trigeminal nerve.

Infranuclear Lesion (Lower Motor Neuron Lesion) or Bell's Palsy

In the most common type of infranuclear paralysis called *Bell's palsy*, the nerve is affected near the stylomastoid foramen. Facial muscles can also be paralysed by interruption of corticonuclear fibres running from the motor cortex to the facial nucleus. This is referred to as supranuclear paralysis.

The effects of paralysis are due to the failure of the muscles concerned to perform their normal actions. Some effects are as follows:

❑ The normal face is more or less symmetrical. When the facial nerve is paralysed on one side, the most noticeable feature is the loss of symmetry.

❑ Normal furrows on the forehead are lost because of paralysis of the occipitofrontalis.

❑ There is drooping of the eyelid, and the palpebral fissure is wider on the paralysed side, because of paralysis of the orbicularis oculi. The conjunctival reflex is lost for the same reason.

❑ There is marked asymmetry of the mouth, because of paralysis of the orbicularis oris and of muscles inserted into the angle of the mouth. This is most obvious when a smile is attempted. As a result of asymmetry, the protruded tongue appears to deviate to one side but is, in fact, in the midline.

❑ During mastication, food tends to accumulate between the cheek and the teeth. This is normally prevented by the buccinator.

Additional effects are observed in injuries to the facial nerve at levels higher than the stylomastoid foramen as follows:

❑ If the injury is proximal to the origin of the chorda tympani, there is loss of sensation of taste on the anterior two-thirds of the tongue.

❑ The transmission of loud sounds to the internal ear is normally dampened by the stapedius muscle. When the lesion is proximal to the origin of the branch to the stapedius, this muscle is paralysed. As a result, even normal sounds appear too loud (*hyperacusis*).

❑ In fractures of the temporal bone or in lesions near the exit of the nerve from the brain, the vestibulocochlear nerve may also be affected (leading to deafness).

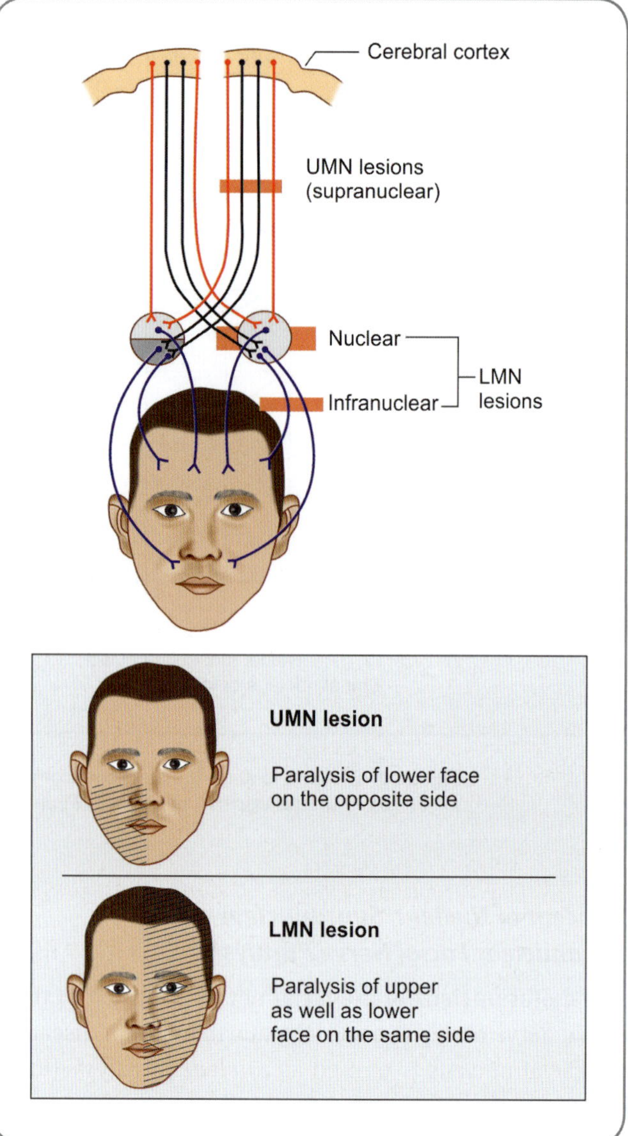

Fig. 21.9: Effects of upper motor neuron and lower motor neuron lesions of the facial nerve

VIII. Vestibulocochlear Nerve

Functional Components

Both the cochlear and vestibular divisions of this nerve are made up of *special somatic afferent* fibres. The fibres of the cochlear nerve are central processes of bipolar cells in the spiral ganglion. The peripheral processes of these neurons supply the organ of Corti. The fibres of the vestibular nerve are central processes of bipolar neurons in the vestibular ganglion. The peripheral processes of these neurons innervate the semicircular ducts, the utricle, and the saccule of the internal ear.

In the past, this nerve has also been called the *auditory* or *statoacoustic nerve*.

⚕ Clinical Correlation

This nerve is responsible for hearing (cochlear part) and for equilibrium (vestibular part). Normally, we test only the cochlear part.

❑ The hearing of the patient can be tested by using a watch. First, place the watch near one ear so that the patient knows what (s)he is expected to hear. Next ask him/her to close his/her eyes and say so when (s)he hears the ticking of the watch. The watch should be held away from the ear and then gradually brought towards it. The distance at which the sounds are first heard should be compared with the other ear.

In doing this test, it must be remembered that loss of hearing can occur from various causes, such as the presence of wax in the ear, or middle ear disease.

Nerve deafness can be distinguished from deafness due to a conduction defect (as in middle ear disease) by noting the following:

○ Sounds can be transmitted to the internal ear through air (normal way) and through bone. Normally, conduction through air is better than that through bone, but in defects of conduction, the sound is better heard through bone.

○ Air conduction and bone conduction can be compared by using a tuning fork. Strike the tuning fork against an object so that it begins to vibrate producing sound. Place the tuning fork near the patient's ear and then immediately put the base of the tuning fork on the mastoid process. Ask the patient where (s)he hears the sound better. This is called **Rinne's test**. In another test, the base of a vibrating tuning fork is placed on the forehead. The sound is heard in both ears but is more clear in the ear with a conduction defect. This is **Weber's test**).

Defects in the vestibular apparatus or in the vestibular nerve are difficult to test, and such cases need to be examined by a specialist.

IX. Glossopharyngeal Nerve

Functional Components

The components of this nerve are as follows:

❑ **Special visceral efferent** fibres arise in the nucleus ambiguus and supply the stylopharyngeus muscle.

❑ **General visceral efferent** fibres (preganglionic) begin from the inferior salivary nucleus and travel to the otic ganglion. Post-ganglionic fibres arising in the ganglion supply the parotid gland.

❑ **General visceral afferent** fibres are peripheral processes of neurons in the inferior ganglion of the nerve. They carry general sensations (touch, pain and temperature) from the pharynx and the posterior part of the tongue to the ganglion. They also carry inputs from the carotid sinus and carotid body. Central processes of the neurons carry these sensations to the nucleus of the solitary tract. Some fibres from the carotid sinus and body reach the paramedian reticular formation of the medulla.

Table 21.4: Important Reflexes mediated by Glossopharyngeal Nerve

Reflex	Afferent Limb	Efferent Limb
Gag reflex	Glossopharyngeal nerve	Vagus nerve
Carotid sinus reflex	Glossopharyngeal nerve	Vagus nerve

❑ **Special visceral afferent** fibres are also peripheral processes of neurons in the inferior ganglion. They carry sensations of taste from the posterior one-third of the tongue to the ganglion. The central processes carry these sensations to the nucleus of the solitary tract. The important reflexes mediated by glossopharyngeal nerve are shown in Table 21.4.

⚕ Clinical Correlation

Testing of this nerve is based on the fact that the nerve carries fibres of taste from the posterior one-third of the tongue and that it provides sensory innervation to the pharynx.

❑ Sensations of taste can be tested by applying substances that are salty (salt), sweet (sugar), sour (lemon), or bitter (quinine) to the posterior one-third of the tongue. The mouth should be rinsed and the tongue dried before the substance is applied.

❑ Touching the pharyngeal mucosa causes reflex constriction of pharyngeal muscles. The glossopharyngeal nerve provides the afferent part of the pathway for gag reflex (Table 21.4).

Lesions of Glossopharyngeal Nerve

The lesion of the glossopharyngeal nerve is associated with the following:

❑ Loss of taste and general sensation from the posterior third of tongue.

❑ Loss of gag reflex, due to interruption of the afferent limb.

❑ Loss of carotid sinus reflex due to interruption of the afferent limb.

❑ Loss of general sensations in pharynx, tonsils, and fauces.

❑ Reduced secretion of saliva, as the nerve is secretomotor to parotid gland.

X. Vagus Nerve

Functional Components

The components of this nerve are as follows (Fig. 21.10):

❑ **Special visceral efferent** fibres are processes of neurons in the nucleus ambiguus and supply the muscles of the pharynx and larynx.

❑ **General visceral efferent** fibres arise in the dorsal (motor) nucleus of the vagus. These are preganglionic parasympathetic fibres. They are distributed to thoracic and abdominal viscera. The postganglionic neurons concerned are situated in ganglia close to or within the walls of the viscera supplied.

❑ **General visceral afferent** fibres are peripheral processes of neurons located in the inferior ganglion of the nerve.

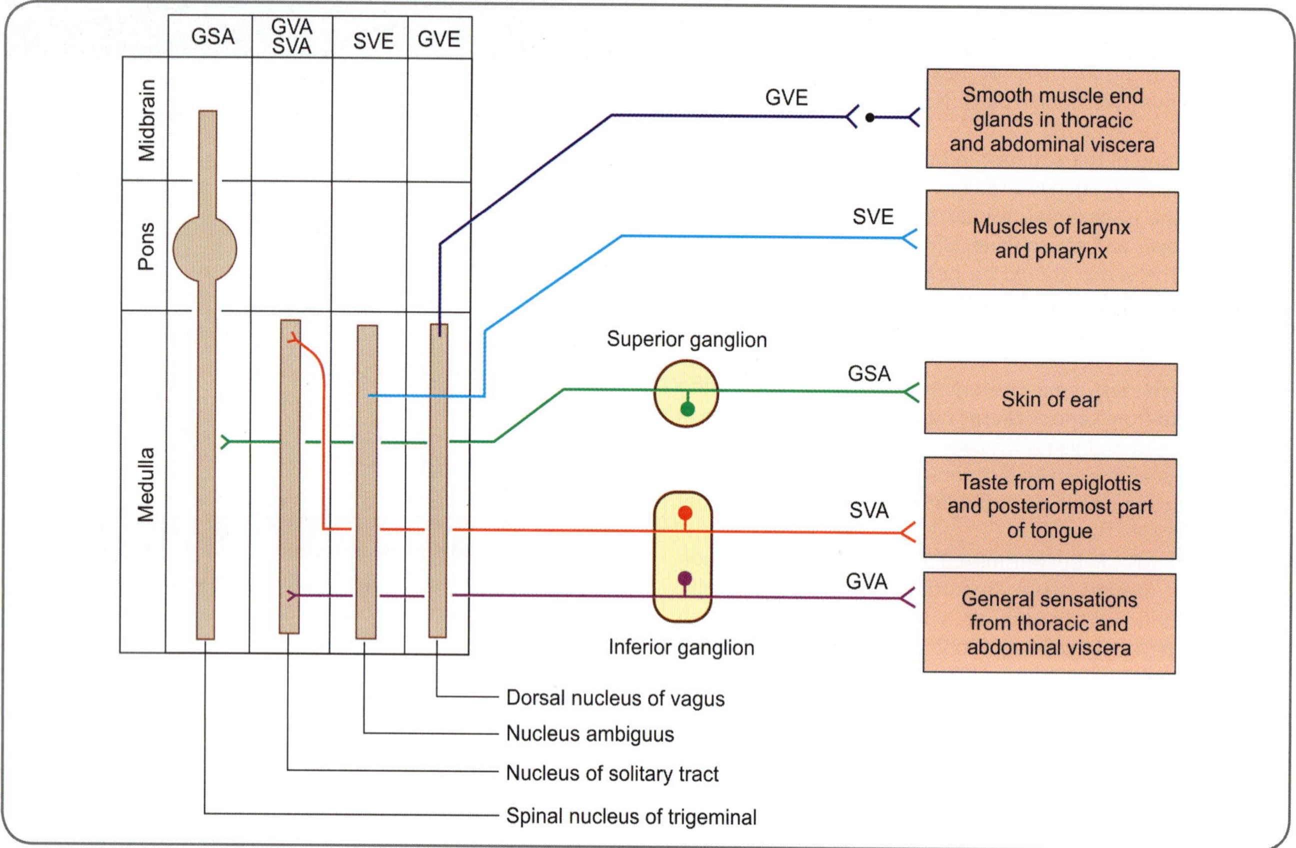

Fig. 21.10: Scheme to show the functional components of the vagus nerve—GSA, general somatic afferent; GVA, general visceral afferent; SVA, special visceral afferent; SVE, special visceral efferent; GVE, general visceral efferent

They bring sensations from the pharynx, larynx, trachea, oesophagus, and abdominal and thoracic viscera. These are conveyed by central processes of the ganglion cells to the nucleus of the solitary tract. According to some authorities, some of these fibres terminate in the dorsal nucleus of the vagus.

- **Special visceral afferent** fibres are also peripheral processes of neurons in the inferior ganglion. They carry sensations of taste from the posteriormost part of the tongue and the epiglottis. The central processes of the neurons concerned terminate in the upper part of the nucleus of the solitary tract.
- **General somatic afferent** fibres are peripheral processes of neurons in the superior ganglion and are distributed to the skin of the external ear. The central processes of the ganglion cells terminate in relation to the spinal nucleus of the trigeminal nerve.

Clinical Correlation

This nerve has an extensive distribution, but testing is based on its motor supply to the soft palate and to the larynx.
- Ask the patient to open the mouth wide and say 'aah'. Observe the movement of the soft palate. In a normal person, the soft palate is elevated. When one vagus nerve is paralysed, the palate is pulled towards the normal side. When the nerve is paralysed on both sides, the soft palate does not move at all.

contd...

Clinical Correlation *contd...*

- In injury to the superior laryngeal nerve, the voice is weak due to paralysis of the cricothyroid muscle. At first, there is hoarseness, but after some time, the opposite cricothyroid compensates for the deficit and hoarseness disappears.
- Injury to the recurrent laryngeal nerve also leads to hoarseness, but this hoarseness is permanent. On examining the larynx through a laryngoscope, it is seen that on the affected side, the vocal fold does not move. It is fixed in a position midway between adduction and abduction. In cases where the recurrent laryngeal nerve is pressed upon by a tumour, it is observed that nerve fibres that supply abductors are lost first.
- In paralysis of both recurrent laryngeal nerves, voice is lost as both vocal folds are immobile.
- It may be remembered that the left recurrent laryngeal nerve runs part of its course in the thorax. It can be involved in bronchial or esophageal carcinoma or by secondary growths in mediastinal lymph nodes.

XI. Accessory nerve

Functional Components

This nerve consists predominantly of **special visceral efferent** fibres, which arise from:
- The nucleus ambiguus to supply striated muscle of the pharynx, the soft palate, and the larynx (along with the vagus) and

❑ The lateral part of the anterior grey column of the upper five or six cervical segments of the spinal cord, to supply the trapezius and sternocleidomastoid muscles.

✦ Clinical Correlation

This nerve is tested as follows:
❑ Put your hands on the right and left shoulders of the patient and ask him/her to elevate (shrug) his/her shoulders. In paralysis, the movement will be weak on one side (due to paralysis of the trapezius).
❑ Ask the patient to turn his/her face to the opposite side (against resistance offered by your hand). In paralysis, the movement is weak on the affected side (due to paralysis of the sternocleidomastoid muscle).

Lesion of the Spinal Accessory Nerve

Unilateral peripheral lesion of spinal accessory nerve leads to paralysis of sternocleidomastoid and trapezius muscles.
❑ Paralysis of sternocleidomastoid leads to *wry neck*, i.e. difficulty in turning of head to the opposite side.
❑ The paralysis of trapezius results in the inability to shrug the shoulder towards the side of injury. As the lower part of trapezius is also supplied by C3 and C4 segments, an injury to the accessory nerve will not result in complete paralysis of trapezius.

Irritation of the Spinal Accessory Nerve

Results in the condition called ***torticollis***. In this condition, there is a spasmodic contraction of sternocleidomastoid and trapezius.

XII. Hypoglossal Nerve

Functional Components

This nerve is made up of the following components:

❑ ***Somatic efferent*** fibres are processes of neurons in the hypoglossal nucleus. They supply the muscles of the tongue.
❑ ***General somatic afferent*** fibres carry proprioceptive impulses from muscles of the tongue. The location of the cells of origin of these fibres is uncertain. In the embryo, a small ganglion is present in relation to the nerve but cannot be identified in the adult. Some unipolar cells can, however, be identified on the roots of the hypoglossal nerve.

Groups of neurons innervating individual muscles of the tongue can be recognised in the hypoglossal nucleus.

✦ Clinical Correlation

This nerve supplies muscles of the tongue. To test the nerve, ask the patient to protrude the tongue. In a normal person, the protruded tongue lies in the midline. If the nerve is paralysed, the tongue deviates to the paralysed side. The explanation for this is as follows.

Protrusion of the tongue is produced by the pull of the right and left genioglossus muscles. The origin of the right and left genioglossus muscles lies anteriorly (on the hyoid bone) and the insertion lies posteriorly (on to the posterior part of the tongue). Each muscle draws the posterior part of the tongue forward and medially. Normally, the medial pull of the two muscles cancels out, but when one muscle is paralysed, it is this medial pull of the intact muscle that causes the tongue to deviate to the opposite side.

Deviation of the tongue should be assessed with reference to the incisor teeth and not to the lips. Remember that, in facial paralysis, the tongue may protrude normally, but may appear to deviate to one side, because of asymmetry of the mouth.

Multiple Choice Questions

1. The cranial nerve whose nuclei of two sides form a single complex lying ventral to the central canal is:
   a. Facial Nerve
   b. Trochlear Nerve
   c. Abducent Nerve
   d. Oculomotor Nerve
2. The spinal nucleus of trigeminal nerve is continuous inferiorly with:
   a. Substantia gelatinosa
   b. Substantia nigra
   c. Nucleus gracilis
   d. Mesencephalic tract
3. Special visceral efferent fibres of the Glossopharyngeal nerve arise from the:
   a. Nucleus ambiguus
   b. Mesencephalic Nucleus
   c. Inferior Salivatory Nucleus
   d. Solitary Nucleus
4. Torticollis is caused by:
   a. Paralysis of sternocleidomastoid
   b. Contraction of trapezius and sternocleidomastoid
   c. Complete paralysis of trapezius
   d. Avulsion of sternocleidomastoid
5. Motor nucleus of trigeminal nerve and Nucleus ambiguus belong to the:
   a. Special Visceral afferent column
   b. General visceral efferent column
   c. Somatic efferent column
   d. Special Visceral efferent column

ANSWERS

1. d   **2.** a   **3.** a   **4.** b   **5.** d

Clinical Problem-solving

Case Study 1: A 27-year-old man complained of pain in his left ear. On a detailed examination, no specific cause could be found out. However, the man was subjected to a dental examination and a caries tooth was located in his lower jaw.
- How would you relate the findings and the complaint?
- What is the phenomenon responsible for such pain?
- Give two other examples of referred pain involving the trigeminal nerve.

Case Study 2: A 38-year-old man had exposure to extreme cold and suffered a Bell's palsy on the right side. You had to examine him.
- If the patient is asked to smile, what effect do you expect?
- How else would you test for the power of the muscles around mouth?
- How would you test for the special visceral afferent component?

(For solutions see Appendix).

Chapter 22

Cerebellum—Gross Anatomy and Internal Structure

Frequently Asked Questions

❑ Discuss the cerebellar cortex with regard to its layers and types of cells.
❑ Write note on: (a) Tentorium cerebelli, (b) Archicerebellum, (c) Cerebellar nuclei, (d) Purkinje cells.
❑ Give a detailed account of the functions of the cerebellum correlating them to cerebellar anatomy.

GROSS ANATOMY

The cerebellum (or small brain) lies in the posterior cranial fossa. In an adult, the weight of the cerebellum is about 150 gm. This is about 10% of the weight of the cerebral hemispheres. Like the cerebrum, the cerebellum has a superficial layer of grey matter, the cerebellar cortex. Because of the presence of numerous fissures, the cerebellar cortex is much more extensive than the size of this part of the brain would suggest. It has been estimated that the surface area of cerebellar cortex is about 50% of the area of the cerebral cortex.

The cerebellum lies behind the pons and the medulla. It is separated from the cerebrum by a fold of dura mater called as ***tentorium cerebelli.*** Anteriorly, the fourth ventricle intervenes between the cerebellum (behind), and the pons and medulla (in front, Fig. 22.1). Part of the cavity of the ventricle extends into the cerebellum as a transverse cleft. This cleft is bounded cranially by the superior (or anterior) medullary velum, a lamina of white matter (Fig. 22.1).

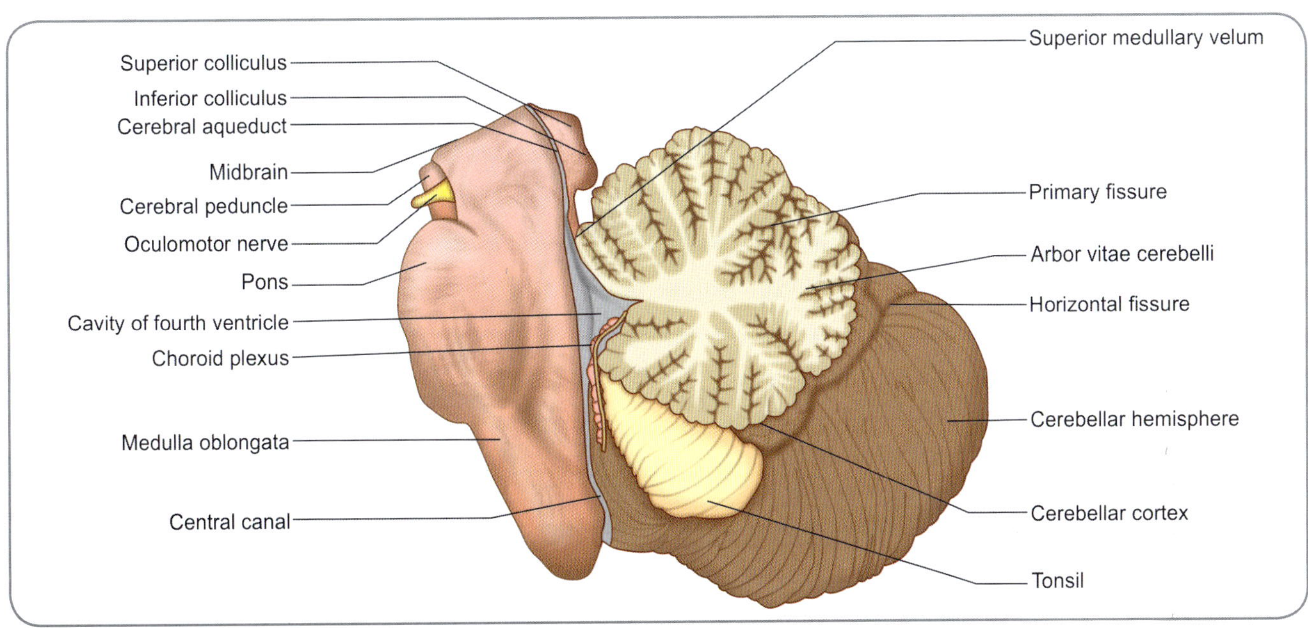

Fig. 22.1: Median sagittal section through brainstem and cerebellum

EXTERNAL FEATURES

Parts of Cerebellum

The cerebellum consists of a part lying near the midline called *vermis* and two lateral *hemispheres.*

Surfaces of Cerebellum

It has two surfaces, *superior* and *inferior.* On the superior aspect, there is no line of distinction between the vermis and the hemispheres. On the inferior aspect, the two hemispheres are separated by a deep depression called the *vallecula.* The vermis lies in the depth of this depression. On each side, the vermis is separated from the corresponding cerebellar hemisphere by a *paramedian sulcus.* Anteriorly and posteriorly, the hemispheres extend beyond the vermis and are separated by anterior and posterior *cerebellar notches.* The falx cerebelli lies in the posterior notch.

Fissures and Lobes of Cerebellum (Figs 22.2 to 22.4)

The surface of the cerebellum is marked by a series of fissures that run more or less parallel to one another. The fissures subdivide the surface of the cerebellum into narrow leaf-like bands or *folia.* The long axis of the majority of folia is more or less transverse. Sections of the cerebellum cut at right angles to this axis have a characteristic tree-like appearance to which the term *arbor-vitae* (tree of life) is applied (Fig. 22.1).

Some of the fissures on the surface of the cerebellum are deeper than others. They divide the cerebellum into *lobes* within which smaller *lobules* may be recognised (Fig. 22.2). The deepest fissures in the cerebellum are:

- The *primary fissure (fissura prima)* running transversely across the superior surface.
- The *posterolateral fissure* seen on the inferior aspect.
- The *horizontal fissure* (Figs 22.2 and 22.4), which divides the cerebellum into upper and lower halves. The parts seen above the horizontal fissure form the superior surface and those below the fissure form the inferior surface of the cerebellum.

The primary and posterolateral fissures divide the cerebellum into three lobes. The part anterior to the primary fissure is the *anterior lobe.* The part between the two fissures is the *posterior lobe* (sometimes called the *middle lobe).* The posterior lobe extends on both superior and inferior surfaces. The remaining part is the *flocculonodular lobe,* present in the inferior surface of the cerebellum. The anterior and posterior lobes together form the *corpus cerebelli,* which constitutes the main mass of the cerebellum, the flocculonodular lobe constitutes a very small part of the cerebellum.

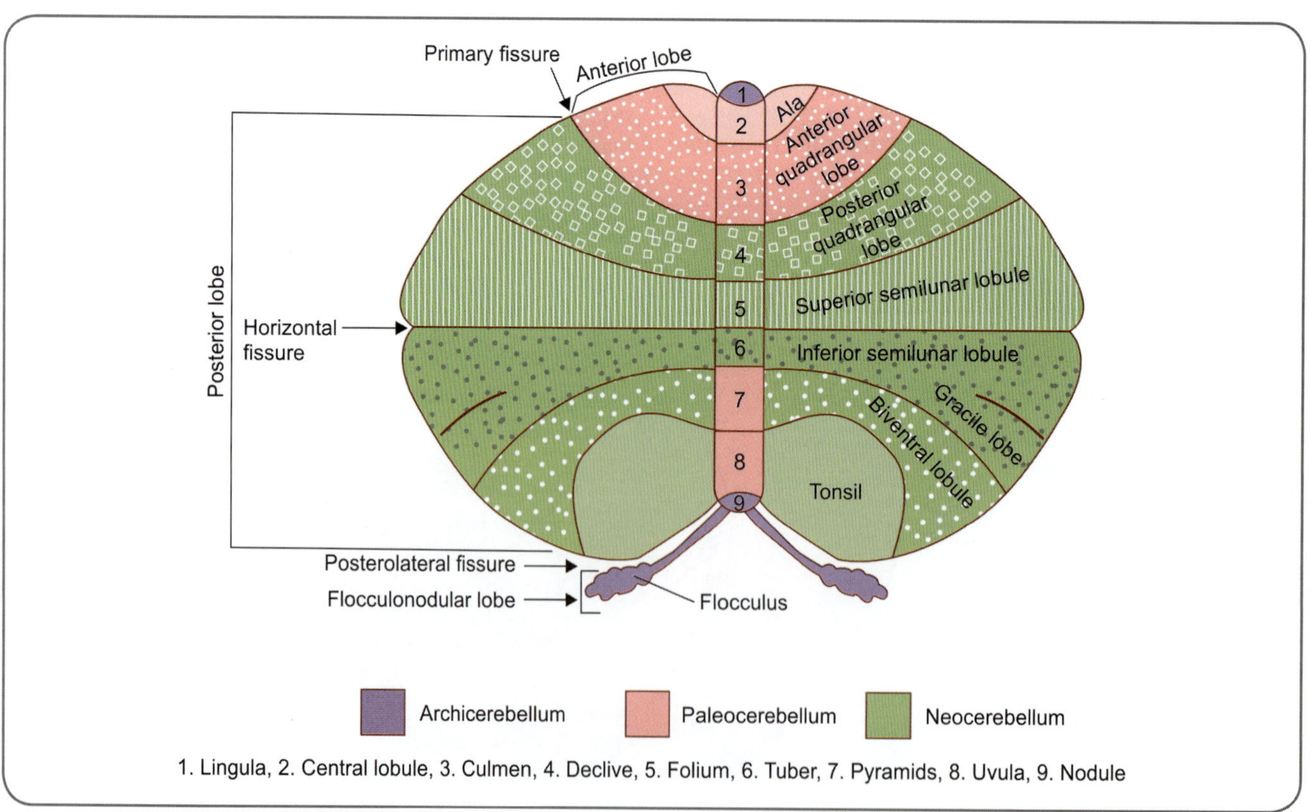

Fig. 22.2: Scheme to show the subdivisions of cerebellum (opened out)

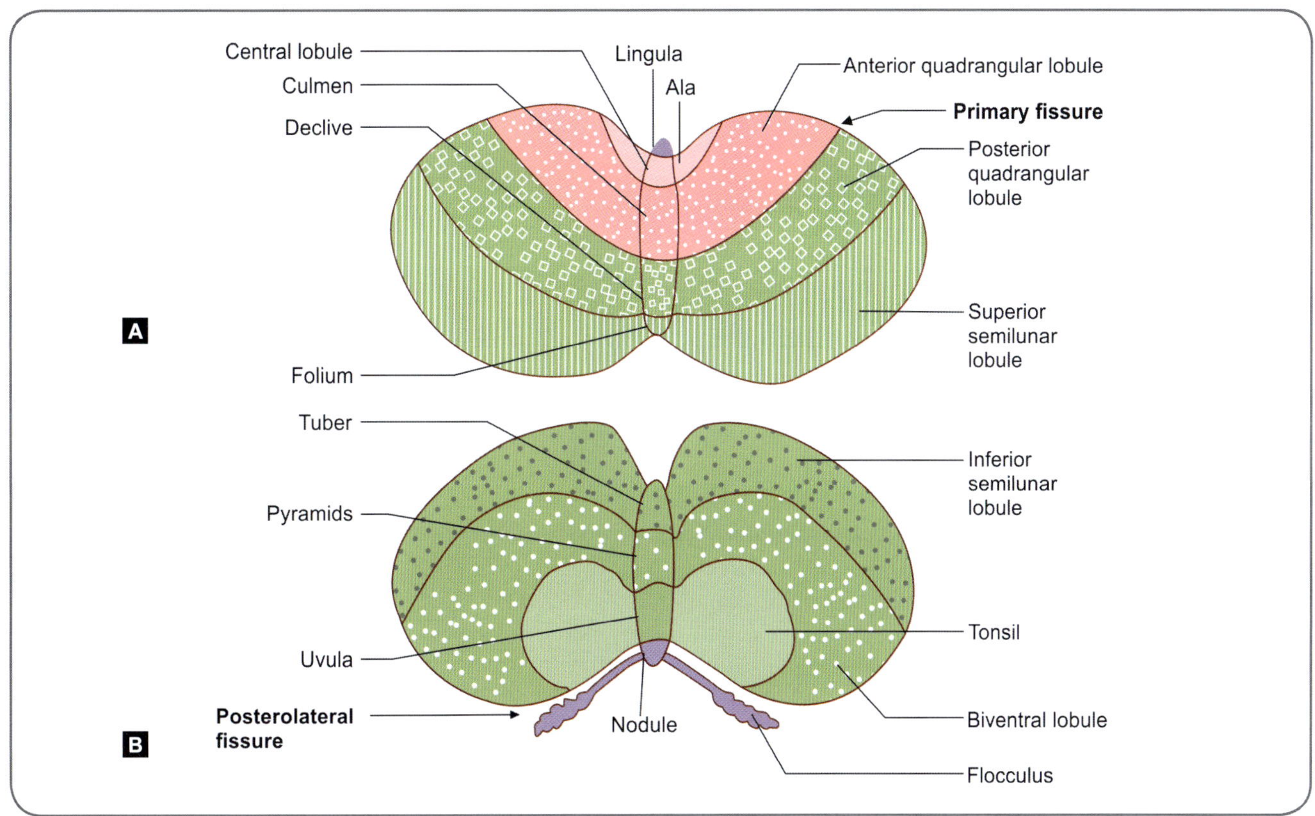

Figs 22.3A and B: Transverse subdivisions of the cerebellum **A.** As seen on superior aspect **B.** As seen on inferior aspect

Note: The primary fissure separates the anterior and posterior lobes. It, therefore, intervenes between the anterior and posterior quadrangular lobules and also separates the culmen and declive. The posterolateral fissure separates the posterior lobe from the flocculonodular lobe and extends into the interval between the nodule and the uvula.

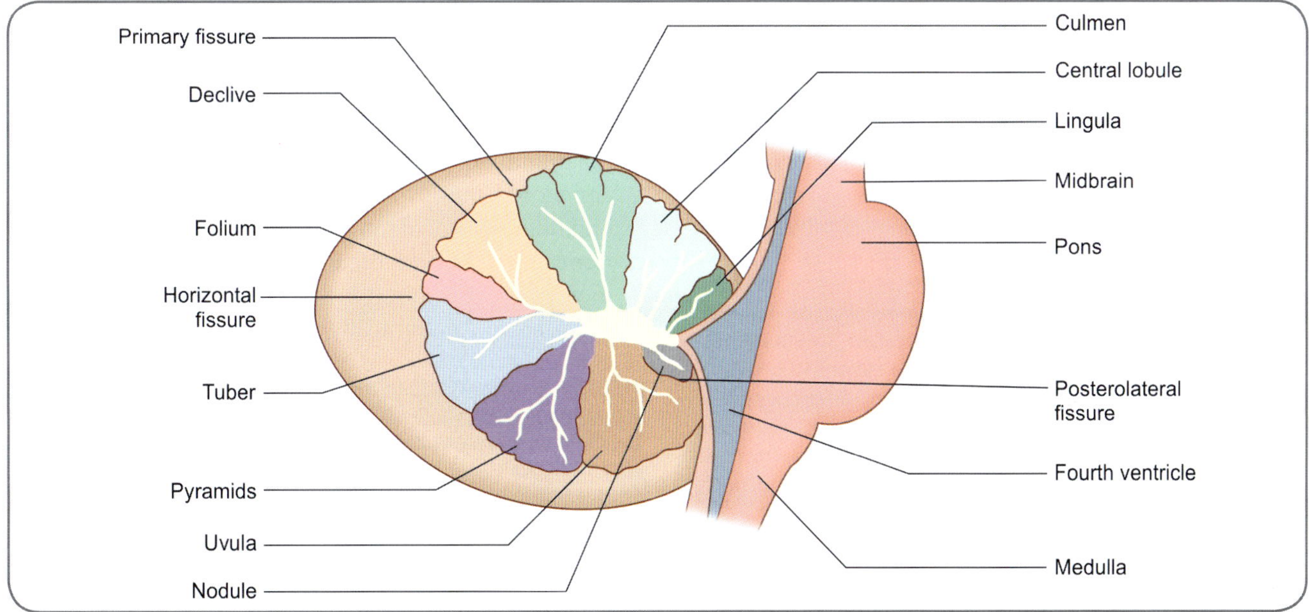

Fig. 22.4: Subdivisions of the vermis of the cerebellum as seen in a median section

Note: The position of horizontal fissure, which divides the cerebellum into upper and lower halves. The parts shown above this fissure are seen on the superior surface of the cerebellum, and those below it on the inferior surface.

Table 22.1: Morphological subdivisions of cerebellum (Fig. 22.2)

Lobes	Surface	Parts of vermis	Parts of hemisphere
Anterior lobe	Superior surface	Lingula Central lobule Culmen	— Ala Anterior quadrangular lobe
Posterior lobe (middle lobe)	Superior surface	Declive Folium	Posterior quadrangular lobe Superior semilunar lobule
	Inferior surface	Tuber Pyramid Uvula	Inferior semilunar lobule Biventral lobule Tonsil
Flocculonodular lobe		Nodule	Flocculus

The vermis is so called because it resembles a worm. Proceeding from above downwards (Fig. 22.2), it consists of the *lingula, central lobule,* and *culmen* (in the anterior lobe); the *declive, folium (or folium vermis), tuber (or tuber vermis), pyramis (or pyramid),* and *uvula* (in the middle lobe); and the *nodule* (in the flocculonodular lobe).

The declive is sometimes referred to as 'simple', as it is part of the simple lobule.

With the exception of the lingula, each subdivision of the vermis is related laterally to a part of the hemisphere (Table 22.1 and Fig. 22.3).

 **Development**

Development of Cerebellum

The cerebellum develops from the metencephalon. It should be rememebered that the alar laminae evert in the metencephalon. The dorsolateral parts of the everted alar laminae on either side, bend inwards and dorsally. These two projections appear as two lips (one on either side) when seen from the dorsal aspect and hence are called the rhombic lips. In the lower part of the metencephalon, the two rhombic lips are apart from each other but in the upper part (since there is no eversion of the laminae in the mesencephalon and the basic tubular pattern continues) the two lips are close to each other. The two rhombic lips are initially in the shape of an inverted 'V'. When the pontine flexure deepens, the rhombic lips get stretched transversely (or flattened cephalocaudally). The transversely placed structure now is called the cerebellar plate. The plate expands dorsally. A small midline swelling is first seen. This will develop into the vermis. The lateral portions further develop and expand to form the lateral hemispheres. The flocculonodular node is first marked off. The other parts develop due to growth, proliferation and expansion.

The cerebellar plate has the germinal, mantle and marginal layers. The germinal layer is on the ventral aspect (lining the ventricular area) and the marginal layer is on the dorsal aspect. Cells from the germinal layer migrate to the dorsal aspect and form an external granular layer on the surface. Other cells from the germinal layer develop and migrate into the mantle layer; they form the Purkinje cells. Cells from the external granule layer migrate into the internal area and form

contd...

 **Development** *contd...*

the granule cells. Basket and stellate cells are formed by cells which differentiate in the mantle zone. Groups of cells from the germinal layer form the deep cerebellar nuclei.

SUBDIVISIONS OF CEREBELLUM

From developmental, phylogenetic, and functional points of view, the cerebellum is often divided into the following (Fig. 22.2 and Table 22.2):

❑ *Archicerebellum:* Phylogenetically, it is the oldest part of cerebellum. Anatomically, it consists of flocculonodular lobe and lingula. The connections of the archicerebellum are predominantly vestibular (hence, called vestibulocerebellum), and it is concerned with the maintenance of body equilibrium.

❑ *Paleocerebellum:* Phylogenetically, it is the next part of cerebellum to arise and is well developed in reptiles and birds. Anatomically, it consists of anterior lobe (except lingula) and pyramids and uvula of the posterior lobe. The paleocerebellum is connected predominantly to the spinal cord (hence, called spinocerebellum). It is concerned mainly with maintenance of muscle tone and finer control of movements.

❑ *Neocerebellum:* It is the most recent part of cerebellum to develop. It is found in mammals only and is largest in humans. Anatomically, it consists of posterior lobe

Table 22.2: Intrinsic neurons of the cerebellar cortex and their location

Intrinsic neuron	Layer of the cerebellar cortex
Outer (external) stellate cells	Molecular layer
Basket cells	Molecular layer
Purkinje cells	Purkinje cell layer
Granule cells	Granular layer
Golgi cells	Granular layer
Brush cells	Granular layer

Note: All the intrinsic neurons of cerebellar cortex are inhibitory except granule cells.

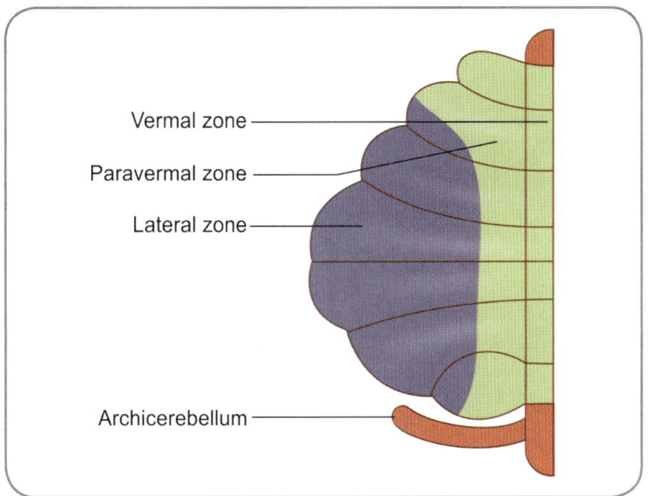

Fig. 22.5: Schematic diagram showing functional subdivisions of the cerebellum

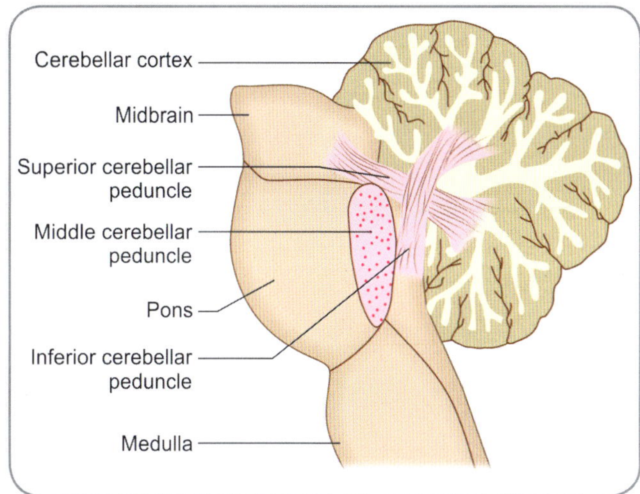

Fig. 22.6: Illustration showing cerebellar peduncles

except pyramids and uvula. The neocerebellum has extensive connections with the cerebral cortex (through pontine nuclei, hence called pontocerebellum). It is usually regarded as being responsible for fine co-ordination of voluntary movements, but its precise role is not known.

From the point of view of its connections, the cerebellar cortex may also be divided into a vermal (vermis), paravermal (or paramedian), and lateral parts—longitudinal parcellation (Fig. 22.5).

CEREBELLAR PEDUNCLES

The cerebellum is attached to the brainstem by three pairs of cerebellar peduncles (Fig. 22.6). The superior cerebellar peduncles connect it to midbrain; the middle peduncles, to the pons; and the inferior peduncles to the spinal cord. The fibres which enter or leave the cerebellum pass through these peduncles.

INTERNAL STRUCTURE

The cerebellum is made up of a thin surface layer of grey matter, the *cerebellar cortex* and a central core of white matter. Embedded within the central core of white matter are masses of grey matter called *intracerebellar nuclei.*

Dissection

The meningeal coverings of the cerebellum are removed.

Study the various parts, surfaces, fissures and notches of the cerebellum. See the lobes of cerebellum. Look out for the three pairs of cerebellar peduncles and trace them to their attachments.

Make a midsagittal section of the cerebellum and divide it into cerebellar hemispheres. See the arbor vitae pattern. Study the deep cerebellar nuclei.

GREY MATTER OF CEREBELLUM

The grey matter of cerebellum is represented by:
- Cerebellar cortex
- Intracerebellar nuclei

Structure of Cerebellar Cortex

Most of the grey matter of the cerebellum is arranged as a thin layer covering the central core of white matter. This layer known as the *cerebellar cortex*.

In striking contrast to the cortex of the cerebral hemispheres, the cerebellar cortex has a uniform structure in all parts of the cerebellum. It is divided into three layers as follows (Fig. 22.7):
1. *Molecular layer* (most superficial),
2. *Purkinje cell layer* and
3. *Granular layer*, which rests on white matter.

The neurons of the cerebellar cortex are of six main types:
1. *Purkinje cells*, forming the layer named after them
2. *Granule cells*, forming the granular layer
3. *Outer (external) stellate cells* lying in the molecular layer
4. *Basket cells*, lying in the molecular layer
5. *Golgi cells*, present in the granular layer
6. *Brush cells* present in the granular layer

Intrinsic neurons of cerebellar cortex and their locations are given in Table 22.2.

Molecular Layer

The molecular layer is the superficial layer of the cortex and situated just below the pia mater.

Two types of cells are found in this layer:
1. *Stellate cells:* These cells situated in the superficial part of the molecular layer.
2. *Basket cells:* These cells situated in the deeper layer

Basket cells and stellate cells are GABAergic. They are inhibitory to Purkinje cells.

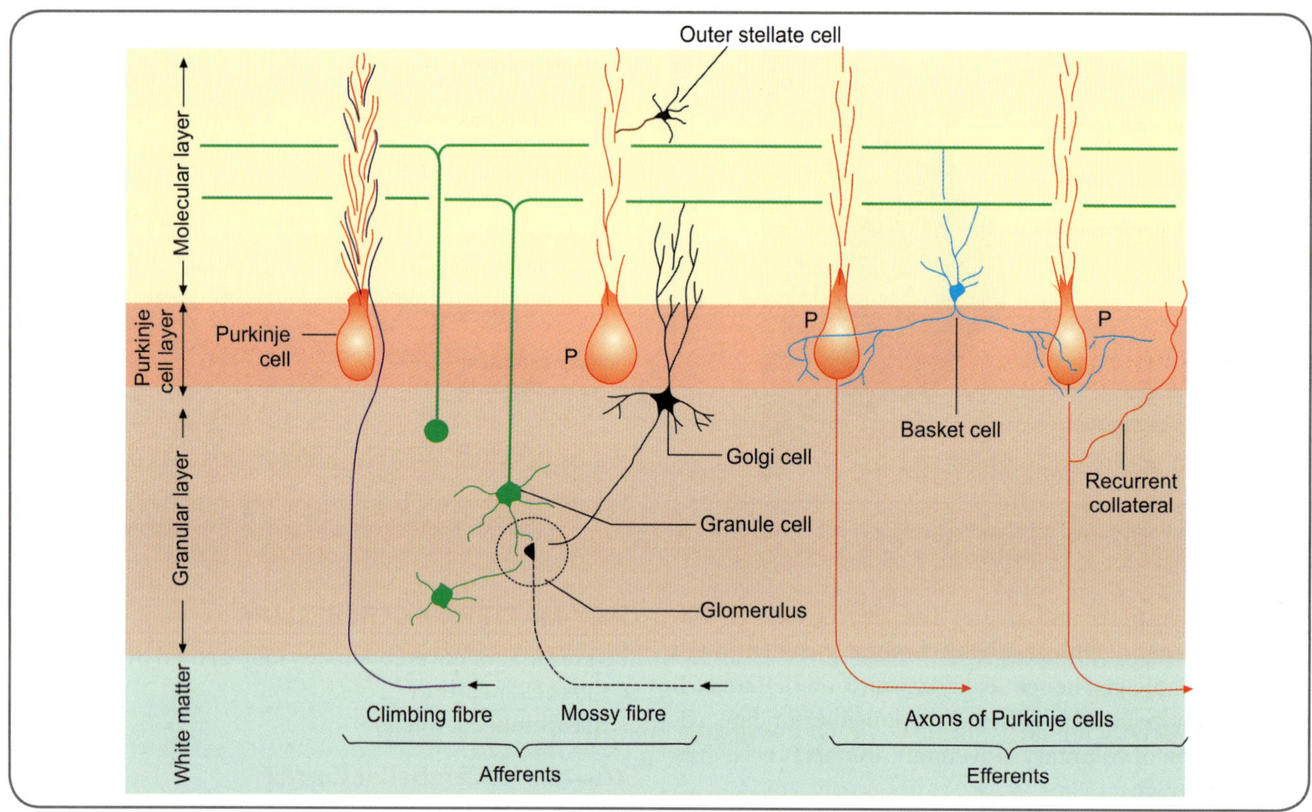

Fig. 22.7: Scheme to show the arrangement of neurons in the cerebellar cortex

Purkinje Cell Layer

The Purkinje cell layer contains flask-shaped cell bodies of Purkinje cells. This layer is unusual in that it contains only one layer of neurons. The Purkinje cells are evenly spaced. A dendrite arises from the 'neck' of the 'flask' and passes upwards into the molecular layer. Here, it divides and subdivides to form an elaborate dendritic tree. The branches of this 'tree' all lie in one plane (like the fins of a fan or like a vine branching against a wall) . This plane is transverse to the long axis of the folium. As a result of this arrangement, the dendritic trees of adjoining Purkinje cells lie in planes more or less parallel to one another.

The axon of each Purkinje cell passes downwards through the granular layer to enter the white matter. As described later, these axons constitute the only efferents of the cerebellar cortex. They end predominantly by synapsing with neurons in cerebellar nuclei. They are inhibitory to these neurons.

Granular Layer

It is the innermost layer and consists of numerous granule cells and a few golgi cells and brush cells.

Granule Cells

These are very small, numerous, spherical neurons that occupy the greater part of the granular layer. The spaces not occupied by them are called *cerebellar islands*. These islands are occupied by special synaptic structures called *glomeruli*.

Each granule cell gives off three to five short dendrites. These end in claw-like endings, which enter the glomeruli where they synapse with the terminals of mossy fibres. The axon of each granule cell enters the molecular layer. Here, it divides into two subdivisions each of which is at right angles to the parent axon (forming a T-junction). These axonal branches of granule cells are called *parallel fibres*.

Golgi Cells

These are large, stellate cells lying in the granular layer (Fig. 22.7), just deep to the Purkinje cells. They are GABAergic inhibitory neurons. Their dendrites enter the molecular layer, where they branch profusely, and synapse with the parallel fibres.

Intracerebellar Nuclei

Embedded within the central core of white matter there are masses of grey matter which constitute the *cerebellar nuclei.* These are located from lateral to medial side as follows (Fig. 22.8):

- The *dentate nucleus* lies in the centre of each cerebellar hemisphere. Cross sections through the nucleus have a striking resemblance to those through the inferior olivary nucleus. Like the latter, it is made up of a thin lamina of grey matter that is folded upon itself so that

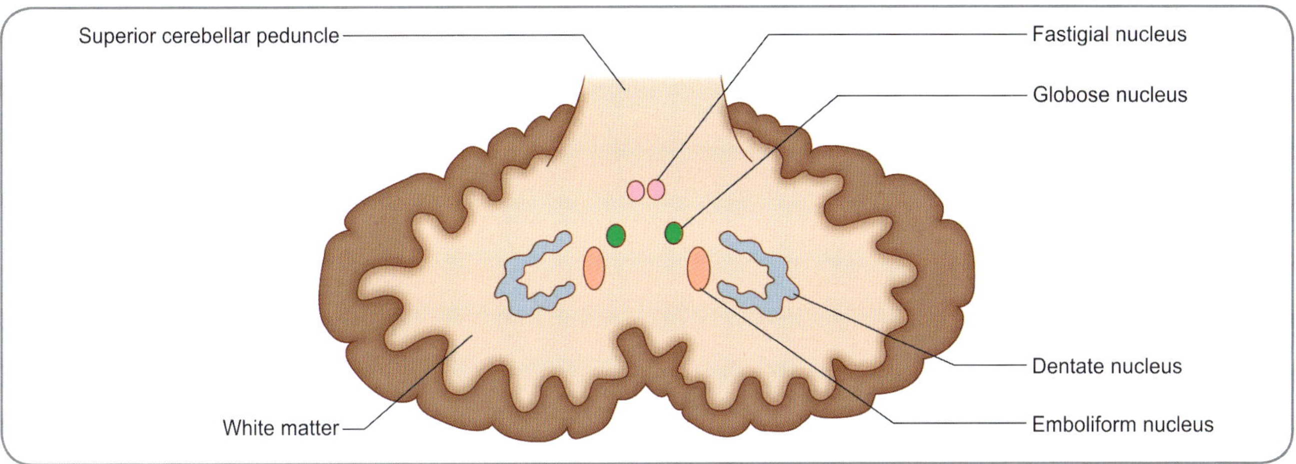

Fig. 22.8: Scheme to show the cerebellar nuclei

it resembles a crumpled purse. Both the nuclei have a hilum directed medially.

❑ The ***emboliform nucleus*** lies on the medial side of the dentate nucleus.

❑ The ***globose nucleus*** lies medial to the emboliform nucleus.

❑ The ***fastigial nucleus*** lies close to the middle line in the anterior part of the superior vermis.

WHITE MATTER OF CEREBELLUM

The central core of each cerebellar hemisphere is formed by the white matter. The peduncles are continued into this white matter. The white matter of the two sides is connected by a thin lamina of fibres that are closely related to the fourth ventricle. The upper part of this lamina forms the superior medullary velum, and its inferior part forms the inferior medullary velum. Both these take part in forming the roof of the fourth ventricle.

The white matter consists of two types of fibres— ***intrinsic*** and ***extrinsic***.

❑ ***Intrinsic fibres:*** Intrinsic fibres remain confined within the cerebellum. They connect different regions of the cerebellum either in the same hemisphere or of the two cerebellar hemispheres.
  ○ Projection fibres connect cerebellar cortex to the cerebellar nuclei.
  ○ Association fibres interconnect different parts of the cerebellar cortex.
  ○ Commissural fibres connect the two cerebellar hemispheres.

❑ ***Extrinsic fibres:*** Extrinsic fibres connect the cerebellum with other parts of the central nervous system, i.e. brain and spinal cord through afferent and efferent fibres. The fibres entering or leaving the cerebellum pass through three thick bundles called the ***cerebellar peduncles: superior, middle, and inferior***.

Afferent Fibres

The afferent fibres to the cerebellar cortex are of two different types:
1. Mossy fibres
2. Climbing fibres

Mossy Fibres

All fibres entering the cerebellum, other than olivo-cerebellar, end as mossy fibres.

They carry most input to cerebellum and end in the granule cell.

Climbing Fibres

These fibres represent terminations of axons reaching the cerebellum from the inferior olivary complex (Fig. 22.9). They pass through the granular layer and the Purkinje cell layer to reach the molecular layer. Each climbing fibre becomes intimately associated with the proximal part of the dendritic tree of one Purkinje cell, and establishes numerous synapses on them (These are called climbing fibres as they appear to ***climb up*** along the Purkinje cell dendrites).

Note: Both mossy and climbing fibres are excitatory.

Efferent Fibres

The efferent fibres from the cerebellar cortex are axons of Purkinje cells, which terminate in the cerebellar (central) nuclei. Axons of the Purkinje cells are inhibitory to cerebellar nuclei.

The fibres from dentate, emboliform, and globose nuclei leave cerebellum through the superior cerebellar peduncle. The fibres from the fastigial nucleus leave the cerebellum through inferior cerebellar peduncle.

The intrinsic neuronal circuit of cerebellum is shown in Figure 22.9.

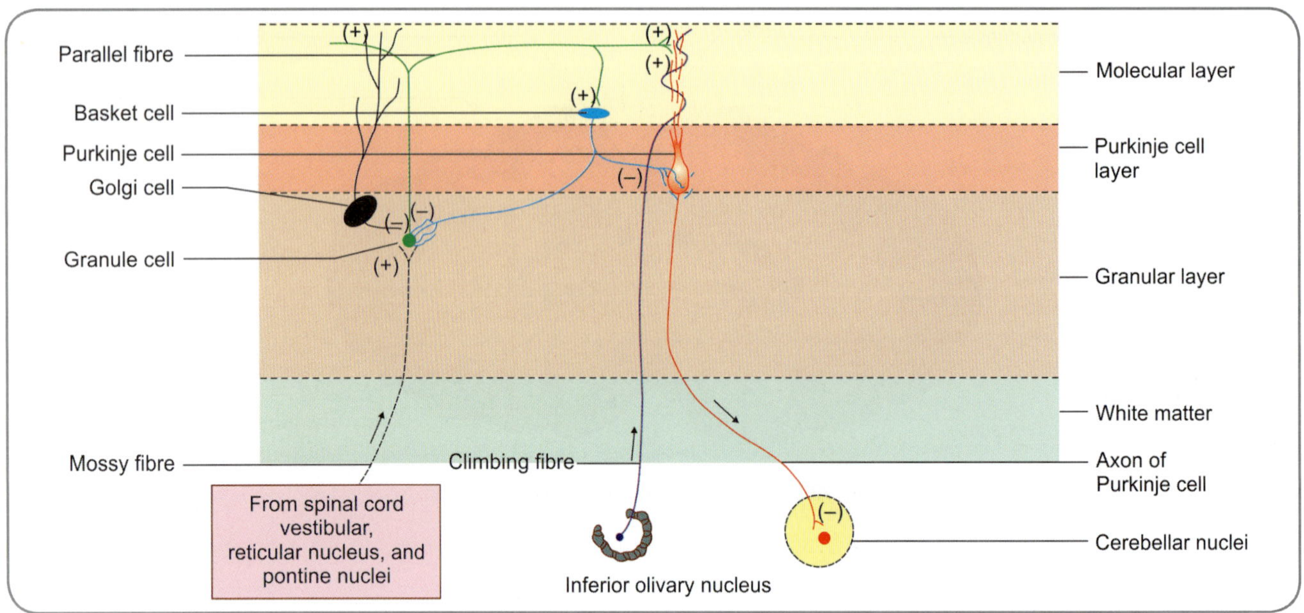

Fig. 22.9: Neuronal circuit of cerebellar cortex

Note:
- All the afferent fibres to the cerebellum (mossy and climbing fibres) are excitatory
- All the cells in the cerebellar cortex are inhibitory except the granule cells, which are excitatory
- The Purkinje cells are the only efferent neurons in the cerebellar cortex, which project to the cerebellar nuclei

FUNCTIONS OF CEREBELLUM

The cerebellum plays an essential role in the control of movement. It is responsible for ensuring that movement takes place smoothly, in the right direction and to the right extent. Cerebellar stimulation modifies movements produced by stimulation of motor areas of the cerebral cortex. The cerebellar cortex is also important for learning of movements (for example, in learning to write).

Through its vestibular and spinal connections, the cerebellum is responsible for maintaining the equilibrium of the body.

These functions are possible because the cerebellum constantly receives proprioceptive information regarding the state of contraction of muscles and of the position of various joints. It also receives information from the eyes, the ears, the vestibular apparatus, the reticular formation, and the cerebral cortex. All this information is integrated and is used to influence movement through motor centres in the brainstem and spinal cord and also through the cerebral cortex.

Recent studies have shown that the importance of the cerebellum may extend beyond the control of motor activity. There is an evidence that the cerebellum influences autonomic functions; through the reticular formation and the thalamus, it influences conduction in ascending sensory pathways and that the cerebral cortex, and cerebellar cortex may cooperate in other complex ways.

ARTERIAL SUPPLY OF CEREBELLUM

The cerebellum is supplied by three pairs of cerebellar arteries:

1. ***Superior cerebellar artery***, a branch of basilar artery supplies the superior surface of the cerebellum.
2. ***Anterior inferior cerebellar artery***, a branch of basilar artery supplies the anterior part of the inferior surface of the cerebellum.
3. ***Posterior inferior cerebellar artery***, a branch of vertebral artery supplies the posterior part of the inferior surface of the cerebellum.

Clinical Correlation

Disorders of Equilibrium

The maintenance of equilibrium and correct posture is dependent on reflex arcs involving various centres including the spinal cord, the cerebellum, and the vestibular nuclei. Afferent impulses for these reflexes are carried by the posterior column tracts (fasciculus gracilis and fasciculus cuneatus), the spinocerebellar tracts, and others. Efferents reach neurons of the ventral grey column (anterior horn cells) through rubrospinal, vestibulospinal, and other extrapyramidal tracts. Interruption of any of these pathways or lesions in the cerebellum, the vestibular nuclei and other centres concerned; can result in various abnormalities involving maintenance of posture and coordination of movements.

contd...

Clinical Correlation *contd...*

Cerebellar Syndrome

The cerebellar lesions due to trauma, haemorrhage, tumours, etc. produce a number of signs and symptoms, which together constitute the **cerebellar syndrome.**

The signs and symptoms produced by cerebellar lesions are as follows:

- ❑ ***Ataxia:*** Inability to maintain the equilibrium of the body, while standing, or while walking, is referred to as ***ataxia.*** This may occur as a result of the interruption of afferent proprioceptive pathways (***sensory ataxia***). Disease of the cerebellum itself, or of efferent pathways, results in more severe disability. Coordination of the activity of different groups of muscles is interfered with, leading to various defects. The person is unable to stand with his/her feet close together—the body sways from side to side and the person may fall. While walking, the patient staggers and is unable to maintain progression in the desired direction. Lack of proprioceptive information can be compensated to a considerable extent by information received through the eyes. The defects mentioned are, therefore, much more pronounced with the eyes closed (***Romberg's sign***).
- ❑ ***Asynergia:*** Lack of coordination of muscles also interferes with purposeful movements (***asynergia***). Movements are jerky and lack precision. For example, the patient finds it difficult to touch his nose with a finger, or to move a finger along a line. There is difficulty in performing movements involving rapid alternating action of opposing groups of muscles (for example, tapping one hand with the other; repeated pronation and supination of the forearm). This phenomenon is called ***dysdiadokokinesis***.
- ❑ ***Dysarthria:*** Incoordination of the muscles responsible for the articulation of words leads to characteristic speech defects (***dysarthria***).
- ❑ ***Nystagmus:*** For the same reason, the eyes are unable to fix the gaze on an object for any length of time. Attempts to bring the gaze back to the same point result in repeated jerky movements of the eyes. This is called ***nystagmus***.
- ❑ ***Hypotonia:*** Apart from incoordination, cerebellar disease is characterised by diminished muscle tone (***hypotonia***).
- ❑ ***Asthenia:*** The muscles are soft, and tire easily. Joints may lack stability (***flail joints***).
- ❑ ***Reflexes:*** Tendon reflexes may be diminished. Alternatively, tapping a tendon may result in oscillating movements of the part concerned like a pendulum.

Multiple Choice Questions

1. The depression separating the two cerebellar hemispheres on the inferior aspect is the:
   a. Vermis
   b. Vallecula
   c. Folium
   d. Declive
2. The middle cerebellar peduncles connect the cerebellum with the:
   a. Cerebral cortex
   b. Medulla
   c. Pons
   d. Midbrain
3. The only efferents of the cerebellar cortex are the:
   a. Axons of the Purkinje cells
   b. Axons of the Basket cells
   c. Dendrites of the Granule cells
   d. Axons of the Golgi cells
4. Climbing fibres are:
   a. Inhibitory in nature
   b. Associated with the dendrites of Purkinje cells
   c. The efferent inputs of cerebellum
   d. Commissural fibres of cerebellum
5. Paleocerebellum is otherwise called:
   a. Vestibulocerebellum
   b. Spinocerebellum
   c. Pontocerebellum
   d. Flocculocerebellum

ANSWERS

1. b   **2.** c   **3.** a   **4.** b   **5.** b

Clinical Problem-solving

Case Study 1: A 67-year-old man had haemorrhage in the cerebellum resulting in cerebellar syndrome.
- ❑ What are the symptoms and signs likely to be seen in this patient?
- ❑ List out any three features of cerebellar lesions and explain their anatomical basis.
- ❑ What would be the status of muscle tone in cerebellar lesions?

Case Study 2: A 57-year-old man with cerebellar disease is not able to fix his gaze on any object. He has jerky movements of his eyes.
- ❑ What is the condition with repetitive jerky movements of eyes called?
- ❑ In what way does the cerebellar disease lead to such a condition?
- ❑ Give other examples of incoordination of muscles.

(For solutions see Appendix).

Chapter 23

Cerebrum, Thalamus and Hypothalamus— Gross Anatomy and Internal Structure

CEREBRUM

The cerebrum is the largest part of the brain. It has an ovoid shape. It consists of two incompletely separated cerebral hemispheres. The outer surface of the cerebral hemisphere is covered with cortex, which is highly folded due to the presence of sulci and gyri. The core of the hemisphere consists of white matter containing a group of nuclei called *basal ganglia*. The cavity inside each hemisphere is called the *lateral ventricle*.

The median longitudinal fissure of cerebrum intervenes between the medial surfaces of the right and left hemispheres. At the bottom of the fissure lies the corpus callosum, which is a connecting bond between the two hemispheres. The contents of the longitudinal fissure are falx cerebri and the accompanying arachnoid mater, pia mater covering the medial surfaces of the hemispheres, anterior cerebral vessels, and the indusium griseum on the superior surface of the corpus callosum.

EXTERNAL FEATURES OF CEREBRAL HEMISPHERE

The cerebral hemisphere has three poles, three surfaces, and three borders.

Poles

When viewed from the lateral aspect, each cerebral hemisphere has the appearance shown in Figure 23.1.

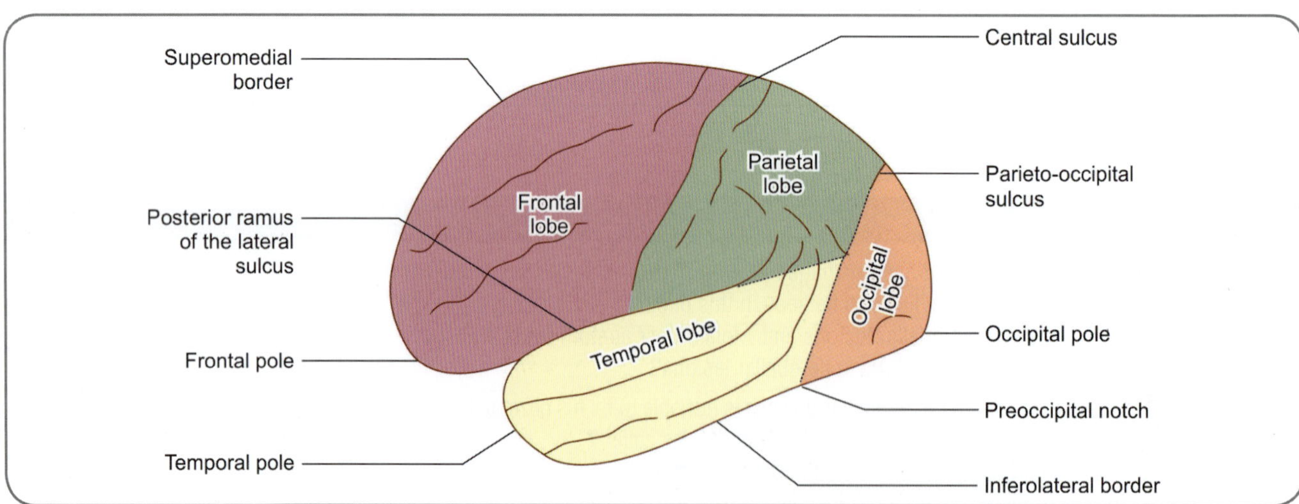

Fig. 23.1: Lateral aspect of the cerebral hemisphere to show borders, poles and lobes

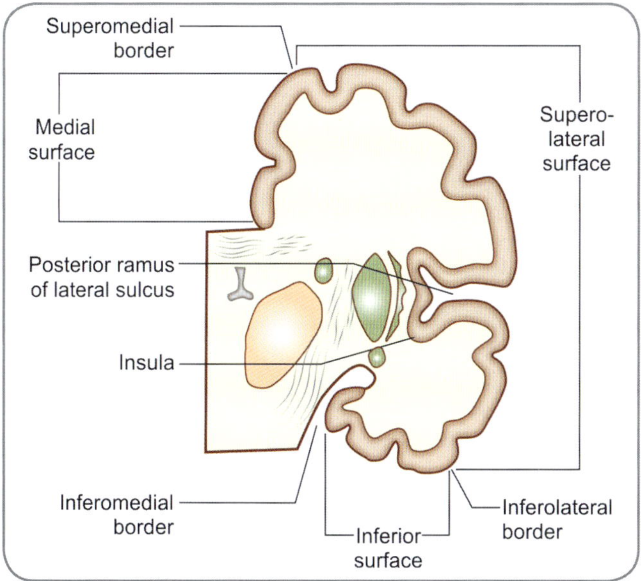

Fig. 23.2: Coronal section through a cerebral hemisphere to show its borders and surfaces

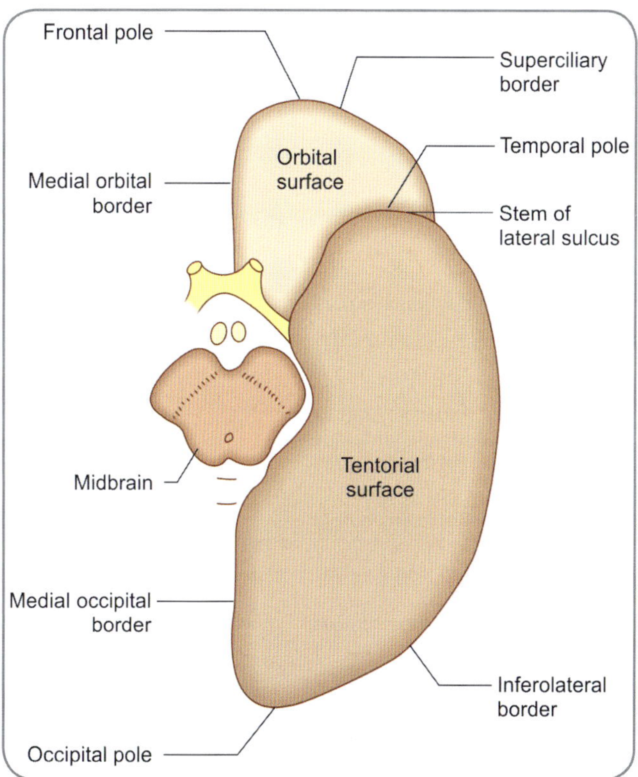

Fig. 23.3: Inferior aspect of a cerebral hemisphere to show its borders, poles and surfaces

Three somewhat pointed ends or poles can be recognised. These are the frontal pole anteriorly, the occipital pole posteriorly, and the temporal pole that lies between the frontal and occipital poles and points forwards and somewhat downwards.

Surfaces

Each cerebral hemisphere has three surfaces—***supero-lateral, medial (or vertical), and inferior*** (Fig. 23.2). The right and left medial surfaces enclose the longitudinal fissure. The inferior surface is subdivided into orbital and tentorial surfaces by the stem of the lateral sulcus (Fig. 23.3).

Borders

A coronal section through the cerebral hemispheres (Fig. 23.2) shows that each hemisphere has three borders: ***superomedial, inferolateral,*** and ***inferomedial.***

The inferomedial border is divided into an anterior part called the ***medial orbital border*** and a posterior part called the ***medial occipital border***. The orbital part of the inferolateral border is called the ***superciliary border*** (as it lies just above the level of the eyebrows).

Sulci and Gyri

The surfaces of the cerebral hemisphere are not smooth. They show a series of grooves or ***sulci***, which are separated by intervening areas that are called ***gyri***.

❑ On the superolateral surface of the hemisphere, there are two prominent sulci (Fig. 23.4). One of these is the ***posterior ramus of the lateral sulcus,*** which begins near the temporal pole and runs backwards and slightly upwards. Its posteriormost part curves sharply upwards. The second sulcus that is used to delimit the lobes is the ***central sulcus.*** It begins on the superomedial margin, a little behind the midpoint between the frontal and occipital poles and runs downwards and forwards to end a little above the posterior ramus of the lateral sulcus.

❑ On the medial surface of the hemisphere, near the occipital pole, there is a sulcus called the ***parieto-occipital sulcus*** (Fig. 23.4). The upper end of this sulcus reaches the superomedial border and a small part of it can be seen on the superolateral surface (Fig. 23.1).

❑ A little anterior to the occipital pole, the inferolateral border shows a slight indentation called the ***preoccipital notch (or preoccipital incisure)***.

Lobes

There are four lobes, namely frontal, parietal, occipital, and temporal, which are well demarcated on the superolateral surface (Fig. 23.1).

To define the limits of various lobes, we have to draw two imaginary lines. The first imaginary line connects the upper end of the parieto-occipital sulcus to the preoccipital notch. The second imaginary line is a backward continuation of the posterior ramus of the lateral sulcus (excluding the posterior upturned part) to meet the first line. We can now define the limits of the various lobes as follows:

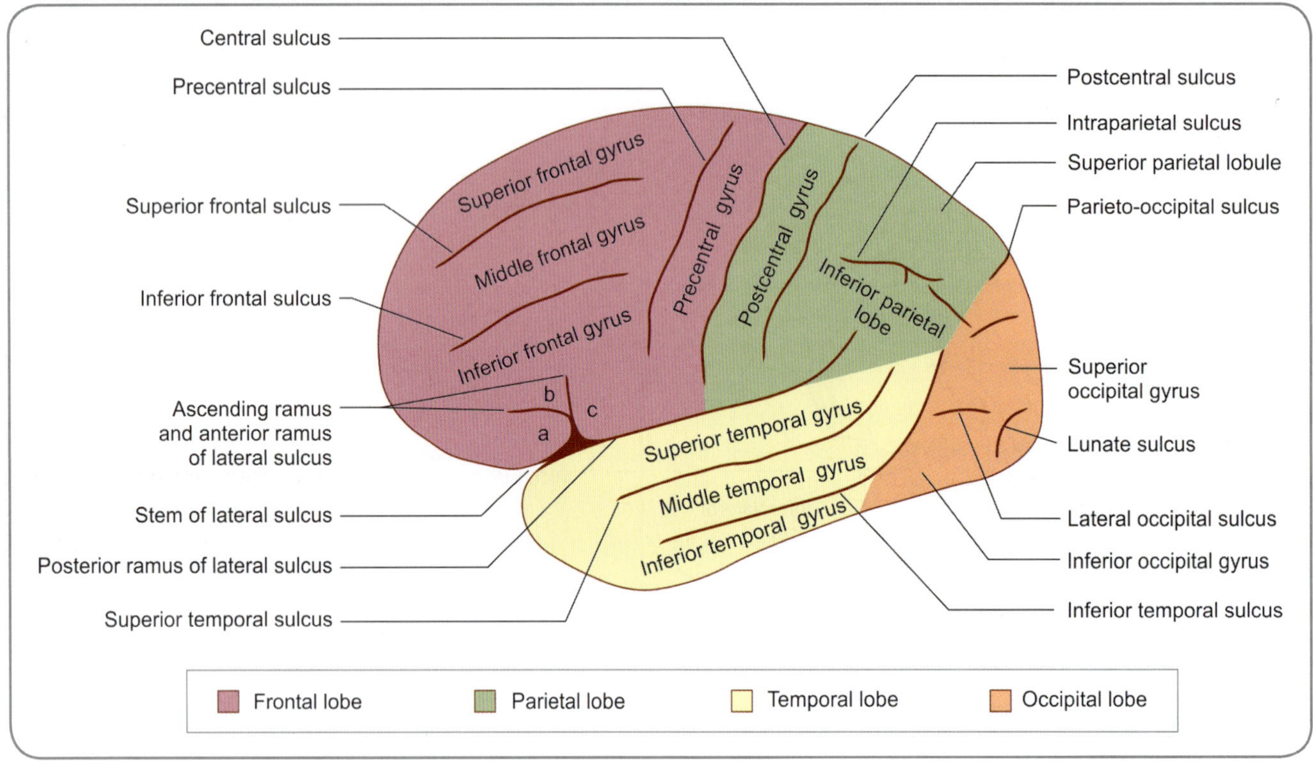

Fig. 23.4: Simplified presentation of sulci and gyri on the superolateral surface of the cerebral hemisphere.
Key: a = pars orbitalis b = pars triangularis c = pars opercularis

- The *frontal lobe* lies anterior to the central sulcus and above the posterior ramus of the lateral sulcus.
- The *parietal lobe* lies behind the central sulcus. It is bounded below by the posterior ramus of the lateral sulcus and by the second imaginary line and behind by the upper part of the first imaginary line.
- The *occipital lobe* is the area lying behind the first imaginary line.
- The *temporal lobe* lies below the posterior ramus of the lateral sulcus and the second imaginary line. It is separated from the occipital lobe by the lower part of the first imaginary line.

Before going onto consider further subdivisions of each of the lobes named above, attention has to be directed to details of some structures already mentioned.

- The upper end of the central sulcus winds round the superomedial border to reach the medial surface. Here its end is surrounded by a gyrus called the *paracentral lobule* (Fig. 23.5). The lower end of the central sulcus is always separated by a small interval from the posterior ramus of the lateral sulcus.
- The lateral sulcus begins on the inferior aspect of the cerebral hemisphere, where it lies between the orbital surface and the anterior part of the temporal lobe. It runs laterally to reach the superolateral surface. On reaching this surface, it divides into three rami (branches). These rami are *anterior (or anterior horizontal), ascending*

(or anterior ascending), and *posterior* (Fig. 23.4). The anterior and ascending rami are short and run into the frontal lobe in the directions indicated by their names. The posterior ramus has already been considered. Unlike most other sulci, the lateral sulcus is very deep. Its walls cover a fairly large area of the surface of the hemisphere called the *insula*.

SUPEROLATERAL SURFACE OF CEREBRAL HEMISPHERE

The subdivisions of the superolateral surface are described below and are shown in Figure 23.4.

Frontal lobe

The frontal lobe is further subdivided as follows—The *precentral sulcus* runs downwards and forwards parallel to and a little anterior to the central sulcus. The area between it and the central sulcus is the *precentral gyrus.* In the region anterior to the precentral gyrus there are two sulci that run in an anteroposterior direction. These are the *superior* and *inferior frontal sulci.* They divide this region into *superior, middle,* and *inferior frontal gyri.* The anterior and ascending rami of the lateral sulcus extend into the inferior frontal gyrus, dividing it into three parts: The part below the anterior ramus is the *pars orbitalis*, that between the anterior and ascending rami is the *pars triangularis*; and the part posterior to the ascending ramus is the *pars opercularis.*

Temporal Lobe

The temporal lobe has two sulci that run parallel to the posterior ramus of the lateral sulcus. They are termed the *superior and inferior temporal sulci.* They divide the superolateral surface of this lobe into *superior, middle, and inferior temporal gyri.*

Parietal Lobe

The parietal lobe shows the following subdivisions. The *postcentral sulcus* runs downwards and forwards parallel to and a little behind the central sulcus. The area between these two sulci is the *postcentral gyrus.* The rest of the parietal lobe is divided into a *superior parietal lobule* and an *inferior parietal lobule* by the *intraparietal sulcus.* The upturned posterior end of the posterior ramus of the lateral sulcus extends into the inferior parietal lobule. The posterior ends of the superior and inferior temporal sulci also turn upwards to enter this lobule. The upturned ends of these three sulci divide the inferior parietal lobule into three parts. The part that arches over the upturned posterior end of the posterior ramus of the lateral sulcus is called the *supramarginal gyrus.* The part that arches over the superior temporal sulcus is called the *angular gyrus.* The part that arches over the posterior end of the inferior temporal sulcus is called the *arcus temporo-occipitalis.*

Occipital Lobe

The occipital lobe shows three rather short sulci. One of these, the *lateral occipital sulcus* lies horizontally and divides the lobe into *superior and inferior occipital gyri.* The *lunate sulcus* runs downwards and slightly forwards just in front of the occipital pole. The vertical strip just in front of it is the *gyrus descendens.* The *transverse occipital sulcus* is located in the uppermost part of the occipital lobe. The upper end of the parieto-occipital sulcus (which just reaches the superolateral surface from the medial surface) is surrounded by the *arcus parieto-occipitalis.* As its name suggests, it belongs partly to the parietal lobe and partly to the occipital lobe.

Insula

In the depth of the stem and posterior ramus of the lateral sulcus, there is a part of the cerebral hemisphere called the *insula* (*insula = hidden*). It is surrounded by a *circular sulcus.* During development of the cerebral hemisphere, this area grows less than surrounding areas, which therefore, come to overlap it and occlude it from surface view. These surrounding areas are called *opercula* (= lids).

MEDIAL SURFACE OF CEREBRAL HEMISPHERE

The features of the medial surface include the sulci and gyri, as well as the corpus callosum and the midline structures below it (Fig. 23.5).

The *corpus callosum* is a prominent arched structure consisting of commissural fibres passing from one hemisphere to the other (Fig. 23.6). It consists of a central part called the *trunk,* a posterior end or *splenium,* and an anterior end or *genu.* A little below the

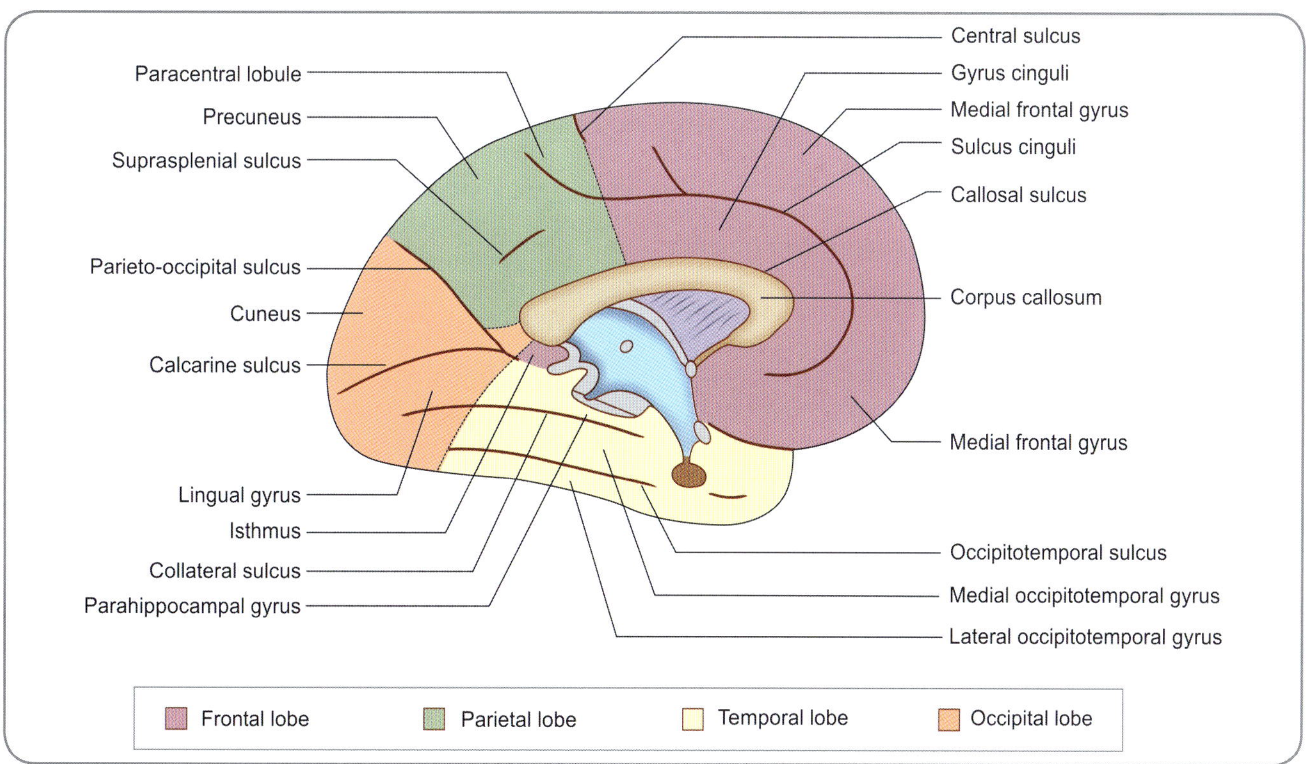

Fig. 23.5: Simplified presentation of sulci and gyri on the medial aspect of the cerebral hemisphere – the corpus callosum and other structures connecting the two hemispheres have been cut across

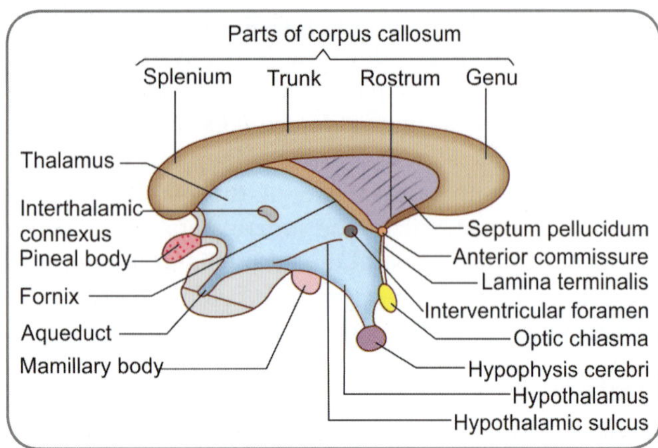

Fig. 23.6: Enlarged view of the central part of Fig. 23.5 to show clearly some structures seen on the medial aspect of the cerebral hemisphere

corpus callosum, there is the third ventricle of the brain. A number of structures can be identified in relation to this ventricle. The ***interventricular foramen,*** through which the third ventricle communicates with the lateral ventricle, can be seen in the upper and anterior part. Posteroinferiorly, the ventricle is continuous with the ***cerebral aqueduct.*** The lateral wall of the ventricle is formed in greater part by a large mass of grey matter called the ***thalamus.*** The right and left thalami are usually interconnected (across the midline) by a strip of grey matter called the ***interthalamic connexus.*** The anteroinferior part of the lateral wall of the third ventricle is formed by a collection of grey matter that constitutes the **hypothalamus.**

Above the thalamus, there is a bundle of fibres called the ***fornix.*** Posteriorly, the fornix is attached to the under surface of the corpus callosum, but anteriorly it disappears from view just in front of the interventricular foramen. Extending between the fornix and the corpus callosum, there is a thin lamina called the ***septum pellucidum (or septum lucidum),*** which separates the right and left lateral ventricles from each other. Removal of the septum pellucidum brings the interior of the lateral ventricle into view.

In the anterior wall of the third ventricle, there are the ***anterior commissure*** and the ***lamina terminalis.*** The anterior commissure is attached to the genu of the corpus callosum through a thin lamina of fibres that constitutes the ***rostrum*** of the corpus callosum. Below, the anterior commissure is continuous with the ***lamina terminalis,*** which is a thin lamina of nervous tissue. The lower end of the lamina terminalis is attached to the optic chiasma. Just in front of the lamina terminalis, there are the ***paraterminal gyrus*** and the ***parolfactory gyrus.*** Posteriorly, the third ventricle is related to the ***pineal body (or pineal gland)*** and inferiorly to the ***hypophysis cerebri.***

Above the corpus callosum (and also in front of and behind it), we see the sulci and gyri of the medial surface of the hemisphere (Fig. 23.5). The most prominent of the sulci is the ***cingulate sulcus,*** which follows a curved course parallel to the upper convex margin of the corpus callosum. Anteriorly, it ends below the rostrum of the corpus callosum. Posteriorly, it turns upwards to reach the superomedial border a little behind the upper end of the central sulcus. The area between the cingulate sulcus and the corpus callosum is called the ***gyrus cinguli.*** It is separated from the corpus callosum by the ***callosal sulcus.***

The part of the medial surface of the hemisphere between the cingulate sulcus and the superomedial border consists of two parts. The smaller posterior part, which is wound around the end of the central sulcus is called the ***paracentral lobule.*** The large anterior part is called the ***medial frontal gyrus.*** These two parts are separated by a short sulcus continuous with the cingulate sulcus.

The part of the medial surface behind the paracentral lobule and the gyrus cinguli shows two major sulci that cut off a triangular area called the ***cuneus.*** The triangle is bounded anteriorly and above by the ***parieto-occipital sulcus,*** inferiorly by the ***calcarine sulcus,*** and posteriorly by the superomedial border of the hemisphere. The calcarine sulcus extends forwards beyond its junction with the parieto-occipital sulcus and ends a little below the splenium of the corpus callosum. The small area separating the splenium from the calcarine sulcus is called the ***isthmus.*** Between the parieto-occipital sulcus and the paracentral lobule, there is a quadrilateral area called the ***precuneus.*** Anteroinferiorly, the precuneus is separated from the posterior part of the gyrus cinguli by the ***suprasplenial (or subparietal) sulcus.***

The precuneus and the posterior part of the paracentral lobule form the medial surface of the parietal lobe.

Although the parieto-occipital and calcarine sulci appear to be continuous with each other on surface view, they are separated by the ***cuneate gyrus*** (or ***cuneolingual gyrus***), which lies in the depth of the area where the two sulci meet. The parts of the calcarine sulcus anterior and posterior to the junction with the parieto-occipital sulcus are separated by a deeply situated ***anterior cuneolingual gyrus.***

INFERIOR SURFACE OF CEREBRAL HEMISPHERE

When the cerebrum is separated from the hindbrain by cutting across the midbrain and is viewed from below, the structures seen are shown in Figure 23.7. Posterior to the midbrain, we see the under surface of the splenium of the corpus callosum. Anterior to the midbrain, there is a depressed area called the ***interpeduncular fossa.*** The fossa is bounded in front by the ***optic chiasma*** and

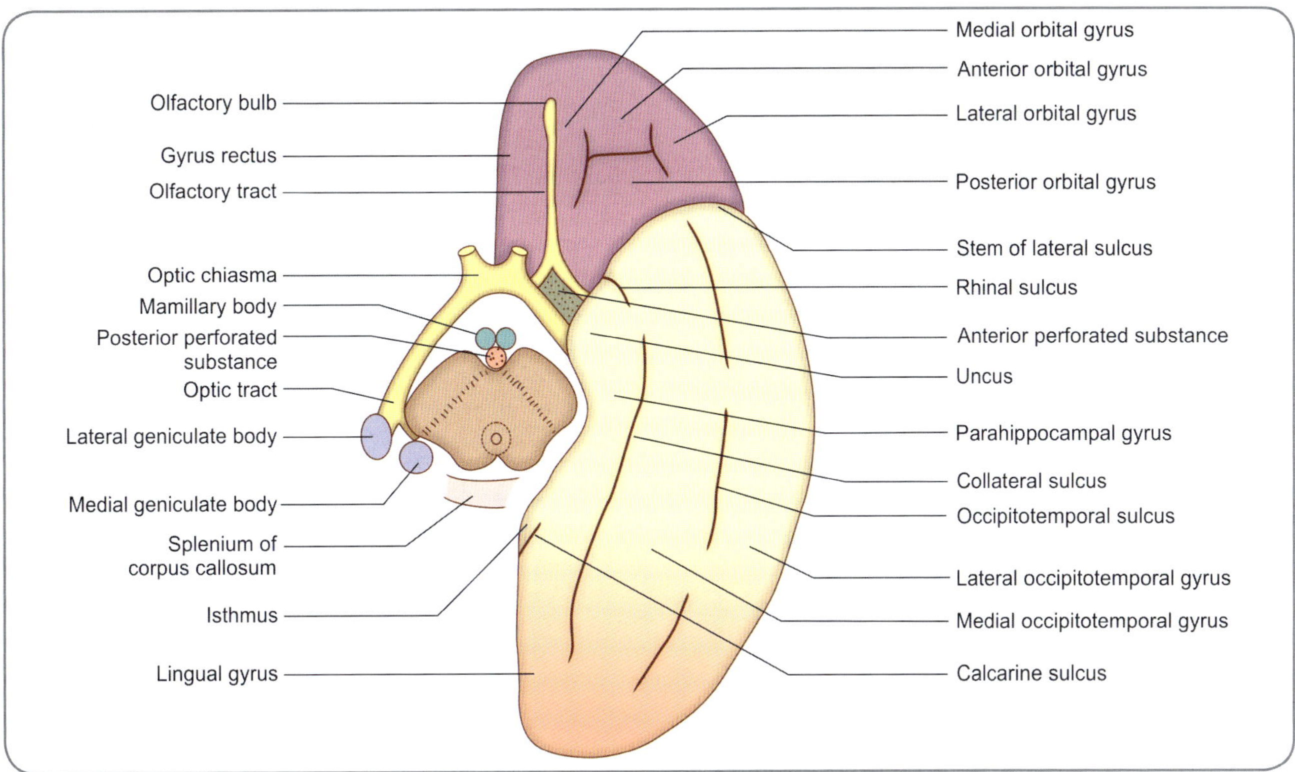

Fig. 23.7: Structures seen on the inferior aspect of the cerebrum—the midbrain has been cut across

on the sides by the right and left *optic tracts*. The optic tracts wind round the sides of the midbrain to terminate on its posterolateral aspect. In this region, two swellings, the *medial and lateral geniculate bodies*, can be seen. Certain structures are seen within the interpeduncular fossa. These are closely related to the floor of the third ventricle (Fig. 23.6). Anterior and medial to the crura of the midbrain, there are two rounded swellings called the *mamillary bodies.* Anterior to these bodies, there is a median elevation called the *tuber cinereum,* to which the infundibulum of the hypophysis cerebri is attached. The triangular interval between the mamillary bodies and the midbrain is pierced by numerous small blood vessels and is called the *posterior perforated substance.* A similar area lying on each side of the optic chiasma is called the *anterior perforated substance*. The anterior perforated substance is bounded anterolaterally by the lateral olfactory stria and posterolaterally by the uncus. The anterior perforated substance is connected to the insula by a band of grey matter, called the *limen insulae,* which lies in the depth of the stem of the lateral sulcus.

In addition to these structures, there are the sulci and gyri on the orbital and tentorial parts of the inferior surface of the each cerebral hemisphere. These parts are separated from each other by the stem of the lateral sulcus.

Sulci and Gyri on Orbital Surface

Close to the medial border of the orbital surface, there is an anteroposterior sulcus called the *olfactory sulcus* because the olfactory bulb and tract lie superficial to it. The area medial to this sulcus is called the *gyrus rectus.* The rest of the orbital surface is divided by an H-shaped *orbital sulcus* into *anterior, posterior, medial,* and *lateral orbital gyri.*

Sulci and Gyri on Tentorial Surface

The tentorial surface is marked by two major sulci that run in an anteroposterior direction. These are the *collateral sulcus,* medially and the *occipitotemporal sulcus,* laterally. The posterior part of the collateral sulcus runs parallel to the calcarine sulcus; the area between them is the *lingual gyrus.* Anteriorly, the lingual gyrus becomes continuous with the *parahippocampal gyrus,* which is related medially to the midbrain and to the interpeduncular fossa. The anterior end of the parahippocampal gyrus is cut off from the curved temporal pole of the hemisphere by a curved *rhinal sulcus.* This part of the parahippocampal gyrus forms a hook-like structure called the *uncus,* details of which are considered later. Posteriorly, the parahippocampal gyrus becomes continuous with the gyrus cinguli through the isthmus (Fig. 23.5). The area between the collateral sulcus and the rhinal sulcus, medially and the occipitotemporal sulcus, laterally is the *medial occipitotemporal gyrus.* The area lateral to the occipitotemporal sulcus is called the *lateral occipitotemporal gyrus.* This gyrus is continuous (around the inferolateral margin of the cerebral hemisphere) with the inferior temporal gyrus.

STRUCTURES WITHIN CEREBRAL HEMISPHERE

The surface of the cerebral hemisphere is covered by a thin layer of grey matter called the *cerebral cortex*. The cortex follows the irregular contour of the sulci and gyri of the hemisphere and extends into the depths of the sulci. As a result of this folding of the cerebral surface, the cerebral cortex acquires a much larger surface area than the size of the hemispheres would otherwise allow.

The greater part of the cerebral hemisphere deep to the cortex is occupied by white matter, within which are embedded certain important masses of grey matter. Immediately lateral to the third ventricle, there are the *thalamus* and *hypothalamus* (and certain smaller masses) derived from the diencephalon. More laterally, there is the *corpus striatum,* which is derived from the telencephalon. It consists of two masses of grey matter, the *caudate nucleus* and the *lentiform nucleus,* which consists of two functionally distinct parts, the putamen and the globus pallidus. A little lateral to the lentiform nucleus, we see the cerebral cortex in the region of the insula. Between the lentiform nucleus and the insula, there is a thin layer of grey matter called the *claustrum.* The caudate nucleus, the lentiform nucleus, the claustrum, and some other masses of grey matter (all of telencephalic origin) are referred to as *basal nuclei* or as *basal ganglia.*

INTERNAL STRUCTURE OF CEREBRAL CORTEX

Due to the presence of a large number of sulci, only about one-third of the total area of cerebral cortex is seen on the surface of the brain. The total area of the cerebral cortex is estimated to be about 2000 cm^2.

Like other masses of grey matter, the cerebral cortex contains the cell bodies of an innumerable number of neurons along with their processes, neuroglia, and blood vessels. The neurons are of various sizes and shapes. They establish extremely intricate connections with each other and with axons reaching the cortex from other masses of grey matter. Despite a very large volume of work on the subject, it is still not possible to explain many functions within the cerebral cortex.

Neurons in Cerebral Cortex

Cortical neurons vary in size, shape of their cell bodies, and lengths, branching patterns, and orientation of their processes. The cerebral cortex consists of many types of nerve cells, but two principal nerve cells are the pyramidal cells and stellate cells, which are described in detail below (Table 23.1 and Fig. 23.8).

Pyramidal Cells

They are the most abundant type of cortical neurons. In contrast, all other neurons in the cortex are referred to as

Table 23.1: Nerve cells in the cerebral cortex	
Pyramidal cells	**Cells of Martinotti**
Stellate cells	Basket cells
Fusiform cells	Chandelier cells
Horizontal cells of cajal	Double bouquet cells

nonpyramidal neurons. About two-thirds of all cortical neurons are pyramidal. Their cell bodies are triangular, with the apex generally directed towards the surface of the cortex. A large dendrite arises from the apex. Other dendrites arise from basal angles. The axon arises from the base of the pyramid. The processes of pyramidal cells extend vertically through the entire thickness of cortex and establish numerous synapses. The axon of a pyramidal cell may terminate in one of the following ways:

❏ It may travel to other regions like the basal ganglia, the brainstem, or the spinal cord. Fibres that leave the cortex commonly give off collaterals that terminate within the cortex.

❏ It may cross to the opposite side (through a commissure) and reach the corresponding region of the opposite hemisphere. Sometimes, it may reach a different region of the opposite hemisphere.

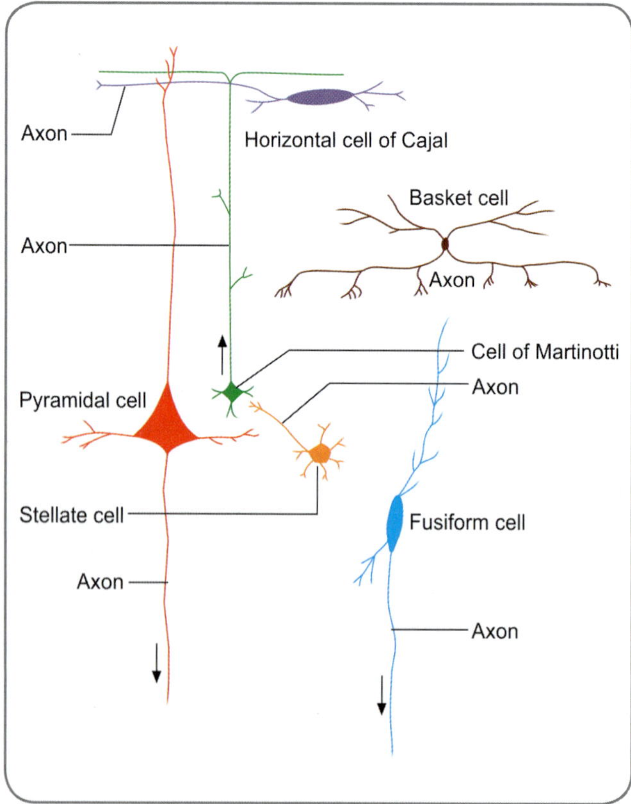

Fig. 23.8: Some of the cell types seen in the cerebral cortex. Other varieties not shown include bipolar cells, Chandelier cells and double bouquet cells

- It may enter the white matter to travel to another part of the cortex.
- It may be short and may terminate within the same area of the cortex.

The neurotransmitter used by pyramidal cells is either glutamate or aspartate.

Stellate Cells

The ***stellate neurons*** are relatively small and multipolar. They form about one-third of the total neuronal population of the cortex. Under low magnifications (and in preparations in which their processes are not demonstrated), these neurons look like granules. They have, therefore, been termed ***granular neurons*** by earlier workers. Stellate cells are of various types depending on their location and pattern of ramification of their processes. Their axons are short and end within the cortex. Their processes extend chiefly in a vertical direction within the cortex, but in some cases, they may be oriented horizontally. Some cells included under the term 'stellate' may be fusiform rather than stellate, with one large process arising at either end.

LAMINAE OF CEREBRAL CORTEX

On the basis of light microscopy (cell bodies displayed by Nissl method and the myelinated fibres stained by Weigert method), the cerebral cortex is described as having six layers or laminae (Fig. 23.9). From the superficial to deep, these laminae are:

- Plexiform or molecular layer
- External granular layer
- Pyramidal cell layer
- Internal granular layer
- Ganglionic layer
- Multiform layer

The plexiform layer is predominantly made up of fibres, although a few cells are present. All the remaining layers contain both stellate and pyramidal neurons, as well as other types of neurons. The external and internal granular layers have predominance of stellate (granular) cells. The prominent neurons in the pyramidal layer and the ganglionic layer are pyramidal neurons. The largest pyramidal cells (giant pyramidal cells of Betz) are found in the ganglionic layer. The multiform layer contains cells of various sizes and shapes.

In addition to the cell bodies of neurons, the cortex contains abundant nerve fibres. Many of these are vertically oriented. Some of these fibres represent afferents entering the cortex. In addition to the vertical fibres, the cortex contains transversely running fibres that form prominent aggregations in certain situations. One such aggregation, present in the internal granular layer is called the ***external band of Baillarger***. Another, present in the ganglionic layer is called the ***internal band of Baillarger***. The space between the cell bodies of neurons is permeated by a dense plexus formed by their processes. This plexus is referred to as the ***neuropil***.

FUNCTIONAL AREAS OF CEREBRAL CORTEX

Some areas of the cerebral cortex can be assigned specific functions. These areas can be defined in terms of known sulci and gyri. However, some areas are commonly referred to by numbers, and it is necessary to know what these numbers mean. Various workers who have studied the microscopic structure of the cerebral cortex, have found that there is a considerable variation from region to region. They have also found that these variations do not necessarily follow the boundaries of sulci and gyri, but often cut across them. Various authors have worked out 'maps' of the cerebral cortex indicating areas of differing structures. The best known scheme is that of Brodmann, who represented different areas by numbers. Although the functional significance of the areas is open to question, areas of the cortex are very frequently referred to by Brodmann numbers. It is, therefore, necessary to be familiar with them. The numbers most commonly referred to are indicated in Figs 23.10A and B.

Primary Motor Area (Area 4 of Brodmann)

The primary motor area is located in the precentral gyrus on the superolateral surface of the hemisphere (Fig. 23.11)

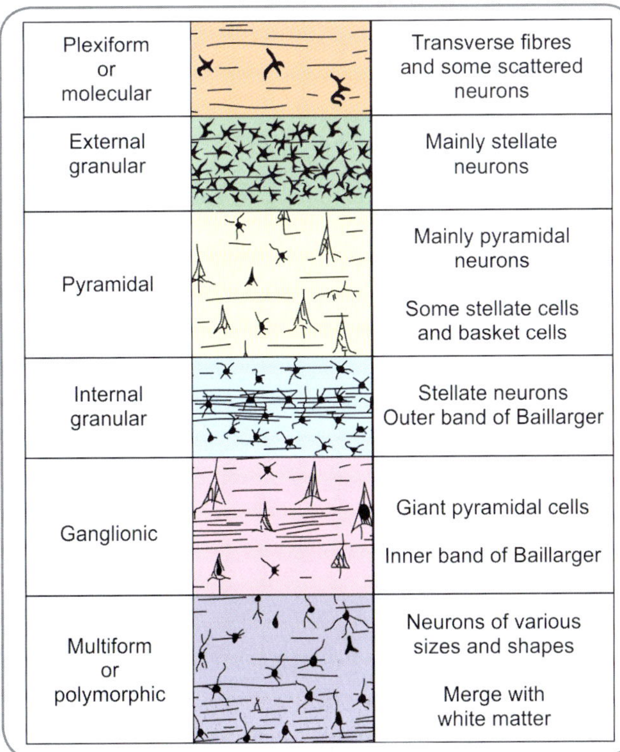

Fig. 23.9: Laminae of cerebral cortex

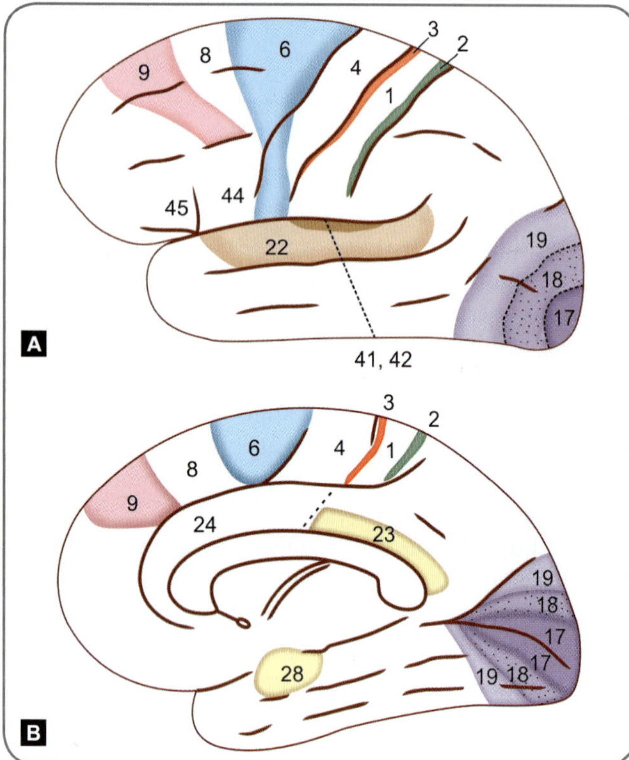

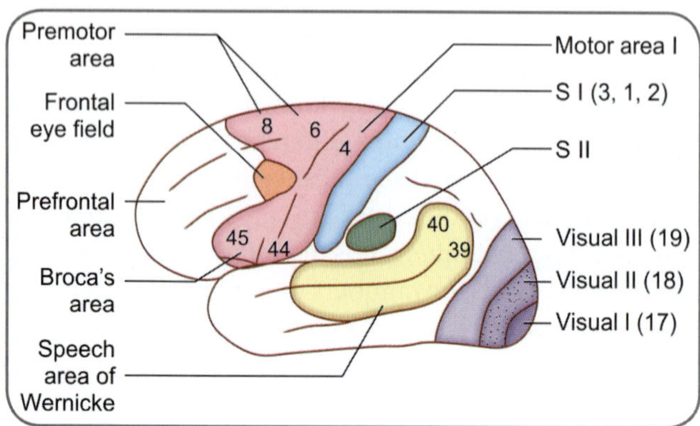

Figs 23.10A and B: A. Location of some of the areas of Brodmann on the superolateral aspect and **B.** Areas of Brodmann on the medial aspect of the cerebral hemisphere

and in the anterior part of the paracentral lobule on the medial surface. It corresponds to area 4 of Brodmann. When these areas are stimulated electrically, movements occur in various parts of the body. Anatomically, these areas give origin to projection fibres that form the corticospinal and corticonuclear tracts (Fig. 23.12).

Specific regions within the area are responsible for movements in specific parts of the body.

Stimulation of the paracentral lobule produces movement in the lower limbs. The trunk and upper limb are represented in the upper part of the precentral gyrus,

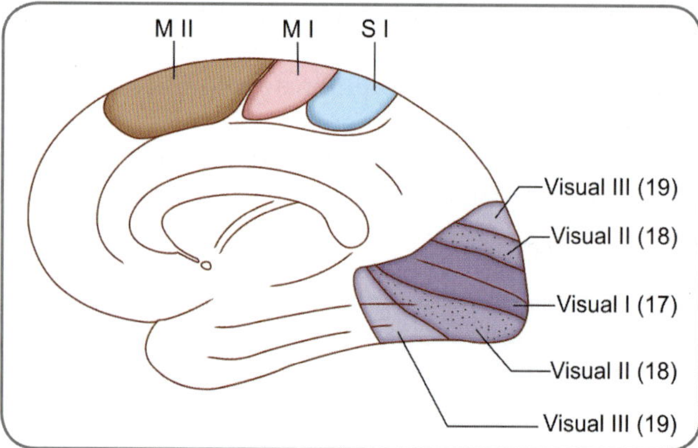

Fig. 23.12: Functional areas on the medial aspect of the cerebral hemisphere (recent concept)

while the face and head are represented in the lower part of the gyrus (i.e. human body is represented in an upside down manner—"***homunculus***") (Fig. 23.13).

Another feature of interest is that the area of cortex representing a part of the body is not proportional to the size of the part, but rather to intricacy of movements in the region. Thus, relatively large areas of cortex represents the hands and lips.

Premotor Area (Areas 6 and 8 of Brodmann)

The premotor area is located just anterior to the motor area. It occupies the posterior parts of the superior, middle and inferior frontal gyri (Fig. 23.11). The part of the premotor area located in the superior and middle frontal gyri corresponds to areas 6 and 8 of Brodmann (Fig. 23.10).

Fig. 23.11: Functional areas on the superolateral aspect of the cerebral hemisphere

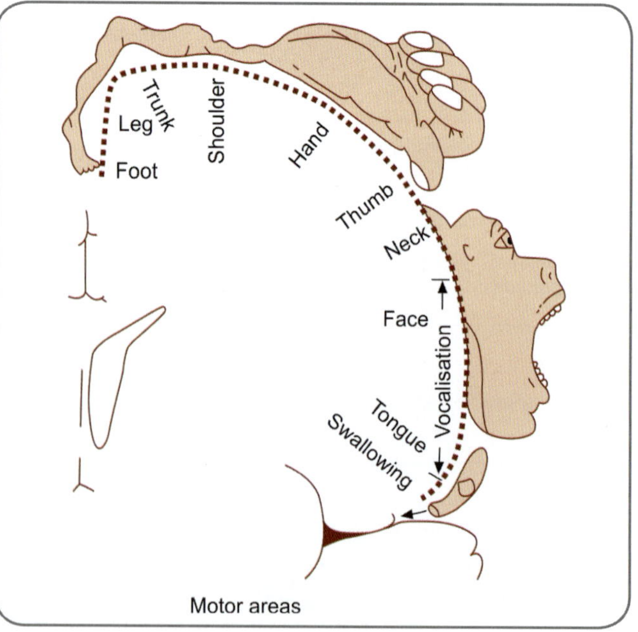

Fig. 23.13: The motor homunculus

Stimulation of the premotor area results in movements, but these are somewhat more intricate than those produced by stimulation of the motor area.

The premotor area appears to be responsible for programming the intended movements and control of movements in progress. It is divisible into a dorsal and a ventral area. The dorsal area is concerned with movements initiated by the individual, while the ventral area is concerned with control of movements that take place in response to external stimulation.

Motor Speech Area

The motor speech area of Broca lies in the inferior frontal gyrus (Brodmann areas 44 and 45, Figs 23.10A and 23.11). Injury to this region results in inability to speak (*aphasia*), even though the muscles concerned are not paralysed. These effects occur only if damage occurs in the left hemisphere in right-handed persons and in the right hemisphere in left-handed persons. In other words, motor control of speech is confined to one hemisphere, which controls the dominant upper limb.

Apart from the motor speech area of Broca, there are two other areas concerned with control of speech. One of these is located in the temporal and parietal lobes (*sensory speech area of Wernicke*), while the other is located in the *supplementary motor area (MII)*.

Frontal Eye Field

The frontal eye field lies in the middle frontal gyrus anterior to the precentral gyrus (Fig. 23.11). It includes parts of areas 6, 8, and 9. Stimulation of this area causes both eyes to move to the opposite side. These are called *conjugate movements.* Movements of the head and dilatation of the pupil may also occur. This area is connected to the cortex of the occipital lobe that is concerned with vision. It is also connected to the thalamus (medial dorsal nucleus).

Note: The frontal eye field and the motor speech area (of Broca) are parts of the premotor area.

Sensory Area

The sensory area is located in the postcentral gyrus and is called the *first somatosensory area (SI)* (Fig. 23.11). It corresponds to areas 3, 1, 2 of Brodmann. It also extends onto the medial surface of the hemisphere where it lies in the posterior part of the paracentral lobule. Responses can be recorded from the sensory area when individual parts of the body are stimulated.

It receives projections from the ventral posteromedial and ventral posterolateral nuclei of the thalamus, conveying impulses received through the medial, spinal, and trigeminal lemnisci.

A definite representation of various parts of the body can be mapped out in the sensory area. It corresponds

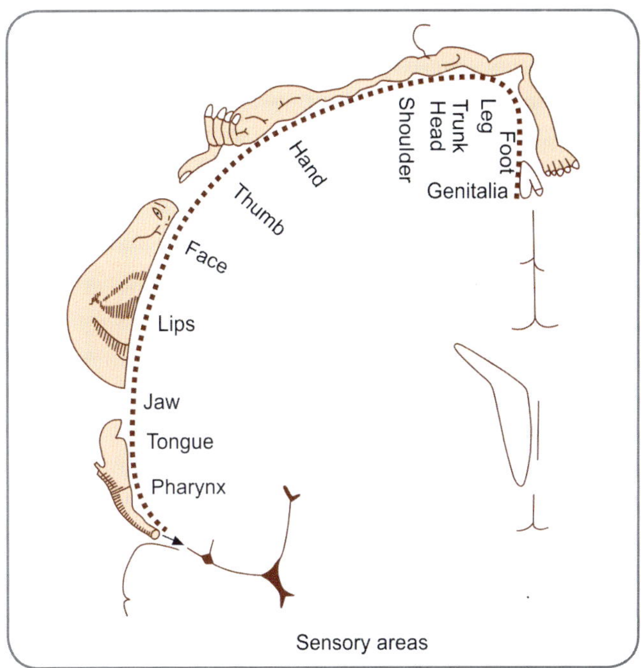

Fig. 23.14: The sensory homunculus

to that in the motor area, in that the body is represented upside down. The area of cortex that receives sensations from a particular part of the body is not proportional to the size of that part but rather to the complexity of sensations received from it. Thus, the digits, the lips, and the tongue have a disproportionately large representation (Fig. 23.14).

It has also been shown that different sensations may be represented in different parts within the area. Unit recordings show that area 2 is concerned mainly with proprioceptive impulses, while area 3 responds only to cutaneous stimuli.

A second area predominantly somatosensory in function (*second somatosensory area or SII*) has been described in relation to the superior lip of the posterior ramus of the lateral sulcus (Fig. 23.11).

The sensory speech area of Wernicke lies in the posterior part of the superior and middle temporal gyri. It extends into areas 39 and 40 of the parietal lobe. This area is responsible for interpretation of speech.

Parts of the superior parietal lobule (areas 5 and 7) help us to recognise shape, size, and texture of objects.

Like area SI, SII receives fibres from the ventral posterior nucleus of the thalamus. It also receives fibres from SI. Descending fibres from SII reach the spinal cord, the nuclei gracilis and cuneatus, and the main trigeminal nucleus. Neurons in SII respond best to intermittent stimulation, for example, vibration. SII may be responsible for perception of pain and temperature.

A small part of the primary sensory area, probably within area 2, serves as *cortical vestibular area.*

Visual Areas

The areas concerned with vision are located in the occipital lobe, mainly on the medial surface, both above and below the calcarine sulcus (area 17). Area 17 extends into the cuneus and into the lingual gyrus (Fig. 23.10). Posteriorly, it may extend onto the superolateral surface, where it is limited (anteriorly) by the lunate sulcus.

It receives fibres of the optic radiation. It is also called the *striate cortex*.

In addition to the striate cortex, additional areas of cortex responding to visual inputs are described. Area 18 (*peristriate area*) is the second visual area, and area 19 (*parastriate area*) is the third visual area.

Areas 18 and 19 are responsible mainly for interpretation of visual impulses reaching area 17, and they are often described as *psychovisual areas*.

Auditory (Acoustic) Area

The acoustic area or the area for hearing is situated in the temporal lobe. It lies in that part of the superior temporal gyrus, which forms the inferior wall of the posterior ramus of the lateral sulcus. In this location, there are two short oblique gyri called the anterior and posterior *transverse temporal gyri* (areas 41, 42, and 52). The acoustic area lies in the anterior transverse temporal gyrus (area 41) and extends to a small extent onto the surface of the hemisphere in the superior temporal gyrus (areas 41 and 42 in Fig. 23.10).

Efferent fibres arising in the acoustic areas project to the medial geniculate body and to the inferior colliculus and possibly also reach motor nuclei of cranial nerves. Some of these efferents may influence the state of contraction of the stapedius and tensor tympani muscles. The acoustic areas are also connected with other parts of the cerebral cortex.

Essentially, the fibres of each lateral lemniscus are bilateral. Hence, the acoustic areas in each cerebral cortex receive fibres from both the right and left cochleae. The close relationship of the acoustic areas to Wernicke's speech area is to be noted. The association is significant in view of the obvious relationship between hearing and speech.

DIENCEPHALON

The diencephalon is the part of the brain between the cerebrum above and midbrain below. It extends from the interventricular foramen to posterior commissure. The hypothalamic sulcus divides the diencephalon into two parts: dorsal (*pars dorsalis*) and ventral part (*pars ventralis*).

1. *Pars dorsalis* consists of the thalamus, metathalamus, and epithalamus.

Table 23.2: Divisions and subdivisions of the diencephalon	
Divisions	**Subdivisions**
Pars dorsalis • Thalamus (dorsal thalamus) • Metathalamus • Epithalamus	 Medial and lateral geniculate bodies Pineal gland (body), habenular nuclei, stria medullaris thalami, habenular commissure, and posterior commissure
Pars ventralis • Subthalamus (ventral thalamus) • Hypothalamus	Subthalamic nucleus and zona incerta

Note: The medial and lateral geniculate bodies are distinct from the other regions of the thalamus and are grouped together as the metathalamus and are integral parts of the dorsal thalamus.

The subthalamic nucleus is included with the basal ganglia to which it is closely related functionally.

2. *Pars ventralis* consists of the hypothalamus and subthalamus.

The cavity of the diencephalon is the third ventricle. The subdivisions of the diencephalon nuclei to be found in each division are summarized in Table 23.2.

THALAMUS (DORSAL THALAMUS)

Anatomically, the thalamus (or dorsal thalamus) is a large egg-shaped mass of grey matter that lies immediately lateral to the third ventricle (Figs 23.15 and 23.16).

External Features

It has two ends (or poles), anterior and posterior; and four surfaces, superior, inferior, medial, and lateral.

The *anterior end (or pole)* lies just behind the interventricular foramen. The *posterior end (or pole)* is called the *pulvinar*. It lies just above and lateral to the superior colliculus. The pulvinar is separated from the geniculate bodies by the *superior brachium quadrigeminum*.

The *medial surface* forms the greater part of the lateral wall of the third ventricle and is lined by ependyma. The medial surfaces of the two thalami are usually interconnected by a mass of grey matter called the *interthalamic adhesion (connexus)* (Fig. 23.15). Inferiorly, the medial surface is separated from the hypothalamus by the *hypothalamic sulcus*. This sulcus runs from the interventricular foramen to the aqueduct (Fig. 23.15).

The *lateral surface* of the thalamus is related to the internal capsule, which separates it from the lentiform nucleus (Fig. 23.15).

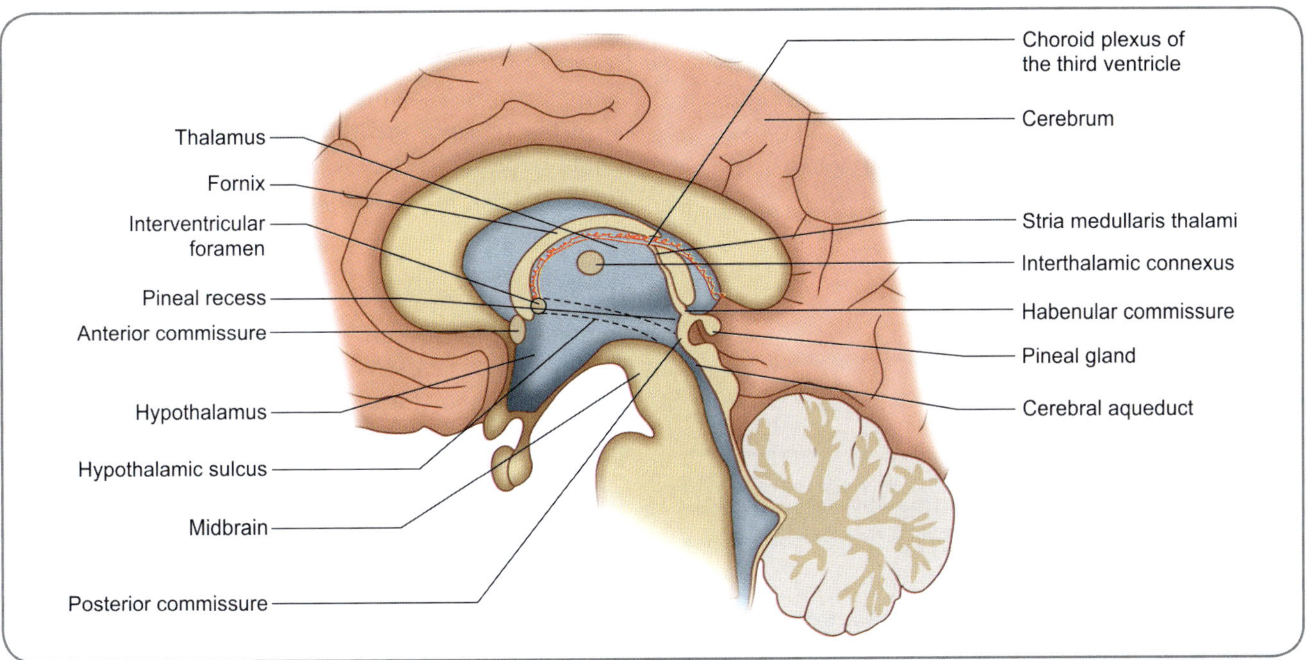

Fig. 23.15: Midsagittal section of the brain showing thalamus, hypothalamus and metathalamus

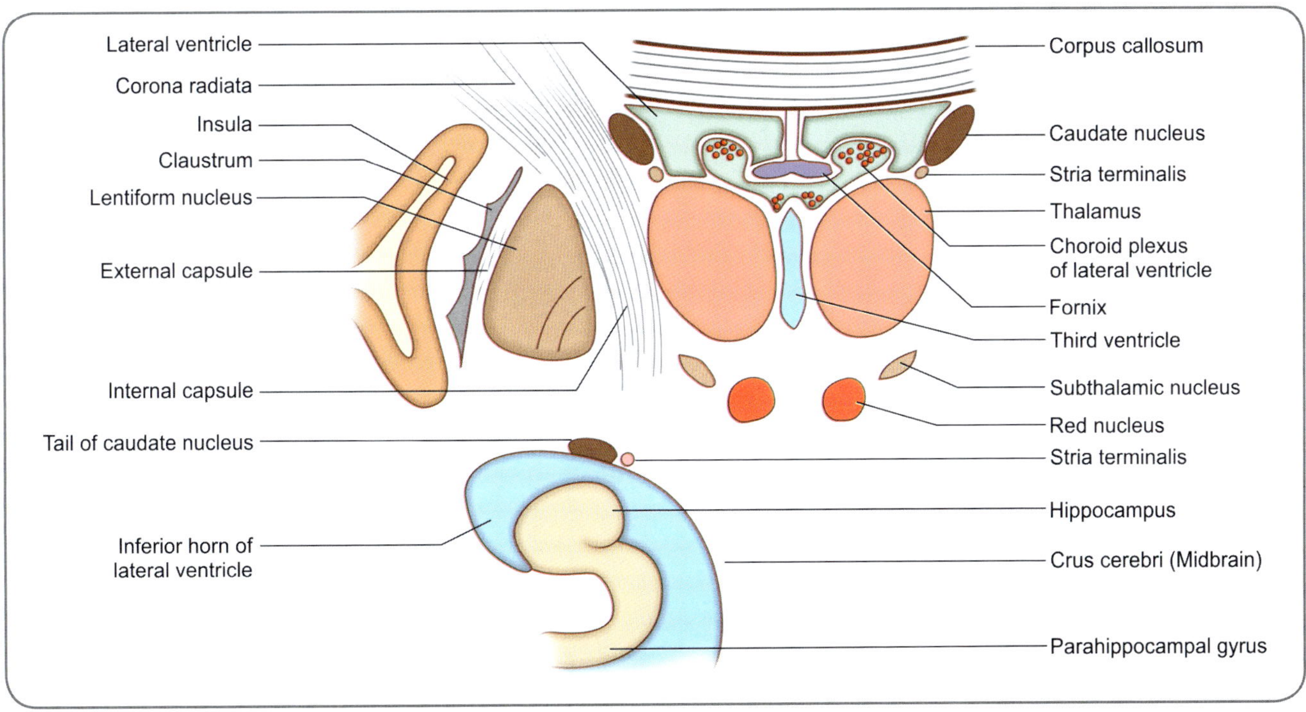

Fig. 23.16: Coronal section through the cerebrum shows structures related to the thalamus

The ***superior (or dorsal) surface*** of the thalamus is related laterally to the caudate nucleus, from which it is separated by a bundle of fibres called the ***stria terminalis***, and by the thalamostriate vein. The thalamus and the caudate nucleus together form the floor of the central part of the lateral ventricle (Fig. 23.16). The medial part of the superior surface of the thalamus is, however, separated from the ventricle by the fornix and by a fold of pia mater called the ***tela choroidea***.

The inferior surface of the thalamus is related to the hypothalamus, anteriorly and the ventral thalamus, posteriorly (Fig. 23.16). The ventral thalamus separates the thalamus from the tegmentum of the midbrain.

At the junction of the medial and superior surfaces of the thalamus, the ependyma of the third ventricle is reflected from the lateral wall to the roof. The line of reflection is marked by a line called the ***taenia thalami***. Underlying it there is a narrow bundle of fibres

413

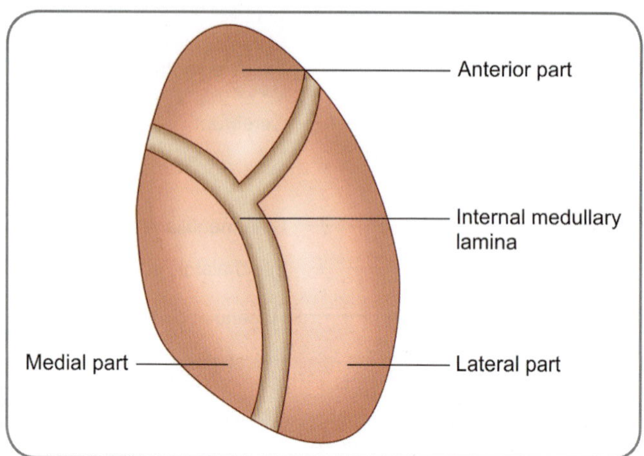

Fig. 23.17: Schematic diagram showing a horizontal section of the thalamus; Note: The Y-shaped internal medullary lamina divides the thalamus into anterior, lateral and medial parts

called the ***stria medullaris thalami*** (not to be confused with the stria medullares present in the floor of the fourth ventricle).

Internal Structure

The thalamus consists mainly of grey matter and only small amount of white matter.

White Matter

The superior surface of thalamus is covered by a thin layer of white matter called the ***stratum zonale*** and its lateral surface, by a similar layer called the ***external medullary lamina.***

Grey Matter

The grey matter of the thalamus is subdivided into three main parts by a Y-shaped sheet of white matter, which is called the ***internal medullary lamina*** (Fig. 23.17). This lamina is placed vertically. It divides the thalamus into lateral, medial, and anterior parts situated between the two limbs of the 'Y'.

Nuclei of Thalamus

A number of nuclei can be distinguished within each of these parts (Figs 23.18A and B and Table 23.3). Only the more important of these are listed below.

Nuclei in the Anterior Part

A number of nuclei can be distinguished, but we shall refer to them collectively as the ***anterior nucleus.***

Table 23.3: Nuclei of thalamus and their functions	
Nucleus	**Functions**
Anterior group	Part of the limbic system concerned with recent memory and emotions
Medial group	Associated with mood and emotional balance
Lateral group	Conveys sensory and motor information and integrates sensory information

Note: Thalamus is a sensory integration and relay station of all the sensory pathways except for the olfactory pathway. The fibres of olfactory pathway are projected directly to the cerebral cortex without being relayed in the thalamus.

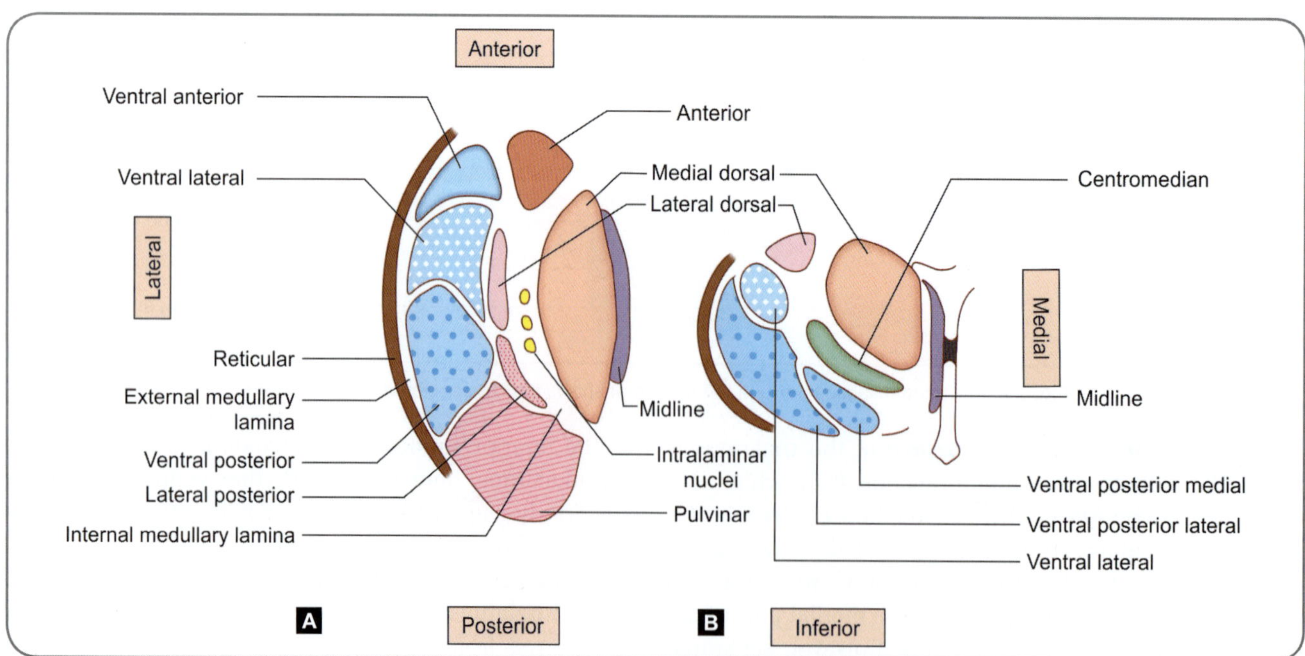

Figs 23.18A and B: Scheme to show the nuclei of the thalamus **A.** Superior aspect **B.** In coronal section

Nuclei in the Medial Part

The largest of these is the **medial dorsal nucleus.** It is divisible into a **magnocellular part** (anteromedial) and a **parvocellular part** (posterolateral).

Nuclei in the Lateral Part

The nuclei in the lateral part can be subdivided into a **ventral group** and a **lateral group.**

The **nuclei in the ventral group** are as follows (in anteroposterior order):

- **Ventral anterior nucleus,**
- **Ventral lateral nucleus** (also called the **ventral intermediate nucleus**),
- **Ventral posterior nucleus**, which is further subdivided into a lateral part, called the **ventral posterolateral nucleus**, and a medial part, called the **ventral posteromedial nucleus** (Fig. 23.18B).

The **nuclei of the lateral group** are as follows (in anteroposterior order):

- **Lateral dorsal nucleus** (or **dorsolateral nucleus**),
- **Lateral posterior nucleus,**
- **Pulvinar.**

Other Thalamic Nuclei

In addition to the above, the thalamus contains the following nuclei:

- The **intralaminar nuclei** are embedded within the internal medullary lamina. There are several nuclei in this group. The most important of these is the **centromedian nucleus** (Fig. 23.18B).
- The **midline nuclei** consist of scattered cells that lie between the medial part of the thalamus and the ependyma of the third ventricle. Several nuclei are recognized.
- The **medial and lateral geniculate bodies** (traditionally described under metathalamus) are now included as part of the thalamus.

The reticular nucleus, earlier described as part of the dorsal thalamus, is now regarded as part of the ventral thalamus.

Connections of the Thalamus—An Overview

See Figure 23.19.

> ### Clinical Correlation
>
> **Thalamic Syndrome**
>
> This occurs due to vascular lesion of the thalamic branch of posterior cerebral artery.
>
> **Characteristic Features**
>
> - Threshold for appreciation of touch pain or temperature is lowered.
> - Sensation that is normal may appear to be exaggerated or unpleasant.
> - There may be spontaneous pain.
> - Emotions may be abnormal.

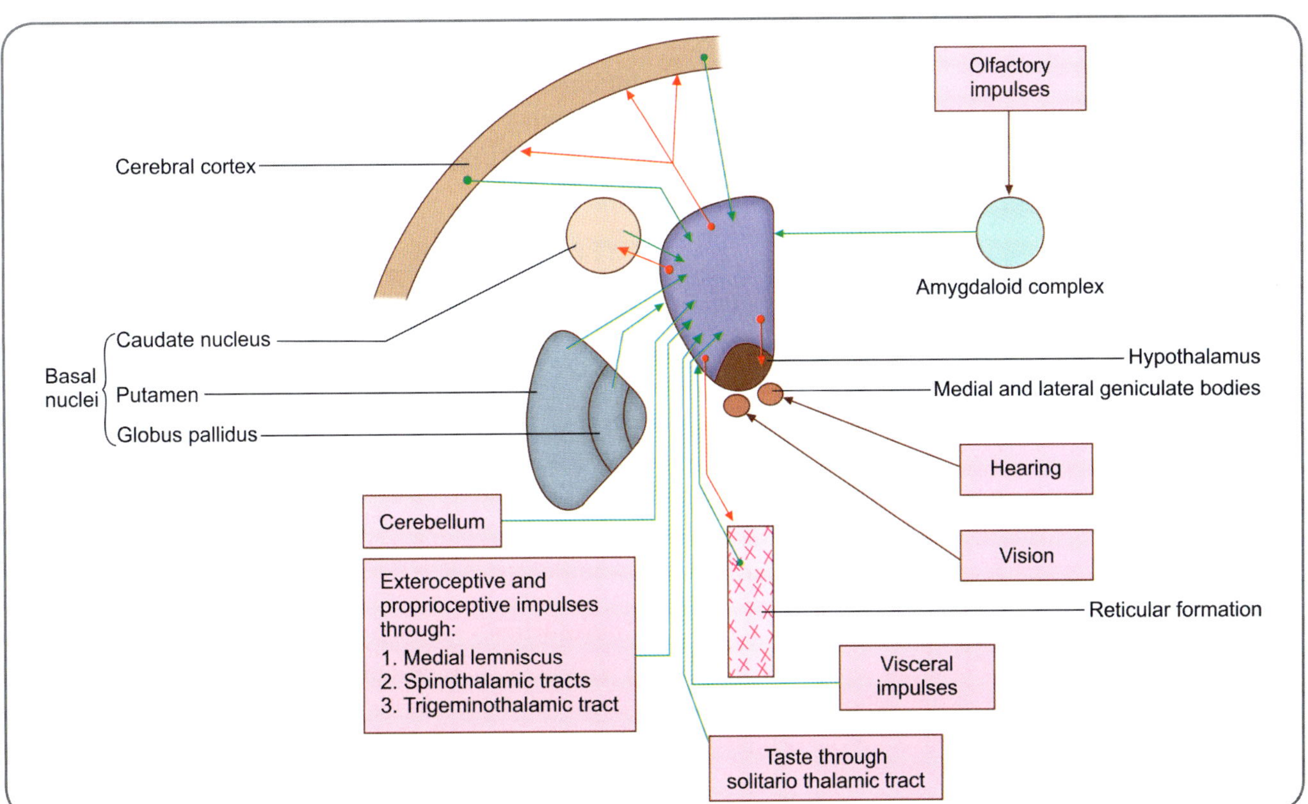

Fig. 23.19: Scheme to show the main connections of thalamus (as a whole)

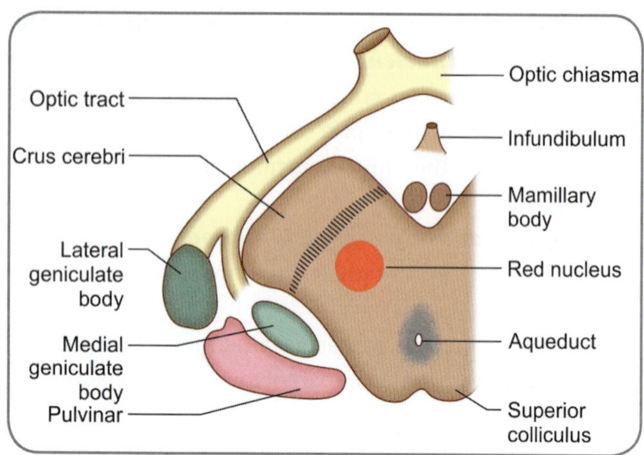

Fig. 23.20: Diagram to show the location of the medial and lateral geniculate bodies

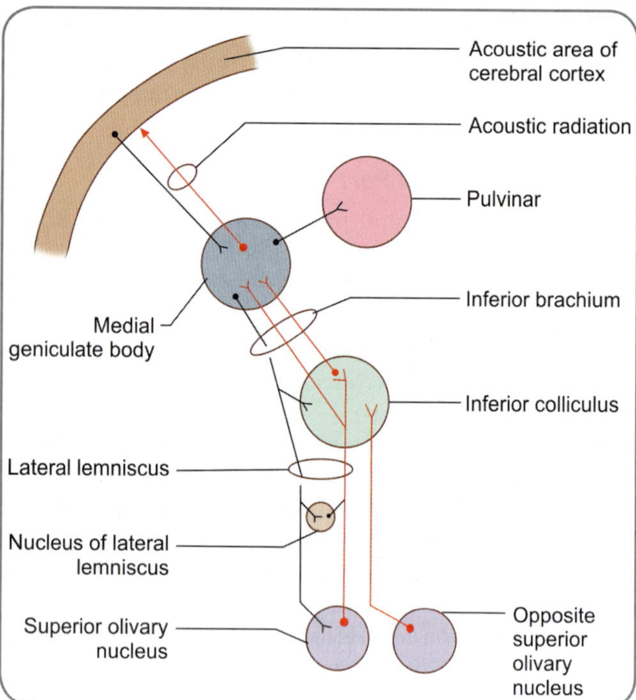

Fig. 23.21: Connections of the medial geniculate body

METATHALAMUS

The metathalamus consists of the medial and lateral geniculate bodies. The medial and lateral geniculate bodies are small oval collections of grey matter situated below the posterior part of the thalamus, lateral to the colliculi of the midbrain (Fig. 23.20). Each mass of grey matter is bent on itself, hence the term **geniculate**. Traditionally, the geniculate bodies have been grouped together under the heading **meta-thalamus**, but because of functional relationships, they are now included in the dorsal thalamus.

Medial Geniculate Body

The medial geniculate body is a relay station on the auditory pathway. Medial, ventral, and dorsal nuclei are described within it.

Connections

The medial geniculate body receives fibres of the lateral lemniscus either directly or after relay in the inferior colliculus (Fig. 23.21). These fibres pass through the brachium of the inferior colliculus. Fibres arising in the medial geniculate body constitute the acoustic radiation. The acoustic radiation passes through the sublentiform part of the internal capsule to reach the acoustic areas of the cerebral cortex.

Each medial geniculate body receives impulses from the cochleae of both sides. It also receives fibres from the auditory area of the cerebral cortex. These fibres form part of the **descending acoustic pathway.**

Different neurons in the ventral nucleus of the medial geniculate body respond to different frequencies of sound (tonotopic organization). The ventral nucleus projects to the primary auditory cortex. The neurons in the dorsal nucleus do not show tonotopic organization. They project to auditory areas around the primary auditory area.

Lateral Geniculate Body

The lateral geniculate body is a relay station on the visual pathway. It is situated on the inferior surface of the pulvinar, anterolateral to the medial geniculate body.

Both anatomical and physiological evidence shows that information from the right and left retinae is brought into alignment in the lateral geniculate body. Integration of information from the two retinae takes place only in the cerebral cortex.

HYPOTHALAMUS

The hypothalamus is a part of the diencephalon. As its name implies, it lies below the thalamus and seperated from it by the hypothalamic sulcus. Most part of hypothalamus is hidden. However, some parts of the hypothalamus can be seen on the external (ventral) surface of the brain. These visible parts of hypothalamus are located in the interpeduncular fossa (Fig. 23.22). On the medial side, it forms the wall of the third ventricle below the level of the hypothalamic sulcus (Fig. 24.5).

Boundaries

- **Laterally,** it is in contact with the internal capsule, and (in the posterior part) with the ventral thalamus (subthalamus).
- **Posteriorly,** the hypothalamus merges with the ventral thalamus and through it, with the tegmentum of the midbrain.

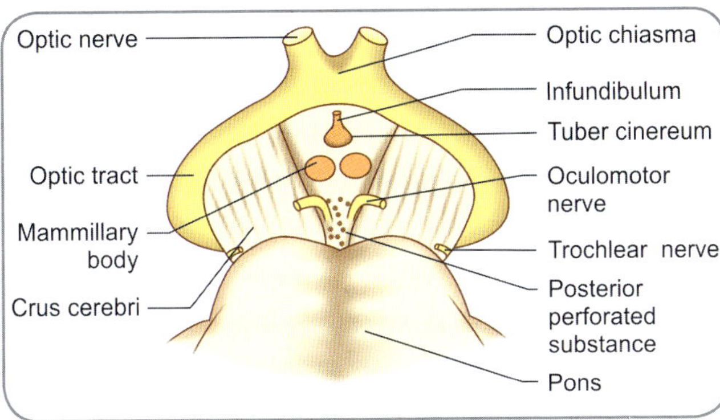

Fig. 23.22: Interpeduncular fossa

❑ *Anteriorly,* it extends up to the lamina terminalis, and merges with certain olfactory structures in the region of the anterior perforated substance.

❑ *Inferiorly,* the hypothalamus is related to structures in the floor of the third ventricle. These are the tuber cinereum, the infundibulum, and the mamillary bodies, which are considered as parts of the hypothalamus.

Subdivisions

For convenience of description the hypothalamus may be subdivided, roughly, into a number of regions. Some authorities divide it (from medial to lateral side) into three *zones,* which are as follows:

1. *Periventricular,*
2. *Intermediate* and
3. *Lateral*.

The periventricular and intermediate zones are often described collectively as the *medial zone* and we will follow this practice here. The column of the fornix lies between the medial and lateral zones. The mamillothalamic tract and the fasciculus retroflexus also lie in this plane.

The hypothalamus is also subdivided anteroposteriorly into four *regions*. These are as follows (Fig. 23.23):

1. The *preoptic region* adjoins the lamina terminalis.
2. The *supraoptic (or chiasmatic) region* lies above the optic chiasma.
3. The *tuberal (or infundibulotuberal) region* includes the infundibulum, the tuber cinereum, and the region above it.
4. The *mamillary (or posterior) region* consists of the mamillary body and the region above it.

The preoptic region differs from the rest of the hypothalamus in being a derivative of the telencephalon. The lamina terminalis also belongs to the telencephalon.

Hypothalamic Nuclei

The entire hypothalamus contains scattered neurons within which some aggregations can be recognised. These aggregations, termed the hypothalamic nuclei, are given in Table 23.4.

Connections of Hypothalamus

See Figure 23.24.

| Table 23.4: Hypothalamic regions and nuclei in them ||
Region	Nucleus
Preoptic region	Preoptic nucleus
Supraoptic region	• Supraoptic nucleus • Anterior nucleus • Paraventricular nucleus
Tuberal region	• Arcuate (infundibular) nucleus • Ventromedial nucleus • Dorsomedial nucleus
Mammillary region	• Posterior nucleus • Mammillary nuclei

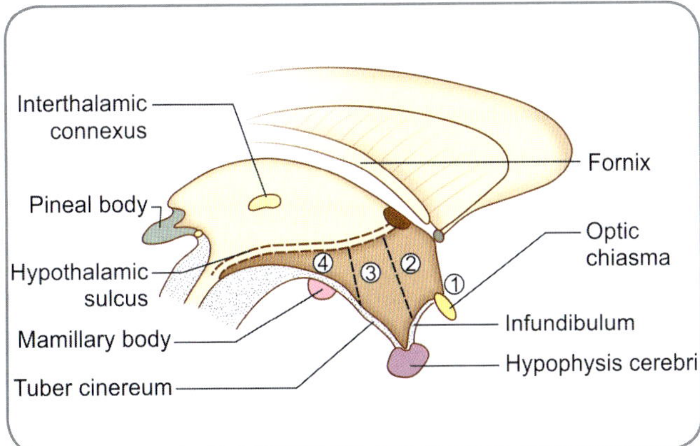

Fig. 23.23: Anteroposterior subdivisions of hypothalamus

Key: 1. Preoptic **2.** Supraoptic **3.** Infundibulotuberal **4.** Mammillary

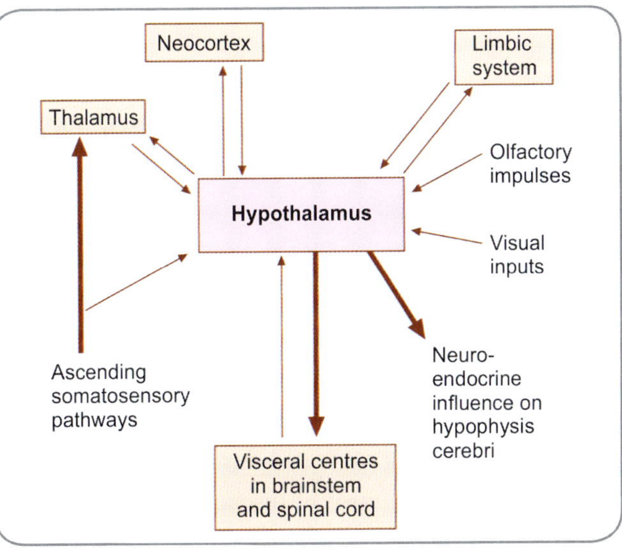

Fig. 23.24: Connections of hypothalamus

Functions

The hypothalamus plays an important role in the control of many functions that are vital for the survival of an animal. In exercising such control, the hypothalamus acts in close coordination with higher centres including the limbic system and the prefrontal cortex and with autonomic centres in the brainstem and spinal cord. The main functions attributed to the hypothalamus are as follows:

- ❑ ***Regulation of eating and drinking behavior:*** The hypothalamus is responsible for feelings of hunger and of satiety, and this determines whether the animal will accept or refuse food. It has been observed that stimulation of the lateral zone of the hypothalamus stimulates hunger, while stimulation of the medial zone produces satiety. The lateral zone is also responsible for thirst and drinking. Based on such studies, a ***feeding centre*** has been described in the lateral hypothalamic nucleus and a ***satiety centre,*** in the ventromedial nucleus.

  Correlate this with the fact that some neurons in the hypothalamus are sensitive to osmolarity (preoptic nucleus), glucose content, and fatty acid content of blood.

- ❑ ***Regulation of sexual activity and reproduction:*** The hypothalamus controls sexual activity, both in the male and female. It also exerts an effect on gametogenesis, ovarian and uterine cycles, and development of secondary sexual characters. These effects are produced by influencing the secretion of gonadotropic hormones by the hypophysis cerebri.

- ❑ ***Control of autonomic activity:*** The hypothalamus exerts an important influence on the activity of the autonomic nervous system and, thus, has considerable effect on cardiovascular, respiratory, and alimentary functions. Sympathetic activity is said to be controlled predominantly by caudal parts of the hypothalamus and parasympathetic activity, by cranial parts, but there is considerable overlap between the regions concerned.

- ❑ ***Emotional behavior:*** The hypothalamus has an important influence on emotions like fear, anger, and pleasure. Stimulation of some areas of the hypothalamus produces sensations of pleasure, while stimulation of other regions produces pain or other unpleasant effects.

- ❑ ***Control of endocrine activity:*** The influence of the hypothalamus in the production of hormones by the pars anterior of the hypophysis cerebri and the elaboration of oxytocin and the antidiuretic hormone by the hypothalamus itself have been described above. Through control of the adenohypophysis, the hypothalamus indirectly influences the thyroid gland, the adrenal cortex, and the gonads.

- ❑ ***Response to stress:*** Through control over the autonomic nervous system and hormones, the hypothalamus plays a complex role in the way a person responds to stress.

- ❑ ***Temperature regulation:*** Some neurons in the preoptic nucleus of the hypothalamus act as a thermostat to control body temperature. When body temperature rises or falls, appropriate mechanisms are brought into play to bring the temperature back to normal.

- ❑ ***Biological clock:*** Several functions of the body show a cyclic variation in activity over the twenty four hours of a day. The most conspicuous of these is the cycle of sleep and waking. Such cycle (called ***circadian rhythms***) are believed to be controlled by the hypothalamus, which is said to function as a biological clock. The suprachiasmatic nucleus is believed to play an important role in this regard. Lesions of the hypothalamus disturb the sleep-waking cycle.

Clinical Correlation

Pressure on the hypothalamus can give rise to the ***hypothalamic syndrome*** that is marked by diabetes insipidus.

EPITHALAMUS

The epithalamus lies in relation to the posterior part of the roof of the third ventricle and in the adjoining part of its lateral wall. The structures included in the epithalamus are as follows:

- ❑ Pineal body,
- ❑ Paraventricular nuclei—anterior and posterior (do not confuse these with the paraventricular nucleus in the hypothalamus),
- ❑ Habenular nuclei: medial and lateral,
- ❑ Stria medullaris thalami and habenular commissure and
- ❑ Posterior commissure.

For details see Authors Textbook of Neuroanatomy.

WHITE MATTER OF CEREBRAL HEMISPHERES

The interior of each cerebral hemisphere consists of a core of white matter, which is composed of myelinated nerve fibres. The fibres of white matter are classified into three types (Figs 23.25 and 23.26):

1. Association fibres,
2. Commissural fibres and
3. Projection fibres.

ASSOCIATION FIBRES

The association fibres connect different parts of the cerebral cortex of the same hemisphere to each other. These are of two types:

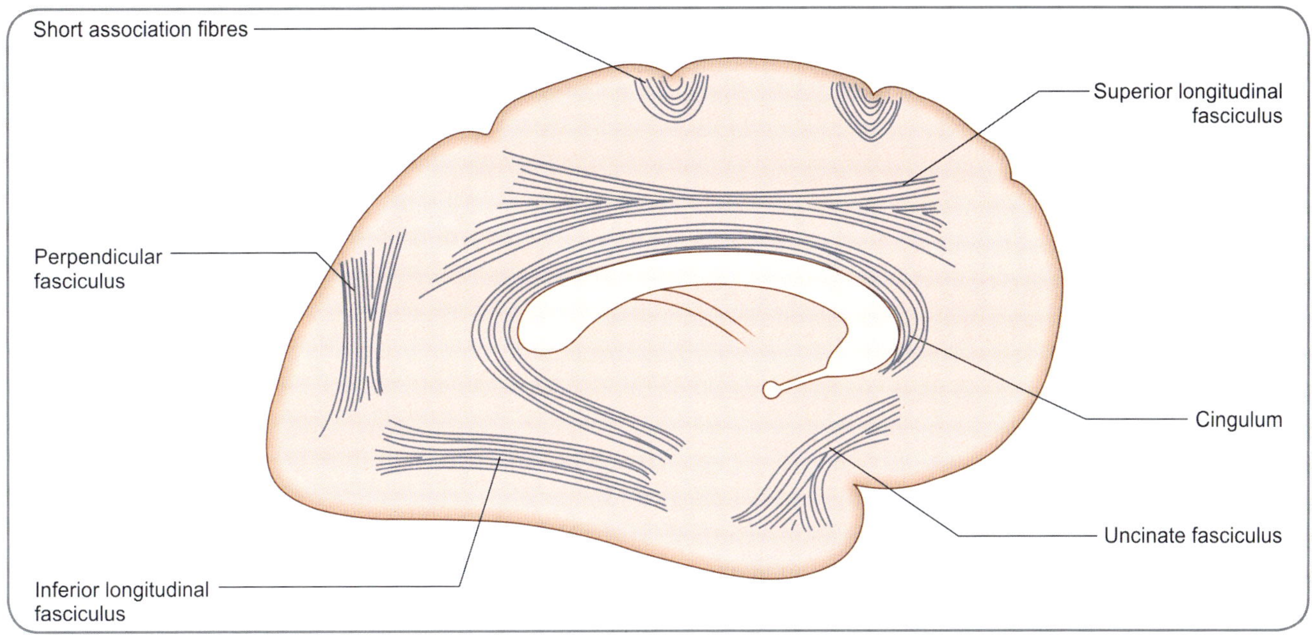

Fig. 23.25: Schematic diagram to show bundles of fibres present within the cerebral hemisphere

1. **Short association fibres,** which connect the adjacent gyri to each other
2. **Long association fibres,** which connect the gyri located at a distance from each other.

Examples of Long Association Fibres

❏ The **cingulum** (girdle-shaped) is located within the cingulate gyrus. It extends from the paraterminal gyrus to the uncus. The cingulum is part of the Papez circuit of the limbic system.

❏ The **uncinate fasciculus** (**arcuate fasciculus**) is a curved fibre bundle. It connects the inferior frontal gyrus and the orbital gyri of the frontal lobe to the cortex of the temporal lobe. Thus, it connects the sensory and motor speech areas to each other in the dominant hemisphere.

❏ The **superior longitudinal fasciculus** is a long bundle that begins in the frontal lobe and arches back via the parietal lobe to the occipital lobe, from where it turns into the temporal lobe. Thus, it connects the occipital lobe to the frontal eye field.

❏ The **inferior longitudinal fasciculus** connects the occipital lobe to the temporal lobe.

COMMISSURAL FIBRES

The commissural fibres cross the midline and connect the identical parts of the two hemispheres.

Note: All fibres crossing from one side of the brain or spinal cord to the opposite side are not commissural fibres. When fibres originating in a mass of grey matter in one half of the CNS end in some other mass of grey matter in the opposite half, they are referred to as **decussating fibres,** and the sites where such crossings take place are referred to as **decussations.**

Examples of Commissural Fibres

❏ The **anterior commissure** connects the right and left temporal lobes. It is of the shape of a **cupid's bow**. It crosses the midline in the upper part of the lamina terminalis anterior to the columns of fornix. Its fibres divide into anterior and posterior bundles. The anterior

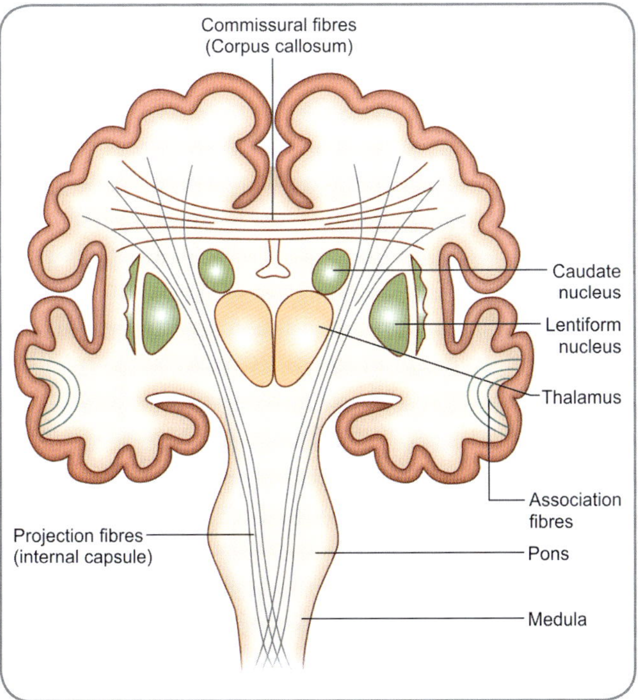

Fig. 23.26: Coronal section of the brain showing association, commissural and projection fibres

bundle connects the anterior perforated substance and the olfactory tracts of the two sides. The posterior bundle, at first, passes through the head of the caudate nucleus and then inclines posteriorly in the lentiform nucleus to enter the middle and inferior gyri of the temporal lobe.

- The **posterior and habenular commissures** are part of epithalamus and are located in the posterior part of the roof of third ventricle.
- The **hippocampal commissure** connects the hippo-campus of the two sides to each other.
- The corpus callosum is the largest commissure of the brain.

PROJECTION FIBRES

The projection fibres connect the cerebral cortex to other regions of central nervous system below it by corticopetal or ascending and corticofugal or descending fibres.

Examples of Projection Fibres

- The **corona radiata** (Fig. 23.16) is a mass of white matter composed of the projection fibres, which converge from the cerebral cortex to the internal capsule and fan out from the internal capsule towards the cortex.
- The **internal capsule** transmits the projection fibres like corticospinal, corticonuclear, and corticopontine fibres. These fibres arise in the cerebral cortex and terminate on the lower neurons (like anterior horn cells, cranial nerve nuclei in brainstem and pontine nuclei). The internal capsule also gives passage to thalamic radiations (comprising to and fro connections between cerebral cortex and thalamic nuclei).
- The **fornix** is composed of projection fibres, which take origin from the hippocampus. The fibres in the fornix are connected to the neurons of the mamillary body.

Corpus Callosum

This is located in the floor of the median longitudinal fissure (Fig. 23.27).

Parts

The corpus callosum consists of four parts. Its anterior end is called the genu, the central part is the **trunk** and the posterior bulbous part forms the **splenium**. The fourth part is the rostrum, which is the prolongation from the genu to the upper end of lamina terminalis (Fig. 23.28).

Relations

- The superior aspect of the corpus callosum is covered with **indusium griseum**, in which medial and lateral longitudinal striae are embedded. The indusium griseum is the rudimentary gray matter of dorsal hippocampos

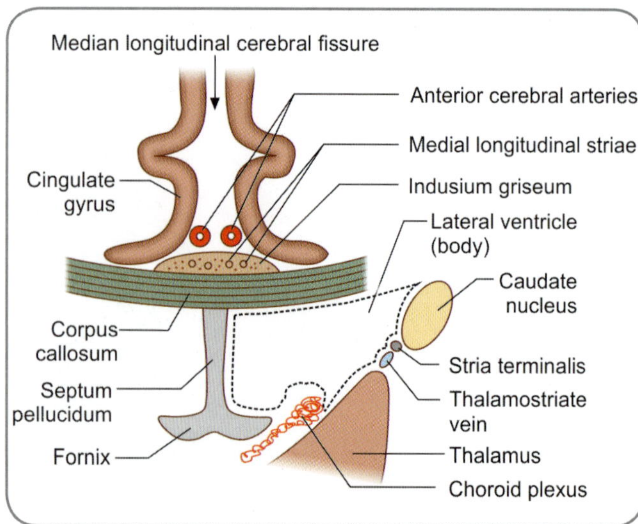

Fig. 23.27: Schematic diagram to show relations of the corpus callosum

- **Transverse fissure** is located between the splenium and the superior colliculi and pineal gland. It gives passage to the tela choroidea of third and lateral ventricles. The posterior choroidal arteries enter the fissure and the internal cerebral veins leave it and unite to form the great cerebral vein beneath the splenium
- The **anterior and superior aspects** of the corpus callosum are in close relation to the anterior cerebral vessels
- The **inferior aspect** of the corpus callosum gives attachment to the septum pellucidum, anteriorly and the fornix, posteriorly
- The **rostrum and genu** form the boundaries of the anterior horn, and the trunk forms the roof of the central part of the lateral ventricle.

Connections

The fibres of the corpus callosum interconnect the corresponding parts of the right and left hemispheres.

- The fibres passing through the rostrum connect the orbital surfaces of the frontal lobes
- The fibres passing through the genu interconnect the two frontal lobes by means of a fork-like bundle of fibres called **forceps minor**

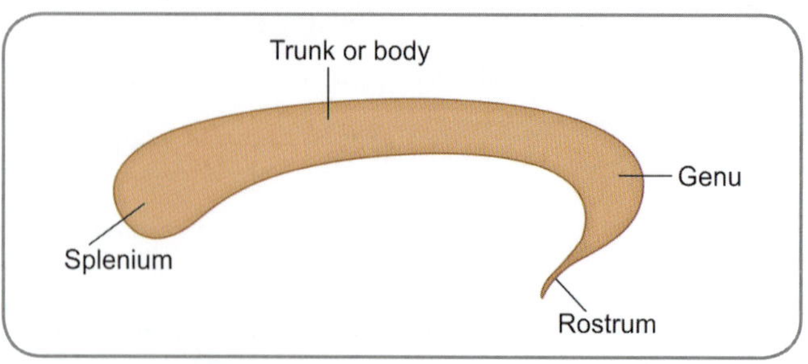

Fig. 23.28: Parts of corpus callosum

❑ The fibres passing through the splenium interconnect the occipital lobes by means of a fork like bundle of fibres called *forceps major*. The forceps major bulges into the medial wall of the posterior horn of lateral ventricle to give rise to an elevation called the *bulb*.

❑ A large number of fibres from the trunk run transversely to intersect with the fibres of the corona radiata. Some fibres of the trunk and adjacent splenium, which do not intersect with corona radiata, are known as the *tapetum*. The tapetum is closely related to the inferior horn and posterior horn of lateral ventricle.

Note: The fibres passing through the corpus callosum generally interconnect corresponding regions of the entire neocortex of the right and left sides. However, some important exceptions are now known.

❑ The greater parts of the visual areas are not interconnected. Only those parts of visual areas that receive impulses from a narrow strip along the vertical meridian of the retina are interconnected. The band of cortex concerned lies at the junction of areas 17 and 18. Similar bands are also present in relation to other visual areas

❑ The parts of the sensorimotor areas (SI and SII) concerned with the hands and feet are not interconnected.

Note: The various regions to which most parts of the cerebral cortex are connected.

Arterial Supply

The rostrum, genu and body of corpus callosum receive bilateral branches from anterior cerebral artery. The splenium receives branches from the posterior cerebral arteries.

⚕ Clinical Correlation

The corpus callosum can be congenitally absent. As the two cerebral hemispheres are not connected, there is a split brain. If one hand is trained to perform an act the other hand may not be able to do so.

INTERNAL CAPSULE

A large number of nerve fibres project from the cerebral cortex to interconnect with subcortical centres in the brainstem and spinal cord and with the thalamus. This compact bundle of fibres is collectively called the *internal capsule*.

Relations

These fibres fan out cranially to form corona radiata and condense caudally to form the crus cerebri of the midbrain. Most of these fibres pass through the interval between the thalamus and caudate nucleus medially and

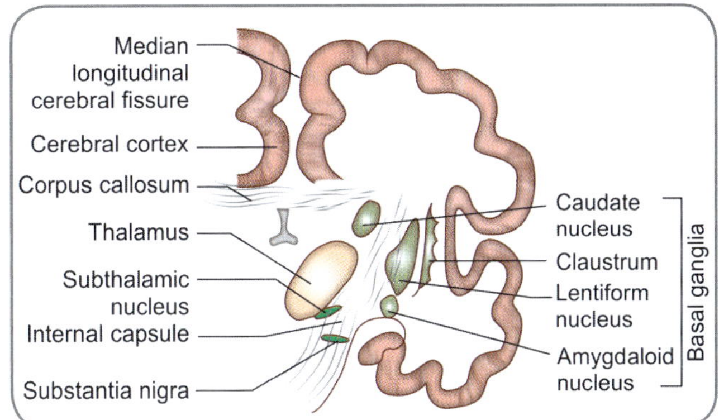

Fig. 23.29: Coronal section through a cerebral hemisphere to show the location of internal capsule

the lentiform nucleus laterally (Fig. 23.29). This region is called the *internal capsule.* Above, the internal capsule is continuous with the corona radiata and below, with the crus cerebri (of the midbrain).

Parts

The internal capsule may be divided into the following parts (Fig. 23.30):

❑ *Anterior limb:* The anterior limb lies between the caudate nucleus medially and the anterior part of the lentiform nucleus laterally.

❑ *Posterior limb:* The posterior limb lies between the thalamus medially and the posterior part of the lentiform nucleus laterally.

❑ *Genu:* In transverse section through the cerebral hemisphere, the anterior and posterior limbs of the internal capsule meet at an angle open outwards. This angle is called the *genu (genu = bend)*.

❑ *Retrolentiform part:* Some fibres of the internal capsule lie behind the posterior end of the lentiform nucleus. They constitute its retrolentiform part.

❑ *Sublentiform part:* Some other fibres pass below the lentiform nucleus (and not medial to it). These fibres constitute the sublentiform part of the internal capsule.

Fibres

The various parts of the internal capsule consist of large number of fibres. The fibres passing through the capsule may be ascending (to the cerebral cortex) or descending (from the cortex). The arrangement of fibres is easily remembered, if it is realized that any group of fibres within the capsule *takes the most direct path* to its destination. Thus, fibres to and from the anterior part of the frontal lobe pass through the anterior limb of the internal capsule. The fibres to and from the posterior part of the frontal lobe and from the greater part of the parietal lobe occupy the genu and posterior limb of the capsule. Fibres to and from the temporal lobe occupy the sublentiform part, while those to

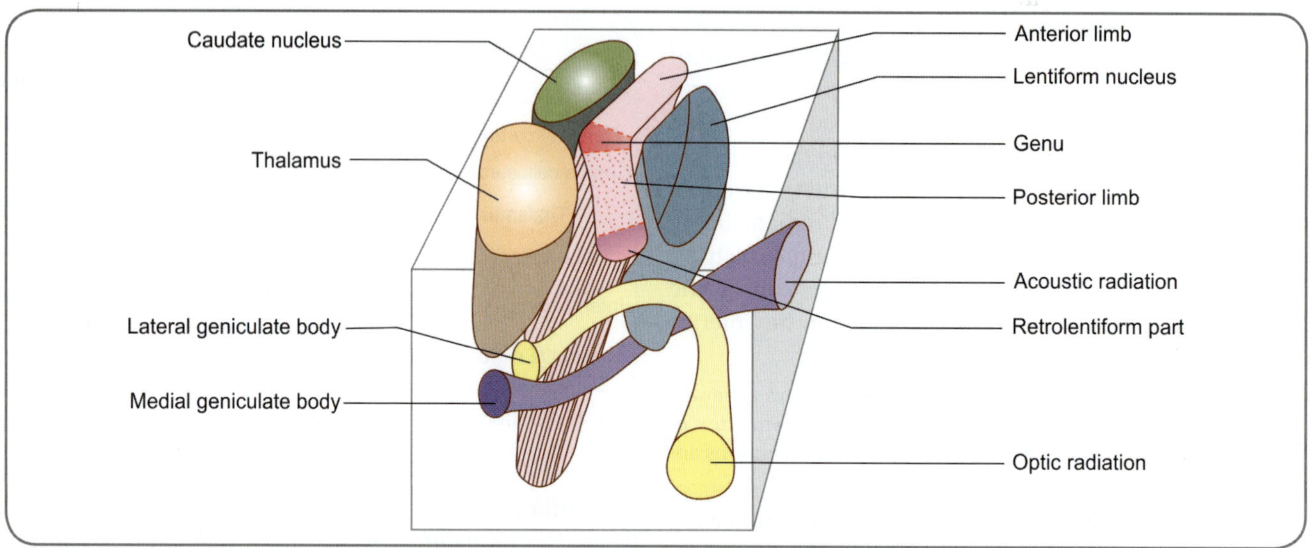

Fig. 23.30: Scheme to show the subdivisions of internal capsule

and from the occipital lobe pass through the retrolentiform part. Some fibres from the lowest parts of the parietal lobe accompany the temporal fibres through the sublentiform part.

Ascending Fibres (Sensory Fibres)

These are *predominantly thalamocortical fibres,* which go from the thalamus to all parts of the cerebral cortex (Figs 23.31A and B).

- Anterior thalamic radiation
- Superior thalamic radiation
- Posterior thalamic radiation
- Inferior thalamic radiation

Descending Fibres (Motor Fibres)

The descending fibres are the projection fibres and are also called centrifugal fibres (Fig. 23.32).

- *Corticopontine fibres:* They originate from all four lobes of cerebral cortex and are named according to the lobe from which they arise:
  - Frontopontine fibres
  - Parietopontine fibres
  - Temporopontine fibres
  - Occipitopontine fibres
- *Pyramidal fibres* (corticospinal and corticonuclear fibres):

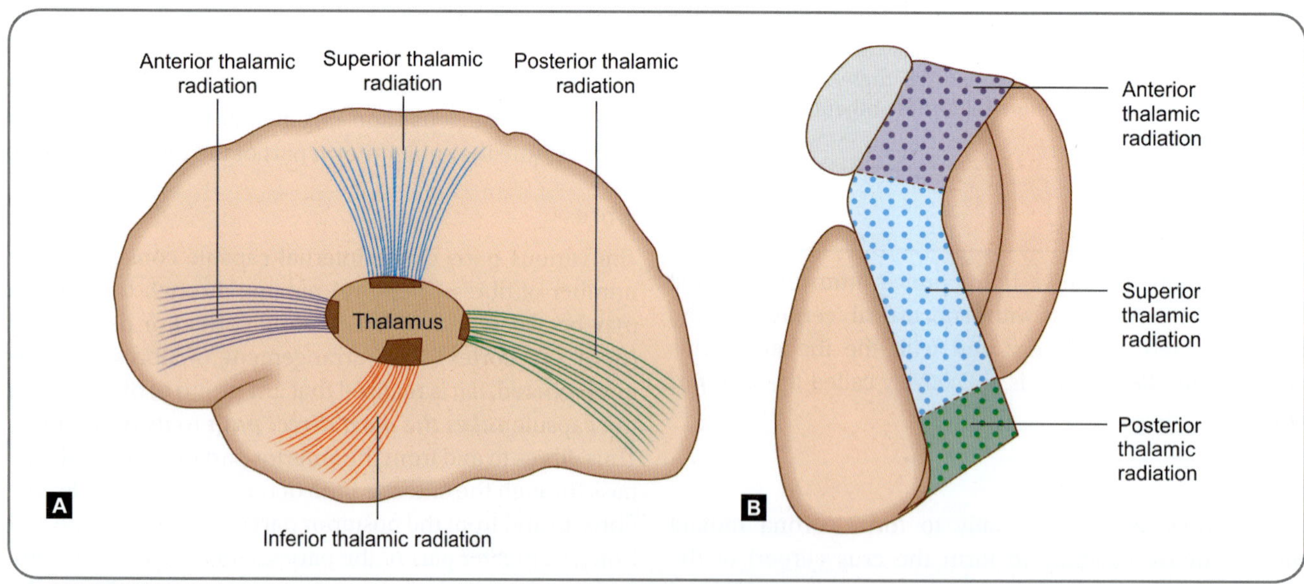

Figs 23.31A and B: Schematic diagram to show ascending fibres passing through the internal capsule **A.** Thalamic radiation **B.** Location of diferent groups of thalamic radiation in different parts of the internal capsule

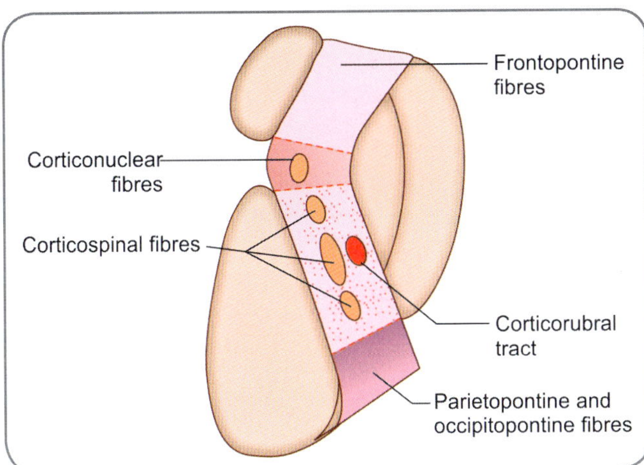

Fig. 23.32: Schematic diagram to show descending fibres passing through the internal capsule

- ○ Corticonuclear fibres
- ○ Corticospinal fibres
- ❑ ***Corticothalamic fibres***
- ❑ ***Extrapyramidal fibres:***
  - ○ Corticostriate fibres
  - ○ Corticorubral fibres
  - ○ Corticoreticular fibres

Arterial Supply

The main arteries supplying the internal capsule are the medial and lateral striate branches of the middle cerebral artery, the recurrent branch of the anterior cerebral, and the anterior choroidal artery. The internal capsule may also receive direct branches from the internal carotid artery and branches from the posterior communicating artery (Fig. 23.33).

- ❑ The ***upper parts*** of the anterior limb, the genu and the posterior limb are supplied by the medial and lateral striate branches of the middle cerebral artery.

One of the lateral striate branches is larger and more frequently ruptured. It is often called ***Charcot's artery of cerebral haemorrhage***. It enters through the anterior perforated substance and supplies the posterior limb of the internal capsule.

- ❑ The ***lower parts*** of these regions are supplied as follows:
  - ○ The lower part of the anterior limb is supplied by the medial striate artery (also called ***recurrent artery of Heubner***), branch of the anterior cerebral artery.
  - ○ The lower part of the genu is supplied by direct branches from the internal carotid and from the posterior communicating artery.
  - ○ The lower part of the posterior limb is supplied by the anterior choroidal artery and striate branch of posterior cerebral artery.
- ❑ The retrolentiform part of the internal capsule is supplied by the anterior choroidal artery.
- ❑ The sublentiform part is probably supplied by the anterior choroidal artery.

✄ Clinical Correlation

The arterial supply of internal capsule is of great clinical significance. Any occlusion of blood vessel supplying the internal capsule can lead to a potentially serious clinical outcome.

A lesion in the internal capsule is most likely to result from thrombosis or rupture of one of the arteries supplying the capsule. The artery most often involved is Charcot's artery of cerebral hemorrhage.

Thrombosis in an artery supplying the internal capsule (cerebral thrombosis) leads to a stroke that results in hemiplegia. The opposite side of the body is affected. As the tracts passing through the internal capsule are closely packed, even a small lesion can cause extensive paralysis. Sensations can also be lost. Reflexes are exaggerated as in a typical upper motor neuron paralysis.

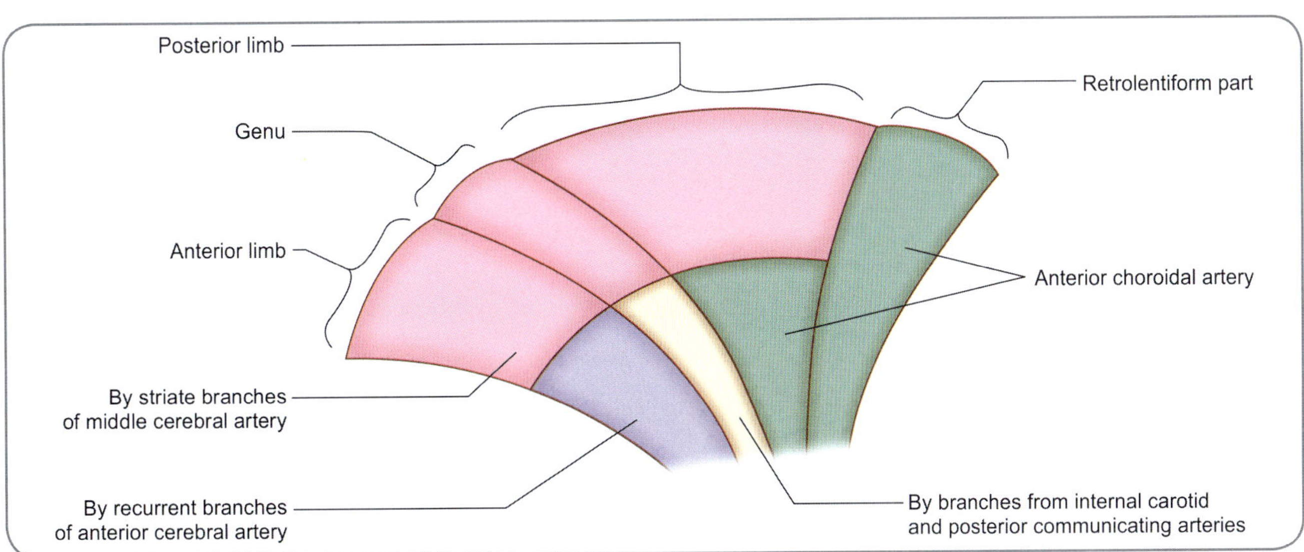

Fig. 23.33: Scheme to show the arterial supply of internal capsule

BASAL GANGLIA

The basal nuclei (or basal ganglia) are large masses of grey matter situated in the cerebral hemispheres. Classically, the following have been included under the definition of basal nuclei (Fig. 23.26). All these are telencephalic in origin.

- *Caudate nucleus,*
- *Lentiform nucleus,* which consists of two functionally distinct parts, the *putamen* and the *globus pallidus,*
- *Amygdaloid nuclear complex* and
- The *claustrum* is often included among the basal ganglia.

The caudate nucleus and the lentiform nucleus together constitute the *corpus striatum.* This consists of two functionally distinct parts. The caudate nucleus and the putamen form one unit called the *striatum,* while the globus pallidus forms the other unit, the *pallidum.*

Masses of grey matter, other than those listed above, are very closely related functionally to the basal nuclei. These are as follows:

- The *subthalamic nucleus* (which is of diencephalic origin) is very closely linked to the basal nuclei and is regarded as belonging to this group.
- The *substantia nigra* (midbrain) is also closely linked, functionally, to the basal nuclei.
- Some masses of grey matter found just below the corpus striatum (near the anterior perforated substance) are described as the *ventral striatum.* The part of the globus pallidus, which lies below the level of the anterior commissure, is designated as the *ventral pallidum.*

CAUDATE NUCLEUS

The caudate nucleus is a C-shaped mass of grey matter (Figs 23.34 and 23.35). It consists of a large head, body, and thin tail. The nucleus is intimately related to the lateral ventricle (Fig. 23.36). The head of the nucleus bulges into the anterior horn of the ventricle and forms the greater part of its floor. The body of the nucleus lies in the floor of the central part and the tail in the roof of the inferior horn of the ventricle. The anterior part of the head of the caudate nucleus is fused, inferiorly, with the lentiform nucleus. This region of fusion is referred to as the *fundus striati.* The fundus striati is continuous, inferiorly, with the anterior perforated substance. The anterior end of the tail of the caudate nucleus ends by becoming continuous with the lentiform nucleus. It lies in close relation to the amygdaloid complex.

The body of the caudate nucleus is related medially to the thalamus and laterally to the internal capsule, which separates it from the lentiform nucleus (Fig. 23.26).

Relationship of the corpus striatum to the internal capsule is shown in Fig. 23.37.

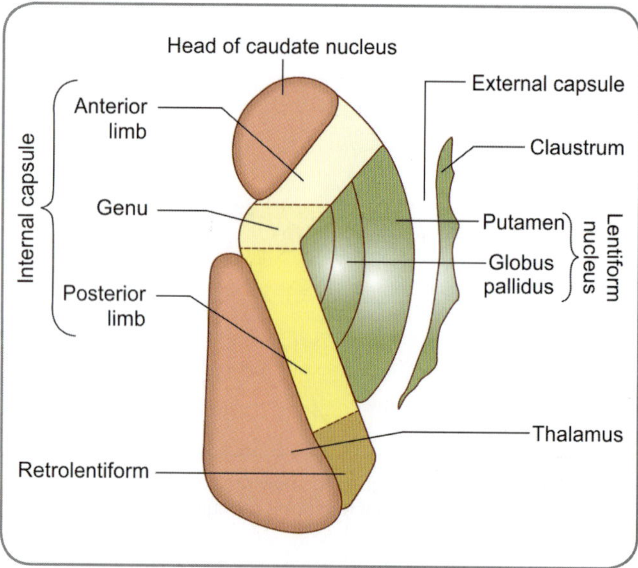

Fig. 23.35: Horizontal section of the cerebral hemisphere showing corpus striatum, thalamus, claustrum and internal capsule

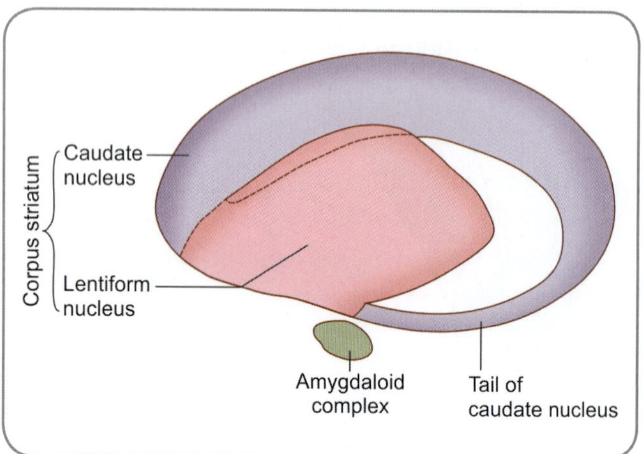

Fig. 23.34: Corpus striatum viewed from the lateral aspect

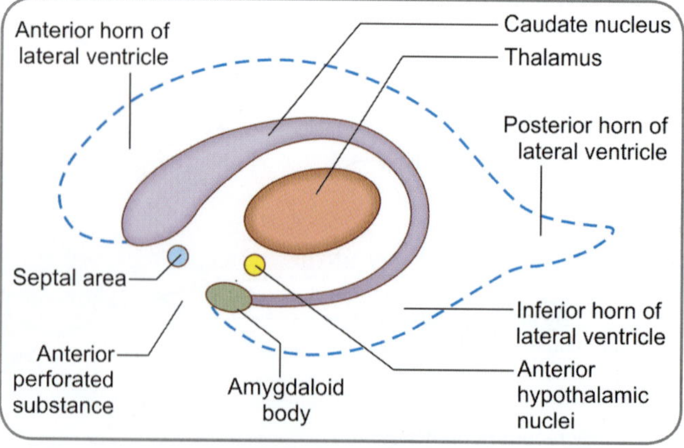

Fig. 23.36: Relationships of caudate nucleus with the cavity of the fourth ventricle and thalamus

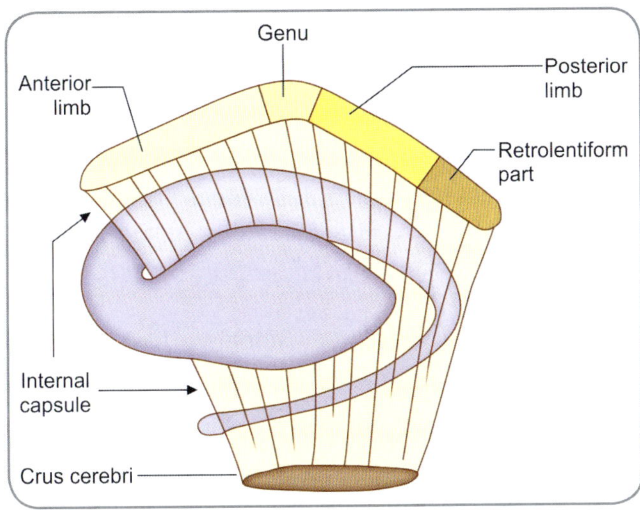

Fig. 23.37: Relationship of the corpus striatum to the internal capsule

LENTIFORM NUCLEUS

The lentiform nucleus appears triangular (or wedge-shaped) in the coronal section.

Relations

The lentiform nucleus lies lateral to the internal capsule. Laterally, it is separated from the claustrum by fibres of the external capsule (Fig. 23.16). (Note that these capsules are so called because they appear, by naked eye, to form a covering for the lentiform nucleus). Superiorly, the lentiform nucleus is related to the corona radiata and inferiorly, to the sublentiform part of the internal capsule.

Parts

It is divided by a thin lamina of white matter into a lateral part, the ***putamen*** and a medial part, the ***globus pallidus.*** The globus pallidus is further subdivided into medial and lateral (or internal and external) segments.

CONNECTIONS OF CORPUS STRIATUM

Afferent Connections

The striatum (caudate and lentiform nuclei) receive afferents from the following:

- The entire cerebral cortex via corticostriate fibres. These fibres are glutamatergic.
- The intralaminar nuclei of the thalamus via thalamo-striate fibres.
- The pars compacta of the substantia nigra via nigrostriate fibres. These fibres are dopaminergic. (Some dopaminergic fibres arise from the retrorubral nuclei lying behind the red nucleus).
- Noradrenergic fibres are received from the locus coeruleus.

- Serotoninergic fibres are received from the raphe nuclei (in the reticular formation of the midbrain).

The afferents from the cerebral cortex and from the thalamus provide the striatum with various modalities of sensory information (other than olfactory).

Efferent Connections

The main output of the striatum is concentrated upon the pallidum and on the substantia nigra (pars reticularis). The outflow from globus pallidus forms four separate bundles:

- ***Fasciculus lenticularis*** arises from the inner segment of the globus pallidus and enters the subthalamic region.
- ***Ansa lenticularis*** arises from both the inner and outer segments of the globus pallidus and enters the subthalamic region where it meets the dentato-rubrothalamic fibres and the fasciculus lenticularis. The union of the three tracts is called the fasciculus thalamicus, which terminates in the ventralis anterior (VA), ventralis lateral (VL), and centromedian nuclei of thalamus.
- ***Subthalamic fasciculus*** consists of reciprocal connections between the globus pallidus and nucleus subthalamicus.
- Some fibres from globus pallidus also pass to the substantia nigra (***pallidonigral fibres***).

Functions of Corpus Striatum

- The corpus striatum mediates enormous number of automatic activities involved in normal motor functions. For example, the maintenance of erect posture when sitting or standing, or swinging of arms during walking
- It helps in smoothening the voluntary motor activity of the body.

Claustrum

This is a thin lamina of grey matter that lies lateral to the lentiform nucleus (Fig. 23.26). It is separated from the latter by fibres of the external capsule. Laterally, it is separated by a thin layer of white matter from the cortex of the insula. Its connections and functions are unknown.

Amygdaloid Complex

This complex (also called the amygdaloid body, amygdala) lies in the temporal lobe of the cerebral hemisphere, close to the temporal pole. It lies deep to the uncus and is related to the anterior end of the inferior horn of the lateral ventricle.

Substantia Nigra

Substantia nigra is a large motor nucleus present in the mid-brain. The nucleus consists of two parts—pars reticulata and pars compacta (Fig. 23.38). Pars reticulata is related functionally to the internal part of globus pallidus.

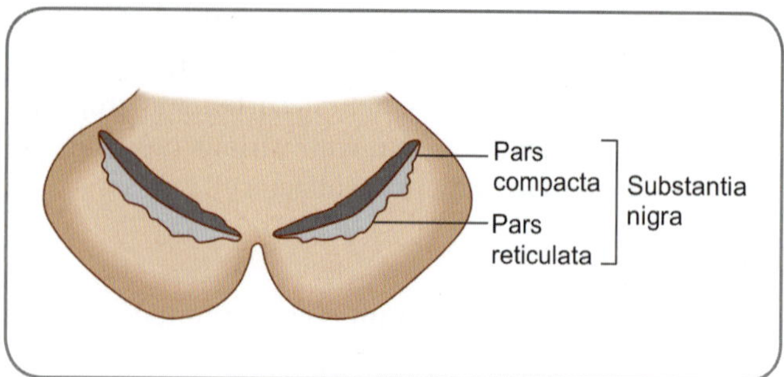

Fig. 23.38: Transverse section of the midbrain showing substantia nigra

Connections of Substantia Nigra

The ***pars compacta*** of the substantia nigra sends a dopaminergic projection to the striatum. A projection from the striatum ends in the pars reticularis of the substantia nigra. This part also receives fibres from the pallidum directly or after relay in the subthalamic nucleus or in the pedunculopontine nucleus (Fig. 23.39).

The ***pars reticularis*** projects to the (middle part) ventral lateral nucleus of the thalamus. These impulses are relayed to cingulate and prefrontal areas of the cerebral cortex. Other efferents of the pars reticularis reach the superior colliculus. They are relayed from there to the reticular formation of the medulla and to the spinal cord. These regions also receive fibres descending from the pedunculopontine nucleus.

In summary,

- The cerebral cortex sends impulses to the corpus striatun through the thalamus, which forms a direct feedback loop.
- The substantia nigra, the subthalamic nucleus, and the pedunculopontine nucleus are integrated with the basal ganglia to form an indirect feedback loop to the cerebral cortex through the thalamus.
- Descending fibres from the basal ganglia influence the superior colliculus, the reticular formation of the medulla, and thus, the motor neurons of the spinal cord.

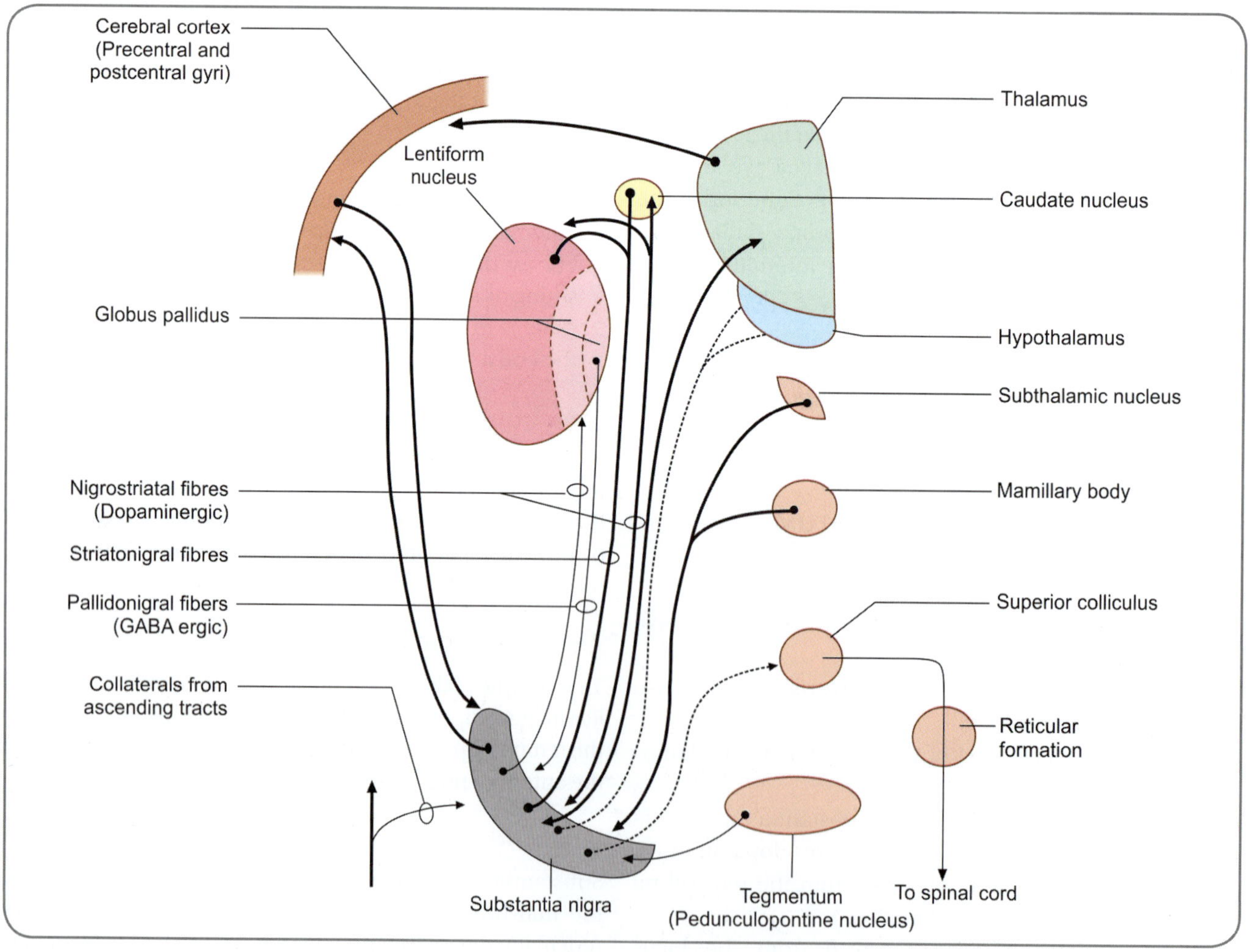

Fig. 23.39: Scheme to show the connections of substantia nigra

Clinical Correlation

The main connections (both afferent and efferent) of substantia nigra are with the striatum (i.e., caudate nucleus and putamen). Dopamine produced by neurons in the substantia nigra (pars compacta) passes along their axons to the striatum (*mesostriatal dopamine system*). Dopamine is much reduced in patients with a disease called *parkinsonism,* in which there is a degeneration of the striatum.

Multiple Choice Questions

1. The orbital and tentorial parts of the inferior surface are separated by:
   a. Stem of lateral sulcus
   b. Inferior temporal sulcus
   c. Whole of lateral sulcus
   d. Intraorbital sulcus
2. The interthalamic connexus is made up of:
   a. Commissural fibres
   b. Association fibres
   c. Projection fibres
   d. Grey matter
3. The primary motor area is in the:
   a. Precentral gyrus
   b. Postcentral gyrus
   c. Middle frontal gyrus
   d. Inferior frontal gyrus
4. Pars ventralis of the diencephalon includes:
   a. Thalamus
   b. Metathalamus
   c. Epithalamus
   d. Hypothalamus
5. Charcot's artery of cerebral haemorrhage is:
   a. Medial striate branch of middle cerebral artery
   b. Lateral striate branch of middle cerebral artery
   c. Medial striate branch of anterior cerebral artery
   d. Lateral striate branch of anterior cerebral artery

ANSWERS

1. a   **2.** d   **3.** a   **4.** d   **5.** b

Clinical Problem-solving

Case Study 1: A 66-year-old man had thromboembolic disease involving arteries to his brain. Motor area and Broca's area of his left cerebral hemisphere were affected. The man presented with right sided paralysis and loss of speech.
- Why was his speech affected?
- What is the relationship of the speech area to the motor area?
- What will be the status of the muscles concerned with speech?

Case Study 2: A 74-year-old man was afflicted by parkinsonism.
- What is the neurotransmitter involved in this disease?
- Where is it produced?
- Which part of the brain undergoes degeneration in this disease?

(For solutions see Appendix).

Chapter **24**

Ventricles of the Brain

- Discuss the lateral ventricles in detail with regard to their parts, walls and choroid plexus. Add a note on the circulation of CSF.
- Write notes on: (a) Fourth Ventricle, (b) Inferior horn of lateral ventricle, (c) Choroid plexuses, (d) Rhomboid fossa.
- Discuss the Rhomboid Fossa.
- Draw a neat and labeled diagram of the rhomboid fossa.
- Give an acoount of circulation of cerebrospinal fluid.
- Write notes on: (a) Foramina of magendie and luschka, (b) Blood—CSF barrier.

The interior of the brain contains a series of cavities (Fig. 24.1). The cerebrum contains a median cavity, the **third ventricle,** and two **lateral ventricles**, one in each hemisphere. Each lateral ventricle opens into the third ventricle through an **interventricular foramen**.

The third ventricle is continuous caudally with the **cerebral aqueduct,** which traverses the midbrain and opens into the fourth ventricle.

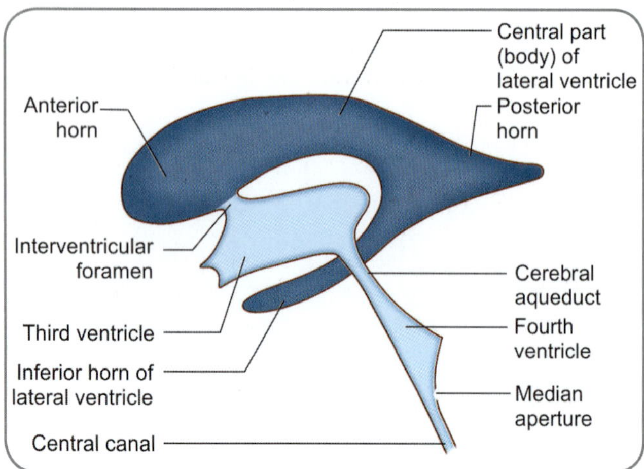

Fig. 24.1: The ventricular system of the brain seen from the lateral side

The **fourth ventricle** is situated dorsal to the pons and medulla and ventral to the cerebellum. It communicates inferiorly, with the **central canal,** which traverses the lower part of the medulla and the spinal cord. The entire ventricular system is lined by an epithelial layer called the **ependyma**.

 **Development**

Developmental Factors

The ventricular system of the brain is developed from the original central canal of the neural tube. Formation of dilatations and flexures, differential growth and expansion of mantle and marginal zones contribute to changes in sizes of the lumen at various levels. The places where the lumen was able to expand came to be known as the ventricles. Eversion of the alar laminae in the metencephalon causes the ventricle to acquire a conical shape and this becomes the fourth ventricle. Lateral expansion of the telencephalon causes the ventricle to be captured within the cerebral hemispheres and these form the two lateral ventricles (one on either side). The alar laminae of the diencephalon proliferate and develop in the lateral walls themselves forming the thalami and the hypothalami. The lumen, therefore, remains narrow between the two lateral walls. This arrangement results in the third ventricle being midline between the thalami and hypothalami. Since the mesencephalon does not undergo any major expansion, the central canal remains almost as it was in the neural tube and forms the cerebral aqueduct.

The roof plate of the fourth ventricle is covered by the vascular mesenchyme of the piamater. The mesenchyme proliferates and invaginates into the roof plate forming the choroid plexus of the fourth ventricle. The roof plate of the third ventricle also has a single layer of ependyma covered by the vascular mesenchyme of piamater. Proliferation of the mesenchyme results in the formation of the choroid plexus which becomes the choroid plexus of the third and lateral ventricles. It is noted that the rudiments of the choroid plexus are seen in the 7th week of intrauterine life. From the 9th to the 17th week, the choroid plexus expands and differentiates

contd...

Development *contd...*

rapidly. From the 17th to the 28th week, brain tissue grows faster than the choroid plexus. From the 29th week onwards, histological changes in the cells of choroid plexus suggest active secretion. However, there is a sequential order in which the choroid plexuses of the ventricles develop and differentiate. The choroid plexus of the fourth ventricle is the first to differentiate, followed by that of the lateral ventricles and lastly by that of the third ventricle.

Cerebrospinal fluid is secreted by the choroid plexuses of the ventricles. Secretion increases gradually during the last few months of development. However, enzymatic reactions responsible for CSF secretion mature postnatally and there is a marked increase in CSF production after the child is born.

The most common cause of hydrocephalus in a new born is aqueductal obstruction or stenosis. CSF from the lateral and third ventricles is not able to pass to the fourth ventricle resulting in expansion of lateral ventricles, compression of brain tissue, widening of the sutures and eventual expansion of the head.

LATERAL VENTRICLES

The lateral ventricles are two cavities, one situated within each cerebral hemisphere. Each ventricle consists of a central part, which gives off three extensions called the anterior, posterior, and inferior horns (Fig. 24.2).

Dissection

Lateral Ventricles: The dissection to expose the lateral ventricle and its horns is difficult. It is preferable to have a senior member by your side when you attempt this dissection. In a midsagittal section, cut upwards from the interventricular foramen in such a way that you reach the superior border. From the superior border extend the incision on to the superolateral surface for a few centimeters. Then turn the incision posteriorly and continue parallel to the superior border but about 2.5 inches lateral to it. Curve behind the postcentral gyrus to reach the posterior ramus of the lateral sulcus. Applying gentle traction on the two pieces of brain thus separated, cut through the white matter now exposed in the medial part of the temporal lobe. The inferior horn of the lateral ventricle will be opened through the roof. Try to study the different parts of the ventricle by gently separating the various parts of brain. If any further incision or cut is required, seek help from the senior member.

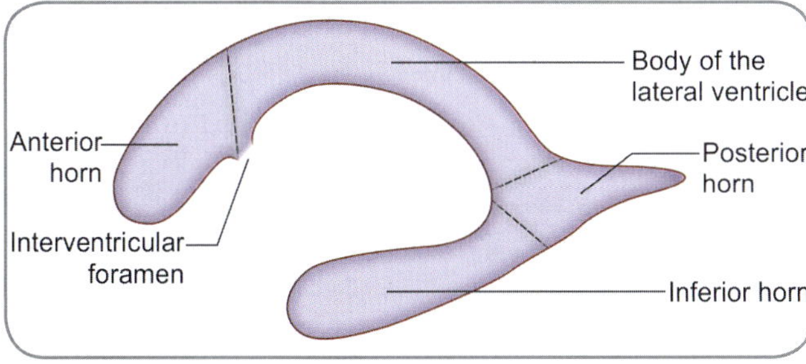

Fig. 24.2: Parts of the lateral ventricle

Central Part

The central part of the lateral ventricle is elongated anteroposteriorly. Anteriorly, it becomes continuous with the anterior horn at the level of the interventricular foramen. Posteriorly, the central part reaches the splenium of the corpus callosum.

The central part is triangular in cross-section (Figs 24.2 and 24.3). It has *a roof, a floor,* and *a medial wall*. The roof and floor meet on the lateral side.

The *roof* is formed by the trunk of the corpus callosum.

The *medial wall* is formed by the septum pellucidum and by the body of the fornix. It is common to the two lateral ventricles.

The *floor* is formed mainly by the superior surface of the thalamus, medially and by the caudate nucleus, laterally. Between these two structures, there are the stria terminalis, laterally and the thalamostriate vein, medially. From Fig. 24.3, it will be seen that there is a space between the fornix and the upper surface of the thalamus. This is the *choroid fissure*.

A fold of pia mater, the *tela choroidea*, invaginates into the ventricle through the fissure and covers part of the thalamus. The tela choroidea is common to the two lateral ventricles and third ventricle. Within each lateral edge of the tela choroidea, there are plexuses of blood vessels that constitute the *choroid plexus* (Fig. 24.3). The tela choroidea and other structures forming the walls of the ventricle are lined by ependyma.

The choroid plexuses are highly vascular structures that are responsible for the formation of cerebrospinal fluid. The surface of each plexus is lined by a membrane formed by fusion of the ventricular ependyma with the pia mater of the tela choroidea.

Anterior Horn

The anterior horn of the lateral ventricle lies anterior to its central part; the two being separated by an imaginary vertical line drawn at the level of the interventricular foramen (Fig. 24.1). This horn is triangular in section. It has a roof, a floor, and a medial wall (Fig. 24.4A). It is closed anteriorly by the genu and rostrum of the corpus callosum.

The *roof* is formed by the most anterior part of the trunk of the corpus callosum. The *floor* is formed mainly by the head of the caudate nucleus. A small part of the floor, near the middle line, is formed by the upper surface of the rostrum of the corpus callosum. The *medial wall* (common to the two sides) is formed by the septum pellucidum. It may be noted that the tela choroidea and the choroid plexus *do not* extend into the anterior horn.

Posterior Horn

The posterior horn of the lateral ventricle extends backwards into the occipital lobe. It has a roof, a lateral wall, and a medial wall (Fig. 24.4B).

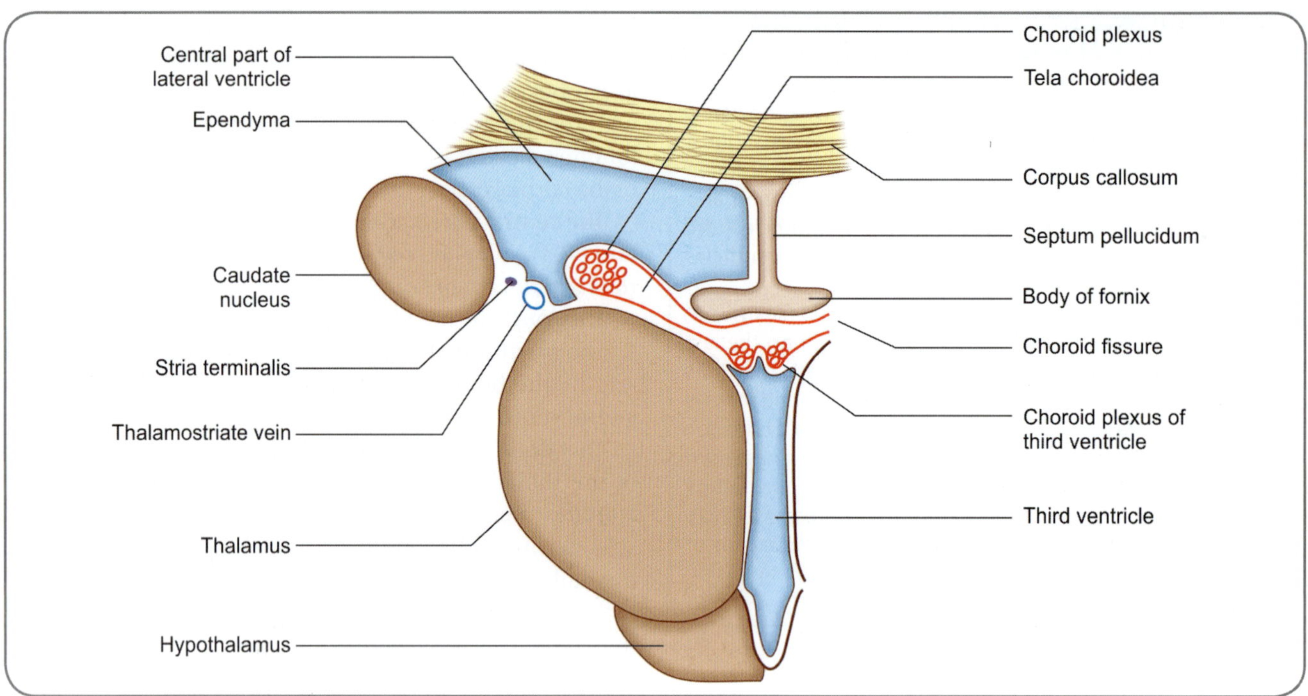

Fig. 24.3: Boundaries of the central part of the lateral ventricle and third ventricle. Note the relationship of the tela choroidea and the choroid plexuses to these ventricles

The **roof** and **lateral wall** are formed by the tapetum (a sheet of fibres from the splenium of the corpus callosum). The **medial wall** shows two elevations. The uppermost of these is the **bulb of the posterior horn**, which is produced by fibres of the forceps major, as they run backwards from the splenium of the corpus callosum. The lower elevation is called the **calcar avis**. It represents white matter '**pushed in**' by formation of the calcarine sulcus.

Inferior Horn

The inferior horn of the lateral ventricle begins at the posterior end of the central part. It runs downwards and forwards into the temporal lobe, its anterior end reaching close to the uncus.

In cross-section, the inferior horn is seen to have a narrow cavity. The cavity is bounded above and laterally by the **roof** and below and medially by the **floor**. Because of this orientation, the lateral part of the roof is sometimes called the **lateral wall** and the medial part of the floor, the **medial wall**.

The lateral part of the roof is formed by fibres of the tapetum. The medial part of the roof is formed by the tail of the caudate nucleus, laterally and the stria terminalis, medially. These structures are continued into the roof of the inferior horn from the floor of the central part. Anteriorly, the tail of the caudate nucleus and the stria terminalis end in relation to the amygdaloid complex, which lies in the most anterior part of the roof.

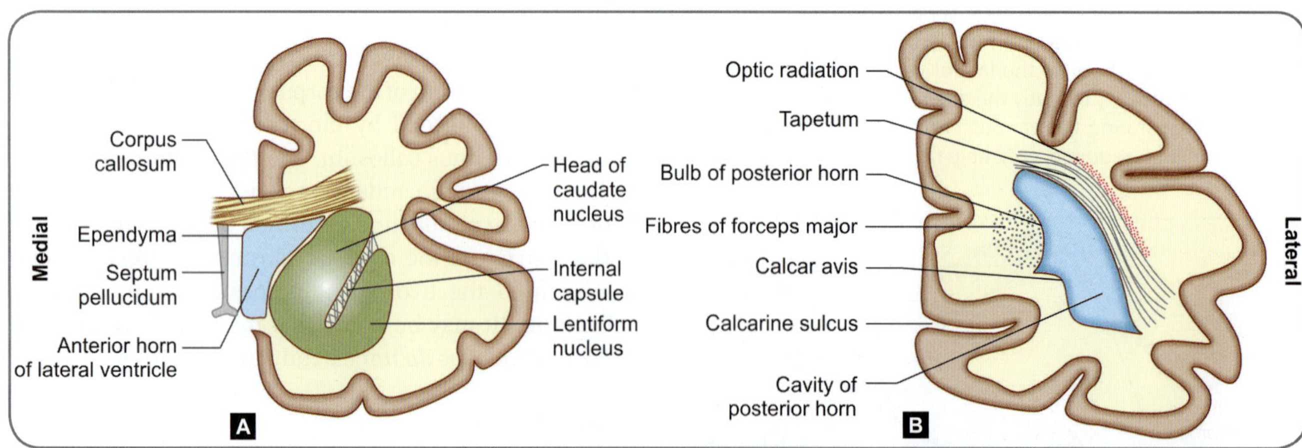

Figs 24.4A and B: A. Coronal section passing through the anterior horn of the lateral ventricle. The roof is formed by the most anterior part of the trunk of the corpus callosum. The floor is formed mainly by the head of the caudate nucleus **B.** Coronal section passing through the posterior horn of the lateral ventricle. The roof and lateral wall of the posterior horn are formed by the tapetum. The medial wall shows two elevations. The uppermost of these is the bulb of the posterior horn, and the lower elevation is called the calcar avis

The floor of the inferior horn is formed mainly by the hippocampus. The fibres of hippocampus form a thin sheet of white matter called ***alveus*** that covers its ventricular surface. The fibres of alveus converge medially to form a ridge called ***fimbria***. In the lateral part of the floor, there is an elevation, the ***collateral eminence***, produced by inward bulging of the white matter by the collateral sulcus.

THIRD VENTRICLE

The third ventricle is the cavity of the diencephalon. It is a median cavity situated between the right and left thalami (Fig. 24.3). It communicates on either side with the lateral ventricle through the interventricular foramen (Figs 24.1 and 24.5). Posteriorly, it continues into the cerebral aqueduct, which connects it to the fourth ventricle. The ventricle has two lateral walls (anterior and posterior), floor, and roof.

Each ***lateral wall*** is marked by the ***hypothalamic sulcus*** (Fig. 24.5), which follows a curved course from the interventricular foramen to the aqueduct. Above the sulcus, the wall is formed by the medial surface of the thalamus. The two thalami are usually connected by a band of grey matter called the ***interthalamic connexus,*** which passes through the ventricle. The lateral wall, below the hypothalamic sulcus, is formed by the medial surface of the hypothalamus. A small part of the lateral wall, above

and behind the thalamus, is formed by the epithalamus. The interventricular foramen is seen on the lateral wall, just behind the column of the fornix.

The ***anterior wall*** of the third ventricle is formed mainly by the lamina terminalis. Its upper part is formed by the anterior commissure and columns of the fornix, as they diverge from each other.

The ***posterior wall*** is formed by the pineal body and posterior commissure.

The ***floor*** is formed by the optic chiasma, tuber cinereum, infundibulum, mamillary bodies, posterior perforated substance, and the tegmentum of the midbrain.

The ***roof*** of the ventricle is formed by the ependyma that stretches across the two thalami (Fig. 24.3). Above the ependyma, there is the tela choroidea. Within the tela choroidea, there are two plexuses of blood vessels (one on either side of the middle line), which bulge downwards into the cavity of the third ventricle. These are ***the choroid plexuses of the third ventricle***.

The cavity of the third ventricle shows a number of prolongations or recesses (Fig. 24.5). The ***infundibular recess*** extends into the infundibulum. The ***optic recess*** lies just above the optic chiasma. The ***pineal recess*** lies between the superior and inferior laminae of the stalk of the pineal body. The ***suprapineal recess*** lies above the pineal body in relation to the epithalamus.

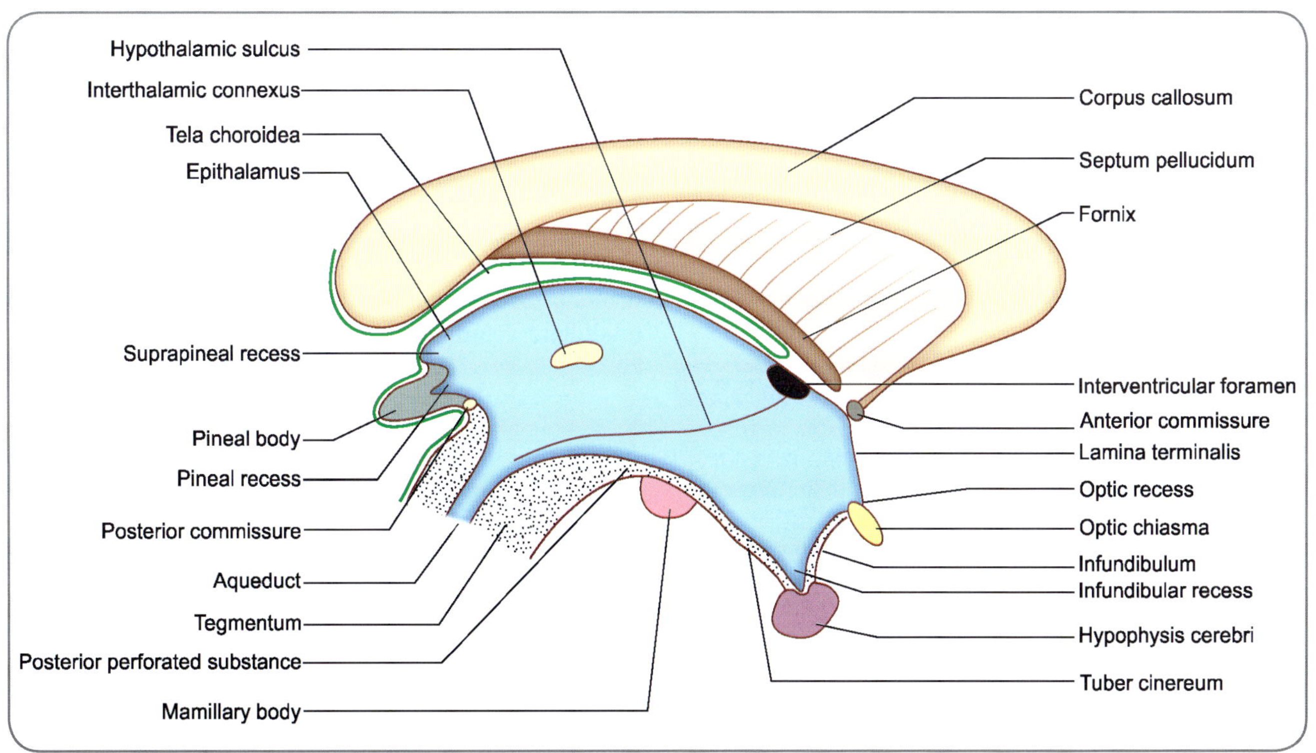

Fig. 24.5: Boundaries and recesses of the third ventricle. Note the mode of formation of the tela choroidea that lies in the roof of the ventricle

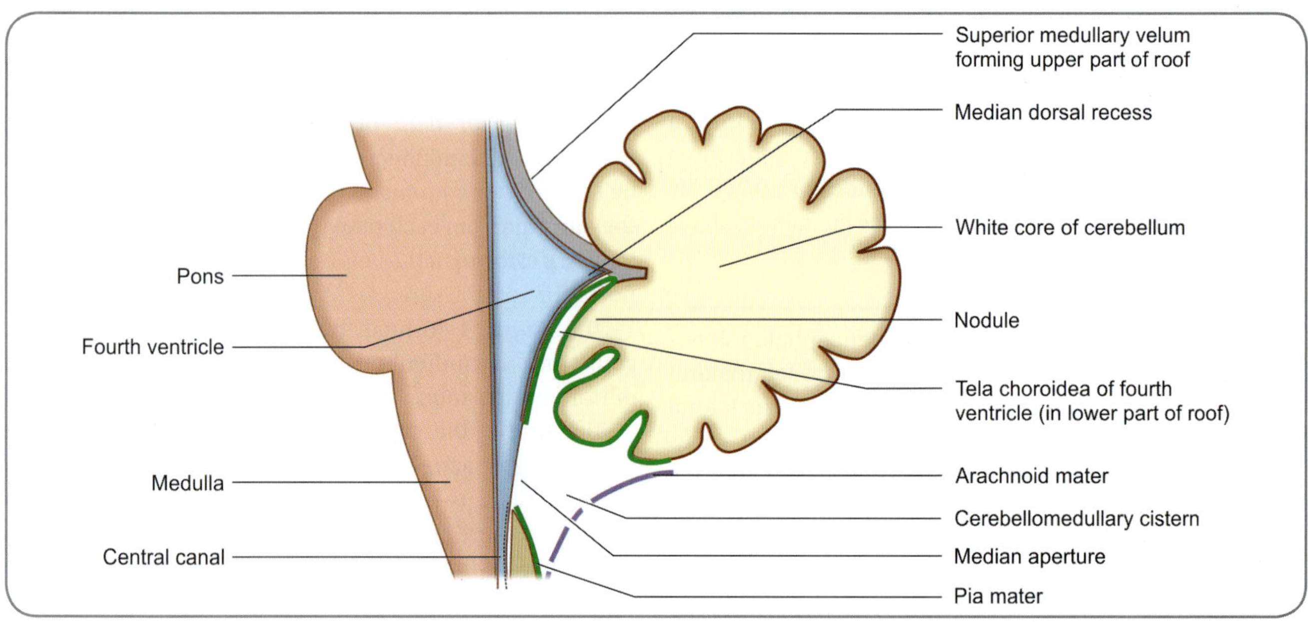

Fig. 24.6: Midsagittal section passing through the fourth ventricle and related structures. The upper part of the ventricle is related to the *superior (or anterior) medullary velum*. When traced inferiorly (and posteriorly), the velum merges into the white matter of the cerebellum. The lower part of the ventricle is related to the nodule. If the tonsil is lifted away, we see that the nodule is continuous laterally with a membrane called the *inferior (or posterior) medullary velum*

Dissection

Third Ventricle: It is difficult to study the third ventricle with all its boundaries *in situ*. Study it in a midsagittal section of the brain. In such a specimen the boundaries should be studied. The third ventricle is a narrow cavity present in the median plane. The thalamaus and the hypothalamus form its lateral boundary. If required, place the complementary opposite section and get a clear picture of the third ventricle.

FOURTH VENTRICLE

The fourth ventricle is a space situated dorsal to the pons and to the upper part of the medulla and ventral to the cerebellum.

For descriptive purposes, the ventricle may be considered as having a cavity, a floor, a roof, and lateral walls.

Dissection

Fourth Ventricle: In a specimen that has the entire brainstem structures, divide the cerebellar hemispheres by cutting through the vermis from the inferior and dorsal aspects. As you proceed gradually and slowly, the arbor vitae pattern of the cerebellum is clearly seen. As you proceed further, the roof of the fourth ventricle comes into view. Locate the foramen of Magendie and the choroid plexus. Separate and hold apart the cerebellar hemispheres so that the floor of the fourth ventricle is well exposed. Study the floor.

Cavity

The cavity of the ventricle is continuous inferiorly with the central canal and superiorly with the cerebral aqueduct.

It communicates with the subarachnoid space through three apertures, one median and two lateral (Figs 24.6 and 24.9A). A number of extensions from the main cavity are described in Fig. 24.7. The largest of these are two *lateral recesses,* one on either side. Each lateral recess passes laterally in the interval between the inferior cerebellar peduncle, ventrally and the peduncle of the flocculus, dorsally (Fig. 24.9A). The lateral extremity of the recess reaches the flocculus. At this extremity, the recess opens into the subarachnoid space as the *lateral aperture*. Another recess present in the middle line is called the *median dorsal recess*. It extends into the white core of the cerebellum and lies just cranial to the nodule (Figs 24.6 and 24.7). Immediately lateral to the nodule, another recess projects dorsally, on either side, above the inferior medullary velum. This is the *lateral dorsal recess* (Fig. 24.7).

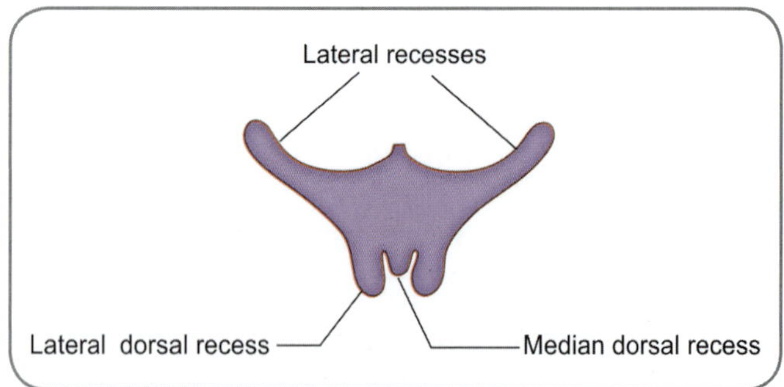

Fig. 24.7: Scheme to show the various recesses of the cavity of the fourth ventricle

Openings in the Fourth Ventricle

The fourth ventricle communicates below (at its inferior angle) with the central part of medulla oblongata. It has three openings in the roof—one median and two lateral, through which it communicates with the subarachnoid space. The median opening (*foramen of Magendie*) is a large opening, in the lower part of the roof. Through this midline opening, it communicates with the cerebellomedullary cistern. Two lateral openings (*foramina of Luschka*), one on each side, lie in the lateral angle of the ventricle between the inferior cerebellar peduncle and flocculus. Through this opening, the cerebrospinal fluid escapes into the subarachnoid space. The choroid plexus of the fourth ventricle also protrudes through this opening.

Floor

Because of its shape, the floor of the fourth ventricle is often called the *rhomboid fossa* (Fig. 24.8). It is divisible into an upper triangular part formed by the posterior surface of the pons, a lower triangular part formed by the upper part of the posterior surface of the medulla, and an intermediate part at the junction of the medulla and pons. The intermediate part is prolonged laterally over the inferior cerebellar peduncle as the floor of the lateral recess. Its surface is marked by the presence of delicate bundles of transversely running fibres. These bundles are the *striae medullares*.

The entire floor is divided into right and left halves by a *median sulcus*. Next to the middle line, there is a longitudinal elevation called the *median eminence*. The eminence is bounded laterally by the *sulcus limitans.* The region lateral to the sulcus limitans is the *vestibular area*, which overlies the vestibular nuclei. The vestibular area lies partly in the pons and partly in the medulla.

The pontine part of the floor shows some features of interest in close relation to the sulcus limitans and the median eminence. The uppermost part of the sulcus limitans overlies an area that is bluish in colour and is called the *locus coeruleus*. Deep to the locus coeruleus, there is the nucleus coeruleus, which extends upwards into the tegmentum of the midbrain. It is regarded as part of the reticular formation.

Somewhat lower down, the sulcus limitans is marked by a depression, the *superior fovea.* At this level, the median eminence shows a swelling, the *facial colliculus*.

The medullary part of the floor also shows some features of interest in relation to the median eminence and the sulcus limitans. The sulcus limitans is marked by a depression, the *inferior fovea.* Descending from the fovea, there is a sulcus that runs obliquely towards the middle line. This sulcus divides the median eminence into two triangles. These are the *hypoglossal triangle*, medially and the *vagal triangle,* laterally. Between the vagal triangle (above) and the gracile tubercle (below), there is a small area called the *area postrema.* Finally, the two terms often used in relation to the medulla must be mentioned. The lowest part of the floor of the fourth ventricle is called the *calamus scriptorius,* because of its resemblance to a nib. Each inferolateral margin of the ventricle is marked by a narrow white ridge or *taenia.* The right and left taeniae meet at the inferior angle of the floor to form a small fold called the *obex.* The term obex is often used to denote the inferior angle itself.

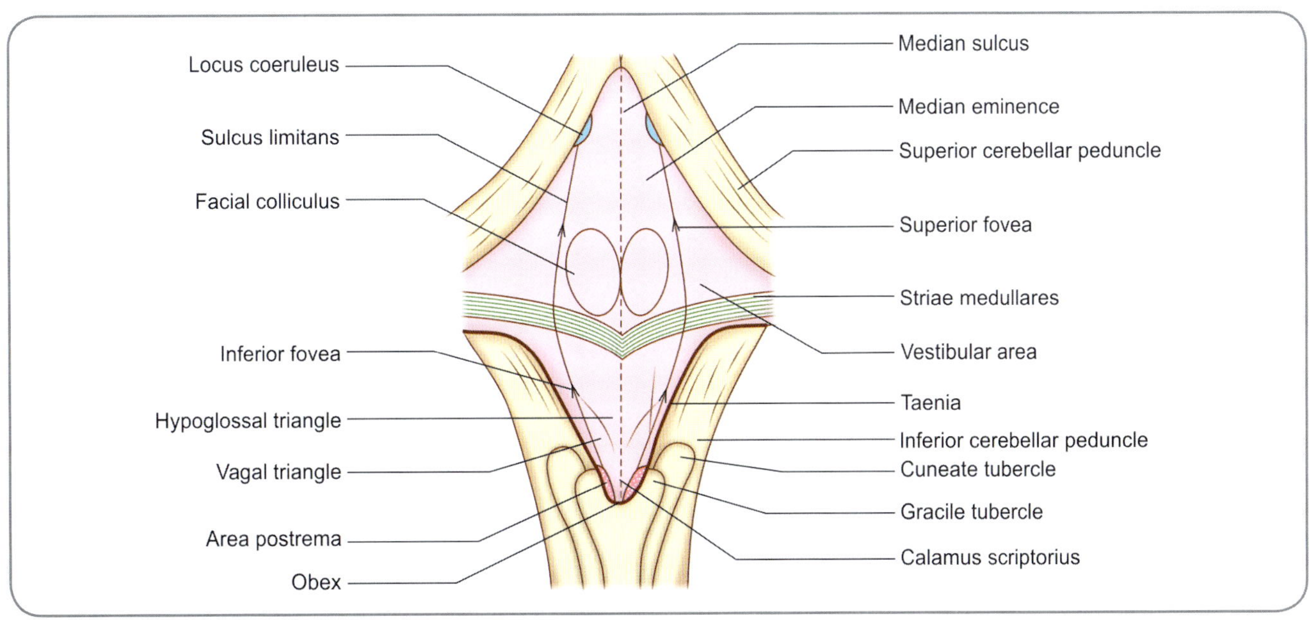

Fig. 24.8: Structures in the floor of the fourth ventricle

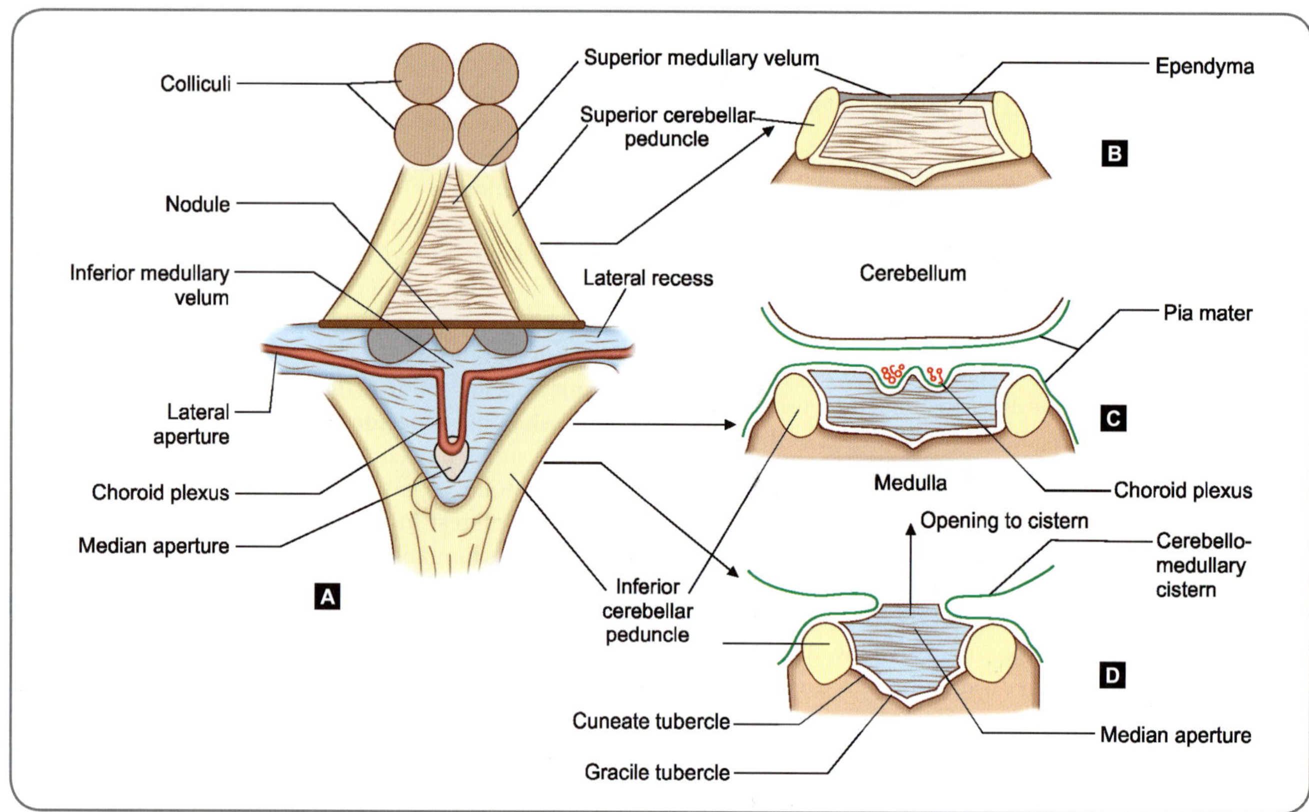

Figs 24.9A to D: Schemes to illustrate the formation of the roof of the fourth ventricle. The bold horizontal line in **A.** Represents the white core of the cerebellum. The part above this line represents the upper part of the roof and the part below the line represents the lower part of the roof **B to D.** Represent transverse sections across the ventricle at the levels indicated

Lateral Walls

The upper part of each lateral wall is formed by the superior cerebellar peduncle (Fig. 24.9B). The lower part is formed by the inferior cerebellar peduncle and by the gracile and cuneate tubercles (Fig. 24.9C and D).

Roof

The roof of the fourth ventricle is tent-shaped and can be divided into upper and lower parts, which meet at an apex (Figs 24.6 and 24.9A). The apex extends into the white core of the cerebellum. The upper part of the roof is formed by the superior cerebellar peduncles and the superior medullary velum (Figs 24.19A and B). The inferior part of the roof is devoid of nervous tissue in most of its extent. It is formed by a membrane consisting of ependyma and a double fold of pia mater, which constitutes the ***tela choroidea of the fourth ventricle*** (Figs 24.9A and C). Laterally, on each side, this membrane reaches and fuses with the inferior cerebellar peduncles. The lower part of the membrane has a large aperture in it. This is the ***median aperture*** of the fourth ventricle through which the ventricle communicates with the subarachnoid space in the region of the cerebellomedullary cistern. In the region of the lateral recess, the membrane is prolonged laterally and helps form the wall of the recess. The inferior medullary velum forms a small part of the roof in the region of the lateral dorsal recess. It may be noted that some authors describe the entire membranous structure, forming the lower part of the roof of the fourth ventricle, as the inferior medullary velum. The nodule is intimately related to the roof of the ventricle in the region of the median dorsal recess.

CEREBROSPINAL FLUID

The cerebrospinal fluid fills the subarachnoid space. It also extends into the ventricles of the brain and the central canal of the spinal cord.

Site of Production

The choroid plexuses of the ventricles of brain secrete the cerebrospinal fluid.

Circulation of Cerebrospinal Fluid

The cerebrospinal fluid formed in each lateral ventricle flows into the third ventricle through the interventricular foramen (Fig. 24.10). From the third ventricle, it passes through the aqueduct into the fourth ventricle. Here, it

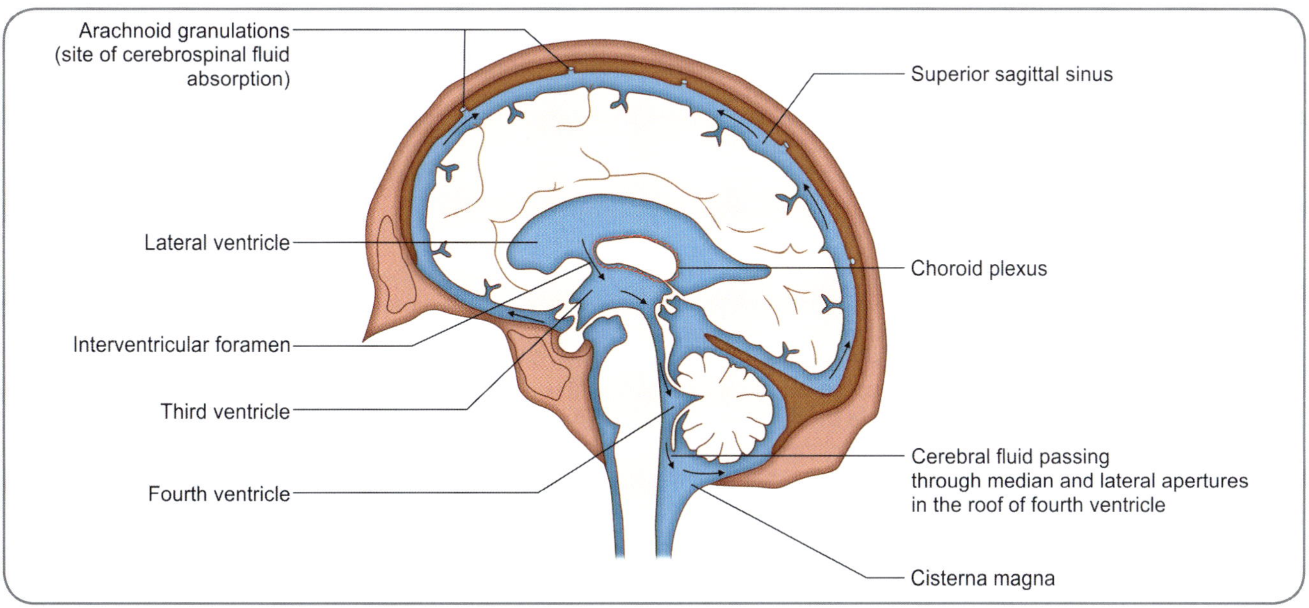

Fig. 24.10: Scheme to show circulation of cerebrospinal fluid

passes through the median and lateral apertures in the roof of this ventricle to enter the part of the subarachnoid space, which forms the cerebellomedullary cistern. From here, the fluid enters other parts of the subarachnoid space. In passing from the posterior cranial fossa into the upper (supratentorial) part of the cranial cavity, the cerebrospinal fluid traverses the narrow interval between the free margin of the tentorium cerebelli and the brainstem. It leaves the subarachnoid space by entering the venous sinuses through arachnoid villi (Fig. 24.10).

Characteristic Features of Cerebrospinal Fluid

- The total volume of cerebrospinal fluid is about 140 ml, of which about 25 ml is in the ventricles. Daily production of cerebrospinal fluid is around 1500 ml, this indicates that the cerebrospinal fluid is constantly replaced
- The normal pressure in supine position in lumbar subarachnoid space is 50–200 mm of water and in sitting position 200–250 mm of water
- Its specific gravity ranges from 1.003 to 1.008
- Glucose level is half of that of blood (40–60 mg%)
- Protein content is very low compared to plasma proteins (20–40 mg%) (Table 24.1).

Table 24.1: Comparison of the composition of CSF and blood plasma

Substance	CSF	Plasma
Protein	20–40 mg%	600 mg%
Glucose	40–60 mg%	100 mg%
Chloride	120 mEq/L	100 mEq/L
Calcium	2.5 mEq/L	4.5 mEq/L

CSF=Cerebrospinal fluid

Blood-cerebrospinal Fluid Barrier

The epithelium and other tissues of the choroid plexuses form an effective barrier between the blood and the cerebrospinal fluid. This blood-cerebrospinal fluid barrier allows selective passage of substances from blood to cerebrospinal fluid, but not in the reverse direction. The arachnoid villi provide a valvular mechanism for flow of cerebrospinal fluid into blood, without permitting backflow of blood into the cerebrospinal fluid.

Functions of Cerebrospinal Fluid

The cerebrospinal fluid provides a fluid cushion, which protects the brain from injury. It probably also helps carry nutrition to the brain and remove waste products.

Clinical Correlation

Lumbar Puncture

Samples of cerebrospinal fluid are often required for help in clinical diagnosis. They are obtained most easily by ***lumbar puncture***. In this procedure, a needle is introduced into the subarachnoid space through the interval between the third and fourth lumbar vertebrae. Under exceptional circumstances, cerebrospinal fluid may be obtained by ***cisternal puncture,*** in which a needle is passed into the cerebellomedullary cistern.

Hydrocephalus

An abnormal increase in the quantity of cerebrospinal fluid can lead to enlargement of the head in children. This condition is called ***hydrocephalus.*** Abnormal pressure of cerebrospinal fluid leads to degeneration of brain tissue. Hydrocephalus may be caused by excessive production of cerebrospinal fluid, by obstruction to its flow, or by impaired absorption through the arachnoid villi. It is classified as ***obstructive,*** when there is obstruction to flow of cerebrospinal fluid from the ventricular system to the subarachnoid space or as ***communicating,*** when such obstruction is not present. Obstruction is most likely to occur where cerebrospinal fluid has to pass through narrow passages, for example, the interventricular foramina, the aqueduct, and the apertures of the fourth ventricle. In each of the above instances, dilatation is confined to cavities proximal to the obstruction. Occasionally, meningitis may lead to obstruction of the narrow interval between the tentorium cerebelli and the brainstem. Meningitis may also lead to hydrocephalus by affecting the arachnoid villi, thus, hampering the reabsorption of cerebrospinal fluid.

Ventriculography

The ventricles of the brain can be studied in living subjects by taking radiographs, after injecting a radiopaque dye into the ventricular system (Fig. 24.11). The procedure is called ventriculography. Parts of the ventricles can also be seen using computed tomography (CT) and magnetic resonanace imaging (MRI) scans.

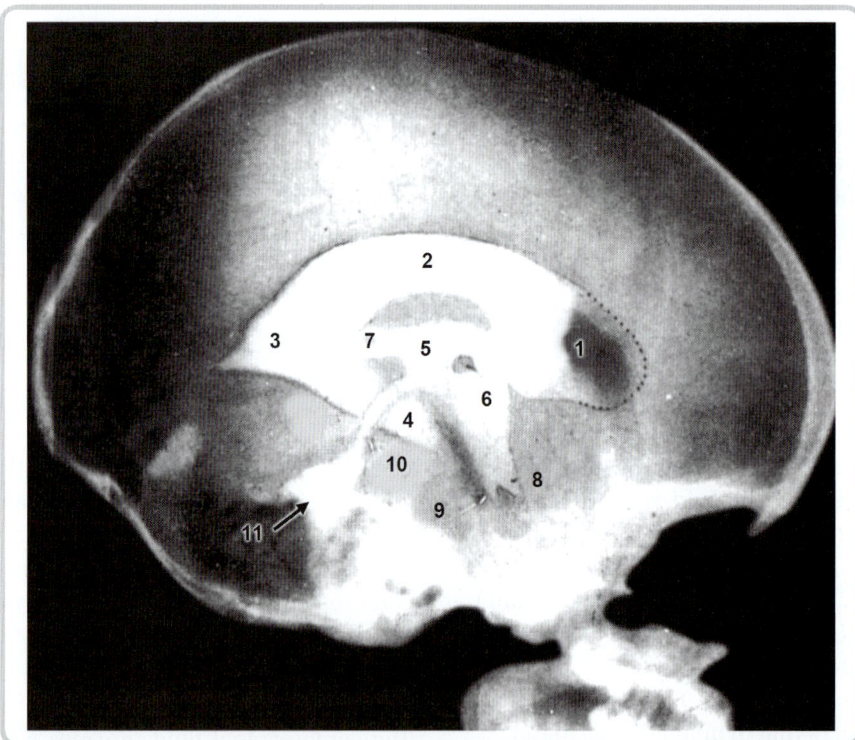

Fig. 24.11: Ventriculogram. Lateral view. A radiograph of the head was taken after injecting a radiopaque dye into the ventricular system. The parts of the lateral ventricle seen are—**1.** Anterior horn (part of which appears dark as air has entered it during injection) **2.** Central part **3.** Posterior horn; and **4.** Inferior horn. In relation to the third ventricle **5.** We can identify the suprapineal recess **7.** The optic recess **8.** And the infundibular recess **9.** The interthalamic connexus is seen as a dark area **6.** Other structures seen are the aqueduct **10.** And the fourth ventricle **11.** Some dye that has reached the cisterna magna through the apertures of the fourth ventricle is also seen

Multiple Choice Questions

1. Alveus of the inferior horn consists of:
   a. Hippocampal cells
   b. Hippocampal fibres
   c. Amygdaloid fibres
   d. Cells of collateral sulcus
2. Vestibular area of the rhomboid fossa:
   a. Overlies the vestibular nucleus
   b. Is medial to the Sulcus limitans
   c. Is completely in pons
   d. Is part of the median emimence
3. Reabsorption of cerebrospinal fluid takes place through the:
   a. Interventricular foramen
   b. Arachnoid villi

   c. Choroid plexuses
   d. Supratentorial notch
4. The anterior wall of the third ventricle is formed by:
   a. Interthalamic connexus
   b. Mammillary bodies
   c. Lamina terminalis
   d. Pineal body
5. The fourth ventricle has:
   a. Two lateral recesses, two lateral dorsal recesses and a median recess
   b. Two lateral recesses and a median recess
   c. Two laterodorsal recesses and a median recess
   d. A median recess and a lateral dorsal recess

ANSWERS

1. b   **2.** a   **3.** b   **4.** c   **5.** a

Clinical Problem-solving

Case Study 1: A 5-year-old boy had an enlargement of his head. He was diagnosed to be having 'hydrocephalus'. On reading his reports, you found that he had 'Obstructive hydrocephalus'.
- What do you understand by the term 'obstructive' here?
- What are the points/areas which could be obstructed to cause such a condition?
- If the arachnoid villi are affected and the condition was only acquired, what possible cause, do you think, lead to this?

Case Study 2: It is necessary that the cerebrospinal fluid is analysed and examined in a particular neurological patient.
- What would be the procedure of your choice for such a study?
- What is the anatomical location at which such procedure can be executed?
- If the said procedure cannot be done, what is the alternate method?

(For solutions see Appendix).

Chapter 25

Special Senses and Their Pathways

Frequently Asked Questions

- ❑ Give a brief account of the olfactory pathway.
- ❑ Write notes on (a) Rods and cones, (b) Pupillary reflex, (c) Primary auditory pathway.
- ❑ Give an account of Geniculocalcarine tract and Visual cortex.

OLFACTORY SYSTEM

The peripheral end organ for smell is the **olfactory epithelium** (Fig. 25.1) that lines the upper and posterior parts of the nasal cavity. Nerve fibres arising in this mucosa collect to form about twenty bundles that together constitute an **olfactory nerve**. The bundles pass through foramina in the cribriform plate of the ethmoid bone to enter the cranial cavity, where they terminate in the **olfactory bulb**.

The olfactory bulb is an elongated oval structure that lies just above the cribriform plate. It is continuous posteriorly with the **olfactory tract**, through which it is connected to the base of the cerebral hemisphere.

When traced posteriorly, the olfactory tract divides into **medial** and **lateral olfactory striae**. The point of bifurcation is expanded and forms the **olfactory trigone.** An **intermediate stria** is sometimes present. The olfactory striae are intimately related to a mass of grey matter called the **anterior perforated substance**. The medial and lateral striae form the anteromedial and anterolateral boundaries of this substance. The intermediate stria extends into the anterior perforated substance and ends in a slight elevation (in the anterior part of the substance) called the **olfactory tubercle**. Posterolaterally, the anterior perforated substance is related to the uncus.

The uncus is a part of the cerebral hemisphere that lies on the tentorial surface, a little behind and medial to the temporal pole. It represents the anterior end of the **parahippocampal gyrus** and is separated from the temporal pole by the rhinal sulcus.

The anterior part of the parahippocampal gyrus, including the uncus, is referred to as the **entorhinal area** (Brodmann area 28). The uncus is subdivided into three parts. From anterior to posterior side, these are the uncinate gyrus, the tail of the dentate gyrus (band of Giacomini), and the intralimbic gyrus.

The **primary olfactory cortex** lies between the anterior perforated substance and the uncus. The **entorhinal area** is often termed **secondary olfactory cortex**. The different parts of olfactory system are shown in Figure 25.2.

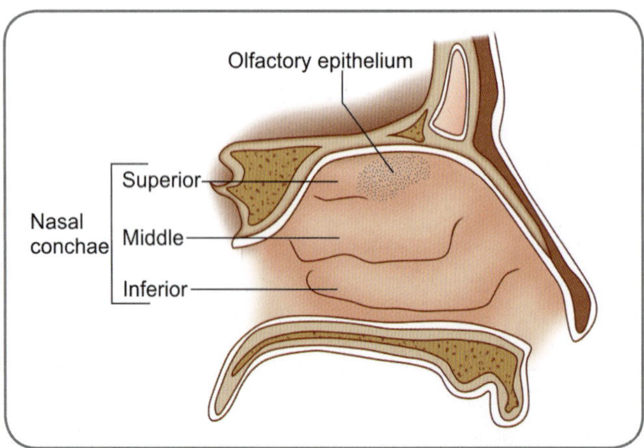

Fig. 25.1: Location of olfactory epithelium in the nasal cavity

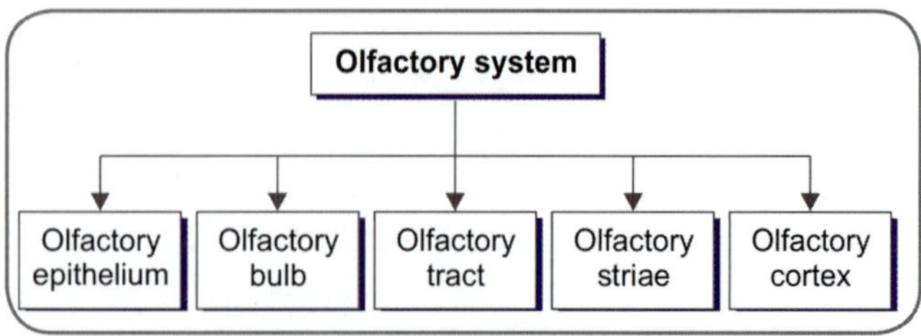

Fig. 25.2: Parts of the olfactory system

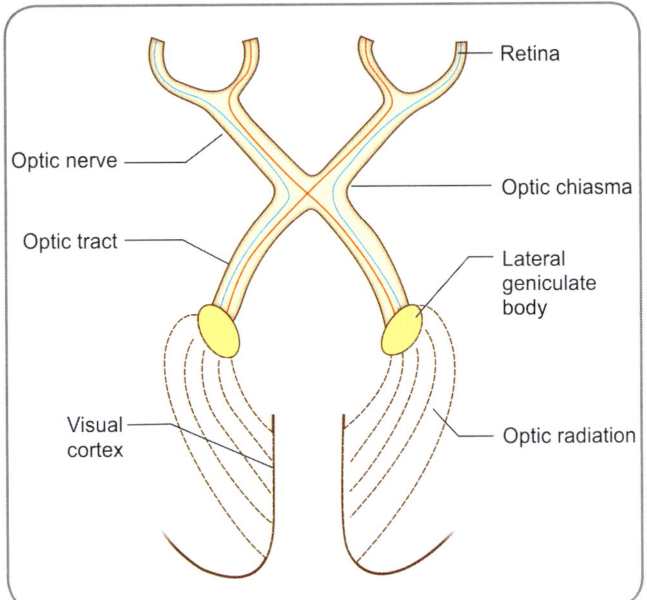

Fig. 25.3: The optic pathway. Note that the fibres from the medial (or nasal) half of each retina cross over to the optic tract of the opposite side

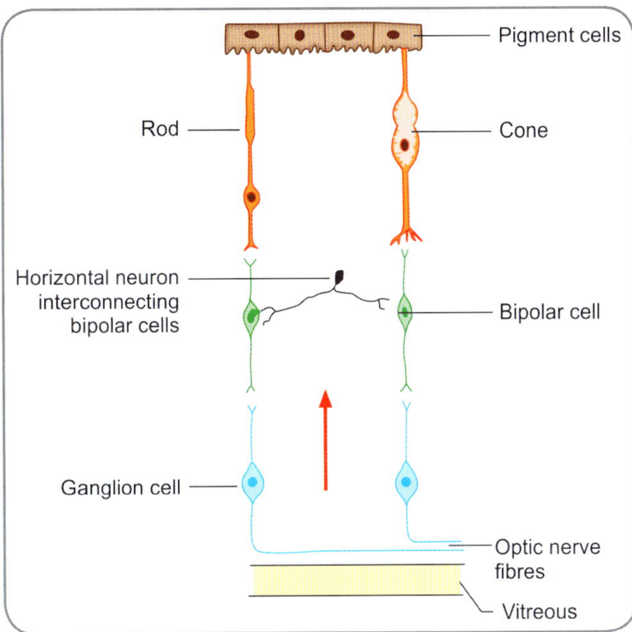

Fig. 25.4: Simplified scheme of the neurons within the retina. The red arrow indicates the direction of light falling on the retina

VISUAL SYSTEM

The peripheral receptors for light are situated in the *retina*. Nerve fibres arising in the retina constitute the *optic nerve*. The right and left optic nerves join to form the *optic chiasma*, in which many of their fibres cross to the opposite side. The uncrossed fibres of the optic nerve, along with the fibres that have crossed over from the opposite side, form the *optic tract*.

The optic tract terminates predominantly in the *lateral geniculate body*. Fibres arising in this body form the *geniculocalcarine tract* or *optic radiation*, which ends in the visual areas of the cerebral cortex (Fig. 25.3).

Retina

The retina has a complex structure. It contains photo-receptors that convert the stimulus of light into nervous impulses. These receptors are of two kinds, *rods* and *cones* (Fig. 25.4). There are about seven million cones in each retina. The rods are far more numerous, around more than a hundred million.

The cones respond best to bright light. They are responsible for sharp vision and for discrimination of colour. They are most numerous in the central region of the retina, which is responsible for sharp vision. This area is about 6 mm in diametre. Within this region, there is a yellowish area called the *macula lutea*. The centre of the macula shows a small depression called the *fovea centralis*.

Medial to the central area, there is a circular area called the *optic disc*. This area is devoid of photoreceptors and is, therefore, called the *blind spot*. The fibres of the optic nerve leave the eyeball through the region of the optic disc.

The macula lies exactly in the optical axis of the eyeball. When any object is viewed critically, its image is formed on the macula. The fovea centralis contains cones only. Rods, on the other hand, predominate in the peripheral parts of the retina. They respond to poor light and specially to movement across the field of vision.

Each rod or cone is a modified sensory neuron. It consists of a cell body, a peripheral process, and a central process.

The central processes of rods and cones synapse with the peripheral processes of *bipolar cells*. The central processes of bipolar cells synapse with dendrites of *ganglion cells*. Axons arising from ganglion cells form the *fibres of the optic nerve*.

Visual Field and Retinal Quadrants

When the head and eyes are maintained in a fixed position and one eye is closed, the area seen by that eye constitutes the *visual field* for that eye. Now, if the other eye is also opened, the area seen is more or less the same as was seen with one eye. In other words, the visual fields of the two eyes overlap to a very great extent. On either side, however, there is a small area seen only by the eye of that side (Fig. 25.5). Although the two eyes view the same area, the relative position of objects within the area

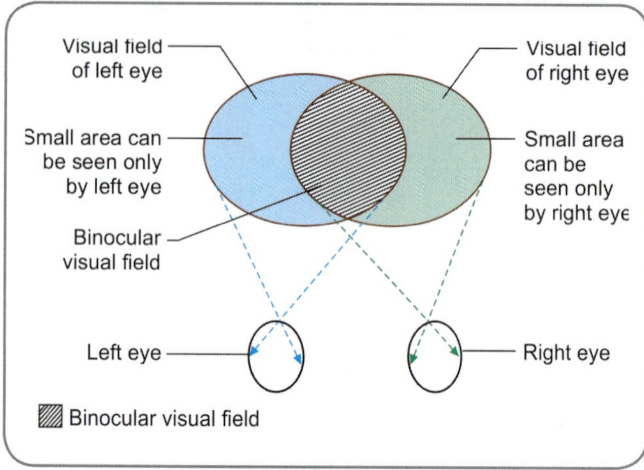

Fig. 25.5: Binocular visual field

crossing in the chiasma. Fibres from the temporal half of each retina enter the optic tract of the same side. Thus, the right optic tract comes to contain fibres from the right halves of both retinae and the left tract from the left halves. In other words, all optic nerve fibres carrying impulses relating to the left half of the field of vision are brought together in the right optic tract and *vice versa*. We have already noted that each optic tract carries these fibres to the lateral geniculate body of the corresponding side.

Lateral Geniculate Body

Lateral geniculate body receives the lateral root of the optic tract. Medially, it is connected to the superior colliculus, and laterally, it gives rise to the optic radiation.

The cells in this body are arranged in six layers. Layers 2, 3, 5 receive ipsilateral fibres, and layers 1, 4, 6 receive contralateral fibres.

Geniculocalcarine Tract and Visual Cortex

Fibres arising from cells of the lateral geniculate body constitute the geniculocalcarine tract or optic radiation. These fibres pass through the retrolentiform part of the internal capsule. The optic radiation ends in the visual areas of the cerebral cortex (areas 17, 18 and 19).

The cortex of each hemisphere receives impulses from the retinal halves of the same side (i.e., from the opposite half of the field of vision). The upper quadrants of the retina are represented above the calcarine sulcus and the lower quadrants below it. The cortical area for the macula is larger than that for peripheral areas. It occupies the posterior part of the visual area. The cortical area for the peripheral part of the retina is situated anterior to the area for the macula.

Neural Pathway for Vision

The first-order sensory neurons carrying visual sensations are bipolar cells of retina. Their dendrites synapse with rods and cones (photoreceptor) and their axons with the dendrites of ganglion cells.

The second-order sensory neurons are the ganglion cells. Axons arising from ganglion cells form the *fibres of the optic nerve*. The right and the left optic nerves join to form the **optic chiasma**, where fibres from nasal half of retina cross to the opposite side and travel through the opposite optic tract to terminate in the opposite lateral geniculate body. The fibres from temporal half of each retina enter the optic tract of the same side to terminate in the ipsilateral geniculate body.

The cell bodies of *third-order sensory neurons* are located in the lateral geniculate body. Their axons form the optic radiations, which project to the visual cortex.

appears somewhat dissimilar to the two eyes, as they view the object from slightly different angles. The difference, though slight, is of considerable importance, as it forms the basis for the perception of depth (***stereoscopic vision***).

For the convenience of description, the visual field is divided into right and left halves. It may also be divided into upper and lower halves so that the visual field can be said to consist of four quadrants. In a similar manner, each retina can also be divided into quadrants. Images of objects in the field of vision are formed on the retina by the lens of the eyeball. As with any convex lens, the image is inverted. If an object is placed in the ***right*** half of the field of vision, its image is formed on the ***left*** half of the retina and ***vice versa***. The two halves of the retina are usually referred to as ***nasal*** (= medial) and ***temporal*** (= lateral) halves. This introduces a complication as the left half of the left eye is the temporal half, while in the case of the right eye, it is the nasal half. Thus, the image of an object placed in the right half of the field of vision falls on the temporal half of the left retina and on the nasal half of the right retina.

Optic Nerve, Optic Chiasma and Optic Tract

The optic nerve is made up of axons of the ganglion cells of the retina. These axons are at first unmyelinated. The fibres from all parts of the retina converge on the optic disc. In this region, the sclera has numerous small apertures and is, therefore, called the ***lamina cribrosa*** (***crib* = sieve**). Bundles of optic nerve fibres pass through these apertures. Each fibre acquires a myelin sheath as soon as it pierces the sclera. The fibres of the nerve arising from the four quadrants of the retina maintain the same relative position within the nerve.

The fibres of the optic nerve arising in the nasal half of each retina enter the optic tract of the opposite side after

Clinical Correlation

Injuries to different parts of the visual pathway can produce various kinds of defects. Loss of vision in one half (right or left) of the visual field is called **hemianopia**. If the same half of the visual field is lost in both eyes, the defect is said to be **homonymous**, and if different halves are lost, the defect is said to be **heteronymous**. Note that the hemianopia is named in relation to the **visual field** and not to the retina.

Injury to the optic nerve will obviously produce total blindness in the concerned eye. Damage to the central part of the optic chiasma (e.g., by pressure from an enlarged hypophysis) interrupts the crossing fibres derived from the nasal halves of the two retinae, resulting in **bitemporal heteronymous hemianopia**. It has been claimed that macular fibres are more susceptible to damage by pressure than peripheral fibres and are affected first. When the lateral part of the chiasma is affected, a **nasal hemianopia** results. This may be unilateral or bilateral. Complete destruction of the optic tract, the lateral geniculate body, the optic radiation, or the visual cortex of one side results in loss of the opposite half of the field of vision. A lesion on the right side leads to **left homonymous hemianopia**.

Partial injury may affect only one quadrant. The resulting condition is called **quadrantic anopia**. Injury to looping fibres of the geniculocalcarine tract may lead to blindness in the upper quadrant, when the lesion is in the temporal lobe and in the lower quadrant, when the lesion is in the parietal lobe.

It may be noted that in lesions of the visual pathway, macular vision is often spared. This is so because of the large size of the macular area and because some areas have a double blood supply (from posterior and middle cerebral arteries).

Lesions confined to areas 18 and 19 (peristriate and parastriate areas) may not result in loss of vision, but the patient may be unable to interpret images seen. This is **visual agnosia**.

Lesions anterior to the lateral geniculate body also interrupt fibres responsible for the **pupillary light reflex**.

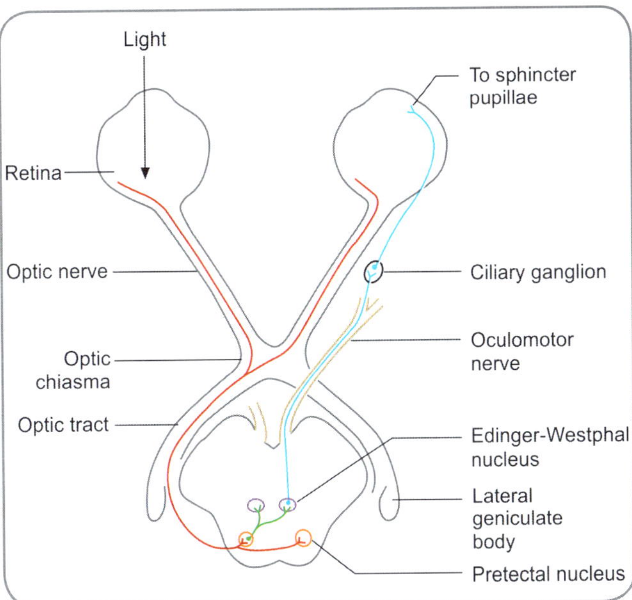

Fig. 25.6: Pathway for the light reflex. Note that all structures shown are bilateral. Some of them are shown only on one side for the sake of clarity

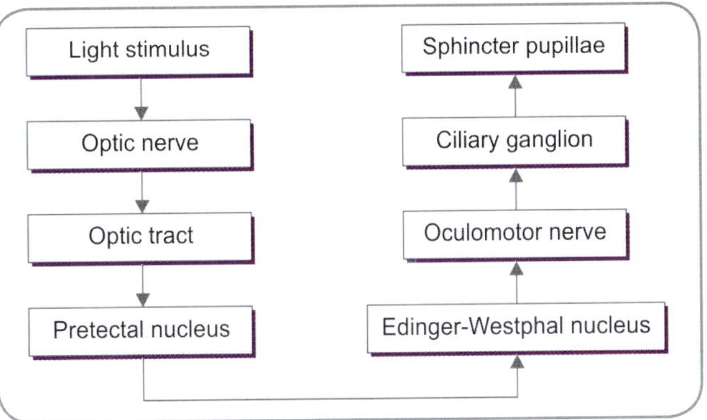

Fig. 25.7: Pathway for light reflex

REFLEXES RELATED TO THE EYEBALL

Pupillary Reflex or Light Reflex

Light thrown on an eye causes the pupil of that eye to contract. This is called the **direct pupillary light reflex**. At the same time, the pupil of the other eye also contracts. This is called the **consensual light reflex.** The pathway for the light reflex is shown in Figures 25.6 and 25.7. Impulses from the retina (of the eye on which light is thrown) travel through the optic nerve, chiasma, and optic tracts. Near the lateral geniculate body, the fibres concerned pass into the midbrain and end in the pretectal nucleus. Axons arising from this nucleus reach the Edinger-Westphal nuclei of both sides. Fibres arising in these nuclei supply the sphincter pupillae after relay in the ciliary ganglion. The consensual reflex may be explained by the following:

- Fibres of each optic nerve enter both optic tracts as a result of partial crossing in the chiasma.
- Fibres from each optic tract end in both pretectal nuclei.
- Fibres from each pretectal nucleus reach the Edinger-Westphal nuclei of both sides.

Clinical Correlation

Constriction of the pupil also takes place when we look at a near object, i.e., during accommodation. The pathway for this accommodation reflex is different from that for the light reflex and is believed to involve the visual cortex. In certain diseases (notably syphlitic infection) of the nervous system, the pupillary light reflex may be abolished, but pupillary constriction in response to accommodation remains. This condition is called Argyll-Robertson pupil. In such cases, the lesion is in the pretectal area.

Corneal Reflex

If the cornea is touched with a small wisp of cotton, it results in closing of both eyes. This is called the **corneal reflex**

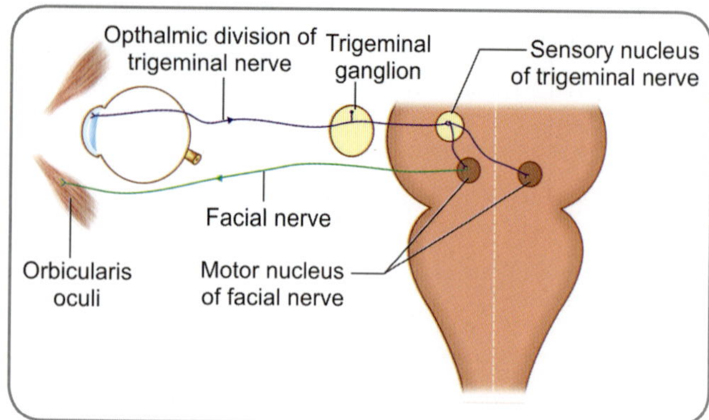

Fig. 25.8: Pathway for corneal reflex

(Fig. 25.8). Impulses from the cornea travel along the branches of the ophthalmic division of the trigeminal nerve to the sensory nuclei of this nerve. Secondary fibres establish connections with the motor nuclei of the facial nerve of both sides. Fibres arising in the facial nuclei and reaching the muscles closing the palpebral aperture complete the reflex arc. In the case of injury to the ophthalmic division of the trigeminal nerve, the corneal reflex cannot be elicited from that side, but a bilateral response is obtained by stimulating the cornea of the normal side. In facial paralysis, the response is seen only on the normal side.

AUDITORY SYSTEM

The internal ear contains the organ of hearing called the cochlea. The cochlea has a central bony core called the modiolus, and a spiral canal runs around it. The organ of Corti, which is the sensory epithelium of hearing, sits on the inner surface of the basilar membrane (Fig. 25.9).

The **first order neurons** of this pathway (**primary auditory pathway**) are located in the **spiral ganglion**,

which consists of two types of cells, type I and type II, present in intimate relationship to the cochlea. These neurons are bipolar (like first order neurons of olfactory and optic nerves, and unlike any other sensory nerves). Their peripheral processes reach the hair cells in the spiral organ of Corti (which is the end organ for hearing). Type I cells innervate inner hair cells. The outer hair cells are innervated by type II cells. The central processes of the neurons form the cochlear nerve, and terminate in the **dorsal and ventral cochlear nuclei**. The neurons in these nuclei are, therefore, second order neurons. Neurons receiving fibres from different parts of the spiral organ are arranged in a definite sequence in the ventral nucleus (tonotopic arrangement).

The axons of the **second order neurons (secondary auditory pathway)** pass medially in the dorsal part of the pons. It has some peculiarities that are as follows:

❑ Most of them cross to the opposite side, (but **some remain uncrossed**), and form the lateral lemniscus. The crossing fibres of the two sides form a conspicuous mass of fibres called the **trapezoid body**.

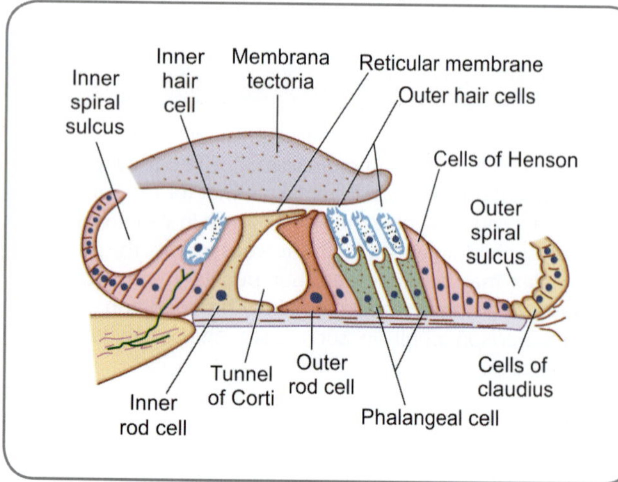

Fig. 25.9: Structure of the inner ear

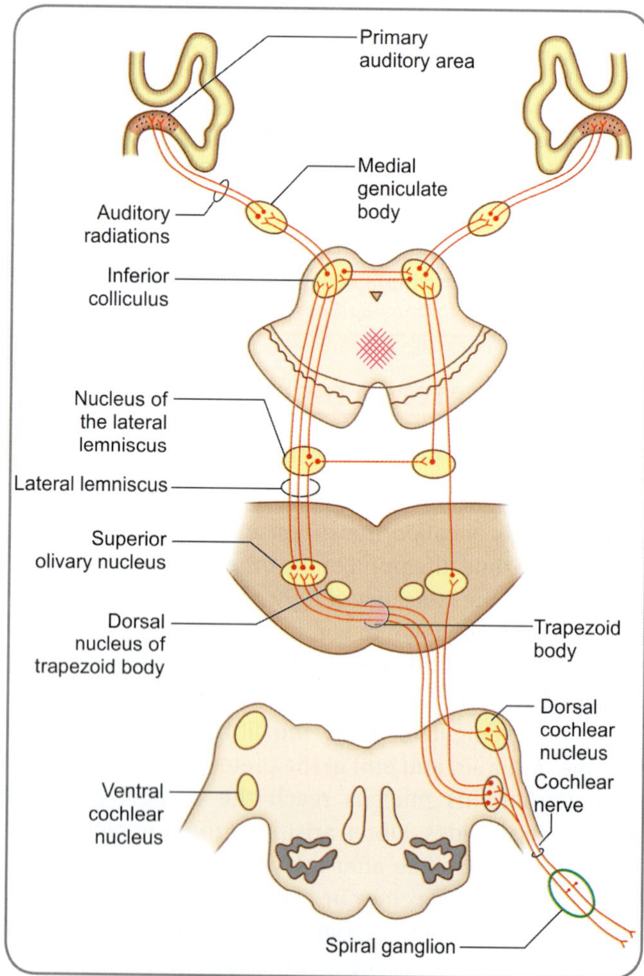

Fig. 25.10: The auditory pathway

❑ The large majority of fibres from the cochlear nuclei terminate in the ***superior olivary complex*** (made up of a number of nuclei). The ***medial superior olivary nucleus*** receives fibres from both cochleae and plays a role in localising the direction of sound (by calculating the time difference in arrival of inputs from the right and left cochleae).

❑ Some cochlear fibres that do not relay in the superior olivary nucleus join the lateral lemniscus after relaying in scattered groups of cells lying within the trapezoid body. These cells constitute the trapezoid nucleus (***nucleus of the trapezoid body***).

❑ Still other cochlear fibres relay in cells that lie within the lemniscus itself, which form the ***nucleus of the lateral lemniscus***.

❑ Most of the fibres of the lateral lemniscus ascend to the midbrain and terminate in the inferior colliculus. Fibres arising in the colliculus enter the inferior brachium to reach the medial geniculate body.

❑ Some fibres of the lateral lemniscus reach the ***medial geniculate body*** without relay in the inferior colliculus. (Fig. 25.10).

Fibres of the third order neurons (tertiary auditory pathway) arising in the medial geniculate body form the auditory radiation which ends tonotopically in the auditory area of the cerebral cortex (***anterior and posterior transverse temporal gyri***, Brodmann's areas 41, 42). Since each lateral lemniscus carries impulses arising in the right and left cochlea, lesions of temporal lobe will not cause complete deafness in either ear.

Multiple Choice Questions

1. The uncus:
   a. Is a part of the cerebellar cortex
   b. Is part of the primary olfactory area
   c. Represents the anterior end of parahippocampal gyrus
   d. Has the uncinate gyrus on the posterior aspect

2. The central processes of rods and cones synapse with:
   a. The dendrites of the bipolar cells
   b. The axons of the bipolar cells
   c. The dendrites of the ganglion cells
   d. The axons of the ganglion cells

3. Contraction of contralateral pupil on throwing light on the ipsilateral eye is:
   a. Direct Pupillary reflex
   b. Indirect Pupillary reflex
   c. Consensual Pupillary reflex
   d. Combined Pupillary reflex

4. Geniculocalcarine tract:
   a. Forms the Optic radiation
   b. Is the second order neuron of the visual pathway
   c. Is the bundle of axons of the ganglion cells
   d. Starts at the Superior Colliculus

5. Cones:
   a. Are responsible for sharp vision
   b. Are more numerous in the periphery of retina
   c. Respond best to poor light
   d. Are not found in the fovea centralis

ANSWERS

1. c   **2.** a   **3.** c   **4.** a   **5.** a

Clinical Problem-solving

Case Study 1: You had to examine a neurological patient. His pupillary lighr reflex was almost nil but his accommodation reflex was normal.
❑ What is this condition called?
❑ Why is there a difference in the two reflexes?
❑ Where, do you think, the lesion is?

Case Study 2: You are asked to test the right corneal reflex in an individual.
❑ What steps would you take?
❑ Trace the afferent and the efferent pathways.
❑ What will be the response if the facial nerve is paralysed on that side?

(For solutions see Appendix).

Blood Supply of the Central Nervous System

The nervous system is richly supplied with blood. Interruption of blood supply even for a short period can result in damage to nervous tissue. Traditionally, it has been taught that lymphatics are not present in nervous tissue, but recently this view has been challenged.

ARTERIES SUPPLYING THE BRAIN

The brain is supplied by branches of the ***internal carotid*** and the ***vertebral*** arteries (Figs 26.1 and 26.2). Each internal carotid artery gives off two major branches to the brain. These are the ***anterior cerebral*** and ***middle cerebral*** arteries.

The two vertebral arteries ascend on the anterolateral aspect of the medulla. At the lower border of the pons, they unite to form the ***basilar*** artery.

Circle of Willis (Circulus Arteriosus)

The circle of Willis is an arterial anastomatic circle present in the interpeduncular fossa. It is polygonal in shape and extends between the superior border of pons and median longitudinal fissure. The arterial circle is an anastomosis between the internal carotid and the vertebrobasilar system of arteries (Fig. 26.2).

Formation

- The anterior communicating artery, which connects the right and left anterior cerebral arteries, forms anterior part of the circle of Willis.
- The anterior cerebral artery forms the anterolateral part on each side.

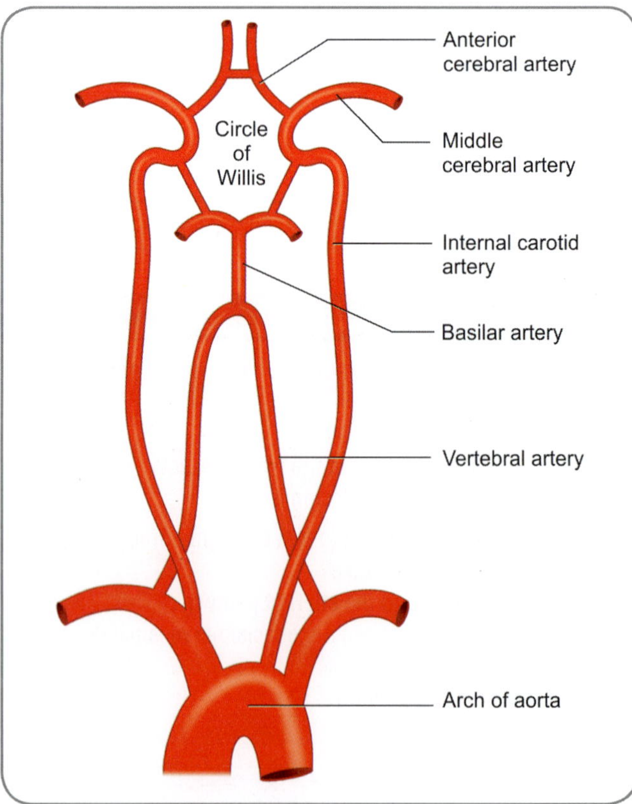

Fig. 26.1: Major arteries supplying the brain

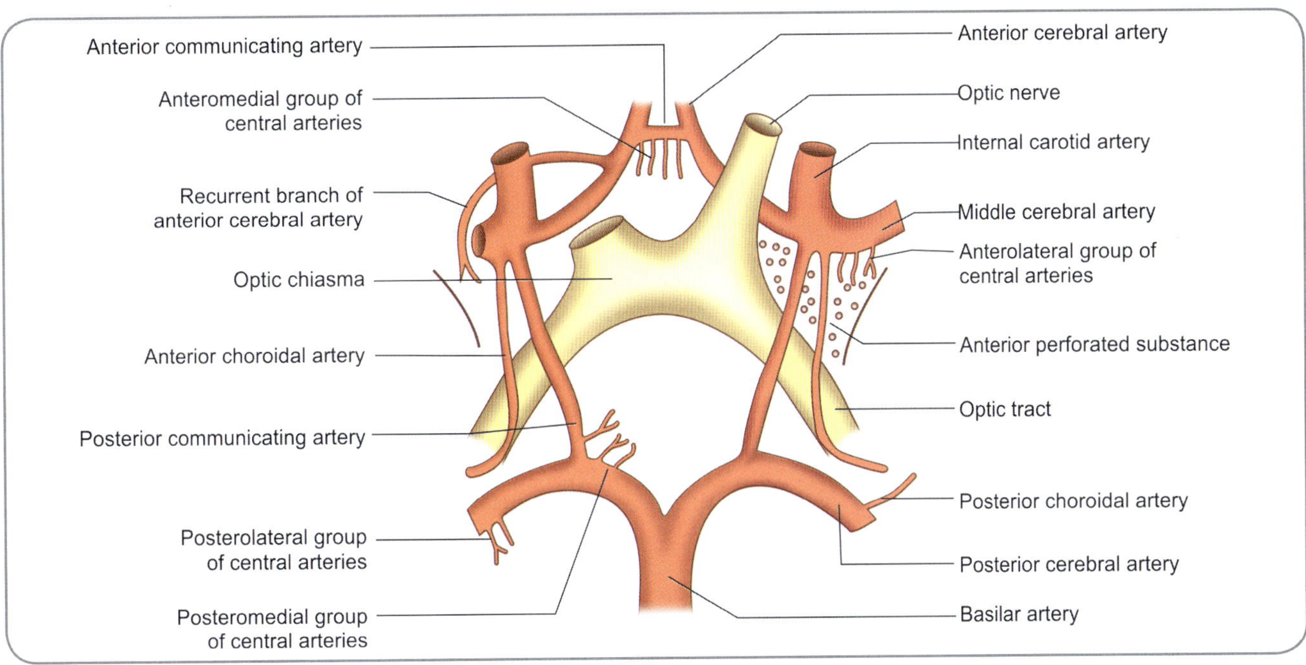

Fig. 26.2: Circulus arteriosus and related structures

❑ The lateral part is formed by the termination of internal carotid artery on each side.

❑ The circle is completed posteriorly by the bifurcation of basilar artery into the right and left posterior cerebral arteries.

❑ Posterolaterally, the posterior communicating artery is the connecting link between the internal carotid and posterior cerebral arteries.

Note: The middle cerebral artery does not take part in the formation of the circle of Willis.

Functional Importance

❑ This arterial circle equalises the pressure of the blood flow to the two sides of the brain, as it is the main collateral channel.

❑ The arterial anastomosis provides an alternative route through which blood entering the internal carotid artery or the basilar artery may be distributed to any part of the cerebral hemisphere. If one of the major arteries forming the circle of Willis is blocked, then it is through this anastomosis that blood can be supplied to the area of blocked artery.

✆ Clinical Correlation

Berry Aneurysm

The berry aneurysm is a localised dilatation on one of the arteries of the circle of Willis due to congenital muscular weakness. The most common sites of berry aneurysm are the junction of anterior cerebral and anterior communicating arteries and at the bifurcation of internal carotid arteries. Rupture of berry aneurysm may cause life-threatening subarachnoid hemorrhage.

VERTEBROBASILAR ARTERIES

❑ Each vertebral artery is a branch of subclavian artery and enters the cranial cavity through foramen magnum (Fig. 26.3). It curves around the ventrolateral aspect of medulla.

❑ The right and left vertebral arteries unite with each other at the lower margin of pons to form the basilar artery (Fig. 26.3).

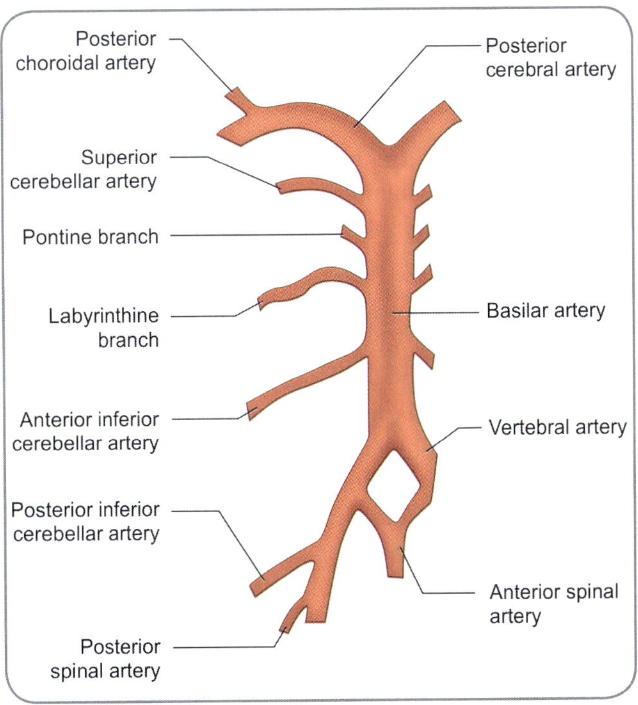

Fig. 26.3: Branches of the vertebral and basilar arteries

Intracranial branches of vertebral artery
- Anterior spinal artery
- Posterior spinal artery
- Posterior inferior cerebellar artery (largest branch of vertebral artery)
- Medullary arteries
- Meningeal branches

- The branches of intracranial part of vertebral artery supply not only the hindbrain but also the spinal cord.
- The basilar artery runs in the median sulcus on the ventral surface of pons and divides into right and left posterior cerebral arteries at upper margin of pons.

Branches of basilar artery
- Superior cerebellar artery
- Anterior inferior cerebellar artery
- Pontine branches
- Labyrinthine artery

In 83% of cases, labyrinthine artery originates from the anterior inferior cerebellar artery (for which reason it is termed as cerebellolabyrinthine artery), but variations in its origin include the lower part of basilar artery.

INTERNAL CAROTID ARTERY

The intracranial course of the artery is subdivided into intrapetrous, intracavernous and supraclinoid parts.

- **Intrapetrous part:** The internal carotid artery ascends in the carotid canal and enters the middle cranial fossa through the upper opening of the foramen lacerum.
- **Intracavernous part:** From foramen lacerum, the artery courses forward in the floor of the cavernous sinus, where it is in close relation to the abducent nerve.
- **Supraclinoid part:** After the artery pierces the dural roof of the cavernous sinus, it bends sharply backward and upward to terminate at the vallecula of cerebrum into middle and anterior cerebral arteries.

Branches

- The intrapetrous part gives origin to superior and inferior caroticotympanic branches, which enter the middle ear.
- The intracavernous part gives branches to the trigeminal ganglion, hypophyseal branches to the pituitary and the meningeal branches.
- The branches of supraclinoid part are ophthalmic artery, posterior communicating artery, anterior choroidal artery, recurrent branch of the anterior cerebral artery and terminal branches (anterior and middle cerebral arteries).

Note: Internal carotid artery gives direct branches to the genu of internal capsule.

Anterior choroidal artery, which arises directly from the internal carotid, runs backward in relation to the optic tract, and enters the inferior horn of the lateral ventricle through the choroid fissure. **Recurrent branch** of the anterior cerebral artery (also called the **artery of Heubner**) runs backward and laterally to enter the anterior perforated substance.

> ### Clinical Correlation
>
> - Thrombosis in the artery of Heubner results in contralateral paralysis of the face and upper extremity.
> - The aneurysm of internal carotid artery may compress the central part of optic chiasma and produce bitemporal hemianopia. Trauma to the internal carotid artery in cavernous sinus leads to the formation of arteriovenous fistula, causing pulsating exophthalmos.
> - The internal carotid artery shows multiple bends, which produce S-shaped shadow called the carotid siphon on an angiogram. The carotid siphon helps in damping down its pulsations in the cranial cavity.

Anterior Cerebral Artery

This artery arises from the internal carotid artery below the anterior perforated substance and lateral to the optic chiasma. It crosses the optic chiasma to reach the median longitudinal fissure. At the anterior end of the longitudinal fissure, the anterior communicating artery connects the right and left anterior cerebral arteries. Inside the longitudinal fissure, the anterior cerebral artery winds round the genu and then runs posteriorly on the superior aspect of the body of corpus callosum.

Distribution

- The orbital branches supply the medial half of orbital surface of the frontal lobe (olfactory bulb, olfactory gyrus and medial olfactory gyri).
- The frontal branches supply the medial surface of the frontal and the parietal lobes including the paracentral lobule, cingulate gyrus and corpus callosum.
- The parietal branches supply the precuneus.

Area of Supply

The anterior cerebral artery supplies the medial part of orbital surface of frontal lobe and the medial surfaces of frontal and parietal lobes. It also supplies 1–2 cm of the superolateral surface (of the frontal and parietal lobes) adjacent to superomedial border. The functional areas that receive blood from anterior cerebral artery are the motor and sensory areas for lower limb and perineum.

> ### ✂ Clinical Correlation
>
> **Effects of Occlusion of Anterior Cerebral Artery**
>
> ❏ Paralysis (or weakness) of muscles of the leg and foot of the opposite side (by involvement of the upper part of the motor area).
> ❏ Loss (or dulling) of sensations from the leg and foot of the opposite side (by involvement of the upper part of the sensory area).
> ❏ Sense of stereognosis is impaired (by involvement of parietal lobe).
> ❏ Personality changes (by involvement of frontal lobe).

Middle Cerebral Artery

It is one of the terminal branches of the internal carotid artery. It turns laterally on the anterior perforated substance to enter the stem of the lateral sulcus, where it divides into four to five cortical branches on the surface of insula.

Distribution

❏ The orbital branches supply the lateral half of orbital surface of frontal lobe.
❏ The frontal branches supply the superolateral surface of frontal lobe (precentral, middle and inferior frontal gyri) excluding the area supplied by anterior cerebral artery.
❏ The parietal branches supply the postcentral gyrus, inferior parietal lobule and superior parietal lobule, excluding the area supplied by the anterior cerebral artery.
❏ The temporal branches supply the lateral surface of the temporal lobe excluding the inferior temporal gyrus but including the temporal pole.

The functional areas that receive blood from the middle cerebral artery are the motor and sensory areas (upper limb, trunk and face), premotor area, frontal eye field, auditory area and the speech centres in the dominant hemisphere.

> ### ✂ Clinical Correlation
>
> **Effects of Occlusion of Middle Cerebral Artery**
>
> ❏ Hemiplegia and loss of sensations on the opposite half of the body. The face and arms are most affected. Foot and leg are spared.
> ❏ Aphasia (by involvement of Broca's and Wernicke's areas), specially if the thrombosis is in the left hemisphere in a right-handed person.
> ❏ Homonymous hemianopia on opposite side (by involvement of optic radiation).
> ❏ Hearing may be affected, but this may be compensated by the opposite hemisphere.

Posterior Cerebral Artery

The right and left posterior cerebral arteries are the terminal branches of basilar artery. Each passes laterally around the crus cerebri of the midbrain, where it receives the posterior communicating artery. It continues along the lateral aspect of the midbrain and enters the supratentorial compartment through the tentorial notch. Then, it courses on the tentorial surface of temporal lobe lying on the parahippocampal gyrus.

Distribution

❏ The posterior temporal branches supply to the inferior surface of temporal lobe, uncus and occipitotemporal and lingual gyri. They also send twigs to inferior temporal gyrus excluding the temporal pole.
❏ The calcarine and parieto-occipital branches supply the medial surface of the occipital lobe, which includes the cuneus. These cortical branches send twigs to superolateral surface of the occipital lobe.

The visual cortex is the important functional area supplied by the posterior cerebral artery. The occipital pole receives blood from the anastomosis between the branches of posterior and middle cerebral arteries. Therefore, in the occlusion of the posterior cerebral artery, there is macular sparing.

> ### ✂ Clinical Correlation
>
> **Effects of Occlusion of Posterior Cerebral Artery**
>
> ❏ The loss of cortical supply results in contralateral homonymous hemianopia with macular sparing. Damage to association cortex of visual area causes visual hallucinations (distortion of colour vision).
> ❏ The part of the visual area responsible for macular vision lies in the region where the territories of supply of the middle and posterior cerebral arteries meet. It may receive a supply from the middle cerebral artery, either directly or through anastomoses with branches of the posterior cerebral artery. This is one explanation for the observation that macular vision is often spared in cases of thrombosis of the posterior cerebral artery. The phenomenon can also be explained by the observation that dye injected into the carotid system (for angiographic studies) often passes into the posterior cerebral artery through the posterior communicating artery.

Note: From the description given above, it will be clear that the main somatic motor and sensory areas are supplied by the middle cerebral artery, except in their uppermost parts (leg areas), which are supplied by the anterior cerebral. The acoustic area is supplied by the middle cerebral artery and the visual area, by the posterior cerebral.

Branching Pattern of the Cerebral Arteries

The cerebral arteries give three types of branches:

1. Cortical
2. Central
3. Choroidal

Cortical Branches

The **cortical branches** ramify on the surface of the cerebral hemispheres and supply the cortex.

They give off branches that run perpendicularly into the substance of the cerebral hemisphere. Some of these are short and end within the grey matter of the cortex. Others are longer and penetrate into the subjacent white matter. While cortical branches may anastomose with each other on the surface of the brain, the perpendicular branches (both long and short) behave as terminal or end arteries. Each branch supplies a limited area of brain tissue and does not anastomose with neighbouring arteries. As a result, blockage of such a branch leads to death (necrosis) of brain tissue in the region of supply.

Central Branches

The central arteries arise in the region of arterial circle of Willis and are end arteries.

They pass deep into the substance of the cerebral hemisphere to supply structures within it and consist of six main groups:

- Anteromedial
- Posteromedial
- Right and left anterolateral
- Right and left posterolateral (Fig. 26.2)

The arteries of the **anteromedial group** arise from the anterior cerebral and anterior communicating arteries. They enter the most medial part of the anterior perforated substance.

The arteries of the **anterolateral group** are the so-called **striate arteries**. They arise mainly from the middle cerebral artery. Some of them arise from the anterior cerebral artery. The anterolateral group of perforating arteries enter the anterior perforated substance and divide into two sets, medial and lateral. The **medial striate arteries** ascend through the lentiform nucleus. They supply this nucleus and also the caudate nucleus and the internal capsule. The **lateral striate arteries (lenticulostriate)** ascend lateral to the lower part of the lentiform nucleus; they then turn medially and pass through the substance of the lentiform nucleus to reach the internal capsule and the caudate nucleus. One of these lateral striate arteries is usually larger than the others. It is called **Charcot's artery**, or **artery of cerebral haemorrhage**.

The **posteromedial group** of central arteries arise from the posterior cerebral and posterior communicating arteries. They enter the interpeduncular region.

The central branches of the **posterolateral group** arise from the posterior cerebral artery, as it winds around the cerebral peduncle.

ARTERIAL SUPPLY OF CEREBRAL CORTEX

The cerebral cortex is supplied by cortical branches of the anterior, middle and posterior cerebral arteries.

Superolateral Surface

The greater part of the superolateral surface is supplied by the middle cerebral artery (Fig. 26.4). The areas **not** supplied by this artery are as follows:

- A strip half to one inch wide along the superomedial border, extending from the frontal pole to the parieto-occipital sulcus is supplied by the anterior cerebral artery.
- The area belonging to the occipital lobe is supplied by the posterior cerebral artery.
- The inferior temporal gyrus (excluding the part adjoining the temporal pole) is also supplied by the posterior cerebral artery.

Medial Surface

The main artery supplying the medial surface is the anterior cerebral artery (Fig. 26.5). The area of this surface belonging to the occipital lobe is supplied by the posterior cerebral artery.

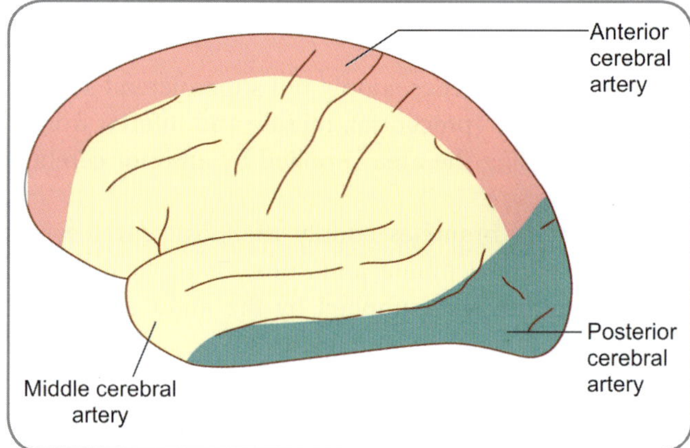

Fig. 26.4: Distribution of the anterior, posterior and middle cerebral arteries on the superolateral surface of the cerebral hemisphere

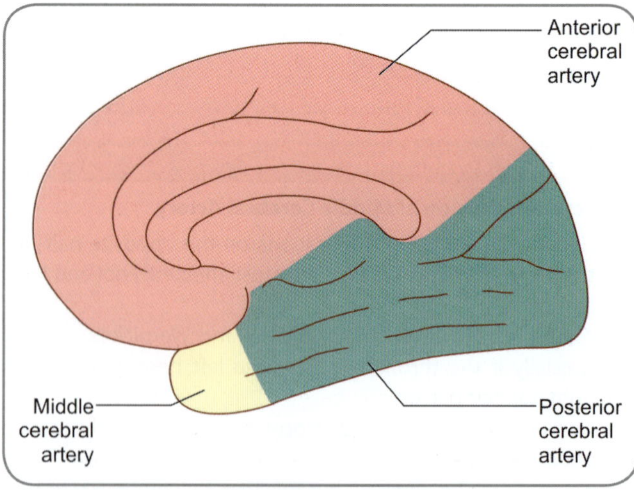

Fig. 26.5: Arteries supplying the medial surface of the cerebral hemisphere

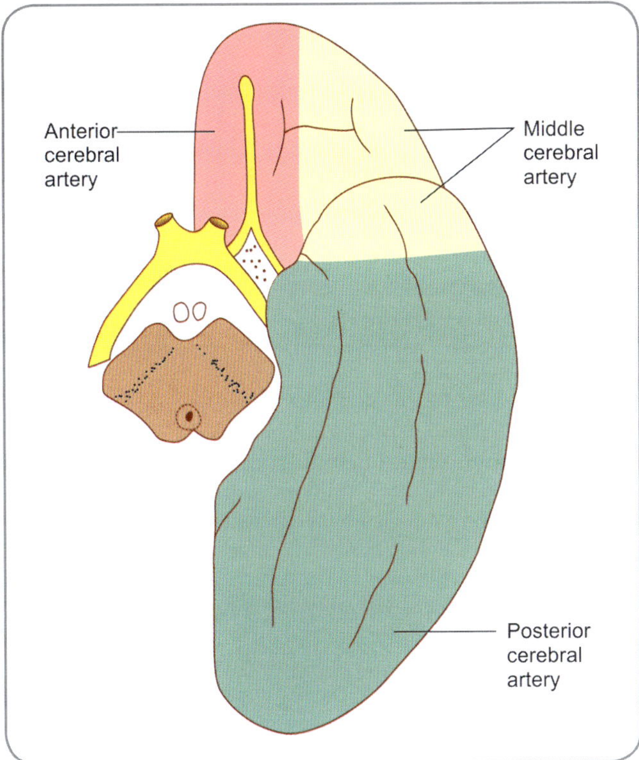

Fig. 26.6: Arteries supplying the orbital and tentorial surfaces of the cerebral hemisphere

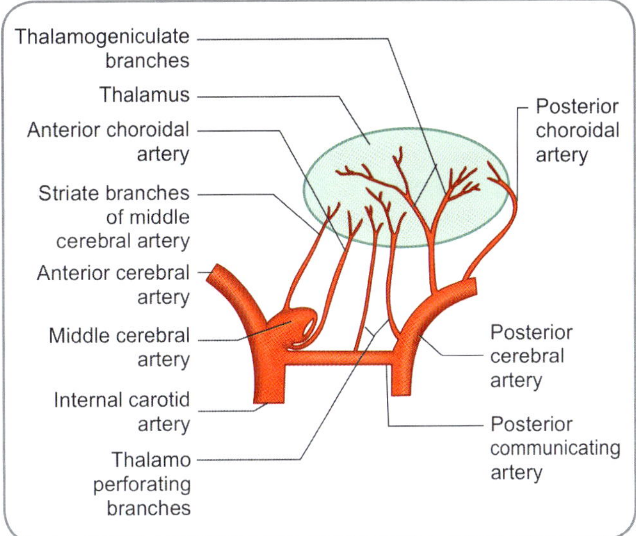

Fig. 26.7: Scheme to show the arteries supplying the thalamus

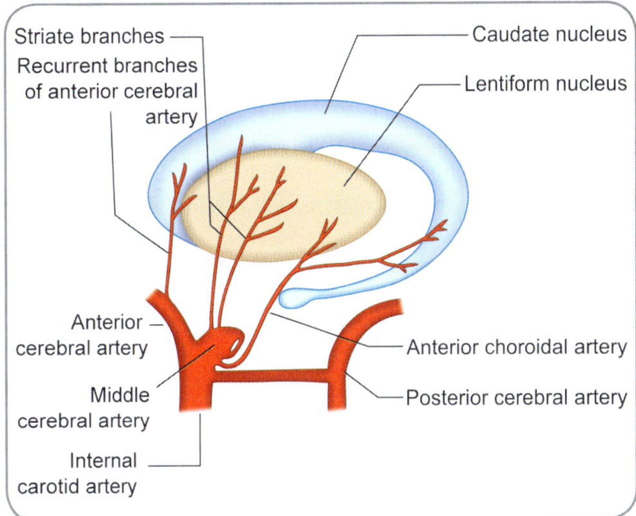

Fig. 26.8: Scheme to show the arteries supplying the corpus striatum

Inferior Surface

The lateral part of the ***orbital surface*** is supplied by the middle cerebral artery and the medial part by the anterior cerebral artery (Fig. 26.6).

The ***tentorial surface*** is supplied by the posterior cerebral artery. The temporal pole is, however, supplied by the middle cerebral artery (Fig. 26.6).

Arteries Supplying the Interior of the Cerebral Hemisphere

Internal Capsule

The main arteries supplying the internal capsule are the medial and lateral striate branches of the middle cerebral artery, the recurrent branch of the anterior cerebral, and the anterior choroidal artery. The internal capsule may also receive direct branches from the internal carotid artery and branches from the posterior communicating artery.

Thalamus

The thalamus is supplied mainly by perforating branches of the posterior cerebral artery (Fig. 26.7). The posteromedial group of branches (also called ***thalamo-perforating*** arteries) supply the medial and anterior parts. The posterolateral group (also called ***thalamogeniculate*** branches) supplies the posterior and lateral parts of the thalamus. The thalamus also receives some branches from the posterior communicating, anterior choroidal, posterior choroidal and middle cerebral arteries.

Hypothalamus

The anterior part of the hypothalamus is supplied by central branches of the anteromedial group (arising from the anterior cerebral artery). The posterior part is supplied by central branches of the posteromedial group (arising from the posterior cerebral and posterior communicating arteries).

Corpus Striatum

The main arterial supply of the ***caudate nucleus*** and ***putamen*** is derived from the medial and lateral striate branches of the middle cerebral artery (Fig. 26.8). In

449

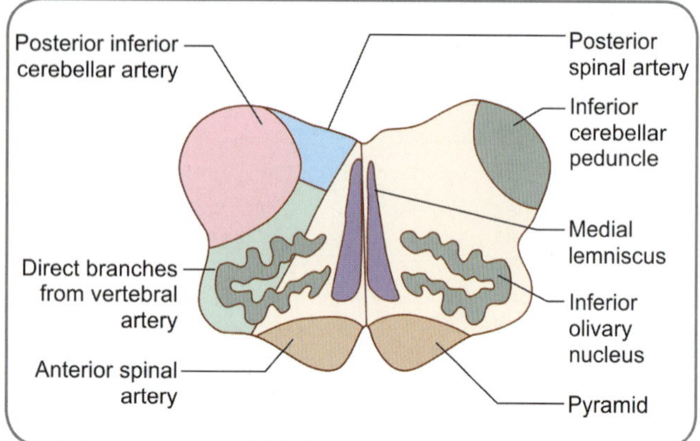

Fig. 26.9: Cross-section passing through the medulla to show the regions supplied by different arteries

addition, their anteriormost parts (including the head of the caudate nucleus) receive their blood supply through the recurrent branch of the anterior cerebral artery and their posterior parts (including the tail of the caudate nucleus) through the anterior choroidal artery.

The main supply of the **globus pallidus** is from the anterior choroidal artery. Its lateral segment also receives blood through the striate arteries. The medialmost part of the globus pallidus receives branches from the posterior communicating artery.

Arterial Supply of the Brainstem

Medulla

The medulla is supplied by various branches of the vertebral arteries. These are the anterior and posterior spinal arteries, the posterior inferior cerebellar artery and small direct branches (Fig. 26.9). The anterior spinal artery supplies a triangular area next to the midline. This area includes the pyramid, the medial lemniscus, and the hypoglossal nucleus. The posterior spinal artery supplies a small area, including the gracile and cuneate nuclei. The posterior inferior cerebellar artery supplies the retro-olivary region. This region contains several important structures, including the spinothalamic tracts, rubrospinal tract, nucleus ambiguus, dorsal vagal nucleus, and descending autonomic fibres. The posterior inferior cerebellar artery also supplies part of the inferior cerebellar peduncle. The rest of the medulla is supplied by direct bulbar branches of the vertebral arteries.

Clinical Correlation

Thrombosis in an artery supplying the medulla produces symptoms depending upon the structures involved. Two characteristic syndromes are the **medial medullary syndrome** produced by thrombosis in the anterior spinal artery and the **lateral medullary syndrome** produced by thrombosis in the posterior inferior cerebellar artery.

contd...

Clinical Correlation *contd...*

Medial Medullary Syndrome

In the medial medullary syndrome, the ventral and medial part of the medulla is damaged. Involvement of corticospinal fibres (pyramids) leads to contralateral hemiplegia. Damage to the hypoglossal nucleus leads to ipsilateral (lower motor neuron type of) paralysis of muscles of the tongue. Involvement of the medial lemniscus leads to loss of the sensation of fine touch, sense of movement, and sense of position.

Lateral Medullary Syndrome

In the lateral medullary syndrome, the structures damaged include the following:

- ❑ Damage to the lateral spinothalamic tract leads to contralateral loss of sensations of pain and temperature.
- ❑ Damage to the spinal nucleus and tract of the trigeminal nerve leads to loss of sensations of pain and temperature over the region supplied by the trigeminal nerve.
- ❑ Damage to the nucleus ambiguus leads to difficulty in swallowing (dysphagia) and in speech (dysarthria).
- ❑ Damage to the cerebellum and to the inferior cerebellar peduncle causes loss of equilibrium (ataxia) and giddiness.
- ❑ Involvement of descending autonomic fibres leads to Horner's syndrome.

Pons

The pons is supplied by branches from the basilar artery. The medial portion of the ventral part of the pons is supplied by **paramedian branches**. The lateral portion of the ventral part is supplied by **short circumferential branches**. The dorsal part of the pons is supplied by **long circumferential branches**. The dorsal part also receives branches from the anterior inferior cerebellar and superior cerebellar arteries. The paramedian branches of the basilar artery may extend into this region from the ventral part of the pons.

Clinical Correlation

Pontine Haemorrhage

Haemorrhage into the pons leads to coma (and is often fatal). The reticular formation, ANS, and the hypothalamus are affected. Apart from coma, the condition is marked by pin-point pupils and hyperpyrexia. Bilateral facial paralysis and paralysis of all four limbs can occur if the haemorrhage is extensive.

Midbrain

The midbrain is supplied mainly by branches of the basilar artery. These are the posterior cerebral and superior cerebellar arteries and direct branches from the basilar artery. Branches are also received from the posterior communicating and anterior choroidal arteries. Branches arising from these vessels may either be **paramedian**, which supply parts near the midline or **circumferential**, which wind round the midbrain to supply lateral and dorsal parts. One of the latter arteries is called the **quadrigeminal artery**. It is the main source of blood to the colliculi.

Cerebellum

The superior surface of the cerebellum is supplied by the **superior cerebellar** branches of the basilar artery. The anterior part of the inferior surface is supplied by the **anterior inferior cerebellar** branches of the same artery. The posterior part of the inferior surface is supplied by the **posterior inferior cerebellar** branch of the vertebral artery.

VENOUS DRAINAGE OF BRAIN

The veins draining the brain open into the dural venous sinuses (Fig. 26.10). These are the **superior sagittal**, **inferior sagittal**, **straight**, **transverse**, **sigmoid**, **cavernous**, **sphenoparietal**, **petrosal**, and **occipital sinuses**. Ultimately, the blood from all these sinuses reaches the sigmoid sinus, which becomes continuous with the **internal jugular vein**. The intracranial venous sinuses communicate with veins outside the skull through emissary veins.

The veins draining the brain are valveless and thin-walled because their walls are devoid of muscles.

Veins of the Cerebral Hemisphere

The veins of the cerebral hemisphere consist of two sets, superficial and deep. The superficial veins lie in the subarachnoid space on the surface of the cerebral hemisphere and drain mainly into the superior sagittal sinus, which ultimately drain into the right internal jugular vein. On the other hand, the deep veins drain mainly into the **great cerebral vein**, which ultimately drain into left internal jugular vein (Table 26.1).

Superficial Cerebral Veins

The superficial veins drain into the neighbouring venous sinuses (Fig. 26.11). The **superior cerebral veins** drain the upper parts of the superolateral and medial surfaces

Table 26.1: Veins of the cerebrum	
Superficial cerebral veins	**Deep cerebral veins**
Superior cerebral veins	Internal cerebral veins
Inferior cerebral veins	Basal veins
Superficial middle cerebral vein	Thalamostriate vein
Superior anastomotic vein	Great cerebral vein
Inferior anastomotic vein	

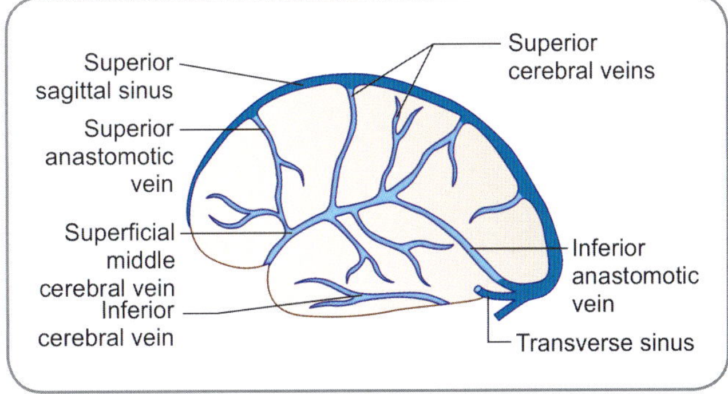

Fig. 26.11: Superior cerebral veins on the superolateral surface of the cerebrum

and end in the superior sagittal sinus. Some veins from the medial surface join the inferior sagittal sinus. **Inferior cerebral veins** drain the lower part of the hemisphere. On the superolateral surface, they drain into the **superficial middle cerebral vein**, which lies superficially along the lateral sulcus and its posterior ramus. The posterior end of this vein is connected to the superior sagittal sinus by the **superior anastomotic vein** and to the transverse sinus by the **inferior anastomotic vein**. The superficial middle cerebral vein terminates in the cavernous sinus. Veins from the inferior surface of the cerebral hemisphere drain into the transverse, superior petrosal, cavernous, and sphenoparietal sinuses. Some may ascend to join the inferior sagittal sinus.

Deep Cerebral Veins

The deep veins of the cerebral hemisphere are the two **internal cerebral veins** that join to form the **great cerebral vein** and the two **basal veins** that wind round the midbrain to end in the great cerebral vein (Fig. 26.12). Each internal cerebral vein begins at the interventricular foramen and runs backwards in the tela choroidea in the roof of the third ventricle. It has numerous tributaries. One of these is the **thalamostriate vein**, which lies in the floor of the lateral ventricle (between the thalamus, medially and the caudate nucleus, laterally). Each basal vein begins near the anterior perforated substance. It is formed by union of the following:

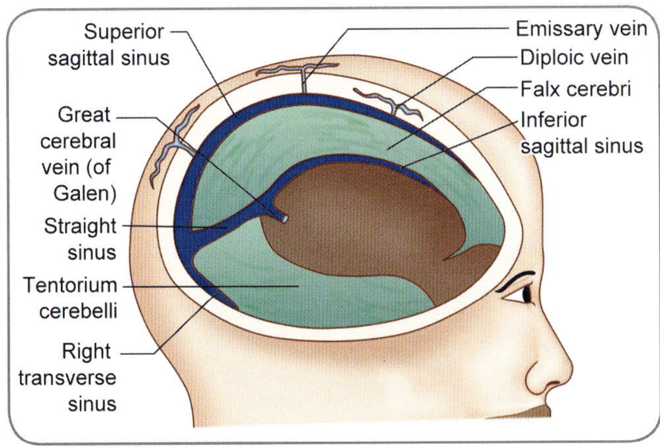

Fig. 26.10: Venous drainage of brain

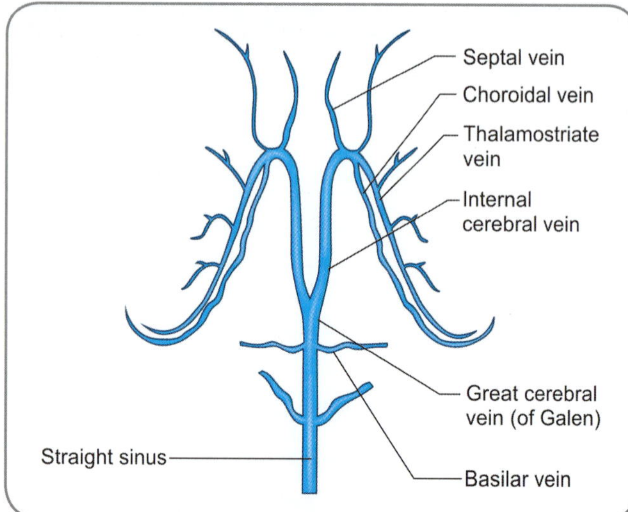

Fig. 26.12: Deep cerebral veins

❑ The ***anterior cerebral vein***, which accompanies the anterior cerebral artery.
❑ The ***deep middle cerebral vein***, which lies deep in the stem and posterior ramus of the lateral sulcus.
❑ Some ***inferior striate veins*** that emerge from the anterior perforated substance.

The great cerebral vein, formed by union of the two internal cerebral veins, passes posteriorly beneath the splenium of the corpus callosum to end in the straight sinus. It receives the basal veins, some veins from the occipital lobes, and some from the corpus callosum.

The deep cerebral veins described above are responsible for draining the thalamus, hypothalamus, corpus striatum, internal capsule, corpus callosum, septum pellucidum and choroid plexuses. Many tributaries of the internal cerebral veins extend beyond the corpus striatum into the white matter of the hemispheres. Here, they establish communications with superficial veins. They can, thus, serve as alternative channels for draining parts of the cerebral cortex.

Venous Drainage of Other Parts of the Brain

The upper part of the ***thalamus*** is drained by the tributaries of the internal cerebral vein (including the thalamostriate vein). The lower part of the thalamus and the hypothalamus are drained by veins that run downwards to end in a plexus of veins present in the interpeduncular fossa. This plexus drains into the cavernous and sphenoparietal sinuses and into the basal veins.

The ***corpus striatum*** and ***internal capsule*** are drained by two sets of striate veins. The ***superior striate veins*** run dorsally and drain into tributaries of the internal cerebral vein. The ***inferior striate veins*** run vertically downwards and emerge on the base of the brain through the anterior perforated substance. Here, they end in the basal vein.

Veins of the Cerebellum and Brainstem

The veins from the upper surface of the **cerebellum** drain into the straight, transverse and superior petrosal venous sinuses. Veins from the inferior surface drain into the right and left sigmoid, inferior petrosal sinuses, occipital sinus and straight sinus.

The veins of the ***midbrain*** drain into the great cerebral vein or into the basal vein. The ***pons*** and ***medulla*** drain into the superior and inferior petrosal sinuses, transverse sinus and occipital sinus. Inferiorly, the veins of the medulla are continuous with the veins of the spinal cord.

BLOOD-BRAIN BARRIER

It has been observed that while some substances can pass from the blood into the brain with ease, others are prevented from doing so. This has given rise to the concept of a selective barrier between blood and the brain.

Structure of Blood-Brain Barrier

Anatomically, the structures that could constitute the barrier are as follows (Fig. 26.13):
❑ Capillary endothelium
❑ Basement membrane of the endothelium
❑ Closely applied to the vessels, there are numerous processes of astrocytes. It has been estimated that these processes cover about 85% of the capillary surface

Areas of the Brain Devoid of Blood-Brain Barrier

Some areas of the brain (and related structures) appear to be devoid of a blood-brain barrier. These include:
❑ Pineal body
❑ Hypophysis cerebri
❑ Choroid plexus
❑ Median eminence (hypothalamus)
❑ Some specialised areas of ependyma in the walls of the third and fourth ventricles

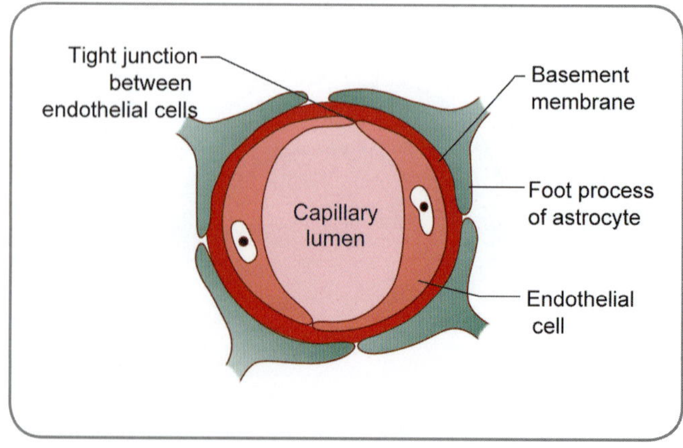

Fig. 26.13: Structure of blood-brain barrier

Clinical Correlation

The blood-brain barrier can break down following ischaemia or infection in the brain. The barrier can also break down in trauma and through the action of toxins. Some drugs, including some antibiotics, can pass through the barrier while others cannot.

In infants, bilirubin can pass through the barrier. There is a danger of encephalitis, if bilirubin levels are high (seen as jaundice in the newborn or kernicterus).

Traditionally, it has been taught that the subarachnoid space is continuous with perivascular spaces (present around blood vessels passing into the brain). However, it has now been shown that the perivascular spaces and the subpial space are completely cut off from the subarachnoid space by pia mater (which is reflected onto arteries as a sleeve). The pia mater, therefore, contributes to the establishment of the blood-brain barrier.

Control of Cerebral Blood Flow

Cerebral blood flow is influenced by sympathetic nerves (which are present around arteries as they pass through the subarachnoid space). Adrenergic nerve fibres within the brain also end on blood vessels.

Blood flow through the brain does not markedly vary with alterations in blood pressure. Blood flow through a part of the brain increases when that part is "active". Such areas can be visualised by using the technique of *positron emission tomography* (PET). Studies using the technique are throwing much light on functions of various areas. PET can be combined with *magnetic resonance imaging* (MRI) to provide accurate localisation of the areas showing altered blood flow.

Multiple Choice Questions

1. The Circle of Willis:
   a. Equalizes pressure of blood flow to the two sides of brain
   b. Has the middle cerebral artery in its lateral part
   c. Is an arteriovenous circle in the interpeduncular fossa
   d. Has no role to play in collateral circulation
2. A branch of the basilar artery is:
   a. Posterior cerebral Artery
   b. Posterior Spinal Artery
   c. Superior Cerebellar Artery
   d. Anterior Choroidal Artery
3. Charcot's artery belongs to:
   a. Medial striate group
   b. Lateral striate group
   c. Thalamogeniculate group
   d. Long circumferential group
4. The part devoid of blood brain barrier is:
   a. Cerebral cortex
   b. Corpus striatum
   c. Midbrain
   d. Pineal body

ANSWERS

1. a   **2.** c   **3.** b   **4.** d

Clinical Problem-solving

Case Study 1: A 54-year-old man has right sided hemiplegia, paralysis of the tongue muscles of the left side and loss of sense of touch on the right side.
- What is the condition affecting this patient?
- What is the cause for this condition?
- Enumerate the anatomical basis of the symptoms seen.

Case Study 2: A 67-year-old man suffered a cerebrovascular accident. Consequently, he had right sided hemiplegia, right sided loss of sensations and aphasia. However, he was able to walk to some extent since his legs and feet were not greatly affected.
- What is the most likely cause for these symptoms?
- What are the other possible symptoms?
- Why are hearing defects not very prominent in conditions of this kind?

(For solutions see Appendix).

Genetics

Chapter 27

Introduction to Genetics

Frequently Asked Questions

- Write short answers on: (a) Classic genetics, (b) Population genetics, (c) Medical genetics, (d) Genetic epidemiology.
- Write very briefly on: (a) History of genetics, (b) Subspecialities of genetics.

AN INTRODUCTION AND OVERVIEW

Genetics is the science that studies about heredity, transmission of hereditary material from one generation to another, factors responsible for such transmission and alterations which are likely to occur in such transmission. The consequences of such alterations and their study are also part of this science.

HISTORY OF GENETICS

Gregor Johann Mendel , an Austrian monk, performed a number of experiments on pea plants and derived at what are called *Mendelian principles*. These principles describe patterns of inheritance. He published the outcome of his work in 1866. Though, at that point of time, many of his postulates were accepted, in the subsequent decades, theories and counter theories arose.

In 1900, Hugo de Vries favoured Mendelian theories. Hailed as rediscovery of Mendel, this led to more research on Mendelian principles. Carl Correns and Erich von Tschermak worked more on these principles, gradually leading genetic workers to apply Mendelian principles to the fruitfly, *Drosophila melanogaster*. Thomas Hunt Morgan did extensive work on Drosophila to be labelled as a 'drosophilist geneticist'. By 1925, Mendelian principles were well established and mendelian models were developed.

Once the genetic patterns of inheritance were established, the attention of scientific workers turned towards the 'gene'. In 1940s and 50s, genes, DNA and chromosomes were researched upon. The double helical structure of the DNA was discovered in 1953, gradually leading to greater concentration on the molecular aspects of genetics. A new subspeciality called *Molecular Genetics* was born. Sequencing of the nucleic acids and proteins started in 1960s; the idea of a *genetic code* was thought of, established and well documented in the 60s. The same decade also saw emphasis on gene expression. In the 1970s, control and manipulation of the gene expression through genetic engineering techniques were made possible. Large scale genetics projects, genome sequencing, cloning projects and other such applicative work were done in the last few decades of the twentieth century.

SUBSPECIALITIES OF GENETICS

With vast research and several breakthroughs, Genetics, as a speciality, has grown far and wide. Many subspecialities have come into existence. Some of these are as follows:

- *Classic genetics*: Study of the visible results of reproductive function and methods. This is the oldest subspeciality. Mendel observed the results of pollination and cross pollination of pea plants and deduced his principles.
- *Molecular genetics*: Study of the structure and functions of genes at a molecular level.
- *Population genetics:* Study of distribution and changes of allele frequency in a population taking into account the factors of recombination, adaptation, speciation and population structure. Forerunners of this subspeciality were J. B. S. Haldane, Sewall Wright and R. A. Fischer.

- ***Quantitative genetics:*** Study of continually varying phenotypes in a population.
- ***Ecological genetics:*** Study of genetics in a natural population. Founder of this subspeciality was E. B. Ford.
- ***Behaviour genetics:*** Study of the role of genetics in the behaviour of the organism, animal or human being.
- ***Conservation genetics:*** Study of application of principles of genetics to conserve and restore biodiversity.
- ***Genomics:*** Study of application of recombinant DNA and DNA sequencing methods to sequence, assemble and analyse the structure and function of genomes.
- ***Human evolutionary genetics:*** Study of the differences between human genomes, their evolutionary past and current effects.
- ***Human mitochondrial genetics:*** Study of the human mitochondrial genome.
- ***Medical genetics:*** Diagnosis and management of human hereditary disorders.
- ***Genetic epidemiology:*** Study of the role of genetic factors in health and disease in families and in population groups with an emphasis on their interplay with environmental factors.

Multiple Choice Questions

1. Mendelian principles describe:
   a. Patterns of inheritance
   b. Genetic code
   c. Manipulation of gene expression
   d. Recombinant technology
2. Scientist who was dubbed the 'drosophilist geneticist' was:
   a. Hugo de Vries
   b. Carl Correns
   c. Thomas Morgan
   d. J.B.S.Haldane
3. Application of recombinant DNA and DNA sequencing: methods are part of
   a. Population genetics
   b. Genetic epidemiology
   c. Genomics
   d. Molecular genetics
4. Who was responsible for rediscovery of Mendel?
   a. E.B.Ford
   b. T.H.Morgan
   c. Sewall Wright
   d. Hugo de Vries
5. The oldest subspeciality of Genetics is:
   a. Molecular genetics
   b. Evolutionary genetics
   c. Human genetics
   d. Classic genetics

ANSWERS

1. a   **2.** c   **3.** c   **4.** d   **5.** d

Chapter 28

Important Terms in Genetics

Frequently Asked Questions

❑ Write short answers on: (a) Deoxyribonucleic acid,
(b) Nucleotide bases, (c) Gene expression.
❑ Write briefly on: (a) Transcription, (b) Genome, (c) Gene pool.

As the science of Genetics progresses day by day, several new discoveries are being made and newer advancements achieved. All these lead to many new words and terms being put into use. A list of the most important terms, which a basic student of Genetics should know, are given under.

❑ ***Adenine:*** One of the four nucleotide bases in DNA or RNA; it pairs with thymine in DNA and uracil in RNA.

❑ ***Allele:*** One of the alternate forms of a single gene occupying a given position or locus in a chromosome.

❑ ***Allelomorph:*** The original full form of the term allele.

❑ ***Chromosome:*** A molecular package for carrying the DNA or hereditary material in a cell.

❑ ***Cytosine:*** One of the four nucleotide bases in DNA or RNA; it pairs with guanine.

❑ ***Deoxyribonucleic acid (DNA):*** A polymer made up of deoxyribo nucleotides. It is usually in the form of a 'double helix' with two paired DNA molecules which have been twisted around each other. Therefore, it appears as a twisted ladder.

❑ ***Gene:*** The unit of heredity in living organisms.

❑ ***Gene expression:*** The process by which the information encoded in a gene is converted into a form that is useful for the cell. If a gene codes for a particular protein, the process by which the protein is synthesized is called 'gene expression' of the encoded information. The steps usually involved in this process are transcription and translation.

❑ ***Gene pool:*** The sum of all the alleles shared by all the members of a single population or population group.

❑ ***Genome:*** The total complement of genetic material contained in an organism or cell.

❑ ***Genotype:*** The complement of alleles in an individual's genome that gives rise to the phenotype of the individual.

❑ ***Guanine:*** One of the four nucleotide bases in DNA or RNA; it pairs with Cytosine.

❑ ***Locus:*** The particular location on/in a chromosome where a particular gene resides or is located.

❑ ***Phenotype:*** The observable physical and behavioural features and traits of an organism; these traits are determined by the genotype of the organism.

❑ ***Ribonucleic acid (RNA):*** A polymer made up of ribonucleotides; though similar to DNA, it is less stable.

❑ ***Thymine:*** One of the four nucleotide bases in DNA; it pairs with adenine.

❑ ***Transcription:*** The first step in gene expression. In this step, an RNA molecule which is complementary to the particular gene code in DNA is synthesized by enzymes called RNA polymerases; this RNA will be called the messenger RNA; it carries the message about the gene code in DNA through its nucleotides and hence the name 'messenger'.

❑ ***Translation:*** The second step in gene expression. In this step, the messenger RNA is read by the ribosomes of the cell and a functional protein is synthesized.

❑ ***Uracil:*** One of the four nucleotide bases in RNA; it pairs with adenine.

Multiple Choice Questions

1. Observable physical features and traits of an organism are collectively called:
   a. Genotype
   b. Genome
   c. Gene code
   d. Phenotype
2. The alternate form of a particular gene in a particular locus is called:
   a. Allele
   b. Nucleotide
   c. Base
   d. Nucleoside
3. The first step in gene expression is:
   a. Coding for a protein
   b. Synthesis of RNA
   c. Reading of RNA
   d. Relocation of RNA
4. In a ribonucleic acid molecule, adenine pairs with:
   a. Guanine
   b. Cytosine
   c. Uracil
   d. Thymine
5. The sum of all alleles shared in a population group is:
   a. Genome
   b. Genotype
   c. Gene expression
   d. Gene pool

ANSWERS

1. d   **2.** a   **3.** b   **4.** c   **5.** d

Chapter 29

Chromosomes

Chromosomes are deeply stained minute structures seen in the nucleus of a dividing cell. They may be rod shaped or V shaped or J shaped. They are formed during cell division by the condensation of chromatin. Chromosomes are composed of deoxyribonucleic acid (DNA) and proteins. DNA molecules are arranged in linear sequences and codes the genetic information inherited from parents.

The number of chromosomes is constant for a species. The ***ploidy*** (Greek.plo=multiple) of a cell denotes the number of sets of chromosomes in it. Each somatic cell of human beings contains 46 chromosomes, which constitute 2 sets or 23 pairs. These cells are ***diploid*** (two+ ploidy; two multiple=2) cells, designated as 2n. Each parent contributes 23 chromosomes, which makes up one member of each pair. The gametes(sperm/ovum) are ***haploid*** (or monoploid) cells (designated as n; Greek. haplous=single), having a single set of 23 chromosomes. The zygote formed after fertilisation therefore has 2 sets or 23 pairs of chromosomes and is diploid.

Out of the 46 chromosomes in the human somatic cells, 44 are autosomes (Greek.auto=self, some=body) and 2 are sex chromosomes (or ***allosomes;*** Greek.allo=other, some=body; or gonosomes; Greek.gonos=offspring). The sex chromosomes are responsible for sex determination and are of 2 types, X and Y. A human female has 44 autosomes and two X chromosomes (44 XX). These two X chromosomes are exactly of the same kind. A human male has 44 autosomes and morphologically dissimilar sex chromosomes, one X and one Y chromosome (44 XY). The X chromosome in male and female are similar to each other.

CELL CYCLE AND CHROMOSOME BEHAVIOUR

Before embarking on a study of chromosomes, it is necessary to consider some details about cell cycle and the relationship of chromosome appearance and behaviour during such cycle. A cell cycle can be defined as the period between one stage of cell division to the same stage of the next division.

Based on early observations, the cell cycle is divided into three parts—(1) nuclear division or mitosis, (2) cytoplasmic division or cytokinesis and (3) resting time or interphase.

Interphase

The cell appears inactive and the nucleus featureless during this phase which made the early cytologists to believe that this is a resting phase. However, the cell is not 'resting'; several important functions of the cell are taking place during this phase. The interphase is subdivided into three phases based on the activities and functions taking place: G1, S and G2.

- ❑ ***G1 phase:*** Gap phase 1 or first gap phase—this is the time immediately after the preceding cell division; genes are active, RNA and protein synthesis occurs and the cell enlarges in size – called the ***major synthetic time***. Rapidly dividing cells spend little time in G1 phase. Slowly dividing cells spend more time here appearing to be present all their life in this phase; such cells are labelled to be in G zero state and to have undergone G arrest.

  Movement to the next phase occurs when the cell is committed to further steps and cell division. This commitment is controlled by a set of genes, collectively called the ***mitotic trigger genes***. Individual genes of this group are called the ***cell division cycle (CDC) gene***. CDC genes interact in varied ways amongst themselves. Faulty or mutated CDC genes may shorten the G1 phase and lead to continuous cell division leading to proliferation of cancerous cells.

- ***S phase:*** From G1, the cell enters into S or synthesis phase. This is the time when DNA synthesis occurs (and hence the name 'synthesis' phase).
- ***G2 phase:*** From S phase, the dividing cell (or a cell committed to further division) enters the G2 (Gap phase 2 or second gap phase). Metabolic activities are completed and the cell readily awaits the next division.

Mitosis

The cell which has entered into this phase, also called the **M phase** or **the phase of nuclear division**, under-goes physical division of its nucleus. Mitosis is further subdivided into four phases (prophase, metaphase, anaphase and telophase) or periods. Chromosomal and non-chromosomal changes and events occur during these phases.

- ***Prophase:*** (Greek.pro=before + phasio=appearance; Greek.prophaino=foreshadow) Chromosomes con-dense from the diffuse state to compact visible bundles. It can be said that prophase actually begins when the chromosomes first become visible as long intertwining strands within the nucleus of the cell. They condense further and become shorter. By late prophase, the chro-mosomes have split into two sister chromatids with a connecting centromere.
- ***Metaphase:*** (Greek.meta=between, phasio=appearance) The condensed chromosomes (which now show sister chromatids with centromeres) are pulled to the central region of the cell called the mitotic plate by the spindle of fibres.
- ***Anaphase:*** (Greek.ana=upward or apart, phasio= appearance) The centromeres of the chromosomes divide in this phase. Infact, anaphase technically begins only when the centromeres show division. The centromeres of all the chromosomes divide at the same time. This division brings about a major change in the status of the chromosome – what was a single chromosome with two sister chromatids and one centromere now becomes two separate progeny chromosomes, each chromosome having its own centromere.

  The progeny chromosomes, pulled by the spindle of fibres, migrate towards opposite poles of the cell. Each (dividing) cell now has two sets of progeny chromosomes; each set is a copy of the set of chromosomes of the parental cell.

- ***Telophase:*** (Greek.telos = end, phasio = appearance) The migrating chromosomes approach the poles. They are subsequently surrounded by new nuclear envelopes thus forming two new sets of chromosomes and two nuclei, preparatory to cytoplasmic division.

Cytokinesis

This is the stage at which a cleavage furrow appears in the middle of the dividing cell, a cell plate is formed and the cell divides into two.

STRUCTURE OF A CHROMOSOME

Chromosomes have certain structural features which distinguish them not only from other structures, but also from other chromosomes. Such features include their size, length, position of centromere and staining patterns. All these features are collectively referred to as ***chromosome morphology***. It can be seen from the aforementioned details of cell cycle that a chromosome and its component parts may be best studied during the metaphase of the cycle.

Centromere and its Location

A metaphase chromosome (this is the phase of the cellular cycle when chromosomes are clearly visible with their component chromatids) consists of two parallel and identical filaments called ***chromatids***, which are held together at a narrowed region called the ***primary constriction*** or ***centromere*** (central part; Greek. meros=part). Each chromatid has two arms, one on each side of the centromere. These are called the ***chromosome arms***. The short arm is denoted as p (from French petite meaning small) and long arm as q (Fig. 29.1).

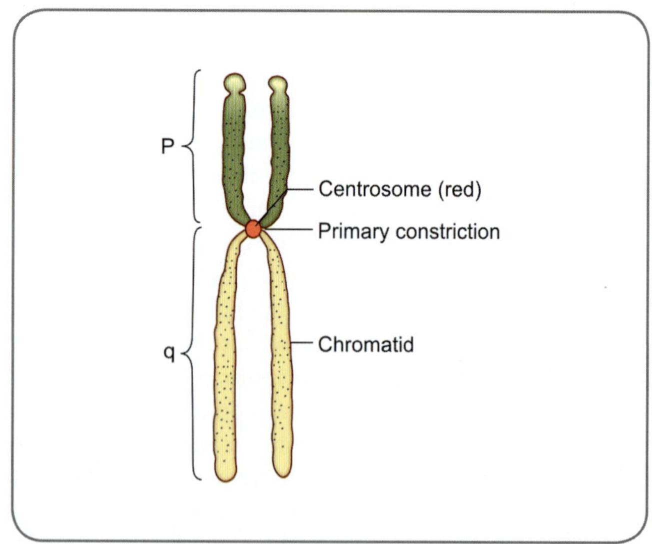

Fig 29.1: Structure of a chromosome

Other Parts of a Chromosome

Some chromosomes have another constriction called the *secondary constriction*. The portion of the chromosome arm separated by the secondary constriction from the rest of the chromosome arm is called the *satellite*.

A location in the chromosome arm closer to the centromere is termed '*proximal*' and a location farther away from the centromere is termed '*distal*'.

Each chromosome arm ends in a structure called the *telomere* (Greek.telos=end, meros=part). The telomere is a zone of repetitive nucleotide sequences; this repetition of the nucleotide sequences gives telomere a polarity that prevents other segments of the chromosome from fusing with it or alteration of the sequence. In vertebrates, the repetitive nucleotide sequence is TTAGGG and in humans, it is estimated that the sequence is repeated 2500 times.

The mitotic spindle of fibres connect to a specialised part of the chromosome called the *kinetochore* (Greek. kinetos=moving, chora=space). The kinetochore is usually located in the centromere and it appears that the spindle is attached to the centromere.

Morphological Grouping of Chromosomes

The morphological grouping of chromosomes is done based on the position of centromere (Figs 29.2A to C) and four groups can be identified.

1. *Metacentric chromosome*: Centromere is situated at or near the centre of the chromosome. In this type, both arms of a chromatid (the chromosome arms) are almost of equal length (Fig. 29.2A).

2. *Submetacentric chromosome*: Centromere lies between the centre and the end of the chromosome; one arm is short and one arm is long (Fig. 29.2B).
3. *Acrocentric chromosome*: Centromere is located near the end; one arm is very short and one arm is very long (Fig. 29.2C).
4. *Telocentric chromosome*: Centromere is at the end of the chromosome; each chromatid has only one arm (Telocentric chromosome is not seen in human beings) (Fig. 29.2D).

CLASSIFICATION OF CHROMOSOMES

For studying and analysing chromosomes, it is necessary to classify them so that appearances of modified/changed/ /altered/abnormal chromosomes can be easily found out. Classification of human chromosomes is done based on certain qualities and morphological features which the chromosomes show. One such feature is the presence of the *centromere and its location*.

Another feature of chromosome morphology is the *size and length* of the chromosome. Some chromosomes are long and even double the length of some other chromosomes which are extremely short.

Utilising these two parameters, attempts to classify the human chromosomes have been made.

Denver Classification

This is a system of identifying and classifying human chromosomes based on their lengths and the position of the centromeres. If the total complement of 46 chromosomes is placed on a tray, it can be seen that they can be pooled into 23 sets because the total of 46 is only a product of 23 pairs (Figs 29.3A and B). If the 23 sets (of which one would be either an XX set or an XY set) are arranged according to their total lengths (or heights), a pattern would emerge. Seven groups have been identified in the order of decreasing length of the chromosomes and are labelled as groups A to G (Table 29.1).

The system has been called the Denver system because it was proposed in a convention of cytoscientists at Denver, Colarado in 1960. It was accepted in a subsequent convention at London in 1963. The chromosomes are numbered in order of descending lengths. The two allosomes are put into two different Denver groups. The 'X' chromosome resembles chromosome No. 6 which is the longest of group C and is therefore placed in group C. The 'Y' chromosome is very small, resembles chromosome 21 and goes into group G.

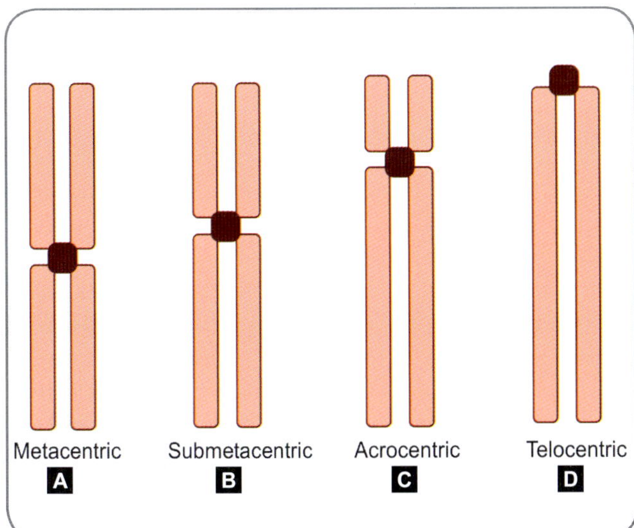

Figs 29.2A to D: Location of centromere and classification of chromosomes

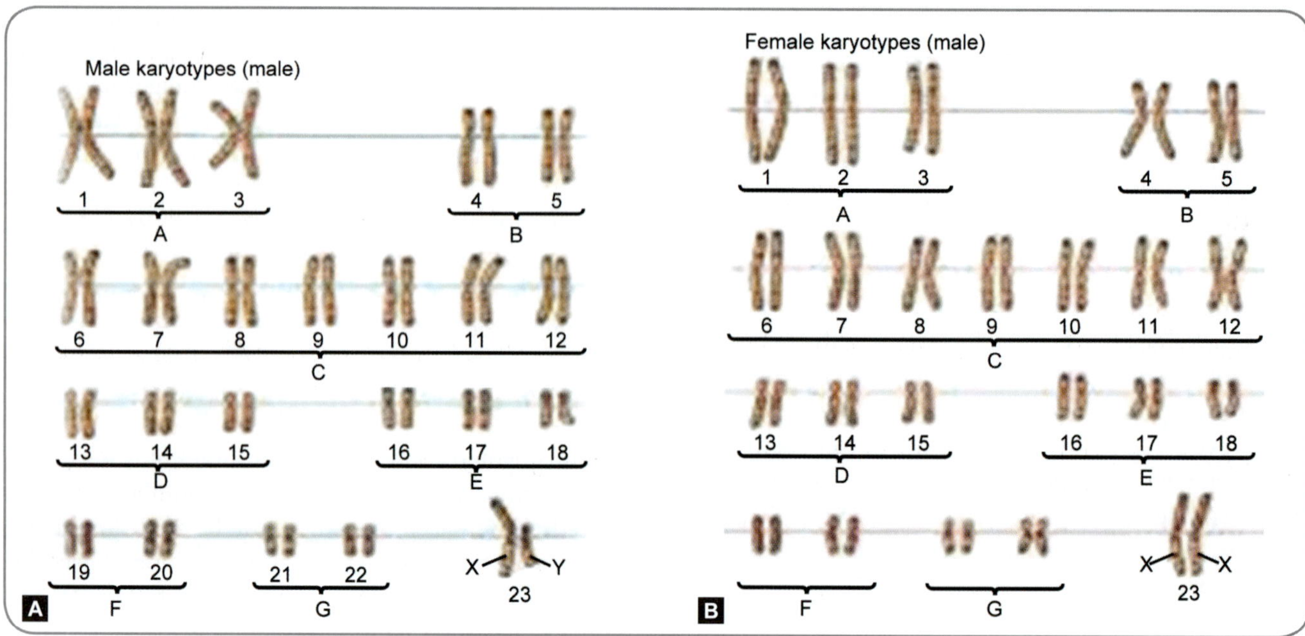

Figs 29.3A and B: Human karyotype **A.** Male **B.** Female

Table 29.1: Seven groups of Human karyotype	
Group	**Chromosome human numbers**
A	1,2,3
B	4,5
C	6,7,8,9,10,11,12 + X
D	13,14,15
E	16,17,18
F	19,20
G	21,22 + Y

Group A has longest metacentric chromosomes; group G has shortest acrocentric chromosomes. Groups D and G have satellites.

Added Information		
More details of Denver Classification		
Group	Chromosome number	Description of chromosomes
A	1, 2, 3	Large chromosomes, Mainly Metacentric
B	4, 5	Large chromosomes, Submetacentric
C	6, 7, 8, 9, 10, 11, 12 And X	Medium sized chromosomes, Mostly Submetacentric; X chromosome closely resembles chromosome number 6

Added Information *contd...*		
D	13, 14, 15	Medium sized chromosomes, Acrocentric, All three have satellites on their short arms
E	16, 17, 18	Small chromosomes, Submetacentric (16 is metacentric)
F	19, 20	Short chromosomes, Metacentric
G	21, 22 And Y	Very short chromosomes, Acrocentric, Satellites in both chromosomes; Y chromosome closely resembles chromosome number 21

Paris Nomenclature

Denver classification came into existence and acceptance by 1963. However, it was soon realised that it was incomplete as a standalone classification. Though chromosomes could be grossly identified, several alterations, mal positions, deletions and abnormalities could not be identified. Meanwhile, chromosomal banding techniques came to be used.

Chromosomes show characteristic patterns of staining when treated with chemicals. Certain chemicals make the chromosomes acquire stripes and bands (areas of staining). Alternate dark and light bands could be seen and this kind of an appearance where a chromosome shows coloured/less coloured/non coloured bands is called *'banding'*. Different chromosomes show different banding patterns and it became possible to know more details about the chromosomes from a study of these patterns.

contd...

Chromosomes treated with quinacrine dyes show bands of fluorescence when exposed to ultraviolet light. These bands are called *Q bands* (named after quinacrine). An inexpensive alternate method of staining led to the G banding. Chromosomes are treated with trypsin and then with Giemsa stain. The bands are called *G bands* (named after Giemsa). France took up to a slightly different method where the reverse of Giemsa banding is used. This is called the *R banding*. The Paris convention of 1971 decided to group chromosomes on the basis of staining properties and banding results.

By this system, the chromosome arms are divided into 1,2 and 3 regions starting from the centromere and each region is subdivided into numbered bands. This classification provides greater accuracy in identifying portions of the individual numbered chromosomes.

ISC Nomenclature

In 1978, a document entitled 'An International System for human Cytogenetic Nomenclature' was made out keeping track of Denver, London and Paris decisions and appropriately incorporating them. This has led to the ISCN and the ISCN is regularly and periodically updated.

KARYOTYPING

Karyotype (Greek.karyos=nucleus) is the chromosomal constitution of a cell. Karyotyping is the process of arranging chromosomes of a cell in a standard manner to study the complete chromosomal complement. 'Karyotype' also means the photomicrograph of the arranged chromosomes. Karyotyping is done to identify and number the chromosomes of a cell and to look for numerical and structural abnormalities of the chromosomes.

The commonly used cells for karyotyping are peripheral blood lymphocytes. 5 ml of venous blood is treated with phytohaemagglutinin (to stimulate cell division) and cultured at 37°c for 3 days. Colchicine is then added to achieve metaphase arrest of the dividing cells. The cells are treated with hypotonic saline (to swell up), fixed and spread on a glass slide by dropping from a height. The metaphase spread thus obtained is stained and microphotographed. From the photomicrograph, individual chromosomes are cut and arranged in descending order of length to obtain the karyotype. With the advent of computerised analysis, some of the manual steps of karyotyping have been replaced by automation.

Chromosome analysis may also be done by chromosome banding and Fluorescent In Situ Hybridisation (FISH) method.

SEX CHROMATIN

Barr and Bertram in 1949 observed a small deeply stained body in the interphase nuclei of female cat neurons. This chromatin body is the *Barr body* (named after the discoverer Murray Barr) or the sex chromatin. It is seen as an ovoid deeply staining body on the inner surface of the nuclear membrane in most cells of human females, though it is mostly studied in a buccal smear. In neutrophils it is seen as a drumstick-shaped appendage of the nucleus.

In 1962, a hypothesis was put forward by Mary F. Lyon, which is called the *Lyon hypothesis* after her name. This hypothesis, which is widely accepted now, states that in somatic cells of female mammals (especially in those where the sex of the individual is determined by the presence of a Y or W chromosome and not the status of diploidy of the X) only one X chromosome is active and the other condensed inactive X chromosome appears in the interphase nuclei as sex chromatin. The process of inactivation is called *lyonisation*. In humans, the number of Barr bodies visible at interphase is always one less than the total number of X chromosomes. Thus, individuals with XX composition (normal females having 46, XX) show one Barr body; those with XY composition (normal males having 46, XY) have none; those with two X chromosomes (men with Klinefelter's syndrome having 47, XXY) show one Barr body; those with three X chromosomes (women with three X chromosomes having 47, XXX) show two Barr bodies.

Added Information

Different species of animals have different chromosome numbers. A list is given below:

Animal	Number of chromosomes
Cat	38
Dog	78
Donkey	62
Cow	60
Horse	64
Rabbit	44
Mouse	40
Rat	42
Elephant	56
Sheep	54
Kingfisher bird	132
Silkworm	56
Chimpanzee and most apes	48

Note: It is stated that two chromosomes united to form chromosome Number 2 in humans during evolution thus reducing the number from 48 to 46; the closest evolutionary relatives of humankind, the apes and chimps have 48.

CHROMOSOMAL ABNORMALITIES AND DISORDERS

Some human diseases and conditions have been traced to abnormalities in chromosome number, shape and size. A few examples are given below:

- ***Down syndrome*** (named after the British physician John Langdin Down who described it in 1866) or DNS syndrome: Trisomy 21—there are three copies of chromosome 21—decreased muscle tone, stocky build, asymmetry of skull, slanting eyes, impaired physical growth, moderate intellectual defects are the usual features.
- ***Edwards syndrome*** (named after John Hilton Edwards who first described it in 1960): Trisomy 18—the second most common autosomal trisomy problem in humans – there are three copies of chromosome 18—motor defects, kidney and visceral disorders, developmental deficiencies, heart defects and arthritic defects leading to characteristically clenched and folded fingers and hands are the features.
- ***Cri du chat syndrome*** (named so because the affected child cries like a cat): Also called Lejeune's syndrome after Jerome Lejeune who first described it in 1963 or 5p minus syndrome–deletion of a portion of the short arm (p arm) of chromosome 5 occurs–wide eyes, small face, head and jaw, laryngeal and tracheal deficiencies leading to the shrill cat like cry, vocal disorders, mental and physical inabilities, hyperactive tantrums and behavioural disorders are features.
- ***Jacobsen syndrome*** or terminal 11 q deletion syndrome – deletion of the terminal portion of chromosome 11 occurs – several physical and mental inabilities may be seen.
- ***Patau syndrome:*** Also called trisomy D or trisomy 13 – there is an additional copy of chromosome 13 or extra parts of the chromosome are seen—shows features similar to Edwards syndrome, but the characteristic folded fingers of Edwards are not seen.
- ***Klinefelter syndrome*** (named after Harry Klinefelter who identified the condition in 1944)—there are two (sometimes more than two) X chromosomes in a male (47, XXY)—sterility, excessively tall, longer arms and legs, shyness, speech difficulties, dyslexia and muscular weakness are the features.
- ***Turner syndrome*** (named after Henry Turner who first described it in 1938): The chromosomal association was proved only in 1964—there is a complete or partial deficiency of X chromosome in the affected female (45, X)—webbed neck, heart and other visceral defects, low set ears, short stature, absence of menstruation, thyroid problems, diabetes, difficulty is spatial visualization resulting in decreased mathematical capabilities are the features.
- ***XYY syndrome:*** There is an extra Y chromosome in the affected male (47, XYY)—taller than usual, aggressive disposition are the features.
- ***XXX syndrome:*** There is an extra X chromosome in the affected female (47, XXX)—taller than usual, Microcephaly, epicanthal folds, muscular weakness, scoliosis, dyslexia, intellectual difficulties may be seen.
- ***Wolf-Hirschhorn syndrome*** (named after Wolf and Hirschhorn who published information about it in 1960s and 70s, though first described by Hirschhorn and Cooper in 1961) or Pitt-Roger-Danks syndrome or deletion Dillan 4 p syndrome – there is complete or partial loss of a portion of the short arm of chromosome 4—growth retardation, facial changes, seizures and immunodeficiency are the features.

Added Information

Chromosome time line

- 1860–1885—Schleiden, Virchow and Butschli recognised the significance of chromosomes;
- 1880s—Walther Flemming identified condensed bodies inside cell nuclei and called them 'chromatin';
- 1888—von Waldeyer-Hartz introduced the term 'chromosome' deriving it from chromatin;
- 1880-1890—Theodor Boveri, by his work, proved that chromosomes are vectors of heredity and inheritance;
- 1885–1900—Wilhelm Roux stated that each chromosome carries a different genetic load;
- 1902—Edmund Beecher Wilson, Theophilus Painter and Walter Sutton put forth the Chromosome theory of Inheritance; This theory, later came to be called the 'Sutton-Boveri theory of inheritance';
- 1923—Theophilus Painter, after viewing processed chromosomes under plain light microscopy and counting 24 sets, stated that the human complement of chromosomes is 24 pairs or 48;
- 1956—Joe Hin Tjio, after modifying certain processing steps and using distilled water in karyotyping techniques, gave the conclusive number as 23 pairs or 46.

Multiple Choice Questions

1. The phase of nuclear division of a cell cycle is called:
   a. Interphase
   b. Cytokinesis
   c. Mitosis
   d. Synthesis phase
2. Splitting of chromosomes into chromatids takes place during:
   a. Prophase
   b. Metaphase
   c. Anaphase
   d. Telophase
3. A satellite of a chromosome is separated from the rest of the chromosome by:
   a. Primary constriction
   b. p arm
   c. Secondary constriction
   d. q arm
4. Nucleotide repetition in the telomere:
   a. Gives polarity
   b. Enhances cell division
   c. Decreases cell division
   d. Gives adhesive power
5. In which phase are the chromosomes arrested for karyotyping?
   a. Prophase
   b. Metaphase
   c. Anaphase
   d. Telophase

ANSWERS

1. c   **2.** a   **3.** c   **4.** a   **5.** b

Clinical Problem-solving

Case Study 1: In the slide analysis class in your genetics laboratory, you are given a slide that shows two Barr bodies in a somatic cell.
- Your friend states that the slide belongs to a normal female. Would you agree with him?
- If you do not agree, what is your reason for disagreement?
- Give your opinion on the slide.

Case Study 2: A 2-year-old child presents with the following: small head and face, wide eyes, behavioural tantrums and a shrill cry. The child has a chromosomal abnormality.
- What disorder do you think does this child suffer from?
- If so, in which chromosome is there the problem?
- What is the cause for the shrill cry?

(For solutions see Appendix).

Chapter 30

DNA and RNA

Frequently Asked Questions

- ❏ Write notes on: (a) DNA, (b) tRNA, (c) mRNA.
- ❏ Write briefly on: (a) Double helix model, (b) Purine and pyrimidine bases (c) rRNA, (d) Nucleosides, (e) Complementary base pairing.

DEOXYRIBONUCLEIC ACID (DNA)

The genetic code is carried by deoxyribonucleic acid molecules. Genes which are the units of heredity are segments of DNA molecules.

Structure of DNA

The **double helix model** of the three dimensional structure of DNA was put forth by James Watson & Francis Crick in 1953. DNA is a double stranded molecule composed of two polynucleotide chains, coiled around each other to form a double helix (Fig. 30.1). Each deoxyribonucleotide has three subunits which are a sugar, a phosphate group and a nitrogenous base. The sugar is a pentose moiety called **deoxyribose**. The base is a nitrogen and carbon containing ring like structure. **Purine bases** (eg: Adenine and Guanine) have 2 rings and **pyrimidine bases** (eg: Cytosine and Thymine) have only one ring. The combination of sugar and base is called a **nucleoside**, which becomes a nucleotide when a phosphate group is attached to the sugar at 5'C position.

Each strand of a double helix is made up of alternate units of sugar and phosphate. Nitrogen bases are attached to the sugars as side groups and point towards the centre of the helix. The two strands of one DNA molecule are held together by hydrogen bonds between the nitrogen bases. The hydrogen bonds are in a specific manner. Adenine binds with thymine by 2 hydrogen bonds (A=T).

Guanine binds with cytosine by 3 hydrogen bonds (G≡C). This manner in which the bases align themselves along the strands is called **complementary base pairing**. The 5'phosphate of one nucleotide bonds with the 3'C of next nucleotide forming a 3'–5' phosphodiester bond. The DNA polynucleotide chain has a free 5'phosphate at one end

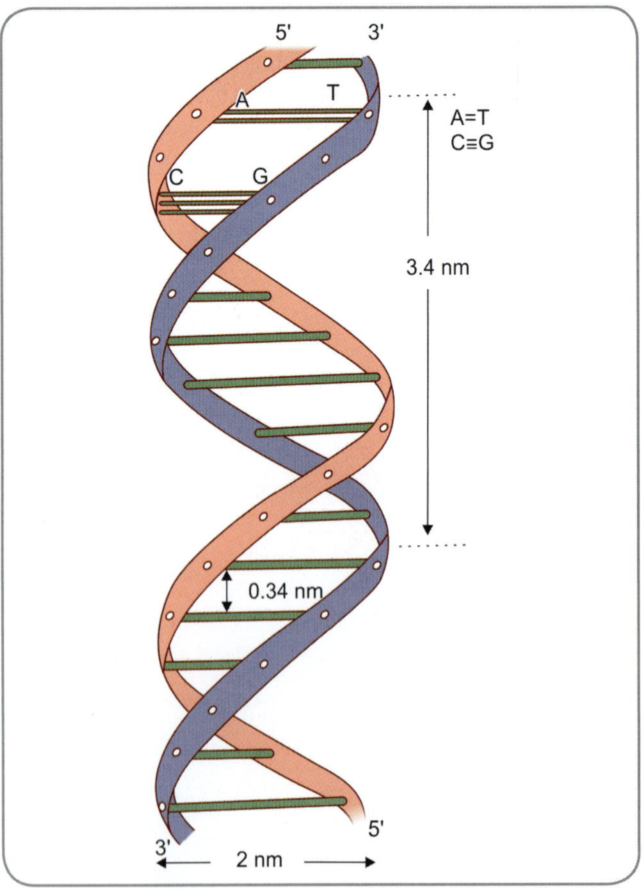

Fig. 30.1: Double helix model of DNA

known as 5'end or tail end or 5'P terminus and a free 3' hydroxyl at the 3'end or head end or 3' OH terminus. Thus a strand of DNA has a polarity running from 5'-3' direction. The arrangement of two strands relative to each other is such that one runs in 5'–3' direction and the other in 3'–5' direction. Therefore the two DNA strands are **antiparallel**.

DNA Packaging

The double helical DNA molecule comprising of billions of base pairs undergoes phenomenal degree of coiling to get accommodated in the tiny nucleus of the cell. This is achieved at various levels by the formation of nucleosomes (where small segments are wound over a histone protein core), solenoid formation and chromatin loops which become highly condensed to form chromosomes.

DNA Replication

During the S phase of cell division the tightly coiled DNA filaments get uncoiled. The two complementary strands of DNA unwind and separate from each other (denaturation of DNA) by breakage of hydrogen bonds. Two new polynucleotide chains are synthesised, each identical to the one from which it has been separated, resulting in the formation of two molecules, each identical to the parent molecule. Thus the mode of replication of DNA is *semiconservative*. The enzyme DNA polymerase mediates the union of nucleotides during DNA synthesis.

Mitochondrial DNA

DNA is also found in the mitochondria of cells. The mitochondrial DNA is double stranded and joined at each end to form a ring.

RIBONUCLEIC ACID (RNA)

Ribonucleic acid is a single stranded polynucleotide molecule, found in the nucleolus and cytoplasm of a cell. RNA is transcribed from DNA. The sugar in RNA is ribose. There are 4 bases—Adenine, Uracil, Cytosine and Guanine. There are three types of RNA—messenger RNA (m RNA), ribosomal RNA (r RNA) and transfer RNA (t RNA).

mRNA

It is synthesised inside the nucleus and carries all the genetic information present on a DNA strand. mRNA acts as a template for protein synthesis during the process of translation.

rRNA

It constitutes 80% of the total RNA content of a cell. rRNA binds with proteins to form macromolecular organelles called ribosomes. Some rRNA molecules exhibit enzymatic activity and are called *ribozymes*.

tRNA

These are the molecules which link the nucleotide sequence of mRNA to the amino acid sequence of a polypeptide during the process of translation. The polynucleotide chain of a tRNA molecule is arranged in the *clover leaf model*, with a bend in the middle of the chain and an arm on each side (Fig. 30.2).

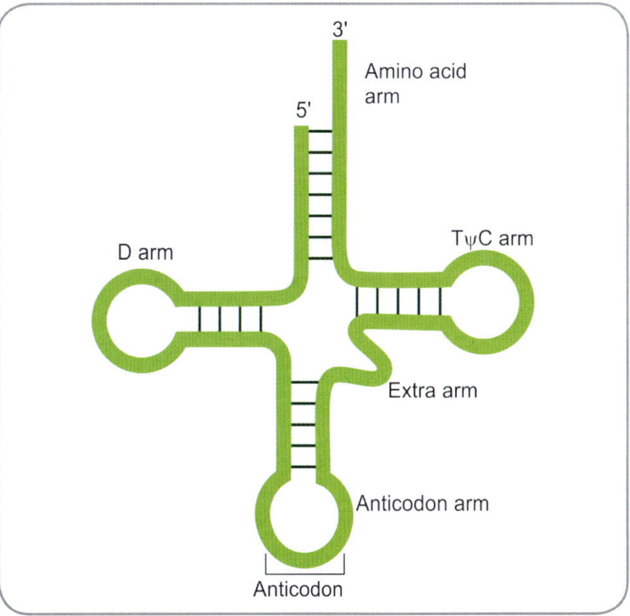

Fig. 30.2: Clover leaf model of tRNA

Multiple Choice Questions

1. The two strands of DNA in a double helix are:
   a. Parallel
   b. Antiparallel
   c. Criss cross
   d. Randomly strung
2. Denaturation of DNA occurs by:
   a. Breakage of hydrogen bonds
   b. Splitting of sugar moiety
   c. Separation of phosphate group
   d. Coiling of a single strand
3. Winding of a small segment of DNA over a histone core is called:
   a. Nucleosome formation
   b. Solenoid formation
   c. Chromatin loop formation
   d. Chromatid formation
4. Replication of DNA is semi-conservative because:
   a. One strand of the parent molecule is retained
   b. One strand of the parent molecule is lost
   c. One molecule of the parent molecule is wound into a helix
   d. One molecule of the parent molecule is only used for replication
5. During translation, the t RNA links up the nucleotide sequence of the mRNA to the:
   a. Aminoacid sequence of the polypeptide
   b. Ribosomal RNA of ribosomes
   c. Enzymes of ribozymes
   d. Template base of the mRNA

ANSWERS

1. b   **2.** a   **3.** a   **4.** a   **5.** a

Chapter 31

Mendelian Laws of Inheritance

Frequently Asked Questions

❑ Write notes on: (a) Transmission genetics, (b) Mendel's first law, (c) Monohybrid cross.
❑ Write briefly on: (a) Dihybrid cross, (b) Mendelian laws, (c) Filial generations

Johann Gregor Mendel (22nd July, 1822—6th January, 1884) was an Austrian monk who did elaborate research on chick pea plants and decoded the principles by which characters are transmitted from one generation to the next. His findings were quite ahead of his time since many facts about the nucleus, chromosomes and cell division were unknown then. He is considered the **Father of Modern Genetics**. Mendelian principles talk about rules of inheritance and so, can be called '*transmission genetics*'.

In 1866, Mendel published a research paper on breeding experiments done using the garden pea plant, *Pisum sativum*. From his results, he developed two basic principles of inheritance.

MONOHYBRID CROSS

Mendel's first set of experiments were on monohybrid crosses. What is a monohybrid cross?

The cross between plants or animals differing in a single pair of contrasting characters is called a *monohybrid cross* (Greek.mono=single, Latin.hybrida=offspring of parents having differing qualities). For example, Mendel crossed pure tall and pure dwarf plants to test the inheritance of the height of stem. Only one character (height of plant) with contrasting features (tall and short) was taken and so, was a monohybrid cross.

F1 generation: The hybrid population obtained from cross-pollination of the pure tall and pure dwarf plants were called the first filial (F1) generation. He observed that all offspring in F1 generation were tall plants. He proposed that the pure tall plant had a pair of factors both of which were T, making the complement TT; and the pure dwarf plant had a pair of factors both of which were t, making the complement tt. When the two plants were crossed, factor T from one parent and factor t from the other united to produce Tt. However, since factor T was dominant over factor t, the appearance of F1 generation was 'tall'.

We can summarize the F1 results by saying

F1 generation Genotype : all Tt; Phenotype : all tall.

Gametes of pure bred tall plants

Gametes of pure bred short plants		t	t
	T	Tt	Tt
	T	Tt	Tt

With the pair of factors of either of the parents segregating within themselves, $2 \times 2 = 4$ filial combinations of these factors are possible. However, since all the gametes of one parent will have T factor and all the gametes of the other parent will have t factor, the combination can only be Tt, thus making all the four possible offspring have Tt.

F2 generation: Mendel then allowed the F1 generation plants to self pollinate and therefore, the F2 generation (second filial generation) was obtained. 75% of the F2 offspring were tall and 25% were dwarf. He observed that the character of dwarfness which disappeared in F1 generation reappeared in F2 generation. Working out the same principles of factors and their combination, Mendel further proposed that TT and Tt appeared tall while tt appeared short. 50% of the gametes of one parent had T factor and the other 50% had t factor. Since both the parents had a combination of Tt factors, this gametic composition was the same for the second parent too. When gametes having T and T united, TT with tall external feature resulted; when gametes having T and t united, Tt with tall external feature resulted; when gametes having t and t united, tt with short external appearance resulted. Thus Mendel was able to show that the F2 generation will have 75% tall and 25% dwarf plants.

We can summarize the F2 results by saying

F2 generation Genotypes—TT:Tt:tt = 1:2:1; Phenotypes—Tall: Short = 3:1.

Gametes of F2 plants

Gametes of F2 plants		T	t
	T	TT	Tt
	t	Tt	tt

F3 generation: Mendel proceeded further with his experiments and obtained a F3 generation by self-pollination of F2 generation plants. He observed that all offspring of dwarf plants were dwarf. Of the tall plants, 1/3rd yielded tall offspring. Of the remaining 2/3rd, tall and dwarf offspring were in the ratio of 3:1.

If the F3 offspring are studied, it can be seen that the feature of dwarfness which had totally disappeared in F1 and had partially surfaced in F2 had almost made a comeback in F3. It indicates that this feature remained hidden somewhere or was hidden by something else. The 'something else' that could have hidden this feature can only be its counterpart, namely tallness. Mendel concluded that tallness was dominant over dwarfness and suppressed the latter. When factor T occurred along with factor t, T was dominant over t and did not allow t to express itself.

In the heterozygous condition (Tt; Greek.heteros= different; condition where the factors are differet from each other) when both factors are present, the feature of tallness dominates over or conceals the feature of shortness. Therefore, the feature of tallness is considered the dominant feature and the gene T is called the ***dominant gene***. On the other hand, the feature of shortness is the recessive feature and gene t is the ***recessive gene***. The recessive feature is expressed only when the gene is present in double dose (tt) and in the absence of the dominant gene (T) (It should be remembered that Mendel did not use the term 'gene' or 'chromosome' because these terms and the entities the terms signify were not known at the time of Mendel. He knew 'something' inside the gamete was responsible for the expression of the characters and called them 'factors'; he termed the characters being studied as 'traits').

Mendel analysed the results of the monohybrid cross. Even as we analyse these results, it can be seen that the two factors T and t, do not blend with each other when they are together; they also tend to separate from each other and are able to bring back the feature they stand for. Mendel suggested that the two factors separate or segregate from each other at the time of formation of gametes. This proposition came to be called the first law of Mendel or ***Mendel's law of segregation***. Law of segregation states that alleles of genes (as we know today that Mendelian factors are alleles of genes) segregate during formation of gametes.

As a sequelae to the first law, the most important principle of Genetics evolved. This is the ***principle of dominance and recessiveness***. Even after the passing of a complete century and with more and more of sophisticated experiments, the principle of dominance and recessiveness which Mendel proposed (and also showed) remains true.

DIHYBRID CROSS

The cross between plants or animals which differ in two pairs of contrasting characters is called a ***dihybrid cross***. Mendel conducted further breeding experiments by cross pollinating pea plants having yellow coloured round seeds (YYRR) and green coloured wrinkled seeds (yyrr) to obtain the F1 generation. He observed that all F1 plants had yellow round seeds.

F1 generation Genotype: all YyRr, Phenotype: all yellow round.

	YR	YR
yr	YyRr	YyRr
yr	YyRr	YyRr

To obtain the F2 generation the F1 plants were allowed to self pollinate. Four phenotypes were observed in F2 generation—yellow round : yellow wrinkled: green round: green wrinkled in the ratio of 9:3:3:1. Mendel deduced that there should be nine combinations of genotypes.

F2 generation Genotype: 9 combinations; Phenotypes: 4 combinations.

	YR	Yr	yR	yr
YR	YYRR	YYRr	YyRR	YyRr
Yr	YYRr	YYrr	YyRr	Yyrr
yR	YyRR	YyRr	yyRR	yyRr
yr	YyRr	Yyrr	yyRr	yyrr

If a genotype-phenotype table is worked out, it will be seen as follows:

Genotype	Phenotype
YYRR	
YyRR	
YyRr	Yellow Round
YYRr	
YYrr	
Yyrr	Yellow Wrinkled
yyRR	
yyRr	Green Round
yyrr	Green Wrinkled

The dihybrid cross revealed two important points:

- As in the monohybrid cross, features which were hidden or masked in the F1 generation appeared in the F2 generation.
- F2 generation also showed appearance of new combinations of features which were not present in the parental generation, examples being yellow wrinkled and green round.

While Mendel had already deduced the first point from the results of the monohybrid cross, the second point gained importance only in the dihybrid cross. The 'factors' separated independent of each other and combined with others in varying combinations. They joined; separated; rejoined but differently; produced an assortment. This led to second law of Mendel or *Mendel's law of independent assortment*.

MENDELIAN LAWS

Therefore, there are two laws of Mendelian principles: *Law of segregation* and *Law of independent Assortment.*

- The *law of segregation* states that each member of one pair of factors (genes) responsible for a particular character, separate from each other during gametogenesis. As a result, a gamete contains only one factor, which remains unaltered and segregates during gametogenesis.
- The *law of independent assortment* states that members of different gene pairs assort independent of one another during gametogenesis. This accounts for the new combinations of characters seen in the offspring.

Added Information

To check on the validity of his deductions and propositions, Mendel did trihybrid cross experiments too. With the results he obtained, he concluded that dominance, allelic segregation and allelic independent assortment were fundamental rules of inheritance for all traits.

- *Back cross:* A cross between an F1 individual and one of the parents is called a back cross. It is done to cross check on the allelic disposition of the heterozygous individual and to confirm the segregation of alleles.
- *Test cross:* A test cross is a specialised type of back cross. An F1 individual is crossed with the homozygous recessive parent. This kind of a cross permits 'testing' of the genotype of the heterozygous individual and so the name.
- *Punnett square:* This is a diagram of squares drawn to analyse the genotypes and phenotypes of a monohybrid and dihybrid crosses. It may be used for higher levels of crossing too. It has been named after the British geneticist Reginald Crundall Punnett who used the method to analyse results of genetic crosses. The gametes of one parent are represented along the y axis and those of the other parent along the X-axis. The combination of a particular gametic union is worked out in the square that is at the junction of the x and the y.

Cross: Aa × Aa

	A	a
A	AA	Aa
a	Aa	aa

A Punnett square

Multiple Choice Questions

1. Mendelian principles form part of transmission genetics because:
   a. They codify passage of characters from one generation to another
   b. They emphasize that characters should be transmitted
   c. They were transmitted from Mendel to other geneticists
   d. Mendel performed his experiments in many generations
2. In the first filial generation of the Mendelian monohybrid cross, the offspring exhibited:
   a. Only the recessive trait
   b. Only the dominant trait
   c. Both the dominant and recessive traits
   d. A new intermediary trait
3. As a result of his dihybrid cross experiments, Mendel proposed the:
   a. Law of Segregation
   b. Law of Dominance
   c. Law of Recessiveness
   d. Law of Independent Assortment

ANSWERS

1. a   **2.** b   **3.** d

Clinical Problem-solving

Case Study 1: You are trying a dihybrid cross on Mendelian principles. One line of purebred plants have red colour and smooth petals. Another line of purebred plants have white colour and wrinkled petals. Red colour and smoothness are the dominant traits. Can you work out Punnett squares for the F2 and F3 generations?

(For solutions see Appendix).

Chapter 32

Alleles

Alleles are alternate forms of the gene located at a specific position (locus) on/in a chromosome. Human beings are diploid; i.e., they (or we) have two sets of chromosomes. Thus, each chromosome has its complementary chromosome in its own pair. If we consider a particular gene located in a chromosome, it can be understood that there will actually be two genes or two copies of the same gene—one in each chromosome of the concerned pair. If we again consider that there can be alternative forms to the gene, then the two copies can be identical or different. If the two copies are different from each other, we can then say that they are two alleles or alternate forms of the same gene.

The term 'allele' is derived from its longer original term 'allelomorph' (other form; Greek.allos=other, allel=each other, morph=form).

IDENTICAL AND NON-IDENTICAL ALLELES (FIG. 32.1)

If both the alleles of a given gene are identical, the organism is called a homozygote and is homozygous with respect to that gene. For example, if the gene for height is taken into consideration, the alleles can both be 'tall' or 'short'.

If the two alleles are different, the organism is a heterozygote and is heterozygous with respect to that gene. If the same characteristic is considered, the organism is a heterozygotic when one allele is 'tall' and the other 'short'.

There are some genes which exist in only one allelic variety, so that all humans are alike for these loci and

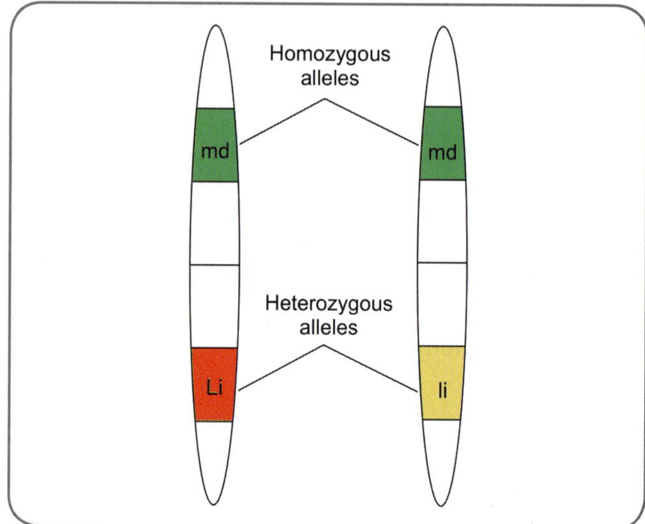

Fig. 32.1: Identical and non identical alleles-at the upper locus, the two alleles are identical-at the-lower locus, the two alleles are non-identical

possess two identical alleles. An example is the gene for the serological properties of man. However, there are many other genes which exist in more than one allelic form. A typical example is the gene for Blood group. This gene is the 'I' gene. People with blood group O possess the allele I^O; people belonging to blood group A possess a different allele I^A. People with blood group B possess the allele I^B.

CHARACTERISTICS OF ALLELES

❑ Alleles segregate from one another in the formation of gametes (Mendel's Law of Segregation).
❑ Alleles of different genes segregate independently (Mendel's Law of Independent Assortment).
❑ Alleles of the same gene can show complete dominance, incomplete dominance, overdominance or

codominance. One allele may also show different types (levels) of dominance.

- Alleles of different genes which produce the same phenotype give a modified dihybrid ratio.
- Mutant allele of one gene can mask an allele of another gene.
- Mutant alleles can show pleiotrophic effects.
- Mutant alleles can show partial penetrance.
- Lethal alleles cause death at particular stages of life cycle.
- Phenotypes produced by conditional alleles are affected by environment.

FUNCTIONAL SIGNIFICANCE OF ALLELES— DOMINANCE AND RECESSIVE EXPRESSIONS

Alleles are expressed; i.e., the trait or characteristic defined by the allele is projected or seen. However, such expression does not occur always.

Some alleles are expressed always—whether in homozygous or heterozygous condition. Such alleles are **dominant alleles** or dominant genes. The trait, so expressed is a dominant trait.

Some alleles are expressed only when in homozygous condition. Such alleles are **recessive alleles** or recessive genes. The trait expressed only in homozygotes for the particular gene is called a recessive trait.

The dominant allele is indicated by an upper case letter and the recessive allele by a lower case letter.

Let us look at an example to study and understand the phenomena of dominance, recessiveness and allelic expression. Human beings have two alleles for determining the earlobe type. There can be two types of earlobes—the attached and the detached types. The dominant allele of this trait (indicated as 'E') is for detached earlobe and the recessive allele (indicated as 'e') is for attached earlobe. One set of genes (and chromosomes) of the diploid human is derived from the male parent (father) and another from the female parent (mother). If there are two alleles, the possibility of the diploid make up can be three— EE (homozygous dominant), Ee (heterozygous) and ee [homozygous recessive (Table 32.1)].

When the individual is EE, it means that he/she has inherited one 'E' from the father and another from the mother. When the individual is Ee, it means that E has come from one parent and e from the other. The combination of alleles that an individual inherits is called the **genotype**—EE or Ee or ee in this case. But what will be seen outside or expressed? The individual will either have an attached earlobe or a detached earlobe. Whatever is the combination of the alleles, there are only two types of

Table 32.1: Genotype and phenotype for two types of earlobes present in humans

Genotype	Phenotype
EE (homozygous dominant)	Detached
Ee (heterozygous dominant)	Detached
ee (homozygous recessive)	Attached

expression—either attached or detached. The expression seen outside is called the **phenotype**.

How do three genotypes give rise to only two phenotypes? The expression is as follows:

It can now be seen that the 'detached earlobe' trait expresses itself whether the allele is single or double. The 'E' allele is able to suppress the 'e' allele when the two of them are together in 'Ee' combination. The 'E' allele is dominant over the 'e' allele. The 'e' allele is able to express itself only if it is in the homozygous state. Thus, three genotypes produce only two phenotypes.

Multiple Choice Questions

1. An allele which can express itself only in the homozygous state is a:
   a. Homozygous allele
   b. Heterozygous allele
   c. Dominant allele
   d. Recessive allele
2. One example for a gene that exists in many allelic forms is:
   a. ABO blood group gene
   b. Earlobe gene
   c. Tallness gene
   d. Serological compatibility gene
3. Phenotypic expression in an individual who is heterozygous to a dominant-recessive genetic complement will be:
   a. Unpredictable
   b. Of the dominant trait
   c. Of the recessive trait
   d. Of the mixed type

ANSWERS

1. d   **2.** a   **3.** b

Clinical Problem-solving

Case Study 1: A father is homozygous for the recessive 'attached' trait of earlobe; the mother is also homozygous for the same trait.
- How many of their children will have the same trait?
- How many of their children will have the 'detached' trait?
- Can you work out the possibilities of incidence of both the traits in the next generation?

(For solutions see Appendix).

Chapter 33

Genotype and Phenotype

Frequently Asked Questions

- Write notes on: (a) Dominance of an allele, (b) Genotype, (c) Phenotype, (d) Codominance, (e) Incomplete dominance.
- Write briefly on the following relationships: (a) Dominance Recessiveness, (b) Genotype-Phenotype.

Genotype is an organism's full hereditary information, even if not expressed. Phenotype is an organism's actually observed properties, such as morphology, development or behaviour.

GENOTYPE

The genotype represents the exact genetic makeup of an organism or the individual—the particular set of genes that the organism or individual possesses.

PHENOTYPE

Phenotype is the outward, physical manifestation of the organism—anything that is part of the observable structure, function or behaviour of the living organism.

ALLELES IN COMBINATION AND EXPRESSION

A gene is expressed when it is used to generate an RNA molecule and then a protein. Different (or alternate) forms of the same gene are called alleles. Two organisms whose genes differ even at one locus are said to have different genotypes. Differences in the genotypes can produce different phenotypes. The mapping of a set of genotypes to a set of phenotypes is referred to as the genotype–phenotype map.

- If two alleles of a given gene are identical, the organism is called a homozygote and is homozygous with respect to that gene. The genotype of the individual is homozygous;

- If two alleles of a given gene are non identical, the organism is called a heterozygote and is heterozygous with respect to that gene. The genotype of the individual is heterozygous.

DOMINANCE AND RECESSIVENESS

The allele that expresses itself whether in homozygous or heterozygous condition is a ***dominant allele***. The allele that expresses itself only when it is in homozygous condition is a recessive allele. The dominant allele masks the ***recessive allele*** in phenotypic expression.

The dominant allele is indicated by an upper case letter and the recessive allele by a lower case letter.

Cleft chin, freckled cheeks, Dimples near the nasolabial furrow, black hair, black or brown iris colour and detached earlobes are some examples of dominant traits. Straight chin, freckle less cheeks, no dimples near the nasolabial furrow, blonde hair, blue iris colour and attached ear lobes are the corresponding recessive traits (Figs 33.1A to C).

INHERITANCE OF GENOTYPE AND PHENOTYPE

When the alleles are plainly dominant and recessive, inheritance of such alleles is also simpler. Let us take a look at a possible situation. We are aware that the child (or the zygote) inherits one set of genes from the father and another from the mother. One of the traits that has a dominant-recessive type of expression is the shape of the face. Round shape (represented by the 'R' allele) is dominant and square shape (represented by the 'r' allele) is recessive.

- Let us think of a situation where the father is homozygous to the square shaped face (rr) and the mother is homozygous to the round shaped face (RR) (Fig. 33.2).

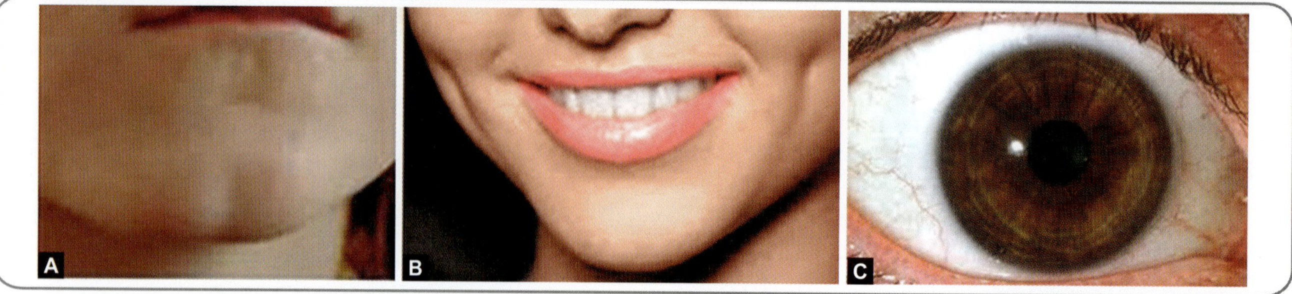

Figs 33.1A to C: Some of the dominant traits **A.** Cleft chin **B.** Dimpled cheeks **C.** Brown iris colour

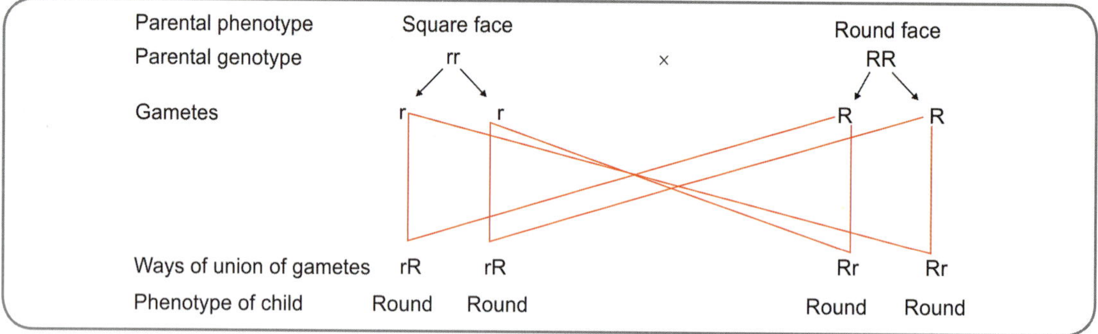

Fig. 33.2: Crossing between homozygous parents' square face (rr) and round face (RR)

We find that, when the father is homozygotic for one allele and the mother homozygotic for another, the children will all be heterozygotic. However, due to the dominant allele masking the recessive allele, all children will have the dominant trait as phenotype.

❑ If the father and mother are both heterozygous (Fig 33.3), what would be the outcome?

We find that, when both the parents are heterozygotic, the children are phenotypically dominant—recessive in the ratio of 3:1. The genotypes are homozygotic dominant–heterozygotic–homozygotic recessive in the ratio of 1:2:1. When both the parents are heterozygotic in single gene dominant–recessive inheritance, the children can have phenotypic expressions of both allelic traits.

When one allele expresses itself over another allele and masks the trait coded for by the second allele, such dominance is spoken of as 'Complete Dominance'.

Codominance

In reality, inheritance and allelic separation are not so simple enough to have straight forward dominant-recessive factors.

When the allelic traits of all alleles are expressed, the situation is defined as that of Codominance. A classic example can be seen in the expression of the ABO blood group alleles. There are three alleles to the ABO group gene; these are the 'I^A', 'I^B' and 'I^O'. I^A is dominant to 'I^O' and so is I^B to 'I^O'. Individuals having genotypes $I^A I^A$ and $I^A I^O$ will be phenotypically A group. Similarly, genotypes

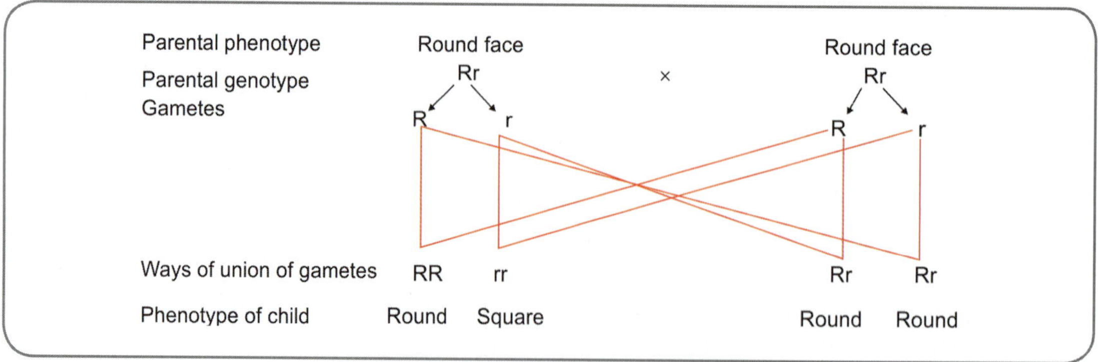

Fig. 33.3: Crossing between heterozygous parents round face (Rr)

I^BI^B and I^BI^O will be B group. But if the genotype is I^AI^B, the individual will have expressions of both alleles and will be of the AB group. Genotype I^OI^O results in O blood group. Thus, alleles I^A and I^B are codominant to each other.

Genotype	AA	AO	BB	BO	AB	OO
Phenotype	A	A	B	B	AB	O

Codominance is therefore defined as a condition in which the alleles of a gene pair in a heterozygote are fully expressed thereby resulting in offspring with phenotypes of both alleles present at the same time. Codominance is the relationship between two alleles of a gene. The contribution of both alleles (genes) are clearly visible and do not overpower each other in the phenotype.

Since codominant alleles also segregate during gamete formation and follow independent assortment principles, the possibility of occurrence of different genotypes in the offspring is manifold. Let us look at an example. What would be the blood group of an offspring when the father is of A group and the mother of B group? The outcome depends on the genotype of both the parents.

Example 1:

Father–genotype AA–phenotype A blood group
Mother–genotype BO–phenotype B blood group
The father is homozygous and the mother heterozygous.

Parental genotype	AA		BO	
Segregation in gametes	A	A	B	O
Possible offspring combinations	AB	AO	AB	AO
Offspring phenotype	AB	A	AB	A

We find that the offspring can either belong to the A group or the AB group. Since the recessive allele I^O (represented by O) or the I^B allele (represented by B) is not available in the father, there is no possibility of the offspring to belong to the O or B groups.

Example 2:

Father–genotype AO–phenotype A blood group
Mother–genotype BO–phenotype B blood group
Both parents are heterozygous.

Parental genotype	AO		BO	
Segregation in gametes	A	O	B	O
Possible offspring combinations	AB	AO	BO	OO
Offspring phenotype	AB	A	B	O

We find that the offspring can phenotypically belong to any of the four blood groups.

Incomplete Dominance

Incomplete dominance is a form of allelic relationship in which one allele for a specific trait is not completely dominant over the other allele. This results in a combined phenotype. The allele is not able to completely express itself in a heterozygous condition and so, the heterozygote has a phenotype which is intermediate to both the homozygotes.

A classic example is seen in cross pollination experiments of certain plants. The snapdragon plants having red flowers were pollinated with plants having white flowers. The results were surprising! The offspring plants did neither have red flowers nor white flowers; they had a different (third) phenotype–pink flowers (colour intermediary between red and white).

Red flower–phenotype red–genotype RR
White flower–phenotype white–genotype WW

Gametes → ↓	W	W
R	RW	RW
R	RW	RW

When the two alleles R and W occur in combination, they produce a phenotype which is not a complete expression of either of them but partial expression of both.

Some examples of incomplete dominance in humans are nose size (large–small), hair texture (curly–wavy), skin colour (dark–light), face shape (round–long), mouth size (large–small) and lip protrusion (well protruded–flat with face).

Multiple Choice Questions

1. Outward physical manifestation of a genetic allele or alleles is called:
   a. Genotype
   b. Phenotype
   c. Dominance
   d. Codominance
2. Which of the following is true about a dominant allele?
   a. A dominant allele can express itself only in a homozygous condition
   b. A dominant allele masks the recessive partner in phenotypic expression
   c. A dominant allele cannot undergo independent assortment
   d. A dominant allele is always dominant in the male
3. When both the genes of an allelic pair express themselves completely and equally, the pattern is called:
   a. Codominance
   b. Intermediate dominance
   c. Incomplete dominance
   d. Complete dominance
4. One of the following is not a relationship between alleles. Identify:
   a. Complete dominance
   b. Incomplete dominance
   c. Codominance
   d. Phenotypic dominance
5. When the father is homozygotic for the dominat allele and the mother homozygotic for the recessive allele, all their children have a phenotypic expression of the dominant trait. The reason for this is:
   a. Only the dominant allele passes into the gametic cells
   b. Male gametes are always dominant over female gametes
   c. Recessive traits will always skip one generation
   d. The dominant allele masks the recessive allele

ANSWERS

1. b   **2.** b   **3.** a   **4.** d   **5.** d

Clinical Problem-solving

Case Study 1: Both a father and a mother belong to the AB group of the ABO blood group system. Their eldest son belongs to the A group and their daughter belongs to the B group.
- Which of their children will belong to the O group?
- What is the percentage of their children likely to have the AO genotype and what percentage, the BO genotype?
- Can you work out the genotypic possibilities of all their children?

Case Study 2: A father is homozygous for large nose size (LL). The mother is heterozygous for the (Ll) alleles with a medium sized nose. The two alleles are incompletely dominant.
- Work out the genotypes and phenotypes of all their possible children.

(For solutions see Appendix).

Chapter 34

Codominance (Features, Significance and Clinical Implications)

Frequently Asked Questions

- ❑ Write notes on: (a) Codominance in ABO blood group system, (b) Haemagglutination.
- ❑ Discuss the genetic basis of the ABO blood group system. Compare it with the MN system.

Codominance is a condition in which the alleles of a gene-pair in a heterozygote are both fully expressed thereby resulting in a phenotype that has traits of both the alleles at the same time. Codominance is the relationship between the two alleles of a gene. The contribution of both alleles (genes) are clearly visible and do not overpower each other in the phenotype.

It can be defined as a condition in which both the alleles of a gene pair in a heterozygote are equally expressed with neither being dominant or recessive to the other.

Examples of Codominance

- ❑ ***ABO blood group system (Fig. 34.1)***: In the ABO Blood group system, the I^A and I^B alleles are codominant in producing the AB blood group phenotype, in which both A and B type antigens are produced.

The genes responsible for the ABO blood group are located in chromosome 9. There are three alleles—I^A, I^B and I^O. I^A causes production of 'A' surface antigens which are found on the surface of the red blood cells. Similarly, allele I^B causes production of 'B' surface antigens. Allele I^O does not cause production of any of these surface antigens. The presence of the surface antigens characterizes the blood group of an individual; those with 'A' antigen belong to 'A' group, those with 'B' antigen to 'B' group and those with no antigen are categorized as 'O' group. The

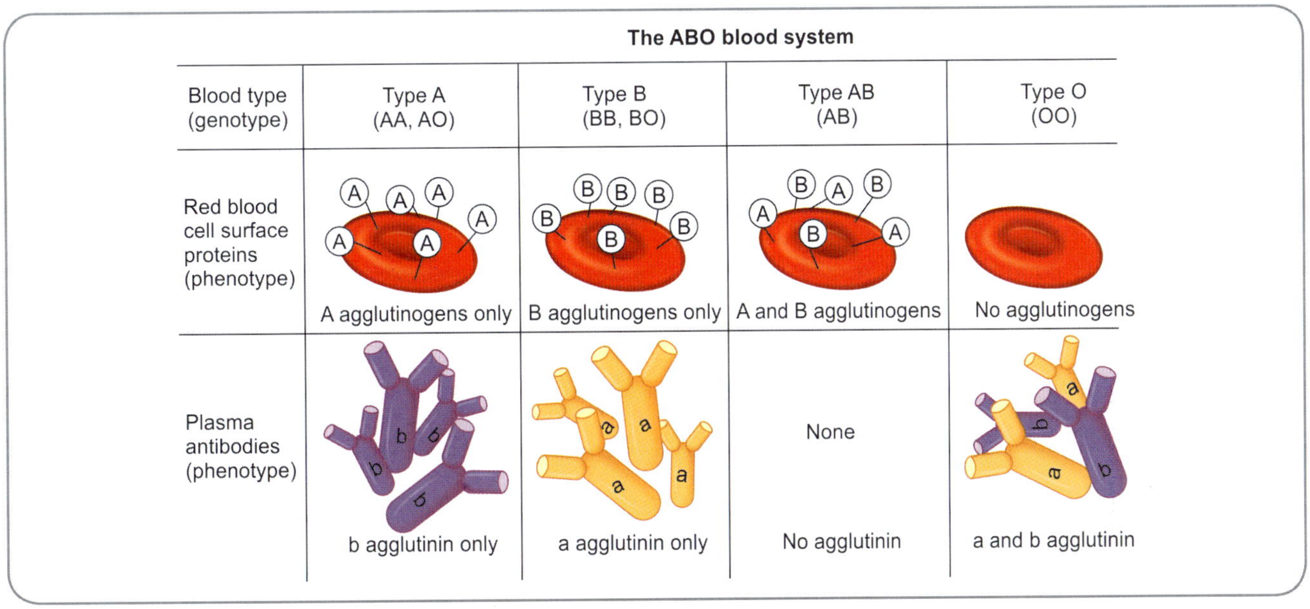

The ABO blood system

Blood type (genotype)	Type A (AA, AO)	Type B (BB, BO)	Type AB (AB)	Type O (OO)
Red blood cell surface proteins (phenotype)	A agglutinogens only	B agglutinogens only	A and B agglutinogens	No agglutinogens
Plasma antibodies (phenotype)	b agglutinin only	a agglutinin only	No agglutinin	a and b agglutinin

Fig. 34.1: ABO blood group system showing antigens and antibodies

Fig. 34.2: Roan colouring in cows

blood plasma of the individuals also has antibodies (preferably called agglutinins) which correspond to the antigen (preferably called agglutinogen) system they possess. Thus, blood group 'A' individuals have anti-B agglutinins, 'B' individuals have anti-A agglutinins and 'O' individuals have both agglutinins.

The presence of both I^A and I^B alleles in an individual causes the expression of both of them and the individual expresses a phenotype of AB blood group. In an individual with AB blood group, the red blood cell has both A and B agglutinogens and the plasma has no agglutinin.

- **MN blood group system:** Within the MN blood groups, there are two antigens M and N (similar to A and B antigens in the ABO system) whose production is determined by a gene in Chromosome 4. This gene has two alleles L^M and L^N. L^M confers the ability to produce the M antigen while L^N confers the ability to produce the N antigen. A heterozygotic individual with both alleles L^M and L^N produces both antigens; the phenotypic expression results in the MN group; the alleles, therefore, are codominant. A homozygote $L^M L^M$ will belong to the M group and a homozgote $L^N L^N$ will belong to the N group.

- **Roan colouring in cows:** This also shows a codominant relationship. Red (R) hair is codominant to white (W). The heterozygous condition (RW) results in a roan colour (red coat with white speckles) (Fig. 34.2).

Detection of Codominant Alleles

Codominant alleles are detected in the laboratory by using Antigen-antibody (agglutinogen–agglutinin) reactions. Antigen is a particular part of a protein or any other molecule; the antibody recognizes it and attaches to it.

The antibody is specific to its antigen. The terms 'antigen' and 'antibody' indicate their relationship. A substance is called an antigen when it is able to 'generate' production of 'antibody' (anti+ gen). A substance is called an antibody when it is 'anti' to another substance (anti+body). However, these terms are now being replaced by the terms agglutinogen and agglutinin (especially if the agglutination involves red cells). These terms also indicate their relationship. A substance is called an agglutinin when it is able to identify another substance and attach to the latter and block it. Thus the agglutinin is able to agglutinate (attach, form a mass and clump together) the agglutinogen. A substance is called an agglutinogen when it is able to generate production of its agglutinins.

This phenomenon of **agglutinogen–agglutinin reaction** is made use of in laboratories to detect the alleles. The commonest and the simplest example of such a detection test is the blood grouping test.

Let us study this in detail. We know from our knowledge about codominance and ABO blood groups that the red cells of a person with blood group A will have the A surface antigen and his/her serum will have anti-B antibodies. Similarly, the red cell of blood group B will have B surface antigens and the serum, anti-A antibodies. A person with surface antigen 'A' cannot have anti-A antibodies and a person with surface antigen 'B' cannot have anti-B antibodies; otherwise, the serum of the same individual will cause clumping of his/her own red cells. Antibodies attach themselves to the antigens; in the case of the anti-A and anti-B antibodies, one antibody molecule is capable of attaching to more than one antigen molecule at the same time. If an anti-A antibody is made to mix with 'A' group red cells, one antibody molecule will attach to the surface antigen molecules in two or more cells. As a result, these two or more cells will come close to each other and adhere to the lattice formed by the antibody molecule. Such coming together and adherence of the cells is called clumping or agglutination. When the involved cell is a red blood cell (as in the blood group system), such clumping is called **haemagglutination** and the terms **agglutinogen** and **agglutinin** are specifically used.

Table 34.1: Determination of blood group		
Agglutination result with		**Blood group**
Antiserum A	**Antiserum B**	
Clumping	No	A
No	Clumping	B
Clumping	Clumping	AB
No	No	O

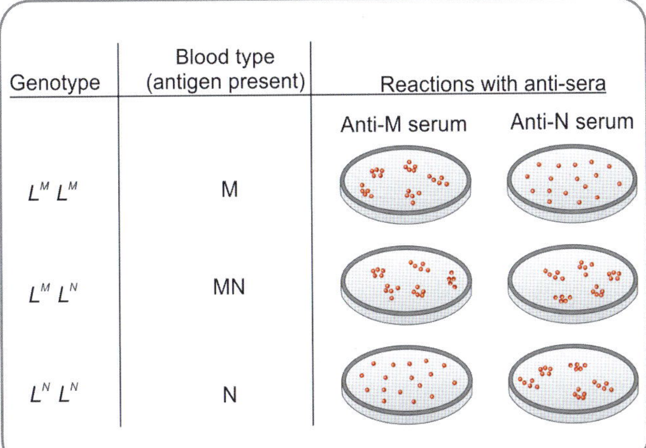

Genotype	Blood type (antigen present)	Reactions with anti-sera	
		Anti-M serum	Anti-N serum
$L^M L^M$	M		
$L^M L^N$	MN		
$L^N L^N$	N		

Fig. 34.3: Antigen–antibody reaction in MN blood type

If we need to find out the group of an individual, the following procedure can be adopted. The blood of that individual is mixed with different sera (plural of serum). The serum containing anti-A antibodies is antiserum A and that containing anti-B antibodies is antiserum B. After mixing with the antisera, the results can be studied (Table 34.1).

The MN antigens and their presence can also be detected in a similar way (Fig. 34.3).

Clinical Correlation

- Analyzing blood samples for codominant alleles is widely used in cases of disputed parentage, particularly of disputed paternity.
- Blood sample analyses are done in transfusion and immunohaematological studies.

Multiple Choice Questions

1. Codominance is the condition when:
   a. The alleles of a gene pair are both expressed
   b. The alleles of a gene pair are both suppressed
   c. The alleles of a gene pair are both dormant
   d. The alleles of a gene pair are both hyperexpressive
2. Blood group of an individual in the ABO system is characterized by:
   a. Surface antigen
   b. Plasma agglutinin
   c. RBC adhesion
   d. Plasma viscosity
3. Determination of one's blood group is done by using:
   a. Karyotyping to locate the chromosome
   b. Genetic mapping to locate the I^A or I^B allele
   c. Agglutinogen-agglutinin reaction
   d. Working out the pedigree chart of the family

ANSWERS

1. a   **2.** a   **3.** c

Clinical Problem-solving

Case Study 1: A father is heterozygotic ($L^M L^N$) for the M and N alleles of the MN Blood group system. The mother is homozygotic ($L^N L^N$) for the N allele.
- Work out the possible genotypes and phenotypes of the children.

(For solutions see Appendix).

Chapter 35

Incomplete Dominance and Overdominance

Frequently Asked Questions

- Write notes on: (a) Incomplete dominance, (b) Heterozygosis.
- Write briefly on the genetic basis of sickle-cell trait and sickle-cell disease.
- Write notes on overdominance and its community uses.

INCOMPLETE DOMINANCE

Incomplete dominance is a form of allelic relationship in which one allele for a specific trait is not completely dominant over the other allele.

This results in a combined (or new) phenotype. Incomplete dominance occurs when one allele is unable to express its full phenotype in a heterozygous individual. This often causes the heterozygote to have a phenotype that is intermediate to both homozygotes.

Examples in human beings include

- A child having wavy hair when one parent has curly hair and the other straight hair;
- A child having medium complexion when one parent is dark skinned and the other light skinned;
- A child having medium level lip protrusion when one parent has large protrusion and the other small protrusion.

In the curly-straight hair example (Table 35.1), all children will have a genotype of Cc, which is a heterozygotic condition. The allele 'C' cannot express itself completely (curliness) in the presence of the allele 'c'; similarly, allele 'c' cannot express itself completely (straightness) in the presence of

Table 35.1: Curly-straight hair
Father—curly hair: CC; Mother—straight hair: cc

Gametes		From mother	
		c	c
From father	C	Cc	Cc
	C	Cc	Cc

allele 'C'. But they influence each other to produce a new phenotype which is neither of them but 'wavy'.

In the dark-light skin example, all children will have a genotype of Dd, which is a heterozygotic condition. The allele 'D' cannot express itself completely (dark skin colour) in the presence of the allele 'd'; similarly, allele 'd' cannot express itself completely (light skin colour) in the presence of allele 'D'. They however influence each other, resulting in a new phenotype 'medium colour' (Table 35.2).

Both the alleles are not in a position to express themselves in a heterozygotic condition. Instead, they try to suppress the other and this mutual influence leads to a third phenotype which is a mix of both traits.

Incomplete dominance differs from codominance in that there is no complete and unmasked expression of the alleles as in codominance.

Table 35.2: Dark-Light skin
Father-dark-skinned: DD; Mother-light-skinned: dd

Gametes		From mother	
		d	d
From father	D	Dd	Dd
	D	Dd	Dd

Clinical Correlation

- Incomplete dominance is responsible for the many variations we find in the characteristics and traits of all animals including human beings.
- Some forms of camptodactyly (the condition where the proximal interphalangeal joint of one or more fingers is bent; clinidactyly (the condition where the proximal interphalangeal joint is bent sideways) and symphalangism (the condition where the proximal interphalangeal joint is fixed; little finger often affected and called 'stiff little finger') show incomplete dominance.

OVERDOMINANCE

Overdominance is a condition where the phenotype of the heterozygote lies outside the phenotypic range of both homozygote parents. The phenotype of the heterozygote is more extreme or better adapted than both homozygote parents. Thus, a trait that shows overdominance sometimes confers a survival advantage in the heterozygote.

It can be said that the heterozygotes have a better or higher degree of fitness than both the homozygotes for the two alleles. The condition is also called *heterozygosis*.

Sickle-cell Trait Alleles

The functional significance of overdominance is shown in the human beings in the presence of the sickle-cell trait alleles.

The human haemoglobin in the adults is predominantly haemoglobin A; this is coded by the allele HB^A. Another allele of the same gene is HB^S. This allele codes for haemoglobin S which has a different aminoacid in a particular place in the haemoglobin chain (valine in position 6 instead of the normal glutamic acid).

❑ When the genotype is $HB^A HB^A$, the individual has normal haemoglobin A and is normal.

❑ When the genotype is $HB^S HB^S$, the individual has the abnormal haemoglobin S. The presence of valine which is highly hydrophobic makes the haemoglobin strands long and adherent. These abnormal haemoglobin strands alter the shape of the red cell in which they exist. Normal human red cells are circular, biconcave and smooth. With haemoglobin S, the red cell becomes long and sickle-shaped. The sickle-shaped red cell is not able to pass through the blood vessels with the same ease as that of the normal red cell; a group of sickle-cells adhere together and block the circulation. Sickling makes the cells brittle, prone to be broken down rapidly and the cells do not survive for more than ten or twenty days; but production of new red cells does not happen at that pace. Blockage of circulation at multiple points in the body and a negative balance in the number of red cells in association with a reduced capacity of the sickle-red cell to carry oxygen results in the highly dangerous sickle-cell anaemia. Individuals with the $HB^S HB^S$ genotype develop this disease and usually die before the reproductive age group is reached.

❑ When the genotype is $HB^A HB^S$, such an individual has what is called a *sickle-cell trait*. The normal allele codes for 50% of the total haemoglobin and so the person does not exhibit the signs and symptoms of the disease. Though he/she has the abnormal haemoglobin and sickle-cells in circulation, normal day to day activities do not suffer. Only when such an individual is exposed to reduced oxygen situations (like climbing a mountain or high-altitude areas) or is dehydrated, he/she has symptoms. However, the person with the heterozygous genotype can transmit the abnormal allele to the next generation. Hence, such people are called *carriers* (carriers of the disease).

People living in the inner parts of Africa, especially sub Saharan Africa are prone to this disease as the incidence of the abnormal allele is very high in this region.

An intriguing situation occurs. The original abnormal allele is supposed to have come into existence by genetic mutation in the African-Arabian areas. However, the allele still occurs in high incidence in certain areas of the globe (Africa, Middle east, India and Mediterranean). If we remember that people afflicted with the sickle-cell disease tend to die early in life and much before the reproductive age group, it is natural to expect that the incidence of the abnormal allele will gradually reduce in the population (because patients will not live long enough to become parents and transmit the gene). Such a situation did not and does not seem to have happened.

On the contrary, the abnormal allele seems to be more in certain areas. When more community research was done, it was noted that the abnormal allele is more prevalent in areas which are endemic to malaria. What then is the relationship?

The heterozygous genotype offers resistance against malaria. The malarial parasite Plasmodium spends most of its life cycle in the red blood cell of the host (affected individual—parasite infested individual). In a person with the sickle-cell trait, entry of the parasite into a red cell causes the cell to immediately sickle and elongate. The cell is thus rendered brittle, ruptures prematurely and is destroyed along with the parasite inside. The parasite is not able to complete further stages of its life cycle and so the disease is curtailed.

So, in areas where malaria is a health hazard (especially the more fatal falciparum malaria), survival rate of an individual increases if he/she carries the sickle-cell trait.

This phenomenon is called *selection for the heterozygote*—the heterozygote is protected from both death due to sickle-cell anaemia and death due to malaria. There is adaptive advantage of the heterozygotic condition. In areas where the incidence of malaria is high, heterozygotes are healthier than individuals who are homozygous for either allele. They neither have anaemia nor malarial symptoms. Under such conditions, the HB^S allele shows overdominance for survival of the host.

Botanical Uses of Overdominance

Certain plants exhibit higher growth potential when they are heterozygous. One such example is the tomato (*Solanum lycopersicum*). Heterozygous genotype for alleles in the Mi locus seems to give protection against certain viral diseases of the plant and to increase yield potential.

Multiple Choice Questions

1. Incomplete dominance occurs because:
   a. The dominant allele is not able to express completely in the presence of the recessive allele
   b. The dominant allele is over expressive and loses its over dosage
   c. The recessive allele gets activated in a heterozygous condition
   d. The two alleles try to enhance the effects of each other
2. In overdominance, the phenotype of the heterozygote lies outside the phenotype range of:
   a. Both dominant and recessive homozygotes
   b. The dominant homozygote
   c. The recessive homozygote
   d. The male parent, whether dominant or recessive
3. What is false about sickle-cell trait?
   a. The individual with the trait is heterozygous
   b. The individual with the trait is a carrier
   c. The individual has no abnormal haemoglobin
   d. The individual has no symptoms under normal conditions

ANSWERS

1. a   **2.** a   **3.** c

Clinical Problem-solving

Case Study 1: A father is homozygous for large lip protrusion (LL). The mother is heterozygous (Ll) with moderate protrusion.
- Work out the possibilities of genotypes and phenotypes of all their children.

(For solutions see Appendix).

Lethal Alleles

Lethal alleles are such alleles which cause an organism to die when they are in a homozygous condition. The gene involved is an essential gene; in other words, such a gene is responsible for vital functions of the body. In the homozygous genotypic condition, the functions performed by the gene are altered in such a way that the organism dies.

To understand the occurence and method of activity of lethal alleles, it is better to learn about the discovery of lethal alleles.

Lucien Julien Cuenot, a French biologist was studying Mendelian principles in the early 1900s; he was performing experiments on the inheritance of coat colour in mice population. He found out that three **mnemons** (his term for **alleles** of the coat colour gene) are responsible for the coat colour. In one of the experiments, he tried a heterozygote cross of yellow mice. Phenotypically the mice were yellow and genotypically they were Yy ('Y' allele being dominant and 'y' allele being recessive). Following Mendelian principles of segregation and assortment,

Heterozygote Cross—expected results

Gametes		From Female Parent	
		Y	y
From Male Parent	Y	YY	Yy
	y	Yy	yy

Genotype YY : Yy : yy = 1 : 2 : 1; **Phenotype** yellow : white = 3 : 1

Heterozygote Cross—result obtained by Cuenot

Gametes		From Female Parent	
		Y	y
From Male Parent	Y	YY	Yy
	y	Yy	yy

Phenotype yellow: white = 2 : 1

Cuenot expected a phenotypic yellow-white ratio of 3:1. But he obtained only a 2:1 ratio.

When Cuenot performed test crosses, he was not able to see the YY genotype at all. Cuenot concluded that the allele 'Y' (dominant for yellow coat colour) should have produced lethal effects on the vital functions of the organism. By the test crosses, he also confirmed that allele 'Y' is the dominant allele and all surviving yellow mice were only heterozygotes.

Subsequent research has proved the existence of lethal alleles. To review Cuenot's discovery, we shall work out the heterozygote cross.

Gametes		From Female Parent	
		Y	y
From Male Parent	Y	YY	Yy
	y	Yy	yy

Genotype YY : Yy : yy
Phenotype dead : yellow : white
So the outcome ratio remainas yellow : white = 2 : 1

LETHAL ALLELES IN HUMANS

Achondroplasia is a disorder where cartilage and bone formation are affected causing the affected individual to be a dwarf. One form of achondroplasia is associated with lethal alleles. The gene responsible is the FGFR3 gene. The normal allele is 'd' and the mutant allele is 'D'.

When the genotype is dd, the individual is normal. When the genotype is Dd, the mutant allele causes disorders of cartilage and bone formation resulting in dwarfism. When the genotype is DD, the alleles produce a lethal effect and the offspring dies before birth or shortly after birth.

What would be the outcome if both the father and the mother are achondroplastic? Let us work out the allelic combinations likely to occur.

Gametes		From Female Parent	
		D	d
From Male Parent D		DD	Dd
d		Dd	dd

Fifty percent of the gametes of an achondroplastic parent will carry the lethal allele. Thus, such a parent has a 50% chance of transmitting dwarfism to each offspring.

When both parents are achondroplastic, of every four pregnancies that the mother has, two children will be achondroplastic (genotype Dd), one child normal (genotype dd) and one child will die (genotype DD).

Recessive Lethality

In the above example of achondroplasia, we find that the lethal effect of the mutant allele occurs only in the homozygous state. When only one mutant allele occurs in the genotype, it is not able to express its lethal effect (Dd— allele D not able to cause lethality) and is being suppressed by its normal allele. So it is recessively lethal.

But the same allele is dominant for achondroplasia. The disease is caused when the genotype is Dd. The allele 'D' acts as a dominant allele to the normal allele 'd' and causes the disease. Achondroplasia is therefore inherited as an autosomal dominant disorder.

An allele is called *recessively lethal* when its lethality is expressed only in the homozygous state.

Other Examples of lethal genes and their effects

- ❑ *Cystic fibrosis:* The involved gene is the CFTR gene (Cystic fibrosis transmembrane conductance regulator gene) in Chromosome 7. The normal allele is dominant to the mutant allele; so in a heterozygote, the normal allele codes for normal secretions in the glands and the individual does not show any disease. In a homozygote for mutant allele, CFTR protein is not produced and the individual is diseased. Thus, cystic fibrosis is inherited as an autosomal recessive disorder.
- ❑ *Tay-Sachs disease:* This is a disease that causes progressive deterioration of nerve cells and affects the mental and neurological abilities of an individual. The gene involved is the HEXA gene in Chromosome 15. Homozygotes for mutant allele (Tay-Sachs allele) exhibit disease symptoms. Heterozygotes with one mutant allele (carrier of the mutant allele) do not have disease symptoms but exhibit some kind of protection to tuberculosis (selective adaptive advantage of heterozygosis).

- ❑ *Congenital ichthyosis:* This is a disease where the various cells of the skin are abnormal and the individual exhibits multiple dermal manifestations. The disease is inherited as an autosomal recessive disorder, meaning, two mutant alleles are required to produce disease symptoms. Heterozygotes appear normal but can transmit the mutant allele as a carrier.

In all the three above mentioned examples, the inheritance of the diseased condition caused by the lethal allele is by autosomal recessiveness. Lethality occurs as the disease progresses and is caused by the homozygous condition for the mutant allele. So, both disease inheritance and lethality inheritance are recessive (compare with achondroplasia, where the disease inheritance is dominant and the lethality inheritance is recessive).

- ❑ *Sickle cell anaemia:* This is a disease where the sickle-shaped red blood cells are destroyed rapidly within the body and circulation suffers due to clogging. Red cells break down within the capillaries themselves. The disease is inherited as an autosomal recessive (two alleles are required to produce the disease); lethality which occurs as a sequelae to the disease is also inherited as an autosomal recessive; but the alleles (normal and mutant) express their effects simultaneously qualifying description as codominant alleles.

Dominant Lethality

This is seen in the Huntington's disease in humans. The gene involved is the Huntingtin gene (also called the HTT gene or HD gene or IT 15 gene) located in chromosome 4. The gene codes for the Huntingtin protein. When the genotype is heterozygous (one normal allele and another mutant allele), the mutant gene alters the Huntingtin protein which in turn leads to neurological degeneration. Once the degeneration sets in, patients deteriorate fast and die. This is a situation where the presence of one lethal mutant allele causes symptoms and so the inheritance is autosomal dominant. The disease leads to lethality and so lethality is also dominantly inherited. However, the lethal allele has not been eliminated from the gene pool because affected persons live past the reproductive period and exhibit symptoms only by the third decade of life.

Types of Lethal Alleles

- ❑ *Embryonic lethal alleles:* These are lethal alleles which cause death in utero or in the gestation period;

❑ *Post natal lethal alleles:* These are lethal alleles which cause disease and death in postnatal life;

❑ *Conditional lethal alleles:* These are lethal alleles which are effective to cause lethality only in certain conditions and not in others. A lethal allele in the Drosophila acts lethal only in deficiency of Vitamin B_6.

Conditional lethal alleles do occur in human beings too, but research studies on them are meagre.

❑ *Synthetic lethal alleles:* These are lethal alleles which cause lethality only in combination with the lethal allele of another gene. If alone, none of these lethal alleles cause lethality and are said to be 'viable'.

Multiple Choice Questions

1. When both parents are achondroplastic and heterozygous for the FgFR3 gene alleles, the probability of the genotype (DD—homozygous dominant) for every four offspring is:
   a. 25%
   b. 50%
   c. 75%
   d. 100%
2. Selective protection advantage of heterozygosis to tuberc ulosis is given by:
   a. CFTR gene
   b. FgFR3 gene
   c. HEXA gene
   d. HTT gene
3. An example for a disease caused by dominant lethality is:
   a. Achondroplasia
   b. Cystic Fibrosis
   c. Tay sachs disease
   d. Huntington chorea

ANSWERS

1. a   2. c   3. d

Clinical Problem-solving

Case Study 1: Both the father and the mother are carriers of the Tay sachs allele.
❑ Work out the Punnett square for the first generation offspring of this family.

(For solutions see Appendix).

Chapter 37

Sex Determination by Chromosomes

Frequently Asked Questions

❑ Write notes on: (a) Sex chromosome, (b) X and Y chromosomes in humans, (c) Sex determination (d) Lyonisation.

SEX CHROMOSOMES

Sex chromosomes are those chromosomes which are involved in the determination of the sex of an organism/individual. They are also involved in the development of sexual characteristics in the organism/individual.

In 1901, Clarence Erwin McClung, who studied the grasshoppers, was the first to suggest that male and female sexes are due to the inheritance of particular chromosomes. He proposed that an accessory chromosome was the determiner of sex. In 1902, further work on grasshoppers revealed that females and males have different numbers of copies of one particular chromosome. This chromosome was called **X chromosome** because its properties and features were not clearly known. The **Y chromosome** was discovered in 1905.

SYSTEM OF SEX CHROMOSOMES

A sex-determination system is a biological system that determines the development of sexual characteristics in an organism. When the presence of a chromosome that was involved in the determination of sex was noted, it was called the **X chromosome**. To differentiate it from the other chromosomes (which were not involved in the process of sex determination), it was also named the allosome (Greek.allos=other, soma=body). Another chromosome, structurally different from the X, but also involved in the sex determination process was subsequently detected. It has been named the 'Y' chromosome (because Y is the next letter to X in the English alphabet).

In mammals, including humans, a normal female has two x chromosomes and a normal male has one X and one Y chromosome. The gametes of a female will have only the X chromosome. Therefore, females belong to the **homogametic sex**. The gametes of a male will be in two groups: one group with the X chromosome and the other with the Y chromosome. Males belong to the **heterogametic sex**.

METHODS OF SEX DETERMINATION

It is not only the presence of the sex chromosome that determines sex, but also sometimes, its absence too. Two methods of sex determination by the presence or absence of sex chromosomes have been found.

Protenor Method or Mode

This method has been so named because of its discovery in the protenor bugs and grasshoppers. The female is the homogametic sex with two X chromosomes and the male is the heterogametic sex with one X chromosome. Sex is determined not by the presence of the X chromosome, but by the ratio of the sex chromosomes to the autosomes. A normal female has two X chromosomes and two sets of autosomes. The ratio of X:A is 1:1. A normal male has one X chromosome and two sets of autosomes. The ratio of X:A is 1:2. A ratio greater than 1:1 will result in extreme female characteristics; an organism with XXX and AA is an example for 'greater than 1:1' and will be a metafemale. A ratio less than 1:2 will result in extreme male characteristics; an organism with X and AAA is an example for 'less than 1:2' and will be a metamale. A ratio of 1:1 will produce the female phenotype; an organism with XXX and AAA will have 1:1 ratio and will be a triploid female. If the ratio is between 1:1 and 1:2, the organism will be an intersex; an organism with XX and AAA will have a ratio of 2:3 and therefore be an intersex.

In the protenor mode, the primary signal for sexual development is the X:A ratio and not the presence of Y chromosome.

Lygaeus Method or Mode

Lygaeus is another variety of bug. In this mode, sex is determined by the presence of a specific chromosome. This specific chromosome is the Y chromosome. This mode of sex determination is seen in most mammals including humans. The primary signal for sex determination is chromosome composition and the presence of Y. The normal female is homogametic with XX composition. The normal male is heterogametic with XY composition. If, due to some anomalous condition, the embryo has only one X chromosome, male sexual characters do not develop. However, again due to some anomaly, even with two X chromosomes, if the embryo has an additional Y chromosome, male sexual characters develop. This development of male characters is not complete because, one Y chromosome is not able to overcome or mask the effects of two X chromosomes. Given these findings, triggering of the development of male sexual characters in the presence of Y irrespective of the status of X chromosomes is a clear indication of the Y signal.

SEX DETERMINATION IN HUMANS

Sex determination in humans depends on the Y chromosome. It has been seen that individuals with a Y chromosome (XY, XXY) develop male characteristics and individuals without a Y (XO, XX, XXX) develop female characteristics. X:A ratio (sex chromosome: autosome ratio) does not have any role to play.

To understand the phenomenon, it is necessary to first study certain genetic steps which happen during development and the female sexual model.

❑ A gene called the DSS gene is found on X chromosome at position Xp 21. This is the Dose Sensitive Sex Reversal (DSS) gene. It represses (or suppresses) the Male Pathway Genes (MPGs).

❑ The tissue which is to develop into a gonad is bipotential in the early embryo. In other words, this tissue has the capability to either become a male gonad or a female gonad (two potentials—bipotential).

❑ If the Male Pathway Genes become active, they direct the bipotential gonadal primordium to develop into a testis.

❑ Female pathway genes (FPGs) are also present; these genes direct the bipotential gonadal primordium to develop into an ovary.

❑ Like how the dose sensitive sex reversal (DSS) gene represses male pathways genes (MPGs), there are also other genes which repress the FPGs.

What happens in an embryo with XX composition?

We know that a normal human female has two X chromosomes. However, early in development, one of the two X chromosomes is made inactive and genetically inert. The other X chromosome which is active has a DSS gene on Xp21. This DSS gene represses the MPGs. A little later in development, FPGs are activated and they direct the bipotential gonad to become an ovary. Subsequent to ovarian development, some other genes which permanently repress the MPGs are triggered. Ovarian development and permanent repression of MPGs result in female sexual development and the individual becomes a female.

What happens in an embryo with XY composition?

The embryo has an X and a Y chromosome. In the short arm of the Y chromosome, next to the terminal zone (called the Ypter) is a gene area called the Sex determining Region on the Y (SRY gene). The SRY gene inactivates the DSS gene. Absence of inhibition by the DSS results in MPGs becoming active. MPGs act on the bipotential gonad and direct it to become a testis. As testis develops, various cells of the testis including the Sertoli cells are formed. By the sixth week of development, Sertoli cells produce the Mullerian Inhibiting Factor (MIF or the antiMullerian hormone gene) that acts on the Mullerian ducts and their further development (MIF actually acts on the FPGs which promote Mullerian development). Since Mullerian (paramesonephric) ducts give rise to the female genital organs, the presence of MIF stops such development. In addition, interstitial cells of the testis produce testosterone which promotes and enhances development of the mesonephric (Wolffian) duct system resulting in male genital organs. It can be clearly seen that the primary signal for male development is from the SRY gene and so, from the Y chromosome.

Role of Y Chromosome

In humans, the Y chromosome has a significant role to play in sexual development. It has already been noted that the Y chromosome gives the primary signal in male development. Female development is more of a routine and regular process and its primary signal is chromosome composition (autosomes plus X) itself. The significance of Y can be better understood by studying two syndromes, namely the Klinefelter syndrome and Turner syndrome.

Individuals with Klinefelter syndrome have normal number of autosomes, two X chromosomes and a Y chromosome. So the total composition is 47 and the sex chromosome pattern is XXY. These individuals develop male sexual characteristics though very often they are sterile. Few female sexual characteristics like enlarged breasts may show out. The overall picture indicates that the SRY gene has acted, DSS has been inhibited and MPGs are allowed to act. Individuals with Turner syndrome have normal number of autosomes, one X chromosome and no Y chromosome. These individuals are physical females externally and develop female characteristics though they may have rudimentary or underdeveloped ovaries. Absence of SRY and presence of one active X has resulted in the development of female characters and DSS inhibition on MPGs. Both these syndromes are evidences of the significance of Y chromosome. However, both developmental processes are not normal because of the imbalance and other genetic factors.

Added Information

- ❑ If the SRY gene is mutated, an embryo with a genetically normal male composition (46, XY) will develop female characteristics and ovaries. There is no SRY, no inhibition on DSS, MPGs are repressed, no inhibition on FPGs which are normally released and female development proceeds. However, the individual may not be a complete female due to other genes.
- ❑ If the SRY gene is translocated on the X chromosome, an embryo with a genetically normal female composition (46, XX) will develop male characteristics and a testis. SRY gives the primary signal and further development proceeds. This individual also will not be a complete male due to other genes.
- ❑ XY embryos with chromosomal duplication of Xp21 (so, there are two Xp21 resulting in two DSS genes and one SRY gene composition) develop into females. This indicates that two DSS genes are able to overcome the effect of one SRY gene.
- ❑ What then happens in the XXY (47, XXY) individual? This individual is usually a Klinefelter syndrome individual and shows male characteristics. If two DSS genes can suppress the effect of one SRY gene, why does it not happen in Klinefelter syndrome? This situation indicates that X chromosome inactivation, in the embryo, occurs before SRY or DSS action. X chromosome inactivation in the early stage leaves the embryo with a single active X chromosome and single DSS. One DSS is not able to suppress the effects of SRY. Presence of SRY triggers the male mechanism.
- ❑ ***Other systems of sex determination***: Several studies have helped researchers identify systems other than the XX/XO (protenor) and XX/XY (lygaeus) systems in sex determination.
  - ○ ***ZW sex determination:*** This is the system found in birds, some reptiles, some fishes and some insects. The male is homogametic (ZZ) and the female is heterogametic (ZW). Since the female is the heterogametic partner, it is the ovum (and not the sperm as in the X-Y system) that decides the sex of the offspring. Z chromosome is larger than the W. There is no relationship between the Z-W chromosomes and the X-Y chromosomes. Z chromosome seems to have some similarity with human autosomal chromosome 9; therefore, it is possible to say that sex chromosomes have been derived from autosomal common ancestor chromosomes.

Added Information *contd...*

- ○ ***Haplodiploidy system:*** This is the system that determines the sex in ants, bees and wasps. The male develops from an unfertilized ovum and is haploid; the female develops from a fertilized ovum and is diploid. This system is also called ***arrhenotoky***.
- ❑ ***Dosage Compensation:*** This is a special compensatory mechanism. We know that the compliment of sex chromosomes is not the same and equal in males and females. The X chromosome has more genes than the Y chromosome. The female with XX will have more genetic material and expression than the male with XY. This imbalance can lead to marked changes within few generations and even endanger the species. To compensate for this imbalance, some method of equalisation has to occur; such equalisation is called dosage compensation (compensating and neutralising the over dosage of certain genetic material). In humans, compensation is done by inactivating one of the two X chromosomes in the female. Murray Llewellyn Barr, a Canadian medical researcher, along with Ewart George Bertram, in 1948 discovered a cell structure which later came to be known as the ***'Barr body'***. This is a darkly stained structure that is attached to the nuclear envelope in the somatic cells of the human female. It consists of one X chromosome that has been inactivated, made heterochromatic and rendered genetically inert. In 1961, Mary Frances Lyon, an English geneticist made the remarkable discovery of X chromosome inactivation. In the human female embryo, at about the 100 cell stage, one of the two X chromosomes is randomly inactivated in the somatic cells. When each of these 100 cells multiplies by mitosis, all the daughter cells of subsequent generations will have only genes of the active chromosome expressed. Which of the two X chromosomes will get inactivated? The selection is at random and different X chromosomes can be inactivated in different cells. However, if an X chromosome is abnormal, it is usually inactivated. It is also true that one X chromosome is left to be active and the others inactivated in the total chromosome composition of the individual. By this, the normal male has no Barr body at all and no X inactivation. The normal female has a single Barr body and one X inactivated. If the individual has XXX (47, XXX), then there are two Barr bodies (and three Barr bodies in case of 48,XXXX etc.). A Klinefelter person has one Barr body (47, XXY) and a Turner person has no Barr body (45, XO). X inactivation ensures that only one X chromosome is active in a somatic cell. The process of X inactivation is also called ***lyonisation in honour of Mary Lyon***.
- ❑ Hermaphrodites have both ovaries and testes. External genitalia are ambiguous and true hermaphrodites are genetic mosaics—some cells are XX, while others are XY.
- ❑ Pseudohermaphrodites have either testicular or ovarian tissue, but not both. Generally the tissue is rudimentary and external genitalia are often ambiguous.
- ❑ Turner Syndrome and Klinefelter syndrome are due to nondysjunction in the male or female parent resulting in gametes without X chromosome.
- ❑ ***XYY Condition:*** It occurs due to nondysjunction of Y chromosome in the Male parent. Phenotypic features are above average height, fertile, sometimes (but not always) retarded.

contd...   *contd...*

Added Information *contd...*

- *Poly-X Females (XXX, XXXX, XXXXX):* Phenotypic features are infantile genitalia, underdeveloped breasts, fertile and sometimes mental retardation.
- Swyer syndrome (also known as 46,XY complete or pure gonadal dysgenesis) is caused by mutations in the *SRY* gene which prevent production of the sex-determining region Y protein or result in the production of a nonfunctioning protein. A foetus whose cells do not produce functional sex-determining region Y protein will develop as a female despite having a Y chromosome.
- In most individuals with 46,XX testicular disorder of sex development, the condition results from an abnormal exchange of genetic material between chromosomes (translocation). This exchange occurs as a random event during the formation of sperm cells in the affected person's father. The *SRY* gene is misplaced in this disorder, almost always onto an X chromosome. A foetus with an X chromosome that carries the *SRY* gene will develop male characteristics despite not having a Y chromosome.
- *SRY* gene mutations that impair, but do not eliminate the function of the sex-determining region Y protein have been identified in a small number of people with 46,XY disorder of sex development or partial gonadal dysgenesis. Affected individuals may have external genitalia that do not look clearly male or clearly female (ambiguous genitalia) or other abnormalities of the genitals and reproductive organs.
- About 10 percent of individuals who have both testicular and ovarian tissue, a condition called *ovotesticular disorder of sex development*, have two X chromosomes with one carrying the *SRY* gene.

Multiple Choice Questions

1. X chromosomes have been so called because:
   a. Of their shape
   b. Their properties were not clearly known then
   c. Of their proximity to the Y chromosome
   d. Of their length
2. Sex is determined in the protenor method:
   a. By the presence of Y chromosomes
   b. By the number of X chromosomes
   c. By the proportion of X chromosomes and autosomes
   d. By the proportion of X and Y chromosomes
3. Male trigger in a human embryo is first given by:
   a. Ypter
   b. Xpter
   c. SRY
   d. MIF

ANSWERS

1. b.   **2.** c   **3.** c

Clinical Problem-solving

Case Study 1: An individual has a total number of 47 chromosomes with an allosomal pattern of XXY. The individual shows male sex characteristics externally.
- Can you give the reason for the male sexual characteristics in this individual?
- What will be the Barr body pattern in this individual?
- To which mode of sex determination does this case belong?

(For solutions see Appendix).

Chapter 38

Twinning—Monozygotic and Dizygotic Twins

A twin is one of the two offsprings produced in the same pregnancy. Twins can either be monozygotic (identical), when they develop from one zygote that splits and forms two embryos, or dizygotic (fraternal) when they develop from two separate eggs which are fertilized by two separate sperms.

Seventy percent of all twins are dizygotic. If the factors of twinning mechanism and gender are taken into consideration, there are five variations of twinning.

- ❑ **Dizygotic:**
  - i. Male-female dizygotic
  - ii. Female-female dizygotic
  - iii. Male-male dizygotic
- ❑ **Monozygotic:**
  - i. Female-female monozygotic
  - ii. Male-male monozygotic

The order of occurrence follows the order given above. Monozygotic twins are genetically identical and are always of the same sex unless a mutation occurs during development.

For easy understanding of the principles and mechanism involved, let us study dizygotic twinning first.

DIZYGOTIC TWINS (FRATERNAL TWINS)

Dizygotic or fraternal twins (also called non-identical twins, dissimilar twins or biovular twins) usually occur when two eggs are independently fertilised by two different sperm cells, both zygotes implant individually in the uterus and each develops its own placenta, amnion and chorionic sac. Since two ova are involved, these twins are called **biovular (or binovular)** and since two zygotes are formed, dizygotic. Sometimes the two placentae are located so close that fusion occurs.

Dizygotic twins have different chromosome profiles. They are usually of different sexes but can be of the same sex. They may have no more resemblance than brothers or sisters of the same age.

Increase in incidence of twin pregnancies are mainly due to fertility drugs and increase in maternal age. Infertility treatments increase the rate of dizygotic twinning, with about 35% of pregnancies from *In Vitro Fertilisation (IVF)* being twin pregnancies. Factors like a mother with larger build, diet with rich dairy products, mother having had several earlier pregnancies also seem to contribute. The role played by dairy products is not direct; use of growth hormone like products in cattle seem to instigate twinning in the mothers consuming milk obtained from such cattle. The incidence of dizygotic twinning is about 6 (in Japanese population) to 12 (in African population) for every 1,000 live births. The incidence in India is near the middle of this range.

MONOZYGOTIC TWINS (IDENTICAL TWINS)

Monozygotic or identical twins are the result of a single fertilised egg splitting into two separate embryos. The two embryos can share a single placenta and can be in the same or different sac. Since identical twins come from the same fertilised egg, they have exactly the same DNA. They have strong resemblance in blood groups, sex and external appearance (Fig. 38.1).

Monozygotic twins develop from a single fertilised ovum. The splitting of the zygote happens at various stages of development and eventually, two separate foetuses develop. Spontaneous division of the zygote into

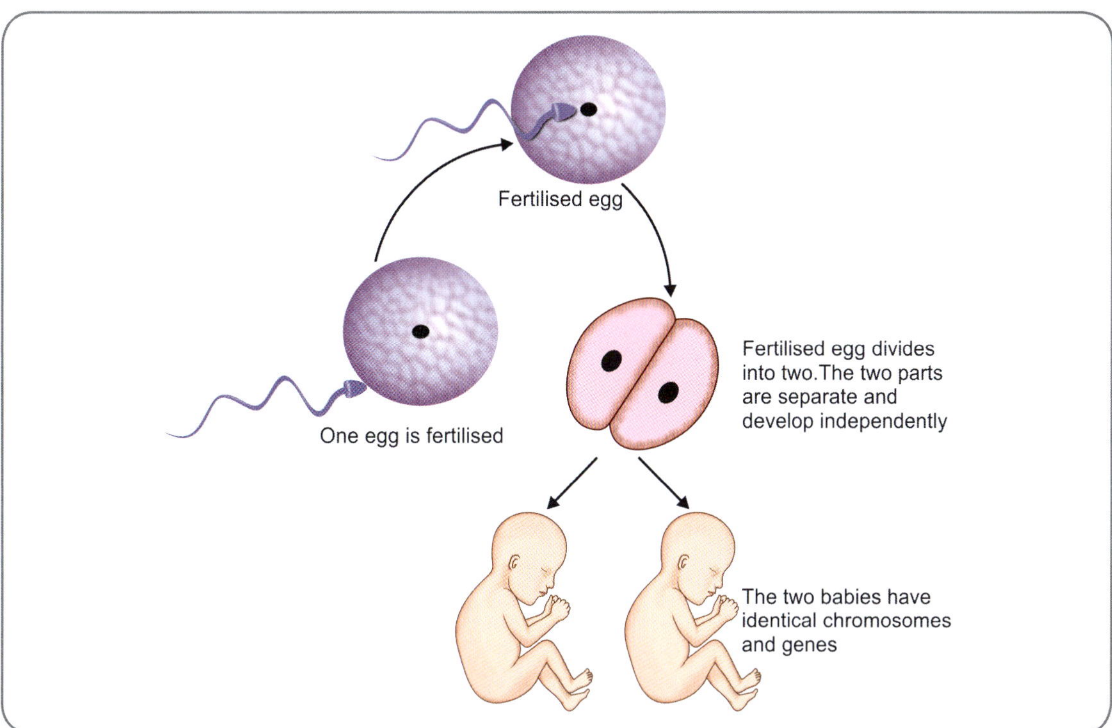

Fig. 38.1: Monozygotic twins

two embryos is not because of heredity but occurs as a spontaneous or random event.

Monozygotic twins can be subclassified by the time of splitting of the zygote (Table 38.1).

❑ ***Dichorionic-diamniotic twins:*** The earliest separation occurs at the two cell stage. The blastocysts develop separately, implant separately and each embryo has its own placenta and chorionic sac. They have two separate choria and amniotic sacs and hence are termed dichorionic-diamniotic twins. They occur in about 25% of monozygotic twins. They have strong resemblance in blood groups, fingerprints, sex and features of external appearance (Fig. 38.2A).

❑ ***Monochorionic-diamniotic twins:*** In most cases, the splitting of the zygote occurs at the early blastocyst stage. The inner cell mass splits into two separate groups of cells within the same blastocyst cavity. The two embryos have a common placenta and a common chorionic cavity but separate amniotic cavities. They are called monochorionic-diamniotic twins and the occurrence is about 60–70% of monozygotic twinning (Fig. 38.2B).

❑ ***Monochorionic-monoamniotic twins:*** In rare cases, separation occurs at the stage of bilaminar germ disc. This results in the formation of two partners with a single placenta and a common chorionic and amniotic sac. They are called monochorionic–monoamniotic twins (Fig. 38.2C).

Table 38.1: Differentiating features between dichorionic-diamniotic, monochorionic-diamniotic and monochorionic-monoamniotic twins		
Type	**Features**	**Probable time of splitting**
Dichorionic-Diamniotic	• Two separate chorionic and amniotic sacs • All the dizygotic twins and 25% of monozygotic twins • Survival good	By third day after fertilisation
Monochorionic-Diamniotic	• One chorionic sac and two amniotic sacs – Shared placenta • About 70% of monozygotic twins • Survival rate around 80% because of risks like twin to twin transfusion	Between fourth and eighth day after fertilization
Monochorionic-Monoamniotic	• Shared chorionic and amniotic sac • About 5% of monozygotic twins • Survival only 50–60% • Risks like twin to twin transfusion and umbilical cord entanglement	After ninth day and mostly before the tenth day

The incidence of monozygotic twinning is about 3 for every 1,000 live births. The likelihood of monozygotic twinning is more uniformly distributed than dizygotic twinning. The reason(s) for the splitting of a zygote is unknown.

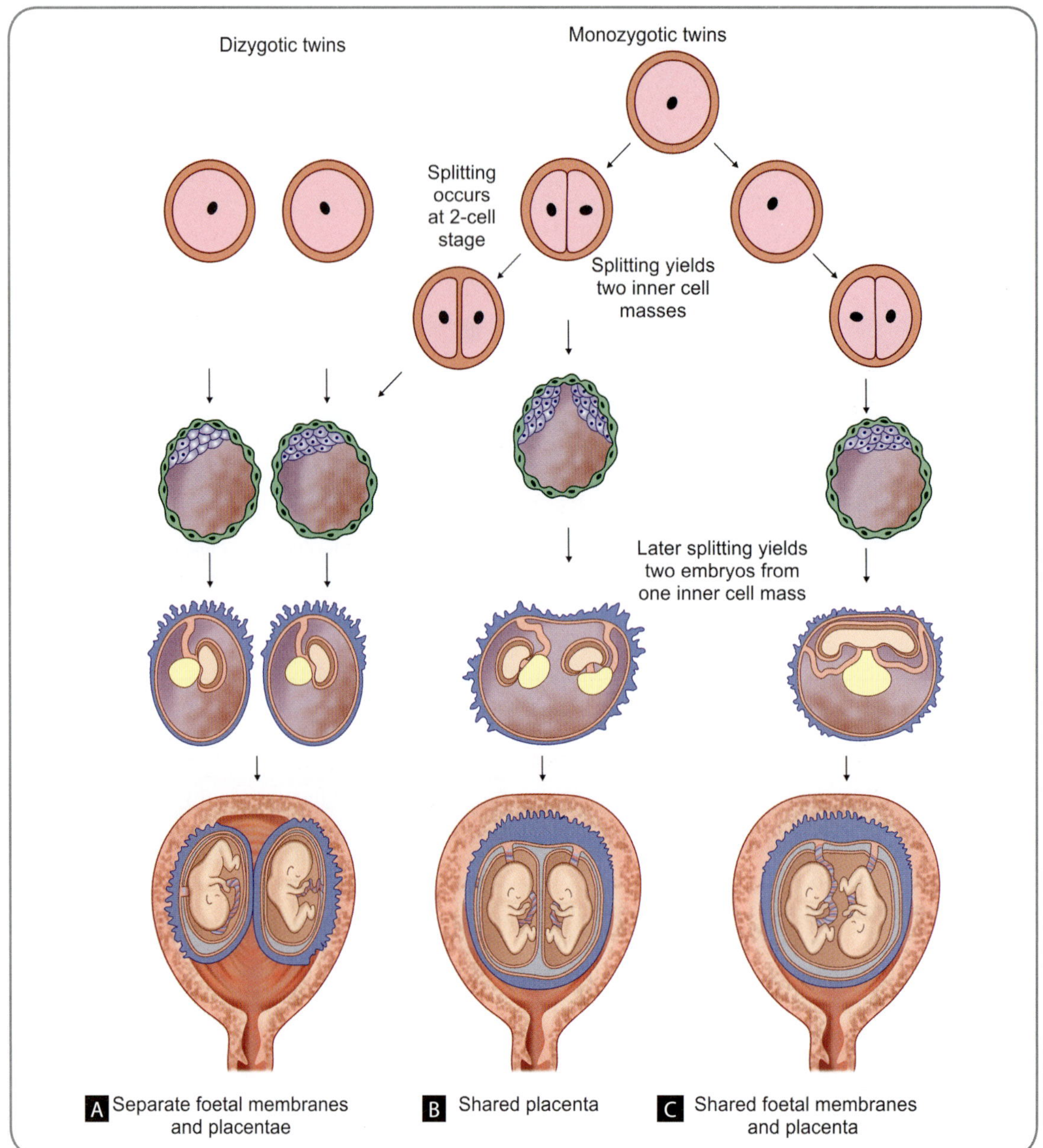

Figs 38.2A to C: Mechanism of twinning **A.** Dizygotic twinning—it also indicates the mechanism in dichorionic–diamniotic twinning **B.** Monochorionic–diamniotic twinning **C.** Monochorionic–monoamniotic twinning

When the division occurs between day 10 and day 12, conjoined twins develop. Conjoined twins are identical twins whose bodies are joined *in utero*. Survival rate is very low because of the complications which can arise due to the sharing of the organs.

Genetic and Epigenetic Similarity in Monozygotic Twins

Monozygotic twins share a common genotype. However, most monozygotic twins are not completely identical. Phenotypic differences have been observed. Variations in susceptibilities to diseases and in anthropomorphic features are common.

One reason for the differences could be mutational changes which are likely to occur during the splitting of the zygote.

The other reason is epigenetic influence. Epigenetic modifications are caused by environmental factors through out the lives of the twins. Such epigenetic modifications are more as age advances. A gene may be switched off or on or partially switched on. Environmental factors trigger such switching on and off. Twins who are 5 years old would have been exposed to less number of epigenetic influences than twins who are 50 years old. Epigenetic discordance is more if members of the twin pair are separated and brought up separately.

Monozygotic twins do not have the same fingerprints. The patterns of arches, loops and whorls of finger prints develop as a result of random stresses that a foetus experiences in the mother's uterus. Even in a small space inside the womb, the two foetuses (members of the twin pair) have contact with different parts of their environment and experience different sets of stresses which cause small variations in their digits, making them unique. Similarity in fingerprints may be present but there will be definitive differences in finer details.

SPECIALISED TYPES OF TWINNING

❑ *Semi-identical twins:* Half-identical or semi-identical twins (also referred to as 'half twins' or 'polar body twins') are the result of a very rare form of twinning in which the twins inherit exactly the same genes from their mother but different genes from their father. These half-identical twins are hypothesized to occur when an unfertilised egg cleaves into two identical attached ova and which are viable for fertilisation. Both ova are then fertilised by different spermatozoa and twinning occurs. Two embryos will be formed, each of which will have different paternal genes and identical maternal genes. This results in a set of twins with identical genes from the mother's side, but different genes from the father's side. Another possibility of the occurrence of half-identical twins is when a single ovum is fertilised by two different spermatozoa and then the zygote splits. There has been a report of only one pair of half-identical twins in the world's medical literature so far and it is difficult to postulate the exact mechanism of such twinning.

❑ *Mirror image twins:* The members of the twin pair have mirror image features. When one member is right handed the other may be left handed. Though the reason for such twinning is not known, it is known to occur when the zygotic split occurs 9 days after fertilisation.

❑ *Chimerism:* A chimera is an ordinary person except that some of his/her parts actually come from the other member of the twin pair or from the mother. A chimera may arise either from monozygotic twin foetuses (where it would be impossible to detect) or from dizygotic foetuses, where identification is possible by chromosomal comparison from various parts of the body. A chimera may be an intersex, composed of cells from a male twin and a female twin.

❑ *Twin-to-twin transfusion syndrome:* Monozygotic twins who share a placenta can develop twin-to-twin transfusion syndrome. Blood from one twin is diverted into the other twin; as a result, the 'donor' twin is small and anaemic and the 'recipient' twin is large and polycythaemic. The lives of both twins are endangered in this condition.

❑ *Parasitic twins:* These are twins which are asymmetrically conjoined. One twin is small, less formed and dependent on the larger twin for survival. Sometimes the parasitic twin becomes an almost indistinguishable part of the other and needs to be treated medically. Even more rare form of parasitic conjoined twinning is the *foetus in foetus*. In this condition, a malformed twin is discovered inside the body of a host twin, either a living child or an adult. In the rare condition called *partial molar twinning,* the single viable twin is endangered when the other zygote becomes cancerous or molar.

Multiple Choice Questions

1. One of the following is different from the others. Identify the odd one:
   a. Fraternal twins
   b. Binovular twins
   c. Dissimilar twins
   d. Identical twins
2. Monochorionic monoamniotic twins occur when splitting of the zygote occurs:
   a. At the 2-cell stage
   b. At the 8-cell stage
   c. At early blastocyst stage
   d. At the bilaminar germ disc stage
3. Epigenetic influence on twins occurs:
   a. During the neonatal period
   b. During childhood
   c. Throughout life
   d. During fourth and fifth decades of life
4. Polar body twins inherit:
   a. Same genes from the father but different genes from the mother
   b. Same genes from the mother but different genes from the father
   c. Same genes from both parents
   d. Different genes from both parents
5. A twin having parts of the body from the other member of the twin pair is:
   a. Mirror image twin
   b. Chimera
   c. Parasitic twin
   d. Transfusion twin

ANSWERS

1. d   **2.** d   **3.** c   **4.** b   **5.** b

Clinical Problem-solving

Case Study 1: Two monozygotic twin brothers who are 50-years-old show more differences in their characteristics than they had when they were 15 years old.

- ❏ What is the possible reason for this increase in differences?
- ❏ Would their fingerprints be the same or different?
- ❏ What is the reason for the fingerprint pattern?

(For solutions see Appendix).

Chapter 39

Inheritance of Genetic Disorders

Frequently Asked Questions

❑ Write notes on: (a) Autosomal dominant genetic disorders, (b) Autosomal recessive genetic disorders, (c) X-linked inheritance.
❑ Give a detailed account of X-linked inheritance.

An inheritance is dominant when one member of the allelic pair is able to express itself with total disregard to the presence of the other member. A trait which is expressed both in heterozygous and homozygous individuals is a *dominant trait*. Recessive inheritance depends on the expression of both partners of an allelic pair. A *recessive trait* is the one that is expressed only in the homozygous condition. Autosomal inheritance is due to genes present on autosomes while sex-linked inheritance is determined by genes present on sex chromosomes (X or Y).

AUTOSOMAL DOMINANT GENETIC DISORDER

An autosomal dominant disorder is inherited when the defective gene is present in an autosome and is dominant; one copy of the defective gene (single allele) is enough to produce the disorder. It can manifest in homozygous and heterozygous conditions. Since autosomal, males and females have equal chances of being affected. The trait or disease is transmitted through all generations. When one parent is affected, there is chance of 50% of the children being affected (Fig. 39.1).

One common example of autosomal dominant inheritance is achondroplasia. This disease is caused by a defective allele of the FGFR 3 gene located in the short arm of chromosome 4 (4p). The normal FGFR 3 gene codes for the fibroblast growth factor receptor 3 protein which is responsible for proper bone growth and ossification. When the gene is defective, it is over active; the resultant

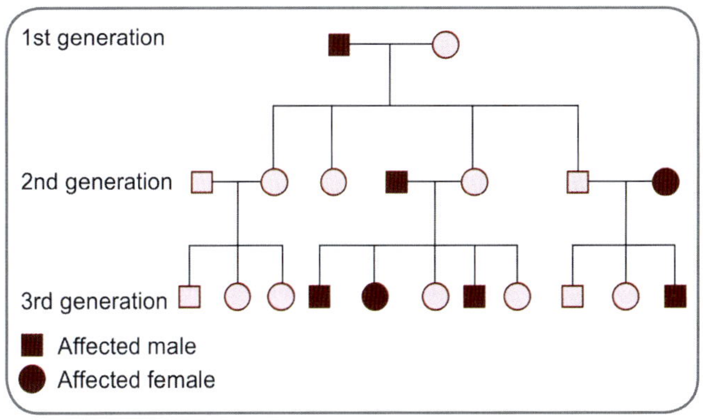

Fig. 39.1: A pedigree chart showing autosomal dominant inheritance; the father of the first generation is affected. Since it is an autosomal dominant inheritance, he is able to transmit it to the next generation where a son and a daughter have been affected - similarly, the disease is transmitted to the next generation.

imbalance in the protein mechanism interferes with bone formation and causes dwarfism that is characteristic of the disease.

An autosomal dominant disease can be transmitted from a father to the son or daughter and by a mother to the son or daughter. As with achondroplasia, a child may be affected when both the parents remain unaffected. Those are situations when a mutation has occurred and made the gene defective.

AUTOSOMAL RECESSIVE GENETIC DISORDER

The genes responsible for such traits are present on autosomes and can manifest only in the homozygous state. A heterozygous individual for an autosomal recessive trait is called a *carrier*. As relatives are more likely to have the same rare gene, offspring of consanguineous marriages have a higher probability inheriting a recessive gene and also manifesting the recessive trait. Since the inheritance

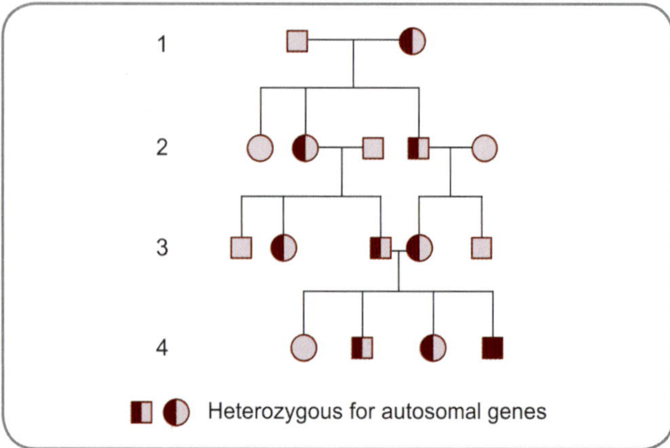

Fig. 39.2: A pedigree chart showing autosomal recessive inheritance; the defective gene has remained masked for three generations—the typical 25-50-25 inheritance can be seen in the 4th generation

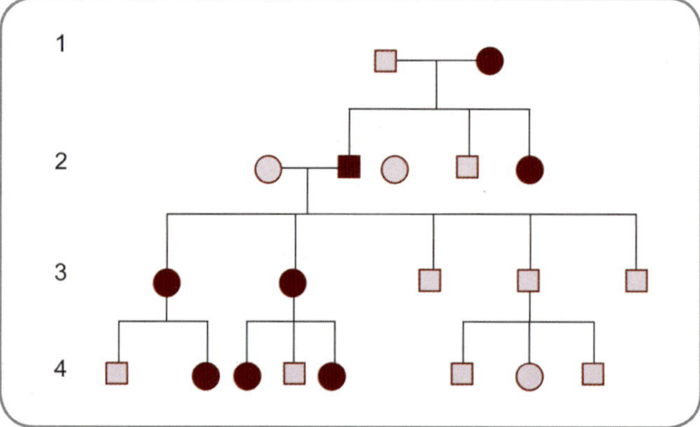

Fig. 39.3: A pedigree chart showing X-linked dominant inheritance —From the pedigree chart, it can be seen that the affected mother of the first generation transmitted it to both son and daughter; the affected father of the second generation transmitted it only to the daughters

is autosomal, males and females have equal chances of being affected. The disease or trait may not manifest in all generations because it is recessive and needs both the alleles to manifest the disease (Fig. 39.2). A normal individual who is a carrier can transmit the gene and individuals of the next generation may be affected.

A typical example of an autosomal recessive disorder is cystic fibrosis. A child with the disease has recessive defective alleles on both chromosomes 7 and is homozygous to the condition. Both the parents would be carriers and heterozygous.

When both parents are carriers for a recessive gene, there is 25% chance of a child inheriting defective genes from both parents and become diseased; there is 50% chance of a child inheriting the defective gene from one of the parents and remain a carrier; there is 25% chance of a child inheriting both normal genes from both parents and remain unaffected.

X-LINKED DOMINANT GENETIC DISORDER

The gene responsible for the concerned trait or disorder is present on the X chromosome and can express itself both in homozygous and heterozygous conditions. In the population, the trait or disorder is seen twice common in females as in males since females possess two X chromosomes whereas males have only one; the possibility of receiving the gene from either of the parents is more in females. An affected male will transmit the disease or trait to all his daughters but not sons. An affected female can transmit the disease or trait to both daughters and sons (Fig. 39.3). An example for this type of inheritance is Vitamin D resistant rickets.

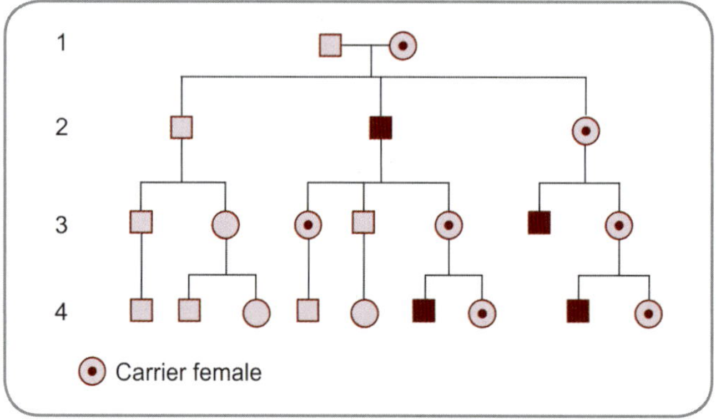

Fig. 39.4: A pedigree chart showing X-linked recessive inheritance—A carrier mother has transmitted the gene to her son and daughter with the daughter becoming a carrier and the son getting affected; the affected male of the second generation transmits the gene to all daughters making them all carriers

X-LINKED RECESSIVE GENETIC DISORDER

The gene responsible for the trait or disorder is present on the X chromosome and can express only in the homozygous condition. The possibility of a homozygous condition occurs only in females because there are two X chromosomes; however, in the population, the possibility of such homozygosity is limited. Most often, only one of the X chromosomes has the defective gene and in such a case, the concerned female becomes a carrier. The trait is expressed by all the males who inherit the gene because they have only one X chromosome (hemizygous condition). As a result, X-linked recessive diseases are often restricted to males and relatively uncommon in females. An affected male will have normal sons and carrier daughters (because his X chromosome will go to all his daughters but not to his sons). A carrier mother will transmit it to 50% percent of her sons and 50% of her daughters (because her affected X chromosome will go to half her gametes) (Fig. 39.4). The daughter of an affected father and a carrier mother will receive two defective X chromosomes from both the parents and will show the disease. By a general rule, a female can be a carrier for the defective gene but not a male. Two common examples of this kind of inheritance are Duchenne muscular dystrophy and Haemophilia.

Y-LINKED INHERITANCE

Traits or disorders inherited by this way are very few in number. In this form of inheritance, only males are affected. All sons of the affected male are affected. Since Y chromosome is absent in females, females do not suffer these diseases or traits and also do not transmit them. One typical example of a Y-linked trait is the occurrence of hairy ears.

Multiple Choice Questions

1. In an autosomal recessive disorder, when an individual is affected:
    a. Both parents are carriers
    b. The father is a carrier
    c. The mother is a carrier
    d. Both parents are affected
2. In an X-linked dominant disorder, an affected father will transmit the disease:
    a. To all his children
    b. To all his daughters
    c. To all his sons
    d. To none of his children
3. The genetic status of a male having an X-linked recessive mutant gene is:
    a. Homozygous
    b. Heterozygous
    c. Gonozygous
    d. Hemizygous

ANSWERS

1. a   **2.** b   **3.** d

Clinical Problem-solving

Case Study 1: A 25-year-old female is afflicted with vitamin D-resistant rickets. She was heterozygous to the gene which is an X-linked dominant gene. Assuming that she marries a normal male, work out the possibilities of the genotypes and phenotypes of their children.

(For solutions see Appendix).

Chapter 40

Mutation and Recombinance

MUTATION

Mutation is any sudden heritable structural change in the genetic material or the process that leads to such a change.

The gene undergoes a structural modification during the process of mutation. As a result of this change, the properties to which the gene codes are also changed and in turn, the phenotypic expression shows alterations. *In toto*, new characteristics appear.

Mutations can produce changes which need not be abnormal. Many of such changes may go unnoticed. Only when the changes brought about are abnormal or harmful, they are noticed.

Mutations may be classified in various ways.
❏ Classification by cause:
  ○ *Spontaneous:* Occurring suddenly and naturally without any human or artificial factor;
  ○ *Induced:* Occurring due to some human or artificial factor.
❏ Classification by type of change:
  ○ *Genome mutation:* The number of the chromosome complement is altered;
  ○ *Chromosome mutation:* Structural changes occur in the chromosome;
  ○ *Gene mutation:* Otherwise called point mutation; changes occur at the molecular level in the DNA.
❏ Classification by location:
  ○ *Somatic mutation:* Occurs in body cells;
  ○ *Germ cell mutation:* Occurs in germ cells.

Factors Producing/Affecting Mutations

Factors affecting mutations are as follows:

❏ *Parental age:* Increasing age of parents increases the incidence of mutations. Fathers of new dominant achondroplastic children are, on an average, slightly older than the other fathers in the population. Similarly, children born to elderly mothers also show mutation effects.
❏ *Parental sex:* Though of limited importance, this factor cannot be completely overruled. Genome mutations are more common in female germ cells. These are mostly due to non-disjunction during meiotic divisions and appear to occur more in the female germ cells. It has been evaluated that most of the trisomies are due to female non-disjunction. Increasing maternal age has a direct influence on such non-disjunctions.
❏ *Autoimmune processes:* Autoimmune processes occurring in both parents may produce mutations.
❏ *Inherent chromosomal properties:* Acrocentric chromosomes show a natural tendency for satellite associations. These can also alter or modify the genetic sequence resulting in mutations.
❏ *Gene size:* Larger genes are more prone to have mutations.
❏ *Genetic hot spots:* Certain nucleotide sequences exhibit increased mutation susceptibility. CG dinucleotides are typical examples and have a mutation rate which is 12 to 13 times higher than that of the other nucleotide sequences.
❏ *Mutagenic agents:* Exposure to certain chemicals and substances produces mutations. Such substances and chemicals are called mutagenic agents. Ionising radiation, gamma and X-rays, neutrons from fast reactors and chemicals like mustard gas, sulphur compounds and high density formaldehyde are known to be mutagenic.

Single Gene Mutations

Many of the ill effects produced by mutations seem to be sequelae of single gene mutations. Single gene mutations are actually point mutations where one base pair in the DNA is altered. The base pair may be substituted, deleted or reversed. In addition, a single base pair may be added to the existing complement. A small change in the base pair sequence will change the amino acid sequence leading to several deleterious effects.

However, not all mutations cause harmful and pronounced effects. Depending upon the effects caused by single gene mutations, they have been grouped as follows:

- ❏ *Silent mutations:* They do not alter amino acid sequences, cause no visible effects.
- ❏ *Mis-sense mutations:* They produce change in a single amino acid sequence.
- ❏ *Non-sense mutations:* They produce one of the stop codons (there are three stop codons in the messenger RNA – UAA, UAG and UGA) and thereby alter the length of the polypeptide chain.
- ❏ *Pseudogene formation:* They produce copies of genes which have some kind of alteration in that the genes cannot be transcribed.
- ❏ *Deletion mutations:* They produce deletion of multiples of three base pairs.
- ❏ *Insertion mutations:* They produce insertion of multiples of three base pairs.
- ❏ *Frame shift mutations:* They produce deletion or insertion of a single base pair.

Of the above mentioned mutations, the last three are dangerous. We know that three base pairs form a codon and code for the amino acids. Alteration of three base pairs by means of either a deletion or insertion will change the entire sequence and result in altered amino acids. The entire protein may be altered as a result of the change in the amino acid. Important amino acids may be deleted or substituted leading to harmful effects. Frame shift mutations are the most dangerous. Here, only one base pair is deleted or added. However, this deletion or insertion will shift the position of the rest of the base pairs. As a result, the entire sequence is altered.

Any of these mutations can produce effects ranging from alterations in enzyme specificity to loss of enzyme activity.

Mutations and Diseases

Several human diseases have been associated to mutations. Human haemoglobin disorders are the most common of single gene disorders. Examples include thalassaemia and sickle cell disease.

RECOMBINANT DNA TECHNOLOGY

Human DNA sequences can be united with DNA molecules of other organisms, especially bacteria. Such 'recombinant DNA' is capable of continuous and infinite duplication under laboratory conditions. Cloning is the process by which a recombinant DNA molecule is produced. Human DNA sequences which have favourable characteristics are chosen and united with bacterial DNA. The ability of the bacterium to generate multiple copies (or clones) of the particular human DNA sequence is the key principle of recombinant DNA technology.

Recombinant DNA technology has a major influence in medical genetics. If favourable DNA sequences can be 'cloned', they can be used in the treatment of several disorders either as 'substitutes' for abnormal / disease producing sequences or as repair agents which set an abnormality 'right'.

Multiple Choice Questions

1. Alteration in the number of chromosomes as a result of mutation is called:
   a. Germinal mutation
   b. Chromosome mutation
   c. Genome mutation
   d. Gonadal mutation
2. Chromosomes which show a natural tendency for satellite association are:
   a. Metacentric
   b. Sub metacentric
   c. Acrocentric
   d. Telocentric
3. Mutations which produce stop codons are:
   a. Mis-sense mutations
   b. Frame shift mutations
   c. Non-sense mutations
   d. Point mutations
4. Recombinant DNA technology involves the principle of generating indefinite number of DNA sequence copies by using:
   a. A complementary DNA from the same species
   b. Gene from a higher animal
   c. X-linked gene
   d. Bacterial gene
5. The major contributing factor for human trisomy disorders is:
   a. Non-disjunction in male cells
   b. Frame shift mutations in male and female cells
   c. Non-disjunction in female cells
   d. Chromosome mutations in female cells

ANSWERS

1. c   2. c   3. c   4. d   5. c

Appendix

DISSECTIONS

INCISIONS FOR FACE AND SCALP

1. Vertical incision from nasion to symphysis menti.
2. Oblique incision from angle of mouth to midpoint of tragus.
3. From one tragus to the other, a curved coronal incision around the scalp.
4. Vertical incision from nasion to incision 3.
5. Curved incision around eye-extended to tragus.
6. Curved incision around mouth.

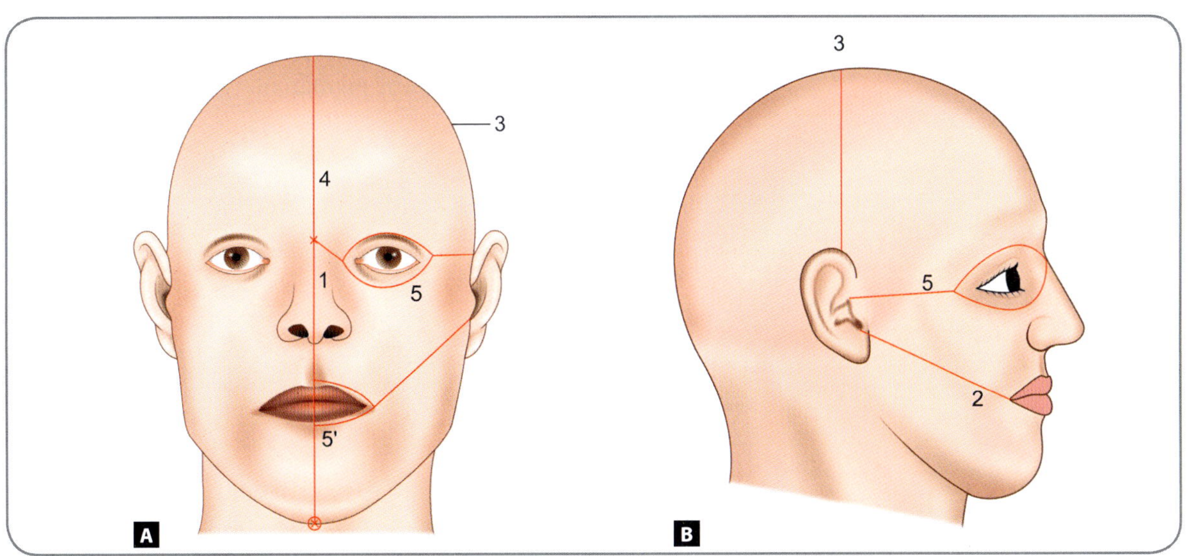

Key: x = nasion; ⊗ = symphysis menti

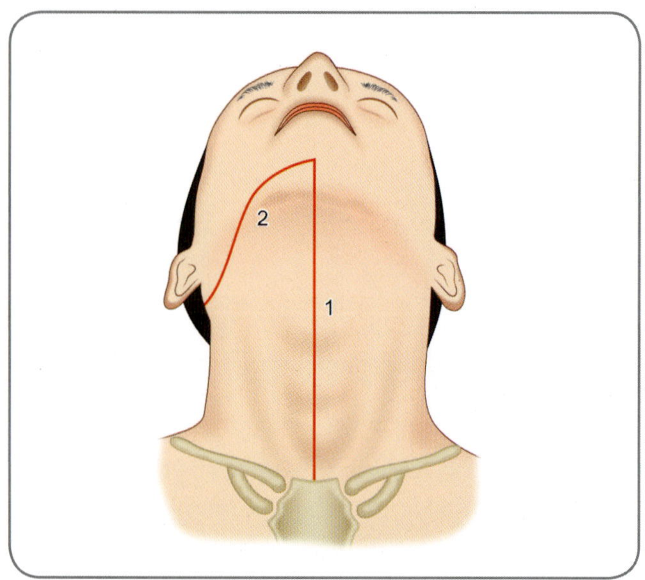

INCISIONS FOR THE NECK—ANTERIOR ASPECT

1. Vertical incision from symphysis menti to jugular notch along midline of neck.
2. Incision from symphysis menti to angle of mandible along mandibular lower border.

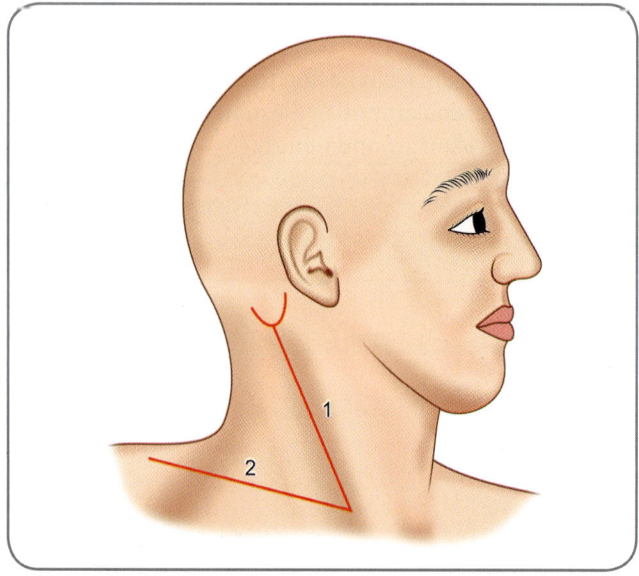

INCISIONS FOR THE NECK—LATERAL AND POSTERIOR ASPECTS

1. Incision from mastoid to jugular notch along anterior border of sternocleidomastoid.
2. Incision from jugular notch to acromion along the clavicle.

SOME FREQUENTLY USED TERMS AND THEIR PLURAL FORMS

Singular	Plural
Viscus	Viscera
Meatus	Meati; meatuses (often used)
Calvaria	Calvariae
Orbit, orbita	Orbits or orbitae
Oculus (eye)	Oculi
Sclera	Sclerae
Choroid, choroidea	Choroideae
Retina	Retinae
Cornea	Corneae
Auricle, auricula	Auriculae
Pinna	Pinnae
Tympanum	Tympana, tympanums (colloquial usage)
Saccule, sacculus	Sacculi
Utricle, utriculus	Utriculi
Vestibule, vestibulus	Vestibuli
Cochlea	Cochleae
Pharynx	Pharynges (pronounced farinjeez)
Larynx	Larynges (pronounced larinjeez)
Epiglottis	Epiglottides (pronounced epiglottideez)
Rectus (as used in muscle name)	Recti
Temporalis (muscle)	Temporales
Occipitofrontalis (muscle)	Occipitofrontales
Scalenus	Scaleni
Trapezius	Trapezii
Orbicularis oculi	Orbiculariae oculi
Rotator (as used in muscle name)	Rotatores
Levator (as used in muscle name)	Levatores
Rotator (as used generally)	Rotators
Levator (as used generally)	Levators
Oesophagus	Oesophagi
Ganglion	Ganglia
Sulcus	Sulci
Gyrus	Gyri
Fontanelle	Fontanelles
Thalamus	Thalami
Hypothalamus	Hypothalami
Cortex	Cortices
Cerebrum	Cerebra

contd…

contd…

Cerebellum	Cerebella
Pons	Pontes
Medulla	Medullae
Uncus	Unci
Insula	Insulae
Fonticulus	Fonticuli
Nucleus	Nuclei
Mandible, mandibula	Mandibulae
Vomer	Vomeria
Maxilla	Maxillae
Concha	Conchae

SYMBOLS COMMONLY USED IN PEDIGREE CHARTS IN GENETIC ANALYSES

Symbols	Meaning
□	Normal male
 ○	Normal female
■	Abnormal/affected male
●	Abnormal/affected female
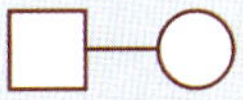 □—○	Mating
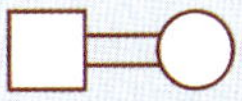 □=○	Consanguineous mating
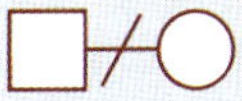 □/○	Divorced
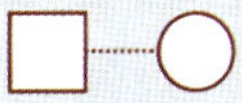 □····○	Illegitimacy
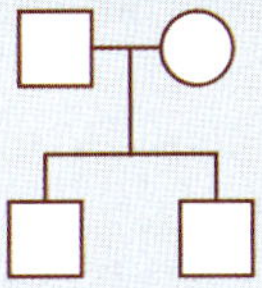	Parents with children—order of birth indicated by sequence from left (first son, second son in this example)
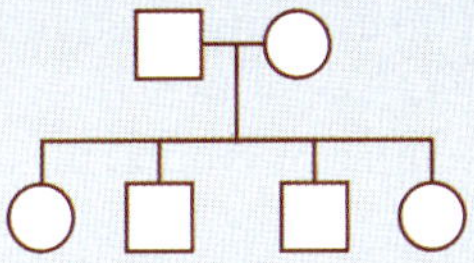	Parents with children (first daughter, second son, third son, fourth daughter in this example)
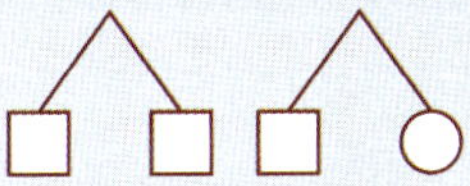	Dizygotic twins
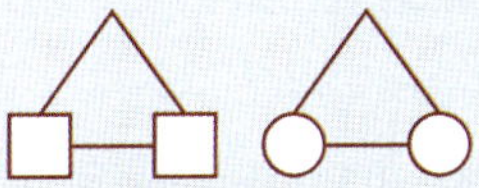	Monozygotic twins
	Propositus and proposita
⊙	X-linked recessive carrier
 ◇	Sex unspecified
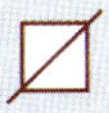 ⊘	Dead
 ① ② 4	Number of children and sex indicated

Solutions to Clinical Problems

CHAPTER 1

Case Study 1

- The ophthalmic division of the trigeminal nerve, especially the external nasal nerve, is affected.
- The dermatomal map of the face has three zones—three broad areas which have posterosuperior wings and these areas are placed one below the other. The upper area also has a central inferior extension that runs over the dorsum of the nose. From above downwards, these areas are the areas subserved by the ophthalmic division, maxillary division and mandibular division of the trigeminal nerve.
- Yes. They can be correlated to the areas developmentally derived from the frontonasal process, maxillary process and mandibular process.

Case Study 2

- The palatine tonsils are most likely to be affected.
- The efferents from this group go to the lower deep cervical nodes or the jugular lymph trunk.

CHAPTER 2

Case Study 1

- The 'soft spots' felt for on a child's head are the fontanelles.
- Information regarding the growth status and hydration status of the child, and presence or absence of intracranial pressure can be obtained by examining the fontanelles.

- Yes. Use of comb or any other foreign object (especially any object that has pointed ends) can cause damage to the soft membranes found in the fontanelles.

Case Study 2

- The fluid could be cerebrospinal fluid. The condition is CSF rhinorrhoea.
- The floor of the anterior cranial fossa should have been involved.
- If CSF leaks happen in the ear, such a condition is called CSF otorrhoea.

CHAPTER 3

Case Study 1

- The provisional diagnosis will be left facial nerve paralysis (lower motor neuron type).
- Sagging down—muscles of face are weakened or paralysed leading to flaccidity and sagging
  - Eversion of left inferior eyelid—paralysis or weakening of orbicularis oculi
  - Widened palpebral fissure on the left side—paralysis or weakening of orbicularis oculi
  - Difficulty in chewing and dribbling of saliva—muscles of the cheek and upper lip area are weakened or paralysed, especially the buccinator.
  - All the above mentioned muscles are supplied by facial nerve and in facial nerve paralysis they are either weakened or paralysed.
- Presence or absence of wrinkles or furrows of the forehead, presence or absence of the nasolabial fold and deviation of the angle of mouth should be looked for. Hyperacusis, any absence of taste sensation over

the anterior two-thirds of the tongue are other features which can be tested for (mainly to ascertain the level of lesion).

Case Study 2

- The probable condition is a stye (or hordeolum).
- If the ciliary gland is involved, the swelling projects on the front of the eyelid. If the tarsal gland is involved, the swelling projects on the eyeball (on the globe).
- The glands of the region are:
  - Tarsal glands – modified sebaceous glands – secrete an oily secretion that lubricates the eyelids;
  - Ciliary glands – small sweat glands or small sebaceous glands opening into the hair follicles of the eye lashes – their secretions are very minimal.

CHAPTER 4

Case Study 1

- The man is probably suffering from dislocation of temporomandibular joint.
- He is not able to close his mouth because the head of mandible would have slipped forwards to the area in front of the articular tubercle. Closure of the mouth is then not possible.
- In yawning, the lateral pterygoid muscle is suddenly and powerfully contracted. This contraction makes the joint unstable and causes the head of mandible to slip forward.

Case Study 2

- The woman is likely to be suffering from submasseteric space infection.
- This space is present between the masseter muscle and the ramus of mandible. Accumulation of pus in this space will not cause any visible swelling because of its internal location.
- Muscle spasm produced by accumulation of pus in the space causes pain and restriction in the opening of the mouth.

CHAPTER 5

Case Study 1

- Yes. It is a normal reaction.
- It is called the 'gag reflex'.
- The afferent pathway of the reflex is through the glossopharyngeal nerve and the efferent pathway is through the glossopharyngeal and vagal nerves.

Case Study 2

- This man has a ranula.
- The sublingual salivary gland is blocked.

- Other salivary glands may be involved. Though major salivary glands may be involved, the condition affects the minor salivary glands more.

CHAPTER 6

Case Study 1

- Neck rigidity is the feature for clinching diagnosis.
- Spasm of neck muscles is the cause for neck rigidity.
- Pia and arachnoid are involved in the inflammation.

CHAPTER 7

Case Study 1

- The vertebral secondaries are likely to have metastasized from a prostatic primary malignancy.
- The vertebral venous plexuses drain into the caval or the lumbar venous systems. Reversal of blood flow can happen in these venous channels when the intra-abdominal pressure is increased or during postural changes. Cancer cells travel from the pelvic area to the vertebrae.
- Under normal conditions blood flows from the vertebral venous plexuses to the caval or lumbar venous systems. The venous channels of these plexuses are devoid of valves. Therefore, reversal of blood flow is possible. When such reversal occurs, malignant cells from the pelvic viscera can travel to the vertebrae and even to the skull.

Case Study 2

- Compression of spinal cord and spinal nerves are the anticipated complications.
- The most common feature would be that of sciatica.

CHAPTER 8

Case Study 1

- The deformity is called torticollis.
- The swelling should have been a fibrous tumour of the sternocleidomastoid that usually occurs just before or after birth.
- Injury to the muscle during child birth or intermittent spasmodic contractions of the muscle produces muscular torticollis and intermittent torticollis respectively.

Case Study 2

- The carotid pulse should be palpated for in the carotid triangle of the neck.
- The transverse process of the C6 vertebra is the bony point.
- It should be palpated for in the lower neck to avoid compression of the carotid sinus and a consequent reduction in blood pressure and heart rate.

CHAPTER 9

Case Study 1

- The possibility of involvement of paranasal sinuses is high.
- The maxillary sinus is most likely to be involved.
- The opening of the maxillary sinus into the nasal cavity is placed close to the roof of the sinus; drainage of the sinus is, therefore, not complete. This leads to chronic infection.

Case Study 2

- The arterioles in the anteroinferior part of the nasal septum form an extensive network and are also superficially placed in that they are covered only by the mucous membrane. Minimal trauma is sufficient to cause damage to these arterioles. Since the arterioles are interconnected by a dense network, injury to even a single arteriole can result in extensive bleeding.
- The anteroinferior part of the nasal septum.
- The anatomical basis is formed by the dense arteriolar network formed by the branches of the anterior ethmoidal, sphenopalatine, greater palatine arteries and the septal branch of superior labial artery.

CHAPTER 10

Case Study 1

- The problem is in the vocal cords; they are injured; scar and nodule formation occurs.
- Repeated friction and trauma result in subepithelial haemorrhages and bruising. Such repeated trauma results in scarring and nodule formation.
- Singing with proper techniques, avoidance of abuse of voice and avoidance of use of force are the advices to be given to the patient.

Case Study 2

- Palatine tonsils are located in the oropharynx. Infection of tonsils can spread to the nasopharynx and to the respiratory tract.
- Ear pain is a consequence of tonsillitis. Infection from the pharynx can spread to the middle ear cavity through the auditory tube.
- Quinsy is abscess in the peritonsillar tissues.

CHAPTER 11

Case Study 1

- Opacity of the lens is called cataract.
- Because of the opacity, focusing power of the lens is lost. Restoration of the focusing power is the principle behind cataract surgery.
- Vision is restored by the use of appropriate glasses (which act as external lenses) or by implanting an artificial lens into the eye (which replaces the original lens).

Case Study 2

- Separation of the pigment and nervous layers of the retina is retinal detachment.
- In the embryo, the two layers are separated by cavity called the intraretinal space. During development, the space gets obliterated. However, the two layers are not firmly fixed to each other. This leads to separation in later life (under appropriate trigger factors).
- History may reveal an episode of trauma that could have happened. Retinal detachment does not usually occur immediately after trauma. History becomes important in such cases because the patient could have forgotten or may not think that it is important to mention about it. Even small degrees of trauma can result in detachment.

CHAPTER 12

Case Study 1

- No, it will not.
- Stapedial surgery will help in conduction defects and not in sensorineural loss.
- The lesion in sensorineural loss is in the pathway from cochlea to brain.

Case Study 2

- Normally, the walls of the auditory tube are in apposition. During plane travel, air pressure on either side of the tympanic membrane is not the same. Air pressure inside the middle ear cavity remains high when compared to the outside and this causes change in the quality of the sound perceived. The individual feels that his ear is blocked.
- Closed position of the auditory tube is the anatomic basis.
- Opening of the mouth actively results in the tube being opened actively. Consumption of sweets which will keep the mouth open for a longer time or ensure repeated opening can help. Swallowing or yawning also can help.

CHAPTER 13

Case Study 1

- The man has acromegaly.
- There is excess secretion of growth hormone.
- He would have had gigantism.

Case Study 2

- The woman has hypothyroidism.
- Her face is bloated because of deposition of mucopolysaccharides.
- Examination of the shin reveals the non-pitting oedema and it can be differentiated from the pitting oedema of other causes.

CHAPTER 14

Case Study 1

- Cavernous sinus thrombosis is a dreaded complication that can occur.
- The facial vein communicates with the cavernous sinus through the deep facial vein and the pterygoid plexus of veins. So, an infection from the face can spread to the cavernous sinus.
- The upper lips and the entire nose are areas which can predispose to such a condition.

Case Study 2

- An arteriovenous communicaton between the internal carotid artery and the cavernous sinus causes such a condition.
- The exophthalmos is pulsating because of the arterial involvement.
- Congestion of the orbital veins results in exophthalmos.

CHAPTER 15

Case Study 1

- The probable diagnosis is facial nerve paralysis.
- Testing for hyperacusis and for taste disturbances in the anterior two-thirds of the tongue will help in ascertaining the level of lesion.
  - Widening of palpebral fissure—due to paralysis of orbicularis oculi;
  - Smoothening of forehead—due to paralysis of occipitofrontalis;
  - Difficulty in chewing—due to paralysis of buccinator;
  - Loss of conjunctival reflex—due to paralysis of orbicularis oculi;
  - Generalised asymmetry of face—paralysis of facial muscles on the affected side.

Case Study 2

- The oculomotor nerve is affected.
- Lateral squint—due to overactivity of lateral rectus.
- Ptosis—due to paralysis of levator palpebrae superioris.
- Fixed and dilated pupil—due to paralysis of sphincter pupillae because of parasympathetic interruption and unopposed action of sympathetic fibres.
- Loss of light and accommodation reflexes—due to paralysis of sphincter pupillae and ciliaris muscles.

CHAPTER 16

Case Study 1

- Anteroposterior radiograph of the skull (occipitofrontal view – Caldwell's view) can be taken.
- No. Another view is required.
- Ocipitomental view – Water's view radiograph will give better visualisation of the maxillary sinus.

Case Study 2

- He should have applied pressure bilaterally and compression of the vagi on both sides would have caused dizziness.
- Eliciting both the carotid pulses at the same time should be avoided and the carotid pulse should also be taken at a lower level in the neck.
- Facial pulse and the superficial temporal pulse can be palpated in the facial region.

SECTION 7 NEUROANATOMY

CHAPTER 17

Case Study 1

- Plantar reflex reaction will be the consequence.
- Reflex arc is the neurological basis.
- The afferent neuron is in the dorsal nerve root ganglion.

CHAPTER 18

Case Study 1

- The man is suffering from herpes zoster.
- The dorsal root ganglion is the structure infected.
- No. The muscles are not affected. They are not affected because only the dorsal root ganglia are involved and not the motor neurons.

Case Study 2

- Yes. The velocity of flow of CSF depends on the pressure of CSF.
- It can be evaluated by counting the rate of the number of drops of CSF flowing out or by connecting the lumbar puncture needle to a manometer.
- Colour, cell content and chemical composition are the other aspects which will help in diagnosis.

CHAPTER 19

Case Study 1

- The neurotransmitter is dopamine.
- It is produced by the neurons of substantia nigra.
- Degeneration is seen in the corpus striatum.

Case Study 2

- The ascending sensory tracts will be involved.
- The spinal nucleus of trigeminal nerve and the nuclei gracilis and cuneatus are affected.
- The spinal nucleus of trigeminal nerve.

CHAPTER 20

Case Study 1

- The internal capsule is likely to be involved.
- If the cerebral cortex is involved, the lesion will be localised. Here the patient has a generalised and widespread paralysis. If the brainstem is involved, cranial nerve nuclei will be involved and those features will be seen along.
- Charcot's artery of cerebral haemorrhage is the artery that is most likely to be involved.

Case Study 2

- The man, most probably, is suffering from thalamic syndrome.
- Sensations are altered because the thalamus is involved.
- The motor pathway is not involved.

CHAPTER 21

Case Study 1

- This is an example of referred pain. Pain arising in a structure supplied by one branch of a nerve may be felt in another area supplied by another branch (or the same branch). Caries tooth produces pain in ear because the lower teeth are supplied by the inferior alveolar nerve, which is a branch of the mandibular nerve and the ear region is supplied by the auriculotemporal nerve which is another branch of the same nerve.
- The phenomenon is referred pain.
- Other examples—(a) Pain in the forehead in frontal sinusitis; frontal sinus is supplied by the supraorbital nerve and skin of forehead also supplied by the supraorbital nerve. (b) Pain in the ear in cancer of the tongue; tongue is supplied by the lingual nerve and ear by the auriculotemporal nerve.

Case Study 2

- The angle of mouth on the paralysed side will not move.
- Other tests are:
  - Filling the mouth with air and then trying to press on it – resistance is less on paralysed side;
  - Filling the mouth with air—it may leak out because of weakened muscles.

- The special visceral afferent component is the taste component. Taste sensations can be tested by applying various substances to the anterior two-thirds of tongue and the taste perception of the individual checked for.

CHAPTER 22

Case Study 1

- Ataxia, asynergia, dysarthria, dysdiadokokinesia, nystagmus, hypotonia, asthenia and diminished reflexes are some of the features likely to be seen.
- *Ataxia:* Coordination between various muscles and/or muscle groups is affected leading to inability in standing. The body sways. Interruption of afferent pathways can also lead to ataxia.
- *Asynergia:* Lack of coordination between muscles causes deficiency in purposeful movements. The patient attempts to do something, but is not able to; tremors occur.
- *Dysarthria:* Lack of coordination causes inability in moving the muscles required for speech.
- Muscle tone will be decreased.

Case Study 2

- The condition with repetitive jerky movements of eye is called nystagmus.
- Due to incoordination of muscles, the eyes are not able to fix the gaze on a particular object. The individual attempts to bring the gaze back to the object. This results in jerky movements of the eyes.
- Ataxia, asynergia and dysarthria are some other examples of muscular incoordination.

CHAPTER 23

Case Study 1

- Speech was affected because the Broca's area on the dominant hemisphere (left hemisphere) was involved.
- The motor speech area is on the inferior frontal gyrus; this area is located inferior to but closely proximal to the motor area, especially that part of the motor area associated with the muscles of the tongue, mouth, larynx and face (area of vocalisation).
- In this case of thromboembolic disease of the cerebral vessels, the muscles concerned are not paralysed.

Case Study 2

- The neurotransmitter involved is dopamine.
- It is produced by neurons in the substantia nigra.
- Degeneration is found in the corpus striatum.

CHAPTER 24

Case Study 1

- The term 'obstructive' indicates that there is some obstruction for the passage of cerebrospinal fluid from the ventricular system to the subarachnoid space.
- The interventricular foramina, the cerebral aqueduct and the foramina of fourth ventricle are the places where the obstruction can occur.
- Meningitis can affect the arachnoid villi and is an acquired condition.

Case Study 2

- Lumbar puncture is the method of choice.
- The needle is introduced into the subarachnoid space at the interval between the third and fourth lumbar vertebrae.
- The alternate method is cisternal puncture.

CHAPTER 25

Case Study 1

- The condition is called Argyll-Robertson pupil.
- The two reflexes are different because the pathways are different. The accommodation reflex is supposed to be routed through the visual cortex.
- The lesion in this condition will be in the pretectal area.

Case Study 2

- The step taken is to touch the cornea with a small thin wisp of cotton.
- The afferent pathway is through the branches of the ophthalmic division of the trigeminal nerve. The efferent pathway is through the facial nerve to the orbicularis oculi muscle.
- There will be no response if the facial nerve is paralysed on that side. However, response can be elicited by stimulating the normal side and producing a bilateral response.

CHAPTER 26

Case Study 1

- The man is affected by medial medullary syndrome.
- The ventral and medial parts of the medulla are affected.
- Involvement of Corticospinal tract—contralateral hemiplegia
  - Hypoglossal nucleus—ipsilateral paralysis of lingual muscles
  - Medial lemniscus—sensory loss and disturbances.

Case Study 2

- Occlusion of middle cerebral artery is the most likely cause.
- Homonymous hemianopia is the likely other feature.
- Hearing disturbance is not pronounced because it is compensated by the other side.

SECTION 8 GENETICS

CHAPTER 29

Case Study 1

- No.
- There are two Barr bodies in this case. In a normal female, there will be only one Barr body.
- This slide shows two Barr bodies. The individual should then have three X chromosomes; could probably be XXX female.

Case Study 2

- Cri-du-chat syndrome.
- Deletion of a portion of the short arm of chromosome 5.
- Laryngeal and tracheal deficiencies result in the shrill cry.

CHAPTER 31

Case Study 1

- Red colour smooth petals (RS) x White colour wrinkled petals (rs)
- F1 generation—genotype – all RrSs; phenotype all Red Smooth

	RS	RS
rs	RrSs	RrSs
rs	RrSs	RrSs

- F2 generation

	RS	Rs	rS	rs
RS	RRSS	RRSs	RrSS	RrSs
Rs	RRSs	RRss	RrSs	Rrss
rS	RrSS	RrSs	rrSS	rrSs
rs	RrSs	Rrss	rrSs	rrss

- F2 generation
  - RRSS, RrSS, RrSs, RRSs—Red Smooth phenotype
  - RRss, Rrss—red wrinkled phenotype
  - rrSS, rrSs—White Smooth phenotype
  - rrss—white wrinkled phenotype
- F3 generation punnett square can be worked out based on the types chosen for cross breeding.

CHAPTER 32

Case Study 1

- All their children will have attached lobes.
- None.
- The appearance or disappearance of a feature will depend on the spouses of their children. Only if the genetic complement of the spouses is available, the pedigree chart can be worked out or the probability analysis done.

CHAPTER 33

Case Study 1

- None. Since both parents are AB, there is no possibility of the gene coding for O group to be present. Since the gene coding for O (I) is a recessive gene, it will be required in the homozygous condition for the blood group O to be expressed. Such homozygous condition for the gene and its phenotypic expression are therefore not possible.
- None. The children can be of any one of the following genotypes: AB, AA, BB.
- As stated above, the genotypes of the children can be AB, AA or BB.

	A	B
A	AA	AB
B	AB	BB

Case Study 2

	L	l
L	LL	Ll
L	LL	Ll

- L – large nose; l – small nose; alleles are incompletely dominant.
- 50% of their children will have genotype LL and so phenotype large nose.
- The other 50% children will have genotype Ll and so phenotype medium sized nose.

CHAPTER 34

Case Study 1

	L^N	L^N
L^M	$L^M L^N$	$L^M L^N$
L^N	$L^N L^N$	$L^N L^N$

- 50% of their children will have genotype $L^M L^N$ and so phenotypically will be of the MN group.

- The other 50% of the children will have genotype $L^N L^N$ and so phenotypically will be of the N group.

CHAPTER 35

Case Study 1

	L	l
L	LL	Ll
L	LL	Ll

- L – large lip protrusion; l – small lip protrusion; Alleles are incompletely dominant.
- 50% of children will have LL and so will have large lip protrusion.
- The other 50% will have Ll and so moderate protrusion.

CHAPTER 36

Case Study 1

- Both father and mother are carriers of Tay-Sachs allele (T).
- T – Tay-Sachs allele; t – normal allele.

	T	t
T	TT	Tt
T	TT	Tt

- 50% of the F1-generation children are likely to have TT genotype and so exhibit symptoms of Tay-Sachs disease. The other 50% will have Tt heterozygous condition and so be carriers.

CHAPTER 37

Case Study 1

- The reason for male sexual characteristics to have shown externally is the Y complement.
- There will be one Barr body.
- This case belongs to the Lygaeus mode of sex determination – human SRY pattern.

CHAPTER 38

Case Study 1

- Epigenetic influence is the reason for the increase in differences.
- The fingerprints will be different.
- The fingerprint pattern is different because of the pressure differences that the two foetuses undergo within the uterus of the mother.

CHAPTER 39

Case Study 1

- ❑ D – dominant X-linked gene; d – normal allele
- ❑ Father – XdY; Mother – XDXd

	XD	Xd
Xd	XDXd	XdXd
Y	XDY	XdY

- ❑ 50% of the children will be daughters XX and the remaining 50% will be sons XY.
- ❑ Of the daughters, 50% (25% of total offspring) will be affected – XDXd; remaining 50% (25% of total offspring) will be normal – XdXd.
- ❑ Of the sons, 50% (25% of total offspring) will be affected – XDY; remaining 50% (25% of total offspring) will be normal – XdY.
- ❑ Therefore, it can be seen that an affected mother (XDXd) can transmit the disease to both sons and daughters.

Index

Page numbers followed by *f* refer to figure and *t* refer to table.

523